Springer Studium Mathematik – Master

Series Editors

Martin Aigner, Berlin, Germany

Heike Faßbender, Braunschweig, Germany

Barbara Gentz, Bielefeld, Germany

Daniel Grieser, Oldenburg, Germany

Peter Gritzmann, Garching, Germany

Jürg Kramer, Berlin, Germany

Volker Mehrmann, Berlin, Germany

Gisbert Wüstholz, Wermatswil, Switzerland

The series "Springer Studium Mathematik" is aimed at students of all areas of mathematics, as well as those studying other subjects involving mathematics, and anyone working in the field of applied mathematics or in teaching. The series is designed for Bachelor's and Master's courses in mathematics, and depending on the courses offered by universities, the books can also be made available in English.

http://www.springer.com/series/13446
http://www.springer.com/series/13893

Die Reihe „Springer Studium Mathematik" richtet sich an Studierende aller mathematischen Studiengänge und an Studierende, die sich mit Mathematik in Verbindung mit einem anderen Studienfach intensiv beschäftigen, wie auch an Personen, die in der Anwendung oder der Vermittlung von Mathematik tätig sind. Sie bietet Studierenden während des gesamten Studiums einen schnellen Zugang zu den wichtigsten mathematischen Teilgebieten entsprechend den gängigen Modulen. Die Reihe vermittelt neben einer soliden Grundausbildung in Mathematik auch fachübergreifende Kompetenzen. Insbesondere im Bachelorstudium möchte die Reihe die Studierenden für die Prinzipien und Arbeitsweisen der Mathematik begeistern. Die Lehr- und Übungsbücher unterstützen bei der Klausurvorbereitung und enthalten neben vielen Beispielen und Übungsaufgaben auch Grundlagen und Hilfen, die beim Übergang von der Schule zur Hochschule am Anfang des Studiums benötigt werden. Weiter begleitet die Reihe die Studierenden im fortgeschrittenen Bachelorstudium und zu Beginn des Masterstudiums bei der Vertiefung und Spezialisierung in einzelnen mathematischen Gebieten mit den passenden Lehrbüchern. Für den Master in Mathematik stellt die Reihe zur fachlichen Expertise Bände zu weiterführenden Themen mit forschungsnahen Einblicken in die moderne Mathematik zur Verfügung. Die Bücher können dem Angebot der Hochschulen entsprechend auch in englischer Sprache abgefasst sein.

More information about this series at http://www.springer.com/series/13893

Ulrich Görtz · Torsten Wedhorn

Algebraic Geometry I: Schemes

With Examples and Exercises

Second Edition

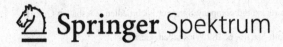

🐎 **Springer** Spektrum

Ulrich Görtz
Fakultät für Mathematik
Universität Duisburg-Essen
Essen, Germany

Torsten Wedhorn
Fachbereich Mathematik
TU Darmstadt
Darmstadt, Germany

ISSN 2509-9310 ISSN 2509-9329 (electronic)
Springer Studium Mathematik – Master
ISBN 978-3-658-30732-5 ISBN 978-3-658-30733-2 (eBook)
https://doi.org/10.1007/978-3-658-30733-2

Planung: Iris Ruhmann
This Springer Spektrum imprint is published by the registered company Springer Fachmedien Wiesbaden GmbH part of Springer Nature.
The registered company address is: Abraham-Lincoln-Str. 46, 65189 Wiesbaden, Germany

Contents

Introduction

Algebraic geometry has its origin in the study of systems of polynomial equations

$$f_1(x_1, \ldots, x_n) = 0,$$
$$\vdots$$
$$f_r(x_1, \ldots, x_n) = 0.$$

Here the $f_i \in k[X_1, \ldots, X_n]$ are polynomials in n variables with coefficients in a field k. The set of solutions is a subset $V(f_1, \ldots, f_r)$ of k^n. Polynomial equations are omnipresent in and outside mathematics, and have been studied since antiquity. The focus of algebraic geometry is studying the geometric structure of their solution sets.

If the polynomials f_i are linear with constant term 0, then $V(f_1, \ldots, f_r)$ is a subvector space of k^n. Its "size" is measured by its dimension and it can be described as the kernel of the linear map $k^n \to k^r$, $x = (x_1, \ldots, x_n) \mapsto (f_1(x), \ldots, f_r(x))$.

For arbitrary polynomials, $V(f_1, \ldots, f_r)$ is in general not a subvector space. To study it, one uses the close connection of geometry and algebra which is a key property of algebraic geometry, and whose first manifestation is the following: If $g = g_1 f_1 + \ldots g_r f_r$ is a linear combination of the f_i (with coefficients $g_i \in k[T_1, \ldots, T_n]$), then we have $V(f_1, \ldots, f_r) = V(g, f_1, \ldots, f_r)$. Thus the set of solutions depends only on the ideal $\mathfrak{a} \subseteq k[T_1, \ldots, T_n]$ generated by the f_i. On the other hand we may consider polynomials $f \in k[T_1, \ldots, T_n]$, view them as polynomial functions $k^n \to k$, and define the ideal $\mathfrak{a}'$ of those f such that f vanishes on $V(f_1, \ldots, f_r)$. It is one of the first main results (Hilbert's Nullstellensatz) that if k is algebraically closed, then $\mathfrak{a}'$ is closely related to $\mathfrak{a}$, more precisely

$$\mathfrak{a}' = \mathrm{rad}(\mathfrak{a}) := \{\, f \in k[T_1, \ldots, T_n] \; ; \; \exists m > 0 : f^m \in \mathfrak{a} \,\}.$$

The quotient $k[T_1, \ldots, T_n]/\mathfrak{a}'$ may be considered as the k-algebra of polynomial functions on the "affine variety" $V(f_1, \ldots, f_r)$. We obtain a close relation between ideals $\mathfrak{a}$ of $k[T_1, \ldots, T_n]$ (or, equivalently, quotient algebras of $k[T_1, \ldots, T_n]$) and affine varieties in k^n – at least if k is algebraically closed. For not algebraically closed fields k this approach is too naive.

Besides this algebraic description, one can endow the sets $V(f_1, \ldots, f_r)$ with a "geometric structure". The only reasonable topology which can be defined purely in algebraic terms, i.e., without appealing to analytic notions as convergence, is the Zariski topology which is the coarsest topology (on k^n, say) such that all zero sets of polynomials are closed. Not surprisingly, it is very coarse and therefore is not sufficient to determine the "geometric structure" of the spaces in question. The right way to remedy this, is to consider each space *together with the entirety of functions on the space*. Similarly as a differentiable (or holomorphic) manifold is determined by its topological structure together with the entirety of differentiable (or holomorphic) functions on all its open subsets, we obtain a satisfactory notion of algebraic geometric objects, "affine varieties over k", by considering closed subsets of k^n together with the entirety of functions on them, which in this case means all functions defined by fractions of polynomials.

© Springer Fachmedien Wiesbaden GmbH, part of Springer Nature 2020
U. Görtz und T. Wedhorn, *Algebraic Geometry I: Schemes*, Springer Studium
Mathematik – Master, https://doi.org/10.1007/978-3-658-30733-2_1

Polynomial equations also arise in number theory, and especially in the last decades algebraic-geometric methods have become extremely fruitful for solving number-theoretic problems. In this case the polynomials have coefficients in $\mathbb{Q}$ or $\mathbb{Z}$ (or more generally in number fields, finite fields, or p-adic rings). One of the most famous examples is Fermat's equation $x^m + y^m = z^m$ with $x, y, z \in \mathbb{Z}$. The proof of Fermat's Last Theorem which asserts that this equation has no solutions for $m \geq 3$, $xyz \neq 0$, by Wiles and Taylor in 1995 relies heavily on modern algebraic geometry.

The unifying approach to study polynomial equations $f_1, \ldots, f_r$ over arbitrary (commutative) rings R is the theory of schemes developed by Grothendieck and his school. It allows to attach to an arbitrary commutative ring A (e.g., $A = R[T_1, \ldots, T_n]/(f_1, \ldots, f_r)$ or $A = \mathbb{Z}[X, Y, Z]/(X^m + Y^m - Z^m)$) a geometric object $\operatorname{Spec} A$ consisting of a topological space X and a datum $\mathcal{O}_X$ of "systems of functions" on this space such that the ring of "globally defined functions" on $\operatorname{Spec} A$ is the ring A itself. Such a pair $(X, \mathcal{O}_X)$ is a so-called locally ringed space. This allows us to view commutative rings as geometric objects called affine schemes. The affine variety $V(f_1, \ldots, f_r) \subseteq k^n$ can be recovered from the affine scheme $\operatorname{Spec} k[T_1, \ldots, T_n]/(f_1, \ldots, f_r)$.

As in elementary geometry some problems only have a satisfying solution if we consider them not in affine space but in projective space. For instance two different lines in the affine plane intersect always in one point except if they are parallel. By adding points at infinity (the "horizon") we obtain the projective plane, where any two different lines intersect in precisely one point. The projective space can be obtained by gluing affine spaces. Vastly generalizing this process we arrive at the central notion of this book: a scheme. It is defined as a locally ringed space that is locally isomorphic to an affine scheme. Note the similarity to the definition of a smooth n-dimensional manifold which is a geometric object that is locally isomorphic – within the right category – to an open subset of $\mathbb{R}^n$.

For schemes geometric notions as dimension or smoothness are defined. As schemes are locally given by commutative rings, many of these notions are defined in terms of rings and ideals. Conversely, every definition or result in commutative algebra has its geometric counterpart in the theory of schemes. Thus algebra and geometry become two aspects of the same theory.

Another reason for the importance of schemes is that it is often possible to parameterize interesting objects by schemes. An example is the Grassmannian which is a scheme that parameterizes subvector spaces of a fixed dimension in a given finite-dimensional vector space. The general concept behind schemes as parameter spaces is the point of view of schemes representing certain functors. This plays an important role in modern algebraic geometry and beyond. It will be one of the main focuses in this book.

Grothendieck's theory of schemes is technically demanding but essential in modern algebraic geometry even for applications in classical complex algebraic geometry. Even more so it is indispensable in arithmetic geometry. Moreover algebraic geometry has also become an important tool with many applications in other fields of mathematics such as topology, representation theory, Lie theory, group theory, string theory, or cryptography.

The goal of this book is to provide its reader with the background in algebraic geometry to go on to current research in algebraic geometry itself, in number theory, or in other fields of mathematics. It strives for the necessary generality to be a stable stepping stone for most of these fields.

There is a wealth of literature on algebraic geometry from which we learned a lot. It is a pleasure to acknowledge the overwhelming influence of the pioneering work of

Grothendieck and Dieudonné ([EGAI],[EGAInew], [EGAII],[EGAIII], [EGAIV]). Other sources are Mumford's red book [Mu1], and the books by Shafarevich [Sh], Hartshorne [Ha3] and Perrin [Per]. Furthermore we list the more specialized books, each with its own focus, by Mumford [Mu2], Griffiths and Harris [GH], Liu [Liu], Harris [Har], Eisenbud and Harris [EH], and Harder [Ha].

Further sources had a more local impact. We followed Kurke [Ku1], [Ku2] and Peskine [Pes] quite closely in our proof of Zariski's main theorem. In our treatment of geometric properties of schemes over a field one of our main references was Jouanolou's book [Jo]. Our main source for determinantal varieties was the book [BV] by Bruns and Vetter. For the example of cubic surfaces we profited much from Beauville [Bea] and [Ge] and for the example of Brauer-Severi varieties from Gille and Szamuely [GS].

Leitfaden

The notion of scheme which is the main object of investigation of the whole book is introduced in Chapter 3, using the affine schemes defined in Chapter 2 as local building blocks. These two chapters are therefore indispensable for all of the book. In Chapter 1 we discuss a precursor of schemes, namely prevarieties (over an algebraically closed field). These prevarieties are much closer to geometric intuition, and on the other hand comprise a large number of interesting schemes. However, besides other defects of this notion, prevarieties are not suitable for discussing arithmetic questions because it is not easily possible to link objects living over base fields of different characteristics. In Chapter 4 we introduce fiber products of schemes which are ubiquitous in all of the remainder. In particular fiber products allow us to view the fibers of morphisms of schemes as schemes, so that we can make precise the philosophy that a morphism $f: X \to S$ of schemes should be seen as a family $(f^{-1}(s))_{s \in S}$ of schemes. For beginners in algebraic geometry, working through all of Chapters 1 to 4 is therefore recommended. For those with a background in classical algebraic geometry, Chapter 1 can probably be skipped, and all readers with some knowledge about schemes should be able to start with Chapter 5 without too many problems.

After this first part of the book, some choices can be made. In Chapter 5, the part on dimension of schemes over a field should be read in any case – not only since it is used at many places, but because the dimension of a scheme is a fundamental notion in algebraic geometry as a whole. The parts on schemes over non-algebraically closed fields, and on base change of the ground field, are more specialized and can be skipped at a first reading. References to the latter can usually be avoided by assuming that the base field in question is perfect or algebraically closed. The part on intersections of plane curves with a proof of Bézout's theorem is one of the first applications of the theory developed so far, but is not strictly necessary for the rest of the text; the only place where it is used again is the discussion of elliptic curves in Chapter 16.

The topic of Chapter 6 are local properties of schemes, in particular the notions of tangent space, smooth, regular and singular points and of normality. We make essential use of the notion of normal scheme in Chapter 12 when we discuss normalizations and Zariski's main theorem.

Chapter 7 provides definitions and results on (quasi-coherent) $\mathscr{O}_X$-modules. Its first part should be read rather selectively because there we collect all constructions of $\mathscr{O}_X$-modules which are used in the rest of the book. The other parts are central for most of the following chapters.

The functor attached to a scheme is introduced in Chapter 4 and discussed in quite some detail in the first part of Chapter 8. This is an essential concept of modern algebraic geometry and is used in many places of the book. So the first part of this chapter is a requisite. The second and third parts on Grassmannians and Brauer-Severi schemes provide examples. While the Brauer-Severi schemes can easily be omitted, if necessary, the example of Grassmannians is of a more fundamental nature, because the projective space and more generally projective bundles are a special case of Grassmannians. Nevertheless, with a little care in Chapter 13, one can replace the definitions of projective bundles in term of Proj schemes (and recover the functorial description).

The first part of Chapter 9 is dedicated to the notions of separated schemes and separated morphisms. Being separated is analogous, in comparison with topological spaces, to being Hausdorff, and not surprisingly is a property which almost all schemes occurring in practice have. In the second part we discuss rational maps, i.e., "morphisms" which are defined only on an open subset of the source. Rational maps and in particular the closely related notion of birational equivalence are a central object of study in algebraic geometry. In the rest of the book, they are relevant in particular in Chapter 11 when we study divisors.

In Chapter 10 we study finiteness notions of schemes. In the noetherian case there are many simplifications, so that we deal with this case first. Next we look at the general case. As it turns out, for quasi-compact and quasi-separated schemes many of the results in the noetherian case have good analogues. In fact, these two properties occur so often that we abbreviate them to *qcqs*. The next two parts of Chapter 10 are dedicated to the question how properties of schemes and morphisms behave under "transition to the limit". More precisely, we study an inductive system of rings $(R_\lambda)_\lambda$, and families $(X_\lambda)_\lambda$, where each X_λ is an R_λ-scheme. This setup is relevant in many different situations; it can often be used to eliminate noetherianness hypotheses, but is also relevant for problems about noetherian schemes. Nevertheless, at a first reading it might be enough to read the first part of Chapter 10 (and make noetherianness assumptions later in the book).

The two main topics of Chapter 11 are vector bundles, and in particular line bundles, and divisors. We look at the close connection between line bundles and divisors. Line bundles are essential in Chapter 13. The study of divisors in the special case of curves is taken up in Chapter 15. The flattening stratification and the classification of vector bundles over the projective line will not be used in the rest of this volume except in some remarks in Chapter 12 and Chapter 16. These two parts may thus be skipped at a first reading.

Next, in Chapter 12, we look at affine, finite and proper morphisms. All three of these are fundamental properties of schemes (and morphisms of schemes) which distinguish interesting classes of schemes. For instance, properness corresponds to the notion of compactness for topological spaces or manifolds. The most important theorem of the chapter is Zariski's main theorem which clarifies the structure of morphisms with finite fibers (so-called quasi-finite morphisms) and has a large number of handy applications. Because the proof is rather involved, it might be appropriate to take the theorem as a black box at first – while the result is used in several places in later chapters, the methods of the proof are not.

Chapter 13 serves to study projective schemes, i.e., closed subschemes of projective space or of projective bundles. From a slightly different point of view, we study, given a scheme X, how X can be embedded into projective space. It turns out that this is controlled by the behavior of line bundles on X. Projective morphisms (i.e., families of projective schemes) are special cases of proper morphisms, so the results of Chapter 12 are used frequently.

The main topic of Chapter 14 is flatness, a notion which encodes that a family of schemes (or modules) varies continuously. For instance, under mild assumptions the dimension of the fibers of a flat morphism is constant. After studying elementary properties of flat morphisms in the first part of the chapter, we prove a number of deep theorems like the valuative criterion and the fiber criterion for flatness in the second part. Here we rely heavily on the local criterion for flatness (see Appendix B). If $X' \to X$ is a flat and surjective morphism, then X inherits many properties of X'; a similar principle applies for morphisms, and to some extent to objects over X' and X, respectively. This principle is called "faithfully flat descent" and is the object of the third part of Chapter 14. As an example, we take up the theory of Brauer-Severi schemes again. The next two parts are dedicated to a more advanced treatment of dimension theory, and in particular to an investigation how the dimensions of the fibers of a morphism vary. Finally, we briefly look at a central example of a scheme parameterizing interesting objects, namely the Hilbert scheme.

A large class of interesting, but relatively well accessible schemes is formed by the 1-dimensional schemes, i.e., by curves. In this case, many of the previously looked at concepts become more concrete and more tangible, and we look at curves in detail in Chapter 15. A particular application are the valuative criteria which characterize separated and proper morphism in a geometrically very tangible way. We also mention, without proof, the theorem of Riemann-Roch, a central result in the theory of curves (and in fact, in a generalized form, also in the theory of much more general schemes).

The final chapter, Chapter 16, contains several examples which are developed in parallel to the advancement of the theory in the main part of the book. Each example is split up into several portions, and for each of them we indicate which of the previous chapters are needed. These examples illustrate most of the concepts introduced in the book. Specifically, we look at determinantal varieties, and at several topics that are linked by their relation to the theory of Hilbert modular surfaces: cubic surfaces, cyclic quotient singularities, and abelian varieties.

Each chapter concludes with exercises. We have marked the easier exercises with the symbol $\Diamond$.

Readers interested only in the noetherian case can omit large parts of Chapter 10, and many reductions in later chapters. The most important facts to keep in mind are that all noetherian schemes are quasi-compact, and that every morphism whose source is noetherian is quasi-compact and quasi-separated. Readers interested only in schemes of finite type over an algebraically closed base field can ignore, in addition, the subtleties of base change by extension fields as detailed in Chapter 5. Over an algebraically closed field, having some property, or having it geometrically, is the same. In a few places it is helpful to assume that the base field is of characteristic 0, but apart from Chapter 16, this does not make a real difference.

Notation

We collect some general notation used throughout the book. By $\subseteq$ we denote an inclusion with equality allowed, and by $\subsetneq$ we denote a proper inclusion; by $\subset$ we denote an inclusion where we do not emphasize that equality must not hold, but where equality never occurs or would not make sense (e.g., $\mathfrak{m} \subset A$ a maximal ideal in a ring).

By Y^c we denote the complement of a subset Y of some bigger set. By $\overline{Y}$ we denote the closure of some subspace Y of a topological space.

By convention, the empty topological space is not connected.

If R is a ring, then we denote by $M_{m \times n}(R)$ the additive group of $(m \times n)$-matrices over R, and by $\mathrm{GL}_n(R)$ the group of invertible $(n \times n)$-matrices over R.

The letters $\mathbb{Z}$, $\mathbb{Q}$, $\mathbb{R}$, $\mathbb{C}$ denote the ring of integers and the fields of rational, real and complex numbers, respectively.

Corrigenda and addenda

Additions and corrections of the text will be posted on the web page

www.algebraic-geometry.de

of this book. We encourage all readers to send us remarks and to give us feedback.

Acknowledgements

We thank all people who sent us their comments about preliminary versions of this text, in particular: Kai Arzdorf, Philipp Hartwig, Andreas Müller, Niko Naumann, Andreas Riedel, Ulrich Schmitt, Otmar Venjakob.

Most of the pictures were produced with the program Surf [Surf].

Preface to the second edition

We received a lot of positive feedback for our book, and are very grateful for that.

In the second edition, we have corrected several serious mistakes and many smaller errors and misprints. Except for the corrections, we did not make substantial changes.

We are very grateful to the many people who notified us of mistakes, large and small, in personal discussions, by email or via the web page listing the errata, in particular:

P. Barik, Alexey Beshenov, Félix Baril Boudreau, Thomas Brazelton, J. Buck, Zhaodong Cai, J. Calabrese, P. Carlucci, Owen Colman, B. Conrad, O. Das, Florian Ebert, A. Elashry, C. Frank, C. Frei, L. Galinat, D. Gerigk, A. Graf, F. Grelak, A. Gross, Vishal Gupta, A. Haase, Yun Hao, U. Hartl, P. Hartwig, B. Heintz, D. Heiss, J. Hilgert, M. Hoyois, Longxi Hu, Yong Hu, H. Iriarte, Alexander Isaev, R. Ishizuka, M. Jarden, Peter Johnson, Shuho Kanda, M. Kaneda, A. Kaučikas, T. Keller, S. Kelly, K. Kidwell, S. Köbele, O. Körner, Mahdi Majidi-Zolbanin, Louis Martini, Akira Masuoka, Sebastian Schlegel Mejia, Nick Mertes, K. Mohri, Shahram Mohsenipour, Laura Brustenga Moncusí, Menachem Dov Mostowicz, A. B. Nguyen, Jesús Martín O., Safak Ozden, M. Pereira, Nathan Pflueger, Lam Pham, Richard Pink, L. Prader, T. Przezdziecki, Caiyong Qiu, Fabian Roll, Matthieu Romagny, Sandeep S, Kannappan Sampath, Immanuel van Santen, J. Scarfy, A. Schiller, Ehsan Shahoseini, Eduardo dos Santos Silva, B. Smithling, A. Steinbach, F. Gispert Sánchez, Viktor Tabakov, Yugo Takanashi, Kuo Tzu-Ang, E. Viehmann, J. Watterlond, Jan Willing, Shaopeng Z, Y. Zaehringer, Victor Zhang, Han Zhou, Yehao Zhou, P. Zsifkovits.

We will continue to collect and publish errata (concerning either of the two editions) at www.algebraic-geometry.de.

1 Prevarieties

Contents

- Affine algebraic sets
- Affine algebraic sets as spaces with functions
- Prevarieties
- Projective varieties

The fundamental topic of algebraic geometry is the study of systems of polynomial equation in several variables. In the end we would like to study polynomial equations with coefficients in an arbitrary ring but as a motivation and a guideline we will assume in this chapter that our ring of coefficients is an algebraically closed field k. In this case the theory has a particularly nice geometric flavor.

If we are given polynomials $f_1, \ldots, f_r \in k[T_1, \ldots, T_n]$, we are interested in "geometric properties" of the set of zeros

$$V(f_1, \ldots, f_r) = \{ (t_1, \ldots, t_n) \in k^n \; ; \; \forall i : \; f_i(t_1, \ldots, t_n) = 0 \} \subseteq k^n.$$

Let us illustrate this by a simple example.

Example 1.1. Consider the polynomial $f = T_2^2 - T_1^2(T_1 + 1) \in k[T_1, T_2]$. To visualize $V(f)$ we show in Figure 1.1 the set of zeros of $T_2^2 - T_1^2(T_1 + 1) \in \mathbb{R}[T_1, T_2]$ in $\mathbb{R}^2$. Of course, this is not an example for our situation as the field $\mathbb{R}$ of real numbers is not algebraically closed (and sometime the visualization obtained in this way may be deceptive, see Exercise 1.8). Nevertheless it is often helpful to look at the "real picture".

In this illustration we see a "one-dimensional" object (the notion of dimension in algebraic geometry will be defined in Chapter 5). Another observation is that the set of zeros looks "locally" in every point except the origin $(0, 0)$ essentially like a real line. But in the origin its local shape is different. We may describe this behavior by saying that at all points outside the origin we can find a unique tangent line, however not in the origin. This corresponds to the distinction of "smooth" and "singular" points, that we will describe in Chapter 6.

Figure 1.1: The solutions in $\mathbb{R}^2$ of the equation $T_2^2 - T_1^2(T_1 + 1) = 0$.

© Springer Fachmedien Wiesbaden GmbH, part of Springer Nature 2020
U. Görtz und T. Wedhorn, *Algebraic Geometry I: Schemes*, Springer Studium
Mathematik – Master, https://doi.org/10.1007/978-3-658-30733-2_2

The theorem of implicit functions implies that the set of zeros of f is diffeomorphic to $\mathbb{R}$ at those points (x_1, x_2) where the Jacobi (1×2)-matrix $\left(\frac{\partial f}{\partial T_1} \ \frac{\partial f}{\partial T_2} \right)$ has rank 1. The partial derivatives of polynomials can be defined over arbitrary base fields and thus this criterion can be formulated algebraically. We will see in Section (6.8) that this is indeed a way to describe which points are "smooth".

Notation

Let k be an algebraically closed field. Occasionally we write $k[\underline{T}]$ instead of $k[T_1, \ldots, T_n]$, the polynomial ring in n variables over k. For $x = (x_1, \ldots, x_n) \in k^n$ and $f \in k[\underline{T}]$ we write $f(x) = f(x_1, \ldots, x_n)$.

Affine algebraic sets

The first step towards geometry is to define a topology on the set of zeros $V(f_1, \ldots, f_r)$. We obtain a very coarse topology which is useful but does not capture all essential geometric properties. The purpose of the following sections will be to endow these topological spaces with additional structure. Here we will see that analytic aids used in differential geometry or complex analysis are replaced by methods of commutative algebra. Over algebraically closed field the connection of geometry and commutative algebra is established by Hilbert's Nullstellensatz 1.7. In its most basic form (Corollary 1.11) it says that attaching to $x = (x_1, \ldots, x_n) \in k^n$ the ideal $(T_1 - x_1, \ldots, T_n - x_n) \subset k[T_1, \ldots, T_n]$ yields a bijection between points in k^n (geometric objects) and maximal ideals in $k[T_1, \ldots, T_n]$ (algebraic objects).

(1.1) The Zariski topology on k^n, the affine space $\mathbb{A}^n(k)$.

Definition 1.2. *Let $M \subseteq k[T_1, \ldots, T_n] = k[\underline{T}]$ be a subset. The set of common zeros of the polynomials in M is denoted by*

$$V(M) = \{ (t_1, \ldots, t_n) \in k^n \ ; \ \forall f \in M : f(t_1, \ldots, t_n) = 0 \}.$$

If M consists of elements f_i, $i \in I$, we also write $V(f_i, i \in I)$ instead of $V(\{f_i ; i \in I\})$.

If $M \subseteq k[\underline{T}]$ is a subset and $\mathfrak{a}$ the ideal generated by M, it is clear that $V(M) = V(\mathfrak{a})$. The Hilbert Basis Theorem (Proposition B.34) implies, that the polynomial ring $k[\underline{T}]$ is a noetherian ring, that is, all ideals are finitely generated. Every generating set M of a finitely generated ideal $\mathfrak{a}$ contains a finite generating set. Hence there exist for every subset $M \subseteq k[\underline{T}]$ finitely many elements $f_1, \ldots, f_r \in M$ such that $V(M) = V(f_1, \ldots, f_r)$.

Another obvious property is that $V(-)$ reverses inclusions: For $M' \subseteq M \subseteq k[\underline{T}]$ we have $V(M') \supseteq V(M)$.

Proposition 1.3. *The sets $V(\mathfrak{a})$, where $\mathfrak{a}$ runs through the set of ideals of $k[\underline{T}]$, are the closed sets of a topology on k^n, called the Zariski topology.*

Proof. The proposition follows from the following more precise assertions.

(1) $\emptyset = V(1)$, $k^n = V(0)$.

(2) For every family $(\mathfrak{a}_i)_{i \in I}$ of ideals $\mathfrak{a}_i \subseteq k[\underline{T}]$ we have

$$\bigcap_{i \in I} V(\mathfrak{a}_i) = V(\sum_{i \in I} \mathfrak{a}_i).$$

(3) For two ideals $\mathfrak{a}, \mathfrak{b} \subseteq k[\underline{T}]$ we have

$$V(\mathfrak{a}) \cup V(\mathfrak{b}) = V(\mathfrak{a} \cap \mathfrak{b}) = V(\mathfrak{ab}).$$

The first point is obvious. Moreover, we have

$$\bigcap_{i \in I} V(\mathfrak{a}_i) = \{x \in k^n; \ \forall i \in I, f \in \mathfrak{a}_i : f(x) = 0\} = V(\bigcup_{i \in I} \mathfrak{a}_i),$$

and this proves the second point because $\sum_{i \in I} \mathfrak{a}_i$ is the ideal generated by the union $\bigcup_i \mathfrak{a}_i$. Third, since $\mathfrak{ab} \subseteq \mathfrak{a} \cap \mathfrak{b} \subseteq \mathfrak{a}, \mathfrak{b}$, it is clear that $V(\mathfrak{a}) \cup V(\mathfrak{b}) \subseteq V(\mathfrak{a} \cap \mathfrak{b}) \subseteq V(\mathfrak{ab})$. If conversely $x \in V(\mathfrak{ab})$ and $x \notin V(\mathfrak{a})$, there exists an $f \in \mathfrak{a}$ with $f(x) \neq 0$, and for all $g \in \mathfrak{b}$ we have $fg \in \mathfrak{ab}$ and hence $f(x)g(x) = (fg)(x) = 0$. Therefore $g(x) = 0$ and hence $x \in V(\mathfrak{b})$. $\qquad\square$

From now on we will consider k^n always as a topological space with the Zariski topology and we will denote this space by $\mathbb{A}^n(k)$. We call this space the *affine space of dimension n (over k)*. The phrase *of dimension n* should be understood as fixed expression for now. Only later we will introduce the notion of dimension (and then of course prove that $\mathbb{A}^n(k)$ really has dimension n).

(1.2) Affine algebraic sets.

Definition 1.4. *Closed subspaces of $\mathbb{A}^n(k)$ are called* affine algebraic sets.

Sets consisting of one point $x = (x_1, \ldots, x_n) \in \mathbb{A}^n(k)$ are closed because $\{x\} = V(\mathfrak{m}_x)$, where $\mathfrak{m}_x = (T_1 - x_1, \ldots, T_n - x_n)$ is the kernel of the evaluation homomorphism $k[\underline{T}] \to k$ which sends f to $f(x)$. As finite unions of closed sets are again closed, we see that all finite subsets of $\mathbb{A}^n(k)$ are closed.

The Zariski topology has the advantage that it can be defined over arbitrary ground fields. On the other hand it is very coarse. Proposition 1.20 will show, that for $n > 0$ it is not Hausdorff. The following examples also show this for $n = 1, 2$.

Example 1.5. For $n = 1$ the polynomial ring $k[T]$ is a principal ideal domain. Therefore the closed subsets are of the form $V(f)$ for a polynomial $f \in k[T]$. As every polynomial $f \neq 0$ has only finitely many zeros, the closed subsets of $\mathbb{A}^1(k)$ are $\mathbb{A}^1(k)$ itself and the finite subsets of $\mathbb{A}^1(k)$.

Example 1.6. To describe the topological space $\mathbb{A}^2(k)$ is more difficult. We have the following list of obvious closed subsets.

- $\mathbb{A}^2(k)$.
- Sets consisting of one point $\{x\} = V(\mathfrak{m}_x)$.
- $V(f)$, $f \in k[T_1, T_2]$ an irreducible polynomial.

We will see later that every closed set is a finite union of sets from this list. In fact the above sets are the closed subsets of the form $V(\mathfrak{p})$ where $\mathfrak{p} \subset k[T_1, T_2]$ is a prime ideal, i.e., the "irreducible" closed subsets (a notion which will be explained in Section (1.5)). This will follow from the fact that in $k[T_1, T_2]$ all non-maximal prime ideals are principal ideals (Proposition 5.31).

(1.3) Hilbert's Nullstellensatz.

As mentioned above, the connection between affine algebraic sets and commutative algebra is established by Hilbert's Nullstellensatz (and its corollaries).

Theorem 1.7. (Hilbert's Nullstellensatz) *Let K be a (not necessarily algebraically closed) field and let A be a finitely generated K-algebra. Then A is Jacobson, that is, for every prime ideal $\mathfrak{p} \subset A$ we have*

$$\mathfrak{p} = \bigcap_{\substack{\mathfrak{m} \supseteq \mathfrak{p} \\ maximal\ ideal}} \mathfrak{m}.$$

If $\mathfrak{m} \subset A$ is a maximal ideal, the field extension $K \subseteq A/\mathfrak{m}$ is finite.

We will base the proof of the theorem on Noether's normalization theorem. Recall that a homomorphism of rings $R \to R'$ is called *integral* if each element of R' is a root of a monic polynomial with coefficients in R. A homomorphism of rings $R \to R'$ is *finite* if it is integral and R' is generated as an R-algebra by finitely many elements (Section (B.10)). We will study this notion in more detail later (see Chapter 12) where we will also obtain a geometric interpretation. Here we remark only that $R \to R'$ is finite if and only if R' is finitely generated as an R-module (Section (B.3)).

Note that this notion of *integral* is not related to the notion of *integral domain*. Below we often use the term *integral K-algebra* (where K is a field) by which we mean a K-algebra which is an integral domain.

Theorem 1.8. (Noether's normalization theorem) *Let K be a field and let $A \neq 0$ be a finitely generated K-algebra. Then there exists an integer $n \geq 0$ and $t_1, \ldots, t_n \in A$ such that the K-algebra homomorphism $K[T_1, \ldots, T_n] \to A$, $T_i \mapsto t_i$ is injective and finite.*

We will not prove this theorem but refer to Theorem B.58. To deduce the Nullstellensatz from Noether's normalization theorem we will first show two lemmas.

Lemma 1.9. *Let A and B be integral domains and let $A \to B$ be an injective integral ring homomorphism. Then A is a field if and only if B is a field.*

Proof. Let A be a field and $b \in B$ nonzero. Then $A[b]$ is an A-vector space of finite dimension. As B is an integral domain, the multiplication $A[b] \to A[b]$ with b is injective. It is clearly A-linear and therefore it is bijective. This shows that b is a unit.

Conversely let B be a field and let $a \in A \setminus \{0\}$. The element $a^{-1} \in B^{\times}$ satisfies a polynomial identity $(a^{-1})^n + \beta_{n-1}(a^{-1})^{n-1} + \cdots + \beta_0 = 0$, $\beta_i \in A$. Therefore we have

$$a^{-1} = -(\beta_{n-1} + \beta_{n-2}a + \cdots + \beta_0 a^{n-1}) \in A. \qquad \square$$

Lemma 1.10. *Let K be a field and let L be a field extension of K that is a finitely generated K-algebra. Then L is a finite extension of K.*

Proof. We apply to L Noether's normalization theorem and obtain a finite injective homomorphism $K[T_1, \ldots, T_n] \to L$ of K-algebras. By Lemma 1.9 we must have $n = 0$ which shows that $K \to L$ is a finite extension. $\square$

Proof. (Hilbert's Nullstellensatz) Lemma 1.10 implies at once the second assertion: If $\mathfrak{m} \subset A$ is a maximal ideal, $A/\mathfrak{m}$ is a field extension of K which is finitely generated as a K-algebra.

For the proof of the first assertion we start with a remark. If L is a finite field extension of K and $\varphi \colon A \to L$ is a K-algebra homomorphism, the image of φ is an integral domain that is finite over K. Thus $\mathrm{Im}\,\varphi$ is a field by Lemma 1.9 and therefore $\mathrm{Ker}\,\varphi$ is a maximal ideal of A.

We now show that A is Jacobson. Let $\mathfrak{p} \subset A$ be a prime ideal. Replacing A by $A/\mathfrak{p}$ it suffices to show that in an integral finitely generated K-algebra the intersection of all maximal ideals is the zero ideal. For $x \neq 0$, $A[x^{-1}]$ is a finitely generated K-algebra $\neq 0$. Let $\mathfrak{n}$ be a maximal ideal of $A[x^{-1}]$, then $L := A[x^{-1}]/\mathfrak{n}$ is a finite extension of K by the second assertion of the Nullstellensatz. The kernel of the composition $\varphi \colon A \to A[x^{-1}] \to L$ is a maximal ideal by the above remark, and it does not contain x. $\square$

If $K = k$ is an algebraically closed field, the Nullstellensatz implies:

Corollary 1.11.
(1) *Let A be a finitely generated k-algebra, $\mathfrak{m} \subset A$ a maximal ideal. Then $A/\mathfrak{m} = k$.*
(2) *Let $\mathfrak{m} \subset k[T_1, \ldots, T_n]$ be a maximal ideal. Then there exists $x = (x_1, \ldots, x_n) \in \mathbb{A}^n(k)$ such that $\mathfrak{m} = \mathfrak{m}_x := (T_1 - x_1, \ldots, T_n - x_n)$.*

Proof. *(1).* As k is algebraically closed, the homomorphism $k \to A \to A/\mathfrak{m}$, which makes $A/\mathfrak{m}$ into a finite field extension of k by the Nullstellensatz, has to be an isomorphism.

(2). Let x_i be the image of T_i by the homomorphism $k[T_1, \ldots, T_n] \to k[\underline{T}]/\mathfrak{m} = k$. Then $\mathfrak{m}$ is a maximal ideal which contains the maximal ideal $\mathfrak{m}_x = (T_1 - x_1, \ldots, T_n - x_n)$. Therefore both are equal. $\square$

(1.4) The correspondence between radical ideals and affine algebraic sets.

To understand the affine algebraic sets $V(\mathfrak{a})$ better, we need the notion of the radical of an ideal: Let A be a ring. Recall that if $\mathfrak{a} \subseteq A$ is an ideal, we call

$$\mathrm{rad}\,\mathfrak{a} := \{ f \in A \;;\; \exists r \in \mathbb{Z}_{\geq 0} : f^r \in \mathfrak{a} \}.$$

the *radical of* $\mathfrak{a}$. It is easy to see that $\mathrm{rad}\,\mathfrak{a}$ is an ideal and that we have $\mathrm{rad}(\mathrm{rad}\,\mathfrak{a}) = \mathrm{rad}\,\mathfrak{a}$. If A is a finitely generated K-algebra for a field K we have

$$(1.4.1) \qquad \mathrm{rad}\,\mathfrak{a} = \bigcap_{\substack{\mathfrak{a} \subseteq \mathfrak{p} \subset A \\ \text{prime ideal}}} \mathfrak{p} = \bigcap_{\substack{\mathfrak{a} \subseteq \mathfrak{m} \subset A \\ \text{maximal ideal}}} \mathfrak{m}.$$

Indeed, the first equality holds in arbitrary commutative rings (B.1.1) and the second equality follows immediately from the Nullstellensatz.

We now study the question when two ideals describe the same closed subset of $\mathbb{A}^n(k)$. Clearly this may happen: As $f^r(x) = 0$ if and only if $f(x) = 0$, we always have the equality $V(\mathfrak{a}) = V(\mathrm{rad}\,\mathfrak{a})$. If $Z \subseteq \mathbb{A}^n(k)$ is a subset, we denote by

$$I(Z) := \{\, f \in k[\underline{T}] \; ; \; \forall x \in Z : f(x) = 0 \,\}$$

the ideal of functions that vanish on Z. For $f \in k[\underline{T}]$ and $x \in \mathbb{A}^n(k)$ we have $f(x) = 0$ if and only if $f \in \mathfrak{m}_x$. Thus we find

(1.4.2)
$$I(Z) = \bigcap_{x \in Z} \mathfrak{m}_x.$$

We have the following consequence of Hilbert's Nullstellensatz.

Proposition 1.12.
(1) *Let $\mathfrak{a} \subseteq k[\underline{T}]$ be an ideal. Then*

$$I(V(\mathfrak{a})) = \operatorname{rad} \mathfrak{a}.$$

(2) *Let $Z \subseteq \mathbb{A}^n(k)$ be a subset and let $\overline{Z}$ be its closure. Then*

$$V(I(Z)) = \overline{Z}.$$

Proof. (1). As $x \in V(\mathfrak{a})$ is equivalent to $\mathfrak{a} \subseteq \mathfrak{m}_x$, we have

$$I(V(\mathfrak{a})) \overset{(1.4.2)}{=} \bigcap_{x \in V(\mathfrak{a})} \mathfrak{m}_x = \bigcap_{\substack{\mathfrak{m} \supseteq \mathfrak{a} \\ \text{maximal ideal}}} \mathfrak{m} \overset{(1.4.1)}{=} \operatorname{rad} \mathfrak{a}.$$

(2). This is a simple assertion for which we do not need the Nullstellensatz. On one hand we have $Z \subseteq V(I(Z))$ and $V(I(Z))$ is closed. This shows $V(I(Z)) \supseteq \overline{Z}$. On the other hand let $V(\mathfrak{a}) \subseteq \mathbb{A}^n(k)$ be a closed subset that contains Z. Then we have $f(x) = 0$ for all $x \in Z$ and $f \in \mathfrak{a}$. This shows $\mathfrak{a} \subseteq I(Z)$ and hence $V(I(Z)) \subseteq V(\mathfrak{a})$. $\qquad\square$

If A is a ring, we call an ideal $\mathfrak{a} \subseteq A$ a *radical ideal* if $\mathfrak{a} = \operatorname{rad}(\mathfrak{a})$. This is equivalent to the property that $A/\mathfrak{a}$ is reduced (i.e., does not contain nilpotent elements $\neq 0$). In particular, every prime ideal is a radical ideal.

The proposition implies:

Corollary 1.13. *The maps*

$$\{radical\ ideals\ \mathfrak{a}\ of\ k[\underline{T}]\} \underset{I(Z) \leftarrow\!\shortmid Z}{\overset{\mathfrak{a} \mapsto V(\mathfrak{a})}{\rightleftarrows}} \{closed\ subsets\ Z\ of\ \mathbb{A}^n(k)\}$$

are mutually inverse bijections, whose restrictions define a bijection

$$\{maximal\ ideals\ of\ k[\underline{T}]\} \leftrightarrow \{points\ of\ \mathbb{A}^n(k)\}.$$

In the following sections we study further properties of the Zariski topology on $\mathbb{A}^n(k)$ and on affine algebraic sets. We will see that these spaces are quite different from Hausdorff spaces for which the notions of irreducible or noetherian spaces introduced below are uninteresting (see Exercise 1.3).

(1.5) Irreducible topological spaces.

Definition 1.14. *A non-empty topological space X is called* irreducible *if X cannot be expressed as the union of two proper closed subsets. A non-empty subset Z of X is called* irreducible *if Z is irreducible when we endow it with the induced topology.*

Proposition 1.15. *Let X be a non-empty topological space. The following assertions are equivalent.*
(i) *X is irreducible.*
(ii) *Any two non-empty open subsets of X have a non-empty intersection.*
(iii) *Every non-empty open subset is dense in X.*
(iv) *Every non-empty open subset is connected.*
(v) *Every non-empty open subset is irreducible.*

Proof. Taking complements the equivalence of (i) and (ii) is immediate. A subset of X is dense if and only if it meets every non-empty open subset of X. This shows that (ii) and (iii) are equivalent. If there exist non-empty open subsets U_1 and U_2 that have an empty intersection, their union is a non-connected open subset. Conversely if U is a non-empty non-connected subset we can write U as the disjoint union of two non-empty open subsets of U (and hence of X). This shows that (iv) and (ii) are equivalent.

Obviously (v) implies (i). Let us show that (iii) implies (v). Let $U \subseteq X$ be open and non-empty. We show that every open non-empty subset $V \subseteq U$ is dense in U (this shows that U is irreducible as we have already seen that (iii) implies (i)). Now V is also open in X and therefore dense in X by (iii). But then V is certainly dense in U. $\square$

Corollary 1.16. *Let $f\colon X \to Y$ be a continuous map of topological spaces. If $Z \subseteq X$ is an irreducible subspace, its image $f(Z)$ is irreducible.*

Proof. If V_1 and V_2 are non-empty open subsets of $f(Z)$, their preimages in Z have a non-empty intersection. This shows that $V_1 \cap V_2 \neq \emptyset$. $\square$

Lemma 1.17. *Let X be a topological space. A subspace $Y \subseteq X$ is irreducible if and only if its closure $\overline{Y}$ is irreducible.*

Proof. By Proposition 1.15 (ii) a subset Z of X is irreducible if and only if for any two open subsets U and V of X with $Z \cap U \neq \emptyset$ and $Z \cap V \neq \emptyset$ we have $Z \cap (U \cap V) \neq \emptyset$. This implies the lemma because an open subset meets Y if and only if it meets $\overline{Y}$. $\square$

If $U \subseteq X$ is an open subset and $Z \subseteq X$ is irreducible and closed, $Z \cap U$ is open in Z and hence, if $Z \cap U \neq \emptyset$, an irreducible closed subset of U whose closure in X is Z. Together with Lemma 1.17 this shows that there are mutually inverse bijective maps

$$\{Y \subseteq U \text{ irreducible closed}\} \leftrightarrow \{Z \subseteq X \text{ irreducible closed with } Z \cap U \neq \emptyset\}$$
(1.5.1)
$$Y \mapsto \overline{Y} \quad (\text{closure in } X)$$
$$Z \cap U \leftarrow\!\shortmid Z$$

Definition 1.18. *A maximal irreducible subset of a topological space X is called an* irreducible component *of X.*

Let X be a topological space. Lemma 1.17 shows that every irreducible component is closed. The set of irreducible subsets of X is ordered inductively, as for every chain of irreducible subsets their union is again irreducible. It is non-empty since every singleton is irreducible. Thus Zorn's lemma implies that every irreducible subset is contained in an irreducible component of X. In particular, every point of X is contained in an irreducible component. This shows that X is the union of its irreducible components.

For later use, we record one more lemma.

Lemma 1.19. *Let X be a topological space and let $X = \bigcup_{i \in I} U_i$ be an open covering of X by connected open subsets U_i.*
(1) *If X is not connected, then there exists a subset $\emptyset \neq J \subsetneq I$ such that for all $j \in J$, $i \in I \setminus J$, $U_j \cap U_i = \emptyset$.*
(2) *If X is connected, I is finite, and all the U_i are irreducible, then X is irreducible.*

Proof. To prove (1), note that if we can write $X = V_1 \cup V_2$ as a disjoint union of open and closed subsets V_1, V_2, then each U_i is contained in either V_1 or V_2, so we can set $J = \{i \in I;\ U_i \subseteq V_1\}$. Now we prove the second part. If $Z \subseteq X$ is an irreducible component and $Z \cap U_i \neq \emptyset$, then $Z \cap U_i$ is dense in Z, so $Z \cap \overline{U_i} = \overline{Z \cap U_i} = Z$. It follows that $Z = \overline{U_i}$ by the maximality of Z and the irreducibility of U_i. In particular, X has only finitely many irreducible components, say $X_1, \ldots, X_n$. Assume $n > 1$. Since the X_i are closed, and X is connected, X_1 must intersect another irreducible component, so we find, say, $x \in X_1 \cap X_2$. Let $i \in I$ with $x \in U_i$. Then $U_i \cap X_1$ is open and hence dense in X_1, and similarly for X_2, so that the closure of U_i in X contains $X_1 \cup X_2$, a contradiction. $\qquad\square$

(1.6) Irreducible affine algebraic sets.

Proposition 1.20. *Let $Z \subseteq \mathbb{A}^n(k)$ be a closed subset. Then Z is irreducible if and only if $I(Z)$ is a prime ideal. In particular $\mathbb{A}^n(k)$ is irreducible.*

Proof. The subset Z is irreducible if and only if it is not union of two proper closed subsets. As every closed subset can be written as intersection of sets of the form $V(f)$, this is equivalent to the property that for any two elements $f, g \in k[T_1, \ldots, T_n]$ with $V(fg) = V(f) \cup V(g) \supseteq Z$ we have $V(f) \supseteq Z$ or $V(g) \supseteq Z$. But this means precisely that for any two polynomials f and g with $fg \in I(Z)$ we have $f \in I(Z)$ or $g \in I(Z)$, that is, that $I(Z)$ is a prime ideal. $\qquad\square$

Remark 1.21. The correspondence of Corollary 1.13 induces a bijection

$$\{\text{irreducible closed subsets of } \mathbb{A}^n(k)\} \leftrightarrow \{\text{prime ideals in } k[T_1, \ldots, T_n]\}.$$

(1.7) Quasi-compact and noetherian topological spaces.

Definition 1.22. *A topological space X is called* quasi-compact *if every open covering of X has a finite subcovering.*

Clearly any closed subspace of a quasi-compact space is again quasi-compact. An open subspace of a quasi-compact space is not necessarily quasi-compact (see however Lemma 1.25 below).

Definition 1.23. *A topological space X is called* noetherian *if every descending chain*

$$X \supseteq Z_1 \supseteq Z_2 \supseteq \cdots$$

of closed subsets of X becomes stationary.

Clearly, X is noetherian if and only if every non-empty set of closed subsets of X has a minimal element with respect to inclusion.

Lemma 1.24. *Let X be a topological space that has a finite covering $X = \bigcup_{i=1}^{r} X_i$ by noetherian subspaces. Then X itself is noetherian.*

Proof. Let $X \supseteq Z_1 \supseteq Z_2 \supseteq \cdots$ be a descending chain of closed subsets of X. Then $(Z_j \cap X_i)_j$ is a descending chain of closed subsets in X_i. Therefore there exists an integer $N_i \geq 1$ such that $Z_j \cap X_i = Z_{N_i} \cap X_i$ for all $j \geq N_i$. For $N = \max\{N_1, \ldots, N_r\}$ we have $Z_j = Z_N$ for all $j \geq N$. $\qquad\square$

Lemma 1.25. *Let X be a noetherian topological space.*
(1) *Every subspace of X is noetherian.*
(2) *Every subset of X is quasi-compact (in particular, X is quasi-compact).*
(3) *Every subset $Z \subseteq X$ has only finitely many irreducible components.*

Proof. *(1).* Let $(Z_i)_i$ be a descending chain of closed subsets of a subspace Y. Then the closures $\overline{Z}_i$ of Z_i in X form a descending chain of closed subsets of X which becomes stationary by hypothesis. As we have $Z_i = Y \cap \overline{Z}_i$, this shows that the chain $(Z_i)_i$ becomes stationary as well. This proves (1).

(2). By (1) it suffices to show that X is quasi-compact. Let $(U_i)_i$ be an open covering of X and let $\mathcal{U}$ be the set of those open subsets of X that are finite unions of the subsets U_i. As X is noetherian, $\mathcal{U}$ has a maximal element V. Clearly $V = X$, otherwise there would exist an U_i such that $V \subsetneq V \cup U_i \in \mathcal{U}$. This shows that $(U_i)_i$ has a finite subcovering.

(3). It suffices to show that every noetherian space X can be written as finite union of irreducible subsets. If the set $\mathcal{M}$ of closed subsets of X that cannot be written as a finite union of irreducible subsets were non-empty, there existed a minimal element $Z \in \mathcal{M}$. The set Z is not irreducible and thus union of two proper closed subsets which do not lie in $\mathcal{M}$. This leads to a contradiction. $\qquad\square$

Proposition 1.26. *Let $X \subseteq \mathbb{A}^n(k)$ be any subspace. Then X is noetherian.*

Proof. By Lemma 1.25 it suffices to show that $\mathbb{A}^n(k)$ is noetherian. But descending chains of closed subsets of $\mathbb{A}^n(k)$ correspond to ascending chains of radical ideals of $k[\underline{T}]$ (Corollary 1.13). As $k[\underline{T}]$ is noetherian by Hilbert's basis theorem, this proves the proposition. $\qquad\square$

By using the correspondence between (irreducible) closed subsets and (prime) radical ideals we obtain from the decomposition of an affine algebraic set into its irreducible components a weak version of the so-called primary decomposition in noetherian rings (e.g., see [AM] Chapter 4 and Chapter 7):

Corollary 1.27. *Let $\mathfrak{a} \subseteq k[T_1, \ldots, T_n]$ be a radical ideal, i.e., $\mathfrak{a} = \mathrm{rad}(\mathfrak{a})$. Then $\mathfrak{a}$ is the intersection of a finite number of prime ideals that do not contain each other. The set of these prime ideals is uniquely determined by $\mathfrak{a}$.*

(1.8) Morphisms of affine algebraic sets.

As affine algebraic sets are zero sets of polynomials, it is only natural to define morphisms between these sets as maps that are given by polynomials, more precisely:

Definition 1.28. *Let $X \subseteq \mathbb{A}^m(k)$ and $Y \subseteq \mathbb{A}^n(k)$ be affine algebraic sets. A morphism $X \to Y$ of affine algebraic sets is a map $f\colon X \to Y$ of the underlying sets such that there exist polynomials $f_1, \ldots, f_n \in k[T_1, \ldots, T_m]$ with $f(x) = (f_1(x), \ldots, f_n(x))$ for all $x \in X$.*

We denote the set of morphisms from X to Y with $\operatorname{Hom}(X, Y)$.

Remark 1.29. The definition shows that a morphism between affine algebraic sets $X \subseteq \mathbb{A}^m(k)$ and $Y \subseteq \mathbb{A}^n(k)$ can always be extended to a morphism $\mathbb{A}^m(k) \to \mathbb{A}^n(k)$ (but not in a unique way unless $X = \mathbb{A}^m(k)$). If $f = (f_1, \ldots, f_n)$ is a tuple of polynomials $f_i \in k[T_1, \ldots, T_m]$ defining a morphism $\mathbb{A}^m(k) \to \mathbb{A}^n(k)$, we obtain a k-algebra homomorphism $\Gamma(f)\colon k[T_1', \ldots, T_n'] \to k[T_1, \ldots, T_m]$ by sending T_i' to f_i. If $V(\mathfrak{a}) \subseteq \mathbb{A}^n(k)$ is a closed subset, then $f^{-1}(V(\mathfrak{a})) = V(\Gamma(f)(\mathfrak{a}))$ is again closed. This shows that morphisms of affine algebraic sets are continuous.

Let $X \subseteq \mathbb{A}^m(k)$, $Y \subseteq \mathbb{A}^n(k)$ and $Z \subseteq \mathbb{A}^r(k)$ be affine algebraic sets and suppose $f\colon X \to Y$ and $g\colon Y \to Z$ are morphisms given by polynomials $f_1, \ldots, f_n \in k[T_1, \ldots, T_m]$ and $g_1, \ldots, g_r \in k[T_1', \ldots, T_n']$. Then we have for $x \in X$:

$$(1.8.1) \qquad g(f(x)) = \big(g_1\big(f_1(x), \ldots, f_n(x)\big), \ldots, g_r\big(f_1(x), \ldots, f_n(x)\big)\big).$$

Therefore $g \circ f$ is given by the polynomials $h_i \in k[T_1, \ldots, T_m]$ $(i = 1, \ldots, r)$ that are obtained from the g_i by replacing the indeterminate T_j' with f_j for $j = 1, \ldots, n$. In particular, $g \circ f$ is again a morphism of affine algebraic sets. We obtain the category of affine algebraic sets.

We give some examples of morphisms of affine algebraic sets.

(1) The map $\mathbb{A}^1(k) \to V(T_1 - T_2^2) \subset \mathbb{A}^2(k)$, $x \mapsto (x^2, x)$ is a morphism of affine algebraic sets. It is even an isomorphism with inverse morphism $(x, y) \mapsto y$. In general a bijective morphism of affine algebraic sets is not an isomorphism (see Exercise 1.12).

(2) The map $\mathbb{A}^1(k) \to V(T_2^2 - T_1^2(T_1 + 1))$, $x \mapsto (x^2 - 1, x(x^2 - 1))$ is a morphism. For $\operatorname{char}(k) \neq 2$ it is not bijective: 1 and -1 are both mapped to the origin $(0, 0)$. In $\operatorname{char}(k) = 2$ it is bijective but not an isomorphism.

(3) We identify the space $M_n(k)$ of $(n \times n)$-matrices with $\mathbb{A}^{n^2}(k)$, thus giving $M_n(k)$ the structure of an affine algebraic set. Then sending a matrix $A \in M_n(k)$ to its determinant $\det(A)$ is a morphism $M_n(k) \to \mathbb{A}^1(k)$ of affine algebraic sets.

(4) For $k = \mathbb{C}$ consider the exponential function $\exp\colon \mathbb{A}^1(\mathbb{C}) \to \mathbb{A}^1(\mathbb{C})$. This is *not* a morphism of algebraic sets (Exercise 1.17).

(1.9) Shortcomings of the notion of affine algebraic sets.

The notion of an affine algebraic set is still not satisfactory. We list three problems:

- Open subsets of affine algebraic sets do not carry the structure of an affine algebraic set in a natural way. In particular we cannot glue affine algebraic sets along open subsets (although this is a "natural operation" for geometric objects).

- Intersections of affine algebraic sets in $\mathbb{A}^n(k)$ are closed and hence again affine algebraic sets. But we cannot distinguish between $V(X) \cap V(Y) \subset \mathbb{A}^2(k)$ and $V(Y) \cap V(X^2 - Y) \subset \mathbb{A}^2(k)$ although the geometric situation seems to be different (we will see similar phenomena later when we study fibers of morphisms).

- Affine algebraic sets seem not to help in studying solutions of polynomial equations in more general rings than algebraically closed fields.

The first problem is due to the fact that affine algebraic sets are necessarily embedded in an affine space. This problem will be solved in the following sections. To deal with the second and the third problem is more difficult and part of the motivation to introduce in Chapter 3 the notion of a scheme.

Affine algebraic sets as spaces with functions

Having defined morphisms between algebraic sets in Section (1.8), we can in particular speak of functions on an affine algebraic set X, i.e., morphisms $X \to \mathbb{A}^1(k)$. These functions form a reduced finitely generated k-algebra $\Gamma(X)$. We will show that this construction yields a contravariant equivalence between the category of affine algebraic sets and the category of reduced finitely generated k-algebras. This is another incarnation of the correspondence of algebraic and of geometric objects.

Next we introduce the algebra of functions $\mathscr{O}_X(U)$ on an open subset U of an irreducible affine algebraic set X. Thus we obtain a topological space X together with a k-algebra of function $\mathscr{O}_X(U)$ for every open subset $U \subseteq X$. This is similar to the language of real smooth manifolds which can also be considered as topological spaces M together with the $\mathbb{R}$-algebras $\mathcal{C}^\infty(U)$ of smooth functions on open subsets $U \subseteq M$. We formalize this concept by introducing the notion of a space with functions. A similar notion ("système local de fonctions") has already been introduced in the Séminaire de Chevalley [Ch]. Although all (real or complex) manifolds, all irreducible algebraic sets, and all prevarieties (defined later in this chapter) are spaces with functions, this concept will be only a stepping stone for us to motivate the notion of ringed spaces that we will need to define schemes. Ringed spaces will be defined in Chapter 2.

Our hypothesis that the algebraic set X is irreducible will not be strictly necessary but it will make the construction of $\mathscr{O}_X$ easier and more explicit. In later chapters, in which we use the languages of schemes, we will get rid of this hypothesis (and several others).

(1.10) The affine coordinate ring.

Let $X \subseteq \mathbb{A}^n(k)$ be a closed subspace. Every polynomial $f \in k[T_1, \ldots, T_n]$ induces a morphism $X \to \mathbb{A}^1(k)$, $x \mapsto f(x)$, of affine algebraic sets. The set $\mathrm{Hom}(X, \mathbb{A}^1(k))$ carries in a natural way the structure of a k-algebra with addition and multiplication

$$(f + g)(x) = f(x) + g(x), \quad (fg)(x) = f(x)g(x).$$

To elements of k we associate the corresponding constant function. The homomorphism $k[\underline{T}] \to \mathrm{Hom}(X, \mathbb{A}^1(k))$ is a surjective homomorphism of k-algebras with kernel $I(X)$.

Definition 1.30. *Let* $X \subseteq \mathbb{A}^n(k)$ *be an affine algebraic set. The k-algebra*

$$\Gamma(X) := k[T_1, \ldots, T_n]/I(X) \cong \mathrm{Hom}(X, \mathbb{A}^1(k))$$

is called the affine coordinate ring *of* X.

For $x = (x_1, \ldots, x_n) \in X$ we denote by $\mathfrak{m}_x$ the ideal

$$\mathfrak{m}_x = \{\, f \in \Gamma(X) \; ; \; f(x) = 0 \,\} \subset \Gamma(X).$$

It is the image of the maximal ideal $(T_1 - x_1, \ldots, T_n - x_n)$ of $\Gamma(\mathbb{A}^n(k)) = k[\underline{T}]$ under the projection $\pi \colon k[\underline{T}] \to \Gamma(X)$. In other words, $\mathfrak{m}_x$ is the kernel of the evaluation homomorphism $\Gamma(X) \to k$, $f \mapsto f(x)$. As the evaluation homomorphism is clearly surjective, $\mathfrak{m}_x$ is a maximal ideal and we find $\Gamma(X)/\mathfrak{m}_x = k$.

If $\mathfrak{a} \subseteq \Gamma(X)$ is an ideal, consider

$$V(\mathfrak{a}) = \{\, x \in X \; ; \; \forall f \in \mathfrak{a} : f(x) = 0 \,\} = V(\pi^{-1}(\mathfrak{a})) \cap X.$$

Thus the $V(\mathfrak{a})$ are precisely the closed subsets of X if we consider X as a subspace of $\mathbb{A}^n(k)$. This topology is again called the *Zariski topology*. For $f \in \Gamma(X)$ we set

$$D(f) := \{\, x \in X \; ; \; f(x) \neq 0 \,\} = X \setminus V(f).$$

These are open subsets of X, called *principal open subsets*.

Lemma 1.31. *The open sets $D(f)$, $f \in \Gamma(X)$, form a basis of the topology (i.e., for every open subset $U \subseteq X$ there exist $f_i \in \Gamma(X)$, $i \in I$, with $U = \bigcup_i D(f_i)$). Finite intersections of principal open subsets are again principal open.*

Proof. Clearly we have $D(f) \cap D(g) = D(fg)$ for $f, g \in \Gamma(X)$. It remains to show the first statement: Every open subset U is a union of principal open subsets. We write $U = X \setminus V(\mathfrak{a})$ for some ideal $\mathfrak{a}$. For generators $f_1, \ldots, f_n$ of this ideal we find $V(\mathfrak{a}) = \bigcap_{i=1}^n V(f_i)$, and hence $U = \bigcup_{i=1}^n D(f_i)$. $\qquad\square$

Proposition 1.32. *Let X be an affine algebraic set. The affine coordinate ring $\Gamma(X)$ is a reduced finitely generated k-algebra. Moreover, X is irreducible if and only if $\Gamma(X)$ is an integral domain.*

Proof. As $\Gamma(X) = k[\underline{T}]/I(X)$, it is a finitely generated k-algebra. As $I(X) = \mathrm{rad}(I(X))$, we find that $\Gamma(X)$ is reduced. Proposition 1.20 shows that X is irreducible if and only if $I(X)$ is a prime ideal, that is, if and only if $\Gamma(X)$ is an integral domain. $\qquad\square$

(1.11) The equivalence between the category of affine algebraic sets and reduced finitely generated algebras.

Let $f \colon X \to Y$ be a morphism of affine algebraic sets. The map

$$\Gamma(f) \colon \mathrm{Hom}(Y, \mathbb{A}^1(k)) \to \mathrm{Hom}(X, \mathbb{A}^1(k)), \quad g \mapsto g \circ f$$

defines a homomorphism of k-algebras. We obtain a functor

Γ: (affine algebraic sets)$^{\text{opp}} \to$ (reduced finitely generated k-algebras).

Proposition 1.33. *The functor Γ induces an equivalence of categories. By restriction one obtains an equivalence of categories*

Γ: (*irreducible affine algebraic sets*)$^{\text{opp}} \to$ (*integral finitely generated k-algebras*).

Proof. We show that Γ is fully faithful, i.e., that for affine algebraic sets $X \subseteq \mathbb{A}^m(k)$, $Y \subseteq \mathbb{A}^n(k)$ the map Γ: $\mathrm{Hom}(X,Y) \to \mathrm{Hom}(\Gamma(Y),\Gamma(X))$ is bijective. We define an inverse map. If $\varphi \colon \Gamma(Y) \to \Gamma(X)$ is given, there exists a k-algebra homomorphism $\tilde{\varphi}$ that makes the following diagram commutative

$$
\begin{array}{ccc}
k[T_1', \ldots, T_n'] & \xrightarrow{\ \tilde{\varphi}\ } & k[T_1, \ldots, T_m] \\
\downarrow & & \downarrow \\
\Gamma(Y) & \xrightarrow{\ \varphi\ } & \Gamma(X).
\end{array}
$$

We define $f \colon X \to Y$ by

$$f(x) := (\tilde{\varphi}(T_1')(x), \ldots, \tilde{\varphi}(T_n')(x))$$

and obtain the desired inverse map.

It remains to show that the functor is essentially surjective, i.e., that for every reduced finitely generated k-algebra A there exists an affine algebraic set X such that $A \cong \Gamma(X)$. By hypothesis, A is isomorphic to $k[T_1, \ldots, T_n]/\mathfrak{a}$, where $\mathfrak{a} \subseteq k[\underline{T}]$ is an ideal with $\mathfrak{a} = \mathrm{rad}\,\mathfrak{a}$. If we set $X = V(\mathfrak{a}) \subseteq \mathbb{A}^n(k)$, we have $\Gamma(X) = k[T_1, \ldots, T_n]/\mathfrak{a}$.

That this equivalence induces an equivalence of the category of irreducible affine algebraic sets with the category of integral finitely generated k-algebras follows from Proposition 1.32. $\qquad\square$

Using the bijective correspondence between points of affine algebraic sets X and maximal ideals of $\Gamma(X)$, we also have the following description of morphisms.

Proposition 1.34. *Let $f \colon X \to Y$ be a morphism of affine algebraic sets and let $\Gamma(f) \colon \Gamma(Y) \to \Gamma(X)$ be the corresponding homomorphism of the affine coordinate rings. Then $\Gamma(f)^{-1}(\mathfrak{m}_x) = \mathfrak{m}_{f(x)}$ for all $x \in X$.*

Proof. This follows from $g(f(x)) = \Gamma(f)(g)(x)$ for $g \in \Gamma(Y) = \mathrm{Hom}(Y, \mathbb{A}^1(k))$. $\qquad\square$

(1.12) Definition of spaces with functions.

We will now define the notion of a *space with functions*. For us this will be the prototype of a "geometric object". It is a special case of a so-called ringed space on which the notion of a scheme will be based.

Definition 1.35. *Let K be a field.*
(1) *A space with functions over K is a topological space X together with a family $\mathcal{O}_X$ of K-subalgebras $\mathcal{O}_X(U) \subseteq \mathrm{Map}(U,K)$ for every open subset $U \subseteq X$ that satisfy the following properties:*

(a) *If $U' \subseteq U \subseteq X$ are open and $f \in \mathscr{O}_X(U)$, the restriction $f_{|U'} \in \mathrm{Map}(U', K)$ is an element of $\mathscr{O}_X(U')$.*

(b) *(Axiom of Gluing) Given open subsets $U_i \subseteq X$, $i \in I$, and $f_i \in \mathscr{O}_X(U_i)$, $i \in I$, with*

$$f_{i|U_i \cap U_j} = f_{j|U_i \cap U_j} \quad \text{for all } i, j \in I,$$

the unique function $f \colon \bigcup_i U_i \to K$ with $f_{|U_i} = f_i$ for all $i \in I$ lies in $\mathscr{O}_X(\bigcup_i U_i)$.
The space with functions $(X, \mathscr{O}_X)$ will often be simply denoted by X.

(2) *A morphism $g \colon (X, \mathscr{O}_X) \to (Y, \mathscr{O}_Y)$ of spaces with functions is a continuous map $g \colon X \to Y$ such that for all open subsets $V \subseteq Y$ and functions $f \in \mathscr{O}_Y(V)$ the function $f \circ g_{|g^{-1}(V)} \colon g^{-1}(V) \to K$ lies in $\mathscr{O}_X(g^{-1}(V))$.*

Clearly spaces with function over K form a category.

Definition 1.36. *Let X be a space with functions and let $U \subseteq X$ be an open subspace. We denote by $(U, \mathscr{O}_{X|U})$ the space U with functions*

$$\mathscr{O}_{X|U}(V) = \mathscr{O}_X(V) \quad \text{for } V \subseteq U \text{ open.}$$

If not stated explicitly otherwise, from now on we will consider only spaces with functions over our fixed algebraically closed field k.

(1.13) The space with functions of an affine algebraic set.

Let $X \subseteq \mathbb{A}^n(k)$ be an irreducible affine algebraic set. It is endowed with the Zariski topology and we want to define for every open subset $U \subseteq X$ a k-algebra of functions $\mathscr{O}_X(U)$ such that $(X, \mathscr{O}_X)$ is a space with functions.

As X is irreducible, the k-algebra $\Gamma(X)$ is a domain, and by definition all the sets $\mathscr{O}_X(U)$ will be k-subalgebras of its field of fractions.

Definition 1.37. *The field of fractions $K(X) := \mathrm{Frac}(\Gamma(X))$ is called the* function field *of X.*

If we consider $\Gamma(X)$ as the set of morphisms $X \to \mathbb{A}^1(k)$, elements of the function field $\frac{f}{g}$, $f, g \in \Gamma(X)$, $g \neq 0$ usually do not define functions on X because the denominator may have zeros on X, but $\frac{f}{g}$ certainly defines a function $D(g) \to \mathbb{A}^1(k)$ (it might be even defined on a bigger open subset of X as there exist representations of the fraction with different denominators). We will use functions of this kind to make X into a space with functions.

Lemma 1.38. *Let X be an irreducible affine algebraic set and let $\frac{f_1}{g_1}$ and $\frac{f_2}{g_2}$ be elements of $K(X)$ ($f_1, f_2, g_1, g_2 \in \Gamma(X)$), such that there exists a non-empty open subset $U \subseteq D(g_1 g_2)$ with:*

$$\forall x \in U : \quad \frac{f_1(x)}{g_1(x)} = \frac{f_2(x)}{g_2(x)}.$$

Then $\frac{f_1}{g_1} = \frac{f_2}{g_2}$ in $K(X)$.

Proof. The closed subset $V(f_1 g_2 - f_2 g_1)$ of X contains the dense subset U and is hence equal to X. That implies that $f_1 g_2 - f_2 g_1 = 0$, because $\Gamma(X)$ is reduced. The lemma follows. $\qquad\square$

Definition 1.39. *Let X be an irreducible affine algebraic set and let $\emptyset \neq U \subseteq X$ be open. We denote by $\mathfrak{m}_x$ the maximal ideal of $\Gamma(X)$ corresponding to $x \in X$ and by $\Gamma(X)_{\mathfrak{m}_x}$ the localization of the affine coordinate ring with respect to $\mathfrak{m}_x$. We define*

$$\mathscr{O}_X(U) = \bigcap_{x \in U} \Gamma(X)_{\mathfrak{m}_x} \subset K(X).$$

We let $\mathscr{O}_X(\emptyset)$ be a singleton.

The localization $\Gamma(X)_{\mathfrak{m}_x}$ can be described in this situation as the union

$$\Gamma(X)_{\mathfrak{m}_x} = \bigcup_{f \in \Gamma(X) \setminus \mathfrak{m}_x} \Gamma(X)_f \subset K(X).$$

To consider $(X, \mathscr{O}_X)$ as space with functions, we first have to explain how to identify elements $f \in \mathscr{O}_X(U)$ with functions $U \to k$. Given $x \in U$ the element f is by definition in $\Gamma(X)_{\mathfrak{m}_x}$ and we may write $f = \frac{g}{h}$ with $g, h \in \Gamma(X)$, $h \notin \mathfrak{m}_x$. But then $h(x) \neq 0$ and we may set $f(x) := \frac{g(x)}{h(x)} \in k$. The value $f(x)$ is well defined and Lemma 1.38 implies that this construction defines an injective map $\mathscr{O}_X(U) \to \mathrm{Map}(U, k)$.

If $\emptyset \neq V \subseteq U \subseteq X$ are open subsets we have $\mathscr{O}_X(U) \subseteq \mathscr{O}_X(V)$ by definition and this inclusion corresponds via the identification with maps $U \to k$ resp. $V \to k$ to the restriction of functions.

To show that $(X, \mathscr{O}_X)$ is a space with functions, we still have to show that we may glue functions together. But this follows immediately from the definition of $\mathscr{O}_X(U)$ as subsets of the function field $K(X)$. We call $(X, \mathscr{O}_X)$ *the space with functions associated with X*. Functions on principal open subsets $D(f)$ can be explicitly described as follows.

Proposition 1.40. *Let $(X, \mathscr{O}_X)$ be the space with functions associated to the irreducible affine algebraic set X and let $f \in \Gamma(X)$. Then there is an equality*

$$\mathscr{O}_X(D(f)) = \Gamma(X)_f$$

(as subsets of $K(X)$). In particular $\mathscr{O}_X(X) = \Gamma(X)$ (taking $f = 1$).

Proof. Clearly we have $\Gamma(X)_f \subseteq \mathscr{O}_X(D(f))$. Let $g \in \mathscr{O}_X(D(f))$ and set

$$\mathfrak{a} = \{ h \in \Gamma(X) \ ; \ hg \in \Gamma(X) \}.$$

Obviously $\mathfrak{a}$ is an ideal of $\Gamma(X)$ and we have to show that $f \in \mathrm{rad}(\mathfrak{a})$. By Hilbert's Nullstellensatz we have $\mathrm{rad}(\mathfrak{a}) = I(V(\mathfrak{a}))$. Therefore it suffices to show $f(x) = 0$ for all $x \in V(\mathfrak{a})$. Let $x \in X$ be a point with $f(x) \neq 0$, i.e., $x \in D(f)$. As $g \in \mathscr{O}_X(D(f))$, we find $g_1, g_2 \in \Gamma(X)$, $g_2 \notin \mathfrak{m}_x$, with $g = \frac{g_1}{g_2}$. Thus $g_2 \in \mathfrak{a}$ and as $g_2(x) \neq 0$ we have $x \notin V(\mathfrak{a})$. $\square$

Remark 1.41. If X is an irreducible affine algebraic set, $U \subseteq X$ open, and $f \in \mathscr{O}_X(U)$, there do not necessarily exist $g, h \in \Gamma(X)$ with $f = \frac{g}{h} \in K(X)$ and $h(x) \neq 0$ for all $x \in U$. Only locally on U we can always find such a representation of f. An example for this situation will be given when we learn dimension theory (Example 5.36). At least, it is easy to see that this problem cannot occur if $\Gamma(X)$ is factorial, e.g. if $X = \mathbb{A}^n(k)$.

Remark 1.42. The proposition shows that we could have defined $(X, \mathscr{O}_X)$ also in another way, namely by setting

$$\mathscr{O}_X(D(f)) = \Gamma(X)_f \quad \text{for } f \in \Gamma(X).$$

As the $D(f)$ for $f \in \Gamma(X)$ form a basis of the topology, the axiom of gluing implies that at most one such space with functions can exist. It would remain to show the existence of such a space (i.e., that for $f, g \in \Gamma(X)$ with $D(f) = D(g)$ we have $\Gamma(X)_f = \Gamma(X)_g$ and that gluing of functions is possible). This is more or less the same as the proof of Proposition 1.40. The way we chose is more comfortable in our situation. For affine schemes we will use the other approach (see Chapter 2).

Remark 1.43. If A is an integral finitely generated k-algebra we may construct the space with functions $(X, \mathscr{O}_X)$ of "the" corresponding irreducible affine algebraic set (uniquely determined up to isomorphism by Proposition 1.33) directly without choosing generators of A. Namely, we obtain X as the set of maximal ideals in A. Closed subsets of X are sets of the form

$$V(\mathfrak{a}) = \{\, \mathfrak{m} \subset A \text{ maximal} \;;\; \mathfrak{m} \supseteq \mathfrak{a} \,\}, \quad \mathfrak{a} \subseteq A \text{ an ideal.}$$

For an open subset $U \subseteq X$ we finally define

$$\mathscr{O}_X(U) = \bigcap_{\mathfrak{m} \in U} A_{\mathfrak{m}} \subset \operatorname{Frac}(A).$$

This defines a space with functions $(X, \mathscr{O}_X)$ which coincides with the space with functions of the irreducible affine algebraic set X corresponding to A. This approach is the point of departure for the definition of schemes.

(1.14) The functor from the category of irreducible affine algebraic sets to the category of spaces with functions.

Proposition 1.44. *Let X, Y be irreducible affine algebraic sets and $f \colon X \to Y$ a map. The following assertions are equivalent.*
(i) *The map f is a morphism of affine algebraic sets.*
(ii) *If $g \in \Gamma(Y)$, then $g \circ f \in \Gamma(X)$.*
(iii) *The map f is a morphism of spaces with functions, i.e., f is continuous and if $U \subseteq Y$ open and $g \in \mathscr{O}_Y(U)$, then $g \circ f_{|f^{-1}(U)} \in \mathscr{O}_X(f^{-1}(U))$.*

Proof. The equivalence of (i) and (ii) has already been proved in Proposition 1.33. Moreover, it is clear that (ii) is implied by (iii) by taking $U = Y$. Let us show that (ii) implies (iii). Let $\varphi \colon \Gamma(Y) \to \Gamma(X)$ be the homomorphism $h \mapsto h \circ f$. For $g \in \Gamma(Y)$ we have

$$f^{-1}(D(g)) = \{\, x \in X \;;\; g(f(x)) \neq 0 \,\} = D(\varphi(g)).$$

As the principal open subsets form a basis of the topology, this shows that f is continuous. The homomorphism φ induces a homomorphism of the localizations $\Gamma(Y)_g \to \Gamma(X)_{\varphi(g)}$. By definition of φ this is the map $\mathscr{O}_Y(D(g)) \to \mathscr{O}_X(D(\varphi(g)))$, $h \mapsto h \circ f$. This shows the claim if U is principal open. As we can obtain functions on arbitrary open subsets of Y by gluing functions on principal open subsets, this proves (iii). $\square$

Altogether we obtain

Theorem 1.45. *The above construction $X \mapsto (X, \mathscr{O}_X)$ defines a fully faithful functor*

$$(\text{Irreducible affine algebraic sets}) \to (\text{Spaces with functions over } k).$$

Prevarieties

We have seen that we can embed the category of irreducible affine algebraic sets into the category of spaces with functions. Of course we do not obtain all spaces with functions in this way. We will now define prevarieties as those connected spaces with functions that can be glued together from finitely many spaces with functions attached to irreducible affine algebraic sets. This is similar to the way a differentiable manifold can be glued from open subsets of $\mathbb{R}^n$ endowed with their differentiable structure (see Remark 1.49).

(1.15) Definition of prevarieties.

We call a space with functions $(X, \mathscr{O}_X)$ *connected*, if the underlying topological space X is connected.

Definition 1.46.
(1) *An* affine variety *is a space with functions that is isomorphic to a space with functions associated to an irreducible affine algebraic set.*
(2) *A* prevariety *is a connected space with functions* $(X, \mathscr{O}_X)$ *with the property that there exists a finite open covering* $X = \bigcup_{i=1}^n U_i$ *such that the space with functions* $(U_i, \mathscr{O}_{X|U_i})$ *is an affine variety for all* $i = 1, \dots, n$.
(3) *A* morphism of prevarieties *is a morphism of spaces with functions.*

We remind the reader that by convention the empty topological space is not connected, whence the empty space with functions is not a prevariety.

We obtain the category of prevarieties. Clearly affine varieties are examples of prevarieties. At this moment we cannot explain why we speak of affine varieties instead of affine prevarieties. Later (in Chapter 9) we will define varieties as "separated" prevarieties and see that affine varieties in the above sense are always "separated".

If X is an affine variety, we often write $\Gamma(X)$ instead of $\mathscr{O}_X(X)$ as we have seen that $\mathscr{O}_X(X)$ is the affine coordinate ring of the corresponding irreducible affine algebraic set.

By Proposition 1.33 and Theorem 1.45 we obtain:

Corollary 1.47. *The following categories are equivalent.*
(i) *The opposite category of the category of integral finitely generated k-algebras.*
(ii) *The category of irreducible affine algebraic sets.*
(iii) *The category of affine varieties.*

We define an *open affine covering of a prevariety* X to be a family of open subspaces with functions $U_i \subseteq X$, $i \in I$ that are affine varieties such that $X = \bigcup_i U_i$.

Proposition 1.48. *Let* $(X, \mathscr{O}_X)$ *be a prevariety. The topological space X is noetherian (in particular quasi-compact) and irreducible.*

Proof. The first assertion follows from Lemma 1.24, the second one from Lemma 1.19. $\square$

Remark 1.49. (Comparison with differential/complex manifolds) In differential geometry (resp. complex geometry) the notion of a differentiable manifold (resp. a complex manifold) is often defined by charts with differentiable (resp. holomorphic) transition maps. This is problematic in our situation because we cannot consider open subsets of affine algebraic sets again as affine algebraic sets. But on the other hand it is possible to use our approach in differential or complex geometry.

If we define for a differentiable manifold X the system $\mathscr{O}_X$ of $\mathbb{R}$-valued functions by $\mathscr{O}_X(U) = C^\infty(U)$ for $U \subseteq X$ open, we obtain a fully faithful functor $X \mapsto (X, \mathscr{O}_X)$ from the category of differentiable manifolds into the category of spaces with functions over $\mathbb{R}$. Thus one could define differentiable manifolds also as those spaces with functions over $\mathbb{R}$ whose underlying topological space is Hausdorff and that have open coverings of those spaces with functions that are attached in the above way to open subsets of $\mathbb{R}^n$. Similarly, using holomorphic functions, one can define complex manifolds.

(1.16) Open Subprevarieties.

We are now able to endow open subsets of affine varieties, and more general of prevarieties, with the structure of a prevariety. Note that in general open subprevarieties of affine varieties are not affine, see Exercise 1.13.

Lemma 1.50. *Let X be an affine variety, $f \in \Gamma(X) = \mathscr{O}_X(X)$, and let $D(f) \subseteq X$ be the corresponding principal open subset. Let $\Gamma(X)_f$ be the localization of $\Gamma(X)$ by f and let $(Y, \mathscr{O}_Y)$ be the affine variety corresponding to this integral finitely generated k-algebra. Then $(D(f), \mathscr{O}_{X|D(f)})$ and $(Y, \mathscr{O}_Y)$ are isomorphic spaces with functions. In particular, $(D(f), \mathscr{O}_{X|D(f)})$ is an affine variety.*

Proof. Let $X \subseteq \mathbb{A}^n(k)$ and $\mathfrak{a} = I(X) \subseteq k[T_1, \ldots, T_n]$ be the corresponding radical ideal. We consider $k[T_1, \ldots, T_n]$ as a subring of $k[T_1, \ldots, T_{n+1}]$ and denote by $\mathfrak{a}' \subseteq k[T_1, \ldots, T_{n+1}]$ the ideal generated by $\mathfrak{a}$ and the polynomial $fT_{n+1} - 1$. Then the affine coordinate ring of Y is $\Gamma(Y) = \Gamma(X)_f \cong k[T_1, \ldots, T_{n+1}]/\mathfrak{a}'$, and we can identify Y with $V(\mathfrak{a}') \subseteq \mathbb{A}^{n+1}(k)$.

The projection $\mathbb{A}^{n+1}(k) \to \mathbb{A}^n(k)$ to the first n coordinates induces a bijective map

$$j \colon Y = \{ (x, x_{n+1}) \in X \times \mathbb{A}^1(k) \; ; \; x_{n+1}f(x) = 1 \} \to D(f) = \{ x \in X \; ; \; f(x) \neq 0 \}.$$

We will show that j is an isomorphism of spaces with functions. As a restriction of a continuous map, j is continuous. It is also open, because for $\frac{g}{f^N} \in \Gamma(Y)$ (with $g \in \Gamma(X)$) we have $j(D(\frac{g}{f^N})) = j(D(gf)) = D(gf)$. Thus j is a homeomorphism.

It remains to show that for all $g \in \Gamma(X)$ the map $\mathscr{O}_X(D(fg)) \to \Gamma(Y)_g$, $s \mapsto s \circ j$, is an isomorphism. But we have $\mathscr{O}_X(D(fg)) = \Gamma(X)_{fg} = \Gamma(Y)_g$, and this identification corresponds to the composition with j. $\qquad\square$

Proposition 1.51. *Let $(X, \mathscr{O}_X)$ be a prevariety and let $U \subseteq X$ be a non-empty open subset. Then $(U, \mathscr{O}_{X|U})$ is a prevariety and the inclusion $U \to X$ is a morphism of prevarieties.*

Proof. As X is irreducible, U is connected (Proposition 1.15). The previous lemma shows that U can be covered by open affine subsets of X. As X is noetherian, U is quasi-compact (Lemma 1.25). Thus a finite covering suffices. $\qquad\square$

The open affine subsets of a prevariety X (i.e., open subsets U of X such that $(U, \mathscr{O}_{X|U})$ is an affine variety) form a basis of the topology of X because this holds by Lemma 1.50 for affine varieties, and X is covered by open affine subvarieties by definition.

(1.17) Function field of a prevariety.

Let X be a prevariety. If $U, V \subseteq X$ are non-empty open affine subvarieties, then $U \cap V$ is open in U and non-empty. We have $\mathscr{O}_X(U) \subseteq \mathscr{O}_X(U \cap V) \subseteq K(U)$ by the definition of functions on U, and therefore $\mathrm{Frac}(\mathscr{O}_X(U \cap V)) = K(U)$. The same argument for V shows $K(U) = K(V)$. Thus the function field of a non-empty open affine subvariety U of X does not depend on U and we denote it by $K(X)$.

Definition 1.52. *The field $K(X)$ is called the* function field *of X.*

Remark 1.53. Let $f \colon X \to Y$ be a morphism of affine varieties. As the corresponding homomorphism $\Gamma(Y) \to \Gamma(X)$ between the affine coordinate rings is not injective in general, it does not induce a homomorphism of function fields $K(Y) \to K(X)$. Thus $K(X)$ is not functorial in X. But if $f \colon X \to Y$ is a morphism of prevarieties whose image contains a non-empty open (and hence dense) subset, f induces a homomorphism $K(Y) \to K(X)$. We will see in Theorem 10.19 that every morphism with dense image satisfies this property (see also Exercise 10.1). Such morphisms will be called dominant.

Proposition 1.54. *Let X be a prevariety and $U \subseteq X$ a non-empty open subset. Then $\mathscr{O}_X(U)$ is a k-subalgebra of the function field $K(X)$. If $U' \subseteq U$ is another non-empty open subset, the restriction map $\mathscr{O}(U) \to \mathscr{O}(U')$ is the inclusion of subalgebras of $K(X)$. If $U, V \subseteq X$ are arbitrary non-empty open subsets, then $\mathscr{O}_X(U \cup V) = \mathscr{O}_X(U) \cap \mathscr{O}_X(V)$.*

Proof. Let $f \colon U \to \mathbb{A}^1(k)$ be an element of $\mathscr{O}_X(U)$. Then its vanishing set $f^{-1}(0) \subseteq U$ is closed as f is continuous and $\{0\} \subset \mathbb{A}^1(k)$ is closed. Therefore if the restriction of f to U' is zero, f is zero because U' is dense in U. This shows that restriction maps are injective. The axiom of gluing implies therefore $\mathscr{O}_X(U \cup V) = \mathscr{O}_X(U) \cap \mathscr{O}_X(V)$ for all open subsets $U, V \subseteq X$. $\qquad\square$

(1.18) Closed subprevarieties.

Let X be a prevariety and let $Z \subseteq X$ be an irreducible closed subset. We want to define on Z the structure of a prevariety. For this we have to define functions on open subsets U of Z. We define:

$$\mathscr{O}'_Z(U) = \{ f \in \mathrm{Map}(U, k) \; ; \; \forall x \in U \colon \exists x \in V \subseteq X \text{ open}, g \in \mathscr{O}_X(V) \colon f_{|U \cap V} = g_{|U \cap V} \}.$$

The definition shows that $(Z, \mathscr{O}'_Z)$ is a space with functions and that $\mathscr{O}'_X = \mathscr{O}_X$. Once we have shown the following lemma, we will always write $\mathscr{O}_Z$ (instead of $\mathscr{O}'_Z$).

Lemma 1.55. *Let $X \subseteq \mathbb{A}^n(k)$ be an irreducible affine algebraic set and let $Z \subseteq X$ be an irreducible closed subset. Then the space with functions $(Z, \mathscr{O}_Z)$ associated to the affine algebraic set Z and the above defined space with functions $(Z, \mathscr{O}'_Z)$ coincide.*

Proof. In both cases Z is endowed with the topology induced by X. As the inclusion $Z \to X$ is a morphism of affine algebraic sets it induces a morphism $(Z, \mathscr{O}_Z) \to (X, \mathscr{O}_X)$. The definition of $\mathscr{O}'_Z$ shows that $\mathscr{O}'_Z(U) \subseteq \mathscr{O}_Z(U)$ for all open subsets $U \subseteq Z$.

Conversely, let $f \in \mathscr{O}_Z(U)$. For $x \in U$ there exists $h \in \Gamma(Z)$ with $x \in D(h) \subseteq U$. The restriction $f_{|D(h)} \in \mathscr{O}_Z(D(h)) = \Gamma(Z)_h$ has the form $f = \frac{g}{h^n}$, $n \geq 0$, $g \in \Gamma(Z)$. We lift g and h to elements in $\tilde{g}, \tilde{h} \in \Gamma(X)$, set $V := D(\tilde{h}) \subseteq X$, and obtain $x \in V$, $\frac{\tilde{g}}{\tilde{h}^n} \in \mathscr{O}_X(D(\tilde{h}))$ and $f_{|U \cap V} = \frac{\tilde{g}}{\tilde{h}^n}_{|U \cap V}$. $\qquad\square$

As a corollary of the lemma we obtain:

Proposition 1.56. *Let X be a prevariety and let $Z \subseteq X$ be an irreducible closed subset. Let $\mathscr{O}_Z$ be the system of functions defined above. Then $(Z, \mathscr{O}_Z)$ is a prevariety. The inclusion $Z \hookrightarrow X$ is a morphism of prevarieties.*

Projective varieties

By far the most important example of prevarieties are projective space $\mathbb{P}^n(k)$ and subvarieties of $\mathbb{P}^n(k)$, called (quasi-)projective varieties. In this subchapter we will define the projective space as a prevariety. Closed subprevarieties of $\mathbb{P}^n(k)$ are vanishing sets of homogeneous polynomials. They are called projective varieties. We will study several examples.

(1.19) Homogeneous polynomials.

To describe the functions on projective space we start with some remarks on homogeneous polynomials. Although in this chapter we will only deal with polynomials with coefficients in k, it will be helpful for later applications to work with more general coefficients. Thus let R be an arbitrary (commutative) ring.

Definition 1.57. *A polynomial $f \in R[X_0, \ldots, X_n]$ is called* homogeneous of degree *$d \in \mathbb{Z}_{\geq 0}$, if f is the sum of monomials of degree d.*

If R is an integral domain with infinitely many elements (e.g., $R = k$), a polynomial $f \in R[X_0, \ldots, X_n]$ is homogeneous of degree d if and only if

$$f(\lambda x_0, \ldots, \lambda x_n) = \lambda^d f(x_0, \ldots, x_n) \quad \text{for all } x_0, \ldots, x_n \in R, \, 0 \neq \lambda \in R$$

(see Exercise 1.20).

The zero polynomial is homogeneous of degree d for all d. We denote by $R[X_0, \ldots, X_n]_d$ the R-submodule of all homogeneous polynomials of degree d. As we can decompose uniquely every polynomial into its homogeneous parts, we have

$$R[X_0, \ldots, X_n] = \bigoplus_{d \geq 0} R[X_0, \ldots, X_n]_d.$$

Lemma 1.58. *Let $i \in \{0, \ldots, n\}$ and $d \geq 0$. There is a bijective R-linear map*

$$\Phi_i = \Phi_i^{(d)} \colon R[X_0, \ldots, X_n]_d \overset{\sim}{\to} \{\, g \in R[T_0, \ldots, \widehat{T_i}, \ldots, T_n] \; ; \; \deg(g) \leq d\,\},$$
$$f \mapsto f(T_0, \ldots, 1, \ldots, T_n).$$

(Elements of a tuple with $\widehat{}$ are omitted.)

Proof. We construct an inverse map. Let g be a polynomial in the right hand side set and let $g = \sum_{j=0}^{d} g_j$ be its decomposition into homogeneous parts (with respect to T_ℓ for $\ell = 0, \ldots, n$, $\ell \neq i$). Define

$$\Psi_i(g) = \sum_{j=0}^{d} X_i^{d-j} g_j(X_0, \ldots, \widehat{X_i}, \ldots, X_n).$$

It is easy to see that Φ_i and Ψ_i are inverse to each other (as both maps are R-linear, it suffices to check this on monomials). $\qquad\square$

The map Φ_i is called *dehomogenization*, the map Ψ_i *homogenization* (*with respect to* X_i). For $f \in R[X_0, \ldots, X_n]_d$ and $g \in R[X_0, \ldots, X_n]_e$ (with $d, e \geq 0$) the product fg is homogeneous of degree $d + e$ and we have

$$(1.19.1) \qquad \Phi_i^{(d)}(f)\Phi_i^{(e)}(g) = \Phi_i^{(d+e)}(fg).$$

If $R = K$ is a field, we will extend homogenization and dehomogenization to fields of fractions as follows. Let $\mathcal{F}$ be the subset of $K(X_0, \ldots, X_n)$ that consists of those elements $\frac{f}{g}$, where $f, g \in K[X_0, \ldots, X_n]$ are homogeneous polynomials of the same degree. It is easy to check that $\mathcal{F}$ is a subfield of $K(X_0, \ldots, X_n)$. By (1.19.1) we have a well defined isomorphism of K-extensions

$$(1.19.2) \qquad \Phi_i \colon \mathcal{F} \overset{\sim}{\to} K(T_0, \ldots, \widehat{T_i}, \ldots, T_n), \qquad \frac{f}{g} \mapsto \frac{\Phi_i(f)}{\Phi_i(g)}.$$

Often, we will identify $K(T_0, \ldots, \widehat{T_i}, \ldots, T_n)$ with the subring $K(\frac{X_0}{X_i}, \ldots, \frac{X_n}{X_i})$ of the field $K(X_0, \ldots, X_n)$. Via this identification the isomorphism (1.19.2) can also be described as follows. Let $\frac{f}{g} \in \mathcal{F}$ with $f, g \in K[X_0, \ldots, X_n]_d$ for some d. Set $\tilde{f} = \frac{f}{X_i^d}$ and $\tilde{g} = \frac{g}{X_i^d}$. Then $\tilde{f}, \tilde{g} \in K[\frac{X_0}{X_i}, \ldots, \frac{X_n}{X_i}]$ and $\Phi_i(\frac{f}{g}) = \frac{\tilde{f}}{\tilde{g}}$.

(1.20) Definition of the projective space $\mathbb{P}^n(k)$.

The projective space $\mathbb{P}^n(k)$ is an extremely important prevariety within algebraic geometry. Many prevarieties of interest are subprevarieties of the projective space. Moreover, the projective space is the correct environment for projective geometry which remedies the "defect" of affine geometry of missing points at infinity. For example, in $\mathbb{A}^2(k)$ there exist lines that do not meet (namely parallel lines) but we will see in Section (1.23) that two different lines in the projective plane always meet in one point.

As a set we define for every field k (not necessarily algebraically closed)

$$(1.20.1) \qquad \mathbb{P}^n(k) = \{\text{lines through the origin in } k^{n+1}\} = (k^{n+1} \setminus \{0\})/k^\times.$$

Here a line through the origin is per definition a 1-dimensional k-subspace and we denote by $(k^{n+1} \setminus \{0\})/k^\times$ the set of equivalence classes in $k^{n+1} \setminus \{0\}$ with respect to the equivalence relation

$$(x_0, \ldots x_n) \sim (x'_0, \ldots, x'_n) \Leftrightarrow \exists \lambda \in k^\times : \forall i : x_i = \lambda x'_i.$$

Then the second equality in (1.20.1) is given by attaching to the equivalence class of $(x_0, \ldots, x_n)$ the 1-dimensional subspace generated by this vector. The equivalence class of a point $(x_0, \ldots, x_n)$ is denoted by $(x_0 : \ldots : x_n)$. We call the x_i the *homogeneous coordinates* on $\mathbb{P}^n(k)$.

To endow $\mathbb{P}^n(k)$ with the structure of a prevariety we will assume from now on that k is algebraically closed. The following observation is essential: For $0 \le i \le n$ we set

$$U_i := \{ (x_0 : \ldots : x_n) \in \mathbb{P}^n(k) \; ; \; x_i \ne 0 \} \subset \mathbb{P}^n(k).$$

This subset is well-defined and the union of the U_i for $0 \le i \le n$ is all of $\mathbb{P}^n(k)$. There are bijections

$$U_i \xrightarrow{\sim} \mathbb{A}^n(k), \quad (x_0 : \ldots : x_n) \mapsto \left(\frac{x_0}{x_i}, \ldots, \frac{\widehat{x_i}}{x_i}, \ldots \frac{x_n}{x_i} \right).$$

Via this bijection we will endow U_i with the structure of a space with functions, isomorphic to $(\mathbb{A}^n(k), \mathscr{O}_{\mathbb{A}^n(k)})$, which we denote by $(U_i, \mathscr{O}_{U_i})$. We want to define on $\mathbb{P}^n(k)$ the structure of a space with functions $(\mathbb{P}^n(k), \mathscr{O}_{\mathbb{P}^n(k)})$ such that U_i becomes an open subset of $\mathbb{P}^n(k)$ and such that $\mathscr{O}_{\mathbb{P}^n(k)|U_i} = \mathscr{O}_{U_i}$ for all $i = 0, \ldots, n$. As $\bigcup_i U_i = \mathbb{P}^n(k)$ there is at most one way to do this:

We define the topology on $\mathbb{P}^n(k)$ by calling a subset $U \subseteq \mathbb{P}^n(k)$ open if $U \cap U_i$ is open in U_i for all i. This defines a topology on $\mathbb{P}^n(k)$ as for all $i \ne j$ the set $U_i \cap U_j = D(T_j) \subseteq U_i$ is open (we use here on $U_i \cong \mathbb{A}^n(k)$ the coordinates $T_0, \ldots, \widehat{T_i}, \ldots, T_n$). With this definition, $(U_i)_{0 \le i \le n}$ is an open covering of $\mathbb{P}^n(k)$.

We still have to define functions on open subsets $U \subseteq \mathbb{P}^n(k)$. We set

$$\mathscr{O}_{\mathbb{P}^n(k)}(U) = \{ f \in \mathrm{Map}(U, k) \; ; \; \forall i \in \{0, \ldots, n\} : f_{|U \cap U_i} \in \mathscr{O}_{U_i}(U \cap U_i) \}.$$

It is clear that this defines the structure of a space with functions on $\mathbb{P}^n(k)$, although we still have to see that $\mathscr{O}_{\mathbb{P}^n(k)|U_i} = \mathscr{O}_{U_i}$ for all i. This follows from the following description of the k-algebras $\mathscr{O}_{\mathbb{P}^n(k)}(U)$ using the inverse of the isomorphism (1.19.2) of the function field $k(T_0, \ldots, \widehat{T_i}, \ldots, T_n)$ of U_i with the subfield $\mathcal{F}$ of $k(X_0, \ldots, X_n)$.

Proposition 1.59. *Let $U \subseteq \mathbb{P}^n(k)$ be open. Then*

$$\mathscr{O}_{\mathbb{P}^n(k)}(U) = \{ f \colon U \to k \; ; \; \forall x \in U \; \exists x \in V \subseteq U \text{ open and}$$

$$g, h \in k[X_0, \ldots, X_n] \text{ homogeneous of the same degree}$$

$$\text{such that } h(v) \ne 0 \text{ and } f(v) = \frac{g(v)}{h(v)} \text{ for all } v \in V \}.$$

Proof. Let $f \in \mathscr{O}_{\mathbb{P}^n(k)}(U)$. As $f_{|U \cap U_i} \in \mathscr{O}_{U_i}(U \cap U_i)$, the function f has locally the form $\frac{\tilde{g}}{\tilde{h}}$ with $\tilde{g}, \tilde{h} \in k[T_0, \ldots, \widehat{T_i}, \ldots, T_n]$. Applying the inverse of (1.19.2) yields the desired form of f.

Conversely, let f be an element of the right hand side. We fix $i \in \{0, \ldots, n\}$. Thus locally on $U \cap U_i$ the function f has the form $\frac{g}{h}$ with $g, h \in k[X_0, \ldots, X_n]_d$ for some d. Once more applying the isomorphism (1.19.2) we obtain that f has locally the form $\frac{\tilde{g}}{\tilde{h}}$ with $\tilde{g}, \tilde{h} \in k[T_0, \ldots, \widehat{T_i}, \ldots, T_n]$. This shows $f_{|U \cap U_i} \in \mathscr{O}_{U_i}(U \cap U_i)$. $\qquad \square$

Corollary 1.60. *Let $i \in \{0, \ldots, n\}$. The bijection $U_i \xrightarrow{\sim} \mathbb{A}^n(k)$ induces an isomorphism*

$$(U_i, \mathscr{O}_{\mathbb{P}^n(k)|U_i}) \xrightarrow{\sim} \mathbb{A}^n(k).$$

of spaces with functions. The space with functions $(\mathbb{P}^n(k), \mathscr{O}_{\mathbb{P}^n(k)})$ is a prevariety.

Proof. The first assertion follows from the proof of Proposition 1.59. This shows that $\mathbb{P}^n(k)$ is a space with functions that has a finite open covering by affine varieties. Moreover, Lemma 1.19 shows that $\mathbb{P}^n(k)$ is irreducible. $\qquad\square$

The function field $K(\mathbb{P}^n(k))$ (Section (1.17)) of $\mathbb{P}^n(k)$ is by its very definition the function field $K(U_i) = k(\frac{X_0}{X_i}, \ldots, \frac{X_n}{X_i})$ of U_i. Using the isomorphism Φ_i (1.19.2), we usually describe $K(\mathbb{P}^n(k))$ as the field

(1.20.2) $\quad K(\mathbb{P}^n(k)) = \{ f/g \; ; \; f, g \in k[X_0, \ldots, X_n] \text{ homog. of the same degree}, g \neq 0 \}.$

For $0 \leq i, j \leq n$ the identification of $K(U_i) \xrightarrow{\sim} K(U_j)$ is given abstractly by $\Phi_j \circ \Phi_i^{-1}$. This can be described explicitly as

$$K(U_i) = k\left(\frac{X_0}{X_i}, \ldots, \frac{X_n}{X_i}\right) \longrightarrow k\left(\frac{X_0}{X_j}, \ldots, \frac{X_n}{X_j}\right) = K(U_j),$$

$$\frac{X_\ell}{X_i} \longmapsto \frac{X_\ell}{X_j}\frac{X_j}{X_i} = \frac{X_\ell}{X_i},$$

i.e., as subfields of $K(X_0, \ldots, X_n)$, all the $K(U_i)$ coincide, and coincide with $K(\mathbb{P}^n(k))$, and the isomorphism induced by our identifications is the identity map. We use these explicit descriptions to prove the following result.

Proposition 1.61. *The only global functions on $\mathbb{P}^n(k)$ are the constant functions, i.e., $\mathscr{O}_{\mathbb{P}^n(k)}(\mathbb{P}^n(k)) = k$. In particular, $\mathbb{P}^n(k)$ is not an affine variety for $n \geq 1$.*

Proof. By Proposition 1.54 we have

$$\mathscr{O}_{\mathbb{P}^n(k)}(\mathbb{P}^n(k)) = \bigcap_{0 \leq i \leq n} \mathscr{O}_{\mathbb{P}^n(k)}(U_i) = \bigcap_{0 \leq i \leq n} k\left[\frac{X_0}{X_i}, \ldots, \frac{X_n}{X_i}\right] = k,$$

where the intersection is taken in $K(\mathbb{P}^n(k))$. The last assertion follows because if $\mathbb{P}^n(k)$ were affine, its set of points would be in bijection to the set of maximal ideals in the ring $k = \mathscr{O}_{\mathbb{P}^n(k)}(\mathbb{P}^n(k))$. This implies that $\mathbb{P}^n(k)$ consists of only one point, so $n = 0$. $\qquad\square$

(1.21) Projective varieties.

Definition 1.62. *A prevariety is called a projective variety if it is isomorphic to a closed subprevariety of a projective space $\mathbb{P}^n(k)$.*

As in the affine case, we speak of projective varieties rather than prevarieties. Similarly, we will talk about subvarieties of projective space, instead of subprevarieties. For an explanation why this is legitimate, we refer to Chapter 9.

For $x = (x_0 : \cdots : x_n) \in \mathbb{P}^n(k)$ and $f \in k[X_0, \ldots, X_n]$ the value $f(x_0, \ldots, x_n)$ obviously depends on the choice of the representative of x and we cannot consider f as a function on $\mathbb{P}^n(k)$. But if f is homogeneous, at least the question whether the value is zero or nonzero is independent of the choice of a representative. Thus we define for homogeneous polynomials $f_1, \ldots, f_m \in k[X_0, \ldots, X_n]$ (not necessarily of the same degree) the vanishing set

$$V_+(f_1, \ldots, f_m) = \{ (x_0 : \ldots : x_n) \in \mathbb{P}^n(k) \; ; \; \forall j : f_j(x_0, \ldots, x_n) = 0 \}.$$

Subsets of the form $V_+(f_1, \ldots, f_m)$ are closed. More precisely we have for $i = 0, \ldots, n$:

$$V_+(f_1, \ldots, f_m) \cap U_i = V(\Phi_i(f_1), \ldots, \Phi_i(f_m)),$$

where Φ_i denotes as usual dehomogenization with respect to X_i. We will see that all closed subsets of the projective space are of this form.

To do this we consider the map

$$f \colon \mathbb{A}^{n+1}(k) \setminus \{0\} \to \mathbb{P}^n(k), \quad (x_0, \ldots, x_n) \mapsto (x_0 : \cdots : x_n).$$

As for all i its restriction $f_{|f^{-1}(U_i)} \colon f^{-1}(U_i) \to U_i$ is a morphism of prevarieties, this also holds for f. If $Z \subseteq \mathbb{P}^n(k)$ is a closed subset, $f^{-1}(Z)$ is a closed subset of $\mathbb{A}^{n+1}(k) \setminus \{0\}$ and we denote by $C(Z)$ its closure in $\mathbb{A}^{n+1}(k)$. Affine algebraic sets $X \subseteq \mathbb{A}^{n+1}(k)$ are called *affine cones* if for all $x \in X$ we have $\lambda x \in X$ for all $\lambda \in k^\times$. Clearly $C(Z)$ is an affine cone in $\mathbb{A}^{n+1}(k)$. It is called the *affine cone of Z*.

Proposition 1.63. *Let $X \subseteq \mathbb{A}^{n+1}(k)$ be an affine algebraic set such that $X \neq \{0\}$. Then the following assertions are equivalent.*
(i) X is an affine cone.
(ii) $I(X)$ is generated by homogeneous polynomials.
(iii) There exists a closed subset $Z \subseteq \mathbb{P}^n(k)$ such that $X = C(Z)$.
If in this case $I(X)$ is generated by homogeneous polynomials $f_1, \ldots, f_m \in k[X_0, \ldots, X_n]$, then $Z = V_+(f_1, \ldots, f_m)$.

Proof. We have already seen that (iii) implies (i). Let us show that (i) implies (ii). To show that $I(X)$ is generated by homogeneous elements, we use that an ideal $\mathfrak{a} \subseteq k[\underline{T}]$ is generated by homogeneous elements if and only if for each $g \in \mathfrak{a}$ its homogeneous components are again in $\mathfrak{a}$. Thus let $g \in I(X)$ and write $g = \sum_d g_d$, where g_d is homogeneous of degree d. As X is an affine cone, we have $g(\lambda x) = 0$ for all $x = (x_0, \ldots, x_n) \in X$ and $\lambda \in k^\times$. If there existed $g_d \notin I(X)$, we would find $x \in X$ such that $g_d(x) \neq 0$. Then $\sum_d g_d(x) T^d \in k[T]$ is not the zero polynomial and there exists a $\lambda \in k^\times$ with

$$0 \neq \sum_d g_d(x) \lambda^d = \sum_d g_d(\lambda x) = g(\lambda x) = 0.$$

Contradiction!

If $I(X)$ is generated by homogeneous polynomials, finitely many suffice, say $f_1, \ldots, f_m$. Then it is clear that for $Z := V_+(f_1, \ldots, f_m)$ we have $X = V(I(X)) = C(Z)$. $\square$

In particular we see that for every closed subset $Z \subseteq \mathbb{P}^n(k)$ there exist homogeneous polynomials $f_1, \ldots, f_r \in k[X_0, \ldots, X_n]$ such that $Z = V_+(f_1, \ldots, f_m)$.

(1.22) Change of coordinates in projective space.

Let $A = (a_{ij})_{i,j=0,\ldots,n} \in \mathrm{GL}_{n+1}(k)$ be an invertible $(n+1) \times (n+1)$-matrix. The map $k^{n+1} \to k^{n+1}$ described by A maps one-dimensional subspaces to one-dimensional subspaces and induces a map $\mathbb{P}^n(k) \to \mathbb{P}^n(k)$. It is given by

$$(x_0 : \cdots : x_n) \mapsto (\sum_{i=0}^{n} a_{0i}x_i : \cdots : \sum_{i=0}^{n} a_{ni}x_i)$$

and we obtain a morphism of prevarieties which we denote by φ_A. For $A, B \in \mathrm{GL}_{n+1}(k)$ we have $\varphi_{AB} = \varphi_A\varphi_B$. In particular, φ_A is an automorphism and we obtain a homomorphism of groups

$$\varphi \colon \mathrm{GL}_{n+1}(k) \to \mathrm{Aut}(\mathbb{P}^n(k)).$$

The automorphism φ_A is called the *change of coordinates described by A*.

The kernel of φ consists of the subgroup $Z := \{\, \lambda I_{n+1} \, ; \, \lambda \in k^\times \,\}$ of scalar matrices. We will see in Section (11.15) that φ is surjective and therefore defines a group isomorphism $\mathrm{PGL}_{n+1}(k) \overset{\sim}{\to} \mathrm{Aut}(\mathbb{P}^n(k))$. Here $\mathrm{PGL}_{n+1}(k) := \mathrm{GL}_{n+1}(k)/Z$ is the so-called projective linear group.

(1.23) Linear Subspaces of the projective space.

For $m \geq -1$ let $\varphi \colon k^{m+1} \to k^{n+1}$ be an injective homomorphism of k-vector spaces. It maps one-dimensional subspaces of k^{m+1} to one-dimensional subspaces of k^{n+1} and we obtain an injective morphism $\iota \colon \mathbb{P}^m(k) \to \mathbb{P}^n(k)$ of prevarieties. This is in fact an isomorphism of $\mathbb{P}^m(k)$ onto a closed subprevariety of $\mathbb{P}^n(k)$: If $A = (a_{ij}) \in M_{\ell \times (n+1)}(k)$ is a matrix such that $\mathrm{Ker}\, A = \mathrm{im}\, \varphi$, then ι defines an isomorphism of $\mathbb{P}^m(k)$ with $V_+(f_1, \ldots, f_\ell)$, where $f_i = \sum_{j=0}^{n} a_{ij}X_j \in k[X_0, \ldots, X_n]$.

Closed subprevarieties of this form are called *linear subspaces of $\mathbb{P}^n(k)$ of dimension m*. They are precisely those closed subprevarieties Z of $\mathbb{P}^n(k)$ for which there exists a subvector space U of dimension $m+1$ of k^{n+1} such that Z consists of the one-dimensional subspaces of k^{n+1} that are contained in U. As $\mathrm{GL}_{n+1}(k)$ acts transitively on the set of subvector spaces of k^{n+1} of dimension $m+1$, the projective linear group $\mathrm{PGL}_{n+1}(k)$ acts transitively by change of coordinates on the set of linear subspaces of $\mathbb{P}^n(k)$ of dimension m.

The only linear subspace of dimension -1 is the empty set, the linear subspaces of dimension 0 are the points. Linear subspaces in $\mathbb{P}^n(k)$ of dimension 1 (resp. 2, resp. $n-1$) are called *lines* (resp. *planes*, resp. *hyperplanes*).

For every two points $p \neq q \in \mathbb{P}^n(k)$ there exists a unique line in $\mathbb{P}^n(k)$ that contains p and q. This is clear, because two different one-dimensional subspaces of k^{n+1} are contained in a unique two-dimensional subspace. We denote this line by $\overline{pq}$.

We also see that two different lines in $\mathbb{P}^2(k)$ always intersect in a unique point: Lines in $\mathbb{P}^2(k)$ correspond to two-dimensional subspaces in k^3 and any two different two-dimensional subspaces in k^3 meet in a unique one-dimensional subspace – which corresponds to a point in $\mathbb{P}^2(k)$. Similar assertions can be made for intersections of linear subspaces in higher-dimensional projective spaces (see Exercise 1.26).

A far reaching generalization for intersections of closed subvarieties of projective spaces that are given by homogeneous polynomials of arbitrary degree is the Theorem of Bézout (see Section (5.15) for a special case and Volume II for the general case).

(1.24) Cones.

Let $H \subset \mathbb{P}^n(k)$ be a hyperplane and let $p \in \mathbb{P}^n(k) \setminus H$ be a point. Let $X \subseteq H$ be a closed subvariety. We define the *cone* $\overline{X,p}$ *of X over p* by

$$\overline{X,p} = \bigcup_{q \in X} \overline{qp}.$$

This is a closed subvariety of $\mathbb{P}^n(k)$: Indeed, after a change of coordinates we may assume $H = V_+(X_n)$ and $p = (0 : \ldots : 0 : 1)$. Then we have

$$X = V_+(f_1, \ldots, f_m) \subseteq \mathbb{P}^{n-1}(k) = H \quad \text{for } f_i \in k[X_0, \ldots, X_{n-1}].$$

Let $\tilde{f}_i$ be the polynomial f_i considered as an element of $k[X_0, \ldots, X_n]$. Then we obtain $\overline{X,p} = V_+(\tilde{f}_1, \ldots, \tilde{f}_m)$.

This construction can be generalized as follows: We say that two linear subspaces Λ and Ψ of $\mathbb{P}^n(k)$ are *complementary* if they are defined by subvector spaces U and V of k^{n+1} that are complements of each other (i.e., we have $\Lambda \cap \Psi = \emptyset$ and the smallest linear subspace of $\mathbb{P}^n(k)$ that contains Λ and Ψ is $\mathbb{P}^n(k)$ itself).

If Λ and Ψ are complementary and if $X \subseteq \Psi$ is a closed subvariety, we define the *cone* $\overline{X,\Lambda}$ *of X over Λ* by

$$(1.24.1) \qquad\qquad \overline{X,\Lambda} = \bigcup_{q \in X} \overline{q,\Lambda},$$

where $\overline{q,\Lambda} = \bigcup_{p \in \Lambda} \overline{qp}$ is the smallest linear subspace that contains q and Λ. Then $\overline{X,\Lambda}$ is a closed subvariety of $\mathbb{P}^n(k)$. This can be shown directly, similarly as above, or by noticing that $\overline{X,\Lambda}$ arises by iterating the first construction for points p_i which span Λ.

In Section (13.17) we will generalize this construction and also show that (1.24.1) still yields a projective variety if we only assume that $X \cap \Lambda = \emptyset$.

(1.25) Morphisms of quasi-projective varieties.

We will now see that morphisms between (open subprevarieties of) projective varieties are given by homogeneous polynomials – just as morphisms of affine varieties are given by polynomials.

Definition 1.64. *A prevariety is called* quasi-projective variety *if it is isomorphic to an open subvariety of a projective variety.*

Projective varieties and affine varieties are clearly quasi-projective. Up to isomorphism quasi-projective varieties are those that are of the form $(Y, \mathscr{O}_Y)$, where $Y \subseteq \mathbb{P}^n(k)$ is a locally closed subspace and where $\mathscr{O}_Y = \mathscr{O}_{X|Y}$ for a closed subvariety X of $\mathbb{P}^n(k)$ such that Y is open in X. The structure of a prevariety depends only on Y and not on the choice of X (although this is not difficult to show, we do not prove it here; once we identified prevarieties with integral schemes of finite type over k, this follows at once from the assertion that for every locally closed subspace of a scheme there exists a unique reduced subscheme structure, see Proposition 3.52).

Proposition 1.65. *Let $Y \subseteq \mathbb{P}^n(k)$ be a quasi-projective variety.*
(1) *Let $f_0, \ldots, f_m \in k[X_0, \ldots, X_n]$ be homogeneous polynomials of the same degree such that for all $y = (y_0 : \ldots : y_n) \in Y$ there exists an index j such that $f_j(y) \neq 0$. Then*

$$h \colon Y \to \mathbb{P}^m(k), \quad y \mapsto (f_0(y) : \ldots : f_m(y))$$

is a morphism of prevarieties. Another family $g_0, \ldots, g_m \in k[X_0, \ldots, X_n]$ as above defines the same morphism h if and only if $f_i(y)g_j(y) = f_j(y)g_i(y)$ for all $y \in Y$ and all $i, j \in \{0, \ldots, m\}$.
(2) *Conversely, let $h \colon Y \to \mathbb{P}^m(k)$ be a morphism of prevarieties. Then there exists for every $y \in Y$ an open neighborhood U of y in Y such that $h_{|U}$ is of the above form.*

Proof. Our hypotheses imply that h is independent of the choice of representative of y, and hence is a well-defined map. Let $U_j = \{ x \in \mathbb{P}^m(k) \; ; \; x_j \neq 0 \}$, as usual. It suffices to show that each component of the restriction

$$h^{-1}(U_j) = \{ y \in Y \; ; \; f_j(y) \neq 0 \} \to U_j \overset{\sim}{\to} \mathbb{A}^m(k), \quad y \mapsto \left(\frac{f_0(y)}{f_j(y)}, \ldots, \frac{\widehat{f_j(y)}}{f_j(y)}, \ldots, \frac{f_m(y)}{f_j(y)} \right)$$

of h is a morphism of prevarieties. But this follows from Proposition 1.59. The second assertion in (1) is clear.

Conversely if $h \colon Y \to \mathbb{P}^m(k)$, $y \mapsto (h_0(y) : \ldots : h_m(y))$, is a morphism, for all $y \in Y$ there exists an open neighborhood U of y such that each component h_j is on U of the form $y \mapsto F_j(y)/G_j(y)$, where F_j and G_j are homogeneous of the same degree. Clearing denominators we see that $h_{|U}$ is as in (1). $\qquad\square$

Example 1.66. (Projections with center in a linear subspace) Let Λ and Ψ be complementary linear subspaces of $\mathbb{P}^n(k)$ of dimensions d and $n - d - 1$, respectively. E.g., if $d = 0$, then Λ is a point and Ψ is a hyperplane not containing Λ. We define a morphism $p_\Lambda \colon \mathbb{P}^n(k) \setminus \Lambda \to \Psi$ as follows. For $x \in \mathbb{P}^n(k) \setminus \Lambda$ let $\overline{x, \Lambda}$ be the $d + 1$-dimensional linear subspace generated by x and Λ. This subspace intersects Ψ in a unique point which we define to be $p_\Lambda(x)$. We call p_Λ a *projection with center* Λ.

Let us show that p_Λ is a morphism of prevarieties. After a change of coordinates, we may assume that

$$\Lambda = \{ (x_0 : \cdots : x_n) \in \mathbb{P}^n_k \; ; \; x_{d+1} = \cdots = x_n = 0 \},$$
$$\Psi = \{ (x_0 : \cdots : x_n) \in \mathbb{P}^n_k \; ; \; x_0 = \cdots = x_d = 0 \}.$$

In this case, $p_\Lambda((x_0 : \cdots : x_n)) = (0 : \cdots : 0 : x_{d+1} : \cdots : x_n)$ and therefore p_Λ is a morphism by Proposition 1.65.

For $y \in \Psi$ the fiber $p_\Lambda^{-1}(y)$ consists of all $x \in \Lambda_y := \overline{y, \Lambda}$ such that $x \notin \Lambda$. Therefore the fiber is an affine space of dimension $\dim \Lambda + 1$ which is openly embedded into the projective space $\Lambda_y = p_\Lambda^{-1}(y) \cup \Lambda$.

Consider the special case that Λ consists of a single point q and let $X \subset \mathbb{P}^n(k)$ be a closed subvariety with $q \notin X$. Then $p_q^{-1}(y) \cap X = \overline{q, y} \cap X$ and this is a proper closed subset of the projective line $\overline{q, y}$ (because it does not contain q) and hence must be finite. Thus we have seen that the restriction $p_{\Lambda|X}$ has finite fibers.

We will generalize this construction in Remark 8.18 and strengthen the last remark in Proposition 13.88.

(1.26) Quadrics.

In this section we assume that $\operatorname{char}(k) \neq 2$.

Definition 1.67. *A* quadric *is a closed subvariety $Q \subseteq \mathbb{P}^n(k)$ of the form $V_+(q)$, where $q \in k[X_0, \ldots, X_n]_2 \setminus \{0\}$ is a non-vanishing homogeneous polynomial of degree 2.*

Let $Q = V_+(q)$ be a quadric and let β be the bilinear form on k^{n+1} corresponding to q, i.e.,

$$\beta(v, w) = \frac{1}{2}(q(v + w) - q(v) - q(w)), \quad v, w \in k^{n+1}.$$

It is an easy argument in bilinear algebra to see that there exists a basis of k^{n+1} such that the matrix of β with respect to this basis is a diagonal matrix with 1 and 0 on its diagonal. By permuting the basis we may assume that the first entries of the diagonal are 1's. Then the change of coordinates induced by the base change matrix yields an isomorphism $Q \overset{\sim}{\to} V_+(X_0^2 + \cdots + X_{r-1}^2)$, where $r \geq 1$ is the rank of β, i.e., the number of 1's. In particular, r is independent of our choice of the basis.

Lemma 1.68. *The polynomial $X_0^2 + \cdots + X_{r-1}^2$ is irreducible if and only if $r > 2$. The closed subspace $V_+(X_0^2 + \cdots + X_{r-1}^2)$ of $\mathbb{P}^n(k)$ is irreducible if and only if $r \neq 2$.*

Proof. The claims are obvious for $r = 1$. For $r = 2$ we have

$$X_0^2 + X_1^2 = (X_0 + \sqrt{-1}X_1)(X_0 - \sqrt{-1}X_1),$$

where $\sqrt{-1} \in k$ is an element whose square is -1. Thus we have

$$V_+(X_0^2 + X_1^2) = V_+(X_0 + \sqrt{-1}X_1) \cup V_+(X_0 - \sqrt{-1}X_1).$$

As $\operatorname{char}(k) \neq 2$, this is a decomposition into different irreducible components. For $r > 2$ it is easy to check that $X_0^2 + \cdots + X_{r-1}^2$ is irreducible (if it were not, we would find a decomposition into two homogeneous polynomials of degree 1; an easy comparison of coefficients then yields a contradiction). Therefore $V_+(X_0^2 + \cdots + X_{r-1}^2)$ is irreducible. $\square$

For the following proposition we will not give a proof here. With some effort we could show the result now, but later (Proposition 6.11) it will follow easily from the general theory, and we use the proposition as one motivation to develop the theory further.

Proposition 1.69. *For $r \neq s$ the quadrics $V_+(X_0^2 + \cdots + X_{r-1}^2)$ and $V_+(X_0^2 + \cdots + X_{s-1}^2)$ are non-isomorphic.*

Linear algebra tells us that there exists no change of coordinates of $\mathbb{P}^n(k)$ that identifies $V_+(X_0^2 + \cdots + X_{r-1}^2)$ with $V_+(X_0^2 + \cdots + X_{s-1}^2)$. As already mentioned above, we will see later (Section (11.15)) that all automorphisms of $\mathbb{P}^n(k)$ are changes of coordinates.

Definition 1.70. *Let $Q \subseteq \mathbb{P}^n(k)$ be a quadric and let $r \geq 1$ be the unique integer such that $Q \cong V_+(X_0^2 + \cdots + X_{r-1}^2)$. Then we say that Q has* dimension $n - 1$ *and* rank r.

Corollary 1.71. *Let Q_1 and Q_2 be quadrics (not necessarily embedded in the same projective space). Then Q_1 and Q_2 are isomorphic as prevarieties if and only if they have the same dimension and the same rank.*

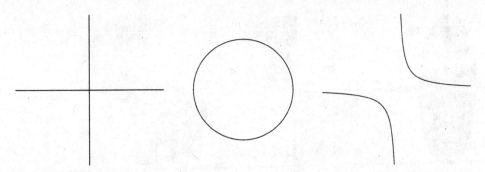

Figure 1.2: The quadric of dimension 1 and rank 2 on the left, and the solution sets in $\mathbb{R}^2$ of the equations $X^2 + Y^2 = 1$ (in the middle), and of $XY = 1$ (on the right). Note that the latter two equations both define quadrics of rank 3, and in particular are isomorphic over $\mathbb{C}$.

Proof. The condition is clearly sufficient. By Proposition 1.69 it suffices to show that two isomorphic quadrics have the same dimension. Let $Q \subseteq \mathbb{P}^n(k)$ be a quadric of rank r. We show that the transcendence degree over k of the function field (of an irreducible component) of Q is always equal to the dimension. As isomorphic prevarieties have isomorphic function field, this shows the corollary. We may assume that $Q = V_+(X_0^2 + \cdots + X_{r-1}^2)$. For $r = 1$ we have $Q = V_+(X_0) \cong \mathbb{P}^{n-1}(k)$, and thus $K(Q) \cong k(T_1, \ldots, T_{n-1})$ and $\mathrm{trdeg}_k K(Q) = n - 1$. For $r = 2$ the two irreducible components Z_1 and Z_2 of Q are given by a linear equation and thus are hyperplanes in $\mathbb{P}^n(k)$. Thus $Z_i \cong \mathbb{P}^{n-1}(k)$ and hence $\mathrm{trdeg}_k(K(Z_i)) = n - 1$.

For $r > 2$ we know that Q is irreducible. We identify $\mathbb{A}^n(k)$ with the open subset U_0 of points $(x_0 : \ldots : x_n) \in \mathbb{P}^n(k)$ with $x_0 \neq 0$. Then for $U := \mathbb{A}^n(k) \cap Q$ we have $U = V(1 + T_1^2 + \cdots + T_r^2) \subset \mathbb{A}^n(k)$ and this is a non-empty open affine subset of Q. We have

$$K(Q) = K(U) = \mathrm{Quot}(\Gamma(U)) = \mathrm{Quot}(k[T_1, \ldots, T_n]/(1 + T_1^2 + \cdots + T_r^2))$$

and again we find $\mathrm{trdeg}_k(K(Q)) = n - 1$. $\qquad\square$

Example 1.72.
(1) in $\mathbb{P}^1(k)$: The quadric of rank 2 consists of two points; in particular it is not irreducible. The quadric of rank 1 consists of a single point.
(2) in $\mathbb{P}^2(k)$: The quadrics of rank 2 and 3 are shown in Figure 1.2 (as usual we show only the "$\mathbb{R}$-valued points", i.e., the solutions of the corresponding equations over $\mathbb{R}$). As a variety it is isomorphic to $\mathbb{P}^1(k)$: We can assume it is given as $Q = V_+(X_0 X_2 - X_1^2)$, and then an isomorphism $\mathbb{P}^1(k) \to Q$ is given by $(x_0 : x_1) \to (x_0^2 : x_0 x_1 : x_1^2)$, cf. Exercise 1.30. The quadric of rank 2 is the union of two different lines, and the quadric of rank 1 is a line.
(3) in $\mathbb{P}^3(k)$: Quadrics of rank 4, 3, and 2 are pictured in Figure 1.3.

A quadric $Q \subset \mathbb{P}^n(k)$ with rank r and dimension d is called *smooth*, if $r = d + 2$, i.e., if the rank of a matrix of q is maximal. In Section (6.8) we will define in general when a prevariety is smooth and see that for quadrics the general definition coincides with the definition given here. We see that if Q is a quadric of rank $r > 2$ and dimension

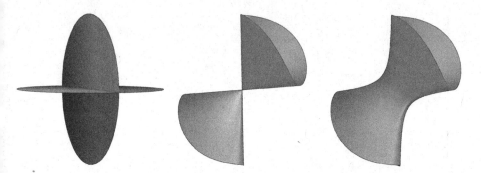

Figure 1.3: Quadrics of dimension 2 and rank 2, 3, and 4.

d, then Q is a cone $\overline{\tilde{Q}, \Lambda}$ of a smooth quadric $\tilde{Q}$ of dimension $r - 2$ with respect to a $d - r + 1$-dimensional subspace Λ.

The cases $r = 1$ and $r = 2$ are "degenerate": The quadric $Q \cong V_+(X_0^2) = V_+(X_0)$ is a hyperplane in $\mathbb{P}^n(k)$. The additional information that Q is given by the square of an irreducible polynomial is not visible for the quadric Q. But in the theory of schemes it will be possible to make a distinction. For $r = 2$ the quadric $Q \cong V_+(X_0^2 + X_1^2)$ is not irreducible and therefore not a prevariety in our sense. But again the theory of schemes will give us a satisfactory tool to deal with "reducible varieties".

Exercises

Throughout, k denotes an algebraically closed field.

Exercise 1.1. (Hilbert's basis theorem) Let A be a noetherian ring. Show that the polynomial ring $A[T]$ is again noetherian.
Hint: Consider for an ideal $\mathfrak{b} \subseteq A[T]$ the chain of ideals $\mathfrak{a}_i \subseteq A$, where $\mathfrak{a}_i$ is the ideal generated by the leading coefficients of all polynomials in $\mathfrak{b}$ of degree $\leq i$.

Exercise 1.2. Show that $I(\mathbb{A}^n(k)) = 0$ without using Hilbert's Nullstellensatz.

Exercise 1.3$\Diamond$. Determine all irreducible Hausdorff spaces. Determine all noetherian Hausdorff spaces. Show that a topological space is noetherian if and only if every open subspace is quasi-compact.

Exercise 1.4$\Diamond$. Show that the underlying topological space X of a prevariety is a T1-space (i.e., for all $x, y \in X$ there exist open neighborhoods U of x and V of y with $y \notin U$ and $x \notin V$).

Exercise 1.5.
(a) Consider the *twisted cubic curve* $C = \{(t, t^2, t^3); \ t \in k\} \subseteq \mathbb{A}^3(k)$. Show that C is an irreducible closed subset of $\mathbb{A}^3(k)$. Find generators for the ideal $I(C)$.
(b) Let $V = V(X^2 - YZ, XZ - X) \subseteq \mathbb{A}^3(k)$. Show that V consists of three irreducible components and determine the corresponding prime ideals.

Exercise 1.6. Let $f \in k[X_1, \ldots, X_n]$ be a non-constant polynomial. Write $f = \prod_{i=1}^{r} f_i^{n_i}$ with irreducible polynomials f_i such that $(f_i) \neq (f_j)$ for all $i \neq j$ and integers $n_i \geq 1$. Show that $\mathrm{rad}(f) = (f_1 \cdots f_r)$ and that the irreducible components of $V(f) \subseteq \mathbb{A}^n(k)$ are the closed subsets $V(f_i)$, $i = 1, \ldots, r$.

Exercise 1.7. Let $f \in k[T_1]$ be a non-constant polynomial. Show that

$$X_1 := V(T_2 - f) \subset \mathbb{A}^2(k)$$

is isomorphic to $\mathbb{A}^1(k)$ and show that $X_2 := V(1 - fT_2) \subset \mathbb{A}^2(k)$ is isomorphic to $\mathbb{A}^1(k) \setminus \{x_1, \ldots, x_n\}$ for some $n \geq 1$. Show that X_1 and X_2 are not isomorphic (look at the invertible elements of their coordinate rings).

Exercise 1.8. Show that the affine algebraic set $V(Y^2 - X^3 + X) \subset \mathbb{A}^2(k)$ is irreducible and in particular connected. Sketch the set $\{(x, y) \in \mathbb{R}^2 ; y^2 = x^3 - x\}$ and show that it is not connected with respect to the analytic topology on $\mathbb{R}^2$.

Exercise 1.9$\diamond$. Describe the union of the n coordinate axes in $\mathbb{A}^n(k)$ as an algebraic set.

Exercise 1.10. Identifying $\mathbb{A}^1(k) \times \mathbb{A}^1(k)$ and $\mathbb{A}^2(k)$ as sets, show that the Zariski topology on $\mathbb{A}^2(k)$ is strictly finer than the product topology.

Exercise 1.11. We identify the space $M_2(k)$ of 2×2-matrices over k with $\mathbb{A}^4(k)$ (with coordinates a, b, c, d). A matrix $A = \begin{pmatrix} a & b \\ c & d \end{pmatrix} \in M_2(k)$ is nilpotent if $A^2 = 0$ or, equivalently, if its determinant and trace are zero. Thus if

$$\mathfrak{a} := (a^2 + bc, d^2 + bc, (a+d)b, (a+d)c), \qquad \mathfrak{b} = (ad - bc, a + d),$$

we have that

$$V(\mathfrak{a}) = V(\mathfrak{b}) = \{A \in M_2(k) ; A \text{ nilpotent}\}.$$

Show that $\mathrm{rad}\,\mathfrak{a} = \mathfrak{b}$ and that $V(\mathfrak{b})$ is an irreducible closed affine cone in $M_2(k) = \mathbb{A}^4(k)$ (the so-called *nilpotent cone*).

Exercise 1.12.
(a) Let char $k \neq 2$ and let $Z_1 = V(U(T-1) - 1)$ and $Z_2 = V(Y^2 - X^2(X+1))$ be closed subsets of $\mathbb{A}^2(k)$. Show that $(t, u) \mapsto (t^2 - 1, t(t^2 - 1))$ defines a bijective morphism $Z_1 \to Z_2$ which is not an isomorphism.
(b) Show that the morphism $\mathbb{A}^1(k) \to V(Y^2 - X^3) \subset \mathbb{A}^2(k)$, $t \mapsto (t^2, t^3)$ is a bijective morphism that is not an isomorphism.

Exercise 1.13. Show that for $n \geq 2$ the open subprevariety $\mathbb{A}^n(k) \setminus \{0\} \subset \mathbb{A}^n(k)$ is not an affine variety. Is $\mathbb{A}^1(k) \setminus \{0\}$ affine?

Exercise 1.14. Let X be a prevariety and let Y be an affine variety. Show that the map

$$\mathrm{Hom}(X, Y) \to \mathrm{Hom}_{(k\text{-Alg})}(\Gamma(Y), \Gamma(X)), \qquad f \mapsto f^* : \varphi \mapsto \varphi \circ f,$$

is bijective. Deduce that $\mathrm{Hom}(X, \mathbb{A}^n(k)) = \Gamma(X)^n$.

Exercise 1.15$\diamond$. Let $n \geq 1$ be an integer. We identify $M_n(k)$ with the affine space $\mathbb{A}^{n^2}(k)$. Show that the subset $\mathrm{GL}_n(k) \subset M_n(k)$ is an open affine subvariety. Describe its coordinate ring $\Gamma(\mathrm{GL}_n(k))$.

Exercise 1.16. Let X be a prevariety. Define on the set of all pairs (U, f), where $\emptyset \neq U \subseteq X$ open and $f \in \Gamma(U)$, an equivalence relation by setting $(U, f) \sim (V, g)$ if there exists $\emptyset \neq W \subseteq U \cap V$ open such that $f_{|W} = g_{|W}$. Define the structure of a field extension of k on the set of equivalence classes and show that this field extension is naturally isomorphic to the function field of X.

Exercise 1.17◊. Let $k = \mathbb{C}$. Show that that Zariski topology on $\mathbb{A}^1(\mathbb{C}) = \mathbb{C}$ is coarser than the complex analytic topology. Show that the exponential function, sine, and cosine, considered as functions $\mathbb{A}^1(\mathbb{C}) \to \mathbb{A}^1(\mathbb{C})$, are not continuous for the Zariski topology (and in particular not morphisms of affine varieties).

Exercise 1.18◊. Let $f \in k[X_0, \ldots, X_n]$ be homogeneous. Show that

$$D_+(f) = \{ x \in \mathbb{P}^n(k) \; ; \; f(x) \neq 0 \}$$

is an open subset of $\mathbb{P}^n(k)$ and that the open subsets of this form are a basis for the topology of $\mathbb{P}^n(k)$.

Exercise 1.19. Formulate and prove a projective analogue of Exercise 1.6.

Exercise 1.20. Let K be a field.
(a) Assume that K is infinite. Show that $f \in K[X_0, \ldots, X_n]$ is homogeneous of degree d if and only if $f(\lambda x_0, \ldots, \lambda x_n) = \lambda^d f(x_0, \ldots, x_n)$ for all $x_0, \ldots, x_n \in k$, $0 \neq \lambda \in K$.
(b) Let $\mathfrak{a} \subseteq K[X_0, \ldots, X_n]$ be an ideal. Show that the following assertions are equivalent.
  (i) The ideal $\mathfrak{a}$ is generated by homogeneous elements.
  (ii) For every $f \in \mathfrak{a}$ all its homogeneous components are again in $\mathfrak{a}$.
  (iii) We have $\mathfrak{a} = \bigoplus_{d \geq 0} \mathfrak{a} \cap K[X_0, \ldots, X_n]_d$.
  An ideal satisfying these equivalent conditions is called *homogeneous*.
(c) Show that intersections, sums, products, and radicals of homogeneous ideals are again homogeneous.
(d) Show that a homogeneous ideal $\mathfrak{p} \subseteq K[X_0, \ldots, X_n]$ is a prime ideal if and only if $fg \in \mathfrak{p}$ implies $f \in \mathfrak{p}$ or $g \in \mathfrak{p}$ for all homogeneous elements f and g.
(e) Show that every homogeneous ideal $\mathfrak{a} \subsetneq K[X_0, \ldots, X_n]$ is contained in the homogeneous ideal $(X_0, \ldots, X_n)$.

Exercise 1.21. Let $\mathfrak{a} \subseteq k[X_0, \ldots, X_n]$ be a homogeneous ideal (Exercise 1.20) and set

$$V_+(\mathfrak{a}) := \{ x \in \mathbb{P}^n(k) \; ; \; f(x) = 0 \text{ for all homogeneous } f \in \mathfrak{a} \}.$$

Show that the maps $\mathfrak{a} \mapsto V_+(\mathfrak{a})$ and $Z \mapsto C(Z)$ define bijections between the following sets.
(1) The set of homogeneous radical ideals $\mathfrak{a} \subseteq k[X_0, \ldots, X_n]$ with $\mathfrak{a} \neq (X_0, \ldots, X_n)$.
(2) The set of closed subspaces Z of $\mathbb{P}^n(k)$.
(3) The set of closed affine cones $C \subseteq \mathbb{A}^{n+1}(k)$ such that $C \neq \{0\}$.
If $Z \subseteq \mathbb{P}^n(k)$ is a closed subset we denote by $I_+(Z)$ the corresponding homogeneous ideal. Show that $I_+(Z) = I(C(Z))$ and deduce that the following assertions are equivalent.
(i) Z is irreducible.
(ii) $I_+(Z)$ is a prime ideal.
(iii) $C(Z)$ is irreducible.

Exercise 1.22. Let L_1 and L_2 be two disjoint lines in $\mathbb{P}^3(k)$.
(a) Show that there exists a change of coordinates such that $L_1 = V_+(X_0, X_1)$ and $L_2 = V_+(X_2, X_3)$.
(b) Let $Z = L_1 \cup L_2$. Determine the homogeneous radical ideal $\mathfrak{a}$ such that $V_+(\mathfrak{a}) = Z$ (Exercise 1.21).

Exercise 1.23. Let $Z \subseteq \mathbb{P}^n(k)$ be a projective variety and let $\mathfrak{p} \subset k[X_0, \ldots, X_n]$ be the corresponding homogeneous prime ideal (Exercise 1.21). Show that the function field $K(Z)$ is isomorphic to the ring of rational functions f/g, where $f, g \in k[X_0, \ldots, X_n]$ are homogeneous of the same degree, $g \notin \mathfrak{p}$, modulo the ideal of f/g with $f \in \mathfrak{p}$.

Exercise 1.24. Let $n \geq 1$ be an integer. We identify $\mathbb{A}^n(k)$ with the open subprevariety $U_0 = \{(x_0 : \cdots : x_n) \in \mathbb{P}^n(k) \; ; \; x_0 \neq 0\}$ of $\mathbb{P}^n(k)$. For $f \in k[T_1, \ldots, T_n]$ of degree d let $\bar{f}$ be its homogenization in $k[X_0, \ldots, X_n]_d$ (with respect to X_0).
Let $X = V(\mathfrak{a}) \subseteq \mathbb{A}^n(k)$ be an affine algebraic set and define $\bar{\mathfrak{a}} \subseteq k[X_0, \ldots, X_n]$ as the ideal generated by the elements $\bar{f}$ for $f \in \mathfrak{a}$. Let $\bar{X}$ be the closure of X in $\mathbb{P}^n(k)$.
(a) Show that X is irreducible if and only if $\bar{X}$ is irreducible.
(b) Show that $\bar{X} = V_+(\bar{\mathfrak{a}})$ (notation of Exercise 1.21).
(c) Find generators $f_1, \ldots, f_r$ for the ideal $I(C)$ of the twisted cubic curve C (Exercise 1.5) such that $\bar{f}_1, \ldots, \bar{f}_r$ do not generate $\overline{I(C)}$.
(d) Show that $X \mapsto \bar{X}$ defines a bijection between the set of non-empty closed subvarieties X of $\mathbb{A}^n(k)$ and closed subvarieties Z of $\mathbb{P}^n(k)$ with $Z \cap \mathbb{A}^n(k) \neq \emptyset$.
The closed subset $\bar{X}$ of $\mathbb{P}^n(k)$ is called the *projective closure of* X.

Exercise 1.25. An *affine subspace* $H \subseteq \mathbb{A}^n(k)$ *of dimension* m is a subset of the form $v + W$, where $v \in k^n$ and $W \subseteq k^n$ is a subvector space of dimension m.
(a) Show that affine subspaces are closed subvarieties of $\mathbb{A}^n(k)$.
(b) Show that attaching to H its projective closure $\bar{H}$ in $\mathbb{P}^n(k)$ (Exercise 1.24) defines an injection of the set of affine subspaces of dimension m of $\mathbb{A}^n(k)$ into the set of linear subspaces of dimension m of $\mathbb{P}^n(k)$. Determine the image of this injection.
(c) Determine those affine algebraic sets in $\mathbb{A}^{n+1}(k)$ that are affine cones of linear subspaces of $\mathbb{P}^n(k)$.

Exercise 1.26. Let Y, Z be linear subspaces of $\mathbb{P}^n(k)$. Show that $Y \cap Z$ is again a linear subspace of dimension $\geq \dim(Y) + \dim(Z) - n$. Deduce that $Y \cap Z$ is always non-empty if $\dim(Y) + \dim(Z) \geq n$.
Conversely let $Y_1, \ldots, Y_r \subseteq \mathbb{P}^n(k)$ be finitely many linear subspaces and let $0 \leq d \leq n$ be an integer such that $\max_i \dim(Y_i) + d < n$. Show that there exists a linear subspace Z of $\mathbb{P}^n(k)$ of dimension d such that $Y_i \cap Z = \emptyset$ for all $i = 1, \ldots, r$. Deduce that for any finite subset $X \subset \mathbb{P}^n(k)$ there exists a hyperplane Z of $\mathbb{P}^n(k)$ such that $X \cap Z = \emptyset$.

Exercise 1.27. Let Y be a quasi-projective variety. Show that every finite subset of Y is contained in an open affine subvariety of Y.
Hint: Exercise 1.26.

Exercise 1.28$\Diamond$. Let $n \geq 0$ be an integer. Let G be the affine variety $\mathrm{GL}_{n+1}(k)$ (Exercise 1.15) and let $L_0 \subset k^{n+1}$ be a fixed one-dimensional subvector space. Show that the map $G \to \mathbb{P}^n(k)$, $g \mapsto g(L_0)$ is a surjective morphism of prevarieties.

Exercise 1.29. Let X be an affine variety.
(a) Show that any morphism $\mathbb{P}^n(k) \to X$ is constant.

(b) Let Z be a prevariety such that every morphism $Z \to \mathbb{P}^1(k)$ has a closed image (we will see in Corollary 13.41 that this is the case if Z is a projective variety). Show that every morphism $Z \to X$ is constant.

Hint: It suffices to consider the case $X = \mathbb{A}^1(k)$.

Exercise 1.30. Let $n, d > 0$ be integers. Let $M_0, \ldots, M_N \in k[X_0, \ldots, X_n]$ be all monomials in $X_0, \ldots, X_n$ of degree d.

(a) Define a k-algebra homomorphism $\theta \colon k[Y_0, \ldots, Y_N] \to k[X_0, \ldots, X_n]$ by $Y_i \mapsto M_i$ and let $\mathfrak{a} = \mathrm{Ker}\,\theta$. Show that $\mathfrak{a}$ is a homogeneous prime ideal (Exercise 1.20). Let $V_+(\mathfrak{a}) \subseteq \mathbb{P}^N(k)$ the projective variety defined by $\mathfrak{a}$ (Exercise 1.21).

(b) Consider the morphism

$$v_d \colon \mathbb{P}^n(k) \to \mathbb{P}^N(k), \quad (x_0 : \cdots : x_n) \mapsto (M_0(x_0, \ldots, x_n) : \cdots : M_N(x_0, \ldots, x_n)),$$

and show that v_d induces an isomorphism $\mathbb{P}^n(k) \cong V_+(\mathfrak{a})$ of prevarieties. Is $V_+(\mathfrak{a})$ a linear subspace of $\mathbb{P}^N(k)$?

(c) Let $f \in k[X_0, \ldots, X_n]$ be homogeneous of degree d. Show that $v_d(V_+(f))$ is the intersection of $V_+(\mathfrak{a})$ and a linear subspace of $\mathbb{P}^N(k)$.

The morphism v_d is called the *d-Uple embedding* or *d-fold Veronese embedding*.

2   Spectrum of a Ring

Contents

In the first chapter we attached to a system of polynomials $f_1, \ldots, f_r \in k[T_1, \ldots, T_n]$ with coefficients in an algebraically closed field k its set of zeros in $\mathbb{A}^n(k) = k^n$. If this set was irreducible we endowed it with the structure of a space with functions. The spaces with functions obtained in this way were called *affine varieties*. This construction gave us an anti-equivalence of categories between finitely generated domains A over k and affine varieties X. Here the points of X corresponded to maximal ideals of A.

In a second step we glued finitely many of these affine varieties along open subsets and called the obtained spaces with functions *prevarieties*. This allowed us to consider sets of zeros of homogeneous polynomials in projective space (projective varieties).

We remark that the reasons that we restricted ourselves to *irreducible* affine varieties and to the gluing of a *finite* number of affine varieties were of purely technical nature. With slightly more work we could have worked with arbitrary closed algebraic subsets to define (a more general notion of) an affine variety and we could have allowed the gluing of infinitely many affine varieties.

But as already explained in the third point of Section (1.9) it is unsatisfactory that our affine varieties depend only on the underlying subset of $\mathbb{A}^n(k)$. Another major drawback is the fact that the theory of varieties works for polynomials with coefficients in an algebraically closed field but not for coefficients in more general rings (as needed in number theory for example).

It would be desirable to associate to arbitrary rings A a geometric object which generalizes the construction of affine varieties out of finitely generated domains over an algebraically closed field. This will be done in the current chapter. In a first step we associate to A a topological space $\operatorname{Spec} A$. As a set this will be the set of prime ideals of A. This differs from the approach in the first chapter where the points of an affine variety corresponded to the maximal ideals. But for arbitrary rings there are "too few" maximal ideals (e.g. for local rings). Moreover, if $\varphi \colon A \to B$ is a ring homomorphism, the inverse image of a maximal ideal in B is not necessarily a maximal ideal of A while the analogous statement for prime ideals is true. Thus working with prime ideals we obtain a functorial construction. This is the content of the first part of this chapter.

But from the topological space $\operatorname{Spec} A$ we certainly cannot get back the ring A: For example for any field K the set $\operatorname{Spec} K$ consists of one element only (in the setting of chapter 1 this ambiguity did not exist, because for any finitely generated algebra A over an algebraically closed field k, which is itself a field, the natural homomorphism $k \to A$ is an isomorphism by Lemma 1.10). Therefore we again endow $\operatorname{Spec} A$ with "functions". More precisely, we define a sheaf $\mathscr{O}_{\operatorname{Spec} A}$ of rings on $\operatorname{Spec} A$. Therefore the second part

© Springer Fachmedien Wiesbaden GmbH, part of Springer Nature 2020
U. Görtz und T. Wedhorn, *Algebraic Geometry I: Schemes*, Springer Studium
Mathematik – Master, https://doi.org/10.1007/978-3-658-30733-2_3

of this chapter will be a short excursion in which we present the necessary notions from the theory of sheaves, which are generalizations of systems of functions. Equipped with this machinery we can construct $\mathscr{O}_{\operatorname{Spec} A}$ in the third part of this chapter. Topological spaces endowed with sheaves of rings are called *ringed spaces*. In fact, $(\operatorname{Spec} A, \mathscr{O}_{\operatorname{Spec} A})$ will be always in the subcategory of so-called *locally ringed spaces*. Locally ringed spaces isomorphic to $(\operatorname{Spec} A, \mathscr{O}_{\operatorname{Spec} A})$ will be called *affine schemes*, and we will show that $A \mapsto (\operatorname{Spec} A, \mathscr{O}_{\operatorname{Spec} A})$ defines an anti-equivalence from the category of rings to the category of affine schemes.

As for affine varieties the next step then will be to define objects obtained by gluing affine schemes. This will be done in the next chapter. In this way we will obtain the basic objects of modern algebraic geometry: schemes.

Spectrum of a ring as a topological space

(2.1) Definition of $\operatorname{Spec} A$ as a topological space.

We start with the following basic definition. Let A be a ring. We set

(2.1.1) $\operatorname{Spec} A := \{\, \mathfrak{p} \subset A \; ; \; \mathfrak{p} \text{ prime ideal} \,\}.$

We will now endow $\operatorname{Spec} A$ with the structure of a topological space. For every subset M of A, we denote by $V(M)$ the set of prime ideals of A containing M. Clearly, if $\mathfrak{a}$ is the ideal generated by M, $V(M) = V(\mathfrak{a})$. For any $f \in A$ we write $V(f)$ instead of $V(\{f\})$.

Lemma 2.1. *The map $\mathfrak{a} \mapsto V(\mathfrak{a})$ is an inclusion reversing map from the set of ideals of A to the set of subsets of $\operatorname{Spec} A$. Moreover, the following relations hold:*
(1) $V(0) = \operatorname{Spec} A$, $V(1) = \emptyset$.
(2) For every family $(\mathfrak{a}_i)_{i \in I}$ of ideals

$$V\Big(\bigcup_{i \in I} \mathfrak{a}_i\Big) = V\Big(\sum_{i \in I} \mathfrak{a}_i\Big) = \bigcap_{i \in I} V(\mathfrak{a}_i).$$

(3) For two ideals $\mathfrak{a}$, $\mathfrak{a}'$

$$V(\mathfrak{a} \cap \mathfrak{a}') = V(\mathfrak{a}\mathfrak{a}') = V(\mathfrak{a}) \cup V(\mathfrak{a}').$$

Proof. Assertions (1) and (2) are obvious, and (3) is simply the fact that a prime ideal contains $\mathfrak{a}$ or $\mathfrak{a}'$ if and only if it contains $\mathfrak{a} \cap \mathfrak{a}'$ or equivalently, if it contains $\mathfrak{a}\mathfrak{a}'$ (Proposition B.2). $\square$

The lemma shows that the subsets $V(\mathfrak{a})$ of $\operatorname{Spec} A$ form the closed sets of a topology on $\operatorname{Spec} A$. This leads us to the following definition.

Definition 2.2. *Let A be a ring. The set $\operatorname{Spec} A$ of all prime ideals of A with the topology whose closed sets are the sets $V(\mathfrak{a})$, where $\mathfrak{a}$ runs through the set of ideals of A, is called the prime spectrum of A or simply the spectrum of A. The topology thus defined is called the Zariski topology on $\operatorname{Spec} A$.*

If x is a point in $\operatorname{Spec} A$, we will often write $\mathfrak{p}_x$ instead of x when we think of x as a prime ideal of A.

Note that the definition of Spec A and the sets $V(\mathfrak{a})$ is analogous to the definitions made in Section (1.1) where we considered the case $A = k[T_1, \ldots, T_n]$ for an algebraically closed field k and where the points in $V(\mathfrak{a})$ corresponded to the maximal ideals of A containing $\mathfrak{a}$. We will explain in Section (3.13), why working with maximal ideals suffices for finitely generated algebras over a field.

Again we have a construction to attach ideals to subsets of Spec A: For every subset Y of Spec A we set

$$(2.1.2) \qquad\qquad I(Y) := \bigcap_{\mathfrak{p} \in Y} \mathfrak{p}.$$

We obtain an inclusion reversing map $Y \mapsto I(Y)$ from the set of subsets of Spec A to the set of ideals of A. Note that $I(\emptyset) = A$. The maps V and I are related as follows.

Proposition 2.3. *Let A be a ring, $\mathfrak{a} \subseteq A$ an ideal, and Y a subset of* Spec A.
(1) $\operatorname{rad}(I(Y)) = I(Y)$.
(2) $I(V(\mathfrak{a})) = \operatorname{rad}(\mathfrak{a})$, $V(I(Y)) = \overline{Y}$, *where $\overline{Y}$ denotes the closure of Y in* Spec A.
(3) *The maps*

$$\{ideals\ \mathfrak{a}\ of\ A\ with\ \mathfrak{a} = \operatorname{rad}(\mathfrak{a})\} \underset{I(Y) \leftarrow Y}{\overset{\mathfrak{a} \mapsto V(\mathfrak{a})}{\rightleftarrows}} \{closed\ subsets\ Y\ of\ \mathrm{Spec}\ A\}$$

are mutually inverse bijections.

Proof. The relation $\mathfrak{a} = \operatorname{rad}(\mathfrak{a})$ means that for $f \in A$, $f^n \in \mathfrak{a}$ implies already $f \in \mathfrak{a}$. This certainly holds for prime ideals and therefore for arbitrary intersections of prime ideals as well. That proves (1). The first assertion of (2) follows from the fact that the radical of an ideal equals the intersection of all prime ideals containing it (B.1.1). A closed set $V(\mathfrak{b})$ (for some ideal $\mathfrak{b}$) contains Y if and only if $\mathfrak{b}$ is contained in all prime ideals that belong to Y. This is equivalent to $\mathfrak{b} \subseteq I(Y)$. Therefore $V(I(Y))$ is the smallest closed subset of Spec A containing Y. This shows the second assertion of (2). Part (3) follows from (2). $\qquad\square$

In particular we see that the closure of a set consisting of only one point $x \in$ Spec A is the set $V(\mathfrak{p}_x)$ of prime ideals containing $\mathfrak{p}_x$.

(2.2) Properties of the topological space Spec A.

Let A be a ring. Let

$$(2.2.1) \qquad\qquad D(f) := D_A(f) := \mathrm{Spec}\ A \setminus V(f)$$

be the open set of prime ideals of A not containing f. Open subsets of Spec A of this form are called *principal open sets of* Spec A. Clearly, $D(0) = \emptyset$, $D(1) = \mathrm{Spec}\ A$, and more generally $D(u) = \mathrm{Spec}\ A$ for every unit $u \in A$. As for a prime ideal $\mathfrak{p}$ and two elements $f, g \in A$ we have $fg \notin \mathfrak{p}$ if and only if $f \notin \mathfrak{p}$ and $g \notin \mathfrak{p}$, we find

$$(2.2.2) \qquad\qquad D(f) \cap D(g) = D(fg).$$

Lemma 2.4. *Let (f_i) be a family of elements in A and let $g \in A$. Then $D(g) \subseteq \bigcup_i D(f_i)$ if and only if there exists an integer $n > 0$ such that g^n is contained in the ideal $\mathfrak{a}$ generated by the f_i.*

Proof. Indeed, $D(g) \subseteq \bigcup_i D(f_i)$ is equivalent to $V(g) \supseteq V(\mathfrak{a})$ which is equivalent to $g \in \mathrm{rad}(\mathfrak{a})$ by Proposition 2.3. $\qquad\square$

Applying this to $g = 1$ it follows that $(D(f_i))_i$ is a covering of $\mathrm{Spec}\,A$ if and only if the ideal generated by the f_i is equal to A.

Proposition 2.5. *Let A be a ring. The principal open subsets $D(f)$ for $f \in A$ form a basis of the topology of $\mathrm{Spec}\,A$. For all $f \in A$ the open sets $D(f)$ are quasi-compact (Definition 1.22). In particular, the space $\mathrm{Spec}\,A$ is quasi-compact.*

Proof. By Lemma 2.1 (2), every closed subset of $\mathrm{Spec}\,A$ is the intersection of closed sets of the form $V(f)$. By taking complements we see that the $D(f)$ form a basis for the topology.

Let $(g_i)_{i \in I}$ be a family of elements of A such that $D(f) \subseteq \bigcup_{i \in I} D(g_i)$. Then we have seen above that there exists an integer $n \geq 1$ such that $f^n = \sum_{i \in I} a_i g_i$, where $a_i \in A$ and $a_i = 0$ for all $i \notin J$, $J \subseteq I$ a suitable finite subset. Hence $D(f) \subseteq \bigcup_{j \in J} D(g_j)$. This proves that $D(f)$ is quasi-compact by the first part of the proposition. $\qquad\square$

Proposition 2.6. *Let A be a ring. A subset Y of $\mathrm{Spec}\,A$ is irreducible if and only if $\mathfrak{p} := I(Y)$ is a prime ideal. In this case $\{\mathfrak{p}\}$ is dense in $\overline{Y}$.*

Proof. Assume that Y is irreducible. Let $f, g \in A$ with $fg \in \mathfrak{p}$. Then

$$Y \subseteq V(fg) = V(f) \cup V(g).$$

As Y is irreducible, $Y \subseteq V(f)$ or $Y \subseteq V(g)$ which implies $f \in \mathfrak{p}$ or $g \in \mathfrak{p}$.

Conversely let $\mathfrak{p}$ be prime. Then by Proposition 2.3, $\overline{Y} = V(\mathfrak{p}) = V(I(\{\mathfrak{p}\})) = \overline{\{\mathfrak{p}\}}$. Therefore $\overline{Y}$ is the closure of the irreducible set $\{\mathfrak{p}\}$ and therefore irreducible. This implies that the dense subset Y is also irreducible (Lemma 1.17). $\qquad\square$

Note that for arbitrary irreducible subsets Y the prime ideal $I(Y)$ is not necessarily a point in Y. But this is clearly true if Y is closed or, more generally, if Y is locally closed (see Exercise 2.8). Together with Proposition 2.3 we obtain:

Corollary 2.7. *The map $\mathfrak{p} \mapsto V(\mathfrak{p}) = \overline{\{\mathfrak{p}\}}$ is a bijection from $\mathrm{Spec}\,A$ onto the set of closed irreducible subsets of $\mathrm{Spec}\,A$. Via this bijection, the minimal prime ideals of A correspond to the irreducible components of $\mathrm{Spec}\,A$.*

We introduce the following notions, which will be used throughout the book, to deal with such topological spaces.

Definition 2.8. *Let X be an arbitrary topological space.*
(1) *A point $x \in X$ is called* closed *if the set $\{x\}$ is closed,*
(2) *We say that a point $\eta \in X$ is a* generic point *if $\overline{\{\eta\}} = X$.*
(3) *Let x and x' be two points of X. We say that x is a* generization *of x' or that x' is a* specialization *of x if $x' \in \overline{\{x\}}$.*
(4) *A point $x \in X$ is called a* maximal point *if its closure $\overline{\{x\}}$ is an irreducible component of X.*

Thus a point $\eta \in X$ is generic if and only if it is a generization of every point of X. As the closure of an irreducible set is again irreducible, the existence of a generic point implies that X is irreducible.

Example 2.9. If $X = \operatorname{Spec} A$ is the spectrum of a ring, the notions introduced in Definition 2.8 have the following algebraic meaning.

(1) A point $x \in X$ is closed if and only if $\mathfrak{p}_x$ is a maximal ideal.

(2) A point $\eta \in X$ is a generic point of X if and only if $\mathfrak{p}_\eta$ is the unique minimal prime ideal. This exists if and only if the nilradical of A is a prime ideal. Thus Proposition 2.6 shows that X is irreducible if and only if its nilradical is a prime ideal.

(3) A point x is a generization of a point x' (in other words, x' is a specialization of x) if and only if $\mathfrak{p}_x \subseteq \mathfrak{p}_{x'}$.

(4) A point $x \in X$ is a maximal point if and only if $\mathfrak{p}_x$ is a minimal prime ideal.

(2.3) The functor $A \mapsto \operatorname{Spec} A$.

We will now show that $A \mapsto \operatorname{Spec} A$ defines a contravariant functor from the category of rings to the category of topological spaces. Let $\varphi \colon A \to B$ be a homomorphism of rings. If $\mathfrak{q}$ is a prime ideal of B, $\varphi^{-1}(\mathfrak{q})$ is a prime ideal of A. Therefore we obtain a map

$$(2.3.1) \qquad {}^a\varphi = \operatorname{Spec} \varphi \colon \operatorname{Spec} B \to \operatorname{Spec} A, \qquad \mathfrak{q} \mapsto \varphi^{-1}(\mathfrak{q}).$$

Proposition 2.10. *Let* $\varphi \colon A \to B$ *be a ring homomorphism.*

(1) *For every subset* $M \subseteq A$, *the relation*

$$ {}^a\varphi^{-1}(V(M)) = V(\varphi(M)) $$

holds. In particular, for $f \in A$,

$$ {}^a\varphi^{-1}(D(f)) = D(\varphi(f)). $$

(2) *For every ideal* $\mathfrak{b}$ *of* B,

$$(2.3.2) \qquad V(\varphi^{-1}(\mathfrak{b})) = \overline{{}^a\varphi(V(\mathfrak{b}))}.$$

Proof. (1). A prime ideal $\mathfrak{q}$ of B contains $\varphi(M)$ if and only if $\varphi^{-1}(\mathfrak{q})$ contains M.

(2). By Proposition 2.3 (2), we can rewrite the right hand side as $V(I({}^a\varphi(V(\mathfrak{b}))))$. But

$$ I({}^a\varphi(V(\mathfrak{b}))) = \bigcap_{\mathfrak{q} \in V(\mathfrak{b})} \varphi^{-1}(\mathfrak{q}) = \varphi^{-1}(\operatorname{rad}(\mathfrak{b})) = \operatorname{rad} \varphi^{-1}(\mathfrak{b}), $$

and the claim follows by applying $V(-)$. $\qquad\square$

The proposition shows in particular that ${}^a\varphi \colon \operatorname{Spec} B \to \operatorname{Spec} A$ is continuous. As ${}^a(\psi \circ \varphi) = {}^a\varphi \circ {}^a\psi$ for any ring homomorphism $\psi \colon B \to C$, we obtain a contravariant functor $A \mapsto \operatorname{Spec} A$ from the category of rings to the category of topological spaces.

Corollary 2.11. *The map* ${}^a\varphi$ *is dominant (i.e., its image is dense in* $\operatorname{Spec} A$) *if and only if every element of* $\operatorname{Ker}(\varphi)$ *is nilpotent.*

Proof. We apply (2.3.2) to $\mathfrak{b} = 0$. $\qquad\square$

Proposition 2.12. *Let A be a ring.*
(1) *Let $\varphi\colon A \to B$ be a surjective homomorphism of rings with kernel $\mathfrak{a}$. Then ${}^a\varphi$ is a homeomorphism of $\operatorname{Spec} B$ onto the closed subset $V(\mathfrak{a})$ of $\operatorname{Spec} A$.*
(2) *Let S be a multiplicative subset of A and let $\varphi\colon A \to S^{-1}A =: B$ be the canonical homomorphism. Then ${}^a\varphi$ is a homeomorphism of $\operatorname{Spec} S^{-1}A$ onto the subspace of $\operatorname{Spec} A$ consisting of prime ideals $\mathfrak{p} \subset A$ with $S \cap \mathfrak{p} = \emptyset$.*

Proof. In both cases it is clear that ${}^a\varphi$ is injective with the stated image. Moreover in both cases a prime ideal $\mathfrak{q}$ of B contains an ideal $\mathfrak{b}$ of B, if and only if $\varphi^{-1}(\mathfrak{q})$ contains $\varphi^{-1}(\mathfrak{b})$. This shows that ${}^a\varphi(V(\mathfrak{b})) = V(\varphi^{-1}(\mathfrak{b})) \cap \operatorname{Im}({}^a\varphi)$. Therefore ${}^a\varphi$ is a homeomorphism onto its image. $\qquad\square$

Remark 2.13. Let A be a ring and let $\mathfrak{p}, \mathfrak{q} \subset A$ be prime ideals. Proposition 2.12 shows that for a prime ideal $\mathfrak{p} \subset A$ the passage from A to $A_\mathfrak{p}$ cuts out all prime ideals except those contained in $\mathfrak{p}$. The passage from A to $A/\mathfrak{q}$ cuts out all prime ideals except those containing $\mathfrak{q}$. Hence if $\mathfrak{q} \subseteq \mathfrak{p}$ localizing with respect to $\mathfrak{p}$ and taking the quotient modulo $\mathfrak{q}$ (in either order as these operations commute) we obtain a ring whose prime ideals are those prime ideals of A that lie between $\mathfrak{q}$ and $\mathfrak{p}$. For $\mathfrak{q} = \mathfrak{p}$ we obtain the field

$$(2.3.3) \qquad\qquad \kappa(\mathfrak{p}) := A_\mathfrak{p}/\mathfrak{p}A_\mathfrak{p} = \operatorname{Frac}(A/\mathfrak{p}),$$

which is called the *residue field at* $\mathfrak{p}$.

(2.4) Examples.

First of all note that $\operatorname{Spec} A = \emptyset$ if and only if $A = \{0\}$. If A is a field or any ring with a single prime ideal, $\operatorname{Spec} A$ consists of a single point. The spectrum of an Artinian ring is finite and discrete (Proposition B.36).

Example 2.14. Let A be a principal ideal domain (e.g. $A = \mathbb{Z}$ or $A = k[T]$ for a field k). In this case, the maximal ideals are of the form (p) for a prime element p of A, and all prime ideals are maximal or the zero ideal. Therefore the closed points of $\operatorname{Spec} A$ correspond to equivalence classes of prime elements $p \in A$, where p and p' are called equivalent if there exists a unit $u \in A^\times$ with $p' = up$ (i.e., p and p' generate the same ideal of A). Let $\eta \in \operatorname{Spec} A$ be the point with $\mathfrak{p}_\eta = \{0\}$. Then the closure of $\{\eta\}$ is $\operatorname{Spec} A$.

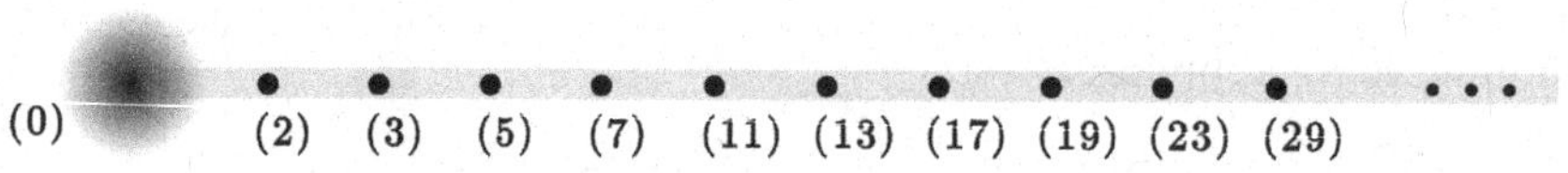

(0) $\quad\quad$ (2) $\quad$ (3) $\quad$ (5) $\quad$ (7) $\quad$ (11) $\quad$ (13) $\quad$ (17) $\quad$ (19) $\quad$ (23) $\quad$ (29) $\quad$ $\cdots$

Figure 2.1: A schematic picture of the spectrum of $\mathbb{Z}$. The closed points correspond to the maximal ideals of $\mathbb{Z}$, and besides them there is the generic point (0) which is dense in $\operatorname{Spec} \mathbb{Z}$.

As A is a principal ideal domain, every closed subset of $\operatorname{Spec} A$ is of the form $V(f)$ for some $f \in A$. Assume $f \neq 0$ (i.e., $V(f) \neq \operatorname{Spec} A$) and let $f = p_1^{e_1} p_2^{e_2} \cdots p_r^{e_r}$ with pairwise non-equivalent prime elements $p_1, \ldots, p_r$ $(r \geq 0)$ and integers $e_i \geq 1$. Then $V(f)$ consists of those closed points which correspond to the prime divisors of f, that is, $V(f) = \{(p_1), \ldots, (p_r)\}$. Therefore the closed subsets $\neq \operatorname{Spec} A$ are the finite sets consisting of closed points.

If $g \neq 0$ is a second element of A, $V(f) \cap V(g) = V(f, g) = V(d)$, where d is a greatest common divisor of f and g. Moreover, $V(f) \cup V(g) = V((f) \cap (g)) = V(e)$, where e is a lowest common multiple of f and g.

If A is a local principal ideal domain, but not a field (i.e., A is a discrete valuation ring), $\operatorname{Spec} A$ consists only of two points η and x, where $\mathfrak{p}_x$ is the maximal ideal and $\mathfrak{p}_\eta = \{0\}$. The only nontrivial open subset of $\operatorname{Spec} A$ is then $\{\eta\}$.

Example 2.15. Let k be an algebraically closed field. We saw in Chapter 1 that there is a contravariant equivalence between the category of finitely generated integral k-algebras A and the category of affine varieties V. If A corresponds to V, the maximal ideals of A are the points of V. Therefore we can consider V as a subset of $\operatorname{Spec} A$. It follows from the definition of the topology on V (see Sections (1.1) and (1.2)) that the variety V carries the topology induced by $\operatorname{Spec} A$.

Example 2.16. Let $A = R[T]$, where R is a principal ideal domain. We assume that R is not a field (otherwise $R[T]$ is a principal ideal domain and this case was already considered in Example 2.14). Let $X = \operatorname{Spec} R[T]$. As R is factorial, $R[T]$ is factorial as well, and the prime elements of $R[T]$ are either of the form p, where p is a prime element of R, or of the form f, where $f \in R[T]$ is a primitive polynomial which is irreducible in $\operatorname{Quot}(R)[T]$ (by Gauß' theorem, e.g., see [La] IV §2, Thm. 2.3).

If $p \in R$ is a prime element, R/pR is a field. By Proposition 2.12, the closure $V(pR[T])$ of $\{pR[T]\}$ is homeomorphic to $\operatorname{Spec}(R/pR)[T]$, and $(R/pR)[T]$ is a principal ideal domain with infinitely many nonequivalent prime elements (cf. Example 2.14). We see that $(pR[T])$ is not a maximal ideal, and the prime ideals in $V(pR[T])$ different from $(pR[T])$ are the maximal ideals generated by p and f where $f \in R[T]$ is a polynomial such that its image in $(R/pR)[T]$ is irreducible.

The situation is more complicated for prime ideals of the form $fR[T]$, where f is a primitive irreducible polynomial. If the leading coefficient of f is a unit in R, it is possible to divide in $R[T]$ by f with unique remainder, and therefore $R[T]/fR[T]$ is finitely generated as R-module (even free of rank $\deg(f)$). This implies that $fR[T]$ is not a maximal ideal, as otherwise R would be a field by Lemma 1.9. For other primitive irreducible polynomials f, $fR[T]$ might be a maximal ideal, namely if R contains only finitely many prime elements (up to equivalence): If $0 \neq a \in R$ is an element which is divisible by all prime elements of R we have, with $f := aT - 1$,

$$R[T]/fR[T] \cong R[a^{-1}] = \operatorname{Quot}(R),$$

which shows that $fR[T]$ is a maximal ideal.

Excursion: Sheaves

It is clear that the topology on the space $\operatorname{Spec} A$ is not sufficiently fine to determine the geometric objects we are looking for, as was already the case with prevarieties. We therefore want to equip this topological space with additional structure. As a guideline, we take the situation of prevarieties: there we defined a system of functions on a prevariety, and found that this additional datum determines the structure up to isomorphisms given by polynomials. The functions made up the affine coordinate ring of the prevariety. In the current situation, in a sense we are working backwards: we start with a ring A, and associate to it the topological space $\operatorname{Spec} A$ of prime ideals in A. *This means that the elements of A should be thought of as the functions we want to consider* on $\operatorname{Spec} A$. Let us discuss this important heuristic a little more precisely.

First, how can we think of elements of A as of functions of $\operatorname{Spec} A$? Strictly speaking, we cannot. However, given $f \in A$ and $x \in \operatorname{Spec} A$, we get an element $f(x) \in \kappa(x)$, where $f(x)$ is the residue class in $\kappa(x) := A_{\mathfrak{p}_x}/\mathfrak{p}_x A_{\mathfrak{p}_x}$ of the image of f in the localization $A_{\mathfrak{p}_x}$. This is completely analogous to the situation with prevarieties. However, we do not get a function with a well defined target, but rather a collection of values $f(x)$ in different targets $\kappa(x)$, $x \in \operatorname{Spec} A$. Nevertheless, this is a useful point of view. For instance, we can interpret $D(f)$ as the set of points where $f(x) \neq 0$, i. e. where the function f does not vanish.

On the other hand, since the elements of A are not, strictly speaking, functions on $\operatorname{Spec} A$, we cannot use the notion of system of functions as in Chapter 1. We need a more flexible construction instead, and it turns out that the key point for working with "functions" is restricting and gluing of functions (rather than evaluating them at points of the source). This leads to the notion of *sheaf*, which we will define and study in the following sections. Although the setting is more abstract now than it was with systems of functions, it is still advisable to think of sections of a sheaf on an open subset (see below) as some kind of functions defined on this open subset.

(2.5) Presheaves and Sheaves.

Definition 2.17. *Let X be a topological space. A* presheaf *$\mathscr{F}$ on X consists of the following data,*
(a) *for every open set U of X a set $\mathscr{F}(U)$,*
(b) *for each pair of open sets $U \subseteq V$ a map $\operatorname{res}^V_U \colon \mathscr{F}(V) \to \mathscr{F}(U)$, called* restriction map,
such that the following conditions hold
(1) *$\operatorname{res}^U_U = \operatorname{id}_{\mathscr{F}(U)}$ for every open set $U \subseteq X$,*
(2) *for $U \subseteq V \subseteq W$ open sets of X, $\operatorname{res}^W_U = \operatorname{res}^V_U \circ \operatorname{res}^W_V$.*

Let $\mathscr{F}_1$ and $\mathscr{F}_2$ be presheaves on X. A morphism of presheaves *$\varphi \colon \mathscr{F}_1 \to \mathscr{F}_2$ is a family of maps $\varphi_U \colon \mathscr{F}_1(U) \to \mathscr{F}_2(U)$ (for all $U \subseteq X$ open), such that for all pairs of open sets $U \subseteq V$ in X the following diagram commutes*

$$
\begin{array}{ccc}
\mathscr{F}_1(V) & \xrightarrow{\ \varphi_V\ } & \mathscr{F}_2(V) \\
{\scriptstyle \operatorname{res}^V_U}\big\downarrow & & \big\downarrow{\scriptstyle \operatorname{res}^V_U} \\
\mathscr{F}_1(U) & \xrightarrow{\ \varphi_U\ } & \mathscr{F}_2(U).
\end{array}
$$

If $U \subseteq V$ are open sets of X and $s \in \mathscr{F}(V)$ we will often write $s_{|U}$ instead of $\operatorname{res}^V_U(s)$. The elements of $\mathscr{F}(U)$ are called *sections* of $\mathscr{F}$ over U. Very often we will also write $\Gamma(U, \mathscr{F})$ instead of $\mathscr{F}(U)$.

We can also describe presheaves as follows. Let (Ouv_X) be the category whose objects are the open sets of X and, for two open sets $U, V \subseteq X$, $\mathrm{Hom}(U, V)$ is empty if $U \not\subseteq V$, and consists of the inclusion map $U \to V$ if $U \subseteq V$ (composition of morphisms being the composition of the inclusion maps). Then a presheaf is the same as a contravariant functor $\mathscr{F}$ from the category (Ouv_X) to the category (Sets) of sets.

By replacing (Sets) in this definition by some other category $\mathcal{C}$ (e.g. the category of abelian groups, the category of rings, the category of R-modules, or the category of R-algebras, R a fixed ring) we obtain the notion of a *presheaf $\mathscr{F}$ with values in $\mathcal{C}$* (e.g. a *presheaf of abelian groups*, a *presheaf of rings*, a *presheaf of R-modules*, or a *presheaf of R-algebras*). This signifies that $\mathscr{F}(U)$ is an object in $\mathcal{C}$ for every open subset U of X and that the restriction maps are morphisms in $\mathcal{C}$. A morphism $\mathscr{F}_1 \to \mathscr{F}_2$ of presheaves with values in $\mathcal{C}$ is then simply a morphism of functors.

Let $\mathscr{F}$ be a presheaf on a topological space X, let U be an open set in X and let $\mathscr{U} = (U_i)_{i \in I}$ be an open covering of U. We define maps (depending on $\mathscr{U}$)

$$\rho\colon \mathscr{F}(U) \to \prod_{i \in I} \mathscr{F}(U_i), \quad s \mapsto (s_{|U_i})_i$$

$$\sigma\colon \prod_{i \in I} \mathscr{F}(U_i) \to \prod_{(i,j) \in I \times I} \mathscr{F}(U_i \cap U_j), \quad (s_i)_i \mapsto (s_{i|U_i \cap U_j})_{(i,j)},$$

$$\sigma'\colon \prod_{i \in I} \mathscr{F}(U_i) \to \prod_{(i,j) \in I \times I} \mathscr{F}(U_i \cap U_j), \quad (s_i)_i \mapsto (s_{j|U_i \cap U_j})_{(i,j)}.$$

Definition 2.18. *The presheaf $\mathscr{F}$ is called a* **sheaf***, if it satisfies for all U and all coverings (U_i) as above the following condition:*
(Sh) *The diagram*

$$\mathscr{F}(U) \xrightarrow{\rho} \prod_{i \in I} \mathscr{F}(U_i) \underset{\sigma'}{\overset{\sigma}{\rightrightarrows}} \prod_{(i,j) \in I \times I} \mathscr{F}(U_i \cap U_j)$$

is exact. This means that the map ρ is injective and that its image is the set of elements $(s_i)_{i \in I} \in \prod_{i \in I} \mathscr{F}(U_i)$ such that $\sigma((s_i)_i) = \sigma'((s_i)_i)$.

In other words, a presheaf $\mathscr{F}$ is a sheaf if and only if for all open sets U in X and every open covering $U = \bigcup_i U_i$ the following two conditions hold:
(Sh1) Let $s, s' \in \mathscr{F}(U)$ with $s_{|U_i} = s'_{|U_i}$ for all i. Then $s = s'$.
(Sh2) Given $s_i \in \mathscr{F}(U_i)$ for all i such that $s_{i|U_i \cap U_j} = s_{j|U_i \cap U_j}$ for all i, j. Then there exists an $s \in \mathscr{F}(U)$ such that $s_{|U_i} = s_i$ (note that s is unique by (Sh1)).
Heuristically, these conditions say that functions are determined by local information, and that functions can be glued. Compare Section (1.12).

A *morphism of sheaves* is a morphism of presheaves. We obtain the category of sheaves on the topological space X, which we denote by $(\mathrm{Sh}(X))$. In the same way we can define the notion of a *sheaf of abelian groups*, a *sheaf of rings*, a *sheaf of R-modules*, or a *sheaf of R-algebras*.

For presheaves of abelian groups (or with values in any abelian category) we can reformulate the definition of a sheaf slightly: Such a presheaf $\mathscr{F}$ is a sheaf if and only if for all open subsets U and all coverings (U_i) of U the sequence of abelian groups

$$0 \to \mathscr{F}(U) \to \prod_i \mathscr{F}(U_i) \quad \to \prod_{i,j} \mathscr{F}(U_i \cap U_j),$$

(2.5.1)

$$s \mapsto (s_{|U_i})_i, \quad (s_i)_i \mapsto (s_{i|U_i \cap U_j} - s_{j|U_i \cap U_j})_{i,j}$$

is exact.

Note that if $\mathscr{F}$ is a sheaf on X, $\mathscr{F}(\emptyset)$ is a set consisting of one element (apply the condition (Sh) to the covering of the empty set with empty index set). In particular, if X consists of one point, a sheaf $\mathscr{F}$ on X is already uniquely determined by $\mathscr{F}(X)$ and sometimes we identify $\mathscr{F}$ with $\mathscr{F}(X)$.

Examples 2.19.
(1) If $\mathscr{F}$ is a presheaf on a topological space X and U is an open subspace of X we obtain a presheaf $\mathscr{F}_{|U}$ on U by setting $\mathscr{F}_{|U}(V) = \mathscr{F}(V)$ for every open subset V in U. If $\mathscr{F}$ is a sheaf, $\mathscr{F}_{|U}$ is a sheaf on U. We call $\mathscr{F}_{|U}$ the *restriction of $\mathscr{F}$ to U*.
(2) Let X and Y be topological spaces. For $U \subseteq X$ open let $\mathscr{F}(U)$ be the set of continuous maps $U \to Y$ and define the restriction maps by the usual restriction of continuous functions. Then $\mathscr{F}$ is a sheaf.
(3) Let K be a field and let $(X, \mathscr{O}_X)$ be a space with functions over K (Section (1.12)). Then $\mathscr{O}_X$ is a sheaf of K-algebras on X.

In particular, if X is a real C^r-manifold ($0 \le r \le \infty$) and if we denote for any open subset U of X by $\mathscr{C}^r_X(U)$ the set of C^r-functions $U \to \mathbb{R}$, $\mathscr{C}^r_X$ is a sheaf of $\mathbb{R}$-algebras on X.
(4) Let X be a topological space and define

$$\mathscr{F}(U) = \{\, f \colon U \to \mathbb{R} \text{ continuous} \; ; \; f(U) \subset \mathbb{R} \text{ bounded} \,\}$$

for all $U \subseteq X$ open. Then $\mathscr{F}$ is a presheaf (the restriction maps being the usual restriction of functions) but it is not a sheaf in general.

If we know the value $\mathscr{F}(U)$ of a sheaf on every element U of some basis $\mathcal{B}$ of the topology on X, we can use the sheaf property to determine $\mathscr{F}(V)$ on an arbitrary open. We simply cover V by elements of $\mathcal{B}$. Here is a more systematic way of saying this:

$$\mathscr{F}(V) \;=\; \{(s_U)_U \in \prod_{\substack{U \in \mathcal{B} \\ U \subseteq V}} \mathscr{F}(U); \text{ for all } U' \subseteq U \text{ both in } \mathcal{B} : s_{U|U'} = s_{U'}\}$$

$$=\; \varprojlim_{\substack{U \in \mathcal{B} \\ U \subseteq V}} \mathscr{F}(U).$$

Using this observation, we see that it suffices to define a sheaf on a basis $\mathcal{B}$ of open sets of the topology of a topological space X: Consider $\mathcal{B}$ as a full subcategory of (Ouv_X). Then a *presheaf on $\mathcal{B}$* is a contravariant functor $\mathscr{F} \colon \mathcal{B} \to (\mathrm{Sets})$. Every such presheaf $\mathscr{F}$ on $\mathcal{B}$ can be extended to a presheaf $\mathscr{F}'$ on X by setting, for V open in X,

(2.5.2) $$\mathscr{F}'(V) = \varprojlim_U \mathscr{F}(U),$$

where U runs through the set of $U \in \mathcal{B}$ with $U \subseteq V$ (ordered by inclusion, the transition maps given by the restriction maps). A morphism of presheaves on $\mathcal{B}$ is again defined as a morphism of functors.

To formulate the sheaf property, first assume that $\mathcal{B}$ is stable under finite intersections. Then we call a presheaf $\mathcal{F}$ on $\mathcal{B}$ a *sheaf* if $\mathcal{F}$ satisfies condition (Sh) of Definition 2.18 for every $U \in \mathcal{B}$ and for every open covering $(U_i)_i$ of U with $U_i \in \mathcal{B}$ for all i.

In general, the intersections $U_i \cap U_j$ in the previous paragraph might not be in $\mathcal{B}$, so we have to cover them by elements of $\mathcal{B}$. We arrive at the following proposition, which is easy to prove.

Proposition 2.20. *The presheaf $\mathcal{F}'$ on X is a sheaf if and only if $\mathcal{F}$ satisfies the following condition: For every $U \in \mathcal{B}$, for every open covering $(U_i)_i$ of U with $U_i \in \mathcal{B}$ for all i, and for every open covering $(U_{ijk})_k$ of $U_i \cap U_j$ with all $U_{ijk} \in \mathcal{B}$, the diagram*

$$\mathcal{F}(U) \xrightarrow{\ \rho\ } \prod_{i \in I} \mathcal{F}(U_i) \underset{\sigma'}{\overset{\sigma}{\rightrightarrows}} \prod_{i,j,k} \mathcal{F}(U_{ijk})$$

is exact (cf. condition (Sh) of Definition 2.18; the maps σ, σ' are defined analogously).

In this case, we say that $\mathcal{F}$ is a *sheaf on $\mathcal{B}$*. Attaching to $\mathcal{F}$ the sheaf $\mathcal{F}'$ on X is clearly functorial in $\mathcal{F}$ and we obtain an equivalence between the category of sheaves on $\mathcal{B}$ and the category of sheaves on X.

Similar results hold for sheaves in a category $\mathcal{C}$ in which projective limits exist, e.g., the category of abelian groups.

(2.6) Stalks of Sheaves.

Let X be a topological space, $\mathcal{F}$ be a presheaf on X, and let $x \in X$ be a point. The system $((\mathcal{F}(U))_U, (\mathrm{res}^V_U)_{V \supseteq U})$ which is indexed by the set of open subsets $U \subseteq X$ with $x \in U$, ordered by reverse containment, is a filtered inductive system.

Definition 2.21. *The inductive limit*

$$\mathcal{F}_x := \varinjlim_{U \ni x} \mathcal{F}(U)$$

is called the stalk *of $\mathcal{F}$ in x.*

In other words, $\mathcal{F}_x$ is the set of equivalence classes of pairs (U, s), where U is an open neighborhood of x and $s \in \mathcal{F}(U)$. Here two such pairs (U_1, s_1) and (U_2, s_2) are equivalent, if there exists an open neighborhood V of x with $V \subseteq U_1 \cap U_2$ such that $s_{1|V} = s_{2|V}$.

For each open neighborhood U of x we have a canonical map

(2.6.1) $$\mathcal{F}(U) \to \mathcal{F}_x, \qquad s \mapsto s_x$$

which sends $s \in \mathcal{F}(U)$ to the class of (U, s) in $\mathcal{F}_x$. We call s_x the *germ of s in x*.

If $\varphi \colon \mathcal{F} \to \mathcal{G}$ is a morphism of presheaves on X, we have an induced map

$$\varphi_x := \varinjlim_{U \ni x} \varphi_U \colon \mathcal{F}_x \to \mathcal{G}_x$$

of the stalks in x. We obtain a functor $\mathcal{F} \mapsto \mathcal{F}_x$ from the category of presheaves on X to the category of sets.

If $\mathcal{F}$ is a presheaf with values in $\mathcal{C}$, where $\mathcal{C}$ is the category of abelian groups, of rings, or any category in which filtered inductive limits exist, then the stalk $\mathcal{F}_x$ is an object in $\mathcal{C}$ and we obtain a functor $\mathcal{F} \mapsto \mathcal{F}_x$ from the category of presheaves on X with values in $\mathcal{C}$ to the category $\mathcal{C}$.

Example 2.22. Let $X = \mathbb{C}$ and $\mathcal{O}_\mathbb{C}$ be the sheaf of holomorphic functions on X (i.e.,for every open set U in $\mathbb{C}$, $\mathcal{O}_\mathbb{C}(U)$ is the set of holomorphic functions $U \to \mathbb{C}$). This is a sheaf of $\mathbb{C}$-algebras. Fix $z_0 \in \mathbb{C}$. Then two holomorphic functions f_1 and f_2 defined in open neighborhoods U_1 and U_2, respectively, of z_0 agree on some open neighborhood $V \subseteq U_1 \cap U_2$ if and only if they have same Taylor expansion around z_0. Therefore

$$\mathcal{O}_{\mathbb{C},z_0} = \left\{ \sum_{n \geq 0} a_n (z - z_0)^n \text{ power series with positive radius of convergence} \right\},$$

and the identity theorem says precisely that for a connected open neighborhood U of z_0, the natural map $\mathcal{O}(U) \to \mathcal{O}_{\mathbb{C},z_0}$ is injective.

Proposition 2.23. *Let X be a topological space, $\mathscr{F}$ and $\mathscr{G}$ presheaves on X, and let $\varphi, \psi \colon \mathscr{F} \to \mathscr{G}$ be two morphisms of presheaves.*
(1) *Assume that $\mathscr{F}$ is a sheaf. Then the induced maps on stalks $\varphi_x \colon \mathscr{F}_x \to \mathscr{G}_x$ are injective for all $x \in X$ if and only if $\varphi_U \colon \mathscr{F}(U) \to \mathscr{G}(U)$ is injective for all open subsets $U \subseteq X$.*
(2) *If $\mathscr{F}$ and $\mathscr{G}$ are both sheaves, the maps φ_x are bijective for all $x \in X$ if and only if φ_U is bijective for all open subsets $U \subseteq X$.*
(3) *If $\mathscr{F}$ and $\mathscr{G}$ are both sheaves, the morphisms φ and ψ are equal if and only if $\varphi_x = \psi_x$ for all $x \in X$.*

Proof. For $U \subseteq X$ open consider the map

$$\mathscr{F}(U) \to \prod_{x \in U} \mathscr{F}_x, \quad s \mapsto (s_x)_{x \in U}.$$

We claim that this map is injective if $\mathscr{F}$ is a sheaf. Indeed let $s, t \in \mathscr{F}(U)$ such that $s_x = t_x$ for all $x \in U$. Then for all $x \in U$ there exists an open neighborhood $V_x \subseteq U$ of x such that $s_{|V_x} = t_{|V_x}$. Clearly, $U = \bigcup_{x \in U} V_x$ and therefore $s = t$ by sheaf condition (Sh1).

Using the commutative diagram

$$
\begin{array}{ccc}
\mathscr{F}(U) & \longrightarrow & \prod_{x \in U} \mathscr{F}_x \\
\varphi_U \downarrow & & \downarrow \prod_x \varphi_x \\
\mathscr{G}(U) & \longrightarrow & \prod_{x \in U} \mathscr{G}_x,
\end{array}
$$

we see that (3) and the necessity of the condition in (1) are implied by the above claim. Moreover, a filtered inductive limit of injective maps is always injective again (as can be checked instantly, see Exercise 2.11), therefore the condition in (1) is also sufficient.

Hence we are done if we show that the bijectivity of φ_x for all $x \in U$ implies the surjectivity of φ_U. Let $t \in \mathscr{G}(U)$. For all $x \in U$ we choose an open neighborhood U^x of x in U and $s^x \in \mathscr{F}(U^x)$ such that $(\varphi_{U^x}(s^x))_x = t_x$. Then there exists an open neighborhood $V^x \subseteq U^x$ of x with $\varphi_{V^x}(s^x_{|V^x}) = t_{|V^x}$. Then $(V^x)_{x \in U}$ is an open covering of U and for $x, y \in U$

$$\varphi_{V^x \cap V^y}(s^x_{|V^x \cap V^y}) = t_{|V^x \cap V^y} = \varphi_{V^x \cap V^y}(s^y_{|V^x \cap V^y}).$$

As we already know that $\varphi_{V^x \cap V^y}$ is injective, this shows $s^x_{|V^x \cap V^y} = s^y_{|V^x \cap V^y}$ and the sheaf condition (Sh2) ensures that we find $s \in \mathscr{F}(U)$ such that $s_{|V^x} = s^x_{|V^x}$ for all $x \in U$. Clearly, we have $\varphi_U(s)_x = t_x$ for all $x \in U$ and hence $\varphi_U(s) = t$.   □

We call a morphism $\varphi\colon \mathscr{F} \to \mathscr{G}$ of sheaves *injective* (resp. *surjective*, resp. *bijective*) if $\varphi_x\colon \mathscr{F}_x \to \mathscr{G}_x$ is injective (resp. surjective, resp. bijective) for all $x \in X$.

If $\varphi\colon \mathscr{F} \to \mathscr{G}$ is a morphism of sheaves, φ is surjective if and only if for all open subsets $U \subseteq X$ and every $t \in \mathscr{G}(U)$ there exist an open covering $U = \bigcup_i U_i$ (depending on t) and sections $s_i \in \mathscr{F}(U_i)$ such that $\varphi_{U_i}(s_i) = t_{|U_i}$, i.e., locally we can find a preimage of t. But the surjectivity of φ does *not* imply that $\varphi_U\colon \mathscr{F}(U) \to \mathscr{G}(U)$ is surjective for all open sets U of X (see Exercise 2.12).

If $\mathscr{F}, \mathscr{G}$ are (pre-)sheaves on X such that $\mathscr{F}(U) \subseteq \mathscr{G}(U)$ for all $U \subseteq X$ open, and such that the restriction maps of $\mathscr{F}$ are induced by those of $\mathscr{G}$, then we call $\mathscr{F}$ a *subsheaf* (or a *subpresheaf*, resp.) of $\mathscr{G}$.

(2.7) Sheaves associated to presheaves.

Proposition 2.24. *Let $\mathscr{F}$ be a presheaf on a topological space X. Then there exists a pair $(\tilde{\mathscr{F}}, \iota_{\mathscr{F}})$, where $\tilde{\mathscr{F}}$ is a sheaf on X and $\iota_{\mathscr{F}}\colon \mathscr{F} \to \tilde{\mathscr{F}}$ is a morphism of presheaves, such that the following holds: If $\mathscr{G}$ is a sheaf on X and $\varphi\colon \mathscr{F} \to \mathscr{G}$ is a morphism of presheaves, then there exists a unique morphism of sheaves $\tilde{\varphi}\colon \tilde{\mathscr{F}} \to \mathscr{G}$ with $\tilde{\varphi} \circ \iota_{\mathscr{F}} = \varphi$. The pair $(\tilde{\mathscr{F}}, \iota_{\mathscr{F}})$ is unique up to unique isomorphism.*

Moreover, the following properties hold:
(1) For all $x \in X$ the map on stalks $\iota_{\mathscr{F},x}\colon \mathscr{F}_x \to \tilde{\mathscr{F}}_x$ is bijective.
(2) For every presheaf $\mathscr{G}$ on X and every morphism of presheaves $\varphi\colon \mathscr{F} \to \mathscr{G}$ there exists a unique morphism $\tilde{\varphi}\colon \tilde{\mathscr{F}} \to \tilde{\mathscr{G}}$ making the diagram

$$(2.7.1)$$

$$
\begin{array}{ccc}
\mathscr{F} & \xrightarrow{\iota_{\mathscr{F}}} & \tilde{\mathscr{F}} \\
\varphi \downarrow & & \downarrow \tilde{\varphi} \\
\mathscr{G} & \xrightarrow{\iota_{\mathscr{G}}} & \tilde{\mathscr{G}}
\end{array}
$$

commutative. In particular, $\mathscr{F} \mapsto \tilde{\mathscr{F}}$ is a functor from the category of presheaves on X to the category of sheaves on X.

The sheaf $\tilde{\mathscr{F}}$ is called the *sheaf associated to* $\mathscr{F}$ or the *sheafification of* $\mathscr{F}$. We can reformulate the first part of the proposition by saying that sheafification is the left adjoint functor to the inclusion functor of the category of sheaves into the category of presheaves.

Proof. For $U \subseteq X$ open, elements of $\tilde{\mathscr{F}}(U)$ are by definition families of elements in the stalks of $\mathscr{F}$ which locally give rise to sections of $\mathscr{F}$. More precisely, we define

$$\tilde{\mathscr{F}}(U) := \Big\{ (s_x) \in \prod_{x \in U} \mathscr{F}_x; \, \forall x \in U\colon \exists \text{ an open neighborhood } W \subseteq U \text{ of } x,$$

$$\text{and } t \in \mathscr{F}(W)\colon \forall w \in W\colon s_w = t_w \Big\}.$$

For $U \subseteq V$ the restriction map $\tilde{\mathscr{F}}(V) \to \tilde{\mathscr{F}}(U)$ is induced by the natural projection $\prod_{x \in V} \mathscr{F}_x \to \prod_{x \in U} \mathscr{F}_x$. Then it is easy to check that $\tilde{\mathscr{F}}$ is a sheaf. For $U \subseteq X$ open, we define $\iota_{\mathscr{F},U}\colon \mathscr{F}(U) \to \tilde{\mathscr{F}}(U)$ by $s \mapsto (s_x)_{x \in U}$. The definition of $\tilde{\mathscr{F}}$ shows that, for $x \in X$, $\tilde{\mathscr{F}}_x = \mathscr{F}_x$ and that $\iota_{\mathscr{F},x}$ is the identity.

Now let $\mathscr{G}$ be a presheaf on X and let $\varphi\colon \mathscr{F} \to \mathscr{G}$ be a morphism. Sending $(s_x)_x \in \tilde{\mathscr{F}}(U)$ to $(\varphi_x(s_x))_x \in \tilde{\mathscr{G}}(U)$ defines a morphism $\tilde{\mathscr{F}} \to \tilde{\mathscr{G}}$. By Proposition 2.23 (3) this is the unique morphism making the diagram (2.7.1) commutative.

If we assume in addition that $\mathscr{G}$ is a sheaf, then the morphism of sheaves $\iota_{\mathscr{G}}\colon \mathscr{G} \to \tilde{\mathscr{G}}$, which is bijective on stalks, is an isomorphism by Proposition 2.23 (2). Composing the morphism $\tilde{\mathscr{F}} \to \tilde{\mathscr{G}}$ with $\iota_{\mathscr{G}}^{-1}$, we obtain the morphism $\tilde{\varphi}\colon \tilde{\mathscr{F}} \to \mathscr{G}$. Finally, the uniqueness of $(\tilde{\mathscr{F}}, \iota_{\mathscr{F}})$ is a formal consequence. □

From this definition and from Proposition 2.23 (2), we get the following characterization of the sheafification: Let $\mathscr{F}$ be a presheaf and $\mathscr{G}$ be a sheaf. Then $\mathscr{G}$ is isomorphic to the sheafification of $\mathscr{F}$ if and only and if there exists a morphism $\iota\colon \mathscr{F} \to \mathscr{G}$ such that ι_x is bijective for all $x \in X$.

Example 2.25. Let E be a set and denote by $\mathscr{F}$ the presheaf such that $\mathscr{F}(U) = E$ for every open set U of X (the restriction maps being the identity). Let $\tilde{\mathscr{F}}$ be the associated sheaf. Then $\tilde{\mathscr{F}}(U)$ is the set of locally constant functions $U \to E$.

The sheaf $\tilde{\mathscr{F}}$ is called the *constant sheaf with value E* and sometimes denoted by $\underline{E}$ or $\underline{E}_X$.

Finally it is clear that if $\mathscr{F}$ is a presheaf of rings, of R-modules, or of R-algebras, its associated sheaf is a sheaf of rings, of R-modules, or of R-algebras.

(2.8) Direct and inverse images of sheaves.

Let $f\colon X \to Y$ be a continuous map of topological spaces. It is a natural question, how we can transport sheaves from X to Y, or the other way around, using f. First, let $\mathscr{F}$ be a presheaf on X. We define a presheaf $f_*\mathscr{F}$ on Y by (for $V \subseteq Y$ open)

$$(f_*\mathscr{F})(V) = \mathscr{F}(f^{-1}(V))$$

the restriction maps given by the restriction maps for $\mathscr{F}$. We call $f_*\mathscr{F}$ the *direct image of $\mathscr{F}$ under f*. Whenever $\varphi\colon \mathscr{F}_1 \to \mathscr{F}_2$ is a morphism of presheaves, the family of maps $f_*(\varphi)_V := \varphi_{f^{-1}(V)}$ for $V \subseteq Y$ open is a morphism $f_*(\varphi)\colon f_*\mathscr{F}_1 \to f_*\mathscr{F}_2$. Therefore f_* is a functor from the category of presheaves on X to the category of presheaves on Y. The following properties are immediate.

Remark 2.26.
(1) If $\mathscr{F}$ is a sheaf on X, $f_*\mathscr{F}$ is a sheaf on Y. Therefore f_* also defines a functor $f_*\colon (\mathrm{Sh}(X)) \to (\mathrm{Sh}(Y))$.
(2) If $g\colon Y \to Z$ is a second continuous map, there exists an identity $g_*(f_*\mathscr{F}) = (g \circ f)_*\mathscr{F}$ which is functorial in $\mathscr{F}$.

We now come to the definition of the inverse image of a presheaf. Again let $f\colon X \to Y$ be a continuous map and let $\mathscr{G}$ be a presheaf on Y. Define a presheaf on X by

(2.8.1) $$U \mapsto \varinjlim_{\substack{V \supseteq f(U), \\ V \subseteq Y \text{ open}}} \mathscr{G}(V),$$

the restriction maps being induced by the restriction maps of $\mathscr{G}$. Momentarily we denote this presheaf by $f^+\mathscr{G}$. Let $f^{-1}\mathscr{G}$ be the sheafification of $f^+\mathscr{G}$. We call $f^{-1}\mathscr{G}$ the *inverse image of $\mathscr{G}$ under f*. Note that even if $\mathscr{G}$ is a sheaf, $f^+\mathscr{G}$ is not a sheaf in general. If f is the inclusion of a subspace X of Y and $\mathscr{G}$ is a sheaf on Y, then we also write $\mathscr{G}_{|X}$ instead of $f^{-1}\mathscr{G}$. If X is an open subspace of Y this definition coincides with the one given in Example 2.19 (1). Although the definition may look complicated at first sight, you should convince yourself that it is the obvious one: Since we can only evaluate $\mathscr{G}$ on open subsets of Y, we cannot talk about sections of $\mathscr{G}$ on $f(U)$. Instead we "approximate" $f(U)$ by open subsets of Y containing it. Compare also the definition of the system of functions on a closed subprevariety, see Section (1.18).

Again the construction of $f^+\mathscr{G}$ and hence of $f^{-1}\mathscr{G}$ is functorial in $\mathscr{G}$. Therefore we obtain a functor f^{-1} from the category of presheaves on Y to the category of sheaves on X.

If x is a point of X and $i\colon \{x\} \to X$ is the inclusion, the definition (2.8.1) shows that

$$i^{-1}\mathscr{F} = \mathscr{F}_x$$

for every presheaf $\mathscr{F}$ on X. It follows that for each presheaf $\mathscr{G}$ on Y we have an identity, functorial in $\mathscr{G}$,

(2.8.2)
$$(f^{-1}\mathscr{G})_x \cong (f^+\mathscr{G})_x = \varinjlim_{x \in U} (f^+\mathscr{G})(U)$$
$$= \varinjlim_{x \in U} \varinjlim_{f(U) \subseteq V} \mathscr{G}(V) = \varinjlim_{f(x) \in V} \mathscr{G}(V) = \mathscr{G}_{f(x)},$$

where the first identification follows from Proposition 2.24 (1).

Now let $g\colon Y \to Z$ be a second continuous map and let $\mathscr{H}$ be a presheaf on Z. Fix an open subset U in X. An open subset $W \subseteq Z$ contains $g(f(U))$ if and only if it contains a subset of the form $g(V)$, where $V \subseteq Y$ is an open set containing $f(U)$. This implies that $f^+(g^+\mathscr{H}) = (g \circ f)^+\mathscr{H}$. Furthermore, (2.8.2) implies that the natural morphism $f^{-1}(g^+\mathscr{H}) \to f^{-1}(g^{-1}\mathscr{H})$ induces isomorphisms on all stalks, and hence is an isomorphism by Proposition 2.23. We deduce an isomorphism

(2.8.3)
$$f^{-1}(g^{-1}\mathscr{H}) \cong (g \circ f)^{-1}\mathscr{H},$$

which is functorial in $\mathscr{H}$.

Direct image and inverse image are functors which are adjoint to each other. More precisely:

Proposition 2.27. *Let $f\colon X \to Y$ be a continuous map, let $\mathscr{F}$ be a sheaf on X and let $\mathscr{G}$ be a presheaf on Y. Then there is a bijection*

$$\mathrm{Hom}_{(\mathrm{Sh}(X))}(f^{-1}\mathscr{G}, \mathscr{F}) \leftrightarrow \mathrm{Hom}_{(\mathrm{PreSh}(Y))}(\mathscr{G}, f_*\mathscr{F}),$$
$$\varphi \mapsto \varphi^\flat,$$
$$\psi^\sharp \leftarrow\!\shortmid \psi$$

which is functorial in $\mathscr{F}$ and $\mathscr{G}$.

Proof. Let $\varphi\colon f^{-1}\mathscr{G} \to \mathscr{F}$ be a morphism of sheaves on X, and let $V \subseteq Y$ be open. Since $f(f^{-1}(V)) \subseteq V$, we have a map $\mathscr{G}(V) \to f^+\mathscr{G}(f^{-1}(V))$, and we define $\varphi_V^\flat$ as the composition

$$\mathscr{G}(V) \to f^+\mathscr{G}(f^{-1}(V)) \longrightarrow f^{-1}\mathscr{G}(f^{-1}(V)) \xrightarrow{\varphi_{f^{-1}(V)}} \mathscr{F}(f^{-1}(V)) = f_*\mathscr{F}(V).$$

Conversely, let $\psi\colon \mathscr{G} \to f_*\mathscr{F}$ be a morphism of presheaves on Y. To define the morphism $\psi^\sharp$ it suffices to define a morphism of presheaves $f^+\mathscr{G} \to \mathscr{F}$, which we call again $\psi^\sharp$. Let U be open in X, and $s \in f^+\mathscr{G}(U)$. If V is some open neighborhood of $f(U)$, U is contained in $f^{-1}(V)$. Let V be such a neighborhood such that there exists $s_V \in \mathscr{G}(V)$ representing s. Then $\psi_V(s_V) \in f_*\mathscr{F}(V) = \mathscr{F}(f^{-1}(V))$. Let $\psi^\sharp_U(s) \in \mathscr{F}(U)$ be the restriction of the section $\psi_V(s_V)$ to U.

Clearly, these two maps are inverse to each other. Moreover, it is straightforward – albeit quite cumbersome – to check that the constructed maps are functorial in $\mathscr{F}$ and $\mathscr{G}$. $\qquad\qquad\qquad\qquad\qquad\qquad\qquad\qquad\qquad\qquad\qquad\qquad\qquad\qquad\qquad\square$

The adjunction between direct image and pull-back and the fact that $(g \circ f)_* = g_* \circ f_*$ for any two morphisms $f\colon X \to Y$, $g\colon Y \to Z$ gives another way to prove (2.8.3).

Remark 2.28. We will almost never use the concrete description of $f^{-1}\mathscr{G}$ in the sequel. Very often we are given f, $\mathscr{F}$, and $\mathscr{G}$ as in the proposition, and a morphism of presheaves $\psi\colon \mathscr{G} \to f_*\mathscr{F}$. Then usually it is sufficient to understand for each $x \in X$ the map

$$\psi^\sharp_x\colon \mathscr{G}_{f(x)} \overset{(2.8.2)}{=\!=} (f^{-1}\mathscr{G})_x \longrightarrow \mathscr{F}_x$$

induced by $\psi^\sharp\colon f^{-1}\mathscr{G} \to \mathscr{F}$ on stalks. The proof of the proposition shows that we can describe this map in terms of ψ as follows: For every open neighborhood $V \subseteq Y$ of $f(x)$, we have maps

$$\mathscr{G}(V) \xrightarrow{\psi_V} \mathscr{F}(f^{-1}(V)) \longrightarrow \mathscr{F}_x,$$

and taking the inductive limit over all V we obtain the map $\psi^\sharp_x\colon \mathscr{G}_{f(x)} \to \mathscr{F}_x$.

Note that if $\mathscr{F}$ is a sheaf of rings (or of R-modules, or of R-algebras) on X, $f_*\mathscr{F}$ is a sheaf on Y with values in the same category. A similar statement holds for the inverse image. Finally, Proposition 2.27 holds (with the same proof) if we consider morphisms of sheaves of rings (or of R-modules, etc.).

(2.9) Locally ringed spaces.

Definition 2.29. *A ringed space is a pair $(X, \mathscr{O}_X)$, where X is a topological space and where $\mathscr{O}_X$ is a sheaf of (commutative) rings on X.*

If $(X, \mathscr{O}_X)$ and $(Y, \mathscr{O}_Y)$ are ringed spaces, we define a morphism of ringed spaces $(X, \mathscr{O}_X) \to (Y, \mathscr{O}_Y)$ as a pair $(f, f^\flat)$, where $f\colon X \to Y$ is a continuous map and where $f^\flat\colon \mathscr{O}_Y \to f_*\mathscr{O}_X$ is a homomorphism of sheaves of rings on Y.

Note that the datum of $f^\flat$ is equivalent to the datum of a homomorphism of sheaves of rings $f^\sharp\colon f^{-1}\mathscr{O}_Y \to \mathscr{O}_X$ on X by Proposition 2.27. Often we simply write f instead of $(f, f^\sharp)$ or $(f, f^\flat)$.

The composition of morphisms of ringed spaces is defined in the obvious way (using Remark 2.26 (2)), and we obtain the category of ringed spaces. We call $\mathscr{O}_X$ the *structure sheaf* of $(X, \mathscr{O}_X)$. Often we simply write X instead of $(X, \mathscr{O}_X)$.

We think of the structure sheaf on X as the system of all "permissible" (within the current context) functions, where permissible might mean continuous, differentiable, holomorphic, given by polynomials, etc. The map $f: X \to Y$ should certainly have the property that composition of a permissible function on an open subset V of Y with f gives rise to a permissible function on $f^{-1}(V)$. Since viewing sections of the structure sheaves as functions is only a heuristic, we cannot actually compose sections with the map f. As a substitute, we request that we are given a map $\mathscr{O}_Y(V) \to \mathscr{O}_X(f^{-1}(V))$ for every open $V \subseteq Y$. These maps must be compatible with restrictions, and constitute the sheaf homomorphism $f^\flat$.

Usually we will work with a subcategory of the category of ringed spaces. To introduce this subcategory, we recall the following notation. If A is a local ring, we denote by $\mathfrak{m}_A$ its maximal ideal and by $\kappa(A) = A/\mathfrak{m}_A$ its residue field. A homomorphism of local rings $\varphi: A \to B$ is called *local*, if $\varphi(\mathfrak{m}_A) \subseteq \mathfrak{m}_B$.

A morphism $(f, f^\flat): X \to Y$ of ringed spaces induces morphisms on the stalks as follows. Let $x \in X$. Let $f^\sharp: f^{-1}\mathscr{O}_Y \to \mathscr{O}_X$ be the morphism corresponding to $f^\flat$ by adjointness. Using the identification $(f^{-1}\mathscr{O}_Y)_x = \mathscr{O}_{Y,f(x)}$ established in (2.8.2), we get

$$f_x^\sharp: \mathscr{O}_{Y,f(x)} \to \mathscr{O}_{X,x}.$$

By Remark 2.28 there is the following more explicit description of this homomorphism: The maps $f_V^\flat: \mathscr{O}_Y(V) \to \mathscr{O}_X(f^{-1}(V))$ for every open neighborhood V of $f(x)$ induce maps $\mathscr{O}_Y(V) \to \mathscr{O}_{X,x}$ and hence a map $f_x^\sharp: \mathscr{O}_{Y,f(x)} = \varinjlim \mathscr{O}_Y(V) \to \mathscr{O}_{X,x}$.

Definition 2.30. *A locally ringed space is a ringed space $(X, \mathscr{O}_X)$ such that for all $x \in X$ the stalk $\mathscr{O}_{X,x}$ is a local ring.*

A morphism of locally ringed spaces $(X, \mathscr{O}_X) \to (Y, \mathscr{O}_Y)$ is a morphism of ringed spaces $(f, f^\flat)$ such that for all $x \in X$ the induced homomorphism on stalks

$$f_x^\sharp: (f^{-1}\mathscr{O}_Y)_x = \mathscr{O}_{Y,f(x)} \to \mathscr{O}_{X,x}$$

is a local ring homomorphism.

The composition of two morphisms of locally ringed spaces is again a morphism of locally ringed spaces. Therefore locally ringed spaces form a category. Note that there exist locally ringed spaces $(X, \mathscr{O}_X)$ and $(Y, \mathscr{O}_Y)$ and morphisms $f: (X, \mathscr{O}_X) \to (Y, \mathscr{O}_Y)$ of ringed spaces which are not morphisms of locally ringed spaces; in other words, the subcategory of locally ringed spaces is not a full subcategory (Exercise 2.18).

Let $(X, \mathscr{O}_X)$ be a locally ringed space and $x \in X$. We call the stalk $\mathscr{O}_{X,x}$ the *local ring of X in x*, denote by $\mathfrak{m}_x$ the maximal ideal of $\mathscr{O}_{X,x}$, and by $\kappa(x) = \mathscr{O}_{X,x}/\mathfrak{m}_x$ the residue field. If U is an open neighborhood of x and $f \in \mathscr{O}_X(U)$, we denote by $f(x) \in \kappa(x)$ the image of f under the canonical homomorphisms $\mathscr{O}_X(U) \to \mathscr{O}_{X,x} \to \kappa(x)$.

Why do we work with *locally* ringed spaces? Although we have not yet defined the structure sheaf on $X = \operatorname{Spec} A$, we can explain this heuristically. We think of sections of the structure sheaf as functions on an open subset, and then the elements of the stalk at a point x are functions defined in some open neighborhood of x. A reasonable property to ask of such functions is that those which do not vanish at x are invertible in some (small) neighborhood of x. Then all elements of the stalk not contained in the ideal of functions vanishing at x are units of the stalk. This shows that the stalk is indeed a local ring, with maximal ideal the ideal of all functions vanishing at x. Now consider a morphism $(f, f^\flat)\colon X \to Y$ of ringed spaces of this nature. The sheaf homomorphism is our replacement for "composition of functions with f". Certainly, if some function on Y vanishes at a point $f(x)$, $x \in X$, then its composition with f must vanish at x. In other words, the maximal ideal of $\mathscr{O}_{Y,f(x)}$ must be mapped into the maximal ideal of $\mathscr{O}_{X,x}$, which is exactly the property we requested above. Since we do not really deal with functions, we do have to put it into our definitions explicitly. In the following two examples, the sections of the structure sheaf are functions, and our philosophy turns into a precise statement.

Example 2.31. Let X be a topological space and consider the sheaf $\mathscr{C}_X$ of $\mathbb{R}$-valued continuous functions on X (i.e., for $U \subseteq X$ open, $\mathscr{C}_X(U)$ is the $\mathbb{R}$-algebra of continuous functions $s\colon U \to \mathbb{R}$). For $x \in X$, $\mathscr{C}_{X,x}$ is the ring of germs $[s]$ of continuous functions s in a neighborhood of x. Let $\mathfrak{m}_x \subset \mathscr{C}_{X,x}$ be the set of germs $[s]$ such that $s(x) = 0$. Clearly, this is a proper ideal.

We claim that this is the unique maximal ideal (and hence $(X, \mathscr{C}_X)$ is a locally ringed space). Indeed, let $[s] \in \mathscr{C}_{X,x} \setminus \mathfrak{m}_x$. For every representative $s \in [s]$ we have $s(x) \neq 0$ and, as s is continuous, there exists an open neighborhood U of x such that $s(u) \neq 0$ for all $u \in U$. Therefore $1/(s_{|U})$ exists. This shows that $\mathscr{C}_{X,x} \setminus \mathfrak{m}_x$ is the group of units in $\mathscr{C}_{X,x}$ which shows our claim. Moreover, the ring homomorphism $\mathscr{C}_{X,x} \to \mathbb{R}$ sending $[s]$ to $s(x)$ is surjective with kernel $\mathfrak{m}_x$ and therefore identifies $\kappa(x)$ with $\mathbb{R}$.

If $f\colon X \to Y$ is a continuous map to another topological space, composition with f defines for all open sets V of Y a homomorphism of $\mathbb{R}$-algebras

$$f_V^\flat\colon \mathscr{C}_Y(V) \to \mathscr{C}_X(f^{-1}(V)) = f_*\mathscr{C}_X(V), \quad t \mapsto t \circ f.$$

The associated homomorphism $f^\sharp$ induces on stalks the map $\mathscr{C}_{Y,f(x)} \to \mathscr{C}_{X,x}$ that sends a germ $[t]$ of continuous functions at $f(x)$ to the germ $[t \circ f]$ of continuous functions at x. Obviously, $f_x^\sharp(\mathfrak{m}_{f(x)}) \subseteq \mathfrak{m}_x$, and $(f, f^\flat)$ is a morphism of locally ringed spaces.

If we interpret the system of functions on a prevariety as a sheaf, we obtain a locally ringed space, as well:

Example 2.32. Let $(X, \mathscr{O}_X)$ be a prevariety over an algebraically closed field k, in the sense of Section (1.15). For $U \subseteq X$ now $\mathscr{O}_X(U)$ consists of certain functions $f\colon U \to \mathbb{A}^1(k)$ which are continuous for the Zariski topology. As in Example 2.31 we see that for $x \in X$ the ideal of germs $[f]$ at x with $f(x) = 0$ is the unique maximal ideal of $\mathscr{O}_{X,x}$ (using that $\mathbb{A}^1(k) \setminus \{0\}$ is open in $\mathbb{A}^1(k)$ for the Zariski topology). Therefore $(X, \mathscr{O}_X)$ is a locally ringed space and the surjective homomorphism $\mathscr{O}_{X,x} \to k$, $[f] \mapsto f(x)$, induces an isomorphism $\kappa(x) \overset{\sim}{\to} k$ for all $x \in X$.

Spectrum of a ring as a locally ringed space

Let A be a ring. We will now endow the topological space $\operatorname{Spec} A$ with the structure of a locally ringed space and obtain a functor $A \mapsto \operatorname{Spec} A$ from the category of rings to the category of locally ringed spaces which we will show to be fully faithful.

(2.10) Structure sheaf on $\operatorname{Spec} A$.

We set $X = \operatorname{Spec}(A)$. We have seen in Proposition 2.5 that the principal open sets $D(f)$ for $f \in A$ form a basis of the topology of X. We will define a presheaf $\mathscr{O}_X$ on this basis (Section (2.5)) and then prove that the sheaf axioms are satisfied with respect to this basis. The basic idea is this: Looking back at the analogy with prevarieties, we certainly want to have $\mathscr{O}_X(X) = A$. More generally, for $f \in A$, we consider the localization A_f of A with respect to the multiplicative set $\{ f^i \; ; \; i \geq 0 \}$. Denote by $\iota_f \colon A \to A_f$ the canonical ring homomorphism $a \mapsto a/1$. By Proposition 2.12, $^a\iota_f$ is a homeomorphism of $\operatorname{Spec} A_f$ onto $D(f)$. So it seems reasonable to set $\mathscr{O}_X(D(f)) = A_f$. (We think of A_f as functions which might have poles along $V(f)$, the set of zeros of f.) Let us check that this is a sensible definition: we must check that $A_f = A_g$ whenever $D(f) = D(g)$, define restriction maps, and check that the sheaf axioms are satisfied on the basis of the topology $(D(f))_{f \in A}$.

Recall from Lemma 2.4 that, for $f, g \in A$, $D(f) \subseteq D(g)$ if and only if there exists an integer $n \geq 1$ such that $f^n \in Ag$ or, equivalently, $g/1 \in (A_f)^\times$. In this case we obtain a unique ring homomorphism $\rho_{f,g} \colon A_g \to A_f$ such that $\rho_{f,g} \circ \iota_g = \iota_f$. Whenever $D(f) \subseteq D(g) \subseteq D(h)$, we have $\rho_{f,g} \circ \rho_{g,h} = \rho_{f,h}$. In particular, if $D(f) = D(g)$, $\rho_{f,g}$ is an isomorphism, which we use to identify A_g and A_f. Therefore we can define

$$(2.10.1) \qquad\qquad \mathscr{O}_X(D(f)) := A_f$$

and obtain a presheaf of rings on the basis $\mathcal{B} := \{ D(f) \; ; \; f \in A \}$ for the topology of $\operatorname{Spec} A$. The restriction maps are the ring homomorphisms $\rho_{f,g}$.

Theorem 2.33. *The presheaf $\mathscr{O}_X$ is a sheaf on $\mathcal{B}$.*

We denote the sheaf of rings on X associated to $\mathscr{O}_X$ again by $\mathscr{O}_X$. For all points $x \in X = \operatorname{Spec}(A)$ we have

$$(2.10.2) \qquad\qquad \mathscr{O}_{X,x} = \varinjlim_{D(f) \ni x} \mathscr{O}_X(D(f)) = \varinjlim_{f \notin \mathfrak{p}_x} A_f = A_{\mathfrak{p}_x}$$

(localization of A in the prime ideal $\mathfrak{p}_x$). In particular, $(X, \mathscr{O}_X)$ is a locally ringed space. We will often simply write $\operatorname{Spec} A$ instead of $(\operatorname{Spec} A, \mathscr{O}_{\operatorname{Spec} A})$.

Proof. Let $D(f)$ be a principal open set and let $D(f) = \bigcup_{i \in I} D(f_i)$ be a covering by principal open sets. We have to show the following two properties.
(1) Let $s \in \mathscr{O}_X(D(f))$ be such that $s_{|D(f_i)} = 0$ for all $i \in I$. Then $s = 0$.
(2) For $i \in I$ let $s_i \in \mathscr{O}_X(D(f_i))$ be such that $s_{i|D(f_i) \cap D(f_j)} = s_{j|D(f_i) \cap D(f_j)}$ for all $i, j \in I$. Then there exists $s \in \mathscr{O}_X(D(f))$ such that $s_{|D(f_i)} = s_i$ for all $i \in I$.

As $D(f)$ is quasi-compact, we can assume that I is finite; this is clear for part (1), and for part (2) we can first glue for a finite subcover and then use part (1) to check that the resulting section s restricts to s_i for all i. Restricting the presheaf $\mathscr{O}_X$ to $D(f)$ and replacing A by A_f we may assume that $f = 1$ and hence $D(f) = X$ to ease the notation. The relation $X = \bigcup_i D(f_i)$ is equivalent to $(f_i \, ; \, i \in I) = A$. As $D(f_i) = D(f_i^n)$ for all integers $n \geq 1$ there exist elements $b_i \in A$ (depending on n) such that

$$(2.10.3) \qquad\qquad \sum_{i \in I} b_i f_i^n = 1.$$

Proof of (1). Let $s = a \in A$ be such that the image of a in A_{f_i} is zero for all i. As I is finite, there exists an integer $n \geq 1$, independent of i, such that $f_i^n a = 0$. By (2.10.3), $a = (\sum_{i \in I} b_i f_i^n) a = 0$.

Proof of (2). As I is finite, we can write $s_i = \frac{a_i}{f_i^n}$ for some n independent of i. By hypothesis, the images of $\frac{a_i}{f_i^n}$ and of $\frac{a_j}{f_j^n}$ in $A_{f_i f_j}$ are equal for all $i, j \in I$. Therefore there exists an integer $m \geq 1$ (which again we can choose independent of i and j) such that $(f_i f_j)^m (f_j^n a_i - f_i^n a_j) = 0$. Replacing a_i by $f_i^m a_i$ and n by $n + m$ (which does not change s_i), we see that

$$(2.10.4) \qquad\qquad\qquad f_j^n a_i = f_i^n a_j$$

for all $i, j \in I$. We set $s := \sum_{j \in I} b_j a_j \in A$, where the b_j are the elements in (2.10.3). Then

$$f_i^n s = f_i^n \sum_{j \in I} b_j a_j = \sum_{j \in I} b_j (f_i^n a_j) \overset{(2.10.4)}{=} (\sum_{j \in I} b_j f_j^n) a_i \overset{(2.10.3)}{=} a_i.$$

This means that the image of s in A_{f_i} is s_i. $\qquad\qquad\qquad\qquad\qquad\qquad\qquad\square$

(2.11) The functor $A \mapsto (\operatorname{Spec} A, \mathscr{O}_{\operatorname{Spec} A})$.

Definition 2.34. *A locally ringed space* $(X, \mathscr{O}_X)$ *is called* affine scheme, *if there exists a ring* A *such that* $(X, \mathscr{O}_X)$ *is isomorphic to* $(\operatorname{Spec} A, \mathscr{O}_{\operatorname{Spec} A})$.

A *morphism of affine schemes* is a morphism of locally ringed spaces. We obtain the category of affine schemes which we denote by (Aff).

Let $\varphi \colon A \to B$ be a homomorphism of rings and set $X = \operatorname{Spec} B$ and $Y = \operatorname{Spec} A$. Let $^a\varphi \colon \operatorname{Spec} B \to \operatorname{Spec} A$ be the associated continuous map (Section (2.3)). We will now define a morphism $(f, f^\flat) \colon X \to Y$ of locally ringed spaces such that $f = {}^a\varphi$ and

$$(2.11.1) \qquad\qquad f_Y^\flat \colon A = \mathscr{O}_Y(Y) \to (f_* \mathscr{O}_X)(Y) = B$$

equals φ.

Set $f = {}^a\varphi$. For $s \in A$, we have $f^{-1}(D(s)) = D(\varphi(s))$ (Proposition 2.10) and we define

$$(2.11.2) \qquad f_{D(s)}^\flat \colon \mathscr{O}_Y(D(s)) = A_s \to B_{\varphi(s)} = (f_* \mathscr{O}_X)(D(s))$$

as the ring homomorphism induced by φ. This ring homomorphism is compatible with restrictions to principal open subsets $D(t) \subseteq D(s)$. As the principal open subsets form a basis of the topology, this defines a homomorphism $f^\flat \colon \mathscr{O}_Y \to f_* \mathscr{O}_X$ of sheaves of rings. Choosing $s = 1$ in (2.11.2) we obtain (2.11.1).

For $x \in X$, the homomorphism

$$f_x^\sharp \colon \mathscr{O}_{Y, f(x)} = A_{\varphi^{-1}(\mathfrak{p}_x)} \to B_{\mathfrak{p}_x} = \mathscr{O}_{X, x}$$

is the homomorphism induced by φ and in particular it is a local ring homomorphism. This finishes the definition of $(f, f^\flat)$.

This morphism $\operatorname{Spec} B \to \operatorname{Spec} A$ of locally ringed spaces associated to φ will be often simply denoted by $\operatorname{Spec}(\varphi)$ or ${}^a\varphi$. It is clear from the definition that, for a second ring homomorphism $\psi\colon B \to C$, we have ${}^a(\psi \circ \varphi) = {}^a\varphi \circ {}^a\psi$. We obtain a contravariant functor

$$\operatorname{Spec}\colon (\mathrm{Ring}) \to (\mathrm{Aff}).$$

Conversely, if $f\colon (X, \mathscr{O}_X) \to (Y, \mathscr{O}_Y)$ is a morphism of ringed spaces, we obtain a ring homomorphism

$$\Gamma(f) := f_Y^\flat \colon \Gamma(Y, \mathscr{O}_Y) = \mathscr{O}_Y(Y) \to \Gamma(X, \mathscr{O}_X) = (f_*\mathscr{O}_X)(Y) = \mathscr{O}_X(X).$$

In this way we get a contravariant functor Γ from the category of ringed spaces to the category of rings. Restricting Γ to the category of affine schemes defines a contravariant functor

$$\Gamma\colon (\mathrm{Aff}) \to (\mathrm{Ring}).$$

Theorem 2.35. *The functors* Spec *and* Γ *define an anti-equivalence between the category of rings and the category of affine schemes.*

Proof. The functor Spec is by definition essentially surjective. Moreover, $\Gamma \circ \operatorname{Spec}$ is clearly isomorphic to $\mathrm{id}_{(\mathrm{Ring})}$. Therefore it suffices to show that for any two rings A and B the maps

$$\operatorname{Hom}_{(\mathrm{Ring})}(A, B) \overset{\operatorname{Spec}}{\underset{\Gamma}{\rightleftarrows}} \operatorname{Hom}_{(\mathrm{Aff})}(\operatorname{Spec} B, \operatorname{Spec} A)$$

are mutually inverse bijections. By (2.11.1), $\Gamma \circ \operatorname{Spec} = \mathrm{id}$. Now let $f\colon \operatorname{Spec} B \to \operatorname{Spec} A$ be a morphism of affine schemes and set $\varphi := \Gamma(f)$. We have to show that ${}^a\varphi = f$. If $\mathfrak{p}_x$ is a prime ideal of B, corresponding to a point $x \in X := \operatorname{Spec} B$, $f_x^\sharp$ is the unique ring homomorphism which makes the diagram

$$(2.11.3) \qquad \begin{array}{ccc} A & \xrightarrow{\ \varphi\ } & B \\ \downarrow & & \downarrow \\ A_{\mathfrak{p}_{f(x)}} & \xrightarrow[\ f_x^\sharp\ ]{} & B_{\mathfrak{p}_x} \end{array}$$

commutative. This shows that $\varphi^{-1}(\mathfrak{p}_x) \subseteq \mathfrak{p}_{f(x)}$. As $f_x^\sharp$ is local, we have equality. This shows that ${}^a\varphi = f$ as continuous maps. Now the definition of ${}^a\varphi^\sharp$ shows that ${}^a\varphi_x^\sharp$ makes (2.11.3) commutative as well and hence ${}^a\varphi_x^\sharp = f_x^\sharp$ for all $x \in X$. This proves ${}^a\varphi^\sharp = f^\sharp$ by Proposition 2.23. $\qquad\square$

(2.12) Examples.

Example 2.36. (Affine Spaces) Let R be a ring. We set $\mathbb{A}_R^n := \operatorname{Spec} R[T_1, \ldots, T_n]$. This is called the *affine space of relative dimension n over R*.

Example 2.37. (Integral Domains) Let A be an integral domain and K its field of fractions. Let $X = \operatorname{Spec} A$. The zero ideal of A is a prime ideal and we denote the corresponding point of X by η. The closure of $\{\eta\}$ consists of X and therefore every non-empty open set of X contains η, i.e., η is a generic point of X. The local ring $\mathscr{O}_{X,\eta}$ is the localization of A by the zero ideal (2.10.2):

$$\mathscr{O}_{X,\eta} = K.$$

For all multiplicative subsets $S \subseteq T$ of A with $0 \notin T$, the canonical ring homomorphism $S^{-1}A \to T^{-1}A$ is injective and we can consider all the localizations $S^{-1}A$ as subrings of K. For all $f \in A$, we have $\mathscr{O}_X(D(f)) = A_f$ by the definition of the structure sheaf. If $U \subseteq X$ is an arbitrary open subset, (2.5.2) shows that $\mathscr{O}_X(U) = \bigcap_f A_f$ where f runs through the set of elements $f \in A$ such that $D(f) \subseteq U$. On the other hand, we have $A_f = \bigcap_{\mathfrak{p}} A_{\mathfrak{p}}$, where $\mathfrak{p}$ runs through the set of prime ideals not containing f, i.e., through the points $x \in D(f)$: Given an element $g \in K$ which lies in the intersection $\bigcap_{\mathfrak{p}} A_{\mathfrak{p}}$, let $\mathfrak{a} = \{h \in A; \ hg \in A\}$. Since $g \in A_{\mathfrak{p}}$ for all $\mathfrak{p}$ with $f \notin \mathfrak{p}$, $\mathfrak{a} \not\subseteq \mathfrak{p}$. In other words, all prime ideals containing $\mathfrak{a}$ also contain f, i.e., $f \in \mathrm{rad}(\mathfrak{a})$. This shows $g \in A_f$, as desired. (Cf. the argument of Proposition 1.40, but note that the Nullstellensatz is not needed in the current setting.) As $A_{\mathfrak{p}_x} = \mathscr{O}_{X,x}$ for every point $x \in X$, we see that for any non-empty open set U of X we have

$$\mathscr{O}_X(U) = \bigcap_{x \in U} \mathscr{O}_{X,x}.$$

Example 2.38. (Principal open subschemes of an affine scheme) Let $X = \operatorname{Spec} A$ be an affine scheme. For $f \in A$ let $j \colon \operatorname{Spec} A_f \to \operatorname{Spec} A$ be the morphism of affine schemes that corresponds to the canonical homomorphism $A \to A_f$. Then j induces a homeomorphism of $\operatorname{Spec} A_f$ onto $D(f)$ by Proposition 2.12. Moreover, for all $x \in D(f)$, $j_x^{\sharp}$ is the canonical isomorphism $A_{\mathfrak{p}_x} \xrightarrow{\sim} (A_f)_{\mathfrak{p}_x}$. Hence we see that $(j, j^{\sharp})$ induces an isomorphism of the affine scheme $\operatorname{Spec} A_f$ with the locally ringed space $(D(f), \mathscr{O}_{X|D(f)})$.

Example 2.39. (Closed subschemes of affine schemes) Let $X = \operatorname{Spec} A$ be an affine scheme. For an ideal $\mathfrak{a}$ of A let $i \colon \operatorname{Spec} A/\mathfrak{a} \to \operatorname{Spec} A$ be the morphism of affine schemes that corresponds to the canonical homomorphism $A \to A/\mathfrak{a}$. Again by Proposition 2.12, i induces a homeomorphism of $\operatorname{Spec} A/\mathfrak{a}$ onto the closed subset $V(\mathfrak{a})$ of $\operatorname{Spec} A$. Moreover, for all $x \in V(\mathfrak{a})$ the morphism $i_x^{\flat}$ is the canonical homomorphism $A_{\mathfrak{p}_x} \to (A/\mathfrak{a})_{\overline{\mathfrak{p}}_x}$ where $\overline{\mathfrak{p}}_x$ is the image of $\mathfrak{p}_x$ in $A/\mathfrak{a}$. We use the homeomorphism $i \colon \operatorname{Spec} A/\mathfrak{a} \to V(\mathfrak{a})$ to equip $V(\mathfrak{a})$ with the structure of a locally ringed space which we again denote by $V(\mathfrak{a})$ and will always identify with $\operatorname{Spec} A/\mathfrak{a}$ via i.

In Chapter 3 we will define the general notion of a closed subscheme and show that every closed subscheme of $\operatorname{Spec} A$ is of the form $V(\mathfrak{a})$ for some ideal $\mathfrak{a} \subseteq A$.

Example 2.40. Let B be a ring, and $\mathfrak{b} \subseteq B$ an ideal. As every prime ideal of B contains $\mathfrak{b}$ if and only if it contains $\mathfrak{b}^n$ for some integer $n \geq 1$, the closed subset $V(\mathfrak{b}^n)$ of $\operatorname{Spec} B$ does not depend on n. But as affine scheme, $\operatorname{Spec}(B/\mathfrak{b}^n) = V(\mathfrak{b}^n)$ depends on n.

We explain the difference between these affine schemes in the case $B = k[T]$ and $\mathfrak{b} = (T)$, where k is an algebraically closed field. The closed points of $\mathbb{A}^1_k = \operatorname{Spec} k[T]$ (i.e., the maximal ideals of $k[T]$) correspond to elements of k, and the point corresponding to $\mathfrak{b}$ is 0. Let $A = k[T]/(T^n)$ and set $X = \operatorname{Spec} A$. Then X consists of a single point x. For the structure sheaf we have $\mathscr{O}_X(X) = \mathscr{O}_{X,x} = A$, the maximal ideal $\mathfrak{m}_x$ is the ideal that is generated by the residue class of T (and hence $\mathfrak{m}_x \neq 0$ for $n > 1$), and $\kappa(x) = k$. As explained in Example 2.39 we should picture X as a closed "subscheme" of $\mathbb{A}^1_k$, that is "concentrated in 0".

As explained in Example 2.15, we can consider $\mathbb{A}^1_k$ as the affine variety $\mathbb{A}^1(k)$, and the k-algebra of functions on $\mathbb{A}^1(k)$ is just $B = k[T]$. The restriction of such a function $f \in k[T]$ to X is given by the canonical homomorphism $k[T] \to k[T]/(T^n)$. For $n = 1$,

$k[T]/(T^n) = k$, and this is the map $f \mapsto f(0)$. But for $n > 1$ we keep the higher order terms of the "Taylor expansion" of f in 0. We suggest to picture X as a slightly fuzzy point, that has an "infinitesimal extension of length $n - 1$ within $\mathbb{A}_k^1$".

We would like to motivate this kind of picture in another way as well. Consider the affine space $\mathbb{A}_k^2 = \operatorname{Spec} k[T, U]$ which we visualize as the plane $\{ (u, t) \; ; \; u, t \in k \}$. The ideals $\mathfrak{a}_1 := (U)$ and $\mathfrak{a}_2 := (U - T^n)$ define closed subvarieties

$$X_1 = \{ (u, t) \in \mathbb{A}^2(k) \; ; \; u = 0 \}, \quad X_2 = \{ (u, t) \in \mathbb{A}^2(k) \; ; \; u = t^n \}.$$

As a set, the intersection of both varieties consists only of the origin $(0, 0)$. But it should play a role that X_1 and X_2 do not intersect transversally for $n > 1$.

As affine scheme we will define the intersection in Section (4.11) as $\operatorname{Spec} k[T, U]/(\mathfrak{a}_1 + \mathfrak{a}_2)$. Using $k[T, U]/\mathfrak{a}_1 \cong k[T]$, $k[T, U]/(\mathfrak{a}_1 + \mathfrak{a}_2)$ is just $k[T]/(T^n)$. Hence the point of view of affine scheme allows us to describe the intersection behavior more precisely.

Exercises

In all exercises, A denotes a ring.

Exercise 2.1$\Diamond$. Let $\mathfrak{a}$ and $\mathfrak{b}$ be ideals of A. Show that the following properties are equivalent:
(i) $V(\mathfrak{a}) \subseteq V(\mathfrak{b})$.
(ii) $\mathfrak{b} \subseteq \operatorname{rad}(\mathfrak{a})$.
(iii) $\operatorname{rad}(\mathfrak{b}) \subseteq \operatorname{rad}(\mathfrak{a})$.

Exercise 2.2$\Diamond$. Let f be an element in A. Show that $D(f) = \emptyset$ if and only if f is nilpotent.

Exercise 2.3. Show that the nilradical of A is equal to the Jacobson radical of A if and only if every non-empty open subset of $\operatorname{Spec} A$ contains a closed point of $\operatorname{Spec} A$.

Exercise 2.4. Let $Z \subseteq X := \operatorname{Spec} A$ be a finite set and let U be an open neighborhood of Z.
(a) Show that there exists an $f \in A$ such that $Z \subseteq D(f) \subseteq U$.
   Hint: Use Proposition B.2 (Section (2)).
(b) Set $\mathscr{O}_{X,Z} := S^{-1}A$, where $S = A \setminus \bigcup_{z \in Z} \mathfrak{p}_z$. Show that $\mathscr{O}_{X,Z} = \varinjlim_{U \supseteq Z \text{ open}} \mathscr{O}_X(U)$.
(c) Show that $\mathscr{O}_{X,Z}$ is a semi-local ring with maximal ideals $S^{-1}\mathfrak{p}_z$ where z runs through those points $x \in Z$ that are not generizations of points $\neq x$ in Z.

Exercise 2.5$\Diamond$. Let $\mathfrak{a} \subseteq A$ be an ideal, let $\mathfrak{p}_1, \ldots, \mathfrak{p}_r$ be prime ideals of A and let $Z_j = V(\mathfrak{p}_j)$, $j = 1, \ldots, r$, the corresponding closed irreducible subset. Assume that for all j there exists an $f_j \in \mathfrak{a}$ such that f_j does not vanish on Z_j (i.e., $Z_j \not\subseteq V(f_j)$). Show that there exists an $f \in \mathfrak{a}$ such that $Z_j \not\subseteq V(f)$ for all j.
Hint: Proposition B.2 (Section (2)).

Exercise 2.6. Show that an open subset U of $\operatorname{Spec} A$ is quasi-compact if and only if it the complement of a closed set of the form $V(\mathfrak{a})$, where $\mathfrak{a}$ is a finitely generated ideal.

Exercise 2.7$\Diamond$**.** Let $\mathfrak{a}$ be an ideal such that $\mathfrak{a}$ is contained in the Jacobson radical of A. Show that the only open set of $X = \operatorname{Spec} A$ that contains $V(\mathfrak{a})$ is X itself. Deduce that, if A is a local ring, the only open set of X containing the unique closed point of X is the space X itself.

Exercise 2.8. Let $X = \operatorname{Spec} A$.
(a) Show that every locally closed irreducible subset of X contains a unique generic point.
(b) Show that every irreducible subset of X contains at most one generic point.
(c) Now let A be a principal ideal domain with infinitely many maximal ideals (e.g. $A = \mathbb{Z}$ or $A = k[T]$, where k is a field). Show that any subset of $\operatorname{Spec} A$ that consists of infinitely many closed points is irreducible but does not contain a generic point.

Exercise 2.9. Let Γ be a totally ordered abelian group. A subgroup Δ of Γ is called *isolated* if $0 \leq \gamma \leq \delta$ and $\delta \in \Delta$ imply $\gamma \in \Delta$.
(a) Let A be a valuation ring, $K = \operatorname{Frac} A$. Show that $\mathfrak{p} \mapsto A_{\mathfrak{p}}$ is an inclusion reversing bijection from $\operatorname{Spec} A$ onto the set of rings B with $A \subseteq B \subseteq K$ and that such a ring B is a valuation ring of K. Its inverse is given by sending B to its maximal ideal (which is contained in A).
(b) Let A be a valuation ring with value group Γ, see Section (B.13). For every isolated subgroup Δ of Γ set $\mathfrak{p}_\Delta := \{ a \in A ; v(a) \notin \Delta \}$. Show that $\Delta \mapsto \mathfrak{p}_\Delta$ defines an order reversing bijection between the set of isolated subgroups of Γ and $\operatorname{Spec} A$ (both totally ordered by inclusion).
(c) Show that the value groups of the valuation ring $A_{\mathfrak{p}_\Delta}$ is isomorphic to Γ/Δ and the value groups of the valuation ring $A/\mathfrak{p}_\Delta$ is isomorphic to Δ.
(d) For an arbitrary totally ordered abelian group Γ let $R = k[\Gamma]$ be the group algebra of Γ over some field k. Write elements $u \in R$ as finite sums $u = \sum_\gamma \alpha_\gamma e^\gamma$. Define a map $v \colon R \setminus \{0\} \to \Gamma$ by sending u to the minimal $\gamma \in \Gamma$ such that $\alpha_\gamma \neq 0$. Show that R is an integral domain and that v can be extended to a valuation v on $K = \operatorname{Frac} R$ with value group Γ. In particular $A := \{ x \in K ; v(x) \geq 0 \}$ is a valuation ring with value group Γ.
(e) Let I be a well-ordered set. Endow $\mathbb{Z}^I$ with the lexicographic order (i.e., we set $(n_i)_{i \in I} < (m_i)_{i \in I}$ if and only if $J := \{ i \in I ; m_i \neq n_i \}$ is non-empty and $n_{i_J} < m_{i_J}$ where i_J is the smallest element of J). Show that $\mathbb{Z}^I$ is a totally ordered abelian group and that for all $k \in I$ the subsets $\Gamma_{\geq k}$ of $(n_i)_i \in \mathbb{Z}^I$ such that $n_i = 0$ for all $i < k$ are isolated subgroups of $\mathbb{Z}^I$.
(f) Deduce that for every cardinal number $\mathfrak{k}$ there are valuation rings whose spectrum has cardinality $\geq \mathfrak{k}$.

Exercise 2.10. Let X be a topological space, $\mathscr{F}$ a sheaf on X and let $s, t \in \mathscr{F}(U)$ be two sections of $\mathscr{F}$ over an open subset $U \subseteq X$. Show that the set of $x \in U$ such that $s_x = t_x$ is open in U.

Exercise 2.11$\Diamond$**.** Let I be a filtered preordered set, fix inductive systems $((X_i)_i, (\alpha_{ji})_{i \leq j})$ and $((Y_i)_i, (\beta_{ji})_{i \leq j})$ of sets, indexed by I, and let X and Y be their respective inductive limits. Let $(u_i \colon X_i \to Y_i)_i$ be a morphism of inductive systems and let $u \colon X \to Y$ be its inductive limit.
(a) Show that if there exists an $i \in I$ such that u_j is injective for all $j \geq i$, u is injective.
(b) Show that if there exists an $i \in I$ such that u_j is surjective for all $j \geq i$, u is surjective.

Exercise 2.12$\Diamond$**.** Let $\mathscr{O}_{\mathbb{C}}$ be the sheaf of holomorphic functions of $\mathbb{C}$.
(a) Show that $(\mathbb{C}, \mathscr{O}_{\mathbb{C}})$ is a locally ringed space. What is $\kappa(z)$ for $z \in \mathbb{C}$?

(b) Let $D: \mathscr{O}_{\mathbb{C}} \to \mathscr{O}_{\mathbb{C}}$ be the morphism of sheaves, which sends $f \in \mathscr{O}_{\mathbb{C}}(U)$ ($U \subseteq \mathbb{C}$ open) to its derivative $f' \in \mathscr{O}_{\mathbb{C}}(U)$. Show that $D_z: \mathscr{O}_{\mathbb{C},z} \to \mathscr{O}_{\mathbb{C},z}$ is surjective for all $z \in \mathbb{C}$. Give an example of an open set U in $\mathbb{C}$ such that D_U is not surjective. Can you characterize the open subsets U, such that D_U is surjective?

Exercise 2.13. Let X be a topological space. A sheaf $\mathscr{F}$ on X is called *locally constant* if every point $x \in X$ has an open neighborhood U such that $\mathscr{F}_{|U}$ is a constant sheaf (Example 2.25).
(a) Let X be irreducible. Show that the following properties for a presheaf $\mathscr{F}$ on X are equivalent.
   (i) The set $\mathscr{F}(\emptyset)$ consists of one element, and for every non-empty open subset $U \subseteq X$, the restriction map $\mathscr{F}(X) \to \mathscr{F}(U)$ is bijective.
   (ii) $\mathscr{F}$ is a constant sheaf on X.
   (iii) $\mathscr{F}$ is a locally constant sheaf on X.
(b) Conversely, let X be a topological space and assume that there exists a sheaf $\mathscr{F}$ such that $\mathscr{F}(X) \to \mathscr{F}(U)$ is bijective for all non-empty open sets $U \subseteq X$ and such that $\mathscr{F}(X)$ contains more than one element. Show that X is irreducible.

Exercise 2.14. Let X be a topological space and $i: Z \to X$ the inclusion of a subspace Z. Let $\mathscr{F}$ be a sheaf on Z. Show the following properties for the stalks $i_*(\mathscr{F})_x$.
(a) For all $x \notin \overline{Z}$, $i_*(\mathscr{F})_x$ is a singleton (i.e., a set consisting of one element).
(b) For all $x \in Z$, $i_*(\mathscr{F})_x = \mathscr{F}_x$.
(c) Now assume that every point in the closure of Z has a fundamental system of open neighborhoods which intersect Z in a connected set and that $\mathscr{F}$ is a constant sheaf with value E, where E is some set. Show that $i_*(\mathscr{F})_x = E$ for all $x \in \overline{Z}$.
Note that the conditions in (c) are automatically satisfied, if $Z = \{x\}$ for some point $x \in X$. Then $i_*(\mathscr{F})$ is called the *skyscraper sheaf in x with value E*.

Exercise 2.15. Let X be a locally compact topological space and let $\mathscr{F}$ be the presheaf of bounded continuous functions on X with values in $\mathbb{R}$ (Example 2.19 (4)). Describe the associated sheaf $\tilde{\mathscr{F}}$.

Exercise 2.16. Let X be a topological space and let $(U_i)_i$ be an open covering of X. For all i let $\mathscr{F}_i$ be a sheaf on U_i. Assume that for each pair (i,j) of indices we are given isomorphisms $\varphi_{ij}: \mathscr{F}_{j|U_i \cap U_j} \xrightarrow{\sim} \mathscr{F}_{i|U_i \cap U_j}$ satisfying for all i,j,k the "cocycle condition" $\varphi_{ik} = \varphi_{ij} \circ \varphi_{jk}$ on $U_i \cap U_j \cap U_k$.
(a) Show that there exists a sheaf $\mathscr{F}$ on X and for all i isomorphisms $\psi_i: \mathscr{F}_i \xrightarrow{\sim} \mathscr{F}_{|U_i}$ such that $\psi_i \circ \varphi_{ij} = \psi_j$ on $U_i \cap U_j$ for all i,j. Show that $\mathscr{F}$ and the ψ_i are uniquely determined up to unique isomorphism by these conditions. Show that an analogous result holds for sheaves with values in an arbitrary category.
   Remark: The sheaf $\mathscr{F}$ is said to be obtained by *gluing the $\mathscr{F}_i$* via the *gluing data φ_{ij}*.
(b) Make sheaves on $(U_i)_i$ and gluing data into a category and show that this category is equivalent to the category of sheaves on X.

Exercise 2.17. Let $(X, \mathscr{O}_X)$ be a locally ringed space.
(a) Let $U \subseteq X$ be an open and closed subset. Show that there exists a unique section $e_U \in \Gamma(X, \mathscr{O}_X)$ such that $e_{U|V} = 1$ for all open subsets V of U and $e_{U|V} = 0$ for all open subsets V of $X \setminus U$. Show that $U \mapsto e_U$ yields a bijection

$$\mathrm{OC}(X) \leftrightarrow \mathrm{Idem}(\Gamma(X, \mathscr{O}_X))$$

from the set of open and closed subsets of X to the set of idempotent elements of the ring $\Gamma(X, \mathscr{O}_X)$.

(b) Show that $e_U e_{U'} = e_{U \cap U'}$ for $U, U' \in \mathrm{OC}(X)$.

(c) Assume that $X \neq \emptyset$. Prove that the following are equivalent:
   (i) X is connected.
   (ii) There exists no idempotent element $e \in \Gamma(X, \mathscr{O}_X)$ with $e \neq 0, 1$.
   (iii) There exists no decomposition $\Gamma(X, \mathscr{O}_X) = R_1 \times R_2$ where R_1, R_2 are non-zero rings.

Exercise 2.18. Let $(X, \mathscr{O}_X)$ be a ringed space such that X consists of one point x (e.g. $X = \operatorname{Spec} K$, where K is a field). We set $B := \mathscr{O}_X(X) = \mathscr{O}_{X,x}$. Clearly, $(X, \mathscr{O}_X)$ is locally ringed if and only if B is a local ring.

(a) Let $(Y, \mathscr{O}_Y)$ be a second ringed space. To every morphism $f \colon (X, \mathscr{O}_X) \to (Y, \mathscr{O}_Y)$ of ringed spaces we associate $(f(x), f_x^\sharp)$. Show that this defines a bijection, functorial in $(Y, \mathscr{O}_Y)$, between the set of morphisms $(X, \mathscr{O}_X) \to (Y, \mathscr{O}_Y)$ and the set of pairs (y, φ) where $y \in Y$ and $\varphi \colon \mathscr{O}_{Y,y} \to B$ is a ring homomorphism.

(b) Show that if B is local and $(Y, \mathscr{O}_Y)$ is a locally ringed space, f is a morphism of locally ringed spaces if and only if $f_x^\sharp$ is a local ring homomorphism.

(c) Give an example of a morphism of ringed spaces between affine schemes which is not a morphism of locally ringed spaces.

Exercise 2.19◊. Let $(X, \mathscr{O}_X)$ be a locally ringed space, and $f \in \mathscr{O}_X(X)$. Define

$$X_f := \{\, x \in X \; ; \; f(x) \neq 0 \,\}.$$

Show that X_f is an open subset of X. What is X_f if X is an affine scheme?

Exercise 2.20. Let A be a local ring. Show that $\operatorname{Spec} A$ is connected.
Hint: Exercise 2.17.

Exercise 2.21. Let A be a ring, $\mathfrak{a} \subset A$ an ideal such that A is $\mathfrak{a}$-adically complete, and let $i \colon \operatorname{Spec} A/\mathfrak{a} \hookrightarrow \operatorname{Spec} A$ be the canonical morphism. Show that $U \mapsto i^{-1}(U)$ yields a bijection from the set of open and closed subsets of $\operatorname{Spec} A$ to the set of open and closed subsets of $\operatorname{Spec} A/\mathfrak{a}$.
Hint: Exercise 2.17.

Exercise 2.22. Let R be a principal ideal domain, and let $f \in R$ be a nonzero element. Describe the affine scheme $X = \operatorname{Spec} R/fR$ (its underlying topological space, the stalks $\mathscr{O}_{X,x}$, and $\mathscr{O}_X(U)$ for every subset U of X) in terms of the decomposition of f into prime factors.

Exercise 2.23. A ring A is called *Boolean* if $a^2 = a$ for all $a \in A$. Let A be a Boolean ring and $X = \operatorname{Spec} A$.

(a) Show that every prime ideal of A is a maximal ideal and that $\kappa(x) = \mathbb{F}_2$ for all $x \in X$. Deduce that $\varphi \mapsto \operatorname{Ker}(\varphi)$ yields a bijection between the set of ring homomorphisms $A \to \mathbb{F}_2$ and $\operatorname{Spec} A$.

(b) Show that X is a compact totally disconnected space and that $A \mapsto \operatorname{Spec} A$ yields an equivalence of the category of Boolean rings (as a full subcategory of all rings) and the category of compact totally disconnected spaces (where the morphisms are continuous maps). A quasi-inverse of $A \mapsto \operatorname{Spec} A$ is given by sending X to the $\mathbb{F}_2$-algebra of continuous maps $X \to \mathbb{F}_2$ (where $\mathbb{F}_2$ is endowed with the discrete topology).

Remark: See also Exercise 10.10.

3   Schemes

Contents

In the current chapter, we will define the notion of *scheme*. In a sense, the remainder of this book is devoted to the study of schemes, so this notion is fundamental for all which follows. Schemes arise by "gluing affine schemes", similarly as prevarieties are obtained by gluing affine varieties. Therefore after the preparations in the previous chapter, the definition is very simple, see Section (3.1). As for varieties we define projective space (Section (3.6)) by gluing copies of affine spaces. This is an example of a scheme which is not affine.

Even though prevarieties are not schemes themselves, we can in a natural way embed the category of prevarieties over some algebraically closed field k as a full subcategory of the category of k-schemes. In Sections (3.8)–(3.13) we will discuss the properties of those schemes which are in the essential image of this embedding, and will explain how we can identify prevarieties over k with "integral schemes of finite type over k".

Finally we will discuss the notion of *subscheme* and in particular of the underlying reduced subscheme of a scheme.

Schemes

(3.1) Definition of Schemes.

In order to define the notion of scheme, we proceed as in Chapter 1 where we defined prevarieties, using affine varieties as building blocks. In the current situation, the local pieces will be affine schemes, i. e. the spectra of rings, seen as locally ringed spaces.

Definition 3.1. *A* scheme *is a locally ringed space* $(X, \mathscr{O}_X)$ *which admits an open covering* $X = \bigcup_{i \in I} U_i$ *such that all locally ringed spaces* $(U_i, \mathscr{O}_X|_{U_i})$ *are affine schemes.*
A morphism of schemes *is a morphism of* locally ringed spaces.

We obtain the category of schemes which we will denote by (Sch). Clearly any affine scheme is a scheme. Usually we denote a scheme $(X, \mathscr{O}_X)$ simply by X.

© Springer Fachmedien Wiesbaden GmbH, part of Springer Nature 2020
U. Görtz und T. Wedhorn, *Algebraic Geometry I: Schemes*, Springer Studium
Mathematik – Master, https://doi.org/10.1007/978-3-658-30733-2_4

Let S be a fixed scheme. The category (Sch/S) of *schemes over S* (or of *S-schemes*) is the category whose objects are the morphisms $X \to S$ of schemes, and whose morphisms $\mathrm{Hom}(X \to S, Y \to S)$ are the morphisms $X \to Y$ of schemes with the property that

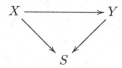

commutes. The morphism $X \to S$ is called the *structural morphism* of the S-scheme X (and often is silently omitted from the notation). The scheme S is also sometimes called the *base scheme*. In case $S = \mathrm{Spec}\,R$ is an affine scheme, one also speaks about *R-schemes* or *schemes over R* instead. For S-schemes X and Y we denote the set of morphisms $X \to Y$ in the category of S-schemes by $\mathrm{Hom}_S(X, Y)$ or by $\mathrm{Hom}_R(X, Y)$, if $S = \mathrm{Spec}\,R$ is affine.

By definition the S-scheme $\mathrm{id}_S \colon S \to S$ is a final object in the category (Sch/S).

We add a remark about the terminology: Originally (in particular in Grothendieck's [EGAI]–[EGAIV]) the objects which we call schemes were called preschemes (and what we will call separated schemes, Definition 9.7, were called schemes). Nowadays (and already in [EGAInew]) only the terminology introduced above is used.

(3.2) Open subschemes.

Proposition and Definition 3.2.
(1) *Let X be a scheme, and $U \subseteq X$ an open subset. Then the locally ringed space $(U, \mathcal{O}_{X|U})$ is a scheme. We call U an* open subscheme *of X. If U is an affine scheme, then U is called an* affine open subscheme.
(2) *Let X be a scheme. The affine open subschemes are a basis of the topology.*

More precisely, in the second part of the proposition we should say that those open subsets which give rise to an affine open subscheme are a basis of the topology.

Proof. By definition the locally ringed space X can be covered by affine schemes, and by Proposition 2.5 each of these affine schemes has a basis of its topology consisting of affine schemes. This yields both parts of the proposition. $\square$

Let $U \subseteq X$ be an open subset, and $j \colon U \to X$ the inclusion. We consider U as an open subscheme of X. If $V \subseteq X$ is open, the restriction map of the structure sheaf $\mathcal{O}_X$ gives us a ring homomorphism

$$\Gamma(V, \mathcal{O}_X) \to \Gamma(V \cap U, \mathcal{O}_X) = \Gamma(j^{-1}(V), \mathcal{O}_{X|U}) = \Gamma(V, j_* \mathcal{O}_{X|U})$$

Altogether, these maps constitute a homomorphism $j^\flat \colon \mathcal{O}_X \to j_* \mathcal{O}_{X|U}$ of sheaves of rings and together with the inclusion $U \subseteq X$ a morphism $U \to X$ of schemes. Whenever we (possibly implicitly) speak about a morphism of schemes from U to X, then this is the one which is meant.

An *affine open covering* of a scheme X is an open covering $X = \bigcup_i U_i$, such that all U_i are affine open subschemes of X.

We will study open (and closed and locally closed) subschemes in more detail in Sections (3.15) – (3.16). Finally, we note the following lemma which will be useful for us. Recall that whenever $X = \operatorname{Spec} A$ is an affine scheme, and $f \in A$ an element of its affine coordinate ring, then $(D(f), \mathscr{O}_{X|D(f)})$ is an affine scheme with coordinate ring A_f. Subschemes of this form are called principal open, see Example 2.38.

Lemma 3.3. *Let X be a scheme, and let U, V be affine open subschemes of X. Then there exists for all $x \in U \cap V$ an open subscheme $W \subseteq U \cap V$ with $W \ni x$ such that W is principal open in the sense of Section (2.2) in U as well as in V.*

Proof. Replacing V by a principal open subset of V containing x, if necessary, we may assume that $V \subseteq U$. Now choose $f \in \Gamma(U, \mathscr{O}_X)$ such that $x \in D(f) \subseteq V$, and let $f|_V$ denote the image of f under the restriction homomorphism $\Gamma(U, \mathscr{O}_X) \to \Gamma(V, \mathscr{O}_X)$. Then $D_U(f) = D_V(f|_V)$. (The sheaf axioms then also imply that $\Gamma(U, \mathscr{O}_X)_f \cong \Gamma(V, \mathscr{O}_X)_{f|_V}$.) $\qquad\square$

(3.3) Morphisms into affine schemes, gluing of morphisms.

Morphisms of an arbitrary scheme (or even an arbitrary locally ringed space) into an affine scheme are easy to understand, as the following proposition shows.

Proposition 3.4. *Let $(X, \mathscr{O}_X)$ be a locally ringed space, $Y = \operatorname{Spec} A$ an affine scheme. Then the natural map*

$$\operatorname{Hom}(X, Y) \longrightarrow \operatorname{Hom}(A, \Gamma(X, \mathscr{O}_X)), \quad (f, f^\flat) \mapsto f_Y^\flat,$$

is a functorial bijection. Here the set on the left side denotes the set of morphisms $X \to Y$ of locally ringed spaces, and the set on the right side denotes the set of ring homomorphisms $A \to \Gamma(X, \mathscr{O}_X)$.

If X is a scheme, the proof is an easy gluing argument that we will give first. After that we will give an independent proof of the general statement.

Proof. First proof if X is a scheme. In this case, there exists an affine open covering $X = \bigcup_i U_i$. We know from Theorem 2.35, that for all U_i the natural map

$$\operatorname{Hom}(U_i, Y) \longrightarrow \operatorname{Hom}(A, \Gamma(U_i, \mathscr{O}_X))$$

is a bijection. For an affine open $V \subseteq U_i \cap U_j$ the diagram

$$
\begin{array}{ccc}
\operatorname{Hom}(U_i, Y) & \longrightarrow & \operatorname{Hom}(A, \Gamma(U_i, \mathscr{O}_X)) \\
\downarrow & & \downarrow \\
\operatorname{Hom}(V, Y) & \longrightarrow & \operatorname{Hom}(A, \Gamma(V, \mathscr{O}_X))
\end{array}
$$

is commutative, since $\Gamma(-)$, i. e. taking global sections, is functorial. The assertion now follows from the very general Proposition 3.5 below about gluing of morphisms.

Second proof in the general case. We first construct a map $\operatorname{Hom}(A, \Gamma(X, \mathscr{O}_X)) \to \operatorname{Hom}(X, Y)$, so let $\varphi \colon A \to \Gamma(X, \mathscr{O}_X)$ be a ring homomorphism. Let us start by defining a map $f \colon X \to Y$ between the underlying sets of X and Y. For $x \in X$, let $\mathfrak{p} := \{\, a \in A \; ; \; \varphi(a)(x) = 0 \text{ in } \kappa(x)\,\}$. Then $A/\mathfrak{p}$ embeds into the field $\kappa(x)$, so $\mathfrak{p}$ is a prime ideal, and we set $f(x) := \mathfrak{p}$.

For $s \in A$, we have $f^{-1}(D(s)) = \{\, x \in X \;;\; \varphi(s)(x) \neq 0 \,\} = \{\, x \in X \;;\; \varphi(s)_x \in \mathscr{O}_{X,x}^\times \,\} =:$
$X_{\varphi(s)}$, an open subset of X (cf. Exercise 2.19). Hence f is continuous. Also note that
$\varphi(s)_{|X_{\varphi(s)}}$ is a unit in $\Gamma(X_{\varphi(s)}, \mathscr{O}_X)$, because this is true locally and the inverse elements
are unique and hence can be glued.

To define a sheaf morphism $\mathscr{O}_Y \to f_* \mathscr{O}_X$, it is enough to give ring homomorphisms
$\mathscr{O}_Y(D(s)) \to \mathscr{O}_X(f^{-1}(D(s)))$, compatible with restriction maps, for all $s \in A$. But
$\mathscr{O}_Y(D(s)) = A_s$, and $\varphi(s)_{|X_s} \in \Gamma(X_{\varphi(s)}, \mathscr{O}_X)^\times = \mathscr{O}_X(f^{-1}(D(s)))^\times$, so there is a unique
such map which is compatible with the given map $\varphi \colon A \to \Gamma(X, \mathscr{O}_X)$. In this way, we
obtain the desired sheaf homomorphism $f^\flat \colon \mathscr{O}_Y \to f_* \mathscr{O}_X$. For $\mathfrak{p} = f(x) \in \operatorname{Spec} A$, we have
$\mathscr{O}_{Y,f(x)} = A_\mathfrak{p}$, and the map induced by $f^\flat$ on stalks is the ring homomorphism $A_\mathfrak{p} \to \mathscr{O}_{X,x}$
induced by φ. For $a \in \mathfrak{p}$, $\varphi(a)$ lies in the maximal ideal of $\mathscr{O}_{X,x}$ by definition of f, which
implies that this is a local ring homomorphism. Altogether we have constructed a map
$\operatorname{Hom}(A, \Gamma(X, \mathscr{O}_X)) \to \operatorname{Hom}(X, Y)$.

It remains to show that the two maps are inverse to each other. By construction, it is
clear that starting from an element $\varphi \in \operatorname{Hom}(A, \Gamma(X, \mathscr{O}_X))$, we get back φ after applying
both maps. Conversely, let $f \in \operatorname{Hom}(X, Y)$, and let $\varphi = f_Y^\flat$. Denote by $g \colon X \to Y$
the morphism constructed from φ as above. Since f is a morphism of locally ringed
spaces, it induces homomorphisms $A \to \kappa(f(x)) \hookrightarrow \kappa(x)$ for every $x \in X$. As a prime
ideal in A, $f(x)$ is just the kernel of $A \to \kappa(f(x))$, so we see that $f = g$ as continuous
maps. To check the equality of the two sheaf morphisms $\mathscr{O}_Y \to f_* \mathscr{O}_X$, it is enough to
consider the maps induced on stalks (Proposition 2.23). But on stalks both f and g
induce homomorphisms $A_\mathfrak{p} \to \mathscr{O}_{X,x}$ (with $\mathfrak{p} = f(x) = g(x)$) which fit into a commutative
diagram with $\varphi \colon A \to \Gamma(X, \mathscr{O}_X)$. Hence they must be equal by the universal property of
the localization.   □

Proposition 3.5. *(Gluing of morphisms) Let X, Y be locally ringed spaces. For every
open subset $U \subseteq X$, let $\operatorname{Hom}(U, Y)$ be the set of morphisms $(U, \mathscr{O}_{X|U}) \to (Y, \mathscr{O}_Y)$ of
locally ringed spaces. Then $U \mapsto \operatorname{Hom}(U, Y)$ is a sheaf of sets on X.*

*In other words: If $X = \bigcup_i U_i$ is an open covering, then a family of morphisms $U_i \to Y$
glues to a morphism $X \to Y$ if and only if the morphisms coincide on intersections
$U_i \cap U_j$, and the resulting morphism $X \to Y$ is uniquely determined.*

Proof. Easy. (Analogously, we can glue morphisms of sets or of topological spaces. It
is then easy to see that one can also define the sheaf homomorphism $\mathscr{O}_Y \to f_* \mathscr{O}_X$ by
gluing.)   □

Because for every ring R there is a unique ring homomorphism $\mathbb{Z} \to R$, we obtain:

Corollary 3.6. *Let X be a locally ringed space. Then there exists a unique morphism
$X \to \operatorname{Spec} \mathbb{Z}$ of locally ringed spaces. In particular, $\operatorname{Spec} \mathbb{Z}$ is a final object in the category
of schemes.*

We also see that $\operatorname{Hom}(X, \operatorname{Spec} \mathbb{Z}[T]) = \Gamma(X, \mathscr{O}_X)$. More generally, for an R-scheme X
we have an identification of R-algebras

(3.3.1) $\operatorname{Hom}_R(X, \mathbb{A}_R^1) = \Gamma(X, \mathscr{O}_X)$.

Thus we may consider global sections of $\mathscr{O}_X$ as morphisms on X with values in the affine
line.

Remark 3.7. We may apply Proposition 3.4 also to $A = \Gamma(X, \mathscr{O}_X)$. Thus for every locally ringed space X there corresponds to $\mathrm{id}_{\Gamma(X, \mathscr{O}_X)}$ a functorial morphism of locally ringed spaces

$$c_X \colon X \to \operatorname{Spec} \Gamma(X, \mathscr{O}_X)$$

which we call *canonical*.

(3.4) Morphisms from $\operatorname{Spec} K$, K a field, into schemes.

Let X be a scheme. Let $x \in X$, and let $U \subseteq X$ be an affine open neighborhood of x, say $U = \operatorname{Spec} A$. Denote by $\mathfrak{p} \subset A$ the prime ideal of A corresponding to x. Then $\mathscr{O}_{X,x} = \mathscr{O}_{U,x} = A_{\mathfrak{p}}$, and the natural homomorphism $A \to A_{\mathfrak{p}}$ gives us a morphism

$$(3.4.1) \qquad j_x \colon \operatorname{Spec} \mathscr{O}_{X,x} = \operatorname{Spec} A_{\mathfrak{p}} \to \operatorname{Spec} A = U \subseteq X$$

of schemes. By Proposition 3.2 (2) this morphism is independent of the choice of U. By Proposition 2.12, j_x is a homeomorphism from $\operatorname{Spec} \mathscr{O}_{X,x}$ onto the subspace Z of all points $x' \in X$ that are generizations of x (Definition 2.8), i.e., Z is the intersection of all open subsets of X which contain x.

Let $\kappa(x) = \mathscr{O}_{X,x}/\mathfrak{m}_x$ be the residue class field of x in X. We obtain a morphism of schemes,

$$(3.4.2) \qquad i_x \colon \operatorname{Spec} \kappa(x) \longrightarrow \operatorname{Spec} \mathscr{O}_{X,x} \longrightarrow X,$$

called *canonical*. The image point of the unique point in $\operatorname{Spec} \kappa(x)$ is x.

Now let K be any field, let $f \colon \operatorname{Spec} K \to X$ be a morphism, and let $x \in X$ be the image point of the unique point p of $\operatorname{Spec} K$. Since f is a morphism of locally ringed spaces, f induces a local homomorphism $\mathscr{O}_{X,x} \to K = \mathscr{O}_{\operatorname{Spec} K, p}$, and hence a homomorphism $\iota \colon \kappa(x) \to K$ between the residue class fields. In other words: The morphism f factors as $f = i_x \circ (\operatorname{Spec} \iota) \colon \operatorname{Spec} K \to \operatorname{Spec} \kappa(x) \to X$.

Proposition 3.8. *The above construction gives rise to a bijection*

$$\operatorname{Hom}(\operatorname{Spec} K, X) \longrightarrow \{(x, \iota);\ x \in X,\ \iota \colon \kappa(x) \to K\}$$

Proof. Conversely, we map an element $(x, \iota \colon \kappa(x) \to K)$ of the right hand side to the morphism

$$\operatorname{Spec} K \xrightarrow{\ \operatorname{Spec} \iota\ } \operatorname{Spec} \kappa(x) \xrightarrow{\ i_x\ } X,$$

and these two maps are inverse to each other. $\qquad\qquad\qquad\qquad\qquad\qquad\square$

More generally, we have an analogous description for $\operatorname{Hom}(\operatorname{Spec} R, X)$, where R is a local ring, see Exercise 3.18.

(3.5) Gluing of schemes, disjoint unions of schemes.

Definition 3.9. *A gluing datum of schemes consists of the following data:*
- *an index set I,*
- *for all $i \in I$ a scheme U_i,*

- *for all $i, j \in I$ an open subset $U_{ij} \subseteq U_i$ (we consider U_{ij} as open subscheme of U_i),*
- *for all $i, j \in I$ an isomorphism $\varphi_{ji} \colon U_{ij} \to U_{ji}$ of schemes,*

such that

(a) *$U_{ii} = U_i$ for all $i \in I$,*

(b) *the* cocycle condition *holds: $\varphi_{kj} \circ \varphi_{ji} = \varphi_{ki}$ on $U_{ij} \cap U_{ik}$, $i, j, k \in I$.*

In the cocycle condition we implicitly assume that in particular $\varphi_{ji}(U_{ij} \cap U_{ik}) \subseteq U_{jk}$, such that the composition is meaningful. For $i = j = k$, the cocycle condition implies that $\varphi_{ii} = \mathrm{id}_{U_i}$ and (for $i = k$) that $\varphi_{ij}^{-1} = \varphi_{ji}$, and that φ_{ji} is an isomorphism $U_{ij} \cap U_{ik} \to U_{ji} \cap U_{jk}$.

Obviously, one can completely analogously define the notion of gluing datum for sets, topological spaces or (locally) ringed spaces. In each of these cases, one can construct from a gluing datum a new object of the concerning category by "gluing" the objects U_i, which satisfies a universal property as explained below. For schemes, we prove this fact in the following proposition.

Proposition 3.10. *Let $((U_i)_{i \in I}, (U_{ij})_{i,j \in I}, (\varphi_{ij})_{i,j \in I})$ be a gluing datum of schemes. Then there exists a scheme X together with morphisms $\psi_i \colon U_i \to X$, such that*

- *for all i the map ψ_i yields an isomorphism from U_i onto an open subscheme of X,*
- *$\psi_j \circ \varphi_{ji} = \psi_i$ on U_{ij} for all i, j,*
- *$X = \bigcup_i \psi_i(U_i)$,*
- *$\psi_i(U_i) \cap \psi_j(U_j) = \psi_i(U_{ij}) = \psi_j(U_{ji})$ for all $i, j \in I$.*

Furthermore, X together with the ψ_i is uniquely determined up to unique isomorphism.

Proposition 3.5 about gluing of morphisms shows that the scheme X in the proposition satisfies the following universal property: If T is a scheme, and for all $i \in I$, $\xi_i \colon U_i \to T$ is a morphism of schemes which induces an isomorphism of U_i with an open subscheme of T_i, such that $\xi_j \circ \varphi_{ji} = \xi_i$ on U_{ij} for all $i, j \in I$, then there exists a unique morphism $\xi \colon X \to T$ with $\xi \circ \psi_i = \xi_i$ for all $i \in I$.

In particular this implies the uniqueness assertion in the proposition. (The uniqueness of course can also easily be obtained directly from Proposition 3.5 about the gluing of morphisms.)

Proof. To define the underlying topological space of X, we start with the disjoint union $\coprod_{i \in I} U_i$ of the (underlying topological spaces of the) U_i and define an equivalence relation $\sim$ on it as follows: points $x_i \in U_i$, $x_j \in U_j$, $i, j \in I$, are equivalent, if and only if $x_i \in U_{ij}$, $x_j \in U_{ji}$ and $x_j = \varphi_{ji}(x_i)$. The cocycle condition implies that $\sim$ is in fact an equivalence relation. As a set, define X to be the set of equivalence classes,

$$X := \coprod_{i \in I} U_i / \sim .$$

The natural maps $\psi_i \colon U_i \to X$ are injective and we have $\psi_i(U_{ij}) = \psi_i(U_i) \cap \psi_j(U_j)$ for all $i, j \in I$. We equip X with the quotient topology, i. e. with the finest topology such that all ψ_i are continuous. That means that a subset $U \subseteq X$ is open if and only if for all i the preimage $\psi_i^{-1}(U)$ is open in U_i. In particular, the $\psi_i(U_i)$ and the $\psi_i(U_{ij}) = \psi_i(U_i) \cap \psi_j(U_j)$ are open in X.

To obtain a locally ringed space, we have to "glue" the structure sheaves on the U_i so as to define a sheaf $\mathscr{O}_X$ of rings on X. The sheaf $\mathscr{O}_X$ is uniquely determined by its sections (and the corresponding restriction maps) on a basis of the topology. It is thus sufficient to define it on those open subsets $U \subseteq X$ which are contained in one of the $\psi_i(U_i)$, and to check that this is well-defined and satisfies the sheaf axioms (Proposition 2.20). For each such U, we fix once and for all an i with $U \subseteq \psi_i(U_i)$, and we set $\mathscr{O}_X(U) = \mathscr{O}_{U_i}(\psi_i^{-1}(U))$. If $U \subseteq \psi_i(U_i) \cap \psi_j(U_j)$, then we identify the rings $\mathscr{O}_{U_i}(\psi_i^{-1}(U))$ and $\mathscr{O}_{U_j}(\psi_j^{-1}(U))$ via φ_{ji}. This allows us to define restriction maps. We obtain a sheaf $\mathscr{O}_X$ of rings on X which is independent of our choices. Since all the U_i are locally ringed spaces, the same is true for X.

Furthermore, with this definition the ψ_i are morphisms of locally ringed spaces; they identify U_i with $(\psi_i(U_i), \mathscr{O}_{X|\psi_i(U_i)})$. Finally all the U_i are schemes by assumption, i. e., they are covered, as locally ringed spaces, by affine schemes, and therefore X is a scheme as well. By construction of X, we have $X = \bigcup \psi_i(U_i)$. □

Example 3.11. (Disjoint union) As a (trivial) special case of this construction we have the disjoint union of schemes (or locally ringed spaces). We simply let $U_{ij} = \emptyset$ for all i, j, so the underlying topological space is indeed the disjoint union of the U_i. The structure sheaf is the "obvious" sheaf. We denote the disjoint union by $\coprod_{i \in I} U_i$.

Example 3.12. Let $X_1, \ldots, X_n$ be affine schemes, $X_i = \operatorname{Spec} A_i$. Then $\coprod_{i=1}^n X_i$ is also an affine scheme, which is isomorphic to $\operatorname{Spec} \prod_{i=1}^n A_i$. The disjoint union of infinitely many (non-empty) affine schemes is not affine, though (see Exercise 3.9).

Example 3.13. (Gluing of two schemes) Note that in case the index set I has only two elements, any two open subsets $U_{12} \subseteq U_1$, $U_{21} \subseteq U_2$ together with an isomorphism $\varphi \colon U_{12} \xrightarrow{\sim} U_{21}$ already yield a gluing datum. Denote by X the scheme obtained by gluing U_1 and U_2 along φ. We can then view U_1 and U_2 as open subschemes of X. The definition of the structure sheaf $\mathscr{O}_X$ in the proof of Proposition 3.10 shows that for every open subset $V \subseteq X$ we have

$$\Gamma(V, \mathscr{O}_X) = \{ (s_1, s_2) \in \Gamma(V \cap U_1, \mathscr{O}_{U_1}) \times \Gamma(V \cap U_2, \mathscr{O}_{U_2}) \; ; \; \varphi^\flat(s_{2|U_{21} \cap V}) = s_{1|U_{12} \cap V} \}.$$

As an example, we consider the "affine line with a double point": Let k be a field, and let $U_1 = U_2 = \mathbb{A}_k^1 = \operatorname{Spec} k[T]$. Fix a closed point $x \in \mathbb{A}_k^1$ and let $U_{12} = U_1 \setminus \{x\}$, $U_{21} = U_2 \setminus \{x\}$. We define a gluing isomorphism $\varphi \colon U_{12} \xrightarrow{\sim} U_{21}$ as the identity morphism. By gluing, we get a scheme X which we should think of as an affine line with the point x doubled. This scheme is not affine (see Exercise 3.26).

Examples of schemes

Again one of the most important examples of a scheme is projective space $\mathbb{P}_R^n$ – now defined over an arbitrary base ring R. Again we may define for a set M of homogeneous polynomials in $R[X_0, \ldots, X_n]$ the vanishing scheme $V_+(M)$. Once we have defined the notion of a closed subscheme in Section (3.15), we will see that $V_+(M)$ is a closed subscheme of $\mathbb{P}_R^n$. In Chapter 13 we will generalize the construction of $\mathbb{P}_R^n$ and of $V_+(M)$ and prove that every closed subscheme of $\mathbb{P}_R^n$ is of the form $V_+(M)$ (Proposition 13.24).

(3.6) Projective Space $\mathbb{P}^n_R$.

Let R be a ring. We define the projective space $\mathbb{P}^n_R$ (over R) by gluing $n+1$ copies of affine space $\mathbb{A}^n_R$. To distinguish between the different copies, we write $U_i = \mathbb{A}^n_R$, $i = 0, \ldots, n$. So U_i is the spectrum of a polynomial ring in n indeterminates over R. It is useful to choose $\frac{X_0}{X_i}, \ldots, \frac{\widehat{X_i}}{X_i}, \ldots, \frac{X_n}{X_i}$ as coordinates (where $\widehat{T_i}$ means that T_i is to be omitted), so we have

$$U_i = \operatorname{Spec} R\left[\frac{X_0}{X_i}, \ldots, \frac{\widehat{X_i}}{X_i}, \ldots, \frac{X_n}{X_i}\right],$$

and we can view all these rings as subrings of the ring $R[X_0, \ldots, X_n, X_0^{-1}, \ldots, X_n^{-1}]$.

We define a gluing datum with index set $\{0, \ldots, n\}$ as follows: For $0 \le i, j \le n$ let $U_{ij} = D_{U_i}(\frac{X_j}{X_i}) \subseteq U_i$ if $i \ne j$, and $U_{ii} = U_i$. Further, let $\varphi_{ii} = \operatorname{id}_{U_i}$ and for $i \ne j$ let

$$\varphi_{ji} \colon U_{ij} \to U_{ji}$$

be the isomorphism defined by the equality

$$R\left[\frac{X_0}{X_i}, \ldots, \frac{\widehat{X_i}}{X_i}, \ldots, \frac{X_n}{X_i}\right]_{\frac{X_j}{X_i}} \longleftarrow R\left[\frac{X_0}{X_j}, \ldots, \frac{\widehat{X_j}}{X_j}, \ldots, \frac{X_n}{X_j}\right]_{\frac{X_i}{X_j}},$$

(as subrings of $R[X_0, \ldots, X_n, X_0^{-1}, \ldots, X_n^{-1}]$) of the affine schemes U_{ij} and U_{ji}. Since the φ_{ij} are defined by equalities, the cocycle condition holds trivially, and we obtain a gluing datum and by Proposition 3.10 a scheme. This scheme is called the *projective space of relative dimension n over R* and it is denoted by $\mathbb{P}^n_R$. We consider the schemes U_i as open subschemes of $\mathbb{P}^n_R$ and denote them also by $D_+(X_i)$.

Remark 3.14. It is easy to see that the canonical ring homomorphism $R \to \Gamma(\mathbb{P}^n_R, \mathscr{O}_{\mathbb{P}^n_R})$ is an isomorphism (Exercise 3.10). This implies that for $n > 0$ the scheme $\mathbb{P}^n_R$ is not affine, since otherwise we would have $\mathbb{P}^n_R = \operatorname{Spec} R$.

Remark 3.15. If k is an algebraically closed field, we had a morphism of prevarieties $\mathbb{A}^{n+1}(k) \setminus \{0\} \to \mathbb{P}^n(k)$ sending $(x_0, \ldots, x_n)$ to $(x_0 : \cdots : x_n)$. An analogous morphism can also be defined for the projective space over an arbitrary ring. Let $R[T_0, \ldots, T_n] \to R$ be the R-algebra homomorphism that sends T_i to 0 for all i. The corresponding scheme morphism $0 \colon \operatorname{Spec} R \to \mathbb{A}^{n+1}_R$ defines an isomorphism of $\operatorname{Spec} R$ onto $V(T_0, \ldots, T_n)$. We denote by $\mathbb{A}^{n+1}_R \setminus \{0\}$ the open complement of $V(T_0, \ldots, T_n)$ in $\mathbb{A}^{n+1}_R$ considered as an open subscheme of $\mathbb{A}^{n+1}_R$. Clearly we have $\mathbb{A}^{n+1}_R \setminus \{0\} = \bigcup_{i=0}^n D(T_i)$. For all $i = 0, \ldots, n$ let p_i be the scheme morphism

$$p_i \colon D(T_i) = \operatorname{Spec} R[T_0, \ldots, T_n, T_i^{-1}] \to D_+(X_i) = \operatorname{Spec} R\left[\frac{X_0}{X_i}, \ldots, \frac{\widehat{X_i}}{X_i}, \ldots, \frac{X_n}{X_i}\right]$$

corresponding to the R-algebra homomorphism given by $X_j/X_i \mapsto T_j/T_i$. Then it is easy to see that the p_i glue together to a scheme morphism

(3.6.1) $p \colon \mathbb{A}^{n+1}_R \setminus \{0\} \to \mathbb{P}^n_R.$

(3.7) Vanishing schemes in projective space.

Let again R be a ring. As in Section (1.19), we view $R[X_0, \dots, X_n]$ as a graded R-algebra. Let $M \subseteq R[X_0, \dots, X_n]$ be a subset of homogeneous polynomials and let $I \subseteq R[X_0, \dots, X_n]$ be the ideal generated by M. Ideals generated by homogeneous elements are called homogeneous ideals. For such an ideal we have $I = \bigoplus_d (I \cap R[X_0, \dots, X_n]_d)$. We want to construct a scheme $V_+(M) = V_+(I)$ which is the analogue of the variety of common zeros of the homogeneous polynomials in I (see Section (1.21)), by gluing affine schemes. If $M = \{f_1, \dots, f_r\}$ is a finite set, we also write $V_+(f_1, \dots, f_r)$ instead of $V_+(M)$.

As in Section (1.20), let $U_i = \operatorname{Spec} R[\frac{X_0}{X_i}, \dots, \widehat{\frac{X_i}{X_i}}, \dots, \frac{X_n}{X_i}]$. By dehomogenizing with respect to X_i (compare Section (1.19)) every homogeneous polynomial in I yields an element in $\Gamma(U_i, \mathscr{O}_{U_i})$. Denote the ideal generated by all these elements by $\Phi_i(I)$. We want to glue the schemes $V_i := \operatorname{Spec} \Gamma(U_i, \mathscr{O}_{U_i})/\Phi_i(I)$ along their open subschemes

$$V_{ij} := D_{V_i}\left(\frac{X_j}{X_i}\right) \subseteq V_i.$$

The gluing isomorphisms which we used to glue the U_i in Section (3.6) restrict to isomorphisms $V_{ji} \xrightarrow{\sim} V_{ij}$, since for a homogeneous polynomial $f \in I$ of degree d we have

$$X_i^d \Phi_i(f) = X_j^d \Phi_j(f),$$

for the dehomogenizations with respect to X_i and X_j, respectively. So if $\frac{X_i}{X_j}$ is invertible, then $\Phi_i(f)$ and $\Phi_j(f)$ differ only by a unit. Hence the images of the ideals $\Phi_i(I)$ and $\Phi_j(I)$ in the coordinate ring of $U_{ij} = D_{U_i}(\frac{X_j}{X_i})$ coincide, and this gives us the desired identification $V_{ij} = V_{ji}$. Since the cocycle condition holds for the U_i and U_{ij}, it is satisfied in this situation, as well. So by gluing we obtain a scheme which we denote by $V_+(I)$, together with a morphism $\iota \colon V_+(I) \to \mathbb{P}_R^n$. The underlying topological space of $V_+(I)$ is a closed subspace of $\mathbb{P}_R^n$, and the morphism ι is a so-called *closed immersion*; see Section (3.15), in particular Example 3.48. We call $V_+(I)$ the *vanishing scheme* of I (or of M).

In this way, we obtain a huge number of examples of schemes: whenever we write down homogeneous polynomials in $X_0, \dots, X_n$ with coefficients in a ring R, we can consider their scheme of common zeros inside $\mathbb{P}_R^n$. We will study these schemes in more detail in Chapter 13. Usually these schemes are not affine (see Corollary 13.77 for a more precise statement).

In Proposition 1.65 we have seen that morphisms between projective varieties can be described by homogeneous polynomials. This is also true over arbitrary rings (see Section (4.14)).

Basic properties of schemes and morphisms of schemes

(3.8) Topological Properties.

Definition 3.16.
(a) *A scheme is called* connected, *if the underlying topological space is connected.*
(b) *A scheme is called* quasi-compact, *if the underlying topological space is quasi-compact, i. e., if every open covering admits a finite subcovering.*
(c) *A scheme is called* irreducible, *if the underlying topological space is irreducible, i. e., if it is non-empty and not equal to the union of two proper closed subsets.*

We have seen in Proposition 2.5 that all affine schemes are quasi-compact. A (trivial) example of a scheme which is not quasi-compact is the disjoint union of infinitely many non-empty schemes. But there also exist connected (and even irreducible) schemes which are not quasi-compact; see Exercises 3.2 and 3.26.

Definition 3.17. *A morphism* $f\colon X \to Y$ *of schemes is called* injective, surjective *or* bijective, *respectively, if the continuous map* $X \to Y$ *of the underlying topological spaces has this property.*

Similarly, f is called open, closed, *or a* homeomorphism, *respectively, if the underlying continuous map has this property.*

Finally, f is called dominant *if $f(X)$ is a dense subspace of Y.*

Note that a homeomorphism of schemes in general is not an isomorphism (see Exercise 3.6 or Exercise 12.21 for examples).

(3.9) Noetherian Schemes.

The notion of a noetherian ring is central in algebra. Of similar importance is its generalization to schemes in algebraic geometry.

Definition 3.18. *A scheme X is called* locally noetherian, *if X admits an affine open cover $X = \bigcup U_i$, such that all the affine coordinate rings $\Gamma(U_i, \mathscr{O}_X)$ are noetherian. If in addition X is quasi-compact, X is called* noetherian.

Because any localization of a noetherian ring is noetherian again, every locally noetherian scheme has a basis of its topology consisting of noetherian affine open subschemes. Because all affine schemes are quasi-compact, in this case the notions "noetherian" and "locally noetherian" coincide.

We also see that all the local rings $\mathscr{O}_{X,x}$ of a locally noetherian scheme X are noetherian. But even for affine schemes X it is not true that if $\mathscr{O}_{X,x}$ is noetherian for all $x \in X$, then X is noetherian (see Exercise 3.21).

Proposition 3.19. *Let $X = \mathrm{Spec}\, A$ be an affine scheme. Then X is noetherian if and only if A is a noetherian ring.*

Proof. By definition, the condition is sufficient. Now assume that X is noetherian. Let $I \subseteq A$ be an ideal; we show that it is finitely generated. By assumption, we can cover $\operatorname{Spec} A$ by finitely many affine open subschemes, which are spectra of noetherian rings. Since any localization of a noetherian ring is noetherian again, we may (using Lemma 3.3) even cover $\operatorname{Spec} A$ by affine open subschemes of the form $D(f_i)$, $f_i \in A$, $i = 1, \ldots, n$, such that all A_{f_i} are noetherian rings. The ideals $J_i := I A_{f_i}$ of A_{f_i} are hence finitely generated, and the claim follows from the following lemma. □

Lemma 3.20. *Let A be a ring, and let $\operatorname{Spec} A = \bigcup_{i \in I} D(f_i)$ be a finite open covering by principal open subsets (i.e., the ideal generated by the f_i is all of A). Let M be an A-module. Then M is finitely generated if and only if for all i, the localization M_{f_i} is a finitely generated A_{f_i}-module.*

Proof. Clearly, the condition is necessary. Now let M_{f_i} be finitely generated over A_{f_i}, for all i, say by elements $m_{ij}/(f_i)^{n_{ij}}$, $j = 1, \ldots, r_i$, $m_{ij} \in M$. Then the submodule $N \subseteq M$ generated by all the m_{ij} is a finitely generated A-module, with the property that $N_{f_i} = M_{f_i}$ for all i. For the A-module M/N we therefore get that all localizations $(M/N)_{\mathfrak{p}} \cong M/N \otimes_A A_{\mathfrak{p}}$ at prime ideals $\mathfrak{p} \in \operatorname{Spec} A$ vanish, which implies that $M/N = 0$, hence $M = N$ is finitely generated. □

Remark 3.21. The underlying topological space of an affine noetherian scheme is clearly noetherian (in the sense of Definition 1.23). Moreover Lemma 1.24 then implies that if X is any noetherian scheme, the underlying topological space of X is noetherian (and in particular has only finitely many irreducible components). The converse statement is false, see Exercise 3.4.

Corollary 3.22. *Let X be a (locally) noetherian scheme and $U \subseteq X$ an open subscheme. Then U is (locally) noetherian.*

Proof. In the locally noetherian case, this is obvious. If X is noetherian, then by Remark 3.21 the underlying topological space is noetherian, hence every open in it is quasi-compact by Lemma 1.25. □

(3.10) Generic Points.

The underlying topological spaces of schemes have usually lots of points that are not closed. In particular they are far from being Hausdorff. Instead of viewing this as a pathology one should think of this as an advantage: There are points x such that their closure $\overline{\{x\}}$ is quite large and this will enable us to reduce the analysis of certain properties to considerations about a single point. We will see examples of this throughout the book.

Let X be a scheme. In Chapter 2 we introduced the following terminology. If Z is a subset of X, then a point $z \in Z$ is called a generic point of Z, if $\{z\}$ is dense in Z. As the closure of an irreducible subset is again irreducible, the subset Z must be irreducible, if it admits a generic point.

In those topological spaces which arise as the underlying spaces of schemes, a much stronger property is satisfied: Every irreducible closed subset has a uniquely determined generic point. This statement is the key point of the following proposition.

Proposition 3.23. *Let X be a scheme. The mapping*

$$X \longrightarrow \{Z \subseteq X; \ Z \ closed, \ irreducible\}$$
$$x \longmapsto \overline{\{x\}}$$

is a bijection, i. e. every irreducible closed subset contains a unique generic point.

Proof. We know already that this property is true for affine schemes (Corollary 2.7). Now let $Z \subseteq X$ be irreducible and closed, and let $U \subseteq X$ be an affine open subset such that $Z \cap U \neq \emptyset$. Then the closure of $Z \cap U$ in X is Z because Z is irreducible. In particular, $Z \cap U$ is irreducible, and the generic point in $Z \cap U$ is a generic point of Z.

If $z \in Z$ is a generic point, then z is contained in every open subset of X which meets Z, hence also in every U as above, and hence the uniqueness statement in the affine case implies the uniqueness of generic points in general. $\qquad\square$

Let X be a scheme. For every point $x \in X$ there exists a maximal point η (i.e., η is the generic point of an irreducible component of X) such that η is a generization of x (i.e., $x \in \overline{\{\eta\}}$). The existence of specializations that are closed points is more subtle. In general it may happen that a non-empty scheme X does not have any closed point, even if X is irreducible (see Exercise 3.14). Clearly this cannot happen if X is affine (because any prime ideal is contained in a maximal ideal), and it is not difficult to deduce that if X is a quasi-compact scheme, then for any $x \in X$ there exists a closed point y of X such that $y \in \overline{\{x\}}$ (Exercise 3.13). In Section (3.12) we will see another important special case where every point has a specialization that is closed, namely schemes locally of finite type over a field. In fact this holds more generally for arbitrary locally noetherian schemes, see Exercise 5.5.

The following purely topological statement is an example for a property that has only to be checked at the generic point.

Proposition 3.24. *Let $f \colon X \to Y$ be an open morphism of schemes and let Y be irreducible with generic point η. Then X is irreducible if and only if the fiber $f^{-1}(\eta)$ (considered as a subspace of X) is irreducible.*

Proof. As f is open, we have $\overline{f^{-1}(\eta)} = f^{-1}(\overline{\{\eta\}}) = f^{-1}(Y) = X$. Thus the claim follows from Lemma 1.17. $\qquad\square$

Although schemes are almost never Hausdorff, they at least satisfy the following weaker separation property.

Proposition 3.25. *Let X be a scheme. Then the underlying topological space of X is a Kolmogorov space (or T_0-space), i. e. for any two distinct points $x, y \in X$ there exists an open subset of X which contains exactly one of the points.*

Proof. Without loss of generality, we may assume that X is affine. Then the points x, y correspond to prime ideals $\mathfrak{p}$, $\mathfrak{q}$ in the affine coordinate ring $\Gamma(X, \mathcal{O}_X)$. We may assume that $\mathfrak{p} \nsubseteq \mathfrak{q}$. Let $f \in \mathfrak{p} \setminus \mathfrak{q}$. Then $D(f)$ is an open subset of X which contains $\mathfrak{q}$, but does not contain $\mathfrak{p}$. $\qquad\square$

We will study in Chapter 9 a property of schemes which in a sense is a substitute for the Hausdorff property, the so-called separatedness.

(3.11) Reduced and integral schemes, function fields.

In this section we generalize the notion of being reduced or an integral domain from rings to schemes.

Definition 3.26.
(a) *A scheme X is called* reduced, *if all local rings $\mathscr{O}_{X,x}$, $x \in X$, are reduced rings.*
(b) *An* integral scheme *is a scheme which is reduced and irreducible.*

Proposition 3.27.
(1) *A scheme X is reduced if and only if for every open subset $U \subseteq X$ the ring $\Gamma(U, \mathscr{O}_X)$ is reduced.*
(2) *A non-empty scheme X is integral if and only if for every open subset $\emptyset \neq U \subseteq X$ the ring $\Gamma(U, \mathscr{O}_X)$ is an integral domain.*
(3) *If X is an integral scheme, then for all $x \in X$ the local ring $\mathscr{O}_{X,x}$ is an integral domain.*

The converse in (3) does not hold (see Exercise 3.16).

Proof. (1). Let X be reduced, let $U \subseteq X$ be open, and consider $f \in \Gamma(U, \mathscr{O}_X)$ such that $f^n = 0$. If we had $f \neq 0$, then there would exist $x \in U$ with $f_x \neq 0$ (in $\mathscr{O}_{X,x}$), but $f_x^n = 0$.

The converse is also easy: Let $\overline{f} \in \mathscr{O}_{X,x}$ be a nilpotent element. Then there exists an open $U \subseteq X$ and a lift $f \in \Gamma(U, \mathscr{O}_X)$ of $\overline{f}$. By shrinking U, if necessary, we may assume that f is nilpotent, and hence $= 0$.

(2). Let X be integral. Because all open subschemes of X are integral, too, it is enough to show that $\Gamma(X, \mathscr{O}_X)$ is a domain. Take $f, g \in \Gamma(X, \mathscr{O}_X)$ such that $fg = 0$. Then $X = V(f) \cup V(g)$, so by the irreducibility we get, say, $X = V(f)$. We want to show that f must then be 0. We can check this locally on X, so we may assume that X is affine. Then f lies in the intersection of all prime ideals, i. e. in the nil-radical of the affine coordinate ring of X. Since X is reduced, by (1) the nil-radical is the zero ideal.

If conversely all $\Gamma(U, \mathscr{O}_X)$ are integral domains, then by (1) X is reduced. If there existed non-empty affine open subsets $U_1, U_2 \subseteq X$ with empty intersection, then the sheaf axioms imply that

$$\Gamma(U_1 \cup U_2, \mathscr{O}_X) = \Gamma(U_1, \mathscr{O}_X) \times \Gamma(U_2, \mathscr{O}_X).$$

But the product on the right hand side obviously contains zero divisors.

(3). This follows from (2), since any (non-zero) localization of a domain is a domain. $\square$

An affine scheme $X = \operatorname{Spec} A$ is integral if and only if A is a domain. The generic point η of X then corresponds to the zero ideal of A, and the local ring $\mathscr{O}_{X,\eta}$ is the localization $A_{(0)}$, which is just the field of fractions of A. This also shows that the local ring at the generic point of an arbitrary integral scheme is a field.

Definition 3.28. *Let X be an integral scheme, and let $\eta \in X$ be its generic point. Then the local ring $\mathscr{O}_{X,\eta}$ is a field, which is called the* function field *of X and denoted by $K(X)$.*

For an integral scheme all "rings of functions" are contained in its function field. More precisely we have:

Proposition 3.29. *Let X be an integral scheme with generic point η and let $K(X)$ be its function field.*
(1) *If $U = \operatorname{Spec} A$ is a non-empty open affine subscheme of X, then $K(X) = \operatorname{Frac}(A)$. If $x \in X$, then $\operatorname{Frac}(\mathscr{O}_{X,x}) = K(X)$.*
(2) *Let $U \subseteq V \subseteq X$ be non-empty open subsets. Then the maps*

$$\Gamma(V, \mathscr{O}_X) \xrightarrow{\ \operatorname{res}^V_U\ } \Gamma(U, \mathscr{O}_X) \xrightarrow{\ f \mapsto f_\eta\ } K(X)$$

are injective.
(3) *For every non-empty open subset $U \subseteq X$ and for every covering $U = \bigcup_i U_i$ by non-empty open subsets U_i we have*

$$\Gamma(U, \mathscr{O}_X) = \bigcap_i \Gamma(U_i, \mathscr{O}_X) = \bigcap_{x \in U} \mathscr{O}_{X,x},$$

where the intersection takes place in $K(X)$.

Proof. For $x \in U = \operatorname{Spec} A \subseteq X$ we have $\eta \in U$ and η corresponds to the zero ideal in the integral domain A. Moreover $\mathscr{O}_{X,x} = A_{\mathfrak{p}_x}$. Hence $K(X) = \mathscr{O}_{U,\eta} = \operatorname{Frac}(A) = \operatorname{Frac}(A_{\mathfrak{p}_x})$. This proves (1).

To show (2) it suffices to prove that for $\emptyset \neq U \subseteq X$ and $f \in \Gamma(U, \mathscr{O}_X)$ with $f_\eta = 0$ we have $f = 0$. As $f = 0$ is equivalent to $f_{|W} = 0$ for all open non-empty affine subschemes $W \subseteq U$, we may assume that $U = \operatorname{Spec} A$ is affine. But in this case the map is simply the canonical inclusion $A \hookrightarrow \operatorname{Frac}(A) = K(X)$.

The first equality in (3) follows from the injectivity of restriction maps and the fact that $\mathscr{O}_X$ is a sheaf. The second equality follows from the analogous assertion for affine integral schemes in Example 2.37. $\qquad\square$

Prevarieties as Schemes

In a sense, schemes provide a generalization of the notion of prevariety which we defined in Chapter 1. However, prevarieties are not schemes – they are missing exactly the generic points of irreducible closed subsets which consist of more than one point. On the other hand, there is a natural way to associate a scheme to any given prevariety. In the case of affine varieties, the obvious way to do this is to associate to an affine variety X the spectrum $\operatorname{Spec} \Gamma(X, \mathscr{O}_X)$ of its coordinate ring. In the following sections, we will deal with the general case. We will obtain a fully faithful functor from the category of prevarieties over an algebraically closed field k to the category of k-schemes. One of our tasks is to analyze which schemes arise in this way. Among the properties all schemes arising from prevarieties have, is being "of finite type" over the base field; this is the content of the next section.

(3.12) Schemes (locally) of finite type over a field.

If A is the coordinate ring of an affine variety over an algebraically closed field k, then A is a finitely generated k-algebra. Let us define a corresponding notion in the case of arbitrary k-schemes (and arbitrary fields k).

Definition 3.30. *Let k be a field, and let $X \to \operatorname{Spec} k$ be a k-scheme. We call X a k-scheme locally of finite type or say that X is locally of finite type over k, if there is an affine open cover $X = \bigcup_{i \in I} U_i$ such that for all i, $U_i = \operatorname{Spec} A_i$ is the spectrum of a finitely generated k-algebra A_i. We say that X is of finite type over k if X is locally of finite type and quasi-compact.*

In Section (10.2) we will define more generally, when a morphism $f \colon X \to Y$ of schemes is called "(locally) of finite type". The definition above is the special case $Y = \operatorname{Spec} k$.

Because every finitely generated k-algebra is noetherian, it follows immediately from the definition that every k-scheme (locally) of finite type is (locally) noetherian.

Proposition 3.31. *Let X be a k-scheme locally of finite type and let $U \subseteq X$ be an open affine subset. Then the k-algebra $\Gamma(U, \mathscr{O}_X)$ is a finitely generated k-algebra.*

Proof. Let $B = \Gamma(U, \mathscr{O}_X)$. Since the localization of a finitely generated k-algebra with respect to a single element is again finitely generated, we see, using Lemma 3.3, that we can cover U by finitely many principal open subsets $D(f_i)$, $f_1, \ldots, f_n \in B$, such that all localizations B_{f_i} are finitely generated k-algebras. The claim now follows from the following lemma (with $A = k$). $\qquad\square$

Lemma 3.32. *Let A be a ring and let B be an A-algebra. Let $f_1, \ldots, f_n \in B$ be elements generating the unit ideal (1), and such that for all i, the localization B_{f_i} is a finitely generated A-algebra. Then B is a finitely generated A-algebra.*

Proof. By assumption, there exist $g_i \in B$ with $\sum_i g_i f_i = 1$. Furthermore, all the B_{f_i} are finitely generated, so we can find finitely many elements b_{ij} which generate B_{f_i} as an A-algebra. We write $b_{ij} = c_{ij}/f_i^m$ with $c_{ij} \in B$ and some $m \geq 0$, which we may assume to be independent of i, j.

Let C be the A-subalgebra of B generated by all elements g_i, f_i, c_{ij}. We will show that $C = B$, which of course implies that B is finitely generated. Let $b \in B$. For N sufficiently large, and all i, we have $f_i^N b \in C$. Because $\sum_i g_i f_i = 1$, we get that the f_i generate the unit ideal in C, so the same is true for $f_1^N, \ldots, f_n^N$ (Lemma 2.4), hence there exist $u_1, \ldots, u_n \in C$ such that $\sum_i u_i f_i^N = 1$. So we obtain $b = (\sum_i u_i f_i^N) b \in C$. $\qquad\square$

Proposition 3.33. *Let k be a field, let X be a k-scheme locally of finite type, and let $x \in X$. Then the following assertions are equivalent.*
(i) *The point $x \in X$ is closed.*
(ii) *The field extension $k \hookrightarrow \kappa(x)$ is finite.*
(iii) *The field extension $k \hookrightarrow \kappa(x)$ is algebraic.*

Proof. If $x \in X$ is a closed point, there exists an open affine neighborhood $U = \operatorname{Spec} A$ of x such that x is closed in U and hence corresponds to a maximal ideal $\mathfrak{m}$ of the finitely generated k-algebra A. By Hilbert's Nullstellensatz (Theorem 1.7) $A/\mathfrak{m} = \kappa(x)$ is a finite extension of k. This proves that (i) implies (ii). The implication "(ii) $\Rightarrow$ (iii)" is clear.

Let us assume that $\kappa(x)$ is an algebraic extension of k and let $U = \operatorname{Spec} A$ be an open affine neighborhood of x. Consider the composition

$$k \hookrightarrow A \to A/\mathfrak{p}_x \hookrightarrow \operatorname{Frac}(A/\mathfrak{p}_x) = \kappa(x).$$

As $\kappa(x)$ is integral over k, the subring $A/\mathfrak{p}_x$ is also integral over k. But by Lemma 1.9 this shows that $A/\mathfrak{p}_x$ is a field. Thus x is closed in U for all open affine neighborhoods U of x. This shows that x is closed in X. $\qquad\square$

Note that in general, it can happen that a point x of some scheme X has got an open neighborhood in which it is closed, without being closed in X (e. g. take x in $U = \{x\}$, where x is the generic point of the spectrum X of a discrete valuation ring, see Example 2.14). However, the proposition shows that for schemes locally of finite type over a field this cannot happen.

We will now see that the set of closed points in a k-scheme locally of finite type is very dense in the following sense.

Definition and Remark 3.34. *A subset Y of a topological space X is called* very dense, *if the following equivalent conditions are satisfied:*
(i) *The map $U \mapsto U \cap Y$ defines a bijection between the set of open subsets of X and the set of open subsets of Y.*
(ii) *The map $F \mapsto F \cap Y$ defines a bijection between the set of closed subsets of X and the set of closed subsets of Y.*
(iii) *For every closed subset $F \subseteq X$, we have $F = \overline{F \cap Y}$.*
(iv) *Every non-empty locally closed subset Z of X contains a point of Y.*

Proof. The equivalence of (i), (ii), and (iii) is clear. We prove "(iii) $\Rightarrow$ (iv)": Write $Z = F \setminus F'$ for closed subsets $F' \subsetneq F$ of X. If we had $(F \cap Y) \setminus (F' \cap Y) = Z \cap Y = \emptyset$, then $F \cap Y = F' \cap Y$ and hence $F = F'$ by (iii).

Let us also prove the implication "(iv) $\Rightarrow$ (ii)": Assume F, F' are closed subsets of X with $F \cap Y = F' \cap Y$, or equivalently $((F \cup F') \setminus (F \cap F')) \cap Y = \emptyset$. Then $(F \cup F') \setminus (F \cap F') = \emptyset$, so $F = F'$. $\qquad\square$

Proposition 3.35. *Let X be a scheme locally of finite type over a field k. Then the subset of closed points of X is very dense in the topological space X.*

Proof. We prove that every non-empty locally closed subset of X contains a closed point. By shrinking the subset in question, we may assume that it is closed in an affine open subset $U = \operatorname{Spec} A$ of X. By our assumption, A is a finitely generated k-algebra. Every closed subset in U has the form $V(\mathfrak{a})$ for an ideal $\mathfrak{a} \subseteq A$, and since $V(\mathfrak{a}) \neq \emptyset$, the ideal $\mathfrak{a}$ is contained in a maximal ideal of A. This shows that $V(\mathfrak{a})$ contains a closed point of $\operatorname{Spec} A$. But Proposition 3.33 shows that in our situation all closed points of $\operatorname{Spec} A$ are also closed in X, and this proves the proposition. $\qquad\square$

As we have seen in Theorem 1.7, in a finitely generated k-algebra every prime ideal is the intersection of maximal ideals. Hence the proposition also follows from Exercise 10.16. As another consequence of Proposition 3.33, we obtain the following corollary, where, in the second term, we write $k = \kappa(x)$ to indicate that the homomorphism $k \to \kappa(x)$ given by the k-scheme structure of X is an isomorphism.

Corollary 3.36. *Let k be algebraically closed and let X be a k-scheme locally of finite type. Then*

$$\{x \in X \ ; \ x \ closed\} = \{x \in X \ ; \ k = \kappa(x)\} = X(k) := \mathrm{Hom}_k(\mathrm{Spec}\,k, X),$$

where the second identity is given by Proposition 3.8. Moreover the set of closed points is very dense in X.

(3.13) Equivalence of the category of integral schemes of finite type over k and prevarieties over k.

We will now define the correspondence between prevarieties and certain k-schemes. So let k be an algebraically closed field. We have already shown that the following categories are equivalent (the equivalence of (i) and (ii) holds by Theorem 2.35 and Proposition 3.31, the equivalence of (ii) and (iii) is Corollary 1.47):
 (i) the category of integral affine schemes of finite type over k
 (ii) the opposite category of the category of integral finitely generated k-algebras
 (iii) the category of affine varieties (in the sense of Definition 1.46)

We extend this equivalence of categories as follows. For a k-scheme X locally of finite type we will identify $X(k) = \mathrm{Hom}_k(\mathrm{Spec}\,k, X)$ with the set of closed points of X (Corollary 3.36). In particular we view $X(k)$ as a very dense subspace of the underlying topological space of X. We define a sheaf of rings by

$$\mathscr{O}_{X(k)} := \alpha^{-1}\mathscr{O}_X,$$

where $\alpha \colon X(k) \to X$ is the inclusion. We obtain a ringed space $(X(k), \mathscr{O}_{X(k)})$.

Theorem 3.37. *The above construction $(X, \mathscr{O}_X) \mapsto (X(k), \mathscr{O}_{X(k)})$ gives rise to an equivalence of the following categories:*
- *the category of integral schemes of finite type over k*
- *the category of prevarieties over k (in the sense of Definition 1.46)*

We will divide the proof into two parts.

Proof. (Part I: The functor $(X, \mathscr{O}_X) \mapsto (X(k), \mathscr{O}_{X(k)})$) We start by showing that the above construction indeed defines a functor from the category of integral k-schemes X of finite type to the category of prevarieties. Let $U \subseteq X$ be an open subset. Let us show that we have inclusions

$$\mathscr{O}_{X(k)}(U \cap X(k)) \hookrightarrow \mathrm{Map}(U \cap X(k), k),$$

such that the restriction maps of the sheaf $\mathscr{O}_{X(k)}$ are given by the restriction of maps. This means that $(X(k), \mathscr{O}_{X(k)})$ is a space with functions.
 Given $f \in \mathscr{O}_{X(k)}(U \cap X(k)) = \mathscr{O}_X(U)$, we associate to it the map

$$U \cap X(k) \longrightarrow k, \quad x \mapsto f(x) := \pi_x(f),$$

where π_x denotes the natural map $\pi_x \colon \mathscr{O}_X(U) \to \mathscr{O}_{X,x} \to \kappa(x) = k$. Here we consider x as a closed point. Then restriction of sections corresponds precisely to restriction of functions. It remains to show that elements $f, g \in \mathscr{O}_{X(k)}(U \cap X(k))$ giving rise to the same function $U \cap X(k) \longrightarrow k$ must coincide. This however can be checked locally on U, so we can assume that U is affine, say $U = \mathrm{Spec}\,A$. Then $\pi_x(f) = \pi_x(g)$ for every closed point $x \in \mathrm{Spec}\,A$, in other words:

$$f - g \in \bigcap_{\substack{\mathfrak{m} \subset A \\ \text{maximal ideal}}} \mathfrak{m} = \text{nil}(A) = 0,$$

because A is a finitely generated reduced k-algebra.

Since we can cover X by finitely many affine schemes, which are the spectrums of integral finitely generated k-algebras, we obtain that the space with functions defined in this way, is indeed a prevariety.

Furthermore the construction is functorial, because any morphism of schemes of finite type over k maps closed points to closed points (Proposition 3.33). □

To define a quasi-inverse of the functor $(X, \mathscr{O}_X) \mapsto (X(k), \mathscr{O}_{X(k)})$ we first start with a general topological construction that will produce from the underlying topological space of a prevariety the underlying topological space of the corresponding scheme.

Given a scheme, every irreducible closed subset of its underlying topological space has got a unique generic point. In some sense, the existence of these points is the only difference between integral schemes of finite type over k and prevarieties over k: in a prevariety, every point is a closed point. To get the equivalence of categories we are aiming at, we would like to construct, for any given prevariety, a locally ringed space whose underlying topological space has the property that every irreducible closed subset contains a unique generic point. Such spaces are also called *sober*. The general topological construction is the following.

Remark 3.38. (Sobrification of topological spaces) Let X be a topological space in which all points are closed. We define a topological space $t(X)$ as follows: As a set, $t(X)$ is the set of all irreducible closed subsets of X.

We now define a topology on $t(X)$: Whenever $Z \subseteq X$ is a closed subset, $t(Z)$ is a subset of $t(X)$. We define the closed subsets of $t(X)$ to be precisely the subsets of the form $t(Z)$, for $Z \subseteq X$ closed. Because $t(\bigcap_i Z_i) = \bigcap t(Z_i)$ and $t(Z_1 \cup Z_2) = t(Z_1) \cup t(Z_2)$ for closed subsets $Z_1, Z_2, Z_i \subseteq X$, we see that these sets in fact are the closed sets of a topology on $t(X)$. If $f : X \to Y$ is a continuous map, then we obtain a continuous map $t(f) : t(X) \to t(Y)$, by mapping each point of $t(X)$, corresponding to an irreducible closed subset of X, to the closure of its image under f, considered as a point of $t(Y)$. All in all, we have defined a functor from the category of topological spaces all of whose points are closed to the category of topological spaces.

Every irreducible closed subset of $t(X)$ is of the form $t(Z)$ for $Z \subseteq X$ closed and irreducible, and has the point $Z \in t(Z)$ as its unique generic point.

Given X, we have a natural continuous map $\alpha_X : X \to t(X)$: it maps each point $x \in X$ to the (irreducible, closed) subset $\overline{\{x\}} \in t(X)$. The mapping $U \mapsto \alpha_X^{-1}(U)$ is a bijection between the set of closed subsets of $t(X)$ and the set of closed subsets of X. So α_X is a homeomorphism from X onto the set of closed points in $t(X)$, and this set is very dense in $t(X)$.

In fact, this construction can be generalized to arbitrary topological spaces. One obtains a functor from the category of topological spaces to the full subcategory of sober topological spaces which is left adjoint to the inclusion functor (see e.g. [EGAInew] $\mathbf{0}_I$ (2.9.2)).

Proof. (Part II: The quasi-inverse of $(X, \mathscr{O}_X) \mapsto (X(k), \mathscr{O}_{X(k)})$) Let X be a prevariety. Let $\alpha_X : X \to t(X)$ be the natural map considered above. We consider the system of functions $\mathscr{O}_X$ we are given on X as a sheaf on X. Then $(t(X), \alpha_{X,*}\mathscr{O}_X)$ is a locally ringed space: In case X is an affine variety with affine coordinate ring A, we can identify the topological space X with the set of maximal ideals of A with the Zariski topology, and then $t(X)$ is homeomorphic to Spec A (Proposition 1.20). Since $\mathcal{O}_X(D(f)) = A_f$ for all $f \in A$, our claim is proved in this case, and the general case follows by considering a covering of X by affine varieties.

Now let $f: X \to Y$ be a morphism of prevarieties. By functoriality, we obtain a continuous map $t(f): t(X) \to t(Y)$ and a sheaf homomorphism $\alpha_{Y,*}\mathcal{O}_Y \to t(f)_*(\alpha_{X,*}\mathcal{O}_X)$. Because the morphism of the "sheaves" on X and Y is given by composition of functions, we get a morphism of locally ringed spaces.

Because affine varieties, as well as affine schemes, are determined uniquely by their affine coordinate ring, it is not hard to see that the two functors are quasi-inverse to each other. $\qquad\square$

Remark 3.39. *Let k be an algebraically closed field, and let X be an integral scheme of finite type over k. Let $X(k)$ be the corresponding prevariety. Then the function fields $K(X)$ and $K(X(k))$ coincide.*

Via this equivalence of categories the k-scheme $\mathbb{A}_k^n$ (resp. $\mathbb{P}_k^n$) corresponds to the prevariety $\mathbb{A}^n(k)$ (resp. $\mathbb{P}^n(k)$). Thus the notation of the functor blends nicely with the notation used in Chapter 1.

In Section (4.2) we will see that although in general a scheme X is (obviously) not determined by its set $X(k)$ of k-valued points for some field k, it is determined by all the sets $X(R) := \operatorname{Hom}(\operatorname{Spec} R, X)$ of "R-valued points", R a ring, together (considered as a functor from the category of rings to the category of sets).

Subschemes and Immersions

(3.14) Open Immersions.

In 3.2 we defined the notion of open subscheme; it has since played an important role. For a scheme X and any open subset $U \subseteq X$, there exists a unique open subscheme whose underlying topological space is U. An *open immersion* is a morphism of schemes which induces an isomorphism between its source and an open subscheme of its target, in other words:

Definition 3.40. *A morphism $j: Y \to X$ of schemes is called an* open immersion, *if the underlying continuous map is a homeomorphism of Y with an open subset U of X, and the sheaf homomorphism $\mathcal{O}_X \to j_*\mathcal{O}_Y$ induces an isomorphism $\mathcal{O}_{X|U} \cong (j_*\mathcal{O}_Y)_{|U}$ (of sheaves on U).*

(3.15) Closed subschemes.

The notion of closed subscheme is a little more involved. This can be seen already in the case of affine schemes: If A is a ring, and $\mathfrak{a}$ an ideal of A, then we can identify the topological space $\operatorname{Spec} A/\mathfrak{a}$ with the closed subspace $V(\mathfrak{a})$ of $\operatorname{Spec} A$. Certainly it is a good start to say that schemes of the form $\operatorname{Spec} A/\mathfrak{a}$ should be the closed subschemes of $\operatorname{Spec} A$. Two such "subschemes" should coincide if and only if the corresponding ideals are equal. This means that unlike in the case of prevarieties, there may be many closed subschemes with the same underlying closed subset (compare Example 2.40). This means that in addition to the closed subset, we also must consider the information given by the structure sheaf. We will show that the following definition of closed subschemes has the desired properties, and that the closed subschemes of an affine scheme $\operatorname{Spec} A$ correspond bijectively to the quotients of A.

Given a ringed space $(X, \mathscr{O}_X)$, we call a subsheaf $\mathscr{J} \subseteq \mathscr{O}_X$ a sheaf of ideals, if for every open subset $U \subseteq X$ the sections $\Gamma(U, \mathscr{J})$ are an ideal in $\Gamma(U, \mathscr{O}_X)$. The quotient sheaf $\mathscr{O}_X/\mathscr{J}$ is defined as the sheaf associated to the presheaf $U \mapsto \mathscr{O}_X(U)/\mathscr{J}(U)$. It is a sheaf of rings. The canonical projection $\mathscr{O}_X \to \mathscr{O}_X/\mathscr{J}$ is surjective (since the stalks of $\mathscr{O}_X \to \mathscr{O}_X/\mathscr{J}$ agree with the stalks of the presheaf above).

Definition 3.41. *Let X be a scheme.*
(1) *A* closed subscheme *of X is given by a closed subset $Z \subseteq X$ and an ideal sheaf $\mathscr{J} \subseteq \mathscr{O}_X$ such that $Z = \{ x \in X \; ; \; (\mathscr{O}_X/\mathscr{J})_x \neq 0 \}$ and $(Z, (\mathscr{O}/\mathscr{J})_{|Z})$ is a scheme.*
(2) *A morphism $i \colon Z \to X$ of schemes is called a* closed immersion, *if the underlying continuous map is a homeomorphism between Z and a closed subset of X, and the sheaf homomorphism $i^\flat \colon \mathscr{O}_X \to i_* \mathscr{O}_Z$ is surjective.*

If $Z \subseteq X$ is a closed subscheme as in (1) with corresponding ideal sheaf $\mathscr{J}$, then Z is determined by $\mathscr{J}$ (in the terminology introduced in Section (7.6), Z is the support of $\mathscr{O}_X/\mathscr{J}$). Writing i for the inclusion $Z \hookrightarrow X$ and $\mathscr{O}_Z = (\mathscr{O}/\mathscr{J})_{|Z}$, we have $i_* \mathscr{O}_Z = \mathscr{O}_X/\mathscr{J}$ and denoting by $i^\flat$ the canonical projection $\mathscr{O}_X \to \mathscr{O}_X/\mathscr{J} = i_* \mathscr{O}_Z$, the morphism $(i, i^\flat)$ is a closed immersion. We can recover $\mathscr{J}$ as the kernel of $i^\flat$. On the other hand, every closed immersion induces an isomorphism of its source with a uniquely determined closed subscheme of its target.

Note that in part (1) of the definition we explicitly require that $(Z, i^{-1}(\mathscr{O}_X/\mathscr{J}))$ be a scheme. This will not be true for an arbitrary sheaf of ideals $\mathscr{J}$. Our aim will be to gain a better understanding about which sheaves of ideals give rise to closed subschemes. (We will ultimately reach this aim in Chapter 7, where we will define the notion of *quasi-coherent* sheaf; we will then see that a sheaf of ideals defines a closed subscheme if and only if it is quasi-coherent. Thus we obtain a bijection between the set of closed subschemes of a scheme X and the set of quasi-coherent ideal sheaves of $\mathscr{O}_X$.)

Given a ring A and an ideal $\mathfrak{a} \subseteq A$, $\operatorname{Spec} A/\mathfrak{a} \subseteq \operatorname{Spec} A$ is a closed subscheme, as we wanted (see Example 2.39). At this point it is not at all clear, however, that all closed subschemes of $\operatorname{Spec} A$ are of this form; we will prove this now.

Theorem 3.42. *Let $X = \operatorname{Spec} A$ be an affine scheme. For every ideal $\mathfrak{a} \subseteq A$ let $V(\mathfrak{a})$ be the corresponding closed subscheme (with the scheme structure induced via the homeomorphism $V(\mathfrak{a}) \cong \operatorname{Spec} A/\mathfrak{a}$). The mapping $\mathfrak{a} \mapsto V(\mathfrak{a})$ is a bijection between the set of ideals of A and the set of closed subschemes of X. In particular, every closed subscheme of an affine scheme is affine.*

Proof. Assume that Z is a closed subscheme of X, and $i \colon Z \subseteq X$ is the inclusion. Then by definition the sheaf homomorphism $\mathscr{O}_X \to i_* \mathscr{O}_Z$ is surjective, and we write

$$I_Z := \operatorname{Ker}(A = \Gamma(X, \mathscr{O}_X) \to \Gamma(X, i_* \mathscr{O}_Z) = \Gamma(Z, \mathscr{O}_Z)).$$

This is an ideal of A. If Z is of the form $V(\mathfrak{a})$, as we want to show, then clearly $I_Z = \mathfrak{a}$. To prove the theorem, it is therefore enough to show that for every closed subscheme Z of $\operatorname{Spec} A$ we have $Z = V(I_Z)$.

By definition, the ring homomorphism $\varphi\colon A \to \Gamma(Z, \mathscr{O}_Z)$ factors through A/I_Z, and hence the inclusion of Z into X factors through $Z \to \operatorname{Spec} A/I_Z$. By replacing A by A/I_Z, we may therefore assume that φ is injective. Under this additional assumption we have to show that the inclusion $Z \hookrightarrow X$ is an isomorphism.

We first show that the underlying continuous map is a homeomorphism. We know that it is injective and closed, so it is enough to show that it is surjective. Let $s \in A$ with $Z \subseteq V(s)$. We claim that for N sufficiently large, $\varphi(s^N) = 0$. If $U \subseteq Z$ is open, such that $(U, \mathscr{O}_{Z|U})$ is affine, we get $U \subseteq V_U(\varphi(s)_{|U})$, hence $\varphi(s)_{|U} \in \Gamma(U, \mathscr{O}_Z)$ is nilpotent. Covering Z by finitely many affine schemes, we obtain that in fact $\varphi(s^N) = 0$. But φ is injective, so $s^N = 0$, which translates to $V(s) = X$. Since Z is closed in X, this shows that $i(Z) = X$.

Let us identify the topological spaces Z and X. It remains to show that the sheaf homomorphism $\mathscr{O}_X \to \mathscr{O}_Z$ is bijective. Since it is surjective by assumption, it is enough to show injectivity. We check this on stalks. For $x \in X$, $\mathscr{O}_{X,x} = A_{\mathfrak{p}_x}$, and we see that it is enough to show that every element of $\operatorname{Ker}(\mathscr{O}_{X,x} \to \mathscr{O}_{Z,x})$ of the form $\frac{g}{1}$ is 0 in $\mathscr{O}_{X,x}$. Given g, we cover $Z = U \cup \bigcup_{i \in I} U_i$ by finitely many open subsets U, U_i, such that:

(1) The schemes $(U, \mathscr{O}_{Z|U})$ and $(U_i, \mathscr{O}_{Z|U_i})$ for all i are affine.

(2) We have $x \in U$ and $\varphi(g)_{|U} = 0$.

Choose $s \in A$ with $x \in D(s) \subseteq U$. If we can show that $\varphi(s^N g) = 0$ for some N, then $s^N g = 0$ because φ is injective, and it follows that $\frac{g}{1} = 0$ in $\mathscr{O}_{X,x}$, as desired, since s is a unit in $\mathscr{O}_{X,x}$. Since $\varphi(g)_{|U} = 0$ by assumption, we have $\varphi(sg)_{|U} = 0$. Now I is finite, so we can search a suitable N for each U_i separately. Because $D_{U_i}(\varphi(s)_{|U_i}) = D(s) \cap U_i \subseteq U \cap U_i$, we obtain $\varphi(g)_{|D_{U_i}(\varphi(s)_{|U_i})} = 0$. In other words, the image of $\varphi(g)$ in the localization $\Gamma(U_i, \mathscr{O}_Z)_{\varphi(s)_{|U_i}}$ is 0, which is precisely what we had to show. $\qquad\square$

(3.16) Subschemes and immersions.

Open and closed subschemes are special cases of the notion of (locally closed) subscheme.

Definition 3.43.
(1) *Let X be a scheme. A subscheme of X is a scheme $(Y, \mathscr{O}_Y)$, such that $Y \subseteq X$ is a locally closed subset, and such that Y is a closed subscheme of the open subscheme $U \subseteq X$, where U is the largest open subset of X which contains Y and in which Y is closed (i. e. U is the complement of $\bar{Y} \setminus Y$). We then have a natural morphism of schemes $Y \to X$.*

(2) *An immersion $i\colon Y \to X$ is a morphism of schemes whose underlying continuous map is a homeomorphism of Y onto a locally closed subset of X, and such that for all $y \in Y$ the ring homomorphism $i_y^\sharp\colon \mathscr{O}_{X,i(y)} \to \mathscr{O}_{Y,y}$ between the local rings is surjective.*

Whenever Y is a subscheme of X, then the natural morphism $Y \hookrightarrow X$ is an immersion. On the other hand, every immersion induces an isomorphism of its source with a unique subscheme of its target. If Y is a subscheme of X, whose underlying subset is closed in X, then Y is a closed subscheme of X. (The corresponding statement for open subschemes is false, cf. Section (3.18) below).

Remark 3.44. Any immersion $i\colon Y \hookrightarrow X$ can be factored into a closed immersion $Y \hookrightarrow U$ followed by an open immersion $U \hookrightarrow X$, where U is the complement of $\overline{i(Y)} \setminus i(Y)$. We will see in Remark 10.31 that under certain (mild) hypotheses it can also be factored into an open immersion $Y \hookrightarrow Z$ followed by a closed immersion $Z \hookrightarrow X$, where the underlying topological space of Z is the closure of $i(Y)$ in X.

Example 3.45. If k is a field, and X is a k-scheme of finite type, then all subschemes of X are of finite type over k. Indeed, if X is affine, then this is obvious for principal open subsets of X; this shows that the statement is true for arbitrary open subschemes of a k-scheme of finite type. On the other hand every closed subscheme of a k-scheme of finite type is again of finite type over k, because the affine coordinate rings are just quotients of the corresponding rings of the larger scheme.

Note that given a morphism $f \colon X \to Y$ and a subscheme $Z \subseteq Y$ with $f(X) \subseteq Z$ (set-theoretically), f will not necessarily factor through Z as a morphism of schemes; see Exercise 3.25. In fact we can define a partial order on subschemes as follows.

Definition 3.46. *Let X be a scheme. For two subschemes Z and Z' of X we say that Z' majorizes Z if the inclusion morphism $Z \to X$ factors through the inclusion morphism $Z' \to X$.*

We sometimes write $Z \leq Z'$ or simply $Z \subseteq Z'$ if Z' majorizes Z. This defines a partial order on the set of subschemes of a scheme X.

We close this section with an easy remark.

Remark 3.47. *Let $\mathbf{P}$ be the property of a morphism of schemes being an "open immersion" (resp. a "closed immersion", resp. an "immersion").*
(1) *The property $\mathbf{P}$ is local on the target, i. e.: If $f \colon Z \to X$ is a morphism of schemes, and $X = \bigcup_i U_i$ is an open covering, then f has $\mathbf{P}$ if and only if for all i the restriction $f^{-1}(U_i) \to U_i$ of f satisfies $\mathbf{P}$.*
(2) *The composition of two morphisms having property $\mathbf{P}$ has again property $\mathbf{P}$.*

Example 3.48. Let R be a ring, and let $I \subseteq R[T_0, \ldots, T_n]$ be a homogeneous ideal. Then the scheme $V_+(I)$ defined in (3.7) is a closed subscheme of $\mathbb{P}_R^n$. This is a direct consequence of part (1) of the previous remark. In Section (13.6) we will see that every closed subscheme of $\mathbb{P}_R^n$ is of this form.

(3.17) Projective and quasi-projective schemes over a field.

Even if, when we defined prevarieties and schemes, one of our goals was to have a definition which is independent of an embedding in a larger space, of course subschemes of a well understood scheme are often easier to handle. In particular, it is often useful if one knows that a certain scheme can be embedded as a subscheme in projective space. In fact, this is the case for many of the schemes which play a role in practice. In Chapter 13 we will study systematically how to embed schemes into projective space.

Definition 3.49. *Let k be a field.*
(1) *A k-scheme X is called* projective, *if there exist $n \geq 0$ and a closed immersion $X \hookrightarrow \mathbb{P}_k^n$.*
(2) *A k-scheme X is called* quasi-projective, *if there exist $n \geq 0$ and an immersion $X \hookrightarrow \mathbb{P}_k^n$.*

As remarked above, the schemes $V_+(I)$, where $I \subseteq k[X_0, \ldots, X_n]$ is a homogeneous ideal, are closed subschemes of projective space, so they are projective schemes. Example 3.45 shows that every quasi-projective k-scheme is of finite type.

Every affine k-scheme X of finite type is quasi-projective: Indeed, let $X = \operatorname{Spec} A$, where $A \cong k[T_1, \ldots, T_n]/\mathfrak{a}$. Therefore there exists a closed immersion $i\colon X \to \mathbb{A}_k^n$. Moreover, projective space $\mathbb{P}_k^n$ is covered by open subschemes which are isomorphic to $\mathbb{A}_k^n$ by construction. In particular we can find an open immersion $j\colon \mathbb{A}_k^n \to \mathbb{P}_k^n$. The composition $j \circ i$ is then an immersion $X \to \mathbb{P}_k^n$ by Remark 3.47.

(3.18) The underlying reduced subscheme of a scheme.

Let X be a scheme. In general, there exist several closed subschemes of X with the same underlying topological space. Among these, there is a smallest one, which is characterized by the fact that it is reduced. To prove this, denote by $\mathcal{N} := \mathcal{N}_X \subset \mathcal{O}_X$ the sheaf of ideals which is the sheaf associated to the presheaf

$$U \mapsto \operatorname{nil}(\Gamma(U, \mathcal{O}_X)), \quad U \subseteq X \text{ open}$$

(here $\operatorname{nil}(R)$ denotes the nilradical of a ring R). Note that the proof of the following proposition shows that $\mathcal{N}_X(U) = \operatorname{nil}(\Gamma(U, \mathcal{O}_X))$ for every affine open subset U of X. This sheaf is called the *nilradical of X*.

Proposition 3.50. *The ringed space $X_{\mathrm{red}} := (X, \mathcal{O}_X/\mathcal{N})$ is a scheme, so it is a closed subscheme of X, and it has the same underlying topological space as X. If $X' \subseteq X$ is any closed subscheme with this property, then the inclusion morphism $X_{\mathrm{red}} \hookrightarrow X$ factors through a closed immersion $X_{\mathrm{red}} \hookrightarrow X'$. Furthermore, X_{red} is reduced.*

If $X = \operatorname{Spec} A$ is affine, $X_{\mathrm{red}} = \operatorname{Spec}(A/\operatorname{nil}(A))$.

We call X_{red} the *underlying reduced subscheme of X*.

Proof. In order to prove that X_{red} is a scheme, it is enough to consider the case that $X = \operatorname{Spec} A$ is affine. The presheaf defined above restricted to the basis of principal open subsets is in fact a sheaf, and for every $f \in A$,

$$\operatorname{nil}(\Gamma(D(f), \mathcal{O}_X)) = \operatorname{nil}(A_f) = \operatorname{nil}(A)A_f$$

is the ideal of A_f generated by the nilradical of A. So in this case, X_{red} is the closed subscheme $\operatorname{Spec} A/\operatorname{nil}(A)$. Obviously this is a reduced scheme.

Now consider X' as in the statement of the proposition. We must show that the sheaf homomorphism $\mathcal{O}_X \to \mathcal{O}_X/\mathcal{N}$ factors through $\mathcal{O}_{X'}$, or in other words, that $\operatorname{Ker}(\mathcal{O}_X \to \mathcal{O}_{X'}) \subseteq \mathcal{N}$. It is enough to show that

$$\operatorname{Ker}(\Gamma(U, \mathcal{O}_X) \to \Gamma(U, \mathcal{O}_{X'})) \subseteq \Gamma(U, \mathcal{N})$$

for every affine open subset $U \subseteq X$, and we may assume that $X = \operatorname{Spec} A$ is affine. Since it is a closed subscheme, X' is affine as well, say $X' = \operatorname{Spec} B$. Our hypothesis that the surjective ring homomorphism $A \to B$ induces a homeomorphism between the spectra, means that its kernel must be contained in every prime ideal, and hence in the nilradical of A. $\square$

Attaching to a scheme X its underlying reduced subscheme X_{red} defines a functor from the category of schemes to the category of reduced schemes:

Proposition 3.51. *For every morphism of schemes $f\colon X \to Y$ there exists a unique morphism of schemes $f_{\mathrm{red}}\colon X_{\mathrm{red}} \to Y_{\mathrm{red}}$ such that*

$$
\begin{array}{ccc}
X_{\mathrm{red}} & \xrightarrow{\ i_X\ } & X \\
{\scriptstyle f_{\mathrm{red}}}\downarrow & & \downarrow{\scriptstyle f} \\
Y_{\mathrm{red}} & \xrightarrow{\ i_Y\ } & Y
\end{array}
$$

commutes, where i_X and i_Y are the canonical inclusion morphisms.

If $g\colon Y \to Z$ is a second morphism of schemes, we have $(g \circ f)_{\mathrm{red}} = g_{\mathrm{red}} \circ f_{\mathrm{red}}$.

Proof. As i_Y is a monomorphism, f_{red} is uniquely determined. To show its existence we therefore may assume that $X = \operatorname{Spec} B$ and $Y = \operatorname{Spec} A$ are affine (Proposition 3.5). Then $f = {}^a\varphi$ for a ring homomorphism $\varphi\colon A \to B$ by Section (2.11). But clearly $\varphi(\mathrm{nil}(A)) \subseteq \mathrm{nil}(B)$ and therefore φ induces a ring homomorphism $\varphi_{\mathrm{red}}\colon A/\mathrm{nil}(A) \to B/\mathrm{nil}(B)$ and we can set $f_{\mathrm{red}} = {}^a(\varphi_{\mathrm{red}})$.

The equality $(g \circ f)_{\mathrm{red}} = g_{\mathrm{red}} \circ f_{\mathrm{red}}$ follows from the uniqueness of $(g \circ f)_{\mathrm{red}}$. $\qquad\square$

Proposition 3.52. *Let X be a scheme and $Z \subseteq X$ a locally closed subset. Then there exists a unique reduced subscheme Z_{red} of X with underlying topological space Z.*

Proof. The previous proposition implies that there exists at most one reduced subscheme with underlying topological space Z. To prove its existence we can replace X by an open neighborhood U of Z such that Z is closed in U. Hence it is enough to consider the case that Z is closed in X.

If $X = \operatorname{Spec} A$ is affine, then Z has the form $V(\mathfrak{a})$, and the closed subscheme we are looking for is $\operatorname{Spec} A/\mathrm{rad}(\mathfrak{a})$. In general, we consider an affine open covering $X = \bigcup U_i$. For every i there is a uniquely determined reduced subscheme Z_i with underlying space $Z \cap U_i$. For all i, j, the scheme $Z_{ij} := (Z \cap U_{ij}, \mathscr{O}_{Z_i|Z\cap U_{ij}})$ (where $U_{ij} = U_i \cap U_j$) is a reduced subscheme with underlying space $Z \cap U_{ij}$. From the analogous property of the Z_{ji}, we obtain $Z_{ji} = Z_{ij}$, and this gives us a gluing datum in the sense of Section (3.5), where the gluing isomorphisms are given by the identity (and thus in particular satisfy the cocycle condition). By gluing the Z_i we construct the reduced subscheme of X with underlying space Z that we are looking for. $\qquad\square$

Proposition 3.52 implies that for any locally closed subspace Z the partially ordered set of subschemes (Definition 3.46) whose underlying topological spaces contain Z has a unique minimal element, namely Z_{red}. It is called the *reduced subscheme with underlying subspace Z*.

Exercises

Exercise 3.1◇. Show that the spectrum of the zero ring is an initial object in the category of schemes.

Exercise 3.2. Prove that there exists a scheme which admits a covering by countably many closed subschemes each of which is isomorphic to the affine line $\mathbb{A}_k^1$ (over an algebraically closed field), indexed by $\mathbb{Z}$, such that the copies of $\mathbb{A}_k^1$ corresponding to i and $i + 1$ intersect in a single point, which is the point 0 when considered as a point in the i-th copy, and the point 1 when considered as an element of the $(i + 1)$-th copy.

Prove that this scheme is connected and locally noetherian, but not quasi-compact.

Exercise 3.3◇. Let k, k' be fields of different characteristic. Let $X \neq \emptyset$ be a k-scheme, and let X' be a k'-scheme. Show that there is no morphism $X \to X'$ of schemes.

Exercise 3.4◇. Give an example of a non-noetherian scheme whose underlying topological space is noetherian (or even consists only of one point).

Exercise 3.5◇. Let p be a prime number, let $\mathbb{F}_p$ be the field with p elements, and let $i_{(p)}\colon \operatorname{Spec}\mathbb{F}_p \to \operatorname{Spec}\mathbb{Z}$ be the canonical morphism. We call a ring A *of characteristic p*, if in A we have $p \cdot 1 = 0$. Prove that for a scheme X the following are equivalent:
 (i) For every open subset $U \subseteq X$, the ring $\Gamma(U, \mathscr{O}_X)$ has characteristic p.
 (ii) The ring $\Gamma(X, \mathscr{O}_X)$ has characteristic p.
 (iii) The scheme morphism $X \to \operatorname{Spec}\mathbb{Z}$ factors through $i_{(p)}$.
A scheme satisfying these equivalent conditions is said to be *of characteristic p*. A scheme X is said to be *of characteristic zero* if the scheme morphism $X \to \operatorname{Spec}\mathbb{Z}$ factors through $\operatorname{Spec}\mathbb{Q} \to \operatorname{Spec}\mathbb{Z}$.

Exercise 3.6◇. Let p be a prime number, and let X be a scheme of characteristic p (Exercise 3.5). Show that there exists a unique morphism $\operatorname{Frob}_X = (f, f^\flat)\colon X \to X$ of schemes such that $f = \operatorname{id}_X$ and that for every open subset $U \subseteq X$, $f_U^\flat$ is given by the ring homomorphism $\Gamma(U, \mathscr{O}_X) \to \Gamma(U, \mathscr{O}_X)$, $a \mapsto a^p$.

Give an example of a scheme X, such that the morphism Frob_X induces an isomorphism on the global sections $\Gamma(X, \mathscr{O}_X)$ without being an isomorphism itself.

The morphism Frob_X is called the *absolute Frobenius morphism of X*.

Exercise 3.7◇. Let X be an irreducible scheme, and let $\eta \in X$ be its generic point. Prove that the intersection of all non-empty open subsets of X is $\{\eta\}$.

Exercise 3.8◇. Let $f\colon X \longrightarrow Y$ be a morphism of integral schemes such that the generic point of Y is in the image of f. Show that f induces an inclusion $K(Y) \longrightarrow K(X)$ of the function fields.

Exercise 3.9◇. Let $(R_i)_{i \in I}$ be a family of rings $R_i \neq \{0\}$.
(a) Assume I is finite. Prove that $\coprod_{i \in I} \operatorname{Spec} R_i = \operatorname{Spec}(\prod_{i \in I} R_i)$.
(b) Assume that I is infinite. Show that $X := \coprod_{i \in I} \operatorname{Spec} R_i$ is not an affine scheme (use that X is not quasi-compact).

Exercise 3.10. With the notation of Section (3.6) set $V_i := U_0 \cup \cdots \cup U_i$ for $i = 0, \ldots, n$. In particular $V_n = \mathbb{P}_R^n$. Show that $\Gamma(V_i, \mathscr{O}_{\mathbb{P}_R^n}) = R$ for $i > 0$ and deduce that V_i is not affine for $i > 0$.

Exercise 3.11. Give an example of a local ring A, such that $\operatorname{Spec} A$ is neither reduced nor irreducible.

Exercise 3.12. Let $f\colon X \to Y$ be a morphism of irreducible schemes, and let η (resp. θ) be the generic point of X (resp. of Y). Show that f is dominant, if and only if $\theta \in f(X)$. In this case $f(\eta) = \theta$.

Exercise 3.13. Let X be a non-empty quasi-compact scheme.
(a) Show that X contains a closed point. Deduce that any point $x \in X$ has a specialization that is a closed point of X.
(b) Assume that X contains exactly one closed point. Prove that X is isomorphic to the spectrum of a local ring.

Exercise 3.14. Let A be a valuation ring such that the maximal ideal of A equals the union of all prime ideals properly contained in it. (In particular, A has infinitely many prime ideals; see Exercise 2.9.) Let $x \in X = \operatorname{Spec} A$ be its closed point. Show that the open subscheme $U := X \setminus \{x\}$ of X does not contain a closed point. Deduce that U is not quasi-compact. Cf. [MO], q/65680, in particular Knaf's comment.

Exercise 3.15. Let X be a locally noetherian scheme. Prove that the set of irreducible components (see Definition 1.18) of X is locally finite (i. e. every point of X has an open neighborhood which meets only finitely many irreducible components of X).

Exercise 3.16. Let X be a scheme.
(a) Consider the following assertions.
  (i) Every connected component of X is irreducible.
  (ii) X is the disjoint union of its irreducible components.
  (iii) For all $x \in X$ the nilradical of $\mathscr{O}_{X,x}$ is a prime ideal. (For instance, this is the case if $\mathscr{O}_{X,x}$ is a domain.)
  Show the implications "(i) $\Rightarrow$ (ii) $\Rightarrow$ (iii)". Show that all assertions are equivalent if the set of irreducible components of X is locally finite (e.g., if X is locally noetherian; see Exercise 3.15).
(b) Let X be connected, and assume that the set of its irreducible component is locally finite. Then X is integral if and only if for all $x \in X$ the local ring $\mathscr{O}_{X,x}$ is a domain.
(c) Let $K_1, \ldots, K_n$ be fields, $n > 1$. Set $X = \operatorname{Spec}(\prod_i K_i)$ and prove that X is not integral, although $\mathscr{O}_{X,x}$ is a field for every $x \in X$.

Exercise 3.17. Let Y be an irreducible scheme with generic point η and let $f \colon X \to Y$ be a morphism of schemes. Then the map $Z \mapsto f^{-1}(\eta) \cap Z$ is a bijective map from the set of irreducible components of X meeting $f^{-1}(\eta)$ onto the set of irreducible components of $f^{-1}(\eta)$, and the generic point of Z is the generic point of $f^{-1}(\eta) \cap Z$.

Exercise 3.18. Let X be a scheme and let R be a local ring. Show that every morphism $\operatorname{Spec} R \to X$ factors through the canonical morphism $j_x \colon \operatorname{Spec} \mathscr{O}_{X,x} \to X$ (see Section (3.4)), where x is the image point of the unique closed point of $\operatorname{Spec} R$. Prove that in this way, one obtains a bijection between $\operatorname{Hom}(\operatorname{Spec} R, X)$ and the set of pairs (x, φ), where $x \in X$ and $\varphi \colon \mathscr{O}_{X,x} \to R$ is a local homomorphism.

Exercise 3.19. Let R be a local ring, and let $n \geq 1$. Show that the set $\operatorname{Hom}(\operatorname{Spec} R, \mathbb{P}^n_R)$ of morphisms from $\operatorname{Spec} R$ to projective space over R can be identified with the set $M/R^\times$, where $M \subset R^{n+1}$ is the subset of tuples where at least one entry is a unit in R, i.e., we can write every "R-valued point" $f \colon \operatorname{Spec} R \to \mathbb{P}^n_R$ (see Section (4.1)) as given by homogeneous coordinates $(x_0 : \cdots : x_n)$, well-determined up to multiplication by a unit in R. What happens without the assumption that R is local?

Exercise 3.20. Let A be a local ring and let $\mathfrak{a} \subsetneq A$ be an ideal such that A is separated and complete for the $\mathfrak{a}$-adic topology. Set $Y := \operatorname{Spec} A$ and $Y_n := \operatorname{Spec} A/\mathfrak{a}^{n+1}$. Let S be a scheme and assume that Y is an S-scheme. Show that for every S-scheme X there is a functorial bijection

$$\mathrm{Hom}_S(Y, X) \overset{\sim}{\to} \varprojlim_{n} \mathrm{Hom}_S(Y_n, X).$$

Hint: Use Exercise 3.18.

Exercise 3.21. Let k be an algebraically closed field and let $(a_i)_{i \in \mathbb{N}}$ be a family of pairwise distinct elements of k. Set

$$A := k[U, T_1, T_2, \dots]/((U - a_i)T_{i+1} - T_i, T_i^2).$$

Show that the nilradical of A is not finitely generated (in particular A is not noetherian) but that $A_\mathfrak{p}$ is noetherian for all prime ideals $\mathfrak{p} \subset A$.

This example is due to J. Rabinoff.

Exercise 3.22. Let R be a ring, $S = \mathrm{Spec}\, R$ and $n \geq 0$ an integer. Show that the following assertions are equivalent.
(i) S is reduced (resp. irreducible, resp. integral).
(ii) $\mathbb{A}_R^n$ is reduced (resp. irreducible, resp. integral).
(iii) $\mathbb{P}_R^n$ is reduced (resp. irreducible, resp. integral).

Exercise 3.23. Let k be a field, $A = k[T_1, T_2, T_3]$, $\mathfrak{p}_1 = (T_1, T_2)$, $\mathfrak{p}_2 = (T_1, T_3)$ and $\mathfrak{a} := \mathfrak{p}_1 \mathfrak{p}_2$. Set $X = \mathbb{A}_k^3 = \mathrm{Spec}\, A$, $Z_i = V(\mathfrak{p}_i)$, $Y = V(\mathfrak{a})$. Show that Z_1 and Z_2 are integral subschemes of X and show that $Y = Z_1 \cup Z_2$ (set-theoretically). Show that Y is not reduced and describe Y_{red}.

Exercise 3.24◊. Prove that every immersion $i \colon Z \to X$ is a monomorphism in the category of schemes.

Exercise 3.25. Let Y be a scheme, and let $i \colon Z \subseteq Y$ be a subscheme. Then a morphism $f \colon X \to Y$ of schemes factors through the subscheme Z if and only if the following conditions are satisfied:
(1) $f(X) \subseteq Z$ (set-theoretically),
(2) $f^\flat \colon \mathscr{O}_Y \to f_* \mathscr{O}_X$ factors through the surjective homomorphism $\mathscr{O}_Y \twoheadrightarrow i_* \mathscr{O}_Z$.
Prove that (1) implies (2) if Z is an open subscheme, or if X is reduced.

Exercise 3.26. Let Y be a scheme, and let U be a non-empty open subscheme. Fix a non-empty index set I. For all $i \in I$ let $U_i := U_{ii} := Y$ and for $i, j \in I$, $i \neq j$, set $U_{ij} := U$, considered as an open subscheme of U_i. For all $i, j \in I$ we define $\varphi_{ij} \colon U_{ji} \to U_{ij}$ as the identity morphism. Check that $((U_i), (U_{ij}), (\varphi_{ij}))$ is a gluing datum. Let X be the scheme obtained by gluing. The U_i can be viewed as open subschemes of X. We assume that Y is integral.
(a) Show that for every open subset V of X and for all $i \in I$, the restriction homomorphism $\Gamma(V, \mathscr{O}_X) \to \Gamma(V \cap U_i, \mathscr{O}_X)$ is an isomorphism (for instance, use Exercise 3.8).
(b) Assume that $U \neq Y$. Conclude from (a), that X is not affine.
(c) Assume that Y is a noetherian scheme, and that $U \neq Y$. Prove that X is integral and locally noetherian. Furthermore, show that X is quasi-compact if and only if I is finite.

Exercise 3.27◊. Describe $\mathbb{A}_{\mathbb{R}}^1$ and $\mathbb{P}_{\mathbb{R}}^1$.

Exercise 3.28◊. Let $n \geq 1$ be an integer and set $X = \mathrm{Spec}\, \mathbb{Q}[S, T]/(S^n + T^n - 1)$. Translate the condition that there exist nonzero integers $x, y, z \in \mathbb{Z}$ with $x^n + y^n = z^n$ into a statement about $X(\mathbb{Q})$.

Exercise 3.29. Let X be a noetherian scheme. Show that the nilradical $\mathcal{N}_X$ is nilpotent (i.e., there exists an integer $k \geq 1$ such that $\Gamma(U, \mathcal{N}_X)^k = 0$ for every open subset $U \subseteq X$; see also Exercise 7.23).

Exercise 3.30. Let k be a field, and let A be a local k-algebra of finite type. Prove that $\operatorname{Spec} A$ consists of a single point, and that A is finite-dimensional as a k-vector space. In particular A is a local Artin ring (why?), and $\kappa(A)/k$ is a finite field extension.

Exercise 3.31. Let X be a scheme.
(a) If X is affine, show that X_{red} is affine.
(b) Assume that X is noetherian. If X_{red} is affine, show that X is affine.
    Hint: Use that $\mathcal{N}_X$ is nilpotent (Exercise 3.29) and reduce to the case $\mathcal{N}_X^2 = 0$. Then show that the canonical morphism $X \to \operatorname{Spec} \Gamma(X, \mathcal{O}_X)$ is an isomorphism.
Remark: The second assertion is also proved in Lemma 12.38 using a criterion of Serre. There it is also explained that the hypothesis that X is noetherian is superfluous.

4   Fiber products

Contents

In this chapter we study one of the central technical tools of algebraic geometry: If S is a scheme and X and Y are S-schemes we define the product $X \times_S Y$ of X and Y over S which is also called fiber product. We do this by defining $X \times_S Y$ as an S-scheme which satisfies a certain universal property (and by proving that such a scheme always exists).

The importance of this construction stems from the fact that different interpretations and special cases of the fiber product allow constructions such as fibers of morphisms, inverse images of subschemes, intersection of subschemes, or the change of the base scheme (e.g. passing from k-schemes to k'-schemes for a field extension $k \hookrightarrow k'$).

To characterize the fiber product by its universal property we start this chapter by considering schemes as functors. This is a point of view that is also very helpful at other occasions and we will see examples of schemes that are defined by their associated functors throughout the book. In Chapter 8 we will study the question whether, conversely, a given functor is defined by a scheme.

Schemes as functors

(4.1) Functors attached to schemes.

The point of origin in algebraic geometry is the goal to understand the set of zeros of polynomial systems of equations. If R is a ring, $f_1, \ldots, f_m$ polynomials in $R[T_1, \ldots, T_n]$ and A an R-algebra, solutions $x \in A^n$ of the equations $f_1(x) = \cdots = f_m(x) = 0$ correspond to homomorphisms $R[T_1, \ldots, T_n]/(f_1, \ldots, f_m) \to A$ of R-algebras and hence to $(\operatorname{Spec} R)$-morphisms $\operatorname{Spec} A \to \operatorname{Spec} R[T_1, \ldots, T_n]/(f_1, \ldots, f_m)$. This observation shows that it is natural to attach to a scheme X the functor

$$h_X \colon (\mathrm{Sch})^{\mathrm{opp}} \to (\mathrm{Sets}),$$
$$T \mapsto h_X(T) := \operatorname{Hom}_{(\mathrm{Sch})}(T, X), \qquad\qquad \text{(on objects)},$$
$$(f \colon T' \to T) \mapsto (\operatorname{Hom}(T, X) \to \operatorname{Hom}(T', X), \ g \mapsto g \circ f), \quad \text{(on morphisms)}.$$

This definition is all the more useful because, as we will see in Section (4.2), the scheme X is determined by the functor h_X.

© Springer Fachmedien Wiesbaden GmbH, part of Springer Nature 2020
U. Görtz und T. Wedhorn, *Algebraic Geometry I: Schemes*, Springer Studium
Mathematik – Master, https://doi.org/10.1007/978-3-658-30733-2_5

The set $\mathrm{Hom}_{(\mathrm{Sch})}(T, X)$ is called the *set of T-valued points of X*. Usually we simply write $X(T)$ instead of $h_X(T) = \mathrm{Hom}_{(\mathrm{Sch})}(T, X)$. If $T = \mathrm{Spec}\, A$ is an affine scheme, we also set $X(A)^{\cdot} := X(\mathrm{Spec}\, A)$. More generally, we might consider an arbitrary functor $F\colon (\mathrm{Sch})^{\mathrm{opp}} \to (\mathrm{Sets})$ as a "geometric object" and we call $F(T)$ the *set of T-valued points of F*.

Let now S be a fixed scheme. Instead of the category (Sch) we also consider the category (Sch/S) of S-schemes (Section (3.1)). Again every S-scheme X provides a functor

$$(\mathrm{Sch}/S) \longrightarrow (\mathrm{Sets}), \qquad T \longmapsto \mathrm{Hom}_S(T, X).$$

Instead of $\mathrm{Hom}_S(T, X)$ we write shorter $X_S(T)$ or even $X(T)$ if it is understood that all schemes are considered as S-schemes. If $S = \mathrm{Spec}\, R$ or $T = \mathrm{Spec}\, A$ is affine, we also write $X_R(T)$ resp. $X_S(A)$ (or even $X_R(A)$, if S and T are both affine).

Example 4.1. Let k be an algebraically closed field and let X be a k-scheme locally of finite type (Section (3.12)). For every k-valued point $x\colon \mathrm{Spec}\, k \to X$ its image $\mathrm{Im}(x)$ is a closed point of the underlying topological space of X. The map $X_k(k) \to X_0$, $x \mapsto \mathrm{Im}(x)$ is a bijection of $X_k(k)$ onto the set of closed points X_0 of X. If X is integral and of finite type, we thus obtain a bijection of $X_k(k)$ onto the associated prevariety (Section (3.13)).

Example 4.2. Consider the affine space: Let $n \geq 0$ and $X = \mathbb{A}^n = \mathrm{Spec}(\mathbb{Z}[T_1, \ldots, T_n])$. Then for every scheme T we have by Section (3.3)

$$\mathrm{Hom}_{(\mathrm{Sch})}(T, \mathbb{A}^n) = \mathrm{Hom}_{(\mathrm{Ring})}(\mathbb{Z}[T_1, \ldots, T_n], \Gamma(T, \mathcal{O}_T)) = \Gamma(T, \mathcal{O}_T)^n,$$
$$\varphi \mapsto (\varphi(T_1), \ldots, \varphi(T_n)).$$

In particular, we have $\mathbb{A}^n(R) = R^n$ for every ring R.

Example 4.3. More generally, let R be a ring and $f_1, \ldots, f_r \in R[T_1, \ldots, T_n]$ polynomials and set $X = \mathrm{Spec}(R[\underline{T}]/(f_1, \ldots, f_r))$. Then for every R-scheme T we have again by Section (3.3)

$$X_R(T) = \mathrm{Hom}_{(R\text{-}\mathrm{Alg})}(R[\underline{T}]/(f_1, \ldots, f_r), \Gamma(T, \mathcal{O}_T))$$
$$= \{\, s \in \Gamma(T, \mathcal{O}_T)^n \;;\; f_1(s) = \cdots = f_r(s) = 0 \,\}.$$

Example 4.4. Set $X = \mathrm{Spec}\, R[U, U^{-1}]$. Then we obtain for every R-scheme T

$$X_R(T) = \mathrm{Hom}_{(R\text{-}\mathrm{Alg})}(R[U, U^{-1}], \Gamma(T, \mathcal{O}_T)) = \Gamma(T, \mathcal{O}_T)^{\times}.$$

In fact, this is a special case of Example 4.3, as we have $R[U, U^{-1}] \cong R[U, V]/(UV - 1)$. In particular we have $X_R(A) = A^{\times}$ for every R-algebra A.

Considering $\Gamma(T, \mathcal{O}_T)^{\times}$ as a group, we obtain a functor from (Sch/R) into the category of groups. Such functors are called group schemes (see Section (4.15) below). The group scheme X above is denoted by $\mathbb{G}_{m,R}$ and is called the *multiplicative group (over $\mathrm{Spec}\, R$)*.

Example 4.5. Let $n \geq 0$ be an integer, R a ring, and let $\tilde{P} = \mathbb{A}_R^n \setminus \{0\}$ be the complement of the zero section (Section (3.6)). Thus $\tilde{P}$ is an open subscheme of $\mathbb{A}_R^n$.

Let A be an R-algebra, let $f\colon \mathrm{Spec}\, A \to \mathbb{A}_R^n$ be an R-morphism, and denote the corresponding R-algebra homomorphism by $\varphi\colon R[T_1, \ldots, T_n] \to A$. Set $a_i = \varphi(T_i) \in A$. Let $\pi\colon R[T_1, \ldots, T_n] \to R$ be the projection mapping each T_i to 0. Then f factors through $\tilde{P}$ if and only if $\mathrm{Ker}(\pi) = (T_1, \ldots, T_n)$ is not contained in $\varphi^{-1}(\mathfrak{p})$ for all $\mathfrak{p} \in \mathrm{Spec}\, A$. Equivalently, there must not exist a prime ideal $\mathfrak{p} \subset A$ that contains $\varphi(\mathrm{Ker}(\pi))$, which is the ideal generated by $a_1, \ldots, a_n$. Thus we have seen that

$$(4.1.1) \qquad \tilde{P}_R(A) = \{\, (a_1, \ldots, a_n) \in A^n \; ; \; a_1, \ldots, a_n \text{ generate the unit ideal in } A \,\}.$$

(4.2) Yoneda Lemma.

Now let $\mathcal{C}$ be an arbitrary category (we will mainly use the examples that $\mathcal{C}$ is the category (Sch$/S$) for some scheme S). As above we define for every object X of $\mathcal{C}$ the functor

$$h_X \colon \mathcal{C}^{\mathrm{opp}} \to (\text{Sets}),$$
$$S \mapsto h_X(S) := \mathrm{Hom}_{\mathcal{C}}(S, X),$$
$$(u \colon S' \to S) \mapsto (h_X(u) \colon h_X(S) \to h_X(S'), x \mapsto x \circ u).$$

If $f \colon X \to Y$ is a morphism in $\mathcal{C}$, for every object S the composition

$$h_f(S) \colon h_X(S) \to h_Y(S), \quad g \mapsto f \circ g$$

defines a morphism $h_f \colon h_X \to h_Y$ of functors. We obtain a (covariant) functor $X \mapsto h_X$ from $\mathcal{C}$ to the category $\widehat{\mathcal{C}}$ of functors $\mathcal{C}^{\mathrm{opp}} \to (\text{Sets})$.

Of central importance will be the Yoneda lemma: Let $F \colon \mathcal{C}^{\mathrm{opp}} \to (\text{Sets})$ be a functor and X be an object of $\mathcal{C}$. Let $\alpha \colon h_X \to F$ be a morphism of functors, in other words, for all objects Y we are given a map $\alpha(Y) \colon h_X(Y) \to F(Y)$, functorially in Y. Then we have $\alpha(X)(\mathrm{id}_X) \in F(X)$.

Lemma 4.6. (Yoneda Lemma) *The map*

$$\mathrm{Hom}_{\widehat{\mathcal{C}}}(h_X, F) \to F(X), \quad \alpha \mapsto \alpha(X)(\mathrm{id}_X)$$

is bijective and functorial in X.

Proof. For $\xi \in F(X)$ we define the map $\alpha_\xi(Y) \colon h_X(Y) \to F(Y)$ by $f \mapsto F(f)(\xi)$ for $f \in h_X(Y) = \mathrm{Hom}_{\mathcal{C}}(Y, X)$. Then $\xi \mapsto \alpha_\xi$ is an inverse map. The functoriality is clear. $\qquad\square$

If we apply the Yoneda lemma to the special case $F = h_Y$ for an object Y of $\mathcal{C}$, we see that the functor $X \mapsto h_X$ induces a bijection

$$(4.2.1) \qquad\qquad \mathrm{Hom}_{\mathcal{C}}(X, Y) \to \mathrm{Hom}_{\widehat{\mathcal{C}}}(h_X, h_Y).$$

In other words, $X \mapsto h_X$ is a fully faithful functor $\mathcal{C} \to \widehat{\mathcal{C}}$. See Exercise 4.1 for an explicit example.

We will apply the Yoneda lemma mainly in the case that $\mathcal{C}$ is the category of S-schemes, where S is a fixed scheme. Then it obtains the following form:

Corollary 4.7. *Let X and Y be S-schemes. Then it is equivalent to give the following data.*
(i) *A morphism of S-schemes from X to Y.*
(ii) *For all S-schemes T a map $f(T) \colon X_S(T) \to Y_S(T)$ of sets which is functorial in T, i.e., for all morphisms $u \colon T' \to T$ of S-schemes the following diagram is commutative*

$$(4.2.2)$$

$$
\begin{CD}
X_S(T) @>{f(T)}>> Y_S(T) \\
@V{X_S(u)}VV @VV{Y_S(u)}V \\
X_S(T') @>{f(T')}>> Y_S(T').
\end{CD}
$$

(iii) *For all affine S-schemes $T = \operatorname{Spec} B$ a map $f(T)\colon X_S(T) \to Y_S(T)$ of sets which is functorial in B.*

Proof. The equivalence of (i) and (ii) is just a reformulation of the Yoneda lemma. The equivalence of (ii) and (iii) follows from the fact that morphisms in the category of schemes can be glued (Proposition 3.5) and that for every scheme the open affine subschemes form a basis of the topology (Proposition 3.2).   □

(4.3) A surjectivity criterion for morphisms of schemes.

By Section (4.2) it is equivalent to give a morphism $f\colon X \to Y$ in a category (Sch/S) or to give maps $f_S(T)\colon X_S(T) \to Y_S(T)$, functorial in T. Often it is helpful to express properties of f in terms of properties of the maps $f_S(T)$ (and vice versa). We give an example here. Other examples we will see throughout the book.

Proposition 4.8. *A morphism of schemes $f\colon X \to Y$ is surjective if and only if for every field K and for every K-valued point $y \in Y(K)$ there exist a field extension L of K and $x \in X(L)$ such that $f(L)(x) = y_L$, where y_L is the image of y under $Y(K) \to Y(L)$.*

For schemes of finite type over a field, this criterion can be sharpened considerably (Exercise 10.6).

Proof. The condition is sufficient: Let y_0 be a point of the underlying topological space of Y, and let $y\colon \operatorname{Spec}(\kappa(y_0)) \to Y$ be the canonical morphism (3.4.2). If $x\colon \operatorname{Spec}(L) \to X$ is an L-valued point of X with $f(L)(x) = y_L$ and $x_0 \in X$ is the image of x, then we have $f(x_0) = y_0$.

The condition is necessary: Let f be surjective, $y \in Y(K)$, and $y_0 \in Y$ be the image of y. There exists a point $x_0 \in X$ with $f(x_0) = y_0$. Consider the corresponding extension $\kappa(y_0) \to \kappa(x_0)$. Choose a field extension L of $\kappa(y_0)$ such that there exist $\kappa(y_0)$-embeddings of $\kappa(x_0)$ and of K into L (e.g., set $L = (\kappa(x_0) \otimes_{\kappa(y_0)} K)/\mathfrak{m}$ where $\mathfrak{m}$ is a maximal ideal of $\kappa(x_0) \otimes_{\kappa(y_0)} K$). The composition $x\colon \operatorname{Spec}(L) \to \operatorname{Spec}(\kappa(x_0)) \to X$ has the desired properties.   □

Remark 4.9. In particular we see that f is surjective if f is surjective on K-valued points for every field K. The converse does not hold: Let $r > 1$ be an integer and let $f_r\colon \mathbb{G}_m \to \mathbb{G}_m$ be given on S-valued points by

$$f_r(S)\colon \mathbb{G}_m(S) = \Gamma(S, \mathscr{O}_S)^\times \longrightarrow \mathbb{G}_m(S), \quad x \to x^r.$$

Then $f_r(K)$ is surjective if and only if for all $x \in K^\times$ there exists an r-th root. In particular, if K is algebraically closed, $f_r(K)$ is surjective, and Proposition 4.8 shows that f_r is surjective. But of course there are fields K such that $f_r(K)$ is not surjective (e.g., $K = \mathbb{R}$ and r even, or $K = \mathbb{Q}$ and $r \geq 2$ arbitrary).

Fiber products of schemes

(4.4) Fiber products in arbitrary categories.

Let $\mathcal{C}$ be a category and let S be a fixed object in $\mathcal{C}$.

Definition 4.10. *For two morphisms $f\colon X \to S$ and $g\colon Y \to S$ in $\mathcal{C}$ we call a triple (Z, p, q) consisting of an object Z in $\mathcal{C}$ and morphisms $p\colon Z \to X$ and $q\colon Z \to Y$ with $f \circ p = g \circ q$ a fiber product of f and g or a fiber product of X and Y over S (with respect to f and g), if for every object T in $\mathcal{C}$ and for all pairs (u, v) of morphisms $u\colon T \to X$ and $v\colon T \to Y$ such that $f \circ u = g \circ v$ there exists a unique morphism $w\colon T \to Z$ such that $p \circ w = u$ and $q \circ w = v$.*

Clearly the fiber product of X and Y over S with respect to f and g is uniquely determined up to unique isomorphism if it exists. We write $X \times_{f,S,g} Y$ or simply $X \times_S Y$ for the object Z. We also call $p\colon X \times_S Y \to X$ the *first projection* and $q\colon X \times_S Y \to Y$ the *second projection*. The morphism w is denoted by $(u, v)_S$. We visualize the universal property of the fiber product by the following diagram

(4.4.1)

$$
\begin{array}{ccc}
 & & T \\
 & \overset{\exists!(u,v)_S}{\searrow} & \quad \overset{\forall u}{\searrow} \\
\forall v & X \times_S Y & \xrightarrow{\ p\ } X \\
 & \downarrow q & \downarrow f \\
 & Y & \xrightarrow[\ g\ ]{} S.
\end{array}
$$

Note that if S is a final object in $\mathcal{C}$, the fiber product $X \times_S Y$ is the categorical product $X \times Y$. Fiber products are special cases of projective limits (Example A.3). In fact it follows formally from the existence of fiber products in the category of schemes (proved in Theorem 4.18 below) that there exist arbitrary finite projective limits in the category of schemes (Exercise 4.3).

Remark 4.11. We can describe the universal property of the fiber product $(X \times_S Y, p, q)$ also as follows. Recall that a morphism $h\colon T \to S$ in $\mathcal{C}$ is called an *S-object*. Sometimes we will simply write T instead of $h\colon T \to S$. The morphism h is called the *structure morphism of T*. Recall that for two S-objects $h\colon T \to S$ and $f\colon X \to S$ in $\mathcal{C}$ we denote by $\mathrm{Hom}_S(T, X)$ the morphisms $w\colon T \to X$ such that $f \circ w = h$. These morphisms are called *S-morphisms*. In this way S-objects and S-morphisms form a category that we will denote by $\mathcal{C}/S$. Usually we write $X_S(T)$ instead of $\mathrm{Hom}_S(T, X)$ and call $X_S(T)$ the *set of T-valued points of X (over S)*. Note that the object id_S in $\mathcal{C}/S$ is a final object.

Then $(X \times_S Y, p, q)$ is the unique (up to unique isomorphism) triple such that for all morphisms $h\colon T \to S$ the map

(4.4.2)
$$
\mathrm{Hom}_S(T, X \times_S Y) \to \mathrm{Hom}_S(T, X) \times \mathrm{Hom}_S(T, Y),
$$
$$
w \mapsto (p \circ w, q \circ w)
$$

is a bijection. In other words, the fiber product of $f\colon X \to S$ and $g\colon Y \to S$ in $\mathcal{C}$ is the same as the product of the S-objects f and g in $\mathcal{C}/S$.

Example 4.12.

(1) In the category of sets (Sets) arbitrary fiber products exist: Let S be a fixed set and let $f\colon X \to S$ and $g\colon Y \to S$ be maps of sets. Then it is immediate that

(4.4.3) $\quad \begin{array}{ccccc} X & \xleftarrow{\;p\;} & X \times_S Y := \{\, (x,y) \in X \times Y \; ; \; f(x) = g(y) \,\} & \xrightarrow{\;q\;} & Y, \\ x & \longleftarrow\!\shortmid & (x,y) & \longmapsto & y \end{array}$

is a fiber product in the category of sets.

(2) Let (Top) be the category of topological spaces (objects are topological spaces, morphisms are continuous maps). If $f\colon X \to S$ and $g\colon Y \to S$ are continuous maps of topological spaces, we can endow the fiber product of the underlying sets $\{\, (x,y) \in X \times Y \; ; \; f(x) = g(y) \,\}$ with the topology induced by the product topology on $X \times Y$. Then it is easy to check that the resulting topological space is a fiber product in the category of topological spaces.

From now on we assume that in the category $\mathcal{C}$ all fiber products exist. We will show in Section (4.5) that this is the case if $\mathcal{C}$ is the category of schemes.

Remark 4.13. The fiber product is functorial in the following sense: If X, Y, X', and Y' are S-objects and $u\colon X \to X'$, $v\colon Y \to Y'$ are S-morphisms, then there exists a unique morphism, denoted $u \times_S v$ (or simply $u \times v$), such that the following diagram is commutative

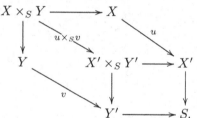

Indeed, we have $u \times_S v = (u \circ p, v \circ q)_S$, where p and q are the projections of $X \times_S Y$.

Recall that the Yoneda lemma implies that it is equivalent to give an S-morphism $f\colon X \to Y$ in $\mathcal{C}/S$ or to give for all S-objects T maps $f_S(T)\colon X_S(T) \to Y_S(T)$ on T-valued points which are functorial in T. The following proposition collects some easy functorial identities of fiber products.

Proposition 4.14. *Let S be an object in $\mathcal{C}$ and let X, Y, and Z be S-objects. There are isomorphisms (functorial in X, Y, and Z), called* canonical,

(4.4.4) $\qquad\qquad X \times_S S \xrightarrow{\sim} X,$

(4.4.5) $\qquad\qquad X \times_S Y \xrightarrow{\sim} Y \times_S X,$

(4.4.6) $\qquad (X \times_S Y) \times_S Z \xrightarrow{\sim} X \times_S (Y \times_S Z),$

given on T-valued points, for $h\colon T \to S$ any S-object, by

$$X_S(T) \times S_S(T) \xrightarrow{\sim} X_S(T), \qquad\qquad (x,h) \mapsto x,$$
$$X_S(T) \times Y_S(T) \xrightarrow{\sim} Y_S(T) \times X_S(T), \qquad\qquad (x,y) \mapsto (y,x),$$
$$(X_S(T) \times Y_S(T)) \times Z_S(T) \xrightarrow{\sim} X_S(T) \times (Y_S(T) \times Z_S(T)), \qquad ((x,y),z) \mapsto (x,(y,z)).$$

A commutative diagram

(4.4.7)

$$\begin{array}{ccc} Z & \xrightarrow{u} & X \\ {\scriptstyle v}\downarrow & & \downarrow{\scriptstyle f} \\ Y & \xrightarrow{g} & S \end{array}$$

in $\mathcal{C}$ is called *cartesian* if the morphism $(u,v)_S\colon Z \to X \times_S Y$ is an isomorphism. We indicate the property of a rectangle being cartesian by putting a square $\square$ in the center. The Yoneda lemma implies:

Remark 4.15. *The commutative diagram (4.4.7) is cartesian, if and only if for all objects T of $\mathcal{C}$ the induced diagram in the category of sets*

$$\begin{array}{ccc} \operatorname{Hom}(T,Z) & \xrightarrow{u(T)} & \operatorname{Hom}(T,X) \\ {\scriptstyle v(T)}\downarrow & & \downarrow{\scriptstyle f(T)} \\ \operatorname{Hom}(T,Y) & \xrightarrow{g(T)} & \operatorname{Hom}(T,S) \end{array}$$

is cartesian (Example 4.12).

Therefore it suffices to prove the following proposition for $\mathcal{C} = (\text{Sets})$, where it follows immediately from the explicit description of the fiber product of sets.

Proposition 4.16. *Let*

(4.4.8)

$$\begin{array}{ccccc} X'' & \xrightarrow{g'} & X' & \xrightarrow{g} & X \\ \downarrow & & \downarrow & \square & \downarrow \\ S'' & \xrightarrow{f'} & S' & \xrightarrow{f} & S \end{array}$$

be a commutative diagram in $\mathcal{C}$ such that the right square is cartesian. Then the left square is cartesian if and only if the entire composed diagram is cartesian.

(4.5) Fiber products of schemes.

We will now prove that fiber products of schemes always exist. To do so, we will reduce to the case of affine schemes. Therefore let us deal with this case first.

Proposition 4.17. *Let $A \leftarrow R \to B$ be homomorphisms of rings, let $S = \operatorname{Spec}(R)$, $X = \operatorname{Spec}(A)$, and $Y = \operatorname{Spec}(B)$. Set $Z = \operatorname{Spec}(A \otimes_R B)$ and let $p\colon Z \to X$ and $q\colon Z \to Y$ be the morphisms of schemes corresponding to the ring homomorphisms*

$$\alpha\colon A \to A \otimes_R B, \quad a \mapsto a \otimes 1,$$
$$\beta\colon B \to A \otimes_R B, \quad b \mapsto 1 \otimes b.$$

Then (Z,p,q) is a fiber product of X and Y over S in the category of schemes, and also in the category of locally ringed spaces.

Proof. Recall that for all locally ringed spaces T and all affine schemes $\operatorname{Spec} C$ there is a functorial bijection $\operatorname{Hom}_{(\mathrm{Sch})}(T, \operatorname{Spec} C) \cong \operatorname{Hom}_{(\mathrm{Ring})}(C, \Gamma(T, \mathscr{O}_T))$ (Proposition 3.4). If $T \to S$ is a morphism of locally ringed spaces, then we therefore have bijections, functorial in T,

$$
\begin{aligned}
\operatorname{Hom}(T, Z) &\cong \operatorname{Hom}_{(R\text{-Alg})}(A \otimes_R B, \Gamma(T, \mathscr{O}_T)) \\
&\cong \operatorname{Hom}_{(R\text{-Alg})}(A, \Gamma(T, \mathscr{O}_T)) \times \operatorname{Hom}_{(R\text{-Alg})}(B, \Gamma(T, \mathscr{O}_T)) \\
&\cong \operatorname{Hom}(T, X) \times \operatorname{Hom}(T, Y),
\end{aligned}
$$

where the second bijection is induced by composition with α and β. This shows the proposition. $\qquad\square$

Theorem 4.18. *Let S be a scheme and let X and Y be two S-schemes. Then the fiber product $X \times_S Y$ exists in the category of schemes, and also satisfies the universal property of the fiber product in the category of locally ringed spaces.*

We can rephrase the theorem as saying that the fiber product of X and Y over S exists in the category of locally ringed spaces, and actually is a scheme (and thus equals the fiber product in the category of schemes).

Proof. When we say in the sequel of this proof that a fiber product of schemes *exists*, we mean that as a short-hand statement to say that *the fiber product exists in the category of schemes, and is also a fiber product in the category of locally ringed spaces*. To ensure this property, we have to do all constructions in the category of schemes, and check the universal property with arbitrary locally ringed spaces as test objects.

The idea of the proof is rather simple: We cover S, X, and Y by open affine subschemes. We have already seen that the fiber product for affine schemes exists. Thus it remains to glue all the fiber products of affine schemes together. To make this more precise, we proceed in several steps. We denote by $x \colon X \to S$ and $y \colon Y \to S$ the structure morphisms.

(i). Let $j \colon U \hookrightarrow X$ be an open subscheme. Assume that $(X \times_S Y, p, q)$ exists. Then we claim that the open subscheme $p^{-1}(U)$ of $X \times_S Y$ together with the restrictions of p and q is the fiber product $U \times_S Y$ in the category of locally ringed spaces. Indeed, if $h \colon T \to p^{-1}(U)$ is a morphism of locally ringed spaces, we obtain morphisms $f := p \circ h \colon T \to U$ and $g := q \circ h \colon T \to Y$ such that $x_{|U} \circ f = y \circ g$. Conversely, let $f \colon T \to U$ and $g \colon T \to Y$ be a pair of morphisms with $x_{|U} \circ f = y \circ g$. As $X \times_S Y$ is a fiber product, there exists a unique morphism $h' \colon T \to X \times_S Y$ such that $p \circ h' = j \circ f$ and $q \circ h' = g$. This shows that h' factors through $p^{-1}(U)$ and thus induces a morphism $h \colon T \to p^{-1}(U)$ such that $f := p \circ h$ and $g := q \circ h$.

(ii). Let $(U_i)_{i \in I}$ be an open covering of X. We claim that if $Z_i := U_i \times_S Y$ exists for all i, then $X \times_S Y$ exists. Indeed, we consider Z_i via the second projection as a Y-scheme. Let $p_i \colon Z_i \to U_i$ be the first projection and set $Z_{ij} := p_i^{-1}(U_i \cap U_j) \subseteq Z_i$. Let $p_{ij} \colon Z_{ij} \to U_i \cap U_j$ be the restriction of p_i. By (i) the schemes Z_{ij} and Z_{ji} are both a fiber product of $U_i \cap U_j$ and Y over S. Therefore there exists a unique isomorphism $\varphi_{ji} \colon Z_{ij} \to Z_{ji}$ of Y-schemes such that $p_{ji} \circ \varphi_{ji} = p_{ij}$. Its inverse isomorphism is φ_{ij}. The uniqueness of these isomorphisms implies that they satisfy the cocycle condition. We denote by Z the scheme obtained by gluing the schemes Z_i along Z_{ij} via the isomorphisms φ_{ij}. The second projections $Z_i \to Y$ glue to a morphism $q \colon Z \to Y$ and the first projections p_i glue to a morphism $p \colon Z \to X$.

It remains to show that (Z, p, q) is the fiber product of X and Y over S. Let $f\colon T \to X$ and $g\colon T \to Y$ be morphisms of locally ringed spaces such that $x \circ f = y \circ g$. Let $f_i\colon T_i := f^{-1}(U_i) \to U_i$ be the restriction of f. Then there exists a unique morphism $h_i\colon T_i \to Z_i$ such that $p_i \circ h_i = f_i$ and $q \circ h_i = g_{|T_i}$. These morphisms glue to a unique morphism $h\colon T \to Z$ such that $p \circ h = f$ and $q \circ h = g$. Therefore (Z, p, q) satisfies the defining property of the fiber product of X and Y over S.

(iii). Let $W \subseteq S$ be an open subset. Assume that $(X \times_S Y, p, q)$ exists. Then we claim that the open subscheme $(x \circ p)^{-1}(W) = (y \circ q)^{-1}(W)$ of $X \times_S Y$ together with the restrictions of p and q is the fiber product $x^{-1}(W) \times_W y^{-1}(W)$. Indeed, similarly as in (i) it is easy to check that $(x \circ p)^{-1}(W)$ satisfies the defining property of $x^{-1}(W) \times_W y^{-1}(W)$.

(iv). Let $(W_i)_{i \in I}$ be an open covering of S, set $X_i := x^{-1}(W_i)$ and $Y_i := y^{-1}(W_i)$. We claim that if $X_i \times_{W_i} Y_i$ exists for all i, then $X \times_S Y$ exists and $(X_i \times_{W_i} Y_i)_i$ is an open covering of $X \times_S Y$. Indeed this is proved in a similar way as (ii) using (iii) instead of (i).

(v). Now we can prove the existence of the fiber product of arbitrary schemes X and Y over S: Covering S by open affine subschemes, (iv) shows that we may assume that S is affine. Covering X by open affine subschemes, (ii) shows that we may assume that X is affine. As the arguments in (ii) are clearly symmetric in X and Y, we finally may also assume that Y is affine. But then we have already shown the existence of the fiber product in Proposition 4.17. $\qquad\square$

If $S = \operatorname{Spec} R$ is affine, we will often write $X \times_R Y$ instead of $X \times_S Y$. If $Y = \operatorname{Spec} B$ is affine, we also write $X \otimes_S B$ or, for $S = \operatorname{Spec} R$ affine, $X \otimes_R B$ instead of $X \times_S Y$.

The proof of Theorem 4.18 shows in particular:

Corollary 4.19. *Let S be a scheme, let X and Y be S-schemes, let $S = \bigcup_i S_i$ be an open covering and denote by X_i (resp. Y_i) the inverse image of S_i in X (resp. in Y). For all i let $X_i = \bigcup_{j \in J_i} X_{ij}$ and $Y_i = \bigcup_{k \in K_i} Y_{ik}$ be open coverings. Then*

$$X \times_S Y = \bigcup_i \bigcup_{j \in J_i, k \in K_i} X_{ij} \times_{S_i} Y_{ik}$$

is an open covering of $X \times_S Y$.

Special cases of the following proposition will be used very often. We start with the following general setting. Let S be a scheme, X and Y two S-schemes, and let $f\colon X' \to X$ be a morphism of S-schemes. Set $g := f \times_S \operatorname{id}_Y$. We obtain a commutative diagram, where all squares are cartesian (Proposition 4.16)

(4.5.1)
$$
\begin{array}{ccccc}
X' \times_S Y & \xrightarrow{\ g\ } & X \times_S Y & \xrightarrow{\ q\ } & Y \\
{\scriptstyle p'}\downarrow & \square & {\scriptstyle p}\downarrow & \square & \downarrow \\
X' & \xrightarrow[\ f\ ]{} & X & \longrightarrow & S
\end{array}
$$

and where p', p, and q are the projections.

Proposition 4.20. *Assume that $f\colon X' \to X$ can be written as the composition of scheme morphisms which satisfy the following condition: each morphism is a homeomorphism onto its image and also satisfies one of the assumptions (I), (II):*

(I) For each point $x' \in X'$, the homomorphism $f_{x'}^\sharp\colon \mathscr{O}_{X, f(x')} \to \mathscr{O}_{X', x'}$ is surjective, and there exists an open affine neighborhood V of $f(x')$ such that $f^{-1}(V)$ is quasi-compact.

(II) For each point $x' \in X'$, the homomorphism $f_{x'}^{\sharp} \colon \mathscr{O}_{X,f(x')} \to \mathscr{O}_{X',x'}$ is bijective.
We set $Z' = X' \times_S Y$ and $Z = X \times_S Y$.
(1) The morphism g is a homeomorphism of Z' onto

(4.5.2) $$g(Z') = p^{-1}(f(X')).$$

(2) For all points $z' \in Z'$ consider the commutative diagram induced on local rings by the left square of (4.5.1)

$$
\begin{array}{ccc}
\mathscr{O}_{Z',z'} & \xleftarrow{\;g_{z'}^{\sharp}\;} & \mathscr{O}_{Z,g(z')} \\[2ex]
\Big\uparrow & & \Big\uparrow{\scriptstyle p_{g(z')}^{\sharp}} \\[2ex]
\mathscr{O}_{X',p'(z')} & \xleftarrow{\;f_{p'(z')}^{\sharp}\;} & \mathscr{O}_{X,p(g(z'))}.
\end{array}
$$

Then the homomorphism $g_{z'}^{\sharp}$ is surjective and its kernel is generated by the image of the kernel of $f_{p'(z')}^{\sharp}$ under $p_{g(z')}^{\sharp}$.

Proof. Because of the transitivity of the fiber product (Proposition 4.16) we may assume that f satisfies assumption (I) or assumption (II).

We first assume that f satisfies (II). Therefore $f^{\sharp} \colon f^{-1}(\mathscr{O}_X) \to \mathscr{O}_{X'}$ is an isomorphism or, in other words, that f yields an isomorphism $(X', \mathscr{O}_{X'}) \xrightarrow{\sim} (f(X'), \mathscr{O}_{X|f(X')})$ of locally ringed spaces. In this case it is easy to check that $(p^{-1}(f(X')), \mathscr{O}_{Z|p^{-1}(f(X'))})$ is a fiber product of X' with Z over X in the category of locally ringed spaces and in particular in the category of schemes. Therefore g gives rise to an isomorphism $(Z', \mathscr{O}_{Z'}) \xrightarrow{\sim} (p^{-1}(f(X')), \mathscr{O}_{Z|p^{-1}(f(X'))})$. This implies all assertions.

Thus from now on we assume that f satisfies assumption (I). All assertions can be checked locally on S, Y and X. Observe that the second assumption in (I) is preserved by replacing X by an affine open $U \subseteq X$: If $x' \in f^{-1}(U)$ and V is an open affine neighborhood of $f(x')$ such that $f^{-1}(V)$ is quasi-compact, there exists a principal open V_1 inside V which is contained in U. Covering $f^{-1}(V)$ by finitely many affine open subschemes, we see that $f^{-1}(V_1)$ is covered by finitely many principal opens inside these (Proposition 2.10), and hence is again quasi-compact. Cf. Proposition/Definition 10.1 for the same argument in a more general context.

We can therefore assume that $S = \operatorname{Spec} R$, $X = \operatorname{Spec} A$ and $Y = \operatorname{Spec} B$ are affine and that X' is quasi-compact. Then f corresponds to an R-algebra homomorphism $\varphi \colon A \to \Gamma(X', \mathscr{O}_{X'})$ and factorizes as

$$f \colon X' \xrightarrow{f_1} \operatorname{Spec}(A/\operatorname{Ker}(\varphi)) \xrightarrow{f_2} X = \operatorname{Spec} A.$$

Then f_2 is a closed immersion and hence surjective on stalks and a homeomorphism onto a closed subspace of X. As f is a homeomorphism onto its image, the same therefore holds for f_1. As f induces surjections on the local rings, the same holds for f_1. We see that it suffices to prove the proposition if f is a closed immersion or if in addition the corresponding R-algebra homomorphism φ is injective.

Let us first consider the case that f is a closed immersion and hence $X' = \operatorname{Spec} A/\mathfrak{a}$ for some ideal $\mathfrak{a} \subseteq A$. Proposition 4.17 shows that Z and Z' are affine as well, and g corresponds to the natural surjective R-algebra homomorphism $A \otimes_R B \to A/\mathfrak{a} \otimes_R B$. Then all assertions are clear.

Therefore we may assume that φ is injective. We claim that for all $x' \in X'$ the surjective ring homomorphism

$$f_{x'}^{\sharp} \colon \mathscr{O}_{X,f(x')} \to \mathscr{O}_{X',x'}$$

is injective and hence bijective (then we are done because we proved all assertions already under assumption (II)).

To prove the claim we start with a general remark (which is a special case of Theorem 7.22 below). For a scheme Z and $t \in \Gamma(Z, \mathscr{O}_Z)$ let Z_t be the open set of $z \in Z$ such that $t(z) \neq 0$. The restriction $\Gamma(Z, \mathscr{O}_Z) \to \Gamma(Z_t, \mathscr{O}_Z)$ defines a homomorphism $\rho_t \colon \Gamma(Z, \mathscr{O}_Z)_t \to \Gamma(Z_t, \mathscr{O}_Z)$. This homomorphism is injective if Z is quasi-compact. Indeed, choose a finite affine covering $(U_i)_i$ of Z and set $C_i = \Gamma(U_i, \mathscr{O}_Z)$. Defining $t_i = t_{|U_i}$ we have $(\prod_i C_i)_t = \prod_i (C_i)_{t_i}$ because the product is finite. We obtain a commutative diagram

$$
\begin{array}{ccc}
\Gamma(Z, \mathscr{O}_Z)_t & \xrightarrow{\rho_t} & \Gamma(Z_t, \mathscr{O}_Z) \\
\downarrow & & \downarrow \\
\prod_i (C_i)_{t_i} & \xrightarrow{\sim} & \prod_i \Gamma((U_i)_{t_i}, \mathscr{O}_Z)
\end{array}
$$

with injective vertical arrows. Moreover, the lower horizontal homomorphism is the product of the isomorphisms $(C_i)_{t_i} \cong \Gamma(D(t_i), \mathscr{O}_{U_i})$ (2.10.1). This shows the injectivity of ρ_t.

To prove the injectivity of $f_{x'}^{\sharp}$ let $\mathfrak{p} \subset A$ be the prime ideal corresponding to $f(x')$. For all $s \in A \setminus \mathfrak{p}$ let $\varphi_s \colon A_s \to \Gamma(X', \mathscr{O}_{X'})_{\varphi(s)}$ be the injective homomorphism obtained from φ by localizing in s and denote by ψ_s the injective composition

$$\psi_s \colon A_s \xrightarrow{\varphi_s} \Gamma(X', \mathscr{O}_{X'})_{\varphi(s)} \xrightarrow{\rho_{\varphi(s)}} \Gamma(X'_{\varphi(s)}, \mathscr{O}_{X'}).$$

Now $X'_{\varphi(s)} = f^{-1}(D(s))$. As the $D(s)$, for $s \in A \setminus \mathfrak{p}$, form a basis of open neighborhoods of $f(x')$ and as f is a homeomorphism onto its image, the $X'_{\varphi(s)}$ form a basis of open neighborhoods of x'. Hence $\varinjlim \Gamma(X'_{\varphi(s)}, \mathscr{O}_{X'}) = \mathscr{O}_{X',x'}$ and we have $\varinjlim \psi_s = f_{x'}^{\sharp}$ which shows the injectivity of $f_{x'}^{\sharp}$. $\qquad \square$

Remark 4.21. The hypotheses of Proposition 4.20 on f are satisfied in the following cases which are the cases of main interest.
(1) f is an immersion of schemes (Section (3.16)).
(2) f is the canonical morphism $\operatorname{Spec} \mathscr{O}_{X,x} \to X$ for some point $x \in X$ (3.4.1).
(3) f is the canonical morphism $\operatorname{Spec} \kappa(x) \to X$ for some point $x \in X$ (3.4.2).
Indeed, in case (1) the morphism f can be written as the composition of a closed immersion (which satisfies assumption (I)) followed by an open immersion (satisfying (II)). In case (2) we can choose an open affine neighborhood U of x and f can be written as the composition of $\operatorname{Spec} \mathscr{O}_{X,x} \to U$ (which satisfies (I)) and the open immersion $U \to X$ (satisfying (II)). In case (3) the morphism f satisfies assumption (I).

(4.6) Examples.

PRODUCTS OF AFFINE SPACES.

Let R be a ring, and $\mathbb{A}^n_R = \mathrm{Spec}(R[T_1, \dots, T_n])$ be the affine space over R. For integers $n, m \geq 0$ one has $R[T_1, \dots, T_n] \otimes_R R[T_{n+1}, \dots, T_{n+m}] \cong R[T_1, \dots, T_{n+m}]$ and therefore the description of fiber products for affine schemes (Proposition 4.17) that

(4.6.1)
$$\mathbb{A}^n_R \times_R \mathbb{A}^m_R \cong \mathbb{A}^{n+m}_R.$$

PRODUCTS OF PREVARIETIES.

Let k be an algebraically closed field and let X be a k-scheme of finite type. In Section (3.13) we have shown that attaching to a k-morphism $x \colon \mathrm{Spec}\, k \to X$ its image defines an identification $X_k(k) = X_0$, where $X_0 \subseteq X$ denotes the subspace consisting of the closed points of X. Moreover we have seen that attaching to a scheme X of finite type over k the ringed space $(X_0, \mathcal{O}_{X|X_0})$ defines an equivalence of the category of integral schemes of finite type over k and the category of prevarieties over k. Let us show that this construction is compatible with products.

Lemma 4.22. *Let k be a field and let X and Y be k-schemes (locally) of finite type. Then $X \times_k Y$ is (locally) of finite type over k.*

This follows from a general result on morphisms of finite type (Proposition 10.7). Alternatively we may see this as follows.

Proof. By definition there exist (finite) open coverings $X = \bigcup_i X_i$ and $Y = \bigcup_j Y_j$, where X_i and Y_j are affine k-schemes of finite type. By Corollary 4.19 the $X_i \times_k Y_j$ form a (finite) open cover of $X \times_k Y$. Therefore we may assume that $X = \mathrm{Spec}\, A$ and $Y = \mathrm{Spec}\, B$, where A and B are finitely generated k-algebras. Then $X \times_k Y$ is the spectrum of $A \otimes_k B$ which is again a k-algebra of finite type. $\qquad\square$

The following lemma follows from the more precise results of Proposition 5.51 below.

Lemma 4.23. *Let k be an algebraically closed field and let X and Y be integral k-schemes. Then $X \times_k Y$ is again integral.*

Let k be an algebraically closed field. Let X and Y be integral k-schemes of finite type. Then the lemmas show that $X \times_k Y$ is also an integral k-scheme of finite type. Let X_0 and Y_0 be the prevarieties corresponding to X and Y, respectively, and let Z_0 be the prevariety corresponding to $X \times_k Y$. Then the universal property of the fiber product (4.4.2) shows

$$Z_0 = (X \times_k Y)_k(k) = X_k(k) \times Y_k(k) = X_0 \times Y_0.$$

If we view the category of prevarieties over k as a full subcategory of the category of k-schemes, we deduce that the fiber product of two prevarieties X_0 and Y_0 is again a prevariety Z_0 and that $Z_0 = X_0 \times Y_0$.

Note that this identity is an identity of sets. The projections $Z_0 \to X_0$ and $Z_0 \to Y_0$ are continuous but the topology on Z_0 is usually finer than the product topology of X_0 and Y_0 (see Exercise 4.11).

Remark/Definition 4.24. (Frobenius morphism) Let p be a prime number and let S be a scheme over $\mathbb{F}_p$ (i.e., for every open subset $U \subseteq S$ and every section $f \in \Gamma(U, \mathscr{O}_S)$ we have $pf = 0$). We denote by $\mathrm{Frob}_S \colon S \to S$ the *absolute Frobenius* of S: Frob_S is the identity on the underlying topological spaces and $\mathrm{Frob}_S^\flat$ is the map $x \mapsto x^p$ on $\Gamma(U, \mathscr{O}_S)$ for all open subsets U of S. Note that $p\Gamma(U, \mathscr{O}_S) = 0$ implies that $\mathrm{Frob}_S^\flat \colon \mathscr{O}_S \to \mathscr{O}_S$ is a homomorphism of sheaves of rings.

Now let $f \colon X \to S$ be an S-scheme. Note that Frob_X is in general not an S-morphism. Instead of the absolute Frobenius we therefore introduce a relative variant. Consider the diagram

(4.6.2)

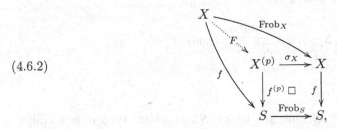

where $X^{(p)}$ is defined by the cartesian square and $F = F_{X/S}$ is the unique morphism making the above diagram commutative. The morphism $F_{X/S}$ is called *the relative Frobenius of X over S*.

We describe this diagram locally: Assume that $S = \operatorname{Spec} R$ and $X = \operatorname{Spec}(A)$ are affine. Via the choice of generators of A as an R-algebra, we can identify A with $R[\underline{T}]/(\underline{f})$ where $\underline{T} = (T_i)_{i \in I}$ is a tuple of indeterminates and $\underline{f} = (f_j)_{j \in J}$ is a tuple of polynomials in $R[\underline{T}]$. Then the diagram (4.6.2) is given by:

(1) $X^{(p)} = \operatorname{Spec}(A^{(p)})$ with $A^{(p)} = R[\underline{T}]/(f_j^{(p)}; j \in J)$, where for any polynomial, say $f = \sum_{\nu \in \mathbb{N}_0^{(I)}} a_\nu \underline{T}^\nu \in R[\underline{T}]$, we set $f^{(p)} = \sum_{\nu \in \mathbb{N}_0^{(I)}} a_\nu^p \underline{T}^\nu$.

(2) The morphism $\sigma_X^* \colon A \to A^{(p)}$ is induced by $R[\underline{T}] \to R[\underline{T}]$, $f \mapsto f^{(p)}$.

(3) The relative Frobenius $F^* = F_{X/S}^*$ is induced by the homomorphism of R-algebras $R[\underline{T}] \to R[\underline{T}]$ which sends an indeterminate T_i to T_i^p.

Base change, Fibers of a morphism

We now study special cases of the fiber product and gain the notion of base change, fibers of a morphism, inverse image of subschemes, or intersections of subschemes.

(4.7) Base change in categories with fiber products.

Let $\mathcal{C}$ be a category in which arbitrary fiber products exist (e.g., the category of schemes), and let $u \colon S' \to S$ be a morphism in $\mathcal{C}$. If $X \to S$ is an S-object, $X \times_S S'$ is an S'-object via the second projection that is sometimes denoted by $u^*(X)$ or by $X_{(S')}$. It is called the *inverse image* or the *base change of X by u*. If $Y \to S$ is a second S-object and $f \colon X \to Y$ an S-morphism, the morphism $f \times_S \mathrm{id}_{S'} \colon X \times_S S' \to Y \times_S S'$ is a morphism of S'-objects that is sometimes denoted by $u^*(f)$ or by $f_{(S')}$ and called the *inverse image* or the *base change of f by u*. We obtain a covariant functor

$$u^* \colon \mathcal{C}/S \to \mathcal{C}/S'$$

from the category of S-objects in $\mathcal{C}$ to the category of S'-objects. This functor is called *base change by u*.

Proposition 4.16 implies that "transitivity of base change" holds: If $u' \colon S'' \to S'$ is another morphism in $\mathcal{C}$, the functors $(u \circ u')^*$ and $(u')^* \circ u^*$ from $\mathcal{C}/S$ to $\mathcal{C}/S''$ are isomorphic.

If $h \colon T \to S'$ is an S'-object, we can consider T as an S-object by composing its structure morphism with u. Let $p \colon X_{(S')} \to X$ be the first projection. We obtain mutually inverse bijections, functorial in T and in X,

(4.7.1)
$$\operatorname{Hom}_{S'}(T, X_{(S')}) \underset{(t,h)_S \mapsfrom t}{\overset{t' \mapsto p \circ t'}{\rightleftarrows}} \operatorname{Hom}_S(T, X).$$

(4.8) Fibers of morphisms.

Let $f \colon X \to S$ be a morphism of schemes and let $s \in S$ be a point. We will now endow the topological fiber $f^{-1}(s)$ with the structure of a scheme.

Definition 4.25. *Let $\operatorname{Spec} \kappa(s) \to S$ be the canonical morphism. Then we call*

$$X_s := X \otimes_S \kappa(s)$$

the fiber of f in s.

Hence X_s is a $\kappa(s)$-scheme (via the second projection). By Proposition 4.20 (applied to S, $X = S$, $X' = \operatorname{Spec} \kappa(s)$, $Y = X$) the underlying topological space of X_s is indeed the subspace $f^{-1}(s)$ of X. In the sequel the notation $f^{-1}(s)$, when understood as a scheme, will always refer to this $\kappa(s)$-scheme.

Thus we have seen that every morphism $f \colon X \to S$ gives rise to a $\kappa(s)$-scheme X_s for every point $s \in S$, in other words, we obtain a family of schemes over fields parameterized by the points of S.

Example 4.26. Let k be an algebraically closed field and set

$$X(k) = \{\, (u, t, s) \in \mathbb{A}^3(k) \ ; \ ut = s \,\}.$$

As $UT - S \in k[U, T, S]$ is irreducible, we may consider $X(k)$ as an affine variety. The associated integral k-scheme of finite type is

$$X = \operatorname{Spec} k[U, T, S]/(UT - S).$$

Let $X \to \mathbb{A}^1_k$ be the projection given on k-valued points by $(u, t, s) \mapsto s$. For each point $s \in \mathbb{A}^1(k)$ the fiber X_s is by definition $X_s = \operatorname{Spec} A_s$, where

$$A_s = k[U, T, S]/(UT - S) \otimes_{k[S]} k[S]/(S - s) = k[U, T]/(UT - s).$$

Note that $UT - s \in k[U, T]$ is irreducible for $s \neq 0$ and reducible for $s = 0$. We see that $X \to \mathbb{A}^1$ defines a family of k-schemes X_s parametrized by $s \in \mathbb{A}^1(k)$ such that X_0 is reducible and X_s is irreducible for all $s \neq 0$.

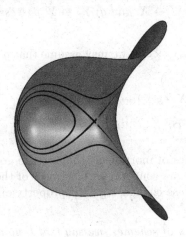

Figure 4.1: The closed subscheme $V(Y^2 - X^2(X+1) - \frac{1}{10}Z) \subset \mathbb{A}^3_\mathbb{R}$. The fibers of the projection $\mathbb{A}^3 \to \mathbb{A}^1$ to the Z-coordinate over $Z = 0$, $Z = 1$ and $Z = -\frac{1}{2}$ are marked.

Example 4.27. For a field k, $a \in k^\times$, we set

$$X := V(Y^2 - X^2(X+1) - aZ) \subset \mathbb{A}^3_k.$$

Let $f\colon X \to \mathbb{A}^1_k = \operatorname{Spec} k[Z]$ be the morphism corresponding to the canonical ring homomorphism $k[Z] \to k[X,Y,Z]/(Y^2 - X^2(X+1) - aZ)$. Let $z \in \mathbb{A}^1(k) = k$, considered as a closed point of $\mathbb{A}^1_k$. We have by definition $X_z = \operatorname{Spec} A_z$, where

$$A_z = k[X,Y]/(Y^2 - X^2(X+1) - az).$$

For $z = 0$ we obtain

$$X_0 = \operatorname{Spec} k[X,Y]/(Y^2 - X^2(X+1))$$

whose $\mathbb{R}$-valued points we have already seen in Chapter 1 (Figure 1.1).

Lemma 4.28. *Let S be a scheme, let $f\colon X \to S$ and $g\colon Y \to S$ be S-schemes and let $p\colon X \times_S Y \to X$ and $q\colon X \times_S Y \to Y$ be the two projections. Let $x \in X$ and $y \in Y$ be points and $\xi\colon \operatorname{Spec} \kappa(x) \to X$ and $\psi\colon \operatorname{Spec} \kappa(y) \to Y$ the canonical morphisms.*
(1) *There exists a point $z \in X \times_S Y$ with $p(z) = x$ and $q(z) = y$ if and only if $f(x) = g(y)$.*
(2) *Assume that the condition in (1) is satisfied and set $s := f(x) = g(y)$. Then*

$$\zeta := \xi \times_S \psi\colon Z := \operatorname{Spec}(\kappa(x) \otimes_{\kappa(s)} \kappa(y)) \to X \times_S Y$$

is a homeomorphism of Z onto the subspace

$$\zeta(Z) = p^{-1}(x) \cap q^{-1}(y).$$

Proof. This follows from the identity $Z = p^{-1}(x) \times_{(X \times_S Y)} q^{-1}(y)$. □

(4.9) Permanence properties of scheme morphisms.

Definition 4.29. *Let $\mathbf{P}$ be a property of morphisms of schemes such that every isomorphism possesses $\mathbf{P}$.*
(1) $\mathbf{P}$ *is called* stable under composition *if for all morphisms $f\colon X \to Y$ and $g\colon Y \to Z$ possessing $\mathbf{P}$, the composition $g \circ f$ also possesses $\mathbf{P}$.*
(2) $\mathbf{P}$ *is called* stable under base change *if for all morphisms $f\colon X \to S$ possessing $\mathbf{P}$ and for all morphisms $S' \to S$, the base change $f_{(S')}\colon X_{(S')} \to S'$ possesses $\mathbf{P}$.*

Proposition 4.30. *Let* **P** *be stable under composition and stable under base change. Then for all schemes* S *and all* S-*morphisms* $f\colon X' \to X$ *and* $g\colon Y' \to Y$ *possessing* **P**, *the fiber product* $f \times_S g$ *also possesses* **P**.

Proof. As $f \times_S g$ is the composition $(f \times_S \mathrm{id}_Y) \circ (\mathrm{id}_{X'} \times_S g)$, we may assume that $g = \mathrm{id}_Y$. We consider X' as an X-scheme via f. Then

$$f \times_S \mathrm{id}_Y \colon X' \times_S Y = X' \times_X (X \times_S Y) \to X \times_S Y$$

is the morphism $f_{(X \times_S Y)}$ and possesses therefore **P**. $\square$

In the category of schemes almost all properties of morphisms that we are going to define will be stable under composition. But there are some properties (most of the time of topological nature) that are not stable under base change, e.g., the property of being injective (see Proposition 4.35 and Exercise 4.16).

Definition 4.31. *If* **P** *is a property of morphisms of schemes, we say that a morphism* $f\colon X \to S$ *of* S-*schemes possesses* **P** *universally if* $f_{(S')}$ *possesses* **P** *for all morphisms* $S' \to S$.

For example, we say that f is *universally injective* if $f_{(S')}$ is injective for all $S' \to S$. In Section (4.10) we will answer the question which of the properties of scheme morphisms defined so far are stable under base change.

Absolute properties of schemes are very often not compatible with base change. E.g., even if k is a field and K and L extension fields of k, $K \otimes_k L$ might be non-noetherian and non-reduced (Exercise 4.18); see also the section in Chapter 5 on extensions of the base field for schemes over a field.

We finally make precise what it means for **P** to be local:

(1) We say that **P** is *local on the target* if for every morphism $f\colon X \to Y$ of schemes and for every open covering $Y = \bigcup_{j \in J} V_j$ the morphism f possesses **P** if and only if $f_{|f^{-1}(V_j)}\colon f^{-1}(V_j) \to V_j$ possesses **P** for all $j \in J$.

(2) We say that **P** is *local on the source* if for every morphism $f\colon X \to Y$ of schemes and for every open covering $X = \bigcup_{i \in I} U_i$ the morphism f possesses **P** if and only if $f_{|U_i}\colon U_i \to Y$ possesses **P** for all $i \in I$.

(4.10) Permanencies of properties of scheme morphisms.

We now study which properties of scheme morphisms introduced so far are stable under composition, base change, local on the target, or local on the source.

Proposition 4.32.

(1) *The following properties of scheme morphisms are stable under composition:* "*injective*", "*surjective*", "*bijective*", "*homeomorphism*", "*open*", "*closed*", "*open immersion*", "*closed immersion*", "*immersion*".

(2) *The following properties of scheme morphisms are stable under base change:* "*surjective*", "*open immersion*", "*closed immersion*", "*immersion*".

(3) *The following properties of scheme morphisms are local on the target:* "*injective*", "*surjective*", "*bijective*", "*homeomorphism*", "*open*", "*closed*", "*open immersion*", "*closed immersion*", "*immersion*".

(4) *The property* "*open*" *is local on the source.*

Proof. The assertions in (1), (3), and (4) are obvious. Proposition 4.20 shows that the properties "open immersion", "closed immersion", and "immersion" are stable under base change (see the discussion at the beginning of Section (4.11)). The property "surjective" is stable under base change by Lemma 4.28. □

The following properties are not stable under base change: "injective", "bijective", and "homeomorphism" (Exercise 4.16), "open" (Exercise 12.23), and "closed" (Exercise 4.21).

Let **P** be a property that is not necessarily stable under base change and let **P'** be the property "universally **P**". Of course, **P'** is stable under base change. Moreover if **P** is stable under composition (resp. local on the target), the same is true for **P'**. In particular, the following corollary follows from Proposition 4.32:

Corollary 4.33. *The following properties of scheme morphisms are stable under composition, stable under base change, and local on the target: "universally injective", "universally bijective", "universal homeomorphism", "universally open", "universally closed".*

Attaching to a scheme X its reduced subscheme X_{red} (Section (3.18)) is compatible with fiber product in the following sense.

Proposition 4.34. *Let S be a scheme and let X and Y be S-schemes. Then the canonical immersions $X_{\mathrm{red}} \to X$ and $Y_{\mathrm{red}} \to Y$ induce an isomorphism*

$$(X_{\mathrm{red}} \times_{S_{\mathrm{red}}} Y_{\mathrm{red}})_{\mathrm{red}} = (X_{\mathrm{red}} \times_S Y_{\mathrm{red}})_{\mathrm{red}} \xrightarrow{\sim} (X \times_S Y)_{\mathrm{red}}.$$

Proof. The equality $X_{\mathrm{red}} \times_{S_{\mathrm{red}}} Y_{\mathrm{red}} = X_{\mathrm{red}} \times_S Y_{\mathrm{red}}$ is clear as $S_{\mathrm{red}} \to S$ is a monomorphism. As $X_{\mathrm{red}} \to X$ and $Y_{\mathrm{red}} \to Y$ are surjective immersions, their fiber product $X_{\mathrm{red}} \times_S Y_{\mathrm{red}} \to X \times_S Y$ is also a surjective immersion (Proposition 4.32) and therefore induces an isomorphism on reduced subschemes. □

Note that the fiber product of reduced schemes (or even the fiber products of spectra of fields) is in general not reduced (Exercise 4.18; see also Proposition 5.49 when this is the case for schemes over a field).

We conclude this subsection with a characterization of universally injective morphisms.

Proposition 4.35. *Let $f\colon X \to Y$ be a morphism of schemes. Then f is universally injective if and only if f is injective and for all $x \in X$ the extension $\kappa(f(x)) \to \kappa(x)$ induced by $f_x^\sharp$ is purely inseparable.*

Proof. Let f be universally injective. Assume that $\kappa(x)$ is not a purely inseparable extension of $\kappa(f(x))$. Then there exist two distinct $\kappa(f(x))$-embeddings of $\kappa(x)$ into an algebraically closed extension K of $\kappa(f(x))$ (Corollary B.102). These embeddings define two distinct Y-morphisms $\mathrm{Spec}\, K \to X$ and hence give rise to two distinct morphisms $\mathrm{Spec}\, K \to X \otimes_Y K$ of K-schemes. Therefore $X \otimes_Y K$ contains at least two points, and the morphism $f_{(K)}\colon X \otimes_Y K \to \mathrm{Spec}\, K$ is not injective.

Now let us prove the converse. First note that f is universally injective if and only if for every morphism $Y' \to Y$ and for every point $y' \in Y'$ with image y in Y the fiber $(X \times_Y Y')_{y'} = X_y \otimes_{\kappa(y)} \mathrm{Spec}\, \kappa(y')$ consists of at most one point. We may assume that X_y is non-empty, so by hypothesis $(X_y)_{\mathrm{red}} = \mathrm{Spec}\, K$, where $K \supset \kappa(y)$ is purely inseparable. We therefore may assume by Proposition 4.34 that $Y = \mathrm{Spec}\, k$ and $X = \mathrm{Spec}\, K$ where $K \supset k$ is purely inseparable and it suffices to show that $\mathrm{Spec}\, K \otimes_k k'$ has only one point for an arbitrary field extension k' of k. This follows from Corollary B.102. □

Sometimes universally injective morphisms are also called *purely inseparable*. The French notion is *morphisme radiciel*.

(4.11) Inverse images and schematic intersections of subschemes.

Let $f\colon X \to Y$ be a morphism of schemes and let $i\colon Z \to Y$ be an immersion. Proposition 4.20 shows that the base change $i_{(X)}\colon Z \times_Y X \to X$ is surjective on stalks and a homeomorphism of $Z \times_Y X$ onto the locally closed subspace $f^{-1}(Z)$ (where we identify $Z = i(Z)$). Therefore $i_{(X)}$ is an immersion. In the sequel we consider $Z \times_Y X$ as a subscheme of X and call it the *inverse image of Z under f*. From now on, $f^{-1}(Z)$, when seen as a scheme, will always mean this subscheme.

Clearly, $f^{-1}(Z)$ is a closed subscheme of X if Z is a closed subscheme of Y. Proposition 4.20 also shows that if Z is an open subscheme of Y, $f^{-1}(Z)$ is an open subscheme of X.

Example 4.36. If $X = \operatorname{Spec} B$ and $Y = \operatorname{Spec} A$ are affine and Z is closed in Y, the morphism f corresponds to a ring homomorphism $\varphi\colon A \to B$ and $Z = V(\mathfrak{a}) = \operatorname{Spec} A/\mathfrak{a}$ for some ideal $\mathfrak{a} \subseteq A$. Then we have an identity of closed subschemes

$$(4.11.1) \qquad\qquad f^{-1}(V(\mathfrak{a})) = V(\varphi(\mathfrak{a})B).$$

As a special case of the inverse image of a subscheme we can define the intersection of two subschemes: Let $i\colon Y \to X$ and $j\colon Z \to X$ be two subschemes. Then we call

$$Y \cap Z := Y \times_X Z = i^{-1}(Z) = j^{-1}(Y)$$

the *(schematic) intersection of Y and Z in X*. The universal property of the fiber product implies the following universal property for $Y \cap Z$: A morphism $T \to X$ factors through $Y \cap Z$ if and only if it factors through Y and through Z.

Example 4.37. If $X = \operatorname{Spec} A$ and $Y = V(\mathfrak{a})$, $Z = V(\mathfrak{b})$, the identity (4.11.1) becomes

$$(4.11.2) \qquad\qquad V(\mathfrak{a}) \cap V(\mathfrak{b}) = V(\mathfrak{a} + \mathfrak{b}).$$

Example 4.38. If R is a ring and if $f_1, \ldots, f_r, g_1, \ldots, g_s \in R[X_0, \ldots, X_n]$ are homogeneous polynomials, we have for the intersection of closed subschemes of $\mathbb{P}_R^n$:

$$V_+(f_1, \ldots, f_r) \cap V_+(g_1, \ldots, g_s) = V_+(f_1, \ldots, f_r, g_1, \ldots, g_s) \subseteq \mathbb{P}_R^n.$$

(4.12) Base change of affine and of projective space.

Affine Space.

Recall that we denote by $\mathbb{A}^n$ the affine space of relative dimension n over $\mathbb{Z}$. For every scheme S we consider $\mathbb{A}_S^n := \mathbb{A}^n \times_{\mathbb{Z}} S$ as an S-scheme via the second projection and call this S-scheme the *affine space of relative dimension n over S*. If $S = \operatorname{Spec} R$ is affine, $\mathbb{A}_S^n = \operatorname{Spec}(\mathbb{Z}[T_1, \ldots, T_n] \otimes_{\mathbb{Z}} R) = \operatorname{Spec} R[T_1, \ldots, T_n]$ is the affine space $\mathbb{A}_R^n$ defined in Example 2.36. For every morphism $S' \to S$ we find

$$(4.12.1) \qquad \mathbb{A}_S^n \times_S S' = \mathbb{A}^n \times_{\mathbb{Z}} S \times_S S' = \mathbb{A}_{S'}^n.$$

For a scheme X "functions on X" are by definition the elements of $\Gamma(X, \mathscr{O}_X)$. We have identifications

$$\Gamma(X, \mathscr{O}_X) = \mathrm{Hom}_{(\mathrm{Ring})}(\mathbb{Z}[T], \Gamma(X, \mathscr{O}_X)) = \mathrm{Hom}_{(\mathrm{Sch})}(X, \mathbb{A}_{\mathbb{Z}}^1),$$

where the first one is given by sending a ring homomorphism $\varphi\colon \mathbb{Z}[T] \to \Gamma(X, \mathscr{O}_X)$ to $\varphi(T)$ and where the second is given by Proposition 3.4. Moreover, if X is an S-scheme we obtain by (4.7.1) an identification

$$(4.12.2) \qquad \Gamma(X, \mathscr{O}_X) = \mathrm{Hom}_S(X, \mathbb{A}_S^1).$$

PROJECTIVE SPACE.

Similar as for the affine space we define $\mathbb{P}_S^n = \mathbb{P}_{\mathbb{Z}}^n \times_{\mathbb{Z}} S$ for every scheme S and call this S-scheme the *projective space of relative dimension n over S*. Then it is easy to see that if $S = \mathrm{Spec}\, R$ is affine, $\mathbb{P}_{\mathrm{Spec}\, R}^n$ is the projective space $\mathbb{P}_R^n$ defined in Section (3.6). For every morphism of schemes $S' \to S$ we have

$$(4.12.3) \qquad \mathbb{P}_{S'}^n \cong \mathbb{P}_S^n \times_S S'.$$

Let R be a ring and let R' be an R-algebra. Let $f_1, \ldots, f_r \in R[T_0, \ldots, T_n]$ be homogeneous polynomials and let $f_1', \ldots, f_r'$ be their images in $R'[T_0, \ldots, T_n]$. Then we have an equality of closed subschemes of $\mathbb{P}_{R'}^n = \mathbb{P}_R^n \otimes_R R'$

$$(4.12.4) \qquad V_+(f_1', \ldots, f_r') = V_+(f_1, \ldots, f_r) \otimes_R R'.$$

(4.13) Morphisms of projective schemes.

In Section (1.25) we have seen that morphisms between projective varieties can be described by homogeneous polynomials. For morphisms between closed subschemes of projective space over an arbitrary ring R a similar description can be given.

Let $\tilde{P}^n = \mathbb{A}_R^{n+1} \setminus \{0\}$ be the complement of the zero section in $\mathbb{A}_R^{n+1}$ (Section (3.6)). Let

$$p = p_n \colon \tilde{P}^n \to \mathbb{P}_R^n$$

be the projection defined in Section (3.6). We choose coordinates $X_0, \ldots, X_n$ on $\mathbb{P}_R^n$. For each i we have

$$(4.13.1) \qquad D(X_i) = p^{-1}(D_+(X_i)) \cong D_+(X_i) \times_R (\mathbb{A}_R^1 \setminus \{0\}).$$

Let $Z \subseteq \mathbb{P}_R^n$ be a closed R-subscheme and set $V_i = D_+(X_i) \cap Z$. Then (4.13.1) shows that

$$(4.13.2) \qquad p^{-1}(V_i) \cong V_i \times_R (\mathbb{A}_R^1 \setminus \{0\}).$$

Let $Z := V_+(I)$ be a closed subscheme of $\mathbb{P}_R^n$, where $I \subseteq R[X_0, \ldots, X_n]$ is a homogeneous ideal. The *affine cone of Z* (defined for varieties in Section (1.21)) is

$$(4.13.3) \qquad C(Z) := \mathrm{Spec}\, R[X_0, \ldots, X_n]/I \subseteq \mathbb{A}_R^{n+1}.$$

Then we have $C(Z) \cap \tilde{P}^n = p^{-1}(Z)$.

Now let $f_0, \ldots, f_m \in R[X_0, \ldots, X_n]$ be homogeneous polynomials of the same degree. They define an R-algebra homomorphism $R[X_0', \ldots, X_m'] \to R[X_0, \ldots, X_n]/I$ by sending X_j' to f_j modulo I. We obtain a morphism of R-schemes $\tilde{f} \colon C(Z) \to \mathbb{A}_R^{m+1}$. This induces for every R-algebra A a map on A-valued points

$$\tilde{f}(A) \colon C(Z)_R(A) = \{\, a = (a_0, \ldots, a_n) \in A^{n+1} \; ; \; \forall g \in I : g(a) = 0 \,\} \to A^{m+1}.$$

By the description of A-valued points of $\tilde{P}^n$ in Example 4.5, the morphism $\tilde{f}$ satisfies $\tilde{f}(C(Z) \cap \tilde{P}^n) \subseteq \tilde{P}^m$ if and only if for all $a = (a_0, \ldots, a_n) \in A^{n+1}$ such that $g(a) = 0$ for all $g \in I$ and such that $a_0, \ldots, a_n$ generate the unit ideal the elements $f_0(a), \ldots, f_m(a) \in A$ generate the unit ideal of A. In this case we obtain a morphism of R-schemes $f \colon Z \to \mathbb{P}_R^m$ that is the unique morphism such that the following diagram commutes

$$
\begin{array}{ccc}
C(Z) \cap \tilde{P}^n & \xrightarrow{\ \tilde{f}\ } & \tilde{P}^m \\
{\scriptstyle p_n}\big\downarrow & & \big\downarrow{\scriptstyle p_m} \\
Z & \xrightarrow{\ f\ } & \mathbb{P}_R^m.
\end{array}
$$

In fact, as above we write $V_i = D_+(X_i) \cap Z$ and identify $p_n^{-1}(V_i) = V_i \times_R (\mathbb{A}_R^1 \setminus \{0\})$. The map $R[T, T^{-1}] \to R$, $T \mapsto 1$, gives us a morphism $\operatorname{Spec} R \to \mathbb{A}_R^1 \setminus \{0\}$ and thus a morphism

$$V_i = V_i \times_R \operatorname{Spec} R \to V_i \times_R (\mathbb{A}_R^1 \setminus \{0\}).$$

Composing this with the morphism

$$V_i \times_R (\mathbb{A}_R^1 \setminus \{0\}) \subseteq C(Z) \cap \tilde{P}^n \xrightarrow{\ \tilde{f}\ } \tilde{P}^m \xrightarrow{\ p_m\ } \mathbb{P}_R^m,$$

we obtain a morphism $V_i \to \mathbb{P}_R^m$. Now one checks that these morphisms for varying i can be glued and give rise to the desired morphism $f \colon Z \to \mathbb{P}_R^m$.

(4.14) Products of projective spaces, Segre embedding.

Let $n, m \geq 1$ be two integers. If R is non-zero, it can be shown that the product $\mathbb{P}_R^n \times_S \mathbb{P}_R^m$ is never isomorphic to $\mathbb{P}_R^N$ for some integer $N \geq 0$ (this is for example an easy corollary of the computation of the Picard groups of these schemes, see Example 11.46; see also Exercise 10.41). But there exists always a natural closed immersion

$$(4.14.1) \qquad\qquad \sigma = \sigma_{n,m} \colon \mathbb{P}_R^n \times_R \mathbb{P}_R^m \hookrightarrow \mathbb{P}_R^{nm+n+m}$$

called the *Segre embedding*. To define σ denote by $\tilde{\sigma} \colon \mathbb{A}_R^{n+1} \times \mathbb{A}_R^{m+1} \to \mathbb{A}_R^{(n+1)(m+1)}$ the morphism of R-schemes that is given on A-valued points by

$$A^{n+1} \times A^{m+1} \ni (x_0, \ldots, x_n, y_0, \ldots, y_m) \mapsto (x_i y_j)_{\substack{0 \leq i \leq n \\ 0 \leq j \leq m}} \in A^{(n+1)(m+1)}.$$

Then $\tilde{\sigma}(\tilde{P}^n \times_R \tilde{P}^m) \subseteq \tilde{P}^{nm+n+m}$ and the Segre embedding σ is the unique morphism of R-schemes that makes the following diagram commutative

$$\begin{array}{ccc}
\tilde{P}^n \times_R \tilde{P}^m & \xrightarrow{\tilde{\sigma}} & \tilde{P}^{nm+n+m} \\
{\scriptstyle p_n \times p_m} \downarrow & & \downarrow {\scriptstyle p_{nm+n+m}} \\
\mathbb{P}^n_R \times_R \mathbb{P}^m_R & \xrightarrow{\sigma} & \mathbb{P}^{nm+n+m}_R .
\end{array}$$

In Remark 8.19 we will describe the Segre embedding on S-valued points.

Proposition 4.39. *The morphism* $\sigma = \sigma_{n,m}$ *is a closed immersion.*

Proof. We choose coordinates X_i, Y_j, and T_{ij} for $i = 0, \ldots, n$, $j = 0, \ldots, m$ on $\mathbb{P}^n_R$, $\mathbb{P}^m_R$, and $\mathbb{P}^{nm+n+m}_R$, respectively. Then we have $\sigma^{-1}(D_+(T_{ij})) = D_+(X_i) \times_R D_+(Y_j)$. As the property of being a closed immersion is local on the target, it suffices to show that the restriction $\sigma^{ij} : D_+(X_i) \times_R D_+(Y_j) \to D_+(T_{ij})$ of σ is a closed immersion. Writing

$$D_+(X_i) = \operatorname{Spec} R\left[\frac{X_{\tilde{i}}}{X_i}, \ 0 \le \tilde{i} \le n\right], \quad D_+(Y_j) = \operatorname{Spec} R\left[\frac{Y_{\tilde{j}}}{Y_j}, \ 0 \le \tilde{j} \le m\right],$$

$$D_+(T_{ij}) = \operatorname{Spec} R\left[\frac{T_{\tilde{i}\tilde{j}}}{T_{ij}}, \ 0 \le \tilde{i} \le n, 0 \le \tilde{j} \le m\right],$$

the morphism σ^{ij} corresponds to the R-algebra homomorphism

$$\frac{T_{\tilde{i}\tilde{j}}}{T_{ij}} \mapsto \frac{X_{\tilde{i}}}{X_i} \otimes \frac{Y_{\tilde{j}}}{Y_j} \in R\left[\frac{X_{\tilde{i}}}{X_i}, \ \tilde{i}\right] \otimes_R R\left[\frac{Y_{\tilde{j}}}{Y_j}, \ \tilde{j}\right].$$

This homomorphism is surjective, as $\frac{X_{\tilde{i}}}{X_i}$ is the image of $\frac{T_{\tilde{i}j}}{T_{ij}}$ and $\frac{Y_{\tilde{j}}}{Y_j}$ is the image of $\frac{T_{i\tilde{j}}}{T_{ij}}$. This shows that σ_{ij} is a closed immersion. $\qquad\square$

Remark 4.40. Bihomogeneous polynomials define closed subschemes of $\mathbb{P}^n_R \times_R \mathbb{P}^m_R$. More precisely, let $M \subset R[X_0, \ldots, X_n, Y_0, \ldots, Y_m]$ be a set of polynomials that are homogeneous in each set of variables $X_0, \ldots, X_n$ and $Y_0, \ldots, Y_m$ and let I be the ideal generated by M. We set

$$U_i = \operatorname{Spec} R\left[\frac{X_0}{X_i}, \ldots, \frac{\widehat{X_i}}{X_i}, \ldots, \frac{X_n}{X_i}\right], \quad W_j = \operatorname{Spec} R\left[\frac{Y_0}{Y_j}, \ldots, \frac{\widehat{Y_j}}{Y_j}, \ldots, \frac{Y_m}{Y_j}\right],$$

for $i = 0, \ldots, n$ and $j = 0, \ldots, m$ such that $(U_i)_i$ (resp. $(W_j)_j$) is an open affine covering of $\mathbb{P}^n_R$ (resp. $\mathbb{P}^m_R$). As in Section (3.7) there exists a (unique) closed subscheme Z of $\mathbb{P}^n_R \times_R \mathbb{P}^m_R$ such that $Z \cap (U_i \times_R W_j) = V(\Phi_{i,j}(I))$ where $\Phi_{i,j}(I)$ is the ideal of dehomogenizations of polynomials in I with respect to the variables X_i and Y_j. With the results of Chapter 13 one can show that all subschemes of $\mathbb{P}^n_R \times_R \mathbb{P}^m_R$ are of this form. We call Z the *vanishing scheme* of M (or of I).

In the same way, closed subschemes of products of any number of projective spaces can be described.

Example 4.41. Let R be a ring, $n \ge 2$ an integer. We will define the closed subscheme C of points in $(\mathbb{P}^n_R)^3$ that are collinear. Collinear points in $\mathbb{P}^n_R$ correspond to lines in $\mathbb{A}^{n+1}_R$ lying in one plane, i.e., lines generated by vectors such that, if we group them together to a matrix, this matrix has rank ≤ 2. We therefore choose coordinates X_{ij} ($i = 0, \ldots, n$) for each copy Y_j of $\mathbb{P}^n_R$ ($j = 1, \ldots, 3$) and let M be the set of 3-minors of the matrix (X_{ij}). Then all elements of M are homogeneous in the X_{ij} for j fixed and we can define C as the vanishing scheme of M.

(4.15) Group schemes.

Let (Grp) be the category of groups and $V\colon$ (Grp) $\to$ (Sets) the forgetful functor. Let S be a scheme and let G be an S-scheme. The following data for G are equivalent by Yoneda's lemma (Section (4.2)).

(i) A factorization of the functor $h_G\colon$ (Sch$/S$)$^{\mathrm{opp}} \to$ (Sets) through the forgetful functor $V\colon$ (Grp) $\to$ (Sets).

(ii) For all S-schemes T the structure of a group on $G_S(T)$ which is functorial in T (i.e., for all S-morphisms $T' \to T$ the associated map $G_S(T) \to G_S(T')$ is a homomorphism of groups).

(iii) Three S-morphisms $m\colon G \times_S G \to G$ (multiplication), $i\colon G \to G$ (inversion), and $e\colon S \to G$ (unit) such that the following diagrams commute.

(4.15.1)

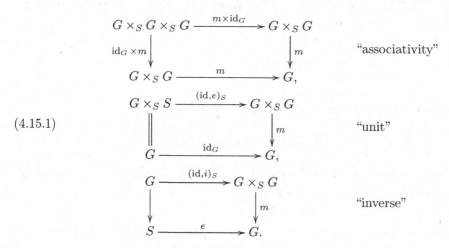

A scheme together with these additional structures is called a *group scheme over* S. As we can glue morphisms (Proposition 3.5) it suffices to give in (ii) functorial group structures on $G_S(R)$, where Spec $R \to S$ is an S-scheme which is affine.

If $f\colon S' \to S$ is a morphism of schemes and (G, m, i, e) is a group scheme over S, then $(G \times_S S', m_{(S')}, i_{(S')}, e_{(S')})$ is a group scheme over S'. For every S'-scheme T we have $(G \times_S S')_{S'}(T) = G_S(T)$, where we consider T as S-scheme via composition with f.

Definition 4.42. *A homomorphism of S-group schemes G and H is a morphism $G \to H$ of S-schemes such that for all S-schemes T the induced map $G(T) \to H(T)$ is a group homomorphism.*

A morphism $f\colon G \to H$ of S-schemes is a homomorphism of group schemes over S if and only if the corresponding morphism $h_G \to h_H$ of functors is a morphism of functors to (Grp). Denoting by m_G, m_H the respective multiplication morphisms, we can also express this condition as $f \circ m_G = m_H \circ (f \times f)$ by the Yoneda lemma.

Example 4.43. The following functors G are group schemes over S.

(1) $S =$ Spec $\mathbb{Z}$ and $G := \mathrm{GL}_n$ with $\mathrm{GL}_n(T) := \mathrm{GL}_n(\Gamma(T, \mathscr{O}_T))$, the group of invertible $(n \times n)$-matrices over $\Gamma(T, \mathscr{O}_T)$, for any scheme T and for a fixed integer $n \geq 1$. The underlying scheme of GL_n is Spec A with $A = \mathbb{Z}[(T_{ij})_{1 \leq i,j \leq n}][\det^{-1}]$, where $\det := \sum_{\sigma \in S_n} \mathrm{sgn}(\sigma) T_{1\sigma(1)} \cdots T_{n\sigma(n)}$ is the determinant of the matrix $(T_{ij})_{i,j}$. This group scheme is called the *general linear group scheme*. We call $\mathbb{G}_m := \mathrm{GL}_1$ the *multiplicative group scheme*.

For an arbitrary scheme S we also define S-group schemes $\mathrm{GL}_{n,S} := \mathrm{GL}_n \times_{\mathbb{Z}} S$ and in particular $\mathbb{G}_{m,S} := \mathbb{G}_m \times_{\mathbb{Z}} S$.

(2) For a group Γ let $G = \underline{\Gamma}_S$ be the associated *constant group scheme*, i.e., $\underline{\Gamma}_S(T)$ is the set of locally constant maps $T \to \Gamma$ for any S-scheme T. The underlying scheme is the disjoint union $\coprod_{\gamma \in \Gamma} S$.

(3) The *additive group scheme* $\mathbb{G}_{a,S}$ over S is defined by $\mathbb{G}_{a,S}(T) = \Gamma(T, \mathscr{O}_T)$ for every S-scheme T. Its underlying S-scheme is $\mathbb{A}^1_S$.

Instead of group schemes one can define *ring schemes* or *schemes of R-algebras* (for R some fixed ring) similarly. Moreover the notion of an action of a group scheme on a scheme is defined in the obvious way:

Definition 4.44. *Let G be an S-group scheme and X be an S-scheme. Then a morphism $a\colon G \times_S X \to X$ of S-schemes is called an* action of G on X *if for all S-schemes T the map $a(T)\colon G(T) \times X(T) \to X(T)$ on T-valued points defines an action of the group $G(T)$ on the set $X(T)$.*

Similarly, we have the notions of subgroup scheme and of kernels:

Definition 4.45.

(1) *Let G be an S-group scheme. A closed subscheme $H \subseteq G$ is called a* subgroup scheme, *if the closed immersion $H \to G$ is a group scheme homomorphism, or equivalently, if for all S-schemes T, $H(T)$ is a subgroup of $G(T)$.*

(2) *Let $f\colon G \to H$ be a homomorphism of S-group schemes. Then the kernel $\operatorname{Ker} f$ of f is the fiber product $G \times_{H,e} S$, where e denotes the unit section of H.*

If $f\colon G \to H$ is a homomorphism of S-group schemes and the unit section e of H is a closed immersion, then $\operatorname{Ker} f$ is a subgroup scheme of G.

Exercises

Exercise 4.1. Let R be a ring, and for every R-algebra A let $\alpha_A\colon A \to A$ be a map of sets such that for every R-algebra homomorphism $\varphi\colon A \to A'$, one has $\varphi \circ \alpha_A = \alpha_{A'} \circ \varphi$. Prove that there exists a polynomial $F \in R[T]$ such that for every A and every $a \in A$, $\alpha_A(a) = F(a)$ in two different ways: using, resp. not using, the Yoneda lemma.

Exercise 4.2$\Diamond$**.** Let S be a scheme and let (X_i) and (Y_j) be two families of S-schemes. Show that

$$\Big(\coprod_i X_i\Big) \times_S \Big(\coprod_j Y_j\Big) = \coprod_{i,j} (X_i \times_S Y_j).$$

Exercise 4.3. Let $\mathcal{C}$ be a category in which a final object and fiber products exist. Show that in $\mathcal{C}$ finite projective limits exist. Deduce that for every scheme S finite projective limits in (Sch/S) exist.

Exercise 4.4$\Diamond$**.** Let $S' \to S$ be a morphism of schemes and let X be an S-scheme considered as a functor $(\mathrm{Sch}/S)^{\mathrm{opp}} \to (\mathrm{Sets})$. Let $X'\colon (\mathrm{Sch}/S')^{\mathrm{opp}} \to (\mathrm{Sets})$ be the restriction of X to $(\mathrm{Sch}/S')^{\mathrm{opp}}$, i.e., $X'(T \to S') := X_S(T \to S' \to S)$. Show that the functor X' is given by the S'-scheme $X \times_S S'$.

Exercise 4.5. Let X be a scheme and let $X = \bigcup_i U_i$ an open covering. Show that for every local ring R we have an equality of R-valued points $X(R) = \bigcup_i U_i(R)$. Give an example where this equality does not hold for a non-local ring R.

Exercise 4.6. Let R be a local ring. Show that $\mathbb{P}^n(R)$ is the set of tuples $(x_0, \ldots, x_n)$ with $x_i \in R$ and some $x_j \in R^\times$ modulo the equivalence relation

$$(x_0, \ldots, x_n) \sim (y_0, \ldots, y_n) \Leftrightarrow \exists \alpha \in R^\times : x_i = \alpha y_i \forall i.$$

Remark: The S-valued points of $\mathbb{P}^n$ for an arbitrary scheme will be determined in Section (8.5).

Exercise 4.7. Let $f \colon X \to S$ and $g \colon Y \to S$ be morphisms in a category $\mathcal{C}$ such that the fiber product $(X \times_S Y, p, q)$ exists. Let $s \colon S \to Y$ be a section of g. Show that $t := (\mathrm{id}_X, s \circ f)_S$ is a section of p and that the following diagram is cartesian

$$\begin{array}{ccc} X & \xrightarrow{\ t\ } & X \times_S Y \\ {\scriptstyle f}\downarrow & & \downarrow{\scriptstyle q} \\ S & \xrightarrow{\ s\ } & Y. \end{array}$$

Exercise 4.8. Let $f \colon X \to X'$ and $g \colon Y \to Y'$ be S-morphisms in a category where fiber products exist. Show that if f and g are monomorphisms, then $f \times_S g$ is a monomorphism.

Exercise 4.9. Let S be an object in a category where fiber products exist. Let X, Y, Z be S-objects, $f \colon X \to Z$, $g \colon Y \to Z$ two S-morphisms. Show that for every morphism $S' \to S$ there is a functorial isomorphism $(X \times_Z Y)_{(S')} = X_{(S')} \times_{Z_{(S')}} Y_{(S')}$.

Exercise 4.10. Let X be a scheme.
(a) Let $f \colon Z \to X$ and $f' \colon Z' \to X$ be (universally) closed morphisms of schemes. Show that the induced morphism $Z \coprod Z' \to X$ is (universally) closed.
(b) Assume that the family $(X_i)_i$ of irreducible components of X is locally finite (e.g., if X is locally noetherian; see Exercise 3.15). Show that the canonical morphism $\coprod_i X_i \to X$ is universally closed and surjective.

Exercise 4.11. For every scheme Z we denote by Z_{top} the underlying topological space of Z and by $Z_0 \subseteq Z_{\mathrm{top}}$ the subspace of closed points. Let S be a scheme and X and Y be two S-schemes. Show that the two projections of $X \times_S Y$ yield a continuous surjection

$$\pi_{\mathrm{top}} \colon (X \times_S Y)_{\mathrm{top}} \to X_{\mathrm{top}} \times_{S_{\mathrm{top}}} Y_{\mathrm{top}},$$

where the right hand side is the fiber product in the category of topological spaces.

Now let $S = \mathrm{Spec}\, k$, where k is a field, and all schemes are assumed to be of finite type over k. For every k-scheme Z we consider $Z_k(k) \subseteq Z_0 \subseteq Z_{\mathrm{top}}$ as subspaces. Show the following assertions.
(a) π_{top} induces a continuous surjective map $\pi_0 \colon (X \times_k Y)_0 \to X_0 \times Y_0$ and a continuous bijective map $\pi(k) \colon (X \times_k Y)_k(k) \to X_k(k) \times Y_k(k)$.
(b) Assume that $X = Y = \mathbb{A}^1_k$. Show that π_0 is bijective if and only if k is separably closed. Show that there is no field k such that π_{top} is bijective or such that π_0 is a homeomorphism. Show that $\pi(k)$ is a homeomorphism if and only if k is finite (in that case $Z_k(k)$ is finite and discrete for any k-scheme Z of finite type).

Exercise 4.12◊. Let S be a scheme and let X and Y be S-schemes. Let $f, g\colon X \to Y$ be two S-morphisms and for all S-schemes T let $f(T)$ and $g(T)$ be the induced maps $X_S(T) \to Y_S(T)$. Show that the following assertions are equivalent.
(i) $f = g$.
(ii) $f(T) = g(T)$ for all S-schemes T.
(iii) $f(X) = g(X)$.
(iv) There exists an open covering $X = \bigcup_i U_i$ such that $f(U_i) = g(U_i)$ for all i.

Exercise 4.13. Let k be a field. Describe the fibers over all points of the following morphisms $\operatorname{Spec} B \to \operatorname{Spec} A$ corresponding in each case to the canonical homomorphism $A \to B$. Which fibers are irreducible or reduced?
(a) $\operatorname{Spec} k[T, U]/(TU - 1) \to \operatorname{Spec} k[T]$.
(b) $\operatorname{Spec} k[T, U]/(T^2 - U^2) \to \operatorname{Spec} k[T]$.
(c) $\operatorname{Spec} k[T, U]/(T^2 + U^2) \to \operatorname{Spec} k[T]$.
(d) $\operatorname{Spec} k[T, U]/(TU) \to \operatorname{Spec} k[T]$.
(e) $\operatorname{Spec} k[T, U, V, W]/((U + T)W, (U + T)(U^3 + U^2 + UV^2 - V^2)) \to \operatorname{Spec} k[T]$.
(f) $\operatorname{Spec} \mathbb{Z}[T] \to \operatorname{Spec} \mathbb{Z}$.
(g) $\operatorname{Spec} \mathbb{Z}[T]/(T^2 + 1) \to \operatorname{Spec} \mathbb{Z}$.
(h) $\operatorname{Spec} \mathbb{C} \to \operatorname{Spec} \mathbb{Z}$.
(i) $\operatorname{Spec} A/\mathfrak{a} \to \operatorname{Spec} A$, where $\mathfrak{a}$ is some ideal of A.

Exercise 4.14. Let k be an algebraically closed field and let $L_1, L_2 \subset \mathbb{P}^n_k$ be two linear subspaces with non-empty intersections. Show that the schematic intersection $L_1 \cap L_2$ is a linear subspace.

Exercise 4.15◊. Let k be a field, $n \geq 1$ an integer and consider in $X = \mathbb{A}^2_k = \operatorname{Spec} k[T, U]$ the closed subschemes $Y_1 := V(T^n + U^n - 1)$ and $Y_2 := V(T + 1)$. Describe the schematic intersection $Z := Y_1 \cap Y_2$. How many points has Z? Is Z irreducible or reduced?

Exercise 4.16◊. Let k be a field, $k \hookrightarrow K$ and $k \hookrightarrow k'$ two field extensions. Set $S = \operatorname{Spec} k$, $X = \operatorname{Spec} K$, and $S' = \operatorname{Spec} k'$. Clearly $f\colon X \to S$ is a homeomorphism. Show that if $K \supset k$ is a finite separable extension of degree n, there exists always a finite separable extension $k' \supset k$ such that $X \times_S S'$ consists of n points. In particular, $f_{(S')}\colon X \times_S S' \to S'$ is not injective if $n > 1$.

Exercise 4.17. Let S be a scheme over $\mathbb{F}_p$. Show that the absolute Frobenius Frob_S is a universal homeomorphism. Let X be an S-scheme. Show that the relative Frobenius $F_{X/S}\colon X \to X^{(p)}$ is a universal homeomorphism.

Exercise 4.18. We keep the notation of Exercise 4.16. Assume that k is a non-perfect field, and that K and k' are perfect closures of k. Show that $A := K \otimes_k k'$ is a ring with a single prime ideal. Show that this prime ideal (equal to the nilradical) is not finitely generated and deduce that A is a non-reduced non-noetherian ring.
Hint: Corollary B.102.

Exercise 4.19. Let k be an algebraically closed field of characteristic $\neq 2$.
(a) Show that the quadric of rank 4 in $\mathbb{P}^3_k$ is isomorphic to $\mathbb{P}^1_k \times_k \mathbb{P}^1_k$.
(b) Show that every quadric in $\mathbb{P}^3$ contains infinitely many lines (i.e., linear subspaces of $\mathbb{P}^3$ of dimension 1).

Exercise 4.20. Let $n, m \geq 0$ be integers. Show that the schemes $\mathbb{P}^n \times \mathbb{P}^m$ and $\mathbb{P}^{n+m}$ are isomorphic if and only if $n = 0$ or $m = 0$.
Hint: Count $(\mathbb{P}^n \times \mathbb{P}^m)(k)$ and $\mathbb{P}^{n+m}(k)$ when k is a finite field.
Remark: See also Example 11.46 for a different proof and Exercise 10.41 for a generalization.

Exercise 4.21. Let k be a field and let $f \colon \mathbb{A}^1_k \to \operatorname{Spec} k$ be the structure morphism. Show that f is closed but that $f_{(\mathbb{A}^1_k)} \colon \mathbb{A}^2_k \to \mathbb{A}^1_k$ is not closed.

Exercise 4.22. Give an example of two group schemes G and H over a field k such that the underlying schemes of G and H are isomorphic but such that G and H are not isomorphic as group schemes.
Hint: Consider the group scheme G such that for all k-algebras R one has

$$G(R) = \{\, (a_{ij}) \in \operatorname{GL}_n(R) \;;\; a_{ij} = 0 \text{ for all } i > j \text{ and } a_{ii} = 1 \text{ for all } i \,\}.$$

5   Schemes over fields

Contents

- Schemes over a field which is not algebraically closed
- Dimension of schemes over a field
- Schemes over fields and extension of the base field
- Intersections of plane curves

A very important special case are schemes that are of finite type over a field. Thus before we progress with the general abstract theory of schemes we focus in this and the next chapter on the case of schemes of finite type over a field (although some of the definitions and results are formulated and proved in greater generality). In fact this is also an important building block for the study of arbitrary morphisms of schemes $f\colon X \to S$ because we have seen how we may attach to each $s \in S$ its fiber $X_s = f^{-1}(s)$ (Section (4.8)). Thus f yields a family of schemes over various fields and we may study f by first studying its fibers and then how these fibers vary.

Schemes over a field which is not algebraically closed

(5.1) Schemes locally of finite type over a field.

In this section let k be an arbitrary field, and let X be a k-scheme locally of finite type (Definition 3.30). A point $x \in X$ is closed if and only if $\kappa(x)$ is a finite extension of k (Proposition 3.33) and the set of closed points is very dense in X (Proposition 3.35). A point $x \in X$ is called *k-rational* if $k \to \kappa(x)$ is an isomorphism. A k-rational point is always closed. Sending a k-morphism Spec $k \to X$ to its image yields a bijection between the set $X(k) = X_k(k)$ of k-valued points of X and the set of k-rational points of X (Proposition 3.8). In the sequel we will often identify k-rational and k-valued points for k-schemes.

If k is algebraically closed, all closed points are k-rational. In other words the very dense subspace of closed points can be identified with the set $X(k)$ of k-valued points of X. If X is of finite type and integral, then X corresponds to a prevariety over k, and we have a good understanding of the topological space X in terms of this prevariety.

For a general field k it is often difficult to decide whether a given k-scheme X has a k-rational point (see Exercise 3.28). The following two examples show that even for quite simple k-schemes the residue fields $\kappa(x)$ might be complicated field extensions of k. Moreover it might happen that there is no k-morphism Spec $k \to X$ (and hence no k-rational point of X).

© Springer Fachmedien Wiesbaden GmbH, part of Springer Nature 2020
U. Görtz und T. Wedhorn, *Algebraic Geometry I: Schemes*, Springer Studium
Mathematik – Master, https://doi.org/10.1007/978-3-658-30733-2_6

Example 5.1. Let $X = \mathbb{A}^1_k = \operatorname{Spec} k[T]$. The points of X are the prime ideals of $k[T]$. As $k[T]$ is a principal ideal domain, the prime ideals are the zero ideal and all principal ideals generated by irreducible polynomials (see Example 2.14). Hence every finite extension field which is generated over k by a single element (in particular every finite separable extension) occurs as a residue class field of a point of X.

Example 5.2. Let $X = \operatorname{Spec} k'$ be the spectrum of a non-trivial extension field k'/k. This gives us an example of a k-scheme which contains no point with residue class field k and which does not admit any k-morphism $\operatorname{Spec} k \to X$. A scheme X of this form is of finite type over k, if and only if the extension k'/k is finite (Lemma 1.10).

(5.2) Points and Galois action.

Let k be a field. Given a k-scheme X and a field extension $k \hookrightarrow K$, we let

$$X(K) := X_k(K) = \operatorname{Hom}_k(\operatorname{Spec} K, X)$$

be the set of *K-valued points of* X (see Section (4.1)). Any K-valued point $x \colon \operatorname{Spec} K \to X$ defines by Proposition 3.8 morphisms $\operatorname{Spec} K \to \operatorname{Spec} \kappa(x) \to X \to \operatorname{Spec} k$ and therefore gives rise to field extensions

$$k \hookrightarrow \kappa(x) \hookrightarrow K.$$

For $K = k$ attaching to $x \colon \operatorname{Spec} k \to X$ its image point in X defines then a bijection

(5.2.1) $X(k) \xrightarrow{\sim} \{\, x \in X \;;\; k \hookrightarrow \kappa(x) \text{ is an isomorphism} \,\}.$

As Example 5.2 shows, $X(k)$ might be empty.

Example 5.3. Let $f_1, \ldots, f_r \in k[X_0, \ldots, X_n]$ be homogeneous polynomials and let $X := V_+(f_1, \ldots, f_r)$. For every field extension $k \hookrightarrow K$ we have

$$X(K) = \{\, x = (x_0 : \ldots : x_n) \in \mathbb{P}^n(K) \;;\; f_1(x) = \cdots = f_r(x) = 0 \,\}$$

Let G be the group of k-automorphisms of K. We obtain an action of G on $X(K)$ by composition of the morphism $x \colon \operatorname{Spec} K \to X$ with $^a\sigma \colon \operatorname{Spec} K \to \operatorname{Spec} K$ for $\sigma \in G$.

Let $X = \operatorname{Spec} A$, where A is a k-algebra. By choosing generators of A as a k-algebra, we can write $A = k[(T_i)_{i \in I}]/((f_j)_{j \in J})$, where $(T_i)_{i \in I}$ is a family of indeterminates, and $(f_j)_{j \in J}$ is a collection of polynomials $f_j \in k[(T_i)_i]$. Then

$$X(K) = \operatorname{Hom}_k(A, K) = \{\, (t_i) \in K^I \;;\; f_j((t_i)_i) = 0 \text{ for all } j \in J \,\},$$

and the group action of G on $X(K)$ defined above is the restriction of the componentwise action of G on K^I. In particular we see that for every subgroup $H \subseteq G$ with fixed field $K^H \subset K$ the set $X(K)^H$ of fix points in $X(K)$ under H coincides with the set $X(K^H)$.

Now consider an arbitrary k-scheme X and an open affine covering $X = \bigcup_\alpha U_\alpha$. Then $X(K) = \bigcup_\alpha U_\alpha(K)$, and $U_\alpha(K)$ is a G-invariant subset of $X(K)$. Therefore again

(5.2.2) $X(K)^H = X(K^H)$

for every subgroup $H \subseteq G$.

In particular, if K is a Galois extension of k, then the set of fix points of the action of $G = \operatorname{Gal}(K/k)$ on $X(K)$ is precisely $X(k)$.

Now let $K = \overline{k}$ be an algebraic closure of k and G be the group of k-automorphisms of $\overline{k}$. For k-schemes locally of finite type we can describe the set of closed points as follows:

Proposition 5.4. *Let X be a k-scheme locally of finite type. Let $\overline{s}$ be the unique point in* $\operatorname{Spec} \overline{k}$. *The map*

$$\alpha\colon X(\overline{k}) \to X, \quad (\overline{x}\colon \operatorname{Spec} \overline{k} \to X) \mapsto \overline{x}(\overline{s})$$

induces a bijection between the set of G-orbits in $X(\overline{k})$ and the set of closed points of X.

Proof. Proposition 3.8 about morphisms from the spectrum of a field into a scheme shows that an element of $X(\overline{k})$ is the same as a point $x \in X$ together with a k-homomorphism $\kappa(x) \to \overline{k}$. Because of Proposition 3.33, x necessarily is a closed point (and for every closed point there exists a corresponding element in $X(\overline{k})$). We obtain a G-equivariant bijection

$$X(\overline{k}) = \{ (x, \iota) \; ; \; x \in X \text{ closed}, \; \iota\colon \kappa(x) \hookrightarrow \overline{k} \},$$

where all the homomorphisms ι are requested to be k-homomorphisms. On the right hand side, G acts via $(x, \iota) \mapsto (x, \sigma \circ \iota)$ for $\sigma \in G$. The proposition follows, since G operates transitively on the set of embeddings $\kappa(x) \hookrightarrow \overline{k}$. $\square$

Dimension of schemes over a field

In this section, we investigate the notion of dimension, where we concentrate on schemes of finite type over a field, which is the situation where this notion works best. A further discussion of the notion of dimension is contained in Chapter 14.

Naively, the dimension of a space should encode the "number of parameters" needed to describe a point of this space. For instance, affine n-space $\mathbb{A}_k^n$ over a field k should certainly have dimension n. If X is a k-scheme of finite type and $X \to \mathbb{A}_k^n$ is a morphism with finite fibers, then X should also have dimension n. Using Noether normalization, one sees that this comes down to defining $\dim X = \operatorname{trdeg}_k K(X)$. This approach leads to a satisfactory theory of dimension for schemes of finite type over a field. For more general schemes, however, we use a different definition, modeled on the fact that the dimension of a vector space is the maximal length of a flag of subspaces. In the context of the Zariski topology, this gives us a viable definition of the dimension of an arbitrary scheme, and we start with this general notion.

(5.3) Definition of dimension.

Definition 5.5. *Let X be a topological space. The* dimension $\dim X$ *of X is the supremum of all lengths of chains*

$$X_0 \supsetneq X_1 \supsetneq \cdots \supsetneq X_l$$

of irreducible closed subsets of X. (The length of a chain as above is l.) If X is a scheme, then its dimension *is by definition the dimension of the underlying topological space. A topological space X is called* equidimensional *(of dimension d), if all irreducible components of X have the same dimension (equal to d).*

So the dimension is $-\infty$ (if and only if $X = \emptyset$), a non-negative integer, or ∞.

For an affine scheme $X = \operatorname{Spec} A$, we have an inclusion reversing bijection between irreducible closed subsets of X and prime ideals of A (Corollary 2.7). Thus we have $\dim X = \dim A$, where

$$\dim A := \sup\{\, l \in \mathbb{N}_0 \; ; \; \exists \mathfrak{p}_0 \subsetneq \mathfrak{p}_1 \subsetneq \cdots \subsetneq \mathfrak{p}_l \text{ chain of prime ideals of } A \,\}$$

is called the *Krull dimension* or simply the *dimension* of the ring A.

Example 5.6. If k is a field, then $\dim \operatorname{Spec} k = 0$. If A is a principal ideal domain (but not a field), then $\dim \operatorname{Spec} A = 1$. In particular, we have $\dim \mathbb{A}_k^1 = 1$ for any field k. If A is any ring and $\mathfrak{p}_0 \subsetneq \cdots \subsetneq \mathfrak{p}_r$ is a chain of prime ideals in A, we obtain a chain of prime ideals in $A[T]$ by $\mathfrak{p}_0 A[T] \subsetneq \cdots \subsetneq \mathfrak{p}_r A[T] \subsetneq \mathfrak{p}_r + (T)$. Therefore we see that

$$(5.3.1) \qquad\qquad\qquad \dim A[T] \geq 1 + \dim A.$$

In Theorem 14.100 we will see that whenever A is a noetherian ring, then $\dim \mathbb{A}_A^n = n + \dim A$. (This statement is not true for an arbitrary ring!) If $A = k$ is a field, we will show in Corollary 5.18 that $\dim \mathbb{A}_k^n = n$ for all n.

We add a word of warning, however. Even for noetherian schemes the notion of dimension is sometimes quite counter-intuitive (see Exercise 5.3, for instance). If one restricts oneself to the case of schemes of finite type over a field, then the theory of dimension works mostly as expected, and is a very useful invariant. We first discuss some simple observations in the general case.

Lemma 5.7. *Let X be a topological space.*
(1) *Let Y be a subspace of X. Then $\dim Y \leq \dim X$. If X is irreducible, $\dim X < \infty$, and $Y \subsetneq X$ is a proper closed subset, then $\dim Y < \dim X$.*
(2) *Let $X = \bigcup_\alpha U_\alpha$ be an open covering. Then*

$$\dim X = \sup_\alpha \dim U_\alpha.$$

(3) *Let I be the set of irreducible components of X. Then*

$$\dim X = \sup_{Y \in I} \dim Y.$$

(4) *Let X be a scheme. Then*
$$\dim X = \sup_{x \in X} \dim \mathcal{O}_{X,x}.$$

Proof. *(1).* If Z is an irreducible closed subset of Y its closure $\bar{Z}$ in X is irreducible and $\bar{Z} \cap Y = Z$. This shows the first assertion of (1). The second follows as every chain inside Y can be enlarged to a chain inside X by adding X.

(2). Let $Z_0 \supsetneq Z_1 \supsetneq \cdots \supsetneq Z_l$ be a chain of irreducible closed subsets of X. Then there exists U_α such that $U_\alpha \cap Z_l \neq \emptyset$. Then $\emptyset \neq U_\alpha \cap Z_i$ is open in Z_i and hence irreducible and closed in U_α. Moreover the closure of $U_\alpha \cap Z_i$ in X is Z_i and thus the $Z_i \cap U_\alpha$ form a chain of length l in U_α. This shows that $\sup_\alpha \dim U_\alpha \geq \dim X$. The converse inequality follows from (1).

Assertion (3) is clear and for the last assertion we may assume that $X = \operatorname{Spec} A$ is affine by (2). But in this case $\dim \mathcal{O}_{X,x}$ is the supremum of the length of chains of prime ideals of A ending in $\mathfrak{p}_x$. $\qquad\square$

Assertion (2) shows that for many questions concerning the dimension, we can restrict to the case of affine schemes and by (3) to the case of irreducible affine schemes. Furthermore the dimension of a scheme depends only on the underlying topological space; in particular, we have $\dim X = \dim X_{\mathrm{red}}$. Thus often we may restrict to the case that X is an affine integral scheme.

The second part of Assertion (1) has the following consequence.

Corollary 5.8. *Let* $i \colon Y \to X$ *be a closed immersion of schemes, where* X *is integral. If* $\dim X = \dim Y < \infty$*, then* i *is an isomorphism.*

For a general morphism of schemes $f \colon X \to Y$ one does not have $\dim X \geq \dim f(X)$ (see Exercise 5.3). However we have:

Proposition 5.9. *Let* $f \colon X \to Y$ *be an open morphism of schemes. Then we have* $\dim X \geq \dim f(X)$.

There is a similar – but simpler – result for closed morphisms of schemes; see Exercise 5.6.

Proof. We may replace Y by the open subscheme $f(X)$ and therefore assume that f is surjective. Then it suffices to show that if $(y_i)_{0 \leq i \leq n}$ is a sequence of points in Y such that $y_{i-1} \in \overline{\{y_i\}}$ for all $i = 1, \ldots, n$, then there exists a sequence $(x_i)_{0 \leq i \leq n}$ of points in X such that $x_{i-1} \in \overline{\{x_i\}}$ for all $i = 1, \ldots, n$ and $f(x_i) = y_i$ for all i. This follows by induction from the following lemma. $\qquad\square$

Lemma 5.10. *Let* $f \colon X \to Y$ *be an open morphism of schemes. For every point* $x \in X$ *and every generization* y' *of* $y := f(x)$ *there exists a generization* $x' \in X$ *of* x *with* $f(x') = y'$.

Proof. We may assume that $X = \operatorname{Spec} B$ and $Y = \operatorname{Spec} A$ are affine. Then the set Z of all generizations of x is $\operatorname{Spec} \mathscr{O}_{X,x} = \bigcap_t D(t)$, where t runs through $B \setminus \mathfrak{p}_x$. As $D(t)$ is an open neighborhood of x, its image $f(D(t))$ is an open neighborhood of y and hence contains y'. We find that, setting $f_t := f_{|D(t)}$, the fiber $f_t^{-1}(y')$ is non-empty for all t.

We now assume that $y' \notin f(Z)$. If g denotes the composition $\operatorname{Spec} \mathscr{O}_{X,x} \to X \to Y$, we therefore have $g^{-1}(y') = \operatorname{Spec}(\mathscr{O}_{X,x} \otimes_A \kappa(y')) = \emptyset$ and hence

$$\mathscr{O}_{X,x} \otimes_A \kappa(y') = \varinjlim_{t \in B \setminus \mathfrak{p}_x} (B_t \otimes_A \kappa(y')) = 0.$$

Therefore $0 = 1$ in the limit, and hence in some $B_t \otimes_A \kappa(y')$. But then we obtain that $f_t^{-1}(y') = \operatorname{Spec}(B_t \otimes_A \kappa(y')) = \emptyset$; contradiction. $\qquad\square$

(5.4) Dimension 0.

Let us describe more precisely what it means for a locally noetherian scheme to have dimension 0. Note that a ring has dimension 0 if and only if every prime ideal is maximal (or equivalently: if every prime ideal is minimal). A noetherian ring is of dimension 0 if any only if it is Artinian, see Proposition B.36. We translate the characterizations of Artinian rings into statements about schemes.

Proposition 5.11. *Let X be a locally noetherian scheme. The following are equivalent:*

(i) $\dim X = 0$

(ii) *The topological space of X carries the discrete topology.*

(iii) *All local rings of X are local Artin rings.*

(iv) *The natural morphism*

$$\coprod_{x \in X} \operatorname{Spec} \mathcal{O}_{X,x} \to X$$

is an isomorphism.

Proof. All properties propagate to open subsets, and can be checked on an open covering. Therefore we may assume that X is noetherian and affine, and the proposition follows from Proposition B.36. □

(5.5) Integral morphisms of affine schemes.

Recall that a ring homomorphism $A \to B$ is called *integral* if every element of B is the zero of a monic polynomial with coefficients in A. The following proposition is a geometric version of Theorem B.56.

Proposition 5.12. *Let $X = \operatorname{Spec} B$ and $Y = \operatorname{Spec} A$ be affine schemes, let $\varphi \colon A \to B$ be an integral ring homomorphism, and let $f \colon X \to Y$ be the corresponding scheme morphism. If $Z = V(\mathfrak{b}) \subseteq X$ is a closed subspace, where $\mathfrak{b} \subseteq B$ is an ideal, then*

$$(5.5.1) \qquad\qquad f(Z) = V(\varphi^{-1}(\mathfrak{b})).$$

In particular, f is closed. Moreover:

(1) *One has $\dim f(Z) = \dim Z$.*

(2) *If φ is in addition injective, then f is surjective.*

Proof. By Proposition 2.10 (2) the image $f(V(\mathfrak{b}))$ is dense in $V(\varphi^{-1}(\mathfrak{b}))$. Replacing A by $A/\varphi^{-1}(\mathfrak{b})$ and B by $B/\mathfrak{b}$ it suffices to show that f is surjective and $\dim X = \dim Y$ if φ is integral and injective. This follows from Theorem B.56: The surjectivity follows from Going Up (2). Going Up also shows that $\dim B \geq \dim A$, and Assertion (1) of loc. cit. implies that $\dim B \leq \dim A$. □

Theorem 5.13. *Let $\varphi \colon A \to B$ be an integral injective ring homomorphism of integral domains, and set $K = \operatorname{Frac}(A)$ and $L = \operatorname{Frac}(B)$. Let $f = {}^a\varphi \colon \operatorname{Spec} B \to \operatorname{Spec} A$ be the associated morphism. Assume that L is a finite extension of K (e.g., if φ is finite) and that A is integrally closed.*

Then the norm map $\mathrm{N}_{L/K} \colon L \to K$ satisfies $\mathrm{N}_{L/K}(B) \subseteq A$. For $b \in B$ we have

$$f(V(b)) = V(\mathrm{N}_{L/K}(b))$$

(equality of sets) and $\dim V(b) = \dim V(\mathrm{N}_{L/K}(b))$.

Proof. Let $b \in B$, and let us show that $a := \mathrm{N}_{L/K}(b)$ is integral over A (and hence lies in A). By assumption b is integral over A, so it is a zero of a monic polynomial $P \in A[T]$. Clearly, the minimal polynomial $\mathrm{minpol}_{L/K,b}$ of b with respect to the field extension L/K divides P, so all zeros (in an algebraic closure of L) of the minimal polynomial are integral over A. But then the same is true for the coefficients, which are polynomial expressions in the zeros, and in particular for the norm a, which is (up to sign) a power of the absolute coefficient of $\mathrm{minpol}_{L/K,b}$.

To complete the proof, it is enough to show that $\mathrm{rad}(bB) \cap A = \mathrm{rad}(aA)$ by Proposition 5.12. Writing $\mathrm{minpol}_{L/K,b} = \sum_{i=0}^{d} a_i T^i$ with $a_d = 1$, we find $a \in a_0 A$. From $a_0 = -\sum_{i>0} a_i b^i$ we see that $a \in bB$. Conversely, if $s \in A$ and $s^n \in bB$, say $s^n = bt$ for some $t \in B$, then

$$s^{n[L:K]} = \mathrm{N}_{L/K}(s^n) = \mathrm{N}_{L/K}(b)\mathrm{N}_{L/K}(t) \in aA. \qquad \square$$

Lemma 5.14. *Let $A \to B$ be a finite ring homomorphism. Then all fibers of the morphism $\mathrm{Spec}\, B \to \mathrm{Spec}\, A$ are finite (as sets).*

Proof. Let $\mathfrak{p} \subset A$ be a prime ideal. We must show that the ring $B \otimes_A \kappa(\mathfrak{p})$ has only finitely many prime ideals. However, this ring is a finite-dimensional $\kappa(\mathfrak{p})$-vector space and thus it is Artinian (every ideal is also a $\kappa(\mathfrak{p})$-subvector space, and every descending chain of subvector spaces of a finite-dimensional vector space becomes stationary) and therefore its spectrum has only finitely many points by Proposition B.36. Alternatively, one can use Proposition 5.11 and Proposition 5.12. $\qquad \square$

(5.6) Dimensions of schemes of finite type over a field.

We fix a field k. We start by recalling a refined version of Noether's normalization theorem (Theorem B.58).

Theorem 5.15. *Let $A \neq 0$ be a finitely generated k-algebra.*
(1) *There exist $t_1, \ldots, t_d \in A$ such that the corresponding k-algebra homomorphism $\varphi \colon k[T_1, \ldots, T_d] \to A$, $T_i \mapsto t_i$, is injective and finite.*
(2) *If $\mathfrak{a}_0 \subseteq \mathfrak{a}_1 \subseteq \cdots \subseteq \mathfrak{a}_r \subsetneq A$ is a chain of ideals in A $(r \geq 0)$, then the t_i in (1) can be chosen such that $\varphi^{-1}(\mathfrak{a}_i) = (T_1, \ldots, T_{h(i)})$ for all $i = 0, \ldots, r$ and suitable $0 \leq h(0) \leq h(1) \leq \cdots \leq h(r) \leq d$.*

Remark 5.16. In more geometric terms, this means the following: Given an affine scheme X of finite type over k, we find a morphism $f \colon X \to \mathbb{A}_k^d$ of k-schemes such that the following assertions hold.
(1) The associated k-algebra homomorphism is finite and injective. In particular, by Proposition 5.12 and Lemma 5.14, f is closed, surjective, and has finite fibers.
(2) Furthermore, given a chain $Z_r \subseteq \cdots \subseteq Z_0$ of closed subschemes of X, we may arrange f such that each Z_i is mapped onto a coordinate hyperplane $V(T_1, \ldots, T_{h(i)})$ in $\mathbb{A}_k^d$.
(3) If $Z_r \subsetneq \cdots \subsetneq Z_0$ is a chain of integral closed subschemes (i.e., $Z_i = V(\mathfrak{a}_i)$ for some *prime* ideal $\mathfrak{a}_i \subset A$), Theorem B.56 (1) shows that we automatically have $h(0) < h(1) < \cdots < h(r)$ for the numbers $h(i)$ obtained from part (2) of the theorem.

Corollary 5.17. *Let $A \neq 0$ be a finitely generated k-algebra, and $d \geq 0$ be an integer. Then $\dim A = d$ if and only if there exists a finite injective k-algebra homomorphism $k[T_1, \ldots, T_d] \hookrightarrow A$.*

Proof. Noether normalization yields a finite injective homomorphism $k[T_1, \ldots, T_d] \hookrightarrow A$. By Proposition 5.12 we have $\dim A = \dim k[T_1, \ldots, T_d]$. By Remark 5.16 (3) we have $\dim A \leq d$ and by (5.3.1) we have $\dim k[T_1, \ldots, T_d] \geq d$. $\qquad\square$

Corollary 5.18. *Let $n \geq 0$ be an integer. Then $\dim \mathbb{A}_k^n = \dim \mathbb{P}_k^n = n$.*

Proof. The equality $\dim \mathbb{A}_k^n = n$ follows immediately from Corollary 5.17. Moreover, as projective space $\mathbb{P}_k^n$ admits an open covering by copies of $\mathbb{A}_k^n$, we have $\dim \mathbb{P}_k^n = n$ by Lemma 5.7 (2). $\qquad\square$

If A is a ring, we call a chain of prime ideals in A *maximal* if it does not admit a refinement. Similarly, we call a chain of closed irreducible subsets of a topological space *maximal*, if it is maximal with respect to refinement.

Theorem 5.19. *Let A be a finitely generated algebra over a field k and set $d = \dim A$. Assume that A is an integral domain. Let $\mathfrak{q}_{h(1)} \subsetneq \cdots \subsetneq \mathfrak{q}_{h(r)}$ be a chain of prime ideals of A such that $\dim V(\mathfrak{q}_{h(i)}) = d - h(i)$.*
(1) There exists a finite injective k-algebra homomorphism $\varphi \colon k[T_1, \ldots, T_d] \to A$ such that $\varphi^{-1}(\mathfrak{q}_{h(i)}) = (T_1, \ldots, T_{h(i)})$ for all $i = 1, \ldots, r$.
(2) For every homomorphism φ as in (1) the chain $(\mathfrak{q}_{h(i)})_i$ can be completed to a chain of prime ideals $\mathfrak{q}_0 \subsetneq \cdots \subsetneq \mathfrak{q}_d$ of A such that $\varphi^{-1}(\mathfrak{q}_j) = (T_1, \ldots, T_j)$ for all $j = 1, \ldots, d$.
In particular, any chain of prime ideals in A can be completed to a maximal chain of prime ideals and all maximal chains have the same length.

Proof. We apply Theorem 5.15 to A and the given chain of prime ideals, and denote by $\varphi \colon k[T_1, \ldots, T_{d'}] \to A$ the resulting homomorphism. By Corollary 5.17 we have $d' = d$. We set $\mathfrak{b}_l = (T_1, \ldots, T_l) \subset k[T_1, \ldots, T_d]$ and thus have $\varphi^{-1}(\mathfrak{q}_{h(i)}) = \mathfrak{b}_{h'(i)}$ for some $0 \leq h'(i) \leq d$. It follows that $h(i) = h'(i)$ because by Proposition 5.12 we have

$$\dim V(\mathfrak{q}_{h(i)}) = \dim V(\mathfrak{b}_{h'(i)}) = \dim k[T_1, \ldots, T_d]/\mathfrak{b}_{h'(i)} = d - h'(i).$$

We now prove (2). Let i be an index such that $h(i+1) > h(i) + 1$. We have to find prime ideals $\mathfrak{q}_j$ for $h(i) < j < h(i+1)$ with $\mathfrak{q}_{h(i)} \subset \mathfrak{q}_{h(i)+1} \subset \cdots \subset \mathfrak{q}_{h(i+1)}$ and $\varphi^{-1}(\mathfrak{q}_j) = \mathfrak{b}_j$. Replacing $k[T_1, \ldots, T_d]$ by $k[T_1, \ldots, T_d]/\mathfrak{b}_{h(i)}$ (this is again a polynomial ring) and A by $A/\mathfrak{q}_{h(i)}$ we may assume that $\mathfrak{q}_{h(i)} = 0$. But then Theorem B.56 (3) shows the existence of the $\mathfrak{q}_j$. $\qquad\square$

Schemes of finite type over a field k of dimension 0 are particularly simple:

Proposition 5.20. *Let X be a non-empty k-scheme of finite type. The following are equivalent:*
(i) $\dim X = 0$.
(ii) The scheme X is affine, the k-vector space $\Gamma(X, \mathcal{O}_X)$ is finite-dimensional, and $\Gamma(X, \mathcal{O}_X) = \prod_x \mathcal{O}_{X,x}$.
(iii) The underlying topological space of X is discrete.
(iv) The underlying topological space of X has only finitely many points.

Proof. If $\dim X = 0$, then (iii) holds by Proposition 5.11, and (iii) implies (iv) because X is quasi-compact. We show that (iv) implies that all points of X are closed (which implies (i)). Let U be the set of points $x \in X$ that are not closed. As X is finite, this is an open subset of X. As the closed points of X are very dense (Proposition 3.35), U has to be empty.

If (ii) holds, X is the spectrum of an Artinian ring and therefore the underlying topological space of X is finite and discrete. Conversely, we show that (i), (iii), and (iv) imply (ii). If X is finite and discrete, X is clearly affine, say $X = \operatorname{Spec} A$, and $A = \prod_x \mathscr{O}_{X,x}$. Corollary 5.17 shows that the structure morphism $X \to \operatorname{Spec} k$ corresponds to a finite homomorphism $k \to A$. $\qquad\square$

Corollary 5.21. *Let X be an integral k-scheme of finite type, such that $\dim X = 0$. Then $X \cong \operatorname{Spec} k'$, where k'/k is a finite field extension.*

The central result on the dimension of schemes over a field is the following theorem.

Theorem 5.22. *Let X be an irreducible k-scheme locally of finite type with generic point η.*
(1) $\dim X = \operatorname{trdeg}_k \kappa(\eta)$.
(2) *Let $x \in X$ be any closed point. Then $\dim \mathscr{O}_{X,x} = \dim X$.*
(3) *Let $f: Y \to X$ be a morphism of k-schemes of finite type such that $f(Y)$ contains the generic point η of X. Then $\dim Y \geq \dim X$. In particular we have $\dim U = \dim X$ for any non-empty open subscheme U of X.*
(4) *Let $f: Y \to X$ be a morphism of k-schemes of finite type with finite fibers. Then $\dim Y \leq \dim X$.*

If X is integral, then $\kappa(\eta)$ is simply the function field of X.

A theorem of Chevalley (see Theorem 10.20) will show that under the hypotheses of the proposition the property that η is contained in $f(Y)$ in (3) is equivalent to the property that $f(Y)$ is dense in X (one can also use Exercise 10.1 or Exercise 3.12 if Y is irreducible).

A morphism f as in (4) is called *quasi-finite*. This notion will be discussed more thoroughly in Section (12.4).

Proof. *(1).* We may assume that X is reduced, and covering X by non-empty open affine subschemes U we may assume that $X = \operatorname{Spec} A$, where A is an integral finitely generated k-algebra. Then we have $\kappa(\eta) = \operatorname{Frac}(A)$. Let $\varphi\colon k[T_1, \ldots, T_d] \to A$ be a finite injective homomorphism as in Corollary 5.17 with $d = \dim A$. Then $\operatorname{Frac}(A)$ is a finite extension of $K := k(T_1, \ldots, T_d)$ and we have $\operatorname{trdeg}_k(\operatorname{Frac}(A)) = \operatorname{trdeg}_k(K) = d$.

(3). By hypothesis there exists $\theta \in Y$ such that $f(\theta) = \eta$. Therefore f induces a k-embedding $\kappa(\eta) \hookrightarrow \kappa(\theta)$. Denote by Z the closure of θ. Then

$$\dim X = \operatorname{trdeg} \kappa(\eta) \leq \operatorname{trdeg} \kappa(\theta) = \dim Z \leq \dim Y.$$

(2). By (3) we may replace X by an open affine neighborhood U of x in X_{red}. Thus again we may assume that $X = \operatorname{Spec} A$, where A is an integral finitely generated k-algebra. Then x corresponds to a maximal ideal $\mathfrak{p}_x$ of A and $\dim(\mathscr{O}_{X,x})$ is the supremum of lengths of chains of prime ideals of A that end in $\mathfrak{p}_x$. But the chain consisting of the single prime ideal $\mathfrak{p}_x$ may be completed to a maximal chain of length $\dim A$ by Theorem 5.19. This proves (2).

(4). Let Z be an irreducible component of Y with generic point θ and set $x := f(\theta)$. We will show that $\mathrm{trdeg}_k \kappa(\theta) \leq \dim X$. Replacing X by an open affine neighborhood U of x and Y by an open affine neighborhood of θ in $f^{-1}(U)$ we may assume that $X = \mathrm{Spec}\, A$ and $Y = \mathrm{Spec}\, B$ are affine. Then B is a k-algebra of finite type and in particular an A-algebra of finite type. The fiber $f^{-1}(x) = \mathrm{Spec}(B \otimes_A \kappa(x))$ is thus a $\kappa(x)$-scheme of finite type with only finitely many points. By Proposition 5.20 the point θ is closed in $f^{-1}(x)$ and therefore $\kappa(\theta)$ is a finite extension of $\kappa(x)$. This shows $\mathrm{trdeg}_k \kappa(\theta) = \mathrm{trdeg}_k \kappa(x) = \dim \overline{\{x\}} \leq \dim X$. $\square$

Corollary 5.23. *Let X be a k-scheme locally of finite type and let $x \in X$ be a closed point. Then $\dim \mathcal{O}_{X,x} = \sup_Z \dim Z$, where Z runs through the (finitely many) irreducible components of X containing x.*

(5.7) Local dimension at a point.

Definition 5.24. *Let X be a topological space and $x \in X$. The* dimension of X at x *is*

$$\dim_x X = \inf_U \dim U,$$

where U runs through all open neighborhoods of x.

Lemma 5.25. *Let X be a topological space.*
(1) Let U be an open neighborhood of x. Then $\dim_x U = \dim_x X$.
(2) One has $\dim X = \sup_{x \in X} \dim_x X$. If X is a quasi-compact scheme and F is the set of closed points in X, then $\dim X = \sup_{x \in F} \dim_x X$.
(3) Let n be an integer. Then $\{\, x \in X \ ;\ \dim_x X \leq n \,\}$ is open in X.

Proof. Recall that we have $\dim Y \leq \dim X$ for every subspace Y of X (Lemma 5.7). This implies (1) and the inequality $\sup_{x \in X} \dim_x X \leq \dim X$ in (2). Let $X_0 \supsetneq \cdots \supsetneq X_l$ be a chain of closed irreducible subsets of X and choose $x \in X_l$. If U is an open neighborhood of x, then $(U \cap X_i)_{0 \leq i \leq l}$ is a chain of irreducible subsets which are closed in U and which are pairwise different because the closure of $U \cap X_i$ in X is X_i. Thus $\dim U \geq l$ which shows $\sup_{x \in X} \dim_x X \geq \dim X$. The second assertions follows because if X is quasi-compact, X_l is also quasi-compact and thus contains a closed point x.

Let $x \in X$ with $\dim_x X = n$. Let U be an open neighborhood of x such that $\dim U = n$. Then for every $y \in U$ we have $\dim_y X = \dim_y U \leq n$ by (1) and (2). This proves (3). $\square$

Proposition 5.26. *Let X be a scheme locally of finite type over a field and let I be the (finite) set of irreducible components of X containing x. Then $\dim_x X = \sup_{Z \in I} \dim Z$. If $x \in X$ is a closed point, then $\dim_x X = \dim \mathcal{O}_{X,x}$.*

Proof. As the set of irreducible components of X is locally finite, $\dim_x X = \inf_U \dim U$ where U runs through those open neighborhoods of x which meet precisely the irreducible components in I. But then $\dim U = \sup_{Z \in I} \dim(Z \cap U) = \sup_{Z \in I} \dim Z$, where the first equality follows from Lemma 5.7 (3) and the second equality from Theorem 5.22 (3). The last assertion follows then from Corollary 5.23. $\square$

(5.8) Codimension of closed subschemes.

Definition 5.27. *Let X be a topological space.*
(1) *Let $Z \subseteq X$ be a closed irreducible subset. The codimension $\mathrm{codim}_X Z$ of Z in X is the supremum of the lengths of chains of irreducible closed subsets $Z_0 \supsetneq Z_1 \supsetneq \cdots \supsetneq Z_l$ such that $Z_l = Z$.*
(2) *Let $Z \subseteq X$ be a closed subset. We say that Z is equi-codimensional (of codimension r), if all irreducible components of Z have the same codimension in X (equal to r).*

If $X = \mathrm{Spec}\, A$ and $Z = V(\mathfrak{p})$ for some prime ideal $\mathfrak{p} \subset A$, the codimension of Z is also called the *height of $\mathfrak{p}$*. It is the supremum of the lengths of chains of prime ideals of A that have $\mathfrak{p}$ as its maximal element. This implies easily that for an arbitrary scheme X and a closed irreducible subset Z with generic point η we have

$$(5.8.1) \qquad \mathrm{codim}_X Z = \dim \mathscr{O}_{X,\eta} = \inf_{z \in Z} \dim \mathscr{O}_{X,z}.$$

This shows that the following definition agrees with the definition given above if Y is closed and irreducible.

Definition 5.28. *Let X be a scheme and let $Y \subseteq X$ be an arbitrary subset. Then*

$$\mathrm{codim}_X(Y) := \inf_{y \in Y} \dim \mathscr{O}_{X,y}$$

is called the codimension of Y in X.

Remark 5.29. Let X be a scheme.
(1) If Y is a closed subset of X, we find

$$\mathrm{codim}_X Y = \inf_Z \mathrm{codim}_X Z,$$

where Z runs through the set of irreducible components of Y.
(2) A closed subset Y of X is of codimension 0 if and only if Y contains an irreducible component of X.

If $Z \subseteq X$ is closed irreducible, we clearly have $\dim Z + \mathrm{codim}_X Z \leq \dim X$. But in general (even for irreducible noetherian schemes) it may happen that this inequality is strict (see Exercise 5.7). The situation is better if X is of finite type over a field k:

Proposition 5.30. *Let X be an irreducible scheme of finite type over a field k. Set $d := \dim X$.*
(1) *All maximal chains of closed irreducible subsets of X have the same length.*
(2) *For all closed subsets Y of X we have*

$$\dim Y + \mathrm{codim}_X Y = \dim X.$$

Proof. (1). If $Z_r \subsetneq \cdots \subsetneq Z_0$ is a maximal chain, then $Z_r = \{x\}$ for some closed point $x \in X$ and we have $r = \dim \mathscr{O}_{X,x}$. Therefore (1) follows from Theorem 5.22 (2).

(2). We first assume that Y is irreducible. Then $\dim Y + \mathrm{codim}_X Y$ is the supremum of the lengths of maximal chains of closed irreducible subsets of X having Y as a member. Thus the claim follows from (1).

In general let Y_i, $1 \leq i \leq n$ be the irreducible components of Y. If $\dim Y_{i_0}$ is maximal for some i_0, then $\mathrm{codim}_X Y_{i_0} = \dim X - \dim Y_{i_0}$ is minimal and we have $\dim Y = \dim Y_{i_0}$ and $\mathrm{codim}_X Y = \mathrm{codim}_X Y_{i_0}$. $\qquad\square$

Note that the hypothesis that X is irreducible is necessary (see Exercise 5.9).

(5.9) Dimension of hypersurfaces.

We work over a fixed field k. In this section we analyze the dimension of closed subschemes defined by a single equation.

In the special case $X = \operatorname{Spec} A$, where A is a unique factorization domain, the situation is particularly simple (and we do not even need that A is of finite type over a field).

Proposition 5.31. *Let $X = \operatorname{Spec} A$, where A is a noetherian unique factorization domain. Let $Z \subset X$ be a reduced closed subscheme of X equi-codimensional of codimension 1. Then there exists $0 \neq f \in A$ with $Z = V(f)$. Conversely, every closed subscheme Z of the form $V(f)$ for an element $0 \neq f \in A$ is equi-codimensional of codimension 1.*

Proof. We first show that an integral subscheme $Z = V(\mathfrak{p})$ for some prime ideal $\mathfrak{p} \subset A$ has codimension 1 if and only if $Z = V(f)$, where $f \in A$ is an irreducible element. Indeed, assume that $\mathfrak{p}$ has height 1 and let $g \in \mathfrak{p}$ be a nonzero element. As $\mathfrak{p}$ is a prime ideal, an irreducible divisor f of g is also contained in $\mathfrak{p}$. Thus we have inclusions of prime ideals $(0) \subsetneq (f) \subseteq \mathfrak{p}$. As $\mathfrak{p}$ has height 1, we have $\mathfrak{p} = (f)$. Conversely let $\mathfrak{p} = (f)$ for some irreducible element f. Then the height of $\mathfrak{p}$ is at least 1 and we find a prime ideal $\mathfrak{q} \subseteq \mathfrak{p}$ of height 1. We have already shown that $\mathfrak{q} = (f')$ for some irreducible element f'. Thus f divides f' and hence $\mathfrak{p} = \mathfrak{q}$.

Now let Z be reduced and equi-codimensional of codimension 1. As A is noetherian, Z has only finitely many irreducible components. Each irreducible component Z_i is of the form $V(f_i)$ for some irreducible element $f_i \in A$ and thus we find $Z = V(\prod_i f_i)$.

Conversely, let $0 \neq f \in A$ and $f = \prod f_i^{e_i}$ be a decomposition in pairwise non-associated irreducible elements f_i with integers $e_i \geq 1$. The irreducible components of $V(f)$ are the $V(f_i^{e_i}) = V(f_i)$ which we have shown to be of codimension 1. □

Note that for the first assertion the hypothesis that Z is reduced is necessary (see Exercise 5.10). In general the situation is more complicated even for schemes of finite type over a field k: it may happen that there are closed integral subschemes Z of codimension 1 of an integral affine k-scheme X of finite type, such that Z cannot be defined as the vanishing scheme of a single equation (Exercise 5.13). On the other hand, every closed subspace of $\mathbb{A}_k^n$ is always the set-theoretic intersection of n hypersurfaces (Exercise 5.14). To study the general situation we start by proving a geometric version of Krull's Hauptidealsatz (principal ideal theorem, cf. Proposition B.63).

Theorem 5.32. *Let X be an integral k-scheme of finite type, and let $f \in \Gamma(X, \mathscr{O}_X)$ be a non-unit, and different from 0 (i. e., $\emptyset \subsetneq V(f) \subsetneq X$). Then $V(f)$ is equi-codimensional of codimension 1 in X.*

Proof. Since $V(f)$ has only finitely many irreducible components $Z_1, \ldots, Z_r$, there exists for each $i = 1, \ldots, r$ an open affine neighborhood U_i of the generic point of Z_i such that $U_i \cap Z_j = \emptyset$ for $j \neq i$. By Theorem 5.22 (3) we have $\dim X = \dim U_i$. Replacing X by U_i and f by $f_{|U_i}$, we therefore may assume that $X = \operatorname{Spec} A$ is affine and that $V(f)$ is irreducible.

Let $\varphi\colon k[T_1,\ldots,T_d] \hookrightarrow A$ be a finite injective k-algebra homomorphism, as given by Noether normalization. Let $K = k(T_1,\ldots,T_d)$, and $L = \mathrm{Frac}(A)$, and let $g = \mathrm{N}_{L/K}(f)$. By Theorem 5.13, $g \in k[T_1,\ldots,T_d]$, and the morphism $X \to \mathbb{A}_k^d$ given by φ induces a surjective morphism $V(f) \to V(g)$ whose associated ring homomorphism is again finite, so that $\dim V(f) = \dim V(g)$ by Proposition 5.12. But on the other hand $\dim V(g) = d - 1$, because in the case $X = \mathbb{A}_k^d$ the theorem follows from the previous proposition. $\qquad\square$

As a generalization, we have the following result.

Corollary 5.33. *Let X be a k-scheme of finite type, and let $f_1,\ldots,f_r \in \Gamma(X,\mathscr{O}_X)$ with $V(f_1,\ldots,f_r) \neq \emptyset$. Then $\mathrm{codim}_X V(f_1,\ldots,f_r) \leq r$.*

Proof. We argue by induction on r. Assume the result is true for $V(f_1,\ldots,f_{r-1})$, and let Z be an irreducible component of the latter scheme. Then f_r, restricted to Z, either vanishes identically, in which case Z is contained in $V(f_1,\ldots,f_r)$, or it does not vanish identically, and we can apply Theorem 5.32. $\qquad\square$

In the following proposition, we look from a different angle at the fact that in the situation of the corollary we can have $\dim V(f_1,\ldots,f_r) > \dim X - r$, namely setting out with a closed subscheme of codimension r. It is not in general possible to obtain it as the set of zeros of r equations. In fact, this is not even possible locally, and the local failure can be seen as a measure of failure of "smoothness" (cf. the discussion of the notion of locally complete intersections in Volume II). Globally, the best we can hope for in general is this:

Proposition 5.34. *Let $X = \mathrm{Spec}\, A$ be an integral affine k-scheme of finite type, and let $Z \subset X$ be an integral closed subscheme of codimension $r > 0$ in X. Then there exist $f_1,\ldots,f_r \in A$ such that Z is an irreducible component of $V(f_1,\ldots,f_r)_{\mathrm{red}}$.*

Proof. We can prove this by induction, but the induction has to be set up carefully. Indeed, let us choose a chain $Z = Z_r \subsetneq Z_{r-1} \subsetneq \cdots \subsetneq Z_1$ of closed irreducible subsets with $\mathrm{codim}\, Z_i = i$. We show that there exist $f_1,\ldots,f_r \in A$ such that for all i, Z_i is an irreducible component of $V(f_1,\ldots,f_i)$ (set-theoretically), and that all irreducible components of $V(f_1,\ldots,f_i)$ have codimension i.

For $r = 1$, take any $f_1 \in A$ which vanishes on Z_1. Theorem 5.32 implies that all irreducible components of $V(f_1)$ have codimension 1, and in particular Z_1 is one of them.

Now let $r > 1$. By induction, we find $f_1,\ldots,f_{r-1}$, such that Z_{r-1} is an irreducible component of $V(f_1,\ldots,f_{r-1})$, and such that the irreducible components $Y_1,\ldots,Y_N$ of $V(f_1,\ldots,f_{r-1})$ all have codimension $r - 1$. By dimension reasons, Z_r does not contain any of the Y_i, so for the corresponding prime ideals we have $I(Z_r) \not\subseteq I(Y_i)$. Because the $I(Y_i)$ are prime, we obtain that $I(Z_r) \not\subseteq \cup_i I(Y_i)$, see Proposition B.2 (2). Now let $f_r \in I(Z_r) \setminus \cup_i I(Y_i)$. Since f_r does not vanish completely on any irreducible component of $V(f_1,\ldots,f_{r-1})$, it follows that $V(f_1,\ldots,f_r)$ is equi-codimensional of codimension r, and in particular $Z_r \subseteq V(f_1,\ldots,f_r)$ is one of its irreducible components. $\qquad\square$

Using Krull's principal ideal theorem (Proposition B.63) as starting point instead of Theorem 5.32 the same proofs as in Proposition 5.34 and Corollary 5.33 show also the following result.

Proposition 5.35. *Let A be a noetherian ring, $X = \operatorname{Spec} A$, let $Z \subseteq X$ be a closed irreducible subspace, and let $r \geq 0$ be an integer. Then the following assertions are equivalent.*

(i) $\operatorname{codim}_X Z \leq r$.

(ii) *There exist elements $f_1, \ldots, f_r \in A$ such that Z is an irreducible component of $V(f_1, \ldots, f_r)$.*

Example 5.36. As an example, we come back to Remark 1.41. Consider $\mathbb{A}_k^4$ with coordinates X, Y, Z, W, and let $C = V(XW - YZ)$, $U = (D(Y) \cup D(W)) \cap C$. So C is the affine cone over a smooth quadric in $\mathbb{P}_k^3$ (at least if $\operatorname{char} k \neq 2$), and U is the complement in C of the affine plane $D := V(W, Y) \subset C$. The scheme C is integral and Theorem 5.32 shows that $\dim(C) = 3$.

Define a function $h \in \Gamma(U, \mathscr{O}_C)$ by gluing the functions $X/Y \in \Gamma(D(Y) \cap C, \mathscr{O}_C)$ and $Z/W \in \Gamma(D(W) \cap C, \mathscr{O}_C)$. Since $XW = YZ$ on C, this can be done. We claim that there do not exist $f, g \in \Gamma(C, \mathscr{O}_C)$ such that $h = f/g$ on all of U. (Of course, h has this property locally on U, as follows from the definition, and as is true for every element of $\Gamma(U, \mathscr{O}_C)$.)

Suppose, to the contrary, that $h = f/g$ on U, where g has no zero in U. This means that $V(g) \subseteq C \setminus U = D$. All irreducible components of $V(g)$ have dimension 2, so either $V(g) = \emptyset$, or $V(g) = D$.

In the first case we would obtain that h extends to a function on all of C, or in other words, h, as an element of $K(C)$, lies in $\Gamma(C, \mathscr{O}_C)$. Since $X = Yh$ on a dense open subset, and hence on all of C, this is a contradiction: evaluating at the point $(1, 0, 0, 0)$ gives $1 = 0$.

In the case that $V(g) = D$ we consider the plane $D' := V(X, Z) \subset C$. We get $\{(0, 0, 0, 0)\} = D \cap D' = V(g) \cap D'$, which means that g gives rise to an element in $\Gamma(D', \mathscr{O}_{D'}) = k[W, Y]$ whose zero set (over an algebraic closure of k) is $\{(0, 0)\}$. Such elements do not exist.

(5.10) Dimension for products and for extensions of the base field.

Proposition 5.37. *Let X, Y be non-empty k-schemes locally of finite type. Then*

$$\dim X \times_k Y = \dim X + \dim Y.$$

Proof. Let $(U_i)_i$ and $(V_j)_j$ be open affine coverings of X and Y, respectively. Then the products $(U_i \times V_j)_{i,j}$ is an open affine covering of $X \times_k Y$ (Corollary 4.19). Thus by Lemma 5.7 (2) we may assume that X and Y are affine. Define $m := \dim X$ and $n := \dim Y$. The Noether normalization theorem gives us finite injective homomorphisms $k[T_1, \ldots, T_m] \to \Gamma(X, \mathscr{O}_X)$ and $k[T_{m+1}, \ldots, T_{m+n}] \to \Gamma(Y, \mathscr{O}_Y)$. Taking the tensor product we obtain a homomorphism

$$k[T_1, \ldots, T_{m+n}] \to \Gamma(X, \mathscr{O}_X) \otimes_k \Gamma(Y, \mathscr{O}_Y) = \Gamma(X \times_k Y, \mathscr{O}_{X \times Y}).$$

This homomorphism is again finite and injective, and the result follows from Corollary 5.17.
$\square$

A similar argument shows:

Proposition 5.38. *Let X be a k-scheme locally of finite type, and let K be a field extension of k. Then $\dim X = \dim X \otimes_k K$.*

Proof. Again we may assume that $X = \operatorname{Spec} A$ is affine of dimension $n \geq 0$ and therefore we find a finite injective homomorphism $k[T_1, \ldots, T_n] \to A$. Tensoring with K we obtain a finite injective homomorphism $K[T_1, \ldots, T_n] \to A \otimes_k K$. This shows that $X \otimes_k K$ has again dimension n by Corollary 5.17. $\square$

This result can be further refined; see Corollary 5.47 below and Exercise 5.12.

(5.11) Dimension of projective varieties.

The above results all yield analogous statements about projective varieties, and in one respect the situation even improves in the projective case: As we will see, if $X \subseteq \mathbb{P}_k^n$ is a projective variety of dimension r, then its intersection with any non-empty closed subscheme $V_+(f_1, \ldots, f_r) \subseteq \mathbb{P}_k^n$ is non-empty. We start with a lemma.

Lemma 5.39. *Let $X \subseteq \mathbb{P}_k^n$ be an integral closed subscheme and let $C(X) \subseteq \mathbb{A}_k^{n+1}$ be the cone over X, i. e., the closure in $\mathbb{A}_k^{n+1}$ of its inverse image under the projection $\mathbb{A}_k^{n+1} \setminus \{0\} \to \mathbb{P}_k^n$. Then $\dim C(X) = \dim X + 1$.*

Proof. This follows from (4.13.2) and Proposition 5.37. $\square$

Proposition 5.40. *Let $X \subseteq \mathbb{P}_k^n$ be an integral closed subscheme of dimension > 0, and let $f \in k[X_0, \ldots, X_n]$ be a homogeneous polynomial such that $V_+(f) \neq \emptyset$ and $X \not\subseteq V_+(f)$. Then $X \cap V_+(f) \neq \emptyset$, and $X \cap V_+(f)$ is equi-codimensional of codimension 1 in X.*

Proof. With the results of the previous section at our disposal, it is enough to show that $X \cap V_+(f) \neq \emptyset$. Let $C(X) \subseteq \mathbb{A}_k^{n+1}$ be the cone over X. Then $\dim C(X) = \dim X + 1 \geq 2$.

Now consider $V(f) \subset \mathbb{A}_k^{n+1}$. Since the origin lies in $C(X) \cap V(f)$, this intersection is non-empty, and hence by Theorem 5.32 has an irreducible component of dimension at least 1. Therefore the origin cannot be the only point in $C(X) \cap V(f)$, and every other point gives rise to a point in $X \cap V_+(f)$. $\square$

By induction, we obtain the following generalization.

Corollary 5.41. *Let $X \subseteq \mathbb{P}_k^n$ be an integral closed subscheme, and let $f_1, \ldots, f_r \in k[X_0, \ldots, X_n]$ be non-constant homogeneous polynomials. Then all irreducible components of $X \cap V_+(f_1, \ldots, f_r)$ have codimension $\leq r$ in X. If $\dim X \geq r$, then the intersection is non-empty.*

There is also an analogous version of Proposition 5.34; see Exercise 5.15.

Corollary 5.42. *Let $Z \subseteq \mathbb{P}_k^n$ be an integral closed subscheme. Then Z is of codimension 1 if and only if $Z = V_+(f)$ for an irreducible homogeneous polynomial f.*

Proof. Apply the reasoning of Proposition 5.31 to $C(Z) \subseteq \mathbb{A}_k^{n+1}$. $\square$

Subschemes of $\mathbb{P}_k^n$ of the form $V_+(f)$ for a homogeneous polynomial $f \in k[X_0, \ldots, X_n]$ of degree $d > 0$ are called *hypersurfaces of degree d*.

Corollary 5.43. *Let $n \geq 2$, and let $X = V_+(f) \subset \mathbb{P}_k^n$ be a hypersurface (f non-constant). Then X is connected.*

Proof. It is enough to show that any two irreducible components of X intersect. But the irreducible components are hypersurfaces themselves, so this follows from Corollary 5.41. $\square$

Schemes over fields and extensions of the base field

We now study the following question. Let k be a field, X a k-scheme and let K be a field extension of k. We set $X_K := X \otimes_k K$. Which properties of X are inherited by X_K and vice versa?

(5.12) Extension of scalars for schemes over a field.

Let k be a field. We use the following result (which we will prove below, see Theorem 14.38).

Proposition 5.44. *Let S be a scheme whose underlying topological space is discrete. Then any morphism $f\colon X \to S$ of schemes is universally open.*

We will use this proposition in the situation where S is a field, and in most cases f will be of finite type. In this case the proof of the proposition simplifies considerably; for instance, one can invoke Theorem 14.35.

Corollary 5.45. *Let X and $Y \neq \emptyset$ be k-schemes and denote by $p\colon X \times_k Y \to X$ the projection.*
(1) *The morphism p is surjective and universally open.*
(2) *The map $Z \mapsto \overline{p(Z)}$ is a well-defined surjective map*

    (5.12.1) $\{$*irreducible components of* $X \times_k Y\} \twoheadrightarrow \{$*irreducible components of* $X\}$.

(3) *The image $p(C)$ of every connected component C of $X \times_k Y$ is contained in a unique connected component of X and we obtain a well-defined surjective map*

    (5.12.2) $\{$*connected components of* $X \times_k Y\} \twoheadrightarrow \{$*connected components of* $X\}$.

(4) *Assume that $X \times_k Y$ has one of the following properties: "irreducible", "connected", "reduced", "integral". Then X has the same property.*

Proof. The structure morphism $Y \to \operatorname{Spec} k$ is universally open by Proposition 5.44 and surjective, so the same is true for its base change p.

We prove (2). To see that the map (5.12.1) is well-defined, we need to show that p sends maximal points (i.e., generic points of irreducible components) in $X \times_k Y$ to maximal points in X. This follows from (1) and Lemma 5.10, or more directly from the fact that the morphism $Y \to \operatorname{Spec} k$, and hence its base change $X \times_k Y \to X$ are faithfully flat; compare Lemma 14.9. At this point, we can however avoid using (1) and the notion of flatness by the following ad hoc argument:

We may assume that $X = \operatorname{Spec} A$ and $Y = \operatorname{Spec} B$ are affine. We set $C = A \otimes_k B$, so $X \times_k Y = \operatorname{Spec} C$. Let $\mathfrak{q} \subset C$ be a minimal prime ideal, let $\mathfrak{p} = \mathfrak{q} \cap A \subset A$ be the image under p. We have to show that $\mathfrak{p}$ is again minimal. As a prime ideal $\mathfrak{r}$ in a ring R is minimal if and only if $R_\mathfrak{r}$ has only one prime ideal, it suffices to show that the homomorphism $A_\mathfrak{p} \to C_\mathfrak{q}$ between the stalks induces a surjective morphism $f \colon \operatorname{Spec} C_\mathfrak{q} \to \operatorname{Spec} A_\mathfrak{p}$. So let $\mathfrak{p}' \subset A_\mathfrak{p}$ be a prime ideal. The fiber of f over this point is the spectrum of $C_\mathfrak{q} \otimes_{A_\mathfrak{p}} \kappa(\mathfrak{p}')$, so we must show that the latter ring is $\neq 0$. We do this in two steps. First, consider the ring $C_\mathfrak{q} \otimes_{A_\mathfrak{p}} A_\mathfrak{p}/\mathfrak{p}' = C_\mathfrak{q}/\mathfrak{p}' C_\mathfrak{q}$. This ring clearly is not zero, because it surjects onto $\kappa(\mathfrak{q})$. On the other hand, we can rewrite these rings as

$$C_\mathfrak{q} \otimes_{A_\mathfrak{p}} \kappa(\mathfrak{p}') = C_\mathfrak{q} \otimes_C C \otimes_A \kappa(\mathfrak{p}') = C_\mathfrak{q} \otimes_C B \otimes_k \kappa(\mathfrak{p}'),$$
$$C_\mathfrak{q} \otimes_{A_\mathfrak{p}} A_\mathfrak{p}/\mathfrak{p}' = C_\mathfrak{q} \otimes_C C \otimes_A A_\mathfrak{p}/\mathfrak{p}' = C_\mathfrak{q} \otimes_C B \otimes_k A_\mathfrak{p}/\mathfrak{p}',$$

and since the injectivity of the map $A_\mathfrak{p}/\mathfrak{p}' \hookrightarrow \operatorname{Frac}(A_\mathfrak{p}/\mathfrak{p}') = \kappa(\mathfrak{p}')$ is preserved by tensor products over the field k, we are done.

The map is surjective because p is surjective. Assertion (3) holds for every continuous surjective map.

The assertion of (4) for the properties "irreducible" and "connected" is clear by the above. To establish the result for reducedness, we may assume that $X = \operatorname{Spec} A$ and $Y = \operatorname{Spec} B$ are affine, so $X \times_k Y = \operatorname{Spec} A \otimes_k B$. But the map $A \to A \otimes_k B$ is injective (because B is free as a k-module), so the claim follows. Together we also get (4) for the property "integral". $\qquad\square$

Corollary 5.46. *Let X be a k-scheme and let $K \supset k$ be a purely inseparable extension. Then the projection $p \colon X_K \to X$ is a universal homeomorphism.*

Proof. By Corollary 5.45 the projection p is universally open and (universally) surjective. By Proposition 4.35 it is universally injective. $\qquad\square$

Corollary 5.45 may be in particular applied if $Y = \operatorname{Spec} K$ for a field extension K of k. We see that if X_K has one of the properties listed in (4), X possesses this property as well. The converse holds only under certain hypotheses, see Corollary 5.56 below.

Corollary 5.47. *Let X be a k-scheme locally of finite type, let K be an extension field of k, let $x \in X$ be a closed point, and let $\overline{x} \in X_K := X \times_{\operatorname{Spec} k} \operatorname{Spec} K$ be a point lying over x. Then*

$$\dim \mathscr{O}_{X,x} = \dim \mathscr{O}_{X_K, \overline{x}}.$$

Proof. Replacing X by an open neighborhood of x, we can remove those irreducible components of X which do not meet x (cf. Corollary 5.23), so we may assume that $\dim X = \dim \mathscr{O}_{X,x}$. Then by Proposition 5.38 we have

$$\dim \mathscr{O}_{X,x} = \dim X = \dim X_K \geq \dim \mathscr{O}_{X_K, \overline{x}}.$$

As $\overline{x}$ is closed in X_K, there exists an open neighborhood $U \subseteq X_K$ of $\overline{x}$ of dimension $\dim \mathscr{O}_{X_K, \overline{x}}$. The projection $p \colon X_K \to X$ is open by Corollary 5.45 (1), so U maps onto an open neighborhood of x, whence $\dim \mathscr{O}_{X,x} \leq \dim p(U) \leq \dim U = \dim \mathscr{O}_{X_K, \overline{x}}$, where the second inequality holds by Proposition 5.9. $\qquad\square$

(5.13) Geometric properties of schemes over fields.

In general it may happen that a k-scheme X has some property **P** (for instance one of the properties in Corollary 5.45 (4)) but there exist field extensions K of k such that the base change X_K does not have **P**. Therefore we make the following definition.

Definition 5.48. *Let* **P** *be one of the following properties of a scheme over a field: "irreducible", "connected", "reduced", or "integral". We say that a k-scheme X possesses* **P** *geometrically if the K-scheme X_K possesses* **P** *for every field extension K of k.*

Thus we say, for example, that X is *geometrically irreducible* if X_K is irreducible for all field extensions K of k. For the next proposition recall that the notion of a separable field extension is defined for an arbitrary (not necessarily algebraic) extension (Definition B.91).

Proposition 5.49. *Let X be a k-scheme. Then the following assertions are equivalent.*
(i) *X is geometrically reduced.*
(ii) *For every reduced k-scheme Y the product $X \times_k Y$ is reduced.*
(iii) *X is reduced and for every maximal point η of X the residue field $\kappa(\eta)$ is a separable extension of k.*
(iv) *There exists a perfect extension Ω of k such that X_Ω is reduced.*
(v) *For every finite purely inseparable extension K of k, X_K is reduced.*

Proof. We may assume that $X = \operatorname{Spec} A$ is affine, where A is a k-algebra. By Corollary 5.45 (4) each assertion implies that X is reduced. Thus we may assume that A is reduced. Let $(\eta_i)_{i \in I}$ be the family of maximal points of X. As A is reduced, the canonical homomorphism

$$(5.13.1) \qquad\qquad A \to \prod_{i \in I} \kappa(\eta_i)$$

is injective and $\kappa(\eta_i)$ is a localization of A for all $i \in I$. Let L be a field extension of k. If $A \otimes_k L$ is reduced, then its localization $\kappa(\eta_i) \otimes_k L$ is reduced. Conversely, we have an injective homomorphism

$$(5.13.2) \qquad A \otimes_k L \hookrightarrow \left(\prod_{i \in I} \kappa(\eta_i) \right) \otimes_k L \hookrightarrow \prod_{i \in I} (\kappa(\eta_i) \otimes_k L).$$

Therefore $A \otimes_k L$ is reduced if and only if $\kappa(\eta_i) \otimes_k L$ is reduced for all $i \in I$. Thus by Proposition B.97 assertion (iii) is equivalent to (i), to (iv), and to (v).

The implication "(ii) $\Rightarrow$ (i)" is trivial. Thus it suffices to show that (iii) implies (ii). For this we may assume that $Y = \operatorname{Spec} B$ is affine. Let $(L_j)_{j \in J}$ be the family of residue fields of maximal points of Y. We obtain an injective homomorphism

$$(5.13.3) \qquad A \otimes_k \left(\prod_{j \in J} L_j \right) \hookrightarrow \left(\prod_{i \in I} \kappa(\eta_i) \right) \otimes_k \left(\prod_{j \in J} L_j \right) \hookrightarrow \prod_{i,j} (\kappa(\eta_i) \otimes_k L_j)$$

Thus first using (5.13.1) for B and then (5.13.3) shows that $A \otimes_k B$ is a k-subalgebra of a product of rings which are reduced (again by Proposition B.97) because all $\kappa(\eta_i)$ are separable over k. $\qquad\square$

Proposition 5.50. *Let X be a k-scheme. Then the following assertions are equivalent.*

(i) *X is geometrically irreducible.*

(ii) *For every irreducible k-scheme Y the product $X \times_k Y$ is irreducible.*

(iii) *X is irreducible and if η is the generic point of X, then k is separably closed in $\kappa(\eta)$.*

(iv) *There exists a separably closed extension Ω of k such that X_Ω is irreducible.*

(v) *For every finite separable extension K of k, X_K is irreducible.*

Proof. By Corollary 5.45 (4) each assertion implies that X is irreducible. Thus we may assume that X is irreducible. Let η be its generic point. For every k-scheme Z the projection $p\colon X \times_k Z \to X$ is open (Corollary 5.45) and thus $X \times_k Z$ is irreducible if and only if $p^{-1}(\eta) = \kappa(\eta) \otimes_k Z$ is irreducible (Proposition 3.24). Applying this remark to $Z = \operatorname{Spec} L$, where L is an arbitrary (resp. a separably closed, resp. a finite separable) extension, we see that (i), (iii), (iv), and (v) are equivalent by Proposition B.101. It remains to show that (iii) implies that $\kappa(\eta) \otimes_k Y$ is irreducible for every irreducible k-scheme Y. But again the projection $q\colon \kappa(\eta) \otimes_k Y \to Y$ is open and Proposition 3.24 shows that $\kappa(\eta) \otimes_k Y$ is irreducible if and only if $\kappa(\eta) \otimes_k \kappa(\theta)$ is irreducible, where θ is the generic point of Y. This again follows from Proposition B.101. $\qquad\square$

Combining Proposition 5.49 and Proposition 5.50 and using Corollary 5.45 we obtain:

Proposition 5.51. *Let X be a k-scheme. Then the following assertions are equivalent.*

(i) *X is geometrically integral.*

(ii) *For every integral k-scheme Y the product $X \times_k Y$ is integral.*

(iii) *X is integral and if η is the generic point of X, then k is algebraically closed in $\kappa(\eta)$ and $\kappa(\eta)$ is separable over k.*

(iv) *There exists an algebraically closed extension Ω of k such that X_Ω is integral.*

(v) *For every finite extension K of k, X_K is integral.*

To characterize the property "geometrically connected" we recall the following purely topological fact (see, e.g., [BouGT] I, 11.3, Prop. 7).

Remark 5.52. Let $f\colon X \to Y$ be a continuous open and surjective map of topological spaces. If Y is connected and if for all $y \in Y$ the fiber $f^{-1}(y)$ is connected, then X is connected.

Proposition 5.53. *Let X be a k-scheme. Then the following assertions are equivalent.*

(i) *X is geometrically connected.*

(ii) *For every connected k-scheme Y the product $X \times_k Y$ is connected.*

(iii) *There exists a separably closed extension Ω of k such that X_Ω is connected.*

If X is quasi-compact, these assertions are also equivalent to

(iv) *For every finite separable extension K of k, X_K is connected.*

We will prove that (iv) implies the other assertions only under the stronger hypothesis that X is of finite type over k. The general case can be proved using the technique of schemes over inductive limits of rings explained in Chapter 10 (see Exercise 10.27 for the idea of a proof if X is quasi-compact and quasi-separated, or [EGAIV] (8.4.5) in general).

Proof. The implications "(ii) $\Rightarrow$ (i) $\Rightarrow$ (iii)" and "(i) $\Rightarrow$ (iv)" are clear. Let us prove that (i) implies (ii). Consider the projection $q\colon X \times_k Y \to Y$ which is open and surjective by Corollary 5.45. By Remark 5.52 it suffices to show $q^{-1}(y) = X \otimes_k \kappa(y)$ is connected for all $y \in Y$. But this holds by hypothesis.

We will show that (iii) implies (i). Let K be an extension of k and let L be a field containing K and Ω. As the canonical morphism $X_L \to X_K$ is surjective, it suffices to show that X_L is connected. Consider the canonical morphism $p\colon X_L \to X_\Omega$ which is open and surjective. Thus again it suffices to show that for all $x \in X_\Omega$ the fiber $p^{-1}(x) = \operatorname{Spec}(\kappa(x) \otimes_\Omega L)$ is connected. But as Ω is separably closed in L all fibers are even irreducible by Proposition 5.50.

It remains to show that (iv) implies (iii) if X is of finite type over k. Let Ω be a separable closure of k. Assume that X_Ω is the union of two closed non-empty disjoint subsets Y and Z. Let $(U_i)_i$ be a finite affine covering of X. Then $U_i = \operatorname{Spec} A_i$, where A_i is a finitely generated k-algebra. Thus $Y \cap (U_i \otimes_k \Omega) = V(\mathfrak{a}_i)$ for an ideal $\mathfrak{a}_i \subseteq A_i \otimes_k \Omega$ generated by finitely many elements f_{ij}. Similarly we find elements g_{il} defining $Z \cap (U_i \otimes_k \Omega)$ in U_i. Let k' be a finite subextension of Ω such that $f_{ij}, g_{il} \in A_i \otimes_k k'$ for all i, j, l. Let $Y' \subseteq X_{k'}$ be the subscheme such that $Y' \cap (U_i \otimes k')$ is defined for all i by the ideal generated by the f_{ij} in $A_i \otimes k'$. Similarly we define $Z' \subseteq X_{k'}$. Then, if $r\colon X_\Omega \to X_{k'}$ is the canonical surjective morphism, we have $r^{-1}(Y') = Y$ and $r^{-1}(Z') = Z$. This shows that the closed subsets Y' and Z' are non-empty and disjoint. Therefore $X_{k'}$ is not connected. $\qquad\square$

Exercise 5.23 shows that if X is a connected k-scheme and k is separably closed in $\kappa(x)$ for *some* point $x \in X$ (e.g., if $X(k) \neq \emptyset$), then X is already geometrically connected.

Corollary 5.54. *Let X be a k-scheme and let K be an algebraically closed extension of k. Then X is geometrically irreducible (resp. geometrically connected, resp. geometrically reduced, resp. geometrically integral) if and only if X_K is irreducible (resp. connected, resp. reduced, resp. integral).*

Remark 5.55. The corollary shows that for a k-scheme X the number of irreducible (resp. connected) components of X_K is independent of the choice of the algebraically closed extension K of k. This number is called the *geometric number of irreducible components* (resp. the *geometric number of connected components*).

Corollary 5.56. *Let X be a k-scheme.*
(1) *If X is reduced and $K \supseteq k$ is a separable extension, then X_K is reduced.*
(2) *If X is irreducible (resp. connected) and $K \supseteq k$ is a field extension such that k is separably closed in K, then X_K is irreducible (resp. connected).*
(3) *If X is integral and $K \supseteq k$ is a separable field extension such that k is algebraically closed in K, then X_K is integral.*

Proof. (1). The k-scheme $\operatorname{Spec} K$ is geometrically reduced and the assertion follows from Proposition 5.49.

(2). The projection $p\colon X_K \to X$ is open and surjective by Corollary 5.45. As X is irreducible (resp. connected), the same holds for X_K if we can show that for all $x \in X$ the fiber $p^{-1}(x)$ is irreducible (resp. connected), see Proposition 3.24 (resp. Remark 5.52). But $\operatorname{Spec} K \to \operatorname{Spec} k$ is geometrically irreducible and therefore $p^{-1}(x) = \operatorname{Spec} \kappa(x) \otimes_k K$ is irreducible (and in particular connected) for all x by Proposition 5.50.

(3). This follows from (1) and (2). $\qquad\square$

Corollary 5.57. *Let k be a perfect field (e.g., if $\mathrm{char}(k) = 0$). Then any reduced k-scheme is geometrically reduced.*

Intersections of plane curves

As an example we study hypersurfaces in $\mathbb{P}^2_k$, i.e., curves in the projective plane. The main result is the Theorem of Bézout 5.61. It says that – roughly speaking – two curves in $\mathbb{P}^2_k$ given by equations of degree d and e, respectively, meet in ed points. Intersections points have to be "counted with multiplicity" and thus we first define intersection numbers for plane curves to motivate this and to make this precise. The proof of Bézout's theorem given here is elementary. In Volume II we will study so-called Hilbert functions of projective schemes and deduce a generalization of Bézout's theorem in higher dimensions.

(5.14) Intersection numbers of plane curves.

Let k be a field.

Definition 5.58. *A plane curve is a closed subscheme C of $\mathbb{P}^2_k$ that is of the form $V_+(f)$, where $0 \neq f \in k[X, Y, T]$ is a non-constant homogeneous polynomial. The degree of f is called the degree of C.*

The degree of a plane curve $C \subset \mathbb{P}^2_k$ may depend on the embedding (rather than only on the isomorphism class of the k-scheme C). In Volume II we will see that if the degree of C is at least 3, it depends only on the isomorphism class of the k-scheme C (more precisely on the arithmetic genus of C). Compare also Section (14.31).

By Proposition 5.40 a plane curve is equidimensional of dimension 1. Let $f = f_1^{e_1} \cdots f_r^{e_r}$ be the decomposition of f into irreducible factors ($e_i \geq 1$ and $(f_i) \neq (f_j)$ for $i \neq j$). Then $V_+(f_i^{e_i})$, $i = 1, \ldots, r$, are the irreducible components of $V_+(f)$. The scheme $V_+(f)$ is reduced if and only if $\mathrm{rad}((f)) = (f)$, that is, if and only if $e_i = 1$ for all i.

We are interested in the (schematic) intersection of two plane curves

$$V_+(f) \cap V_+(g) = V_+(f, g) \subseteq \mathbb{P}^2_k.$$

We start with the following lemma.

Lemma 5.59. *Let $0 \neq f, g \in k[X, Y, T]$ be non-constant homogeneous polynomials. Then $\dim V_+(f, g) = 0$ if and only if f and g have no common factor.*

Proof. As $V_+(f)$ and $V_+(g)$ are both of dimension 1, we have $\dim(V_+(f, g)) \leq 1$. By Proposition 5.40 we know that $V_+(f, g) \neq \emptyset$. Now a homogeneous polynomial h of positive degree is a common factor of f and g if and only if the intersection $V_+(f, g)$ contains the plane curve $V_+(h)$. Therefore the existence of such an h implies $\dim(V_+(f, g)) = 1$. Conversely, if $\dim V_+(f, g) = 1$, Corollary 5.42 shows that there exists a closed integral subscheme $Z \subseteq V_+(f, g)$ of the form $Z = V_+(h)$ for an irreducible factor h of f and g. $\square$

Thus if f and g do not have a common factor, then Proposition 5.20 implies that $Z := V_+(f,g)$ is the spectrum of a finite-dimensional k-algebra, consists of finitely many points $z_1, \ldots, z_n$, and as a scheme it is the disjoint union $\coprod_{i=1}^n \operatorname{Spec} \mathscr{O}_{Z,z_i}$, where $\mathscr{O}_{Z,z_i}$ is a finite-dimensional local k-algebra. One or both of the following phenomena may occur:

(1) $\mathscr{O}_{Z,z_i}$ is non-reduced.
(2) The residue field of $\mathscr{O}_{Z,z_i}$ is a finite nontrivial extension of k.

The geometric interpretation of (1) is that $V_+(f)$ and $V_+(g)$ do not intersect "transversally" in z_i. We will give a precise definition of transversal intersection in Volume II (see also Exercise 6.7). Here we illustrate this remark just with two examples:

Let $f = YT - X^2$ and $g = Y$. Then $(0:0:1) \in Z(k)$ defines a point $z \in V_+(f,g)$ and this is the only point of Z. By dehomogenizing with respect to T we have

$$\Gamma(Z, \mathscr{O}_Z) = \mathscr{O}_{Z,z} \cong k[U,V]/(V - U^2, V) = k[U]/(U^2).$$

This agrees with the picture that $V_+(f)$ and $V_+(g)$ are tangent of order 2 to each other. Note that we have $\dim_k(\Gamma(Z, \mathscr{O}_Z)) = 2$.

As a second example we choose $f = Y^2T - X^2(X + T)$ and $g = Y$. As $V_+(T) \cap Z = \emptyset$, we consider $Z = Z \cap D_+(T)$, that is we dehomogenize again by T. Then

$$\Gamma(Z, \mathscr{O}_Z) \cong k[U]/(U^2(U + 1)) = k[U]/(U^2) \times k[U]/(U + 1).$$

Again this agrees with our picture that $V_+(g)$ meets $V_+(f)$ transversally in $(-1:0:1)$ (corresponding to $\dim_k k[U]/(U+1) = 1$) and meets two "branches" of $V_+(f)$ in $(0:0:1)$ (corresponding to $\dim_k k[U]/(U^2) = 2$). In this case we have $\dim_k(\Gamma(Z, \mathscr{O}_Z)) = 3$.

To illustrate (2) consider $k = \mathbb{R}$, $f = X^2 + Y^2 + T^2$ and $g = Y$. Then $V_+(f)$ and hence Z has no $\mathbb{R}$-valued points, but Z has the $\mathbb{C}$-valued points $(i:0:1)$ and $(-i:0:1)$, where i is a square root of -1. Both points lie in the same orbit with respect to the action of the Galois group $\operatorname{Gal}(\mathbb{C}/\mathbb{R})$, and therefore define a single closed point z of Z (Proposition 5.4). Dehomogenizing with respect to T we find $\Gamma(Z, \mathscr{O}_Z) = \mathscr{O}_{Z,z} \cong \mathbb{R}[U]/(U^2 + 1) \cong \mathbb{C}$ and $\dim_{\mathbb{R}}(\Gamma(Z, \mathscr{O}_Z)) = 2$.

These examples make the following definition plausible.

Definition 5.60. *Let $C, D \subset \mathbb{P}^2_k$ be two plane curves such that $Z := C \cap D$ is a k-scheme of dimension 0. Then we call $i(C, D) := \dim_k(\Gamma(Z, \mathscr{O}_Z))$ the* intersection number *of C and D. For $z \in Z$ we call $i_z(C, D) := \dim_k(\mathscr{O}_{Z,z})$ the* intersection number *of C and D at z.*

As explained above, we have $i(C, D) = \sum_{z \in C \cap D} i_z(C, D)$.

(5.15) Bézout's theorem.

The aim of the section is to show the following theorem.

Theorem 5.61. *(Theorem of Bézout) Let k be a field. Let $C = V_+(f)$ and $D = V_+(g)$ be plane curves in $\mathbb{P}^2_k$ given by polynomials without a common factor. Then*

$$i(C, D) = (\deg f)(\deg g).$$

In particular, the intersection $C \cap D$ is non-empty and consists of a finite number of closed points.

The proof will occupy the rest of the section. We start with an easy remark. Let R be a factorial ring and let $f, g \in R$ be elements that do not have a common divisor. Then for $a, b \in R$ we have

$$(5.15.1) \qquad af + bg = 0 \Leftrightarrow \exists t \in R : a = gt, \ b = -ft.$$

Indeed, the existence of t is clearly sufficient. Conversely let $af + bg = 0$. As f and g do not have a common divisor, g divides a and there exists $t \in R$ such that $a = gt$. Then $af + bg = 0$ shows that $b = -ft$.

The following easy lemma shows that we may replace k by some field extension.

Lemma 5.62. *Let K be a field extension of k and set $C_K := C \otimes_k K$ and $D_K = D \otimes_k K$. Then $C_K = V_+(f_K) \subset \mathbb{P}^2_K$, where f_K is the polynomial f considered as a homogeneous polynomial with coefficients in K. Similarly, $D_K = V_+(g_K) \subset \mathbb{P}^2_K$. We have*

$$i(C, D) = i(C_K, D_K).$$

Proof. Everything but the last equality is clear. We have

$$C_K \cap D_K := C_K \times_{\mathbb{P}^2_K} D_K = (C \times_{\mathbb{P}^2_k} D) \otimes_k K = (C \cap D) \otimes_k K = (C \cap D)_K.$$

Setting $A := \Gamma(C \cap D, \mathcal{O}_{C \cap D})$ we therefore have $A \otimes_k K = \Gamma((C \cap D)_K, \mathcal{O}_{(C \cap D)_K})$ and hence

$$i(C_K, D_K) = \dim_K(A \otimes_k K) = \dim_k A = i(C, D). \qquad \square$$

To prove Bézout's theorem we may therefore assume that k is algebraically closed. As $Z := C \cap D$ is finite, we find a hyperplane (= line) L in $\mathbb{P}^2_k$ such that $L \cap Z = \emptyset$. Thus we may choose coordinates X, Y, and T on $\mathbb{P}^2_k$ such that $Z \cap V_+(T) = \emptyset$.

We set $S := k[X, Y, T]$ and denote the degrees of $f, g \in S$ by $n := \deg(f)$, $m := \deg(g)$. Then $S = \bigoplus_d S_d$ is a graded k-algebra and $\mathfrak{a} := (f, g)$ is an ideal of S which is generated by homogeneous elements. Therefore $B := S/\mathfrak{a}$ is graded as well, that is $B = \bigoplus B_d$. Note that $\dim_k S_d = \binom{d+2}{2}$. In particular B_d is a finite-dimensional k-vector space.

Lemma 5.63. *For $d \geq n + m$ we have $\dim_k B_d = nm$.*

Proof. We claim that the sequence

$$(5.15.2) \qquad 0 \to S_{d-n-m} \xrightarrow{\mu} S_{d-n} \oplus S_{d-m} \xrightarrow{\nu} S_d \longrightarrow B_d \to 0$$

with $\mu(s) = (gs, -fs)$ and $\nu(s', s'') = fs' + gs''$ is exact. Clearly μ is injective and $\mathrm{Coker}(\nu) = B_d$. Moreover we have $\mathrm{Im}(\mu) = \mathrm{Ker}(\nu)$ by (5.15.1) and this shows our claim.

Thus we have $\dim_k B_d = \dim_k S_d - \dim_k S_{d-n} - \dim_k S_{d-m} + \dim_k S_{d-n-m}$ and an easy calculation using $\dim_k S_d = (d+2)(d+1)/2$ proves the lemma. $\qquad \square$

In the language that will be introduced in Volume II this lemma shows that the Hilbert polynomial of Z is constant with absolute coefficient nm.

Proof. (Bézout's theorem) Let

$$\Phi \colon S = k[X, Y, T] \to k[X, Y], \qquad h \mapsto \tilde{h} = h(X, Y, 1),$$

be the dehomogenization map with respect to T. As $Z \subset D_+(T)$ we have $Z = \operatorname{Spec} A$ with $A = k[X,Y]/(\tilde{f}, \tilde{g})$. Then the map Φ induces a surjective k-algebra homomorphism $B = S/(f,g) \to A$ and in particular a k-linear map $v_d \colon B_d \to A$. To prove Bézout's theorem we have to show that $\dim_k A = nm$. Thus in view of the above Lemma it suffices to show:

Claim: The k-linear map v_d is an isomorphism for $d \geq n + m$.

(i). We first show that the multiplication with T on B is injective. For $h \in S$ we denote the image of h in $k[X,Y,T]/(T)$ by h_0. Now let $h \in S$ such that $Th \in (f,g)$, that is, $Th = af + bg$. We have to show that $h \in (f,g)$. As $V_+(f,g) \cap V_+(T) = \emptyset$, the images f_0 and g_0 are still without a common divisor. Thus (5.15.1) shows that there exists $t_0 \in k[X,Y,T]/(T)$ such that $a_0 = g_0 t_0$ and $b_0 = -f_0 t_0$. Lifting t_0 to an element $t \in k[X,Y,T]$ we find $a = gt + Ta'$ and $b = -ft + Tb'$. This implies $h = a'f + b'g \in (f,g)$.

Note that (i) together with Lemma 5.63 shows that the multiplication with T induces an isomorphism $B_d \xrightarrow{\sim} B_{d+1}$ for $d \geq n + m$.

(ii). We now show that v_d is injective. Let $h \in S_d$ such that $\Phi(h) = \tilde{a}\tilde{f} + \tilde{b}\tilde{g}$ for $\tilde{a}, \tilde{b} \in k[X,Y]$. Let $a, b \in S_{d'}$ be the homogenization of $\tilde{a}$ and $\tilde{b}$ with respect to T for some $d' \geq d$. Then we find $T^\alpha h = T^\beta af + T^\gamma bg \in (f,g)$. By (i) we have $h \in (f,g)$ and this shows the injectivity of v_d.

(iii). It remains to show that v_d is surjective. Let $\tilde{h} \in A$ be arbitrary. Lifting $\tilde{h}$ to $k[X,Y]$ and homogenization with respect to T yields an element $h \in S_e$. We may assume $e \geq d$. Let $\bar{h} \in B_e$ be its image. As remarked above, multiplication by T^{e-d} is an isomorphism $B_d \to B_e$, so there exists an element $\bar{h}' \in B_d$ such that $T^{e-d}\bar{h}' = \bar{h}$. Then we have $v_d(\bar{h}') = \tilde{h}$ and v_d is surjective. $\qquad\square$

Exercises

Exercise 5.1$\lozenge$. Let p be a prime number, $q = p^r$ for an integer $r \geq 1$, and let $\mathbb{F}_q$ be a finite field with q elements. Describe $\mathbb{A}^1_{\mathbb{F}_q}$.

Exercise 5.2. Let k be a field, k^{sep} a separable closure, $\Gamma := \operatorname{Gal}(k^{\mathrm{sep}}/k)$, and let X be a k-scheme locally of finite type. Show that for all $x \in X(k^{\mathrm{sep}})$ the Γ-orbit of x in $X(k^{\mathrm{sep}})$ is finite.

Remark: The finiteness of the Γ-orbits is equivalent to the continuity of the action

$$\Gamma \times X(k^{\mathrm{sep}}) \to X(k^{\mathrm{sep}}),$$

where Γ is endowed with its profinite topology and $X(k^{\mathrm{sep}})$ is endowed with the discrete topology.

Exercise 5.3$\lozenge$. Let A be a discrete valuation ring and let $Y = \operatorname{Spec} A$. Let K be the field of fractions of A and k its residue class field. The canonical homomorphisms $i \colon A \to K$ and $\pi \colon A \to k$ yield a ring homomorphism $\varphi \colon A \to K \times k$, $a \mapsto (i(a), \pi(a))$. Let $X = \operatorname{Spec}(K \times k)$ and let $f \colon X \to Y$ be the morphism of schemes corresponding to φ.

Prove that $K \times k$ is an A-algebra of finite type, that f is bijective, and that $\dim(X) = 0$ and $\dim(Y) = 1$.

Exercise 5.4. Let X be a locally noetherian scheme and let $x \in X$ a point such that $\overline{\{x\}}$ is locally closed in X. Show that $\dim \overline{\{x\}} \leq 1$.
Hint: Use the lemma of Artin-Tate (Corollary B.65).

Exercise 5.5. Let X be a locally noetherian scheme. Show that any closed subset Z of X contains a point that is closed in X. Deduce that for every $x \in X$ there exists a specialization of x that is closed in X.
Hint: Show first that there exists a point $w \in Z$ such that $\overline{\{w\}}$ is locally closed in X. Then use Exercise 5.4.

Exercise 5.6. Let $f\colon X \to Y$ be a morphism of schemes.
(a) Show that if f is closed and surjective, then we have $\dim X \geq \dim Y$.
(b) Show that if f is injective, then we have $\dim X \leq \dim Y$.

Exercise 5.7. Let A be a discrete valuation ring with uniformizing element π. Let $X = \operatorname{Spec} A[T]$. Define ideals $\mathfrak{m}_1 = (\pi T - 1)$ and $\mathfrak{m}_2 = (T, \pi)$ of $A[T]$. Show that these ideals are maximal in $A[T]$. In particular $Z_i := V(\mathfrak{m}_i)$ has dimension 0. Show that $\operatorname{codim}_X Z_1 = 1$ and $\operatorname{codim}_X Z_2 = 2$.

Exercise 5.8◇. Let X be an equidimensional scheme of dimension d of finite type over a field.
(a) Let $Z_{d_0} \subsetneq \cdots \subsetneq Z_{d_r} \subseteq X$ be a chain of closed integral subschemes Z_{d_i} such that $\dim Z_{d_i} = d_i$. Show that this chain can be completed to a chain $Z_0 \subsetneq Z_1 \subsetneq \cdots \subsetneq Z_d$ of closed integral subschemes of X such that $\dim Z_j = j$ for all $j = 0, \ldots, d$.
(b) For any two closed irreducible subsets $Y \subseteq Z \subseteq X$ show that

$$\operatorname{codim}_X Y = \operatorname{codim}_Z Y + \operatorname{codim}_X Z.$$

Exercise 5.9◇. Give an example of a non-irreducible scheme X of finite type over a field k and an irreducible closed subset Y of X such that $\operatorname{codim}_X Y + \dim Y < \dim X$.

Exercise 5.10. Let $X = \mathbb{A}_k^2$ with coordinates T, U. Show that $Z = V(TU, T^2)$ is closed irreducible of codimension 1 in X but that there is no $f \in k[T, U]$ such that $Z = V(f)$.

Exercise 5.11. Let k be a field with $\operatorname{char}(k) \neq 2$, $n \geq 2$ an integer, and let $\operatorname{SO}_{n,k}$ be the group scheme over k such that $\operatorname{SO}_{n,k}(R) = \{A \in \operatorname{GL}_n(R) \; ; \; {}^t\!AA = I_n, \det(A) = 1\}$ for all k-algebras R. Let $\mathfrak{so}_{n,k}$ be the k-scheme such that $\mathfrak{so}_{n,k}(R) = \{A \in M_n(R) \; ; \; {}^t\!A = -A\}$.
(a) Show that $\operatorname{SO}_{n,k}$ is a scheme of finite type over k and that $\mathfrak{so}_{n,k} \cong \mathbb{A}_k^{n(n-1)/2}$.
(b) Show that $A \mapsto (I_n + A)^{-1}(I_n - A)$ defines an isomorphism of an open dense subscheme of $\mathfrak{so}_{n,k}$ onto an open dense subscheme of $\operatorname{SO}_{n,k}$.
(c) Deduce that $\operatorname{SO}_{n,k}$ is irreducible and of dimension $n(n-1)/2$.
Remark: $\mathfrak{so}_{n,k}(k)$ is the tangent space at $I_n \in \operatorname{SO}_n(k)$; see Exercise 6.5.

Exercise 5.12. Let X be a k-scheme locally of finite type, let $Z \subseteq X$ be an irreducible component. Let $K \supseteq k$ be a field extension, let $p\colon X \otimes_k K \to X$ be the projection and let $Z' \subseteq X \otimes_k K$ be an irreducible component such that $\overline{p(Z')} = Z$. Then $\dim Z' = \dim Z$.

Exercise 5.13. Let k be a field, $A = k[a, b, c, d]/(ad - bc)$, $X = \operatorname{Spec} A$ and $Z := V(a, b)$. Show that X is an integral scheme, Z is a closed integral subscheme of codimension 1, that Z is an irreducible component of $V(a)$, and that Z itself cannot be written as $V(f)$ for some $f \in A$.

Exercise 5.14. Let k be a field, $n \geq 0$ an integer, and X a closed reduced subscheme of $\mathbb{A}_k^n$. Show that there exist $f_1, \ldots, f_n \in k[T_1, \ldots, T_n]$ such that $X = V(f_1, \ldots, f_n)_{\mathrm{red}}$.
Hint: The following steps show the following more general result.

    (*) Let R be a noetherian ring of finite dimension d and let $\mathfrak{a} \subset R[T]$ be an ideal. Then there exist $f_1, \ldots, f_{d+1} \in \mathfrak{a}$ such that $\mathrm{rad}(\mathfrak{a}) = \mathrm{rad}(f_1, \ldots, f_{d+1})$.

The result can be shown by induction on d along the following steps.
(a) Show that it suffices to prove (*) if R is reduced. From now on let R be reduced.
(b) Let S be the multiplicative set of regular elements in R and $K := \mathrm{Frac}\, R = S^{-1}R$. Show that $K[T]$ is a finite product of principal ideal domains and deduce that every ideal in $K[T]$ is a principal ideal.
(c) Let $f \in \mathfrak{a}$ be a generator of $S^{-1}\mathfrak{a}$ and $g_1, \ldots, g_t$ generators of $\mathfrak{a}$. Show that there exists a regular element $r \in R$, $h_j \in R[T]$ and integers $n_j \geq 1$ such that $rg_j^{n_j} = h_j f$. Show that $V(\mathfrak{a}) \subseteq V(f) \subseteq V(r) \cup V(\mathfrak{a})$ (as sets). Deduce the result if r is a unit.
(d) Apply the induction hypothesis on $R/(r)$.

Exercise 5.15.
(a) Let $X \subseteq \mathbb{P}_k^n$ be a closed integral subscheme, where k is a field. Let Y be a non-empty integral closed subscheme of X, with $\mathrm{codim}_X Y = r$. Show that there exist non-constant homogeneous polynomials $f_1, \ldots, f_r \in k[X_0, \ldots, X_n]$ such that Y is an irreducible component of $(X \cap V_+(f_1, \ldots, f_r))_{\mathrm{red}}$.
(b) Let X be as in part (a), and let $r = \dim X + 1$. Prove that there exist non-constant homogeneous polynomials $f_1, \ldots, f_r \in k[X_0, \ldots, X_n]$ such that $X \cap V_+(f_1, \ldots, f_r)$ is empty.

Exercise 5.16. Let k be a field, $\mathrm{char}(k) \neq 2$, let $a \in k$ be an element that is not a square in k, and $X = \mathrm{Spec}\, k[T, U]/(T^2 - aU^2)$.
(a) Show that X is integral, geometrically reduced and geometrically connected.
(b) Show that X is not geometrically integral.

Exercise 5.17. Let k be a field, let X and Y be k-schemes locally of finite type, and let $f, g \colon X \to Y$ be two k-morphisms. Assume that X is geometrically reduced over k. Show that $f = g$ if and only if there exists an algebraically closed extension Ω of k such that f and g induce the same map $X_k(\Omega) \to Y_k(\Omega)$ on Ω-valued points.

Exercise 5.18◊. Let X be a scheme over a field k. Show that X is geometrically connected (resp. geometrically irreducible) if and only if X_{red} is.

Exercise 5.19. Let k be a field and let X and Y be k-schemes. If X and Y are geometrically reduced (resp. geometrically irreducible, resp. geometrically integral, resp. geometrically connected), then show that $X \times_k Y$ has the same property.

Exercise 5.20. Let k be a field, X a k-scheme, $x \in X$.
(a) Show that the following assertions are equivalent.
    (i) For every extension K of k and for every point $x' \in X_K$ over x the local ring $\mathcal{O}_{X_K, x'}$ is reduced.
    (ii) $\mathrm{Spec}\, \mathcal{O}_{X, x}$ is a geometrically reduced k-scheme.
    (iii) There exists a perfect extension Ω and a point $x' \in X_\Omega$ over x such that $\mathcal{O}_{X_\Omega, x'}$ is reduced.
    (iv) $\mathcal{O}_{X, x}$ is reduced and for every irreducible component Z of X containing x the residue field $\kappa(\eta_Z)$ in the generic point of Z is a separable extension k.
    (v) For every finite purely inseparable extension k' of k the local ring $\mathcal{O}_{X_{k'}, x'}$ in the unique point x' over x is reduced.

If these equivalent conditions are satisfied, X is called *geometrically reduced at x*.
(b) Show that if X is locally noetherian, then the conditions in (a) are equivalent to the condition that there exists an open neighborhood U of x such that U is geometrically reduced over k.

Exercise 5.21. Let k be a field, X a locally noetherian k-scheme, $x \in X$. Then X is called *geometrically integral at x* if for every extension K of k and for every point $x' \in X_K$ over x the local ring $\mathscr{O}_{X_K, x'}$ is an integral domain. If X is geometrically integral at all points $x \in X$, we call X *geometrically locally integral*. Now assume that X is of finite type over k and show that the following assertions are equivalent.
 (i) X is geometrically locally integral.
(ii) X is geometrically reduced and the geometric number of irreducible components is equal to the geometric number of connected components.

Exercise 5.22. Let $A \to B$ be an integral ring homomorphism. Show that the corresponding morphism of schemes $\operatorname{Spec} B \to \operatorname{Spec} A$ is universally closed. Deduce that if K is an algebraic extension of a field k, then the projection $X_K \to X$ is universally open, universally closed, and surjective.

Exercise 5.23. Let k be a field and let X be a connected k-scheme.
(a) Assume that there exists a non-empty geometrically connected k-scheme Y and a k-morphism $Y \to X$. Show that X is geometrically connected.
 Hint: Use Exercise 5.22 to show that $X_\Omega \to X$ is open, closed, and surjective for a separable closure Ω of k and show that there are no nontrivial open and closed subsets of X_Ω.
(b) Assume that there exists a point $x \in X$ such that k is separably closed in $\kappa(x)$ (e.g., if $X(k) \neq \emptyset$). Show that X is geometrically connected.
 Hint: Use that $\operatorname{Spec} \kappa(x)$ is a geometrically irreducible k-scheme.

6   Local Properties of Schemes

Contents

- The tangent space
- Smooth morphisms
- Regular schemes
- Normal schemes

Consider a scheme X of finite type over an algebraically closed field k. If X is reduced then "locally" around almost all closed points X looks like affine space. Compare Figure 1.1: zooming in sufficiently, this is true for the pictured curve at all points except for the point where it self-intersects. However, while in differential geometry this can be used as the definition of a manifold, the Zariski topology is too coarse to capture appropriately what should be meant by "local". Instead, one should look at whether X can be "well approximated by a linear space".

To make this precise, we define the (absolute) tangent space $T_x X$ of a scheme X at a point x. If X is embedded in an affine or projective space, then we can imagine $T_x X$ as the linear subspace generated by all tangents to X in x. Although the definition of $T_x(X)$ is purely algebraic, it can be described similarly as in differential geometry: If X is (locally around x) the vanishing scheme of equations $f_1, \ldots, f_r$ in $\mathbb{A}_k^n$, then $T_x(X)$ is (after possibly extending the base field) the kernel of the Jacobi-matrix defined by $f_1, \ldots, f_r$. Alternatively, $T_x(X)$ can also be described as space of "derivatives of small curves through the point x". A locally noetherian scheme X where $\dim T_x(X) = \dim \mathscr{O}_{X,x}$ is called *regular*.

If X is locally at x the vanishing scheme of equations $f_1, \ldots, f_r$ in $\mathbb{A}_k^n$ ($r \le n$) such that the rank of the Jacobi matrix of the f_i at x is r, then we call X *smooth* at x over k. The notion of smoothness is a relative one and depends on the base field k. In fact we will define "smoothness" for an arbitrary morphism of schemes. It is connected to the notion of relative tangent space $T_x(X/k)$ which is introduced in Section (6.6) and which behaves better under base change and in families. On the other hand we will see, that its calculation can be reduced to the calculation of an absolute tangent space. In this chapter we will mainly consider smooth schemes over a field. General smooth morphisms will be studied in Volume II.

In Theorem 6.28 we will link the notions of regularity and smoothness. In particular we will see that they are equivalent if k is algebraically closed (or, more generally, if k is perfect).

For singular, i.e., (possibly) non-smooth, schemes there is a whole arsenal of notions to describe their singularities. Maybe the most important one is normality. It corresponds to the algebraic concept of an integrally closed ring and we will study it at the end of the chapter. Its importance stems from the fact, that for all (integral) schemes there is a rather simple process to "normalize" these schemes. This will be studied in Chapter 12.

© Springer Fachmedien Wiesbaden GmbH, part of Springer Nature 2020
U. Görtz und T. Wedhorn, *Algebraic Geometry I: Schemes*, Springer Studium
Mathematik – Master, https://doi.org/10.1007/978-3-658-30733-2_7

The tangent space

(6.1) Formal derivatives.

If R is a ring, and $f = \sum_{i=0}^{d} a_i T^i \in R[T]$ is a polynomial, we define the *formal derivative*

$$\frac{\partial f}{\partial T} = \sum_{i=1}^{d} i a_i T^{i-1}.$$

If $f \in R[T_1, \ldots, T_n]$, we define the "partial derivative" $\partial f/\partial T_i$ by viewing the polynomial ring as $R[T_1, \ldots, \widehat{T_i}, \ldots, T_n][T_i]$ and applying the previous definition with the ground ring $R[T_1, \ldots, \widehat{T_i}, \ldots, T_n]$. Then $\partial/\partial T_i \colon R[\underline{T}] \to R[\underline{T}]$ is an R-derivation of $R[\underline{T}]$, i.e. $\partial/\partial T_i$ is R-linear and

$$\frac{\partial fg}{\partial T_i} = f \frac{\partial g}{\partial T_i} + g \frac{\partial f}{\partial T_i}$$

for all $f, g \in R[\underline{T}]$ ("Leibniz rule").

Let $S \subset R[T_1, \ldots, T_n]$ be a multiplicative set. Then we can extend the definition to the localization $S^{-1} R[T_1, \ldots, T_n]$ (using the customary rules for derivatives of fractions). In particular, if $R = k$ is a field, we can take partial formal derivatives of elements of $k(T_1, \ldots, T_n)$. An analogous definition applies to formal power series instead of polynomials.

Lemma 6.1. *Let R be a ring, and let $f \in R[T_0, \ldots, T_n]$ be homogeneous of degree d. Then the partial derivatives satisfy the* Euler relation

$$\sum_{j=0}^{n} \frac{\partial f}{\partial T_j} \cdot T_j = d \cdot f.$$

Proof. Because of linearity, it is enough to check the statement when f is a monomial, and that is easy. $\qquad\square$

(6.2) Zariski's definition of the tangent space.

Let X be a scheme. We want to define the tangent space of X in a point x; it is a $\kappa(x)$-vector space. Heuristically, thinking of X as embedded in some ambient space, we would like to obtain the "vector space generated by all lines tangent to X in x". This heuristic is not a good starting point, though, for instance because the tangent space would possibly depend on the embedding. It will turn out however that the notion we define fits well into this picture, if X is embedded in an affine space. Let $\mathfrak{m}_x$ denote the maximal ideal of the local ring $\mathscr{O}_{X,x}$.

Definition 6.2. *Let X be a scheme, and let $x \in X$. Then $\mathfrak{m}_x/\mathfrak{m}_x^2$ is a vector space over $\mathscr{O}_{X,x}/\mathfrak{m}_x = \kappa(x)$, and the (Zariski, or absolute) tangent space of X at x is by definition the dual vector space*

$$T_x X = (\mathfrak{m}_x/\mathfrak{m}_x^2)^*.$$

Let us point out right away that this notion behaves best if X is a scheme over a field k, and $x \in X$ is a point with residue field k, which is the setting which we will see almost exclusively below. On the other hand, if for instance η is the generic point of any integral scheme X, we have $\mathfrak{m}_\eta = 0$, so the $K(X)$-vector space $(\mathfrak{m}_\eta/\mathfrak{m}_\eta^2)^*$ does not contain any information about X.

In Section (6.6) we will introduce the notion of relative tangent space, which behaves better for non-rational points and in families. But we will also see how to obtain the relative tangent space from an absolute tangent space. For now, we are concerned only with the absolute tangent space in the sense above, and usually simply call it the tangent space of X at x.

Remark 6.3. Let X be a scheme and $x \in X$.
(1) Nakayama's lemma shows that if $\mathfrak{m}_x$ is finitely generated, $\dim_{\kappa(x)} T_x X$ is the cardinality of any minimal generating set of $\mathfrak{m}_x$. In particular $\dim_{\kappa(x)} T_x X$ is finite if X is locally noetherian.
(2) If $U \subseteq X$ is an open neighborhood of x, we clearly have $T_x X = T_x U$.
(3) The tangent space is functorial in (X, x) in the following sense. Let $f \colon X \to Y$ be a morphism of schemes and let $x \in X$ be a point such that $\dim_{\kappa(f(x))} T_{f(x)} Y$ is finite. Then the local homomorphism $f_x^\sharp \colon \mathcal{O}_{Y,f(x)} \to \mathcal{O}_{X,x}$ induces a $\kappa(x)$-linear map $\mathfrak{m}_{f(x)}/\mathfrak{m}_{f(x)}^2 \otimes_{\kappa(f(x))} \kappa(x) \to \mathfrak{m}_x/\mathfrak{m}_x^2$. If the extension $\kappa(x)/\kappa(f(x))$ is finite or $T_{f(x)} Y$ is a finite-dimensional $\kappa(f(x))$-vector space, then dualizing we obtain an induced map on tangent spaces

(6.2.1) $$df_x \colon T_x X \to T_{f(x)} Y \otimes_{\kappa(f(x))} \kappa(x).$$

This construction is compatible with composition of morphisms in the obvious way.

(6.3) Tangent spaces of affine schemes over a field.

Let us investigate (and at the same time motivate) the definition of the tangent space in detail in the case where X is a scheme over a field k, and x is a k-valued point of X. We start with the situation for affine spaces.

Example 6.4. Let k be a field. We first compute the tangent spaces of k-valued points of $\mathbb{A}_k^n$. So let $x \in \mathbb{A}_k^n(k) = k^n$, say $x = (x_1, \dots, x_n)$. The maximal ideal in $k[T_1, \dots, T_n]$ corresponding to x is $(T_1 - x_1, \dots, T_n - x_n)$. The elements $T_i - x_i$ yield a basis of the k-vector space $\mathfrak{m}_x/\mathfrak{m}_x^2$. We can describe the resulting isomorphism $k^n \overset{\sim}{\to} T_x \mathbb{A}_k^n$ explicitly by

$$(v_1, \dots, v_n) \mapsto \left(\mathfrak{m}_x/\mathfrak{m}_x^2 \to k, \ \overline{g} \mapsto \sum v_i \frac{\partial g}{\partial T_i}(x) \right).$$

Now let $f_1, \dots, f_r \in k[T_1, \dots, T_n]$, and let $f \colon \mathbb{A}_k^n \to \mathbb{A}_k^r$ be the map given by the f_i. Let $x = (x_1, \dots, x_n) \in k^n = \mathbb{A}^n(k)$. Then the induced map $df_x \colon T_x \mathbb{A}_k^n \to T_{f(x)} \mathbb{A}^r$ is given, using the identifications of the tangent spaces with k^n and k^r, resp., as above, by the matrix

$$\left(\frac{\partial f_i}{\partial T_j}(x) \right)_{\substack{i=1,\dots,r, \\ j=1,\dots,n}}.$$

This is checked easily by using "Taylor expansions" of the f_i around $x = (x_1, \dots, x_n)$, i. e., by writing the f_i as polynomials in $T_j - x_j$.

Whenever we are given polynomials $f_1, \ldots, f_r \in R[T_1, \ldots, T_n]$ over some ring R, we denote by

$$(6.3.1) \qquad J_{f_1,\ldots,f_r} := \left(\frac{\partial f_i}{\partial T_j} \right)_{i,j} \in M_{r \times n}(R[T_1, \ldots, T_n])$$

the *Jacobian matrix* of the f_i.

Example 6.5. As in the previous example, let k be a field. Let $X = V(f_1, \ldots, f_r) \subseteq \mathbb{A}_k^n$ be a closed subscheme, $f_i \in k[T_1, \ldots, T_n]$. Fix a k-valued point x of X (in other words, a (closed) point with residue field $\kappa(x) = k$). The natural morphisms $X \hookrightarrow \mathbb{A}_k^n \to \mathbb{A}_k^r$ (the second one being given by the f_i) yield homomorphisms

$$\mathscr{O}_{\mathbb{A}_k^r, f(x)} \to \mathscr{O}_{\mathbb{A}_k^n, x} \to \mathscr{O}_{\mathbb{A}_k^n, x}/(f_1, \ldots, f_r) = \mathscr{O}_{X, x}$$

and taking the maximal ideals of these local rings modulo their squares, and dualizing, we get an exact sequence

$$0 \to T_x X \to T_x \mathbb{A}_k^n \to T_{f(x)} \mathbb{A}_k^r$$

of k-vector spaces. Together with the computation in the previous example, we see that we can identify $T_x X$ with the subspace

$$T_x X = \mathrm{Ker}(J_{f_1,\ldots,f_r}(x)) \subseteq T_x \mathbb{A}_k^n = k^n.$$

This is the description which is customarily used in differential geometry for submanifolds of affine space.

If X is any scheme locally of finite type over k and x is a k-valued point of X, we can choose an open affine neighborhood $U = \mathrm{Spec}\, A$ of x. Then $U \cong V(f_1, \ldots, f_r)$ for suitable polynomials $f_i \in k[T_1, \ldots, T_n]$ and we can compute $T_x X = T_x U$ as above.

Example 6.6. The one basic piece of computation which is still missing is computing the map induced on tangent spaces by a morphism given by rational polynomials.

Again, let k be a field. Let $f_1, \ldots, f_r, g_1, \ldots, g_r \in k[T_1, \ldots, T_n]$. Consider the open subset $X = D(g_1 \cdots g_r) \subseteq \mathbb{A}_k^n$, and consider the morphism $h \colon X \to \mathbb{A}_k^r$ given on R-valued points, R a k-algebra, by

$$(x_1, \ldots, x_n) \mapsto \left(\frac{f_1(x)}{g_1(x)}, \ldots, \frac{f_r(x)}{g_r(x)} \right).$$

For $x \in X(k)$, it induces a homomorphism $dh_x \colon k^n = T_x \mathbb{A}_k^n = T_x X \to T_{h(x)} \mathbb{A}_k^r = k^r$. To compute it, we identify X with the closed subset $V(g_1 T_{n+1} - 1, \ldots, g_r T_{n+r} - 1) \subseteq \mathbb{A}_k^{n+r}$. We can then decompose h as $X \hookrightarrow \mathbb{A}_k^{n+r} \to \mathbb{A}_k^r$, where the second morphism is given by

$$(x_1, \ldots, x_{n+r}) \mapsto (f_1(x_1, \ldots, x_n) x_{n+1}, \ldots, f_r(x_1, \ldots, x_n) x_{n+r}).$$

Taking the tangent space is compatible with composition of morphisms, and it is a straightforward computation to show that dh_x is given by the $(r \times n)$-matrix

$$\left(\frac{\partial (f_i/g_i)}{\partial T_j}(x) \right)_{i,j}.$$

(6.4) Tangent space as set of $k[\varepsilon]$-valued points.

Often k-schemes are not defined via equations but by their R-valued points for k-algebras R. Then the following interpretation of the tangent space in functorial terms is often helpful (see e.g., Exercise 6.4 or Section (8.9)). It is the algebraic version of the definition of the tangent spaces in differential geometry for abstract manifolds as the space of derivatives of small curves through the point x.

We start with the definition of the algebraic version of an "infinitesimally small curve". For a field k, we denote the ring $k[T]/(T^2)$ by $k[\varepsilon]$, where ε denotes the residue class of T, i. e., $\varepsilon^2 = 0$. This ring is called the *ring of dual numbers (over k)*. We think of $\operatorname{Spec} k[\varepsilon]$ as an infinitesimal small subcurve of $\mathbb{A}^1_k$; see Example 2.40.

Now let X be a k-scheme. Recall the notation $X(k[\varepsilon]) = \operatorname{Hom}_k(\operatorname{Spec} k[\varepsilon], X)$. Given a k-valued point x of X, we denote by $X(k[\varepsilon])_x \subset X(k[\varepsilon])$ the preimage of x under the projection $X(k[\varepsilon]) \to X(k)$.

If $f\colon \operatorname{Spec} k[\varepsilon] \to X$ is an element of $X(k[\varepsilon])_x$, then the induced morphism $\mathscr{O}_{X,x} \to k[\varepsilon]$ maps $\mathfrak{m}_x$ into $\varepsilon k \cong k$, and we obtain a map $\mathfrak{m}_x/\mathfrak{m}_x^2 \to k$, i. e., an element of $T_x X$.

Proposition 6.7. *Let X be a k-scheme, and $x \in X(k)$. The map*

$$X(k[\varepsilon])_x \to T_x X$$

constructed above is a bijection which is functorial in (X, x).

Proof. Take a homomorphism $t\colon \mathfrak{m}_x/\mathfrak{m}_x^2 \to k$, and consider the induced map $\mathfrak{m}_x \to \varepsilon k$, $m \mapsto \varepsilon t(m \bmod \mathfrak{m}^2)$. Since by assumption $\mathscr{O}_{X,x}/\mathfrak{m}_x = k$, we can extend this map in a unique way to a k-algebra homomorphism $\mathscr{O}_{X,x} \to k[\varepsilon]$. This construction is inverse to the one described before. □

Remark 6.8. We can also express the structure of k-vector space on $X(k[\varepsilon])_x$ induced by the above bijection in functorial terms. We set $k[\varepsilon_1, \varepsilon_2] := k[T_1, T_2]/(T_1^2, T_2^2, T_1 T_2)$, where ε_i is the residue class of T_i. Then we have a natural map $p\colon k[\varepsilon_1, \varepsilon_2] \to k[\varepsilon]$, $\varepsilon_1 \mapsto \varepsilon$, $\varepsilon_2 \mapsto \varepsilon$. Now let $v_1, v_2 \in T_x X$ be tangent vectors which we view, using the above bijection, as morphisms $v_i\colon \operatorname{Spec} k[\varepsilon] \to X$ with image $\{x\}$. The v_i correspond to morphisms $\mathscr{O}_{X,x} \to k[\varepsilon]$, $s \mapsto s(x) + \varepsilon \dot{v}_i(s)$, and we obtain a map $\alpha\colon \mathscr{O}_{X,x} \to k[\varepsilon_1, \varepsilon_2]$, $s \mapsto s(x) + \varepsilon_1 \dot{v}_1(s) + \varepsilon_2 \dot{v}_2(s)$. Then the sum $v_1 + v_2$ corresponds to the $k[\varepsilon]$-valued point

$$\operatorname{Spec} k[\varepsilon] \to \operatorname{Spec} k[\varepsilon_1, \varepsilon_2] \to X,$$

where the first morphism is the one induced by p, and the second one is induced by α. The multiplication by scalars a in k is given as follows: Let $m_a\colon k[\varepsilon] \to k[\varepsilon]$ the k-algebra homomorphism that sends ε to $a\varepsilon$. If $v \in T_x X$ corresponds to $v\colon \operatorname{Spec} k[\varepsilon] \to X$ then av corresponds to the composition

$$\operatorname{Spec} k[\varepsilon] \xrightarrow{\operatorname{Spec} m_a} \operatorname{Spec} k[\varepsilon] \xrightarrow{v} X.$$

As an immediate application we obtain that the formation of the tangent space is compatible with products:

Proposition 6.9. *Let k be a field, let X, Y be k-schemes, and let $x \in X(k)$, $y \in Y(k)$ be k-valued points. Then there is a natural isomorphism*

$$T_{(x,y)}(X \times_k Y) \xrightarrow{\sim} T_x X \oplus T_y Y.$$

Proof. This is an easy consequence of the description of the tangent space as the space of $k[\varepsilon]$-valued points and the universal property of fiber products (4.4.2). $\qquad\square$

This result can be easily generalized to fiber products of k-schemes (Exercise 6.6).

(6.5) Computation of the tangent spaces of projective schemes.

Let us apply Example 6.6 in order to look at the tangent spaces of points in projective space. We continue to work over our fixed field k. Clearly, since $\mathbb{P}_k^n$ can be covered by affine spaces of dimension n over k, all tangent spaces $T_x\mathbb{P}_k^n$ of points $x \in \mathbb{P}_k^n(k)$ are n-dimensional k-vector spaces. However, this reasoning does not give us a natural choice of basis, and hence no natural identification with k^n. To obtain a more intrinsic description, we consider the morphism $\mathbb{A}_k^{n+1} \setminus \{0\} \to \mathbb{P}_k^n$ (see (3.6.1)). Let us fix a point $\dot{x} = (x_0,\ldots,x_n) \in \mathbb{A}_k^{n+1}(k) \setminus \{0\}$, and denote by x its image in $\mathbb{P}_k^n(k)$. We obtain a homomorphism

$$k^{n+1} = T_{\dot{x}}\mathbb{A}_k^{n+1} \to T_x\mathbb{P}_k^n.$$

Let us compute its kernel. There exists i with $x_i \neq 0$. To simplify the notation, we assume that $i = 0$. We denote by $U_0 \subset \mathbb{P}_k^n$ the corresponding open chart of $\mathbb{P}_k^n$ of points with 0-th homogeneous coordinate $\neq 0$, and we identify $U_0 = \mathbb{A}_k^n$ as usual. We use $T_0,\ldots,T_n$ as coordinates on $\mathbb{A}_k^{n+1}$, and consider the principal open subset $D(T_0)$. We can compute the kernel we are interested in terms of the map

$$D(T_0) \to U_0 = \mathbb{A}_k^n, \quad (x_0,\ldots,x_n) \mapsto (x_0^{-1}x_1,\ldots,x_0^{-1}x_n),$$

which means that we are in the situation of Example 6.6. The map induced on the tangent spaces is given by the matrix

$$x_0^{-1}\begin{pmatrix} -x_1/x_0 & 1 & & \\ \vdots & & \ddots & \\ -x_n/x_0 & & & 1 \end{pmatrix} \in M_{n\times(n+1)}(k).$$

Therefore the kernel of the homomorphism $k^{n+1} \to T_x\mathbb{P}_k^n$ is the line $k \cdot (x_0,\ldots,x_n)$. Since $T_x\mathbb{P}_k^n$ is n-dimensional, this implies in particular that the map in question is surjective, and altogether we have proved the first part of

Proposition 6.10. *Let k be a field.*
(1) *Let $x = (x_0 : \cdots : x_n) \in \mathbb{P}_k^n(k)$. Then we have a natural identification*

$$T_x\mathbb{P}_k^n = k^{n+1}/k(x_0,\ldots,x_n).$$

(2) *Let $X = V_+(f_1,\ldots,f_r) \subseteq \mathbb{P}_k^n$ be a closed subscheme, given by homogeneous polynomials f_i, and let $x \in X(k)$. Then*

$$T_xX = \left(\mathrm{Ker}\left(\frac{\partial f_i}{\partial T_j}(x)\right)_{i,j}\right)/kx.$$

Proof. We write $C(X) = V(f_1,\ldots,f_r) \subseteq \mathbb{A}_k^{n+1}$, and obtain a cartesian diagram (Section (4.14))

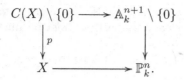

Fix a point $\dot{x} \in C(X) \setminus \{0\}$ mapping to x, and consider the corresponding homomorphism $dp_{\dot{x}} \colon T_{\dot{x}} C(X) \to T_x X$. Because of Lemma 6.1, we have $\sum_{j=0}^{n} \frac{\partial f_i}{\partial T_j}(x) x_j = \deg(f_i) f_i(x) = 0$ for all i, so the line $k\dot{x}$ is contained in the kernel. Now to prove our claim, it is enough to show that the homomorphism $dp_{\dot{x}}$ is surjective. This can be checked locally on X, so by (4.13.1) we are reduced to a product situation, and Proposition 6.9 shows that the homomorphism on the tangent spaces is the projection onto a direct summand.   □

Let us apply the proposition to continue our investigation of quadrics; see Section (1.26).

Proposition 6.11. *Let k be an algebraically closed field of characteristic $\neq 2$, and let Q be a quadric over k, i. e., Q is isomorphic to a closed subscheme of the form $V_+(X_0^2 + \cdots + X_{r-1}^2) \subset \mathbb{P}_k^n$ for some $n \geq 1$ and $1 \leq r \leq n+1$.*
(1) The scheme Q is reduced if and only if $r > 1$.
(2) The scheme Q is irreducible if and only if $r \neq 2$.
(3) Assume that Q is reduced. Then for all closed points $x \in Q$, we have

$$n - 1 \leq \dim T_x Q \leq n,$$

and there exist x with $\dim T_x Q = n - 1$. If $r = n+1$, then the set of closed points $x \in Q$ where $\dim T_x Q = n$ is empty; otherwise it is the set of closed points of a linear subspace of $\mathbb{P}_k^n$ of dimension $n - r$.
In particular, this proves that quadrics of different rank or different dimension cannot be isomorphic; cf. Proposition 1.69, Corollary 1.71.

Proof. Parts (1) and (2) are clear; note that they were discussed without using the language of schemes in Section (1.26).

Part (3) easily follows from Proposition 6.10, which shows that the tangent space in a closed point $x = (x_0 : \cdots : x_n)$ of Q is given by

$$(\mathrm{Ker}(2x_0, \ldots, 2x_{r-1}, 0, \ldots, 0) \colon k^{n+1} \to k)/kx.   □$$

(6.6) The relative tangent space at a k-valued point.

The Zariski tangent space is a good concept for k-rational points of a scheme over a field k. For arbitrary points the following generalization of the tangent space at a k-valued point is more useful. Consider a commutative diagram of schemes, where K is a field

(6.6.1)

$$\begin{array}{ccc} \mathrm{Spec}\, K & \xrightarrow{\;\xi\;} & X \\ & \searrow & \downarrow \\ & & S. \end{array}$$

Define $T_\xi(X/S)$ as the set of S-morphisms $t\colon \operatorname{Spec} K[\varepsilon] \to X$ such that the composition of t with $\operatorname{Spec} K \to \operatorname{Spec} K[\varepsilon]$ is equal to ξ. As in Section (6.4) we can endow $T_\xi(X/S)$ with the structure of a K-vector space. We call this K-vector space $T_\xi(X/S)$ the *(relative) tangent space of X in ξ over S*. If $x \in X$ is a point and $\xi\colon \operatorname{Spec} \kappa(x) \to X$ is the canonical morphism we also write $T_x(X/S)$ instead of $T_\xi(X/S)$. If $S = \operatorname{Spec} R$ is affine, we also write $T_\xi(X/R)$ or $T_x(X/R)$.

Note that $T_\xi(X/S)$ depends on S. If k is a field, $S = \operatorname{Spec} k = \operatorname{Spec} K$, and x is a k-valued point, we have $T_x(X/k) = T_x X$. But in general one has to distinguish between the relative and the absolute tangent space: Even for a k-scheme X of finite type it may happen that the $\kappa(x)$-vector spaces $T_x X$ and $T_x(X/k)$ are not isomorphic for a point $x \in X$ that is not k-rational (see Exercise 6.3).

Remark 6.12.

(1) The relative tangent space can be defined in terms of the absolute tangent space as follows. Consider again the diagram (6.6.1) above. By definition of the fiber product, the morphism ξ corresponds to a K-valued point $\bar{x}$ of the k-scheme $X \times_S \operatorname{Spec} K$. In a similar way, the S-morphisms $\operatorname{Spec} K[\varepsilon] \to X$ correspond to K-morphisms $\operatorname{Spec} K[\varepsilon] \to X \times_S \operatorname{Spec} K$. Via Proposition 6.7 we obtain an identification of K-vector spaces
$$T_\xi(X/S) = T_{\bar{x}}(X \times_S \operatorname{Spec} K).$$

(2) The relative tangent space is functorial in ξ in the following sense. In the situation of (6.6.1), let $\iota\colon K \hookrightarrow L$ be a field extension corresponding to the morphism $p\colon \operatorname{Spec} L \to \operatorname{Spec} K$. Then composition with $\iota \otimes \operatorname{id}_{K[\varepsilon]}\colon K[\varepsilon] \to L[\varepsilon]$ induces an isomorphism of L-vector spaces

(6.6.2) $$T_\xi(X/S) \otimes_K L \xrightarrow{\sim} T_{\xi \circ p}(X/S).$$

(3) Consider the following special case. Let X be a scheme over a field k, $x \in X$ a point, $k \hookrightarrow k'$ a field extension, and let $x' \in X' := X \otimes_k k'$ be a point that projects to x. Then we have

(6.6.3) $$T_{x'}(X'/k') = T_\xi(X/k) \cong T_x(X/k) \otimes_{\kappa(x)} \kappa(x'),$$

where ξ is the composition $\operatorname{Spec} \kappa(x') \to \operatorname{Spec} \kappa(x) \to X$.

The dimension of the relative tangent space is upper semi-continuous:

Proposition 6.13. *Let k be a field and let X be a k-scheme locally of finite type. Then for each integer d the set $\{\, x \in X \;;\; \dim_{\kappa(x)} T_x(X/k) \geq d \,\}$ is closed in X.*

Proof. The question is local on X and we can therefore assume that $X = \operatorname{Spec} A$ with $A = k[T_1, \ldots, T_n]/(f_1, \ldots, f_r)$. For each $x \in X$ we have
$$T_x(X/k) = T_x(X \otimes_k \kappa(x)) = \operatorname{Ker}(J_{f_1, \ldots, f_r}(x)) \subseteq \kappa(x)^n.$$

Thus we see that $\{\, x \in X \;;\; \dim T_x(X/k) \geq d \,\}$ is the closed subspace of zeros of the ideal of A generated by the images of $(n - d + 1) \times (n - d + 1)$-minors of the Jacobian matrix $J_{f_1, \ldots, f_r}$. $\square$

(6.7) The projective tangent space.

Let k be a field, let X be a closed subscheme of $\mathbb{P}^N_k$, and let $x\colon \operatorname{Spec} K \to X$ be a K-valued point, where K is some field extension of k. We can consider the (relative) tangent space $T_x(X/k)$ of X at x as a K-subvector space of $T_x(\mathbb{P}^N_k/k) = T_x(\mathbb{P}^N_K)$. Then there is a unique linear subspace $\Lambda \subseteq \mathbb{P}^N_K$ such that $T_x(\Lambda) = T_x(X/k)$: If $X = V_+(f_1,\dots,f_r)$ for homogeneous polynomials $f_i \in k[T_0,\dots,T_N]$, then Λ is the subspace of $\mathbb{P}^N_k$ corresponding to the linear subspace of K^{N+1} given by the kernel of the matrix $\big((\partial f_i/\partial T_j)(x)\big)_{i,j}$ viewed as linear map $K^{N+1} \to K^r$ (Proposition 6.10). We call Λ the *projective tangent space* of X at x and denote it by $T_x(X \subset \mathbb{P}^N/k)$. Note that it depends on the embedding of X into $\mathbb{P}^N$. Its dimension as a linear subspace of $\mathbb{P}^N_K$ agrees with $\dim_K T_x(X/k)$ again by Proposition 6.10.

Smooth morphisms

(6.8) Definition of smoothness.

In this section, we define the notion of smooth morphism, and in particular the notion of smooth k-schemes (for a field k). Heuristically, smoothness should mean that the scheme in question "locally" looks like affine space. However, the Zariski topology is not sufficiently fine to appropriately make sense of this. Instead one has to use the étale topology; this will be explained in Volume II. Here we define smoothness by the condition that locally, the scheme in question is defined by equations $f_1,\dots,f_r$ in some affine space which behave as coordinate functions $T_1,\dots,T_r$, at least if we only consider their first derivatives. This notion of smoothness is the same as the one used in differential geometry.

Definition 6.14. *Let $f\colon X \to Y$ be a morphism of schemes and $d \geq 0$ be an integer.*
(1) *We say that f is* smooth of relative dimension d *at $x \in X$, if there exist affine open neighborhoods U of x and $V = \operatorname{Spec} R$ of $f(x)$ such that $f(U) \subset V$, and an open immersion*
$$U \hookrightarrow \operatorname{Spec} R[T_1,\dots,T_n]/(f_1,\dots,f_{n-d})$$
of R-schemes for suitable n and f_i, such that the Jacobian matrix
$$J_{f_1,\dots,f_{n-d}}(x) = \left(\frac{\partial f_i}{\partial T_j}(x)\right)_{i,j} \in M_{(n-d)\times n}(\kappa(x))$$
has rank $n-d$.
(2) *We say that $f\colon X \to Y$ is* smooth, *or that X is* smooth over Y, *(of relative dimension d), if it is smooth (of relative dimension d) at all points $x \in X$.*

Recall for this definition that we denote for $g \in R[T_1,\dots,T_n]$ (e.g., $g = \frac{\partial f_i}{\partial T_j}$) and for $x \in \mathbb{A}^n_R$ (or x in a subscheme U of $\mathbb{A}^n_R$) by $g(x) \in \kappa(x)$ the image of g in $\mathcal{O}_{\mathbb{A}^n_R,x}/\mathfrak{m}_x$. In part (1), requiring that the open immersion $U \hookrightarrow \operatorname{Spec} R[T_1,\dots,T_n]/(f_1,\dots,f_{n-d})$ be a morphism of R-schemes amounts precisely to saying that its composition with the projection to $\operatorname{Spec} R$ equals the restriction of f to U. If $R = k$ is a field and $\mathfrak{p}_x = (T_1 - a_1,\dots,T_n - a_n)$ for $a = (a_1,\dots,a_n) \in k^n$, then $g(x) \in \kappa(x) = k$ is simply the usual evaluation of g in a.

One of the obvious problems of our definition of smoothness is that it is hard to prove that a certain morphism is not smooth. To this end, it is desirable to show that, in a suitable sense, the validity of the condition is independent of the choice of representation as a subscheme of some $\mathbb{A}_R^n$. We will come back to this question in a special situation in the section about regular rings, see Corollary 6.31, and in full generality in Volume II.

Proposition 6.15.
(1) Let $f\colon X \to Y$ be a morphism of schemes, and let $x \in X$ be a smooth point of relative dimension d for f. There exists an open neighborhood U of x, such that f is smooth of relative dimension d in every point of U.
(2) Smoothness is local on the source and on the target, in the following sense: given a morphism $f\colon X \to Y$, an open subset $U \subseteq X$, and $x \in U$, then f is smooth at x if and only if the restriction $U \to Y$ is smooth at x. If $V \subseteq Y$ is open and $x \in f^{-1}(V)$, then f is smooth at x if and only if the restriction $f^{-1}(V) \to V$ is smooth at x.
(3) If $f\colon X \to Y$ is smooth of relative dimension d at $x \in X$, and $Y' \to Y$ is a morphism, then the morphism $X \times_Y Y' \to Y'$ obtained by base change is smooth of relative dimension d at all points of $X \times_Y Y'$ which project to x.
(4) Let $f\colon X \to Y$ and $g\colon Y \to Z$ be morphisms, and assume that f is smooth at $x \in X$, and that g is smooth at $f(x) \in Y$. Then $g \circ f$ is smooth at x.
(5) Open immersions are smooth of relative dimension 0.

For a morphism $f\colon X \to Y$ we call the open subscheme

$$(6.8.1) \qquad X_{\mathrm{sm}} := X_{\mathrm{sm}}(f) := \{\, x \in X \;;\; f \text{ is smooth at } x \,\} \subseteq X$$

the *smooth locus* of X with respect to f.

Proof. Since the rank condition in the definition of smoothness can be phrased by saying that there exists a $r \times r$ minor of the Jacobian matrix which does not vanish at x (i. e. is $\neq 0$ in $\kappa(x)$), it is an open condition. This proves the first point. The assertions (2), (3), and (5) are clear. We skip the proof of Assertion (4) (it is not used in Volume I and we will give a proof in Volume II). $\qquad \square$

A smooth morphism of relative dimension 0 is also called *étale*. We will study étale morphisms in more detail in Volume II.

In particular, whenever $f\colon X \to Y$ is a morphism of schemes which is smooth at x of relative dimension d, then the fiber $X_{f(x)} = X \times_Y \operatorname{Spec} \kappa(f(x))$ is smooth over $\kappa(f(x))$ at x. As Theorem 6.28 below shows, in this case $\dim \mathscr{O}_{X_{f(x)},x} = d$, i.e. the maximal dimension of the irreducible components of $X_{f(x)}$ containing x is d (Corollary 5.23). This justifies the term "of relative dimension d".

In most of this chapter, we will only consider the special case $Y = \operatorname{Spec} k$, k a field. Note that for a k-scheme X, whenever $x \in X$ is a smooth point, the neighborhood U of x in the definition above is of finite type over k. Therefore, every smooth k-scheme is locally of finite type over k.

Examples 6.16.
(1) Let S be a scheme. Then $\mathbb{A}_S^n$ and $\mathbb{P}_S^n$ are smooth of relative dimension n over S.

(2) Let k be a field, let X be a k-scheme locally of finite type, and let $x \in X(k)$. Then we will see in Theorem 6.28 that X is smooth at x if and only if $\dim T_x X = \dim_x X$. This shows that in this case the criterion defining smoothness is independent of the choice of embedding of a neighborhood of x into affine space. We also see that for $X = V_+(f_1, \ldots, f_r) \subseteq \mathbb{P}^n_k$, f_i homogeneous, and $x \in X(k)$, we can check whether X is smooth at x looking at the Jacobian matrix of the homogeneous equations defining X (use Proposition 6.10). See also Exercise 6.13.

(3) In particular: Let k be an algebraically closed field of characteristic $\neq 2$, and let Q be a quadric of dimension n and rank $n + 1$ over k. Then Q is smooth over k.

(4) Let k be an algebraically closed field of characteristic $\neq 2$, and let $f \in k[X]$ be a polynomial. Then $V(Y^2 - f(X)) \subset \mathbb{A}^2_k$ is smooth if and only if f has no multiple zeros. (To see that $V(Y^2 - f(X))$ is not smooth when f has multiple zeros, use Theorem 6.28, cf. part (1) above.)

(6.9) Existence of smooth points for schemes over a field.

Lemma 6.17. *Let X and Y be k-schemes locally of finite type. Let $x \in X$ and $y \in Y$ and let $\varphi \colon \mathcal{O}_{Y,y} \overset{\sim}{\to} \mathcal{O}_{X,x}$ be an isomorphism of k-algebras. Then there exists an open neighborhood U of x in X and V of y in Y and an isomorphism $h \colon U \overset{\sim}{\to} V$ of k-schemes with $h(x) = y$ such that $h^\sharp_x = \varphi$.*

This is a special case of a more general result of extending morphisms from stalks to open neighborhoods, see Proposition 10.52. Here we will give a quick proof in the case that X and Y are integral (the only case we need in this chapter).

Proof. (if X and Y are integral) We may assume that $X = \operatorname{Spec} B$ and $Y = \operatorname{Spec} A$ are affine. Let $\mathfrak{p} = \mathfrak{p}_y \subset A$ and $\mathfrak{q} = \mathfrak{p}_x \subset B$ be the prime ideals corresponding to y and x. By hypothesis there exists an isomorphism $\varphi \colon A_\mathfrak{p} \overset{\sim}{\to} B_\mathfrak{q}$. We denote the induced isomorphism $\operatorname{Frac}(A) \overset{\sim}{\to} \operatorname{Frac}(B)$ again by φ.

Since A and B are finitely generated, we can find elements $f \in A$, $g' \in B$ such that $\varphi(A) \subseteq B_{g'} \subseteq \varphi(A_f)$, and since $\varphi(A) \subset \varphi(A_\mathfrak{p}) = B_\mathfrak{q} \supset B$, we can even choose $f \in A \setminus \mathfrak{p}$, $g' \in B \setminus \mathfrak{q}$. But then for suitable $n > 0$, $g := (g')^n \varphi(f)$ lies in $B \setminus \mathfrak{q}$, too, and $\varphi(A_f) = B_g$. Thus φ yields an isomorphism $h \colon U := D(g) \overset{\sim}{\to} V := D(f)$. $\qquad\square$

Proposition 6.18. *Let k be a field and let X be an integral k-scheme of finite type of dimension $d \geq 0$. Assume that its function field $K(X)$ is a separable extension of k (e.g., if k is perfect). Then there exists a dense open subscheme $U \subseteq X$ such that U is isomorphic to a dense open subscheme of $\operatorname{Spec} k[T_1, \ldots, T_d, T]/(g)$, where g is a monic separable irreducible polynomial $g \in k(T_1, \ldots, T_d)[T]$ with coefficients in $k[T_1, \ldots, T_d]$.*

Proof. The function field $K(X)$ is a finitely generated separable extension of k, i. e., we can find a transcendence basis $T_1, \ldots, T_d$ of $K(X)$ over k such that $K(X) \supseteq k(T_1, \ldots, T_d)$ is finite separable; see Proposition B.97. In particular, the latter extension is generated by a single element $\alpha \in K(X)$ (Proposition B.98). Replacing α by $f\alpha$ for a suitable $f \in k[T_1, \ldots, T_d]$, we may assume that the minimal polynomial $g \in k(T_1, \ldots, T_d)[T]$ of α over $k(T_1, \ldots, T_d)$ has coefficients in $k[T_1, \ldots, T_d]$. Set $B = k[T_1, \ldots, T_d, T]/(g)$ and $Y = \operatorname{Spec} B$. Then X and Y have isomorphic function fields and the claim follows from Lemma 6.17, applied to the generic points. $\qquad\square$

Theorem 6.19. *Let k be a perfect field, and let X be a non-empty reduced k-scheme which is locally of finite type. Then the smooth locus X_{sm} of X over k is open and dense.*

Proof. We know already that X_{sm} is open in X (Proposition 6.15 (1)). To prove density it suffices to show that for every irreducible component Z of X there exists a non-empty open affine subscheme U of X which is contained in Z such that U_{sm} is dense in U. The set of irreducible components of X is locally finite because X is locally noetherian. Thus there exists a non-empty open subset U of Z that does not meet any other irreducible component of X and hence is also open in X. Thus we may assume that X is integral. By Proposition 6.18 we may then assume that $X = \mathrm{Spec}\, k[T_1, \ldots, T_d, T_{d+1}]/(g)$ with g irreducible and separable as a polynomial in T_{d+1} with coefficients in the field $k(T_1, \ldots, T_d)$.

Consider the partial derivatives $\partial g/\partial T_i \in k[T_1, \ldots, T_{d+1}]$. We must show that they are not all divisible by the irreducible polynomial g. Because of degree reasons, this just means that they do not all vanish. But since g is irreducible and separable as a polynomial in T_{d+1}, we have $\partial g/\partial T_{d+1} \neq 0$. $\qquad\square$

Remark 6.20. The proof shows that instead of assuming that k is perfect it suffices to make one of the following weaker assumptions (which are equivalent by Proposition 5.49):
(i) For each irreducible component Z of X the function field $K(Z)$ is a separable extension of k.
(ii) The k-scheme X is geometrically reduced.

A similar idea shows the following result.

Proposition 6.21. *Let X be a scheme locally of finite type over a field k. Assume that X_{red} is geometrically reduced. Then the set of closed points $x \in X$ such that $\kappa(x)$ is separable over k contains an open dense subset of X.*

Proof. Replacing X by X_{red} we may assume that X is geometrically reduced. Let Y be the closed subset of points in X which are contained in at least two irreducible components. Replacing X by the open and dense subscheme $X \setminus Y$ we may assume that every connected component of X is irreducible. By proving the theorem for each component we may assume that X is integral.

As X is geometrically reduced, its function field is a separable extension of k (Proposition 5.49). By Proposition 6.18 we may assume that $X = \mathrm{Spec}(B)$, where $B = k[T_1, \ldots, T_{d+1}]/(g)$ for a separable monic irreducible polynomial $g \in k(T_1, \ldots, T_d)[T_{d+1}]$ with coefficients in $k[T_1, \ldots, T_d]$.

We have obtained a morphism $f \colon \mathrm{Spec}(B) \to \mathbb{A}_k^d = \mathrm{Spec}(k[T_1, \ldots, T_d])$. The subset of points $z \in \mathbb{A}_k^d$ such that the image $\overline{g}$ of g in $\kappa(z)[T_{d+1}]$ is non-separable is the vanishing locus of the discriminant of the polynomial g, cf. Section (B.20) and Exercise 6.26, hence a Zariski closed subset. Since g is separable over $k(T_1, \ldots, T_d)$ it does not contain the generic point, so its complement V is open and dense. Whenever $z \in V$, the fiber $\mathrm{Spec}(\kappa(z)[T_{d+1}]/(\overline{g}))$ of the above morphism over z is the spectrum of a product of separable extensions of k (since $\overline{g}$ might not be irreducible over $\kappa(z)$, we might have more than one factor). In particular, for all $x \in f^{-1}(V)$, the extension $\kappa(x)/k$ is separable. As X is integral, $f^{-1}(V)$ is (open and) dense. This shows the proposition. $\qquad\square$

(6.10) Complete local rings.

In the analytic world, we have the inverse function theorem which shows that under a condition analogous to our smoothness condition above, zero sets are locally isomorphic to $\mathbb{R}^n$ (or $\mathbb{C}^n$). In our setting, we cannot expect the same result to hold because the Zariski topology is far too coarse. We cannot work with convergent power series either, because we do not have a notion of convergence. At least, Proposition 6.23 below shows that the corresponding result holds *formally*, i. e., for formal power series, where we simply do not require convergence: the complete local ring at a smooth k-valued point is a power series ring.

Lemma 6.22. *Let R be a ring, $n \geq 1$. An R-algebra homomorphism*

$$\varphi \colon R[\![Y_1, \ldots, Y_n]\!] \to R[\![X_1, \ldots, X_n]\!] \text{ with } \varphi(Y_j) \in (X_1, \ldots, X_n) \text{ for all } j$$

is an isomorphism if the Jacobian matrix

$$\left(\frac{\partial \varphi(Y_i)}{\partial X_j}(0) \right)_{i,j} \in M_{n \times n}(R)$$

is invertible over R.

Proof. The ring $R[\![Y_1, \ldots, Y_n]\!]$ is complete with respect to the $(Y_1, \ldots, Y_n)$-adic topology (Example B.47 (1)). Therefore by Proposition B.49 the homomorphism φ is an isomorphism if and only if the induced homomorphism $\operatorname{gr} \varphi \colon R[Y_1, \ldots, Y_n] \to R[X_1, \ldots, X_n]$ on the associated graded rings (Example B.48) is an isomorphism. But

$$(\operatorname{gr} \varphi)(Y_j) = \sum_{i=1}^{n} \frac{\partial \varphi(Y_j)}{\partial X_i}(0) \cdot X_i. \qquad \square$$

Proposition 6.23. *Let k be a field, let X be a k-scheme, and let $x \in X(k)$ be a point which is smooth of relative dimension d over k. Then the completion $\widehat{\mathscr{O}}_{X,x}$ of the local ring $\mathscr{O}_{X,x}$ (with respect to its maximal ideal) is isomorphic to a power series ring over k in d indeterminates.*

Proof. By the definition of smoothness, and because we can compute the local ring in an arbitrary open neighborhood of x, we may assume that X is the closed subscheme of $\mathbb{A}_k^n$ defined by polynomials $f_1, \ldots, f_r$ with $r = n - d$, such that the Jacobian matrix $J := J_{f_1, \ldots, f_r}(x)$ has rank r. By a change of coordinates in $\mathbb{A}_k^n$ we may furthermore assume that x is the origin in $\mathbb{A}_k^n$ (and hence $f_i \in (T_1, \ldots, T_n)$ for all i). By renumbering the T_i we can assume that the $(r \times r)$-minor given by the first r columns of J does not vanish at x. Using Lemma 6.22, we see that the homomorphism $k[\![U_1, \ldots, U_n]\!] \to k[\![T_1, \ldots, T_n]\!] \cong \widehat{\mathscr{O}}_{\mathbb{A}_k^n, x}$ given by

$$U_i \mapsto \begin{cases} f_i & 1 \leq i \leq r \\ T_i & r < i \leq n \end{cases}$$

is an isomorphism. But this means that

$$\widehat{\mathscr{O}}_{X,x} \cong \widehat{\mathscr{O}}_{\mathbb{A}_k^n, x}/(f_1, \ldots, f_r) \cong k[\![U_{r+1}, \ldots, U_n]\!]$$

is isomorphic to a ring of formal power series over k in $d = n - r$ indeterminates. $\qquad \square$

Regular schemes

We now study the notion of regular schemes, which is based on the commutative algebra notion of a regular (local) ring. For schemes over a field it mostly can (should) be replaced by the notion of smoothness. However, it does serve an important purpose in the arithmetic setting. Let us explain this. It is often important to consider morphisms $f\colon X \to Y$ of k-schemes where X is smooth, but f is not necessarily smooth. If k is a perfect field, we can replace the smoothness condition by requiring that X is regular. A typical analogue in the arithmetic setting would be a morphism $f\colon X \to \operatorname{Spec}\mathbb{Z}$. Continuing the analogy, we would not want to require f to be smooth; asking that X is regular, however, is often useful (and cannot be replaced, in this context, by a smoothness condition on X because there exists no ground field); see Exercise 6.16 for examples.

(6.11) Regular schemes.

It is straightforward to transfer the definition of regularity from rings (Definition B.76) to schemes:

Definition 6.24. *Let X be a locally noetherian scheme. A point $x \in X$ is called regular, if the local ring $\mathscr{O}_{X,x}$ is regular. The scheme X is called regular if every point is regular.*

Remark 6.25.
(1) Let A be a noetherian ring. Then Proposition B.77 (1) shows that A is regular if and only if $\operatorname{Spec} A$ is regular.
(2) We can also express regularity in terms of the tangent space: A point $x \in X$ is regular if and only if $T_x X$ has dimension (as $\kappa(x)$-vector space) $\dim \mathscr{O}_{X,x}$.
(3) Let X be a locally noetherian scheme. Then every point of X specializes to a closed point (this is easy to see e. g., if X is quasi-compact, or locally of finite type over a field; but in fact holds for an arbitrary locally noetherian scheme, see Exercise 5.5). Then Proposition B.77 (1) shows that X is regular if all of its closed points are regular.
(4) The question, whether for a locally noetherian scheme X the *regular locus*

$$(6.11.1) \qquad\qquad X_{\mathrm{reg}} := \{\, x \in X \; ; \; \mathscr{O}_{X,x} \text{ is regular}\,\}$$

is open in X, is rather delicate in general. We will state a sufficient criterion in Section (12.12) which in particular implies that X_{reg} is open if X is locally of finite type over a field k. Over a perfect field k a point x of X is regular if and only if X is smooth at x (see Corollary 6.32 and Remark 6.33) and the openness of X_{reg} follows from the openness of the smooth locus X_{sm} (6.8.1).

(6.12) Regular and smooth schemes over a field.

Our next goal is to relate the notions of smoothness (over a field) and of regularity. We begin by showing that smooth points are regular.

Lemma 6.26. *Let k be a field, and X a k-scheme, locally of finite type. Let $x \in X$ be a point such that X is smooth at x of relative dimension d over k. Then the local ring $\mathscr{O}_{X,x}$ is regular of dimension $\leq d$. If x is closed, then $\dim \mathscr{O}_{X,x} = d$.*

Proof. Let U be an open neighborhood of x such that U is smooth over k. As the closed points of X are very dense, there exists a closed point x' of X with $x' \in U \cap \overline{\{x\}}$. If we have shown that $\mathscr{O}_{X,x'}$ is regular, then $\mathscr{O}_{X,x}$, being a localization of $\mathscr{O}_{X,x'}$, is also regular by Proposition B.77 (1) and $\dim \mathscr{O}_{X,x} \leq \dim \mathscr{O}_{X,x'}$. Thus we may assume that x is a closed point.

We embed a neighborhood of x into an affine space as in the definition of smoothness and denote by $y \in \mathbb{A}_k^n$ the image of x. Then $\mathscr{O}_{X,x} \cong \mathscr{O}_{\mathbb{A}_k^n,y}/(f_1,\dots,f_{n-d})$, with polynomials $f_j \in k[T_1,\dots,T_n]$ such that the matrix $((\partial f_i/\partial T_j)(y))_{i,j}$ has rank $n-d$. By Example 6.5 this means that the images of $f_1,\dots,f_{n-d}$ in $(T_y\mathbb{A}_k^n)^* = \mathfrak{m}_y/\mathfrak{m}_y^2$ are linearly independent (where $\mathfrak{m}_y$ is the maximal ideal of $\mathscr{O}_{\mathbb{A}_k^n,y}$). In fact, if $\kappa(y) = k$, then this follows immediately. The general case is reduced to this case as follows. Consider the base change $X \otimes_k \kappa(y) \subseteq \mathbb{A}_{\kappa(y)}^n$. The morphism $X \otimes_k \kappa(y) \to X$ is surjective, and we fix a point y' lying over y, with corresponding maximal ideal $\mathfrak{m}_{y'} \subset \kappa(y)[T_1,\dots,T_n]$. The rank of the Jacobian matrix of the polynomials f_j is the same, regardless of whether we consider it over k or over $\kappa(y)$. Now consider the $\kappa(y)$-linear map $\mathfrak{m}_y/\mathfrak{m}_y^2 \to \mathfrak{m}_{y'}/\mathfrak{m}_{y'}^2$. By the previous case, the images of the f_j in $\mathfrak{m}_{y'}/\mathfrak{m}_{y'}^2$ form a linearly independent system, and it follows that the same holds in $\mathfrak{m}_y/\mathfrak{m}_y^2$, as desired. As $\mathscr{O}_{\mathbb{A}_k^n,y}$ is regular, Proposition B.77 (3) implies that $\mathscr{O}_{X,x}$ is regular of dimension d. $\square$

Lemma 6.27. *Let k be a field, and $X = V(g_1,\dots,g_s) \subseteq \mathbb{A}_k^n$. Let $x \in X$ be a closed point, such that*

$$\mathrm{rk}_{\kappa(x)}\, J_{g_1,\dots,g_s}(x) = n - \dim \mathscr{O}_{X,x}.$$

Then x is a smooth point of X.

Proof. Write $d = \dim \mathscr{O}_{X,x}$. We may assume, after renumbering the g_i if necessary, that the first $n-d$ rows of the Jacobian matrix above are linearly independent. Let $Y = V(g_1,\dots,g_{n-d})$, so $x \in X \subseteq Y \subseteq \mathbb{A}_k^n$. Observe that Y is smooth of relative dimension d over k at x. By Lemma 6.26, the local ring $\mathscr{O}_{Y,x}$ of x in Y is regular of dimension d, so that the natural map $\mathscr{O}_{Y,x} \to \mathscr{O}_{X,x}$ is a surjection of d-dimensional rings, where $\mathscr{O}_{Y,x}$ is an integral domain. It is therefore an isomorphism by Corollary 5.8, and we conclude using Lemma 6.17. $\square$

Combining the lemmas, we get a rather complete picture of the relationship between regularity and smoothness for schemes over a field.

Theorem 6.28. *Let k be a field, X a k-scheme locally of finite type, and let $x \in X$ be a closed point. Let $d \geq 0$ be an integer. We fix an algebraically closed extension K of k, and write $X_K = X \otimes_k K$. The following are equivalent:*
(i) *The k-scheme X is smooth of relative dimension d at x.*
(ii) *For every point $\bar{x} \in X_K$ lying over x, X_K is smooth of relative dimension d in $\bar{x}$.*
(iii) *For every point $\bar{x} \in X_K$ lying over x, the completed local ring $\widehat{\mathscr{O}}_{X_K,\bar{x}}$ is isomorphic to a ring of formal power series $K[[T_1,\dots,T_d]]$ over K.*
(iv) *For every point $\bar{x} \in X_K$ lying over x, the local ring $\mathscr{O}_{X_K,\bar{x}}$ is regular and has dimension d.*
(v) *The equalities $\dim_{\kappa(x)} T_x(X/k) = \dim \mathscr{O}_{X,x} = d$ hold.*
If these conditions are satisfied, then
(vi) *The local ring $\mathscr{O}_{X,x}$ is regular and has dimension d.*
Furthermore, if $\kappa(x) = k$, then the final condition implies the other ones.

Proof. We have already seen (Lemma 6.26) that (i) implies (vi). By Proposition 6.15 (3), (i) implies (ii). The residue field $\kappa(x)$ is a finite extension of k. In particular there exists a k-embedding $\kappa(x) \hookrightarrow K$, and every point $\overline{x} \in X_K$ lying over x has residue field K. Therefore (ii) implies (iii) by Proposition 6.23. Using Proposition B.77 we see that $K[[T_1, \ldots, T_d]]$ is regular of dimension d and that therefore $\mathscr{O}_{X_K, \overline{x}}$ is regular (and of of dimension d by Proposition B.67). Thus we get that (iii) implies (iv). Note that Lemma 6.26 proves directly that (ii) implies (iv), without using the notion of completion.

Now let us show that every point $x \in X$ with residue class field k and regular local ring $\mathscr{O}_{X,x}$ is a smooth point of X. This shows the final statement of the proposition. Write $d = \dim \mathscr{O}_{X,x} = \dim_k T_x X$. Choose an affine open neighborhood U of x, and embed U as a closed subscheme $V(g_1, \ldots, g_s)$ in some affine space $\mathbb{A}_k^n$. We then have

$$\mathrm{rk}_{\kappa(x)} \, J_{g_1, \ldots, g_s}(x) = n - d,$$

see Example 6.5. Lemma 6.27 implies that x is a smooth point of X.

Recall that $\dim \mathscr{O}_{X,x} = \dim \mathscr{O}_{X_K, \overline{x}}$ by Corollary 5.47. Thus (iv) and (v) are equivalent because we have $\dim_{\kappa(x)} T_x(X/k) = \dim_K T_{\overline{x}}(X_K)$ by (6.6.2).

It remains to show that (iv) implies (i). Fix a point $\overline{x}$ of X_K lying over x. By what we have just seen, $\overline{x}$ is a smooth point of X_K. We may replace X by an open neighborhood of x, and can then assume that $X = V(g_1, \ldots, g_s) \subseteq \mathbb{A}_k^n$, for suitable polynomials g_i. By Lemma 6.27, it is enough to show that

$$\mathrm{rk}_{\kappa(x)} \, J_{g_1, \ldots, g_s}(x) = n - d,$$

since $\dim \mathscr{O}_{X,x} = \dim \mathscr{O}_{X_K, \overline{x}} = d$. But we have

$$\mathrm{rk}_{\kappa(x)} \, J_{g_1, \ldots, g_s}(x) = \mathrm{rk}_K \, J_{g_1, \ldots, g_s}(\overline{x}),$$

because the rank of a matrix is independent of the field over which we view it, and the latter term is equal to $n - d$ (by the same reasoning as above, since $\overline{x}$ is a regular K-valued point of X_K). $\qquad\square$

Using Theorem 5.22 we obtain the following corollary.

Corollary 6.29. *Let X be an irreducible scheme locally of finite type over a field k and let $x \in X(k)$ be a k-valued point (considered as a closed point of X). Then X is smooth over k at x if and only if $\dim_k T_x X = \dim X$.*

Remark 6.30.
(1) As the proof of Theorem 6.28 shows, the statements are also equivalent to statements (ii'), (iii'), (iv'), where the requirement that the given condition holds for every $\overline{x}$ over x is replaced by asking that there exists a point $\overline{x}$ over x for which it is true.
(2) Theorem 6.28 in particular implies that for every field extension L of k the scheme X is smooth over k if and only if $X \otimes_k L$ is smooth over L: The condition is necessary by Proposition 6.15 (3). Conversely, if $X \otimes_k L$ is smooth, then $X \otimes_k K$ is smooth for an algebraic closure K of L and hence X is smooth.

   This assertion is a special case of the assertion that smoothness is stable under faithfully flat descent; see Section (14.11) and Appendix C.
(3) We will prove in Volume II that the regularity of $\mathscr{O}_{X,x}$ implies that X is smooth at x, if we only assume that the extension $\kappa(x)/k$ is separable. As Example 6.34 below shows, in general the regularity of $\mathscr{O}_{X,x}$ is not sufficient for x being a smooth point.

The proof of Theorem 6.28 shows that the smoothness criterion is independent of the choice of embedding in a strong sense; we have:

Corollary 6.31. *Let k be a field, $X = V(g_1, \ldots, g_s) \subseteq \mathbb{A}_k^n$, and let $x \in X$ be a smooth closed point. Let $d = \dim \mathscr{O}_{X,x}$. Then* $\operatorname{rk}\left(\frac{\partial g_i}{\partial T_j}(x)\right)_{\substack{i=1,\ldots,s \\ j=1,\ldots,n}} = n - d$; *in particular, $s \geq n - d$. After possibly renumbering the g_i, such that* $\operatorname{rk}\left(\frac{\partial g_i}{\partial T_j}(x)\right)_{\substack{i=1,\ldots,n-d \\ j=1,\ldots,n}} = n - d$, *there exists an open neighborhood U of x in X such that the map $U \hookrightarrow V(g_1, \ldots, g_{n-d})$ is an open immersion.*

We will come back to this point in Volume II, when we discuss smoothness over arbitrary base schemes and related properties in more detail.

Theorem 6.28 (together with the fact that smoothness is stable under base change, Proposition 6.15 (3)) implies:

Corollary 6.32. *For a k-scheme X locally of finite type the following assertions are equivalent.*
(i) *X is smooth over k.*
(ii) *X is geometrically regular (i.e., $X \otimes_k L$ is regular for every field extension L of k).*
(iii) *There exists an algebraically closed extension K of k such that $X \otimes_k K$ is regular.*

Remark 6.33. The corollary shows that a scheme locally of finite type over an algebraically closed field k is regular if and only if X is smooth over k. In fact, one can show that this also holds if k is any perfect field; see Exercise 6.19, where the property "geometrically regular" is studied in more detail (or Volume II).

Whenever purely inseparable extensions come into the play, the notions of smoothness and regularity differ:

Example 6.34. Let p be a prime number. The morphism $\operatorname{Spec} \mathbb{F}_p(T) \to \operatorname{Spec} \mathbb{F}_p(T^p)$ is not smooth. To see this, we first do a base change with respect to the same morphism, and get $\operatorname{Spec} \mathbb{F}_p(T)[X]/(X^p) \cong \operatorname{Spec}(\mathbb{F}_p(T) \otimes_{\mathbb{F}_p(T^p)} \mathbb{F}_p(T)) \to \operatorname{Spec} \mathbb{F}_p(T)$. Since the source is not even reduced, it is not regular, so the resulting morphism is not smooth. Since smoothness is stable under base change, our claim is proved. On the other hand, $\operatorname{Spec} \mathbb{F}_p(T)$ is clearly regular.

More generally, it can be shown that a field extension is smooth if and only if it is finite and separable; see Exercise 6.12 (or Volume II).

Example 6.34 shows that there exist regular k-schemes X of finite type and finite field extensions K of k such that $X \otimes_k K$ is not regular. As explained above, this can only happen if K is not separable over k. Conversely, using the notion of flatness and Proposition B.77 (5) one shows easily (see Proposition 14.59 for a more general result and its proof):

Proposition 6.35. *Let k be a field, let X be a k-scheme locally of finite type, and let K be a field extension of k. If $X \otimes_k K$ is regular, then X is regular.*

There exist regular geometrically integral schemes of finite type over k such that X is not smooth (Exercise 6.22).

Normal schemes

The notions of smoothness and regularity discussed above are central properties in algebraic geometry, however, often they are too strict a requirement. In this section we will discuss the property of schemes of being *normal*, which is implied by regularity, but satisfied by a much larger class of schemes.

(6.13) Normal schemes.

Definition 6.36. *A scheme X is* normal *at a point $x \in X$, if the local ring $\mathscr{O}_{X,x}$ is a normal domain. A scheme X is* normal *if it is normal at all points.*

Remark 6.37. Even the property that all local rings of X are domains is quite useful; see Exercise 3.16 which shows that in a locally noetherian scheme with the property that all the local rings are domains, every connected component is already irreducible. If a point in an arbitrary scheme lies on more than one irreducible component, then its local ring will have more than one minimal prime ideal and hence cannot be an integral domain. In particular, such a point is not normal.

Lemma 6.38.
(1) *Let X be a locally noetherian normal scheme, and let $U \subseteq X$ be a connected open subset. Then $\Gamma(U, \mathscr{O}_X)$ is a normal domain.*
(2) *A quasi-compact scheme X is normal if for all closed points $x \in X$ the local ring $\mathscr{O}_{X,x}$ is normal.*
(3) *Let X be a scheme which admits an affine open cover $X = \bigcup U_i$, such that all $\Gamma(U_i, \mathscr{O}_X)$ are normal domains. Then X is normal.*

The proof below shows that (2) also holds for locally noetherian schemes by Exercise 5.5.

Proof. To show (1) we may assume that $X = U$ (and in particular that X is connected). As explained in Remark 6.37, X is even integral. Let $K(X)$ be the function field of X, and let $s \in K(X)$ be an element which is integral over $A := \Gamma(X, \mathscr{O}_X)$. In particular, s is integral over all subrings of $K(X)$ containing A, and the normality assumption implies that s lies in all stalks $\mathscr{O}_{X,x}$, $x \in X$. Using Proposition 3.29, we get that $s \in A$. Parts (2) and (3) follow immediately from the fact that every localization of a normal domain is normal again. In (2) we use in addition that every closed subset of a quasi-compact scheme contains a closed point. $\square$

By Remark B.78 there is the following result.

Corollary 6.39. *Let X be a locally noetherian scheme. If $x \in X$ is a regular point, then x is normal in X. In particular, if X is regular, then X is normal.*

On the other hand, normal rings of dimension 1 are discrete valuation rings, and hence in particular are regular (Proposition B.73 (3)). Therefore we have

Proposition 6.40. *Let X be a locally noetherian scheme, and let $x \in X$ be a normal point such that $\dim \mathscr{O}_{X,x} \leq 1$. Then x is a regular point.*

We call a locally noetherian scheme X *regular in codimension* 1, if for all $x \in X$ with $\dim \mathcal{O}_{X,x} = 1$ the local ring $\mathcal{O}_{X,x}$ is regular (or equivalently, normal). In other words, all local rings of X satisfy condition (R_1) (Definition B.79). So normal schemes are regular in codimension 1. As it turns out, for schemes defined by one equation in a regular scheme, one can prove the following converse.

Proposition 6.41. *Let $X = \operatorname{Spec} A$ be a regular affine scheme, and consider a closed integral subscheme $Z = V(f) = \operatorname{Spec} A/(f)$ of X defined by a single element $f \in A$. Then Z is normal if and only if Z is regular in codimension 1.*

Proof. We give a proof using more advanced notions of commutative algebra. Using Serre's criterion for normality, see Proposition B.81, it is enough to check condition (S_2) for the ring $A/(f)$. However, f is a regular element of the regular ring A, so the quotient $A/(f)$ is Cohen-Macaulay (Proposition B.86), i. e., satisfies conditions (S_k) for all k. $\square$

Remark 6.42. The proof uses Serre's criterion for normality (Proposition B.81) which can be formulated as follows. A locally noetherian scheme is normal if and only if the following assertions hold.
(1) For all $x \in X$ with $\dim \mathcal{O}_{X,x} \leq 1$ the local ring $\mathcal{O}_{X,x}$ is regular.
(2) For all $x \in X$ with $\dim \mathcal{O}_{X,x} \geq 2$ there exists in the maximal ideal $\mathfrak{m}_x$ a regular sequence (Definition B.60) of length 2.
The proof shows in fact it would be enough to assume that A is Cohen-Macaulay (instead of being regular), and that we have a similar result for any subscheme defined by a regular sequence (instead of a single element) using Proposition B.86.

Example 6.43. The proposition shows us many examples of normal, non-regular varieties. For instance, every quadric of rank at least 3 is normal.

Similarly as we defined the smooth or the regular locus above, we call for a locally noetherian scheme X the set

$$(6.13.1) \qquad X_{\mathrm{norm}} := \{ x \in X \; ; \; \mathcal{O}_{X,x} \text{ is normal} \}$$

the *normal locus* of X. It is stable under generization. We will see later (Section (12.12)) that under mild assumptions the normal locus is even an open subset of X (for instance, if X is a scheme of finite type over a field).

Remark 6.44. Let k be a field and let X be a k-scheme locally of finite type. Example 6.34 shows that even if X is normal in general there exist extensions K of k such that $X \otimes_k K$ is not normal (although this can never happen if K is a separable extension of k by Exercise 6.18). But conversely, if $X \otimes_k K$ is normal for some field extension K, then X is normal. This follows from Proposition B.81 and Proposition B.82; see also Proposition 14.59 for a proof a more general result.

(6.14) Geometric concept of normality, Hartogs's theorem.

We mention two further properties of normal schemes here which also exhibit that normality, despite its algebraic definition, is a property which is geometrically tangible.

An important theorem about normal varieties is the analogue of Hartogs's theorem. Recall that Hartogs's theorem says that whenever $n > 1$, $U \subseteq \mathbb{C}^n$ is open, $x \in U$ a point, and $f \colon U \setminus \{x\} \to \mathbb{C}$ is a holomorphic function, then f extends to a holomorphic function on all of U. In the algebraic setting, we have

Theorem 6.45. *Let X be a locally noetherian normal scheme, and let $U \subseteq X$ be an open subset with $\mathrm{codim}_X(X \setminus U) \geq 2$. Then the restriction map $\Gamma(X, \mathscr{O}_X) \to \Gamma(U, \mathscr{O}_X)$ is an isomorphism. In other words: every function $f \in \Gamma(U, \mathscr{O}_X)$ on U extends uniquely to X.*

Proof. We may assume that $X = \mathrm{Spec}\, A$ is affine, where A is a normal integral domain. For every non-empty open set V of X we may consider $\Gamma(V, \mathscr{O}_X)$ as a subring of the function field $K(X) = \mathrm{Frac}\, A$ such that the restriction maps are given by inclusions of rings (Example 2.37). Let $Z \subset X$ be an irreducible closed subset of codimension 1. By hypothesis, U intersects Z non-trivially, so in particular it contains the generic point η of Z. In other words, the subring $\Gamma(U, \mathscr{O}_X)$ of the function field $K(X)$ is contained in the stalk $\mathscr{O}_{X,\eta}$. But by Proposition B.73 (3), $A = \Gamma(X, \mathscr{O}_X)$ is the intersection of all the stalks $\mathscr{O}_{X,\eta}$, where η is a prime ideal of height 1; in other words, where η is the generic point of an irreducible closed subset of codimension 1. $\square$

Another important property of normal points of schemes (of finite type over a field, say) is that "only one branch emanates from the point". Here the notion of branch, which we deliberately leave vague, should be thought of as a more local concept than irreducible components. In the analytic setting (over the complex numbers), a normal point x inside a "complex analytic space" has the following property: There exist arbitrarily small neighborhoods U of x, such that $U \cap X_{\mathrm{sm}}$ is connected. It is instructive to check that the subset $\{ (x,y) \in \mathbb{C}^2 \; ; \; y^2 = x^2(x+1) \}$ does not have this property at the origin.

In an algebraic context, the most important theorem which captures this property of normal varieties is the following result that will be discussed later in more detail (see Theorem 12.50 and Theorem 12.51). For now we only state the following version without proof.

Theorem 6.46. *Let k be a field, and let X be a k-scheme, locally of finite type over k. If $x \in X$ is normal, then the completion $\widehat{\mathscr{O}}_{X,x}$ is a normal domain.*

Even the fact that $\widehat{\mathscr{O}}_{X,x}$ is a domain tells us something about the geometry of X: Similarly as in the discussion in Section (6.10) above, this statement should be understood as saying that even "very locally" around x, X is irreducible. Similarly as above, it is instructive to check that for $X = V(Y^2 - X^2(X+1)) \subset \mathbb{A}_k^2$, $x = (0,0)$, the complete local ring $\widehat{\mathscr{O}}_{X,x}$ is not a domain, although X itself (and in particular $\mathscr{O}_{X,x}$) is integral.

Exercises

Exercise 6.1$\lozenge$**.** Let $i \colon Y \to X$ be an immersion and let $y \in Y$. Show that the homomorphism $di_y \colon T_y Y \to T_{i(y)} X$ is injective. Show that di_y is bijective, if i is an open immersion.

Exercise 6.2. Let k be an algebraically closed field. Give examples of k-schemes X_1, X_2, X_3, X_4, such that for all four schemes there exists a morphism of k-schemes $\mathbb{A}_k^1 \to X_i$ which is a homeomorphism, and such that
(a) for all closed points x_1 of X_1, we have: $\dim T_{x_1}(X_1) = 1$,
(b) for all except exactly one closed point x_2 of X_2, we have: $\dim T_{x_2}(X_2) = 1$, and X_2 is reduced.

(c) for all except exactly one closed point x_3 of X_3 we have: $\dim T_{x_3}(X_3) = 1$, and X_3 is not reduced.

(d) for all closed points x_4 of X_4, we have: $\dim T_{x_4}(X_4) > 1$.

Exercise 6.3. Let k be a field and let $K \supseteq k$ be a finite field extension. Set $X = \operatorname{Spec} K$, let $x \in X$ be the unique point and let $\xi \colon \operatorname{Spec} K \to X$ be the identity. Show that $T_x X = 0$, but that $T_\xi(X/\operatorname{Spec} k) = 0$ if and only if K is a separable extension of k.

Exercise 6.4. Let k be a field, and let $G = \operatorname{GL}_{n,k} = D(\det) \subset \mathbb{A}_k^{n^2}$. (We think of the points of $\mathbb{A}_k^{n^2}$ as $(n \times n)$-matrices.) Denote by $e \in \operatorname{GL}_n(k)$ the unit matrix. We identify $T_e \operatorname{GL}_{n,k} = k^{n^2}$ with the space $M_n(k)$ of $(n \times n)$-matrices.

(a) Denote by $m \colon G \times G \to G$ the multiplication map. Show that the homomorphism $dm_{(e,e)} \colon T_e G \times T_e G = T_{(e,e)} G \times G \to T_e G$ is given by $(v, w) \mapsto v + w$.

(b) Show that the k-linear map $d\det_e \colon M_n(k) \to k$ induced by $\det \colon \operatorname{GL}_{n,k} \to \operatorname{GL}_{1,k}$ is the trace.

(c) Let $i \colon G \to G$ be the inverse. Show that $di_e \colon T_e G \to T_e G$ is given by $v \mapsto -v$.

(d) Let $f \colon G \to G$ be the morphism $A \mapsto {}^t A^{-1}$. Show that df_e is given by $v \mapsto -{}^t v$.

Exercise 6.5. We keep the notation of Exercise 6.4. Determine the tangent spaces at e as subspaces of $T_e \operatorname{GL}_{n,k} = M_n(k)$ of the following subschemes H of $\operatorname{GL}_{n,k}$, given on R-valued points for a k-algebra R:

(a) $H = \operatorname{SL}_{n,k}$, with $\operatorname{SL}_{n,k}(R) = \operatorname{SL}_n(R)$.

(b) $H = \operatorname{Sym}_{T,k}$, where $T \in \operatorname{GL}_n(k)$, with $\operatorname{Sym}_{T,k}(R) = \{ A \in \operatorname{GL}_n(R) \; ; \; {}^t A T A = T \}$.

(c) $H = \operatorname{SSym}_{T,k}$, with $\operatorname{SSym}_{T,k}(R) = \{ A \in \operatorname{SL}_n(R) \; ; \; {}^t A T A = T \}$.

Exercise 6.6. Let k be a field and let $f \colon X \to S$ and $g \colon Y \to S$ be morphisms of k-schemes. Let $x \in X(k)$ and $y \in Y(k)$ be k-valued points such that their images in $S(k)$ are equal. We thus obtain a k-valued point $z := (x, y)_S \in (X \times_S Y)(k)$. Show that

$$T_z(X \times_S Y) = \{ (t, u) \in T_x X \times T_y Y \; ; \; df_x(t) = dg_y(u) \}.$$

Deduce that if X and Y are subschemes of a scheme S, we have for a k-valued point z of the (schematic) intersection:

$$T_z(X \cap Y) = T_z(X) \cap T_z(Y).$$

Generalize these assertions to the relative tangent space.

Exercise 6.7. Let k be a field, let X be a k-scheme, and let Y_1 and Y_2 be closed subschemes of X and let $Y_1 \cap Y_2$ be their schematic intersection. Let $x \in (Y_1 \cap Y_2)(k)$ be a k-valued point and assume that X, Y_1, and Y_2 are smooth at x over k of relative dimension d, $d - c_1$, and $d - c_2$, respectively. Show that the following assertions are equivalent.

(i) $Y_1 \cap Y_2$ is smooth of relative dimension $d - (c_1 + c_2)$.

(ii) $T_x Y_1 + T_x Y_2 = T_x X$.

If these equivalent conditions are satisfied, we say that Y_1 and Y_2 *intersect transversally*.

Exercise 6.8. Let k be a field and X, Y schemes locally of finite type over k. Let $p \colon X \times_k Y \to X$ and $q \colon X \times_k Y \to Y$ be the projections. Let $z \in X \times_k Y$ and set $x := p(z)$ and $y := q(z)$. Show that z is a smooth point of $X \times_k Y$ if and only if x is a smooth point of X and y is a smooth point of Y.

Exercise 6.9. Let k be a field, $f \in k[T_0, \ldots, T_n]$ be a nonzero homogeneous polynomial of degree $d \geq 1$, and set $f_i := \frac{\partial f}{\partial T_i}$. Define $X := V_+(f)$ and $Z := V_+(f_0, \ldots, f_n)$. Let $X(k)_{\text{sing}}$ be the set of $x \in X(k)$ that are not smooth points of the k-scheme X.
(a) Show that $X(k)_{\text{sing}} = X(k) \cap Z(k)$.
(b) Assume that $\text{char}(k) = 0$ or that $\text{char}(k)$ is prime to d. Show that $X(k)_{\text{sing}} = Z(k)$.

Exercise 6.10. Let k be a field. If $f \in k[T_1, \ldots, T_n]$ and $f = f_r + f_{r+1} + \cdots + f_d$ is its decomposition into homogeneous polynomials f_i of degree i with $f_r \neq 0$, then $f^* := f_r$ is called the *leading term* of f. For an ideal $\mathfrak{a}$ of $k[T_1, \ldots, T_n]$ let $\mathfrak{a}^*$ be the ideal generated by f^* for $f \in \mathfrak{a}$.
Now let X be a scheme locally of finite type over k. Let $x \in X(k)$ and let $\mathfrak{m} = \mathfrak{m}_x$ be the maximal ideal at x. Then $A := \bigoplus_{d \geq 0} \mathfrak{m}^d / \mathfrak{m}^{d+1}$ is a k-algebra. The affine scheme $\text{TC}_x(X) := \text{Spec} \, A$ is called the *tangent cone* of X at x.
(a) For every open affine neighborhood U of x there exists a closed immersion $i \colon U \hookrightarrow \mathbb{A}_k^n$ for some $n \geq 0$ with $i(x) = 0$ and that i yields an isomorphism $U \xrightarrow{\sim} V(\mathfrak{a})$ for an ideal $\mathfrak{a} \subseteq k[T_1, \ldots, T_n]$ (why?). Show that $\text{TC}_x(X) \cong \text{Spec} \, k[T_1, \ldots, T_n]/\mathfrak{a}^*$.
(b) Assume that x is contained in a single irreducible component Z of X. Show that every irreducible component of $\text{TC}_x(X)$ has dimension $\dim Z$. In general, let $Z_1, \ldots, Z_m$ be the irreducible components of X containing x. Show that $\dim \text{TC}_x(X) = \sup_i \dim Z_i$.

Exercise 6.11. Let k be an algebraically closed field and let $f \in k[T, U]$ be a polynomial with $f(0,0) = 0$. Set $X = \text{Spec} \, k[T, U]/(f)$, $x = (0,0) \in X(k)$ and let $\text{TC}_x(X)$ be the tangent cone of X at x (Exercise 6.10).
(a) Show that $\text{TC}_x(X) = \text{Spec} \, k[T, U]/f^*$, where f^* is the leading term of f. The degree of f^* is called the *multiplicity* of X at x and denoted by $\mu_x(X)$. If $\mu_x(X) = 2$ (resp. $= 3$), then x is called a *double point* (resp. *triple point*) of X.
(b) Show that f^* is the product $f^* = \ell_1 \cdots \ell_r$ of homogeneous polynomials ℓ_i of degree 1. Deduce that $\text{TC}_x(X)$ is (set-theoretically) a union of lines.
(c) Show that x is a smooth point of X if and only if $\mu_x(X) = 1$.
(d) Assume that $\mu_x(X) = 2$. Show that $\text{TC}_x(X)$ either has two reduced irreducible components (then x is called a *node* of X) or that $\text{TC}_x(X)$ is irreducible but not reduced (then x is called a *cusp* of X).
(e) Determine $\text{TC}_x(X)$ and "sketch" X and $\text{TC}_x(X)$ for $f = U^2 - T^3$ (a cusp), for $f = T^3 + T^2 - U^2$ (a node in characteristic $\neq 2$), and for $f = T^4 - T^2 U + U^3$.

Exercise 6.12. Let k be a field
(a) Let $K \supseteq k$ be a field extension and $f \colon \text{Spec} \, K \to \text{Spec} \, k$ the corresponding morphism of schemes. Show that f is smooth if and only if K is a finite separable field extension.
(b) Show that for a non-empty k-scheme X the following assertions are equivalent:
  (i) X is smooth of relative dimension 0 over k (i.e., X is étale over k).
  (ii) X is smooth over k and $\dim X = 0$.
  (iii) $X \cong \text{Spec}(K_1 \times \cdots \times K_n)$, where K_i are finite separable field extensions of k.

Exercise 6.13. Let k be a field.
(a) Let X be a closed subscheme of $\mathbb{A}_k^n$, say $X = V(f_1, \ldots, f_r)$ with $f_i \in k[T_1, \ldots, T_n]$, let $x \in X(k)$, and $d := \dim_x X = \dim \mathcal{O}_{X,x}$. Show that X is smooth at x over k if and only if $\text{rk} \, J_{f_1, \ldots, f_r}(x) = n - d$.

(b) Let X be a closed subscheme of $\mathbb{P}^n_k$, say $X = V_+(f_1, \ldots, f_r)$ with homogeneous polynomials $f_i \in k[T_0, T_1, \ldots, T_n]$, let $x = (x_0 : \ldots : x_n) \in X(k) \subseteq \mathbb{P}^n(k)$, and $d := \dim_x X = \dim \mathscr{O}_{X,x}$. Show that $\operatorname{rk} J_{f_1, \ldots, f_r}(x_0, \ldots, x_n)$ does not depend on the choice of a representative $(x_0, \ldots, x_n)$ of x. Show that X is smooth at x over k if and only if $\operatorname{rk} J_{f_1, \ldots, f_r}(x_0, \ldots, x_n) = n - d$.

Exercise 6.14. Let k be a field and let $X = V_+(T_0^d + \cdots + T_n^d) \subseteq \mathbb{P}^n_k$ be the *Fermat hypersurface of degree d*. Show that if $p := \operatorname{char}(k)$ does not divide d, then X is smooth over k. If p divides d, show that X is not reduced, determine X_{red} and show that X_{red} is smooth over k.

Exercise 6.15. Let Y be an integral scheme with generic point η and let $f \colon X \to Y$ be a morphism. Let X be reduced (resp. irreducible, resp. normal, resp. regular). Show that the generic fiber X_η has the same property.

Exercise 6.16. Show that the following morphisms are non-smooth morphisms between regular schemes.
(a) $\operatorname{Spec} \mathbb{Z}[T]/(T^2 + 1) \to \operatorname{Spec} \mathbb{Z}$.
(b) $\operatorname{Spec} R[T_1, \ldots, T_n]/(T_1 T_2 \cdots T_n - \pi) \to \operatorname{Spec} R$, where R is a discrete valuation ring, $\pi \in R$ a uniformizing element, and $n \geq 2$.
(c) $\mathbb{A}^m_k \to \mathbb{A}^n_k$, $(a_1, \ldots, a_m) \mapsto (a_1, \ldots, a_m, 0, \ldots, 0)$, where k is a field and $0 \leq m < n$.

Exercise 6.17. Let k be a field and let K and L be field extensions of k such that one of the field extensions is finitely generated.
(a) Show that $K \otimes_k L$ is noetherian.
(b) Show that $K \otimes_k L$ is regular if one of the field extensions is separable.

Exercise 6.18. Let k be a field and let $K \supseteq k$ be a separable field extension. Let X be a normal (resp. regular) scheme locally of finite type over k. Show that $X \otimes_k K$ is again normal (resp. regular).
Hint: Use Proposition B.77 (5) (resp. Proposition B.73 (6)) and Exercise 6.17.

Exercise 6.19. Let X be a locally noetherian scheme over a field k. Show that the following assertions are equivalent.
(i) For every finite field extension $K \supseteq k$ the scheme $X \otimes_k K$ is normal (resp. regular).
(ii) For every finitely generated extension $K \supseteq k$ the scheme $X \otimes_k K$ is normal (resp. regular).
(iii) For every finite inseparable field extension $K \supseteq k$ the scheme $X \otimes_k K$ is normal (resp. regular).
If X is locally of finite type over k, show that the assertions (i)–(iii) are further equivalent to:
(iv) For every field extension $K \supseteq k$ the scheme $X \otimes_k K$ is normal (resp. regular).
(v) There exists a perfect extension $K \supseteq k$ such that $X \otimes_k K$ is normal (resp. regular).
If X satisfies the equivalent conditions (i)–(iii), X is called *geometrically normal* (resp. *geometrically regular*).
Deduce that if k is perfect (e.g., if $\operatorname{char}(k) = 0$), every locally noetherian normal (resp. regular) k-scheme is geometrically normal (resp. geometrically regular).
Hint: Exercise 6.18.

Exercise 6.20$\Diamond$. Let k be a field and X a normal k-scheme of finite type. Show that X is geometrically connected if and only if X is geometrically irreducible.
Hint: Use Exercise 6.18.

Exercise 6.21. Let X be a geometrically normal scheme (Exercise 6.19) of finite type over a field k and let X_{sing} be the closed set of $x \in X$ such that X is not smooth at x over k. Show that $\text{codim}_X X_{\text{sing}} \geq 2$.

Exercise 6.22. Let k be a non-perfect field of characteristic $p > 2$. Let $a \in k$ be an element that is not a p-th power and let $f = T^2 - U^p + a \in k[T, U]$. We define $C := \text{Spec}\, k[T, U]/(f)$.
(a) Show that C is integral, normal and $\dim C = 1$. Deduce that C is regular. Let $K = K(C)$ be the function field of C.
(b) Show that K is separable over k and that k is algebraically closed in K. Deduce that C is geometrically integral.
(c) Let $k' = k[a^{1/p}]$. Show that $C \otimes_k k'$ is not regular and deduce that C is not smooth.

Exercise 6.23. Let k be a field of characteristic $\neq 2$.
(a) Let $X = V(f) \subset \mathbb{A}_k^5$ with $f = T_1^2 + \cdots + T_5^2$ and $x \in X(k)$ the zero point. Show that $\mathscr{O}_{X,x}$ is factorial but not regular.
   Hint: Use Proposition B.75 (3).
(b) Let $X = V(f) \subset \mathbb{A}_k^3$ with $f = T_1 T_2 - T_3^2$ and $x \in X(k)$ the zero point. Show that $\mathscr{O}_{X,x}$ is normal but not factorial.

Exercise 6.24. Let k be a field and let $C = V_+(f) \subset \mathbb{P}_k^2$ be a plane curve (Section (5.14)). Let $c \in C$ be a k-rational point. Show that the following assertions are equivalent.
(i) C is smooth at c.
(ii) There exists a line $L \subset \mathbb{P}_k^2$ such that $i_c(L, C) = 1$ (where $i_c(L, C)$ is the intersection number (Definition 5.60)).

Exercise 6.25. Let k be an algebraically closed field. Let C be a plane cubic over k (i.e., C is a k-scheme of the form $V_+(f) \subset \mathbb{P}_k^2$ with f homogeneous of degree 3). Assume that C is reduced (i.e., no irreducible component of f occurs with multiplicity > 1).
(a) Assume that C is integral (i.e., f is irreducible). Show that C has at most one non-smooth point.
   Hint: Use Exercise 6.24.
(b) Assume that C has three non-smooth points. Show that C is the union of three lines.

Exercise 6.26. Let R be a ring, and let $f \in R[X]$ be a monic polynomial. Let $X = \text{Spec}\, R[X]/(f)$, a closed subscheme of $\mathbb{A}_R^1$, and consider the morphism $p \colon X \to \text{Spec}\, R$ which is the restriction of the natural projection $\mathbb{A}_R^1 \to \text{Spec}\, R$. Let $\text{disc}(f)$ denote the discriminant of f, cf. Section (B.20).
(a) For $x \in \text{Spec}\, R$, show that $x \notin V(\text{disc}(f))$ if and only if the fiber $p^{-1}(x)$ is the spectrum of a product of separable field extensions of $\kappa(x)$.
(b) For x with algebraically closed residue class field, conclude that $x \notin V(\text{disc}(f))$ if and only if the topological space $p^{-1}(x)$ has $\deg(f)$ points.
(c) Show that, for all x, the condition in part (a) is also equivalent to $p^{-1}(x)$ being geometrically reduced. In particular, whenever $x \notin V(\text{disc}(f))$, then $p^{-1}(x)$ is reduced.

7   Quasi-coherent modules

Contents

In Chapters 2 and 3 we associated to each ring a geometric object, namely an affine scheme, and defined schemes as "globalizations" of affine schemes. In this chapter we will show how to attach to a module M over a ring A a module $\tilde{M}$ over the sheaf of rings $\mathscr{O}_{\operatorname{Spec} A}$ and prove that this defines an equivalence of the category of A-modules with the category of so-called quasi-coherent $\mathscr{O}_{\operatorname{Spec} A}$-modules. Then we globalize, and an $\mathscr{O}_X$-module on a scheme X will be called quasi-coherent if its restriction to each open affine subscheme is quasi-coherent.

We start with an arbitrary ringed space $(X, \mathscr{O}_X)$ and begin with an excursion on $\mathscr{O}_X$-modules which are the "sheaf version" of modules over a ring. For easy reference we collect all general results on $\mathscr{O}_X$-modules in this section. Thus the reading of this section might be somewhat tedious. We suggest to start only with the definition of $\mathscr{O}_X$-modules and then to come back to the other definitions later when they are needed.

The next part introduces the notion of a quasi-coherent $\mathscr{O}_X$-module on a scheme X and shows the results mentioned above. We prove that attaching to an A-module M the quasi-coherent $\mathscr{O}_{\operatorname{Spec} A}$-module $\tilde{M}$ is compatible with all kinds of constructions for modules (such as direct sums, exact sequences, tensor products).

In the last part we define certain properties for $\mathscr{O}_X$-modules $\mathscr{F}$ (to be flat, of finite type, of finite presentation, etc.) and show that if $X = \operatorname{Spec} A$ is affine and $\mathscr{F} = \tilde{M}$ is quasi-coherent these notions are equivalent to the similar notions for the A-module M. We prove that for an arbitrary locally ringed space X, an $\mathscr{O}_X$-module of finite type $\mathscr{F}$, and an integer $r \geq 0$, the locus where $\mathscr{F}$ can be generated by r elements is open in X. Moreover we will show that two $\mathscr{O}_X$-modules $\mathscr{F}$ and $\mathscr{G}$ of finite presentation have isomorphic stalks at a point $x \in X$ if and only if there exists an open neighborhood U of x such that their restrictions to U are isomorphic.

Excursion: $\mathscr{O}_X$-modules

(7.1) Definition of $\mathscr{O}_X$-modules.

Let X be a topological space, and let $\mathscr{F}$ and $\mathscr{F}'$ be two presheaves on X. We define a presheaf $\mathscr{F} \times \mathscr{F}'$ on X by setting

© Springer Fachmedien Wiesbaden GmbH, part of Springer Nature 2020
U. Görtz und T. Wedhorn, *Algebraic Geometry I: Schemes*, Springer Studium
Mathematik – Master, https://doi.org/10.1007/978-3-658-30733-2_8

$$(\mathscr{F} \times \mathscr{F}')(U) := \mathscr{F}(U) \times \mathscr{F}'(U),$$

where the restriction maps are the products of the restriction maps for $\mathscr{F}$ and $\mathscr{F}'$. Clearly this is a sheaf if $\mathscr{F}$ and $\mathscr{F}'$ are sheaves.

From now on let $(X, \mathscr{O}_X)$ be a ringed space, Section (2.9).

Definition 7.1. *An $\mathscr{O}_X$-module is a sheaf $\mathscr{F}$ on X together with two morphisms (addition and scalar multiplication) of sheaves*

$$\mathscr{F} \times \mathscr{F} \to \mathscr{F}, \quad (s, s') \mapsto s + s' \qquad \text{for } s, s' \in \mathscr{F}(U),\ U \subseteq X \text{ open},$$
$$\mathscr{O}_X \times \mathscr{F} \to \mathscr{F}, \quad (a, s) \mapsto as \qquad \text{for } a \in \mathscr{O}_X(U),\ s \in \mathscr{F}(U),\ U \subseteq X \text{ open}$$

such that addition and scalar multiplication by $\mathscr{O}_X(U)$ define on $\mathscr{F}(U)$ the structure of an $\mathscr{O}_X(U)$-module.

If $\mathscr{F}_1$ and $\mathscr{F}_2$ are $\mathscr{O}_X$-modules, a morphism of sheaves $w\colon \mathscr{F}_1 \to \mathscr{F}_2$ is called a homomorphism of $\mathscr{O}_X$-modules, if for each open subset $U \subseteq X$ the map $\mathscr{F}_1(U) \to \mathscr{F}_2(U)$ is an $\mathscr{O}_X(U)$-module homomorphism, i. e.

$$w_U(s + s') = w_U(s) + w_U(s'),$$
$$w_U(as) = a w_U(s)$$

for all $s, s' \in \mathscr{F}(U)$ and $a \in \mathscr{O}_X(U)$.

The composition of two homomorphisms of $\mathscr{O}_X$-modules is again a homomorphism of $\mathscr{O}_X$-modules. We obtain the category of $\mathscr{O}_X$-modules which we denote by $(\mathscr{O}_X\text{-Mod})$. The $\mathscr{O}_X$-module $\mathscr{F}$ such that $\mathscr{F}(U) = \{0\}$ for all open sets $U \subseteq X$ is called the zero module and simply denoted by 0.

Examples 7.2.
(1) Let X be a topological space and let $\underline{\mathbb{Z}}$ be the constant sheaf of rings on X with value $\mathbb{Z}$ (Example 2.25). Then a $\underline{\mathbb{Z}}$-module is simply a sheaf of abelian groups on X.
(2) Let A be a ring. Let X be a space that consists of a single point and let $\mathscr{O}_X$ be the sheaf of rings with $\mathscr{O}_X(X) = A$. Then an $\mathscr{O}_X$-module $\mathscr{F}$ is just an A-module M (by attaching $M = \mathscr{F}(X)$ to $\mathscr{F}$).
(3) Let X be a real C^∞-manifold and denote by $\mathscr{C}_X^\infty$ the sheaf of rings of smooth functions on X. For $i \geq 0$ let Ω_X^i be the sheaf of smooth differential forms of degree i on X. Then Ω_X^i is a $\mathscr{C}_X^\infty$-module.

Let $\mathscr{F}$ be an $\mathscr{O}_X$-module and $x \in X$. The $\mathscr{O}_X(U)$-module structures on $\mathscr{F}(U)$, where U is an open neighborhood of x, induce on the stalk $\mathscr{F}_x$ an $\mathscr{O}_{X,x}$-module structure. If $w\colon \mathscr{F} \to \mathscr{F}'$ is a homomorphism of $\mathscr{O}_X$-modules, $w_x\colon \mathscr{F}_x \to \mathscr{F}'_x$ is a homomorphism of $\mathscr{O}_{X,x}$-modules.

If $(X, \mathscr{O}_X)$ is a locally ringed space we call the $\kappa(x)$-vector space

$$(7.1.1) \qquad \mathscr{F}(x) := \mathscr{F}_x / \mathfrak{m}_x \mathscr{F}_x = \mathscr{F}_x \otimes_{\mathscr{O}_{X,x}} \kappa(x)$$

the *fiber of $\mathscr{F}$ in x*. If s is a section of $\mathscr{F}$ over an open neighborhood U of x, we denote by $s(x)$ the image of the germ $s_x \in \mathscr{F}_x$ in $\mathscr{F}(x)$.

Let $\mathscr{F}$ be an $\mathscr{O}_X$-module. We now discuss a number of constructions of $\mathscr{O}_X$-modules. As a general principle, given a functor on usual R-modules, we apply it to all sets of sections $\mathscr{F}(U)$. With functors that are well behaved with respect to products, and which are left exact, the result will be a sheaf: in this way we get kernels and products, for instance. In general, we will only obtain a presheaf, and take the associated sheaf as the final result. Examples are given by the image and the cokernel of a homomorphism, the direct sum, and the tensor product of $\mathscr{O}_X$-modules. We also endow the set of homomorphisms between $\mathscr{O}_X$-modules with the structure of a $\Gamma(X, \mathscr{O}_X)$-module and show that the category of $\mathscr{O}_X$-modules is an *abelian category*.

(7.2) Submodules and quotient modules.

SUBMODULES.

An $\mathscr{O}_X$-*submodule of* $\mathscr{F}$ is an $\mathscr{O}_X$-module $\mathscr{G}$ such that $\mathscr{G}(U)$ is a subset of $\mathscr{F}(U)$ for all open sets $U \subseteq X$ and such that the inclusions $\iota_U \colon \mathscr{G}(U) \hookrightarrow \mathscr{F}(U)$ form a homomorphism $\iota \colon \mathscr{G} \to \mathscr{F}$ of $\mathscr{O}_X$-modules.

The $\mathscr{O}_X$-submodules of $\mathscr{O}_X$ are called *ideals of* $\mathscr{O}_X$.

QUOTIENT MODULES.

If $\mathscr{G}$ is an $\mathscr{O}_X$-submodule of an $\mathscr{O}_X$-module, the presheaf

$$U \mapsto \mathscr{F}(U)/\mathscr{G}(U)$$

is not a sheaf in general (compare Exercise 2.12). The associated sheaf together with the addition and scalar multiplication induced from $\mathscr{F}$ is again an $\mathscr{O}_X$-module which is denoted by $\mathscr{F}/\mathscr{G}$ and called the *quotient of* $\mathscr{F}$ *by* $\mathscr{G}$. The canonical homomorphisms $\mathscr{F}(U) \to \mathscr{F}(U)/\mathscr{G}(U)$ induce a homomorphism $\mathscr{F} \to \mathscr{F}/\mathscr{G}$ of $\mathscr{O}_X$-modules which is called the *canonical projection* or *canonical homomorphism*. If $x \in X$ is a point,

$$\varinjlim_{U \ni x} \mathscr{F}(U)/\mathscr{G}(U) = \mathscr{F}_x/\mathscr{G}_x,$$

where U runs through the open neighborhoods of x. As sheafification does not change the stalk, we obtain

(7.2.1) $$(\mathscr{F}/\mathscr{G})_x = \mathscr{F}_x/\mathscr{G}_x$$

for all $x \in X$.

(7.3) Kernel, image, cokernel, exact sequences, modules of homomorphisms for $\mathscr{O}_X$-modules.

We recall from Section (2.6) that a morphism $w \colon \mathscr{F} \to \mathscr{F}'$ of sheaves on X is called injective (resp. surjective, resp. bijective), if the induced map on stalks $w_x \colon \mathscr{F}_x \to \mathscr{F}'_x$ is injective (resp. surjective, resp. bijective) for all $x \in X$. We have already seen in Proposition 2.23 that w is injective (resp. bijective) if and only if $w_U \colon \mathscr{F}(U) \to \mathscr{F}'(U)$ is injective (resp. bijective) for all open subsets $U \subseteq X$. The analogous statement for "surjective" does not hold.

Now let $w \colon \mathscr{F} \to \mathscr{F}'$ be a homomorphism of $\mathscr{O}_X$-modules.

KERNELS.

The presheaf
$$U \mapsto \mathrm{Ker}(w_U \colon \mathscr{F}(U) \to \mathscr{F}'(U))$$

is a sheaf and therefore an $\mathscr{O}_X$-submodule of $\mathscr{F}$ called the *kernel of w* and denoted by $\mathrm{Ker}(w)$. It is easily checked that

(7.3.1) $$\mathrm{Ker}(w)_x = \mathrm{Ker}(w_x)$$

for all $x \in X$. Therefore w is injective if and only $\mathrm{Ker}(w) = 0$.

IMAGES.

The presheaf
$$U \mapsto \mathrm{Im}(w_U \colon \mathscr{F}(U) \to \mathscr{F}'(U))$$

is not a sheaf in general (see Exercise 2.12). The associated sheaf is an $\mathscr{O}_X$-module denoted by $\mathrm{Im}(w)$ and called the *image of w*. It is an $\mathscr{O}_X$-submodule of $\mathscr{F}'$. Again we have

(7.3.2) $$\mathrm{Im}(w)_x = \mathrm{Im}(w_x)$$

for all $x \in X$. Hence w is surjective if and only if $\mathrm{Im}(w) = \mathscr{F}'$.

COKERNELS.

The presheaf
$$U \mapsto \mathrm{Coker}(w_U \colon \mathscr{F}(U) \to \mathscr{F}'(U))$$

is not a sheaf in general. The associated sheaf is an $\mathscr{O}_X$-module denoted by $\mathrm{Coker}(w)$ and called the *cokernel of w*. The canonical homomorphisms $\mathscr{F}'(U) \to \mathrm{Coker}(w_U)$ define a homomorphism of $\mathscr{O}_X$-modules $\mathscr{F}' \to \mathrm{Coker}(w)$, called *canonical*. Again

(7.3.3) $$\mathrm{Coker}(w)_x = \mathrm{Coker}(w_x)$$

for all $x \in X$. Clearly, $\mathrm{Coker}(w) = \mathscr{F}'/\mathrm{Im}(w)$.

As the formation of quotients, kernels, and images is compatible with the formation of stalks, every homomorphism $w \colon \mathscr{F} \to \mathscr{F}'$ of $\mathscr{O}_X$-modules induces an isomorphism

(7.3.4) $$\mathscr{F}/\mathrm{Ker}(w) \xrightarrow{\sim} \mathrm{Im}(w).$$

EXACT SEQUENCES.

Let

(7.3.5) $$\mathscr{F} \xrightarrow{\ w\ } \mathscr{F}' \xrightarrow{\ w'\ } \mathscr{F}''$$

be a sequence of two homomorphisms of $\mathscr{O}_X$-modules. Then (7.3.5) is called *exact* if the following two equivalent conditions are satisfied.
(i) $\mathrm{Im}(w) = \mathrm{Ker}(w')$
(ii) The induced sequence $\mathscr{F}_x \to \mathscr{F}'_x \to \mathscr{F}''_x$ of $\mathscr{O}_{X,x}$-modules is exact for all $x \in X$.
A sequence
$$\cdots \to \mathscr{F}_{i-1} \to \mathscr{F}_i \to \mathscr{F}_{i+1} \to \mathscr{F}_{i+2} \to \cdots$$

is *exact*, if $\mathscr{F}_{i-1} \to \mathscr{F}_i \to \mathscr{F}_{i+1}$ is exact for all i.

THE MODULE OF HOMOMORPHISMS OF $\mathscr{O}_X$-MODULES.

Let $\mathscr{F}$ and $\mathscr{F}'$ be two $\mathscr{O}_X$-modules. If w_1 and w_2 are two homomorphisms $\mathscr{F} \to \mathscr{F}'$, we can define their sum in the obvious way:

$$(w_1 + w_2)_U := w_{1,U} + w_{2,U} \colon \mathscr{F}(U) \to \mathscr{F}'(U).$$

Let $a \in \Gamma(X, \mathscr{O}_X)$ and $w \colon \mathscr{F} \to \mathscr{F}'$ a homomorphism. Then we define a homomorphism $aw \colon \mathscr{F} \to \mathscr{F}'$ by $(aw)_U := (a_{|U})w_U$. In this way we endow the set of homomorphisms $\mathscr{F} \to \mathscr{F}'$ of $\mathscr{O}_X$-modules with the structure of a $\Gamma(X, \mathscr{O}_X)$-module. We denote this $\Gamma(X, \mathscr{O}_X)$-module by

(7.3.6) $$\mathrm{Hom}_{\mathscr{O}_X}(\mathscr{F}, \mathscr{F}').$$

The following proposition can be shown as for modules over a ring.

Proposition 7.3.
(1) *A sequence* $0 \to \mathscr{F}' \to \mathscr{F} \to \mathscr{F}''$ *of* $\mathscr{O}_X$-*modules is exact if and only if for all open subsets* $U \subseteq X$ *and for all* $\mathscr{O}_U$-*modules* $\mathscr{G}$ *the sequence*

$$0 \to \mathrm{Hom}_{\mathscr{O}_U}(\mathscr{G}, \mathscr{F}'_{|U}) \to \mathrm{Hom}_{\mathscr{O}_U}(\mathscr{G}, \mathscr{F}_{|U}) \to \mathrm{Hom}_{\mathscr{O}_U}(\mathscr{G}, \mathscr{F}''_{|U})$$

of $\Gamma(U, \mathscr{O}_X)$-*modules is exact.*
(2) *A sequence* $\mathscr{F}' \to \mathscr{F} \to \mathscr{F}'' \to 0$ *of* $\mathscr{O}_X$-*modules is exact if and only if for all open subsets* $U \subseteq X$ *and all* $\mathscr{O}_U$-*modules* $\mathscr{G}$ *the sequence*

$$0 \to \mathrm{Hom}_{\mathscr{O}_U}(\mathscr{F}''_{|U}, \mathscr{G}) \to \mathrm{Hom}_{\mathscr{O}_U}(\mathscr{F}_{|U}, \mathscr{G}) \to \mathrm{Hom}_{\mathscr{O}_U}(\mathscr{F}'_{|U}, \mathscr{G})$$

of $\Gamma(U, \mathscr{O}_X)$-*modules is exact.*
In particular, the Hom-*functor is left exact in both entries.*

(7.4) Basic constructions of $\mathscr{O}_X$-modules.

As for modules over a ring we can define direct sums, products, tensor products, etc. In the entire section, $(X, \mathscr{O}_X)$ will denote a ringed space.

DIRECT SUMS AND PRODUCTS.

Let $(\mathscr{F}_i)_{i \in I}$ be a family of $\mathscr{O}_X$-modules. Consider the presheaves

$$U \mapsto \bigoplus_{i \in I} \mathscr{F}_i(U), \qquad U \mapsto \prod_{i \in I} \mathscr{F}_i(U),$$

where the restriction maps are induced by the restriction maps of the $\mathscr{F}_i$. The product presheaf is a sheaf. In the case of the direct sum, we take the associated sheaf. Defining addition and scalar multiplication componentwise, we obtain $\mathscr{O}_X$-modules that are denoted $\bigoplus_i \mathscr{F}_i$ (resp. $\prod_i \mathscr{F}_i$) and called *direct sum* (resp. *product*) of the family $(\mathscr{F}_i)$. Clearly we have a natural injective homomorphism $\bigoplus_i \mathscr{F}_i \to \prod_i \mathscr{F}_i$ which is an isomorphism if I is finite.

If for all $i \in I$, $\mathscr{F}_i = \mathscr{F}$ for some $\mathscr{O}_X$-module $\mathscr{F}$, we also write $\mathscr{F}^{(I)}$ (resp. $\mathscr{F}^I$) instead of $\bigoplus_i \mathscr{F}_i$ (resp. $\prod_i \mathscr{F}_i$).

Let $x \in X$ be a point. For every open neighborhood U of x the maps $\mathscr{F}_i(U) \to \mathscr{F}_{i,x}$ sending a section s_i to its germ $s_{i,x}$ define a map $(\bigoplus_i \mathscr{F}_i)(U) \to \bigoplus_i \mathscr{F}_{i,x}$. Forming the inductive limit over all open neighborhoods U of x, we obtain an isomorphism of $\mathscr{O}_{X,x}$-modules

$$(7.4.1) \qquad \left(\bigoplus_{i \in I} \mathscr{F}_i\right)_x \xrightarrow{\sim} \bigoplus_{i \in I} \mathscr{F}_{i,x}$$

because direct sums and inductive limits commute with each other.

Similarly we obtain an $\mathscr{O}_{X,x}$-module homomorphism

$$(7.4.2) \qquad \left(\prod_{i \in I} \mathscr{F}_i\right)_x \to \prod_{i \in I} \mathscr{F}_{i,x}.$$

This homomorphism is in general not an isomorphism.

The structure of an abelian group defined on $\operatorname{Hom}_{\mathscr{O}_X}(\mathscr{F}, \mathscr{G})$ for all $\mathscr{O}_X$-modules $\mathscr{F}$ and $\mathscr{G}$ defines on the category of $\mathscr{O}_X$-modules the structure of an additive category (Section (A.4)) in which even infinite direct sums and infinite products exist. Moreover, it is immediate that the kernels and cokernels defined in Section (7.3) are kernels and cokernels, respectively, in the categorical sense. Together with (7.3.4) this shows:

Proposition 7.4. *The additive category of $\mathscr{O}_X$-modules is an abelian category.*

We denote the abelian category of $\mathscr{O}_X$-modules by $(\mathscr{O}_X\text{-Mod})$.

SUMS AND INTERSECTIONS OF SUBMODULES.

Let $\mathscr{F}$ be an $\mathscr{O}_X$-module and let $(\mathscr{F}_i)_{i \in I}$ be a family of $\mathscr{O}_X$-submodules. The *sum* $\sum \mathscr{F}_i$ of the family $(\mathscr{F}_i)_i$ is the $\mathscr{O}_X$-submodule of $\mathscr{F}$ defined as the image of the canonical homomorphism

$$\bigoplus_{i \in I} \mathscr{F}_i \to \mathscr{F}.$$

It is the sheaf associated to the presheaf $U \mapsto \sum_i \mathscr{F}_i(U)$. Combining (7.4.1) and (7.3.2) we obtain for all $x \in X$ an isomorphism of $\mathscr{O}_{X,x}$-modules

$$(7.4.3) \qquad \left(\sum_{i \in I} \mathscr{F}_i\right)_x \xrightarrow{\sim} \sum_{i \in I} \mathscr{F}_{i,x}.$$

The *intersection* $\bigcap_i \mathscr{F}_i$ of the family $(\mathscr{F}_i)$ is the $\mathscr{O}_X$-submodule of $\mathscr{F}$ defined as the kernel of the canonical homomorphism

$$\mathscr{F} \to \prod_{i \in I} \mathscr{F}/\mathscr{F}_i.$$

It is the sheaf $U \mapsto \bigcap_i \mathscr{F}_i(U)$. The canonical homomorphism of $\mathscr{O}_{X,x}$-modules

$$(7.4.4) \qquad \left(\bigcap_{i \in I} \mathscr{F}_i\right)_x \to \bigcap_{i \in I} \mathscr{F}_{i,x}.$$

is in general not an isomorphism. If I is finite, then $\bigcap_i \mathscr{F}_i$ can also be described as the kernel of the canonical homomorphism $\mathscr{F} \to \bigoplus_{i \in I} \mathscr{F}/\mathscr{F}_i$. In this case (7.2.1), (7.3.1), and (7.4.1) show that (7.4.4) is an isomorphism

MODULES GENERATED BY SECTIONS.

Let $\mathscr{F}$ be an $\mathscr{O}_X$-module. The map

$$(7.4.5) \qquad \operatorname{Hom}_{\mathscr{O}_X}(\mathscr{O}_X, \mathscr{F}) \to \Gamma(X, \mathscr{F}), \quad w \mapsto w_X(1)$$

is an isomorphism of $\Gamma(X, \mathscr{O}_X)$-modules which is functorial in $\mathscr{F}$. Indeed, it is clear that the map is $\Gamma(X, \mathscr{O}_X)$-linear and functorial in $\mathscr{F}$. If $s \in \Gamma(X, \mathscr{F})$, there is a unique homomorphism $w \colon \mathscr{O}_X \to \mathscr{F}$ such that $w_U(1) = s_{|U} \in \Gamma(U, \mathscr{F})$ for all open sets $U \subseteq X$. This defines an inverse map.

If I is any index set, we obtain a functorial isomorphism of $\Gamma(X, \mathscr{O}_X)$-modules

$$(7.4.6) \qquad \operatorname{Hom}_{\mathscr{O}_X}(\mathscr{O}_X^{(I)}, \mathscr{F}) = \operatorname{Hom}_{\mathscr{O}_X}(\mathscr{O}_X, \mathscr{F})^I \xrightarrow{\sim} \Gamma(X, \mathscr{F})^I.$$

Definition 7.5. *Let $(s_i)_{i \in I}$ be a family of sections $s_i \in \Gamma(X, \mathscr{F})$. We say that $\mathscr{F}$ is generated by the family (s_i), if the corresponding homomorphism $\mathscr{O}_X^{(I)} \to \mathscr{F}$ is surjective.*

In other words, $\mathscr{F}$ is generated by $(s_i)_i$, if for all $x \in X$ the $\mathscr{O}_{X,x}$-module $\mathscr{F}_x$ is generated by the family $(s_{i,x})_i$.

TENSOR PRODUCTS.

Let $\mathscr{F}$ and $\mathscr{G}$ be two $\mathscr{O}_X$-modules. The presheaf

$$U \mapsto \mathscr{F}(U) \otimes_{\mathscr{O}_X(U)} \mathscr{G}(U),$$

where the restriction maps are the tensor products of the restriction maps of $\mathscr{F}$ and $\mathscr{G}$, is in general not a sheaf (see Exercise 7.11). If we define addition and scalar multiplication in the obvious way on this presheaf, its sheafification is an $\mathscr{O}_X$-module that is called the *tensor product of $\mathscr{F}$ and $\mathscr{G}$* and denoted by $\mathscr{F} \otimes_{\mathscr{O}_X} \mathscr{G}$. Similarly as for R-modules, the tensor product can be characterized by a universal property. For $s \in \mathscr{F}(U)$, $t \in \mathscr{G}(U)$, we denote again by $s \otimes t$ the image of $s \otimes t \in \mathscr{F}(U) \otimes_{\mathscr{O}_X(U)} \mathscr{G}(U)$ in $(\mathscr{F} \otimes_{\mathscr{O}_X} \mathscr{G})(U)$.

For all $x \in X$ the canonical homomorphism of $\mathscr{O}_{X,x}$-modules

$$(7.4.7) \qquad (\mathscr{F} \otimes_{\mathscr{O}_X} \mathscr{G})_x \to \mathscr{F}_x \otimes_{\mathscr{O}_{X,x}} \mathscr{G}_x$$

is an isomorphism because inductive limits commute with tensor products.

If $n \geq 1$ is an integer we set

$$(7.4.8) \qquad \mathscr{F}^{\otimes n} := \underbrace{\mathscr{F} \otimes_{\mathscr{O}_X} \cdots \otimes_{\mathscr{O}_X} \mathscr{F}}_{n \text{ terms}}.$$

We also define $\mathscr{F}^{\otimes 0} := \mathscr{O}_X$. If $s \in \mathscr{F}(U)$ is a section, $U \subseteq X$ open, we denote by $s^{\otimes n}$ the image of $(s, \dots, s)$ under the canonical homomorphism

$$\mathscr{F}(U)^n \to \mathscr{F}(U) \otimes_{\mathscr{O}_X(U)} \cdots \otimes_{\mathscr{O}_X(U)} \mathscr{F}(U) \to \mathscr{F}^{\otimes n}(U).$$

Let $\mathscr{I} \subseteq \mathscr{O}_X$ be an ideal. Then the image of the multiplication $\mathscr{I} \otimes_{\mathscr{O}_X} \mathscr{F} \to \mathscr{F}$ is denoted by $\mathscr{I}\mathscr{F}$. It is an $\mathscr{O}_X$-submodule of $\mathscr{F}$. For all $x \in X$ we have $(\mathscr{I}\mathscr{F})_x = \mathscr{I}_x \mathscr{F}_x$.

THE $\mathscr{O}_X$-MODULE OF HOMOMORPHISMS.

Let $\mathscr{F}$ and $\mathscr{G}$ be two $\mathscr{O}_X$-modules. The presheaf

$$U \mapsto \mathrm{Hom}_{\mathscr{O}_{X|U}}(\mathscr{F}_{|U}, \mathscr{G}_{|U})$$

with the obvious restriction maps is a sheaf. As explained in Section (7.1), the right hand side is a $\Gamma(U, \mathscr{O}_X)$-module. Therefore this sheaf has the structure of an $\mathscr{O}_X$-module, and we denote this $\mathscr{O}_X$-module by $\mathscr{H}om_{\mathscr{O}_X}(\mathscr{F}, \mathscr{G})$.

For all $x \in X$ there is a canonical homomorphism of $\mathscr{O}_{X,x}$-modules

$$(7.4.9) \qquad \mathscr{H}om_{\mathscr{O}_X}(\mathscr{F}, \mathscr{G})_x \to \mathrm{Hom}_{\mathscr{O}_{X,x}}(\mathscr{F}_x, \mathscr{G}_x),$$

which in general is neither injective nor surjective (see however Proposition 7.27 below).

The $\mathscr{O}_X$-module

$$(7.4.10) \qquad \mathscr{F}^{\vee} := \mathscr{H}om_{\mathscr{O}_X}(\mathscr{F}, \mathscr{O}_X)$$

is called the *dual $\mathscr{O}_X$-module of $\mathscr{F}$*.

We deduce from (7.4.6) a canonical isomorphism of $\mathscr{O}_X$-modules

$$(7.4.11) \qquad \mathscr{H}om_{\mathscr{O}_X}(\mathscr{O}_X^{(I)}, \mathscr{F}) \xrightarrow{\sim} \mathscr{F}^I.$$

Finally, if $\mathscr{F}$ is any $\mathscr{O}_X$-module, sending a section $s \in \mathscr{O}_X(U)$, $U \subseteq X$ open, to the scalar multiplication on $\mathscr{F}_{|U}$ with s defines a homomorphism of $\mathscr{O}_X$-modules

$$(7.4.12) \qquad \mathscr{O}_X \to \mathscr{H}om_{\mathscr{O}_X}(\mathscr{F}, \mathscr{F}).$$

INDUCTIVE LIMITS OF $\mathscr{O}_X$-MODULES.

Let I be a filtered partially preordered set and let $((\mathscr{F}_i)_{i \in I}, (u_{ji})_{i \leq j})$ be an inductive system of $\mathscr{O}_X$-modules. Consider the presheaf

$$U \mapsto \varinjlim_i \mathscr{F}_i(U),$$

where the restriction maps are the inductive limits of the restriction maps of the $\mathscr{F}_i$. This presheaf is in general not a sheaf. If we define addition and scalar multiplication in the obvious way on this presheaf, its sheafification is an $\mathscr{O}_X$-module that is called the *inductive limit of $((\mathscr{F}_i)_{i \in I}, (u_{ji})_{i \leq j})$* and denoted by $\varinjlim_i \mathscr{F}_i$. This sheaf is an inductive limit in the category of $\mathscr{O}_X$-modules: If $\mathscr{G}$ is an $\mathscr{O}_X$-module we have an isomorphism, functorial in $\mathscr{G}$,

$$(7.4.13) \qquad \mathrm{Hom}_{\mathscr{O}_X}(\varinjlim_i \mathscr{F}_i, \mathscr{G}) \xrightarrow{\sim} \varprojlim_i \mathrm{Hom}_{\mathscr{O}_X}(\mathscr{F}_i, \mathscr{G}).$$

As any two filtered inductive limits commute with each other, we obtain for all $x \in X$ an isomorphism of $\mathscr{O}_{X,x}$-modules

$$(7.4.14) \qquad \varinjlim_i \mathscr{F}_{i,x} \xrightarrow{\sim} (\varinjlim_i \mathscr{F}_i)_x.$$

(7.5) Finite locally free $\mathscr{O}_X$-modules.

LOCALLY FREE $\mathscr{O}_X$-MODULES.

Let $(X, \mathscr{O}_X)$ be a ringed space.

Definition 7.6. *An $\mathscr{O}_X$-module $\mathscr{F}$ is called* locally free, *if for all $x \in X$ there exists an open neighborhood U of x such that $\mathscr{F}_{|U}$ is isomorphic to an $\mathscr{O}_{X|U}$-module of the form $\mathscr{O}_X^{(I)}|_U$ for a set I (depending on x). If I is finite for all U, we say that $\mathscr{F}$ is* finite locally free *or* locally free of finite type.

If $\mathscr{F}_{|U} \cong \mathscr{O}_X^{(I)}|_U$, it follows from (7.4.1) that $\mathscr{F}_x$ is isomorphic to $\mathscr{O}_{X,x}^{(I)}$ as an $\mathscr{O}_{X,x}$-module for all $x \in U$. In particular we see that the cardinality of I depends only on $\mathscr{F}$ and on x, and we call this cardinal number the *rank of $\mathscr{F}$ at x* and denote it by $\mathrm{rk}_x(\mathscr{F})$. Clearly, the function $x \mapsto \mathrm{rk}_x(\mathscr{F})$ is locally constant on X. This function is called the *rank of $\mathscr{F}$* and denoted by $\mathrm{rk}(\mathscr{F})$.

If $\mathscr{F}$ is a finite locally free $\mathscr{O}_X$-module of rank n (hence n is a locally constant function $X \to \mathbb{N}_0$), its dual $\mathscr{F}^\vee$ is again locally free of rank n. More generally, if $\mathscr{E}$ is a finite locally free $\mathscr{O}_X$-module of rank m, the $\mathscr{O}_X$-module $\mathscr{H}om_{\mathscr{O}_X}(\mathscr{F}, \mathscr{E})$ is locally free of rank nm.

MODULES OF HOMOMORPHISMS AND THE TENSOR PRODUCT.

Let $\mathscr{F}$, $\mathscr{G}$ and $\mathscr{H}$ be $\mathscr{O}_X$-modules. As in the case of modules over a ring it is easy to see that there is a functorial isomorphism of $\Gamma(X, \mathscr{O}_X)$-modules

$$(7.5.1) \qquad \mathrm{Hom}_{\mathscr{O}_X}(\mathscr{F} \otimes_{\mathscr{O}_X} \mathscr{G}, \mathscr{H}) \xrightarrow{\sim} \mathrm{Hom}_{\mathscr{O}_X}(\mathscr{F}, \mathscr{H}om_{\mathscr{O}_X}(\mathscr{G}, \mathscr{H}))$$

and hence a functorial isomorphism of $\mathscr{O}_X$-modules

$$(7.5.2) \qquad \mathscr{H}om_{\mathscr{O}_X}(\mathscr{F} \otimes_{\mathscr{O}_X} \mathscr{G}, \mathscr{H}) \xrightarrow{\sim} \mathscr{H}om_{\mathscr{O}_X}(\mathscr{F}, \mathscr{H}om_{\mathscr{O}_X}(\mathscr{G}, \mathscr{H})).$$

There is also a functorial homomorphism of $\mathscr{O}_X$-modules

$$(7.5.3) \qquad \mathscr{H}om_{\mathscr{O}_X}(\mathscr{F}, \mathscr{G}) \otimes_{\mathscr{O}_X} \mathscr{H} \longrightarrow \mathscr{H}om_{\mathscr{O}_X}(\mathscr{F}, \mathscr{G} \otimes_{\mathscr{O}_X} \mathscr{H})$$

which is defined as follows. For each open set $U \subseteq X$, we send a pair (w, t), where $w \in \Gamma(U, \mathscr{H}om_{\mathscr{O}_X}(\mathscr{F}, \mathscr{G})) = \mathrm{Hom}_{\mathscr{O}_U}(\mathscr{F}_{|U}, \mathscr{G}_{|U})$ and $t \in \Gamma(U, \mathscr{H})$, to the homomorphism $\mathscr{F}_{|U} \to (\mathscr{G} \otimes_{\mathscr{O}_X} \mathscr{H})_{|U} = \mathscr{G}_{|U} \otimes_{\mathscr{O}_U} \mathscr{H}_{|U}$ that attaches to a section $s \in \Gamma(V, \mathscr{F})$ ($V \subseteq U$ open) the image of $w_V(s) \otimes t_{|V}$ in $\Gamma(V, \mathscr{G} \otimes_{\mathscr{O}_X} \mathscr{H})$. In general, this homomorphism is neither injective or surjective.

Proposition 7.7. *The homomorphism (7.5.3) is an isomorphism if $\mathscr{F}$ or $\mathscr{H}$ is finite locally free.*

Proof. As a morphism of sheaves is bijective if and only if it is bijective on stalks, this is a local question, i. e. it suffices that each $x \in X$ has an open neighborhood U such that the restriction of (7.5.3) to U is an isomorphism. Therefore we may assume that $\mathscr{F} = \mathscr{O}_X^n$ or that $\mathscr{H} = \mathscr{O}_X^n$ (for some integer $n \geq 0$).

If $\mathscr{F} = \mathscr{O}_X^n$, the homomorphism (7.5.3) is the composition of the functorial isomorphisms

$$\mathscr{H}om_{\mathscr{O}_X}(\mathscr{F}, \mathscr{G}) \otimes_{\mathscr{O}_X} \mathscr{H} \xrightarrow{\sim} \mathscr{H}om_{\mathscr{O}_X}(\mathscr{O}_X, \mathscr{G})^n \otimes_{\mathscr{O}_X} \mathscr{H} \xrightarrow{\sim} \mathscr{G}^n \otimes_{\mathscr{O}_X} \mathscr{H}$$

$$\xrightarrow{\sim} (\mathscr{G} \otimes_{\mathscr{O}_X} \mathscr{H})^n \xrightarrow{\sim} \mathscr{H}om_{\mathscr{O}_X}(\mathscr{O}_X, \mathscr{G} \otimes_{\mathscr{O}_X} \mathscr{H})^n$$

$$\xrightarrow{\sim} \mathscr{H}om_{\mathscr{O}_X}(\mathscr{F}, \mathscr{G} \otimes_{\mathscr{O}_X} \mathscr{H})$$

and therefore an isomorphism. The case that $\mathcal{H} = \mathcal{O}_X^n$ is proved similarly. $\qquad\square$

Setting $\mathcal{G} = \mathcal{O}_X$ in (7.5.3), we obtain a functorial homomorphism

$$(7.5.4) \qquad \mathcal{F}^\vee \otimes_{\mathcal{O}_X} \mathcal{H} \longrightarrow \mathcal{H}om_{\mathcal{O}_X}(\mathcal{F}, \mathcal{H})$$

that is an isomorphism if $\mathcal{F}$ or $\mathcal{H}$ is finite locally free. If $\mathcal{F}$ is finite locally free, we can precompose this isomorphism with (7.5.2) and obtain a functorial isomorphism of $\mathcal{O}_X$-modules

$$(7.5.5) \qquad \mathcal{H}om_{\mathcal{O}_X}(\mathcal{G} \otimes_{\mathcal{O}_X} \mathcal{F}, \mathcal{H}) \xrightarrow{\sim} \mathcal{H}om_{\mathcal{O}_X}(\mathcal{G}, \mathcal{F}^\vee \otimes_{\mathcal{O}_X} \mathcal{H}).$$

INVERTIBLE $\mathcal{O}_X$-MODULES.

Let $\mathcal{L}$ be a locally free $\mathcal{O}_X$-module of rank 1. Then its dual (7.4.10) is again locally free of rank 1. By Proposition 7.7 there is a functorial isomorphism

$$(7.5.6) \qquad \mathcal{L}^\vee \otimes_{\mathcal{O}_X} \mathcal{L} \xrightarrow{\sim} \mathcal{H}om_{\mathcal{O}_X}(\mathcal{L}, \mathcal{L}).$$

We claim that for locally free $\mathcal{O}_X$-modules $\mathcal{L}$ of rank 1 the canonical homomorphism

$$(7.5.7) \qquad \iota \colon \mathcal{O}_X \to \mathcal{H}om_{\mathcal{O}_X}(\mathcal{L}, \mathcal{L})$$

defined in (7.4.12) is an isomorphism. Indeed, this is a local question and we can therefore assume that $\mathcal{L} = \mathcal{O}_X$ and in this case the isomorphism (7.4.11) is an inverse isomorphism of ι. Combining (7.5.7) and (7.5.6) we obtain an isomorphism of $\mathcal{O}_X$-modules

$$(7.5.8) \qquad \mathcal{L}^\vee \otimes_{\mathcal{O}_X} \mathcal{L} \xrightarrow{\sim} \mathcal{O}_X.$$

We say that an $\mathcal{O}_X$-module $\mathcal{L}$ is *invertible* if $\mathcal{L}$ is locally free of rank 1 and we set $\mathcal{L}^{\otimes -1} := \mathcal{L}^\vee$. More generally, we set $\mathcal{L}^{\otimes -n} := (\mathcal{L}^\vee)^{\otimes n}$ for every integer $n \geq 1$. Then for all integers $n, m \in \mathbb{Z}$ there is an isomorphism of $\mathcal{O}_X$-modules

$$(7.5.9) \qquad \mathcal{L}^{\otimes n} \otimes_{\mathcal{O}_X} \mathcal{L}^{\otimes m} \xrightarrow{\sim} \mathcal{L}^{\otimes(n+m)}$$

which is functorial in $\mathcal{L}$. We see that the set of isomorphism classes of invertible $\mathcal{O}_X$-modules forms a group, the so-called Picard group $\mathrm{Pic}(X)$ of X. We will come back to this notion in Chapter 11.

(7.6) Support of an $\mathcal{O}_X$-module.

Let $\mathcal{F}$ be an $\mathcal{O}_X$-module. Then

$$(7.6.1) \qquad \mathrm{Supp}(\mathcal{F}) := \{\, x \in X \; ; \; \mathcal{F}_x \neq 0 \,\}$$

is called the *support of $\mathcal{F}$*. This is not necessarily a closed subset of X (see Exercise 7.3). However we will see in Corollary 7.32 that $\mathrm{Supp}(\mathcal{F})$ is closed if $\mathcal{F}$ is of finite type (see Definition 7.25 below).

For every section s of $\mathcal{F}$ over an open subset U, the *support of s* is the set of $x \in U$ such that $s_x \neq 0$. This is always a closed subset of U (see Exercise 2.10).

(7.7) $\mathcal{O}_X$-algebras.

Again let $(X, \mathcal{O}_X)$ be an arbitrary ringed space. An $\mathcal{O}_X$-*algebra* is an $\mathcal{O}_X$-module $\mathscr{A}$ together with an $\mathcal{O}_X$-bilinear multiplication

$$\mathscr{A} \times \mathscr{A} \to \mathscr{A}, \quad (a, a') \mapsto aa' \qquad \text{for } a, a' \in \mathscr{A}(U), \, U \subseteq X \text{ open}$$

such that this multiplication defines on $\mathscr{A}(U)$ the structure of an $\mathcal{O}_X(U)$-algebra for all open subsets $U \subseteq X$. We call $\mathscr{A}$ *commutative*, if $\mathscr{A}(U)$ is a commutative $\mathcal{O}_X(U)$-algebra.

As for algebras over a ring, an $\mathcal{O}_X$-algebra $\mathscr{A}$ is the same as a sheaf of non-necessarily commutative rings together with a morphism $\alpha \colon \mathcal{O}_X \to \mathscr{A}$ of sheaves of rings such that the image of $\alpha(U)$ is contained in the center of the ring $\mathscr{A}(U)$ for all open sets $U \subseteq X$.

(7.8) Direct and inverse image of $\mathcal{O}_X$-modules.

Let $f \colon (X, \mathcal{O}_X) \to (Y, \mathcal{O}_Y)$ be a morphism of ringed spaces. In other words we are given a continuous map $f \colon X \to Y$ and a morphism $f^\flat \colon \mathcal{O}_Y \to f_*(\mathcal{O}_X)$ of sheaves of rings on Y (or, equivalently, a morphism $f^\sharp \colon f^{-1}\mathcal{O}_Y \to \mathcal{O}_X$ of sheaves of rings on X), see Section (2.9).

DIRECT IMAGE.

Note that if $\mathscr{F}$ and $\mathscr{F}'$ are sheaves on X, and $V \subseteq Y$ is an open set, we have

$$f_*(\mathscr{F} \times \mathscr{F}')(V) = (\mathscr{F} \times \mathscr{F}')(f^{-1}(V)) = \mathscr{F}(f^{-1}(V)) \times \mathscr{F}'(f^{-1}(V))$$
$$= f_*(\mathscr{F})(V) \times f_*(\mathscr{F}')(V)$$

and hence

(7.8.1) $$f_*(\mathscr{F} \times \mathscr{F}') = f_*(\mathscr{F}) \times f_*(\mathscr{F}').$$

Now let $\mathscr{F}$ be an $\mathcal{O}_X$-module. Then addition and scalar multiplication by $\mathcal{O}_X$ define by (7.8.1) via functoriality morphisms of sheaves

$$f_*(\mathscr{F}) \times f_*(\mathscr{F}) \to f_*(\mathscr{F}), \qquad f_*(\mathcal{O}_X) \times f_*(\mathscr{F}) \to f_*(\mathscr{F})$$

and in this way $f_*(\mathscr{F})$ is an $f_*(\mathcal{O}_X)$-module. Via $f^\flat$ we obtain the structure of an $\mathcal{O}_Y$-module on $f_*(\mathscr{F})$. This $\mathcal{O}_Y$-module is called the *direct image of $\mathscr{F}$ under f*.

Remark 7.8. This construction is clearly functorial in $\mathscr{F}$ and we obtain a functor f_* from the category of $\mathcal{O}_X$-modules to the category of $\mathcal{O}_Y$-modules. It is easy to see that this functor is left exact (this also follows formally from Proposition 7.11 below which shows that f_* is right adjoint to another functor).

If $g \colon (Y, \mathcal{O}_Y) \to (Z, \mathcal{O}_Z)$ is a second morphism of ringed spaces, there is an identity

(7.8.2) $$(g \circ f)_*(\mathscr{F}) = g_*(f_*(\mathscr{F}))$$

of $\mathcal{O}_Z$-modules which is functorial in $\mathscr{F}$.

Let $\mathscr{F}$ and $\mathscr{F}'$ be $\mathcal{O}_X$-modules. For all open subsets V of Y the canonical map

$$\mathscr{F}(f^{-1}(V)) \times \mathscr{F}'(f^{-1}(V)) \to (\mathscr{F} \otimes_{\mathcal{O}_X} \mathscr{F}')(f^{-1}(V))$$

is $\mathscr{O}_X(f^{-1}(V))$-bilinear and thus in particular $\mathscr{O}_Y(V)$-bilinear. Therefore we obtain a homomorphism of $\mathscr{O}_Y$-modules

$$(7.8.3) \qquad f_*(\mathscr{F}) \otimes_{\mathscr{O}_Y} f_*(\mathscr{F}') \to f_*(\mathscr{F} \otimes_{\mathscr{O}_X} \mathscr{F}')$$

which is functorial in $\mathscr{F}$ and $\mathscr{F}'$. In general this homomorphism is neither injective nor surjective.

INVERSE IMAGE.

Recall that for a sheaf $\mathscr{G}$ on Y we defined in Section (2.8) $f^{-1}\mathscr{G}$ as the sheaf attached to the presheaf

$$f^+\mathscr{G} : U \mapsto \varinjlim_{V \supset f(U)} \mathscr{G}(V).$$

If $\mathscr{G}'$ is a second sheaf on Y, we clearly have $f^+(\mathscr{G} \times \mathscr{G}') = f^+(\mathscr{G}) \times f^+(\mathscr{G}')$ and hence

$$(7.8.4) \qquad f^{-1}(\mathscr{G} \times \mathscr{G}') = f^{-1}(\mathscr{G}) \times f^{-1}(\mathscr{G}').$$

Now let $\mathscr{G}$ be an $\mathscr{O}_Y$-module. Using (7.8.4), addition and scalar multiplication on $\mathscr{G}$ define on $f^{-1}\mathscr{G}$ the structure of an $f^{-1}\mathscr{O}_Y$-module. Via $f^\sharp \colon f^{-1}\mathscr{O}_Y \to \mathscr{O}_X$, $\mathscr{O}_X$ is an $f^{-1}\mathscr{O}_Y$-algebra. Therefore

$$(7.8.5) \qquad f^*\mathscr{G} := \mathscr{O}_X \otimes_{f^{-1}\mathscr{O}_Y} f^{-1}\mathscr{G}$$

is endowed with the structure of an $\mathscr{O}_X$-module which we call the *inverse image of $\mathscr{G}$ under f*.

For every $x \in X$ there is a functorial isomorphism of stalks $f^{-1}(\mathscr{G})_x \cong \mathscr{G}_{f(x)}$ by (2.8.2) and hence by (7.4.7) a functorial isomorphism of $\mathscr{O}_{X,x}$-modules

$$(7.8.6) \qquad (f^*\mathscr{G})_x \cong \mathscr{O}_{X,x} \otimes_{\mathscr{O}_{Y,f(x)}} \mathscr{G}_{f(x)}.$$

Remark 7.9. Clearly $\mathscr{G} \mapsto f^{-1}(\mathscr{G}) \mapsto f^*(\mathscr{G})$ is functorial in $\mathscr{G}$, and we obtain a functor f^* from the category of $\mathscr{O}_Y$-modules to the category of $\mathscr{O}_X$-modules. By (7.8.6), f^* is right exact (again this also follows formally from Proposition 7.11 below which shows that f^* is left adjoint to f_*).

Remark 7.10. If $f \colon (X, \mathscr{O}_X) \to (Y, \mathscr{O}_Y)$ is a morphism of ringed spaces, then

$$f^*\mathscr{O}_Y = \mathscr{O}_X.$$

Let $\mathscr{G}'$ be a second $\mathscr{O}_Y$-module. The definition of $f^{-1}\mathscr{G}$ shows that there is a natural functorial homomorphism

$$(7.8.7) \qquad f^{-1}(\mathscr{G}) \otimes_{f^{-1}\mathscr{O}_Y} f^{-1}(\mathscr{G}') \to f^{-1}(\mathscr{G} \otimes_{\mathscr{O}_Y} \mathscr{G}').$$

By (2.8.2) and (7.4.7) this is an isomorphism on stalks and thus an isomorphism. This isomorphism induces by the definition of f^* an isomorphism

$$(7.8.8) \qquad f^*(\mathscr{G}) \otimes_{\mathscr{O}_X} f^*(\mathscr{G}') \xrightarrow{\sim} f^*(\mathscr{G} \otimes_{\mathscr{O}_Y} \mathscr{G}'),$$

functorial in $\mathscr{G}$ and $\mathscr{G}'$.

If $g \colon (Y, \mathscr{O}_Y) \to (Z, \mathscr{O}_Z)$ is a second morphism of ringed spaces and $\mathscr{H}$ an $\mathscr{O}_Z$-module, the isomorphism (7.8.7) defines an isomorphism, functorial in $\mathscr{H}$,

$$(7.8.9) \qquad\qquad (g \circ f)^* \mathscr{H} \cong f^*(g^* \mathscr{H}).$$

The existence of this isomorphism follows also from the isomorphism of functors (7.8.2) and from the following Proposition.

Proposition 7.11. *For every $\mathscr{O}_X$-module $\mathscr{F}$ and every $\mathscr{O}_Y$-module $\mathscr{G}$ there is an isomorphism of $\Gamma(Y, \mathscr{O}_Y)$-modules*

$$\operatorname{Hom}_{\mathscr{O}_X}(f^*\mathscr{G}, \mathscr{F}) \xrightarrow{\sim} \operatorname{Hom}_{\mathscr{O}_Y}(\mathscr{G}, f_*\mathscr{F})$$

which is functorial in $\mathscr{F}$ and $\mathscr{G}$.

Here the left hand side, being an $\Gamma(X, \mathscr{O}_X)$-module, is considered as a $\Gamma(Y, \mathscr{O}_Y)$-module via the ring homomorphism $f_Y^\flat \colon \Gamma(Y, \mathscr{O}_Y) \to \Gamma(X, \mathscr{O}_X)$.

Proof. It is straightforward to check – albeit a bit tedious – that there are functorial isomorphisms

$$\operatorname{Hom}_{\mathscr{O}_X}(f^*\mathscr{G}, \mathscr{F}) \xrightarrow{\sim} \operatorname{Hom}_{f^{-1}\mathscr{O}_Y}(f^{-1}\mathscr{G}, \mathscr{F}) \xrightarrow{\sim} \operatorname{Hom}_{\mathscr{O}_Y}(\mathscr{G}, f_*\mathscr{F}),$$

the second isomorphism being a variant of Proposition 2.27. $\qquad\qquad\qquad\qquad\qquad\square$

In particular, $\operatorname{id}_{f_*\mathscr{G}}$ and $\operatorname{id}_{f_*\mathscr{F}}$ correspond to homomorphisms

$$(7.8.10) \qquad\qquad \mathscr{G} \to f_*(f^*\mathscr{G}), \qquad f^*(f_*\mathscr{F}) \to \mathscr{F}$$

which we call *canonical*. These homomorphisms are in general neither injective nor surjective. If $\mathscr{G} = \mathscr{O}_Y$, then $\mathscr{G} \to f_*(f^*\mathscr{G})$ is the homomorphism $f^\flat \colon \mathscr{O}_Y \to f_*\mathscr{O}_X$ using Remark 7.10.

Using the canonical homomorphism $\mathscr{G} \to f_*(f^*\mathscr{G})$ we define for every open subset $V \subseteq Y$ a map

$$(7.8.11) \qquad f^* = f_V^* \colon \Gamma(V, \mathscr{G}) \to \Gamma(V, f_*(f^*\mathscr{G})) = \Gamma(f^{-1}(V), f^*\mathscr{G})$$

which we call the *pull back of sections under f*. If $\mathscr{G} = \mathscr{O}_Y$, then $f_V^* = f_V^\flat$ and we obtain the usual pullback of functions.

Quasi-coherent modules on a scheme

In this part of the chapter we attach to every module M over a ring A a module $\tilde{M}$ over the ringed space $X = \operatorname{Spec} A$. We obtain a fully faithful functor from the category of A-modules to the category of $\mathscr{O}_X$-modules which is compatible with those constructions that commute with localization (e.g., forming kernels and cokernels, direct sums, filtered inductive limits; but not infinite products or infinite intersections of submodules).

The notion of a quasi-coherent $\mathscr{O}_X$-module is introduced for an arbitrary ringed space $(X, \mathscr{O}_X)$. Then we show that for $X = \operatorname{Spec} A$ an $\mathscr{O}_X$-module $\mathscr{F}$ is quasi-coherent if and only if $\mathscr{F} \cong \tilde{M}$ for some A-module M (Theorem 7.16). The key property of quasi-coherent $\mathscr{O}_X$-modules $\mathscr{F}$ used in the proof is the fact that for $f \in A$ any section u of $\mathscr{F}$ over the principal open subset $D(f)$ can be extended essentially uniquely to X after multiplying u with some positive power of f. This useful property holds in fact much more generally (Theorem 7.22).

Finally we will discuss direct and inverse images of quasi-coherent modules.

(7.9) The $\mathscr{O}_{\operatorname{Spec} A}$-module $\tilde{M}$ attached to an A-module M.

Let A be a ring and let $X = \operatorname{Spec} A$ be the associated affine scheme. Let M be an A-module. We define an $\mathscr{O}_X$-module $\tilde{M}$ similarly as we defined the structure sheaf $\mathscr{O}_X$ in Section (2.10): Recall that $\{ D(f) \ ; \ f \in A \}$ is a basis for the topology of $\operatorname{Spec} A$. We define

$$\Gamma(D(f), \tilde{M}) := M_f,$$

where M_f is the localization of M with respect the multiplicative set $\{ f^i \ ; \ i \geq 0 \}$. As in Section (2.10) it is easy to see that this is well-defined and that we obtain a presheaf $\tilde{M}$ on the basis $\{ D(f) \ ; \ f \in A \}$. Verbatim the same arguments as in the proof of Theorem 2.33 show:

Theorem 7.12. *The presheaf $\tilde{M}$ is a sheaf on $\{ D(f) \ ; \ f \in A \}$.*

We denote the attached sheaf on X again by $\tilde{M}$. For each $f \in A$, $M_f = \Gamma(D(f), \tilde{M})$ is a module over the ring $A_f = \Gamma(D(f), \mathscr{O}_X)$, and it is easy to check that this defines the structure of an $\mathscr{O}_X$-module on $\tilde{M}$. If we consider A as a module over itself, then $\tilde{A} = \mathscr{O}_X$.

As for the structure sheaf $\mathscr{O}_X$ in (2.10.2), we have for every point $x \in X$

$$(7.9.1) \qquad \tilde{M}_x = \varinjlim_{D(f) \ni x} \Gamma(D(f), \tilde{M}) = \varinjlim_{f \notin \mathfrak{p}_x} M_f = M_{\mathfrak{p}_x},$$

where $M_{\mathfrak{p}_x}$ denotes the localization of M in the prime ideal $\mathfrak{p}_x$.

If N is a second A-module and $u \colon M \to N$ a homomorphism of A-modules, u induces a homomorphism $u_f \colon M_f \to N_f$ of A_f-modules for all $f \in A$. For all $f, g \in A$ with $D(f) \subseteq D(g)$ we have a commutative diagram

$$
\begin{array}{ccc}
M_g & \xrightarrow{\ u_g\ } & N_g \\
\downarrow & & \downarrow \\
M_f & \xrightarrow{\ u_f\ } & N_f,
\end{array}
$$

where the vertical maps are the canonical homomorphisms. Therefore u induces a homomorphism of $\mathscr{O}_X$-modules

$$\tilde{u} \colon \tilde{M} \to \tilde{N}.$$

In this way we obtain a functor $M \mapsto \tilde{M}$ from the category of A-modules to the category of $\mathscr{O}_X$-modules.

Conversely, a homomorphism $w \colon \mathscr{F} \to \mathscr{G}$ of $\mathscr{O}_X$-modules induces on global sections a homomorphism of modules over the ring $A = \Gamma(X, \mathscr{O}_X)$

$$w_X \colon \Gamma(X, \mathscr{F}) \to \Gamma(X, \mathscr{G})$$

and we obtain a functor $\Gamma \colon \mathscr{F} \mapsto \Gamma(X, \mathscr{F})$ from the category of $\mathscr{O}_X$-modules to the category of A-modules.

Proposition 7.13. *Let A be a ring, $X = \operatorname{Spec} A$. Then for all A-modules M and N the maps*

$$\operatorname{Hom}_A(M, N) \underset{\Gamma}{\overset{u \mapsto \tilde{u}}{\rightleftarrows}} \operatorname{Hom}_{\mathscr{O}_X}(\tilde{M}, \tilde{N})$$

are mutually inverse. In particular, the functor $M \mapsto \tilde{M}$ is fully faithful.

Proof. The definition of $\tilde{u}$ shows that $\Gamma(\tilde{u}) = u$. Conversely let $w \colon \tilde{M} \to \tilde{N}$ be a homomorphism of $\mathscr{O}_X$-modules and set $u = \Gamma(w) = w_X$. For all $f \in A$, the A_f-module homomorphism $w_{D(f)}$ makes the diagram

$$\begin{array}{ccc} M & \xrightarrow{\ u\ } & N \\ \downarrow & & \downarrow \\ M_f & \xrightarrow{\ w_{D(f)}\ } & N_f, \end{array}$$

commutative. Hence $w_{D(f)} = u_f$ for all $f \in A$ and therefore $\tilde{u} = w$. $\qquad\square$

By setting $M = A$ in Proposition 7.13, it follows that an A-module N is zero if and only if $\tilde{N}$ is zero.

Proposition 7.14. *Let A be a ring, $X = \operatorname{Spec} A$.*
(1) *Let*

(7.9.2) $$M \xrightarrow{\ u\ } N \xrightarrow{\ v\ } P$$

be a sequence of A-modules. Then this sequence is exact if and only if the sequence

(7.9.3) $$\tilde{M} \xrightarrow{\ \tilde{u}\ } \tilde{N} \xrightarrow{\ \tilde{v}\ } \tilde{P}$$

is an exact sequence of $\mathscr{O}_X$-modules.
(2) *Let $u \colon M \to N$ be a homomorphism of A-modules. Then*

$$\operatorname{Ker}(u)^{\sim} = \operatorname{Ker}(\tilde{u}), \qquad \operatorname{Im}(u)^{\sim} = \operatorname{Im}(\tilde{u}), \qquad \operatorname{Coker}(u)^{\sim} = \operatorname{Coker}(\tilde{u}).$$

In particular, u is injective (resp. surjective, resp. bijective) if and only if $\tilde{u}$ is.
(3) *Let $(M_i)_{i \in I}$ be a family of A-modules. Then*

$$\bigoplus_{i \in I} \tilde{M}_i = \Big(\bigoplus_{i \in I} M_i \Big)^{\sim}.$$

(4) *Let M be the filtered inductive limit of an inductive system of A-modules M_λ. Then $\tilde{M}$ is the inductive limit of the inductive system $\tilde{M}_\lambda$ of $\mathscr{O}_X$-modules.*

Proof. The sequence (7.9.2) is exact if and only if for all $x \in X$ the induced sequence $M_{\mathfrak{p}_x} \to N_{\mathfrak{p}_x} \to P_{\mathfrak{p}_x}$ is exact, which is equivalent to the exactness of (7.9.3) by (7.9.1).

Assertion (2) follows immediately from (1). To prove Assertion (3) set $M := \bigoplus_i M_i$. The inclusions $M_i \to M$ define homomorphisms $\tilde{M}_i \to \tilde{M}$ and hence a homomorphism $\bigoplus_i \tilde{M}_i \to \tilde{M}$. As localization commutes with direct sums, this homomorphism is bijective on stalks and hence an isomorphism. The same proof shows (4). $\qquad\square$

(7.10) Quasi-coherent modules.

Let A be a ring. We will now identify the essential image of the functor $M \mapsto \tilde{M}$. Note that if M is any A-module, there exists an exact sequence of A-modules

$$A^{(J)} \to A^{(I)} \to M \to 0$$

for some index sets J and I. Therefore by Proposition 7.14, the following condition for an $\mathscr{O}_X$-module (for $X = \operatorname{Spec} A$) is clearly necessary for being of the form $\tilde{M}$.

Definition 7.15. *Let* $(X, \mathscr{O}_X)$ *be a ringed space. An* $\mathscr{O}_X$*-module* $\mathscr{F}$ *is called* quasi-coherent *if for all* $x \in X$ *there exists an open neighborhood* U *of* x *and an exact sequence of* $\mathscr{O}_{X|U}$*-modules of the form*

$$\mathscr{O}_X^{(J)}|_U \longrightarrow \mathscr{O}_X^{(I)}|_U \longrightarrow \mathscr{F}|_U \longrightarrow 0,$$

where I *and* J *are arbitrary index sets (depending on* x*).*

We will now show that for $X = \operatorname{Spec} A$ an $\mathscr{O}_X$-module $\mathscr{F}$ is isomorphic to $\tilde{M}$ for some A-module M if and only if $\mathscr{F}$ is quasi-coherent. We introduce the following notation. Let $(X, \mathscr{O}_X)$ be a locally ringed space and let $f \in \Gamma(X, \mathscr{O}_X)$ be a global section. We define

$$X_f := \{\, x \in X \ ; \ f_x \text{ is invertible in } \mathscr{O}_{X,x} \,\}.$$

Note that f_x is invertible in $\mathscr{O}_{X,x}$ if and only if the residue class $f(x)$ of f in $\kappa(x)$ is $\neq 0$. The set X_f is easily seen to be an open subset of X. The image of f under the restriction homomorphism $\Gamma(X, \mathscr{O}_X) \to \Gamma(X_f, \mathscr{O}_X)$ is invertible. Therefore for every $\mathscr{O}_X$-module $\mathscr{F}$ the restriction homomorphism $\Gamma(X, \mathscr{F}) \to \Gamma(X_f, \mathscr{F})$ induces a homomorphism of $\Gamma(X, \mathscr{O}_X)$-modules

$$(7.10.1) \qquad\qquad \Gamma(X, \mathscr{F})_f \longrightarrow \Gamma(X_f, \mathscr{F}),$$

where the left hand side is the localization of the $\Gamma(X, \mathscr{O}_X)$-module $\Gamma(X, \mathscr{F})$ by f.

If $X = \operatorname{Spec} A$ is an affine scheme and $f \in A = \Gamma(X, \mathscr{O}_X)$, X_f is simply the principal open set $D(f)$. If $\mathscr{F} = \tilde{M}$ for an A-module M, the homomorphism (7.10.1) is the identity morphism of $M_f = \Gamma(D(f), \tilde{M})$ which holds by definition of $\tilde{M}$.

Theorem 7.16. *Let* X *be a scheme and let* $\mathscr{F}$ *be an* $\mathscr{O}_X$*-module. Then the following assertions are equivalent.*

(i) *For every open affine subset* $U = \operatorname{Spec} A$ *of* X *there exists an* A*-module* M *such that* $\mathscr{F}|_U \cong \tilde{M}$.

(ii) *There exists an open affine covering* $(U_i)_i$ *of* X, $U_i = \operatorname{Spec} A_i$, *and for each* i *an* A_i*-module* M_i *such that* $\mathscr{F}|_{U_i} \cong \tilde{M}_i$ *for all* i.

(iii) *The* $\mathscr{O}_X$*-module* $\mathscr{F}$ *is quasi-coherent.*

(iv) *For every open affine subset* $U = \operatorname{Spec} A$ *of* X *and every* $f \in A$ *the homomorphism* (7.10.1)

$$\Gamma(U, \mathscr{F})_f \to \Gamma(D(f), \mathscr{F})$$

is an isomorphism.

Proof. The implication "(i) $\Rightarrow$ (ii)" is clear, and the implication "(ii) $\Rightarrow$ (iii)" follows from the remarks above. To show "(iv) $\Rightarrow$ (i)" we may assume that $X = U = \operatorname{Spec} A$ is affine. Setting $M := \Gamma(X, \mathscr{F})$, (iv) implies that there is an isomorphism

$$\Gamma(D(f), \tilde{M}) = M_f \overset{\sim}{\to} \Gamma(D(f), \mathscr{F})$$

for all $f \in A$, compatible with restriction to $D(g)$ for $g \in A$ with $D(g) \subseteq D(f)$. This proves $\mathscr{F} \cong \tilde{M}$.

To show the implication "(iii) $\Rightarrow$ (iv)" we can assume that $X = \operatorname{Spec} A$ is affine and that $U = X$. We have already seen above that if $\mathscr{F}$ is of the form $\tilde{M}$ for an A-module M, condition (iv) holds. As X is quasi-compact, the hypothesis implies that there exist finitely many $g_i \in A$ such that $X = \bigcup_i D(g_i)$ and such that $\mathscr{F}_{|D(g_i)}$ is isomorphic to the cokernel of a homomorphism $\tilde{A}_{g_i}^{(J)} \to \tilde{A}_{g_i}^{(I)}$. By Proposition 7.13 this homomorphism is of the form $\tilde{u}$ for some homomorphism $u : A_{g_i}^{(J)} \to A_{g_i}^{(I)}$ and hence

$$\mathscr{F}_{|D(g_i)} \cong \operatorname{Coker}(\tilde{u}) = \operatorname{Coker}(u)^{\sim}$$

by Proposition 7.14.

This shows that $\mathscr{F}_{|D(g_i)}$ satisfies condition (iv). The same argument proves condition (iv) for $\mathscr{F}_{|D(g_i) \cap D(g_j)} = \mathscr{F}_{|D(g_i g_j)}$ as well.

Let us denote the image of f in A_{g_i} and in $A_{g_i g_j}$ again by f. As $\mathscr{F}$ is a sheaf, we have a commutative diagram with exact rows

$$
\begin{array}{ccccccc}
0 & \longrightarrow & \Gamma(X, \mathscr{F})_f & \longrightarrow & \prod_i \Gamma(D(g_i), \mathscr{F})_f & \longrightarrow & \prod_{i,j} \Gamma(D(g_i g_j), \mathscr{F})_f \\
& & \downarrow{\scriptstyle\alpha} & & \downarrow{\scriptstyle\alpha'} & & \downarrow{\scriptstyle\alpha''} \\
0 & \longrightarrow & \Gamma(D(f), \mathscr{F}) & \longrightarrow & \prod_i \Gamma(D(f g_i), \mathscr{F}) & \longrightarrow & \prod_{i,j} \Gamma(D(f g_i g_j), \mathscr{F}).
\end{array}
$$

Here the upper row is obtained by localizing the exact sequence (2.5.1), and we use that localization commutes with finite products. As we have seen, α' and α'' are isomorphisms, hence α is an isomorphism by the Five Lemma (Proposition B.5). $\square$

Corollary 7.17. *Let A be a ring, $X = \operatorname{Spec} A$. The functor $M \mapsto \tilde{M}$ induces an equivalence of the category of A-modules with the category of quasi-coherent $\mathscr{O}_X$-modules.*

Remark 7.18. Let X be a ringed space. An $\mathscr{O}_X$-algebra is called *quasi-coherent* if its underlying $\mathscr{O}_X$-module is quasi-coherent.

If $X = \operatorname{Spec} A$ is an affine scheme, the equivalence in Corollary 7.17 immediately implies that the functor $B \mapsto \tilde{B}$ induces an equivalence of the category of A-algebras with the category of quasi-coherent $\mathscr{O}_X$-algebras. Moreover, an A-algebra B is commutative if and only if $\tilde{B}$ is commutative.

Corollary 7.19. *Let X be a scheme.*
(1) *Let $u : \mathscr{F} \to \mathscr{G}$ be a homomorphism of quasi-coherent $\mathscr{O}_X$-modules. Then $\operatorname{Ker}(u)$, $\operatorname{Coker}(u)$, and $\operatorname{Im}(u)$ are quasi-coherent $\mathscr{O}_X$-modules.*
(2) *The direct sum of quasi-coherent $\mathscr{O}_X$-modules is again quasi-coherent.*
(3) *Let $(\mathscr{F}_i')_{i \in I}$ be a family of quasi-coherent submodules of a quasi-coherent $\mathscr{O}_X$-module $\mathscr{F}$. Then their sum $\sum_i \mathscr{F}_i'$ is quasi-coherent. If I is finite, then their intersection $\bigcap_i \mathscr{F}_i'$ is quasi-coherent.*
(4) *Let $\mathscr{F}$ and $\mathscr{G}$ be quasi-coherent $\mathscr{O}_X$-modules. The tensor product $\mathscr{F} \otimes_{\mathscr{O}_X} \mathscr{G}$ is quasi-coherent, and for every open affine subset $U \subseteq X$ we have*

(7.10.2) $\Gamma(U, \mathscr{F} \otimes_{\mathscr{O}_X} \mathscr{G}) = \Gamma(U, \mathscr{F}) \otimes_{\Gamma(U, \mathscr{O}_X)} \Gamma(U, \mathscr{G})$.

Proof. We may assume that $X = \operatorname{Spec} A$ is affine. Then $\mathscr{F} \cong \tilde{M}$ and $\mathscr{G} \cong \tilde{N}$, where M and N are A-modules. Then (1) and (2) follow from Proposition 7.14. Now $\sum_i \mathscr{F}_i'$ is the image of $\bigoplus \mathscr{F}_i' \to \mathscr{F}$ and, if I is finite, $\bigcap_i \mathscr{F}_i'$ is the kernel of $\mathscr{F} \to \bigoplus_i \mathscr{F}/\mathscr{F}_i'$. Thus (1) and (2) imply (3).

To prove (4) it suffices to show that there is an isomorphism

$$(7.10.3) \qquad \tilde{M} \otimes_{\tilde{A}} \tilde{N} \cong (M \otimes_A N)^{\sim},$$

functorial in M and N. The sheaf $\tilde{M} \otimes_{\tilde{A}} \tilde{N}$ is attached to the presheaf

$$U \mapsto \mathscr{H}(U) := \Gamma(U, \tilde{M}) \otimes_{\Gamma(U, \tilde{A})} \Gamma(U, \tilde{N}),$$

defined for principal open subsets $U = D(f)$, $f \in A$. There are functorial isomorphisms

$$\mathscr{H}(D(f)) \cong M_f \otimes_{A_f} N_f \cong (M \otimes_A N)_f \cong \Gamma(D(f), (M \otimes_A N)^{\sim})$$

which are compatible with restriction from $D(f)$ to $D(g) \subseteq D(f)$. This defines the desired isomorphism (7.10.3). $\qquad \square$

Arbitrary intersections of quasi-coherent submodules are not necessarily quasi-coherent (Exercise 7.12).

Remark 7.20. The corollary shows in particular that on a scheme X the category of quasi-coherent $\mathscr{O}_X$-modules is an abelian category.

In Corollary 12.34 we will see that given an affine scheme $X = \operatorname{Spec} A$ and an exact sequence $0 \to \mathscr{F}' \to \mathscr{F} \to \mathscr{F}'' \to 0$ of $\mathscr{O}_X$-modules with $\mathscr{F}'$ quasi-coherent, then the induced sequence of A-modules

$$0 \to \Gamma(X, \mathscr{F}') \to \Gamma(X, \mathscr{F}) \to \Gamma(X, \mathscr{F}'') \to 0$$

is exact.

(7.11) Extending sections of quasi-coherent modules.

The implication "(iii) $\Rightarrow$ (iv)" of Theorem 7.16 is a special case of a very useful theorem about the extension of sections of quasi-coherent modules. We will generalize it in two ways: We replace $\mathscr{O}_X$ by an arbitrary invertible $\mathscr{O}_X$-module $\mathscr{L}$ (Section (7.5)) and we relax the condition of X to be affine.

Let $(X, \mathscr{O}_X)$ be a locally ringed space and let $\mathscr{L}$ be an invertible $\mathscr{O}_X$-module. Let $s \in \Gamma(X, \mathscr{L})$ be a global section. By definition there exists for every point $x \in X$ an open neighborhood U of x and an isomorphism $w \colon \mathscr{L}_{|U} \overset{\sim}{\to} \mathscr{O}_{X|U}$. We say that s is *invertible in* x, if $w_x(s_x) \in \mathscr{O}_{X,x}^{\times}$. Using the notion of fiber introduced above (7.1.1), we can express this as $s(x) \neq 0 \in \mathscr{L}(x)$. In particular, this property does not depend on the choice of w. Set

$$(7.11.1) \qquad X_s(\mathscr{L}) := \{\, x \in X \; ; \; s \text{ is invertible in } x \,\}.$$

For $\mathscr{L} = \mathscr{O}_X$ we have $X_s(\mathscr{O}_X) = X_s$. We claim that $X_s(\mathscr{L})$ is an open set in X. Indeed, a subset $W \subseteq X$ is open if and only if there exists an open covering $X = \bigcup_i U_i$ such that $W \cap U_i$ is open in U_i for all i. To show that $X_s(\mathscr{L})$ is open in X we can therefore assume that $\mathscr{L} = \mathscr{O}_X$. If $x \in X_s$ there exists a $t_x \in \mathscr{O}_{X,x}$ such that $s_x t_x = 1$. Choose an open neighborhood V of x such that there exists $t \in \Gamma(V, \mathscr{O}_X)$ whose germ at x is t_x. By shrinking V, we can assume that $(s_{|V})t = 1$ and therefore $V \subseteq X_s$.

Definition 7.21. *A scheme is called* quasi-separated *if for every two affine open subsets* $U, V \subseteq X$ *the intersection* $U \cap V$ *is quasi-compact.*

For a more thorough discussion of this notion we refer to Section (10.7). Here we note only that by Corollary 3.22 every locally noetherian scheme is quasi-separated. Moreover, every affine scheme X is quasi-separated: As U and V are quasi-compact, we can choose finite coverings $U = \bigcup_k D(f_k)$ and $V = \bigcup_l D(g_l)$ by principal open subsets of X. As $D(g) \cap D(g') = D(gg')$, $U \cap V = \bigcup_{k,l} D(f_k g_l)$ is a finite union of quasi-compact sets and thus quasi-compact. (Proposition 9.15 shows that $U \cap V$ is always affine.)

Therefore the following theorem applies in particular if X is affine or if X is a noetherian scheme.

Theorem 7.22. *Let X be a quasi-compact and quasi-separated scheme, let $\mathscr{L}$ be an invertible $\mathscr{O}_X$-module, and let $s \in \Gamma(X, \mathscr{L})$ be a global section. Let $\mathscr{F}$ be a quasi-coherent $\mathscr{O}_X$-module.*
(1) *Let $t \in \Gamma(X, \mathscr{F})$ be a global section such that $t_{|X_s} = 0$. Then there exists an integer $n > 0$ such that $t \otimes s^{\otimes n} = 0 \in \Gamma(X, \mathscr{F} \otimes \mathscr{L}^{\otimes n})$.*
(2) *For every section $t' \in \Gamma(X_s, \mathscr{F})$ there exist $n > 0$ and a section $t \in \Gamma(X, \mathscr{F} \otimes \mathscr{L}^{\otimes n})$ such that $t_{|X_s} = t' \otimes s^{\otimes n}$.*

The proof will show that (1) also holds if X is only assumed to be quasi-compact but not necessarily quasi-separated.

Proof. We can write X as a finite union of open affine subsets U_i such that we can fix an isomorphism $w_i \colon \mathscr{L}_{|U_i} \xrightarrow{\sim} \mathscr{O}_{U_i}$.

To show (1) we can therefore assume that X is affine and that $\mathscr{L} = \mathscr{O}_X$. Then (1) is equivalent to the injectivity of the natural map $\Gamma(X, \mathscr{F})_s \to \Gamma(D(s), \mathscr{F})$ which has been proved to be an isomorphism in Theorem 7.16.

We prove (2). Let s_i be the image of $s_{|U_i}$ in $\mathscr{O}_X(U_i)$ under w_i. We first remark that the surjectivity of $\Gamma(U_i, \mathscr{F} \otimes \mathscr{L}^{\otimes n})_{s_i} \cong \Gamma(U_i, \mathscr{F})_{s_i} \to \Gamma(U_i \cap X_s, \mathscr{F})$ proved in Theorem 7.16 shows that there exists an integer $l > 0$ (independent of i), such that $(t' \otimes s^{\otimes l})_{|U_i \cap X_s}$ can be extended to a section $t_i \in \Gamma(U_i, \mathscr{F} \otimes \mathscr{L}^{\otimes l})$. If we denote by $t_{i|j}$ the restriction of t_i to $U_i \cap U_j$, the restriction of $t_{i|j} - t_{j|i}$ to $X_s \cap U_i \cap U_j$ is zero. As X is quasi-separated, $U_i \cap U_j$ is quasi-compact and quasi-separated. By (1) there exists an integer $m > 0$ (which we can choose to be independent of i and j) such that $(t_{i|j} - t_{j|i}) \otimes s^{\otimes m} = 0$. Therefore there exists a section $t \in \Gamma(X, \mathscr{F} \otimes \mathscr{L}^{\otimes(l+m)})$ such that $t_{|U_i} = t_i \otimes s^{\otimes m}$ and hence $t_{|X_s} = t' \otimes s^{\otimes(m+l)}$. $\square$

(7.12) Direct and inverse image of quasi-coherent modules.

Remark 7.23. Let $f \colon (X, \mathscr{O}_X) \to (Y, \mathscr{O}_Y)$ be a morphism of ringed spaces. If $\mathscr{G}$ is a quasi-coherent module on $\mathscr{O}_Y$ it follows from $f^*(\mathscr{O}_Y) = \mathscr{O}_X$ and from the fact that f^* is right exact and commutes with direct sums that $f^*(\mathscr{G})$ is a quasi-coherent $\mathscr{O}_X$-module.

If $\mathscr{F}$ is a quasi-coherent $\mathscr{O}_X$-module, it is not true in general that $f_*(\mathscr{F})$ is again quasi-coherent, even if $(X, \mathscr{O}_X)$ and $(Y, \mathscr{O}_Y)$ are schemes (see Exercise 10.14 for such an example). However this is true if $(X, \mathscr{O}_X)$ and $(Y, \mathscr{O}_Y)$ are affine schemes. This follows from the following more precise description of $f_*(\mathscr{F})$ and $f^*(\mathscr{G})$ in this case.

Proposition 7.24. *Let* $f\colon X = \mathrm{Spec}(B) \to Y = \mathrm{Spec}(A)$ *be a morphism of affine schemes and let* $\varphi\colon A \to B$ *be the corresponding ring homomorphism.*

(1) *Let N be a B-module and let $\varphi_*(N)$ be the restriction of scalars to A, i.e., $\varphi_*(N) = N$ considered as an A-module via φ. Then there is a functorial isomorphism of $\mathcal{O}_Y$-modules*

$$f_*(\tilde{N}) \cong \varphi_*(N)^{\sim}.$$

(2) *Let M be an A-module. Then there is a functorial isomorphism of $\mathcal{O}_X$-modules*

$$f^*(\tilde{M}) \cong (B \otimes_A M)^{\sim}.$$

Proof. For all $g \in A$ we have $f^{-1}(D(g)) = D(\varphi(g))$ and therefore we obtain

$$\Gamma(D(g), f_*(\tilde{N})) = \Gamma(D(\varphi(g)), \tilde{N}) = N_{\varphi(g)} = (\varphi_*(N))_g = \Gamma(D(g), \varphi_*(N)^{\sim}).$$

These identifications are compatible with restriction maps for $D(g') \subseteq D(g)$ and functorial in N. This proves (1). To show (2), note that we already know that $f^*(\tilde{M})$ is a quasi-coherent $\mathcal{O}_X$-module. Now for all quasi-coherent $\mathcal{O}_X$-modules $\mathcal{F}$ we have identifications, functorial in M and $\mathcal{F}$,

$$\mathrm{Hom}_{\mathcal{O}_X}(f^*\tilde{M}, \mathcal{F}) \overset{7.11}{=} \mathrm{Hom}_{\mathcal{O}_Y}(\tilde{M}, f_*\mathcal{F})$$
$$\overset{(1)}{=} \mathrm{Hom}_{\mathcal{O}_Y}(\tilde{M}, \varphi_*(\Gamma(X, \mathcal{F}))^{\sim}) \overset{7.13}{=} \mathrm{Hom}_A(M, \varphi_*(\Gamma(X, \mathcal{F})))$$
$$= \mathrm{Hom}_B(B \otimes_A M, \Gamma(X, \mathcal{F})) \overset{7.13}{=} \mathrm{Hom}_{\mathcal{O}_X}((B \otimes_A M)^{\sim}, \mathcal{F}).$$

This proves (2) by the Yoneda lemma (Section (4.2)): Using the identity morphisms for $\mathcal{F} = f^*\tilde{M}$ and $\mathcal{F} = (B \otimes_A M)^{\sim}$, we obtain homomorphisms $f^*\tilde{M} \to (B \otimes_A M)^{\sim}$ and $(B \otimes_A M)^{\sim} \to f^*\tilde{M}$, and the functoriality in $\mathcal{F}$ implies that they are inverse to each other. $\square$

In Corollary 10.27 we will see that $f_*(\mathcal{F})$ is quasi-coherent for quasi-coherent $\mathcal{O}_X$-modules $\mathcal{F}$ if f is a morphism of schemes satisfying certain mild finiteness conditions (more precisely, if f is quasi-compact and quasi-separated) which for instance are satisfied whenever X is noetherian.

(7.13) Example: Invertible sheaves on Dedekind schemes.

A very important example of an $\mathcal{O}_X$-module are those formed by functions on a scheme X which have poles or zeros of prescribed orders. We will study this in general in Chapter 11. Here we discuss this construction only for the simple (but still very interesting) case of so-called Dedekind schemes.

Recall (Proposition B.87) that a *Dedekind ring* is a noetherian integral domain A such that for each maximal ideal $\mathfrak{m}$ of A the local ring $A_{\mathfrak{m}}$ is a principal ideal domain. In other words, a Dedekind ring is a noetherian regular (or equivalently normal) domain of dimension ≤ 1.

A noetherian integral scheme X is called a *Dedekind scheme* if $\Gamma(U, \mathcal{O}_X)$ is a Dedekind ring for every open affine subscheme $U \subseteq X$. In other words, a Dedekind scheme is a noetherian integral regular scheme of dimension ≤ 1. If X is an integral scheme which has a finite open covering $X = \bigcup_{i=1}^n U_i$ by Dedekind schemes U_i, then X is a Dedekind scheme.

Examples for Dedekind schemes are of course affine schemes $X = \operatorname{Spec} A$ where A is a Dedekind ring, e.g., if A is the integral closure of $\mathbb{Z}$ in a finite field extension of $\mathbb{Q}$. If k is a field, the affine line $\mathbb{A}^1_k = \operatorname{Spec}(k[T])$ is a Dedekind scheme. Moreover, as the projective line $\mathbb{P}^1_k$ has an open covering by two copies of the affine line, $\mathbb{P}^1_k$ is a Dedekind scheme. More generally any regular integral curve C over k (i.e., C is a regular integral k-scheme of finite type with $\dim C = 1$) is a Dedekind scheme.

From now on we denote by X a Dedekind scheme of dimension 1. Then $\mathscr{O}_{X,x}$ is a discrete valuation ring for every closed point $x \in X$. We denote by X_0 its set of closed points. Let $K := K(X) = \mathscr{O}_{X,\eta}$ be the function field. As explained in Proposition 3.29, we may consider $\mathscr{O}_{X,x}$ (for $x \in X$) and $\Gamma(U, \mathscr{O}_X)$ (for $U \subseteq X$ non-empty open) as subrings of $K(X)$. If U is affine, K is the field of fractions of $\Gamma(U, \mathscr{O}_X)$. We denote by $v_x \colon K^\times \to \mathbb{Z}$ the normalized valuation and by π_x a uniformizing element of the discrete valuation ring $\mathscr{O}_{X,x}$ (Section (B.13)). As usual, we set $v_x(0) := \infty$. Then $\mathscr{O}_{X,x} = \{ a \in K \; ; \; v_x(a) \geq 0 \}$. By Proposition 3.29 we have for every open set $U \subseteq X$

$$\Gamma(U, \mathscr{O}_X) = \bigcap_{x \in X_0 \cap U} \mathscr{O}_{X,x} = \{ a \in K \; ; \; v_x(a) \geq 0 \text{ for all } x \in X_0 \cap U \}.$$

Let $D := (n_x)_{x \in X_0} \in \operatorname{Div}(X) := \mathbb{Z}^{(X_0)}$ be a tuple of integers n_x (for $x \in X_0$) such that $n_x = 0$ for almost all $x \in X_0$. Such a D is called a *divisor on X* (see Section (11.11) for the general definition of a divisor and Example 11.47 for the connection to the definition given here). We attach to D an $\mathscr{O}_X$-module $\mathscr{L}_D$ as follows: If $U \subseteq X$ is open, we set

$$\Gamma(U, \mathscr{L}_D) := \{ s \in K \; ; \; v_x(s) \geq n_x \text{ for all } x \in X_0 \cap U \}$$

For $V \subseteq U$ we define the restriction maps to be the inclusions $\Gamma(U, \mathscr{L}_D) \hookrightarrow \Gamma(V, \mathscr{L}_D)$ and this defines a sheaf $\mathscr{L}_D$ of abelian groups on X. Multiplication within K defines a scalar multiplication $\Gamma(U, \mathscr{O}_X) \times \Gamma(U, \mathscr{L}_D) \to \Gamma(U, \mathscr{L}_D)$ that makes $\mathscr{L}_D$ into an $\mathscr{O}_X$-module.

We think of elements in $\Gamma(U, \mathscr{L}_D)$ as "meromorphic functions on U" that have at x a zero of order at least n_x (or, for $n_x < 0$, a pole of order at most $-n_x$). The notion of a meromorphic function will be made precise in Section (11.10).

If $D' := (n'_x)_{x \in X_0} \in \operatorname{Div}(X)$ is a second divisor with $n'_x \leq n_x$ for all $x \in X_0$, we have a natural inclusion

(7.13.1) $\mathscr{L}_D \hookrightarrow \mathscr{L}_{D'}.$

We claim that $\mathscr{L}_D$ is locally free of rank 1, i.e. an invertible $\mathscr{O}_X$-module (in Theorem 11.40 we will see that conversely every invertible $\mathscr{O}_X$-module $\mathscr{L}$ is of the form $\mathscr{L}_D$ for some divisor D). Let $\operatorname{Supp}(D)$ be the set of $x \in X_0$ such that $n_x \neq 0$. For each $x \in \operatorname{Supp}(D)$ let U_x be an open affine neighborhood of x which does not contain any other point of $\operatorname{Supp}(D)$ and such that there exists a section $s_x \in \Gamma(U_x, \mathscr{O}_X)$ whose germ at x is the chosen uniformizing element $\pi_x \in \mathscr{O}_{X,x}$. By shrinking U_x we can further assume that $v_y(s_x) = 0$ for all $y \in (U_x \cap X_0) \setminus \{x\}$. We also set $V := X \setminus \operatorname{Supp}(D)$. Then V and the U_x for $x \in \operatorname{Supp}(D)$ form an open covering of X. By definition we have $\mathscr{L}_{D|V} \cong \mathscr{O}_{X|V}$ and over U_x multiplication by $s_x^{n_x} \in K$ defines an isomorphism

$$\mathscr{O}_{X|U_x} \xrightarrow{\sim} \mathscr{L}_{D|U_x}.$$

For two divisors $D, D' \in \operatorname{Div}(X)$ we have $\mathscr{L}_D \otimes_{\mathscr{O}_X} \mathscr{L}_{D'} = \mathscr{L}_{D+D'}$. For every $f \in K^\times$ set $\operatorname{div}(f) = (v_x(f))_{x \in X_0} \in \operatorname{Div}(X)$ (such divisors are called *principal divisors*; see

Definition 11.26 for the general definition). The multiplication by f defines an isomorphism $\mathscr{L}_D \overset{\sim}{\to} \mathscr{L}_{D+\mathrm{div}(f)}$. In Proposition 11.28 we will see that, conversely, if two invertible sheaves $\mathscr{L}_D$ and $\mathscr{L}_{D'}$ are isomorphic, there exists an $f \in K^\times$ such that $D + \mathrm{div}(f) = D'$. In Exercise 7.10 the case $X = \mathbb{P}^1_k$ is studied in more detail.

Properties of quasi-coherent modules

We now define analogues of properties **P** of modules M over a ring A for $\mathscr{O}_X$-modules over a scheme (or even a ringed space) $(X, \mathscr{O}_X)$. Of course, all these properties are defined in such a way that if $X = \operatorname{Spec} A$, then M has property **P** if and only if the corresponding quasi-coherent $\mathscr{O}_X$-module $\tilde{M}$ has this property – although sometimes this requires a proof.

(7.14) Modules of finite type and of finite presentation.

Definition 7.25. *Let* $(X, \mathscr{O}_X)$ *be a ringed space. An* $\mathscr{O}_X$*-module* $\mathscr{F}$ *is called of finite type (resp. of finite presentation) if for all* $x \in X$ *there exists an open neighborhood* U *of* x *and an exact sequence of* $\mathscr{O}_{X|U}$*-modules of the form*

$$\mathscr{O}^n_{X|U} \longrightarrow \mathscr{F}_{|U} \longrightarrow 0$$

(resp. of the form

$$\mathscr{O}^m_{X|U} \longrightarrow \mathscr{O}^n_{X|U} \longrightarrow \mathscr{F}_{|U} \longrightarrow 0),$$

where $n, m \geq 0$ *are integers (dependent on* x*).*

In other words, $\mathscr{F}$ is of finite type if $\mathscr{F}$ is locally generated by finitely many sections. Clearly, every $\mathscr{O}_X$-module of finite presentation is quasi-coherent and of finite type.

Proposition 7.26. *Let* $X = \operatorname{Spec} A$ *be an affine scheme. An* A*-module* M *is of finite type (resp. of finite presentation) if and only if* $\tilde{M}$ *is an* $\mathscr{O}_X$*-module of finite type (resp. of finite presentation).*

Proof. The condition is clearly necessary as $M \mapsto \tilde{M}$ is an exact functor. Conversely, assume that $\tilde{M}$ is an $\mathscr{O}_X$-module of finite type (resp. of finite presentation). As the principal open subsets $D(f)$ form a basis of the topology of X and as X is quasi-compact, we may assume that there exists a finite open covering $X = \bigcup_i D(f_i)$ such that $\tilde{M}_{|D(f_i)} = (M_{f_i})^\sim$ is generated by a finite number of global sections (resp. admits a finite presentation). Therefore M_{f_i} is a finitely generated A_{f_i}-module (resp. an A_{f_i}-module of finite presentation) for all i by Proposition 7.14. Hence M is a finitely generated A-module by Lemma 3.20 (resp. an A-module of finite presentation by Proposition B.28; see also Proposition 14.48 below). $\qquad\square$

Finitely generated modules over noetherian rings are of finite presentation. Hence if X is a locally noetherian scheme, a quasi-coherent $\mathscr{O}_X$-module is of finite type if and only if it is of finite presentation. (Note however that even when X is noetherian, in general there exist $\mathscr{O}_X$-modules which are of finite type, but not of finite presentation: Just take the quotient of $\mathscr{O}_X$ by an ideal sheaf which is not quasi-coherent.)

For modules of finite presentation, properties on the stalks in some point $x \in X$ can often be extended to properties on an open neighborhood of x. The key proposition for this is the following.

Proposition 7.27. *Let* $(X, \mathscr{O}_X)$ *be a ringed space and let* $\mathscr{F}$ *be an* $\mathscr{O}_X$-*module of finite presentation.*

(1) *For all* $x \in X$ *and for each* $\mathscr{O}_X$-*module* $\mathscr{G}$, *the canonical homomorphism of* $\mathscr{O}_{X,x}$-*modules*

$$\mathscr{H}om_{\mathscr{O}_X}(\mathscr{F}, \mathscr{G})_x \to \mathrm{Hom}_{\mathscr{O}_{X,x}}(\mathscr{F}_x, \mathscr{G}_x)$$

is bijective.

(2) *Let* $\mathscr{F}$ *and* $\mathscr{G}$ *be two* $\mathscr{O}_X$-*modules of finite presentation. Let* $x \in X$ *be a point and let* $\theta \colon \mathscr{F}_x \overset{\sim}{\to} \mathscr{G}_x$ *be an isomorphism of* $\mathscr{O}_{X,x}$-*modules. Then there exists an open neighborhood* U *of* x *and an isomorphism* $u \colon \mathscr{F}_{|U} \overset{\sim}{\to} \mathscr{G}_{|U}$ *of* $\mathscr{O}_U$-*modules such that* $u_x = \theta$.

Proof. The proof of (1) uses a standard technique for proving assertions on modules of finite presentation: Denote by $\mathcal{F}$ and $\mathcal{F}'$ the following contravariant functors from the category of $\mathscr{O}_X$-modules to the category of $\mathscr{O}_{X,x}$-modules

$$\mathcal{F} \colon \mathscr{F} \mapsto \mathscr{H}om_{\mathscr{O}_X}(\mathscr{F}, \mathscr{G})_x$$
$$\mathcal{F}' \colon \mathscr{F} \mapsto \mathrm{Hom}_{\mathscr{O}_{X,x}}(\mathscr{F}_x, \mathscr{G}_x).$$

Both functors are left exact and commute with finite direct sums. We are given a morphism of functors $\mathcal{F} \to \mathcal{F}'$ and we want to show that $\mathcal{F}(\mathscr{F}) \to \mathcal{F}'(\mathscr{F})$ is an isomorphism if $\mathscr{F}$ is of finite presentation.

By replacing X by a sufficiently small open neighborhood of x we may assume that there exists a finite presentation $\mathscr{O}_X^m \to \mathscr{O}_X^n \to \mathscr{F} \to 0$. Therefore we obtain a commutative diagram with exact rows

$$
\begin{array}{ccccccc}
0 & \longrightarrow & \mathcal{F}(\mathscr{F}) & \longrightarrow & \mathcal{F}(\mathscr{O}_X)^n & \longrightarrow & \mathcal{F}(\mathscr{O}_X)^m \\
& & \downarrow & & \downarrow & & \downarrow \\
0 & \longrightarrow & \mathcal{F}'(\mathscr{F}) & \longrightarrow & \mathcal{F}'(\mathscr{O}_X)^n & \longrightarrow & \mathcal{F}'(\mathscr{O}_X)^m.
\end{array}
$$

By the Five Lemma it suffices to show that $\mathcal{F}(\mathscr{O}_X) \to \mathcal{F}'(\mathscr{O}_X)$ is an isomorphism which is obvious.

Now (2) is a direct corollary of (1): Let $\theta \colon \mathscr{F}_x \to \mathscr{G}_x$ and $\eta \colon \mathscr{G}_x \to \mathscr{F}_x$ be mutually inverse isomorphisms of $\mathscr{O}_{X,x}$-modules. By (1) there exist open neighborhoods U and V of x and homomorphisms $u \colon \mathscr{F}_{|U} \to \mathscr{G}_{|U}$ and $v \colon \mathscr{G}_{|V} \to \mathscr{F}_{|V}$ such that $u_x = \theta$ and $v_x = \eta$. As $\theta \circ \eta = \mathrm{id}_{\mathscr{G}_x}$ and $\eta \circ \theta = \mathrm{id}_{\mathscr{F}_x}$, we can find an open neighborhood $W \subseteq U \cap V$ of x such that $u_{|W} \circ v_{|W} = \mathrm{id}_{\mathscr{G}_{|W}}$ and $v_{|W} \circ u_{|W} = \mathrm{id}_{\mathscr{F}_{|W}}$. $\qquad\square$

As for modules over a ring we obtain also the following characterization of modules of finite presentation (Proposition B.8 (2)).

Proposition 7.28. *Let* $(X, \mathscr{O}_X)$ *be a ringed space and* $\mathscr{F}$ *be an* $\mathscr{O}_X$-*module of finite type. Then* $\mathscr{F}$ *is of finite presentation if and only if for each open set* $U \subseteq X$ *and for each exact sequence of* $\mathscr{O}_U$-*modules*

$$0 \to \mathscr{F}' \to \mathscr{G} \to \mathscr{F}_{|U} \to 0,$$

where $\mathscr{G}$ is of finite type, $\mathscr{F}'$ is an $\mathscr{O}_U$-module of finite type.

Proposition 7.29. *Let X be a scheme, let $\mathscr{F}$ be an $\mathscr{O}_X$-module of finite presentation and let $\mathscr{G}$ be a quasi-coherent $\mathscr{O}_X$-module. Then the $\mathscr{O}_X$-module $\mathscr{H}om_{\mathscr{O}_X}(\mathscr{F},\mathscr{G})$ is again quasi-coherent.*

If $X = \operatorname{Spec} A$ is affine and M and N are A-modules such that $\tilde{M} = \mathscr{F}$ (of finite presentation) and $\tilde{N} = \mathscr{G}$, then there is an isomorphism $\operatorname{Hom}_A(M,N)^\sim \xrightarrow{\sim} \mathscr{H}om_{\mathscr{O}_X}(\tilde{M},\tilde{N})$ of $\mathscr{O}_X$-modules which is functorial in M and N.

Proof. It suffices to show the second assertion. For $f \in A$ we have functorial homomorphisms of A_f-modules

$$(*) \quad \begin{aligned} \operatorname{Hom}_A(M,N)^\sim(D(f)) &= \operatorname{Hom}_A(M,N)_f \\ &\longrightarrow \operatorname{Hom}_A(M,N_f) = \operatorname{Hom}_{A_f}(M_f,N_f) = \mathscr{H}om_{\mathscr{O}_X}(\tilde{M},\tilde{N})(D(f)) \end{aligned}$$

compatible with restrictions of principal open subsets $D(g) \subseteq D(f)$. Here the arrow is defined by $u/f^r \mapsto (m \mapsto u(m)/f^r)$ and the last equality holds because of the equivalence between the categories of A_f-modules and of quasi-coherent $\mathscr{O}_{D(f)}$-modules.

If $\mathscr{F}$ is of finite presentation, then by Proposition 7.26 there exists a finite presentation $A^m \to A^n \to M \to 0$. To show that $(*)$ is an isomorphism we can now argue as in the proof of Proposition 7.27 and therefore may assume that $M = A$. Then the arrow in $(*)$ is the identity $N_f \xrightarrow{\sim} N_f$. $\square$

(7.15) Support of a module of finite type.

One of the key properties of modules of finite type is the following proposition and its corollaries.

Proposition 7.30. *Let $(X, \mathscr{O}_X)$ be a ringed space and let $\mathscr{F}$ be an $\mathscr{O}_X$-module of finite type. Let $x \in X$ be a point and let $s_i \in \Gamma(U, \mathscr{F})$ for $i = 1, \ldots, n$ be sections over some open neighborhood of x such that the germs $(s_i)_x$ generate the stalk $\mathscr{F}_x$. Then there exists an open neighborhood $V \subseteq U$, such that the $s_{i|V}$ generate $\mathscr{F}_{|V}$.*

Proof. Let $U' \subseteq U$ be an open neighborhood of x such that $\mathscr{F}_{|U'}$ is generated by sections $t_j \in \Gamma(U', \mathscr{F})$ for $j = 1, \ldots, m$. As the $(s_i)_x$ generate $\mathscr{F}_x$, there exist sections a_{ij} of $\mathscr{O}_X$ over an open neighborhood $U'' \subseteq U'$ of x such that $(t_j)_x = \sum_i (a_{ij})_x (s_i)_x$ for all j. Therefore there exists an open neighborhood $V \subseteq U''$ of x such that $(t_j)_y = \sum_i (a_{ij})_y (s_i)_y$ for all j and all $y \in V$. Hence the $(s_i)_y$ generate $\mathscr{F}_y$ for all $y \in V$. $\square$

Corollary 7.31. *Let $(X, \mathscr{O}_X)$ be a ringed space. For every $\mathscr{O}_X$-module $\mathscr{F}$ of finite type and any integer $r \geq 0$ the subset*

$$X_r := \{\, x \in X \;;\; \mathscr{F}_x \text{ can be generated by } r \text{ elements as } \mathscr{O}_{X,x}\text{-module}\,\}$$

is open in X.

Note that if $(X, \mathscr{O}_X)$ is a locally ringed space, we have

$$X_r = \{\, x \in X \;;\; \dim_{\kappa(x)} \mathscr{F}(x) \leq r \,\}$$

by the Lemma of Nakayama (Proposition B.3 (3)).

As X_0 is the complement of the support of $\mathscr{F}$ (7.6.1), we obtain in particular:

Corollary 7.32. *Let* $(X, \mathscr{O}_X)$ *be a ringed space and let* $\mathscr{F}$ *be an* $\mathscr{O}_X$-*module of finite type. Then* $\mathrm{Supp}(\mathscr{F})$ *is closed in* X.

(7.16) Closed immersions revisited.

Let X be a scheme. For any subset Y of X we denote by $i_Y \colon Y \hookrightarrow X$ the inclusion map. In Section (3.15) we defined a closed subscheme of X to be a scheme $(Z, \mathscr{O}_Z)$, such that Z is a closed subset X and $(i_Z)_* \mathscr{O}_Z \cong \mathscr{O}_X/\mathscr{J}$, where $\mathscr{J}$ is a (necessarily unique) ideal of $\mathscr{O}_X$. Note that $i_Z^{-1}(i_Z)_* \mathscr{O}_Z = \mathscr{O}_Z$ and hence $\mathscr{O}_Z \cong i_Z^{-1}(\mathscr{O}_X/\mathscr{J})$ if $(Z, \mathscr{O}_Z)$ is a closed subscheme. Moreover, $Z = \mathrm{Supp}(\mathscr{O}_X/\mathscr{J})$ in this case. The following proposition characterizes the ideals of $\mathscr{O}_X$ that define closed subschemes.

Proposition 7.33. *Let* X *be a scheme, let* $\mathscr{J}$ *be an ideal of* $\mathscr{O}_X$, *and set*

$$Z = \mathrm{Supp}(\mathscr{O}_X/\mathscr{J}), \qquad \mathscr{O}_Z = i_Z^{-1}(\mathscr{O}_X/\mathscr{J}).$$

Then Z *is a closed subset of* X, *and* $(Z, \mathscr{O}_Z)$ *is a closed subscheme of* X *if and only if* $\mathscr{J}$ *is a quasi-coherent* $\mathscr{O}_X$-*module.*

Proof. As $\mathscr{O}_X/\mathscr{J}$ is an $\mathscr{O}_X$-module of finite type, its support is closed by Corollary 7.32. Because the properties of being a scheme and of being quasi-coherent can both be checked locally, we may assume that $X = \mathrm{Spec}\, A$ is an affine scheme. Now $\mathscr{J}$ is quasi-coherent if and only if there exists an ideal $\mathfrak{a}$ of A such that $\mathscr{J} = \tilde{\mathfrak{a}}$ (Theorem 7.16). In this case $Z = V(\mathfrak{a})$ and $\mathscr{O}_Z = (A/\mathfrak{a})^{\sim}$ and hence $Z = \mathrm{Spec}(A/\mathfrak{a})$. Conversely, if Z is a closed subscheme of $X = \mathrm{Spec}(A)$, we have already seen in Theorem 3.42 that there exists an ideal $\mathfrak{a}$ such that $Z = \mathrm{Spec}(A/\mathfrak{a})$. Then $\mathscr{J} = \mathrm{Ker}(\tilde{A} \to (A/\mathfrak{a})^{\sim})$ and hence $\mathscr{J} = \tilde{\mathfrak{a}}$ by Proposition 7.14. $\square$

Corollary 7.34. *Let* X *be a scheme. Attaching to a quasi-coherent ideal* $\mathscr{J}$ *the closed subscheme* $(Z := \mathrm{Supp}(\mathscr{O}_X/\mathscr{J}), i_Z^{-1}(\mathscr{O}_X/\mathscr{J}))$ *defines a bijection between the set of quasi-coherent ideals of* $\mathscr{O}_X$ *and the set of closed subschemes of* X. *An inverse bijection is given by attaching to a closed subscheme* $(Z, \mathscr{O}_Z)$ *the kernel of* $\mathscr{O}_X \to (i_Z)_* \mathscr{O}_Z$.

The closed subscheme of X corresponding to a quasi-coherent ideal $\mathscr{J} \subseteq \mathscr{O}_X$ is denoted by $V(\mathscr{J})$. It is called the *vanishing scheme* of $\mathscr{J}$.

(7.17) The annihilator of an $\mathscr{O}_X$-**module.**

If X is a scheme and $\mathscr{F}$ is quasi-coherent of finite type, there is the following possibility to endow the closed subset $\mathrm{Supp}(\mathscr{F})$ with a natural subscheme structure: We will call the kernel of the canonical homomorphism $\mathscr{O}_X \to \mathscr{H}om_{\mathscr{O}_X}(\mathscr{F}, \mathscr{F})$ (7.4.12) the *annihilator of* $\mathscr{F}$ and denote it by $\mathrm{Ann}(\mathscr{F})$.

Proposition 7.35. *Let* X *be a scheme and let* $\mathscr{F}$ *be a quasi-coherent* $\mathscr{O}_X$-*module of finite type. Then* $\mathrm{Ann}(\mathscr{F})$ *is a quasi-coherent ideal of* $\mathscr{O}_X$, *for every open affine subset* $U \subseteq X$ *we have* $\Gamma(U, \mathrm{Ann}(\mathscr{F})) = \mathrm{Ann}\, \Gamma(U, \mathscr{F})$, *and the underlying topological space of* $V(\mathrm{Ann}(\mathscr{F}))$ *is* $\mathrm{Supp}(\mathscr{F})$.

Proof. As this is a local question, we may assume that $X = \mathrm{Spec}\, A$ is affine and thus $\mathscr{F} = \tilde{M}$, where M is an A-module, generated by a finite number of elements $t_1, \ldots, t_n$.

We first show that $\mathrm{Ann}(\mathscr{F})$ is quasi-coherent. As finite intersections of quasi-coherent submodules are again quasi-coherent (Corollary 7.19), we even may assume that M is generated by a single element t. But then $\mathrm{Ann}(\mathscr{F})$ is the kernel of the homomorphism of quasi-coherent $\mathscr{O}_X$-modules $\mathscr{O}_X \to \mathscr{F}$ that is given by multiplication with t and hence quasi-coherent as well.

Moreover, we also see that $\mathrm{Ann}(\mathscr{F})$ is the quasi-coherent ideal corresponding to the ideal $\bigcap_i \mathrm{Ann}(t_i) = \mathrm{Ann}(M)$. If $\mathfrak{p} \subset A$ is a prime ideal, we have

$$\mathfrak{p} \in \mathrm{Supp}(\mathscr{F}) \Leftrightarrow M_{\mathfrak{p}} \neq 0 \Leftrightarrow \exists t_i : \mathrm{Ann}(t_i) \subseteq \mathfrak{p} \Leftrightarrow \mathfrak{p} \supseteq \bigcap_i \mathrm{Ann}(t_i) \Leftrightarrow \mathfrak{p} \in V(\mathrm{Ann}(M)). \quad \square$$

Remark 7.36. Under the hypotheses of Proposition 7.35 let $\mathscr{I}$ be any quasi-coherent ideal with $\mathscr{I} \subseteq \mathrm{Ann}(\mathscr{F})$ and let $i \colon V(\mathscr{I}) \to X$ be the corresponding closed immersion. Then we have $\mathscr{I}\mathscr{F} = 0$ and therefore $\mathscr{F}$ is an $\mathscr{O}_X/\mathscr{I}$-module. This shows that the canonical homomorphism $\mathscr{F} \to i_*(i^*\mathscr{F})$ is an isomorphism.

Remark 7.37. Let X be a scheme, $\mathscr{F}$ be a quasi-coherent $\mathscr{O}_X$-module and let $\mathscr{E}, \mathscr{E}' \subseteq \mathscr{F}$ be quasi-coherent $\mathscr{O}_X$-submodules. Then let $(\mathscr{E} : \mathscr{E}')$ be the ideal of $\mathscr{O}_X$ defined by

$$\Gamma(U, (\mathscr{E} : \mathscr{E}')) := \{\, a \in \Gamma(U, \mathscr{O}_X) \; ; \; \forall V \subseteq U \text{ open}, \, m' \in \Gamma(V, \mathscr{E}') : (a_{|V})m' \in \Gamma(V, \mathscr{E}) \,\}.$$

For instance we have $(0 : \mathscr{F}) = \mathrm{Ann}(\mathscr{F})$. A similar proof as in Proposition 7.35 shows that $(\mathscr{E} : \mathscr{E}')$ is a quasi-coherent ideal of $\mathscr{O}_X$ if $\mathscr{E}'$ is of finite type.

(7.18) Flat and finite locally free modules.

Let $f \colon X \to Y$ be a morphism of ringed spaces and let $\mathscr{F}$ be an $\mathscr{O}_X$-module. For each $x \in X$, the $\mathscr{O}_{X,x}$-module $\mathscr{F}_x$ is endowed via the homomorphism $f_x^\sharp \colon \mathscr{O}_{Y,f(x)} \to \mathscr{O}_{X,x}$ with the structure of an $\mathscr{O}_{Y,f(x)}$-module. For the notions of flat and faithfully flat modules over a ring, see Section (B.4). The notion of a flat morphism of schemes introduced below is of great importance, because it expresses algebraically the property of a "family varying in a continuous way". Here we will address only some basic results on flat modules. For a more thorough discussion see Chapter 14.

Definition 7.38.
(1) The $\mathscr{O}_X$-module $\mathscr{F}$ is called flat over Y at x or f-flat at x if $\mathscr{F}_x$ is a flat $\mathscr{O}_{Y,f(x)}$-module. It is called flat over Y or f-flat if $\mathscr{F}$ is flat over Y at all points $x \in X$.
(2) If $X = Y$ and $f = \mathrm{id}_X$, we simply say that $\mathscr{F}$ is flat at x if it is id_X-flat at x, i.e. if $\mathscr{F}_x$ is a flat $\mathscr{O}_{X,x}$-module. Similarly, $\mathscr{F}$ is called flat, if $\mathscr{F}_x$ is a flat $\mathscr{O}_{X,x}$-module for all $x \in X$.
(3) We say that f is flat, or that X is flat over Y, if $\mathscr{O}_X$ is flat over Y.

Remark 7.39. Let $\varphi \colon A \to B$ be a homomorphism of rings, set $Y = \mathrm{Spec}\, A$ and $X = \mathrm{Spec}\, B$ and let $f \colon X \to Y$ the morphism of schemes corresponding to φ. Then a B-module M is flat over A if and only if $M_{\mathfrak{p}_x}$ is a flat $A_{\varphi^{-1}(\mathfrak{p}_x)}$-module for all points $x \in \mathrm{Spec}\, B$ (Proposition B.27). By definition this is equivalent to the assertion that the quasi-coherent $\mathscr{O}_X$-module $\tilde{M}$ is flat over Y.

In particular an A-module M is flat if and only if the corresponding quasi-coherent $\mathscr{O}_Y$-module $\tilde{M}$ is flat.

If $\mathscr{F}$ is f-flat at x, for all open neighborhoods V of $y = f(x)$, the functor

$$(\mathscr{O}_V\text{-Mod}) \to (\mathscr{O}_{X,x}\text{-Mod}),$$
$$\mathscr{G} \mapsto (f^*\mathscr{G} \otimes_{\mathscr{O}_X} \mathscr{F})_x \overset{(7.4.7)}{=} (f^*\mathscr{G})_x \otimes_{\mathscr{O}_{X,x}} \mathscr{F}_x \overset{(7.8.6)}{=} \mathscr{G}_y \otimes_{\mathscr{O}_{Y,y}} \mathscr{F}_x$$

is exact. In particular, if $\mathscr{F}$ is flat over Y, then the functor

(7.18.1) $(\mathscr{O}_Y\text{-Mod}) \to (\mathscr{O}_X\text{-Mod}), \qquad \mathscr{G} \mapsto f^*\mathscr{G} \otimes_{\mathscr{O}_X} \mathscr{F}$

is exact.

Proposition 7.40. *Let $f\colon X \to Y$ be a morphism of schemes and let*

(7.18.2) $0 \to \mathscr{F}' \to \mathscr{F} \to \mathscr{F}'' \to 0$

be an exact sequence of quasi-coherent $\mathscr{O}_X$-modules. Assume that $\mathscr{F}''$ is flat over Y.
(1) *For every morphism $g\colon Y' \to Y$ and every quasi-coherent $\mathscr{O}_{Y'}$-module $\mathscr{G}'$ the sequence*

$$0 \to \mathscr{F}' \boxtimes_Y \mathscr{G}' \to \mathscr{F} \boxtimes_Y \mathscr{G}' \to \mathscr{F}'' \boxtimes_Y \mathscr{G}' \to 0$$

of $\mathscr{O}_{X\times_Y Y'}$-modules is exact (here we set $\mathscr{H} \boxtimes_Y \mathscr{G}' := p^*\mathscr{H} \otimes_{\mathscr{O}_{X\times_Y Y'}} q^*\mathscr{G}'$ for an $\mathscr{O}_X$-module $\mathscr{H}$, where $p\colon X \times_Y Y' \to X$ and $q\colon X \times_Y Y' \to Y'$ are the projections).
(2) *The $\mathscr{O}_X$-module $\mathscr{F}$ is flat over Y if and only if $\mathscr{F}'$ is flat over Y.*

Proof. We may assume that $Y = \operatorname{Spec} A$, $X = \operatorname{Spec} B$, and Y' are affine. Let us write $M := \Gamma(X, \mathscr{F})$, $M' := \Gamma(X, \mathscr{F}')$, $M'' := \Gamma(X, \mathscr{F}'')$. Then (7.18.2) corresponds to an exact sequence $0 \to M' \to M \to M'' \to 0$ of B-modules (Proposition 7.14) and M'' is a flat A-module by hypothesis. Thus the proposition follows from the analogous assertions for modules over a ring (Proposition B.16).   □

Clearly for every ringed space $(X, \mathscr{O}_X)$, locally free $\mathscr{O}_X$-modules are flat. There is also the following converse.

Proposition 7.41. *Let $(X, \mathscr{O}_X)$ be a locally ringed space and let $\mathscr{F}$ be an $\mathscr{O}_X$-module. Then the following assertions are equivalent.*
(i) *$\mathscr{F}$ is locally free of finite type.*
(ii) *$\mathscr{F}$ is of finite presentation and $\mathscr{F}_x$ is a free $\mathscr{O}_{X,x}$-module for all $x \in X$.*
(iii) *$\mathscr{F}$ is flat and of finite presentation.*

Proof. The conditions (i) and (ii) are equivalent by Proposition 7.27, and (ii) clearly implies (iii). Conversely, if $\mathscr{F}$ is flat and of finite presentation, the $\mathscr{O}_{X,x}$-module $\mathscr{F}_x$ is flat and of finite presentation and hence free by Proposition B.21.   □

By Proposition B.29 we obtain the following corollary.

Corollary 7.42. *Let $X = \operatorname{Spec} A$ be an affine scheme and let M be an A-module. Then the following assertions are equivalent.*
(i) *$\tilde{M}$ is a locally free $\mathscr{O}_X$-module of finite type.*
(ii) *M is a finitely generated projective A-module.*
(iii) *M is a flat A-module of finite presentation.*

There exist projective A-modules M (not finitely generated) such that $\tilde{M}$ is not locally free (Exercise 7.19).

Lemma 7.43. *Let X be a scheme, let $Z \subseteq X$ be a finite set of points and let $U = \operatorname{Spec} A$ be an open affine neighborhood of Z. Let $\mathscr{E}$ be a finite locally free $\mathscr{O}_X$-module of constant rank r. Then there exists an $s \in A$ such that $D(s) \supset Z$ and $\mathscr{E}_{|D(s)} \cong \mathscr{O}_{D(s)}^r$.*

Proof. We may assume $X = U$ and thus $\mathscr{E} = \tilde{M}$ for an A-module M. For every $z \in Z$ choose a specialization z' of z which is a closed point of X. Every open neighborhood of the set $\{ z' \; ; \; z \in Z \}$ then is an open neighborhood of Z. Thus we may assume that Z consists of finitely many maximal ideals $\mathfrak{m}_1, \ldots, \mathfrak{m}_n \subset A$. Let S be the complement of $\bigcup_i \mathfrak{m}_i$, which is a multiplicative subset of A. We set $A' = S^{-1}A$ and $M' = S^{-1}M$. Then A' is a semi-local ring. If $\mathfrak{r}$ denotes its radical, $A'/\mathfrak{r}$ is a product of fields. As the rank of $\mathscr{E}$ is constant, $M'/\mathfrak{r}M'$ is free. Let $m_1, \ldots, m_r \in M'$ be elements whose images in $M'/\mathfrak{r}M'$ are a basis. The corresponding homomorphism $A'^r \to M'$ of A'-modules is surjective by Nakayama's lemma and hence an isomorphism because M' is locally free of rank r. By Proposition 7.27 there exists an $s \in S$ and an isomorphism $A_s^r \xrightarrow{\sim} M_s$. In other words $\mathscr{O}_{D(s)}^r \cong \mathscr{E}_{|D(s)}$. $\qquad\square$

Remark 7.44. The property of being projective (but not necessarily of finite type) can be globalized to schemes as follows. Let X be a scheme. A quasi-coherent $\mathscr{O}_X$-module $\mathscr{F}$ is called *locally projective* if for all $x \in X$ there exists an open affine neighborhood U of x such that $\mathscr{F}_{|U}$ is isomorphic to a direct summand of $\mathscr{O}_U^{(I)}$ for some index set I. It is then a highly non-trivial result that a quasi-coherent $\mathscr{O}_X$-module is locally projective if and only if for all open affine subschemes $U = \operatorname{Spec} A \subseteq X$ the restriction $\mathscr{F}_{|U}$ is isomorphic to $\tilde{P}$, where P is a projective A-module (see [RG], 2nd part, 3.1, see also [Pey], [St] 058B, 05A5).

(7.19) Coherent modules.

Definition 7.45. *Let $(X, \mathscr{O}_X)$ be a ringed space. An $\mathscr{O}_X$-module $\mathscr{F}$ is called* coherent *if $\mathscr{F}$ is of finite type and if for every open subset $U \subseteq X$, every integer $n \geq 0$, and for every homomorphism $w: \mathscr{O}_{X|U}^n \to \mathscr{F}_{|U}$ the kernel of w is of finite type.*

By Proposition 7.28 every coherent $\mathscr{O}_X$-module is of finite presentation and in particular quasi-coherent. The converse is not true: note that we do not require w above to be surjective (in fact, there exist affine schemes X such that $\mathscr{O}_{X,x}$ is noetherian for every point $x \in X$ and such that $\mathscr{O}_X$ is not a coherent $\mathscr{O}_X$-module; e.g., see [Gl] Chapter 2, Section 4).

The notion of coherence is important in the analytic setting, as well. For instance it is a basic result in complex analysis that the structure sheaf $\mathscr{O}_X$ of a complex analytic space X is coherent (Theorem of Oka; e.g., see [Re] Theorem 7.4). The same holds for the structure sheaf of p-adic analytic spaces in the sense of Berkovich ([Du] Lemme 0.1). In the sequel, we will use the notion of a coherent $\mathscr{O}_X$-module almost always for locally noetherian schemes X. In this case several of the finiteness conditions on $\mathscr{O}_X$-modules are equivalent:

Proposition 7.46. *Let X be a locally noetherian scheme and let $\mathscr{F}$ be an $\mathscr{O}_X$-module. Then the following assertions are equivalent:*
(i) *$\mathscr{F}$ is coherent.*
(ii) *$\mathscr{F}$ is of finite presentation.*

(iii) $\mathscr{F}$ is of finite type and quasi-coherent.

Proof. The implications "(i) $\Rightarrow$ (ii) $\Rightarrow$ (iii)" are clear. To prove that (iii) implies (i) we may assume that X is affine and that we are given a homomorphism $w\colon \mathscr{O}_X^n \to \mathscr{F}$. We have to show that $\mathrm{Ker}(w)$ is of finite type. By Proposition 3.19, $X = \mathrm{Spec}\, A$, where A is a noetherian ring. Then $\mathscr{F} \cong \tilde{M}$ for a finitely generated A-module M (Theorem 7.16 and Proposition 7.26) and w is of the form $\tilde{u}$ for some homomorphism $u\colon A^n \to M$ by Proposition 7.13. As A is noetherian, the kernel of u is finitely generated. By Proposition 7.14, $\mathrm{Ker}(w) = \mathrm{Ker}(u)^\sim$ and therefore $\mathrm{Ker}(w)$ is of finite type. $\qquad\square$

Corollary 7.47. *Let X be a locally noetherian scheme. Let $0 \to \mathscr{F}' \to \mathscr{F} \to \mathscr{F}'' \to 0$ be an exact sequence of quasi-coherent $\mathscr{O}_X$-modules. Then $\mathscr{F}$ is coherent if and only if $\mathscr{F}'$ and $\mathscr{F}''$ are coherent.*

Proof. We may assume that $X = \mathrm{Spec}\, A$ is affine. Then A is noetherian and the assertions follow from the analogous assertions for modules over a noetherian ring (Proposition B.32 and Proposition B.33). $\qquad\square$

In particular we see that on a locally noetherian scheme X the category of coherent $\mathscr{O}_X$-modules is an abelian category.

(7.20) Exterior powers, determinant, and trace.

We recall the notion of exteriors powers. Let R be a ring, let M be an R-module and let $r \geq 1$ be an integer. For any R-module N we call a multilinear map $\alpha\colon M^r \to N$ *alternating* if $\alpha(m_1, \ldots, m_r) = 0$ whenever there exist indices $1 \leq i \neq j \leq r$ such that $m_i = m_j$. The *r-th exterior power* is an R-module $\bigwedge^r M = \bigwedge_R^r M$ together with an alternating map $\varpi\colon M^r \to \bigwedge^r M$ such that for all R-modules N and for all alternating maps $\alpha\colon M^r \to N$ there exists a unique R-linear map $u\colon \bigwedge^r M \to N$ such that $u \circ \varpi = \alpha$. Clearly, the pair $(\bigwedge^r M, \varpi)$ is unique up to unique isomorphism. To show its existence we denote by L the submodule of

$$M^{\otimes r} := \underbrace{M \otimes_R \cdots \otimes_R M}_{r \text{ times}}$$

that is generated by elements of the form $m_1 \otimes \cdots \otimes m_r$ with $m_i \in M$ and such that there exist indices $1 \leq i \neq j \leq r$ with $m_i = m_j$. Then $M^{\otimes r}/L$ together with the map $\varpi\colon M^r \to M^{\otimes r}/L$ that sends $(m_1, \ldots, m_r)$ to the image of $m_1 \otimes \cdots \otimes m_r$ is an r-th exterior power of M. For $m_1, \ldots, m_r \in M$ we set

$$m_1 \wedge \cdots \wedge m_r := \varpi(m_1, \ldots, m_r) \in \bigwedge^r M.$$

Elements of this form generate $\bigwedge^r M$. We have $\bigwedge^1 M = M$ and we set $\bigwedge^0 M := R$. The construction of $\bigwedge^r M$ is functorial in M: If $u\colon M \to N$ is a homomorphism of R-modules, then there exists a unique R-linear homomorphism $\bigwedge^r(u)\colon \bigwedge^r(M) \to \bigwedge^r(N)$ that sends $m_1 \wedge \cdots \wedge m_r$ to $u(m_1) \wedge \cdots \wedge u(m_r)$.

If $\varphi\colon R \to R'$ is a ring homomorphism, the construction of $\bigwedge^r M$ shows that there is an isomorphism of R'-modules

$$(7.20.1) \qquad (\bigwedge_R^r M) \otimes_R R' \xrightarrow{\sim} \bigwedge_{R'}^r (M \otimes_R R')$$

which is functorial in M.

As tensor products commute with filtered inductive limits, it is easy to see that the same holds for exterior powers: Let $((M_i)_i, (u_{ji})_{i \leq j})$ be a filtered inductive system of R-modules. Via functoriality of the r-th exterior power we obtain an filtered inductive system $((\bigwedge^r M_i), (\bigwedge^r u_{ji}))$ and the R-linear homomorphisms $\bigwedge^r M_i \to \bigwedge^r \varinjlim M_i$ yield an isomorphism of R-modules

$$(7.20.2) \qquad \varinjlim_i \bigwedge^r M_i \xrightarrow{\sim} \bigwedge^r \varinjlim_i M_i.$$

We now collect some results about exterior powers of free modules. Let R be a ring and let $A = (a_{ij}) \in M_{n \times m}(R)$ be a matrix. Let $J = \{j_1 < \cdots < j_s\} \subseteq \{1, \ldots, m\}$ and $I = \{i_1 < \cdots < i_t\} \subseteq \{1, \ldots, n\}$ be subsets. We denote by

$$(7.20.3) \qquad A_{I,J} = (a_{i_\lambda, j_\mu})_{\substack{1 \leq \lambda \leq t \\ 1 \leq \mu \leq s}}$$

the submatrix of A consisting only of rows (resp. columns) numbered by elements in I (resp. in J).

Now let M be an R-module, and $(e_1, \ldots, e_m)$ a tuple of elements of M. For every finite subset $J = \{j_1 < \cdots < j_r\} \subseteq \{1, \ldots, m\}$ we set

$$e_J := e_{j_1} \wedge e_{j_2} \wedge \cdots \wedge e_{j_r} \in \bigwedge^r (M).$$

We write $e_\emptyset := 1 \in \bigwedge^0(M) = R$. Finally we denote by $\mathcal{F}_r(m)$ the set of subsets J of $\{1, \ldots, m\}$ such that $\#J = r$ and by $\mathcal{F}(m)$ the power set of $\{1, \ldots, m\}$. Recall the following result (e.g. [BouAI] Chapter 3, §7.8 Theorem 1 and §8.5 Proposition 10).

Proposition 7.48. *Let M be a free R-module of rank m, let $(e_1, \ldots, e_m)$ be a basis of M, and let $r \geq 0$ be an integer.*
(1) Then the elements e_J for $J \in \mathcal{F}_r(m)$ form a basis of $\bigwedge^r(M)$.
(2) Let N be a free R-module with basis $(f_1, \ldots, f_n)$, let $u \colon M \to N$ be a linear map, and let $A \in M_{n \times m}(R)$ be the matrix of u with respect to the given bases of M and N. Then the matrix of $\bigwedge^r(u)$ with respect to the bases $(e_J)_{J \in \mathcal{F}_r(m)}$ of $\bigwedge^r(M)$ and $(f_I)_{I \in \mathcal{F}_r(n)}$ of $\bigwedge^r(N)$ is the matrix

$$(\det(A_{I,J}))_{I \in \mathcal{F}_r(n), J \in \mathcal{F}_r(m)}.$$

We also need the following formula for an endomorphism u of a free R-module of rank n and for all scalars $a, b \in R$ (see [BouAI] Chapter 3, §8.5 Proposition 11):

$$(7.20.4) \qquad \det(a \operatorname{id}_M + bu) = \sum_{r=0}^n \operatorname{tr}(\bigwedge^r(u)) a^{n-r} b^r.$$

We will now globalize the exterior power to ringed spaces. Let $(X, \mathcal{O}_X)$ be a ringed space and let $\mathcal{F}$ be an $\mathcal{O}_X$-module and let $r \geq 0$ be an integer. The sheaf attached to the presheaf $U \mapsto \bigwedge_{\Gamma(U, \mathcal{O}_X)}^r \Gamma(U, \mathcal{F})$ is an $\mathcal{O}_X$-module denoted by $\bigwedge_{\mathcal{O}_X}^r \mathcal{F}$ or simply $\bigwedge^r \mathcal{F}$. It is called the *r-th exterior power of* $\mathcal{F}$. Clearly $\bigwedge^r \mathcal{F}$ is a covariant functor in $\mathcal{F}$. As filtered inductive limits commute with exterior powers (7.20.2), there is for each point $x \in X$ a functorial isomorphism of $\mathcal{O}_{X,x}$-modules

$$(7.20.5) \qquad\qquad (\overset{r}{\bigwedge} \mathcal{F})_x \xrightarrow{\sim} \overset{r}{\bigwedge} \mathcal{F}_x.$$

For quasi-coherent modules the exterior power is again quasi-coherent:

Proposition 7.49. *Let S be a scheme, let $\mathcal{F}$ be a quasi-coherent $\mathcal{O}_S$-module, and let $r \geq 0$ be an integer.*
(1) *The r-th exterior product $\bigwedge^r(\mathcal{F})$ is quasi-coherent.*
(2) *If $S = \operatorname{Spec} R$ is affine and $\mathcal{F} = \tilde{M}$ for some R-module M, then the quasi-coherent $\mathcal{O}_S$-module corresponding to $\bigwedge^r(M)$ is $\bigwedge^r(\mathcal{F})$.*
(3) *For every morphism $f \colon T \to S$ of schemes there is an isomorphism of quasi-coherent $\mathcal{O}_T$-modules, functorial in $\mathcal{F}$,*

$$(7.20.6) \qquad\qquad f^* \overset{r}{\bigwedge}(\mathcal{F}) \xrightarrow{\sim} \overset{r}{\bigwedge}(f^*\mathcal{F}).$$

Proof. Let us show (2). We apply (7.20.1) to the localization $R' = R_s$ for an element $s \in R$. Therefore on the basis of principal open subsets of $\operatorname{Spec} R$ the presheaf

$$D(s) \mapsto \overset{r}{\underset{R_s}{\bigwedge}} \Gamma(D(s), \mathcal{F}) = \overset{r}{\underset{R_s}{\bigwedge}} M_s = (\overset{r}{\underset{R}{\bigwedge}} M) \otimes_R R_s$$

is the quasi-coherent $\mathcal{O}_S$-module associated to $\bigwedge_R^r M$. This shows (2). As quasi-coherence can be shown locally, (1) follows from (2). Finally (3) is implied by (7.20.1). $\qquad\square$

Now assume that $\mathcal{F}$ is a locally free $\mathcal{O}_X$-module of rank n. As the r-th exterior power of a free module of rank n over any ring is free of rank $\binom{n}{r}$, it is immediate that $\bigwedge^r \mathcal{F}$ is a locally free $\mathcal{O}_X$-module of rank $\binom{n}{r}$. In particular, $\det(\mathcal{F}) := \bigwedge^n \mathcal{F}$ is an invertible $\mathcal{O}_X$-module which we call the *determinant of* $\mathcal{F}$.

Let us define determinant and trace of an endomorphism of $\mathcal{F}$. Consider the morphism of sheaves

$$\det \colon \mathscr{H}om_{\mathcal{O}_X}(\mathcal{F}, \mathcal{F}) \to \mathscr{H}om_{\mathcal{O}_X}(\det(\mathcal{F}), \det(\mathcal{F})) \cong \mathcal{O}_X,$$

$$(7.20.7) \qquad\qquad u \mapsto \det(u) := \overset{n}{\bigwedge}(u),$$

where the isomorphism is the one given by (7.5.7) and where u is an endomorphism of $\mathcal{F}_{|U}$ for some open subset $U \subseteq X$. We call $\det(u)$ the *determinant of u*. If $X = \operatorname{Spec} R$, $\mathcal{F} = \widetilde{R^n}$, and $u \colon \mathcal{F} \to \mathcal{F}$ an endomorphism, then u corresponds to an endomorphism w of R^n and $\det(u) = \det(w) \in \Gamma(X, \mathcal{O}_X) = R$ by Proposition 7.48.

The trace is defined as follows.

$$\operatorname{tr} \colon \mathscr{H}om_{\mathcal{O}_X}(\mathcal{F}, \mathcal{F}) \cong \mathcal{F}^\vee \otimes_{\mathcal{O}_X} \mathcal{F} \to \mathcal{O}_X,$$

$$(7.20.8) \qquad\qquad \lambda \otimes s \mapsto \lambda(s),$$

where the isomorphism is the one given by (7.5.4) and where λ and s are sections of $\mathscr{F}^\vee$ and of $\mathscr{F}$, respectively, over some open subset $U \subseteq X$. If u is an endomorphism of $\mathscr{F}_{|U}$, we call its image $\operatorname{tr}(u) \in \Gamma(U, \mathscr{O}_X)$ the *trace of u*. Again this coincides with the usual notion of trace if X is affine and $\mathscr{F}$ is free (see Exercise 7.31).

Exercises

Exercise 7.1. Give an example of a scheme X and a family $(\mathscr{F}_i)_i$ of $\mathscr{O}_X$-modules such that the presheaf $U \mapsto \bigoplus_{i \in I} \mathscr{F}_i(U)$ is not a sheaf.

Exercise 7.2. Let $(X, \mathscr{O}_X)$ be a locally ringed space and let $\mathscr{F}$ be an $\mathscr{O}_X$-module of finite type.
(a) Show that $\operatorname{Supp}(\mathscr{F}) = \{\, x \in X \; ; \; \mathscr{F}(x) \neq 0 \,\}$.
(b) If $\mathscr{G}$ is another $\mathscr{O}_X$-module of finite type, show that

$$\operatorname{Supp}(\mathscr{F} \otimes_{\mathscr{O}_X} \mathscr{G}) = \operatorname{Supp}(\mathscr{F}) \cap \operatorname{Supp}(\mathscr{G}).$$

(c) Let $f: (X', \mathscr{O}_{X'}) \to (X, \mathscr{O}_X)$ be a morphism of locally ringed spaces. Show that

$$\operatorname{Supp}(f^*(\mathscr{F})) = f^{-1}(\operatorname{Supp}(\mathscr{F})).$$

Exercise 7.3. Let A be a ring and let $Y \subseteq X = \operatorname{Spec} A$ be a subset of closed points of X. Let $M = \bigoplus_{y \in Y} \kappa(y)$. Show that $\operatorname{Supp}(\tilde{M}) = Y$. Give an example of a ring A and a subset Y of closed points of $X = \operatorname{Spec} A$ such that Y is not closed in X.

Exercise 7.4. Let $(X, \mathscr{O}_X)$ be a locally ringed space, let $\mathscr{F}$ be an $\mathscr{O}_X$-module of finite type and let $s_1, \ldots, s_r \in \Gamma(X, \mathscr{F})$.
(a) Show that $\{\, x \in X \; ; \; s_1(x), \ldots s_r(x) \text{ generate } \mathscr{F}(x) \,\}$ is open in X.
(b) Show that $\{\, x \in X \; ; \; s_1(x), \ldots s_r(x) \in \mathscr{F}(x) \text{ is linearly independent} \,\}$ is open in X if $\mathscr{F}$ is a finite locally free $\mathscr{O}_X$-module.

Exercise 7.5$\diamondsuit$. Let $(X, \mathscr{O}_X)$ be a ringed space and let $\mathscr{F}$ be an $\mathscr{O}_X$-module of finite presentation.
(a) Fix an integer $r \geq 0$. Show that

$$Y_r := \{\, x \in X \; ; \; \mathscr{F}_x \text{ is a free } \mathscr{O}_{X,x}\text{-module of rank } r \,\}$$

is an open subset of X and that $\mathscr{F}_{|Y_r}$ is a locally free $\mathscr{O}_{Y_r}$-module of rank r.
(b) Now let X be an integral scheme and let $\mathscr{F}$ be an $\mathscr{O}_X$-module of finite presentation. Show that there exists an open dense subset U of X and an integer $n \geq 0$ such that $\mathscr{F}_{|U} \cong \mathscr{O}_X^n{}_{|U}$.

Exercise 7.6$\diamondsuit$. Let $(X, \mathscr{O}_X)$ be a ringed space, let $x \in X$ be a point, and let $\mathscr{F}$ and $\mathscr{G}$ be $\mathscr{O}_X$-modules. Assume that $\mathscr{F}$ is of finite type.
(a) Let $w: \mathscr{G} \to \mathscr{F}$ be a homomorphism of $\mathscr{O}_X$-modules such that w_x is surjective. Show that there exists an open neighborhood U of x such that $w_{|U}$ is surjective.
(b) Let $w: \mathscr{F} \to \mathscr{G}$ be a homomorphism of $\mathscr{O}_X$-modules such that $w_x = 0$. Show that there exists an open neighborhood U of x such that $w_{|U} = 0$.

Exercise 7.7. Let A be an integral domain with field of fractions K, set $X = \operatorname{Spec} A$ and let $\eta \in X$ be the generic point. Let $x \in X$ be a point and let $\mathscr{F}^x$ be the skyscraper sheaf with value K concentrated in the point x (see Exercise 2.14). Show that $\mathscr{F}^x$ is an $\mathscr{O}_X$-module.
(a) Show that $\mathscr{F}^x$ is quasi-coherent if and only if $x = \eta$.
(b) Now assume that $x \neq \eta$. Show that there is a canonical injective homomorphism of $\mathscr{O}_X$-modules $\mathscr{F}^x \to \mathscr{F}^\eta$ and let $\mathscr{G}$ be its cokernel. Show that there exists no open neighborhood U of x such that $\mathscr{G}_{|U}$ is generated by its global sections.

Exercise 7.8◊. Let X be a quasi-compact and quasi-separated scheme, let $\mathscr{F}$ be a quasi-coherent $\mathscr{O}_X$-module, and let $f \in \Gamma(X, \mathscr{O}_X)$. Show that the homomorphism of $\Gamma(X, \mathscr{O}_X)_f$-modules $\Gamma(X, \mathscr{F})_f \to \Gamma(X_f, \mathscr{F})$ (7.10.1) is an isomorphism.

Exercise 7.9. Let X be a locally noetherian scheme. Show that

$$\{\, x \in X \;;\; \mathscr{O}_{X,x} \text{ is reduced}\,\}$$

is an open subset of X.
Hint: Use that the nilradical of X is an $\mathscr{O}_X$-module of finite type.

Exercise 7.10. Let k be an algebraically closed field and let $X = \mathbb{P}^1_k$ be the projective line over k. Its function field is $k(T)$, the field of fractions of the polynomial ring $k[T]$. We use the notation of Section (7.13).
(a) For each monic irreducible polynomials $p \in k[T]$ let v_p be the p-adic valuation on $k(T)$. Define also the discrete valuation v_∞ on $k(T)$ by $v_\infty(f/g) = \deg(g) - \deg(f)$, if $f, g \in k[T]$ are nonzero polynomials. Show that these valuations are the normalized discrete valuations corresponding to the closed points X_0 of X.
(b) For a divisor $D = (n_x) \in \operatorname{Div}(X)$ we call $\deg(D) := \sum_{x \in X_0} n_x$ its *degree*. Show that every principal divisor on X has degree 0.
(c) Show that for two divisors $D, D' \in \operatorname{Div}(X)$ the invertible sheaves $\mathscr{L}_D$ and $\mathscr{L}_{D'}$ are isomorphic if and only if $\deg(D) = \deg(D')$. For all $n \in \mathbb{Z}$ we denote by $\mathscr{O}_X(n)$ the isomorphism class of an invertible sheaf attached to a divisor of degree $-n$.
(d) Show that $\Gamma(\mathbb{P}^1_k, \mathscr{O}_X(n))$ is a k-vector space of dimension $n + 1$, if $n \geq 0$, and of dimension 0, if $n < 0$.
Remark: Divisors and their degrees on general 1-dimensional k-schemes will be studied in Section (15.9).

Exercise 7.11. With the notation of Exercise 7.10 show $\mathscr{O}_X(n) \otimes_{\mathscr{O}_X} \mathscr{O}_X(m) \cong \mathscr{O}_X(n+m)$ for all $n, m \in \mathbb{Z}$. Find $n, m \in \mathbb{Z}$ such that

$$\Gamma(\mathbb{P}^1_k, \mathscr{O}_X(n) \otimes_{\mathscr{O}_X} \mathscr{O}_X(m)) \not\cong \Gamma(\mathbb{P}^1_k, \mathscr{O}_X(n)) \otimes_{\Gamma(\mathbb{P}^1_k, \mathscr{O}_X)} \Gamma(\mathbb{P}^1_k, \mathscr{O}_X(m)).$$

Exercise 7.12. Let A be a discrete valuation ring with uniformizing element π. Set $X = \operatorname{Spec} A$ and for $n \geq 1$ let $\mathscr{F}_n = \tilde{\mathfrak{a}}_n$ be the quasi-coherent $\mathscr{O}_X$-submodule of $\mathscr{O}_X$ corresponding to the ideal $\mathfrak{a}_n = (\pi^n)$. Show that $\bigcap_n \mathscr{F}_n$ is not quasi-coherent. This example is due to P. Hartwig.

Exercise 7.13◊. Let X be a scheme, let $\mathscr{I} \subseteq \mathscr{O}_X$ be a quasi-coherent ideal, and let $\mathscr{F}$ be a quasi-coherent $\mathscr{O}_X$-module. Show that $\mathscr{I}\mathscr{F}$ is a quasi-coherent $\mathscr{O}_X$-module.

Exercise 7.14. Let X be a scheme and let $Z \subseteq X$ be a closed subspace. Define an ideal $\mathscr{I}_Z$ of $\mathscr{O}_X$ by

$$\mathscr{I}_Z(U) := \{\, f \in \mathscr{O}_X(U) \;;\; f(x) = 0 \text{ for } x \in U \cap Z \,\}.$$

Show that $\mathscr{I}_Z$ is quasi-coherent and that the corresponding closed subscheme is Z_{red}.

Exercise 7.15. Let $(X, \mathcal{O}_X)$ be a ringed space, let $\mathcal{E}$ be a locally free $\mathcal{O}_X$-module of finite rank, and let $u \in \operatorname{End}_{\mathcal{O}_X}(\mathcal{E})$ be an endomorphism. Show that u is an automorphism if and only if $\det(u) \in \Gamma(X, \mathcal{O}_X)$ is a unit.

Exercise 7.16. Let $(X, \mathcal{O}_X)$ be a ringed space and let $\mathcal{E}$ be a locally free $\mathcal{O}_X$-module of constant rank $n \geq 0$. For $i = 1, \dots, n$ define

$$\sigma_i \colon \mathcal{H}om_{\mathcal{O}_X}(\mathcal{F}, \mathcal{F}) \to \mathcal{O}_X, \qquad u \mapsto \operatorname{tr}(\overset{i}{\bigwedge}(u))$$

for a section $u \in \Gamma(U, \mathcal{H}om_{\mathcal{O}_X}(\mathcal{F}, \mathcal{F}))$, $U \subseteq X$ open. Then

$$\operatorname{charpol}_u := T^n - \sigma_1(u)T^{n-1} + \cdots + (-1)^n \sigma_n(u) \in \Gamma(U, \mathcal{O}_X)[T]$$

is called the *characteristic polynomial of u*.
(a) Show that if $X = U = \operatorname{Spec} A$ and $\mathcal{E} = (A^n)^\sim$, then $\operatorname{charpol}_u$ is the usual characteristic polynomial of $u \in \operatorname{End}_A(A^n)$.
(b) Show that $\operatorname{charpol}_u(u) = 0$.

Exercise 7.17. Let X be a scheme, $\mathcal{E}$ be a locally free $\mathcal{O}_X$-module of constant rank $n \geq 0$, and let $u \colon \mathcal{E} \to \mathcal{E}$ be an $\mathcal{O}_X$-linear endomorphism. Let $0 \leq r \leq n$ be an integer.
(a) Show that the following assertions are equivalent.
   (i) There exists u-stable decomposition $\mathcal{E} = \mathcal{E}_{\mathrm{nil}}(u) \oplus \mathcal{E}_{\mathrm{iso}}(u)$ of $\mathcal{O}_X$-modules, such that the characteristic polynomial of $u_{|\mathcal{E}_{\mathrm{nil}}}$ is T^r and such that $u_{|\mathcal{E}_{\mathrm{iso}}}$ is an automorphism of $\mathcal{E}_{\mathrm{iso}}$.
   (ii) There exists an open affine covering $X = \bigcup U_i$, $U_i = \operatorname{Spec} A_i$, such that

$$\operatorname{charpol}_{u_{|U_i}} = T^r(T^{n-r} + c_1 T^{n-r-1} + \cdots + c_{n-r}) \in A_i[T]$$

   with $c_{n-r} \in A_i^\times$.
(b) Show that $\mathcal{E}_{\mathrm{nil}}(u)$ and $\mathcal{E}_{\mathrm{iso}}(u)$ are uniquely determined if they exist and that they are v-stable for every $\mathcal{O}_X$-linear endomorphism v of $\mathcal{E}$ with $v \circ u = u \circ v$.
(c) Show that for every morphism of schemes $f \colon X' \to X$ one has $\mathcal{E}_{\mathrm{nil}}(f^*u) = f^*\mathcal{E}_{\mathrm{nil}}(u)$ and $\mathcal{E}_{\mathrm{iso}}(f^*u) = f^*\mathcal{E}_{\mathrm{iso}}(u)$.
The decomposition $\mathcal{E} = \mathcal{E}_{\mathrm{nil}}(u) \oplus \mathcal{E}_{\mathrm{iso}}(u)$ is called *Fitting decomposition*.

Exercise 7.18. Let X be a scheme and let $0 \neq \mathcal{I} \subseteq \mathcal{O}_X$ be a quasi-coherent ideal that is locally free as an $\mathcal{O}_X$-module. Show that $\mathcal{I}$ is an invertible $\mathcal{O}_X$-module.

Exercise 7.19. Let k be a field, let I be an infinite set, set $A = k^I$ and $X = \operatorname{Spec} A$. Show that $\mathfrak{a} := k^{(I)}$ is a projective ideal of A such that $\tilde{\mathfrak{a}}$ is not a locally free $\mathcal{O}_X$-module. *Hint:* Exercise 7.18.

Exercise 7.20. Let $f \colon (X, \mathcal{O}_X) \to (Y, \mathcal{O}_Y)$ be a morphism of ringed spaces, and let $\mathcal{G}$ and $\mathcal{G}'$ be two $\mathcal{O}_Y$-modules. Define a natural homomorphism of $\mathcal{O}_X$-modules

$$\alpha \colon f^* \mathcal{H}om_{\mathcal{O}_Y}(\mathcal{G}, \mathcal{G}') \to \mathcal{H}om_{\mathcal{O}_X}(f^*\mathcal{G}, f^*\mathcal{G}'),$$

functorial in $\mathcal{G}$ and $\mathcal{G}'$. Show that α is an isomorphism in each of the following two cases.
(a) $\mathcal{G}$ is locally free of finite rank.
(b) $\mathcal{G}$ is of finite presentation and f is flat.

Exercise 7.21. Let X be a quasi-compact scheme, $\mathscr{F}$ a quasi-coherent $\mathscr{O}_X$-module of finite type, and let $\mathscr{I} \subseteq \mathscr{O}_X$ be a quasi-coherent ideal of finite type. We assume that $\mathrm{Supp}(\mathscr{F}) \subseteq \mathrm{Supp}(\mathscr{O}/\mathscr{I})$ (as closed subsets). Show that there exists an $n > 0$ such that $\mathscr{I}^n \mathscr{F} = 0$.

Exercise 7.22◊. Let X be a scheme and let $\mathscr{F}$ be a quasi-coherent $\mathscr{O}_X$-module of finite type. Show that any surjective $\mathscr{O}_X$-module endomorphism $\mathscr{F} \to \mathscr{F}$ is bijective.
Hint: Use Corollary B.4.

Exercise 7.23◊. Let X be a scheme, let $\mathscr{I} \subseteq \mathscr{O}_X$ be a quasi-coherent ideal and let $k \geq 1$ be an integer. Show that the following assertions are equivalent.
(i) $\mathscr{I}^k = 0$.
(ii) There exists an open affine covering $(U_i)_i$ of X such that $\Gamma(U_i, \mathscr{I})^k = 0$.
(iii) For all $x \in X$ we have $\mathscr{I}_x^k = 0$.

Exercise 7.24. Let $(X, \mathscr{O}_X)$ be a ringed space.
(a) Let $0 \to \mathscr{F}' \to \mathscr{F} \to \mathscr{F}'' \to 0$ be an exact sequence of $\mathscr{O}_X$-modules. Show that if two of the three modules are coherent, the third one is also coherent.
(b) Let $u\colon \mathscr{F} \to \mathscr{G}$ be a homomorphism of coherent $\mathscr{O}_X$-modules. Show that $\mathrm{Ker}(u)$, $\mathrm{Im}(u)$, and $\mathrm{Coker}(u)$ are coherent.
(c) Let $\mathscr{F}$ and $\mathscr{G}$ be two coherent $\mathscr{O}_X$-modules. Show that $\mathscr{F} \otimes_{\mathscr{O}_X} \mathscr{G}$ and $\mathscr{H}om_{\mathscr{O}_X}(\mathscr{F}, \mathscr{G})$ are coherent.

Exercise 7.25. Let $(X, \mathscr{O}_X)$ be a ringed space such that $\mathscr{O}_X$ is a coherent $\mathscr{O}_X$-module. Show that an $\mathscr{O}_X$-module $\mathscr{F}$ is of finite presentation if and only if $\mathscr{F}$ is coherent.
Hint: Use Exercise 7.24.

Exercise 7.26. Let $(X, \mathscr{O}_X)$ be a ringed space such that $\mathscr{O}_X$ is a coherent $\mathscr{O}_X$-module and let $x \in X$ be a point.
(a) Let $\mathscr{F}$ be a coherent $\mathscr{O}_X$-module and let $E \subseteq \mathscr{F}_x$ be an $\mathscr{O}_{X,x}$-submodule of finite type. Show that there exist an open neighborhood U of x and a coherent $\mathscr{O}_U$-submodule $\mathscr{E}$ of $\mathscr{F}_{|U}$ such that $\mathscr{E}_x = E$.
(b) Let M be an $\mathscr{O}_{X,x}$-module of finite presentation. Show that there exist an open neighborhood U of x and a coherent $\mathscr{O}_U$-module $\mathscr{F}$ such that $\mathscr{F}_x \cong M$.

Exercise 7.27. Let $(X, \mathscr{O}_X)$ be a ringed space, let $\mathscr{E}, \mathscr{F}, \mathscr{H}$ be $\mathscr{O}_X$-modules. Assume that $\mathscr{H}$ is finite locally free.
(a) Show that there are functorial isomorphisms of $\Gamma(X, \mathscr{O}_X)$-modules

$$\mathrm{Hom}_{\mathscr{O}_X}(\mathscr{E} \otimes_{\mathscr{O}_X} \mathscr{F}, \mathscr{H}) \xrightarrow{\sim} \mathrm{Hom}_{\mathscr{O}_X}(\mathscr{E}, \mathscr{F}^\vee \otimes_{\mathscr{O}_X} \mathscr{H}) \xrightarrow{\sim} \mathrm{Hom}_{\mathscr{O}_X}(\mathscr{F}, \mathscr{E}^\vee \otimes_{\mathscr{O}_X} \mathscr{H}).$$

An $\mathscr{O}_X$-linear homomorphism $\beta\colon \mathscr{E} \otimes_{\mathscr{O}_X} \mathscr{F} \to \mathscr{H}$ is called a *pairing*. If $\mathscr{H} = \mathscr{O}_X$, then β is called a *bilinear form*. A pairing is called *perfect* if the corresponding $\mathscr{O}_X$-module homomorphisms $s_\beta\colon \mathscr{E} \to \mathscr{F}^\vee \otimes_{\mathscr{O}_X} \mathscr{H}$ and $r_\beta\colon \mathscr{F} \to \mathscr{E}^\vee \otimes_{\mathscr{O}_X} \mathscr{H}$ are both isomorphisms.
(b) Show that if $\mathscr{E}$ and $\mathscr{F}$ are finite locally free, then r_β is an isomorphism if and only if s_β is an isomorphism.

Exercise 7.28. Let $(X, \mathscr{O}_X)$ be a ringed space, $n \geq 1$ an integer, and let $\mathscr{F}$ be a locally free $\mathscr{O}_X$-module of rank n. Show that the wedge product yields for all $1 \leq r \leq n$ a perfect pairing (Exercise 7.27)

$$\bigwedge^{r} \mathscr{F} \otimes_{\mathscr{O}_X} \bigwedge^{n-r} \mathscr{F} \xrightarrow{\cdot} \bigwedge^{n} \mathscr{F} = \det(\mathscr{F}).$$

In particular we obtain an isomorphism of $\mathscr{O}_X$-modules

$$\bigwedge^{r} \mathscr{F} \xrightarrow{\sim} (\bigwedge^{n-r} \mathscr{F})^{\vee} \otimes_{\mathscr{O}_X} \det(\mathscr{F}).$$

Exercise 7.29. Let $(X, \mathscr{O}_X)$ be a ringed space and let $0 \to \mathscr{F}' \to \mathscr{F} \to \mathscr{F}'' \to 0$ be an exact sequence of finite locally free $\mathscr{O}_X$-modules. Show that $\det \mathscr{F} \cong \det \mathscr{F}' \otimes \det \mathscr{F}''$.

Exercise 7.30. Let $(X, \mathscr{O}_X)$ be a ringed space, let $\mathscr{E}$ be a finite locally free $\mathscr{O}_X$-module, and let $\omega \colon \mathscr{E} \otimes_{\mathscr{O}_X} \mathscr{E} \to \mathscr{O}_X$ be an alternating bilinear form (Exercise 7.27). It is called *symplectic* if ω is perfect (Exercise 7.27).

Show that if ω is a symplectic form on $\mathscr{E}$, then there exists for all $x \in X$ an open neighborhood U of x and integer $n \geq 0$ and an isomorphism of $\mathscr{O}_U$-modules $\mathscr{E}_{|U} \xrightarrow{\sim} \mathscr{O}_U^{2n}$ which identifies ω with the alternating form on the free module $\mathscr{O}_U^{2n}$ given by the matrix

$$J := \begin{pmatrix} 0 & I_n \\ -I_n & 0 \end{pmatrix},$$

where I_n is the $(n \times n)$-identity matrix.

Hint: Use that every finitely generated free module M over any ring with a symplectic form ω admits a basis such that ω is given by J with respect to this basis.

Remark: If $(X, \mathscr{O}_X)$ is a real manifold and $\mathscr{E}$ is the tangent bundle of X, then this result is called *Theorem of Darboux*.

Exercise 7.31. Let R be a ring, $n \geq 1$ an integer, and let $(e_1, \ldots, e_n)$ be the standard basis of R^n. Show that under the isomorphism (7.5.4)

$$(R^n)^{\vee} \otimes_R R^n \xrightarrow{\sim} \mathrm{End}_R(R^n) = M_n(R)$$

the base vector $e_i^{\vee} \otimes e_j$ is sent to the matrix all of whose coefficients are zero except for the (i, j)-th coefficient which is 1. Deduce that for any matrix $A = (a_{ij}) \in M_n(R)$ with corresponding endomorphism u of R^n the trace of u in the sense of (7.20.8) is equal to $\mathrm{tr}(A) = \sum_i a_{ii}$.

8   Representable Functors

Contents

In Chapter 4 we attached to a scheme X a contravariant functor h_X from the category
of schemes to the category of sets. The Yoneda lemma 4.6 tells us that we obtain an
embedding of the category of schemes into the category of such functors and thus we
can consider schemes also as functors. Functors F that lie in the essential image of this
embedding are called *representable*. We say that a scheme X *represents* F if $h_X \cong F$. It
is one of the central problems within algebraic geometry to study functors that classify
certain interesting objects and to decide whether they are representable, i.e., whether
they are "geometric objects". For general functors F and G it may be difficult to envisage
them as geometric objects. But it makes sense to say that a morphism $f \colon F \to G$ is
"geometric" (called representable), even if F and G are not necessarily representable. Thus
we may speak of immersions or of open coverings of functors. We will show that a functor
that is a sheaf for the Zariski topology and has an open covering by representable functors
is itself representable.

In the second part of this chapter we will define the important example of the Grassmannian functor and illustrate the abstract notions of the first part. The Grassmannian
scheme $\mathrm{Grass}_{d,n}$ represents a functor F such that $F(\operatorname{Spec} k)$ is the set of subvector spaces
of dimension d in k^n for all fields k. To make this precise we will first have to define what
the correct analogue for "subvector space of dimension d" is over an arbitrary scheme
S. In the special case $d = 1$ we will see that $\mathrm{Grass}_{1,n} \cong \mathbb{P}^{n-1}$. As we will have defined
S-valued points of $\mathrm{Grass}_{d,n}$, we will in particular obtain a description of the S-valued
points of $\mathbb{P}^{n-1}$.

A variant of the Grassmannian then leads to Brauer-Severi schemes. These schemes
are forms of the projective space, that is, they are schemes over a field k that become
isomorphic to projective space after base change to a field extension. In this chapter we
will explain how to construct Brauer-Severi schemes. As an application of descent theory
we will prove in Section (14.23) that this construction yields all Brauer-Severi schemes.

Representable Functors

(8.1) Representable Functors.

The Yoneda lemma (Section (4.2)) shows that we can embed a category $\mathcal{C}$ into the category
$\widehat{\mathcal{C}}$ of functors $F \colon \mathcal{C}^{\mathrm{opp}} \to (\mathrm{Sets})$ via the functor $X \mapsto h_X$.

© Springer Fachmedien Wiesbaden GmbH, part of Springer Nature 2020
U. Görtz und T. Wedhorn, *Algebraic Geometry I: Schemes*, Springer Studium
Mathematik – Master, https://doi.org/10.1007/978-3-658-30733-2_9

Definition 8.1. *A functor* $F\colon \mathcal{C}^{\mathrm{opp}} \to$ (Sets) *is called* representable *if there exist an object X and an isomorphism* $\xi\colon h_X \overset{\sim}{\to} F$.

The pair (X, ξ) is then uniquely determined up to unique isomorphism. In other words, F is representable if it lies in the essential image of the functor $X \mapsto h_X$. In the sequel we will often not distinguish between an object X and the representing functor h_X.

Example 8.2. (Affine Space) Let $n \geq 0$ be an integer. The functor $(\mathrm{Sch})^{\mathrm{opp}} \to$ (Sets) that sends a scheme S to $\Gamma(S, \mathcal{O}_S)^n$ is represented by $\mathbb{A}^n$ (Section (4.1)).

Example 8.3. (Fiber products of functors) In the category $\widehat{\mathcal{C}}$ fiber products always exist: Let F, G, and H be contravariant functors from $\mathcal{C}$ to (Sets), and let $F \to H$ and $G \to H$ be morphisms of functors. We set for every object $T \in \mathcal{C}$

$$(F \times_H G)(T) := F(T) \times_{H(T)} G(T),$$

where the right hand side denotes the fiber product in the category of sets (Example 4.12). Then $(F \times_H G)(T)$ is clearly functorial in T and we obtain a functor $F \times_H G \in \widehat{\mathcal{C}}$. The projections $p(T)\colon (F \times_H G)(T) \to F(T)$ and $q(T)\colon (F \times_H G)(T) \to G(T)$ define morphisms of functors $p\colon F \times_H G \to F$ and $q\colon F \times_H G \to G$. Then $(F \times_H G, p, q)$ is the fiber product of F and G over H in $\widehat{\mathcal{C}}$.

If F, G, and H are representable, say $F \cong h_X$, $G \cong h_Y$, and $H \cong h_S$, the fiber product $F \times_H G$ in $\widehat{\mathcal{C}}$ is representable by an object Z if and only if $X \times_S Y$ exists in $\mathcal{C}$ and in this case $Z = X \times_S Y$. Indeed, the universal property of $(X \times_S Y, p, q)$ says that for every object T in $\mathcal{C}$ the morphisms p and q yield a bijection (automatically functorial in T)

$$
\begin{aligned}
h_{X \times_S Y}(T) &= \mathrm{Hom}(T, X \times_S Y) \overset{\sim}{\to} \mathrm{Hom}(T, X) \times_{\mathrm{Hom}(T,S)} \mathrm{Hom}(T, Y) \\
&= h_X(T) \times_{h_S(T)} h_Y(T).
\end{aligned}
$$

Let us give less formal examples of representable functors which will be useful at several occasions.

Proposition 8.4. *Let S be a scheme and let $v\colon \mathcal{E} \to \mathcal{F}$ be a homomorphism of quasi-coherent $\mathcal{O}_S$-modules.*
(1) *Let $\mathcal{F}$ be of finite type. Then the locus where v is surjective is open, i.e., the functor*

$$F\colon (\mathrm{Sch}/S)^{\mathrm{opp}} \to (\mathrm{Sets}), \qquad F(T) = \{\, f \in \mathrm{Hom}_S(T, S) \;;\; f^*(v) \text{ is surjective}\,\}$$

is represented by an open subscheme of S.
(2) *Let $\mathcal{F}$ be finite locally free. Then the locus $v = 0$ is closed, i.e., the functor*

$$F\colon (\mathrm{Sch}/S)^{\mathrm{opp}} \to (\mathrm{Sets}), \qquad F(T) = \{\, f \in \mathrm{Hom}_S(T, S) \;;\; f^*(v) = 0\,\}$$

is represented by a closed subscheme of S.

Proof. Assertion (1) means that there exists an open subscheme $U \subseteq S$ such that a morphism $f\colon T \to S$ factors through U if and only if $f^*(v)\colon f^*\mathcal{E} \to f^*\mathcal{F}$ is surjective, i.e., if and only if $\mathrm{Coker}(f^*(v)) = 0$. As $\mathrm{Coker}(v)$ is of finite type, its support is closed (Corollary 7.32). As f^* is right exact, we have $\mathrm{Coker}(f^*(v)) = f^*(\mathrm{Coker}(v))$. Therefore the open subscheme $U = S \setminus \mathrm{Supp}(\mathrm{Coker}(v))$ has the desired property. Cf. Exercise 7.2 (c).

If $\mathscr{F}$ is finite locally free, we have $\mathscr{H}om(\mathscr{E},\mathscr{F}) = \mathscr{E}^\vee \otimes \mathscr{F} = (\mathscr{E} \otimes \mathscr{F}^\vee)^\vee$ by (7.5.4). Thus v corresponds to a homomorphism $\tilde{v}\colon \mathscr{E} \otimes \mathscr{F}^\vee \to \mathscr{O}_S$ of quasi-coherent $\mathscr{O}_S$-modules. Let $\mathscr{I}$ be the image of $\tilde{v}$. Then $\mathscr{I}$ is a quasi-coherent ideal of $\mathscr{O}_S$. If $f\colon T \to S$ is a morphism, then $f^*(v) = 0$ if and only if $f^*(\tilde{v}) = 0$. But $f^*(\tilde{v}) = 0$ if and only if the ideal generated by $\mathscr{I}$ in $\mathscr{O}_T$ is zero, which happens if and only if f factors through the closed subscheme $V(\mathscr{I})$. $\qquad\square$

Once we have introduced the flattening stratification in Section (11.8) for modules it is not difficult to show further assertions in this direction, see Exercise 11.10.

Also note the following variant: Let S be a scheme, let X be an S-scheme, and let $v\colon \mathscr{E} \to \mathscr{F}$ be a homomorphism of quasi-coherent $\mathscr{O}_X$-modules with $\mathscr{F}$ of finite type. Then the functor F' on S-schemes with $F'(T) = \{\, f \in \mathrm{Hom}_S(T, X)\,;\; f^*(v) \text{ surjective}\,\}$ is representable by an open subscheme of X. (Apply the original proposition to X, v to obtain an open subscheme U of X, and observe that the S-scheme U represents the functor F'.)

Often the most interesting functors are those that classify interesting objects. As examples, we list the following questions:
(1) Is there a functor $F\colon (\mathrm{Sch})^{\mathrm{opp}} \to (\mathrm{Sets})$ such that $F(\mathrm{Spec}\, k)$ is the set of isomorphism classes of schemes of finite type over k for every field k? Can such an F be representable?

The answer to the first question is "Yes" (one could define $F(S)$ as the set of isomorphism classes of S-schemes of finite presentation, see Section (10.9) below). The answer to the second question is "No" (see Exercise 8.3).
(2) Given a scheme X over a field k. Is there a functor $H\colon (\mathrm{Sch}/k)^{\mathrm{opp}} \to (\mathrm{Sets})$ such that $H(\mathrm{Spec}\, K)$ is the set of closed subschemes of X_K for every field extension K of k? Is it representable?

Again the first answer is "Yes", and the second answer is also very often "Yes", e.g., if X is quasi-projective over k (see the discussion of the Hilbert scheme in Section (14.32)).
(3) Consider in (2) the special case $X = \mathbb{P}^n_k$ and fix $0 \le d \le n$. Is there a representable functor $G\colon (\mathrm{Sch}/k)^{\mathrm{opp}} \to (\mathrm{Sets})$ such that $G(\mathrm{Spec}\, K)$ is the set of linear subspaces of $\mathbb{P}^n_K$ of dimension d for every field extension K of k?

Again the answer is "Yes". This functor can be identified with the Grassmannian $\mathrm{Grass}_{n+1,d+1}$ considered in Section (8.4), see Section (8.8) (and also Exercise 8.10).

(8.2) Representable morphisms.

Let F and G be two functors $(\mathrm{Sch})^{\mathrm{opp}} \to (\mathrm{Sets})$. For certain morphisms $f\colon F \to G$ of functors it is possible to make sense of the assertion that f possesses a property $\mathbf{P}$, where $\mathbf{P}$ is a property of morphisms of schemes:

Let X be a scheme and let $g\colon X \to G$ be a morphism in $\widehat{(\mathrm{Sch})}$ (recall that we do not distinguish explicitly between a scheme X and the attached functor h_X). Let $F \times_G X$ be the fiber product in the category $\widehat{(\mathrm{Sch})}$ (Example 8.3). Denote by

$$p\colon F \times_G X \to F, \qquad q\colon F \times_G X \to X,$$

the two projections.

Definition 8.5. *A morphism* $f\colon F \to G$ *of functors in* $\widehat{(\mathrm{Sch})}$ *is called* representable *if for all schemes X and all morphisms $g\colon X \to G$ in* $\widehat{(\mathrm{Sch})}$ *the functor $F \times_G X$ is representable.*

Let Z be a scheme and let $\zeta\colon Z \overset{\sim}{\to} F \times_G X$ be an isomorphism. By the Yoneda lemma 4.6 the composition of ζ with the second projection $F \times_G X \to X$ is given by a unique scheme morphism $Z \to X$ which is independent of the choice of (Z, ζ) up to composition with an isomorphism. Thus the following definition makes sense.

Definition 8.6. *Let* $\mathbf{P}$ *be a property of morphisms of schemes such that the composition of a morphism possessing $\mathbf{P}$ with an isomorphism from the right or from the left has again property $\mathbf{P}$ (this will hold for all properties of morphisms of schemes that we will be considering). We say that a representable morphism $f\colon F \to G$ of functors in* $\widehat{(\mathrm{Sch})}$ *possesses the property* $\mathbf{P}$ *if for all schemes X and for every morphism $g\colon X \to G$ the second projection $f_{(X)}\colon F \times_G X \to X$ possesses* $\mathbf{P}$.

Remark 8.7. If $F \cong h_Y$ and $G \cong h_S$ are representable, we have by definition of the universal property of the fiber product that $h_Y \times_{h_S} h_X \cong h_{X \times_S Y}$, where $X \times_S Y$ is the fiber product in the category of schemes, which exists by Theorem 4.18. Thus the fiber product of three representable functors is again representable. In particular, every morphism of representable functors is itself representable.

In general it is not correct that if a scheme morphism $f\colon X \to Y$ possesses a property $\mathbf{P}$, the corresponding morphism of functors $h_X \to h_Y$ also possesses $\mathbf{P}$. But this is true for all properties that are stable under base change; see Section (4.7).

(8.3) Zariski sheaves, Zariski coverings of functors.

Often we are given for a fixed scheme S a functor $F\colon (\mathrm{Sch}/S)^{\mathrm{opp}} \to (\mathrm{Sets})$ and we would like to know whether this functor is representable by an S-scheme. In general this is a difficult problem. We will now give a first criterion for the representability of a functor which we will use to prove the representability of the Grassmannian in Section (8.4).

Let $F\colon (\mathrm{Sch}/S)^{\mathrm{opp}} \to (\mathrm{Sets})$ be a functor. If $j\colon U \to X$ is an open immersion of S-schemes and $\xi \in F(X)$ we write, like for sections of presheaves, simply $\xi_{|U}$ instead of $F(j)(\xi)$.

We say that F is a *sheaf for the Zariski topology* or shorter a *Zariski sheaf* (on (Sch/S)) if the usual sheaf axioms are satisfied, that is, for every S-scheme X and for every open covering $X = \bigcup_{i \in I} U_i$ we have:

(Sh) Given $\xi_i \in F(U_i)$ for all $i \in I$ such that $\xi_{i|(U_i \cap U_j)} = \xi_{j|(U_i \cap U_j)}$ for all $i, j \in I$, there exists a unique element $\xi \in F(X)$ such that $\xi_{|U_i} = \xi_i$ for all $i \in I$.

Then Proposition 3.5 on gluing of morphisms can be reformulated as:

Proposition 8.8. *Every representable functor $F\colon (\mathrm{Sch}/S)^{\mathrm{opp}} \to (\mathrm{Sets})$ is a sheaf for the Zariski topology.*

Later on we will treat faithfully flat descent and we will see that representable functors are sheaves for much finer "coverings" than coverings in the Zariski topology (see Section (14.17)).

We will now show that every Zariski sheaf that has a Zariski covering by representable functors is itself representable. Let us make this more precise: Let S be a fixed scheme and let $F\colon (\mathrm{Sch}/S)^{\mathrm{opp}} \to (\mathrm{Sets})$ be a contravariant functor. An *open subfunctor F' of F*

is a representable morphism $f\colon F' \to F$ that is an open immersion (Definition 8.6). By definition this means that for every S-scheme X and for every S-morphism $g\colon X \to F$ the second projection $f_{(X)}\colon F' \times_F X \to X$ is an open immersion of schemes.

A family $(f_i\colon F_i \to F)_{i \in I}$ of open subfunctors is called a *Zariski open covering* of F, if for every S-scheme X and every S-morphism $g\colon X \to F$ the images of the $(f_i)_{(X)}$ form a covering of X.

Theorem 8.9. *Let $F\colon (\mathrm{Sch}/S)^{\mathrm{opp}} \to (\mathrm{Sets})$ be a functor such that*
(a) F is a sheaf for the Zariski topology,
(b) F has a Zariski open covering $(f_i\colon F_i \to F)_{i \in I}$ by representable functors F_i.
Then F is representable.

Proof. Let the F_i be represented by S-schemes X_i. We will show that we may glue the schemes X_i to a scheme X that represents F. The morphisms $F_i \to F$ are representable by open immersions and open immersions are monomorphisms. Therefore the Yoneda lemma implies that for all S-schemes T the maps $F_i(T) \to F(T)$ are injective. For all $i, j \in I$ and all T we can therefore identify $(F_i \times_F F_j)(T)$ with $F_i(T) \cap F_j(T) \subseteq F(T)$. In virtue of this identification, the functors $F_i \times_F F_j$ and $F_j \times_F F_i$ are equal. Let $X_{\{i,j\}}$ be a scheme that represents this functor. Likewise we identify for $i, j, k \in I$ the functors $F_i \times_F F_j \times_F F_k$, $F_j \times_F F_i \times_F F_k$ etc. and write $F_{\{i,j,k\}}$ for them.

The morphisms $X_{\{i,j\}} \to X_i$, induced by the projections $F_i \times_F F_j \to F_i$, are open immersions and we denote their images by U_{ij}. These immersions induce isomorphisms $\psi_{i,j}\colon X_{\{i,j\}} \xrightarrow{\sim} U_{ij}$. We set $\varphi_{ji} = \psi_{j,i} \circ \psi_{i,j}^{-1}\colon U_{ij} \cong U_{ji}$ and claim that the tuple $((X_i)_{i \in I}, (U_{ij}), (\varphi_{ij}))$ is a gluing datum of schemes (Section (3.5)).

We have to check the cocycle condition, that is,

(8.3.1) $$\varphi_{kj} \circ \varphi_{ji} = \varphi_{ki} \quad \text{on } U_{ij} \cap U_{ik}$$

for all $i, j, k \in I$. For the open subscheme $U_{ij} \cap U_{ik}$ of X_i and for all T we then have $(U_{ij} \cap U_{ik})(T) = U_{ij}(T) \cap U_{ik}(T) = F_{\{i,j,k\}}(T)$ and there is a commutative diagram

$$
\begin{array}{ccc}
U_{ij} \cap U_{ik} & \xrightarrow{\;\varphi_{ji}|_{U_{ij} \cap U_{ik}}\;} & U_{ji} \cap U_{jk} \\
\Big\downarrow{\cong} & & \Big\downarrow{\cong} \\
F_{\{i,j,k\}} & \xrightarrow{\quad\mathrm{id}\quad} & F_{\{i,j,k\}}.
\end{array}
$$

Therefore it suffices to show that the equality (8.3.1) holds for the corresponding morphisms between the functors $F_{\{i,j,k\}}$. But this is obvious as these morphisms are by construction the identity morphisms.

Let X be the S-scheme that results from gluing the X_i with respect to this gluing datum. As F is a Zariski sheaf, the open immersions $f_i\colon X_i \to F$ can be glued to an S-isomorphism $f\colon X \to F$ and thus F is represented by X. $\qquad\square$

The example of the Grassmannian

(8.4) The Grassmannian functor and its representability.

Let $1 \leq d \leq n$ be integers. We would like to find a scheme that classifies d-dimensional subspaces within an n-dimensional vector space. Thus we want to define a representable functor $\mathrm{Grass}_{d,n} \colon (\mathrm{Sch})^{\mathrm{opp}} \to (\mathrm{Sets})$ such that we have for every field k:

$$\mathrm{Grass}_{d,n}(\mathrm{Spec}\, k) = \{\, U \subseteq k^n \;;\; U \text{ is } d\text{-dimensional } k\text{-subspace of } k^n \,\}.$$

What is the correct replacement for "d-dimensional subspace" when we want to define $\mathrm{Grass}_{d,n}(S)$, where S is an arbitrary scheme? We will need the following result.

Proposition 8.10. *Let S be a scheme and let $\iota \colon \mathscr{U} \to \mathscr{E}$ be a homomorphism of $\mathscr{O}_S$-modules. Let $\mathscr{U}$ be of finite type and $\mathscr{E}$ finite locally free. Then the following assertions are equivalent.*
(i) *For every open affine subset U of S there exists a homomorphism $\pi \colon \mathscr{E}_{|U} \to \mathscr{U}_{|U}$ such that $\pi \circ \iota_{|U} = \mathrm{id}$.*
(ii) *The homomorphism ι is injective and the $\mathscr{O}_S$-module $\mathscr{E}/\iota(\mathscr{U})$ is locally free.*
(iii) *For all $s \in S$ the homomorphism $\iota \otimes \mathrm{id}_{\kappa(s)} \colon \mathscr{U}(s) \to \mathscr{E}(s)$ of $\kappa(s)$-vector spaces is injective.*
(iv) *For every scheme morphism $f \colon T \to S$ the homomorphism $f^*(\iota) \colon f^*(\mathscr{U}) \to f^*(\mathscr{E})$ is injective.*
(v) *$\mathscr{U}$ is a finite locally free $\mathscr{O}_S$-module and the dual $\iota^\vee \colon \mathscr{E}^\vee \to \mathscr{U}^\vee$ is surjective.*

Proof. For all implications we may assume that $S = \mathrm{Spec}\, R$ is affine. We recall that Corollary 7.42 shows that if $\mathscr{F} \cong \tilde{N}$ is a quasi-coherent $\mathscr{O}_S$-module, $\mathscr{F}$ is finite locally free if and only if N is a finitely generated projective R-module. In particular we have $\mathscr{E} \cong \tilde{M}$, where M is a finitely generated projective R-module.

"(ii) $\Rightarrow$ (i)". We may assume that $U = S$. The exact sequence

$$0 \to \mathscr{U} \longrightarrow \mathscr{E} \longrightarrow \mathscr{E}/\iota(\mathscr{U}) \to 0$$

shows that $\mathscr{U}$ is quasi-coherent by Proposition 7.14. It corresponds to an exact sequence $0 \to N \xrightarrow{i} M \longrightarrow P \to 0$ of R-modules, where P is projective. Therefore there exists a homomorphism $r \colon M \to N$ such that $r \circ i = \mathrm{id}_N$ and we may set $\pi = \tilde{r}$.

"(i) $\Rightarrow$ (iv) $\Rightarrow$ (iii)". Assertion (i) implies clearly (iv), and (iii) is the special case $T = \mathrm{Spec}(\kappa(s)) \to S$ of (iv).

"(iii) $\Rightarrow$ (ii)". For $s \in S$ we set $A := \mathscr{O}_{S,s}$ and denote by $\mathfrak{m}$ its maximal ideal. We write $N := \mathscr{U}_s$, $M := \mathscr{E}_s$, and $i := \iota_s$. Thus we are given a homomorphism $i \colon N \to M$ of A-modules such that the homomorphism $i_0 := i \otimes \mathrm{id}_{A/\mathfrak{m}} \colon N/\mathfrak{m}N \to M/\mathfrak{m}M$ of $(A/\mathfrak{m})$-vector spaces is injective. We claim that i is injective and that $i(N)$ is a direct summand of M.

Indeed, let r_0 be a left inverse of i_0. The surjection $M \longrightarrow M/\mathfrak{m}M \xrightarrow{r_0} N/\mathfrak{m}N$ can be factorized into $M \xrightarrow{r'} N \longrightarrow N/\mathfrak{m}N$, because M is projective. Now $r' \circ i$ is an endomorphism of N which is modulo $\mathfrak{m}$ the identity. Hence $r' \circ i$ is surjective by the Lemma of Nakayama. But surjective endomorphisms of finitely generated modules are bijective by Corollary B.4. Thus if we set $r := (r' \circ i)^{-1} \circ r' \colon M \to N$, we have $r \circ i = \mathrm{id}_N$. This shows our claim.

The claim proves that ι is injective and that all stalks $\mathscr{E}/\iota(\mathscr{U})_s$ are free $\mathscr{O}_{S,s}$-modules. As $\mathscr{U}$ is of finite type, $\mathscr{E}/\iota(\mathscr{U})$ is of finite presentation and thus locally free (Proposition 7.41).

"$(iii) \Leftrightarrow (v)$". As (iii) implies (ii) and (i), we know already that (iii) implies that $\mathscr{U} \cong \iota(\mathscr{U})$ is locally a direct summand of $\mathscr{E}$ and hence $\mathscr{U}$ is locally free. Therefore we have $(\iota^\vee)^\vee = \iota$ and $(\iota \otimes \mathrm{id}_{\kappa(s)})^\vee = \iota^\vee \otimes \mathrm{id}_{\kappa(s)}$. This shows that (iii) is equivalent to the surjectivity of $\iota^\vee \otimes \mathrm{id}_{\kappa(s)}$ for all $s \in S$ and hence by Nakayama's Lemma to the surjectivity of $\iota^\vee$.   □

If ι is the inclusion of an $\mathscr{O}_S$-submodule $\mathscr{U}$ of $\mathscr{E}$, it satisfies these equivalent conditions if and only if for all $s \in S$ there exists an open neighborhood V of s such that $\mathscr{U}_{|V}$ is a direct summand of $\mathscr{E}_{|V}$ (and in that case $\mathscr{U}_{|U}$ is a direct summand of $\mathscr{E}_{|U}$ for every open affine subscheme U of S). In this case $\mathscr{U}$ is called *locally a direct summand of $\mathscr{E}$*.

Remark 8.11. The proof shows that the implications "(ii) $\Rightarrow$ (i) $\Rightarrow$ (iv) $\Rightarrow$ (iii)" also hold if $\mathscr{U}$ is an arbitrary $\mathscr{O}_S$-module and $\mathscr{E}$ is only assumed to be quasi-coherent.

Corollary 8.12. *Let S be a scheme and let $\pi\colon \mathscr{E} \to \mathscr{F}$ be a homomorphism of finite locally free $\mathscr{O}_S$-modules of the same rank. Then the following assertions are equivalent.*
(i) *π is an isomorphism.*
(ii) *For all $s \in S$ the homomorphism $\pi \otimes \mathrm{id}_{\kappa(s)}\colon \mathscr{E}(s) \to \mathscr{F}(s)$ is surjective.*
(iii) *$\det(\pi)\colon \det(\mathscr{E}) \to \det(\mathscr{F})$ is bijective.*

Proof. The equivalence of (i) and (ii) follows from Proposition 8.10 applied to $\iota := \pi^\vee$. The equivalence of (i) and (iii) can be checked on stalks and thus follows from the analogous property of homomorphisms between free modules of the same rank.   □

This corollary will be generalized in Lemma 16.17 below.

We would like to define $\mathrm{Grass}_{d,n}(S)$ as a set of "certain" $\mathscr{O}_S$-submodules $\mathscr{U} \subseteq \mathscr{O}_S^n$. This will be made more precise now. We first demand that these submodules $\mathscr{U}$ are of finite type. In order to make $\mathrm{Grass}_{d,n}$ into a functor we also have to define for every morphism of schemes $f\colon T \to S$ a map $\mathrm{Grass}_{d,n}(f)\colon \mathrm{Grass}_{d,n}(S) \to \mathrm{Grass}_{d,n}(T)$. We set $\mathrm{Grass}_{d,n}(f)(\mathscr{U}) := f^*(\mathscr{U})$ for $\mathscr{U} \in \mathrm{Grass}_{d,n}(S)$. The inclusion $\mathscr{U} \hookrightarrow \mathscr{O}_S^n$ yields via functoriality a homomorphism of $\mathscr{O}_T$-modules $f^*(\mathscr{U}) \to f^*(\mathscr{O}_S^n) = \mathscr{O}_T^n$. By the proposition above this homomorphism is injective for all f if and only if $\mathscr{O}_S^n/\mathscr{U}$ is a locally free $\mathscr{O}_S$-module. If this is the case, $\mathscr{U}$ is a locally free $\mathscr{O}_S$-module. This suggests that the correct replacement for "d-dimensional subspace" is the condition that $\mathscr{O}_S/\mathscr{U}$ is locally free of rank $n - d$. Therefore we define

$$\mathrm{Grass}_{d,n}(S) = \{\, \mathscr{U} \subseteq \mathscr{O}_S^n \,;\; \mathscr{O}_S^n/\mathscr{U} \text{ is a locally free } \mathscr{O}_S\text{-module of rank } n - d \,\}.$$

We want to prove that the functor $\mathrm{Grass}_{d,n}$ is representable. As $\mathscr{O}_S$-submodules can be glued (as all sheaves), $\mathrm{Grass}_{d,n}$ is a sheaf for the Zariski topology. Therefore it suffices to show that $\mathrm{Grass}_{d,n}$ has a covering by open representable subfunctors which we are going to define now. For every subset $I \subseteq \{1, \ldots, n\}$ with $n - d$ elements we define a subfunctor of $\mathrm{Grass}_{d,n}$ by

$$(8.4.1) \quad \mathrm{Grass}_{d,n}^I(S) := \{\, \mathscr{U} \in \mathrm{Grass}_{d,n}(S) \,;\; \mathscr{O}_S^I \hookrightarrow \mathscr{O}_S^n \twoheadrightarrow \mathscr{O}_S^n/\mathscr{U} \text{ is an isomorphism} \,\}.$$

Here the first arrow is the homomorphism of $\mathscr{O}_S$-modules $u^I \colon \mathscr{O}_S^I \to \mathscr{O}_S^{\{1,\dots,n\}} = \mathscr{O}_S^n$, induced by the inclusion $I \hookrightarrow \{1,\dots,n\}$. In other words, $\mathrm{Grass}_{d,n}^I(S)$ consists of those $\mathscr{U} \in \mathrm{Grass}_{d,n}(S)$ such that $\mathscr{U} \oplus \mathscr{O}_S^I = \mathscr{O}_S^n$. The inclusion $\mathrm{Grass}_{d,n}^I(S) \hookrightarrow \mathrm{Grass}_{d,n}(S)$ defines a morphism of functors

$$\iota^I \colon \mathrm{Grass}_{d,n}^I \to \mathrm{Grass}_{d,n} .$$

Lemma 8.13.
(1) *The morphism ι^I is representable and an open immersion.*
(2) *The functor $\mathrm{Grass}_{d,n}^I$ is isomorphic to $\mathbb{A}^{d(n-d)}$, in particular it is representable.*

Proof. (1). Let X be a scheme and let $g \colon X \to \mathrm{Grass}_{d,n}$ be a morphism of functors, that is, g is an X-valued point of $\mathrm{Grass}_{d,n}$ and thus corresponds to an $\mathscr{O}_X$-submodule $\mathscr{U}$ of $\mathscr{O}_X^n$ such that $\mathscr{O}_X^n/\mathscr{U}$ is locally free of rank $n - d$. Let S be an arbitrary scheme. By definition we have

$$(\mathrm{Grass}_{d,n}^I \times_{\mathrm{Grass}_{d,n}} X)(S) = \{\, f \in X(S) = \mathrm{Hom}(S, X) \;;\; f^*(\mathscr{U}) \in \mathrm{Grass}_{d,n}^I(S) \,\}.$$

Therefore we have to show that there exists an open subscheme U of X with the following property: A morphism of schemes $f \colon S \to X$ factors through U if and only if the composition $v_f \colon \mathscr{O}_S^I \to \mathscr{O}_S^n \to \mathscr{O}_S^n/f^*(\mathscr{U})$ is an isomorphism. As source and target of v_f are locally free $\mathscr{O}_S$-modules of the same rank, v_f is an isomorphism if and only if v_f is surjective (Corollary 8.12), and (1) follows from Proposition 8.4 (1).

(2). Let S be a scheme and $\mathscr{U} \in \mathrm{Grass}_{d,n}^I(S)$. Then $w \colon \mathscr{O}_S^I \to \mathscr{O}_S^n/\mathscr{U}$ is by definition an isomorphism. The kernel of the composition

$$u_{\mathscr{U}} \colon \mathscr{O}_S^n \twoheadrightarrow \mathscr{O}_S^n/\mathscr{U} \xrightarrow{w^{-1}} \mathscr{O}_S^I$$

is $\mathscr{U}$ and we have $u_{\mathscr{U}} \circ u^I = \mathrm{id}_{\mathscr{O}_S^I}$. Conversely, given a homomorphism $u \colon \mathscr{O}_S^n \to \mathscr{O}_S^I$ with $u \circ u^I = \mathrm{id}$, we find $\mathrm{Ker}(u) \in \mathrm{Grass}_{d,n}^I(S)$. Therefore the map

$$(8.4.2) \quad F(S) := \{\, u \in \mathrm{Hom}_{\mathscr{O}_S}(\mathscr{O}_S^n, \mathscr{O}_S^I) \;;\; u \circ u^I = \mathrm{id} \,\} \to \mathrm{Grass}_{d,n}^I(S), \quad u \mapsto \mathrm{Ker}(u)$$

is bijective. It is functorial in S and we obtain an isomorphism of functors $F \xrightarrow{\sim} \mathrm{Grass}_{d,n}^I$. Setting $J := \{1,\dots,n\} \setminus I$, the map

$$(8.4.3) \qquad F(S) \to \mathrm{Hom}_{\mathscr{O}_S}(\mathscr{O}_S^J, \mathscr{O}_S^I) = \Gamma(S, \mathscr{O}_S)^{J \times I} \cong \mathbb{A}^{d(n-d)}(S), \qquad u \mapsto u|_{\mathscr{O}_S^J}$$

is bijective and functorial in S and therefore we obtain an isomorphism $F \xrightarrow{\sim} \mathbb{A}^{d(n-d)}$. $\square$

Now we use this lemma and the representability criterion Theorem 8.9 to prove:

Proposition 8.14. *Let $n \geq d \geq 1$ be integers. Then the functor $\mathrm{Grass}_{d,n}$ is representable.*

The representing scheme is called *Grassmannian* and is also denoted by $\mathrm{Grass}_{d,n}$.

Proof. By Theorem 8.9 it suffices to show that the family $(\iota^I \colon \mathrm{Grass}_{d,n}^I \to \mathrm{Grass}_{d,n})_I$ of open subfunctors (where I runs through the set of subsets of $\{1,\dots,n\}$ with $(n - d)$ elements) is a Zariski covering of $\mathrm{Grass}_{d,n}$.

Let X be a scheme and let $g\colon X \to \mathrm{Grass}_{d,n}$ be a morphism of functors corresponding via the Yoneda lemma to an element $\mathscr{U} \in \mathrm{Grass}_{d,n}(X)$. In Lemma 8.13 we have seen that $U^I := \mathrm{Grass}_{d,n}^I \times_{\mathrm{Grass}_{d,n}} X$ is representable by an open subscheme of X which again we denote by U^I. We have to show that the morphism $f\colon \coprod_I U^I \to X$, induced by the open immersion $U^I \to X$, is surjective. For this it suffices to show that f is surjective on K-valued points, where K is an arbitrary field (Proposition 4.8).

Let $x\colon \mathrm{Spec}(K) \to X$ be a K-valued point of X. By composition with g we obtain a K-valued point of $\mathrm{Grass}_{d,n}$ which corresponds to a d-dimensional subvector space U of K^n. By definition, x lies in the image of $U^I(K) \hookrightarrow X(K)$ if and only if K^I is a complement of U. But linear algebra tells us that we can complete any basis of U by a part of the standard basis to a basis of K^n and thus there exists a subset I of $\{1, \ldots, n\}$ with $n - d$ elements such that K^I is a complement of U. $\square$

In Lemma 8.13 we have seen that $\mathrm{Grass}_{d,n}^I \cong \mathbb{A}^{d(n-d)}$ for all I. We therefore obtain the following corollary.

Corollary 8.15. *The scheme $\mathrm{Grass}_{d,n}$ has a finite open covering by schemes that are isomorphic to $\mathbb{A}^{d(n-d)}$. In particular $\mathrm{Grass}_{d,n}$ is smooth over $\mathrm{Spec}\,\mathbb{Z}$ of relative dimension $d(n-d)$.*

(8.5) Projective Space as a Grassmannian.

We now show that the projective space is the special case $d = 1$ of the Grassmannian. Note that $\mathrm{Grass}_{1,n}(S)$ is the set of locally direct summands $\mathscr{L}$ of $\mathscr{O}^n$ of rank 1. In particular we find that $\mathrm{Grass}_{1,n}(k) = \mathbb{P}^{n-1}(k)$ for every algebraically closed field k. In fact, we will see now that $\mathrm{Grass}_{1,n} = \mathbb{P}^{n-1}$.

If $\mathscr{L}$ is a locally free submodule of $\mathscr{O}^n$ of rank 1, then locally on S there exists a section of $\Gamma(S, \mathscr{O}^n)$ that generates $\mathscr{L}$. This means that for every point $s \in S$ there exists an open affine neighborhood U of s such that $\mathscr{L}_{|U}$ is the submodule of $\mathscr{O}_U^n$ generated by a section $x = (x_0, \ldots, x_{n-1}) \in \Gamma(U, \mathscr{O}_U)^n$. Moreover, $\mathscr{L}_{|U}$ is a direct summand of $\mathscr{O}_U^n$ if and only if the ideal generated by $x_0, \ldots, x_{n-1}$ in $\Gamma(U, \mathscr{O}_U)$ is the unit ideal.

For $i \in \{0, \ldots, n-1\}$ set $I_i := \{1, \ldots, n\} \setminus \{i+1\}$ and denote by $U_i := \mathrm{Grass}_{1,n}^{I_i}$ the open subscheme of $\mathrm{Grass}_{1,n}$ defined in (8.4.1). Then $U_i(S)$ consists of those $\mathscr{L} \in \mathrm{Grass}_{1,n}(S)$ that are locally on S generated by $(x_0, \ldots, x_{n-1}) \in \Gamma(S, \mathscr{O}_S)^n$ with $x_i \in \Gamma(S, \mathscr{O}_S)^\times$. The isomorphism $U_i \xrightarrow{\sim} \mathbb{A}^{n-1}$ constructed in the proof of Lemma 8.13 is given locally on S by

$$U_i \ni \langle x_0, \ldots, x_{n-1} \rangle \longmapsto \left(\frac{x_0}{x_i}, \ldots, \widehat{\frac{x_i}{x_i}}, \ldots, \frac{x_{n-1}}{x_i} \right).$$

The definition of $\mathbb{P}^{n-1}$ in Section (3.6) shows that the isomorphisms

$$U_i \xrightarrow{\sim} \mathbb{A}^{n-1} \xrightarrow{\sim} \mathbb{P}^{n-1} \setminus V_+(X_i) \subset \mathbb{P}^{n-1}$$

glue to an isomorphism

(8.5.1) $\mathrm{Grass}_{1,n} \xrightarrow{\sim} \mathbb{P}^{n-1}$.

Remark 8.16. The canonical surjective morphism $p\colon \mathbb{A}^n \setminus \{0\} \to \mathbb{P}^{n-1}$ in (3.6.1) is given on R-valued points (R some ring) by sending $(x_0, \ldots, x_{n-1}) \in (\mathbb{A}^n \setminus \{0\})(R)$ (see Example 4.5) to the free direct summand of R^n generated by $(x_0, \ldots, x_{n-1})$. Conversely, if

$U \subset R^n$ is a free direct summand of rank 1 generated by an element $(x_0, \ldots, x_{n-1}) \in R^n$, then the ideal in R generated by the x_i is the unit ideal. In particular we see that p is surjective on R-valued points if and only if every direct summand of R^n of rank 1 is a *free* R-module (e.g., if R is a local ring).

(8.6) The Grassmannian of a quasi-coherent module.

Instead of considering submodules of $\mathscr{O}_S^n$ we can also look at the following more general situation. Let S be a scheme, $\mathscr{E}$ a quasi-coherent $\mathscr{O}_S$-module, and $e \geq 0$ an integer. For every S-scheme $h\colon T \to S$ denote by $\mathrm{Grass}^e(\mathscr{E})(T)$ the set of $\mathscr{O}_T$-submodules $\mathscr{U} \subseteq h^*(\mathscr{E})$ such that $h^*(\mathscr{E})/\mathscr{U}$ is a locally free $\mathscr{O}_T$-module of rank e. Every morphism $f\colon T' \to T$ of S-schemes yields a map

$$\mathrm{Grass}^e(\mathscr{E})(T) \longrightarrow \mathrm{Grass}^e(\mathscr{E})(T'), \qquad \mathscr{U} \longmapsto f^*(\mathscr{U})$$

by Remark 8.11. Therefore we obtain a contravariant functor $\mathrm{Grass}^e(\mathscr{E})$ from the category of S-schemes to the category (Sets). Setting $\mathscr{E} = \mathscr{O}_S^n$ and $e = n - d$ we obtain the special case $\mathrm{Grass}^e(\mathscr{E}) = \mathrm{Grass}_{d,n} \times_{\mathbb{Z}} S$. Every surjection $v\colon \mathscr{E}_1 \to \mathscr{E}_2$ of $\mathscr{O}_S$-modules induces a morphism $i_v\colon \mathrm{Grass}^e(\mathscr{E}_2) \to \mathrm{Grass}^e(\mathscr{E}_1)$.

Proposition 8.17.
(1) Let $\mathscr{E}$ be a quasi-coherent $\mathscr{O}_S$-module and $e \geq 0$ an integer. Then $\mathrm{Grass}^e(\mathscr{E})$ is representable by an S-scheme.
(2) If $v\colon \mathscr{E}_1 \to \mathscr{E}_2$ is a surjection of quasi-coherent $\mathscr{O}_S$-modules, the induced morphism $i_v\colon \mathrm{Grass}^e(\mathscr{E}_2) \to \mathrm{Grass}^e(\mathscr{E}_1)$ is a closed immersion.

The representing S-scheme is called *Grassmannian of quotients of $\mathscr{E}$ of rank e* and it is again denoted by $\mathrm{Grass}^e(\mathscr{E})$.

We will only give the proof of (1) if $\mathscr{E}$ is of finite type. For a sketch of the proof in the general case we refer to Exercise 8.11 where also the open subfunctors $\mathrm{Grass}_{d,n}^I$ are generalized. See also [EGAInew] 9.7 for a different proof.

Proof. (if $\mathscr{E}$ is of finite type) It is a consequence of Proposition 8.4 (2) that i_v in part (2) is representable and a closed immersion (for this part of the proof it is not necessary to assume that $\mathscr{E}$ is of finite type). In fact, let $h\colon X \to S$ be an S-scheme. Given $X \to \mathrm{Grass}^e(\mathscr{E}_1)$, let $\mathscr{V}_X$ be the corresponding element of $\mathrm{Grass}^e(\mathscr{E}_1)(X)$, and apply Proposition 8.4 (2) to the composite $\ker(h^*(v)) \to h^*(\mathscr{E}_1)/\mathscr{V}_X$. Moreover, as $\mathrm{Grass}^e(\mathscr{E})$ is a Zariski-sheaf, the question of the representability of $\mathrm{Grass}^e(\mathscr{E})$ is local on S (Theorem 8.9) and we may assume that $S = \mathrm{Spec}\, R$ is affine and that $\mathscr{E} = \tilde{E}$ for a finitely generated R-module E. Let $u\colon R^n \twoheadrightarrow E$ be a surjection of R-modules. As $i_{\tilde{u}}\colon \mathrm{Grass}^e(\mathscr{E}) \to \mathrm{Grass}^e(\mathscr{O}_S^n) = \mathrm{Grass}_{n-e,n} \times_{\mathbb{Z}} S$ is representable by a closed immersion, $\mathrm{Grass}^e(\mathscr{E})$ is representable by an S-scheme which is isomorphic to a closed subscheme of $\mathrm{Grass}_{n-e,n} \times_{\mathbb{Z}} S$. $\square$

If $\mathscr{E}$ is a finite locally free $\mathscr{O}_S$-module of rank n, dualizing yields an isomorphism

(8.6.1)
$$\mathrm{Grass}^e(\mathscr{E}) \overset{\sim}{\to} \mathrm{Grass}^{n-e}(\mathscr{E}^\vee),$$
$$(\iota\colon \mathscr{U} \hookrightarrow h^*\mathscr{E}) \mapsto (\mathrm{Coker}(\iota)^\vee \hookrightarrow h^*\mathscr{E}^\vee) \quad \text{on } (h\colon T \to S)\text{-valued points.}$$

(8.7) Base change of Grassmannians.

Let S be a scheme, $\mathscr{E}$ a quasi-coherent $\mathscr{O}_S$-module and $e \geq 0$ an integer. Let $u\colon S' \to S$ be a morphism of schemes. By composing with u we can consider every S'-scheme $h'\colon T \to S'$ as an S-scheme and we find identities, functorial in T,

$$\operatorname{Hom}_{S'}(T, \operatorname{Grass}^e(\mathscr{E}) \times_S S') \overset{(4.7.1)}{=} \operatorname{Hom}_S(T, \operatorname{Grass}^e(\mathscr{E})) = \operatorname{Hom}_{S'}(T, \operatorname{Grass}^e(u^*\mathscr{E})),$$

where the second equality follows from the definition of the Grassmannian functor and the isomorphism $(u \circ h')^*\mathscr{E} = (h')^*(u^*\mathscr{E})$ (7.8.9). By the Yoneda lemma we obtain an isomorphism of S'-schemes

(8.7.1) $$u^* \operatorname{Grass}^e(\mathscr{E}) \cong \operatorname{Grass}^e(u^*\mathscr{E}).$$

For every scheme S we also set $\operatorname{Grass}_{d,n,S} := \operatorname{Grass}_{d,n} \times_{\mathbb{Z}} S$. Then

$$\operatorname{Grass}_{d,n,S} = \operatorname{Grass}^{n-d}(\mathscr{O}_S^n) \cong \operatorname{Grass}^d((\mathscr{O}_S^n)^\vee),$$

where the second identity is given by (8.6.1).

(8.8) Projective bundles and linear subbundles.

Let S be a scheme and $\mathscr{E}$ be a quasi-coherent $\mathscr{O}_S$-module. As a generalization of the projective space we define the *projective bundle defined by $\mathscr{E}$* as

(8.8.1) $$\mathbb{P}(\mathscr{E}) := \operatorname{Grass}^1(\mathscr{E}).$$

Thus the S-scheme $\mathbb{P}(\mathscr{E})$ represents the functor that attaches to every S-scheme $h\colon T \to S$ the set $\mathbb{P}(\mathscr{E})(T)$ of equivalence classes of surjections $h^*\mathscr{E} \twoheadrightarrow \mathscr{L}$ where $\mathscr{L}$ is a locally free $\mathscr{O}_T$-module of rank 1. In particular there is a universal surjection $p^*\mathscr{E} \twoheadrightarrow \mathscr{L}_{\mathrm{univ}}$ corresponding to $\operatorname{id}_{\mathbb{P}(\mathscr{E})}$, where $p\colon \mathbb{P}(\mathscr{E}) \to S$ is the structure morphism.

   In the special case $S = \operatorname{Spec}\mathbb{Z}$ and $\mathscr{E} = (\mathscr{O}_{\operatorname{Spec}\mathbb{Z}}^{n+1})^\vee$ the dualizing isomorphism (8.6.1) yields an identification

(8.8.2) $$\mathbb{P}((\mathscr{O}_{\operatorname{Spec}\mathbb{Z}}^{n+1})^\vee) = \mathbb{P}^n.$$

   By (8.7.1) we have for every morphism $u\colon S' \to S$ an isomorphism of S'-schemes

(8.8.3) $$u^*\mathbb{P}(\mathscr{E}) = \mathbb{P}(u^*\mathscr{E})$$

   In Section (13.8) we will show that $\mathbb{P}(\mathscr{E})$ can also be described quite differently (using the projective spectrum of the symmetric algebra attached to $\mathscr{E}$).

In Section (1.23) we defined the notion of a linear subspace of $\mathbb{P}^n(k)$ (k an algebraically closed field). This is generalized to projective bundles as follows. Assume that $\mathscr{E}$ is a finite locally free $\mathscr{O}_S$-module. We call a closed subscheme Z of $\mathbb{P}(\mathscr{E})$ a *linear subbundle* if there exists a surjective homomorphism $\mathscr{E} \twoheadrightarrow \mathscr{F}$ onto a finite locally free $\mathscr{O}_S$-module $\mathscr{F}$ such that the associated closed immersion $\mathbb{P}(\mathscr{F}) \to \mathbb{P}(\mathscr{E})$ induces an isomorphism $\mathbb{P}(\mathscr{F}) \overset{\sim}{\to} Z$. If $\mathscr{F}$ is of rank $m + 1$, we say that the corresponding linear subbundle is *of rank m*.

   We see that a linear subbundle of $\mathbb{P}(\mathscr{E})$ of rank m is the same as an S-valued point of $\operatorname{Grass}^{m+1}(\mathscr{E})$. Therefore we define

(8.8.4) $$\operatorname{LinSub}_m(\mathbb{P}(\mathscr{E})) := \operatorname{Grass}^{m+1}(\mathscr{E})$$

and consider $\operatorname{LinSub}_m(\mathbb{P}(\mathscr{E}))$ as the scheme parameterizing linear subbundles of rank m.

Remark 8.18. (Projections with center in a linear subspace) Now fix a surjective homomorphism $q\colon \mathscr{E} \twoheadrightarrow \mathscr{F}$ as above and identify $\mathbb{P}(\mathscr{F})$ with the associated linear subbundle of $\mathbb{P}(\mathscr{E})$. Let $\mathscr{G} := \operatorname{Ker}(q)$. This is locally a direct summand of $\mathscr{E}$ and in particular again finite locally free (Proposition 8.10). Generalizing Example 1.66 we will define a morphism of S-schemes

$$\pi_q\colon \mathbb{P}(\mathscr{E}) \setminus \mathbb{P}(\mathscr{F}) \to \mathbb{P}(\mathscr{G})$$

which is called *projection with center* q. First we will identify the T-valued points of $U := \mathbb{P}(\mathscr{E}) \setminus \mathbb{P}(\mathscr{F})$. Define a subfunctor $\mathbb{P}(\mathscr{E})_q$ of $\mathbb{P}(\mathscr{E})$ by

$$\mathbb{P}(\mathscr{E})_q(T) = \{\, [h^*\mathscr{E} \twoheadrightarrow \mathscr{L}] \in \mathbb{P}(\mathscr{E})(T) \;;\; h^*\mathscr{G} \hookrightarrow h^*\mathscr{E} \twoheadrightarrow \mathscr{L} \text{ is surjective}\,\}$$

By Proposition 8.4 it is representable by the open subscheme U' of $\mathbb{P}(\mathscr{E})$ which is the complement of the support of the cokernel of the composition $p^*\mathscr{G} \to p^*\mathscr{E} \twoheadrightarrow \mathscr{L}_{\mathrm{univ}}$. Moreover, it is clear that for any field k the open subschemes U and U' have the same k-valued points. Thus they have the same underlying topological spaces and hence are equal. Therefore we may define π_q by sending every $(h\colon T \to S)$-valued point of $\mathbb{P}(\mathscr{E})_q$ to the composition $h^*\mathscr{G} \hookrightarrow h^*\mathscr{E} \twoheadrightarrow \mathscr{L}$ which is a T-valued point of $\mathbb{P}(\mathscr{G})$.

If T is an affine S-scheme, we can choose a complement of $h^*\mathscr{G}$ in $h^*\mathscr{E}$ and extend any surjection $h^*\mathscr{G} \twoheadrightarrow \mathscr{L}$ to a surjection $h^*\mathscr{E} \twoheadrightarrow \mathscr{L}$. This shows that $\pi_q(T)$ is surjective. In particular π_q is surjective (Proposition 4.8).

Any surjection $\mathscr{F} \to \mathscr{F}'$ of $\mathscr{O}_S$-modules (corresponding to a linear subbundle $\mathbb{P}(\mathscr{F}')$ of $\mathbb{P}(\mathscr{F})$) yields a factorization of π_q as follows. Let q' be the composition $\mathscr{E} \twoheadrightarrow \mathscr{F} \twoheadrightarrow \mathscr{F}'$ and let $\mathscr{G}'$ be its kernel. We denote by q'' the canonical projection $\mathscr{G}' \twoheadrightarrow \mathscr{G}'/\mathscr{G}$. Note that we have an exact sequence $0 \to \mathscr{G}'/\mathscr{G} \to \mathscr{F} \to \mathscr{F}' \to 0$. Therefore $\pi_{q'}^{-1}(\mathbb{P}(\mathscr{G}'/\mathscr{G})) = \mathbb{P}(\mathscr{F})$ and π_q is the composition

$$\pi_q\colon \mathbb{P}(\mathscr{E}) \setminus \mathbb{P}(\mathscr{F}) \xrightarrow{\pi_{q'}|\mathbb{P}(\mathscr{E})\setminus\mathbb{P}(\mathscr{F})} \mathbb{P}(\mathscr{G}') \setminus \mathbb{P}(\mathscr{G}'/\mathscr{G}) \xrightarrow{\pi_{q''}} \mathbb{P}(\mathscr{G}).$$

In Section (13.17) we will give a more geometric interpretation of π_q.

Remark 8.19. (Segre embedding revisited) The Segre embedding for products of projective spaces (Section (4.14)) can be generalized and defined via the representing functors as follows. Let $\mathscr{E}$ and $\mathscr{F}$ be two locally free $\mathscr{O}_S$-modules of finite rank. For every S-scheme $h\colon T \to S$ we have $h^*\mathscr{E} \otimes_{\mathscr{O}_T} h^*(\mathscr{F}) = h^*(\mathscr{E} \otimes_{\mathscr{O}_S} \mathscr{F})$ (7.8.8). We define

$$(8.8.5) \quad \begin{aligned} &\sigma(T)\colon \mathbb{P}(\mathscr{E})(T) \times \mathbb{P}(\mathscr{F})(T) \to \mathbb{P}(\mathscr{E} \otimes_{\mathscr{O}_S} \mathscr{F})(T), \\ &(u\colon h^*\mathscr{E} \twoheadrightarrow \mathscr{L}, \ v\colon h^*\mathscr{F} \twoheadrightarrow \mathscr{M}) \mapsto (u \otimes v\colon h^*(\mathscr{E} \otimes \mathscr{F}) \twoheadrightarrow \mathscr{L} \otimes \mathscr{M}). \end{aligned}$$

This map is clearly functorial in T and therefore defines a morphism of S-schemes

$$(8.8.6) \qquad \sigma = \sigma_{\mathscr{E},\mathscr{F}}\colon \mathbb{P}(\mathscr{E}) \times_S \mathbb{P}(\mathscr{F}) \to \mathbb{P}(\mathscr{E} \otimes_{\mathscr{O}_S} \mathscr{F})$$

which is called the *Segre embedding*. For $\mathscr{E} = (\mathscr{O}_S^{n+1})^\vee$ and $\mathscr{F} = (\mathscr{O}_S^{m+1})^\vee$ we have $\mathbb{P}(\mathscr{E}) = \mathbb{P}_S^n$, $\mathbb{P}(\mathscr{F}) = \mathbb{P}_S^m$, and $\mathbb{P}(\mathscr{E} \otimes_{\mathscr{O}_S} \mathscr{F}) = \mathbb{P}_S^{nm+n+m}$, and we obtain the usual Segre embedding

$$\sigma_{n,m}\colon \mathbb{P}_S^n \times_S \mathbb{P}_S^m \to \mathbb{P}_S^{nm+n+m}$$

defined in Section (4.14). It follows immediately from the definition that the Segre embedding is compatible with base change: For every morphism $f\colon S' \to S$ we have

$$(8.8.7) \qquad (\sigma_{\mathscr{E},\mathscr{F}})_{(S')} = \sigma_{f^*\mathscr{E},f^*\mathscr{F}}.$$

Proposition 8.20. *The morphism* $\sigma_{\mathscr{E},\mathscr{F}}$ *is a closed immersion.*

Proof. The property of being a closed immersion is local on the target and in particular local on S. Replacing S by a sufficiently small open affine subset we may assume by (8.8.7) that $S = \operatorname{Spec} R$ and that $\mathscr{E} \cong (\mathscr{O}_S^{n+1})^{\vee}$ and $\mathscr{F} \cong (\mathscr{O}_S^{m+1})^{\vee}$. Therefore the claim follows from the fact that the Segre embedding $\sigma_{n,m} \colon \mathbb{P}_R^n \times_R \mathbb{P}_R^m \to \mathbb{P}_R^{nm+n+m}$ is a closed immersion (Proposition 4.39). $\qquad\square$

The construction of $\sigma_{\mathscr{E},\mathscr{F}}$ and Proposition 8.20 can be generalized to arbitrary quasi-coherent $\mathscr{O}_S$-modules $\mathscr{E}$ and $\mathscr{F}$.

Remark 8.21. Let S be a scheme, let $\mathscr{E}$ and $\mathscr{F}$ be quasi-coherent $\mathscr{O}_S$-modules, and let $e \geq 1$ be an integer. Then we may consider the disjoint union of $\operatorname{Grass}^e(\mathscr{E})$ and $\operatorname{Grass}^e(\mathscr{F})$ as a closed subscheme of $\operatorname{Grass}^e(\mathscr{E} \oplus \mathscr{F})$: The canonical surjections $\mathscr{E} \oplus \mathscr{F} \to \mathscr{E}$ and $\mathscr{E} \oplus \mathscr{F} \to \mathscr{F}$ yield closed immersions $\operatorname{Grass}^e(\mathscr{E}) \hookrightarrow \operatorname{Grass}^e(\mathscr{E} \oplus \mathscr{F})$ (Proposition 8.17). We have to show that there is no field k and no k-valued point $x \colon \operatorname{Spec} k \to \operatorname{Grass}^e(\mathscr{E} \oplus \mathscr{F})$ such that x factors through $\operatorname{Grass}^e(\mathscr{E})$ and through $\operatorname{Grass}^e(\mathscr{F})$. But such a point x would correspond to a subvector space U of $x^*\mathscr{E} \oplus x^*\mathscr{F}$ of codimension e which is contained and of codimension e in $x^*\mathscr{E}$ and in $x^*\mathscr{F}$. This is absurd.

In particular, there is a closed embedding of S-schemes

$$(8.8.8) \qquad \mathbb{P}(\mathscr{E}) \amalg \mathbb{P}(\mathscr{F}) \hookrightarrow \mathbb{P}(\mathscr{E} \oplus \mathscr{F}).$$

(8.9) The tangent space of the Grassmannian.

In Section (6.6) we defined the relative tangent space of a scheme in terms of the functor it represents. We illustrate this in the case of the Grassmannian.

Example 8.22. Let k be a field, $x \colon \operatorname{Spec} k \to X := \operatorname{Grass}_{d,n}$ be a k-valued point. It corresponds to a sub-vector space $U_x \subseteq V := k^n$ of dimension d. The tangent space $T_x(X/\mathbb{Z})$ can then be identified with the set of free $k[\varepsilon]$-submodules $U \subseteq V \otimes k[\varepsilon]$ of rank d with $U \otimes k = U_x$. Note that we have $V \otimes k[\varepsilon] = V \oplus \varepsilon V$ as k-vector spaces.

Every $U \in T_x(X/\mathbb{Z})$ is the image of a $k[\varepsilon]$-linear map $U_x \otimes_k k[\varepsilon] \to V \otimes k[\varepsilon]$ that is the inclusion modulo ε. Therefore for $w \in \operatorname{Hom}_k(U_x, V)$ let $\tilde{w} \colon U_x \otimes_k k[\varepsilon] \to V \otimes k[\varepsilon]$ be the unique $k[\varepsilon]$-linear homomorphism such that $\tilde{w}_{|U_x} = \iota \oplus \varepsilon w$, where $\iota \colon U_x \to V$ is the inclusion. Then the image of $\tilde{w}$ is an element U_w of $T_x(X/\mathbb{Z})$ and $w \mapsto U_w$ defines a surjective map $\pi \colon \operatorname{Hom}_k(U_x, V) \to T_x(X/\mathbb{Z})$ that is readily checked to be k-linear. Its kernel consists of those k-linear maps w such that $w(U_x) \subseteq U_x$. Therefore π induces an isomorphism

$$(8.9.1) \qquad \operatorname{Hom}_k(U_x, V/U_x) \xrightarrow{\sim} T_x(\operatorname{Grass}_{d,n}/\mathbb{Z}).$$

In particular we see that its dimension is $d(n-d)$ (which we already knew by applying Corollary 6.29 to the scheme $\operatorname{Grass}_{d,n} \otimes_{\mathbb{Z}} k$ which is smooth of relative dimension $d(n-d)$ over k by Corollary 8.15).

(8.10) The Plücker embedding.

For $n \geq d \geq 1$ the d-th exterior power $\bigwedge_{\mathscr{O}_S}^d \mathscr{O}_S^n$ is a free $\mathscr{O}_S$-module of rank $\binom{n}{d}$. If $\mathscr{U} \subseteq \mathscr{O}_S^n$ is a locally direct summand of rank d, $\bigwedge_{\mathscr{O}_S}^d \mathscr{U}$ is a locally direct summand of $\bigwedge_{\mathscr{O}_S}^d \mathscr{O}_S^n$ of rank 1. Therefore we obtain a morphism of functors

$$\varpi\colon \operatorname{Grass}_{d,n} \longrightarrow \operatorname{Grass}^{\binom{n}{d}-1}(\bigwedge^{d}(\mathscr{O}^{n}_{\operatorname{Spec}(\mathbb{Z})})) \cong \mathbb{P}^{\binom{n}{d}-1}, \qquad \mathscr{U} \mapsto \bigwedge^{d}\mathscr{U}.$$

Both sides are representable functors and therefore ϖ is a morphism of schemes, called the *Plücker embedding*.

Proposition 8.23. *The Plücker embedding is a closed immersion.*

Proof. We identify $\mathscr{O}^{n}_{\operatorname{Spec}(\mathbb{Z})}$ with its dual using the standard basis. Then dualizing (8.6.1) defines an isomorphism $\operatorname{Grass}_{d,n} \xrightarrow{\sim} G := \operatorname{Grass}_{n-d,n}$ and

$$\operatorname{Grass}^{1}(\bigwedge^{d}\mathscr{O}^{n}_{\operatorname{Spec}(\mathbb{Z})}) \xrightarrow{\sim} P := \mathbb{P}(\bigwedge^{d}\mathscr{O}^{n}_{\operatorname{Spec}(\mathbb{Z})}).$$

Via these isomorphism the Plücker embedding is given on S-valued points by

$$\pi(S)\colon G(S) \longrightarrow P(S), \qquad (\mathscr{O}^{n}_{S} \twoheadrightarrow \mathscr{O}^{n}_{S}/\mathscr{U}) \mapsto (\bigwedge^{d}\mathscr{O}^{n}_{S} \twoheadrightarrow \bigwedge^{d}\mathscr{O}^{n}_{S}/\mathscr{U}).$$

For a subset $J \subseteq \{1,\dots,n\}$ with d elements we set $G^{J} := \operatorname{Grass}^{J}_{n-d,n}$. We also define a subfunctor P^{J} of P by

$$P^{J}(S) = \{(\bigwedge^{d}\mathscr{O}^{n}_{S} \twoheadrightarrow \mathscr{L}) \in P(S);\; \bigwedge^{d}\mathscr{O}^{J}_{S} \hookrightarrow \bigwedge^{d}\mathscr{O}^{n}_{S} \to \mathscr{L} \text{ is an isomorphism}\}.$$

In Section (8.4) we have seen that $(G^{J})_{J}$ is an open covering of G. Similarly, $(P^{J})_{J}$ is an open covering of P. As a homomorphism u between locally free modules of rank d is an isomorphism if and only if $\bigwedge^{d}(u)$ is an isomorphism (Corollary 8.12), we see that $\pi(S)^{-1}(P^{J}(S)) = G^{J}(S)$. The property to be a closed immersion is local on the base and therefore it suffices to show that the restriction $\pi^{J}\colon G^{J} \to P^{J}$ of π is a closed immersion.

We set $I := \{1,\dots,n\} \setminus J$. By (8.4.2) and (8.4.3) we have a bijection, functorial in S,

$$G^{J}(S) \xrightarrow{\sim} G'^{J}(S) := \operatorname{Hom}_{\mathscr{O}_{S}}(\mathscr{O}^{I}_{S}, \mathscr{O}^{J}_{S}).$$

Likewise we can identify $P^{J}(S)$ with $\operatorname{Hom}_{\mathscr{O}_{S}}(\mathscr{E}, \bigwedge^{d}\mathscr{O}^{J}_{S})$ if $\mathscr{E}$ is a complement of $\bigwedge^{d}\mathscr{O}^{J}_{S}$ in $\bigwedge^{d}\mathscr{O}^{n}_{S}$. As we have

$$\bigwedge^{d}\mathscr{O}^{n}_{S} = \bigoplus_{i=0}^{d}\mathscr{E}_{i}, \qquad \text{where } \mathscr{E}_{i} := (\bigwedge^{d-i}\mathscr{O}^{J}_{S} \otimes \bigwedge^{i}\mathscr{O}^{I}_{S}),$$

we may choose $\mathscr{E} := \bigoplus_{i=1}^{d}\mathscr{E}_{i}$ and can identify $P^{J}(S)$ with

$$P'^{J}(S) := \bigoplus_{i=1}^{d}\operatorname{Hom}_{\mathscr{O}_{S}}(\mathscr{E}_{i}, \bigwedge^{d}\mathscr{O}^{J}_{S}),$$

functorial in S. Because the $\mathscr{O}_{S}$-modules $\mathscr{E}_{i}$ are free, it follows that $P'^{J} \cong \mathbb{A}^{N}$, where $N = \operatorname{rk}(\mathscr{E}_{1}) + \cdots + \operatorname{rk}(\mathscr{E}_{d})$.

Via the identification of G^{J} with G'^{J} and of P^{J} with P'^{J}, the morphism π is given by $G'^{J}(S) \ni u \mapsto \pi(u) =: (u_{i})_{1 \leq i \leq d} \in P'^{J}(S)$ with

$$u_i \colon \mathscr{E}_i \longrightarrow \overset{d}{\bigwedge} \mathscr{O}_S^J, \quad x \otimes y \mapsto x \wedge \left(\overset{i}{\bigwedge}(u)\right)(y)$$

for $x \in \Gamma(U, \bigwedge^{d-i} \mathscr{O}_S^J)$ and $y \in \Gamma(U, \bigwedge^i \mathscr{O}_S^I)$, U open in S.

Now u and the first component u_1 of $\pi(u)$ determine each other because u corresponds to u_1 via the isomorphism

$$\alpha \colon \mathrm{Hom}_{\mathscr{O}_S}(\mathscr{O}_S^I, \mathscr{O}_S^J) \xrightarrow{\sim} \mathrm{Hom}_{\mathscr{O}_S}\left(\overset{d-1}{\bigwedge} \mathscr{O}_S^J \otimes \mathscr{O}_S^I, \overset{d}{\bigwedge} \mathscr{O}_S^J\right),$$

$$v \longmapsto (x \otimes y \mapsto x \wedge v(y)).$$

Therefore we can identify $G'^J(S)$ with those tuples $(u_i) \in P'^J(S)$ such that

$$u_i(x \otimes y) = x \wedge \left(\overset{i}{\bigwedge} \alpha^{-1}(u_1)\right)(y) \quad \text{for all } i = 2, \dots, d.$$

This shows that $G'^J(S) \cong \Gamma(S, \mathscr{O}_S)^{(n-d)d}$ is the vanishing set in $P'^J(S) \cong \Gamma(S, \mathscr{O}_S)^N$ of certain polynomials which have integer coefficients independent of S (we have to check the equality only for standard basis vectors due to the linearity in x and in y). In other words, G'^J is a closed subscheme of P'^J. $\square$

The theorem implies that $\mathrm{Grass}_{d,n}$ is isomorphic to a closed subscheme of $\mathbb{P}^{\binom{n}{d}-1}$. Using the terminology that we will introduce in Chapter 13, this shows that $\mathrm{Grass}_{d,n}$ is a projective $\mathbb{Z}$-scheme.

Remark 8.24. Let $\mathscr{E}$ be an arbitrary quasi-coherent $\mathscr{O}_S$-module and let $e \geq 0$ be an integer. Let $f \colon T \to S$ be an S-scheme. For each T-valued point $\mathscr{U} = \mathrm{Ker}(f^*\mathscr{E} \twoheadrightarrow \mathscr{Q})$ of $\mathrm{Grass}^e(\mathscr{E})$ we define $\varpi_{\mathscr{E}}(T)(\mathscr{U}) \in \mathbb{P}(\bigwedge^e \mathscr{E})(T)$ as the kernel of $\bigwedge^e f^*\mathscr{E} \to \bigwedge^e \mathscr{Q}$. We obtain a morphism

$$\varpi_{\mathscr{E}} \colon \mathrm{Grass}^e(\mathscr{E}) \to \mathbb{P}(\overset{e}{\bigwedge} \mathscr{E}).$$

It is not difficult to modify the argument in the proof of Proposition 8.23 in order to show that $\varpi_{\mathscr{E}}$ is a closed immersion.

Brauer-Severi schemes

(8.11) Brauer-Severi schemes.

Let k be a field. It is a common phenomenon that, given k-schemes X and Y, there is a field extension K of k such that X_K and Y_K are isomorphic as K-schemes without X and Y being isomorphic k-schemes. In that case we say that X and Y are (K/k)-*forms* of each other. This will be studied in detail in Section (14.22).

Here is an explicit example: the $\mathbb{R}$-schemes $V_+(X_0^2 + X_1^2 + X_2^2)$ and $V_+(X_0^2 + X_1^2 - X_2^2)$ in $\mathbb{P}_{\mathbb{R}}^2$ are not isomorphic (the first one does not have an $\mathbb{R}$-valued point), but over $\mathbb{C}$ they are isomorphic. In fact, over $\mathbb{C}$ they are isomorphic to $\mathbb{P}_{\mathbb{C}}^1$ (see Example 1.72), so this is an instance of the following important class of varieties (see also Exercise 8.16):

Definition 8.25. *A* Brauer-Severi scheme *over k is a k-scheme X such that there exist a field extension K of k and an integer $n \geq 0$ such that the K-schemes $X \otimes_k K$ and $\mathbb{P}^n_K$ are isomorphic.*

Clearly $\mathbb{P}^n_k$ is a Brauer-Severi scheme (use $K = k$). Also note that $\mathbb{P}^n_K \otimes_K K' = \mathbb{P}^n_{K'}$ for every field extension K'/K. By Proposition 5.51 X is geometrically integral because $\mathbb{P}^n_K$ is integral for every field K. (Therefore one often speaks of Brauer-Severi varieties.) We will see in Section (14.23) that every Brauer-Severi scheme X is smooth and of finite type over k and that there exists a finite Galois extension K of k such that $X \otimes_k K \cong \mathbb{P}^n_K$.

(8.12) Morita equivalence.

We will construct Brauer-Severi schemes by the functors which they represent. For that we need a general useful equivalence of categories: the *Morita equivalence*. Let $(S, \mathscr{O}_S)$ be a ringed space. For every $\mathscr{O}_S$-module $\mathscr{E}$ we have defined in Section (7.4) the sheaf of endomorphisms $\mathscr{E}nd(\mathscr{E}) := \mathscr{H}om_{\mathscr{O}_S}(\mathscr{E}, \mathscr{E})$. It is naturally a (usually non-commutative) $\mathscr{O}_S$-algebra (Section (7.7)). The maps

$$\mathscr{E}nd(\mathscr{E})(U) \times \mathscr{E}(U) \to \mathscr{E}(U), \qquad (v, s) \mapsto v_U(s),$$
$$\mathscr{E}^{\vee}(U) \times \mathscr{E}nd(\mathscr{E})(U) \to \mathscr{E}^{\vee}(U), \qquad (\lambda, v) \mapsto \lambda \circ v,$$

for $U \subseteq S$ open, make $\mathscr{E}$ into a left $\mathscr{E}nd(\mathscr{E})$-module and $\mathscr{E}^{\vee}$ into a right $\mathscr{E}nd(\mathscr{E})$-module. We obtain functors

$$F \colon (\mathscr{O}_S\text{-Mod}) \to (\mathscr{E}nd(\mathscr{E})\text{-LeftMod}), \qquad \mathscr{F} \mapsto \mathscr{E} \otimes_{\mathscr{O}_S} \mathscr{F},$$
$$G \colon (\mathscr{E}nd(\mathscr{E})\text{-LeftMod}) \to (\mathscr{O}_S\text{-Mod}), \qquad \mathscr{H} \mapsto \mathscr{E}^{\vee} \otimes_{\mathscr{E}nd(\mathscr{E})} \mathscr{H}.$$

Proposition 8.26. *Assume that $\mathscr{E}$ is a finite locally free $\mathscr{O}_S$-module such that $\mathscr{E}_s \neq 0$ for all $s \in S$. Then the functors F and G are quasi-inverse to each other.*

Proof. As $\mathscr{E}$ is finite locally free, we have $\mathscr{E} \otimes_{\mathscr{O}_S} \mathscr{E}^{\vee} \cong \mathscr{E}nd(\mathscr{E})$ (7.5.4) and therefore $F \circ G \cong \mathrm{id}$. To show that $G \circ F \cong \mathrm{id}$, it suffices to show that the evaluation homomorphism $\mathrm{ev} \colon \mathscr{E}^{\vee} \otimes_{\mathscr{E}nd(\mathscr{E})} \mathscr{E} \to \mathscr{O}_S$ is an isomorphism. This is a local question and can be checked on stalks. Therefore it suffices to show that for a ring R and an integer $n > 0$ the evaluation map $\mathrm{ev} \colon (R^n)^{\vee} \otimes_{M_n(R)} R^n \to R$ is an isomorphism. Identifying $(R^n)^{\vee}$ with the module $M_{1 \times n}(R)$ of $(1 \times n)$-matrices and R^n with the module $M_{n \times 1}(R)$ of $(n \times 1)$-matrices, ev is given by multiplication

$$m \colon M_{1 \times n} \otimes_{M_n(R)} M_{n \times 1}(R) \to R, \qquad A \otimes B \mapsto AB.$$

Clearly m is surjective.

Let $\sum_i A_i \otimes B_i \in \mathrm{Ker}(m)$, that is, $\sum_i A_i B_i = 0$. Fix $D \in M_{1 \times n}(R)$ and $E \in M_{n \times 1}(R)$ such that $DE = 1$. Then

$$\sum_i A_i \otimes B_i = (\sum_i A_i \otimes B_i)DE = \sum_i A_i \otimes (B_i D)E$$

$$= \sum_i A_i(B_i D) \otimes E = (\sum_i A_i B_i)(D \otimes E) = 0. \qquad \square$$

Note that the proof is purely formal and even works if $\mathscr{O}_S$ is a sheaf of not necessarily commutative rings. It is also possible to show the converse of Proposition 8.26: If F and G are quasi-inverse functors, the $\mathscr{O}_S$-module $\mathscr{E}$ is finite locally free with $\mathscr{E}_s \neq 0$ for all $s \in S$.

If $\mathscr{E} = \mathscr{O}_S^n$ we can identify $\mathscr{End}(\mathscr{E})$ with the sheaf of $(n \times n)$-matrices $M_n(\mathscr{O}_S)$ and we obtain an equivalence of the category of $\mathscr{O}_S$-modules and the category of left $M_n(\mathscr{O}_S)$-modules.

(8.13) Construction of Brauer-Severi schemes.

If A is a not necessarily commutative k-algebra (e.g., $A = \mathrm{End}(V)$ for a k-vector space V) and $f \colon S \to \mathrm{Spec}\, k$ is a k-scheme, we often simply write A_S instead of $f^*\tilde{A}$, where $\tilde{A}$ is the quasi-coherent $\mathscr{O}_{\mathrm{Spec}\, k}$-algebra corresponding to A (Remark 7.18). We fix an integer $e \geq 0$ and define a subfunctor $\mathcal{F}^e(A) \colon (\mathrm{Sch}/k)^{\mathrm{opp}} \to (\mathrm{Sets})$ of $\mathrm{Grass}^e(A)$ by sending a k-scheme S to the set of left ideals $\mathscr{I} \subseteq A_S$ such that $A_S/\mathscr{I}$ is a locally free $\mathscr{O}_S$-module of rank e. This functor is representable by a closed subscheme of $\mathrm{Grass}^e(A)$:

Lemma 8.27. *The inclusion* $\mathcal{F}^e(A) \to \mathrm{Grass}^e(A)$ *is representable and a closed immersion.*

Proof. Let S be a k-scheme. We have to show that if $\mathscr{U} \subseteq A_S$ is a submodule such that $A_S/\mathscr{U}$ is finite locally free, there exists a closed subscheme Z of S such that a morphism $f \colon T \to S$ factors through Z if and only if $f^*(\mathscr{U}) \subseteq A_T$ is a left ideal. But $f^*(\mathscr{U})$ is a left ideal if and only if the composition

$$A_T \times f^*\mathscr{U} \hookrightarrow A_T \times A_T \xrightarrow{\mathrm{mult}} A_T \twoheadrightarrow A_T/f^*\mathscr{U}$$

is zero. But the locus where a homomorphism to a finite locally free module is zero, is a closed subscheme by Proposition 8.4 (2). $\qquad\square$

A similar proof shows a much more general version of this lemma; see Exercise 8.14.

We recall (e.g., [We] IX, §1) that a finite-dimensional k-algebra A is called *central and simple* if A satisfies the following equivalent conditions.

(i) The center of A is k and A has no nontrivial two-sided ideals.

(ii) There exists a field extension K of k such that the K-algebra $A \otimes_k K$ is isomorphic to $\mathrm{End}_K(V)$ for some finite-dimensional K-vector space $V \neq 0$.

(iii) There exists a finite Galois extension K of k such that the K-algebra $A \otimes_k K$ is isomorphic to $\mathrm{End}_K(V)$ for some finite-dimensional K-vector space $V \neq 0$.

(iv) There exists a division algebra D with $\mathrm{Cent}(D) = k$ and an integer $r \geq 1$ such that $A \cong M_r(D)$.

Moreover, the isomorphism class of the division algebra D in (iv) is uniquely determined by A as $D = \mathrm{End}_A(M)$ where M is any simple left A-module. It is called the *Brauer class of A* and we denote it by $[A]$. Two central simple k-algebras A and B are called *Brauer equivalent* if $[A] = [B]$.

From now on let A be a central simple k-algebra and K a field extension of k such that $A_K := A \otimes_k K \cong \mathrm{End}_K(V)$. Then $\dim_k(A) = n^2$, where $n = \dim_K(V)$ is called the *degree of A*. The degree of the division algebra D in (iv) is called the *index of A*. Clearly the index of A always divides the degree of A.

For every integer $e \geq 0$ we set $\mathcal{BS}^e(A) := \mathcal{F}^{ne}(A)$. Therefore for every k-algebra R the R-valued points of $\mathcal{BS}^e(A)$ are the left ideals $I \subseteq A \otimes_k R$ such that $(A \otimes_k R)/I$ is finite projective of rank ne.

We claim that $\mathcal{BS}^e(A)$ is a form of $\mathrm{Grass}^e(V)$. More precisely we have

$$(8.13.1) \qquad \mathcal{BS}^e(A) \otimes_k K \cong \mathrm{Grass}^e(V_K).$$

Indeed, we have $\mathcal{BS}^e(A) \otimes_k K = \mathcal{BS}^e(A_K) \cong \mathcal{BS}^e(\mathrm{End}(V_K))$. Let S be a K-scheme, denote by $V_S = (V_K)_S$ the base change, and let

$$\mathscr{I} \subseteq A_S \cong \mathscr{E}nd_{\mathcal{O}_S}(V_S) = V_S^\vee \otimes V_S$$

be a left ideal. By Morita equivalence, $\mathscr{I}$ corresponds to an $\mathcal{O}_S$-submodule $\mathscr{U}$ of V_S such that $V_S/\mathscr{U}$ is locally free of rank e. This defines the functorial bijection between $\mathrm{Grass}^e(V_K)(S)$ and $\mathcal{BS}^e(A_K)(S)$ and hence an isomorphism (8.13.1).

We call $\mathcal{BS}^e(A)$ the e-th *Brauer-Severi scheme attached to* A. It is a closed subscheme of $\mathrm{Grass}^{ne}(A)$. For every field extension L of k we have $\mathcal{BS}^e(A) \otimes_k L = \mathcal{BS}^e(A_L)$ because both sides have the same S-valued points for every L-scheme S.

We also write $\mathcal{BS}(A)$ instead of $\mathcal{BS}^{n-1}(A)$ and call it the *Brauer-Severi scheme attached to* A. This is a Brauer-Severi scheme in the sense of Definition 8.25 as we have $\mathcal{BS}(A) \otimes_k K \cong \mathbb{P}_K^{n-1}$ by (8.13.1). In particular we obtain $\mathcal{BS}(M_n(k)) = \mathbb{P}_k^{n-1}$.

For each fixed integer $n \geq 1$ we obtain a map

$$(8.13.2) \qquad \left\{ \begin{matrix} \text{central simple } k\text{-algebras} \\ \text{of degree } n \end{matrix} \right\} \longrightarrow \left\{ \begin{matrix} \text{isomorphism classes of Brauer-Severi} \\ \text{varieties over } k \text{ that are forms of } \mathbb{P}_k^{n-1} \end{matrix} \right\},$$

$$A \longmapsto \mathcal{BS}(A).$$

As an application of descent theory we will see in Section (14.23) that this map is a bijection. Here we remark only that for a finite-dimensional division algebra D with $\mathrm{Cent}(D) = k$ there exist no nontrivial left ideals. This shows that $\mathcal{BS}^e(D)(k) = \emptyset$ for $0 < e < n$, where n is the index of D (see Exercise 8.15 for a more precise version). In particular we see that $\mathcal{BS}^e(D) \ncong \mathrm{Grass}^e(k^n)$ even though $\mathcal{BS}^e(D)_K \cong \mathrm{Grass}^e(k^n)_K$.

Exercises

Exercise 8.1. Let (Aff) be the category of affine schemes. A contravariant functor $F: (\mathrm{Aff})^{\mathrm{opp}} \to (\mathrm{Sets})$ is called a *Zariski sheaf on* (Aff) if for all affine schemes $X = \mathrm{Spec}\, A$ and for every finite open covering $(U_i)_{1 \leq i \leq n}$ of X with $U_i = D(f_i)$ for $f_i \in A$ the sheaf axiom (Sh) in Section (8.3) holds. By restriction of a functor on (Sch) to (Aff) we obtain a functor from the category $\mathrm{Sh}_{/(\mathrm{Sch})}$ of Zariski sheaves on (Sch) to the category $\mathrm{Sh}_{/(\mathrm{Aff})}$ of Zariski sheaves on (Aff). Show that this induces an equivalence of categories of $\mathrm{Sh}_{/(\mathrm{Sch})}$ and $\mathrm{Sh}_{/(\mathrm{Aff})}$.

Exercise 8.2$\Diamond$.
(a) Let F in $\widehat{(\mathrm{Sch})}$ be a representable functor and let $f: T \to S$ be an epimorphism (Section (A.1)) of schemes. Show that $F(f): F(S) \to F(T)$ is injective.
(b) Let $f: T \to S$ be a scheme morphism such that f is surjective and $f^\flat: \mathcal{O}_S \to f_*(\mathcal{O}_T)$ is injective. Show that f is an epimorphism in the category of schemes.

(c) Let $(U_i)_{i \in I}$ be an open covering of a scheme S. Show that the morphism $\coprod_i U_i \to S$, induced by the immersions $U_i \hookrightarrow S$, is an epimorphism.

(d) Let K be a field and A an K-algebra. Show that the corresponding morphism $\operatorname{Spec} A \to \operatorname{Spec} K$ is an epimorphism.

Remark: It follows from the results in descent theory of schemes (Section (14.17)) that every faithfully flat quasi-compact morphism of schemes is an epimorphism.

Exercise 8.3. Does there exist a scheme X such that $X(k)$ is the set of isomorphism classes of k-schemes of finite type for every field k?
Hint: Use Exercise 8.2.

Exercise 8.4. One motivation for the introduction of schemes was the "geometric globalization" of rings, i.e., we wanted to define "an object" whose "local geometry" is determined by a ring. For that we made rings into locally ringed spaces (affine schemes) and defined schemes as locally ringed spaces that are locally isomorphic to affine schemes. But one might argue that there is no natural reason to work with locally ringed spaces. It might be more natural to embed the category of rings into a category of functors via the Yoneda lemma and define schemes as those functors that are locally isomorphic to a functor given by a ring. In fact that yields the same result:

(a) Define for a (covariant) functor $F \colon (\text{Ring}) \to (\text{Sets})$ when F is called representable or a Zariski sheaf (cf. Exercise 8.1). Define the notion of a (not necessarily finite) open covering of F.

(b) Show that the category of covariant functors $F \colon (\text{Ring}) \to (\text{Sets})$ that are Zariski sheaves and have an open covering by representable functors is equivalent to the category of schemes.

(c) Let (MfStandard) be the category of open subsets of $\mathbb{R}^n$ for n non-fixed, where morphisms are differentiable maps. Identify the category of (not necessarily Hausdorff, locally finite-dimensional) differentiable manifolds with a globalization of (MfStandard) using contravariant functors $F \colon (\text{MfStandard})^{\text{opp}} \to (\text{Sets})$.

Exercise 8.5◇. Let S be a scheme, let $\mathscr{E}$ be a finite locally free $\mathcal{O}_S$-module, and let $\mathscr{G} \subseteq \mathscr{H} \subseteq \mathscr{E}$ be $\mathcal{O}_S$-submodules. Assume that $\mathscr{H}$ is locally a direct summand of $\mathscr{E}$. Show that $\mathscr{G}$ is locally a direct summand of $\mathscr{E}$ if and only if $\mathscr{G}$ is locally a direct summand of $\mathscr{H}$.

Exercise 8.6. Let S be a scheme and let $\mathscr{F}$ be a quasi-coherent $\mathcal{O}_S$-module. Define functors

$$\underline{End}(\mathscr{F}) \colon (\text{Sch}/S)^{\text{opp}} \to (\text{Sets}), \qquad (h \colon T \to S) \mapsto \operatorname{End}_{\mathcal{O}_T}(h^*\mathscr{F}),$$
$$\underline{Aut}(\mathscr{F}) \colon (\text{Sch}/S)^{\text{opp}} \to (\text{Sets}), \qquad (h \colon T \to S) \mapsto \operatorname{Aut}_{\mathcal{O}_T}(h^*\mathscr{F}).$$

(a) Assume that $\mathscr{F}$ is of finite type. Show that the inclusion of functors $\underline{Aut}(\mathscr{F}) \hookrightarrow \underline{End}(\mathscr{F})$ is representable and an open immersion.

(b) Show that $\underline{End}(\mathscr{F})$ and $\underline{Aut}(\mathscr{F})$ are representable if $\mathscr{F}$ is finite locally free.
Hint: To show (a) use Exercise 7.22.
Remark: If $\mathscr{F}$ is of finite presentation, (b) has a converse, see [Nau].

Exercise 8.7. Let S be a scheme, $\mathscr{E}$ be a locally free $\mathscr{O}_S$-module of constant rank $n \geq 0$, and let $0 \leq r \leq n$ be an integer. Show that there exists a subscheme $\mathscr{N}_r$ of the S-scheme $\underline{\mathrm{End}}(\mathscr{E})$ (Exercise 8.6) such that for an S-scheme $h \colon T \to S$ an S-morphism $T \to \underline{\mathrm{End}}(\mathscr{E})$, corresponding to an endomorphism u of $h^*\mathscr{E}$, factors through $\mathscr{N}_r$ if and only if the Fitting decomposition of u (Exercise 7.17) exists and $(h^*\mathscr{E})_{\mathrm{nil}}(u)$ is locally free of rank r. Show that $\mathscr{N}_0$ is the open subscheme $\underline{\mathrm{Aut}}(\mathscr{E})$ and that $\mathscr{N}_n$ is a closed subscheme.

$\mathscr{N}_n$ is the called the *scheme of nilpotent endomorphisms* or the *nilpotent cone in* $\underline{\mathrm{End}}(\mathscr{E})$.

Hint: Let $\sigma \colon \underline{\mathrm{End}}(\mathscr{E}) \to \mathbb{A}_S^n$ be the morphism of S-schemes which is given on T-valued points (T an S-scheme) by $u \mapsto (\mathrm{tr}(\bigwedge^i(u)))_{1 \leq i \leq n}$ (cf. Exercise 7.16). Let $Z_r \subset \mathbb{A}_S^n$ be the subscheme $(\mathbb{A}_S^1 \setminus \{0\})^{n-r} \times \{0\}^r$ and show that $\mathscr{N}_r = \sigma^{-1}(Z_r)$.

Exercise 8.8. Let S be a scheme, let $\mathscr{F}$ be a quasi-coherent $\mathscr{O}_S$-module, and let $\mathscr{G} \subseteq \mathscr{F}$ be a quasi-coherent $\mathscr{O}_X$-submodule such that $\mathscr{F}/\mathscr{G}$ is finite locally free. Show that the locus where an endomorphism $u \colon \mathscr{F} \to \mathscr{F}$ satisfies $u(\mathscr{G}) \subseteq \mathscr{G}$ is a closed subscheme of S (i.e., there exists a closed subscheme Z of S such that a morphism $h \colon T \to S$ factors through Z if and only if $h^*(u)(h^*\mathscr{G}) \subseteq h^*\mathscr{G}$).

Exercise 8.9. For a finite locally free $\mathscr{O}_S$-module $\mathscr{E}$ let $G = W(\mathscr{E})$ be the associated *vector group*, i.e., for every S-scheme $h \colon T \to S$ we set $W(\mathscr{E})(T) := \Gamma(T, h^*\mathscr{E})$, considered as a group via the addition. Show that $W(\mathscr{E})$ is representable by a group scheme.

Exercise 8.10◇. Let $n \geq d \geq 1$ be integers.
(a) Let k be an algebraically closed field. Show that the k-scheme $\mathrm{Grass}_{d,n} \otimes_{\mathbb{Z}} k$ is an integral regular projective k-scheme. The points of the corresponding projective variety correspond to the set of d-dimensional subspaces of k^n.
(b) Show that $\mathrm{Grass}_{d,n}(k)$ can be also considered as the set of $(d-1)$-dimensional linear subspaces of $\mathbb{P}^{n-1}(k)$ (see Section (1.23)).

Exercise 8.11. Let S be a scheme, let $\mathscr{E}$ be a quasi-coherent $\mathscr{O}_S$-module, and let $e \geq 0$ be an integer. Show that $\mathrm{Grass}^e(\mathscr{E})$ is representable. The following steps might be useful.
(a) Let $\mathscr{G} \subseteq \mathscr{E}$ be a quasi-coherent submodule. Define a subfunctor $\mathrm{Grass}^e(\mathscr{E})_{\mathscr{G}}$ of $\mathrm{Grass}^e(\mathscr{E})$ by

$$\mathrm{Grass}^e(\mathscr{E})_{\mathscr{G}}(T) := \{\, \mathscr{U} \in \mathrm{Grass}^e(\mathscr{E})(T) \; ; \; f^*\mathscr{G} \to f^*\mathscr{E} \twoheadrightarrow f^*\mathscr{E}/\mathscr{U} \text{ is surjective} \,\}$$

for any S-scheme $f \colon T \to S$. Show that the morphism $\mathrm{Grass}^e(\mathscr{E})_{\mathscr{G}} \to \mathrm{Grass}^e(\mathscr{E})$ of functors is representable and an open immersion.
(b) Check that $\mathrm{Grass}^e(\mathscr{E})$ is a sheaf for the Zariski topology and deduce that it suffices to show the representability of $\mathrm{Grass}^e(\mathscr{E})$ if $S = \mathrm{Spec}\, R$ is affine and $\mathscr{E} = \tilde{E}$ for an R-module E.
(c) Let I be the set of finitely generated R-submodules $G \subseteq E$. Show that the open subfunctors $(\mathrm{Grass}^e(\mathscr{E})_{\tilde{G}})_{G \in I}$ form an open covering of $\mathrm{Grass}^e(\mathscr{E})$ and deduce that $\mathrm{Grass}^e(\mathscr{E})$ is representable.

Exercise 8.12. Let k be an algebraically closed field. Let $n \geq d \geq 1$ be integers and let $T := \mathcal{F}_d(n)$ be the set of subsets $J \subseteq \{1, \ldots, n\}$ with d elements. Denote by $\mathbb{P}((k^T)^\vee)$ the projective space of lines through the origin in k^T.
(a) For every $(n \times d)$-matrix $A \in \mathrm{M}_{n \times d}(k) \cong \mathbb{A}^{nd}(k)$ let U_A be the subspace of k^n generated by the column vectors of A. Show that $V := \{\, A \in \mathrm{M}_{n \times d}(k) \; ; \; A \text{ has rank } d \,\}$ is an open subvariety of $\mathrm{M}_{n \times d}(k)$, and that the map $V \to \mathrm{Grass}_{d,n}(k)$, $A \mapsto U_A$ is a morphism of prevarieties.

(b) For $J \in \mathcal{F}_d(n)$ and $A \in M_{n \times d}(k)$ let $A_J \in M_{d \times d}(k)$ be the matrix that consists only of the j-th rows of A for $j \in J$. Show that the morphism $V \to \mathbb{A}^T(k)$, $A \mapsto (\det(A_J))_{J \in T}$ induces the Plücker embedding $\varpi(k) \colon \mathrm{Grass}_{d,n}(k) \to \mathbb{P}((k^T)^\vee)$.

(c) Let x_J for $J \in \mathcal{F}_d(n)$ be the coordinates on $\mathbb{P}(k^T)$. Let $(j_1, \ldots, j_d)$ be a tuple of integers $j_i \in \{1, \ldots, n\}$. If the j_i are pairwise different, we set $x_{j_1, \ldots, j_d} := \varepsilon x_J$, where $J = \{j_1, \ldots, j_d\}$ and $\varepsilon = \varepsilon(j_1, \ldots, j_d)$ is the sign of the permutation σ of $\{1, \ldots, d\}$ such that $j_{\sigma(1)} < \cdots < j_{\sigma(d)}$. Otherwise we set $x_{j_1, \ldots, j_d} := 0$. Show that the image of the Plücker embedding is the closed subvariety of $\mathbb{P}((k^T)^\vee)$ that is given by the quadratic homogeneous polynomials

$$\sum_{\alpha=1}^{d+1} (-1)^\alpha x_{i_1, \ldots, i_{d-1}, j_\alpha} x_{j_1, \ldots, \hat{j}_\alpha, \ldots, j_{d+1}} = 0,$$

for all sequences $i_1 < \cdots < i_{d-1}$ and $j_1 < \cdots < j_{d+1}$ of integers in $\{1, \ldots, n\}$.

(d) Deduce that the Plücker embedding identifies the variety $\mathrm{Grass}_{2,4}(k)$ with the quadric

$$V_+(x_{\{1,2\}} x_{\{3,4\}} - x_{\{1,3\}} x_{\{2,4\}} + x_{\{1,4\}} x_{\{2,3\}})$$

in $\mathbb{P}((k^{\mathcal{F}_2(4)})^\vee) \cong \mathbb{P}^5(k)$.

Exercise 8.13. Let k be an algebraically closed field and let $\mathcal{G}$ be the Grassmannian variety $\mathrm{Grass}_{2,4}(k)$ over k.

(a) Let $p \in \mathbb{P}^3(k)$ be a point and $H \subset \mathbb{P}^3(k)$ be a plane that contains p. Let $\Sigma_{p,H} \subset \mathcal{G}$ be the set of lines $\mathbb{P}^3(k)$ that lie in H and that meet p. Show that $\Sigma_{p,H}$ is mapped by the Plücker embedding $\varpi \colon \mathcal{G} \hookrightarrow \mathbb{P}^5(k)$ to a line in $\mathbb{P}^5(k)$ and that every line in the image of ϖ arises in that way.

(b) For $p \in \mathbb{P}^3(k)$ let $\Sigma_p \subset \mathcal{G}$ be the set of lines in $\mathbb{P}^3(k)$ that contain p. For every plane $H \subset \mathbb{P}^3(k)$ let $\Sigma_H \subseteq \mathcal{G}$ be the set of lines that are contained in H. Show that Σ_p and Σ_H are mapped by the Plücker embedding to planes in $\mathbb{P}^5(k)$ and that every plane in $\mathbb{P}^5(k)$ that is contained in the image of ϖ is the image of some Σ_p or Σ_H.

Exercise 8.14. Let S be a scheme, let $\mathscr{A}$ be a quasi-coherent $\mathcal{O}_S$-algebra (not necessarily commutative), and let $\mathscr{E}$ be a quasi-coherent $\mathcal{O}_S$-module with an $\mathscr{A}$-action (i.e., a homomorphism of $\mathcal{O}_S$-algebras $\rho \colon \mathscr{A} \to \mathcal{E}nd_{\mathcal{O}_S}(\mathscr{E})$). For each integer $e \geq 1$ define a functor $\mathrm{Grass}^e_{\mathscr{A}}(\mathscr{E})$ by sending an S-schemes $f \colon T \to S$ to the set of $f^*(\mathscr{A})$-submodules $\mathscr{U} \subseteq f^*\mathscr{E}$ such that $f^*\mathscr{E}/\mathscr{U}$ is a locally free $\mathcal{O}_T$-module of rank e. Show that $\mathrm{Grass}^e_{\mathscr{A}}(\mathscr{E})$ is representable by a closed subscheme of $G := \mathrm{Grass}^e(\mathscr{E})$ and that the quasi-coherent ideal $\mathscr{I} \subseteq \mathcal{O}_G$ defining $\mathrm{Grass}^e_{\mathscr{A}}(\mathscr{E})$ is of finite type if $\mathscr{A}$ is locally finitely generated (as an $\mathcal{O}_S$-algebra).

Exercise 8.15. Let k be a field, A a finite-dimensional central simple k-algebra of degree n, and $1 \leq e \leq n - 1$ an integer. Show that $\mathcal{BS}^e(A)$ has a K-valued point for a field extension K of k if and only if the index of A_K divides e. Deduce

$$\mathcal{BS}(A)(k) \neq \emptyset \Leftrightarrow A \cong M_n(k) \Leftrightarrow \mathcal{BS}(A) \cong \mathbb{P}^{n-1}_k.$$

Exercise 8.16. Let k be a field of characteristic not 2. For $a, b \in k^\times$ define (a, b) as the 4-dimensional k-algebra with basis $1, i, j, k$, where the multiplication is determined by $i^2 = a$, $j^2 = b$, $k = ij = -ji$. (The Hamilton quaternions are the special case $k = \mathbb{R}$ and $a = b = -1$.)

(a) Show that (a, b) is a central simple k-algebra which is either isomorphic to $M_2(k)$ or a division algebra. Conversely, every 4-dimensional central simple k-algebra is isomorphic to (a, b) for some $a, b \in k^\times$.

(b) Let $C(a, b) \subset \mathbb{P}^2_k$ be the *associated cone*, that is, the closed subscheme given by the homogeneous equation $ax_0^2 + bx_1^2 - x_2^2$. Show that if two quaternion k-algebras (a, b) and (a', b') are isomorphic, their associated cones $C(a, b)$ and $C(a', b')$ are isomorphic. (The converse is also true, see Theorem 14.95).

(c) Show that $C(a, b)$ is a form of $\mathbb{P}^1_k$ and in particular a Brauer-Severi scheme. Show that $C(a, b) \cong \mathbb{P}^1_k \Leftrightarrow (a, b) \cong M_2(k) \Leftrightarrow b$ is a norm of the extension $k[\sqrt{a}] \supseteq k$.

9 Separated morphisms

Contents

- Diagonal of scheme morphisms and separated morphisms
- Rational maps and function fields

Recall that a topological space X is Hausdorff if and only if the following equivalent conditions are satisfied.

(i) The diagonal $\{(x,x) \; ; \; x \in X\}$ is closed in $X \times X$ (with respect to the product topology).

(ii) For every topological space Y and every continuous map $f\colon Y \to X$ its graph $\{(y, f(y)) \; ; \; y \in Y\}$ is closed in $Y \times X$.

(iii) For every topological space Y and any two continuous maps $f, g\colon Y \to X$ the equalizer $\{y \in Y \; ; \; f(y) = g(y)\}$ is closed in Y.

Now the underlying topological spaces of schemes are rarely Hausdorff but the analogues of the properties (i)–(iii) can be used to define an analogue of the Hausdorff property for schemes. Since the topology on fiber products of schemes is (usually) not the product topology, this gives rise to the different (and in fact very useful) notion of a separated scheme. We will start in this chapter with the definition of diagonal, graph, and equalizer for morphisms of schemes and then define the notion of a separated morphism in analogy to (i)–(iii). Almost all schemes and morphisms encountered in practice will turn out to be separated, but in particular if one uses gluing constructions, then this might not be obvious.

If $f, g\colon X \to Y$ are morphisms of S-schemes and $U \subseteq X$ is an open dense subscheme such that $f_{|U} = g_{|U}$, then – in contrast to the analogous statement for topological spaces – this does not imply that $f = g$, even if Y is separated over S. It only implies that f and g coincide on a closed subscheme Z of X whose underlying topological space is the same as that of X (Corollary 9.9). If X is reduced, then one has necessarily $Z = X$ but for non-reduced schemes we will define the stronger notion of a *schematically dense* open subscheme U. For locally noetherian schemes, schematic density can be expressed in terms of associated prime ideals.

The second part of this chapter deals with *rational maps* $X \dashrightarrow Y$, i.e., morphisms to Y that are only defined on some schematically dense open subscheme. This leads us to the notion of birational equivalence of schemes X and Y which means that there exist schematically dense open subschemes $U \subseteq X$ and $V \subseteq Y$ such that $U \cong V$. For integral schemes X and Y of finite type over a field k we will prove that X and Y are birationally equivalent if and only if their functions fields are isomorphic.

© Springer Fachmedien Wiesbaden GmbH, part of Springer Nature 2020
U. Görtz und T. Wedhorn, *Algebraic Geometry I: Schemes*, Springer Studium
Mathematik – Master, https://doi.org/10.1007/978-3-658-30733-2_10

Diagonal of scheme morphisms and separated morphisms

(9.1) Diagonals, graphs, and equalizers in arbitrary categories.

Let $\mathcal{C}$ be a category where arbitrary fiber products exist. Let S be an object of $\mathcal{C}$. Recall that if X and T are S-objects, we write $X_S(T)$ for the set of S-morphisms $T \to X$.

Definition 9.1.
(1) *Let* $u\colon X \to S$ *be an S-object. The morphism*

$$(9.1.1) \qquad \Delta_{X/S} := \Delta_u := (\mathrm{id}_X, \mathrm{id}_X)_S \colon X \to X \times_S X$$

is called the diagonal (morphism) *of X over S.*

(2) *Let* $f\colon X \to Y$ *be a morphism of S-objects. The morphism*

$$(9.1.2) \qquad \Gamma_f := (\mathrm{id}_X, f)_S \colon X \to X \times_S Y$$

is called the graph (morphism) *of f.*

(3) *Let* $f, g\colon X \to Y$ *be two S-morphisms. An S-object K together with an S-morphism $i\colon K \to X$ is called* equalizer *(or* difference kernel*) of f and g if for all S-objects T the map $i(T)$ yields a bijection*

$$K_S(T) \overset{\sim}{\to} \{\, x \in X_S(T) \; ; \; f(T)(x) = g(T)(x) \,\}$$

We denote the equalizer of f and g by $\mathrm{Eq}(f, g)_S$ *or simply* $\mathrm{Eq}(f, g)$ *and call the morphism* $i\colon \mathrm{Eq}(f, g) \to X$ *the* canonical morphism.

Thus the equalizer of f and g represents the contravariant functor which sends an S-object T to the set $\{\, x \in X_S(T) \; ; \; f(T)(x) = g(T)(x) \,\}$. We will see in Proposition 9.3 that the equalizer can be described as a fiber product, in particular it always exists in $\mathcal{C}$. For all S-objects T the canonical morphism $\mathrm{Eq}(f, g) \to X$ is injective on T-valued points, in other words, it is a monomorphism in the category of S-objects.

Example 9.2. If $\mathcal{C}$ is the category of sets, the diagonal of an S-object $u\colon X \to S$ and graph of an S-map $f\colon X \to Y$ to an S-object $v\colon Y \to S$ are given by the usual diagonal and graph

$$\Delta_u\colon X \to X \times_S X = \{\, (x, x') \in X \times X \; ; \; u(x) = u(x') \,\}, \qquad x \mapsto (x, x);$$
$$\Gamma_f\colon X \to X \times_S Y = \{\, (x, y) \in X \times Y \; ; \; u(x) = v(y) \,\}, \qquad x \mapsto (x, f(x)).$$

If $g\colon X \to Y$ is a second S-morphism, we have

$$\mathrm{Eq}(f, g) = \{\, x \in X \; ; \; f(x) = g(x) \,\}.$$

If $p\colon X \times_S Y \to X$ is the first projection, we have $p \circ \Gamma_f = \mathrm{id}_X$. In particular Γ_f and $\Delta_{X/S} = \Gamma_{\mathrm{id}_X}$ are monomorphisms. Diagonal, graph, and equalizer are related as follows.

Proposition 9.3. *Let* $u\colon X \to S$, $v\colon Y \to S$ *be S-objects, let* $p\colon X \times_S Y \to X$ *and* $q\colon X \times_S Y \to Y$ *be the projections, and* $f, g\colon X \to Y$ *two S-morphisms.*

(1)

(9.1.3) $\Delta_{X/S} = \Gamma_{\mathrm{id}_X}$, $\Gamma_f = (\mathrm{can}\colon \mathrm{Eq}(\, X \times_S Y \overset{q}{\underset{f \circ p}{\rightrightarrows}} Y\,) \to X \times_S Y)$.

(2) *All rectangles of the following diagram are cartesian.*

(9.1.4)

$$
\begin{array}{ccccc}
\mathrm{Eq}(f,g) & \xrightarrow{\ \mathrm{can}\ } & X & \xrightarrow{\ f\ } & Y \\
{\scriptstyle \mathrm{can}}\downarrow & \square & {\scriptstyle \Gamma_f}\downarrow & \square & \downarrow{\scriptstyle \Delta_{Y/S}} \\
X & \xrightarrow[\ \Gamma_g\ ]{} & X \times_S Y & \xrightarrow[\ f \times \mathrm{id}_Y\ ]{} & Y \times_S Y.
\end{array}
$$

(3) *Let* $s\colon S \to X$ *be a section of* u *(i.e.,* $u \circ s = \mathrm{id}_S$*). The following diagram is cartesian.*

(9.1.5)

$$
\begin{array}{ccc}
S & \xrightarrow{\ s\ } & X \\
{\scriptstyle s}\downarrow & \square & \downarrow{\scriptstyle \Gamma_{s \circ u}} \\
X & \xrightarrow[\ \Delta_{X/S}\ ]{} & X \times_S X.
\end{array}
$$

Proof. Using the Yoneda lemma it suffices to treat the case that $\mathcal{C}$ is the category of sets (Remark 4.15), where the claims follow from the explicit description given in Example 9.2. $\square$

(9.2) Diagonal, graph, and equalizers for morphisms of schemes.

In the category of schemes arbitrary fiber products exist (Theorem 4.18). Therefore diagonal, graph, and equalizer exist. For affine schemes diagonal and graph are described as follows.

Proposition 9.4. *Let* $S = \mathrm{Spec}\, R$ *be an affine scheme, let* $X = \mathrm{Spec}\, B \to S$ *and* $Y = \mathrm{Spec}\, A \to S$ *be affine* S-*schemes and let* $f\colon X \to Y$ *be an* S-*morphism corresponding to an* R-*algebra morphism* $\varphi\colon A \to B$*. Then the diagonal morphism* $\Delta_{X/S}$ *and graph morphism* Γ_f *correspond to the following surjective ring homomorphisms.*

$$
\begin{aligned}
\Delta_{B/R}\colon B \otimes_R B \to B, & \qquad\qquad b \otimes b' \mapsto bb', \\
\Gamma_\varphi\colon A \otimes_R B \to B, & \qquad\qquad a \otimes b \mapsto \varphi(a)b.
\end{aligned}
$$

In particular $\Delta_{X/S}$ *and* Γ_f *are closed immersions.*

In general, $\Delta_{X/S}$ and Γ_f are still immersions but not necessarily closed. To show this we first remark that if S is a scheme, X an S-scheme and $Z, Z' \subseteq X$ are subschemes, we may consider $Z \times_S Z'$ as a subscheme of $X \times_S X$ (immersions are stable under base change and composition) and we have an equality of subschemes of X,

(9.2.1) $Z \cap Z' = \Delta_{X/S}^{-1}(Z \times_S Z')$.

Proposition 9.5. *Let* S *be a scheme, let* X *and* Y *be* S-*schemes, and let* $f, g\colon X \to Y$ *be morphisms of* S-*schemes. Then* $\Delta_{X/S}$, Γ_f, *and the canonical morphism* $\mathrm{Eq}(f,g) \to X$ *are immersions.*

Proof. As being an immersion is stable under base change, the cartesian diagram (9.1.4) shows that it suffices to prove that $\Delta_{X/S}$ is an immersion. As the question is local on the target, we may assume that S is affine. If $X = \bigcup_{i \in I} U_i$ is an open covering, the open subschemes $U_i \times_S U_i$ form an open covering of $\Delta_{X/S}(X)$. Thus by (9.2.1) we may assume that X is also affine. But in this case we have already seen that $\Delta_{X/S}$ is an immersion. $\square$

Therefore $\Delta_{X/S}$ yields an isomorphism of X onto a subscheme of $X \times_S X$ which we call the *diagonal* of $X \times_S X$. Similarly, we call the subscheme of $X \times_S Y$ attached to the immersion Γ_f the *graph of f*. Finally, we will usually consider $\mathrm{Eq}(f, g)$ as a subscheme of X.

Remark 9.6.
(1) A subscheme Γ of $X \times_S Y$ is the graph of an S-morphism $f \colon X \to Y$ if and only if the restriction of the first projection $p \colon X \times_S Y \to X$ to Γ is an isomorphism (in this case $f = q \circ (p_{|\Gamma})^{-1}$, where q is the second projection).
(2) Let $p, q \colon X \times_S X \to X$ be the projections, and $\Delta \subseteq X \times_S X$ the diagonal. Then we have an inclusion of sets $\Delta \subseteq \{ z \in X \times_S X \; ; \; p(z) = q(z) \}$ but this inclusion is in general not an equality (Exercise 9.4).

(9.3) Separated morphisms and separated schemes.

In analogy of the notion of a Hausdorff space we define now the notion of a separated morphism.

Definition and Proposition 9.7. *A morphism of schemes $v \colon Y \to S$ is called* separated *if the following equivalent conditions are satisfied.*
(i) *The diagonal morphism $\Delta_{Y/S}$ is a closed immersion.*
(ii) *For every S-scheme X and for any two S-morphisms $f, g \colon X \to Y$ the equalizer $\mathrm{Eq}(f, g) \subseteq X$ is a closed subscheme of X.*
(iii) *For every S-scheme X and for any S-morphism $f \colon X \to Y$ its graph Γ_f is a closed immersion.*
In this case we also say that Y is separated *over S. A scheme Y is called* separated *if it is separated over $\mathbb{Z}$.*

Proof. The equivalence of (i), (ii), and (iii) follows from Proposition 9.3 and the fact that being a closed immersion is stable under base change. $\square$

Remark 9.8. Proposition 9.4 shows that every morphism between affine schemes is separated. In particular every affine scheme is separated.

Corollary 9.9. *Let S be a scheme, X and Y two S-schemes. Assume that Y is separated over S. Let $U \subseteq X$ be an open dense subscheme, and $f, g \colon X \to Y$ two S-morphisms such that $f_{|U} = g_{|U}$. Then we have $f_{|X_{\mathrm{red}}} = g_{|X_{\mathrm{red}}}$.*

Proof. By hypothesis, U is majorized by $\mathrm{Eq}(f, g)$ (Definition 3.46). As Y is separated over S, $\mathrm{Eq}(f, g)$ is a closed subscheme of X. As U is dense in X, the underlying topological space of $\mathrm{Eq}(f, g)$ has to be X. This implies $X_{\mathrm{red}} \subseteq \mathrm{Eq}(f, g)$. $\square$

Example 9.10. Let U be a scheme and let $V \subseteq U$ be an open subscheme. Let X be the scheme obtained from gluing two copies of U along V (Example 3.13). Assume that V is not closed in U (e.g., if X is the affine line with a double point in loc. cit.). We claim that X is not separated. Indeed, let $j, j' \colon U \to X$ be the open immersions whose images are the two copies of U. Then $\mathrm{Eq}(j, j') = V$ and therefore X cannot be separated.

Remark 9.11. Let **P** be a property of morphisms of schemes that is stable under composition and stable under base change. Moreover, we assume that every immersion (resp. closed immersion) possesses **P**.
(1) Then for every commutative diagram of schemes

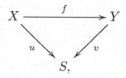

where u possesses **P** (resp. where u possesses **P** and v is separated), f also possesses **P**. Indeed, f can be written as the composition

$$X \xrightarrow{\Gamma_f} X \times_S Y \xrightarrow{q} Y,$$

where $q = u_{(Y)}$ is the second projection.
(2) If $f \colon X \to Y$ is a morphism that possesses **P**, then f_{red} possesses **P**.

    Indeed, let $i_X \colon X_{\mathrm{red}} \to X$ and $i_Y \colon Y_{\mathrm{red}} \to Y$ be the canonical closed immersions. As f possesses **P**, $f \circ i_X = i_Y \circ f_{\mathrm{red}}$ possesses **P**. Therefore f_{red} possesses **P** by (1).

Example 9.12. The Remark 9.11 can be applied to **P** being the property to be a (closed) immersion. Thus we see that if a composition $g \circ f$ is an immersion (resp. is a closed immersion and g is separated), then f is an immersion (resp. a closed immersion).

    In particular, if $g \colon X \to S$ is a morphism (resp. a separated morphism), any section of g (i.e., any morphism $i \colon S \to X$ with $g \circ i = \mathrm{id}_S$) is an immersion (resp. a closed immersion).

Proposition 9.13.
(1) *Every monomorphism of schemes (and in particular every immersion) is separated.*
(2) *The property of being separated is stable under composition, stable under base change, and local on the target.*
(3) *If the composition $X \to Y \to Z$ of two morphisms is separated, $X \to Y$ is separated.*
(4) *A morphism $f \colon X \to Y$ is separated if and only if $f_{\mathrm{red}} \colon X_{\mathrm{red}} \to Y_{\mathrm{red}}$ is separated.*

Proof. (1). If f is a monomorphism (i.e., injective on T-valued points for all schemes T), Δ_f is an isomorphism (i.e., bijective on T-valued points for all T). In particular, Δ_f is a closed immersion.

    (2). Let $f \colon X \to Y$ and $g \colon Y \to Z$ be two separated morphisms of schemes, $p, q \colon X \times_Y X \to X$ the two projections. The following diagram is commutative and the square on the right hand side is cartesian (easily checked in the category of sets)

(9.3.1)
$$X \xrightarrow{\Delta_f} X \times_Y X \xrightarrow{f \circ p = f \circ q} Y$$

$$\Delta_{g \circ f} \searrow \quad \downarrow (p,q)_Z \quad \square \quad \downarrow \Delta_g$$

$$X \times_Z X \xrightarrow{f \times f} Y \times_Z Y.$$

As Δ_g is a closed immersion, $(p,q)_Z$ is a closed immersion and therefore its composition $\Delta_{g \circ f}$ with the closed immersion Δ_f is a closed immersion. This shows that $g \circ f$ is separated.

If $f \colon X \to S$ is a separated morphism and $S' \to S$ a morphism, we have

$$(\Delta_f)_{(S')} = \Delta_{f_{(S')}} \colon X \times_S S' \to (X \times_S S') \times_{S'} (X \times_S S') = (X \times_S X) \times_S S'.$$

As Δ_f is a closed immersion, $\Delta_{f_{(S')}}$ is a closed immersion. This shows that being separated is stable under base change. Finally, being separated is local on the target, as follows now easily from the fact that the same is true for closed immersions.

(3). Assertion (3) follows via Remark 9.11 from (1) and (2).

(4). Let $f \colon X \to S$ be a morphism and let $i \colon X_{\mathrm{red}} \to X$ be the canonical immersion. Then i is a surjective immersion and thus a universal homeomorphism. Identifying $X_{\mathrm{red}} \times_{S_{\mathrm{red}}} X_{\mathrm{red}}$ with $X_{\mathrm{red}} \times_S X_{\mathrm{red}}$ we have $\Delta_f \circ i = (i \times_S i) \circ \Delta_{f_{\mathrm{red}}}$. Therefore Δ_f is a closed immersion if and only if $\Delta_{f_{\mathrm{red}}}$ is a closed immersion. $\qquad\square$

Example 9.14. Let S be a scheme. As $\mathbb{A}_{\mathbb{Z}}^n$ is an affine scheme and therefore separated (over $\mathbb{Z}$), the affine space $\mathbb{A}_S^n = \mathbb{A}_{\mathbb{Z}}^n \times_{\mathbb{Z}} S$ is separated over S (being separated is stable under base change). More generally, every subscheme of $\mathbb{A}_S^n$ is separated over S.

Proposition 9.15. *Let $S = \operatorname{Spec} R$ be an affine scheme and let X be an S-scheme. Then the following assertions are equivalent.*
(i) *X is separated.*
(ii) *For every two open affine sets $U, V \subseteq X$ the intersection $U \cap V$ is affine and*

$$\rho_{U,V} \colon \Gamma(U, \mathscr{O}_X) \otimes_R \Gamma(V, \mathscr{O}_X) \to \Gamma(U \cap V, \mathscr{O}_X), \qquad (s,t) \mapsto s_{|U \cap V} \cdot t_{|U \cap V}$$

is surjective.
(iii) *There exists an open affine covering $X = \bigcup_i U_i$ such that $U_i \cap U_j$ is affine and $\rho_{U_i, U_j} \colon \Gamma(U_i, \mathscr{O}_X) \otimes_R \Gamma(U_j, \mathscr{O}_X) \to \Gamma(U_i \cap U_j, \mathscr{O}_X)$ is surjective for all i, j.*

Proof. For all open subschemes $U, V \subseteq X$ we have $\Delta_{X/S}^{-1}(U \times_S V) = U \cap V$ by (9.2.1). As being a closed immersion is local on the target, $\Delta_{X/S}$ is a closed immersion if and only if its restriction $U \cap V \to U \times_S V$ is a closed immersions for all pairs (U, V) of open subschemes or, equivalently, for all pairs (U_i, U_j) for $i, j \in I$ if $X = \bigcup U_i$ is an open covering (note that $X \times_S X = \bigcup_{i,j} U_i \times_S U_j$ by Corollary 4.19).

But if $U = \operatorname{Spec} A$ and $V = \operatorname{Spec} B$ are affine, $U \times_S V = \operatorname{Spec}(A \otimes_R B)$ is affine as well. Thus the restriction of $\Delta_{X/S}$ to a morphism $U \cap V \to U \times_S V$ is a closed immersion if and only if $U \cap V$ is affine and the induced homomorphism is surjective (Theorem 3.42). $\qquad\square$

(9.4) Examples of separated schemes.

Example 9.16. (Projective spaces and the Grassmannian) Let S be a scheme and let $\mathscr{E}$ be a finite locally free $\mathscr{O}_S$-module. Then $\mathbb{P}(\mathscr{E})$ is separated over S and in particular $\mathbb{P}_S^n$ is separated over S for all integers $n \geq 0$. Indeed, the question is local on S and we may assume that $S = \operatorname{Spec} R$ is affine and that $\mathscr{E}$ is a finite free $\mathscr{O}_S$-module. Thus we have $\mathbb{P}(\mathscr{E}) \cong \mathbb{P}_R^n$ for some n. Let $\mathbb{P}_R^n = \bigcup_{i=0}^n U_i$ be the standard open affine covering. Then the explicit description in Section (3.6) shows that condition (iii) of Proposition 9.15 is satisfied for this covering.

More generally, let $e \geq 0$ be an integer and let $\mathscr{E}$ be a quasi-coherent $\mathscr{O}_S$-module of finite type. Then the Grassmannian $\mathrm{Grass}^e(\mathscr{E})$ is separated. Indeed, we may assume that $S = \mathrm{Spec}\, R$ is affine. Then $\mathscr{E} = \tilde{M}$, where M is a finitely generated R-module. Let $u \colon R^n \twoheadrightarrow M$ be a surjection of R-modules for some $n \geq 0$. By Proposition 8.17, u yields a closed immersion

$$\mathrm{Grass}^e(\mathscr{E}) \hookrightarrow \mathrm{Grass}^e((R^n)^{\sim}) = \mathrm{Grass}_{n-e,n} \otimes_{\mathbb{Z}} R.$$

As immersions are separated, it suffices to show that $\mathrm{Grass}_{n-e,n}$ is a separated scheme. The Plücker embedding $\varpi \colon \mathrm{Grass}_{n-e,n} \to \mathbb{P}^{\binom{n}{e}-1}$ is a closed immersion (Section (8.10)). As $\mathbb{P}^{\binom{n}{e}-1}$ is separated, $\mathrm{Grass}_{n-e,n}$ is separated.

In fact, it can be shown that $\mathrm{Grass}^e(\mathscr{E})$ is separated over S for every quasi-coherent $\mathscr{O}_S$-module $\mathscr{E}$ ([EGAInew] (9.7.7)).

Example 9.17. (Varieties) Let k be an algebraically closed field and let X be a prevariety over k (which we view as an integral k-scheme of finite type via Theorem 3.37). We call X a *variety* if X is separated over k. As affine schemes are separated, a prevariety that is affine is a variety. Moreover, Example 9.16 shows that $\mathbb{P}^n(k)$ is a variety. As subschemes of separated schemes are separated (Proposition 9.13) we see that any prevariety that is quasi-projective is indeed a variety. That explains why we did not speak of affine (or quasi-projective, or projective) *prevarieties*.

Rational maps and function fields

Sometimes it is useful to consider, instead of morphisms, "morphisms" which are not everywhere defined but only on an open dense set. As an example from complex geometry, meromorphic functions on a complex manifold X are defined only on the open dense subset that is the complement of their poles. Below we discuss this concept in the setting of algebraic geometry. We start by discussing the scheme-theoretic version of the notion "open and dense".

(9.5) Schematically dense open subschemes.

If Y is a Hausdorff topological space, and $f, g \colon X \to Y$ are continuous maps of topological spaces such that there exists a dense subset $U \subseteq X$ with $f_{|U} = g_{|U}$, then we have $f = g$. The naive generalization to schemes does not hold: If Y is separated over a scheme S, X is an S-scheme, and $f, g \colon X \to Y$ are S-morphisms coinciding on an open dense subscheme U we saw in Corollary 9.9 that $f_{|X_{\mathrm{red}}} = g_{|X_{\mathrm{red}}}$. But in general we do not have $f = g$ (Exercise 9.10). The reason is that X is not necessarily the only closed subscheme Z of X such that Z majorizes U. Therefore we make the following definition.

Definition 9.18. *Let X be a scheme. An open subscheme U of X is called schematically dense in X if for every open subscheme V of X the only closed subscheme of V that majorizes $U \cap V$ is V itself.*

An open immersion $j \colon Y \to X$ is called schematically dominant *if the open subscheme $j(Y)$ is schematically dense in X.*

Then we indeed have:

Proposition 9.19. *Let S be a scheme, let X be an S-scheme, and let $j : U \hookrightarrow X$ be an open subscheme of X. Then the following assertions are equivalent.*

(i) *U is schematically dense in X.*

(ii) *The homomorphism $j^\flat \colon \mathscr{O}_X \to j_* \mathscr{O}_U$ is injective.*

(iii) *For every open subscheme V of X, for every separated S-scheme Y, and for any two S-morphisms $f, g \colon V \to Y$ such that $f_{|U \cap V} = g_{|U \cap V}$ we have $f = g$.*

Proof. "(ii) $\Rightarrow$ (i)". Let $i \colon Z \hookrightarrow V$ be a closed subscheme of an open subscheme V defined by the quasi-coherent ideal $\mathscr{J} \subseteq \mathscr{O}_V$. Assume that Z majorizes $U \cap V$. As $j^\flat_{|V}$ is injective, $i^\flat \colon \mathscr{O}_V \to i_*(\mathscr{O}_Z) = \mathscr{O}_V / \mathscr{J}$ is injective. This implies $\mathscr{J} = 0$.

"(i) $\Rightarrow$ (iii)". Eq(f, g) is a closed subscheme majorizing $U \cap V$.

"(iii) $\Rightarrow$ (ii)". Let $V \subseteq X$ be an open subset. We have to show that the restriction $\rho \colon \Gamma(V, \mathscr{O}_X) \to \Gamma(V \cap U, \mathscr{O}_X)$ is injective. This follows by applying (iii) to $Y = \mathbb{A}^1_S$ and using that $\operatorname{Hom}_S(W, \mathbb{A}^1_S) = \Gamma(W, \mathscr{O}_X)$ (4.12.2) for every open subscheme W of X. $\square$

Remark 9.20.

(1) Corollary 9.9 shows that if X is reduced and U is an open dense subscheme of X, then U is schematically dense in X. On the other hand, a schematically dense open subscheme $U \subseteq X$ does not have to be dense in general; see [St] 01RC for an example. This is true, however, if the open immersion $U \hookrightarrow X$ is "quasi-compact", e.g., if X is locally noetherian (cf. Definition 10.1, Proposition 10.30, Remark 10.31).

(2) If U and U' are schematically dense open subschemes of a scheme X, their intersection $U \cap U'$ is again a schematically dense open in X.

(3) If U is schematically dense in X and W is schematically dense in U, W is schematically dense in X.

(4) If $X = \bigcup_j V_j$ is an open covering, an open subset U of X is schematically dense in X if and only if $U \cap V_j$ is schematically dense in V_j for all j.

For locally noetherian schemes X, we can express whether an open $U \subseteq X$ is schematically dense in X in terms of associated points. Recall from Section (B.11) that if A is a ring, a prime ideal $\mathfrak{p}$ of A is called *associated* if there exists an element $a \in A$ such that $\mathfrak{p} = \operatorname{Ann}(a) = \{ b \in A ; ba = 0 \}$. The set of associated prime ideals of A is denoted by $\operatorname{Ass} A$.

Definition 9.21. *If X is a locally noetherian scheme, we call a point $x \in X$ associated if $\mathfrak{m}_x$ is an associated prime ideal of $\mathscr{O}_{X,x}$. The set of associated points of X is denoted by $\operatorname{Ass}(X)$.*

It follows from Equation (B.11.1) that if $X = \operatorname{Spec} A$ is an affine scheme, we have $\operatorname{Ass}(X) = \operatorname{Ass}(A)$. Every maximal point of X is an associated point. The closure $\overline{\{x\}}$ of an associated point x is called an *associated component*. It is called an *embedded component* if x is not a maximal point. If all associated points are maximal points, we say that X has *no embedded components*.

The connection between associated points and schematically dense subsets is given by the following proposition.

Proposition 9.22. *Let X be a locally noetherian scheme and let $U \subseteq X$ be an open subset. Then U is schematically dense in X if and only if U contains $\operatorname{Ass}(X)$.*

Proof. By definition we have $\text{Ass}(V) = \text{Ass}(X) \cap V$ for every open subscheme V of X. Moreover, the property of being schematically dense may be checked locally (Remark 9.20 (4)). Therefore it suffices to prove the following more precise lemma.   □

Lemma 9.23. *Let A be a noetherian ring, let $U \subseteq X = \text{Spec } A$ be an open subset, and let $\mathfrak{a} \subseteq A$ be any ideal such that $U = X \setminus V(\mathfrak{a})$. Then the following assertions are equivalent.*
(i) *U is schematically dense in X.*
(ii) *U contains a principal open subset $D(t)$, where $t \in A$ is not a zero divisor.*
(iii) *$\mathfrak{a}$ contains an element $t \in A$ that is not a zero divisor.*
(iv) *$\text{Ann}(\mathfrak{a}) := \{\, s \in A \; ; \; s\mathfrak{a} = 0 \,\}$ is zero.*
(v) *U contains $\text{Ass}(A)$.*

Proof. Clearly (ii) and (iii) are equivalent.

"(ii) $\Rightarrow$ (i)". It suffices to show that $D(t)$ is schematically dense in X. Let $V \subseteq X$ be open and let $s \in \Gamma(V, \mathscr{O}_X)$ be such that $s_{|V \cap D(t)} = 0$. Then for every open affine subset $W \subseteq V$ there exists an $n \geq 1$ such that $t^n s_{|W} = 0$ (Theorem 7.22). As $t_{|W}$ is not a zero divisor, we see that $s_{|W} = 0$ for all W and hence $s = 0$.

"(i) $\Rightarrow$ (iv)". Let $s \in \text{Ann}(\mathfrak{a})$ and $x \in U$. As $\mathfrak{p}_x \notin V(\mathfrak{a})$, there exists an $a \in \mathfrak{a} \setminus \mathfrak{p}_x$. Then $a_x \in \mathscr{O}_{X,x}^\times$ and $sa = 0$. This shows that $s_x = 0$ for all $x \in U$ and hence $s_{|U} = 0$. As U is schematically dense, this implies $s = 0$.

"(iv) $\Rightarrow$ (v)". If we had $\text{Ass}(A) \cap V(\mathfrak{a}) \neq \emptyset$, we would find $\mathfrak{p} \in \text{Ass}(A)$ with $\mathfrak{p} \supseteq \mathfrak{a}$. But $\mathfrak{p} = \text{Ann}(s)$ for some $0 \neq s \in A$ and hence we get $s \in \text{Ann}(\mathfrak{a})$.

"(v) $\Rightarrow$ (iii)". Assume that all elements of $\mathfrak{a}$ are zero divisors. As A is noetherian, $\mathfrak{a}$ would be contained in the finite union of all associated prime ideals (Proposition B.59) and hence in one of them (Proposition B.2 (2)). This contradicts $V(\mathfrak{a}) \cap \text{Ass}(A) = \emptyset$.   □

Remark 9.24. The proof shows that the implications "(iii) $\Leftrightarrow$ (ii) $\Rightarrow$ (i) $\Rightarrow$ (iv) $\Rightarrow$ (v)" hold also for non-noetherian rings.

Remark 9.25. For schemes flat over a base scheme (Definition 7.38; flat morphism will be studied in detail in Chapter 14), schematic density can be checked on fibers (see [EGAIV] (11.10.9)): Let S be a scheme, let X be a locally noetherian S-scheme which is flat over S, and let $U \subseteq X$ be an open subscheme. Then the following assertions are equivalent.
 (i) For all $s \in S$ the fiber U_s is a schematically dense open subscheme of the fiber X_s.
(ii) For every morphism $g \colon S' \to S$ the open scheme $U \times_S S'$ of $X \times_S S'$ is schematically dense.
The hypothesis that X is locally noetherian can be replaced by the hypothesis that X is locally of finite presentation over S in the sense of Definition 10.33 below (see [EGAIV] (11.10.10)).

(9.6) Rational maps and rational functions.

Let X and Y be schemes. Let $\mathcal{R}(X, Y)$ be the set of pairs $(U, \tilde{f})$ where $U \subseteq X$ is an open and schematically dense subscheme of X (Definition 9.18) and where $\tilde{f} \colon U \to Y$ is a morphism of schemes. We call $(U, \tilde{f})$, $(V, \tilde{g}) \in \mathcal{R}(X, Y)$ equivalent if there exists a schematically dense open subset $W \subseteq U \cap V$ such that $\tilde{f}_{|W} = \tilde{g}_{|W}$. Clearly this is an equivalence relation on $\mathcal{R}(X, Y)$.

Definition 9.26. *An equivalence class in* $\mathcal{R}(X, Y)$ *is called a* rational map *from* X *to* Y.

Note that we deviate here from the terminology used in EGA: What we call a rational map is called a *strict rational map* or *pseudo-morphism* in [EGAIV] 20.2. In the definition of rational map in [EGAInew] 8.1 the property of being *open and schematically dense* that we use in the definition of $\mathcal{R}(X, Y)$ is replaced by *open and dense*.

We denote a rational map f from X to Y by $f \colon X \dashrightarrow Y$. If X and Y are schemes over a scheme S, we say that a rational map f from X to Y is a *rational S-map* if there exists a representative in f which is an S-morphism. If S is a separated scheme, it is not difficult to see that for every representative $(U, \tilde{f})$ of a rational S-map f the morphism $\tilde{f}$ is an S-morphism (Exercise 9.17). The set of all rational S-maps $X \dashrightarrow Y$ is denoted by $\mathrm{Rat}_S(X, Y)$. The equivalence class of an S-morphism $f \colon X \to Y$ is a rational S-map. Therefore we obtain a map $\mathrm{Hom}_S(X, Y) \to \mathrm{Rat}_S(X, Y)$. Proposition 9.19 shows that this map is injective, if Y is separated over S.

Let W be an open subscheme of an S-scheme X and let f be a rational S-map from X to Y. We choose a representative $(U, \tilde{f})$ of f. Then $U \cap W$ is open and schematically dense in W and we denote by $f_{|W}$ the rational equivalence class of $(U \cap W, \tilde{f}_{|U \cap W})$. Then $f_{|W}$ is a rational S-map $W \dashrightarrow Y$ which does not depend on our choice of $(U, \tilde{f})$ and which is called the *restriction of f to W*.

For a rational S-map $f \colon X \dashrightarrow Y$ let $\mathrm{dom}(f) = \mathrm{dom}_S(f)$ be the set of points of $x \in X$ such that there exists a representative $(U, \tilde{f})$ of f such that $\tilde{f}$ is an S-morphism and such that $x \in U$. We call $\mathrm{dom}(f)$ the *domain of definition* of f. Clearly, $\mathrm{dom}(f)$ is an open schematically dense subset of X. Often rational S-maps $f \colon X \dashrightarrow Y$ are the same as S-morphisms $\mathrm{dom}(f) \to Y$:

Proposition 9.27. *Let S be a scheme, X and Y two S-schemes. Assume that Y is separated over S. Let $f \colon X \dashrightarrow Y$ be a rational S-map. Then there exists a unique S-morphism $f_0 \colon \mathrm{dom}(f) \to Y$ in the class f. It is the unique representative of f that cannot be extended to a larger schematically dense open subscheme.*

Proof. As morphisms can be glued together (Proposition 3.5), it suffices to show that if $(U, \tilde{f})$ and $(U', \tilde{f}')$ are representatives of f, then we have $\tilde{f}_{|U \cap U'} = \tilde{f}'_{|U \cap U'}$. This follows from Proposition 9.19. $\qquad\square$

We will now define rational functions on an S-scheme X. Recall (4.12.2) that we have $\Gamma(X, \mathscr{O}_X) = \mathrm{Hom}_S(X, \mathbb{A}^1_S)$. This leads us to the following definition.

Definition 9.28. *Let X be an S-scheme. A* rational S-function *is a rational S-map* $X \dashrightarrow \mathbb{A}^1_S$.

We denote the set of rational S-functions by $R(X)$. As explained above we may consider $R(X)$ also as the set of equivalence classes of pairs $(U, \tilde{s})$, where U is an open schematically dense subset of X and $\tilde{s} \in \Gamma(U, \mathscr{O}_X)$ is a function on U. In other words, we have

$$(9.6.1) \qquad\qquad R(X) = \varinjlim_U \Gamma(U, \mathscr{O}_X),$$

where U runs through the directed set of open schematically dense subschemes of X.

Remark 9.29. If X is integral, an open set U is schematically dense in X if and only if it is non-empty, i.e., if and only if it is an open neighborhood of the generic point η of X. Therefore we have in this case

$$R(X) = \mathcal{O}_{X,\eta} = K(X).$$

Example 9.30. Let k be an algebraically closed field. We view the closed points of $\mathbb{P}^1_k$ as the lines in k^2 containing the origin, where we, a little unusually, identify the point $(\lambda : \mu)$ with the line given by the equation $\lambda x = \mu y$. Let $\tilde{X}$ be the "incidence scheme" of pairs (p, ℓ), where $\ell \subseteq k^2$ is a line containing p (and the origin), i.e.,

$$\tilde{X}(k) = \{((x,y),(\lambda : \mu)) \in \mathbb{A}^2(k) \times \mathbb{P}^1(k); \ \mu x = \lambda y\}.$$

For $p \in \mathbb{A}^2(k) \setminus \{(0,0)\}$, there is a unique line through the origin containing p, and in this way we obtain a morphism given on k-valued points by

$$\mathbb{A}^2_k \setminus \{(0,0)\} \to \tilde{X}(k), \quad (x,y) \to ((x,y),(x:y)).$$

This gives us a rational map $\mathbb{A}^2_k \dashrightarrow \tilde{X}$. It is checked easily that the domain of definition of this map is $\mathbb{A}^2_k \setminus \{(0,0)\}$. Geometrically, the reason is that there is no natural choice of a line through the origin in k^2. This phenomenon of *indeterminacy* is the most important obstacle to extending rational maps. Compare Section (13.18).

(9.7) Birational equivalence.

Let X and Y be S-schemes.

Definition 9.31. *A rational S-map $f\colon X \dashrightarrow Y$ is called* birational *if there exists a representative $(U, \tilde{f})$ of f such that $\tilde{f}$ induces an S-isomorphism from U onto an open schematically dense subscheme V of Y.*

We say that X and Y are birationally equivalent *if there exists a birational S-map $X \dashrightarrow Y$.*

A morphism $f\colon X \to Y$ of S-schemes is called *birational* if its rational equivalence class is birational. In other words, if f induces an isomorphism of an open schematically dense subscheme U of X onto an open schematically dense V of Y.

If X and Y are integral, then a birational morphism induces an isomorphism of function fields $K(Y) \xrightarrow{\sim} K(X)$. In Proposition 10.52 we will see that the converse holds if X and Y are "locally of finite presentation over S", see Definition 10.33. For schemes of finite type over a field we will show in Proposition 9.35 a more precise result.

If X is an integral S-scheme with generic point η, every non-empty open set is schematically dense and contains η, so composition with the canonical morphism $\operatorname{Spec} K(X) \to X$ yields a well-defined map

(9.7.1) $\rho\colon \operatorname{Rat}_S(X,Y) \to \operatorname{Hom}_S(\operatorname{Spec} K(X), Y).$

If $f\colon X \dashrightarrow Y$ is a rational S-map and $(U, \tilde{f})$ is a representative, the point $\tilde{f}(\eta) \in Y$ is independent of the choice of the representative and we set $f(\eta) := \tilde{f}(\eta)$. The description of $K(X)$-valued points of Y in Proposition 3.8 shows that for each fixed point $y \in Y$ the map ρ induces a map

(9.7.2) $\rho_y\colon \{f \in \operatorname{Rat}_S(X,Y) \ ; \ f(\eta) = y\} \to \operatorname{Hom}_S(\operatorname{Spec} K(X), \operatorname{Spec} \kappa(y)).$

The case that Y is integral and $y = \vartheta$ is the generic point is particularly interesting. In this case we have:

Lemma 9.32. *Let X and Y be integral S-schemes with generic points η and ϑ, respectively. Let $f\colon X \dashrightarrow Y$ be a rational S-map. Then the following assertions are equivalent.*
(i) *There exists a representative $(U, \tilde{f})$ of f such that $\tilde{f}(U)$ is dense in Y.*
(ii) *$f(\eta) = \vartheta$.*

Proof. Clearly (ii) implies (i). Conversely, let $(U, \tilde{f})$ be a representative of f such that $\tilde{f}(U)$ is dense in Y. If V is an arbitrary non-empty open subset of Y, $V \cap \tilde{f}(U) \neq \emptyset$ and thus $\tilde{f}^{-1}(V)$ is a non-empty open subset of X and therefore contains η. This implies $\eta \in \bigcap_V \tilde{f}^{-1}(V) = \tilde{f}^{-1}(\vartheta)$. $\qquad\square$

For arbitrary S-schemes X and Y, we call a rational S-map $f\colon X \dashrightarrow Y$ *dominant* if f satisfies condition (i) of Lemma 9.32.

In Proposition 10.52 we will see that the map (9.7.1) (and thus (9.7.2) as well) is a bijection if $Y \to S$ satisfies a certain finiteness condition (namely to be locally of finite presentation, Definition 10.33). This finiteness condition is in particular satisfied if $S = \operatorname{Spec} k$ for a field k and Y is locally of finite type over k. For now we restrict ourselves to the following result that follows from Lemma 6.17.

Lemma 9.33. *Let k be a field, let X and Y be integral k-schemes of finite type. Then the map*

$$\{f \in \operatorname{Rat}_k(X, Y)\; ;\; f\ dominant\} \to \operatorname{Hom}_k(\operatorname{Spec} K(X), \operatorname{Spec} K(Y))$$
$$= \operatorname{Hom}_k(K(Y), K(X))$$

is bijective.

In particular, a rational map $f\colon X \dashrightarrow Y$ between integral k-schemes of finite type is birational if and only if f induces an isomorphism on function fields $K(Y) \to K(X)$. This implies that integral schemes of finite type over k are up to birational equivalence already determined by their function fields.

Remark 9.34. Let $f\colon X \dashrightarrow Y$ and $g\colon Y \dashrightarrow Z$ be dominant rational k-maps of integral k-schemes of finite type. Their composition $g \circ f$ is defined as follows. Let $(U, \tilde{f})$ and $(V, \tilde{g})$ be representatives of f and g, respectively. The inverse image $\tilde{f}^{-1}(V)$ is open and dense because it contains the generic point of X by Lemma 9.32. Replacing U by $U' := U \cap \tilde{f}^{-1}(V)$ and $\tilde{f}$ by $\tilde{f}' := \tilde{f}_{|U'}$, which also represent f, we define $g \circ f$ as the class of $\tilde{g} \circ \tilde{f}'$. It is easy to check that $g \circ f$ does not depend on the choice of $(U, \tilde{f})$ and $(V, \tilde{g})$. This composition makes integral k-schemes of finite type together with dominant rational k-maps into a category.

Proposition 9.35. *Attaching to an integral k-scheme X its function field $K(X)$ induces an equivalence between the following two categories.*
(1) *The category whose objects are integral k-schemes of finite type over k and whose morphisms are dominant rational k-maps $f\colon X \dashrightarrow Y$.*
(2) *The category of finitely generated field extensions of k*

Proof. By Lemma 9.33 the functor is fully faithful. It remains to show that it is essentially surjective. Let K be a field extension of k generated by $t_1, \dots, t_n \in K$ and let A be the sub-k-algebra of K generated by $t_1, \dots, t_n$. Then $X := \operatorname{Spec} A$ is an integral k-scheme of finite type such that $K(X) = K$. $\qquad\square$

In particular we see that two integral k-schemes of finite type are birationally equivalent if and only if their function fields are isomorphic extensions of k.

Remark 9.36. Let X be an integral k-scheme of finite type of dimension d. Assume that k is perfect (or, more generally, that X is geometrically reduced). Then Proposition 6.18 shows that X is birationally equivalent to a hypersurface in $\mathbb{A}_k^{d+1}$ (or in $\mathbb{P}_k^{d+1}$).

Projective integral k-schemes X of finite type that are birational to projective space $\mathbb{P}_k^d$ itself are called *rational*. We will see in Section (15.7) that if X is normal and of dimension 1, then X is rational if and only if $X \cong \mathbb{P}_k^1$. For schemes of higher dimension this does not hold (Exercise 9.18).

Exercises

Exercise 9.1◇. Let $\mathcal{C}$ be a category where arbitrary fiber products exist.
(a) Show that a morphism $f\colon X \to S$ is a monomorphism if and only if Δ_f is an isomorphism.
(b) Show that if $f\colon X \to S$ is a monomorphism and $S' \to S$ is a morphism, $f_{(S')}$ is a monomorphism.

Exercise 9.2. Let S be a scheme, let X and Y be S-schemes and let p and q be the projections of $X \times_S Y$. Let $u\colon S \to T$ a morphism. Show that $(p,q)_T\colon X \times_S Y \to X \times_T Y$ is an immersion. Show that it is a closed immersion (resp. an isomorphism) if u is separated (resp. if u is an monomorphism).

Exercise 9.3. Let S be a scheme, let $f\colon X \to Y$ be a morphism of S-schemes, let X' be a subscheme of X, let Y' be a subscheme of Y and consider $X' \times_S Y'$ as a subscheme of $X \times_S Y$. Show the equality $\Gamma_f^{-1}(X' \times_S Y') = X' \cap f^{-1}(Y')$ of subschemes of X.

Exercise 9.4. Let S be a scheme, let X be an S-scheme, let $\Delta \subseteq X \times_S X$ be the diagonal, and let Z be the subset $\{ z \in X \times_S X \; ; \; p(z) = q(z) \}$, where $p, q\colon X \times_S X \to X$ are the projections. Show that $\Delta \subseteq Z$ (as sets). Now let $S = \operatorname{Spec} k$, where k is a field, let $X = \operatorname{Spec} K$, where K is a field extension of k. Show that $\Delta = Z$ if and only if $k \to K$ is purely inseparable.

Exercise 9.5. Let $f\colon X \to Y$ and $g\colon Y \to Z$ be scheme morphisms such that f is surjective and universally closed and such that $g \circ f$ is separated. Show that g is separated.

Exercise 9.6. Show that a monomorphism $f\colon X \to S$ of schemes is separated, purely inseparable and for $x \in X$ the induced homomorphism $\kappa(f(x)) \to \kappa(x)$ is an isomorphism.

Exercise 9.7. Let S be a scheme and let $f, g\colon X \to Y$ be morphism of S-schemes. Show that $x \in X$ lies in the subscheme $\operatorname{Eq}(f,g)$ if and only if $f(x) = g(x)$ and the homomorphisms $\kappa(f(x)) = \kappa(g(x)) \hookrightarrow \kappa(x)$ induced by f and g coincide.

Exercise 9.8◇. Let S be a scheme, and let X and Y be S-schemes. Assume that Y is separated over S and that X is integral with generic point η.
(a) Let $f, g\colon X \to Y$ be two S-morphisms such that $f(\eta) = g(\eta) =: \vartheta$ and that the homomorphisms $\kappa(\vartheta) \hookrightarrow \kappa(\eta)$ induced by f and g coincide. Show that $f = g$. *Hint*: Exercise 9.7.

(b) Let $S = X$ be integral with generic point η. Show that two sections $s, t \colon S \to Y$ of $Y \to S$ are equal if and only if $s(\eta) = t(\eta)$.

Exercise 9.9. Let $f \colon X \to Y$ be a morphism of schemes. Show that the following assertions are equivalent.
(i) f is purely inseparable.
(ii) For every field K the induced map $f(K) \colon X(K) \to Y(K)$ is injective.
(iii) For every field K there exists an algebraically closed extension K' of K such that the induced map $f(K') \colon X(K') \to Y(K')$ is injective.
(iv) The diagonal $\Delta_f \colon X \to X \times_Y X$ is surjective.
Deduce that every purely inseparable morphism is separated and that the property "purely inseparable" is stable under composition, stable under base change and local on the target.

Exercise 9.10◊. Let k be a field, $A := k[X,Y]/(XY, Y^2)$. Define two k-algebra homomorphisms $\varphi_i \colon k[T]/(T^2) \to A$ for $i = 1, 2$ by $\varphi_1(T) = 0$ and $\varphi_2(T) = Y$. Let $f_i \colon \operatorname{Spec} A \to \operatorname{Spec} k[T]/(T^2)$ be the associated morphisms of schemes. Show that there exists an open dense subset $U \subseteq \operatorname{Spec} A$ such that $f_{1|U} = f_{2|U}$ although $f_1 \neq f_2$. Determine $\operatorname{Eq}(f_1, f_2)$.

Exercise 9.11. Show that a morphism of schemes $f \colon X \to S$ is separated if and only if for every S-scheme S' every section of the base change $f_{(S')} \colon X \times_S S' \to S'$ is a closed immersion.

Exercise 9.12. Let $f \colon X \to Y$ be a morphism of schemes and assume that the family of irreducible components of X is locally finite. Show that f is separated if and only if $f_{|Z_{\mathrm{red}}}$ is separated for every irreducible component Z of X.
Hint: Use Exercise 4.10 and Exercise 9.5.

Exercise 9.13. Let $f \colon X \to S$ be a morphism of schemes with S separated. Show that for all affine open subschemes $V \subseteq S$ and $U \subseteq X$ the intersection $U \cap f^{-1}(V)$ is affine.
Hint: Consider X and S as schemes over $\mathbb{Z}$ and use Exercise 9.3.

Exercise 9.14. Let k be a field. Describe the schematically dense open subschemes of $\operatorname{Spec} A$ for the following rings A:
(a) $A = k[T, U]/(T^2)$,
(b) $A = k[T, U]/(T^2, TU)$,
(c) $A = B \oplus \kappa(\mathfrak{p})$, where B is a reduced noetherian ring, $\mathfrak{p}$ is a prime ideal of B and the multiplication on A is given by $(b, x)(b', x') = (bb', bx' + b'x)$.

Exercise 9.15. Describe all associated components of the scheme Y defined in Exercise 3.23.

Exercise 9.16. Let X be a locally noetherian scheme. Prove that X is reduced if and only if X is generically reduced (i.e., $\mathscr{O}_{X,x}$ is reduced for all maximal points x of X) and X has no embedded components.

Exercise 9.17. Let S be a separated scheme and let X and Y be S-schemes. Show that for every representative $(U, \tilde{f})$ of a rational S-map $f \colon X \dashrightarrow Y$ the morphism $\tilde{f}$ is an S-morphism.

Exercise 9.18. Let k be a field and $n, m \geq 1$ be integers. Show that $\mathbb{P}^n_k \times_k \mathbb{P}^m_k$ is birational equivalent to $\mathbb{P}^{n+m}_k$.
Remark: $\mathbb{P}^n_k \times_k \mathbb{P}^m_k$ is never isomorphic to $\mathbb{P}^{n+m}_k$ (see Example 11.46 or Exercise 10.41).

10 Finiteness Conditions

Contents

In this chapter we generalize the notion and properties of "schemes locally of finite type over a field" to arbitrary morphisms $X \to Y$ of schemes (Section (10.2)). Roughly speaking that means that locally on X and Y the morphism corresponds to a ring homomorphism $A \to B$ that makes B into a finitely generated A-algebra, i.e. $B \cong A[T_1, \ldots, T_n]/\mathfrak{a}$ for an ideal $\mathfrak{a} \subseteq A[T_1, \ldots, T_n]$. If Y is locally noetherian, A will be noetherian, and the ideal $\mathfrak{a}$ will be finitely generated by Hilbert's basis theorem. Thus B will be given over A by finitely many generators and finitely many relations. In this case many of the definitions simplify considerably and we will therefore deal with this case first. One of the main results will be Chevalley's theorem that the image of a morphism of finite type of noetherian schemes is constructible (i.e., a finite union of locally closed subspaces).

In the non-noetherian case the property of being defined by finitely many genera-tors and by finitely many relations leads to the notion "locally of finite presentation" (Section (10.9)). Moreover to exclude the pathology that intersections of quasi-compact open subsets are not quasi-compact in general we also have to introduce the notion of quasi-separatedness (which is automatic for locally noetherian schemes), see (10.7). In fact, the properties "quasi-compact" and "quasi-separated" will occur so often together in the sequel that we coin the non-standard abbreviation "qcqs". Keeping this technical difficulties in mind, the theory then runs along the same lines as in the noetherian case. In particular we will prove a non-noetherian version of Chevalley's theorem for a suitably generalized notion of constructibility.

If R is a ring which is the filtered inductive limit of rings R_λ and if X is a scheme that is defined by finitely many generators and relations over $S = \operatorname{Spec} R$ (more precisely, that is "of finite presentation" over S), then X will already be defined over some R_λ. Moreover, for most properties of the S-scheme X we can find a model of X over some R_λ that has the same property. This technique will allow us to extend results from schemes over the local ring at a point to some open neighborhood and to make reductions from arbitrary field extensions to finite field extensions, from non-noetherian schemes to noetherian schemes (or even schemes of finite type over $\mathbb{Z}$), or from schemes over arbitrary fields to schemes over finite fields.

In the last part we study the question about the constructibility of the locus, where a certain property of the fibers of a morphism $X \to S$ is satisfied. We will prove results only for a few examples and refer to Appendix E and the references there for a more exhaustive list of results.

© Springer Fachmedien Wiesbaden GmbH, part of Springer Nature 2020
U. Görtz und T. Wedhorn, *Algebraic Geometry I: Schemes*, Springer Studium Mathematik – Master, https://doi.org/10.1007/978-3-658-30733-2_11

Finiteness conditions (noetherian case)

(10.1) Quasi-compact morphisms.

We first introduce a relative notion of quasi-compactness.

Proposition and Definition 10.1. *A morphism $f : X \to Y$ of schemes is called* quasi-compact, *if the following equivalent conditions are satisfied:*
(i) *For every quasi-compact open subset $V \subseteq Y$ the preimage $f^{-1}(V)$ is quasi-compact.*
(ii) *There exists a covering $Y = \bigcup_i V_i$ by affine open subsets V_i such that for every i, the inverse image $f^{-1}(V_i)$ is quasi-compact.*

Proof. We only have to show that the second condition is sufficient. By assumption, for every i the preimage $f^{-1}(V_i)$ is quasi-compact, so it is a finite union of affine schemes, say $f^{-1}(V_i) = \bigcup_{j=1}^{n_i} U_{ij}$. For $s \in \Gamma(V_i, \mathscr{O}_Y)$, denote its image in $\Gamma(U_{ij}, \mathscr{O}_X)$ by t_j. Then $f^{-1}(D_{V_i}(s))$ is the union of the $D_{U_{ij}}(t_j)$, and in particular is quasi-compact, too. Now if $V \subseteq Y$ is an arbitrary quasi-compact open subset, we can cover it by finitely many distinguished open subsets of the V_i, and it follows that $f^{-1}(V)$ is quasi-compact. $\qquad\square$

Remark 10.2.
(1) If Y is an affine scheme, then a morphism $f : X \to Y$ is quasi-compact if and only if X is quasi-compact. In particular X is quasi-compact if and only if the unique morphism $X \to \operatorname{Spec} \mathbb{Z}$ is quasi-compact.
(2) Every closed immersion is quasi-compact.
(3) If X is noetherian, then every morphism $f : X \to Y$ is quasi-compact, because every open subset of X is quasi-compact (Lemma 1.25).
(4) There exist quasi-compact schemes X, Y and a morphism $X \to Y$ which is not quasi-compact (Exercise 10.13). In particular, in part (ii) of the proposition, it is not sufficient to consider a covering by arbitrary quasi-compact open subsets $V_i \subseteq Y$ (although Proposition 10.3 below shows that this problem can arise only if Y is not separated).

Proposition 10.3.
(1) *The property "quasi-compact" of morphisms of schemes is stable under composition and under base change, and is local on the target.*
(2) *Let $f : X \to Y$ and $g : Y \to Z$ be morphisms of schemes such that g is separated. If $g \circ f$ is quasi-compact, then f is quasi-compact.*

Proof. It is clear that the property of being quasi-compact is stable under composition and local on the target. Let us check that it is stable under base change. So suppose that $f : X \to Y$ is a quasi-compact morphism, and let $g : Y' \to Y$ be an arbitrary morphism of schemes. Since quasi-compactness is local on the target, we may replace Y by an affine open subset, and correspondingly replace X and Y' by its inverse images. So we may assume that Y is affine and that X is quasi-compact, say X is the union of finitely many affine open subsets U_i.

We have to check that for every open affine $V' \subseteq Y'$ the inverse image in $X \times_Y Y'$ is quasi-compact. This inverse image is homeomorphic to $X \times_Y V'$, and we can cover this fiber product by the finitely many affine schemes $U_i \times_Y V'$, so it is indeed quasi-compact.

Assertion (2) follows from (1) and Remark 9.11. $\square$

Remark 10.4. The conclusion in (2) also holds (with the same argument) if g is only assumed to be quasi-separated (see Definition 10.22 below), in particular whenever Y locally noetherian (Exercise 10.12).

(10.2) Morphisms (locally) of finite type.

We now generalize the notion of schemes *(locally) of finite type over a field*, as introduced in Section (3.12), to more general base schemes.

Proposition and Definition 10.5. *A morphism $f\colon X \to Y$ of schemes is called* locally of finite type, *if the following equivalent conditions are satisfied:*
(i) *For every affine open subscheme V of Y and every affine open subscheme U of $f^{-1}(V)$, the $\Gamma(V, \mathscr{O}_Y)$-algebra $\Gamma(U, \mathscr{O}_X)$ is finitely generated.*
(ii) *There exist a covering $Y = \bigcup_i V_i$ by open affine subschemes $V_i \cong \operatorname{Spec} A_i$, and for each i a covering $f^{-1}(V_i) = \bigcup_j U_{ij}$ by open affine subschemes $U_{ij} \cong \operatorname{Spec} B_{ij}$ of X, such that for all i, j the A_i-algebra B_{ij} is finitely generated.*
We also say that X is locally of finite type over Y, *or that X is a Y-scheme locally of finite type in this case.*

Proof. Assume that condition (ii) is satisfied, and let $V \subseteq Y$ and $U \subseteq f^{-1}(V)$ be open affine. For $s \in \Gamma(V_i, \mathscr{O}_Y)$, we can cover $f^{-1}(D_{V_i}(s))$ by principal open subsets of the U_{ij}, and their affine coordinate rings will be of finite type over the localizations $A_{i,s}$. This means that we may assume that V is the union of (finitely many) V_i. To simplify the notation, we replace Y by V, and write $Y = \operatorname{Spec} A$. Using Lemma 3.3 and replacing the covering $Y = \bigcup V_i$ by a finer covering, we reduce to the case that $V_i = D(s_i) \subseteq Y$, $s_i \in A$.

Replacing the U_{ij} by principal open subsets, we may also assume that U is covered by some of the U_{ij}, so we can replace X by U. We write $X = \operatorname{Spec} B$, and denote by $\varphi\colon A \to B$ the ring homomorphism corresponding to f. Since $V_i = D(s_i)$, we have $f^{-1}(V_i) = D_X(\varphi(s_i))$, and using Lemma 3.32, we see that $\Gamma(D_X(\varphi(s_i)), \mathscr{O}_X) = B_{\varphi(s_i)}$ is a finitely generated A_{s_i}-algebra. We choose, for each i, finitely many elements $b_{ij} \in B$ whose images in $B_{\varphi(s_i)}$ generate $B_{\varphi(s_i)}$ as an A_{s_i}-algebra.

Let $B' \subset B$ be the sub-A-algebra generated by all the s_i and b_{ij}, and let $b \in B$. By definition of B', for each i there exists an N such that $\varphi(s_i)^N b \in B'$. But there are only finitely many i, so we find an N which works for all i simultaneously. Since the s_i generate the unit ideal of A, so do the s_i^N, and the $\varphi(s_i)^N$ generate the unit ideal of B. This implies that $b \in B'$. So $B = B'$ is a finitely generated A-algebra. $\square$

Definition 10.6. *A morphism $f\colon X \to Y$ is called* of finite type, *if f is locally of finite type and quasi-compact.*

As usual, the following standard properties are easy to check.

Proposition 10.7.
(1) *Every immersion is locally of finite type. Every quasi-compact immersion $i\colon Y \to X$ is of finite type. In particular i is of finite type whenever i is a closed immersion or Y is noetherian.*

(2) *The properties "locally of finite type" and "of finite type" of morphisms of schemes are stable under composition and under base change, and local on the target. The property "locally of finite type" is also local on the source.*

(3) *Let $f\colon X \to Y$ and $g\colon Y \to Z$ be morphisms of schemes. If $g \circ f$ is locally of finite type (resp. of finite type and g is separated), then f is locally of finite type (resp. of finite type).*

In (3) it suffices to assume that $g \circ f$ is of finite type and g is quasi-separated (Definition 10.22) to conclude that f is of finite type.

The property "of finite type" cannot be checked on fibers. There exist morphisms $f\colon X \to Y$ of affine schemes such that every fiber X_y is of finite type over $\kappa(y)$ and such that f is not of finite type (Exercise 10.9).

Example 10.8.
(1) Let R be a ring. Then $\mathbb{A}_R^n$ is clearly of finite type over R. Thus $\mathbb{P}_R^n$ (having a finite covering of open subschemes isomorphic to $\mathbb{A}_R^n$) is of finite type over R. This shows that all closed subschemes of $\mathbb{P}_R^n$ and of $\mathbb{A}_R^n$ are of finite type over R. If R is noetherian, any subscheme of $\mathbb{P}_R^n$ is of finite type over R.

(2) Let S be a scheme, let $\mathscr{E}$ be a quasi-coherent $\mathscr{O}_S$-module of finite type and let $e \geq 0$ be an integer. Then $\mathrm{Grass}^e(\mathscr{E})$ is of finite type over S. Indeed, the question is local on S and we may assume that S is affine. Then we can choose a surjection $\mathscr{O}_S^n \twoheadrightarrow \mathscr{E}$ which yields a closed immersion $\mathrm{Grass}^e(\mathscr{E}) \hookrightarrow \mathrm{Grass}_{n-e,n} \times_{\mathbb{Z}} S$. Composing with the Plücker embedding (Section (8.10)) we obtain a closed immersion of $\mathrm{Grass}^e(\mathscr{E})$ into some $\mathbb{P}_R^N$ which shows the claim by (1). Alternatively we can use the covering by affine spaces produced in the proof of the representability.

(3) Choosing $e = 1$ in (2) shows in particular that the projective bundle $\mathbb{P}(\mathscr{E})$ is of finite type over S if $\mathscr{E}$ is a quasi-coherent $\mathscr{O}_S$-module of finite type.

Proposition 10.9. *Let $f\colon X \to Y$ be a morphism of schemes which is (locally) of finite type. If Y is (locally) noetherian, then X is (locally) noetherian.*

Proof. This statement reduces to the affine case by tracing through the definitions, and then follows from Hilbert's Basissatz (Proposition B.34). $\qquad\square$

(10.3) Quasi-coherence of direct images of quasi-coherent modules.

Proposition 10.10. *Let $f\colon X \to Y$ be a quasi-compact, separated morphism, and let $\mathscr{F}$ be a quasi-coherent $\mathscr{O}_X$-module. Then the direct image $f_*\mathscr{F}$ is a quasi-coherent $\mathscr{O}_Y$-module.*

Proof. We may assume that Y is affine, and hence that X is quasi-compact, so that X is a finite union of open affine subschemes, say $X = \bigcup_{i=1}^n X_i$. Because f is separated, all the intersections $X_i \cap X_j$ are also affine (Proposition 9.15). We write $\mathscr{F}_i = \mathscr{F}_{|X_i}$, $\mathscr{F}_{ij} = \mathscr{F}_{|X_i \cap X_j}$, and $\mathscr{F}'_i = (f_{|X_i})_*\mathscr{F}_i$, $\mathscr{F}'_{ij} = (f_{|X_i \cap X_j})_*\mathscr{F}_{ij}$. We know by Proposition 7.24 that the $\mathscr{F}'_i$, $\mathscr{F}'_{ij}$ are quasi-coherent, so the *finite* products

$$\mathscr{G} = \prod_i \mathscr{F}'_i, \quad \mathscr{H} = \prod_{i,j} \mathscr{F}'_{ij}$$

are quasi-coherent, as well (Corollary 7.19 (2)). Now for each open $U \subseteq X$, we have an exact sequence

$$0 \to \mathscr{F}(U) \to \prod_i \mathscr{F}(U \cap X_i) \to \prod_{i,j} \mathscr{F}(U \cap X_i \cap X_j),$$

and in particular, for $V \subseteq Y$ open, we obtain an exact sequence

(10.3.1) $$0 \to (f_* \mathscr{F})(V) \to \prod_i \mathscr{F}_i'(V) \to \prod_{i,j} \mathscr{F}_{ij}'(V),$$

because $(f_* \mathscr{F})(V) = \mathscr{F}(f^{-1}(V))$, $\mathscr{F}_i'(V) = \mathscr{F}_{|X_i}((f_{|X_i})^{-1}(V)) = \mathscr{F}(f^{-1}(V) \cap X_i)$, and similarly for $\mathscr{F}_{ij}'$. Altogether, we get an exact sequence

(10.3.2) $$0 \to f_* \mathscr{F} \to \mathscr{G} \to \mathscr{H},$$

which shows by Corollary 7.19 (1) that $f_* \mathscr{F}$ is quasi-coherent, as desired. $\square$

In Corollary 10.27 below we will show a more general version of the proposition which will in particular show that for all *noetherian* schemes X and for all morphisms $f \colon X \to Y$ the direct image of any quasi-coherent $\mathscr{O}_X$-module is quasi-coherent again.

(10.4) Noetherian Induction.

Recall the principle of noetherian induction: Let I be a partially ordered set such that every non-empty subset $J \subseteq I$ has a minimal element (equivalently, every descending chain in I becomes stationary). Let $P \subseteq I$ be a subset such that for all $i \in I$:

$$\{ j \in I \;;\; j < i \} \subseteq P \Rightarrow i \in P.$$

Then $P = I$.

For instance, we can take I as the set of closed subsets of a noetherian topological space ordered by inclusion. In fact, we have already applied this principle (implicitly) when we proved that a noetherian topological space has only finitely many irreducible components, see the proof of Lemma 1.25 (3). There is a refined version of this principle for noetherian schemes:

Proposition 10.11. *Let X be a noetherian scheme, and let* **P** *be a property of closed subschemes of X. Assume that whenever $Y \subseteq X$ is a closed subscheme such that all proper closed subschemes of Y possess the property* **P**, *then Y has the property* **P**. *Then every closed subscheme of X has the property* **P**.

Proof. By the principle of noetherian induction it suffices to show that every descending chain of closed subschemes of a noetherian scheme X becomes stationary. Since X is quasi-compact, we can cover it by finitely many affine open subschemes, and we may hence assume that X is itself affine. In the affine case, however, there is an inclusion-reversing bijection between closed subschemes of X and ideals of the ring $\Gamma(X, \mathscr{O}_X)$ which is noetherian by Proposition 3.19. Our assertion is proved. $\square$

We will apply this principle in several places below, see for example the proof of Chevalley's theorem, Theorem 10.20.

(10.5) Constructible sets in noetherian schemes.

Let X be a noetherian topological space.

Definition 10.12. *A subset $C \subseteq X$ is called* constructible, *if C is a finite union of locally closed subsets of X.*

The following proposition gives an elegant description of the set of all constructible subsets.

Proposition 10.13. *The set of constructible subsets of X is the smallest subset of the power set of X which contains all open subsets, and is closed under taking complements and under finite unions (and hence under finite intersections).*

Proof. Clearly the smallest set with the properties stated above contains all constructible sets. On the other hand, finite unions of constructible subsets are obviously constructible. So it is enough to show that the complement of a constructible set is constructible again. But if $C = \bigcup_{i \in I} U_i \cap Z_i$ is a finite union of locally closed subsets, $U_i \subseteq X$ open, $Z_i \subseteq X$ closed, then, denoting complements by $-^c$,

$$C^c = \bigcap_i U_i^c \cup Z_i^c$$

is the union of all sets of the form $\left(\bigcap_{i \in J} U_i^c\right) \cap \left(\bigcap_{i \in I \setminus J} Z_i^c\right)$, $J \subseteq I$. All those sets are locally closed, whence C^c is constructible. $\square$

Recall that a subset E of a topological space X is called *nowhere dense*, if the closure of E has empty interior, or in other words, if the complement of the closure of E is dense in X. If X is irreducible, then $E \subseteq X$ is nowhere dense if and only if the closure of E is $\neq X$, or equivalently if $X \setminus E$ contains a non-empty open subset of X.

Proposition 10.14. *Let X be a noetherian topological space. A subset $E \subseteq X$ is constructible if and only if for every closed irreducible subset $Y \subseteq X$, $E \cap Y$ contains a non-empty open subset of Y or is nowhere dense in Y.*

Proof. If E is constructible in X, then $E \cap Y$ is constructible in Y, so it is a finite union of locally closed subsets of Y. If one of these locally closed subsets is dense, then it is open in Y. If none is dense, then their union is not dense, and is hence nowhere dense, because Y is irreducible.

To show the converse, we apply the principle of noetherian induction to the set of closed subsets $Y \subseteq X$ such that $E \cap Y$ is constructible (in Y, or equivalently in X). So we may assume that for every proper closed subset $Y \subset X$, the intersection $E \cap Y$ is constructible. If X is not irreducible, then let X_i be the irreducible components of X. The X_i are closed in X, so by induction hypothesis the $E \cap X_i$ are constructible, hence so is their union E. Finally, let us assume that X is irreducible. Then either E is nowhere dense in X, i. e. it is contained in a proper closed subset $Y \subset X$, and we are done by again applying the induction hypothesis. Or E contains a non-empty open subset U of X. Since $E \setminus U = E \cap (X \setminus U)$ is constructible, so is $E = U \cup (E \setminus U)$. $\square$

Remark 10.15. Let X be a noetherian scheme and let X_0 be a very dense subset of X (Definition 3.34). As intersecting with X_0 yields a bijection of the set of open (resp. closed) subsets of X with the set of open (resp. closed) subsets of X_0, we also obtain a bijection

$$\{\text{constructible subsets of } X\} \leftrightarrow \{\text{constructible subsets of } X_0\}, \qquad C \mapsto C \cap X_0.$$

The most important special case for this situation is the case that X is a Jacobson scheme, i.e., a scheme such that the set of closed points X_0 in X is very dense. This is the case if X is a scheme of finite type over a field (Proposition 3.35). More generally, this holds if X is of finite type over a Jacobson ring A (Exercise 10.16 and Exercise 10.17), e.g., $A = \mathbb{Z}$.

Lemma 10.16. *Let X be a noetherian scheme, and let C be a constructible subset. Then there exists an affine scheme X' and a morphism $f \colon X' \to X$ of finite type such that $f(X') = C$.*

Proof. We write C as a finite union of locally closed subsets in X. Assembling X' as a suitable disjoint union, we may assume that C is itself locally closed, say $C = U \cap Z$ with U open, Z closed in X. Similarly, we can pass to a finite open affine covering of U and hence assume that U is affine. But then $U \cap Z$, seen as a closed subscheme of U, is affine, as well, and the morphism $U \cap Z \to X$ is an immersion, and in particular of finite type, since X is noetherian.   □

Below we will prove some kind of converse to the lemma, namely Chevalley's Theorem, Theorem 10.20, which asserts that every morphism of finite type between noetherian schemes has a constructible image.

Lemma 10.17. *Let X be a noetherian scheme, and let $C \subseteq X$ be a subset.*
(1) C is closed if and only if C is constructible and stable under specialization.
(2) C is open if and only if C is constructible and stable under generization.

Proof. Let $C \subseteq X$ be constructible and stable under specialization. We have to show that C is closed. To do so, we can work on an open covering of X, so we may assume that $X = \operatorname{Spec} A$ is affine. By Lemma 10.16, we can write C as the image of a morphism $f \colon \operatorname{Spec} B \to X$.

Now let $x \in X \setminus C$. We must show that some open neighborhood of x in X is contained in $X \setminus C$. On the other hand, by assumption we have that C is stable under specialization, so $X \setminus C$ is stable under generization, and we get that $\operatorname{Spec} \mathscr{O}_{X,x} \subseteq X \setminus C$. In other words, $f^{-1}(\operatorname{Spec} \mathscr{O}_{X,x}) = \emptyset$, which we can also express as $B \otimes_A \mathscr{O}_{X,x} = 0$.

Since $\mathscr{O}_{X,x} = \varinjlim_s \Gamma(D(s), \mathscr{O}_X) = \varinjlim_s A_s$, where the limit runs over all s with $s(x) \neq 0$ (in $\kappa(x)$), we obtain that $\varinjlim B \otimes_A A_s = 0$. This means that $1 = 0$ in the inductive limit, so $1 = 0$ holds for suitable s. We get that $B \otimes_A A_s = 0$ for some s, and translating this back, we have $D(s) \subseteq X \setminus C$, which proves (1).

Now (2) follows from (1) by taking complements.   □

(10.6) Images of constructible sets.

Let $f \colon X \to Y$ be a morphism of schemes. For arbitrary morphisms the image can be anything: Let $Z \subseteq Y$ be any subset. Then the canonical morphism $\coprod_{z \in Z} \operatorname{Spec} \kappa(z) \to Y$ has image Z. And even if X and Y are both of finite type over a field, the image $f(X)$ is in general not locally closed (see Exercise 10.4). But we now prove Chevalley's theorem (in the noetherian case; see Theorem 10.70 below for a more general version), which states that under suitable finiteness conditions the image is always constructible. This is a fundamental result.

Lemma 10.18. *Let R be an integral domain, and let $R \hookrightarrow A$ be an injective ring homomorphism such that A is a finitely generated R-algebra. Then there exist $0 \neq s \in R$ and a finite injective homomorphism $R_s[T_1, \ldots T_n] \to A_s$.*

Proof. Let $S = R \setminus \{0\}$, so that $K := S^{-1}R$ is the field of fractions of R. Since $S^{-1}A$ is a finitely generated K-algebra, we can apply Noether normalization (Theorem 1.8), and find elements $y_1, \ldots, y_n \in S^{-1}A$ which are algebraically independent over K, and such that $S^{-1}A$ is finite over $K[y_1, \ldots, y_n]$. Replacing R, and correspondingly A, by a localization, we may assume that the y_i all lie in A. Choose finitely many generators of A as an R-algebra. They are zeros of monic polynomials with coefficients in $K[y_1, \ldots, y_n]$, and there exists an $s \in R \setminus \{0\}$ such that all the polynomials involved have coefficients in $R_s[y_1, \ldots, y_n]$, so that the inclusion $R_s[y_1, \ldots, y_n] \subset A_s$ is a finite injective ring homomorphism. $\square$

The key ingredient of the proof of Chevalley's theorem, which is also interesting in its own right, is:

Theorem 10.19. *Let X and Y be noetherian schemes, and let $f : X \to Y$ be a dominant morphism of finite type. Then $f(X)$ contains an open dense subset of Y.*

Proof. Let us first assume that Y is irreducible. Since we are only interested in the topological properties of the image of f, we may assume that X and Y are reduced. Replacing X and Y by non-empty affine open subsets, we may assume that both are affine. Then f corresponds to a ring homomorphism $\varphi : A \to B$, where A is an integral domain. As f is dominant, φ is injective (Corollary 2.11). Lemma 10.18 shows that there exists $s \in A \setminus \{0\}$ such that the induced morphism $\varphi_s : A_s \to B_s$ factors into $A_s \to A_s[T_1, \ldots, T_n] \hookrightarrow B_s$ where the second homomorphism is finite and injective. Thus φ_s induces a surjection $\operatorname{Spec} B_s \to \operatorname{Spec} A_s$ (Proposition 5.12) and we find $D(s) \subseteq f(X)$.

If Y is not irreducible, denote by $Y_1, \ldots, Y_n$ the irreducible components of Y and let f_i be the restriction $f^{-1}(Y_i) \to Y_i$ of f. We have already shown that the image of f_i contains an open dense subset V_i of Y_i. Replacing V_i by $V_i \setminus \bigcup_{j \neq i} Y_j$ we may assume that V_i is also open in Y and then $\bigcup_i V_i$ is an open dense subset of Y which is contained in $f(X)$. $\square$

Now we can prove Chevalley's theorem about images of morphisms.

Theorem 10.20. *(Chevalley's theorem) Let Y be a noetherian scheme, let $f : X \to Y$ be a morphism of finite type. Then for every constructible subset $C \subseteq X$, the image $f(C)$ is a constructible subset of Y. In particular, $f(X)$ is a constructible subset of Y.*

Proof. By Lemma 10.16 we may assume that $C = X$, and by Proposition 10.14, it is enough to show that for every irreducible closed subset $Z \subseteq Y$ whose generic point lies in $f(X)$, $f(X) \cap Z$ contains an open subset of Z. This follows from Theorem 10.19, applied to the restriction $f^{-1}(Z) \to Z$. $\square$

Corollary 10.21. *Let $f : X \to Y$ be a morphism of finite type between noetherian schemes. The following assertions are equivalent:*
(i) *The morphism f is open.*
(ii) *For every $x \in X$, $f(\operatorname{Spec} \mathscr{O}_{X,x}) = \operatorname{Spec} \mathscr{O}_{Y,f(y)}$ (where we consider the spectra of the stalks as subsets of X and Y, resp., in the natural way).*

(iii) *For every $x \in X$ and every generalization y' of $y := f(x)$, there exists a generalization x' of x with $f(x') = y'$.*

Proof. Clearly, conditions (ii) and (iii) are equivalent. We have already proved that (i) implies (iii); see Lemma 5.10. So it remains to show that (iii) implies (i). Since condition (iii) is still satisfied after replacing X by an open subset, it is enough to show that $f(X)$ is open. By Theorem 10.20, we know that $f(X)$ is constructible, hence by Lemma 10.17 (2), it is open. $\qquad\square$

See Corollary 10.72 below for a generalization for non-noetherian schemes.

Finiteness conditions in the non-noetherian case

To deal with non-noetherian schemes the properties "quasi-compact" and "locally of finite type" for morphisms of schemes are often not sufficient to get analogous results as in the noetherian case. The notion "quasi-compact" has to be supplemented by "quasi-separated", and the notion "locally of finite type" has to be replaced by "locally of finite presentation". For morphisms of noetherian schemes these new notions are equivalent to the old ones.

(10.7) Quasi-separated morphisms.

Definition 10.22. *A morphism $f\colon X \to Y$ is called* quasi-separated, *if the diagonal morphism $\Delta_{X/Y}\colon X \to X \times_Y X$ is quasi-compact.*

As usual, we call a scheme X *quasi-separated*, if the morphism $X \to \operatorname{Spec}\mathbb{Z}$ is quasi-separated. Since closed immersions are quasi-compact, every separated morphism is quasi-separated.

Proposition 10.23. *Let $f\colon X \to Y$ be a morphism of schemes. The following are equivalent:*
(i) *The morphism f is quasi-separated.*
(ii) *For every affine open $V \subseteq Y$, and all affine open subsets $U_1, U_2 \subseteq f^{-1}(V)$, the intersection $U_1 \cap U_2$ is quasi-compact.*

Proof. For V, U_1, U_2 as in (ii), the inverse image of $U_1 \times_V U_2$ under the diagonal morphism $\Delta_{X/Y}$ is just the intersection $U_1 \cap U_2$. So if f is quasi-separated, then the intersection is quasi-compact. The converse follows, because we can cover the fiber product $X \times_Y X$ by its (affine open) subsets of the form $U_1 \times_V U_2$. $\qquad\square$

This immediately implies that every locally noetherian scheme is quasi-separated, or more generally:

Corollary 10.24. *Let X be a locally noetherian scheme. Then all morphisms $f\colon X \to Y$ are quasi-separated.*

Again the following permanence properties of "quasi-separated" are easy to check.

Proposition 10.25.
(1) *The property "quasi-separated" is stable under composition, stable under base change and local on the target.*
(2) *Let $f\colon X \to Y$ and $g\colon Y \to Z$ be morphisms of schemes such that $g \circ f$ is quasi-separated. Then f is quasi-separated. If f is quasi-compact and surjective, then g is quasi-separated.*

A very common finiteness condition on a scheme X will be that X is quasi-compact and quasi-separated, in other words that X is quasi-compact and the intersection of any two quasi-compact open subsets is again quasi-compact. This condition is satisfied if X is noetherian or if X is affine. One reason for the importance of this finiteness condition is the following principle.

Lemma 10.26. *Let X be a quasi-compact and quasi-separated scheme and let $\mathbf{P}(U)$ be a property of quasi-compact open subschemes U of X such that:*
(a) *$\mathbf{P}(U)$ holds for every open affine subscheme U of X.*
(b) *If $U \subseteq X$ is a quasi-compact open subscheme and $U = \bigcup_{i=1}^{n} U_i$ is a finite open affine covering of U such that $\mathbf{P}(U_i \cap U_j)$ holds for all $i, j = 1, \ldots, n$, then $\mathbf{P}(U)$ holds.*
Then $\mathbf{P}(U)$ holds for all quasi-compact open subschemes U of X.

Proof. Let $U \subseteq X$ be a quasi-compact and separated open subscheme and $U = \bigcup_i U_i$ be a finite open affine covering. Then $U_i \cap U_j$ is also affine for all i, j by Proposition 9.15. Thus (a) and (b) imply that $\mathbf{P}(U)$ holds.

Now let U be an arbitrary quasi-compact open subscheme and let $U = \bigcup_i V_i$ be a finite open affine covering. Then the intersections $V_i \cap V_j$ are quasi-compact (because X is quasi-separated) and separated (as subschemes of the separated scheme V_i). Thus we have just seen that $\mathbf{P}(V_i \cap V_j)$ holds for all i, j and we are done by (b). $\square$

One application of this lemma is the following generalization of Proposition 10.10:

Corollary 10.27. *Let $f\colon X \to Y$ be a quasi-compact quasi-separated morphism of schemes. If $\mathscr{F}$ is a quasi-coherent $\mathscr{O}_X$-module, $f_*\mathscr{F}$ is a quasi-coherent $\mathscr{O}_Y$-module.*

Proof. We may assume that Y is affine. Then X is quasi-compact and quasi-separated. We want to use Lemma 10.26 where we define the property $\mathbf{P}(U)$ to hold ($U \subseteq X$ open quasi-compact) if for every quasi-coherent $\mathscr{O}_U$-module $\mathscr{G}$ its direct image $(f_{|U})_*\mathscr{G}$ is again quasi-coherent. Then $\mathbf{P}(U)$ holds if U is affine (Proposition 7.24). The exact sequence (10.3.2) shows that $\mathbf{P}$ satisfies also the second assumption of Lemma 10.26. This proves the Corollary. $\square$

In particular assume under the hypotheses of the corollary that Y is affine. Then $f_*\mathscr{F}$ is the quasi-coherent module associated to its global sections. Thus we find

$$(10.7.1) \qquad f_*\mathscr{F} = \Gamma(X, \mathscr{F})^{\sim},$$

where we consider the $\Gamma(X, \mathscr{O}_X)$-module $\Gamma(X, \mathscr{F})$ as an A-module via the homomorphism $f_V^\flat\colon A = \Gamma(Y, \mathscr{O}_Y) \to \Gamma(X, \mathscr{O}_X)$.

Because of the frequent use of the term "quasi-compact and quasi-separated", we introduce the following abbreviation:

Abbreviation 10.28. We use *qcqs* as an abbreviation for quasi-compact and quasi-separated:
(1) A qcqs scheme X is a quasi-compact and quasi-separated scheme. In other words, X is quasi-compact and the intersection of any two quasi-compact open subsets is again quasi-compact.
(2) A qcqs morphism $f\colon X \to Y$ is a quasi-compact and quasi-separated morphism. In other words, for any open affine subset $V \subseteq Y$ the inverse image $f^{-1}(V)$ is a qcqs scheme.

Every noetherian scheme and every affine scheme is qcqs. Every morphism with a noetherian source is qcqs by Corollary 10.24.

(10.8) Schematic image.

Definition and Lemma 10.29. *Let $f\colon X \to Y$ be a morphism of schemes. Then there exists a unique closed subscheme $\mathrm{Im}(f)$ of Y satisfying the following conditions.*
(a) *f factorizes through the inclusion $\mathrm{Im}(f) \hookrightarrow Y$.*
(b) *If f factorizes through the inclusion $Z \hookrightarrow Y$ of a closed subscheme Z of Y, then Z majorizes $\mathrm{Im}(f)$ (Definition 3.46).*
We call $\mathrm{Im}(f)$ the schematic image of f and denote the corresponding quasi-coherent ideal of $\mathscr{O}_Y$ by $\mathscr{I}_f$.

Proof. For each closed subscheme Z such that f factorizes through Z let $\mathscr{I}_Z$ be the corresponding quasi-coherent ideal of $\mathscr{O}_Y$. The sum $\sum_Z \mathscr{I}_Z$ is a quasi-coherent ideal by Corollary 7.19 which defines $\mathrm{Im}(f)$. $\qquad\square$

The underlying topological space of $\mathrm{Im}(f)$ contains the closure of $f(X)$, but in general we do not have equality (Exercise 10.18). However we have the following result (which was communicated to us by P. Hartwig).

Proposition 10.30. *Let $f\colon X \to Y$ be a quasi-compact morphism of schemes. Then the kernel $\mathscr{K}_f$ of $f^\flat\colon \mathscr{O}_Y \to f_*\mathscr{O}_X$ is a quasi-coherent ideal, $\mathrm{Im}(f)$ is the closed subscheme of Y defined by $\mathscr{K}_f$, and the underlying topological space of $\mathrm{Im}(f)$ is the closure of $f(X)$ in Y.*

Proof. Let $i\colon Z \to Y$ be a closed subscheme and let $\mathscr{I}_Z$ be the corresponding quasi-coherent ideal of $\mathscr{O}_Y$. Then f factorizes through Z if and only if $f(X) \subseteq Z$ and the homomorphism $f^\flat\colon \mathscr{O}_Y \to f_*\mathscr{O}_X$ factorizes through the surjection $i^\flat\colon \mathscr{O}_Y \to i_*\mathscr{O}_Z$, i.e., if and only if $\mathscr{I}_Z \subseteq \mathscr{K}_f$.

This shows that $\mathscr{K}_f$ equals the ideal sheaf defining $\mathrm{Im}(f)$ if and only if $\mathscr{K}_f$ is a quasi-coherent ideal of $\mathscr{O}_Y$. As $\mathrm{Supp}(\mathscr{O}_Y/\mathscr{K}_f) = \overline{f(X)}$ we see that the underlying topological space of $\mathrm{Im}(f)$ equals $\overline{f(X)}$ in this case.

It remains to show that $\mathscr{K}_f$ is quasi-coherent if f is quasi-compact. We may assume that Y is affine. Let $X = \bigcup_i U_i$ be a finite open affine covering and let $j_i\colon U_i \to X$ be the inclusion. Applying f_* to the injective homomorphism $\mathscr{O}_X \hookrightarrow \bigoplus_i (j_i)_*\mathscr{O}_{U_i}$ we obtain an injection $f_*\mathscr{O}_X \hookrightarrow \mathscr{G} := \bigoplus_i (f \circ j_i)_*\mathscr{O}_{U_i}$. As $f \circ j_i$ is a morphism of affine schemes, $(f \circ j_i)_*\mathscr{O}_{U_i}$ and hence $\mathscr{G}$ are quasi-coherent. Thus $\mathscr{K}_f$ is also the kernel of the homomorphism $\mathscr{O}_Y \to \mathscr{G}$ of quasi-coherent $\mathscr{O}_Y$-modules and thus is quasi-coherent itself. $\qquad\square$

Remark 10.31. Let $i\colon X \to Y$ be a quasi-compact immersion. Then the induced morphism $j\colon X \to \bar{X} := \operatorname{Im}(i)$ is an open schematically dominant immersion because $j^\flat\colon \mathscr{O}_{\bar{X}} = (\mathscr{O}_Y/\mathscr{K}_i)_{|\bar{X}} \to j_*\mathscr{O}_X$ is by definition of $\mathscr{K}_i$ injective. In this case we call $\operatorname{Im}(i)$ the *schematic closure* of X in Y.

Remark 10.32. Let $f\colon X \to Y$ be a morphism of schemes where X is reduced. Let Y' be the closure of $f(X)$ endowed with its reduced subscheme structure. Then f factors through $Y' \hookrightarrow Y$ and Y' is the smallest subscheme having $\overline{f(X)}$ as underlying topological space (Section (3.18)). Thus Y' is the schematic image of f in this case.

(10.9) Morphisms (locally) of finite presentation.

In a non-noetherian situation, it is often not enough to assume that morphisms are of finite type: even if we can embed X as a closed subscheme of an affine space $\mathbb{A}_Y^n$ (which can be done locally if X is a Y-scheme of finite type), it is possible that infinitely many equations are needed to define the closed subscheme X. The more useful finiteness notion is the notion of morphisms that are "(locally) of finite presentation". It will be clear from the definitions that in the noetherian case these notions are equivalent to the notions "(locally) of finite type".

Recall that an A-algebra B is called *of finite presentation* if the following two equivalent conditions are satisfied (see Proposition B.11).
(i) $B \cong A[T_1, \ldots, T_n]/\mathfrak{a}$ for some $n \geq 0$ and some *finitely generated* ideal $\mathfrak{a}$.
(ii) B is a finitely generated A-algebra and for every finitely generated A-algebra B' and every surjective A-algebra homomorphism $\varphi\colon B' \to B$ the kernel $\operatorname{Ker}(\varphi)$ is a finitely generated ideal of B'.

Proposition and Definition 10.33. *A morphism $f\colon X \to Y$ of schemes is called* locally of finite presentation, *if the following equivalent conditions are satisfied:*
(i) *For every affine open subset $V \subseteq Y$ and every affine open $U \subseteq f^{-1}(V)$ the $\Gamma(V, \mathscr{O}_Y)$-algebra $\Gamma(U, \mathscr{O}_X)$ is of finite presentation.*
(ii) *There exist a covering $Y = \bigcup_i V_i$ by open affine subsets $V_i \cong \operatorname{Spec} A_i$, and for each i a covering $f^{-1}(V_i) = \bigcup_j U_{ij}$ by open affine subschemes $U_{ij} \cong \operatorname{Spec} B_{ij}$ of X, such that for all i, j the A_i-algebra B_{ij} is of finite presentation.*

Proof. We leave the proof, which is similar to the proof of Proposition 10.5, to the reader, see Exercise 10.20. $\qquad\square$

Definition 10.34. *A morphism $f\colon X \to Y$ of schemes is called* of finite presentation, *if it is locally of finite presentation and qcqs.*

We list the following standard properties which are not difficult to show using Proposition B.11:

Proposition 10.35.
(1) *An open immersion is locally of finite presentation. It is of finite presentation if it is quasi-compact.*

A closed immersion $i\colon Y \to X$ is of finite presentation if and only if the quasi-coherent ideal of $\mathcal{O}_X$ corresponding to i is an $\mathcal{O}_X$-module of finite type.

(2) The properties "locally of finite presentation" and "of finite presentation" of morphisms of schemes are stable under composition and under base change, and local on the target. The property "locally of finite presentation" is also local on the source.

(3) Let $f\colon X \to Y$ and $g\colon Y \to Z$ be morphisms of schemes such that g is locally of finite type (resp. quasi-separated and locally of finite type). If $g \circ f$ is locally of finite presentation (resp. of finite presentation), then f is locally of finite presentation (resp. of finite presentation).

Remark 10.36. Let Y be a locally noetherian scheme. Then a morphism of schemes $f\colon X \to Y$ is (locally) of finite presentation if and only if it is (locally) of finite type.

Indeed, the condition is certainly necessary. Conversely, if f is (locally) of finite type, then X is (locally) noetherian (Proposition 10.9). Thus f is quasi-separated (Corollary 10.24) and hence (locally) of finite presentation.

Example 10.37. Let R be a ring and let $f_1, \ldots, f_r \in R[T_0, \ldots, T_n]$ be homogeneous polynomials. Then the closed subscheme $V_+(f_1, \ldots, f_r)$ of $\mathbb{P}^n_R$ is an R-scheme of finite presentation.

Example 10.38. Let S be a scheme, let $\mathscr{E}$ be an $\mathcal{O}_S$-module of finite presentation, and let $e \geq 0$ be an integer. Then $\mathrm{Grass}^e(\mathscr{E})$ and in particular $\mathbb{P}(\mathscr{E})$ is an S-scheme of finite presentation.

(10.10) Constructible sets in arbitrary schemes.

In the non-noetherian case, the right notion of constructible subset is slightly more complicated.

Definition 10.39. Let X be a topological space. A subset $C \subseteq X$ is called
(1) retro-compact, if for every quasi-compact open subset $U \subseteq X$, the intersection $C \cap U$ is quasi-compact,
(2) globally constructible, if it is a finite union of subsets of the form $U \cap (X \setminus V)$, where U and V are retro-compact open subsets of X,
(3) constructible, if every point $x \in X$ has an open neighborhood U such that $C \cap U$ is globally constructible in U.

Here we use the terminology of [EGAInew], which differs from that of [EGAIV] §1.

Remark 10.40. Let X be a topological space.
(1) A finite union of retro-compact subsets is again retro-compact.
(2) Every globally constructible subset is retro-compact: If U is a quasi-compact open and if V and V' are retro-compact open subsets, $U \cap V \cap (X \setminus V')$ is closed in the quasi-compact space $U \cap V$ and hence quasi-compact. Thus the claim follows from (1).
(3) If X is quasi-compact, then the notions "globally constructible" and "constructible" coincide.
(4) If X is noetherian, then every open subset is retro-compact, and the notions of "globally constructible" and "constructible" coincide with the notion of "constructible" as defined in Definition 10.12; cf. Exercise 10.28.

The following proposition can be shown in the same way as its noetherian analogue Proposition 10.13.

Proposition 10.41. *Let X be a topological space. The set of globally constructible subsets of X is the smallest subset of the power set of X which contains all retro-compact open subsets of X and is stable under finite intersections, finite unions, and taking complements.*

Corollary 10.42. *Let X be a scheme.*
(1) *Finite intersections, finite unions, and complements of constructible subsets of X are again constructible.*
(2) *Let $C \subseteq X$ be a constructible set and let $Y \subseteq X$ be a closed irreducible subspace. Then either $C \cap Y$ or $Y \setminus C$ contains an open dense subset of Y.*

Proof. Assertion (1) follows from Proposition 10.41 and the fact that in a scheme every point has an open quasi-compact neighborhood in which the notions "globally constructible" and "constructible" coincide.

To prove (2) we may assume that $X = Y$ and that C contains the generic point η of X (otherwise replace C by its complement). We claim that C contains an open dense subset. Replacing X by an open affine neighborhood of η, we may assume C is globally constructible. In this case we conclude by the same argument as in the proof of Proposition 10.14. $\qquad\square$

Proposition 10.43. *Let $f\colon X \to Y$ be a morphism of schemes, and let $E \subseteq Y$ be a (globally) constructible subset. Then the inverse image $f^{-1}(E)$ is (globally) constructible in X.*

Proof. We first assume that E is globally constructible. In this case the claim follows from the following observation: if $V \subseteq Y$ is open, retro-compact, i. e. the morphism $V \to Y$ is a quasi-compact open immersion, then the same is true for the morphism $f^{-1}(V) = V \times_Y X \to X$, see Proposition 10.3.

Now assume that E is constructible, let $x \in X$, and let $V \subseteq Y$ be an open neighborhood of $f(x)$ such that $E \cap V$ is constructible in V. By the first case, $f^{-1}(E) \cap f^{-1}(V)$ is constructible in $f^{-1}(V)$, and the latter is an open neighborhood of x. $\qquad\square$

For qcqs schemes the notion of constructibility simplifies significantly, as the following proposition shows.

Proposition 10.44. *Let X be a qcqs scheme. An open subset $U \subseteq X$ is constructible if and only if it is quasi-compact. A subset C is constructible if and only if it is the union of finitely many subsets of the form $V \setminus V' = V \cap (X \setminus V')$ where V and V' are quasi-compact open subsets.*

Proof. It suffices to show the first assertions. As X is quasi-separated, U is retro-compact if and only if U is quasi-compact. Therefore U is constructible if U is quasi-compact. Conversely, if U is constructible, then U is globally constructible because X is quasi-compact and has a basis of retro-compact open subsets. Therefore U is retro-compact. Cf. Remark 10.40. $\qquad\square$

Similarly as in the noetherian case (see Lemma 10.16), we have:

Proposition 10.45. *Let X be a qcqs scheme, and let $C \subseteq X$ be a constructible subset. Then there exists an affine scheme X' and a morphism $f\colon X' \to X$ of finite presentation such that $f(X') = C$.*

Proof. We leave the proof as an exercise, see Exercise 10.43.   □

Remark 10.46. The same proof as for Lemma 10.17 then shows that if X is an arbitrary scheme, a constructible subset C of X is open (resp. closed) if and only if it is stable under generization (resp. specialization).

(10.11) Extending quasi-coherent modules.

Let X be a scheme, let $U \subseteq X$ be an open subset and let $\mathscr{F}'$ be a quasi-coherent $\mathscr{O}_U$-module. Assume that the open immersion $j\colon U \hookrightarrow X$ is quasi-compact (e.g., if X is locally noetherian or if X is quasi-separated and U is quasi-compact). Then $j_*\mathscr{F}'$ is a quasi-coherent $\mathscr{O}_X$-module (Proposition 10.10) which extends $\mathscr{F}'$ (i.e., $(j_*\mathscr{F}')_{|U} \cong \mathscr{F}'$).

Now assume that $\mathscr{F}'$ is an $\mathscr{O}_U$-module of finite type. Nevertheless $j_*\mathscr{F}'$ will rarely be of finite type. Thus we now study the question whether it is possible to extend $\mathscr{F}'$ to a quasi-coherent $\mathscr{O}_X$-module $\mathscr{F}$ of finite type. In fact we will construct $\mathscr{F}$ as an $\mathscr{O}_X$-submodule of $j_*\mathscr{F}'$. We start with the following lemma.

Lemma 10.47. *Let X be a quasi-compact scheme and let $u\colon \mathscr{G} \to \mathscr{F}$ be a homomorphism of quasi-coherent $\mathscr{O}_X$-modules. Assume that $\mathscr{F}$ is of finite type and that $\mathscr{G}$ is the filtered inductive limit of quasi-coherent $\mathscr{O}_X$-modules $\mathscr{G}_\lambda$. Then u is surjective if and only if there exists an index λ_0 such that the homomorphism $\mathscr{G}_\lambda \to \mathscr{F}$ is surjective for all $\lambda \geq \lambda_0$.*

Proof. The condition is clearly sufficient. To prove the converse, we may assume that $X = \operatorname{Spec} R$ is affine because we can cover X by finitely many affine schemes. Then $\mathscr{F} \cong \tilde{M}$, where M is an R-module generated by a finite set E. If $u\colon N = \varinjlim N_\lambda \to M$ is a surjective homomorphism of R-modules, for each $m \in E$ there exists an index $\lambda(m)$ such that m is contained in the image of $N_{\lambda(m)} \to M$, and we can take λ_0 to be an index which is $\geq \lambda(m)$ for all $m \in E$.   □

Proposition 10.48. *Let X be a qcqs scheme, let $U \subseteq X$ be a quasi-compact open subscheme, and let $\mathscr{G}$ be a quasi-coherent $\mathscr{O}_X$-module. For all quasi-coherent $\mathscr{O}_U$-submodules $\mathscr{F}' \subseteq \mathscr{G}_{|U}$ of finite type there exists a quasi-coherent $\mathscr{O}_X$-submodule $\mathscr{F} \subseteq \mathscr{G}$ of finite type such that $\mathscr{F}_{|U} = \mathscr{F}'$.*

Proof. We first show that it is always possible to extend $\mathscr{F}'$ to some $\mathscr{O}_X$-submodule of $\mathscr{G}$. Denote the open immersion $U \hookrightarrow X$ by j. Define an $\mathscr{O}_X$-submodule $\overline{\mathscr{F}}$ of $\mathscr{G}$ as follows

$$\Gamma(V, \overline{\mathscr{F}}) := \{\, s \in \Gamma(V, \mathscr{G}) \,;\, s_{|U \cap V} \in \Gamma(U \cap V, \mathscr{F}') \,\}, \qquad V \subseteq X \text{ open.}$$

In other words, $\overline{\mathscr{F}}$ is the inverse image of the $\mathscr{O}_X$-submodule $j_*\mathscr{F}'$ of $j_*j^*\mathscr{G}$ under the canonical morphism $\mathscr{G} \to j_*j^*\mathscr{G}$. In particular $\overline{\mathscr{F}}$ is a quasi-coherent $\mathscr{O}_X$-submodule of $\mathscr{G}$. We have $\overline{\mathscr{F}}_{|U} = \mathscr{F}'$.

We now prove the proposition in the special case that $X = \operatorname{Spec} A$ is affine. Then $\overline{\mathscr{F}} = \tilde{N}$ for some A-submodule N of $\Gamma(X, \mathscr{G})$. We write N as the filtered inductive limit of its finitely generated A-submodules N_λ. Set $\mathscr{F}_\lambda := \tilde{N}_\lambda$ which is a quasi-coherent $\mathscr{O}_X$-submodule of $\mathscr{G}$ of finite type. Then $\mathscr{F}' = \overline{\mathscr{F}}_{|U}$ is the inductive limit of the $\mathscr{F}_{\lambda|U}$ and by Lemma 10.47 there exists an index λ such that $\mathscr{F}_{\lambda|U} = \mathscr{F}'$.

If X is not necessarily affine, let $(V_\alpha)_{1 \le \alpha \le m}$ be a finite open affine covering of X, set $W_\alpha := \bigcup_{\beta \le \alpha} V_\beta$ and $U_\alpha := U \cup W_\alpha$ for all $\alpha = 0, \ldots, m$. The U_α are qcqs and the set $\mathcal{U} := \{ U_\alpha \ ; \ \alpha = 0, \ldots, m \}$ is totally ordered by inclusion: $\mathcal{U} = \{ U = U_0 = U^0 \subsetneq U^1 \subsetneq \cdots \subsetneq U^n = U_m = X \}$. We set $\mathscr{F}_0 := \mathscr{F}'$ and show by induction on n that there exists a family $(\mathscr{F}_i)_{1 \le i \le n}$ of $\mathcal{O}_{U^i}$-submodules $\mathscr{F}_i$ of $\mathscr{G}_{|U^i}$ such that $\mathscr{F}_{i|U^j} \cong \mathscr{F}_j$ for all $0 \le j < i$.

Replacing U by U^{n-1} and using the induction hypothesis, we see that it suffices to prove the proposition if $X = U \cup V$, where V is open affine. As we can glue $\mathcal{O}_X$-modules, it suffices to find a quasi-coherent $\mathcal{O}_V$-submodule $\mathscr{F}''$ of finite type of $\mathscr{G}_{|V}$ such that $\mathscr{F}''_{|U \cap V} \cong \mathscr{F}'_{|U \cap V}$. As X is quasi-separated, $U \cap V$ is quasi-compact, and the existence of $\mathscr{F}''$ follows from the special case. $\qquad \square$

Corollary 10.49. *Let X be a qcqs scheme, let $U \subseteq X$ be a quasi-compact open subscheme (e.g., if X is noetherian and $U \subseteq X$ is arbitrary open), and let $\mathscr{F}'$ be a quasi-coherent $\mathcal{O}_U$-module of finite type. Then there exists a quasi-coherent $\mathcal{O}_X$-module $\mathscr{F}$ of finite type such that $\mathscr{F}_{|U} \cong \mathscr{F}'$.*

Under the hypotheses of the corollary one can also show that if $\mathscr{F}'$ is an $\mathcal{O}_U$-module of finite presentation, then there exists an $\mathcal{O}_X$-module $\mathscr{F}$ of finite presentation such that $\mathscr{F}_{|U} \cong \mathscr{F}'$ ([EGAInew] (6.9.11)).

Proof. The open immersion $j \colon U \to X$ is quasi-compact and thus we can apply Proposition 10.48 with the quasi-coherent $\mathcal{O}_X$-module $\mathscr{G} := j_* \mathscr{F}'$. $\qquad \square$

Corollary 10.50. *Let X be a qcqs scheme.*
(1) *Every quasi-coherent $\mathcal{O}_X$-module is the filtered inductive limit of its quasi-coherent submodules of finite type.*
(2) *Every quasi-coherent $\mathcal{O}_X$-module is a filtered inductive limit of $\mathcal{O}_X$-modules of finite presentation.*

Note that if X is noetherian (and thus qcqs), then every quasi-coherent $\mathcal{O}_X$-module of finite type is of finite presentation (Proposition 7.46), and the second assertion follows from the first one.

Proof. We will prove only Assertion (1); for Assertion (2) we refer to [EGAInew] (6.9.12). The claim is clear if X is affine because the analogous assertion for a module over a ring holds. In general we find a finite open affine covering $X = \bigcup_i U_i$. Let $\mathscr{G}$ be a quasi-coherent $\mathcal{O}_X$-module. Then $\mathscr{G}_{|U_i}$ is the limit of its quasi-coherent submodules $\mathscr{F}'_{i,\lambda}$ of finite type for all i. Each $\mathcal{O}_{U_i}$-submodule $\mathscr{F}'_{i,\lambda}$ can be extended to a quasi-coherent submodule $\mathscr{F}_{i,\lambda}$ of $\mathscr{G}$ of finite type by Proposition 10.48. All finite sums of submodules of the form $\mathscr{F}_{i,\lambda}$ are again quasi-coherent $\mathcal{O}_X$-submodules of $\mathscr{G}$ of finite type and $\mathscr{F}$ is the inductive limit of these finite sums. $\qquad \square$

Corollary 10.51. *Let X be a qcqs scheme. Every quasi-coherent $\mathcal{O}_X$-algebra $\mathscr{B}$ is the filtered inductive limit of its quasi-coherent $\mathcal{O}_X$-subalgebras of finite type.*

Proof. The $\mathcal{O}_X$-module $\mathscr{B}$ is the filtered inductive limit of its quasi-coherent $\mathcal{O}_X$-submodules of finite type by Corollary 10.50. Thus it suffices to show that the $\mathcal{O}_X$-subalgebra $\mathscr{C}$ generated by such a submodule $\mathscr{E}$ is quasi-coherent and of finite type. To do this we may assume that $X = \operatorname{Spec} A$ and hence $\mathscr{B} = \tilde{B}$ for an A-algebra B and $\mathscr{E} = \tilde{E}$

for a finitely generated A-submodule E of B. The A-subalgebra C of B generated by E is finitely generated. Moreover, C is the sum of the images of the multiplication homomorphisms $E^{\otimes n} \to B$ for $n \geq 0$. Thus $\mathscr{C} = \tilde{C}$ is quasi-coherent and finitely generated by Corollary 7.19. $\qquad\square$

(10.12) Extending morphisms.

As a warm-up exercise for the next sections we prove a result which relates homomorphisms between the local rings of two schemes to scheme morphisms defined on a suitable open neighborhood of the point in question. The following proposition will also follow from the more general results about schemes over inductive limits of rings below (see Theorem 10.63), but since it is of great importance, we will prove it separately here.

We call a morphism $f \colon Y \to S$ *locally of finite type at a point* $y \in Y$, if there exist affine open neighborhoods V of $f(y)$ and U of y such that $\Gamma(U, \mathscr{O}_Y)$ is a $\Gamma(V, \mathscr{O}_S)$-algebra of finite type. Similarly, we can speak of morphisms *locally of finite presentation at a point*.

Proposition 10.52. *Let S be a scheme, let X and Y be S-schemes, and let $x \in X$, $y \in Y$ be points lying over the same point $s \in S$.*
(1) *Suppose that Y is locally of finite type over S at y. Let $f, f' \colon X \to Y$ be S-morphisms with $f(x) = f'(x) = y$ and such that the induced morphisms $\mathscr{O}_{Y,y} \to \mathscr{O}_{X,x}$ coincide. Then f and f' coincide on an open neighborhood of x.*
(2) *Suppose that Y is locally of finite presentation over S at y. Let $\varphi_x \colon \mathscr{O}_{Y,y} \to \mathscr{O}_{X,x}$ be a local $\mathscr{O}_{S,s}$-homomorphism. Then there exists an open neighborhood U of x and an S-morphism $f \colon U \to Y$ with $f(x) = y$ and such that the homomorphism $\mathscr{O}_{Y,y} \to \mathscr{O}_{X,x}$ induced by f is φ_x.*
(3) *If in (2) in addition X is locally of finite type over S at x, then one can find U and f such that f is of finite type.*
(4) *If in (2) in addition X is locally of finite presentation over S at x and φ_x is an isomorphism, then one can find U and f such that f is an isomorphism of U onto an open neighborhood of y.*

Proof. We can, for all four parts of the proposition, replace S, Y and X by suitable affine open subschemes, and may therefore assume that $S = \operatorname{Spec} R$, $X = \operatorname{Spec} B$ and $Y = \operatorname{Spec} A$ are affine. The points x and y correspond to prime ideals $\mathfrak{q} \subset B$ and $\mathfrak{p} \subset A$

We first prove (1). The morphisms f, f' correspond to R-algebra homomorphisms φ, $\varphi' \colon A \to B$ with $\mathfrak{p} = \varphi^{-1}(\mathfrak{q}) = \varphi'^{-1}(\mathfrak{q})$, and we assume that the morphisms $A_{\mathfrak{p}} \to B_{\mathfrak{q}}$ induced by φ, φ' coincide.

Now by assumption A is a finitely generated R-algebra, and we fix generators $a_1, \dots, a_n$. We then know that for every i, the images of $\varphi(a_i)$ and $\varphi'(a_i)$ in $B_{\mathfrak{q}}$ are the same. Hence they coincide in some localization B_t, $t \in B \setminus \mathfrak{q}$, so the restrictions of f and f' to $D_X(t)$ coincide.

Let us prove (2). By assumption, A is isomorphic to $R[T_1, \dots, T_n]/(f_1, \dots, f_m)$. We can lift the images of the T_i in $B_{\mathfrak{q}}$ to some B_t, and choosing $t \in B \setminus \mathfrak{q}$ suitably, we can achieve that all the f_j map to 0 in B_t, so that the homomorphism $A \to A_{\mathfrak{p}} \to B_{\mathfrak{q}}$ factors through $\varphi \colon A \to B_t$. It is clear that the corresponding morphism $D(t) \to Y$ maps x to y and induces the morphism φ_x on the stalks.

To prove (3), we continue to use the notation of the proof of (2), and observe that, if B is a finitely generated R-algebra, then a fortiori B_t is a finitely generated A-algebra.

Finally, to prove (4) we apply (2) to φ_x and φ_x^{-1} to get morphisms $f\colon U \to Y$ and $g\colon V \to X$ extending φ_x and φ_x^{-1}, respectively. Then (1) implies that after possibly shrinking U and V, we have $f \circ g = \mathrm{id}_V$ and $g \circ f = \mathrm{id}_U$. $\qquad\square$

Schemes over inductive limits of rings

In the vast majority of cases, objects arising in algebraic geometry are "defined by finitely many equations"; in particular, only "finitely many indeterminates" and "finitely many coefficients" occur. Technically, these properties are captured by the notion of morphisms of finite presentation. Therefore, even if the base is a very big ring (say the field $\mathbb{C}$ of complex numbers), to describe a $\mathbb{C}$-scheme of finite presentation we need only work with a finitely generated subfield. The purpose of the present section is to make this observation more precise, and to develop these techniques in a more general situation.

The standard reference for this kind of result is [EGAIV] vol. 3, §8, §9, which contains a very comprehensive treatment.

Because one of the important applications of the results below is the elimination of noetherianness hypotheses, we choose not to restrict to the noetherian case. However, even in the noetherian case these techniques are of great importance, so for the reader who is only interested in this case, we add the reminder that for a noetherian scheme S,

(1) every open subset of S is quasi-compact, and in particular S is quasi-separated,
(2) every morphism $X \to S$ of finite type is of finite presentation, and X is then again noetherian,
(3) the notion of constructibility simplifies, see Section (10.5) and the remarks in Section (10.10).

(10.13) Introduction and notation.

To avoid repetition, we collect the relevant notation in one place:

1. Let Λ be a filtered partially ordered set with a unique minimal element 0.
2. Let $(R_\lambda)_\lambda$ be an inductive system of rings indexed by Λ and with transition maps $\sigma_{\lambda\mu}\colon R_\lambda \to R_\mu$. Set $R := \varinjlim R_\lambda$ and let $\sigma_\lambda\colon R_\lambda \to R$ be the natural maps.
   Set $S_\lambda := \operatorname{Spec} R_\lambda$, $S := \operatorname{Spec} R$, and let $s_{\lambda\mu}\colon S_\mu \to S_\lambda$, $s_\lambda\colon S \to S_\lambda$ be the morphisms associated with the σ's.
3. Let X_0 be an S_0-scheme, $X_\lambda := X_0 \times_{S_0} S_\lambda$, $X := X_0 \times_{S_0} S$, $x_{\lambda\mu}\colon X_\mu \to X_\lambda$, $x_\lambda\colon X \to X_\lambda$ be the morphisms obtained by base change from the $s_{\lambda\mu}$, s_λ.
4. If X_0 is affine, we set $A_\lambda := \Gamma(X_\lambda, \mathscr{O}_{X_\lambda})$, $A := \Gamma(X, \mathscr{O}_X)$, and denote the ring homomorphisms corresponding to the $x_{\lambda\mu}$, x_λ by $\xi_{\lambda\mu}$, ξ_λ.
5. Similarly, for an S_0-scheme Y_0, we have Y_λ, Y, $y_{\lambda\mu}$, y_λ, and if Y_0 is affine, B_λ, B, $v_{\lambda\mu}$, v_λ.
6. For a morphism $f_0\colon X_0 \to Y_0$ of S_0-schemes, we denote by $f_\lambda\colon X_\lambda \to Y_\lambda$, $f\colon X \to Y$ the morphisms obtained by base change.

The basic goal is to compare properties of X with properties of (some or all) X_λ. More precisely:

(EXIST) In the setting 1., 2. let X be an S-scheme of finite presentation. Then there exist an index λ and an S_λ-scheme X_λ such that $X_\lambda \times_{S_\lambda} S \cong X$. More generally, given an S-morphism $f\colon X \to Y$ of S-schemes of finite presentation, there exist an index λ and an S_λ-morphism $f_\lambda\colon X_\lambda \to Y_\lambda$ of S_λ-schemes of finite presentation such that $f_\lambda \times_{S_\lambda} \mathrm{id}_S = f$. This is shown in Corollary 10.67.

(PROP) Throughout the book we will see plenty of examples of properties $\mathbf{P}$ such that the following assertions holds: In the setting 1.–3., 5., 6. assume that X (resp. that f) has the property $\mathbf{P}$. Then there exists an index λ such that X_λ (resp. that f_λ) has the property $\mathbf{P}$.

We will then say that $\mathbf{P}$ is *compatible with inductive limits of rings*.

Very often the results proved in this section are used as follows. In the setting 1.–2. let $f\colon X \to Y$ be a morphism of R-schemes that are of finite presentation over R. Assume that f possesses a certain property $\mathbf{P}$ that is compatible with inductive limits of rings and that we want to show that f has a property $\mathbf{Q}$ that is stable under base change. Then there exist an index λ and a morphism $f_\lambda\colon X_\lambda \to Y_\lambda$ of R_λ-schemes of finite presentation such that $f_\lambda \times \mathrm{id}_S = f$ and such that f_λ possesses $\mathbf{P}$. If we can show that $\mathbf{P}$ for f_λ implies $\mathbf{Q}$ for f_λ, then we have shown the same implication for f.

Let us give some frequently encountered examples where this technique is applied to morphisms $f\colon X \to Y$ of S-schemes (or to an S-scheme $X \to S$).

(1) If T is a scheme, $t \in T$, then $R := \mathscr{O}_{T,t}$ is the inductive limit of the coordinate rings R_λ of suitable open affine subschemes of T. Therefore we can exploit the techniques to be developed in order to relate properties over the local ring at t to properties over some open neighborhood of t. (See Proposition 10.95 for an example where the generic point is "approximated" by non-empty open subsets to prove the constancy of the fiber dimension near generic points.)

(2) Any ring R can be written as the inductive limit of its finitely generated $\mathbb{Z}$-subalgebras R_λ. In this way one can often reduce questions over arbitrary rings to the noetherian case (see Theorem 10.69).

More generally, let R_0 be a ring and let R be an R_0-algebra. Then R is isomorphic to a filtered inductive limit of R_0-algebras of finite presentation (see Exercise 10.21 or [EGAInew] $\mathbf{0}_I$ (6.3.8)).

(3) There are other applications of (2): The reduction of properties of schemes over fields of characteristic 0 to properties of schemes over fields of positive characteristic (see Remark 10.90 below for details).

(4) Another variant of (2) is the case that $R = L$ is a field extension of a field K. Then we can write L as the inductive limit of its subfields which are finitely generated over K (see Proposition 10.78, for instance.)

Note that in previous chapters we have already used ad hoc arguments of this flavor, see for instance Proposition 5.9 or Lemma 6.17.

(10.14) The spectrum of an inductive limit.

We work in the setting of Section (10.13), 1., 2.

Proposition 10.53. *The scheme S (together with the s_λ) is the projective limit in the category of schemes of the projective system $(S_\lambda, s_{\lambda\mu})$.*

Proof. Given a scheme X, we have to show that the natural map

$$\operatorname{Hom}(X, S) \to \varprojlim \operatorname{Hom}(X, S_\lambda)$$

is bijective. Since the S_λ and S are affine, we can rewrite these sets of homomorphisms as $\operatorname{Hom}(R, \Gamma(X, \mathscr{O}_X))$ and $\operatorname{Hom}(R_\lambda, \Gamma(X, \mathscr{O}_X))$ (see Proposition 3.4), and the assertion follows from the definitions. $\qquad \square$

Lemma 10.54. *The scheme S is empty if and only if for some λ, $S_\lambda = \emptyset$ (and hence $S_\mu = \emptyset$ for all $\mu \geq \lambda$).*

Proof. If the inductive limit $A = \varinjlim A_\lambda$ is 0, then $1 = 0$ in this ring, so $1 = 0$ in some A_λ. $\qquad \square$

(10.15) Topological properties of limits.

We work in the setting of Section (10.13) 1.–3.

Proposition 10.55. *The topological space underlying X is the projective limit of the system of topological spaces X_λ in the category of topological spaces.*

Proof. We denote by T the projective limit in the category of topological spaces, and by $t_\lambda \colon T \to X_\lambda$ and by $\psi \colon X \to T$ the continuous maps obtained from the universal property of $\varprojlim$. We will show that ψ is a homeomorphism by proving that the open sets in X are precisely the sets of the form $\psi^{-1}(V)$, $V \subseteq T$ open, and that ψ is bijective.

For the first step, it is in fact enough to show that every open subset of X is a union of sets $\psi^{-1}(V)$, $V \subseteq T$ open. Consider an open affine subset $U_0' \subseteq X_0$ and let $U' \subseteq X$ (resp. $U_\lambda' \subseteq X_\lambda$) be its inverse image, an affine open inside X (resp. inside X_λ). Let $U = D(s)$ for $s \in \Gamma(U', \mathscr{O}_{U'})$. Every open subset of X can be written as a union of subsets of this form. But since $\Gamma(U', \mathscr{O}_{U'}) = \varinjlim \Gamma(U_\lambda', \mathscr{O}_{U_\lambda'})$, we find λ so that we can lift s to $s_\lambda \in \Gamma(U_\lambda', \mathscr{O}_{U_\lambda'})$, and since localization is compatible with tensor products, we see that $D(s) = D_{U_\lambda'}(s_\lambda) \times_{S_\lambda} S$. Topologically, this means that $D(s) = x_\lambda^{-1}(D(s_\lambda)) = \psi^{-1}(t_\lambda^{-1}(D(s_\lambda)))$, and by definition of the topology on T, $t_\lambda^{-1}(D(s_\lambda))$ is open in T.

Now it remains to prove that ψ is bijective. For any two distinct points of X, we can find an open subset which contains exactly one of them (Proposition 3.25), so from the above it follows that ψ is injective. The surjectivity can be checked on an open covering of T; we take one induced by a covering of X_0 by open affine subsets, and can hence assume that all X_λ and X are affine. We denote by A_λ, $A = \varinjlim A_\lambda$ the corresponding rings, and by $\xi_{\lambda\mu}$ the transition maps. A point of T then corresponds to a system $\mathfrak{p}_\lambda \subset A_\lambda$ of prime ideals with $\xi_{\lambda\mu}^{-1}(\mathfrak{p}_\mu) = \mathfrak{p}_\lambda$. Then the $\mathfrak{p}_\lambda$ form an inductive system, and $\mathfrak{p} := \varinjlim \mathfrak{p}_\lambda$ is an ideal of A (by the exactness of $\varinjlim$), which is easily checked to be a prime ideal. The image of $\mathfrak{p}$ under ψ is the point $(\mathfrak{p}_\lambda)_\lambda$, and thus the surjectivity is proved. $\qquad \square$

Proposition 10.56. *Assume that X_0 is qcqs. Let $E_0, F_0 \subseteq X_0$ be constructible subsets, and set $E_\lambda = x_{0\lambda}^{-1}(E_0)$, $E = x_0^{-1}(E_0)$, and similarly for F_λ, F. The following are equivalent:*

(i) $E \subseteq F$
(ii) *There exists λ such that $E_\lambda \subseteq F_\lambda$.*

Proof. Clearly the second condition implies the first one. So let us assume that $E \subseteq F$. We set $G_0 = E_0 \setminus F_0$, $G_\lambda = x_{0\lambda}^{-1}(G_0) = E_\lambda \setminus F_\lambda$, $G = \dot{x}_\lambda^{-1}(G_0) = E \setminus F$. Then our assumption says that $G = \emptyset$, and we have to show that $G_\lambda = \emptyset$ for some λ.

Because G_0 is again constructible, by Proposition 10.45 there exists an affine scheme X_0' and a morphism $f_0 \colon X_0' \to X_0$ with $f(X_0') = G_0$. We set $X_\lambda' = X_0' \times_{X_0} X_\lambda = X_0' \times_{S_0} S_\lambda$, $X' = X_0' \times_{X_0} X$, and similarly for f_λ, f, and obtain $G_\lambda = f_\lambda(X_\lambda')$, $G = f(X')$ (see Lemma 4.28). So we must have $X' = \emptyset$, and Lemma 10.54 shows that $X_\lambda' = \emptyset$ for some λ, as desired. $\qquad \square$

For a topological space X, we denote by $\mathfrak{C}(X)$, $\mathfrak{Oc}(X)$, and $\mathfrak{Fc}(X)$ the sets of constructible, open constructible, and closed constructible subsets, resp. If X_0 is a qcqs scheme, then X is qcqs. In this case, an open subset of X is constructible if and only if it is quasi-compact (see Proposition 10.44). Since taking inverse images under morphisms preserves all these properties (Proposition 10.43), in the setting above we have natural maps $\mathfrak{C}(X_\lambda) \to \mathfrak{C}(X)$, and hence $\varinjlim \mathfrak{C}(X_\lambda) \to \mathfrak{C}(X)$, and similarly for $\mathfrak{Oc}(X)$ and $\mathfrak{Fc}(X)$.

Theorem 10.57. *Assume that X_0 is qcqs. Then the following maps are bijections:*
(1) $\varinjlim \mathfrak{C}(X_\lambda) \to \mathfrak{C}(X)$
(2) $\varinjlim \mathfrak{Oc}(X_\lambda) \to \mathfrak{Oc}(X)$
(3) $\varinjlim \mathfrak{Fc}(X_\lambda) \to \mathfrak{Fc}(X)$

Proof. The injectivity follows from Proposition 10.56. Let us prove the surjectivity. Because every constructible subset is a union of subsets of the form $U \cap V^c$ for open quasi-compact subsets, it is enough to prove the surjectivity in case (2), from which (3) follows by taking complements. But we have seen in the proof of Proposition 10.55 that we can cover any open subset U of X by principal open subsets $D(s_i)$ that are inverse images of a principal open subset D_i of some X_{λ_i}. If U is quasi-compact, finitely many suffice. Let λ be an index with $\lambda \geq \lambda_i$ for all i. Then $\bigcup_i x_{\lambda_i \lambda}^{-1}(D_i)$ is an open quasi-compact subset of X_λ whose inverse image in X is U. $\qquad \square$

(10.16) Modules of finite presentation and limits.

We will now prove a generalization of Proposition 7.27. We work in the setting of Section (10.13) 1., 2. In addition, let $\mathscr{F}_0$ and $\mathscr{G}_0$ be $\mathscr{O}_{S_0}$-modules, and for each λ, set $\mathscr{F}_\lambda := s_{0\lambda}^* \mathscr{F}_0$. Let $\mathscr{F} = s_\lambda^* \mathscr{F}_\lambda$, and similarly for $\mathscr{G}$.

The functoriality of the pull-back $s_{\lambda\mu}^*$ gives us a natural homomorphism

$$u_{\mathscr{F},\mathscr{G}} \colon \varinjlim \operatorname{Hom}_{S_\lambda}(\mathscr{F}_\lambda, \mathscr{G}_\lambda) \to \operatorname{Hom}_S(\mathscr{F}, \mathscr{G}).$$

Theorem 10.58. *Suppose that for some λ, $\mathscr{F}_\lambda$ is quasi-coherent and of finite type, and that $\mathscr{G}_\lambda$ is quasi-coherent. Then the homomorphism $u_{\mathscr{F},\mathscr{G}}$ is injective.*

If furthermore $\mathscr{F}_\lambda$ is of finite presentation, then $u_{\mathscr{F},\mathscr{G}}$ is bijective.

Proof. Because of the quasi-coherence condition, everything is readily translated into a statement about modules over rings (recall that the S_λ are affine by assumption), so that it is enough to prove the following lemma. $\qquad \square$

Lemma 10.59. *Let $(R_\lambda)_\lambda$ be an inductive system of rings, let M_0 and N_0 be R_0-modules, and write $R = \varinjlim R_\lambda$; $M_\lambda = M_0 \otimes_{R_0} R_\lambda$, $M = M_0 \otimes_{R_0} R$, and similarly for N_λ, N. If M_0 is finitely generated as an R_0-module, then the natural homomorphism*

$$\varinjlim \mathrm{Hom}_{R_\lambda}(M_\lambda, N_\lambda) \to \mathrm{Hom}_R(M, N)$$

is injective. If M_0 is of finite presentation, then it is bijective.

Proof. We rewrite
$$\mathrm{Hom}_{R_\lambda}(M_\lambda, N_\lambda) = \mathrm{Hom}_{R_0}(M_0, N_\lambda)$$
and
$$\mathrm{Hom}_R(M, N) = \mathrm{Hom}_{R_0}(M_0, N).$$

It is clear that the statement of the lemma is true if $M_0 \cong R_0^n$ for some integer $n \geq 0$. Now fix a presentation $R_0^n \to M_0 \to 0$ (or $R_0^m \to R_0^n \to M_0 \to 0$ in the second case). The assertion then follows from the left-exactness of the Hom functor and the exactness of $\varinjlim$. $\qquad\square$

Theorem 10.60. *In the setting of Section (10.13) 1. and 2. let $\mathscr{F}$ be an $\mathscr{O}_S$-module of finite presentation.*
(1) *There exist $\lambda \in \Lambda$ and an $\mathscr{O}_{S_\lambda}$-module $\mathscr{F}_\lambda$ of finite presentation such that $s_\lambda^* \mathscr{F}_\lambda$ is isomorphic to $\mathscr{F}$.*
(2) *The $\mathscr{O}_S$-module $\mathscr{F}$ is finite locally free if and only if there exists a $\mu \geq \lambda$ such that $\mathscr{F}_\mu := s_{\lambda\mu}^* \mathscr{F}_\lambda$ is finite locally free.*

Using a gluing argument is it not difficult to prove a much more general result, see Exercise 10.33.

Proof. Again, we are in an affine situation, so $\mathscr{F} = \widetilde{M}$ for some R-module of finite presentation. Let $R^m \to R^n \to M \to 0$ be a presentation of M. Clearly the homomorphism $f \colon R^m \to R^n$ (which we can imagine as given by a matrix with entries in R) is induced from a homomorphism $f_\lambda \colon R_\lambda^m \to R_\lambda^n$ for suitable λ. Then $\mathscr{F}_\lambda = \widetilde{\mathrm{Coker}\, f_\lambda}$ has the desired property because of right-exactness of $\otimes$.

As $s_\mu^* \mathscr{F}_\mu \cong \mathscr{F}$, the condition is clearly sufficient. Conversely if $\mathscr{F}$ is locally free, then M is projective and thus a direct summand of a finitely generated free R-module. Thus we may assume f as above with $m = n$ and $f^2 = f$. Considering this as an identity of matrices, we see that there exists a $\mu \geq \lambda$ such that for $f_\mu := f_\lambda \otimes \mathrm{id}_{R_\mu}$ we have $f_\mu^2 = f_\mu$. $\qquad\square$

(10.17) Morphisms of schemes and limits.

We work in the setting of Section (10.13) 1.–3., 5., 6. We assume that X_0 is quasi-compact.

Lemma 10.61. *Let X_0 be quasi-compact, and let $U_0 \subseteq X_0$ be an open subset with $U := x_0^{-1}(U_0) = X$. Then $U_\lambda := x_{0\lambda}^{-1}(U_0) = X_\lambda$ for some $\lambda \in \Lambda$.*

Proof. Since X_0 is quasi-compact, and our index set is filtered, we can check this on a finite affine open covering, and hence may assume that X_0 is affine. Then X_λ and X are affine. We set $Z_0 = X_0 \setminus U_0$, and view it as a reduced closed subscheme of X_0 which is affine. By Proposition 4.20 and the remark following it, we have $Z_\lambda := Z_0 \times_{X_0} X_\lambda = X_\lambda \setminus U_\lambda$, and $Z := Z_0 \times_{X_0} X = X \setminus U = \emptyset$ (as topological spaces). Lemma 10.54 now shows that $Z_\lambda = \emptyset$ for some λ, i. e. $U_\lambda = X_\lambda$. $\square$

We need another lemma about inductive limits of algebras; we use the notation of Section (10.13) 1., 2.

Lemma 10.62. *Let A_0 be an R_0-algebra of finite presentation, and let B_0 be an R_0-algebra. Set $A_\lambda = A_0 \otimes_{R_0} R_\lambda$, $A = A_0 \otimes_{R_0} R$, and similarly for B_λ, B. Then the canonical homomorphism*

$$\varinjlim_\lambda \mathrm{Hom}_{(R_\lambda\text{-Alg})}(A_\lambda, B_\lambda) \to \mathrm{Hom}_{(R\text{-Alg})}(A, B)$$

is bijective.

Proof. Using the universal property of the tensor product, we can rewrite the Hom-sets on the left, resp. right, hand side as $\mathrm{Hom}_{R_0}(A_0, B_\lambda)$ and $\mathrm{Hom}_{R_0}(A_0, B)$, and therefore in the sequel we consider the homomorphism

$$\varinjlim_\lambda \mathrm{Hom}_{R_0}(A_0, B_\lambda) \to \mathrm{Hom}_{R_0}(A_0, B).$$

Let us first show that it is injective. So suppose that $f_\lambda, g_\lambda \colon A_0 \to B_\lambda$ are homomorphisms of R_0-algebras that induce the same homomorphism $f = g \colon A_0 \to B_\lambda \to B$. Let $a_1, \ldots, a_n$ be generators of A_0 as an R_0-algebra. Since $f(a_i) = g(a_i)$ in $B = \varinjlim_{\mu \geq \lambda} B_\mu$, there exists $\mu \geq \lambda$ with $f_\mu(a_i) = g_\mu(a_i)$, and since there are only finitely many a_i, and the index set is filtered, we can find μ which works simultaneously for all $i = 1, \ldots, n$. But then $f_\mu = g_\mu$, and we are done.

Now we prove surjectivity. We write $A_0 = R_0[T_1, \ldots, T_n]/(f_1, \ldots, f_m)$. Assume that $f \colon A_0 \to B$ is an R_0-homomorphism. If μ is sufficiently large, such that $f(T_i)$ is in the image of B_μ for each i, then we can define a homomorphism $g \colon R[T_1, \ldots, T_n] \to B_\mu$ such that the square

$$\begin{array}{ccc} R[T_1, \ldots, T_n] & \stackrel{g}{\longrightarrow} & B_\mu \\ \downarrow & & \downarrow \\ A_0 & \stackrel{f}{\longrightarrow} & B \end{array}$$

is commutative. Since $g(f_i)$ maps to 0 in B, for each i, we find $\nu \geq \mu$ such that all $g(f_i)$ are 0 in B_ν. But then the homomorphism $R[T_1, \ldots, T_n] \to B_\nu$ factors through $f_\nu \colon A_0 \to B_\nu$, and this is the preimage of f that we were looking for. $\square$

Now we can prove an analogous theorem for schemes which are not necessarily affine. We work in the setting of Section (10.13) 1.–3., 5.; Point 6. is not in force anymore.

Theorem 10.63. *Suppose that X_0 is qcqs, and that Y_0 is locally of finite presentation over S_0. Then the natural morphism*

$$(10.17.1) \qquad \varinjlim \operatorname{Hom}_{S_\lambda}(X_\lambda, Y_\lambda) \to \operatorname{Hom}_S(X, Y)$$

is bijective.

Proof. We first prove injectivity, so assume that morphisms $f_\lambda, g_\lambda \colon X_\lambda \to Y_\lambda$ which induce the same morphism $f = g \colon X \to Y$ are given. Since X is quasi-compact, so is $f(X)$, and it follows that $f(X)$ is contained in a finite union $\bigcup_{i=1}^n V_i$ of open affine subsets of Y, and we may assume that these V_i arise by base change from a covering $Y_\lambda = \bigcup_{i=1}^n V_{i,\lambda}$. In particular, we have a covering $Y_\mu = \bigcup_{i=1}^n V_{i,\mu}$ for every $\mu \geq \lambda$.

We would like to show that every point $x \in X_\lambda$ admits an affine open neighborhood $W(x)$ such that $f_\lambda(W(x))$, $g_\lambda(W(x))$ are contained in the same $V_{i,\lambda}$. Since X is quasi-compact, finitely many of these $W(x)$ will cover X, and we can apply the lemma above to get the injectivity result. We will show that this is true after possibly replacing λ by some $\mu \geq \lambda$, which of course is good enough for our purposes.

We write $U'_{i,\lambda} = f_\lambda^{-1}(V_{i,\lambda})$, $U''_{i,\lambda} = g_\lambda^{-1}(V_{i,\lambda})$, $U_{i,\lambda} = U'_{i,\lambda} \cap U''_{i,\lambda}$. The inverse image of $\bigcup_i U_{i,\lambda}$ in X is equal to X, so by Lemma 10.61 we conclude that for some $\mu \geq \lambda$ we have $\bigcup_i U_{i,\mu} = X_\mu$. So every point of X_μ has an open affine neighborhood which is entirely contained in one of the $U_{i,\mu}$, and any such neighborhood is mapped to $V_{i,\mu}$ by both f_μ and g_μ.

Now let us show the surjectivity, so consider an S-morphism $f \colon X \to Y$. Choose V_i, $V_{i,\lambda}$ as in the first part. We fix finite open affine coverings $f^{-1}(V_i) = \bigcup_j U_{ij}$. By Theorem 10.57 we may assume that the U_{ij} arise by base change from open subsets $U_{ij,\lambda} \subseteq X_\lambda$. Using the lemma above, we find λ and morphisms $U_{ij,\lambda} \to Y_\lambda$ which (after the base change $\times_{S_\lambda} S$) are compatible with f. Since the intersections $U_{ij,\lambda} \cap U_{i'j',\lambda}$ are quasi-compact, the injectivity statement which we proved already is available for them. Therefore, after passing to a suitable index $\mu \geq \lambda$ we can glue these morphisms and obtain the desired preimage $f_\mu \colon X_\mu \to Y_\mu$ of f. $\qquad\square$

Note that the proof shows that (10.17.1) is still injective if we only assume that X_0 is quasi-compact and that Y_0 is locally of finite type.

Corollary 10.64. *Suppose that in the situation of Theorem 10.63 both X_0 and Y_0 are of finite presentation over S_0.*
(1) *Let $f_0 \colon X_0 \to Y_0$ be an S_0-morphism. Then the induced morphism $f \colon X \to Y$ is an isomorphism if and only if there exists λ such that f_λ is an isomorphism.*
(2) *The S-schemes X and Y are isomorphic if and only if there exists λ such that X_λ and Y_λ are isomorphic.*

Proof. Under these assumptions, we can apply the theorem to morphisms $X \to Y$ and $Y \to X$, and the assertions follow easily. $\qquad\square$

Now that we have investigated the behavior of morphisms with respect to taking limits, we look at the analogous question for schemes. The affine case is Lemma B.10 which we restate here.

Lemma 10.65. *Let R be a ring, and let $(A_\lambda)_{\lambda \in \Lambda}$ be a filtered inductive system of R-algebras. Let $A = \varinjlim A_\lambda$. Furthermore, let B be an A-algebra of finite presentation. Then there exists $\lambda \in \Lambda$ and an A_λ-algebra B_λ of finite presentation such that $B \cong B_\lambda \otimes_{A_\lambda} A$.*

Theorem 10.66. *Let X be an S-scheme of finite presentation. Then there exists λ, and an S_λ-scheme X_λ of finite presentation such that X is isomorphic to $X_\lambda \times_{S_\lambda} S$.*

Proof. Since X is of finite presentation, there exists a finite open affine covering $(X_i)_i$ of X such that each $\Gamma(X_i, \mathscr{O}_X)$ is a $\Gamma(S, \mathscr{O}_S)$-algebra of finite presentation. Now we use Lemma 10.65: Passing to a suitable λ, we find $\Gamma(S_\lambda, \mathscr{O}_{S_\lambda})$-algebras B_i of finite presentation such that $B_i \otimes_{\Gamma(S_\lambda, \mathscr{O}_{S_\lambda})} \Gamma(S, \mathscr{O}_S) = \Gamma(X_i, \mathscr{O}_X)$.

We would like to glue the B_i to an S_λ-scheme whose base change to S is isomorphic to X. We achieve this by constructing a suitable gluing datum, see Section (3.5), using our previous results. At most steps, we have to pass to some $\mu \geq \lambda$, and we usually do so silently, i. e. using λ to denote the new index! We write $Z_{i,\lambda} = \operatorname{Spec} B_i$, $Z_i = Z_{i,\lambda} \times_{S_\lambda} S = X_i$.

First, since $X_i \cap X_j$ is quasi-compact and hence constructible in the (quasi-separated!) X, by Theorem 10.57, the open subset $X_i \cap X_j \subseteq X_i = Z_i$ comes from an open subset $Z_{ij\lambda} \subset Z_{i\lambda}$ (we repeat that by abuse of notation, we keep the notation λ for the (suitably increased) index). Then $Z_{ij} := Z_{ij\lambda} \times_{S_\lambda} S = X_i \cap X_j = Z_{ji}$, and by Corollary 10.64 this isomorphism comes from an isomorphism $Z_{ij\lambda} \xrightarrow{\sim} Z_{ji\lambda}$. Since there are only finitely many indices i, j, we may assume that a single λ works for all i, j. Finally, since all these isomorphisms satisfy the cocycle condition after base change to S, the cocycle condition is satisfied when we pass to a sufficiently high index λ. Therefore the $Z_{i\lambda}$ glue to a scheme X_λ, such that $X_\lambda \times_{S_\lambda} S$ is isomorphic to X. From this construction it is also easily checked that X_λ is of finite presentation over S_λ. $\square$

Combining Theorem 10.66 and Theorem 10.63 we obtain the following result.

Corollary 10.67. *Let X and Y be S-schemes of finite presentation and let $f\colon X \to Y$ be a morphism of S-schemes. Then there exists λ, and S_λ-schemes X_λ and Y_λ of finite presentation and a morphism $f_\lambda\colon X_\lambda \to Y_\lambda$ such that $X \cong X_\lambda \times_{S_\lambda} S$, $Y \cong Y_\lambda \times_{S_\lambda} S$, and $f = f_\lambda \times_{S_\lambda} \operatorname{id}_S$.*

Remark 10.68. To simplify the exposition we chose to present results and proofs only in the case where S_0, S_λ, and S are affine. Most of the results above and in particular Corollary 10.67 can be generalized to the case where we replace the hypotheses 2. of Section (10.13) by the following more general assumption (see [EGAIV] §8, §9).

$2'$. S_0 is qcqs (e.g. S_0 noetherian), and $s_{0\lambda}\colon S_\lambda \to S_0$ is an affine morphism (i.e., there exists an open affine covering (U_i) of S_0 such that $s_{0\lambda}^{-1}(U_i)$ is affine for all i; see Section (12.1) for details) for all λ.

(10.18) Elimination of noetherian hypothesis.

We apply the results of the previous section to the elimination of hypotheses about noetherianness.

Theorem 10.69. *Let R be a ring, and let $f\colon X \to Y$ be a morphism of R-schemes. The following are equivalent:*
(i) The schemes X and Y are of finite presentation over R.

(ii) *There exist a noetherian ring R_0, schemes X_0 and Y_0 of finite type over R_0, a morphism $f_0 \colon X_0 \to Y_0$ of R_0-schemes, and a ring homomorphism $R_0 \to R$, such that $X \cong X_0 \otimes_{R_0} R$, $Y \cong Y_0 \otimes_{R_0} R$, and $f = f_0 \otimes \mathrm{id}_R$.*

In fact, in (ii) one can even assume that R_0 is a $\mathbb{Z}$-algebra of finite type.

Proof. Use Corollary 10.67 and the fact that every ring is the filtered union of its finitely generated $\mathbb{Z}$-subalgebras. $\qquad\square$

As an application of these techniques, we prove that Chevalley's theorem also holds (under suitable assumptions) in the non-noetherian case.

Theorem 10.70. *Let $f \colon X \to Y$ be a morphism which is quasi-compact and locally of finite presentation. Then $f(X)$ is a constructible subset of Y.*

Proof. By checking this on an open affine covering of Y, we reduce to the case that $Y = \operatorname{Spec} A$ is affine. Since X is then quasi-compact, and finite unions of constructible sets are again constructible, we may assume that X is affine, as well, say $X = \operatorname{Spec} B$, and in particular that X is of finite presentation over Y. Now write A as the inductive limit of its $\mathbb{Z}$-subalgebras of finite type. By Lemma 10.65 we may assume that f is the base change of a morphism $f_0 \colon \operatorname{Spec} B_0 \to \operatorname{Spec} A_0$ of finite presentation, where A_0 is a finitely generated $\mathbb{Z}$-algebra, and in particular is noetherian. Now $f(X)$ is the inverse image of $f_0(\operatorname{Spec} B_0)$ under the canonical projection, and since the latter is constructible by Theorem 10.20, we are done by Proposition 10.43. $\qquad\square$

We also note the following immediate corollary:

Corollary 10.71.
(1) *Let $f \colon X \to Y$ be a morphism which is of finite presentation. Let $C \subseteq X$ be constructible. Then $f(C)$ is a constructible subset of Y.*
(2) *Let $f \colon X \to Y$ be a morphism which is locally of finite presentation. Then $f(X)$ is a union of constructible subsets of Y.*

Proof. The claim of the first point can be checked locally on Y, so that we can assume that Y is affine, and hence that X is qcqs. Using Proposition 10.45, we can write C as the image of a morphism of finite presentation. Applying the theorem to the composition, we are done. The second point is clear. $\qquad\square$

As an application, we can characterize open morphisms:

Corollary 10.72. *Let $f \colon X \to Y$ be a morphism locally of finite presentation. The following are equivalent:*
(i) *The morphism f is open.*
(ii) *For every $x \in X$, $f(\operatorname{Spec} \mathscr{O}_{X,x}) = \operatorname{Spec} \mathscr{O}_{Y,f(x)}$ (where we consider the spectra of the stalks as subsets of X and Y, resp., in the natural way).*
(iii) *For every $x \in X$ and every generization y' of $y := f(x)$, there exists a generization x' of x with $f(x') = y'$.*

Proof. With the above results about the non-noetherian case at hand, the proof of the corollary is basically the same as the proof of Corollary 10.21. Clearly, conditions (ii) and (iii) are equivalent. We have already proved that (i) implies (iii); see Lemma 5.10. So it remains to show that (iii) implies (i).

Since condition (iii) is still satisfied after replacing X by an open subset, it is enough to show that $f(X)$ is open. We may check this on open coverings of Y and X, so we may assume that Y and X are affine, so that f is of finite presentation. By Theorem 10.70, we know that $f(X)$ is constructible. Using Proposition 10.45, and the method of the proof of Lemma 10.17, we see that it is open.   □

Remark 10.73. Given an arbitrary scheme S and a morphism $f\colon X \to Y$ of S-schemes of finite presentation, Theorem 10.69 allows often to reduce to the noetherian case for questions that are local on S. Here we use that if $S = \operatorname{Spec} R$ is affine, then S is the filtered projective limit of noetherian affine schemes. For assertions that are global on S one can use Remark 10.68 together with the following result (see [TT] Theorem C9 and Theorem C7):

Let R_0 be a noetherian ring (e.g., $R_0 = \mathbb{Z}$) and let S be an R_0-scheme. Then S is qcqs if and only if S is the limit of a projective system (S_λ) of R_0-schemes of finite type where all transition morphisms $s_{\lambda\mu}\colon S_\mu \to S_\lambda$ are affine and such that $\mathscr{O}_{S_\lambda} \to s_{\lambda\mu*}\mathscr{O}_{S_\mu}$ is injective. Moreover, S is separated if and only if there exists an index λ_0 such that S_λ is separated for all $\lambda \geq \lambda_0$.

To use the above elimination techniques for hypotheses about noetherianness it is necessary to assume that the morphisms involved are of finite presentation. Sometimes it is possible to weaken this assumption by the following result (see [Co] 4.3 and A1 for a proof):

Theorem 10.74. *Let S be a qcqs scheme and let X be a quasi-separated S-scheme of finite type. Then there exists a closed S-immersion $X \hookrightarrow X_1$, where X_1 is of finite presentation over S. If X is separated over S, one can choose X_1 to be separated over S. If X is finite over S (see Definition 12.9 below), then one can choose X_1 to be finite over S.*

(10.19) Properties of morphisms and inductive limits.

We work in the setting of Section (10.13) 1.–3., 5., 6. We give some examples of properties of morphisms of schemes which are compatible with inductive limits of rings as defined in Section (10.13).

Proposition 10.75. *Let X_0, Y_0 be S_0-schemes of finite presentation, and let $f_0\colon X_0 \to Y_0$ be a morphism between them. Let $\mathbf{P}$ denote the property of being*
(1) an immersion (resp. an open immersion, resp. a closed immersion), or
(2) surjective.
Then f has the property $\mathbf{P}$ if and only if there exists λ such that f_λ has the property $\mathbf{P}$.

Proof. Since all the properties are stable under base change, if f_λ has the property $\mathbf{P}$ for some λ, then so has f, and our task is to prove the converse in each case.

Note that f_0 (and therefore f) is of finite presentation (Proposition 10.35). All the properties above can be checked on an affine open covering of Y_0. Therefore we may assume that Y_0 is affine. Then Y is affine, too.

First assume that f is an open immersion, and denote by $V \subseteq Y$ the open subscheme $f(X)$. By Corollary 10.71 (1), $f(X)$ is constructible, so we can apply Theorem 10.57 and find that there exists λ and an open subset V_λ whose inverse image in Y is V. But if we consider V_λ as an open subscheme of Y_λ, then again the inverse image in Y, as a scheme, is V. Now f by assumption induces an isomorphism $X \xrightarrow{\sim} V$, and it follows from Corollary 10.64 that $X_\mu \cong V_\mu$ for some μ, and by Theorem 10.63 we get that by possibly increasing μ this isomorphism, composed with the inclusion $V_\mu \subseteq Y_\mu$, induces the morphism f_μ.

Now assume that f is a closed immersion, and let $\mathfrak{a} \subseteq \Gamma(Y, \mathscr{O}_Y) =: A$ be the ideal corresponding to f. Since f is of finite presentation over S, the ideal $\mathfrak{a}$ is finitely generated, say by elements $a_1, \dots, a_r \in A$. Each a_i has a preimage in some ring A_{λ_i}, where $A_\lambda = \Gamma(Y_\lambda, \mathscr{O}_{Y_\lambda})$. Therefore we find an index λ and a finitely generated ideal $\mathfrak{a}_\lambda$ of A_λ such that $\mathfrak{a}_\lambda A = \mathfrak{a}$. We obtain a closed immersion $f'_\lambda \colon \operatorname{Spec} A_\lambda/\mathfrak{a}_\lambda \to Y_\lambda = \operatorname{Spec} A_\lambda$ such that $f'_\lambda \times \operatorname{id}_S = f$. Using the same argument as in the case of an open immersion, we arrive at the desired conclusion.

The case that f is an immersion then follows by writing f as the composition of closed immersion followed by an open immersion.

Finally, we assume that f is surjective. The image $Z_0 := f(X_0)$ is constructible in Y_0 by Corollary 10.71 (1). Since we have $f(X_\lambda) = Z_\lambda := y_{0\lambda}^{-1}(Z_0)$, $f(X) = Z := y_0^{-1}(Z_0)$, and by assumption $Z = Y$, it follows from Theorem 10.57 (1) that $Z_\lambda = Y_\lambda$ for some λ. $\qquad \square$

In fact, most of the properties we define for morphisms of schemes are compatible with inductive limits of rings, see Appendix C.

(10.20) Popescu's theorem.

We have seen in Theorem 10.66 that (in the usual setting) an S-scheme X of finite presentation can be "approximated", i. e. can be expressed as a base change $X_\lambda \times_{S_\lambda} S$. As Proposition 10.75 shows, one can even hope to do this in a way which preserves certain properties of X, or rather of morphisms of S-schemes. In this section we give an outlook on a related result which concerns the compatibility of smoothness and limits. This is Popescu's theorem ([Po], see also [Sp]), an important and highly non-trivial result which generalizes results of Néron and is also related to Artin's approximation theorem.

We call a morphism $\operatorname{Spec} B \to \operatorname{Spec} A$ of noetherian affine schemes *regular* if it is flat (i. e. B is a flat A-module) and if all the fibers $\operatorname{Spec} B \otimes_A \kappa(s)$, $s \in \operatorname{Spec} A$, are geometrically regular (see Exercise 6.19).

Theorem 10.76. *A morphism $\operatorname{Spec} B \to \operatorname{Spec} A$ of noetherian affine schemes is regular if and only if B is isomorphic to a filtered inductive limit of smooth A-algebras.*

Note that in the situation of the theorem, in general B is not equal to the inductive limit of its smooth A-*subalgebras*, cf. [Sp] Problem 1.3, Section 10. As we will see in Volume II, the notion of regular morphism is very close to the notion of smooth morphism. In fact, if B is an A-algebra of finite presentation, then $\operatorname{Spec} B \to \operatorname{Spec} A$ is regular if and only if it is smooth. Because of this, it is not hard to see that the morphism $\operatorname{Spec} B \to \operatorname{Spec} A$ is regular in case B is a filtered inductive limit of smooth A-algebras. The converse, however, is a deep result. If $A = k$ and $B = K$ are fields, then $\operatorname{Spec} K \to \operatorname{Spec} k$ is regular if and only if $k \to K$ is a separable extension (Proposition B.97).

In [An] André has shown that the following class of morphisms is regular (and thus Popescu's theorem can be applied).

Theorem 10.77. *Let $\varphi \colon A \to B$ be a local homomorphism of local noetherian rings. Let $\mathfrak{r}$ be the Jacobson radical of B. Suppose that the following conditions are satisfied.*
(a) *A is quasi-excellent (see Definition 12.49 and Theorem 12.51 below).*

(b) *For every A-algebra C and every ideal $I \subset C$ with $I^2 = 0$ and for every A-algebra homomorphism $\bar{\psi} \colon B \to C/I$ with $\bar{\psi}(\mathfrak{r}^n) = 0$ for some $n \geq 1$ there exists an A-algebra homomorphism $\psi \colon B \to C$ such that its composition with the canonical homomorphism $C \to C/I$ is $\bar{\psi}$.*

Then the morphism $\operatorname{Spec} B \to \operatorname{Spec} A$ corresponding to φ is regular.

The property (b) means that B with the $\mathfrak{r}$-adic topology is formally smooth over A (cf. [Mat] §28).

(10.21) The principle of the finite extension.

Proposition 10.78. *Let k be a field, and let $\mathbf{P}$ be a property of k-algebras such that*
(a) *if $R \to R'$ is a homomorphism of k-algebras and if $\mathbf{P}$ holds for R, then $\mathbf{P}$ holds for R',*
(b) *if $(R_\lambda)_\lambda$ is an inductive system of k-algebras, and $\mathbf{P}$ holds for $R = \varinjlim R_\lambda$, then it holds for some R_λ,*
(c) *there exists a k-algebra R such that $\mathbf{P}$ holds for R.*
Then there exists a finite extension K/k such that $\mathbf{P}$ holds for K, and in particular, $\mathbf{P}$ holds for any algebraically closed extension field of k.

Proof. Let R be a k-algebra as in (c). Writing it as the inductive limit of its k-subalgebras of finite type and using (b), we see that $\mathbf{P}$ holds for some finitely generated k-algebra A. By Hilbert's Nullstellensatz (Theorem 1.7), every quotient of A by a maximal ideal is a finite extension of k, so the claim follows from (a). $\qquad\square$

Corollary 10.79. *Let k be a field, and let X, Y be k-schemes of finite type, such that for some k-algebra R, there exists a morphism $X_R \to Y_R$ of R-schemes. Then there exists a finite extension K of k such that there exists a K-morphism $X_K \to Y_K$, and furthermore, whenever Ω is an algebraically closed extension field of k, then there exists a morphism $X_\Omega \to Y_\Omega$.*

Moreover, let $\mathbf{Q}$ be a property of scheme morphisms that is stable under base change, under composition and compatible with inductive limits of rings and assume that the morphism $X_R \to Y_R$ has $\mathbf{Q}$. Then there exists a finite extension K of k and a K-morphism $X_K \to Y_K$ that has property $\mathbf{Q}$.

Proof. We apply the proposition above to the property $\mathbf{P}$ of a k-algebra R such that $\mathbf{P}$ holds if and only if there exists a morphism of R-schemes $X_R \to Y_R$ (and has property $\mathbf{Q}$). Then the properties (a) and (c) are obviously satisfied, and (b) is satisfied by Theorem 10.63. (Note that we may assume that $R_0 = k$, so that $X_0 = X$ and $Y_0 = Y$, being of finite type over k, satisfy the hypothesis of the theorem.) $\qquad\square$

Corollary 10.80. *Let k be a field, let X, Y be k-schemes of finite type, such that for some k-algebra R, there exists an isomorphism $X_R \xrightarrow{\sim} Y_R$. Then there exists a finite extension K of k such that there exists an isomorphism $X_K \xrightarrow{\sim} Y_K$, and furthermore, whenever Ω is an algebraically closed extension field of k, then X_Ω and Y_Ω are isomorphic.*

Constructible properties

Let $\mathbf{P}(Z, k)$ be a property of a k-scheme Z and let $\mathbf{Q}(Z, z, k)$ be a property of a point z of a k-scheme Z. We will now study the question whether for a morphism $f\colon X \to S$ of finite presentation the sets

$$\{\, s \in S \;;\; \mathbf{P}(X_s, \kappa(s)) \text{ holds}\,\},$$
$$\{\, x \in X \;;\; \mathbf{Q}(X_{f(x)}, x, \kappa(f(x))) \text{ holds}\,\}$$

are constructible. In fact, there are plenty of variants of such questions as studying the locus where the fibers of a quasi-coherent $\mathscr{O}_X$-module or of a morphism $X \to Y$ of S-schemes have a certain property. In this chapter we will look only at a few selected examples. We refer to Appendix E and the references therein for a more exhaustive list.

(10.22) Generic freeness.

We start with proving a result about generic freeness of modules which is interesting in itself and that we will need in the subsequent sections.

Lemma 10.81. *Let S be an integral scheme and let $\mathscr{F}$ be an $\mathscr{O}_S$-module of finite presentation. Then there exists a dense open subset U of S such that $\mathscr{F}_{|U} \cong \mathscr{O}_U^n$ for some integer $n \geq 0$.*

Proof. This follows from Proposition 7.27: If $\eta \in S$ is the generic point, then $\mathscr{O}_{S,\eta}$ is the function field and hence $\mathscr{F}_\eta \cong (\mathscr{O}_S^n)_\eta$. $\qquad\square$

See also Exercise 10.37, and Section (11.8). This lemma can be vastly generalized. We start with another lemma:

Lemma 10.82. *Let R be a noetherian ring, let A be an R-algebra which is generated by a single element $x \in A$. Let M be a finitely generated A-module, and let $N \subseteq M$ be a finitely generated R-submodule such that $AN = M$. Then there exists a (not necessarily finite) filtration*

$$M/N \supseteq \cdots \supseteq M_{i+1} \supseteq M_i \supseteq \cdots \supseteq M_0 = 0$$

by finitely generated sub-R-modules, such that there exists an integer $j \geq 0$ such that $M_{i+1}/M_i \cong M_{j+1}/M_j$ for all $i \geq j$.

Proof. We let

$$(10.22.1) \qquad\qquad M_i = \left(\sum_{j=0}^{i} x^j N \right) /N,$$

$$(10.22.2) \qquad\qquad N_i = \mathrm{Ker}(x^{i+1}\colon N \to M/M_i).$$

(We denote multiplication by x^{i+1} by x^{i+1}.)

Clearly, the M_i are an ascending chain of R-submodules of M/N whose union is all of M/N. Furthermore, the N_i are an ascending chain of R-submodules of N, which must terminate because N is finitely generated over the noetherian ring R. Because $M_{i+1}/M_i \cong N/N_i$ for every i, the claim follows. $\qquad\square$

Theorem 10.83. *Let R be a noetherian integral domain, let A be a finitely generated R-algebra, and let M be a finitely generated A-module. Then there exists $s \in R \setminus \{0\}$, such that the localization M_s is a free R_s-module.*

Proof. Let $x_1, \ldots, x_n \in A$ be generators of A as an R-algebra. We prove the proposition by induction on n. If $n = 0$, i. e. $A = R$, then we are done by Lemma 10.81. Otherwise let A' be the R-subalgebra of A generated by $x_1, \ldots x_{n-1}$.

Let $e_1, \ldots, e_r$ be generators of M as an A-module, and set $N = \sum A' e_i$. We can apply the previous lemma to A', A, M and N, and get a filtration of M/N by finitely generated A'-submodules with only finitely many subquotients (up to isomorphism). Adding N as a filtration step, we obtain a filtration of M with the same property. By the induction hypothesis, all the subquotients of this filtration are free over a suitable localization of R, and over this localization the filtration splits (Proposition B.15). This shows the proposition. □

We may use this result to prove the Theorem of generic flatness:

Theorem 10.84. *Let $f\colon X \to Y$ be a quasi-compact morphism locally of finite presentation, and assume that Y is integral. Let $\mathscr{F}$ be a quasi-coherent $\mathscr{O}_X$-module of finite presentation. Then there exists a dense open subset $U \subseteq Y$ such that the restriction $\mathscr{F}_{|f^{-1}(U)}$ is flat over U.*

Proof. The question is local on Y, so we assume that $Y = \operatorname{Spec} A$ is affine, where A is an integral domain. Since f is quasi-compact, we find a finite open affine covering $X = \bigcup_i U_i$. If we find dense open subsets U of Y as in the theorem for each of the restrictions $U_i \to Y$, their intersection will satisfy the desired conclusion with respect to f, so we may also assume that $X = \operatorname{Spec} B$ is affine, and then B is an A-algebra of finite presentation, and $\mathscr{F}$ is the quasi-coherent $\mathscr{O}_X$-module associated with the B-module $M = \Gamma(X, \mathscr{F})$ of finite presentation.

Using the technique of eliminating the noetherianness hypothesis (Theorem 10.69 and Theorem 10.60), we may assume that the situation arises by base change for $A_0 \to A$, where A_0 is a noetherian subring of A, from an analogous situation over A_0. Over A_0, the conclusion we are looking for follows directly from Theorem 10.83, and since flatness is stable under base change, we are done. □

Applying the theorem to the structure sheaf $\mathscr{O}_X$, we get

Corollary 10.85. *Let $f\colon X \to Y$ be a morphism of finite type and locally of finite presentation, and assume that Y is integral. Then there exists a dense open subset $U \subseteq Y$ such that the restriction $f_{|f^{-1}(U)}\colon f^{-1}(U) \to U$ is flat.*

Proposition 10.86. *Let S be a noetherian scheme and let $f\colon X \to S$ be a morphism of finite type. Let $\mathbf{P}$ be a property of morphisms of schemes that is compatible with inductive limits of rings. We assume that for all $s \in S$ the morphism $f^{-1}(s) \to \operatorname{Spec} \kappa(s)$ satisfies $\mathbf{P}$. Then S is the (set-theoretical) disjoint union of finitely many affine subschemes S_i such that the restriction $f^{-1}(S_i) \to S_i$ of f is flat and satisfies property $\mathbf{P}$.*

Proof. By noetherian induction we may assume that the theorem holds for the restriction $f^{-1}(S') \to S'$ for every proper closed subscheme S' of S. Moreover, we may assume that

S is reduced. Let $\eta \in S$ be a maximal point. Let V be an open neighborhood of η that does not meet any other irreducible component. Then V is an integral subscheme of S. By Corollary 10.85 there exists an open affine neighborhood $U \subseteq V$ of η such that the restriction $f^{-1}(U) \to U$ is flat. Writing $\mathscr{O}_{S,\eta}$ as the inductive limit of the coordinate rings of the open affine neighborhoods of η in U we find an affine open neighborhood S_0 of η in U such that the restriction $f^{-1}(S_0) \to S_0$ has property $\mathbf{P}$ (because $\mathbf{P}$ is compatible with inductive limits). Application of the induction hypothesis to the complement $S \setminus S_0$ (endowed with its reduced subscheme structure) now proves the proposition. $\qquad \square$

The same result holds for any qcqs scheme S, if we assume in addition that f is of finite presentation and that $\mathbf{P}$ is stable under faithfully flat descent (see Definition 14.54 below).

(10.23) Constructible properties.

We now come to the question, whether for a scheme S and an S-scheme X of finite presentation the set of $s \in S$, such that the fiber X_s has a certain property $\mathbf{P}$, is constructible. One might call such a property $\mathbf{P}$ "constructible" but usually it is difficult to check directly whether a given property $\mathbf{P}$ is constructible in this naive sense. Thus we have the following definition which makes it usually easier to check whether a property is constructible.

Definition 10.87. *Let $\mathbf{P} = \mathbf{P}(X, k)$ be a property of schemes X of finite type over a field k. The property $\mathbf{P}$ is called* constructible, *if*
(a) *For every field extension k' of k and every k-scheme X of finite type, $\mathbf{P}(X, k)$ is equivalent to $\mathbf{P}(X \otimes_k k', k')$.*
(b) *For every integral noetherian scheme S with generic point η, and every morphism $u \colon X \to S$ of finite type, writing $E = \{ s \in S \ ; \ \mathbf{P}(X_s, \kappa(s)) \ holds \}$, we have that either E or $S \setminus E$ contains a non-empty open subset of S. (Here X_s denotes the scheme-theoretic fiber of u over s, see Section (4.8).)*

There are obvious variants of this definition for properties of pairs (X, k) as in Definition 10.87 together with additional data (e.g., an $\mathscr{O}_X$-module $\mathscr{F}$ or a subscheme Z of X) which we do not make explicit.

Condition (a) is satisfied if $\mathbf{P}(X, k)$ is of the form "the morphism $X \to \operatorname{Spec} k$ has property $\mathbf{P}'$", where $\mathbf{P}'$ is a property of morphisms of schemes that is stable under base change (Section (4.9)) and under faithfully flat descent (see Definition 14.54 below). This is in fact true for most of the properties on morphisms of schemes that we define (see Appendix C).

The following proposition shows that if a property is constructible, then it is also constructible in the naive sense discussed above.

Proposition 10.88. *Let $\mathbf{P}$ be a constructible property as in the definition. Let $X \to S$ be a morphism of finite presentation. Then $\{ s \in S \ ; \ \mathbf{P}(X_s, \kappa(s)) \ holds \}$ is a constructible subset of S.*

Proof. We may assume that $S = \operatorname{Spec} A$ is affine, because constructibility is a local property. By Theorem 10.69, there exists a subring $A_0 \subseteq A$ which is a $\mathbb{Z}$-algebra of finite type, and an A_0-scheme X_0 of finite type such that $X_0 \otimes_{A_0} A \cong X$. Let $p \colon S \to S_0$ be the natural morphism, and let $E_0 = \{ s \in S_0; \ \mathbf{P}((X_0)_s, \kappa(s)) \}$. From property (a) of the definition of constructible properties, and the transitivity of fiber products, it follows that $E = p^{-1}(E_0)$, and by Proposition 10.43 it is enough to show that E_0 is constructible in S_0. Thus we may assume that S is noetherian.

Now we use Proposition 10.14. If $Z \subseteq S$ is an irreducible closed subset, we consider it as an integral closed subscheme, and use Definition 10.87 for the base change $X \times_S Z \to Z$. Because of the transitivity of the fiber product, the fibers over points in Z are the same in both cases, and the result follows. $\square$

For an example of a constructible property, see Proposition 10.96. One can show that the properties "geometrically reduced" and "geometrically irreducible" are constructible (see Appendix E for a more exhaustive lists); see also Exercise 10.42.

Remark 10.89. Let k be a field, let X and S be schemes of finite type over k and let $f \colon X \to S$ be a k-morphism. Let $\mathbf{P}$ be a constructible property. Assume that we are given a constructible subset C of S (e.g., $C = S$) and that we want to show that for all $s \in C$ the $\kappa(s)$-scheme $f^{-1}(s)$ has property $\mathbf{P}$. Then it follows from Remark 10.15 that it suffices to show that $f^{-1}(s)$ has property $\mathbf{P}$ for every point $s \in C$ that is closed in S.

More generally, this is true whenever k is a Jacobson ring (Exercise 10.16).

Remark 10.90. For schemes over finite fields there are a number of tools available that do not exist for schemes over arbitrary fields, e.g, there is the additional structure of the Frobenius morphism (Remark/Definition 4.24) and there are techniques to count points. Thus the following principle (and variants thereof) is sometimes useful (see Exercise 10.41 for an example).

Let $\mathbf{P}(X \to S)$ and $\mathbf{Q}(X \to S)$ be properties of a scheme X of finite type over an affine noetherian integral scheme S. Let K be an arbitrary field. Assume that the following conditions hold.
(a) $\mathbf{P}$ is stable under base change of the form $\operatorname{Spec} \kappa(s) \to S$, where $s \in S$ is a closed point. $\mathbf{P}$ is compatible with inductive limits of rings.
(b) $\mathbf{Q}$ is stable under base change of the form $\operatorname{Spec} k' \to S = \operatorname{Spec} k$, where k' is a field extension of a field k. For all fields k the property $\mathbf{Q}(X \to \operatorname{Spec} k)$ is constructible.
(c) $\mathbf{P}(X \to \operatorname{Spec} k)$ implies $\mathbf{Q}(X \to \operatorname{Spec} k)$ for all finite fields k.
Then for every K-scheme X of finite type $\mathbf{P}(X \to \operatorname{Spec} K)$ implies $\mathbf{Q}(X \to \operatorname{Spec} K)$.

Indeed, if K is a field of characteristic $p > 0$ (resp. of characteristic 0), we may write K as the filtered union of finitely generated $\mathbb{F}_p$-subalgebras (resp. $\mathbb{Z}$-subalgebras) R_λ. By (a) and by Theorem 10.66 there exists an index λ and an R_λ-scheme X_λ of finite type such that $\mathbf{P}(X_\lambda \to \operatorname{Spec} R_\lambda)$ holds. As K is an extension of the field of fractions of R_λ, it suffices to prove that the generic fiber of X_λ has property $\mathbf{Q}$. For each closed point $s \in \operatorname{Spec} R_\lambda$ the residue field $\kappa(s)$ is a finite field. As $\mathbf{P}$ is stable under base change, $\mathbf{P}$ holds for the fiber $X_{\lambda,s} \to \operatorname{Spec} \kappa(s)$. Then $\mathbf{Q}$ holds for $X_{\lambda,s} \to \operatorname{Spec} \kappa(s)$ by Assumption (c). But the set E of $s \in \operatorname{Spec} R_\lambda$ where $\mathbf{Q}(X_{\lambda,s} \to \operatorname{Spec} \kappa(s))$ holds is constructible by Proposition 10.88. As the set of closed points in $\operatorname{Spec} R_\lambda$ is very dense by Proposition 3.35 (resp. Exercise 10.17), we conclude that E contains the generic point.

The argument shows that the Assumption (c) can be weakened:
(1) Let K be of characteristic p. Then it suffices that (c) is satisfied for finite fields of characteristic p.
(2) Let K be of characteristic 0 and let $\mathcal{P}$ be a given infinite set of prime numbers. Then it suffices to show that (c) is satisfied for finite fields of characteristic p with $p \in \mathcal{P}$.

Indeed, let R be a finitely generated $\mathbb{Z}$-subalgebra of a field of characteristic zero and $E \subseteq S := \operatorname{Spec} R$ a constructible set that contains every closed point $s \in S$ such that $\operatorname{char}(\kappa(s)) \in \mathcal{P}$. We have to show that E contains the generic point η of S. By Proposition 10.14 it suffices to show that E is dense in S. Let $\pi \colon S \to \operatorname{Spec} \mathbb{Z}$ be the unique morphism. By assumption we have $\overline{E} \supseteq \pi^{-1}(\mathcal{P})$, where we consider $\mathcal{P}$ as a subset of $\operatorname{Spec} \mathbb{Z}$. (If $\pi(\mathfrak{p}) = (p)$ with $p \in \mathcal{P}$, then $\pi(\mathfrak{m}) = (p)$ for all closed points $\mathfrak{m}$ which are specializations of $\mathfrak{p}$, as π is continuous. Since all such closed points belong to E and R is a finitely-generated $\mathbb{Z}$-algebra and, in particular, Jacobson, we conclude that $\mathfrak{p} \in \overline{E}$.) As R is a torsion-free module over the principal ideal domain $\mathbb{Z}$, π is flat (Proposition B.89). In Theorem 14.35 below we will see that this implies that π is open. Thus we have $\overline{E} \supseteq \overline{\pi^{-1}(\mathcal{P})} = \pi^{-1}(\overline{\mathcal{P}}) = S$.

By similar arguments one can often show a kind of converse, in the sense that properties which hold in characteristic 0 hold for almost all prime characteristics.

(10.24) Constructibility of properties of modules.

We consider the following situation: $f \colon X \to S$ is a morphism of schemes, and $\mathcal{F}$ is a quasi-coherent $\mathcal{O}_X$-module. Then for $s \in S$, we denote by $\mathcal{F}_s$ the restriction of $\mathcal{F}$ to the fiber $X_s = f^{-1}(s)$, i. e. $\mathcal{F}_s := g^* \mathcal{F}$, where $g \colon X_s \to X$ is the natural morphism.

Recall that over a (locally) noetherian scheme X, an $\mathcal{O}_X$-module $\mathcal{F}$ is *coherent*, if and only if it is quasi-coherent and of finite type, or equivalently, if it is of finite presentation; see Proposition 7.46.

Proposition 10.91. *Let S be a noetherian scheme, let X be an S-scheme of finite type, and let $\mathcal{F}$ be a coherent $\mathcal{O}_X$-module. Then the set*

$$E = \{ s \in S \; ; \; \mathcal{F}_s \neq 0 \}$$

is constructible.

In fact, a more general version of the proposition holds: Defining the notion of constructible property of triples $(X, \mathcal{F}, k)$, in analogy to Definition 10.87, the properties that $\mathcal{F} \neq 0$ (or $= 0$) are constructible. Thus E is constructible for an arbitrary scheme S if X is of finite presentation over S and $\mathcal{F}$ is an $\mathcal{O}_X$-module of finite presentation.

Proof. The support of $\mathcal{F}$ is closed, so in particular constructible, and therefore its image in S is also constructible by Chevalley's Theorem 10.20. But this image is precisely the set E of the proposition. $\square$

Next we consider exactness properties of sequences of $\mathcal{O}_X$-module homomorphisms.

Lemma 10.92. *Let S be a noetherian integral scheme with generic point η, let $f \colon X \to S$ be a morphism of finite type, and let $\mathcal{F}, \mathcal{G}, \mathcal{H}$ be coherent $\mathcal{O}_X$-modules. Furthermore, let $u \colon \mathcal{F} \to \mathcal{G}$, $v \colon \mathcal{G} \to \mathcal{H}$ be homomorphisms such that the sequence*

$$\mathcal{F}_\eta \to \mathcal{G}_\eta \to \mathcal{H}_\eta$$

is exact. Then there exists a non-empty open subset $U \subseteq S$, such that for every $s \in U$, the sequence

$$\mathcal{F}_s \to \mathcal{G}_s \to \mathcal{H}_s$$

is exact.

Proof. We first show that there exists a non-empty open subset $U' \subseteq S$ such that the sequence

$$\mathscr{F}_{|f^{-1}(U')} \to \mathscr{G}_{|f^{-1}(U')} \to \mathscr{H}_{|f^{-1}(U')}$$

is exact. To check this, we may assume for the moment that $S = \operatorname{Spec} R$ and $X = \operatorname{Spec} A$ are affine, so that the sequence $\mathscr{F} \to \mathscr{G} \to \mathscr{H}$ corresponds to a sequence $L \to M \to N$ of A-modules. By assumption this sequence becomes exact after tensoring with $\otimes_R T^{-1}R$, where $T = R \setminus \{0\}$, or equivalently after tensoring with $\otimes_A T^{-1}A$. We can also express this in terms of direct limits, e. g. $L \otimes_A T^{-1}A = \varinjlim_{t \in T} L \otimes_A A_t$. Since L is finitely generated, it is clear that the composition $L \to N$ is zero after tensoring with A_t for some $t \in T$. Replacing R by R_t we may assume that $L \to N$ is zero. Then $\operatorname{Im}(L \to M) \subseteq \operatorname{Ker}(M \to N)$, and since the quotient is zero after tensoring with $\otimes_A T^{-1}A$, and is finitely generated, we see that it is zero after tensoring with $\otimes_A A_t$ for a suitable $t \in T$. This proves our claim, and replacing S by a non-empty open subset, we may therefore assume that the sequence $\mathscr{F} \to \mathscr{G} \to \mathscr{H}$ is exact.

It is enough to prove that $(\operatorname{Ker} v)_s = \operatorname{Ker}(v_s)$, and $(\operatorname{Im} u)_s = \operatorname{Im}(u_s)$ for all s in a non-empty open U. Splitting the sequence $\mathscr{F} \to \mathscr{G} \to \mathscr{H}$ into suitable short exact sequences, we may thus even assume that

$$0 \to \mathscr{F} \to \mathscr{G} \to \mathscr{H} \to 0$$

is exact. Now we apply Theorem 10.84: for a suitable non-empty open $U \subseteq S$, $\mathscr{H}_{|f^{-1}(U)}$ is flat over U, so that the sequence above remains flat after tensoring by $\otimes_{\mathscr{O}_U} \kappa(s)$ for every $s \in U$ (Proposition 7.40). That is what we had to show. □

Proposition 10.93. *Let S be a noetherian scheme and let $f \colon X \to S$ be a morphism of finite type. Let $\mathscr{F}, \mathscr{G}, \mathscr{H}$ be coherent $\mathscr{O}_X$-modules. Furthermore, let $u \colon \mathscr{F} \to \mathscr{G}$, $v \colon \mathscr{G} \to \mathscr{H}$ be homomorphisms of $\mathscr{O}_X$-modules. Then the set*

$$E = \big\{ s \in S \; ; \; \mathscr{F}_s \to \mathscr{G}_s \to \mathscr{H}_s \text{ is exact} \big\}$$

is a constructible subset of S.

Proof. We use Proposition 10.14, so we assume that in addition S is noetherian and integral with generic point η. If $\eta \in E$, then it follows from the lemma above that E contains a non-empty open subset of S. It remains to handle the case that $\eta \notin E$. We have to show that $S \setminus E$ contains a non-empty open subset of S.

Let us write $w = v \circ u$, and first suppose that $w_\eta = v_\eta \circ u_\eta \neq 0$. The kernel $\mathscr{N} := \operatorname{Ker} w$ is again coherent, and we can apply the lemma to the sequence $0 \to \mathscr{N} \to \mathscr{F} \to \mathscr{H}$. This shows, that replacing S by a suitable non-empty open subset, we may assume that $\operatorname{Ker}(w_s) = \mathscr{N}_s$ for all $s \in S$. Then we also have $\mathscr{F}_s / \mathscr{N}_s = (\mathscr{F}/\mathscr{N})_s$, because taking tensor products is right exact. Therefore, $w_s \neq 0$ if and only if $(\mathscr{F}/\mathscr{N})_s \neq 0$, and by Proposition 10.91 the set of $s \in S$ with $w_s \neq 0$ (and in particular $S \setminus E$) contains a non-empty open subset.

Finally suppose that $w_\eta = 0$, so that the exactness fails because $\operatorname{Im} u_\eta \subsetneq \operatorname{Ker} v_\eta$. Similarly as above, by shrinking S we may assume that $\operatorname{Im}(u_s) = (\operatorname{Im} u)_s$, $\operatorname{Ker}(v_s) = (\operatorname{Ker} v)_s$ for all $s \in S$. Then the claim follows from Proposition 10.91, applied to $\operatorname{Ker} v / \operatorname{Im} u$. □

The proposition remains true if S is not necessarily noetherian, but f is of finite presentation and if $\mathscr{F}$, $\mathscr{G}$, and $\mathscr{H}$ are $\mathscr{O}_X$-modules of finite presentation. The previous lemma, which is still the key point, also holds in this more general context, because it can be reduced to the noetherian case using standard techniques, similar to those developed above. See [EGAIV] 9.4.

Our next goal is to prove that the property of being of dimension d for a fixed d is constructible. We start with the following lemma whose assertion is interesting in itself.

Lemma 10.94. *Let $S = \operatorname{Spec} A$ be an integral noetherian scheme with generic point η, let $X = \operatorname{Spec} B$ be an affine S-scheme of finite type, and let $\mathscr{F}$ be a coherent $\mathscr{O}_X$-module. Assume that for all $s \in S$, the fiber X_s is integral, and that $\mathscr{F}_\eta$ is a torsion-free $\mathscr{O}_{X_\eta}$-module. Then for all s in a non-empty open subset of S, the $\mathscr{O}_{X_s}$-module $\mathscr{F}_s$ is torsion-free.*

Proof. Let $M = \Gamma(\mathscr{F}, X)$, and $M_s = \Gamma(X_s, \mathscr{F}_s) = M \otimes_A \kappa(s)$ for $s \in S$, and similarly define B_s, $s \in S$. Since M_η is a torsion-free B_η-module, the natural homomorphism $M \to M \otimes_{B_\eta} \operatorname{Frac}(B_\eta) \cong \operatorname{Frac}(B_\eta)^n$ is injective. The latter is the union of free B_η-modules, and because M is finitely generated, we can embed M_η into a free module B_η^n. By Theorem 10.58, the homomorphism $M_\eta \to B_\eta^n$ can be extended to an open subset of S, and by Proposition 10.93, it is still injective on the fibers over a possibly smaller open subset of S. $\qquad\square$

Note that it is not difficult to show this lemma if S is an arbitrary locally noetherian scheme and X is of finite type over S. We are here mainly interested in the following application for which the version proved above suffices.

Proposition 10.95. *Let $f\colon X \to S$ be a dominant morphism of finite type between irreducible noetherian schemes. Let η denote the generic point of S. Then there exists a non-empty open subset $U \subseteq S$, such that for every $s \in U$, all the irreducible components of the fiber X_s have dimension $\dim X_\eta$.*

Proof. We may assume that $S = \operatorname{Spec} R$ is affine, and replacing f by f_{red}, we may assume that R is a domain and that X is integral. Note furthermore that X_η is irreducible, because it contains the generic point of X.

Let $W \subseteq X_\eta$ be an affine open subscheme, and let $V \subseteq X$ be an open subset such that $V \cap X_\eta = W$. We claim that for all s in a suitable non-empty open subset of S, V_s is dense in X_s. We can check this property on a (finite) open affine covering of X, and hence assume for the moment that X is affine, say $X = \operatorname{Spec} A$. Let $t \in A$ with $D(t) \subseteq V$. Clearly it is enough to prove the desired density statement for $D(t)$ rather than for V. Since X_η is irreducible, $D(t)_\eta$ is dense in X_η, or in other words, multiplication by t induces an injection $\mathscr{O}_{X,\eta} \to \mathscr{O}_{X,\eta}$. This means that multiplication induces injective homomorphisms $\mathscr{O}_{X,s} \to \mathscr{O}_{X,s}$ for all s in a non-empty open subset (use Proposition 10.93), and for all those s, $D(t)_s$ is dense X_s. The claim is proved, and therefore we may replace X by V, and hence assume that $W = X_\eta$.

This means that X_η is affine, and of finite type over $\kappa(\eta)$, so there is a closed immersion $X_\eta \hookrightarrow \mathbb{A}^N_{\kappa(\eta)}$ into some affine space over $\kappa(\eta)$. Using Theorem 10.63 and Proposition 10.75, we see that after shrinking S further, we find a closed immersion $X \hookrightarrow \mathbb{A}^N_S$, so in particular X is affine, say $X = \operatorname{Spec} A$.

Now Lemma 10.18 gives us, after possibly making S even smaller, a finite injective ring homomorphism $R[T_1, \ldots, T_n] \to A$, where $n = \dim X_\eta$. Since A is a domain, this homomorphism makes A into a finitely generated torsion-free $R[T_1, \ldots, T_n]$-module. Therefore $A \otimes_R \kappa(\eta)$ is a finitely generated torsion-free $\kappa(\eta)[T_1, \ldots, T_n]$-module, and by Lemma 10.94 there exists a non-empty open subset $U \subseteq S$, such that for all $s \in U$, $A \otimes_R \kappa(s)$ is a finitely generated torsion-free $\kappa(s)[T_1, \ldots, T_n]$-module. For all these s, in particular the homomorphism $\kappa(s)[T_1, \ldots, T_n] \to A \otimes_R \kappa(s)$ is injective.

Now let $Z \subseteq X_s = \operatorname{Spec} A \otimes_R \kappa(s)$ be an irreducible component. Then we have $Z = \operatorname{Spec}(A \otimes_R \kappa(s))/\mathfrak{p}$, where $\mathfrak{p} \subset A \otimes_R \kappa(s)$ is a minimal prime ideal. Therefore $\mathfrak{p}$ consists entirely of zero-divisors (Proposition B.59), and since $A \otimes_R \kappa(s)$ is torsion-free over the polynomial ring, the homomorphism $\kappa(s)[T_1, \ldots, T_n] \to A \otimes_R \kappa(s)/\mathfrak{p}$ is still injective (and, of course, finite). By Corollary 5.17, we have $\dim Z = n$, as desired.   □

Now we can also easily give an example of a constructible property of k-schemes in the sense of Definition 10.87.

Proposition 10.96. *The following properties of k-schemes X are constructible:*

(1) X *is empty.*

(2) X *is finite.*

(3) $\dim X$ *is contained in a fixed subset* $\Phi \subseteq \mathbb{Z} \cup \{-\infty\}$.

Proof. The first and second properties are special cases of the third one (taking $\Phi = \{-\infty\}$, and $\Phi = \{-\infty, 0\}$, respectively – see Section (5.4)), so it is enough to prove the third assertion.

By Proposition 5.38, the first property of Definition 10.87 is satisfied, so it remains to prove the second one. So assume that S is noetherian and integral with generic point η, and that X is an S-scheme of finite type. By noetherian induction we see that it is enough to prove that there exists a non-empty open subset $U \subseteq S$ such that $\dim X_u = \dim X_\eta$ for all $u \in U$.

Denote by X_i, $i \in I$ the finitely many irreducible components of X. To each X_i, considered as an integral scheme, we can apply Proposition 10.95 and obtain an open subscheme $U_i \subseteq S$ where all fibers of $X_i \to S$ have the same dimension. As we have $\dim X_s = \sup_i \dim(X_i)_s$, we can set $U := \bigcap_i U_i$.   □

(10.25) Constructibility of the local dimension of fibers.

Let $f \colon X \to S$ be a morphism. Often it is interesting to know that the subset of all points $x \in X$ which satisfy a certain property inside the fiber $f^{-1}(f(x))$ is constructible. Recall that we defined in Section (5.7) the dimension of X at a point x.

Theorem 10.97. *Let $f \colon X \to S$ be a morphism locally of finite presentation, and let $\Phi \subset \mathbb{Z} \cup \{\pm\infty\}$ be a finite subset. Then the set*

$$\{ x \in X \; ; \; \dim_x(X_{f(x)}) \in \Phi \}$$

is a constructible subset of X.

Proof. We will give the proof only in the case that S is noetherian. The general case can be reduced to the noetherian case with some additional work, using the methods above; see Exercise 10.44.

Since the question is local on X, we may also assume that X is of finite type over S, and in particular that X is noetherian, as well. Note that all the fibers X_s, $s \in S$, of f are schemes of finite type over a field (namely $\kappa(s)$), and in particular the dimension $\dim_x X_s$ ($x \in X_s$) is just the supremum of the dimensions of those irreducible components of X_s that contain x.

To prove the theorem, we apply the criterion given in Proposition 10.14. Let Y be a closed irreducible subset of X, and let ξ be its generic point. It is then enough to show that Y contains a non-empty open subset V such that for all $v \in V$, the dimensions $\dim_v X_{f(v)}$ coincide. As $f(Y)$ is contained in the closure of $f(\xi) =: \eta$, we may replace S by this closure (with the reduced scheme structure), and hence may assume that S is an integral scheme with generic point η.

Let X_i, $i \in I$, be the irreducible components of X which contain ξ (and therefore contain Y). Looking at generic points, one sees that the intersections $X_i \cap X_\eta$ are precisely the irreducible components of the fiber X_η. Since we are looking for an open neighborhood of ξ, we may discard all those irreducible components which do not contain ξ, so we assume that $X = \bigcup_i X_i$.

Since X_i is closed and hence constructible, so is its image $f(X_i \cap Y)$ by Chevalley's Theorem 10.20. Since it contains η, it contains a non-empty open subset of S (Proposition 10.14). By shrinking S, we may therefore assume that all the fibers X_s meet all the components X_i.

Now applying Proposition 10.95 to X_i, we obtain a non-empty open subset $U \subseteq S$, such that for every $s \in U$ all irreducible components of $X_i \cap X_s$ have dimension $\dim X_i \cap X_\eta$. Since I is finite, we may assume that this holds for all i simultaneously. Then for $v \in V := Y \cap f^{-1}(U)$, we have, since $Y \subseteq \bigcap_i X_i$,

$$\dim_v X_{f(v)} = \sup_i \dim X_i \cap X_\eta = \dim X_\eta,$$

independently of v. $\qquad\qquad\square$

Exercises

Exercise 10.1. Let $f \colon X \to Y$ be a quasi-compact morphism. Show that the following assertions are equivalent.
(i) f is dominant.
(ii) For every maximal point y of Y the fiber $f^{-1}(y)$ is non-empty.
(iii) For every maximal point y of Y the fiber $f^{-1}(y)$ contains a maximal point of X.
Hint: To show that (i) implies (iii) reduce to the case that X and Y are affine and reduced. Then use Corollary 2.11.

Exercise 10.2. Let $f \colon X \to Y$ be a quasi-compact morphism. Show that f is closed if and only if the image of f is stable under specialization (i.e., if $y \in f(X)$ and y' is a specialization of y, then $y' \in f(X)$).
Hint: To show that the condition is sufficient reduce to the case that X and Y are irreducible and use Exercise 10.1.

Exercise 10.3. Let $f\colon X \to Y$ be a surjective morphism.

(a) Show that the following properties are equivalent.

   (i) Every quasi-compact open subset of Y is the image of a quasi-compact open subset of X.

   (ii) There exists an open affine covering $(V_i)_i$ of Y such that each V_i is the image of a quasi-compact open subset of X.

   (iii) For all $x \in X$ there exists a quasi-compact open neighborhood U of x in X such that $f(U)$ is open and affine in Y.

   We call a morphism satisfying these properties *locally quasi-compact surjective*.

(b) Show that the property "locally quasi-compact surjective" is local on the target, stable under composition and stable under base change.

(c) Show that every quasi-compact surjective morphism is locally quasi-compact surjective. Show that every surjective open morphism is locally quasi-compact surjective.

Exercise 10.4◇. Let k be an algebraically closed field and let $f\colon \mathbb{A}_k^2 \to \mathbb{A}_k^2$ be the morphism given on k-valued points by $(x, y) \mapsto (x, xy)$. Describe the image of f and show that it is not locally closed. Write it as a finite union of locally closed subsets.

Exercise 10.5. Let k be an algebraically closed field. Show that the subset D of diagonalizable matrices in $M_n(k)$ is constructible and contains an open dense subset.

Exercise 10.6. Let k be a field and let $f\colon X \to Y$ be a k-morphism of schemes of finite type over k. Show that the following assertions are equivalent.

(i) f is surjective.

(ii) Every closed point of Y is in the set-theoretic image of f.

(iii) There exists an algebraically closed extension K of k such that f induces a surjection $X_k(K) \to Y_k(K)$ on K-valued points.

(iv) For every algebraically closed extension K of k the morphism f induces a surjection $X_k(K) \to Y_k(K)$ on K-valued points.

Exercise 10.7. Let S be a locally noetherian scheme, let k be a field, let $f\colon \operatorname{Spec} k \to S$ be a morphism of schemes, and let $s \in S$ be its image point. Show that f is of finite type if and only if $[k : \kappa(s)]$ is finite and $\{s\}$ is locally closed in S (which implies that $\dim \overline{\{s\}} \leq 1$ by Exercise 5.4). Give an example where f is of finite type and $\{s\}$ is not closed in S.

Exercise 10.8. Let X be a noetherian integral scheme with function field $K(X)$. Show that the canonical morphism $\operatorname{Spec} K(X) \to X$ is of finite type if and only if $\dim X \leq 1$ and X consists of finitely many points.
Hint: Exercise 10.7.

Exercise 10.9. Let R be a discrete valuation ring and let K be its field of fractions. Set $A := \{ f \in K[T] \; ; \; f(0) \in R \}$.

(a) Show that A is a R-subalgebra of $K[T]$ which is not noetherian. Deduce that the corresponding morphism of schemes $f\colon X := \operatorname{Spec} A \to S := \operatorname{Spec} R$ is not of finite type.

(b) Show that $f^{-1}(s) \cong \operatorname{Spec} \kappa(s)$ and $f^{-1}(\eta) \cong \mathbb{A}_K^1$, where s (resp. η) is the closed (resp. the generic) point of S.

(c) Let x be the unique point in $f^{-1}(s)$. Show that $\operatorname{Spec} \mathcal{O}_{X,x}$ consists of 3 points and has dimension 2.

(d) Sketch X.

Exercise 10.10. Let k be a fixed field. Let X be a compact totally disconnected space (e.g., the underlying topological space of the ring of p-adic integers $\mathbb{Z}_p$) and for $U \subseteq X$ open let $\mathscr{O}_X(U)$ be the k-algebra of locally constant functions $U \to k$. Show that $(X, \mathscr{O}_X)$ is a locally ringed space which is isomorphic to Spec A, where $A = \Gamma(X, \mathscr{O}_X)$.
Hint: Use that X can be written as projective limit (in the category of topological spaces) of finite discrete spaces (see, e.g., [NSW] (1.1.1)).

Exercise 10.11. Give an example of an affine scheme X and an open subset $U \subset X$ such that U is not quasi-compact. Let Y be the scheme which is obtained by gluing two copies of X along U via id_U. Show that Y is quasi-compact but not quasi-separated.
Hint: Exercise 10.10.

Exercise 10.12. Let $f \colon X \to Y$ and $g \colon Y \to Z$ be morphisms of schemes such that $g \circ f$ is quasi-compact and g is quasi-separated. Show that f is quasi-compact.

Exercise 10.13. Give an example of a morphism between quasi-compact schemes which is not quasi-compact.
Hint: Exercise 10.11.

Exercise 10.14. This exercise gives an example of an open immersion $j \colon U \to X$ such that $j_* \mathscr{O}_U$ is not quasi-coherent. The following example is taken from [AHK] (note that the example given in [EGAInew] (6.7.3) contains a mistake).

Let k be a field and let A be the subring of $k[T]^{\mathbb{N}}$ consisting of sequences $(f_n)_n$ of polynomials such that there exists an integer $N \geq 1$ with $f_{n+1} = f_n$ for all $n \geq N$. Let $\mathfrak{a} \subset A$ be the ideal of sequences (f_n) such that there exists an $N \geq 1$ such that $f_n = 0$ for all $n \geq N$. Set $X := \operatorname{Spec} A$, $U := X \setminus V(\mathfrak{a})$, and let $j \colon U \to X$ be the inclusion. Show that the $\mathscr{O}_X$-module $j_* \mathscr{O}_U$ is not quasi-coherent.
Hint: Let $g = (T, T, T, \ldots) \in A$. Show that the image of the canonical homomorphism $\Gamma(X, j_* \mathscr{O}_U)_g \to \Gamma(D(g), j_* \mathscr{O}_U)$ does not contain the element $(1/T^n)_n \in \Gamma(D(g), j_* \mathscr{O}_U) = (k[T]_T)^{\mathbb{N}}$.

Exercise 10.15. Show that for a ring A the following assertions are equivalent.
 (i) Every prime ideal in A is the intersection of (in general infinitely many) maximal ideals.
 (ii) For every ideal $\mathfrak{a}$ of A the nilradical of $A/\mathfrak{a}$ is equal to the Jacobson radical of $A/\mathfrak{a}$.
 (iii) The subset of closed points of $X = \operatorname{Spec} A$ is very dense in X.
 (iv) For every finitely generated A-algebra B the nilradical of B is equal to the Jacobson radical of B.
 (v) Every finitely generated A-algebra which is a field is a finite A-algebra.
Hint: Use Lemma 10.18 and Exercise 2.3.
 A ring A satisfying these conditions is called a *Jacobson ring*.

Exercise 10.16. A scheme X is called a *Jacobson scheme*, if the subset of closed points is very dense in X.
(a) Show that for a scheme X the following are equivalent:
  (i) X is a Jacobson scheme.
  (ii) For every affine open subset $U \subseteq X$ the ring $\Gamma(U, \mathscr{O}_X)$ is a Jacobson ring (Exercise 10.15).
  (iii) There exists an affine open cover $X = \bigcup_i U_i$, such that for all i the ring $\Gamma(U_i, \mathscr{O}_X)$ is a Jacobson ring.

(b) Show that a Dedekind ring A is a Jacobson ring if and only if A has infinitely many prime ideals (e.g., this is the case for $A = \mathbb{Z}$).
   Hint: Use Proposition B.87.
It can be shown ([EGAIV] (10.4.5) and (10.5.3)) that a noetherian scheme of dimension ≤ 1 is Jacobson if and only if each irreducible component consists either of a single or of infinitely many points.

Exercise 10.17. Let S be a Jacobson scheme (Exercise 10.16) and let $f\colon X \to S$ be a morphism locally of finite type. Show that X is a Jacobson scheme and that for every closed point $x \in X$ its image $f(x)$ is closed in S and $\kappa(x)$ is a finite extension of $\kappa(f(x))$.

Exercise 10.18. Let R be a discrete valuation ring, $\mathfrak{m}$ its maximal ideal, and define $X := \coprod_{n \geq 1} \operatorname{Spec} R/\mathfrak{m}^n$. Let $f\colon X \to S := \operatorname{Spec} R$ be the canonical morphism. Show that $f(X)$ consists only of the closed point of $\operatorname{Spec} R$ and that the schematic image $\operatorname{Im}(f)$ is equal to S.
   Hint: Show that $\Gamma(S, \mathscr{K}_f) = 0$ and $\Gamma(\{\eta\}, \mathscr{K}_f) = \operatorname{Frac} R$ (where $\eta \in S$ is the generic point and $\mathscr{K}_f := \operatorname{Ker}(f^\flat\colon \mathcal{O}_S \to f_*\mathcal{O}_X)$).

Exercise 10.19◊. Let $f\colon X \to Y$ be a surjective morphism of finite presentation and let $C \subseteq Y$ be a subset. Show that C is constructible in Y if and only if $f^{-1}(C)$ is constructible in X.
   Remark: It can be shown ([EGAInew] (7.2.10)) that the same result holds for surjective morphisms f that are locally of finite presentation *or* quasi-compact.

Exercise 10.20. Prove Proposition 10.33.

Exercise 10.21. Let R_0 be a ring and let R be an R_0-algebra. Show that R is isomorphic to a filtered inductive limit of R_0-algebras of finite presentation.

Exercise 10.22. Let R be a ring, let $((B_\lambda), (\varphi_{\mu\lambda}))$ be a filtered inductive system of R-algebras, and let B be its inductive limit. Let Y be an R-scheme locally of finite type (resp. locally of finite presentation). Show that the canonical map

$$\varinjlim \operatorname{Hom}_R(\operatorname{Spec} B_\lambda, Y) \to \operatorname{Hom}_R(\operatorname{Spec} B, Y)$$

is injective (resp. bijective).

Exercise 10.23. Let S be a scheme, let X and Y be S-schemes locally of finite presentation, and let $f, g\colon X \to Y$ be two morphisms of S-schemes. Show that $f = g$ if and only if for all $x \in X$ the induced maps $f(\mathcal{O}_{X,x})$ and $g(\mathcal{O}_{X,x})$ on $\mathcal{O}_{X,x}$-valued points are equal.
   Hint: One can assume that S is affine and then use Exercise 4.12 and Exercise 10.22.
   Remark: If S is locally noetherian, the result can be further strengthened (Exercise 14.15).

Exercise 10.24. Let S be a locally noetherian scheme, let X and Y be S-schemes locally of finite type. Let Y be separated over S. Assume that for all $x \in X$ there is given a map $f_x\colon X(\mathcal{O}_{X,x}) \to Y(\mathcal{O}_{X,x})$ such that for every generization x' of x the diagram

$$
\begin{array}{ccc}
X(\mathcal{O}_{X,x}) & \xrightarrow{\ f_x\ } & Y(\mathcal{O}_{X,x}) \\
\downarrow & & \downarrow \\
X(\mathcal{O}_{X,x'}) & \xrightarrow{\ f_{x'}\ } & Y(\mathcal{O}_{X,x'})
\end{array}
$$

is commutative, where the vertical maps are induced by the localization $\mathcal{O}_{X,x} \to \mathcal{O}_{X,x'}$. Show that there exists a unique morphism $f\colon X \to Y$ such that $f(\mathcal{O}_{X,x}) = f_x$.

Exercise 10.25. Assume the setting of Section (10.13) 1.–3. and assume that X_0 is qcqs.
(a) Show that if X is not connected, then there exists an index λ such that X_μ is not connected for all $\mu \geq \lambda$.
(b) Assume that the underlying topological space of X is noetherian. Show that if X is not irreducible, then there exists an index λ such that X_μ is not irreducible for all $\mu \geq \lambda$.

Exercise 10.26. Assume the setting of Section (10.13) 1.–3. and assume that X_0 is qcqs and that $x_{\lambda\mu} \colon X_\mu \to X_\lambda$ is dominant for all $\mu \geq \lambda$.
(a) Show that X is connected if and only if there exists an index λ such that X_μ is connected for all $\mu \geq \lambda$.
(b) Assume that the underlying topological space of X is noetherian. Show that X is irreducible if and only if there exists λ such that X_μ is irreducible for all $\mu \geq \lambda$.

Exercise 10.27. Let X be a qcqs scheme over a field k. Show that X is geometrically connected if (and only if) $X \otimes_k K$ is connected for all finite separable extensions K of k. *Hint:* Show that $X \otimes_k k^{\text{sep}}$ is connected, where k^{sep} is a separable closure of k, by writing k^{sep} as filtered union of finite separable extensions and use Exercise 10.26.

Exercise 10.28.
(a) Let X be a quasi-compact topological space. Show that the notions of "globally constructible" and "constructible" coincide.
(b) Let X be a noetherian topological space. Show that a subset of X is constructible in the sense of Definition 10.12 if and only if it is constructible in the sense of Definition 10.39.

Exercise 10.29. Let X be a qcqs scheme and let $U \subseteq X$ be a quasi-compact open subscheme. Show that there exists a quasi-coherent ideal $\mathscr{I} \subseteq \mathscr{O}_X$ of finite type, such that the underlying topological space of $V(\mathscr{I})$ is $X \setminus U$.
Hint: Reduce to the case that X is affine and use Exercise 2.6.

Exercise 10.30. A morphism $f \colon X \to Y$ of schemes is called a *local immersion* if for all $x \in X$ there exists an open neighborhood U of x such that $f_{|U} \colon U \to Y$ is an immersion.
(a) Let S be a scheme, let $f \colon X \to Y$ be a morphism of S-schemes that are locally of finite presentation over S, and let $x \in X$. Show that there exists an open neighborhood U of x such that $f_{|U}$ is an immersion if and only if the homomorphism $f_x^\# \colon \mathscr{O}_{Y,f(x)} \to \mathscr{O}_{X,x}$ is surjective.
(b) Let k be an algebraically closed field, define $A := k[T,U]/((T - U^2 + U)T)$ and $B := k[T,U]/(TU)$. Let $f \colon \operatorname{Spec} B \to \operatorname{Spec} A$ the morphism of k-schemes sending (t, u) to $(u - t, t^2 - t)$ (on k-valued points). Show that f is a local immersion but not injective (and hence not an immersion).
Remark: Note that $\operatorname{Spec} B$ is not irreducible. In Lemma 14.18 it will be shown that a separable local immersion with irreducible source is an immersion.

Exercise 10.31. Let S be a noetherian scheme, let X and Y be S-schemes of finite type over S, and let $f \colon X \to Y$ be a monomorphism of S-schemes. Assume that X is irreducible.
(a) Show that there exists an open and dense subscheme U of X such that $f_{|U} \colon U \to Y$ is an immersion.

Hint: Use Exercise 10.30 (a) applied to the generic point of X and that every monomorphism of a non-empty scheme into the spectrum of a field is an isomorphism by Exercise 9.6.

(b) Show that there exists an open and dense subscheme V of Y such that $f^{-1}(V)$ is non-empty and such that $f_{|f^{-1}(V)}$ is an immersion.

Exercise 10.32. Assume the setting of Section (10.13) 1.–3. and assume that X_0 is qcqs. Let $\mathscr{F}_0$ and $\mathscr{G}_0$ be quasi-coherent $\mathscr{O}_{X_0}$-modules and set $\mathscr{F}_\lambda := x_{0\lambda}^* \mathscr{F}_0$, $\mathscr{G}_\lambda := x_{0\lambda}^* \mathscr{G}_0$, $\mathscr{F} := x_0^* \mathscr{F}_0$, and $\mathscr{G} := x_0^* \mathscr{G}_0$.
(a) Show that there exists a homomorphism of $\Gamma(X_0, \mathscr{O}_{X_0})$-modules, functorial in $\mathscr{F}_0$ and $\mathscr{G}_0$,

$$\varinjlim_\lambda \operatorname{Hom}_{\mathscr{O}_{X_\lambda}}(\mathscr{F}_\lambda, \mathscr{G}_\lambda) \to \operatorname{Hom}_{\mathscr{O}_X}(\mathscr{F}, \mathscr{G})$$

which is injective (resp. an isomorphism) if $\mathscr{F}_0$ is of finite type (resp. of finite presentation).
(b) Let $u_\lambda \colon \mathscr{F}_\lambda \to \mathscr{G}_\lambda$ be a homomorphism of $\mathscr{O}_{X_\lambda}$-modules and assume that $\mathscr{F}_\lambda$ is of finite type and that $\mathscr{G}_\lambda$ is of finite presentation. Show that the homomorphism $u := x_\lambda^*(u_\lambda) \colon \mathscr{F} \to \mathscr{G}$ of $\mathscr{O}_X$-modules is an isomorphism if and only if there exists an index $\mu \geq \lambda$ such that $u_\mu := x_{\lambda\mu}^* u_\lambda \colon \mathscr{F}_\mu \to \mathscr{G}_\mu$ is an isomorphism.

Exercise 10.33. Assume the setting of Section (10.13) 1.–3. and assume that X_0 is qcqs. Let $\mathscr{F}$ be an $\mathscr{O}_X$-module of finite presentation.
(a) Show that there exists an index λ and an $\mathscr{O}_{X_\lambda}$-module $\mathscr{F}_\lambda$ of finite presentation such that $x_\lambda^* \mathscr{F}_\lambda \cong \mathscr{F}$.
Hint: If X_0 is affine, use Theorem 10.60. In general use Exercise 10.32 and a gluing argument.
(b) Show that $\mathscr{F}$ is locally free of rank n if and only if there exists an index $\mu \geq \lambda$ such that $\mathscr{F}_\mu := x_{\lambda\mu}^* \mathscr{F}_\lambda$ is locally free of rank n.

Exercise 10.34. Let R be a ring, let (A_λ) be a filtered inductive system of R-algebras, and let A be its inductive limit. Set $S = \operatorname{Spec} R$, $X = \operatorname{Spec} A$, $X_\lambda = \operatorname{Spec} A_\lambda$, and let $f \colon X \to S$ and $f_\lambda \colon X_\lambda \to S$ be the structure morphisms. Show that $f(X) = \bigcap_\lambda f_\lambda(X_\lambda)$.

Exercise 10.35. Using the notation of Section (10.13) 1.–3., 5., 6. assume that X_0 and Y_0 are S_0-schemes of finite presentation. Let $\mathbf{P}$ be one of the following properties: "separated", "purely inseparable", "open", "universally open". Show that f possesses $\mathbf{P}$ if and only if there exists a λ such that f_μ has $\mathbf{P}$ for all $\mu \geq \lambda$.
Hint: For "purely inseparable" use Exercise 9.9.

Exercise 10.36. Let $f \colon X \to Y$ be a morphism of schemes. Show that f is universally open if and only if for all $n \geq 0$ the base change $f_{\mathbb{A}_Y^n} \colon X \times_Y \mathbb{A}_Y^n \to \mathbb{A}_Y^n$ is open.
Hint: Use Exercise 10.35.

Exercise 10.37.
(a) Show that in Lemma 10.81 the assumption that M is finitely generated cannot be omitted.
(b) Give a proof of Lemma 10.81 in the spirit of the proof of Lemma 10.82.

Exercise 10.38. Using the notation of Section (10.13) 1.–3. let

$$\mathscr{F}_0 \to \mathscr{G}_0 \to \mathscr{H}_0 \to 0$$

be a sequence of quasi-coherent $\mathcal{O}_{X_0}$-modules, and assume that $\mathscr{F}_0$, $\mathscr{G}_0$ are of finite type, and that $\mathscr{H}_0$ is of finite presentation. Then the pull-back of this sequence to X is exact if and only if the pull-back to some X_λ is exact.

Exercise 10.39. Show that the property of morphisms "smooth" is compatible with inductive limits of rings.

Exercise 10.40. Let Y be an integral scheme, let $f\colon X \to Y$ be a morphism of finite presentation. Assume that $K(Y)$ is perfect (e.g., if $\operatorname{char}(K(Y)) = 0$).
(a) Assume that f is dominant and that X is integral. Show that there exists an open dense subscheme U of X such that $f_{|U}$ is smooth.
   Hint: Use Exercise 6.15 to show that $f^{-1}(\eta)$ is geometrically reduced, where η is the generic point of Y. Shrink X such that $f^{-1}(\eta)$ is smooth over $K(Y)$. Then use Exercise 10.39.
(b) Assume that X is regular. Show that there exists an open dense subscheme V of Y such that the restriction $f^{-1}(V) \to V$ is smooth.
   Hint: Use Exercise 6.15 to show that $f^{-1}(\eta)$ is smooth over $K(Y)$. Then use Exercise 10.39.
Remark: One often says that f is *generically smooth*.

Exercise 10.41.
(a) Let X and Y be two $\mathbb{Z}$-schemes of finite type and let $N \geq 1$ be an integer. Show that there exists a non-empty scheme S and an isomorphism $X \times_\mathbb{Z} S \overset{\sim}{\to} Y \times_\mathbb{Z} S$ if and only if there exist a finite field k with $\#k > N$ and an isomorphism $X \otimes_\mathbb{Z} k \overset{\sim}{\to} Y \otimes_\mathbb{Z} k$.
   Hint: Reduce to the case $S = \operatorname{Spec} K$ for a field K and use the technique introduced in Remark 10.90.
(b) Let $n, m \geq 1$ be integers. Show that there exists no non-empty scheme S such that $\mathbb{P}^n_S \times_S \mathbb{P}^m_S$ is isomorphic to $\mathbb{P}^{n+m}_S$.
   Hint: Count the k-valued points of $\mathbb{P}^n \times \mathbb{P}^m$ and $\mathbb{P}^{n+m}$ for finite fields k.
(c) Show that there exists no non-empty scheme S such that $\operatorname{Grass}_{2,4} \times_\mathbb{Z} S$ is isomorphic to $\mathbb{P}^r_S$ for some $r \geq 1$.
   Hint: Again count points.
   Remark: This argument can also be used to show the analogous assertion for $X = \operatorname{Grass}_{d,n}$ where $n \geq 4$ and $2 \leq d \leq n - 2$.

Exercise 10.42.
(a) Prove that the properties "irreducible", "reduced", and "connected" all violate the condition in Definition 10.87 (a).
(b) Let $f\colon \operatorname{Spec} \mathbb{R}[T, U]/(U^2 - T) \to S := \operatorname{Spec} \mathbb{R}[T]$ be the canonical morphism. Show that
$$\{\, s \in S \;;\; f^{-1}(s) \text{ is irreducible} \,\} = \{\, s \in S \;;\; f^{-1}(s) \text{ is connected} \,\}$$
is not constructible in S. Deduce that the properties "irreducible" and "connected" both violate the condition in Definition 10.87 (b).
(c) Let k be a non-perfect field of characteristic p and let $f\colon \operatorname{Spec} k[T, U]/(U^p - T) \to S := \operatorname{Spec} k[T]$ be the canonical morphism. Show that $\{\, s \in S \;;\; f^{-1}(s) \text{ is reduced} \,\}$ is not constructible in S.

Exercise 10.43. Prove Proposition 10.45.

Exercise 10.44. Carry through the reduction to the noetherian case needed at the beginning of the proof of Theorem 10.97.

11   Vector bundles

Contents

- – Vector bundles and locally free modules
- – Flattening stratification for modules
- – Divisors
- – Vector bundles on $\mathbb{P}^1$.

A central object in differential geometry is the tangent bundle of a manifold M. It can be described as a manifold T together with a projection $T \to M$ such that the fiber over each point $m \in M$ is in a natural way identified with the tangent space of M in m. In particular, T is locally on M isomorphic to $M \times \mathbb{R}^n$ and such that the transition maps between two such local descriptions are given by a linear map. These objects are called vector bundles (of rank n). Practically the same definition can be used in algebraic geometry to obtain the notion of a (geometric) vector bundle over a scheme X. But in fact we will see that we have already come across vector bundles – only in a different guise: Given a quasi-coherent $\mathscr{O}_X$-module $\mathscr{E}$ we will define the attached *quasi-coherent bundle* $\mathbb{V}(\mathscr{E})$ that is a scheme over X, and this construction yields an equivalence between the category of finite locally free $\mathscr{O}_X$-modules and the category of vector bundles over X. Its inverse is given by attaching to a vector bundle V over X (the dual of) the $\mathscr{O}_X$-module of sections of V (a construction which is also standard in differential geometry).

As all vector bundles of rank n are locally trivial, their isomorphism classes are given by the gluing data of locally trivial vector bundles. We will explain how this can be encoded in non-abelian cohomology either defined via torsors or via Čech cohomology (which we show to yield the same cohomology sets).

In the second part we will show that for an arbitrary quasi-coherent $\mathscr{O}_X$-module $\mathscr{F}$ of finite type over a scheme X and an integer $r \geq 0$ we can define the "flattening stratification" of X with respect to $\mathscr{F}$. This is a decomposition into (locally closed) subschemes $F_{=r}(\mathscr{F})$ which are – roughly speaking – the loci where $\mathscr{F}$ is finite locally free of rank r. In the part on determinantal schemes in Chapter 16 we will also define closed subschemes $F_{\geq r}(\mathscr{F})$ which are the loci where $\mathscr{F}$ is at least of rank r and such that $F_{=r}(\mathscr{F}) = F_{\geq r}(\mathscr{F}) \setminus F_{\geq r+1}(\mathscr{F})$.

The third part of this chapter is devoted to "configurations of poles and zeros" (called *divisors*) on a scheme X and the definition of corresponding $\mathscr{O}_X$-modules of "meromorphic functions" that have zeros (resp. poles) of prescribed orders. These $\mathscr{O}_X$-module will (under very mild conditions) be line bundles (i.e., vector bundles of rank 1). In the fourth part we use the techniques developed so far to classify vector bundles on $\mathbb{P}^1_k$ for a field k.

© Springer Fachmedien Wiesbaden GmbH, part of Springer Nature 2020
U. Görtz und T. Wedhorn, *Algebraic Geometry I: Schemes*, Springer Studium
Mathematik – Master, https://doi.org/10.1007/978-3-658-30733-2_12

Vector bundles and locally free modules

Our first goal in this chapter is to identify geometric vector bundles over a scheme X (which are schemes that look locally like an affine space over X, Definition 11.6) and finite locally free $\mathscr{O}_X$-modules. We will start by explaining two constructions – of the symmetric algebra of an $\mathscr{O}_X$-module and of the spectrum of a quasi-coherent $\mathscr{O}_X$-algebra – which will be useful at other occasions as well.

(11.1) The symmetric algebra of an $\mathscr{O}_X$-module.

Let us briefly recall the tensor algebra and the symmetric algebra of a module. Let A be a ring and let M be an A-module. The *tensor algebra of M* is the (in general non-commutative) A-algebra

$$T(M) := T_A(M) := \bigoplus_{n \geq 0} T^n(M), \qquad \text{with } T^n(M) := M^{\otimes n} := \underbrace{M \otimes_A \cdots \otimes_A M}_{n \text{ times}},$$

where the product is given by

$$(m_1 \otimes \cdots \otimes m_n, \, m_1' \otimes \cdots \otimes m_{n'}') \mapsto m_1 \otimes \cdots \otimes m_n \otimes m_1' \otimes \cdots \otimes m_{n'}'.$$

This is a graded A-algebra (i.e., $(T^n M)(T^m M) \subseteq T^{n+m} M$, see Section (13.1) below for more details on graded algebras) with $T^0 M = A$ and $T^1 M = M$.

The *symmetric algebra of M* is the quotient $T(M)/I$, where I is the two-sided ideal of $T(M)$ generated by the elements $m \otimes m' - m' \otimes m$ for $m, m' \in M$. It is denoted by $\mathrm{Sym}_A(M)$ or simply by $\mathrm{Sym}(M)$. This is a commutative A-algebra. As I is generated by homogeneous elements, $\mathrm{Sym}(M)$ is a graded A-algebra, $\mathrm{Sym}(M) = \bigoplus_{n \geq 0} \mathrm{Sym}^n(M)$. We have $\mathrm{Sym}^0(M) = A$ and $\mathrm{Sym}^1(M) = M$. The A-algebra $\mathrm{Sym}(M)$ and the A-linear map $\iota \colon M = \mathrm{Sym}^1(M) \hookrightarrow \mathrm{Sym}(M)$ satisfy the following universal property. For every commutative A-algebra B, composition with ι yields a bijection

$$(11.1.1) \qquad \mathrm{Hom}_{(A\text{-Alg})}(\mathrm{Sym}(M), B) \xrightarrow{\sim} \mathrm{Hom}_{(A\text{-Mod})}(M, B), \qquad \varphi \mapsto \varphi \circ \iota.$$

If $u \colon M \to N$ is an A-linear map, applying (11.1.1) to $B = \mathrm{Sym}(N)$, we see that u induces an A-algebra homomorphism $\mathrm{Sym}(u) \colon \mathrm{Sym}(M) \to \mathrm{Sym}(N)$ which is easily seen to be graded. Thus we obtain a functor Sym from the category of A-modules into the category of commutative graded A-algebras. Since (11.1.1) is functorial in M and B, Sym as a functor to the category of commutative A-algebras is left adjoint to the forgetful functor $(A\text{-Alg}) \to (A\text{-Mod})$.

If $\varphi \colon A \to B$ is a ring homomorphism, then (11.1.1) implies that there is an isomorphism of B-algebras

$$(11.1.2) \qquad \mathrm{Sym}_A(M) \otimes_A B \xrightarrow{\sim} \mathrm{Sym}_B(M \otimes_A B)$$

which is functorial in M and compatible with the grading. Another immediate corollary of (11.1.1) is an isomorphism of graded A-algebras, functorial in A-modules M and M',

(11.1.3) $\mathrm{Sym}(M \oplus M') \overset{\sim}{\to} \mathrm{Sym}(M) \otimes_A \mathrm{Sym}(M')$.

If M is a free A-module with basis $(x_1, \ldots, x_r)$, an easy induction on r using (11.1.3) shows that there is a unique isomorphism of graded A-algebras

(11.1.4) $A[T_1, \ldots, T_r] \overset{\sim}{\to} \mathrm{Sym}(M)$,   with $T_i \mapsto x_i$.

In particular, $\mathrm{Sym}^n(M)$ is a free A-module of rank $\binom{r+n-1}{n}$.

We now globalize the construction of the symmetric algebra to schemes (or even to ringed spaces). Let $(X, \mathscr{O}_X)$ be a ringed space and $\mathscr{E}$ an $\mathscr{O}_X$-module. The sheaf associated to the presheaf $U \mapsto \mathrm{Sym}_{\Gamma(U, \mathscr{O}_X)}(\Gamma(U, \mathscr{E}))$ on X is a commutative graded $\mathscr{O}_X$-algebra

$$\mathrm{Sym}(\mathscr{E}) = \bigoplus_{n \geq 0} \mathrm{Sym}^n(\mathscr{E}),$$

called the *symmetric algebra of $\mathscr{E}$*. Again $\mathscr{E} \mapsto \mathrm{Sym}(\mathscr{E})$ defines a functor from the category of $\mathscr{O}_X$-modules into the category of commutative $\mathscr{O}_X$-algebras which is left adjoint to the forgetful functor, that is, for every commutative $\mathscr{O}_X$-algebra $\mathscr{A}$ we have bijections which are functorial in $\mathscr{A}$ and in $\mathscr{E}$

(11.1.5) $\mathrm{Hom}_{(\mathscr{O}_X\text{-Alg})}(\mathrm{Sym}(\mathscr{E}), \mathscr{A}) \overset{\sim}{\to} \mathrm{Hom}_{\mathscr{O}_X}(\mathscr{E}, \mathscr{A})$.

Let $X = \mathrm{Spec}\, A$ be an affine scheme, M an A-module, and let $\mathscr{E} = \tilde{M}$ be the associated quasi-coherent $\mathscr{O}_X$-module. Then (11.1.2) applied to $B = A_f$ for elements $f \in A$ shows that $\mathrm{Sym}_A(M)_f = \mathrm{Sym}_{A_f}(M_f) = \Gamma(D(f), \mathrm{Sym}(\mathscr{E}))$. Therefore $\mathrm{Sym}(\mathscr{E})$ is the quasi-coherent $\mathscr{O}_X$-algebra associated to $\mathrm{Sym}(M)$.

This shows that if X is an arbitrary scheme and $\mathscr{E}$ is a quasi-coherent $\mathscr{O}_X$-module, $\mathrm{Sym}(\mathscr{E})$ is a quasi-coherent $\mathscr{O}_X$-algebra. If $f \colon T \to X$ is a morphism of schemes, then (11.1.2) yields

(11.1.6) $f^* \mathrm{Sym}(\mathscr{E}) \cong \mathrm{Sym}(f^* \mathscr{E})$.

(11.2) Spectrum of quasi-coherent $\mathscr{O}_X$-algebras.

If A is a ring, $X = \mathrm{Spec}\, A$, and B an A-algebra, we have seen in Proposition 3.4 that for every X-scheme T there exists a bijection

(11.2.1) $\mathrm{Hom}_X(T, \mathrm{Spec}\, B) \overset{\sim}{\to} \mathrm{Hom}_{(A\text{-Alg})}(B, \Gamma(T, \mathscr{O}_T))$

which is functorial in T and in B. In other words, the A-scheme $\mathrm{Spec}\, B$ represents the functor $T \mapsto \mathrm{Hom}_{(A\text{-Alg})}(B, \Gamma(T, \mathscr{O}_T))$. We will now globalize this construction.

Proposition 11.1. *Let X be a scheme and let $\mathscr{B}$ be a quasi-coherent $\mathscr{O}_X$-algebra. Then there exists an X-scheme $\mathrm{Spec}(\mathscr{B})$ such that for all X-schemes $f \colon T \to X$ there are bijections, functorial in T,*

(11.2.2) $\mathrm{Hom}_X(T, \mathrm{Spec}(\mathscr{B})) \overset{\sim}{\to} \mathrm{Hom}_{(\mathscr{O}_X\text{-Alg})}(\mathscr{B}, f_* \mathscr{O}_T)$.

In other words, $\mathrm{Spec}(\mathscr{B})$ represents the functor

$$(\mathrm{Sch}/X)^{\mathrm{opp}} \to (\mathrm{Sets}), \qquad (f \colon T \to X) \mapsto \mathrm{Hom}_{(\mathscr{O}_X\text{-Alg})}(\mathscr{B}, f_* \mathscr{O}_T).$$

Proof. For an X-scheme $f: T \to X$ we set $F(T) := \mathrm{Hom}_{(\mathscr{O}_X\text{-Alg})}(\mathscr{B}, f_*\mathscr{O}_T)$. As f_* and Hom are left exact, F is a sheaf for the Zariski topology, so by Theorem 8.9 we may assume that $X = \mathrm{Spec}\, A$ is affine. But in this case (11.2.1) shows that F is represented by $\mathrm{Spec}\, B$, where $B = \Gamma(X, \mathscr{B})$. $\qquad\square$

We denote the representing object by $\mathrm{Spec}\,\mathscr{B}$ and call it the *spectrum of* $\mathscr{B}$. Let $h\colon \mathrm{Spec}\,\mathscr{B} \to X$ denote its structure morphism. Then for every open affine subset $U = \mathrm{Spec}\, A$ of X the construction shows that there exists an isomorphism of affine U-schemes $h^{-1}(U) \cong \mathrm{Spec}\,\Gamma(U, \mathscr{B})$. This also shows

$$(11.2.3) \qquad\qquad h_*\mathscr{O}_{\mathrm{Spec}\,\mathscr{B}} = \mathscr{B}.$$

Let $\mathscr{B}'$ be another quasi-coherent $\mathscr{O}_X$-algebra. Then applying (11.2.2) to $T = \mathrm{Spec}\,\mathscr{B}'$ and using (11.2.3) we obtain a functorial isomorphism

$$(11.2.4) \qquad \mathrm{Hom}_{(\mathscr{O}_X\text{-Alg})}(\mathscr{B}, \mathscr{B}') \cong \mathrm{Hom}_X(\mathrm{Spec}\,\mathscr{B}', \mathrm{Spec}\,\mathscr{B}).$$

So $\mathscr{B} \mapsto \mathrm{Spec}\,\mathscr{B}$ is a fully faithful functor from the category of quasi-coherent $\mathscr{O}_X$-algebras into the category of X-schemes. In Corollary 12.2 we will see that its essential image consists of those X-schemes that are "affine over X", i. e. those X-schemes X' such that for every open affine in X, the inverse image in X' is again affine.

If $g\colon X' \to X$ is a morphism of schemes, there is an isomorphism of X'-schemes, functorial in $\mathscr{B}$,

$$(11.2.5) \qquad\qquad \mathrm{Spec}\, g^*\mathscr{B} \cong \mathrm{Spec}\,\mathscr{B} \times_X X'.$$

Indeed, we have $\mathrm{Hom}_{X'}(T', \mathrm{Spec}\,\mathscr{B} \times_X X') = \mathrm{Hom}_X(T', \mathrm{Spec}\,\mathscr{B})$ for every X'-scheme $f'\colon T' \to X'$. Therefore both X'-schemes represent the same functor on X'-schemes given by

$$(f'\colon T' \to X') \mapsto \mathrm{Hom}_{(\mathscr{O}_{X'}\text{-Alg})}(g^*\mathscr{B}, f'_*\mathscr{O}_{T'}) = \mathrm{Hom}_{(\mathscr{O}_X\text{-Alg})}(\mathscr{B}, g_*(f'_*\mathscr{O}_{T'})).$$

(11.3) Quasi-coherent bundles.

Let X be a scheme.

Definition and Remark 11.2. For every quasi-coherent $\mathscr{O}_X$-module $\mathscr{E}$ we set

$$(11.3.1) \qquad\qquad \mathbb{V}(\mathscr{E}) := \mathrm{Spec}(\mathrm{Sym}(\mathscr{E})).$$

We obtain a contravariant functor $\mathscr{E} \mapsto \mathbb{V}(\mathscr{E})$ from the category of quasi-coherent $\mathscr{O}_X$-modules into the category of X-schemes.

If $h\colon \mathbb{V}(\mathscr{E}) \to X$ is the structure morphism, we have by (11.2.3)

$$(11.3.2) \qquad\qquad h_*\mathscr{O}_{\mathbb{V}(\mathscr{E})} = \mathrm{Sym}(\mathscr{E}) = \bigoplus_n \mathrm{Sym}^n(\mathscr{E})$$

and in particular $h_*\mathscr{O}_{\mathbb{V}(\mathscr{E})}$ is a graded $\mathscr{O}_X$-algebra with $(h_*\mathscr{O}_{\mathbb{V}(\mathscr{E})})^1 = \mathrm{Sym}^1\mathscr{E} = \mathscr{E}$.

For two quasi-coherent $\mathscr{O}_X$-modules $\mathscr{E}$ and $\mathscr{F}$ we obtain by functoriality of $\mathbb{V}(-)$ a map $\mathrm{Hom}_{\mathscr{O}_X}(\mathscr{F}, \mathscr{E}) \to \mathrm{Hom}_X(\mathbb{V}(\mathscr{E}), \mathbb{V}(\mathscr{F}))$. It is easy to see (by restricting to open affine subschemes) that this map is injective. We call a morphism of X-schemes $\mathbb{V}(\mathscr{E}) \to \mathbb{V}(\mathscr{F})$ *linear*, if it is induced by an $\mathscr{O}_X$-linear homomorphism $\mathscr{F} \to \mathscr{E}$.

An X-morphism $f\colon \mathbb{V}(\mathscr{E}) \to \mathbb{V}(\mathscr{F})$ induces a homomorphism $\varphi\colon \mathrm{Sym}(\mathscr{F}) \to \mathrm{Sym}(\mathscr{E})$ of $\mathscr{O}_X$-algebras by (11.3.2). The morphism f is linear if and only if φ preserves the grading (i.e., φ induces homomorphisms of $\mathscr{O}_X$-modules $\varphi_n\colon \mathrm{Sym}^n(\mathscr{F}) \to \mathrm{Sym}^n(\mathscr{E})$). In this case one has automatically $\varphi = \mathrm{Sym}(\varphi_1)$ since $\mathscr{F} = \mathrm{Sym}^1\mathscr{F}$ generates $\mathrm{Sym}(\mathscr{F})$ as an $\mathscr{O}_X$-algebra.

Therefore the functor $\mathscr{E} \mapsto \mathbb{V}(\mathscr{E})$ yields a contravariant equivalence between the category of quasi-coherent $\mathscr{O}_X$-modules and the category of *quasi-coherent bundles over X* which is defined as the category whose objects are pairs consisting of

(a) an X-scheme $h\colon V \to X$ such that $V \cong \mathrm{Spec}\,\mathscr{A}$ for some quasi-coherent $\mathscr{O}_X$-algebra $\mathscr{A}$ (by Section (12.1) below these are those morphisms h that are affine)

(b) and a grading of quasi-coherent $\mathscr{O}_X$-algebras $h_*\mathscr{O}_V = \bigoplus_{n\geq 0}(h_*\mathscr{O}_V)_n$ such that the unique homomorphism $\mathrm{Sym}((h_*\mathscr{O})_1) \to h_*\mathscr{O}_V$ that is the identity in degree 1 is an isomorphism.

A morphism $(h\colon V \to X, h_*\mathscr{O}_V = \bigoplus_n (h_*\mathscr{O}_V)_n) \to (h'\colon V' \to X, h'_*\mathscr{O}_{V'} = \bigoplus_n (h'_*\mathscr{O}_{V'})_n)$ of quasi-coherent bundles is a morphism $f\colon V \to V'$ of X schemes such that the homomorphism $\mathscr{O}_{V'} \to f_*\mathscr{O}_V$ induces a homomorphism $h'_*\mathscr{O}_{V'} \to h'_*f_*\mathscr{O}_V = h_*\mathscr{O}_V$ which preserves the grading, i.e., maps $(h'_*\mathscr{O}_{V'})_n$ into $(h_*\mathscr{O}_V)_n$ for all n.

If $h\colon V \to X$ and $h'\colon V' \to X$ are quasi-coherent bundles over X, then a morphism of X-schemes $f\colon V \to V'$ is called *linear* if it is a morphism of quasi-coherent bundles. If one chooses isomorphisms $V \cong \mathbb{V}(\mathscr{E})$ and $V' \cong \mathbb{V}(\mathscr{E}')$ of quasi-coherent bundles, then the notion of linearity corresponds to the one defined above.

Remark 11.3. The formation of $\mathbb{V}(\mathscr{E})$ is compatible with base change: If $g\colon X' \to X$ is a scheme morphism, the identities (11.1.6) and (11.2.5) show that there is an isomorphism of quasi-coherent bundles over X', functorial in $\mathscr{E}$,

$$(11.3.3) \qquad\qquad \mathbb{V}(g^*\mathscr{E}) \overset{\sim}{\to} \mathbb{V}(\mathscr{E}) \times_X X'.$$

This implies in particular that for every open $U \subseteq X$ we have $\mathbb{V}(\mathscr{E}_{|U}) = \mathbb{V}(\mathscr{E})_{|U}$.

Proposition 11.4. *Let $\mathscr{E}$ be a quasi-coherent $\mathscr{O}_X$-module. Then for every X-scheme $h\colon T \to X$ there is a bijection, functorial in T,*

$$(11.3.4) \qquad\qquad \mathrm{Hom}_X(T, \mathbb{V}(\mathscr{E})) \overset{\sim}{\to} \Gamma(T, (h^*\mathscr{E})^{\vee}).$$

In other words, the X-scheme $\mathbb{V}(\mathscr{E})$ represents the functor $T \mapsto \Gamma(T, (h^*\mathscr{E})^{\vee})$.

Proof. This follows from the existence of identities, functorial in $h\colon T \to X$,

$$\mathrm{Hom}_X(T, \mathbb{V}(\mathscr{E})) \overset{(11.2.2)}{=} \mathrm{Hom}_{(\mathscr{O}_X\text{-Alg})}(\mathrm{Sym}(\mathscr{E}), h_*\mathscr{O}_T) \overset{(11.1.5)}{=} \mathrm{Hom}_{\mathscr{O}_X}(\mathscr{E}, h_*\mathscr{O}_T)$$
$$= \mathrm{Hom}_{\mathscr{O}_T}(h^*\mathscr{E}, \mathscr{O}_T) = \Gamma(T, (h^*\mathscr{E})^{\vee}). \qquad \square$$

The zero element in $\Gamma(X, \mathscr{E}^{\vee})$ corresponds via (11.3.4) to a section $z\colon X \to \mathbb{V}(\mathscr{E})$ of the structure morphism $f\colon \mathbb{V}(\mathscr{E}) \to X$, i.e., $f \circ z = \mathrm{id}_X$. It is called the *zero section of* $\mathbb{V}(\mathscr{E})$. As $\mathbb{V}(\mathscr{E}) \to X$ is locally on X a morphism between affine schemes, it is separated. Therefore the zero section is a closed immersion (Example 9.12).

Example 11.5. Consider the special case $\mathscr{E} = (\mathscr{O}_X^n)^{\vee}$ which we often identify with $\mathscr{O}_X^n$ using the standard basis. Then we have a functorial bijection $\mathrm{Hom}_X(T, \mathbb{V}((\mathscr{O}_X^n)^{\vee})) \overset{\sim}{\to} \Gamma(T, \mathscr{O}_T)^n = \mathbb{A}_X^n(T)$ which shows

(11.3.5) $$\mathbb{V}((\mathscr{O}_X^n)^\vee) = \mathbb{A}_X^n.$$

In this case, the zero section is the morphism corresponding to the $\mathscr{O}_X$-algebra homomorphism $\mathscr{O}_X[T_1, \ldots, T_n] \to \mathscr{O}_X$ with $T_i \mapsto 0$ for all i. Geometrically, every point $x \in X$ is mapped to the origin in the fiber $\mathbb{A}_{\kappa(x)}^n$.

Linear endomorphisms of $\mathbb{A}_X^n$ are those endomorphisms of X-schemes that are locally for $X = \operatorname{Spec} A$ given by an A-algebra automorphism φ of $A[T_1, \ldots, T_n]$ such that $\varphi(T_i) = \sum_j a_{ji} T_j$ for suitable $a_{ji} \in A$.

(11.4) Vector bundles and locally free modules.

Let X be a scheme. For an X-scheme $f \colon Y \to X$ and for $U \subseteq X$ open we simply write $Y_{|U}$ for the U-scheme $f^{-1}(U)$. Let $n \geq 0$ be an integer and let $\mathscr{E}$ be a locally free $\mathscr{O}_X$-module of rank n. We will now relate the quasi-coherent bundle $\mathbb{V}(\mathscr{E})$ with a more classical notion of a vector bundle.

By definition, an $\mathscr{O}_X$-module $\mathscr{E}$ is locally free of rank n if and only if it is locally isomorphic to $\mathscr{O}_X^n$. Therefore the equivalence of categories in Remark 11.2 yields a contravariant equivalence between the category of locally free $\mathscr{O}_X$-modules of rank n and the category of quasi-coherent bundles $h \colon V \to X$ such that there exists an open covering $X = \bigcup_i U_i$ and linear isomorphisms of U_i-schemes $c_i \colon V_{|U_i} \xrightarrow{\sim} \mathbb{A}_{U_i}^n$.

The structure of a quasi-coherent bundle on V, i.e., the grading on the $\mathscr{O}_X$-algebra $h_* \mathscr{O}_V$, can be obtained by transport of structure from $\mathbb{A}_{U_i}^n = \operatorname{Spec} \mathscr{O}_{U_i}[T_1, \ldots, T_r]$ via the c_i. Therefore we can describe quasi-coherent bundles attached to locally free modules of rank n also as follows.

Definition 11.6. *Let $n \geq 0$ be an integer. A (geometric) vector bundle of rank n over X is a pair consisting of an X-scheme V and an equivalence class of families (U_i, c_i) where $(U_i)_i$ is an open covering of X and $c_i \colon V_{|U_i} \xrightarrow{\sim} \mathbb{A}_{U_i}^n$ are isomorphisms of U_i-schemes such that for all i, j the automorphisms $c_i \circ c_j^{-1}$ of $\mathbb{A}_{U_i \cap U_j}^n$ are linear.*

Such a family (U_i, c_i) is called a vector bundle atlas, *and two vector bundle atlases (U_i, c_i) and (U_j', c_j') are called equivalent if the covering of X consisting of all the U_i and U_j' together with the isomorphisms c_i and c_j' also form a vector bundle atlas.*

A morphism $(V, [U_i, c_i]) \to (V', [U_j', c_j'])$ of geometric vector bundles over X is an X-morphism $f \colon V \to V'$ such that $c_j' \circ f \circ c_i^{-1} \colon \mathbb{A}_{U_i \cap U_j'}^n \to \mathbb{A}_{U_i \cap U_j'}^n$ is linear for all i, j.

Proposition 11.7. *The functor $\mathscr{E} \mapsto \mathbb{V}(\mathscr{E})$ yields a contravariant equivalence of the category of locally free $\mathscr{O}_X$-modules of rank n and the category of geometric vector bundles of rank n over X.*

Since for a locally free $\mathscr{O}_X$-module $\mathscr{E}$ of finite rank, we have $\mathscr{E}^{\vee\vee} = \mathscr{E}$, Proposition 11.4 shows a quasi-inverse of the functor $\mathscr{E} \mapsto \mathbb{V}(\mathscr{E})$ is given by $V \mapsto \mathscr{E}_V$, where $\mathscr{E}_V$ is the $\mathscr{O}_X$-module defined as follows.

Let $(h \colon V \to X, [U_i, c_i])$ be a geometric vector bundle over X. Define a presheaf $\mathscr{S}(V/X)$ on X by attaching to an open subset $U \subseteq X$ the set of sections of V over U, that is, the set of morphisms $s \colon U \to V_{|U}$ such that $f \circ s = \operatorname{id}_U$. The restriction maps of $\mathscr{S}(V/X)$ are given by the restriction of scheme morphisms. As scheme morphisms can be glued (Proposition 3.5), $\mathscr{S}(V/X)$ is a sheaf on X. To endow $\mathscr{S}(V/X)$ with the structure of a locally free $\mathscr{O}_X$-module, we may work locally and can assume that $V = \mathbb{V}((\mathscr{O}_X^n)^\vee)$ via our chosen vector bundle atlas (U_i, c_i). Then Proposition 11.4 shows that $\mathscr{S}(V/X) = \mathscr{O}_X^n$ which has an obvious structure of a (locally) free $\mathscr{O}_X$-module of rank n. Then we set

$$\mathscr{E}_V := \mathscr{S}(V/X)^\vee.$$

In the sequel we will use the terms "vector bundles over X" and "finite locally free $\mathscr{O}_X$-modules" synonymously and will usually consider them as $\mathscr{O}_X$-modules. Similarly we will call invertible $\mathscr{O}_X$-modules (that is, locally free $\mathscr{O}_X$-modules of rank 1) also *line bundles*.

All properties that can be checked locally on the target and are satisfied by affine space also hold for geometric vector bundles, for instance:

Remark 11.8. If $p\colon V \to X$ is a geometric vector bundle over X, then the morphism p is smooth and separated.

Example 11.9. (Universal bundle over the Grassmannian) Let S be a scheme, let $\mathscr{E}$ be a locally free $\mathscr{O}_S$-module of finite type, and let $e \geq 0$ be an integer. The S-scheme $X = \mathrm{Grass}^e(\mathscr{E})$ (Section (8.4)) represents the functor that attaches to each S-scheme $h\colon T \to S$ the set of isomorphism classes of surjections $\mathscr{E}_T := h^*\mathscr{E} \twoheadrightarrow \mathscr{Q}$, where $\mathscr{Q}$ is a locally free $\mathscr{O}_T$-module of rank e. By dualization (Proposition 8.10) we may also describe the T-valued points of X as

$$(11.4.1) \quad X(T) = \{\, \mathscr{U} \subseteq \mathscr{E}_T^\vee \text{ submodule} \; ; \; \mathscr{U} \text{ is locally a direct summand of rank } e\,\}.$$

Define a subfunctor $Z = Z_{\mathscr{E}}^e\colon (\mathrm{Sch}/S)^{\mathrm{opp}} \to (\mathrm{Sets})$ of $\mathbb{V}(\mathscr{E}) \times X$ by

$$Z(T) = Z_{\mathscr{E}}^e(T) := \{\, (s, \mathscr{U}) \in \Gamma(T, \mathscr{E}_T^\vee) \times X(T) \; ; \; s \in \Gamma(T, \mathscr{U}) \,\}.$$

We have a morphism of functors $Z \to X$ by sending a T-valued point $(s, \mathscr{U})$ to $\mathscr{U} \in X(T)$.

As an example assume that $S = \mathrm{Spec}\, k$ is the spectrum of a field and $\Gamma(S, \mathscr{E}) = V^\vee$ for a finite-dimensional k-vector space V. For every field extension K of k the set $Z(K)$ consists of pairs (s, U), where $U \subseteq V \otimes_k K$ is an e-dimensional subspace and where $s \in U$.

Now setting $T = X$ in (11.4.1), the identity $\mathrm{id}_X \in X(X)$ corresponds to a locally free $\mathscr{O}_X$-submodule $\mathscr{U}_{\mathrm{univ}} \subseteq \mathscr{E}_X^\vee$ of rank e such that the identity (11.4.1) is given by sending an X-scheme $f\colon T \to X$ to $\mathscr{U} := f^*\mathscr{U}_{\mathrm{univ}} \subseteq f^*\mathscr{E}_X^\vee = \mathscr{E}_T^\vee$. In particular we have $\Gamma(T, f^*\mathscr{U}_{\mathrm{univ}}) = \Gamma(T, \mathscr{U})$. This shows that we have an identity of functors over X

$$(11.4.2) \qquad\qquad Z_{\mathscr{E}}^e = \mathbb{V}(\mathscr{U}_{\mathrm{univ}}^\vee).$$

In particular $Z_{\mathscr{E}}^e$ is representable by an X-scheme and is endowed with the structure of a vector bundle of rank e over $X = \mathrm{Grass}^e(\mathscr{E})$.

(11.5) Excursion: Torsors and non-abelian cohomology.

Isomorphism classes of vector bundles can also be described via non-abelian cohomology which also will play an important role in descent theory in Chapter 14. These cohomology sets are not groups in general, but only *pointed sets*, that is, pairs (H, e) consisting of a set H and an element $e \in H$. A homomorphism $(H, e) \to (H', e')$ of pointed sets is a map $\gamma\colon H \to H'$ with $\gamma(e) = e'$. A sequence

$$(H_1, e_1) \xrightarrow{\gamma_1} (H_2, e_2) \xrightarrow{\gamma_2} (H_3, e_3)$$

of pointed sets is called *exact* if $\mathrm{Im}(\gamma_1) = \mathrm{Ker}(\gamma_2) := \{\, h_2 \in H_2 \; ; \; \gamma_2(h_2) = e_3 \,\}$. Note that a morphism of pointed sets with trivial kernel is not necessarily injective.

We start our excursion to non-abelian cohomology by giving the general definition of a torsor on a topological space.

Torsors.

Let X be a topological space, and let G be a sheaf of groups on X. If T is a sheaf (of sets) on X, we say that G *acts on* T or that T *is a* G-*sheaf* if there is given a morphisms of sheaves $G \times T \to T$ such that for every open subset $U \subseteq X$ the map $G(U) \times T(U) \to T(U)$ is a left action of the group $G(U)$ on the set $T(U)$. A *morphism of* G-*sheaves* is a morphism of sheaves $\varphi \colon T \to T'$ such that $\varphi_U \colon T(U) \to T'(U)$ is $G(U)$-equivariant for all open subsets $U \subseteq X$. We obtain the category of G-sheaves.

Definition 11.10. *A* G-*sheaf* T *is called a* G-torsor *(for the Zariski topology) if it satisfies the following two properties.*
(a) *The group* $G(U)$ *acts simply transitively on* $T(U)$ *for every open subset* $U \subseteq X$.
(b) *There exists an open covering* $\mathcal{U} = (U_i)_i$ *of* X *such that* $T(U_i) \neq \emptyset$ *for all* i.

Recall that the action of a group G on a set T is called simply transitive, if for every two elements $t_1, t_2 \in T$, there exists a unique $g \in G$ with $gt_1 = t_2$. In particular, $T = \emptyset$ is allowed.

One example for a G-torsor is the sheaf G itself on which G acts by left multiplication. This torsor is called the *trivial* G-*torsor*. A G-torsor T is isomorphic to the trivial G-torsor if and only if $T(X) \neq \emptyset$ because any $t \in T(X)$ yields an isomorphism $G(U) \to T(U)$, $g \mapsto gt_{|U}$.

We denote by $H^1(X, G)$ the set of isomorphism classes of G-torsors. This is a pointed set, where the distinguished element is the isomorphism class of the trivial G-torsor. Later we will study G-torsors for a finer (Grothendieck) topology than the Zariski topology. If we want to stress that we consider G-torsors for the Zariski topology we write $H^1_{\mathrm{Zar}}(X, G)$ instead of $H^1(X, G)$.

Non-abelian Čech cohomology.

We may also give a more elementary description of $H^1(X, G)$ in terms of cocycles which is often advantageous for concrete calculations. To ease the notation, for two sections $s \in \Gamma(U, G)$ and $t \in \Gamma(V, G)$ we will often write $st \in \Gamma(U \cap V, G)$ instead of $s_{|U \cap V} t_{|U \cap V}$, and $s = t$ instead of $s_{|U \cap V} = t_{|U \cap V}$.

Fix an open covering $\mathcal{U} = (U_i)_{i \in I}$ of X. A *Čech 1-cocycle of* G *on* $\mathcal{U}$ is a tuple $\theta = (g_{ij})_{i,j \in I}$, where $g_{ij} \in G(U_i \cap U_j)$, such that the cocycle condition

$$(11.5.1) \qquad g_{kj} g_{ji} = g_{ki}$$

holds for all i, j, k. This implies $g_{ii} = 1$ and $g_{ij} = g_{ji}^{-1}$ for all $i, j \in I$.

Two Čech 1-cocycles θ and θ' on $\mathcal{U}$ are called *cohomologous* if there exist $h_i \in G(U_i)$ for all i such that we have

$$h_i g_{ij} = g'_{ij} h_j$$

for all $i, j \in I$. This is easily checked to be an equivalence relation on the set of Čech 1-cocycles of G on $\mathcal{U}$. The equivalence classes are called *cohomology classes*, and the set of cohomology classes of Čech 1-cocycles on $\mathcal{U}$ is called the *(first) Čech cohomology of* G *on* $\mathcal{U}$ and is denoted by $\check{H}^1(\mathcal{U}, G)$. This is a pointed set in which the distinguished element is the cohomology class of the cocycle (g_{ij}) with $g_{ij} = 1$ for all i, j.

We say that a covering $\mathcal{V} = (V_j)_{j \in J}$ of X is a *refinement* of a covering $\mathcal{U} = (U_i)_{i \in I}$ if there exists a map $\tau \colon J \to I$ such that $V_j \subseteq U_{\tau(j)}$ for all $j \in J$. If $(g_{ii'})$ is a Čech 1-cocycle on $\mathcal{U}$, then the tuple $\tau^*(g)_{jj'} = g_{\tau(j)\tau(j')|V_j \cap V_{j'}}$ is a Čech 1-cocycle on $\mathcal{V}$. It is easy to see that this construction induces a map

$$\tau^*\colon \check{H}^1(\mathcal{U},G) \to \check{H}^1(\mathcal{V},G).$$

This map is independent of the choice of τ (this requires a careful although straightforward argument; e.g., see [Se1] I, §3, Prop. 3). Two coverings $\mathcal{U}$ and $\mathcal{V}$ are called *equivalent* if each one is a refinement of the other one. In this case we use the isomorphisms described above to identify $\check{H}^1(\mathcal{U},G) = \check{H}^1(\mathcal{V},G)$.

Definition 11.11. *Let G be a sheaf of groups on a topological space. The pointed set*

$$\check{H}^1(X,G) := \varinjlim_{\mathcal{U}} \check{H}^1(\mathcal{U},G),$$

where $\mathcal{U}$ runs through the set of equivalence classes of open coverings of X, is called the (first) *Čech cohomology of G on X.*

Remark 11.12. If G is a sheaf of abelian groups, the set of Čech 1-cocycles of G on $\mathcal{U}$ forms an abelian group with respect to componentwise multiplication $(g_{ij})(g'_{ij}) := (g_{ij}g'_{ij})$. The equivalence relation of being cohomologous is compatible with the group structure and therefore $\check{H}^1(\mathcal{U},G)$ is an abelian group. For every refinement $\mathcal{V}$ of $\mathcal{U}$ the map $\check{H}^1(\mathcal{U},G) \to \check{H}^1(\mathcal{V},G)$ is clearly a homomorphism of abelian groups. Therefore $\check{H}^1(X,G)$ is an abelian group.

Now let $(X,\mathcal{O}_X)$ be a ringed space and $G = \mathcal{F}$ be an $\mathcal{O}_X$-module. We define the groups $\check{H}^1(\mathcal{U},\mathcal{F})$ and $\check{H}^1(X,\mathcal{F})$ by forgetting the module structure and viewing $\mathcal{F}$ as a sheaf of abelian groups. Every global section $a \in \Gamma(X,\mathcal{O}_X)$ acts after restriction via multiplication on $\mathcal{F}(U)$ for all open subsets $U \subseteq X$ and thus acts on $\check{H}^1(\mathcal{U},\mathcal{F})$ and $\check{H}^1(X,\mathcal{F})$, making these abelian groups into $\Gamma(X,\mathcal{O}_X)$-modules.

TORSORS AND ČECH COHOMOLOGY.

We will now construct an isomorphism

$$(11.5.2) \qquad\qquad H^1(X,G) \cong \check{H}^1(X,G)$$

of pointed sets.

Let T be a G-torsor and let $\mathcal{U} = (U_i)_{i \in I}$ be an open covering of X that trivializes T, i.e., $T(U_i) \ne \emptyset$ for all i. Set $U_{ij} = U_i \cap U_j$ for all $i,j \in I$. Choose elements $t_i \in T(U_i)$. As G acts simply transitively, there exists a unique element $g_{ij} \in G(U_{ij})$ such that $g_{ij}t_i = t_i$. We have $g_{kj}g_{ji}t_i = t_k = g_{ki}t_i$ and thus $g_{kj}g_{ji} = g_{ki}$. For a different choice of elements t_i we obtain a cohomologous 1-cocycle. We obtain a map of pointed sets

$$c_{G,\mathcal{U}}\colon H^1(\mathcal{U},G) := \{\, T \in H^1(X,G)\ ;\ T \text{ is trivialized by } \mathcal{U}\,\} \to \check{H}^1(\mathcal{U},G).$$

By taking inductive limits we obtain a map of pointed sets

$$c_G\colon H^1(X,G) \to \check{H}^1(X,G).$$

Proposition 11.13. *The maps $c_{G,\mathcal{U}}$ are isomorphisms of pointed sets. In particular, c_G is an isomorphism.*

Proof. We define an inverse of $c_{G,\mathcal{U}}$ as follows. Let (g_{ij}) be a representative of a 1-cocycle in $\check{H}^1(\mathcal{U},G)$. For $V \subseteq X$ open we set

$$(11.5.3) \qquad T(V) = \{ (t_i) \in \prod_i G(U_i \cap V) \; ; \; t_i t_j^{-1} = g_{ij} \}.$$

Endowed with the obvious restriction maps, T is a sheaf. We define a G-action on T via $g \cdot (t_i)_i = (t_i g^{-1})_i$. For a fixed $k \in I$ and for $V \subseteq U_k$ the map $G(V) \to T(V)$, $g \mapsto (g_{ik} g^{-1})_i$, defines an isomorphism of $G_{|U_k}$-sheaves $G_{|U_k} \to T_{|U_k}$ whose inverse is given by $(t_i) \mapsto t_k^{-1}$. Thus T is a G-torsor which is trivialized by $\mathcal{U}$. If (g_{ij}) is replaced by a cohomologous cocycle $(g_{ij}') = (h_i g_{ij} h_j^{-1})$ with associated G-torsor T', then $(t_i) \mapsto (h_i t_i)_i$, defines an isomorphism $T \xrightarrow{\sim} T'$ of G-torsors. $\qquad \square$

FIRST TERM SEQUENCE OF COHOMOLOGY.

Let $\varphi \colon G \to G'$ be a homomorphism of sheaves of groups. Sending a Čech cocycle (g_{ij}) of G to $(\varphi(g_{ij}))$ defines a homomorphism of pointed sets

$$(11.5.4) \qquad \check{H}^1(\varphi) \colon \check{H}^1(X, G) \to \check{H}^1(X, G').$$

This map can also be defined in terms of torsors (Exercise 11.3).

Now let

$$(11.5.5) \qquad 1 \longrightarrow G' \xrightarrow{\varphi} G \xrightarrow{\psi} G'' \longrightarrow 1$$

be an exact sequence of sheaves of groups. Define a *connecting map*

$$(11.5.6) \qquad \delta \colon G''(X) \to \check{H}^1(X, G')$$

as follows. For $g'' \in G''(X)$ let $\mathcal{U} = (U_i)_i$ be an open covering of X such that there exist $g_i \in G(U_i)$ whose image in $G''(U_i)$ is $g''_{|U_i}$. For all i, j let $g_{ij}' \in G'(U_{ij})$ be the unique element that is mapped to $g_i g_j^{-1} \in G(U_i \cap U_j)$. Then (g_{ij}') is a Čech cocycle on $\mathcal{U}$. A different choice of elements g_i yields a cohomologous cocycle. Therefore its class $\delta(g'')$ in $\check{H}^1(X, G)$ is well defined. It is clear that δ is a morphism of pointed sets. The following result follows immediately from the definitions.

Proposition 11.14. *The following sequence of pointed sets is exact*

$$(11.5.7) \quad 1 \to G'(X) \to G(X) \to G''(X) \xrightarrow{\delta} \check{H}^1(X, G') \to \check{H}^1(X, G) \to \check{H}^1(X, G'').$$

Moreover, we have:
(1) *Assume that G' is a subgroup sheaf of the center of G (in particular it is a sheaf of abelian groups). Then δ is a homomorphism of groups, and $\check{H}^1(\varphi)$ induces an injection of the group $\mathrm{Coker}(\delta)$ into the pointed set $\check{H}^1(X, G)$.*
(2) *Assume that G', G, and G'' are abelian sheaves. Then the sequence (11.5.7) is an exact sequence of abelian groups.*

In Volume II we will see that if G is a sheaf of abelian groups, we can also identify $\check{H}^1(X, G)$ and hence $H^1(X, G)$ with the first derived functor of $G \mapsto \Gamma(X, G)$. If the sheaves G', G, and G'' in (11.5.5) are abelian, the sequence (11.5.7) is the beginning of the long exact cohomology sequence.

If X is an affine scheme and if $\mathscr{F}$ is a quasi-coherent $\mathscr{O}_X$-module, we will see in Proposition 12.32 that $H^1(X, \mathscr{F}) = 0$.

(11.6) Vector bundles and GL_n-torsors.

We now link isomorphism classes of vector bundles and non-abelian cohomology. The general philosophy is the following. Assume that we are given a category of sheaves E on a space X that are locally isomorphic to some standard sheaf SS of that category. For all E the sheaf of isomorphisms $\mathscr{I}\!som(SS, E)$ is a G-torsor, where $G = \mathscr{A}ut(SS)$ is the sheaf of automorphisms of SS. We obtain a map α from the set of isomorphism classes of such sheaves E to $H^1(X, G)$. The fact that sheaves can be glued together with $H^1(X, G) = \check{H}^1(X, G)$ then implies that α is a bijection. We will not make this precise in this generality (which is possible, see [Gi]). Instead we will give within this book several examples that will illuminate this philosophy.

The first example is that of vector bundles of rank n over a ringed space $(X, \mathscr{O}_X)$, where $n \geq 0$ is an integer. These are defined as those $\mathscr{O}_X$-modules that are locally isomorphic to the "standard module" $\mathscr{O}_X^n$. The group of automorphisms of $\mathscr{O}_X^n$ is the sheaf in groups

$$(11.6.1) \qquad \mathrm{GL}_n(\mathscr{O}_X)\colon U \mapsto \mathrm{Aut}_{\mathscr{O}_U}(\mathscr{O}_{X|U}^n) = \mathrm{GL}_n(\Gamma(U, \mathscr{O}_U)).$$

According to our philosophy we attach to every locally free $\mathscr{O}_X$-module $\mathscr{E}$ of rank n the sheaf $\mathscr{I}\!som_{\mathscr{O}_X}(\mathscr{O}_X^n, \mathscr{E})$. Then $\mathrm{GL}_n(\mathscr{O}_X)(U)$ acts on $\mathscr{I}\!som(\mathscr{O}_X^n, \mathscr{E})(U)$ by $(g, u) \mapsto u \circ g^{-1}$ and this action is simply transitive. There exists an open covering $X = \bigcup_i U_i$ such that $\mathscr{E}_{|U_i} \cong \mathscr{O}_{X|U_i}^n$ for all i. In other words, $\mathscr{I}\!som(\mathscr{O}_X^n, \mathscr{E})(U_i) \neq \emptyset$ for all i. Thus we see that $\mathscr{I}\!som(\mathscr{O}_X^n, \mathscr{E})$ is a $\mathrm{GL}_n(\mathscr{O}_X)$-torsor. We obtain a map of pointed sets

$$(11.6.2) \qquad \alpha\colon \left\{ \begin{array}{c} \text{isomorphism classes of} \\ \text{locally free } \mathscr{O}_X\text{-modules of rank } n \end{array} \right\} \longrightarrow H^1(X, \mathrm{GL}_n(\mathscr{O}_X)).$$

Proposition 11.15. *The map* (11.6.2) *is bijective.*

Proof. We identify $H^1(X, \mathrm{GL}_n(\mathscr{O}_X))$ with $\check{H}^1(X, \mathrm{GL}_n(\mathscr{O}_X))$ (Proposition 11.13) and define an inverse to α as follows. Let $\Theta \in \check{H}^1(X, \mathrm{GL}_n(\mathscr{O}_X))$ be represented by a Čech cocycle (g_{ij}) on an open covering $(U_i)_i$ of X. We set $\mathscr{S}_i := \mathscr{O}_{U_i}^n$ and glue these modules using $g_{ij}\colon \mathscr{S}_{j|U_i \cap U_j} \overset{\sim}{\to} \mathscr{S}_{i|U_i \cap U_j}$. The cocycle condition $g_{ik} = g_{ij}g_{jk}$ ensures that there exists an $\mathscr{O}_X$-module $\mathscr{E} = \mathscr{E}_\Theta$ and isomorphisms $t_i\colon \mathscr{E}_{|U_i} \overset{\sim}{\to} \mathscr{S}_i$ such that $g_{ij} = t_i \circ t_j^{-1}$ after restricting to $U_i \cap U_j$. The isomorphism class of $\mathscr{E}_\Theta$ does not depend on the choice of $(U_i)_i$ and not on the choice of the representing cocycle (g_{ij}). Using the explicit definition of a torsor attached to a 1-cocycle (11.5.3) it is immediate that this defines an inverse to α. $\square$

Remark 11.16. The proof (together with Proposition 11.13) shows that for every open covering $\mathcal{U} = (U_i)_i$ of X we have a bijection

$$\left\{ \begin{array}{c} \text{isomorphism classes of locally free} \\ \mathscr{O}_X\text{-modules } \mathscr{E} \text{ of rank } n \\ \text{such that } \mathscr{E}_{|U_i} \cong \mathscr{O}_{U_i}^n \text{ for all } i \end{array} \right\} \longrightarrow \check{H}^1(\mathcal{U}, \mathrm{GL}_n(\mathscr{O}_X)).$$

(11.7) Line bundles and the Picard group.

Let $(X, \mathscr{O}_X)$ be a ringed space. Recall that we use the notions "line bundle over X", "locally free $\mathscr{O}_X$-module of rank 1", and "invertible $\mathscr{O}_X$-module" synonymously. Let $\mathrm{Pic}(X)$

be the set of isomorphism classes of invertible $\mathscr{O}_X$-modules. For two invertible $\mathscr{O}_X$-modules $\mathscr{L}$ and $\mathscr{M}$ we define their product to be the tensor product $\mathscr{L} \otimes_{\mathscr{O}_X} \mathscr{M}$. This induces a multiplication on $\mathrm{Pic}(X)$ which is associative and commutative. The (isomorphism class of the) structure sheaf $\mathscr{O}_X$ is a neutral element. Moreover we have seen in (7.5.8) that $\mathscr{L}^\vee \otimes \mathscr{L} \cong \mathscr{O}_X$ for every invertible module $\mathscr{L}$. Thus $\mathrm{Pic}(X)$ becomes an abelian group which is called the *Picard group of* $(X, \mathscr{O}_X)$.

If $f \colon (X, \mathscr{O}_X) \to (Y, \mathscr{O}_Y)$ is a morphism of ringed spaces and $\mathscr{M}$ is an invertible $\mathscr{O}_Y$-module, $f^*(\mathscr{M})$ is an invertible $\mathscr{O}_X$-module. We obtain a map $f^* \colon \mathrm{Pic}(Y) \to \mathrm{Pic}(X)$ which is a homomorphism of groups by (7.8.8). As $(f \circ g)^* \cong g^* \circ f^*$ by (7.8.9), attaching $\mathrm{Pic}(X)$ to $(X, \mathscr{O}_X)$ yields a contravariant functor from the category of ringed spaces to the category of abelian groups.

Proposition 11.15 for $n = 1$ yields an isomorphism of pointed sets

$$(11.7.1) \qquad \mathrm{Pic}(X) \overset{\sim}{\to} H^1(X, \mathscr{O}_X^\times),$$

which is easily seen to be an isomorphism of abelian groups.

Examples 11.17. (Affine schemes with trivial Picard groups)
If $X = \mathrm{Spec}\, A$ is an affine scheme, we also write $\mathrm{Pic}(A)$ instead of $\mathrm{Pic}(X)$. For an A-module M the associated $\mathscr{O}_X$-module $\tilde{M}$ is invertible if and only if M is projective and the localization $M_\mathfrak{m}$ is a free $A_\mathfrak{m}$-module of rank 1 for all maximal ideals $\mathfrak{m}$ (Corollary 7.42).

In particular we see that $\mathrm{Pic}(A) = 0$ in the following two cases.
(1) A is a local ring.
(2) A is a factorial ring (see Example 11.44 below).
In particular Gauß' Theorem (Proposition B.75 (1)) implies that $\mathrm{Pic}(\mathbb{A}_R^n) = 0$ whenever R is factorial (e.g., if R is a field or $R = \mathbb{Z}$).

Flattening stratification for modules

(11.8) Flattening stratification.

Let X be a scheme and let $\mathscr{F}$ be an $\mathscr{O}_X$-module of finite type. The function

$$(11.8.1) \qquad \mathrm{rk}(\mathscr{F}) \colon X \to \mathbb{N}_0, \qquad x \mapsto \mathrm{rk}_x(\mathscr{F}) := \dim_{\kappa(x)} \mathscr{F}(x)$$

is called the *rank of* $\mathscr{F}$.

If $\mathscr{F}$ is a locally free, $\mathrm{rk}(\mathscr{F})$ is a locally constant function. The converse is not necessarily true as the following example shows: Let $X = \mathrm{Spec}\, A$ where A is a local Artin ring (thus X has only one point) and $\mathscr{F} = \tilde{M}$, where M is a finitely generated A-module that is not free (e.g., $A = k[T]/(T^2)$ for a field k and $M = k$). But we have the following theorem.

Theorem 11.18. *Let $\mathscr{F}$ be a quasi-coherent $\mathscr{O}_X$-module of finite type, let $r \geq 0$ be an integer. Then there exists a unique subscheme $F_{=r}(\mathscr{F})$ of X such that a scheme morphism $f \colon T \to X$ factors through $F_{=r}(\mathscr{F})$ if and only if $f^*\mathscr{F}$ is locally free of rank r.*

Thus we may say that $F_{=r}(\mathscr{F})$ is the locus where $\mathscr{F}$ is locally free of rank r. A point $x \in X$ with canonical morphism $i_x \colon \mathrm{Spec}(\kappa(x)) \to X$ lies in $F_{=r}(\mathscr{F})$ if and only if $i_x^*\mathscr{F}$ is (locally) free of rank r. Therefore we see that the underlying set of $F_{=r}(\mathscr{F})$ is $\{ x \in X \; ; \; \mathrm{rk}_x(\mathscr{F}) = r \}$. In particular, X is the disjoint union (as a set) of the subschemes $F_{=r}(\mathscr{F})$ for $r \geq 0$. The family $(F_{=r}(\mathscr{F}))_r$ is called the *flattening stratification of* $\mathscr{F}$.

Proof. The uniqueness is clear. Moreover, if we have shown that $F_{=r}(\mathscr{F})$ exists and $U \subseteq X$ is an open subscheme, $F_{=r}(\mathscr{F}_{|U})$ exists and is equal to $F_{=r}(\mathscr{F}) \cap U$. From these two facts it follows formally that if $X = \bigcup_i U_i$ is an open covering and $F_{=r}(\mathscr{F}_{|U_i})$ exists for all i, the subscheme $F_{=r}(\mathscr{F})$ exists, too. Indeed we have $F_{=r}(\mathscr{F}_{|U_i}) \cap U_j = F_{=r}(\mathscr{F}_{|U_i \cap U_j}) = F_{=r}(\mathscr{F}_{|U_j}) \cap U_i$ for all i, j and therefore the subschemes $F_{=r}(\mathscr{F}_{|U_i})$ can be glued to a subscheme $F_{=r}(\mathscr{F})$ which has the desired properties.

Fix an integer $r \geq 0$. By Corollary 7.31 the set $X_r = \{ x \in X ; \operatorname{rk}_x(\mathscr{F}) \leq r \}$ is open in X. We consider X_r as an open subscheme. If $F_{=r}(\mathscr{F})$ exists, it will be a subscheme of X_r. Therefore we may replace X by X_r and can assume that $\operatorname{rk}_x(\mathscr{F}) \leq r$ for all $x \in X$. Under this additional assumption we will show that $F_{=r}(\mathscr{F})$ is a closed subscheme of X.

As we have seen, it suffices to construct for every $x \in X$ an open neighborhood U such that $F_{=r}(\mathscr{F}_{|U})$ exists. If x is a point with $r' := \operatorname{rk}_x(\mathscr{F}) < r$, then $U = X_{r'}$ is an open neighborhood of x and we can define $F_{=r}(\mathscr{F}_{|U})$ as the empty scheme. Therefore we may assume that x is a point with $\operatorname{rk}_x(\mathscr{F}) = r$. By Nakayama's Lemma $\mathscr{F}_x$ is generated by r elements $s_{1,x}, \ldots, s_{r,x}$. By Proposition 7.30 we may extend $s_{1,x}, \ldots, s_{r,x}$ to sections $s_1, \ldots, s_r$ of $\mathscr{F}$ over some open affine neighborhood U of x that generate $\mathscr{F}_{|U}$. Therefore we may assume that $X = \operatorname{Spec} A$ is affine and that there exists an $\mathscr{O}_X$-linear surjection $u \colon \mathscr{O}_X^r \twoheadrightarrow \mathscr{F}$ that induces an isomorphism $\kappa(x)^r \to \mathscr{F}(x)$ at some point $x \in X$. By choosing generators of $\Gamma(X, \operatorname{Ker}(u))$ we obtain an exact sequence

$$\mathscr{O}_X^{(I)} \xrightarrow{v} \mathscr{O}_X^r \xrightarrow{u} \mathscr{F} \longrightarrow 0.$$

Now v is given by elements $a_{ji} \in A$ for $i \in I$ and $j = 1, \ldots, r$ and we define $F_{=r}(\mathscr{F})$ to be the closed subscheme defined by the ideal generated by these elements. In other words, $F_{=r}(\mathscr{F})$ is the locus where v is zero (Proposition 8.4). A morphism $f \colon T \to X$ factors through $F_{=r}(\mathscr{F})$ if and only if $f^*(v) = 0$. In this case $f^*(u)$ is an isomorphism and $f^*\mathscr{F}$ is locally free. Conversely, if $f^*\mathscr{F}$ is locally free of rank r, the surjection $f^*(u)$ is bijective by Corollary 8.12. This implies that $f^*(v) = 0$. $\square$

Clearly, set-theoretically X is the union of the locally closed subsets $F_{=r}(\mathscr{F})$. If $\mathscr{F}$ is of finite presentation, we can choose the index set I in the proof to be finite and then the proof shows that the immersion $F_{=r}(\mathscr{F}) \to X$ is locally of finite presentation. In Section (16.9) we will see that there is a canonical way to endow $\bigcup_{i \geq r} F_{=i}(\mathscr{F})$ with the structure of a closed subscheme.

As a corollary of Theorem 11.18 we obtain:

Corollary 11.19. *Let X be a reduced scheme and let $\mathscr{F}$ be a quasi-coherent $\mathscr{O}_X$-module of finite type. Then $\mathscr{F}$ is locally free if and only if $\operatorname{rk}(\mathscr{F})$ is a locally constant function.*

Proof. We only have to show that the condition is sufficient and we may assume that $\operatorname{rk}(\mathscr{F})$ is constant equal to some integer r. Then $F_{=r}(\mathscr{F})$ is a subscheme of X with the same underlying topological space as X. Since X is reduced, $F_{=r}(\mathscr{F}) = X$ and therefore $\mathscr{F}$ is locally free. $\square$

This corollary may be further generalized (Exercise 11.9).

Divisors

It is a classical and important question to determine the sets of zeros and poles of rational ("meromorphic") functions on a given variety or scheme. Mainly, one is interested in determining the relationship between the local and the global structure of the given scheme in this regard. So given zero/pole configurations on an open covering, we ask whether these configurations are induced from a global rational function.

We mention in passing that this type of question is also very important in complex analysis. The theorem of Mittag-Leffler and the Weierstrass product theorem are classical results in this direction.

There are two ways of making precise the notion of "configuration" used here. The more geometric one is to consider formal $\mathbb{Z}$-linear combinations of irreducible closed subsets of codimension one, where the coefficient says whether we want to see the corresponding subset as a zero locus of some (positive) multiplicity, or as the locus of poles of some order (negative multiplicity). This idea leads to the notion of Weil divisor, see Section (11.13). However, this notion works well only if suitable assumptions on the underlying scheme X are made. The notion which works better in the general case and which we therefore treat first, is the notion of Cartier divisor. Here we just define a "configuration" as an equivalence class of rational functions where we call rational functions f, g on some open subset U equivalent if $f = ug$ for some $u \in \Gamma(U, \mathscr{O}_X)^\times$ – clearly since u is a unit, f and g should be thought of having the same zero/pole configuration in this case.

If X is an integral scheme, then many of the definitions simplify considerably, and therefore we treat this case separately as a first step.

(11.9) Divisors on integral schemes.

Let X be an integral scheme. We denote by $\mathscr{K}_X$ the constant sheaf with value the function field $K(X)$ of X. In other words, for every non-empty open $U \subseteq X$, we have $\mathscr{K}_X(U) = K(X)$. For every point $x \in X$, $\mathscr{K}_{X,x} = K(X)$.

Since any non-empty open subset $U \subseteq X$ is schematically dense, the ring $R(U)$ of rational functions coincides with $K(X)$ by Remark 9.29. In particular, the sheaves $U \mapsto R(U)$ and $\mathscr{K}_X$ are equal.

Definition 11.20. *A* Cartier divisor *D on the integral scheme X is given by a tuple (U_i, f_i) where the U_i form an open covering of X and where $f_i \in K(X)^\times$ are elements with $f_i f_j^{-1} \in \Gamma(U_i \cap U_j, \mathscr{O}_X^\times)$ for all i, j. Two tuples (U_i, f_i), (V_i, g_i) give rise to the same Cartier divisor, if $f_i g_j^{-1} \in \Gamma(U_i \cap V_j, \mathscr{O}_X^\times)$ for all i, j.*

The set of Cartier divisors is denoted by $\mathrm{Div}(X)$. It is an abelian group: in fact, given divisors D, E represented by families (U_i, f_i), (V_i, g_i), we define the sum $D + E$ as the divisor given by $(U_i \cap V_j, f_i g_j)$.

A Cartier divisor is called principal, if it is equal to a divisor given by (X, f). Two divisors D, E are called linearly equivalent, if their difference $D - E$ is a principal divisor.

We denote by $\mathrm{DivCl}(X)$ the quotient of $\mathrm{Div}(X)$ by the subgroup of principal divisors. We obtain an exact sequence

$$1 \to \Gamma(X, \mathscr{O}_X)^\times \to K(X)^\times \to \mathrm{Div}(X) \to \mathrm{DivCl}(X) \to 0.$$

There is a close relationship between divisors and line bundles. To a Cartier divisor D we attach the line bundle $\mathscr{O}_X(D)$, given by

$$\Gamma(V, \mathscr{O}_X(D)) = \{ f \in K(X) \; ; \; f_i f \in \Gamma(U_i \cap V, \mathscr{O}_X) \text{ for all } i \}$$

for $V \subseteq X$ open. Over U_i, $\mathscr{O}_X(D)$ is isomorphic to the free $\mathscr{O}_{U_i}$-submodule of rank 1 of $\mathscr{K}_{U_i}$ generated by f_i^{-1}. In Proposition 11.29 we will see that:

Proposition 11.21. *The map* $\mathrm{Div}(X) \to \mathrm{Pic}(X)$, $D \mapsto \mathscr{O}_X(D)$, *induces an isomorphism* $\mathrm{DivCl}(X) \xrightarrow{\sim} \mathrm{Pic}(X)$ *of abelian groups.*

Thinking of a divisor as an object which locally models the sets of zeros and poles of meromorphic functions, we would like to understand divisors as geometric objects of their own. The notion one introduces first in this regard is the notion of support of a divisor D, a closed subset of X. We define

$$\mathrm{Supp}(D) = \{ x \in X \; ; \; (f_i)_x \notin \mathscr{O}_{X,x}^\times \text{ for some (or all) } i \text{ with } x \in U_i \}.$$

Under suitable assumptions on X (like normality or regularity), one can extend this to viewing Cartier divisors as "subschemes of codimension 1 with multiplicities attached to the irreducible components", i. e. as so-called Weil divisors. See Section (11.13).

(11.10) Sheaves of meromorphic functions and rational functions.

For arbitrary schemes we do not have a function field in general. Instead of working with fraction fields we use more generally the total fraction ring $\mathrm{Frac}(A)$ of a ring A. For that recall that $a \in A$ is called *regular* if it is not a zero divisor in A, in other words, multiplication $x \mapsto ax$ is an injective homomorphism $A \to A$. Then the *total fraction ring of* A is defined as $\mathrm{Frac}(A) := R^{-1}A$, where R is the multiplicative set of regular elements in A.

More generally, let M be an A-module. Then an element $m \in M$ is called *regular* if the A-linear map $A \to M$, $a \mapsto am$ is injective. This notion can be globalized as follows.

Definition and Remark 11.22. *Let X be a scheme and let $\mathscr{F}$ be a quasi-coherent $\mathscr{O}_X$-module. Then a section $s \in \Gamma(X, \mathscr{F})$ is called* regular *if the following equivalent assertions hold.*
(i) *The homomorphism of $\mathscr{O}_X$-modules $\mathscr{O}_X \to \mathscr{F}$, $f \mapsto fs$, is injective.*
(ii) *For all $x \in X$ the image of s in $\mathscr{F}_x$ is a regular element of the $\mathscr{O}_{X,x}$-module $\mathscr{F}_x$.*
(iii) *For any open affine $U = \mathrm{Spec}\, A \subseteq X$, the restriction $f_{|U}$ is a regular element of the A-module $\Gamma(U, \mathscr{F})$.*

The equivalence is immediate because injectivity can be checked on stalks for arbitrary sheaves (by definition) and on sections over open affine subschemes for quasi-coherent modules (Proposition 7.14).

Let X be a scheme. For $U \subseteq X$ open we let S_U be the subset of regular sections of $\Gamma(U, \mathscr{O}_X)$. This is a multiplicative subset of $\Gamma(U, \mathscr{O}_X)$. We define the sheaf $\mathscr{K}_X$ of $\mathscr{O}_X$-algebras as the sheaf associated to the presheaf

$$(11.10.1) \qquad\qquad \mathscr{K}_X' : U \mapsto S_U^{-1}\Gamma(U, \mathscr{O}_X).$$

This sheaf is called the *sheaf of meromorphic functions* on X. A section of this sheaf is locally the quotient f/s for a function f and a regular function s, and we therefore call such a section sometimes a *meromorphic function*. Note that the presheaf $\mathscr{K}'_X$ in general is not a sheaf, cf. [Kl].

If X is integral, then $\Gamma(U, \mathscr{O}_X)$ is an integral domain for every non-empty open subset $U \subseteq X$ and S_U is the set of non-zero elements of $\Gamma(U, \mathscr{O}_X)$. Hence in this case $\mathscr{K}_X$ is the constant sheaf with value the function field $K(X)$ defined in the beginning of Section (11.9).

As every element of S_U is a non-zero-divisor of $\Gamma(U, \mathscr{O}_X)$, the homomorphism $\mathscr{O}_X \to \mathscr{K}_X$ is injective. This allows us to identify $\mathscr{O}_X$ with a subsheaf of rings of $\mathscr{K}_X$. We will see below in Remark 11.25 that if X is a locally noetherian scheme, we have $\Gamma(U, \mathscr{K}_X) = \mathrm{Frac}(\Gamma(U, \mathscr{O}_X))$ for every open affine subscheme U. We mention that $\mathscr{K}_X$ is not always quasi-coherent (even if X is noetherian), although it is quasi-coherent if X is noetherian and reduced (Exercise 11.12) or if X is integral.

Meromorphic functions $f \in \Gamma(X, \mathscr{K}_X)$ can also be considered as rational functions on X as follows. Let $\mathrm{dom}(f)$ be the set of $x \in X$ such that $f_x \in \mathscr{O}_{X,x} \subseteq \mathscr{K}_{X,x}$. It is called the *domain of definition of f*.

Lemma 11.23. *For every $f \in \Gamma(X, \mathscr{K}_X)$ its domain of definition $\mathrm{dom}(f)$ is a schematically dense open subscheme of X.*

Proof. For $x \in \mathrm{dom}(f)$ there exists an open neighborhood W of x and $g \in \Gamma(W, \mathscr{O}_X)$ such that $g_x = f_x$. Hence we can find an open neighborhood $W' \subseteq W$ of x such that $g_{|W'} = f_{|W'}$. This shows that $\mathrm{dom}(f)$ is open in X.

For every $x \in X$ there exists an open affine neighborhood $U = U_x$ of x and a regular element $g = g_x \in \Gamma(U, \mathscr{O}_X)$ such that $gf_{|U} \in \Gamma(U, \mathscr{O}_X)$. Now $D(g_x)$ is schematically dense in U by Remark 9.24. As $D(g_x) \subseteq \mathrm{dom}(f) \cap U_x$, we see that $\bigcup_x D(g_x)$ is a schematically dense open subset of X which is contained in $\mathrm{dom}(f)$. $\qquad\square$

For $f \in \Gamma(X, \mathscr{K}_X)$ the restriction $f_{|\mathrm{dom}(f)}$ defines an element in $\Gamma(\mathrm{dom}(f), \mathscr{O}_X)$ and the lemma shows that we obtain a rational function on X (Definition 9.28). For all open subsets $U \subseteq X$ we therefore have a homomorphism $\alpha_U \colon \Gamma(U, \mathscr{K}_X) \to R(U)$, where $R(U)$ denotes the ring of rational functions on U. If $\mathscr{R}_X$ denotes the sheaf $U \mapsto R(U)$ on X, the α_U define a homomorphism

$$(11.10.2) \qquad\qquad \alpha \colon \mathscr{K}_X \to \mathscr{R}_X$$

of $\mathscr{O}_X$-algebras. As the canonical map $\Gamma(V, \mathscr{O}_X) \to R(U)$ is injective for every schematically dense open subset V of U, the homomorphism α is injective. Very often, the notions of meromorphic and of rational functions coincide:

Proposition 11.24. *If X is integral or a locally noetherian scheme, then (11.10.2) is an isomorphism.*

Proof. In case X is integral, this is clear; cf. Section (11.9).

Now assume that X is locally noetherian. We have to show that α_U is bijective for all open affine subsets $U = \mathrm{Spec}\, A$ of X. We have homomorphisms

$$\mathrm{Frac}(A) = \mathscr{K}'_X(U) \longrightarrow \mathscr{K}_X(U) \xrightarrow{\alpha_U} R(U),$$

where $\mathscr{K}'_X$ is the presheaf defined in (11.10.1). As we already know that α is injective, it suffices to show that $\mathrm{Frac}(A) \to R(U)$ is bijective. By definition we have

$$\text{Frac}(A) = \varinjlim_{t} \Gamma(D(t), \mathscr{O}_X), \qquad R(U) = \varinjlim_{V} \Gamma(V, \mathscr{O}_X),$$

where t runs through the set of regular elements of A and where V runs through the set of schematically dense open subsets of U. But by Lemma 9.23 an open subset V is schematically dense if and only if there exists a regular element $t \in A$ such that $D(t) \subseteq V$. This proves the proposition. $\qquad\qquad\qquad\qquad\qquad\qquad\qquad\qquad\qquad\qquad\qquad\qquad\square$

Thus for locally noetherian schemes we will often use the notions "meromorphic function" and "rational function" synonymously.

Remark 11.25. The proof of Proposition 11.24 shows in particular that for a noetherian affine scheme $X = \text{Spec } A$ we have $\Gamma(X, \mathscr{K}_X) = \text{Frac}(A)$. This implies that for a locally noetherian scheme X and a point $x \in X$ we have $\mathscr{K}_{X,x} = \text{Frac}(\mathscr{O}_{X,x})$.

(11.11) Cartier Divisors.

Definition 11.26. *Let X be a scheme.*
(1) *A* Cartier divisor *or simply a* divisor *on X is an element of the group*

$$\text{Div}(X) := \Gamma(X, \mathscr{K}_X^\times / \mathscr{O}_X^\times).$$

The group law is noted additively.
(2) *A divisor is called* principal *if it is in the image of the canonical homomorphism $\Gamma(X, \mathscr{K}_X^\times) \to \text{Div}(X)$. The principal divisor defined by $f \in \Gamma(X, \mathscr{K}_X^\times)$ is denoted by $\text{div}(f)$.*
(3) *Two divisors D_1 and D_2 are called* linearly equivalent, *denoted $D_1 \sim D_2$, if $D_1 - D_2$ is principal.*
(4) *A divisor D is called* effective *if it is in the submonoid $\text{Div}_+(X) = \Gamma(X, (\mathscr{K}_X^\times \cap \mathscr{O}_X)/\mathscr{O}_X^\times)$. In this case we write $D \geq 0$. For two divisors $D, D' \in \text{Div}(X)$ we write $D \geq D'$ if $D - D' \geq 0$.*

We can describe divisors more concretely. For $f \in \Gamma(U, \mathscr{K}_X)$ and $g \in \Gamma(V, \mathscr{K}_X)$ we write simply fg instead of $f_{|U \cap V} g_{|U \cap V}$. Then a divisor is given by a tuple $(U_i, f_i)_i$, where $X = \bigcup_i U_i$ is an open covering and where $f_i \in \Gamma(U_i, \mathscr{K}_X^\times)$ is a meromorphic function that is the quotient of two regular functions such that $f_i f_j^{-1} \in \Gamma(U_i \cap U_j, \mathscr{O}_X^\times)$ for all i, j.

Let D and E be divisors given by tuples $(U_i, f_i)_i$ and $(V_j, g_j)_j$, respectively. Then $D = E$ if and only if $f_i g_j^{-1} \in \Gamma(U_i \cap V_j, \mathscr{O}_X^\times)$ for all i, j. The sum $D + E$ is given by the tuple $(U_i \cap V_j, f_i g_j)_{i,j}$ and the inverse $-D$ is given by (U_i, f_i^{-1}). The divisor D is effective if and only if $f_i \in \Gamma(U_i, \mathscr{O}_X)$ for all i. A divisor D is principal if and only if it can be given by (X, f).

The principal divisors form a subgroup $\text{Div}_{\text{princ}}(X)$ of $\text{Div}(X)$ and the quotient

(11.11.1) $$\text{DivCl}(X) := \text{Div}(X)/\text{Div}_{\text{princ}}(X)$$

is called the group of *divisor classes* on X. We obtain an exact sequence

(11.11.2) $$1 \to \Gamma(X, \mathscr{O}_X)^\times \to \Gamma(X, \mathscr{K}_X)^\times \to \text{Div}(X) \to \text{DivCl}(X) \to 0.$$

(11.12) Divisors and line bundles.

Let X be a scheme. We will now attach to a divisor D on X an $\mathscr{O}_X$-submodule $\mathscr{O}_X(D)$ of $\mathscr{K}_X$ which is an invertible $\mathscr{O}_X$-module and then examine the question when two such invertible modules are isomorphic and if all invertible $\mathscr{O}_X$-modules (up to isomorphism) can be obtained in this way.

DIVISORS AND INVERTIBLE FRACTIONAL IDEALS.

We call an invertible $\mathscr{O}_X$-submodule of $\mathscr{K}_X$ an *invertible fractional ideal* of $\mathscr{O}_X$. If $\mathscr{I}_1$ and $\mathscr{I}_2$ are two invertible fractional ideals, we define their product to be the invertible submodule $\mathscr{I}_1\mathscr{I}_2$ of $\mathscr{K}_X$.

Let D be a divisor on X, represented by $(U_i, f_i)_i$. As $f_i f_j^{-1} \in \Gamma(U_i \cap U_j, \mathscr{O}_X)^\times$ for all i, j, there exists a (unique) invertible fractional ideal $\mathscr{I}_X(D)$ such that

$$\mathscr{I}_X(D)_{|U_i} = \mathscr{O}_{U_i} f_i.$$

Moreover $\mathscr{I}_X(D)$ does not depend on the choice of the representation $(U_i, f_i)_i$. Conversely, if $\mathscr{I}$ is an invertible fractional ideal, there exists an open covering $X = \bigcup_i U_i$ and sections $f_i \in \Gamma(U_i, \mathscr{K}_X^\times)$ such that $\mathscr{I}_{|U_i} = f_i \mathscr{O}_{U_i}$. We obtain an isomorphism of abelian groups

$$(11.12.1) \qquad \mathrm{Div}(X) \xrightarrow{\sim} \{\text{invertible fractional ideals on } X\}, \qquad D \mapsto \mathscr{I}_X(D).$$

For two divisors $D_1, D_2 \in \mathrm{Div}(X)$ we have $D_1 \leq D_2$ if and only if $\mathscr{I}_X(D_1) \supseteq \mathscr{I}_X(D_2)$. We set

$$\mathscr{O}_X(D) := \mathscr{I}_X(D)^{-1},$$

i. e. for an open subset $U \subseteq X$ its sections are given by

$$\Gamma(U, \mathscr{O}_X(D)) = \{\, f \in \mathscr{K}_X(U) \; ; \; f \mathscr{I}_X(D)(U) \subseteq \mathscr{O}_X(U) \,\}.$$

Thus if D is represented by $(U_i, f_i)_i$, we have $\mathscr{O}_X(D)_{|U_i} = \mathscr{O}_{U_i} f_i^{-1}$. We obtain

$$(11.12.2) \qquad D_1 \leq D_2 \Leftrightarrow \mathscr{O}_X(D_1) \subseteq \mathscr{O}_X(D_2).$$

For every $f \in \Gamma(X, \mathscr{K}_X^\times)$ we find

$$(11.12.3) \qquad f \in \Gamma(X, \mathscr{O}_X(D)) \Leftrightarrow \mathrm{div}(f) \geq -D.$$

One checks that $D \mapsto \mathscr{O}_X(D)$ is a homomorphism of groups, i. e.

$$(11.12.4) \qquad \begin{aligned} &\mathscr{O}_X(0) = \mathscr{O}_X, \\ &\mathscr{O}_X(D_1 + D_2) = \mathscr{O}_X(D_1)\mathscr{O}_X(D_2) \cong \mathscr{O}_X(D_1) \otimes_{\mathscr{O}_X} \mathscr{O}_X(D_2), \end{aligned}$$

where the product is taken in $\mathscr{K}_X$.

Remark 11.27. A divisor D is effective if and only if $\mathscr{I}_X(D)$ is an ideal of $\mathscr{O}_X$. The corresponding closed subscheme of X is often denoted by D as well. If D is represented by $(U_i, f_i)_i$, then $D \cap U_i = V(f_i)$ (as subschemes of U_i). By definition there is an exact sequence of $\mathscr{O}_X$-modules

$$(11.12.5) \qquad 0 \to \mathscr{O}_X(-D) \to \mathscr{O}_X \to \mathscr{O}_D \to 0.$$

Conversely, assume that Y is a closed subscheme of X such that there exists an open affine covering (U_i) of X and regular elements $f_i \in \Gamma(U_i, \mathscr{O}_X)$ such that $Y \cap U_i = V(f_i)$ for all i (such subschemes will be studied in more detail in Volume II where they will be called *regularly immersed of codimension* 1). Then $(U_i, f_i)_i$ represents an effective Cartier divisor. We obtain a bijective correspondence between regularly immersed closed subschemes of codimension 1 and effective Cartier divisors and will often use both notions synonymously.

INVERTIBLE FRACTIONAL IDEALS AND LINE BUNDLES.

We have just seen that attaching to a divisor D the invertible submodule $\mathscr{O}_X(D)$ of $\mathscr{K}_X$ yields a group isomorphism of $\mathrm{Div}(X)$ with the group of invertible fractional ideals of X. We now study the following two questions.
(1) For two divisors D_1 and D_2, when are two invertible fractional ideals $\mathscr{O}_X(D_1)$ and $\mathscr{O}_X(D_2)$ isomorphic as $\mathscr{O}_X$-modules?
(2) Which invertible $\mathscr{O}_X$-modules are isomorphic to an invertible fractional ideal?
The first question is answered by the following proposition.

Proposition 11.28. *Let D_1 and D_2 be divisors. Then the $\mathscr{O}_X$-modules $\mathscr{O}_X(D_1)$ and $\mathscr{O}_X(D_2)$ are isomorphic if and only if D_1 and D_2 are linearly equivalent.*

Proof. We have to show that a divisor D is principal if and only if $\mathscr{O}_X(D) \cong \mathscr{O}_X$. The condition is clearly necessary. Conversely, if $u \colon \mathscr{O}_X \xrightarrow{\sim} \mathscr{O}_X(D)$ is an isomorphism, the image of $1 \in \Gamma(X, \mathscr{O}_X)$ is an element $f \in \Gamma(X, \mathscr{O}_X(D)) \subseteq \Gamma(X, \mathscr{K}_X)$. Then $\mathscr{O}_X(D) = \mathscr{O}_X f$. As $\mathscr{O}_X(D)$ is invertible, we have $f \in \Gamma(X, \mathscr{K}_X)^\times$ and $D = \mathrm{div}(f^{-1})$. $\qquad\square$

Proposition 11.28 can be also expressed by saying that the following is an exact sequence of abelian groups.

(11.12.6)
$$1 \to \Gamma(X, \mathscr{O}_X)^\times \to \Gamma(X, \mathscr{K}_X)^\times \to \mathrm{Div}(X) \xrightarrow{\delta} \mathrm{Pic}(X),$$
$$D \longmapsto \mathscr{O}_X(D).$$

It is easy to see (Exercise 11.13) that this exact sequence is the beginning of the exact cohomology sequence attached to the short exact sequence of abelian sheaves

(11.12.7)
$$1 \to \mathscr{O}_X^\times \to \mathscr{K}_X^\times \to \mathscr{K}_X^\times / \mathscr{O}_X^\times \to 1.$$

The second question is about the image of δ in $\mathrm{Pic}(X)$. It consists of those invertible $\mathscr{O}_X$-modules $\mathscr{L}$ such that there exists an embedding $\mathscr{L} \hookrightarrow \mathscr{K}_X$ of $\mathscr{O}_X$-modules. In general, δ is not surjective, even for noetherian schemes; see [Ha1], I.1.3, for an example of Kleiman. But very often this will be the case:

Proposition 11.29. *Let X be a scheme. We assume that*
(1) *X is integral, or*
(2) *X is a locally noetherian scheme that contains an affine open subscheme U that is schematically dense in X.*
Then the homomorphism $\mathrm{Div}(X) \to \mathrm{Pic}(X)$, $D \mapsto \mathscr{O}_X(D)$ is surjective and therefore induces an isomorphism
$$\mathrm{DivCl}(X) \xrightarrow{\sim} \mathrm{Pic}(X).$$

Proof. First step. Let U be an affine open subscheme that is schematically dense in X (under hypothesis (1) we can take for U any non-empty open affine subscheme). Let $\mathscr{L}$ be an invertible $\mathcal{O}_X$-module. We first claim that every embedding $s\colon \mathscr{L}_{|U} \hookrightarrow \mathscr{K}_{X|U}$ can be extended to a unique embedding $\tilde{s}\colon \mathscr{L} \hookrightarrow \mathscr{K}_X$. The open subsets $V \subseteq X$ with $\mathcal{O}_V \cong \mathscr{L}_{|V}$ form a basis for the topology on X and $U \cap V$ is schematically dense in V for all V. Therefore we may assume that $\mathscr{L} = \mathcal{O}_X$. Then s can be considered as a section $t \in \Gamma(U, \mathscr{K}_X)$. The injectivity of s is equivalent to $t \in \Gamma(U, \mathscr{K}_X)^\times$. By Proposition 11.24, t may be considered as a rational function on U. But a rational function on U is the same as a rational function on X and therefore restriction defines an isomorphism $\Gamma(X, \mathscr{K}_X) \overset{\sim}{\to} \Gamma(U, \mathscr{K}_X)$. This shows our claim.

Second step. Thus our hypothesis implies that we may assume that $X = \operatorname{Spec} A$ is affine, where A is a domain or a noetherian ring. Let $\mathscr{L}$ be an invertible $\mathcal{O}_X$-module. Then $\mathscr{L} = \tilde{M}$ where M is a projective A-module of rank 1. Let $S \subset A$ be the set of regular elements. We have $\Gamma(X, \mathscr{K}_X) = S^{-1}A$ (using Remark 11.25 in the noetherian case). We have to show that there exists a linear injective homomorphism $M \to S^{-1}A$. We claim that $S^{-1}A$ is a semi-local ring. This is clear, if A is an integral domain. If A is noetherian, S is the complement of the union of the finitely many associated ideals of A (Proposition B.59) which implies our claim. Thus the finite projective $S^{-1}A$-module $S^{-1}M$ is free (Proposition B.21) of rank 1 and we can choose $M \hookrightarrow S^{-1}M \cong S^{-1}A$. $\qquad\square$

There are plenty of situations where Proposition 11.29 may be applied. We give two examples (the second follows from results in Chapter 13).

Corollary 11.30. *Assume that X satisfies one of the following conditions.*
(1) X is noetherian and reduced.
(2) X is a subscheme of the projective space $\mathbb{P}^n_R$, where R is a noetherian ring.
Then $D \mapsto \mathcal{O}_X(D)$ induces an isomorphism $\operatorname{DivCl}(X) \overset{\sim}{\to} \operatorname{Pic}(X)$.

Proof. (1). As X is noetherian, there are only finitely many irreducible components $Z_1, \dots, Z_r$. Choose within each irreducible component Z_i a non-empty open affine subset $U_i \subseteq Z_i$ that does not meet any of the other irreducible components. Then the union $U := \bigcup_i U_i = \coprod_i U_i$ is affine and dense in X. As X is reduced, U is also schematically dense (Section (9.5)).

(2). By Proposition 9.22 it suffices to show that there exists an open affine subset that contains $\operatorname{Ass}(X)$. As X is noetherian, $\operatorname{Ass}(X)$ is finite. But under the hypothesis (2), *any* finite subset of X is contained in an open affine subset (Proposition 13.49). $\qquad\square$

Remark 11.31. Using the language introduced in Chapter 13, the proof of (2) shows that $\operatorname{DivCl}(X) \cong \operatorname{Pic}(X)$ whenever X is a noetherian scheme such that there exists an ample line bundle on X.

SUPPORT OF A DIVISOR.

The support of a divisor is simply the support as a section of the abelian sheaf $\mathscr{K}_X^\times / \mathcal{O}_X^\times$:

Definition 11.32. *The support of a divisor D on a scheme X is the subspace*

$$\operatorname{Supp}(D) := \{\, x \in X \;;\; D_x \neq 1 \,\}.$$

Here D_x denotes the image of $D \in \Gamma(X, \mathscr{K}_X^\times / \mathscr{O}_X^\times)$ in the stalk $(\mathscr{K}_X^\times / \mathscr{O}_X^\times)_x$. In other words, if D is represented by $(U_i, f_i)_i$, then a point $x \in X$ lies in $\mathrm{Supp}(D)$ if and only if $(f_i)_x \notin \mathscr{O}_{X,x}^\times$ for all (or, equivalently, for one) index i with $U_i \ni x$. As the support of any section of an abelian sheaf is closed (Section (7.6)), $\mathrm{Supp}(D)$ is closed in X.

Remark 11.33. Let D be a divisor on a scheme X, represented by $(U_i, f_i)_i$.
(1) If D is an effective divisor, then $\mathrm{Supp}(D)$ is the underlying closed subspace of the closed subscheme corresponding to D (Remark 11.27).
(2) Let V be any open subset of X. Then $\Gamma(V, \mathscr{O}_X(D))$ consists of those sections s in $\Gamma(V, \mathscr{K}_X)$ such that $sf_i \in \Gamma(U_i \cap V, \mathscr{O}_X)$ for all i by (11.12.3). Therefore if U_{eff} is the locus of points where D is effective (i.e., the open subset of $x \in X$ such that $(f_i)_x \in \mathscr{O}_{X,x}$ for those i with $U_i \ni x$), then $1 \in \Gamma(U_{\mathrm{eff}}, \mathscr{K}_X)$ is a section $s_D \in \Gamma(U_{\mathrm{eff}}, \mathscr{O}_X(D))$. This section is called the *canonical section of (the line bundle attached to) D*.
(3) If U is the complement of $\mathrm{Supp}(D)$ in X, then $U \subseteq U_{\mathrm{eff}}$ and $s_{D|U}$ defines an isomorphism $\mathscr{O}_X(D)_{|U} \xrightarrow{\sim} \mathscr{O}_{X|U}$.

Proposition 11.34. *Let X be a scheme, let $\mathscr{L}$ be a line bundle on X, and let $R_{\mathscr{L}}$ be the set of $s \in \Gamma(X, \mathscr{L})$ that are regular. Then there is a natural bijection*

$$\left\{ \begin{array}{c} \textit{effective Cartier divisors } D \\ \textit{such that } \mathscr{O}_X(D) \cong \mathscr{L} \end{array} \right\} \leftrightarrow R_{\mathscr{L}} / \sim,$$

where $s \sim s'$ if there exists an $a \in \Gamma(X, \mathscr{O}_X^\times)$ with $s' = as$.

In Section (13.13) we will give a more geometric interpretation of this result.

Proof. For an effective divisor D and an isomorphism $\alpha \colon \mathscr{O}_X(D) \xrightarrow{\sim} \mathscr{L}$ the corresponding section $s \in R_{\mathscr{L}}$ is the image of the canonical section s_D under α. This yields a well defined map from the left hand side to the right hand side. Conversely let $s \in R_{\mathscr{L}}$. Choose an open covering (U_i) of X and isomorphisms $\mathscr{L}_{|U_i} \cong \mathscr{O}_{X|U_i}$. Then the images of $s_{|U_i}$ define local equations f_i for an effective divisor D, which depends only on the equivalence class of s. The section s then defines a monomorphism $\mathscr{O}_X \hookrightarrow \mathscr{L}$ which extends by definition to an isomorphism $\mathscr{O}_X(D) \xrightarrow{\sim} \mathscr{L}$. $\square$

Lemma 11.35. *Let X be a locally noetherian scheme and let D be a divisor on X. Then* $\mathrm{codim}_X(\mathrm{Supp}(D)) \geq 1$, *i.e., for all $z \in \mathrm{Supp}(D)$ we have $\dim(\mathscr{O}_{X,z}) \geq 1$.*

Proof. Let $\eta \in X$ be a point with $\dim(\mathscr{O}_{X,\eta}) = 0$. Then $\mathscr{O}_{X,\eta}$ is a local Artin ring and thus every regular element in $\mathscr{O}_{X,\eta}$ is invertible. Thus $\mathscr{O}_{X,\eta} = \mathrm{Frac}(\mathscr{O}_{X,\eta}) = \mathscr{K}_{X,\eta}$, where the second identity holds by Remark 11.25. This shows that the germ $D_\eta \in \mathscr{K}_{X,\eta}^\times / \mathscr{O}_{X,\eta}^\times = 1$ is trivial. $\square$

The lemma can be made more precise: For every maximal point η of $\mathrm{Supp}(D)$ we have $\mathrm{depth}(\mathscr{O}_{X,\eta}) = 1$ (Exercise 11.14). In particular, the complement of $\mathrm{Supp}\, D$ is schematically dense in X if X is locally noetherian.

(11.13) Cartier divisors and Weil divisors.

We will now consider Cartier divisors more geometrically, namely as so-called Weil divisors. Let X be a noetherian scheme. Let $C \subseteq X$ be a closed irreducible subset with generic point ξ. We sometimes call $\mathscr{O}_{X,C} := \mathscr{O}_{X,\xi}$ the *local ring of X at C*. Recall that

$$(11.13.1) \qquad\qquad \operatorname{codim}_X C = \dim \mathscr{O}_{X,C}$$

is called the codimension of C in X, see Section (5.8). More generally, for an arbitrary closed subset Z of X we have $\operatorname{codim}_X(Z) = \inf_C \operatorname{codim}_X(C)$, where C runs through the set of irreducible components of Z (Remark 5.29). Thus Z is of codimension zero if and only if Z contains an irreducible component of X.

For an integer $k \geq 0$ we denote by X^k the set of closed integral subschemes of codimension k. The free abelian group $\mathbb{Z}^{(X^k)}$ is denoted by $Z^k(X)$. Thus elements of $Z^k(X)$ are finite linear combinations $\sum_C n_C[C]$, where C runs through X^k.

Definition 11.36. *An element of $Z^1(X)$ is called a Weil divisor. The elements of X^1 are called prime Weil divisors. We denote by $Z^1_+(X)$ the set of $\sum_{C \in X^1} n_C[C] \in Z^1(X)$ such that $n_C \geq 0$ for all C, and call its elements effective Weil divisors.*

We will now define a group homomorphism

$$(11.13.2) \qquad\qquad \operatorname{cyc} \colon \operatorname{Div}(X) \to Z^1(X).$$

For this we will have to define for a prime Weil divisor C the order $\operatorname{ord}_C(f)$ of a meromorphic function $f \in \Gamma(U, \mathscr{K}_X)$ along C. Thus assume that U contains the generic point of C. If $\mathscr{O}_{X,C}$ is a discrete valuation ring (e.g., if X is normal; see Proposition 11.39 below), the germ f_C is a nonzero element in $K := \operatorname{Frac}(\mathscr{O}_{X,C})$ and we set

$$\operatorname{ord}_C(f) := v_C(f_C),$$

where $v_C(f)$ is the normalized discrete valuation of $f \in K$ given by $\mathscr{O}_{X,C}$.

In general, $\mathscr{O}_{X,C}$ is only a local noetherian ring of dimension 1 and we use the following lemma to define the order of an element $f \in \operatorname{Frac}(\mathscr{O}_{X,C})^\times$.

Lemma 11.37. *Let A be a noetherian local ring of dimension 1. Write $f \in \operatorname{Frac}(A)^\times$ as $f = ab^{-1}$ for regular elements $a, b \in A$ and set*

$$\operatorname{ord}_A(f) := \lg_A(A/(a)) - \lg_A(A/(b)).$$

Then $\operatorname{ord} \colon \operatorname{Frac}(A)^\times \to \mathbb{Z}$ is a well defined group homomorphism and $A^\times \subseteq \operatorname{Ker}(\operatorname{ord})$.

Proof. Let $a \in A$ be a regular element. Then $\dim A/(a) = 0$ (Corollary B.64) and thus $A/(a)$ is a local Artin ring and hence of finite length (Proposition B.36). If $b \in A$ is a second regular element, multiplication with b yields an isomorphism $A/(a) \cong bA/(ab)$. The exact sequence $0 \to bA/(ab) \to A/(ab) \to A/(b) \to 0$ therefore shows that $\lg A/(ab) = \lg A/(a) + \lg A/(b)$. This proves that ord is a well defined group homomorphism $\operatorname{Frac}(A)^\times \to \mathbb{Z}$. The fact that $\operatorname{ord}(u) = 0$ for all units $u \in A^\times$ is then clear. $\qquad\square$

To define cyc (11.13.2) let $D \in \operatorname{Div}(X)$ be a Cartier divisor represented by (U_i, f_i). For $C \in X^1$ a prime Weil divisor we choose an index i such that the generic point η_C of C is contained in U_i. Let $f \in \mathscr{K}_{X,\eta_C} = \operatorname{Frac}(\mathscr{O}_{X,C})$ be the germ of f_i at η_C. Then f does not depend on the choice of the presentation (U_i, f_i) or on i up to a unit of $\mathscr{O}_{X,C}$. Therefore

(11.13.3) $\mathrm{ord}_C(D) := \mathrm{ord}_{\mathscr{O}_{X,C}}(f) \in \mathbb{Z}$

depends only on D and C. The integer $\mathrm{ord}_C(D)$ is called the *vanishing order of D at C*. If $\mathrm{ord}_C(D) \geq 0$, we also say that D has a *zero of order* $\mathrm{ord}_C(D)$ *in C*, and if $\mathrm{ord}_C(D) < 0$, we say that D has a *pole of order* $- \mathrm{ord}_C(D)$ *in C*.

If C is not contained in the support of D, then we have $\mathrm{ord}_C(D) = 0$. By Lemma 11.35 one has $\mathrm{codim}_X(\mathrm{Supp}\, D) \geq 1$. Therefore every $C \in X^1$ with $C \subseteq \mathrm{Supp}\, D$ is an irreducible component. As X and hence $\mathrm{Supp}\, D$ is a noetherian topological space there are (for a given divisor D) only finitely many $C \in X^1$ such that $\mathrm{ord}_C(D) \neq 0$. Now we define

(11.13.4) $\mathrm{cyc}\colon \mathrm{Div}(X) \to Z^1(X), \qquad D \mapsto \sum_{C \in X^1} \mathrm{ord}_C(D)[C].$

Lemma 11.37 shows that cyc is a group homomorphism.

If $f \in \Gamma(X, \mathscr{K}_X)^\times$ is an invertible rational function, we call $\mathrm{ord}_C(f) := \mathrm{ord}_C(\mathrm{div}(f))$ the *vanishing order of f at C* and write $\mathrm{cyc}(f)$ instead of $\mathrm{cyc}(\mathrm{div}(f))$. A Weil divisor is called *principal* if it is of the form $\mathrm{cyc}(f)$ for some $f \in \Gamma(X, \mathscr{K}_X)^\times$. The principal Weil divisors form a subgroup $Z^1_{\mathrm{princ}}(X)$ of $Z^1(X)$. We denote the quotient by

(11.13.5) $\mathrm{Cl}(X) := Z^1(X)/Z^1_{\mathrm{princ}}(X).$

The homomorphism cyc induces a homomorphism $\mathrm{DivCl}(X) \to \mathrm{Cl}(X)$.

In general cyc is neither injective nor surjective (Exercise 11.18). To formulate a positive result we introduce the following notion.

Definition 11.38. *A scheme X is called* locally factorial *if for all $x \in X$ the local ring $\mathscr{O}_{X,x}$ is factorial.*

Proposition 11.39. *Let X be a locally noetherian scheme. Consider the following assertions.*
(i) *X is regular.*
(ii) *X is locally factorial.*
(iii) *For all $x \in X$ the local ring $\mathscr{O}_{X,x}$ is an integral domain and every closed integral subscheme $C \subset X$ of codimension 1 is regularly immersed (i.e., for all $x \in C$ the closed subscheme $C \cap \mathrm{Spec}\, \mathscr{O}_{X,x}$ of $\mathrm{Spec}\, \mathscr{O}_{X,x}$ is of the form $V(f)$ for some regular element $f \in \mathscr{O}_{X,x}$).*
(iv) *X is normal.*
(v) *For every closed integral subscheme $C \subset X$ of codimension 1 the local ring $\mathscr{O}_{X,C}$ is a discrete valuation ring.*
Then we have the implications "(i) $\Rightarrow$ (ii) $\Leftrightarrow$ (iii) $\Rightarrow$ (iv) $\Rightarrow$ (v)".

Proof. This is just the geometric version of Remark B.78 and Proposition B.75 (2). $\square$

Theorem 11.40. *Let X be a noetherian scheme.*
(1) *If X is normal, the homomorphism $\mathrm{cyc}\colon \mathrm{Div}(X) \to Z^1(X)$ is injective. In particular it induces an injective homomorphism $\mathrm{DivCl}(X) \hookrightarrow \mathrm{Cl}(X)$.*
(2) *If X is locally factorial, the homomorphism $\mathrm{cyc}\colon \mathrm{Div}(X) \to Z^1(X)$ is bijective. In particular it induces an isomorphism $\mathrm{DivCl}(X) \xrightarrow{\sim} \mathrm{Cl}(X)$.*

Proof. (1). As we have $\mathrm{Div}_+(X) \cap (-\mathrm{Div}_+(X)) = 0$ and $Z^1_+(X) \cap (-Z^1_+(X)) = 0$, it suffices to show that $\mathrm{cyc}^{-1}(Z^1_+(X)) = \mathrm{Div}_+(X)$. Let D be a Cartier divisor such that $\mathrm{cyc}(D) \in Z^1_+(X)$. To prove that $D \in \mathrm{Div}_+(X)$ is a local question on X and we may assume that $X = \mathrm{Spec}\, A$ is affine, where A is an integrally closed domain, and that D is a principal divisor given by an element $f \in K(X) = \mathrm{Quot}(A)$. By hypothesis we have $f \in A_{\mathfrak{p}}$ for every prime ideal $\mathfrak{p}$ of height one. But as A is integrally closed, this implies $f \in A$ by Proposition B.73 (3).

(2). We construct an inverse map d. Let Z be a prime Weil divisor given by the quasi-coherent ideal $\mathscr{I}_Z \subset \mathscr{O}_X$. Let $x \in X$ be a point. By Proposition 11.39, $\mathscr{I}_{Z,x}$ is generated by one element $f_x \in \mathscr{O}_{X,x}$. By Proposition 7.27 there exists an open neighborhood U of x and a section $f \in \Gamma(U, \mathscr{O}_X)$ whose germ in x is f_x such that $\mathscr{I}_{|U} = \mathscr{O}_U f$. This shows that we find an open affine covering $(U_i)_i$ of X and $f_i \in \Gamma(U_i, \mathscr{O}_X)$ such that $\mathscr{I}_{|U_i} = \mathscr{O}_{U_i} f_i$. As two principal ideals (f) and (f') are equal if and only if there exists a unit u with $f' = fu$, the tuple $(U_i, f_i)_i$ defines an effective Cartier divisor $D = d(Z)$ on X. Clearly we have $\mathrm{cyc}(d(Z)) = Z$. Extending d by linearity we obtain a homomorphism $d \colon Z^1(X) \to \mathrm{Div}(X)$ such that $\mathrm{cyc} \circ d = \mathrm{id}$. By (1) this implies that d is an inverse. $\square$

Theorem 11.40 also has a converse (Exercise 11.17).

Remark 11.41. Let X be an integral noetherian locally factorial scheme with function field $K(X)$. Combining Theorem 11.40 and Section (11.12) we have an isomorphism of abelian groups

$$(11.13.6) \qquad Z^1(X) \xrightarrow{\sim} \{\text{invertible fractional ideals of } \mathscr{O}_X\}$$

which induces by Corollary 11.30 (1) an isomorphism

$$(11.13.7) \qquad \mathrm{Cl}(X) \xrightarrow{\sim} \mathrm{Pic}(X).$$

The isomorphism (11.13.6) can be described as follows. Let $\sum_C n_C[C] \in Z^1(X)$. Then the corresponding invertible fractional ideal $\mathscr{L}$ is given by (for $U \subseteq X$ open)

$$\Gamma(U, \mathscr{L}) := \{\, f \in K(X) \,;\, v_C(f) \geq -n_C \text{ for all } C \in X^1 \text{ with } C \cap U \neq \emptyset \,\}.$$

Here v_C is the normalized discrete valuation given by the discrete valuation ring $\mathscr{O}_{X,C}$ on $\mathrm{Frac}\, \mathscr{O}_{X,C} = K(X)$. Thus $\Gamma(X, \mathscr{L})$ consists of those meromorphic functions whose order at C is at least $-n_C$.

If $U \subseteq X$ is an open subscheme of a noetherian scheme X, we define a restriction homomorphism $Z^1(X) \to Z^1(U)$ by $\sum_C n_C[C] \mapsto \sum_C n_C[C \cap U]$, where the second sum is only indexed by those $C \in X^1$ such that $C \cap U \neq \emptyset$. For those C we have $\mathscr{O}_{U,C \cap U} = \mathscr{O}_{X,C}$. This shows that if D is a Cartier divisor, the restriction of $\mathrm{cyc}(D)$ to U is equal to $\mathrm{cyc}(D_{|U})$. As the restriction of principal Cartier divisors are again principal Cartier divisors, this homomorphism induces a restriction homomorphism $\mathrm{Cl}(X) \to \mathrm{Cl}(U)$. If $U \subseteq X$ is schematically dense, this restriction is surjective and one can describe its kernel:

Proposition 11.42. *Let X be a noetherian scheme and let Z be a closed subscheme that does not contain an irreducible component of X. Let $Z_1, \ldots, Z_r$ be the irreducible components of Z that have codimension 1 in X and set $U = X \setminus Z$. Suppose that U is schematically dense in X. Then we have an exact sequence*

$$\bigoplus_{i=1}^{r} \mathbb{Z} \cdot [Z_i] \to \mathrm{Cl}(X) \to \mathrm{Cl}(U) \to 0.$$

The hypotheses on Z imply that U is always dense in X. Hence U is automatically schematically dense in X if X is reduced (Remark 9.20 (1)) or, more generally, if no embedded component of X is contained in Z (Proposition 9.22).

Proof. If C_U is a prime Weil divisor on U, its closure in X is a prime Weil divisor on X. This shows that $Z^1(X) \to Z^1(U)$ is surjective, in particular $\mathrm{Cl}(X) \to \mathrm{Cl}(U)$ is surjective. The kernel of $Z^1(X) \to Z^1(U)$ is equal to $\bigoplus_{i=1}^{r} \mathbb{Z}[Z_i]$ and in particular $[Z_i]$ is in the kernel of $\mathrm{Cl}(X) \to \mathrm{Cl}(U)$. Conversely, let $\sum_C n_C[C] \in Z^1(X)$ such that $\sum_C n_C[C \cap U]$ is the divisor of an invertible meromorphic function f_U on U. As U is schematically dense, f_U can be uniquely extended to a meromorphic function f on X (Proposition 11.24). Then $\mathrm{div}(f) - \sum_C n_C[C]$ is a linear combination of the Weil prime divisors $[Z_i]$. This shows that the divisor class of $\sum_C n_C[C]$ is in the image of $\bigoplus_{i=1}^{r} \mathbb{Z} \cdot [Z_i]$. $\qquad\square$

Under the hypotheses of Proposition 11.42 we obtain a commutative diagram

(11.13.8)
$$
\begin{array}{ccc}
\mathrm{Pic}(X) \xleftarrow{\mathscr{O}_X(\ )} \mathrm{DivCl}(X) \xrightarrow{\mathrm{cyc}} \mathrm{Cl}(X) \\
\downarrow \qquad\qquad \downarrow \qquad\qquad \downarrow \\
\mathrm{Pic}(U) \xleftarrow[\mathscr{O}_U(\ )]{} \mathrm{DivCl}(U) \xrightarrow{\mathrm{cyc}} \mathrm{Cl}(U).
\end{array}
$$

where the vertical arrows are the restriction maps. If X is locally factorial, then the horizontal maps are isomorphisms by Corollary 11.30 and Theorem 11.40. Therefore we obtain the following corollary.

Corollary 11.43. *Let X be a noetherian locally factorial scheme (e.g., if X is regular), and let $Z \subset X$ be a closed subset of codimension at least 1. Let $U = X \setminus Z$ be the complement. Then $\mathrm{Pic}(X) \to \mathrm{Pic}(U)$, $\mathscr{L} \mapsto \mathscr{L}_{|U}$ is surjective. If $\mathrm{codim}_X(Z) \geq 2$ this map is an isomorphism.*

(11.14) Examples for divisor class groups.

Example 11.44. (Factorial rings)
Let A be a noetherian integrally closed domain. Recall that A is factorial if and only if all prime ideals of height 1 are principal ideals (Proposition B.75 (2)), i.e., every prime Weil divisor on $X = \mathrm{Spec}\, A$ is principal. In other words, A is factorial if and only if $\mathrm{Cl}(X) = 0$. In this case we also have

$$\mathrm{DivCl}(X) = \mathrm{Pic}(X) = 0$$

by Theorem 11.40 and Corollary 11.30. In particular, for every field k we find

(11.14.1) $\mathrm{Pic}(\mathbb{A}_k^n) = \mathrm{DivCl}(\mathbb{A}_k^n) = \mathrm{Cl}(\mathbb{A}_k^n) = 0.$

In fact, Exercise 11.24 shows that $\mathrm{Pic}(\mathrm{Spec}\, A) = 0$ for any factorial (not necessarily noetherian) ring.

Example 11.45. (Projective space)

Let k be a field, $n \geq 1$ be an integer and set $S := k[T_0, \ldots, T_n]$. If $f \in S$ is an irreducible homogeneous polynomial, $V_+(f)$ is a prime Weil divisor (as this holds for its affine cone $V(f) \subset \mathbb{A}_k^{n+1}$). Define a subgroup of $k(T_0, \ldots, T_n)^\times$ by

$$\mathcal{R} := \{\, f = g/h \;;\; g, h \in S \text{ nonzero homogeneous}\,\}$$

For $f = g/h \in \mathcal{R}$ we set $\deg(f) := \deg(g) - \deg(h)$ and obtain a surjective homomorphism $\deg \colon \mathcal{R} \to \mathbb{Z}$.

As S is factorial, we may write any $f \in \mathcal{R}$ as product $f = f_1 f_2 \ldots f_r$, where $f_i \in S$ are irreducible homogeneous polynomials. Moreover, the f_i are uniquely determined up to units and up to order. Thus we obtain a map

$$Z \colon \mathcal{R} \to Z^1(\mathbb{P}_k^n), \qquad f \mapsto \sum_i [V_+(f_i)]$$

which is clearly a homomorphism of groups.

The nonzero rational functions on $\mathbb{P}_k^n$ are the elements $f \in \mathcal{R}$ that can be written as a quotient $f = g/h$, where $g, h \in S$ are homogeneous of the same degree. In other words

$$K(\mathbb{P}_k^n)^\times = \operatorname{Ker}(\deg \colon \mathcal{R} \to \mathbb{Z}).$$

By Corollary 5.42 every integral closed subscheme of codimension 1 is of the form $V_+(f)$ for an irreducible homogeneous polynomial f. This shows that Z is surjective and induces therefore isomorphisms of groups

$$(11.14.2) \qquad \mathcal{R}/K(\mathbb{P}_k^n)^\times \overset{\sim}{\to} \operatorname{Cl}(\mathbb{P}_k^n) \overset{\sim}{\to} \mathbb{Z},$$

where the second isomorphism is induced by $[V_+(f)] \mapsto \deg(f)$.

As $\mathbb{P}_k^n$ is regular (and in particular locally factorial), we find

$$(11.14.3) \qquad \operatorname{Div}(\mathbb{P}_k^n) = Z^1(\mathbb{P}_k^n) = \mathcal{R}/k^\times, \quad \operatorname{Pic}(\mathbb{P}_k^n) = \operatorname{DivCl}(\mathbb{P}_k^n) = \operatorname{Cl}(\mathbb{P}_k^n) \cong \mathbb{Z}.$$

For $f \in S$ irreducible and homogeneous, the Cartier divisor corresponding to the prime Weil divisor $V_+(f)$ is represented by $(D_+(T_i), f/T_i^{\deg f})_i$.

For $d \in \mathbb{Z}$ fix some $f \in \mathcal{R}$ of degree d. We denote by $\mathscr{O}_{\mathbb{P}_k^n}(d)$ the invertible $\mathscr{O}_{\mathbb{P}_k^n}$-submodule of $\mathscr{K}_{\mathbb{P}_k^n}$ whose restriction to $D_+(T_i)$ is equal to $\mathscr{O}_{D_+(T_i)} T_i^d / f \subset \mathscr{K}_{D_+(T_i)}$. Then via $\operatorname{Pic}(\mathbb{P}_k^n) = H^1(\mathbb{P}_k^n, \mathscr{O}_{\mathbb{P}_k^n}^\times)$, the isomorphism class of the invertible module $\mathscr{O}_{\mathbb{P}_k^n}(d)$ corresponds to the Čech cohomology class represented by the cocycle $(T_j^d/T_i^d)_{i,j}$ on the open covering $(D_+(T_i))_i$. In particular it is independent of the choice of f. Thus we have

$$(11.14.4) \qquad \begin{aligned} \Gamma(D_+(T_i), \mathscr{O}_{\mathbb{P}_k^n}(d)) &= T_i^d k[T_0/T_i, \ldots, T_n/T_i] \\ &= \{\, f/T_i^m \in \operatorname{Frac}(S) \;;\; m \geq 0,\ f \in k[T_0, \ldots, T_n]_{d+m}\,\}. \end{aligned}$$

We claim that

$$(11.14.5) \qquad \Gamma(\mathbb{P}_k^n, \mathscr{O}_{\mathbb{P}_k^n}(d)) = k[T_0, \ldots, T_n]_d,$$

where $k[T_0, \ldots, T_n]_d$ denotes the subspace of homogeneous polynomials of degree d. Indeed,

$$\Gamma(\mathbb{P}_k^n, \mathscr{O}_{\mathbb{P}_k^n}(d)) = \bigcap_{i=0}^{n} \Gamma(D_+(T_i), \mathscr{O}_{\mathbb{P}_k^n}(d))$$

and $k[T_0, \ldots, T_n]_d$ is contained in this intersection. Conversely let $s \in \Gamma(\mathbb{P}_k^n, \mathscr{O}_{\mathbb{P}_k^n}(d))$. For all $i = 0, \ldots, n$ we can write $s = f_i/T_i^{m_i}$ where $m_i \geq 0$ and f_i is homogeneous of degree $d + m_i$. We may assume that T_i does not divide f_i. Thus for all $i \neq j$ we find $T_i^{m_i} f_j = T_j^{m_j} f_i$. As T_i does not divide f_i, we have $m_i = 0$ for all i and thus $s = f_i$ is homogeneous of degree d.

In particular, $\Gamma(\mathbb{P}_k^n, \mathscr{O}_{\mathbb{P}_k^n}(d))$ is a finite-dimensional vector space and

$$(11.14.6) \qquad \dim \Gamma(\mathbb{P}_k^n, \mathscr{O}_{\mathbb{P}_k^n}(d)) = \begin{cases} 0, & \text{if } d < 0; \\ \binom{n+d}{n}, & \text{if } d \geq 0. \end{cases}$$

The line bundle $\mathscr{O}_{\mathbb{P}_k^n}(1)$ can also be described as the universal quotient bundle on $\mathbb{P}_k^n$ (see Exercise 11.23, and Section (13.8) for a systematic treatment).

Example 11.46. Let k be a field, $n_1, \ldots, n_r \geq 1$ be integers and $X = \mathbb{P}_k^{n_1} \times_k \cdots \times_k \mathbb{P}_k^{n_r}$. This is a regular integral scheme. Choose coordinates T_{si} ($s = 1, \ldots, r$, $i = 0, \ldots, n_s$). Set $S := k[(T_{ij})]$ and let $\mathcal{R}$ be the subgroup of $\mathrm{Frac}(S)^\times$ consisting of fractions g/h, where $g, h \in S$ are nonzero and homogeneous separately in each set of variables $\underline{T}_s := (T_{si})_{0 \leq i \leq n_s}$. As in Example 11.45 we obtain for all $s = 1, \ldots, r$ surjective homomorphisms of groups $\deg_s \colon \mathcal{R} \to \mathbb{Z}$ and thus a surjective homomorphism $\deg \colon \mathcal{R} \to \mathbb{Z}^r$. Again, the set of nonzero elements of the function field of X is the kernel of $\deg$. Therefore $\deg$ yields an isomorphism

$$(11.14.7) \qquad \mathrm{Pic}(X) = \mathrm{DivCl}(X) = \mathrm{Cl}(X) \overset{\sim}{\to} \mathbb{Z}^r.$$

In particular this shows that a nontrivial product of projective spaces is never isomorphic to projective space.

Example 11.47. (Divisors on Dedekind schemes)
We continue the example given in Section (7.13). Thus X is a Dedekind scheme that is not the spectrum of a field, X_0 is the set of closed points of X, and for all $x \in X_0$ we denote by v_x the normalized valuation on the function field $K(X)$ given by the discrete valuation ring $\mathscr{O}_{X,x}$. Clearly we have $Z^1(X) = \mathbb{Z}^{(X_0)}$. As X is regular, we find $\mathrm{Div}(X) = Z^1(X)$ as already suggested by the notation of Section (7.13). For $D \in \mathrm{Div}(X)$ the invertible sheaf $\mathscr{L}_D$ defined in loc. cit. is the sheaf $\mathscr{I}_X(D) = \mathscr{O}_X(D)^{-1}$ defined in Section (11.12).

If $X = \mathrm{Spec}\, O_K$, where O_K is the ring of integers in a number field K (that is, the integral closure of $\mathbb{Z}$ in the finite extension K of $\mathbb{Q}$), $\mathrm{Cl}(O_K) = \mathrm{Pic}(O_K)$ is the divisor class group studied in algebraic number theory. The exact sequence (11.11.2) becomes the exact sequence

$$1 \longrightarrow O_K^\times \longrightarrow K^\times \longrightarrow \mathrm{Div}(O_K) \longrightarrow \mathrm{Pic}(O_K) \longrightarrow 1.$$

In this case it is a basic result from number theory that the divisor class group $\mathrm{Pic}(O_K)$ is a finite group (e.g. [Neu] I, §6) and that $O_K^\times \cong \mu(K) \times \mathbb{Z}^{r+s-1}$, where $\mu(K)$ is the finite cyclic group of roots in unity in K and where r (resp. s) is the number of real (resp. half the number of complex) embeddings of K (e.g. [Neu] I, §7).

For divisors on not necessarily regular 1-dimensional schemes we refer to Proposition 15.25 below and to Exercise 11.18.

(11.15) The automorphism group of $\mathbb{P}_k^n$.

Using the results of Example 11.45, we now compute the automorphism group of projective space $\mathbb{P}_k^n$ over a field k. As we have seen before in Section (1.22), every invertible matrix $A \in \mathrm{GL}_{n+1}(k)$ induces an automorphism of $\mathbb{P}_k^n$. Since scalar matrices induce the identity morphism, we obtain a map

$$(11.15.1) \qquad \mathrm{PGL}_{n+1}(k) := \mathrm{GL}_{n+1}(k)/(k^\times \cdot E_n) \to \mathrm{Aut}_k(\mathbb{P}_k^n).$$

This is a group homomorphism, and it is clearly injective.

Proposition 11.48. *Let k be a field, and $n \geq 0$. The natural map* (11.15.1) *is a group isomorphism*

$$\mathrm{PGL}_{n+1}(k) \xrightarrow{\sim} \mathrm{Aut}(\mathbb{P}_k^n).$$

Proof. We will show that the map is an isomorphism by constructing an inverse. Set $P := \mathbb{P}_k^n$. In Example 11.45 we have seen that $\mathrm{Pic}(\mathbb{P}_k^n) \cong \mathbb{Z}$, and the sheaf $\mathscr{O}_P(1)$ can be characterized as the unique generator of $\mathrm{Pic}(\mathbb{P}_k^n)$ whose space of global sections is non-trivial. In particular, the pull-back of $\mathscr{O}_P(1)$ under any automorphism of P is isomorphic to $\mathscr{O}_P(1)$, and we obtain a map

$$(11.15.2) \qquad \mathrm{Aut}(P) \to \mathrm{GL}(\Gamma(P, \mathscr{O}_P(1)))/\mathrm{Aut}_k(\mathscr{O}_P(1)).$$

Now $\mathrm{Hom}(\mathscr{O}_P(1), \mathscr{O}_P(1)) = \mathrm{Hom}(\mathscr{O}_P, \mathscr{O}_P) = \Gamma(P, \mathscr{O}_P) = k$ and therefore we obtain $\mathrm{Aut}_k(\mathscr{O}_P(1)) = k^\times \cdot \mathrm{id}$. Choose homogeneous coordinates $X_0, \ldots, X_n$ on $P = \mathbb{P}_k^n$ and identify $\Gamma(P, \mathscr{O}_P(1))$ with $k[X_0, \ldots, X_n]_1 = \bigoplus_{i=0}^n kX_i$ by (11.14.5). Thus we can identify the right hand side of (11.15.2) with $\mathrm{PGL}_{n+1}(k)$, and it is a straightforward computation to check that this map is the inverse of the map above. $\qquad\square$

(11.16) Inverse image of divisors.

In general, although we can always form the inverse image of a line bundle along a morphism of schemes, the inverse image of a Cartier divisor on Y along a morphism $f : X \to Y$ of schemes is not well-defined (e.g., think of the case that the image of X is contained in the support of D).

The key question, given $f : X \to Y$, is whether the homomorphism $\mathscr{O}_Y \to f_*\mathscr{O}_X$ given by f extends to a homomorphism $\mathscr{K}_Y \to f_*\mathscr{K}_X$. Assume for a moment that this is the case. We obtain a homomorphism $\mathscr{K}_Y^\times/\mathscr{O}_Y^\times \to f_*\mathscr{K}_X^\times/f_*\mathscr{O}_X^\times \to f_*(\mathscr{K}_X^\times/\mathscr{O}_X^\times)$, and hence a homomorphism

$$(11.16.1) \qquad f^* : \mathrm{Div}(Y) = \Gamma(Y, \mathscr{K}_Y^\times/\mathscr{O}_Y^\times) \to \Gamma(X, \mathscr{K}_X^\times/\mathscr{O}_X^\times) = \mathrm{Div}(X).$$

Definition 11.49. *Let $f : X \to Y$ be a morphism of schemes such that the homomorphism $f^\flat : \mathscr{O}_Y \to f_*\mathscr{O}_X$ given by f extends to a homomorphism $\mathscr{K}_Y \to f_*\mathscr{K}_X$. Let D be a Cartier divisor on Y. Then its image f^*D under the map* (11.16.1) *is called the* pullback *of D or the* inverse image *of D.*

Clearly, f^* is a group homomorphism, i.e., $f^*(D_1 + D_2) = f^*(D_1) + f^*(D_2)$. One easily checks that $f^*\mathscr{O}(D) \cong \mathscr{O}(f^*D)$, i.e., that pullback of divisors is compatible with the pullback morphism $f^* : \mathrm{Pic}(Y) \to \mathrm{Pic}(X)$.

The following proposition gives two criteria when it is possible to form the inverse image of divisors.

Proposition 11.50. *Let $f\colon X \to Y$ be a morphism of schemes, and assume that one of the following conditions holds:*
(a) *f is flat, or*
(b) *X is reduced and locally noetherian, and each irreducible component of X dominates an irreducible component of Y.*
Then the homomorphism $f^\flat\colon \mathscr{O}_Y \to f_*\mathscr{O}_X$ extends to a homomorphism $\mathscr{K}_Y \to f_*\mathscr{K}_X$.

One can show that the proposition still holds if in (b) the assumption that X is locally noetherian is replaced by the assumption that the set of connected components of X is locally finite (see [EGAIV] (21.4.5) (iii)).

Proof. Recall that $\mathscr{K}_X$ is the sheaf associated with the presheaf $U \mapsto S_U^{-1}\Gamma(U, \mathscr{O}_X)$, where $S_U \subseteq \Gamma(U, \mathscr{O}_X)$ is the multiplicative subset of regular sections (Definition 11.22). It is enough to show that the homomorphism $\mathscr{O}_Y \to f_*\mathscr{O}_X$ extends to a morphism between these presheaves, i.e., that for every open subset $V \subseteq Y$, the homomorphism $f_V^\flat\colon \Gamma(V, \mathscr{O}_Y) \to \Gamma(f^{-1}(V), \mathscr{O}_X)$ induces a homomorphism between the corresponding rings of total fractions. This means that we have to show that $f_V^\flat$ maps regular sections to regular sections. As being a regular section can be checked on stalks, it suffices to show that $f_x^\sharp\colon \mathscr{O}_{Y, f(x)} \to \mathscr{O}_{X,x}$ maps regular elements to regular elements for all $x \in X$.

This is clear in case f is flat, since then $f_x^\sharp$ is a flat ring homomorphism. Now assume that (b) holds. By assumption, $\mathscr{O}_{X,x}$ is reduced, so that the regular elements are precisely those which are not contained in a minimal prime ideal (cf. Proposition B.59). Since every irreducible component of X dominates an irreducible component of Y, the inverse image of a minimal prime ideal under $f_x^\sharp$ is again a minimal prime ideal. The claim follows. $\square$

For general morphisms one can still define the pullback of Cartier divisors in a suitable subgroup of $\mathrm{Div}(Y)$ – roughly speaking, the subgroup where the above procedure works.

Corollary 11.51. *Let $f\colon X \to Y$ be a morphism of schemes satisfying one of the hypotheses of Proposition 11.50. Let $Z \subset Y$ be an effective Cartier divisor. Then the inverse image $f^{-1}(Z)$ (as a subscheme) is an effective Cartier divisor in X and this divisor is the inverse image $f^*(Z)$ (as a divisor).*

Proof. We may assume that $Y = \mathrm{Spec}\, A$ and $X = \mathrm{Spec}\, B$ are affine and that $Z = V(t)$ for a regular element $t \in A$. Let $\varphi\colon A \to B$ be the homomorphism corresponding to f. We have seen in the proof of Proposition 11.50 that $\varphi(t)$ is regular in B. So $f^{-1}(Z) = V(\varphi(t))$ is an effective Cartier divisor which by definition equals the inverse image $f^*(Z)$. $\square$

Vector bundles on $\mathbb{P}^1$

(11.17) Vector bundles on $\mathbb{P}^1$.

We now study vector bundles on $\mathbb{P}^1_k$, where k is a field. This case is particularly simple because we can cover $\mathbb{P}^1_k$ by two open affine subschemes U_0 and U_1, each isomorphic to $\mathbb{A}^1_k$. We will identify the function field of $\mathbb{P}^1_k$ with $k(T)$ such that $U_0 = \mathrm{Spec}\, k[T]$ and $U_1 = \mathrm{Spec}\, k[T^{-1}]$. To shorten notation we set

$$R^+ := k[T], \quad R^- := k[T^{-1}], \quad R^\pm := k[T, T^{-1}].$$

Vector bundles over U_i ($i = 0, 1$) correspond to finitely generated projective R^+-modules (resp. R^--modules) by Corollary 7.42. These rings are principal ideal domains, and any projective module over a principal ideal domain is free (Proposition B.89). Thus Remark 11.16 shows that isomorphism classes of vector bundles over $\mathbb{P}^1_k$ of a fixed rank n are in bijection to elements in $\check{H}^1(\mathcal{U}, G)$, where $\mathcal{U} = (U_0, U_1)$ and $G = \mathrm{GL}_n(\mathscr{O}_{\mathbb{P}^1_k})$.

As $\mathcal{U}$ consists of only two open sets, a Čech 1-cocycle of G over $\mathcal{U}$ is simply given by a single element $g = g_{01} \in G(U_0 \cap U_1) = \mathrm{GL}_n(R^{\pm})$. And two such Čech 1-cocycles g and g' are cohomologous if and only if there exist $h^+ \in \mathrm{GL}_n(R^+)$ and $h^- \in \mathrm{GL}_n(R^-)$ such that $g' = h^+ g h^-$. Thus we have a bijection

$$(11.17.1) \qquad \left\{ \begin{matrix} \text{isomorphism classes of vector} \\ \text{bundles of rank } n \text{ on } \mathbb{P}^1_k \end{matrix} \right\} \longleftrightarrow \mathrm{GL}_n(R^+) \backslash \mathrm{GL}_n(R^{\pm}) / \mathrm{GL}_n(R^-)$$

Elements in this double quotient can be described as follows.

Lemma 11.52. *Let $(\mathbb{Z}^n)_+$ be the set of $\mathbf{d} = (d_i)_i \in \mathbb{Z}^n$ with $d_1 \geq d_2 \geq \cdots \geq d_n$. For each $\mathbf{d} \in (\mathbb{Z}^n)_+$ let $T^{\mathbf{d}} \in \mathrm{GL}_n(R^{\pm})$ be the diagonal matrix with entries $T^{d_1}, \ldots, T^{d_n}$. Then the following map is surjective*

$$\tau \colon (\mathbb{Z}^n)_+ \to \mathrm{GL}_n(R^+) \backslash \mathrm{GL}_n(R^{\pm}) / \mathrm{GL}_n(R^-),$$

$$\mathbf{d} \mapsto \mathrm{GL}_n(R^+) T^{\mathbf{d}} \, \mathrm{GL}_n(R^-).$$

The uniqueness part of Theorem 11.53 below will show that the map τ is bijective (which also can be proved directly).

Proof. We have to show that any matrix in $A = (a_{ij}) \in \mathrm{GL}_n(R^{\pm})$ can be transformed to a matrix of the form $T^{\mathbf{d}}$ by invertible row operations over R^+ and invertible column operations over R^-.

For $a = \sum_{i \in \mathbb{Z}} \alpha_i T^i \in R^{\pm}$ let $v_T(a) := \inf\{ i \; ; \; \alpha_i \neq 0 \}$ be its T-adic valuation. Using invertible row operation over R^+, we can reduce the first column to $^t(a, 0, \ldots, 0)$, where $v_T(a) = \min_i v_T(a_{i1})$. By induction on n this shows that we can reduce A to an upper triangular matrix. Then any diagonal entry a_{ii} has to be a unit in $R^{\pm}$ and it is thus of the form $\alpha_i T^{d_i}$ for some $\alpha_i \in k^{\times}$ and $d_i \in \mathbb{Z}$. By multiplication with a diagonal matrix with entries in $k^{\times}$ we may assume that $\alpha_i = 1$. We therefore may assume that A is of the form

$$(11.17.2) \qquad \begin{pmatrix} T^{d_1} & a_{12} & a_{13} & \cdots & & a_{1n} \\ & T^{d_2} & a_{23} & \cdots & & a_{2n} \\ & & \ddots & \ddots & & \vdots \\ & & & \ddots & & a_{n-1,n} \\ & & & & & T^{d_n} \end{pmatrix}.$$

Let Φ be the set of pairs (i, j) of integers $1 \leq i < j \leq n$. For such a pair $\alpha = (i_0, j_0)$ let $G_{\alpha} \subseteq \mathrm{GL}_n$ be the subgroup of matrices (a_{ij}) with $a_{ij} = 0$ for $i \neq i_0$ or $j \neq j_0$ and $i \neq j$ and $a_{ii} = 1$ for $i \neq i_0, j_0$. Hence $G_{\alpha} \cong \mathrm{GL}_2$. We endow Φ with the linear order

$$(1, 2) < (2, 3) < \cdots < (n-1, n) < (1, 3) < \cdots < (n-2, n) < \cdots < (1, n)$$

We will show that for all $\alpha = (i, j) \in \Phi$ and any matrix A of the form (11.17.2) with $a_{i',j'} = 0$ for $(i', j') < (i, j)$, we can use left multiplication with elements in $G_\alpha(R^+)$ and right multiplication with elements in $G_\alpha(R^-)$ to obtain a matrix of the form (11.17.2) with $a_{i',j'} = 0$ for $(i', j') \leq (i, j)$. By induction, it follows that we can transform A to a diagonal matrix with powers of T on the diagonal, and by conjugating by a suitable permutation we can permute the entries so that the exponents are in descending order.

The induction hypothesis means that all non-diagonal entries below and to the left of a_{ij} are zero. Thus we can assume that $n = 2$ and hence

$$A = \begin{pmatrix} T^d & a \\ & T^e \end{pmatrix}.$$

Adding R^+-multiples of the second row to the first row and adding R^--multiples of the first column to the second column, we may assume that $a = \sum_{i=d+1}^{e-1} \alpha_i T^i$. Thus we are done if $d \geq e$ and it suffices to prove the claim by ascending induction on $e - d$.

We may assume $a = \alpha_v T^v + \cdots + \alpha_{e-1} T^{e-1}$ with $\alpha_v \neq 0$ and $d < v < e$. Then T^{e-v} and aT^{-v} are polynomials in R^+ that are prime to each other. So we find $p, q \in R^+$ with

(11.17.3)						$qaT^{-v} + pT^{e-v} = 1.$

Then $B := \begin{pmatrix} q & p \\ -T^{e-v} & aT^{-v} \end{pmatrix} \in \mathrm{GL}_2(R^+)$ and we have

$$B \begin{pmatrix} T^d & a \\ & T^e \end{pmatrix} \begin{pmatrix} 0 & -1 \\ 1 & 0 \end{pmatrix} = \begin{pmatrix} qT^d & qa + pT^e \\ -T^{d+e-v} & 0 \end{pmatrix} \begin{pmatrix} 0 & -1 \\ 1 & 0 \end{pmatrix} = \begin{pmatrix} T^v & -qT^d \\ 0 & T^{d+e-v} \end{pmatrix}$$

As $(d + e - v) - v < d + e - 2d = e - d$, we can now apply the induction hypothesis. $\square$

We now use this lemma to show the following classification of vector bundles on $\mathbb{P}^1_k$.

Theorem 11.53. *For every vector bundle $\mathcal{E}$ of rank n on $\mathbb{P}^1_k$ there exist uniquely determined integers $d_1 \geq \cdots \geq d_n$ such that*

(11.17.4)						$\mathcal{E} \cong \bigoplus_{i=1}^{n} \mathcal{O}_{\mathbb{P}^1_k}(d_i).$

In Volume II we will give another proof of this theorem using cohomological methods.

Proof. In Example 11.45 we have seen that for $n = 1$ the matrix (T^d) corresponds to the line bundle $\mathcal{O}_{\mathbb{P}^1_k}(d)$. Thus the existence of such a decomposition follows from the previous lemma using (11.17.1).

To show the uniqueness, we set $X = \mathbb{P}^1_k$. For integers $d, e \in \mathbb{Z}$ we have

$$\mathrm{Hom}_{\mathcal{O}_X}(\mathcal{O}_X(d), \mathcal{O}_X(e)) = \Gamma(X, \mathcal{O}_X(d)^\vee \otimes_{\mathcal{O}_X} \mathcal{O}_X(e)) = \Gamma(X, \mathcal{O}_X(e - d)).$$

Thus by (11.14.6) we have $\mathrm{Hom}_{\mathcal{O}_X}(\mathcal{O}_X(d), \mathcal{O}_X(e)) = 0$ if and only if $e < d$. Hence given $\mathcal{E} = \bigoplus_{i=1}^{n} \mathcal{O}_{\mathbb{P}^1_k}(d_i)$ we see that for every integer λ the sub-vector bundle $\mathcal{E}^\lambda := \bigoplus_{d_i \geq \lambda} \mathcal{O}_X(d_i)$ is unique. In particular, we see that for every λ the number $\#\{i \; ; \; d_i \geq \lambda\}$ is uniquely determined by $\mathcal{E}$. This shows that $(d_1 \geq \cdots \geq d_n)$ is unique. $\square$

Remark 11.54. (Harder-Narasimhan filtration) The individual summands in (11.17.4) are not unique (for $n > 1$). But the proof shows that we obtain a unique filtration

(11.17.5)
$$\mathscr{E} \supseteq \cdots \supseteq \mathscr{E}^\lambda \supseteq \mathscr{E}^{\lambda+1} \supseteq \cdots$$

where $\lambda \in \mathbb{Z}$, $\mathscr{E}^\lambda = \mathscr{E}$ for small λ and $\mathscr{E}^\lambda = 0$ for large λ, and such that all subquotients $\mathscr{E}^\lambda/\mathscr{E}^{\lambda+1}$ have the form $\mathscr{O}(\lambda)^{N_\lambda}$. This filtration is functorial in $\mathscr{E}$.

For other curves (schemes of dimension 1 over a field) a vector bundle is not necessarily a direct sum of line bundles. But if C is a smooth projective scheme over k equidimensional of dimension 1, then for every vector bundle $\mathscr{E}$ on C there exists an analogue of the filtration (11.17.5) which is called the Harder-Narasimhan filtration (see [HN]).

For k algebraically closed and $n > 1$ there exist in general vector bundles on $\mathbb{P}^n_k$ that are non-split, i.e., that are not direct sums of line bundles. Horrocks and Mumford [HoMu] constructed a vector bundle of rank 2 on $\mathbb{P}^4_k$ which is non-split. In general, it is not known whether for given n and r there exist non-split vector bundles of rank r on $\mathbb{P}^n_k$. Hartshorne conjectured in [Ha2] that every vector bundle of rank 2 on $\mathbb{P}^n_k$ for $n \geq 7$ splits. This conjecture is still open. Horrocks [Ho] proved that for $n \geq 3$ a vector bundle on $\mathbb{P}^n_k$ splits if and only if its restriction to any linear hyperplane $H \cong \mathbb{P}^{n-1}_k$ in $\mathbb{P}^n_k$ is split. This reduces the question whether a vector bundle on $\mathbb{P}^n_k$ is split to a question on vector bundles on $\mathbb{P}^2_k$.

Exercises

Exercise 11.1◇**.** Let X be a scheme.
(a) Show that a homomorphism $\varphi\colon \mathscr{B} \to \mathscr{B}'$ of $\mathscr{O}_X$-algebras is surjective if and only if the associated morphism of X-schemes $\operatorname{Spec} \varphi\colon \operatorname{Spec} \mathscr{B}' \to \operatorname{Spec} \mathscr{B}$ is a closed immersion.
(b) Show that a homomorphism $u\colon \mathscr{E} \to \mathscr{F}$ of quasi-coherent $\mathscr{O}_X$-modules is surjective if and only if the associated morphism of X-schemes $\mathbb{V}(u)\colon \mathbb{V}(\mathscr{F}) \to \mathbb{V}(\mathscr{E})$ is a closed immersion.

Exercise 11.2. Let X be a scheme and let $\mathscr{E}$ and $\mathscr{F}$ be quasi-coherent $\mathscr{O}_X$-modules.
(a) Let $\mathscr{E}$ be of finite type (resp. of finite presentation). Show that $\mathbb{V}(\mathscr{E})$ is an X-scheme of finite type (resp. of finite presentation).
(b) Show that $\mathbb{V}(\mathscr{E} \oplus \mathscr{F}) \cong \mathbb{V}(\mathscr{E}) \times_X \mathbb{V}(\mathscr{F})$.

Exercise 11.3. Let X be a topological space and G be a sheaf of groups on X. Let T (resp. T') be a sheaf of sets on X which is endowed with a right action $T \times G \to T$ (resp. a left action $G \times T' \to T'$) of G. Define a left G-action of $T \times T'$ by $g \cdot (t, t') := (tg^{-1}, gt')$ for $g \in G(U)$, $t \in T(U)$, $t' \in T'(U)$, $U \subseteq X$ open. The *contracted product* $T \times^G T'$ is defined as the sheaf associated to the presheaf whose value on U is the set of $G(U)$-orbits of $T(U) \times T'(U)$. Let $\varphi\colon G \to G'$ be a homomorphism of sheaves of groups on X and let G'_r be the sheaf G' with the right G-action given by $g' \cdot g := g'\varphi(g)$ on local sections.
(a) Show that if T is a G-torsor, then $G'_r \times^G T$ is a G'-torsor (where G' acts from the left on $G'_r \times^G T$ via the first factor). We obtain a map $H^1(\varphi)\colon H^1(X, G) \to H^1(X, G')$ of pointed sets.
(b) Show that the following diagram is commutative

$$H^1(X,G) \xrightarrow{H^1(\varphi)} H^1(X,G')$$

$$c_G \downarrow \qquad\qquad \downarrow c_{G'}$$

$$\check{H}^1(X,G) \xrightarrow{\check{H}^1(\varphi)} \check{H}^1(X,G').$$

Exercise 11.4. Let $(X,\mathscr{O}_X)$ be a ringed space and let $\det\colon \mathrm{GL}_n(\mathscr{O}_X) \to \mathscr{O}_X^\times$ be the determinant. Show that $\check{H}^1(\det)$ corresponds via the identification (11.6.2) to the map induced by $\mathscr{E} \mapsto \bigwedge^n \mathscr{E}$. Deduce that there is an isomorphism of pointed sets between $H^1(X,\mathrm{SL}_n(\mathscr{O}_X))$ and isomorphism classes of locally free $\mathscr{O}_X$-modules $\mathscr{E}$ of rank n such that $\bigwedge^n \mathscr{E} \cong \mathscr{O}_X$.

Exercise 11.5◇. Let $X = \operatorname{Spec} A$ where A is a local ring. Show that $H^1(X,\mathrm{GL}_n)$ and $H^1(X,\mathrm{SL}_n)$ (Exercise 11.4) are trivial.

Exercise 11.6. Let k be a field, $n \geq 1$. Choose homogeneous coordinates $T_0,\ldots,T_n$ on $\mathbb{P}_k^n$, and let $\mathcal{U} = (D_+(T_i))_i$ be the standard covering of $\mathbb{P}_k^n$. Show that for $d \in \mathbb{Z}$ one has

$$\dim_k H^1(\mathcal{U},\mathscr{O}_{\mathbb{P}_k^n}(d)) = \begin{cases} 0, & \text{if } n > 1 \text{ or } d \geq -1; \\ -d-1, & \text{if } n = 1 \text{ and } d \leq -2. \end{cases}$$

Remark: In Volume II we will see that $H^1(\mathcal{U},\mathscr{O}_{\mathbb{P}_k^n}(d)) = H^1(\mathbb{P}_k^n,\mathscr{O}_{\mathbb{P}_k^n}(d))$.

Exercise 11.7. Let X be a scheme of characteristic p and let $\mathrm{Frob}_X\colon X \to X$ be the absolute Frobenius.
(a) Let $\mathscr{E}$ be a locally free $\mathscr{O}_X$-module of rank n given by a Čech cocycle (g_{ij}). Show that $\mathrm{Frob}_X^* \mathscr{E}$ is given by the Čech cocycle $(g_{ij}^{(p)})$ (here for a matrix $A = (a_{kl})$ we set $A^{(p)} := (a_{kl}^p)$).
(b) Let $\mathscr{L}$ be an invertible $\mathscr{O}_X$-module. Show that $\mathrm{Frob}_X^* \mathscr{L} \cong \mathscr{L}^{\otimes p}$.

Exercise 11.8. Let S be a scheme, let $\mathscr{E}$ be a quasi-coherent $\mathscr{O}_S$-module, and let $\mathscr{F}$ be a finite locally free $\mathscr{O}_S$-module.
(a) Show that the functor $\underline{\mathrm{Hom}}(\mathscr{E},\mathscr{F})$ (16.7.1) is representable by the quasi-coherent bundle $\mathbb{V}(\mathscr{E} \otimes \mathscr{F}^\vee)$.
(b) Show that the structure morphism $h\colon \underline{\mathrm{Hom}}(\mathscr{E},\mathscr{F}) \to S$ is of finite type (resp. of finite presentation) if $\mathscr{E}$ is an $\mathscr{O}_S$-module of finite type (resp. of finite presentation).
Hint: This can be checked locally on S. Use Exercise 11.2.

Exercise 11.9. Let X be a locally noetherian scheme, let $\mathscr{F}$ be an $\mathscr{O}_X$-module of finite type and let $n \geq 0$ be an integer. Show that $\mathscr{F}$ is locally free of rank n if and only if $\mathrm{rk}(\mathscr{F})$ is locally constant and $\mathscr{F}_x$ is a free $\mathscr{O}_{X,x}$-module of rank n for all $x \in \mathrm{Ass}(X)$.

Exercise 11.10. Let S be a scheme and let $v\colon \mathscr{E} \to \mathscr{F}$ be a homomorphism of quasi-coherent $\mathscr{O}_S$-modules. Assume that $\mathscr{F}$ is finite locally free.
(a) Let $d \geq 0$ be an integer. Show that the locus where $\mathrm{Im}(v) \subseteq \mathscr{F}$ is locally a direct summand of rank d is a subscheme of S, i.e., there exists a subscheme Y of S such that a morphism $f\colon T \to S$ factorizes through Y if and only if the image of $f^*(v)$ is locally a direct summand of $f^*\mathscr{F}$ of rank d. Show that the immersion $Y \hookrightarrow S$ is of finite presentation if $\mathscr{E}$ is of finite type.

(b) Assume that $\mathscr{E}$ is of finite type. Show that the locus where v is an isomorphism is an open subscheme of S, i.e., there exists an open subscheme $U \subseteq S$ such that a morphism $f \colon T \to S$ factorizes through U if and only if f^*v is an isomorphism.
Hint: Use Proposition 8.4 and the flattening stratification or use Exercise 11.8.

Exercise 11.11◇. Let X be a noetherian reduced scheme and let $\eta_1, \ldots, \eta_r$ be its maximal points. Show that $\Gamma(U, \mathscr{K}_X) = \prod_{\eta_i \in U} \kappa(\eta_i)$ for every open subset $U \subseteq X$.

Exercise 11.12. Let X be a locally noetherian scheme.
(a) Show that $\mathscr{K}_X$ is a quasi-coherent $\mathcal{O}_X$-module if X is reduced.
(b) Let B be a local noetherian ring with $\dim(B) \geq 2$ and residue field k. Let $A = B \oplus k$ with multiplication $(b, x) \cdot (b', x') := (bb', bx' + b'x)$, and let $X = \operatorname{Spec} A$. Show that $\Gamma(X, \mathscr{K}_X) = A$ and deduce that $\mathscr{K}_X$ is not a quasi-coherent $\mathcal{O}_X$-module.

Exercise 11.13. Show that for the exact sequence (11.12.7) the connecting map δ (11.5.6) can be identified with the homomorphism

$$\operatorname{Div}(X) = \Gamma(X, \mathscr{K}_X^\times / \mathcal{O}_X^\times) \to \operatorname{Pic}(X) = H^1(X, \mathcal{O}_X^\times), \qquad D \mapsto \mathcal{O}_X(D).$$

Exercise 11.14. Let X be a locally noetherian scheme. Let D and D' be Cartier divisors on X.
(a) Show that $D = D'$ (resp. $D \geq D'$) if and only if for all $x \in X$ with $\operatorname{depth}(\mathcal{O}_{X,x}) = 1$ we have $D_x = D'_x$ (resp. $D_x \geq D'_x$). (See Definition B.61 for the notion of depth of a local ring.)
(b) Show that for every irreducible component C of $\operatorname{Supp}(D)$ we have $\operatorname{depth} \mathcal{O}_{X,C} = 1$.

Exercise 11.15. Let X be a scheme, let D_1 and D_2 be divisors on X, assume that D_1 is effective, and let $i \colon D_1 \to X$ be the inclusion of the corresponding closed subscheme. Set $D := D_1 + D_2$. Show that there is an exact sequence of $\mathcal{O}_X$-modules

$$1 \to \mathcal{O}_X(D_2) \to \mathcal{O}_X(D) \to i_*(i^*(\mathcal{O}_X(D))) \to 0$$

Exercise 11.16. Let X be a noetherian scheme. For a Weil divisor $E = \sum_C n_C [C]$ on X we define the *support of E*, denoted by $\operatorname{Supp}(E)$, as the union of those $C \in X^1$ such that $n_C \neq 0$. Show that for every Cartier divisor D we have $\operatorname{Supp}(\operatorname{cyc}(D)) \subseteq \operatorname{Supp}(D)$ and that we have equality if D is effective or if X is locally factorial. Give an example, where $\operatorname{Supp}(\operatorname{cyc}(D)) \subsetneq \operatorname{Supp}(D)$.

Exercise 11.17. Let X be a reduced noetherian scheme. Show that the following assertions are equivalent.
(i) X is normal and $\operatorname{cyc} \colon \operatorname{Div}(X) \to Z^1(X)$ is bijective.
(ii) X is locally factorial.
(iii) Every closed integral subscheme Z of X of codimension 1 is regularly immersed.

Exercise 11.18. Let A be a local noetherian ring of dimension 1, and let $\mathfrak{m} \subset A$ be the maximal ideal and let $\mathfrak{p}_1, \ldots, \mathfrak{p}_r$ be the minimal prime ideals of A.
(a) Show that $\operatorname{Pic}(A) = 0$, $Z^1(\operatorname{Spec} A) \cong \mathbb{Z}^r$, $\operatorname{Div}(A) = \operatorname{Frac}(A)^\times / A^\times$, and $\operatorname{DivCl}(A) = 0$. Describe $\operatorname{cyc} \colon \operatorname{Div}(A) \to Z^1(\operatorname{Spec} A)$ in this case.
(b) Show that we have $\mathfrak{m} \in \operatorname{Ass}(A) \Leftrightarrow A = \operatorname{Frac}(A) \Leftrightarrow \operatorname{Div}(A) = 0$. Deduce that cyc is not surjective. Give an example of a ring, where $\mathfrak{m} \in \operatorname{Ass}(A)$.

(c) Now let $\mathfrak{m} \notin \mathrm{Ass}(A)$. Let R be the set of regular elements in A, let $\mathrm{nil}(A)$ be the nilradical of A and set $A_{\mathrm{red}} = A/\mathrm{nil}(A)$. Show that there is an exact sequence $0 \to R^{-1}\mathrm{nil}(A) \to \mathrm{Frac}(A) \to \mathrm{Frac}(A_{\mathrm{red}}) \to 0$ (use that the hypothesis implies that R is the complement of the union of the minimal prime ideals of A). Deduce an exact sequence

$$1 \to 1 + \mathrm{nil}(A) \to 1 + R^{-1}\mathrm{nil}(A) \to \mathrm{Div}(A) \to \mathrm{Div}(A_{\mathrm{red}}) \to 0.$$

(d) Assume that A is reduced and let C be the cokernel of the injection $A \hookrightarrow \prod_i A/\mathfrak{p}_i$. Show that there is an exact sequence

$$1 \to C \to \mathrm{Div}(A) \to \bigoplus_{i=1}^{r} \mathrm{Div}(A/\mathfrak{p}_i) \to 0.$$

(e) Assume that A is an integral domain with field of fractions K and let A' be the integral closure of A in K. Show that there is an exact sequence

$$1 \to A'^{\times}/A^{\times} \to \mathrm{Div}(A) \to \mathrm{Div}(A') \to 0.$$

Note that A' is a discrete valuation ring and thus $\mathrm{Div}(A') = Z^1(\mathrm{Spec}\, A') \cong \mathbb{Z}$.

Exercise 11.19. Let k be a field, set $X = \mathrm{Spec}\, k[S,T,U]/(UT - S^2)$ and let C be the closed subscheme $V(S,T)$ of X.
(a) Show that $C \cong \mathbb{A}_k^1$ is a prime Weil divisor on X and that $\mathrm{Cl}(X)$ is a group of order 2 which is generated by the class of C.
(b) Show that $\mathrm{DivCl}(X) = 0$ and deduce that $\mathrm{cyc}\colon \mathrm{Div}(X) \to Z^1(X)$ is not surjective.

Exercise 11.20$\Diamond$. Let k be a field, let $f \in k[T_0, \ldots, T_n]$ be an irreducible homogeneous polynomial of degree $d > 0$, and set $U = \mathbb{P}_k^n \setminus V_+(f)$. Show that $\mathrm{Pic}(U) = \mathrm{Cl}(U) \cong \mathbb{Z}/d\mathbb{Z}$.

Exercise 11.21. Let X be a noetherian integral scheme and let $\pi\colon \mathbb{A}_X^n \to X$ be the structure morphism.
(a) Show that if Z is a prime Weil divisor of X its inverse image $\pi^{-1}(Z)$ is a prime Weil divisor of $\mathbb{A}_X^n$. Thus we obtain an injective homomorphism $Z^1(X) \to Z^1(\mathbb{A}_X^n)$. Show that this homomorphism induces a homomorphism $\pi^*\colon \mathrm{Cl}(X) \to \mathrm{Cl}(\mathbb{A}_X^n)$.
(b) Prove that π^* is bijective.

Exercise 11.22. Let X be a scheme and $n \geq 0$ an integer. An X-scheme $f\colon Y \to X$ is called *affine bundle over X* if there exist an open covering $X = \bigcup U_i$ and isomorphisms of U_i-schemes $f^{-1}(U_i) \xrightarrow{\sim} \mathbb{A}_{U_i}^n$. Now let X be reduced and noetherian.
(a) Show that taking the inverse image of prime Weil divisors is a surjective homomorphism $f^*\colon \mathrm{Cl}(X) \to \mathrm{Cl}(Y)$ (use Exercise 11.21 and noetherian induction on X).
(b) Now assume that X is locally factorial and that f has a section $s\colon X \to Y$ (e.g., if Y is a vector bundle over X). Show that Y is locally factorial and that f^* is an isomorphism.
Remark: One can show that f^* is an isomorphism for vector bundles without the assumption that X is locally factorial ([Fu2] Theorem 3.3).

Exercise 11.23. Let V be a k-vector space of dimension $n + 1$ (also considered as a quasi-coherent module on $\mathrm{Spec}\, k$) and let Z be the line bundle over $\mathbb{P}(V) = \mathrm{Grass}^1(V)$ defined in Example 11.9 whose corresponding invertible sheaf $\mathscr{Q}$ on $\mathbb{P}(V)$ is the universal quotient $p^*(V) \to \mathscr{Q}$, where $p\colon \mathbb{P}(V) \to \mathrm{Spec}\, k$ is the structure morphism. Show that for $V = (k^{n+1})^{\vee}$ (and thus $\mathbb{P}(V) = \mathbb{P}_k^n$) one has $\mathscr{Q} \cong \mathcal{O}_{\mathbb{P}_k^n}(1)$.

Exercise 11.24. Let A be a factorial ring. Show that $\text{Pic}(\text{Spec } A) = 0$.

Hint: Show first that every ideal of A that is an invertible A-module is a principal ideal. Then use Proposition 11.29 to show that every projective A-module of rank 1 is isomorphic to an ideal of A.

12   Affine and proper morphisms

Contents

- Affine morphisms
- Finite and quasi-finite morphisms
- Serre's and Chevalley's criteria to be affine
- Normalization
- Proper morphisms
- Zariski's main theorem

In this chapter, we will study properties of morphisms of schemes which distinguish important subclasses of morphisms. The emphasis in this chapter is on properties that are *not* local on the source. We start with a relative version of being affine and then study finite and quasi-finite morphisms.

We prove two criteria for a scheme to be affine and explain the important technique of normalization which corresponds to the algebraic construction of integral closure.

The notion of a *proper* morphism is the analogue in algebraic geometry of the notion of a *compact* manifold in differential geometry and is of similar importance. The last part of this chapter is devoted to the statement and the proof of different versions of a central result in algebraic geometry: Zariski's main theorem.

Affine morphisms

We now define a relative notion of affineness. We will see that a morphism $X \to S$ of schemes is affine if and only if $X = \operatorname{Spec} \mathscr{A}$, where $\mathscr{A}$ is a quasi-coherent $\mathscr{O}_S$-algebra (see Section (11.2)).

(12.1) Affine Morphisms and spectra of quasi-coherent algebras.

Recall that there is a bijective correspondence between morphisms from X into an affine scheme $\operatorname{Spec} A$ and ring homomorphisms $A \to \Gamma(X, \mathscr{O}_X)$ (Proposition 3.4). In particular, for every scheme X one has a morphism $X \to \operatorname{Spec} \Gamma(X, \mathscr{O}_X)$ corresponding to $\operatorname{id}_{\Gamma(X, \mathscr{O}_X)}$. We generalize this observation to the relative situation.

Let $f \colon X \to Y$ be a qcqs morphism of schemes. Then the $\mathscr{O}_Y$-algebra $f_* \mathscr{O}_X$ is quasi-coherent by Corollary 10.27. Applying Proposition 11.1 to this quasi-coherent algebra, we obtain a canonical morphism of Y-schemes

© Springer Fachmedien Wiesbaden GmbH, part of Springer Nature 2020
U. Görtz und T. Wedhorn, *Algebraic Geometry I: Schemes*, Springer Studium
Mathematik – Master, https://doi.org/10.1007/978-3-658-30733-2_13

(12.1.1) $$c_X : X \to \operatorname{Spec} f_* \mathscr{O}_X$$

corresponding to the identity of $f_* \mathscr{O}_X$ via (11.2.2).

Proposition and Definition 12.1. *For a morphism of schemes $f : X \to Y$ the following assertions are equivalent.*
(i) *There exists an open affine covering $(V_i)_i$ of Y such that $f^{-1}(V_i)$ is an affine scheme for all i.*
(ii) *For every open affine subscheme V of Y its inverse image $f^{-1}(V)$ is affine.*
(iii) *The morphism f is quasi-compact and separated and the morphism c_X is an isomorphism.*
If these properties are satisfied, f is called affine *or X is called* affine over Y.

Proof. Clearly, (ii) implies (i). We have already seen in Section (11.2) that if $X \cong \operatorname{Spec} \mathscr{B}$ for a quasi-coherent $\mathscr{O}_Y$-algebra, we have for every open affine subscheme V of Y an isomorphism $f^{-1}(V) \cong \operatorname{Spec} \Gamma(V, \mathscr{B})$. Thus (iii) implies (ii).

It remains to prove "(i) $\Rightarrow$ (iii)". If f satisfies (i), it is quasi-compact and separated because both properties are local on the target and hold for morphisms between affine schemes (Remark 10.2, Proposition 9.4). The question whether the Y-morphism c_X is an isomorphism is local on Y and we may assume that $Y = \operatorname{Spec} A$ is an affine scheme such that $X = f^{-1}(Y) = \operatorname{Spec} B$ is affine. Then it is clear that c_X is an isomorphism. $\square$

We have already seen in (11.2.4) that the functor $\mathscr{B} \mapsto \operatorname{Spec} \mathscr{B}$ from the category of quasi-coherent $\mathscr{O}_Y$-algebras to the category of Y-schemes is fully faithful. Proposition 12.1 together with (11.2.3) gives us its essential image:

Corollary 12.2. *Let Y be a scheme. The functor $\mathscr{B} \mapsto \operatorname{Spec} \mathscr{B}$ yields a contravariant equivalence of the category of quasi-coherent $\mathscr{O}_Y$-algebras and the category of Y-schemes that are affine over Y. A quasi-inverse is given by $(f : X \to Y) \mapsto f_* \mathscr{O}_X$.*

Using Proposition 12.1 it is immediate that if $f : X \to Y$ is a morphism and Y is an affine scheme, then X is an affine scheme if and only if f is affine.

Proposition 12.3.
(1) *Every closed immersion is affine.*
(2) *The property of scheme morphisms to be affine is local on the target, stable under composition, and stable under base change (Section (4.9)).*
(3) *If $f : X \to Y$ is a morphism of S-schemes such that X is affine over S and such that Y is separated over S, then f is affine.*
(4) *For morphisms of finite presentation, the property of being affine is compatible with inductive limits of rings in the sense of Section (10.13).*

Taking $S = \operatorname{Spec} \mathbb{Z}$ in (3), we see in particular that every morphism of an affine scheme into a separated scheme is affine.

Proof. Assertion (1) has already been shown in Theorem 3.42. By definition the property of being affine is local on the target. Stability under composition is clear using (i) of Proposition 12.1. Stability under base change follows from (11.2.5) using (iii) of Proposition 12.1 and that the properties to be quasi-compact and to be separated are stable under base change. Moreover, (3) follows formally from (1) and (2) by Remark 9.11.

To show (4) it suffices to remark that a morphism $f\colon X \to Y$ of finite presentation is affine if and only if it can be factorized locally on Y into a closed immersion $X \hookrightarrow \mathbb{A}^n_Y$ followed by the canonical morphism $\mathbb{A}^n_Y \to Y$, and this property is compatible with inductive limits by Theorem 10.63 and Proposition 10.75. □

Examples 12.4.

(1) If S is a scheme, any closed subscheme of $\mathbb{A}^n_S$ is affine over S. On the other hand, closed subschemes of $\mathbb{P}^n_S$ are rarely affine over S (more precisely, they are affine over S if and only if they are finite over S in the sense of Definition 12.9 below; to see this combine Corollary 13.41 and Corollary 12.89 below).

Projective space $\mathbb{P}^n_S$ itself is never affine over S if $n \geq 1$ and S is non-empty. Indeed, assume that $\mathbb{P}^n_S$ is affine over S. Choose any morphism $\operatorname{Spec} k \to S$ where k is an algebraically closed field. Then the base change $\mathbb{P}^n_k = \mathbb{P}^n_S \times_S \operatorname{Spec} k$ is affine over k (Proposition 12.3). But we have seen in Proposition 1.61 that the structure morphism $\mathbb{P}^n_k \to \operatorname{Spec} k$ induces an isomorphism on global sections $k \to \Gamma(\mathbb{P}^n_k, \mathscr{O}_{\mathbb{P}^n_k})$. If $\mathbb{P}^n_k$ is affine, we have $\mathbb{P}^n_k \cong \operatorname{Spec} k$ and hence $n = 0$.

(2) If $\mathscr{E}$ is a finite locally free $\mathscr{O}_S$-module, the corresponding vector bundle $\mathbb{V}(\mathscr{E})$ (Section (11.4)) is affine over S because it is the spectrum of the quasi-coherent $\mathscr{O}_S$-algebra $\operatorname{Sym}(\mathscr{E})$.

(3) An open immersion is not necessarily affine. Take for example a noetherian normal affine scheme X and a closed non-empty subset $Z \subset X$ such that $\operatorname{codim}_X(Z) \geq 2$. Let U be the open subscheme $X \setminus Z$. Then the inclusion morphism $U \to X$ is not affine. Indeed, otherwise U would be affine as well. But by Hartogs's theorem (Theorem 6.45) the restriction $\Gamma(X, \mathscr{O}_X) \to \Gamma(U, \mathscr{O}_X)$ is an isomorphism which, because U and X are affine, yields that $U \hookrightarrow X$ is an isomorphism; contradiction. See also Exercise 12.18.

(4) On the other hand, let X be a scheme, $\mathscr{L}$ a line bundle on X, and $s \in \Gamma(X, \mathscr{L})$ a global section. Let $X_s(\mathscr{L})$ be the open subscheme of points $x \in X$ where s is invertible (Section (7.11)). Then the inclusion $j\colon X_s(\mathscr{L}) \to X$ is affine. Indeed, it suffices to show this locally on X. Therefore we may assume that $X = \operatorname{Spec} A$ is affine and that $\mathscr{L} = \mathscr{O}_X$. Then $X_s(\mathscr{L}) = D(s) = \operatorname{Spec}(A_s)$ and j is a morphism between affine schemes and hence affine.

As the category of quasi-coherent modules over an affine scheme $\operatorname{Spec} A$ and the category of A-modules are equivalent, one easily gets the following global analogue.

Proposition 12.5. *Let $f\colon X \to Y$ be an affine morphism. Then the functor $\mathscr{F} \mapsto f_*\mathscr{F}$ yields an equivalence of the category of quasi-coherent $\mathscr{O}_X$-modules and the category of quasi-coherent $f_*\mathscr{O}_X$-modules.*

(12.2) Base change of direct images of quasi-coherent modules.

Consider a commutative diagram of schemes

(12.2.1)
$$\begin{array}{ccc} X' & \overset{g'}{\longrightarrow} & X \\ {\scriptstyle f'}\downarrow & & \downarrow{\scriptstyle f} \\ Y' & \underset{g}{\longrightarrow} & Y \end{array}$$

Let $\mathscr{F}$ be an $\mathscr{O}_X$-module. We will construct a homomorphism of $\mathscr{O}_{Y'}$-modules

(12.2.2) $$g^*(f_* \mathscr{F}) \to f'_*(g'^* \mathscr{F})$$

which is functorial in $\mathscr{F}$ as follows. As the functor g^* is left adjoint to g_* (Proposition 7.11), it suffices to construct a functorial homomorphism $v \colon f_* \mathscr{F} \to g_* f'_*(g'^* \mathscr{F}) = f_* g'_*(g'^* \mathscr{F})$. But again by adjointness the identity of $g'^* \mathscr{F}$ corresponds to a morphism $\mathscr{F} \to g'_*(g'^* \mathscr{F})$ and applying the functor f_* we obtain v.

If $\mathscr{G}'$ is an $\mathscr{O}_{Y'}$-module we may tensor (12.2.2) with the canonical homomorphism $\mathscr{G}' \to f'_*(f'^* \mathscr{G}')$. We obtain a homomorphism of $\mathscr{O}_{Y'}$-modules, functorial in $\mathscr{F}$ and $\mathscr{G}'$,

(12.2.3) $$\mathscr{G}' \otimes_{\mathscr{O}_{Y'}} g^*(f_* \mathscr{F}) \to f'_*(f'^* \mathscr{G}') \otimes_{\mathscr{O}_{Y'}} f'_*(g'^* \mathscr{F}) \to f'_*(f'^* \mathscr{G}' \otimes_{\mathscr{O}_{X'}} g'^* \mathscr{F}),$$

where the second homomorphism is (7.8.3). For $\mathscr{G}' = \mathscr{O}_{Y'}$, the homomorphism (12.2.3) specializes to (12.2.2).

Assume that $Y = \operatorname{Spec} A$, $X = \operatorname{Spec} B$, and $Y' = \operatorname{Spec} A'$ are affine, that the diagram (12.2.1) is cartesian (and hence $X' = \operatorname{Spec}(B' \otimes_B A)$), and that $\mathscr{F} = \tilde{N}$ and $\mathscr{G}' = \tilde{M}'$ are quasi-coherent. Then the pull back corresponds to extension of scalars and the push forward to restriction of scalars (Proposition 7.24). In this case, (12.2.3) is the isomorphism which corresponds to the canonical isomorphism of A'-modules

$$M' \otimes_{A'} (A' \otimes_A N) \xrightarrow{\sim} M' \otimes_A N \xrightarrow{\sim} (M' \otimes_A B) \otimes_B N$$
$$\xrightarrow{\sim} (M' \otimes_{A'} (A' \otimes_A B)) \otimes_{(A' \otimes_A B)} \otimes (A' \otimes_A B) \otimes_B N$$

As the question whether (12.2.3) is an isomorphism is local on Y and Y' this implies the following result if f is affine.

Proposition 12.6. *Let the diagram (12.2.1) be cartesian and let $\mathscr{F}$ be a quasi-coherent $\mathscr{O}_X$-module and $\mathscr{G}'$ a quasi-coherent $\mathscr{O}_{Y'}$-module. Let one of the following assumptions be satisfied.*
(1) f is affine, or
(2) f is qcqs and $\mathscr{G}'$ is flat over Y.
Then the homomorphism (12.2.3) is an isomorphism. In particular, (12.2.2) is an isomorphism if f is affine or if f is qcqs and g is flat.

Proof. It remains to show the proposition if assumption (2) is satisfied. We may assume that $Y = \operatorname{Spec} A$ and $Y' = \operatorname{Spec} A'$ are affine. If X is also affine, we are done by (1). In general, X is qcqs. To show (2) we will use Lemma 10.26. Let $(U_i)_i$ be a finite open affine covering of X and set $\mathscr{F}_i := \mathscr{F}_{|U_i}$, $\mathscr{F}_{ij} := \mathscr{F}_{|U_i \cap U_j}$, $f_i := f_{|U_i}$, and $f_{ij} := f_{|U_i \cap U_j}$. From the exact sequence (10.3.1) we obtain a sequence

$$0 \to \mathscr{G}' \otimes g^* f_* \mathscr{F} \to \mathscr{G}' \otimes \bigoplus_i g^*(f_i)_* \mathscr{F}_i \to \mathscr{G}' \otimes \bigoplus_{i,j} g^*(f_{ij})_* \mathscr{F}_{ij}$$

which is still exact because $\mathscr{G}'$ is flat over Y (7.18.1). On the other hand, we define $U'_i := g'^{-1}(U_i)$, $U'_{ij} := g'^{-1}(U_i \cap U_j)$, and sheaves $\mathscr{H}' := f'^* \mathscr{G}' \otimes_{\mathscr{O}_{X'}} g'^* \mathscr{F}$, $\mathscr{H}'_i := \mathscr{H}'_{|U'_i}$, and $\mathscr{H}'_{ij} := \mathscr{H}'_{|U'_{ij}}$. Let f'_i (resp. f'_{ij}) be the restriction of f' to U'_i (resp. to U'_{ij}).

The base change morphism for $\mathscr{F}_i$ and the cartesian diagram

$$\begin{array}{ccc} U'_i & \longrightarrow & U_i \\ \downarrow & & \downarrow \\ Y' & \longrightarrow & Y \end{array}$$

is a morphism $\mathscr{G}' \otimes g^*(f_i)_*\mathscr{F}_i \to (f'_i)_*\mathscr{H}'_i$ which on global sections is given by $\mathscr{G}'(Y') \otimes_{A'} (\mathscr{F}(U_i) \otimes_A A') \to (f'^*\mathscr{G}' \otimes_{\mathscr{O}_{X'}} g'^*\mathscr{F})(U'_i)$. Since all the schemes in the above diagram are affine, it is an isomorphism by the first part of the proposition.

We obtain a diagram with exact rows

$$
\begin{array}{ccccccc}
0 & \longrightarrow & \mathscr{G}' \otimes g^* f_*\mathscr{F} & \longrightarrow & \bigoplus_i(\mathscr{G}' \otimes g^*(f_i)_*\mathscr{F}_i) & \longrightarrow & \bigoplus_{i,j}(\mathscr{G}' \otimes g^*(f_{ij})_*\mathscr{F}_{ij}) \\
 & & \downarrow{\scriptstyle u} & & \downarrow{\scriptstyle u'} & & \downarrow{\scriptstyle u''} \\
0 & \longrightarrow & f'_*\mathscr{H}' & \longrightarrow & \bigoplus_i(f'_i)_*\mathscr{H}'_i & \longrightarrow & \bigoplus_{ij}(f'_{ij})_*\mathscr{H}'_{ij}.
\end{array}
$$

This diagram is commutative: In fact, since these are sheaves on the affine scheme Y', it is enough to check the commutativity on global sections, where it follows from the description above (and analogous descriptions for the first and last columns).

To prove the proposition, we use Lemma 10.26. This means that we may assume that u' and u'' are isomorphisms. But then u is an isomorphism by the five lemma. $\quad\square$

If X is noetherian, any morphism $f: X \to Y$ is qcqs. Thus we obtain the following corollary.

Corollary 12.7. *Assume that the diagram (12.2.1) is cartesian. Then for every quasi-coherent $\mathscr{O}_X$-module $\mathscr{F}$ the canonical homomorphism $g^*(f_*\mathscr{F}) \to f'_*(g'^*\mathscr{F})$ is an isomorphism if f is affine or if g is flat and X is noetherian.*

Corollary 12.8. *Let X be a qcqs scheme, let $Y = \operatorname{Spec} A$ be an affine scheme, and let $f: X \to Y$ be a morphism of schemes. Let A' be an A-algebra, set $X' = X \otimes_A A'$ and let $p: X' \to X$ be the projection. If A' is a flat A-algebra, then there exists for every quasi-coherent $\mathscr{O}_X$-module $\mathscr{F}$ a functorial isomorphism*

$$\Gamma(X, \mathscr{F}) \otimes_A A' \xrightarrow{\sim} \Gamma(X', p^*\mathscr{F}).$$

Proof. As $f_*\mathscr{F}$ is quasi-coherent by Corollary 10.27, we find that $f_*\mathscr{F}$ is the $\mathscr{O}_Y$-module corresponding to the A-module $\Gamma(Y, f_*\mathscr{F}) = \Gamma(X, \mathscr{F})$. The same argument shows that $f'_*(p^*\mathscr{F}) = \Gamma(X', p^*\mathscr{F})^\sim$, where $f': X' \to \operatorname{Spec} A'$ is the second projection. Thus the corollary follows from Proposition 12.6. $\quad\square$

In general, even if the diagram (12.2.1) is a cartesian diagram of noetherian schemes, the homomorphism (12.2.2) is not an isomorphism. To understand this failure is an important question and we will study ways to deal with this problem and their applications in detail in Volume II using cohomological methods.

Finite and quasi-finite morphisms

(12.3) Integral and finite morphisms.

Recall from Section (B.10) that an A-algebra B is called *integral over* A if all elements $b \in B$ are integral (i.e., all elements b are zeros of a monic polynomial in $A[T]$). Proposition B.53 shows that an A-algebra B is finite (i.e., finitely generated as an A-module) if and only if B is integral over A and finitely generated as an A-algebra.

The property of ring homomorphisms to be finite or integral is generalized to morphisms of schemes as follows.

Proposition and Definition 12.9. *For a morphism of schemes $f: X \to Y$ the following assertions are equivalent.*

(i) *There exists an open affine covering $(V_i)_i$ of Y such that $f^{-1}(V_i)$ is an affine scheme and the induced homomorphism $\Gamma(V_i, \mathscr{O}_Y) \to \Gamma(f^{-1}(V_i), \mathscr{O}_X)$ is a finite (resp. integral) ring homomorphism for all i.*

(ii) *For every open affine subscheme V of Y its inverse image $f^{-1}(V)$ is affine and the induced homomorphism $\Gamma(V, \mathscr{O}_Y) \to \Gamma(f^{-1}(V), \mathscr{O}_X)$ is a finite (resp. integral) ring homomorphism.*

If these properties are satisfied, f is called finite *(resp.* integral*) or X is called* finite over *Y (resp.* integral over *Y).*

Proof. Clearly (ii) implies (i). To prove the converse we may assume that $Y = \operatorname{Spec} A$ is affine. Then X is affine by Proposition 12.1, say $X = \operatorname{Spec} B$. Since any localization of a finite (resp. integral) algebra is finite (resp. integral) again, we may (using Lemma 3.3) cover $\operatorname{Spec} A$ by affine open subschemes of the form $D(f_i)$, $f_i \in A$, $i = 1, \ldots, n$, such that all B_{f_i} is a finite (resp. integral) A_{f_i}-algebra. If all B_{f_i} are finite over A_{f_i}, Lemma 3.20 shows that B is a finite A-algebra. Almost the same proof as in Lemma 3.20 shows the analogous assertion for the property "integral". $\qquad\square$

The notion of integral homomorphism should not be confused with the notion of integral scheme.

Remark 12.10.

(1) Integral and finite morphisms are affine.

(2) A scheme morphism $f: X \to Y$ is finite if and only if f is of finite type and integral.

(3) Let Y be a scheme. By Proposition 7.26 and Corollary 12.2 there is an equivalence between the category of finite morphisms $f: X \to Y$ and quasi-coherent $\mathscr{O}_Y$-algebras $\mathscr{A}$ that are $\mathscr{O}_Y$-modules of finite type given by

$$(f: X \to Y) \mapsto f_*\mathscr{O}_X, \qquad \mathscr{A} \mapsto \operatorname{Spec}(\mathscr{A}).$$

(4) Noether normalization (see Theorem 5.15 and also Proposition 5.12) shows that, if k is a field, for every affine k-scheme X of finite type there exists a finite surjective morphism $X \twoheadrightarrow \mathbb{A}_k^n$, where $n = \dim X$.

The following permanence properties are either clear or easy to check using Proposition B.54.

Proposition 12.11.

(1) *Any closed immersion is finite (and in particular integral).*

(2) *The property of being finite (resp. integral) is stable under composition, stable under base change and local on the target (Section (4.9)).*

(3) *If $f: X \to Y$ is a morphism of S-schemes such that X is finite (resp. integral) over S and such that Y is separated over S, then f is finite (resp. integral).*

(4) *For morphisms of finite presentation, the property of being finite is compatible with inductive limits of rings in the sense of Section (10.13).*

We also record the following result which follows by reduction to the affine case from Proposition 5.12.

Proposition 12.12. *Let* $f\colon X \to Y$ *be an integral morphism (e.g., if f is finite) and let* $Z \subseteq X$ *be a closed set. Then $f(Z)$ is closed in Y and one has* $\dim Z = \dim f(Z)$.

Proposition 12.13. *Let* $f\colon X \to Y$ *be a finite morphism. Then* $\mathscr{F} \mapsto f_*\mathscr{F}$ *yields an equivalence between the category of finite locally free $\mathscr{O}_X$-modules and finite locally free $f_*\mathscr{O}_X$-modules. If $\mathscr{F}$ has rank r, then $f_*\mathscr{F}$ has rank r (as $f_*\mathscr{O}_X$-module).*

Proof. By Proposition 12.5 it suffices to show that an $\mathscr{O}_X$-module $\mathscr{F}$ is finite locally free of rank r if and only if the $f_*\mathscr{O}_X$-module $f_*\mathscr{F}$ is finite locally free of rank r. The condition is clearly sufficient. Conversely, let $\mathscr{F}$ be finite locally free of rank r. We have to show that there exists an open covering $(V_j)_j$ of Y such that $\mathscr{F}_{|f^{-1}(V_j)}$ is a free $\mathscr{O}_{f^{-1}(V_j)}$-module of rank r. We may assume that Y is affine and hence that X is affine. Let $y \in Y$ be a point. If V runs through a fundamental system of open neighborhoods of y, then $f^{-1}(V)$ runs through a fundamental system of open neighborhoods of $f^{-1}(y)$ because f is closed (Proposition 12.12). Now $f^{-1}(y)$ is a finite $\kappa(y)$-scheme and hence consists only of finitely many points (Proposition 5.20). Thus the claim follows from Lemma 7.43.   □

This proposition can be further generalized, see Exercise 12.10.

(12.4) Quasi-finite morphisms.

Definition 12.14. *A scheme morphism $f\colon X \to Y$ is called* quasi-finite *if it is of finite type and if for all $y \in Y$ the fiber $f^{-1}(y)$ consists of only finitely many points.*

Remark 12.15.
(1) A morphism $f\colon X \to Y$ of finite type is quasi-finite if and only if for all $y \in Y$ the fibers $f^{-1}(y)$ are $\kappa(y)$-schemes that satisfy the equivalent properties of Proposition 5.20.
(2) As being finite is stable under base change, every finite morphism has finite fibers. Therefore finite morphisms are quasi-finite. The converse does not hold in general. E.g., any quasi-compact open immersion is clearly quasi-finite. But it is finite if and only if it is also a closed immersion because we have already seen that finite morphisms are always closed (Proposition 12.12).
(3) It is one of the versions of Zariski's main theorem that any quasi-finite and separated morphism $f\colon X \to Y$ is locally on Y the composition of an open immersion followed by a finite morphism (see Corollary 12.85 below).

Remark 12.16. Let k be a field, let X and Y be schemes of finite type over k and let $f\colon X \to Y$ be a morphism of k-schemes. Let K be an algebraically closed extension of k. Then f is quasi-finite if and only if the map on K-valued points $f(K)\colon X(K) \to Y(K)$ has finite fibers (as a map of sets).

Indeed, first note that f is of finite type since X, Y are of finite type over k (Proposition 10.7 (3)). Define $X_y := f^{-1}(y)$ for $y \in Y$. The set $Y_0 := \{\, y \in Y \ ;\ X_y \text{ is finite}\,\} = \{\, y \in Y \ ;\ \dim X_y = 0\,\}$ is constructible by Proposition 10.96. Therefore $Y_0 = Y$ if and only if Y_0 contains every closed point of Y (Remark 10.15). If $y \in Y$ is a closed point, $\kappa(y)$ is a finite extension of k and there exists a k-embedding $\kappa(y) \hookrightarrow K$. We have $\dim X_y = \dim X_y \otimes_{\kappa(y)} K$ (Proposition 5.38) and $\dim X_y \otimes_{\kappa(y)} K = 0$ if and only if $X_y(K) = (X_y \otimes_{\kappa(y)} K)(K)$ is a finite set (Proposition 5.20).

We leave the (easy) proof of the following permanence properties to the reader (or refer to [EGAInew] (6.11.5) and [EGAIV] (8.10.5)).

Proposition 12.17.

(1) *Any quasi-compact immersion is quasi-finite.*

(2) *The property of being quasi-finite is stable under composition, stable under base change and local on the target (Section (4.9)).*

(3) *If $f: X \to Y$ is a morphism of S-schemes such that X is quasi-finite over S and such that Y is quasi-separated over S. Then f is quasi-finite.*

(4) *For morphisms of finite presentation, the property of being quasi-finite is compatible with inductive limits of rings in the sense of Section (10.13).*

(12.5) Ramification and inertia index.

Let k be a field and let $X = \operatorname{Spec} A$ be a finite k-scheme. Then X satisfies the equivalent properties of Proposition 5.20. Thus we have

$$(12.5.1) \qquad A = \prod_{x \in X} \mathscr{O}_{X,x},$$

where $\mathscr{O}_{X,x}$ is a local finite-dimensional k-algebra. We will attach several numerical invariants to A.

Recall that for any finite field extension $L \supseteq K$ we denote by $[L : K]_{\text{sep}}$ its separability degree and by $[L : K]_{\text{insep}}$ its inseparability degree. Thus $[L : K]_{\text{sep}}[L : K]_{\text{insep}}$ is the degree $[L : K]$ of the extension $L \supseteq K$.

We attach to $x \in X$ the following numerical invariants.

$$e_x := e_x(X) := \lg_{\mathscr{O}_{X,x}}(\mathscr{O}_{X,x})$$
$$f'_x := f'_x(X) := [\kappa(x) : k]_{\text{insep}},$$
$$f''_x := f''_x(X) := [\kappa(x) : k]_{\text{sep}},$$
$$f_x := f_x(X) := f'_x f''_x = [\kappa(x) : k].$$

These are positive integers. Let us denote by $\mathfrak{m}$ the maximal ideal of $\mathscr{O}_{X,x}$. We have $\mathscr{O}_{X,x} \cong \bigoplus_i \mathfrak{m}^i/\mathfrak{m}^{i+1}$ as k-vector spaces. Furthermore, $\mathfrak{m}^i/\mathfrak{m}^{i+1}$ is a $\kappa(x)$-vector space whose $\kappa(x)$-dimension is the same as its $\mathscr{O}_{X,x}$-length. The additivity of the length in short exact sequences implies that

$$(12.5.2) \qquad \dim_k \mathscr{O}_{X,x} = e_x f_x$$

and therefore by (12.5.1) we find

$$(12.5.3) \qquad \dim_k A = \sum_{x \in X} e_x f_x.$$

We have $e_x = 1$ if and only if $\mathscr{O}_{X,x} = \kappa(x)$. Thus $e_x f'_x = 1$ means that $\mathscr{O}_{X,x}$ is a finite separable field extension of k.

Let $\pi: X \to Y$ be a quasi-finite morphism and let $y \in Y$ be a point. Then $X_y := \pi^{-1}(y)$ is a finite $\kappa(y)$-scheme (possibly empty). For any $x \in X_y$ we set $e_{x/y}(\pi) := e_{x/y} := e_x(X_y)$ and define $f'_{x/y}$, $f''_{x/y}$, and $f_{x/y}$ similarly. We call π *unramified in x* if $e_{x/y} f'_{x/y} = 1$, i.e., if $\mathscr{O}_{X_y,x}$ is a separable field extension of $\kappa(y)$. Unramified morphisms will be studied in more detail in Volume II.

The invariants $f''_{x/y}$ and the property of being unramified are stable under base change in the following sense.

Proposition 12.18. *Consider a cartesian diagram of schemes*

$$
\begin{array}{ccc}
X' & \xrightarrow{\;\alpha'\;} & X \\
{\scriptstyle\pi'}\downarrow & & \downarrow{\scriptstyle\pi} \\
Y' & \xrightarrow{\;\alpha\;} & Y
\end{array}
$$

where π is quasi-finite. Let $y' \in Y'$ and let $y \in Y$ be its image under α. For $x \in X$ with $\pi(x) = y$ the following assertions hold.
(1)

$$
f''_{x/y} = \sum_{x' \in \pi'^{-1}(y') \cap \alpha'^{-1}(x)} f''_{x'/y'}
$$

(2) *The morphism π is unramified in x if and only if π' is unramified at all points x' of $\pi'^{-1}(y') \cap \alpha'^{-1}(x)$.*

Proof. As the invariants depend only on the fibers, we may assume that $Y = \operatorname{Spec} k$ and that $Y' = \operatorname{Spec} k'$ for a field extension $k' \supseteq k$. Then $X = \operatorname{Spec} A$ for a finite k-algebra A. Moreover we can replace A by $\mathscr{O}_{X,x}$ and thus may assume that A is local with residue field $\kappa(x)$. As we have $\dim_k A = \dim_{k'} A \otimes_k k'$ the identity (12.5.3) shows that it suffices to prove the second equality. As X and $\operatorname{Spec} \kappa(x)$ have the same underlying topological space, $\pi'^{-1}(y') \cap \alpha'^{-1}(x)$ and X' also have the same underlying topological space. Let Ω an algebraically closed extension of k'. We have

$$
f''_{x/y} = [\kappa(x) : k]_{\mathrm{sep}} = \#\operatorname{Hom}_k(A, \Omega) = \#\operatorname{Hom}_{k'}(A \otimes_k k', \Omega)
$$
$$
= \sum_{x' \in X'} [\kappa(x') : k']_{\mathrm{sep}} = \sum_{x' \in X'} f''_{x'/y'}. \qquad \square
$$

A similar statement as in Proposition 12.18 for the invariants $e_{x/y}$ and $f_{x/y}$ does not hold (Exercise 12.11).

The integer $e_{x/y}$ is called the *ramification index of x over y* and $f_{x/y}$ is called the *inertia index of x over y*.

(12.6) Finite locally free morphisms.

Proposition 12.19. *For a morphism $f \colon X \to Y$ the following assertions are equivalent.*
(i) *f is affine and $f_* \mathscr{O}_X$ is a finite locally free $\mathscr{O}_Y$-module.*
(ii) *f is finite, flat, and of finite presentation.*
(iii) *For all open affine subsets $V = \operatorname{Spec} A$ of Y the inverse image $f^{-1}(V)$ is affine, say $f^{-1}(V) = \operatorname{Spec} B$, and the A-algebra B is a finite projective A-module.*

A morphism f satisfying these equivalent properties is called *finite locally free*. If Y is locally noetherian, f is finite locally free if and only if f is finite and flat.

Proof. An affine morphism f is finite, flat, and of finite presentation if and only if for all open affine subsets $V = \operatorname{Spec} A$ of Y the A-algebra $B := \Gamma(f^{-1}(V), \mathscr{O}_X) = \Gamma(V, f_* \mathscr{O}_X)$ is finite, flat and of finite presentation (as A-algebra). But a finite algebra is of finite presentation as an algebra if and only if it is of finite presentation as a module (Proposition B.13). Thus the equivalence of all three statements follows from Corollary 7.42. $\qquad \square$

If $f : X \to Y$ is finite locally free, $f_* \mathcal{O}_X$ is a locally free $\mathcal{O}_Y$-module and therefore the function

$$(12.6.1) \qquad Y \to \mathbb{N}_0, \quad y \mapsto \dim_{\kappa(y)} (f_* \mathcal{O}_X)(y)$$

is locally constant on Y. It is called the *degree of f* or sometimes the *rank of f*. It is denoted by $\deg f$. (Recall that by $\mathscr{F}(y)$ we denote the fiber $\mathscr{F}_y \otimes_{\mathcal{O}_{Y,y}} \kappa(y)$ of an $\mathcal{O}_Y$-module $\mathscr{F}$ at a point $y \in Y$ (7.1.1)).

If f is finite, the formation of $f_* \mathcal{O}_X$ commutes with base change by Proposition 12.6. Thus we obtain from Theorem 11.18 the following result.

Corollary 12.20. *Let $f : X \to Y$ be a finite morphism. Then Y is the disjoint union (as a set) of subschemes Y_r for $r \in \mathbb{N}_0$ such that the restriction $f^{-1}(Y_r) \to Y_r$ is finite locally free of degree r.*

Moreover Y_r is uniquely determined by the property that a scheme morphism $Y' \to Y$ factors through Y_r if and only if the base change $f_{(Y')}$ is finite locally free of degree r.

The ramification index and inertia index defined in Section (12.5) are related to the degree of a finite locally free morphism as follows.

Proposition 12.21. *Let $\pi : X \to Y$ be a finite locally free morphism of constant degree. Then for all $y \in Y$ we have*

$$\deg \pi = \dim_{\kappa(y)} \Gamma(X_y, \mathcal{O}_{X_y}) = \sum_{x \in \pi^{-1}(y)} e_{x/y} f_{x/y}.$$

Proof. We may replace Y by $\operatorname{Spec} \mathcal{O}_{Y,y}$ and assume that $Y = \operatorname{Spec} A$ is affine and local and that $\mathfrak{p}_y = \mathfrak{m}$ is the maximal ideal of A. Then $X = \operatorname{Spec} B$, where B is a finite free A-module of rank $r := \deg \pi$. Set $k := \kappa(y) = A/\mathfrak{m}$. We have $X_y = \operatorname{Spec} B/\mathfrak{m}B$ and $\dim_k B/\mathfrak{m}B = r$ which shows the first equality. The second equality has already been shown in Section (12.5). $\qquad \square$

If $\pi : X \to Y$ is a finite locally free morphism of integral schemes, then its degree is necessarily constant ≥ 1. In particular, π is surjective, so the generic point η_X of X is mapped to the generic point η_Y of Y. As all fibers of a finite morphism are discrete, we find $\pi^{-1}(\eta_Y) = \{\eta_X\}$. If we apply Proposition 12.21 to the generic point of Y and we obtain $\deg \pi = [K(X) : K(Y)]$. Thus we see that in this case we have for an arbitrary point $y \in Y$ the so-called fundamental equality

$$(12.6.2) \qquad [K(X) : K(Y)] = \sum_{x \in \pi^{-1}(y)} e_{x/y} f_{x/y}.$$

Proposition 12.22. *Let $\pi : X \to Y$ and $\varpi : Y \to Z$ be finite locally free morphisms of constant degree.*
(1) The composition $\varpi \circ \pi$ is finite locally free of degree $(\deg \pi)(\deg \varpi)$.
(2) Let $x \in X$ be a point, set $y := \pi(x)$, $z := \varpi(y)$. Then

$$e_{x/z} = e_{y/z} e_{x/y}, \quad f'_{x/z} = f'_{y/z} f'_{x/y}, \quad f''_{x/z} = f''_{y/z} f''_{x/y}$$

Proof. Assertion (1) is clear. The equalities for f' and f'' are simply the multiplicativity of the separability (resp. inseparability) degree for the extensions $\kappa(z) \hookrightarrow \kappa(y) \hookrightarrow \kappa(x)$. To show the equality for e, we may assume that $Z = \operatorname{Spec} k$ for a field k, that $Y = \operatorname{Spec} \mathscr{O}_{Y,y}$ and that $X = \operatorname{Spec} \mathscr{O}_{X,x}$. We have $\deg \varpi = e_{y/z} f_{y/z}$ by (12.5.3). As π is finite locally free, we also find $\deg \pi = e_{x/y} f_{x/y}$ by Proposition 12.21. Thus the multiplicativity of e follows from the one for the degree and the inertia index f. $\square$

Remark 12.23. The proof shows that if π and ϖ are arbitrary quasi-finite morphisms, the multiplicativity for f' and f'' still holds, but in general we have only $e_{x/z} \leq e_{y/z} e_{x/y}$.

Example 12.24. We give an interpretation of the invariants $e_{x/y}$ and $f_{x/y}$ if $\pi \colon X \to Y$ is a finite locally free morphism of Dedekind schemes. We may assume that $Y = \operatorname{Spec} A$ is affine. Then $X = \operatorname{Spec} B$, and A and B are Dedekind rings. Fix a closed point $y \in Y$ with corresponding maximal ideal $\mathfrak{p} \subset A$. We have $\pi^{-1}(y) = \operatorname{Spec} B/\mathfrak{p}B$. As B is a Dedekind ring, we have

$$\mathfrak{p}B = \mathfrak{q}_1^{e_1} \ldots \mathfrak{q}_r^{e_r},$$

where the $\mathfrak{q}_i \subset B$ are maximal ideals of B which are pairwise prime to each other and where $e_i = e_{\mathfrak{q}_i/\mathfrak{p}}$ are positive integers. Moreover this decomposition is unique up to order. By the Chinese remainder theorem we have

$$B/\mathfrak{p}B = \prod_{i=1}^{r} B/\mathfrak{q}_i^{e_i}.$$

Thus the points in $\pi^{-1}(y)$ are the prime ideals $x_i = \mathfrak{q}_i$ and we have

$$e_{x_i/y} = e_i, \qquad f_{x_i/y} = [B/\mathfrak{q}_i : A/\mathfrak{p}].$$

This shows that in this special case (12.6.2) is the "fundamental equality" in algebraic number theory

$$[\operatorname{Frac}(B) : \operatorname{Frac}(A)] = \sum_{\mathfrak{q}|\mathfrak{p}} e_{\mathfrak{q}/\mathfrak{p}} f_{\mathfrak{q}/\mathfrak{p}}$$

Remark 12.25. (Norm of a line bundle) Let $f \colon X \to Y$ be a finite locally free morphism of schemes of rank > 0. We will define a homomorphism of groups, called norm,

$$N_{X/Y} \colon \operatorname{Pic} X \to \operatorname{Pic} Y.$$

By hypothesis $\mathscr{B} := f_* \mathscr{O}_X$ is a finite locally free $\mathscr{O}_Y$-algebra.

We first define a morphism $N_{\mathscr{B}/\mathscr{O}_Y} \colon \mathscr{B} \to \mathscr{O}_Y$ as usual: For $V \subseteq Y$ open and for $b \in \Gamma(V, \mathscr{B})$ let $m_b \colon \Gamma(V, \mathscr{B}) \to \Gamma(V, \mathscr{B})$ be the multiplication with b. As $\mathscr{B}$ is a finite locally free $\mathscr{O}_Y$-module we can define $N_{\mathscr{B}/\mathscr{O}_Y}(b) := \det(m_b) \in \Gamma(V, \mathscr{O}_Y)$ see (7.20.7). This defines the *norm homomorphism*

$$N_{\mathscr{B}/\mathscr{O}_Y} \colon \mathscr{B} \to \mathscr{O}_Y.$$

It is multiplicative, i.e., $N_{\mathscr{B}/\mathscr{O}_Y}(bb') = N_{\mathscr{B}/\mathscr{O}_Y}(b) N_{\mathscr{B}/\mathscr{O}_Y}(b')$ for $b, b' \in \Gamma(V, \mathscr{B})$. In particular, it induces a homomorphism of abelian sheaves $\mathscr{B}^\times \to \mathscr{O}_Y^\times$.

Now let $\mathscr{L}$ be an invertible $\mathscr{O}_X$-module. Then $f_*\mathscr{L}$ is an invertible $\mathscr{B}$-module (Proposition 12.13) and there exists an open covering $\mathcal{V} = (V_i)_i$ of Y such that $f_*\mathscr{L}$ is given by a Čech 1-cocycle (g_{ij}) of $\mathscr{B}^\times$ on $\mathcal{V}$ (applying Section (11.6) for the ringed space $(Y, \mathscr{B})$). Thus $g_{ij} \in \Gamma(V_i \cap V_j, \mathscr{B}^\times)$ for all i, j. Then $(N_{\mathscr{B}/\mathscr{O}_Y}(g_{ij}))_{ij}$ is a Čech 1-cocycle of $\mathscr{O}_Y^\times$ on $\mathcal{V}$ which defines an invertible $\mathscr{O}_Y$-module which we denote by $N_{X/Y}(\mathscr{L})$. This induces a well-defined group homomorphism $N_{X/Y}\colon \operatorname{Pic} X \to \operatorname{Pic} Y$. It is the composition of the isomorphism $H^1(X, \mathscr{O}_X^\times) \cong H^1(Y, \mathscr{B}^\times)$ given by Proposition 12.13 and the homomorphism $H^1(Y, \mathscr{B}^\times) \to H^1(Y, \mathscr{O}_Y^\times)$ induced by functoriality of $H^1(Y, \cdot)$ by $N_{\mathscr{B}/\mathscr{O}_Y}$ (11.5.4). We call $N_{X/Y}(\mathscr{L})$ the *norm of $\mathscr{L}$ under f*.

If $\mathscr{M}$ is an invertible $\mathscr{O}_Y$-module, then the definition of $N_{X/Y}$ shows that

$$(12.6.3) \qquad N_{X/Y}(f^*\mathscr{M}) \cong \mathscr{M}^{\otimes n},$$

where n is the rank of f.

If $u\colon \mathscr{L}' \to \mathscr{L}$ is a homomorphism of invertible $\mathscr{O}_X$-modules, then its direct image $v := f_*(u)\colon f_*\mathscr{L}' \to f_*\mathscr{L}$ is a homomorphism of invertible $\mathscr{B}$-modules. Denote by $T' := \mathscr{I}som_{\mathscr{B}}(\mathscr{B}, f_*\mathscr{L}')$ and $T := \mathscr{I}som_{\mathscr{B}}(\mathscr{B}, f_*\mathscr{L})$ the corresponding $\mathscr{B}^\times$-torsors and let $\mathcal{V} = (V_i)_i$ be an open covering of Y which trivializes T and T'. For all i choose $t_i' \in T'(V_i)$ and $t_i \in T(V_i)$. Then the homomorphism v corresponds to a family $(h_i)_i$ with $h_i \in \Gamma(V_i, \mathscr{B})$ by $h_i := t_i^{-1} \circ v_{|V_i} \circ t_i'$. The family $(N_{\mathscr{B}/\mathscr{O}_Y}(h_i))_i$ then defines a homomorphism $N_{X/Y}(u)\colon N_{X/Y}(\mathscr{L}') \to N_{X/Y}(\mathscr{L})$. As the norm is multiplicative, different choices of $(t_i)_i$ or $(t_i')_i$ yield the same homomorphism. We obtain a map

$$(12.6.4) \qquad N_{X/Y}\colon \operatorname{Hom}_{\mathscr{O}_X}(\mathscr{L}', \mathscr{L}) \to \operatorname{Hom}_{\mathscr{O}_Y}(N_{X/Y}(\mathscr{L}'), N_{X/Y}(\mathscr{L})).$$

For $\mathscr{L}' = \mathscr{O}_X$ we obtain in particular a map

$$(12.6.5) \qquad N_{X/Y}\colon \Gamma(X, \mathscr{L}) \to \Gamma(Y, N_{X/Y}(\mathscr{L})).$$

Proposition 12.26. *Let $f\colon X \to Y$ be finite locally free of degree > 0 and let $u\colon \mathscr{L}' \to \mathscr{L}$ be homomorphism of invertible $\mathscr{O}_X$-modules. Then u is an isomorphism if and only if $N_{X/Y}(u)\colon N_{X/Y}(\mathscr{L}') \to N_{X/Y}(\mathscr{L})$ is an isomorphism.*

Proof. By Proposition 12.13, u is an isomorphism if and only if $v := f_*(u)$ is an isomorphism. With the notation above, v is an isomorphism if and only if $h_i \in \Gamma(V_i, \mathscr{B})^\times$ for all i. This is equivalent to $N_{\mathscr{B}/\mathscr{O}_Y}(h_i) \in \Gamma(V_i, \mathscr{O}_Y)^\times$ for all i (an endomorphism is an isomorphism if and only if its determinant is an isomorphism) which is the case if and only if $N_{X/Y}(u)$ is an isomorphism. $\square$

If X and Y are integral and Y is normal, then one can define $N_{X/Y}$ even if f is only finite and surjective (Exercise 12.25).

(12.7) Invariants under a finite group.

Let X be a scheme and let G be a group of automorphisms of X. A *quotient of X by G* is a pair (Y, p), where Y is a scheme and $p\colon X \to Y$ is a morphism of schemes with $p \circ g = p$ for all $g \in G$ such that for every scheme morphism $f\colon X \to Z$ with $f \circ g = f$ for all g there exists a unique morphism $\bar{f}\colon Y \to Z$ such that $\bar{f} \circ p = f$. Clearly if such a quotient exists, it is unique up to unique isomorphism. In this case we write X/G instead of Y.

We prove the existence of such a quotient and its properties in a special but interesting case. First note that when $X = \operatorname{Spec} A$ is affine, then a scheme automorphism of X is the same as a ring automorphism of A. To obtain a left action of G on A, we define, for each $g \in G$, $g\colon A \to A$ to be the automorphism corresponding to $g^{-1}\colon X \to X$.

Proposition 12.27. *Let $X = \operatorname{Spec} A$ be an affine scheme and let G be a finite group of automorphisms of X. Let $A^G = \{\, a \in A \; ; \; g(a) = a \;\forall g \in G \,\}$ the ring of invariants and let $p\colon X \to Y := \operatorname{Spec}(A^G)$ be the morphism corresponding to the inclusion $A^G \hookrightarrow A$.*
(1) The pair (Y, p) is a quotient of X by G (therefore we write $Y = X/G$).
(2) For $x, x' \in X$ we have $p(x) = p(x')$ if and only if there exists a $g \in G$ with $g(x) = x'$. The homomorphism $p^\flat\colon \mathscr{O}_{X/G} \to p_* \mathscr{O}_X$ induces an isomorphism $\mathscr{O}_{X/G} \overset{\sim}{\to} (p_* \mathscr{O}_X)^G$.
(3) The morphism p is integral, surjective, and has finite fibers.
(4) Let X be of finite type over a noetherian ring R and let G act on X by R-linear automorphisms. Then p is finite and X/G is of finite type over R.

In general it may happen that p is not finite. We start with a preliminary remark.

Remark 12.28. Assume that A is an R-algebra for a ring R and that G acts by R-linear automorphisms (e.g., $R = A^G$). Consider the R-linear map

$$u\colon A \to \prod_{g \in G} A, \qquad a \mapsto (a - g(a))_{g \in G}.$$

Then $\operatorname{Ker}(u) = A^G$. Let B be any R-algebra. Then G acts on $B \otimes_R A$ via the second factor and we obtain a canonical ring homomorphism

$$(12.7.1) \qquad\qquad \iota_B\colon B \otimes_R A^G \to (B \otimes_R A)^G, \quad b \otimes a \mapsto b \otimes a$$

which is simply the canonical homomorphism $B \otimes \operatorname{Ker}(u) \to \operatorname{Ker}(\operatorname{id}_B \otimes u)$. Thus ι_B is an isomorphism if B is a flat R-algebra.

Proof. (of Proposition 12.27) Define a map $N\colon A \to A^G$ by $N(a) = \prod_{g \in G} g(a)$. We first prove (2). Let $\mathfrak{p}, \mathfrak{p}' \subset A$ be two prime ideals such there exists a $g \in G$ with $g(\mathfrak{p}) = \mathfrak{p}'$. Then clearly $\mathfrak{p} \cap A^G = \mathfrak{p}' \cap A^G$. Conversely, if $\mathfrak{p} \cap A^G = \mathfrak{p}' \cap A^G$, then we have $N(a) \in \mathfrak{p}'$ for all $a \in \mathfrak{p}$. As $\mathfrak{p}'$ is a prime ideal, this shows that $\mathfrak{p}$ is contained in the union of the finitely many prime ideals $g(\mathfrak{p}')$ for $g \in G$. By prime ideal avoidance (Proposition B.2 (2)) there exists a $g \in G$ such that $\mathfrak{p} \subseteq g(\mathfrak{p}')$. By symmetry we conclude that $\mathfrak{p}$ and $\mathfrak{p}'$ lie in the same G-orbit. Moreover, by the remark above (with $R = A^G$ and $B = (A^G)_f$) we have $(A^G)_f = (A_f)^G$ for every $f \in A^G$ which proves the isomorphism $\mathscr{O}_Y \overset{\sim}{\to} (p_* \mathscr{O}_X)^G$.

Now (2) immediately implies (1) and that each fiber of p has at most $\#G$ elements. Moreover, every element $a \in A$ is a zero of the monic polynomial $\chi_a(T) = \prod_g (T - g(a))$ which has coefficients in A^G. This shows that p is integral and for every ideal $\mathfrak{a} \subseteq A$ we have $p(V(\mathfrak{a})) = V(\mathfrak{a} \cap A^G)$ by Proposition 5.12. The case $\mathfrak{a} = 0$ yields the surjectivity of p.

It remains to prove (4). As A is a finitely generated R-algebra, A is certainly a finitely generated A^G-algebra. Thus p is integral and of finite type and hence finite (we did not use that R is noetherian for this part of (4)). To prove that A^G is a finitely generated R-algebra let $a_1, \ldots, a_n$ be generators of the R-algebra A. As A is integral over A^G we find monic polynomials $P_i \in A^G[T]$ such that $P_i(a_i) = 0$ for all i. Let B be the R-algebra generated by all coefficients of the P_i. Then A is integral and of finite type over B and thus finite over B. As R is noetherian, B is noetherian as well, and the B-submodule A^G

of A is therefore finitely generated. This shows that A^G is of finite type over B and hence over R. $\qquad\square$

Example 12.29. Let R be a ring. The symmetric group S_n on n letters acts on the n-dimensional affine space $\mathbb{A}^n_R$ over R by permuting the coordinates. Correspondingly, the action on the affine coordinate ring $R[T_1, \dots, T_n]$ is given by permuting the indeterminates. Recall the following facts from the theory of symmetric functions (e.g. [BouAII] IV, 6.1, Theorem 1).

(1) The invariant ring $R[T_1, \dots, T_n]^{S_n}$ is the R-subalgebra generated by the elementary symmetric polynomials $\sigma_1 = T_1 + \cdots + T_n, \dots, \sigma_n = T_1 \cdots T_n$.

(2) The polynomial ring $R[T_1, \dots, T_n]$ is a free $R[T_1, \dots, T_n]^{S_n}$-module of rank $n!$.

(3) The σ_i are algebraically independent over R, i. e. the R-algebra homomorphism $R[X_1, \dots, X_n] \to R[T_1, \dots, T_n]^{S_n}$ that sends X_i to σ_i is an isomorphism, and hence the σ_i define an isomorphism

$$(12.7.2) \qquad\qquad \sigma \colon \mathbb{A}^n_R/S_n \xrightarrow{\sim} \mathbb{A}^n_R.$$

Another interpretation of this isomorphism is the following. Let A be any R-algebra. The quotient map $p \colon \mathbb{A}^n_R \twoheadrightarrow \mathbb{A}^n_R/S_n$ defines a map

$$p_A \colon A^n = \mathbb{A}^n_R(A) \to (\mathbb{A}^n_R/S_n)(A)$$

on A-valued points (we will see that the map p_A is usually not surjective).

For $(t_1, \dots, t_n) \in A^n$ we have

$$(T + t_1) \cdots (T + t_n) = T^n + \sigma_1(t_1, \dots, t_n)T^{n-1} + \cdots + \sigma_n(t_1, \dots, t_n) \in A[T].$$

Since this expression does not depend on the order of the t_i (almost by definition), we see that p_A factors through the projection $A^n \to A^n/S_n$. This also follows directly from the universal property of the quotient morphism p.

We may describe the A-valued points of $\mathbb{A}^n_R/S_n$ as follows. Let $\mathcal{C}_A$ be the set of isomorphism classes of finite A-algebras that are of the form $A[X]/(f)$ where $f \in A[X]$ is a monic polynomial of degree ≥ 1. Then $(\mathbb{A}^n_R/S_n)(A)$ is the set of equivalence classes of tuples $(A', (t'_1, \dots, t'_n))$ with $A' \in \mathcal{C}_A$ and $(t'_1, \dots, t'_n) \in A'^n$ such that $\prod_i(T + t'_i) \in A[T]$, where $(A', (t'_1, \dots, t'_n))$ and $(A'', (t''_1, \dots, t''_n))$ are equivalent if there exist $B \in \mathcal{C}_A$, A-algebra homomorphisms $\varphi' \colon A' \to B$ and $\varphi'' \colon A'' \to B$ and $\pi \in S_n$ such that $\varphi''(t''_i) = \varphi'(t'_{\pi(i)})$ for all $i = 1, \dots, n$.

Let us check the validity of this description. A pair $(A', (t'_1, \dots, t'_n))$ defines an A'-valued point of $\mathbb{A}^n_R$, and hence an element $[A', (t'_\bullet)] \in (\mathbb{A}^n_R/S_n)(A')$. The image of this point under σ is

$$(a_1, \dots, a_n) \in (A')^n, \quad \text{where } (T + t'_1) \cdots (T + t'_n) = T^n + a_1 T^{n-1} + \cdots + a_n.$$

Now under our assumption the tuple $(a_1, \dots, a_n)$ actually lies in $A^n \subseteq (A')^n$, and this shows that $[A', (t'_\bullet)]$ lies in the subset $(\mathbb{A}^n_R/S_n)(A) \subseteq (\mathbb{A}^n_R/S_n)(A')$. It is not hard to check that this map is a bijection, as we claimed.

Then σ is described on A-valued points by sending $(A', (t'_1, \dots, t'_n))$ to the coefficients of the polynomial $\prod_i(T + t'_i)$. Its inverse σ^{-1} may be described by sending $(a_1, \dots, a_n) \in A^n$ to the unordered set of zeros of the polynomial $T^n - a_1 T^{n-1} + \cdots + (-1)^n a_n$ in some finite A-algebra.

Application 12.30. Let $n \geq 1$ be an integer and let M_n be the scheme of $(n \times n)$-matrices (isomorphic to $\mathbb{A}_{\mathbb{Z}}^{n^2}$). Define a morphism $\chi\colon M_n \to \mathbb{A}^n$ on R-valued points (R some ring) as follows. It sends a matrix $A \in M_n(R)$ to $(a_1, \ldots, a_n)$ if $T^n - a_1 T^{n-1} + \cdots + (-1)^n a_n$ is the characteristic polynomial of A (a_i can also be described as $\mathrm{tr}(\bigwedge^i(A))$, where we consider A as an endomorphism of the R-module R^n, see Exercise 7.16). Now let

$$(12.7.3) \qquad\qquad \varepsilon\colon M_n \to \mathbb{A}^n/S_n$$

be the composition of χ followed by the inverse of the isomorphism σ (12.7.2). Then ε can be interpreted as the morphism that sends a matrix to the unordered tuple of its eigenvalues (with multiplicity).

Remark 12.31. Whenever k is a field and $G \subset \mathrm{GL}_n(k)$ is a finite subgroup, we can consider the action of G on $\mathbb{A}_k^n = \mathrm{Spec}\, k[T_1, \ldots, T_n]$. We mention here the Theorem of Shephard-Todd and Chevalley which says that if $\mathrm{char}\, k$ does not divide the order of G, the following are equivalent:
  (i) The group G is generated by pseudo-reflections, i.e., by elements $s \in \mathrm{GL}_n(k)$ such that $\mathrm{rk}(\mathrm{id} - s) = 1$.
  (ii) The k-algebra $k[T_1, \ldots, T_n]^G$ is a regular ring.
  (iii) The k-algebra $k[T_1, \ldots, T_n]^G$ is a polynomial ring (in n variables) over k.
See [BH] Theorem 6.4.12, [BouLie] V.5.5, Theorem 4.

In general the question whether quotients of schemes by groups exist is a difficult question. One might argue that it is more natural to ask for the existence of a quotient in a larger category than the category of schemes, the category of algebraic spaces for example. On the other hand, it is also interesting to consider variants of the notion of quotient defined above. Moreover one should pose the question more generally by considering not only actions of abstract groups but of group schemes. We will not elaborate on this topic.

Serre's and Chevalley's criteria to be affine

(12.8) Serre's affineness criterion.

We now prove a theorem due to Serre which states that a (qcqs) scheme X is affine if and only if the first cohomology $H^1(X, \mathscr{F}) = \check{H}^1(X, \mathscr{F})$ as defined in Section (11.5) vanishes for all quasi-coherent modules $\mathscr{F}$.

Proposition 12.32. *Let $X = \mathrm{Spec}\, A$ be an affine scheme and let $\mathscr{F}$ be a quasi-coherent $\mathscr{O}_X$-module. Then $H^1(X, \mathscr{F}) = 0$.*

Proof. As $\mathscr{F}$ is quasi-coherent, it is of the form $\mathscr{F} = \tilde{M}$ for an A-module M. As the principal open subsets form a basis of the topology of X, it suffices to show that $\check{H}^1(\mathcal{U}, \mathscr{F}) = 0$, where $\mathcal{U} = (D(f_i))_i$ is a finite covering of X by principal open subsets. Then $\check{H}^1(\mathcal{U}, \mathscr{F}) = 0$ is equivalent to the following lemma for $p = 1$. $\qquad\square$

Lemma 12.33. *Let A be a ring and let $(f_i)_{i \in I}$ be a finite family of elements $f_i \in A$ which generate the unit ideal in A (equivalently, $X := \operatorname{Spec} A = \bigcup_i U_i$ with $U_i := D(f_i)$). For every integer $p \geq 0$ and for $\mathbf{i} = (i_0, \ldots, i_p) \in I^{p+1}$ we set $f_{\mathbf{i}} = f_{i_0} f_{i_1} \cdots f_{i_p}$. Thus we have*

$$D(f_{\mathbf{i}}) = U_{\mathbf{i}} := U_{i_0} \cap \cdots \cap U_{i_p}.$$

Let M be an A-module and let $\mathscr{F} = \tilde{M}$ the corresponding quasi-coherent $\mathscr{O}_X$-module. Let $m = (m_{\mathbf{i}})_{\mathbf{i} \in I^{p+1}}$ be a family of elements $m_{\mathbf{i}} \in \mathscr{F}(U_{\mathbf{i}}) = M_{f_{\mathbf{i}}}$ such that

$$d(m)_{i_0 \ldots i_{p+1}} := \sum_{r=0}^{p+1} (-1)^r m_{i_0 \ldots \hat{i}_r \ldots i_{p+1}} |_{U_{i_0 \ldots i_{p+1}}} = 0$$

for all $(i_0, \ldots, i_{p+1}) \in I^{p+2}$. Then there exists a family $n = (n_{\mathbf{j}})_{\mathbf{j} \in I^p}$ of $n_{\mathbf{j}} \in \mathscr{F}(U_{\mathbf{j}})$ such that $d(n) = m$.

Proof. For $\mathbf{i} = (i_0, \ldots, i_p)$ we may write $m_{\mathbf{i}} = x_{\mathbf{i}}/f_{\mathbf{i}}^k$ for some integer $k > 0$ and $x_{\mathbf{i}} \in M$. As $d(m)_{j i_0 \ldots i_p} = 0$ for all $j \in I$, we find

$$\left(\frac{x_{\mathbf{i}}}{f_{\mathbf{i}}^k} + \sum_{r=0}^p (-1)^{r+1} \frac{x_{j i_0 \ldots \hat{i}_r \ldots i_p} f_{i_r}^k}{f_j^k f_{\mathbf{i}}^k} \right)_{|U_{j\mathbf{i}}} = 0.$$

Hence there exists an integer $l \geq 1$ such that for all $j \in I$ and all $\mathbf{i} \in I^{p+1}$ we have the following equality of elements in $M_{f_{\mathbf{i}}}$:

$$(12.8.1) \qquad f_j^{k+l} m_{\mathbf{i}} = \frac{f_j^{k+l} x_{\mathbf{i}}}{f_{\mathbf{i}}^k} = \sum_{r=0}^p (-1)^r \frac{f_j^l f_{i_r}^k x_{j i_0 \ldots \hat{i}_r \ldots i_p}}{f_{\mathbf{i}}^k}.$$

As the open subsets $D(f_j^{k+l})$ for $j \in I$ cover $\operatorname{Spec} A$, there exist elements $h_j \in A$ such that $\sum_{j \in I} h_j f_j^{k+l} = 1$. We now set

$$n_{i_0 \ldots i_{p-1}} := \sum_{j \in I} h_j f_j^l \frac{x_{j i_0 \ldots i_{p-1}}}{f_{i_0 \ldots i_{p-1}}^k} \in M_{f_{i_0 \ldots i_{p-1}}}.$$

Then we find $d(n) = m$ because we have

$$
\begin{aligned}
d(n)_{i_0 \ldots i_p} &= \sum_{r=0}^p (-1)^r (n_{i_0 \ldots \hat{i}_r \ldots i_p})_{|U_{\mathbf{i}}} \\
&= \sum_{r=0}^p (-1)^r \sum_{j \in I} h_j \frac{f_j^l f_{i_r}^k x_{j i_0 \ldots \hat{i}_r \ldots i_p}}{f_{\mathbf{i}}^k} \\
&= \sum_{j \in I} h_j f_j^{k+l} m_{\mathbf{i}} = m_{\mathbf{i}}.
\end{aligned}
$$

where the third equality is given by (12.8.1). $\square$

In Volume II we will define higher Čech cohomology groups for abelian sheaves. Then Lemma 12.33 simply says that $\check{H}^i(X, \mathscr{F}) = 0$ for $i > 0$, for all affine schemes X, and for all quasi-coherent $\mathscr{O}_X$-modules $\mathscr{F}$.

Using the exact cohomology sequence (Proposition 11.14), we obtain:

Corollary 12.34. *Let X be an affine scheme and let $0 \to \mathscr{F}' \to \mathscr{F} \to \mathscr{F}'' \to 0$ be an exact sequence of $\mathscr{O}_X$-modules, where $\mathscr{F}'$ is quasi-coherent. Then the sequence of $\Gamma(X, \mathscr{O}_X)$-modules $0 \to \Gamma(X, \mathscr{F}') \to \Gamma(X, \mathscr{F}) \to \Gamma(X, \mathscr{F}'') \to 0$ is exact.*

Serre's criterion is a converse to Proposition 12.32.

Theorem 12.35. *Let X be a qcqs scheme. Then the following assertions are equivalent.*

(i) *X is affine.*

(ii) *There exists a family of elements $f_i \in A := \Gamma(X, \mathscr{O}_X)$ such that the open subscheme $X_{f_i} = \{ x \in X ; f_i(x) \neq 0 \}$ of X is affine for all i and the ideal generated by the f_i in A is the unit ideal.*

(iii) *$H^1(X, \mathscr{F}) = 0$ for every quasi-coherent $\mathscr{O}_X$-module $\mathscr{F}$.*

(iv) *$H^1(X, \mathscr{I}) = 0$ for every quasi-coherent ideal $\mathscr{I}$ of $\mathscr{O}_X$.*

Proof. We add an auxiliary assertion:

(v) $H^1(X, \mathscr{F}) = 0$ for every quasi-coherent submodule $\mathscr{F}$ of $\mathscr{O}_X^n$ for some integer n.

"(ii) $\Rightarrow$ (i)". We have $X = \bigcup_i X_{f_i}$ because the f_i generate the unit ideal. Let $g \colon X \to \operatorname{Spec} A$ be the canonical scheme morphism corresponding to the identity of A via Proposition 3.4. Then $g^{-1}(D(f_i)) = X_{f_i}$ and it suffices to show that the restriction of g to a morphism $g_i \colon X_{f_i} \to D(f_i)$ is an isomorphism. As X is qcqs, we have $\Gamma(X_{f_i}, \mathscr{O}_X) = \Gamma(X, \mathscr{O}_X)_{f_i}$ (Theorem 7.22). Therefore g_i induces an isomorphism on global sections and thus is an isomorphism as X_{f_i} and $D(f_i)$ are affine.

"(i) $\Rightarrow$ (iii) $\Rightarrow$ (v) $\Rightarrow$ (iv)". The first implication follows from Proposition 12.32 and the other implications are clear.

"(iv) $\Rightarrow$ (v)". Let $\mathscr{F}$ be a submodule of $\mathscr{O}_X^n$. For all $k \leq n$ consider $\mathscr{O}_X^k$ as submodule of $\mathscr{O}_X^n$ and set $\mathscr{F}_k = \mathscr{F} \cap \mathscr{O}_X^k$. Then $\mathscr{F}_k$ is quasi-coherent (Corollary 7.19) and $\mathscr{F}_k/\mathscr{F}_{k-1}$ is isomorphic to a quasi-coherent ideal of $\mathscr{O}_X$. The cohomology sequence yields an exact sequence $H^1(X, \mathscr{F}_{k-1}) \to H^1(X, \mathscr{F}_k) \to H^1(X, \mathscr{F}_k/\mathscr{F}_{k-1}) = 0$ and by induction on k we obtain $H^1(X, \mathscr{F}) = 0$.

"(v) $\Rightarrow$ (ii)". We first show that there exists a finite covering of X by open affine subsets which are of the form X_{f_i} for some $f_i \in A := \Gamma(X, \mathscr{O}_X)$. Indeed, if $x \in X$ is a point, its closure $Z = \overline{\{x\}}$ is quasi-compact because X is quasi-compact. Thus Z contains a point that is closed in Z and hence closed in X. Hence if $(U_i)_i$ is a family of open subsets whose union contains every closed point, then the family $(U_i)_i$ is a covering of X.

Therefore it suffices to show the following claim: For every closed point $x \in X$ and for every open affine neighborhood U of x there exists $f \in A$ such that $x \in X_f \subseteq U$ (because then $X_f = D_U(f_{|U})$ is affine). To prove the claim let $\mathscr{I}$ and $\mathscr{I}'$ be the quasi-coherent ideals of $\mathscr{O}_X$ defining the closed reduced subscheme whose underlying subset is $X \setminus U$ resp. $(X \setminus U) \cup \{x\}$. Then $\mathscr{I}' \subset \mathscr{I}$ and $\mathscr{I}/\mathscr{I}' = \kappa(x)$ (considered as a sheaf with support on the single point x). As $H^1(X, \mathscr{I}') = 0$, the homomorphism $\Gamma(X, \mathscr{I}) \to \kappa(x)$ is surjective. Thus there exists an $f \in \Gamma(X, \mathscr{I})$ with $f(x) \neq 0$ and hence $x \in X_f \subseteq U$.

It remains to prove that if $X = \bigcup X_{f_i}$ for elements $f_1, \ldots, f_n \in A$, then these elements generate the unit ideal. But the global sections f_i define a homomorphism of $\mathscr{O}_X$-modules $u \colon \mathscr{O}_X^n \to \mathscr{O}_X$ whose restriction to the X_{f_i} is surjective. Therefore u is surjective. By hypothesis, $H^1(X, \operatorname{Ker}(u)) = 0$ and thus u is also surjective on global sections. $\square$

Remark 12.36. If X is noetherian, every quasi-coherent ideal $\mathscr{I}$ of $\mathscr{O}_X$ is coherent (Proposition 7.46). Therefore a noetherian scheme is affine if and only if $H^1(X, \mathscr{I}) = 0$ for every coherent ideal $\mathscr{I}$ of $\mathscr{O}_X$.

Using the Leray spectral sequence for f_* (see Volume II) we obtain the following entirely formal consequence (see also Exercise 12.14 for a proof without the language of spectral sequences).

Corollary 12.37. *Let $f\colon X \to Y$ be an affine scheme morphism. Then for every quasi-coherent $\mathscr{O}_X$-module there is an isomorphism, functorial in $\mathscr{F}$,*

$$(12.8.2) \qquad H^1(Y, f_*\mathscr{F}) \overset{\sim}{\to} H^1(X, \mathscr{F}).$$

(12.9) Chevalley's affineness criterion.

A second important criterion for a scheme to be affine is Chevalley's theorem. We start with the following lemma.

Lemma 12.38. *Let X be a scheme and let X_0 be a closed subscheme which has the same underlying topological space. Then X is affine if and only if X_0 is affine.*

We will prove the lemma only under the additional hypothesis that the quasi-coherent ideal $\mathscr{I} \subset \mathscr{O}_X$ defining X_0 is nilpotent. This is for example always the case if X is noetherian (because then the nilradical $\mathscr{N}_X$ is nilpotent and $\mathscr{I} \subseteq \mathscr{N}_X$). For the proof of Chevalley's theorem we will use the lemma only in the case that X is noetherian. The general case follows then from a non-noetherian variant of Chevalley's theorem; see Corollary 12.40 below.

Proof. If X is affine, any closed subscheme, in particular X_0, is affine. Therefore let X_0 be affine. Let $n \geq 1$ be an integer such that $\mathscr{I}^n = 0$ and set $X_k = V(\mathscr{I}^{k+1})$ for $k \geq 1$. Then $X_{n-1} = X$ and X_{k-1} is a closed subscheme of X_k defined by a quasi-coherent ideal of square zero. Using induction on k we may assume that $\mathscr{I}^2 = 0$. Let i be the closed immersion $X_0 \to X$.

As the schemes X and X_0 have the same underlying topological space and as X_0 is quasi-compact and separated (being affine), the same holds for X (being separated depends only on the underlying reduced subscheme by Proposition 9.13 (4)). Hence we may apply Serre's criterion and it suffices to show that $H^1(X, \mathscr{F}) = 0$ for every quasi-coherent $\mathscr{O}_X$-module. Note that if $\mathscr{G}$ is any sheaf of groups, we have $H^1(X, \mathscr{G}) = H^1(X_0, \mathscr{G})$ because X and X_0 have the same underlying topological spaces (but of course in general an $\mathscr{O}_X$-module will not be an $\mathscr{O}_{X_0}$-module).

Consider the exact sequence $0 \to \mathscr{I}\mathscr{F} \to \mathscr{F} \to \mathscr{F}/\mathscr{I}\mathscr{F} \to 0$ which yields an exact sequence

$$(12.9.1) \qquad H^1(X, \mathscr{I}\mathscr{F}) \to H^1(X, \mathscr{F}) \to H^1(X, \mathscr{F}/\mathscr{I}\mathscr{F}).$$

Then $\mathscr{F}/\mathscr{I}\mathscr{F}$ is also a quasi-coherent module over $\mathscr{O}_{X_0} = \mathscr{O}_X/\mathscr{I}$. Moreover, $i^*(\mathscr{I}\mathscr{F}) = \mathscr{I}\mathscr{F}/\mathscr{I}^2\mathscr{F} = \mathscr{I}\mathscr{F}$, and we can consider also $\mathscr{I}\mathscr{F}$ as quasi-coherent $\mathscr{O}_{X_0}$-module. Thus we find $H^1(X_0, \mathscr{I}\mathscr{F}) = H^1(X_0, \mathscr{F}/\mathscr{I}\mathscr{F}) = 0$ and hence $H^1(X, \mathscr{F}) = 0$. $\qquad\square$

Using this lemma and Serre's affineness criterion, we can prove Chevalley's theorem:

Theorem 12.39. *Let Y be a noetherian scheme and let $f: X \to Y$ be a finite surjective morphism. Then X is affine if and only if Y is affine.*

Proof. Clearly, if Y is affine, X is affine as well, because f is affine. We have to show the converse. Hence let X be affine. If $Y' \subseteq Y$ is a closed subscheme, $f_{(Y')}: f^{-1}(Y') \to Y'$ is again finite and surjective (as both properties are stable under base change) and $f^{-1}(Y')$ is again affine (being a closed subscheme of X). Thus by noetherian induction (Proposition 10.11) we may assume that every proper closed subscheme of Y is affine. By Lemma 12.38 we may also assume that Y is reduced. Under these assumptions we will show that $H^1(Y, \mathscr{F}) = 0$ for every coherent ideal $\mathscr{F}$ of $\mathscr{O}_Y$ which proves that Y is affine by Serre's criterion Theorem 12.35.

(i). Let $\mathscr{F}$ be a coherent $\mathscr{O}_Y$-module such that $\mathrm{Supp}(\mathscr{F}) \neq Y$. Then $Y' := V(\mathrm{Ann}(\mathscr{F}))$ is a proper closed subscheme of Y (Proposition 7.35) and hence it is affine by induction hypothesis. Denote by $i: Y' \to Y$ the closed immersion. We have $\mathscr{F} \cong i_*(i^*\mathscr{F})$ by Remark 7.36 and we find

$$(12.9.2) \qquad H^1(Y, \mathscr{F}) = H^1(Y, i_*(i^*\mathscr{F})) \overset{(12.8.2)}{=} H^1(Y', i^*\mathscr{F}) = 0.$$

(ii). Let Y be non-irreducible. We show that $H^1(Y, \mathscr{F}) = 0$ for every coherent $\mathscr{O}_Y$-module $\mathscr{F}$. Let $Z \subset Y$ be an irreducible component (considered as an integral closed subscheme), let η be its generic point, and let $i: Z \to Y$ be its inclusion. For a coherent $\mathscr{O}_Y$-module $\mathscr{F}$ consider the canonical homomorphism $u: \mathscr{F} \to i_*(i^*\mathscr{F})$ of coherent $\mathscr{O}_Y$-modules and the induced exact sequence

$$H^1(Y, \mathrm{Ker}(u)) \to H^1(Y, \mathscr{F}) \to H^1(Y, \mathrm{Im}(u)).$$

The stalk u_η is an isomorphism and thus $\mathrm{Supp}(\mathrm{Ker}(u))$ is a proper closed subspace of Y. Therefore $H^1(Y, \mathrm{Ker}(u)) = 0$ by (12.9.2). As $\mathrm{Supp}(\mathrm{Im}(u)) \subseteq \mathrm{Supp}(i_*(i^*\mathscr{F})) \subseteq Z$, we find that $\mathrm{Supp}(\mathrm{Im}(u))$ is a proper subset as well and hence again $H^1(Y, \mathrm{Im}(u)) = 0$ by step (i). This proves $H^1(Y, \mathscr{F}) = 0$.

(iii). Thus we may assume that Y is integral. As f is surjective, there exists an irreducible component X' of X such that $f(X')$ contains the generic point η of Y. As f is closed (being finite), the restriction of f to X'_{red} is still surjective and finite. Thus we may assume that X is integral as well.

As f is finite, $\mathscr{B} := f_*\mathscr{O}_X$ is a finite $\mathscr{O}_Y$-algebra (i.e., of finite type as an $\mathscr{O}_Y$-module). The localization of $\mathscr{B}$ in the generic point η of Y is a vector space over the function field $K(Y)$ of Y. Therefore by Proposition 7.27 there exists a non-empty open affine subset $V = \mathrm{Spec}\, A \subseteq Y$ such that $\mathscr{B}_{|V}$ is a finite free $\mathscr{O}_V$-module. Thus $U := f^{-1}(V)$ is a non-empty open affine subset such that $B := \Gamma(U, \mathscr{O}_X) = \Gamma(V, \mathscr{B})$ is a free A-module. Let $(b_1, \dots, b_n)$ be an A-basis of B. As X is affine, there exists an element $0 \neq g \in \tilde{B} := \Gamma(X, \mathscr{O}_X)$ such that $s_i := gb_i \in \tilde{B}$ for all i. These elements $s_i \in \tilde{B} = \Gamma(Y, \mathscr{B})$ define a homomorphism $u: \mathscr{O}_Y^n \to \mathscr{B} = f_*\mathscr{O}_X$ of $\mathscr{O}_Y$-modules which induces an isomorphism $\mathscr{O}_{Y,\eta}^n \to \mathscr{B}_\eta$. Again by Proposition 7.27 we see that there exists a non-empty open subset $W \subseteq Y$ such that $u_{|W}$ is an isomorphism. For every coherent $\mathscr{O}_Y$-module $\mathscr{F}$ composition with u yields a homomorphism of $\mathscr{O}_Y$-modules

$$v: \mathscr{G} := \mathscr{H}\!om_{\mathscr{O}_Y}(f_*\mathscr{O}_X, \mathscr{F}) \to \mathscr{H}\!om_{\mathscr{O}_Y}(\mathscr{O}_Y^n, \mathscr{F}) = \mathscr{F}^n$$

whose restriction to W is an isomorphism.

By Serre's criterion it suffices to prove $H^1(Y, \mathscr{F})^n = H^1(Y, \mathscr{F}^n) = 0$ for every coherent ideal $\mathscr{F} \subseteq \mathscr{O}_Y$. As Y is integral, all restriction maps for $\mathscr{F}$ are injective. Therefore v is injective because $v_{|W}$ is an isomorphism. Set $\mathscr{K} = \operatorname{Coker}(v)$. As $v_{|W}$ is an isomorphism, the annihilator of $\mathscr{K}$ is a quasi-coherent ideal which defines a proper closed subscheme Z of Y. By induction hypothesis, Z is affine and therefore $H^1(Z, \mathscr{K}) = H^1(Y, \mathscr{K}) = 0$. The exact cohomology sequence yields $H^1(Y, \mathscr{G}) \to H^1(Y, \mathscr{F}^n) \to H^1(Y, \mathscr{K}) = 0$. Thus it suffices to show that $H^1(Y, \mathscr{G}) = 0$.

Now $\mathscr{G}$ is a coherent $f_*\mathscr{O}_X$-module via the first factor. By Proposition 12.5 there exists a quasi-coherent $\mathscr{O}_X$-module $\mathscr{H}$ such that $f_*\mathscr{H} \cong \mathscr{G}$. But by Corollary 12.37 we have $H^1(Y, f_*\mathscr{H}) = H^1(X, \mathscr{H})$ and $H^1(X, \mathscr{H}) = 0$ by Proposition 12.32. $\qquad\square$

Using Remark 10.73 and the fact that any finite Y-scheme can be embedded into a finite Y-scheme that is of finite presentation over Y, one can deduce from Theorem 12.39 the following non-noetherian generalization of Chevalley's theorem. We refer to the appendix of [Co] for the details.

Corollary 12.40. *Let* $f \colon X \to Y$ *be a finite surjective morphism of schemes. Then* X *is affine if and only if* Y *is affine.*

Normalization

(12.10) Integral closure in a quasi-coherent algebra.

Let A be a ring and let B be an A-algebra. Recall that the set of all elements in B that are integral over A form an A-subalgebra A' of B (Proposition B.53 (2)) which is called the *integral closure of A in B*.

We will now globalize this notion. Let X be a scheme and let $\mathscr{B}$ be a quasi-coherent $\mathscr{O}_X$-algebra. Define a presheaf $\mathscr{A}'$ of $\mathscr{O}_X$-subalgebras of $\mathscr{B}$ by

$$\Gamma(U, \mathscr{A}') := \{ b \in \Gamma(U, \mathscr{B}) \ ; \ b \text{ is integral over } \Gamma(U, \mathscr{O}_X) \}.$$

As being integral can be checked locally for the Zariski topology (combine Proposition B.53 (1) and Lemma 3.20), $\mathscr{A}'$ is a sheaf and hence an $\mathscr{O}_X$-algebra. If $X = \operatorname{Spec} A$ is affine and $f \in A$ we have $\Gamma(D(f), \mathscr{A}') = \Gamma(X, \mathscr{A}')_f$ because forming the integral closure is compatible with localization (Proposition B.55). This shows that $\mathscr{A}'$ is a quasi-coherent $\mathscr{O}_X$-algebra (Theorem 7.16) and we can form the affine X-scheme $\operatorname{Spec} \mathscr{A}'$ (Section (11.2)).

Definition 12.41. *The X-scheme* $\operatorname{Spec} \mathscr{A}'$ *is called the* integral closure of X in $\mathscr{B}$.

(12.11) Normalization of schemes.

Let X be an integral scheme with function field $K(X)$ and let L be a field extension of $K(X)$. The constant sheaf L_X with value L on X is a quasi-coherent $\mathscr{O}_X$-algebra. The corresponding affine X-scheme (Corollary 12.2) is the canonical morphism $\operatorname{Spec} L \to X$.

Definition 12.42. *The integral closure of X in L_X is called the* normalization of X in *L. The normalization of X in $K(X)$ is called the* normalization of X.

Proposition 12.43. *Let X be an integral scheme, let L be an algebraic extension of $K(X)$, and let $\pi\colon X' \to X$ be the normalization of X in L.*
(1) *The scheme X' is integral and normal, and $K(X') = L$.*
(2) *The morphism π is integral and surjective and $\dim X' = \dim X$.*
(3) *Let $U \subseteq X$ be a non-empty open subscheme. Then the restriction $\pi^{-1}(U) \to U$ is the normalization of U in L.*
(4) *Assume that $X = \operatorname{Spec} A$ is affine and let $S \subset A$ be the multiplicative set of nonzero elements. Then $X' = \operatorname{Spec} A'$ with*

$$A' = \{\, b \in L \; ; \; b \text{ is integral over } A \,\}$$

and $S^{-1}A' = L$.

Proof. The equality $X' = \operatorname{Spec} A'$ in Assertion (4) is by definition. Moreover $S^{-1}A'$ is the integral closure of $S^{-1}A = \operatorname{Frac}(A)$ in L (Proposition B.55). As L is algebraic over $\operatorname{Frac}(A)$, we find $S^{-1}A' = L$. Assertion (3) holds because the formation of integral closure is compatible with localization (again by Proposition B.55). Because of (3) we may assume that $X = \operatorname{Spec} A$ is affine to show (1) and (2). Then (1) and the integrality of π follow at once from (4). Moreover, the image of π contains the generic point of X. By Proposition 12.12, we have $\dim X' = \dim X$ and π is closed. Hence it is surjective. $\square$

The normalization of X can be characterized as follows.

Proposition 12.44. *Let $\pi\colon X' \to X$ be a dominant morphism of integral schemes. Assume that X' is normal. Then the following assertions are equivalent.*
(i) *The morphism $\pi\colon X' \to X$ is the normalization of X.*
(ii) *The morphism π is integral and induces an isomorphism $K(X) \to K(X')$.*
(iii) *For every integral and normal scheme Y and every dominant morphism $f\colon Y \to X$ there exists a unique morphism $f'\colon Y \to X'$ such that $\pi \circ f' = f$.*

Proof. The equivalence of (i) and (ii) follows from Proposition 12.43. As a pair (X', π) satisfying (iii) is unique up to unique isomorphism, a morphism π satisfying (iii) is the normalization if we have shown that (i) implies (iii). Because of the uniqueness assertion, this is a local question on X and we may assume that $X = \operatorname{Spec} A$ is affine. Let A' be the integral closure of A in $\operatorname{Frac} A$ (and hence $X' = \operatorname{Spec} A'$). As f is dominant and A is reduced, the corresponding homomorphism $\varphi\colon A \to \Gamma(Y, \mathscr{O}_Y)$ is injective. By Lemma 6.38, $\Gamma(Y, \mathscr{O}_Y)$ is integrally closed in its fraction field which is an extension of $\operatorname{Frac} A$. Therefore φ factorizes uniquely through A'. $\square$

The universal property for the normalization also yields a characterization for a scheme to be normal.

Corollary 12.45. *Let Y be an integral scheme. Then Y is normal if and only if for every dominant morphism $f\colon Y \to X$ to an integral scheme X with normalization $\pi\colon X' \to X$ there exists a morphism $f'\colon Y \to X'$ with $\pi \circ f' = f$.*

Proof. We have already seen that the condition is necessary. Conversely, we can apply the universal property to $f = \mathrm{id}_Y$ and we obtain a section $s\colon Y \to Y'$ of the normalization $\pi\colon Y' \to Y$. As π is separated, s is a closed immersion by Example 9.12. But the image of s also contains the generic point of Y', thus s yields an isomorphism of Y onto a closed subscheme Z of Y' with the same underlying topological space as Y'. As Y' is reduced, we necessarily have $Z = Y'$. $\qquad\square$

Remark 12.46. Let X be an integral scheme, let $\pi\colon X' \to X$ be its normalization, and let $U \subseteq X$ be an open subscheme that is normal. Then the restriction $\pi^{-1}(U) \to U$ is an isomorphism.

In particular we can apply this to the subset U of points $x \in X$ where $\mathscr{O}_{X,x}$ is normal if that subset is open (which is often the case; see Section (12.12) below).

Example 12.47. Let k be a field. Consider the "cusp" $X := \operatorname{Spec} k[T,U]/(T^2 - U^3)$. The homomorphism $k[V^2, V^3] \to k[T,U]/(T^2 - U^3)$ with $V^2 \mapsto U$ and $V^3 \mapsto T$ is an isomorphism. Thus the normalization $\pi\colon X' \to X$ corresponds to the inclusion $k[V^2, V^3] \hookrightarrow k[V]$, and in particular X is not normal.

It is easy to see that π is injective and that for all $x' \in X'$ the induced homomorphism on residue fields $\kappa(\pi(x')) \to \kappa(x')$ is an isomorphism. Therefore π is universally injective (Proposition 4.35). As π is also surjective and universally closed, it is a finite birational universal homeomorphism which induces isomorphisms on all residue fields. But π is not an isomorphism.

Example 12.48. Let A be an integral domain, $X = \operatorname{Spec} A$, and let G be a finite group of automorphisms of X. Then $X/G = \operatorname{Spec} A^G$ (Section (12.7)) and A^G is an integral domain because it is a subring of A. We claim that if X is normal, then X/G is normal.

We use Corollary 12.45 to prove the claim. Denote by $p\colon X \to X/G$ the canonical morphism. Let $f\colon X/G \to Y$ be a dominant morphism, where Y is an integral scheme, and let $\pi\colon Y' \to Y$ be its normalization. We have to find a morphism $f'\colon X/G \to Y'$ such that $\pi \circ f' = f$. As X is normal, there exists a unique morphism $\tilde{f}\colon X \to Y'$ such that $\pi \circ \tilde{f} = f \circ p$. For all $g \in G$ the composition $\tilde{f} \circ g$ also satisfies $\pi \circ \tilde{f} \circ g = f \circ p$ because $p \circ g = p$. Therefore we find $\tilde{f} \circ g = \tilde{f}$ for all $g \in G$. Thus the universal property of the quotient shows that there exists a unique morphism $f'\colon X/G \to Y'$ such that $f' \circ p = \tilde{f}$. Applying the uniqueness statement in the definition of quotient again, we obtain $\pi \circ f' = f$.

(12.12) Finiteness of normalization and quasi-excellent schemes.

If X is an integral scheme, its normalization $\pi\colon X' \to X$ is not necessarily a finite morphism: Nagata gave an example of a local noetherian integral domain such that its integral closure is not noetherian (and in particular not finite over A) (see [Na1] A1, Example 5). In addition, even if A is integrally closed it may happen that its normalization in a finite extension of its field of fractions is not finite over A: Again Nagata gave an example where A even is a discrete valuation ring (see [Na1] A1, Example 3.3).

But for many applications the following class of schemes is sufficient.

Definition 12.49. *A locally noetherian scheme X is called* quasi-excellent *if it satisfies the following two conditions.*

(a) *For all $x \in X$ every fiber of the canonical morphism $\operatorname{Spec} \hat{\mathscr{O}}_{X,x} \to \operatorname{Spec} \mathscr{O}_{X,x}$ is geometrically regular.*
(b) *For every morphism $Y \to X$ of finite type the set Y_{reg} of points $y \in Y$, where $\mathscr{O}_{Y,y}$ is regular, is open in Y.*
The scheme X is called excellent *if it satisfies in addition the following condition.*
(c) *For every morphism $Y \to X$ of finite type and for every pair of closed irreducible subsets $Z \subseteq Z' \subseteq Y$, every maximal chain $Z = Z_0 \subsetneq Z_1 \subsetneq \cdots \subsetneq Z_r = Z'$ of closed irreducible subsets has the same length.*

Schemes satisfying condition (c) are called universally catenary. This notion will be studied in more detail in Section (14.25). Quasi-excellent schemes have the following properties (in particular the normalization is always finite):

Theorem 12.50. *Let X be a quasi-excellent scheme.*
(1) *The set X_{norm} of points $x \in X$, where $\mathscr{O}_{X,x}$ is normal, is open in X.*
(2) *Let X be integral and let L be a finite extension of its function field. Then the normalization of X in L is finite over X.*
(3) *Let $x \in X$ be a point. Then the local ring $\mathscr{O}_{X,x}$ is integrally closed (resp. satisfies condition (R_k), resp. satisfies condition (S_k)) if and only if the completion $\hat{\mathscr{O}}_{X,x}$ has the same property.*

The last assertions does not hold in general: Again Nagata gave in [Na1] A1 examples of local integrally closed domains whose completion is not reduced or irreducible (Example 6/7 of loc. cit.). The following result shows that the class of excellent schemes is quite large.

Theorem 12.51. *The following assertions hold.*
(1) *If X is an excellent scheme and if $f\colon X' \to X$ is a morphism locally of finite type, then X' is an excellent scheme.*
(2) *Assume that R is a complete local noetherian ring (e.g., if R is a field) or a Dedekind ring whose field of fractions has characteristic zero (e.g., $R = \mathbb{Z}$). Then $\operatorname{Spec} R$ is excellent.*

This shows that every scheme locally of finite type over a ring R as in (2) of Theorem 12.51 is excellent. In particular we have:

Corollary 12.52. *Let k be a field.*
(1) *Let X be an integral scheme of finite type over k. For every finite extension L of the function field $K(X)$ the normalization of X in L is finite over X.*
(2) *Let X be a scheme locally of finite type over k. Then X_{reg} and X_{norm} are open subsets of X.*

We will prove neither Theorem 12.50 nor Theorem 12.51 but refer to [EGAIV] Section 7.8.

Here we will only directly prove the first assertion of Corollary 12.52. We start with the following general finiteness result for normal schemes and for separable field extensions.

Proposition 12.53. *Let X be a normal noetherian integral scheme and let L be a finite separable extension of $K(X)$. Then the normalization of X in L is finite over X.*

Proof. We may assume that $X = \operatorname{Spec} A$ is affine. Then we can apply Proposition B.57. $\square$

Proof. (of Corollary 12.52 (1)) Again we may assume that $X = \operatorname{Spec} A$ is affine. Define $K := K(X) = \operatorname{Frac} A$. Let B be the integral closure of A in L. We have to show that B is a finite A-module. We start with two preliminary remarks.

(i). Let N/L be a finite extension, and let C be the integral closure C of A in N. If C is finite over A, then the ring B, which is an A-submodule of C, is also a finite A-algebra because A is noetherian. Thus we can always replace L by a finite extension.

(ii). If K' is a subextension of L and A' is the integral closure of A in K', the integral closure of A' in L is B. Thus if we know that A' is finite over A and B is finite over A', we see that B is finite over A.

By Noether normalization there is a finite injective homomorphism $k[T_1, \ldots, T_n] \hookrightarrow A$. The integral closure B of A in L is also the integral closure of $k[T_1, \ldots, T_n]$ in L, and if B is finite over $k[T_1, \ldots, T_n]$ it is also finite over A. Thus we may assume that $A = k[T_1, \ldots, T_n]$ and hence $K = k(T_1, \ldots, T_n)$. By (i) we may replace L by its normal hull over K and we may assume that L is a normal extension of K. The subfield K' of elements in L that are fixed by all K-automorphisms of L is a purely inseparable extension of K and L is a Galois extension of K'. Using Proposition 12.53 and (ii) we may assume that L is a purely inseparable extension of K.

If $\operatorname{char}(k) = 0$, we are done. Thus let $\operatorname{char}(k) = p > 0$. As L is a finite purely inseparable extension, there exists a power q of p such that L is obtained from K by adjoining finitely many q-th roots of elements $f_i \in K$. By (i) we may replace L by $k'(T_1^{1/q}, \ldots, T_n^{1/q})$ where k' is the finite extension of k obtained by adjoining all q-th roots of all coefficients of the f_i. Then the integral closure of A in L is $k'[T_1^{1/q}, \ldots, T_n^{1/q}]$ which is finite over A. $\qquad\square$

Once we know that the normalization is always finite for integral schemes of finite type over a field k, it is not difficult to deduce that X_{norm} is open for schemes locally of finite type over k; see Exercise 12.17.

For integral noetherian schemes of dimension 1 it still may happen that their normalization is not finite over X, but at least it will be always noetherian again. More precisely, we have the following result.

Theorem 12.54. *(Krull-Akizuki) Let X be an integral noetherian scheme such that $\dim X = 1$, let L be a finite extension of $K(X)$, and let $\pi\colon X' \to X$ be the normalization of X in L. Then X' is a Dedekind scheme (Section (7.13)) and for all proper closed subschemes $Z \subsetneq X$ the restriction $\pi^{-1}(Z) \to Z$ is a finite morphism.*

Proof. From the general properties of the normalization it is clear that X' is an integral normal scheme of dimension 1. It remains to show that X' is noetherian (then X' will be regular by Proposition 6.40 and hence a Dedekind scheme) and that $\pi^{-1}(Z) \to Z$ is finite. To show this we may assume that $X = \operatorname{Spec} A$ is affine. But then the result follows from the Theorem of Krull-Akizuki for rings (Proposition B.90). $\qquad\square$

Proper morphisms

We have seen in Chapter 9 that separatedness is a good analogue, in the world of schemes, of the Hausdorff property of topological spaces. Now we seek to define an analogue of the notion of compactness. Certainly affine space $\mathbb{A}^n$ for $n > 0$ should not be compact, so we are looking for a stronger condition than being quasi-compact (and separated). As in the case of separatedness, it turns out that a good notion is obtained by carrying over an equivalent characterization of compactness for locally compact topological spaces.

More precisely, let $f\colon X \to Y$ be a continuous map of topological spaces, where X is Hausdorff and Y is locally compact. Then $f^{-1}(Z)$ is compact for any compact subset $Z \subseteq Y$ if and only if for all topological spaces Y' the product $f \times \mathrm{id}_{Y'}\colon X \times Y' \to Y \times Y'$ is closed. In particular, taking the one-point space as Y, we can characterize compactness of X. For a proof of this result, see [BouGT] I, 10.3, Proposition 7. Now we imitate this definition – adding some suitable finiteness conditions.

(12.13) Definition of proper morphisms.

Recall (Definition 4.31) that a morphism $f\colon X \to Y$ of schemes is called *universally closed* if for any morphism $Y' \to Y$ its base change $f_{(Y')}\colon X \times_Y Y' \to Y'$ is closed. Now the algebraic geometric analogue of being compact is the following fundamental notion.

Definition 12.55. *A morphism of schemes $f\colon X \to Y$ is called* proper *(or X is called proper over Y) if f is of finite type, separated and universally closed.*

We will see in Corollary 13.101 that a separated morphism $f\colon X \to Y$ of finite type is proper if and only if the base change $f_{(\mathbb{A}_Y^n)}\colon X \times_Y \mathbb{A}_Y^n \to \mathbb{A}_Y^n$ is closed for all n.

Examples 12.56.
(1) Possibly the most important example of a proper morphism is the structure morphism $\mathbb{P}_Y^n \to Y$ of the projective space (we will prove that it is proper in Corollary 13.42 below).
(2) An integral morphism $f\colon X \to Y$ is separated and universally closed. Indeed, it is separated because it is affine. As being integral is stable under base change, it suffices to show that integral morphisms are closed which we have seen in Proposition 12.12.
    The converse also holds: A scheme morphism is integral if and only if it is affine and universally closed (see Exercise 12.19).
(3) Any finite morphism is proper. This follows from (2) because a morphism is finite if (and only if) it is integral and of finite type. In Corollary 12.89 below we will see that a morphism is finite if and only if it is proper and affine.
(4) Let k be a field and $n \geq 1$ be an integer. Then the structure morphism $\mathbb{A}_k^n \to \operatorname{Spec} k$ is separated, of finite type, and closed. But it is not proper over k: Consider the base change $q\colon \mathbb{A}_k^n \times_k \mathbb{A}_k^1 \to \mathbb{A}_k^1$. Writing $\mathbb{A}_k^n = \operatorname{Spec} k[T_1, \ldots, T_n]$ and $\mathbb{A}_k^1 = \operatorname{Spec} k[U]$, consider the closed subscheme $Z = V(UT_1 - 1) \subset \mathbb{A}_k^n \times_k \mathbb{A}_k^1$. Then $q(Z) = \mathbb{A}_k^1 \setminus \{0\}$ which is not closed in $\mathbb{A}_k^1$.

Remark 12.57. We will often use the following characterization of closed morphisms. A continuous map $f\colon X \to Y$ of topological spaces is closed if and only if for all $y \in Y$ and all open neighborhoods U of $f^{-1}(y)$ in X there exists an open neighborhood V of y in Y such that $f^{-1}(V) \subseteq U$.

The property of being proper satisfies the usual permanence properties.

Proposition 12.58.
(1) *Any closed immersion is proper.*
(2) *The property of being proper is stable under composition, stable under base change and local on the target (Section (4.9)).*
(3) *If $f\colon X \to Y$ is a morphism of S-schemes such that X is proper over S and such that Y is separated over S. Then f is proper.*
(4) *For morphisms of finite presentation, the property of being proper is compatible with inductive limits of rings in the sense of Section (10.13).*
(5) *If f is a morphism of finite type, then f is proper if and only if f_{red} is proper.*

For the proof of (4) we refer to [EGAIV] (8.10.5) (it uses the Lemma of Chow, see Theorem 13.100). The remaining assertions are easy to show:

Proof. The first assertion is clear and (2) follows from the fact that the properties "universally closed", "separated", and "of finite type" are stable under composition, stable under base change and local on the target. Then (3) follows formally by Remark 9.11. The property of being universally closed depends only on the underlying topological space. Together with Proposition 9.13 this shows (5). □

Images of proper schemes are again proper:

Proposition 12.59. *Let S be a scheme, and let $f\colon X \to Y$ be a surjective morphism of S-schemes. Let X be proper over S and Y separated and of finite type over S. Then Y is proper over S.*

Proof. We have to show that Y is universally closed over S. Let $S' \to S$ be a morphism of schemes. Consider the base change $X \times_S S' \to Y \times_S S' \to S'$. The composition is closed and the first morphism is surjective ("surjective" is stable under base change). Therefore the second morphism is closed. □

Recall that every rational S-map $f\colon X \dashrightarrow Y$ to a separated S-scheme Y has a unique largest domain of definition $\mathrm{dom}(f)$ (Proposition 9.27). The following proposition shows that in many important cases, the domain of definition is "very large".

Theorem 12.60. *Let S be a scheme, let Y be a proper S-scheme of finite presentation, and let X be a normal noetherian S-scheme. Let $f\colon X \dashrightarrow Y$ be a rational S-map. Then*

$$\mathrm{codim}_X(X \setminus \mathrm{dom}(f)) \geq 2.$$

The proof will show that it suffices to assume that X is a noetherian S-scheme which satisfies Serre's condition (R_1).

Proof. We have to show that $U := \mathrm{dom}(f)$ contains all points $x \in X$ with $\dim \mathscr{O}_{X,x} \leq 1$. As U is open and dense, it contains all points with $\dim \mathscr{O}_{X,x} = 0$. Assume $\dim \mathscr{O}_{X,x} = 1$. As X is normal, $\mathscr{O}_{X,x}$ is a discrete valuation ring (Proposition B.73 (3)). We define $Z := \mathrm{Spec}\, \mathscr{O}_{X,x}$. At least the generic point of Z lies in U and we denote by f' the rational map $Z \to X \dashrightarrow Y$. If we can show that $\mathrm{dom}(f') = Z$, then f is defined in an open neighborhood of x because Y is of finite presentation over S (Proposition 10.52). Thus we are done if we show the following lemma. □

Lemma 12.61. *Let X, Y be schemes. Let A be a valuation ring, $S := \operatorname{Spec} A$, $\eta \in S$ its generic point, and $K := \operatorname{Frac} A = \kappa(\eta)$, and fix a morphism $S \to Y$ so that S (and hence $\operatorname{Spec} K$) are Y-schemes. Let $f \colon X \to Y$ be a morphism. If f is universally closed (resp. separated), the canonical map $\rho \colon \operatorname{Hom}_Y(S, X) \to \operatorname{Hom}_Y(\operatorname{Spec} K, X)$ is surjective (resp. injective).*

Proof. If f is separated, ρ is injective because for any two morphisms $g, g' \in \operatorname{Hom}_Y(S, X)$ with $\rho(g) = \rho(g')$ the subscheme $\operatorname{Eq}(g, g')$ contains $\{\eta\}$, and is closed (by the Definition 9.7 of "separated"). As S is reduced, this shows $\operatorname{Eq}(g, g') = S$ and hence $g = g'$.

Now let f be universally closed, and let $\alpha \colon \operatorname{Spec} K \to X$ be a Y-morphism. The desired extension $S \to X$ is the same as a section of the projection $X \times_Y S \to S$ which extends the morphism $\operatorname{Spec} K \to X \times_Y S$ induced by α.. Let x be the image point of $\operatorname{Spec} K$ in $X \times_Y S$, and let X' be the closure of $\{x\}$, seen as a reduced subscheme of $X \times_Y S$. Since f is universally closed, the morphism $X' \to S$ is surjective, so there exists $x' \in X'$ mapping to the closed point of S.

We have induced ring homomorphisms

$$A \to \mathscr{O}_{X',x'} \to \mathscr{O}_{X',x} \to K.$$

Since X' is integral, the stalk $\mathscr{O}_{X',x}$ is a field, and since it contains A, it must be equal to K. We find that $\mathscr{O}_{X',x'}$ is a local ring dominating A, hence $A = \mathscr{O}_{X',x'}$ (Proposition B.69). Therefore $S \xrightarrow{\sim} \operatorname{Spec} \mathscr{O}_{X',x'} \to X' \to X \times_Y S$ is a section as we wanted. $\qquad\square$

Later, in Section (15.3), we will prove a kind of converse to this lemma, the so-called valuative criteria for separatedness and properness.

Corollary 12.62. *Let A be a valuation ring, K its field of fractions. Let $f \colon X \to \operatorname{Spec} A$ be a proper morphism. Then the canonical map $X(A) \to X(K)$ is bijective.*

(12.14) Coherence of direct images and Stein factorization.

One of the central properties of proper morphism $X \to Y$ is that direct images of coherent modules are again coherent; see Theorem 12.68 below for the precise statement. This result will be proved in Volume II using cohomological methods. Here we will give a proof only in the case where $Y = \operatorname{Spec} k$ for a field k. We start with a lemma which will also be useful in the general case.

Lemma 12.63. (Lemme de dévissage) *Let X be a noetherian scheme and let $\mathcal{K}$ be a subset of the set of isomorphism classes of coherent $\mathscr{O}_X$-modules satisfying the following properties.*
(a) *Let $0 \to \mathscr{F}' \to \mathscr{F} \to \mathscr{F}'' \to 0$ be a short exact sequence of coherent sheaves on X. Then if the isomorphism classes of $\mathscr{F}$ and $\mathscr{F}''$ (resp. of $\mathscr{F}'$ and $\mathscr{F}''$) are in $\mathcal{K}$, then the third is contained in $\mathcal{K}$.*
(b) *Let $\mathscr{F}$ be a coherent $\mathscr{O}_X$-module and $d \geq 1$ an integer. If the isomorphism class of $\mathscr{F}^d$ is in $\mathcal{K}$, then the class of $\mathscr{F}$ is in $\mathcal{K}$.*
Assume that for every closed integral subscheme $i \colon Z \to X$ with generic point η there exists a coherent $\mathscr{O}_Z$-module $\mathscr{H}$ such that $\mathscr{H}_\eta \neq 0$ and such that $i_* \mathscr{H}$ is in $\mathcal{K}$. Then $\mathcal{K}$ contains the isomorphism class of every coherent $\mathscr{O}_X$-module.

Proof. We will simply say that $\mathscr{F}$ is in $\mathcal{K}$ if its isomorphism class is in $\mathcal{K}$. Let Z be a closed subscheme of X defined by a quasi-coherent ideal $\mathscr{I}$. We say that a coherent $\mathscr{O}_X$-module $\mathscr{G}$ has support on Z if $\mathscr{I}\mathscr{G} = 0$. By noetherian induction we may assume that all coherent $\mathscr{O}_X$-modules $\mathscr{G}$ with support on a proper closed subscheme $Z \subsetneq X$ are in $\mathcal{K}$. Let $\mathscr{F}$ be a coherent $\mathscr{O}_X$-module and let us show that $\mathscr{F}$ is in $\mathcal{K}$.

(i). Let $u\colon \mathscr{G}' \to \mathscr{G}$ be a homomorphism of coherent $\mathscr{O}_X$-modules. Assume that $\mathscr{G}$, $\operatorname{Ker}(u)$, and $\operatorname{Coker}(u)$ are in $\mathcal{K}$. Then $\mathscr{G}'$ is in $\mathcal{K}$. Indeed, applying (a) to the exact sequence $0 \to \operatorname{Im}(u) \to \mathscr{G} \to \operatorname{Coker}(u) \to 0$ shows that $\operatorname{Im}(u)$ is in $\mathcal{K}$. Using the exact sequence $0 \to \operatorname{Ker}(u) \to \mathscr{G}' \to \operatorname{Im}(u) \to 0$ then shows that $\mathscr{G}'$ is in $\mathcal{K}$.

Moreover, (a) also implies that if $\mathscr{G}$ and $\mathscr{G}'$ are in $\mathcal{K}$, then $\mathscr{G} \oplus \mathscr{G}'$ is in $\mathcal{K}$: Use the exact sequence $0 \to \mathscr{G} \to \mathscr{G} \oplus \mathscr{G}' \to \mathscr{G}' \to 0$.

(ii). Assume that X is not reduced and let $0 \neq \mathscr{I} \subset \mathscr{O}_X$ be a quasi-coherent nilpotent ideal. Let $n > 1$ be minimal with the property $\mathscr{I}^n = 0$. Then $\mathscr{I}\mathscr{F}$ is annihilated by $\mathscr{I}^{n-1}$ and $\mathscr{F}/\mathscr{I}\mathscr{F}$ is annihilated by $\mathscr{I}$. Thus both modules are in $\mathcal{K}$ by the noetherian induction hypothesis. Hence $\mathscr{F}$ is in $\mathcal{K}$ by (a). Thus we can assume that X is reduced.

Assume that X is not irreducible. Let $(X_k)_k$ be the family of irreducible components of X endowed with their reduced scheme structures and denote $i_k\colon X_k \to X$ the inclusion. Let $u_k\colon \mathscr{F} \to \mathscr{F}_k := i_{k*}i_k^*\mathscr{F}$ be the canonical homomorphisms, and let $u\colon \mathscr{F} \to \mathscr{G} := \bigoplus_k \mathscr{F}_k$ be the induced homomorphism. By induction hypothesis all $\mathscr{F}_k$ and hence $\mathscr{G}$ are in $\mathcal{K}$. Moreover the restriction of u to $X_k \setminus \bigcap_{l \neq k} X_l$ is an isomorphism for all k. Thus $\operatorname{Ker}(u)$ and $\operatorname{Coker}(u)$ are in $\mathcal{K}$ by induction hypothesis. Hence $\mathscr{F}$ is in $\mathcal{K}$ by (i).

(iii). Thus we may assume that X is integral. Let $\mathscr{F}$ be a coherent $\mathscr{O}_X$-module. To show that $\mathscr{F}$ is in $\mathcal{K}$ we may assume that $\operatorname{Supp}\mathscr{F} = X$ by induction hypothesis. Let η be the generic point of X. Let $\mathscr{H}$ be a coherent $\mathscr{O}_X$-module in $\mathcal{K}$ with $\mathscr{H}_\eta \neq 0$ which exists by hypothesis. There exist integers $d, e \geq 1$ such that the $K(X)$-vector spaces $\mathscr{F}_\eta^d$ and $\mathscr{H}_\eta^e$ are isomorphic. By (b) it suffices to show that $\mathscr{F}^d$ is in $\mathcal{K}$. Thus replacing $\mathscr{H}$ by $\mathscr{H}^e$ we may assume that $\mathscr{F}_\eta \cong \mathscr{H}_\eta$. Then there exists an open non-empty affine subset $U \subseteq X$ and an isomorphism $v\colon \mathscr{F}_{|U} \overset{\sim}{\to} \mathscr{H}_{|U}$ (Proposition 7.27). Let $j\colon U \to X$ be the inclusion. Let w be the composition $\mathscr{F} \to j_*\mathscr{F}_{|U} \overset{\sim}{\to} j_*\mathscr{H}_{|U}$ which is a homomorphism of quasi-coherent $\mathscr{O}_X$-modules (Corollary 10.27). Let $\mathscr{G}$ be the sum of the coherent module $\operatorname{Im}(w)$ and the image of the canonical homomorphism $\mathscr{H} \to j_*\mathscr{H}_{|U}$. This is a coherent $\mathscr{O}_X$-module. We have $\mathscr{G}_{|U} = \mathscr{H}_{|U}$. Thus by induction hypothesis $\mathscr{G}/\mathscr{H}$ is in $\mathcal{K}$. As $\mathscr{H}$ is in $\mathcal{K}$, we see that $\mathscr{G}$ is in $\mathcal{K}$. Let u be the composition

$$u\colon \mathscr{F} \overset{w}{\longrightarrow} \operatorname{Im}(w) \hookrightarrow \mathscr{G}.$$

The restriction of u to U is the isomorphism v. Thus by induction hypothesis $\operatorname{Ker}(u)$ and $\operatorname{Coker}(u)$ are in $\mathcal{K}$. Hence $\mathscr{F}$ is in $\mathcal{K}$ by (i). $\qquad\square$

Corollary 12.64. *Let $f\colon X \to Y$ be a morphism of noetherian schemes. Assume that for every closed integral subscheme $i\colon Z \to X$ with generic point η there exists a coherent $\mathscr{O}_Z$-module $\mathscr{H}$ such that $\mathscr{H}_\eta \neq 0$ and such that $f_*i_*\mathscr{H}$ is a coherent $\mathscr{O}_Y$-module. Then $f_*\mathscr{F}$ is coherent for every coherent $\mathscr{O}_X$-module $\mathscr{F}$.*

Proof. Let $\mathcal{K}$ be the isomorphism classes of coherent $\mathscr{O}_X$-modules such that $f_*\mathscr{F}$ is coherent. We have to show that $\mathcal{K}$ satisfies conditions (a) and (b) of the Lemme de dévissage. Condition (b) is clear. The morphism f is quasi-separated by Corollary 10.24 and quasi-compact. Thus by Corollary 10.27 we know that $f_*\mathscr{G}$ is at least quasi-coherent for all quasi-coherent $\mathscr{O}_X$-modules $\mathscr{G}$. Let $0 \to \mathscr{F}' \to \mathscr{F} \to \mathscr{F}'' \to 0$ be an exact sequence of coherent $\mathscr{O}_X$-modules. As f_* is left exact, the sequence $0 \to f_*\mathscr{F}' \to f_*\mathscr{F} \to f_*\mathscr{F}''$ is exact. Thus if $f_*\mathscr{F}$ and $f_*\mathscr{F}''$ are coherent, $f_*\mathscr{F}'$ is coherent. Now let $f_*\mathscr{F}''$ and $f_*\mathscr{F}'$ be coherent. Then the image of $f_*\mathscr{F} \to f_*\mathscr{F}''$ is a quasi-coherent submodule of $f_*\mathscr{F}''$ and hence coherent (Corollary 7.47). As every extension of coherent modules is again coherent, this shows that $f_*\mathscr{F}$ is coherent. $\qquad\square$

Theorem 12.65. *Let k be a field and let X be a proper k-scheme. Then $\Gamma(X, \mathscr{F})$ is a finite-dimensional k-vector space for every coherent $\mathscr{O}_X$-module $\mathscr{F}$.*

Proof. By Corollary 12.8 we may assume that k is algebraically closed. By Corollary 12.64 it suffices to prove the following result (which is then applied to every integral closed subscheme Z of X and to $\mathscr{H} = \mathscr{O}_Z$). $\qquad\square$

Proposition 12.66. *Let k be a field and let X be a proper geometrically connected and geometrically reduced k-scheme. Then $\Gamma(X, \mathscr{O}_X) = k$.*

Proof. Again by Corollary 12.8 we may assume that k is an algebraically closed field. Let $s \in \Gamma(X, \mathscr{O}_X)$ be a global section which we consider as a k-morphism $s\colon X \to \mathbb{A}_k^1$. We have to show that s factors through a k-valued point of $\mathbb{A}_k^1$. As X is proper, its image $Z = s(X)$ is closed. Consider Z as a closed reduced subscheme of $\mathbb{A}_k^1$. As X is reduced, s factors through $Z \hookrightarrow \mathbb{A}_k^1$. We will show that $Z = \operatorname{Spec} k$. By embedding $\mathbb{A}_k^1$ into $\mathbb{P}_k^1$ we may consider s also as a morphism $X \to \mathbb{P}_k^1$. Thus Z is also closed in $\mathbb{P}_k^1$ and thus must be a finite k-scheme. As X is connected, Z is connected. As Z is reduced, we must have $Z = \operatorname{Spec} k'$ for a finite field extension k' of k. As k is algebraically closed, $k' = k$. $\qquad\square$

If Y is an affine k-scheme, we can embed it into affine space $\mathbb{A}_k^{(I)}$ (I some index set). We have just seen that for every morphism $X \to Y$ every component of $X \to Y \hookrightarrow \mathbb{A}_k^{(I)}$ is constant. This shows the following result.

Corollary 12.67. *Let k be a field, let X be a proper geometrically connected and geometrically reduced k-scheme and let Y be an affine k-scheme. Then every morphism $X \to Y$ of k-schemes factors through a k-valued point of Y.*

As mentioned above, we will prove in Volume II the following general result.

Theorem 12.68. *Let Y be a locally noetherian scheme and let $f\colon X \to Y$ be a proper morphism. Then for every coherent $\mathscr{O}_X$-module $\mathscr{F}$ its direct image $f_*\mathscr{F}$ is again coherent.*

Under the hypotheses of Theorem 12.68 it follows that $f_*\mathscr{O}_X$ is a coherent $\mathscr{O}_Y$-algebra. We obtain a factorization

$$(12.14.1) \qquad\qquad X \xrightarrow{f'} \operatorname{Spec} f_*\mathscr{O}_X \xrightarrow{g} Y$$

of f, where g is a finite morphism (Remark 12.10 (3)) and where f' corresponds to the identity morphism of $f_*\mathscr{O}_X$ via Proposition 11.1. This factorization is called the *Stein factorization*. It has the following property which we will prove in Volume II.

Theorem 12.69. *Let Y be a locally noetherian scheme, let $f\colon X \to Y$ be a proper morphism, and let $f = g \circ f'$ be its Stein factorization. Then g is finite and f' is proper with geometrically connected fibers.*

In particular, if $\mathscr{O}_Y \to f_* \mathscr{O}_X$ is an isomorphism, all fibers of f are geometrically connected.

The difficult assertion of this theorem is the connectedness of the fibers of f'. See Exercise 12.29 for a situation where the assumption $\mathscr{O}_Y \cong f_* \mathscr{O}_X$ is satisfied.

(12.15) Compactification of schemes.

In topology it is often useful to have for every locally compact topological space an embedding into a compact topological space. In algebraic geometry we have the following result which is essentially due to Nagata.

Theorem 12.70. *Let S be a qcqs scheme (e.g., if S is noetherian) and let $f: X \to S$ be a separated morphism of finite type. Then the S-scheme X is isomorphic to an open schematically dense subscheme of a proper S-scheme $\bar{X}$.*

In analogy to the language in topology, such an S-scheme $\bar{X}$ is sometimes called a *compactification of X*.

We will not give a proof here of this difficult result. If X is quasi-projective over S (i.e., there exists an immersion of S-schemes $i: X \hookrightarrow \mathbb{P}(\mathscr{E})$ for some finite locally free $\mathscr{O}_S$-module; see Definition 13.68), we can simply define $\bar{X}$ as the schematic closure of $i(X)$ in $\mathbb{P}(\mathscr{E})$. As a closed subscheme of $\mathbb{P}(\mathscr{E})$ this is proper over S because $\mathbb{P}(\mathscr{E})$ is proper over S by Corollary 13.42 below. In particular we see that locally the existence of such a compactification is clear: If $S = \operatorname{Spec} R$ and $X = \operatorname{Spec} A$ are affine, we can write A as a quotient of $R[T_1, \ldots, T_n]$ and we find a closed immersion $X \hookrightarrow \mathbb{A}^n_S$. Embedding $\mathbb{A}^n_S$ into $\mathbb{P}^n_S$, we see that there exists an immersion $X \hookrightarrow \mathbb{P}^n_S$.

Nagata proved this theorem in [Na2] and [Na3] for noetherian schemes without using the language of schemes. A modern exposition and proof of this result was given by Conrad [Co] based on notes of Deligne (see also [Lü]). The proof in the general case requires a very careful gluing of local compactifications which have to be modified using blow-ups (see Section (13.19)). Although difficult, the proof is quite elementary and does not use any methods which are not explained in this Volume.

If X has good properties (e.g., being smooth over S) one would like to have a compactification $\bar{X}$ that has similar good properties. Moreover it would be desirable that the "boundary" $\bar{X} \setminus X$ is as simple as possible. For the property of smoothness the existence of a smooth compactification is not known in general – even if $S = \operatorname{Spec} k$ for a field k. However, if $\operatorname{char}(k) = 0$, then one can use Hironaka's result on resolution of singularities; see Section (13.23).

From Nagata's compactification theorem Temkin [Te2] has deduced the following factorization result. This deduction is based on an approximation theorem similar to Remark 10.73. Again his proof uses only methods explained in this Volume.

Theorem 12.71. *Let Y be a qcqs scheme and let $f: X \to Y$ be a separated quasi-compact morphism. Then there exists a factorization*

$$X \xrightarrow{h} Z \xrightarrow{g} Y,$$

where h is affine and g is proper.

Conversely, using the results of Chapter 13 it is not difficult to deduce Nagata's compactification Theorem 12.70 from Temkin's result (Exercise 13.21). As Temkin gives also a proof of Theorem 12.71 that does not use Nagata compactification (but the theory of Riemann-Zariski spaces - which is beyond the scope of this book), he obtains a new proof of Theorem 12.70.

Zariski's main theorem

We now come to a difficult but extremely important result: Zariski's main theorem. There are several versions of it. We state and prove a very general version and deduce several corollaries which are also often referred to as Zariski's main theorem.

(12.16) Statement and proof of Zariski's main theorem.

We call a point x of a topological space X *isolated in* X if $\{x\}$ is open and closed in X. Note that this definition differs from the usual definition of an isolated point in topology which defines x to be isolated if $\{x\}$ is open in X. If X is a scheme locally of finite type over a field, then there is no difference:

Lemma 12.72. *Let X be a scheme locally of finite type over a field. The following assertions are equivalent for a point $x \in X$.*
(i) *x is isolated in X.*
(ii) *$\{x\}$ is open in X.*
(iii) *$\dim_x X = 0$ (i.e., every irreducible component of X containing x has dimension 0, see Proposition 5.26).*

Proof. Indeed, every open subset of X contains a closed point (Proposition 3.35). This shows that (i) and (ii) are equivalent. The equivalence of (i) and (iii) follows from Proposition 5.20.   □

If $f\colon X \to Y$ is a morphism of schemes, we denote by

$$\mathrm{Isol}(f) := \mathrm{Isol}(X/Y) := \{\, x \in X \ ; \ x \text{ is isolated in } f^{-1}(f(x)) \,\}$$

the set of points that are isolated in their fiber. If $g\colon Y \to Z$ is a second morphism, then

(12.16.1) $\mathrm{Isol}(g \circ f) \subseteq \mathrm{Isol}(f).$

Let $\varphi\colon A \to B$ be a ring homomorphism of finite type. We say that a prime ideal $\mathfrak{q} \subset B$ is *isolated over* A if the corresponding point of $\mathrm{Spec}\, B$ lies in $\mathrm{Isol}(B/A) := \mathrm{Isol}(\mathrm{Spec}\, B/ \mathrm{Spec}\, A)$, i. e. if it is maximal and minimal among the prime ideals $\mathfrak{q}'$ of B such that $\varphi^{-1}(\mathfrak{q}') = \varphi^{-1}(\mathfrak{q})$.

Zariski's main theorem is the following result.

Theorem 12.73. *Let Y be a qcqs scheme and $f\colon X \to Y$ a separated morphism of finite type. Then $\mathrm{Isol}(f)$ is open in X and for every quasi-compact open subset $V' \subseteq \mathrm{Isol}(f)$ there exists a factorization*

$$X \xrightarrow{h} Y' \xrightarrow{g} Y$$

of f, such that g is finite and $h_{|V'}: V' \to Y'$ is a quasi-compact open immersion with $h^{-1}(h(V')) = V'$.

Often $\mathrm{Isol}(f)$ itself is quasi-compact, and then we may take $V' = \mathrm{Isol}(f)$. This is for instance the case if X is noetherian (then every open subset is quasi-compact) or if f is of finite presentation (see Remark 14.114 below).

The proof of Zariski's main theorem 12.73 consists of two parts. Part I is the proof of a local version. Then Part II is to deduce the general statement from this local version.

I. PROOF OF A LOCAL VERSION OF ZARISKI'S MAIN THEOREM.

We consider the following situation. Let B be a ring and let $A \subseteq B$ be a subring. Set $Y = \mathrm{Spec}\, A$, $X = \mathrm{Spec}\, B$ and let $f: X \to Y$ be the morphism corresponding to the inclusion $A \hookrightarrow B$. We start with a Lemma.

Lemma 12.74. *Assume that A is integrally closed in B and that $B = A[b]$ for an element $b \in B$. Let $y \in Y$ be a point and $\mathfrak{p} \subset A$ be the corresponding prime ideal. Then the fiber $f^{-1}(y) = \mathrm{Spec}\, B_{\mathfrak{p}}/\mathfrak{p}B_{\mathfrak{p}}$ has one of the following three forms:*
(i) $f^{-1}(y) = \emptyset$.
(ii) *$f^{-1}(y)$ consists of a single point. Then there exists an open neighborhood V of y in Y such that f induces an isomorphism $f^{-1}(V) \xrightarrow{\sim} V$.*
(iii) *We have $(A/\mathfrak{p})[T] \cong B/\mathfrak{p}B$ (and in particular $f^{-1}(y) \cong \mathbb{A}^1_{\kappa(y)}$ and for the generic point $x \in f^{-1}(y)$ the residue field $\kappa(x)$ is transcendental over $\kappa(y)$).*

Proof. Let $\varepsilon: A[T] \twoheadrightarrow B$ be the A-algebra homomorphism that sends T to b.

First step. We first show that $\mathrm{Ker}(\varepsilon)$ is generated by linear polynomials (i.e., by polynomials of the form $cT + d$ with $c, d \in A$). Let

$$g = a_n T^n + a_{n-1} T^{n-1} + \cdots + a_0 \in \mathrm{Ker}(\varepsilon)$$

with $a_n \neq 0$. We show that g is a linear combination (with coefficients in $A[T]$) of linear polynomials contained in $\mathrm{Ker}(\varepsilon)$ by induction on n. We may assume that $n \geq 2$. Multiplying with a_n^{n-1} and evaluating in b we obtain

$$(a_n b)^n + a_{n-1}(a_n b)^{n-1} + a_{n-2}a_n(a_n b)^{n-2} + \cdots + a_0 a_n^{n-1} = 0.$$

Thus $a := a_n b \in A$ because A is integrally closed in B and $a_n T - a$ is a linear polynomial contained in $\mathrm{Ker}(\varepsilon)$. We obtain

$$g = T^{n-1}(a_n T - a) + h,$$

where $h \in \mathrm{Ker}(\varepsilon)$ with degree $< n$. Thus our claim follows from the induction hypothesis.

Second step. Let $\bar{\varepsilon}$ be the composition of ε with $B \to B/\mathfrak{p}B$. Then $\mathfrak{p}A[T] \subseteq \mathrm{Ker}(\bar{\varepsilon})$. Thus case (iii) is equivalent to $\mathrm{Ker}(\bar{\varepsilon}) \subseteq \mathfrak{p}A[T]$ and also equivalent to $\mathrm{Ker}(\varepsilon) \subseteq \mathfrak{p}A[T]$ because $\mathrm{Ker}(\bar{\varepsilon}) = \mathrm{Ker}(\varepsilon) + \mathfrak{p}A[T]$.

If we are not in case (iii), then by the first step we find elements $c, d \in A$ with $cb = d$ and $c \notin \mathfrak{p}$ or $d \notin \mathfrak{p}$. If $c \notin \mathfrak{p}$, then $A_c = B_c$ and $V := D(c)$ is an open neighborhood as in case (ii). If $d \notin \mathfrak{p}$ and $c \in \mathfrak{p}$, then b is a unit in B_d and $b^{-1} \in \mathfrak{p}B_d$. Thus $\mathfrak{p}B_d = B_d$ and $f^{-1}(y) = \mathrm{Spec}\, B_{\mathfrak{p}}/\mathfrak{p}B_{\mathfrak{p}} = \emptyset$. $\square$

Let $\mathrm{LocIsom}(B/A)$ be the set of prime ideals $\mathfrak{q} \subset B$ such that there exists $s \in A$ with $s \notin \mathfrak{q}$ and $B_s = A_s$. If $Y = \mathrm{Spec}\,A$, $X = \mathrm{Spec}\,B$ and $f \colon X \to Y$ is the morphism corresponding to $\varphi \colon A \to B$, then $\mathrm{LocIsom}(B/A)$ consists of the points $x \in X$ such that there exists an open neighborhood V of $f(x)$ such that f induces an isomorphism $f^{-1}(V) \xrightarrow{\sim} V$. We start by listing some basic properties of the sets LocIsom and Isol.

Lemma 12.75. *Assume that B is a finitely generated algebra over a subring A.*
(1) *For every A-subalgebra A' of B that is integral over A one has*

$$\mathrm{LocIsom}(B/A) \subseteq \mathrm{LocIsom}(B/A') \subseteq \mathrm{Isol}(B/A) \subseteq \mathrm{Isol}(B/A')$$

(2) $\mathrm{LocIsom}(B/A)$ *is open in* $\mathrm{Spec}\,B$.
(3) *If $S \subseteq A$ is multiplicatively closed, then*

$$\mathrm{Isol}(S^{-1}B/S^{-1}A) = \mathrm{Isol}(B/A) \cap \mathrm{Spec}\,S^{-1}B,$$
$$\mathrm{LocIsom}(S^{-1}B/S^{-1}A) = \mathrm{LocIsom}(B/A) \cap \mathrm{Spec}\,S^{-1}B$$

(4) *Assume that there exists $b \in B$ such that the extension $A[b] \to B$ is integral. Then $\mathrm{Isol}(B/A)$ is stable under generization.*
(5) *For every A-subalgebra A' of B that is integral over A one has*

$$\mathrm{LocIsom}(B/A') = \bigcup_{A''} \mathrm{LocIsom}(B/A''),$$

where the union is over all $A \subseteq A'' \subseteq A'$ that are finite A-algebras.

Proof. The first and the third inclusion in (1) are clear.

Let us next show (5). If $\mathfrak{q} \in \mathrm{LocIsom}(B/A')$, then there is an element $f \in A' \setminus \mathfrak{q}$ such that $A'_f = B_f$. As B_f is finitely generated over A, there is a finitely generated A-subalgebra A'' of A' such that $A''_f = B_f$. Then A'' is integral and finitely generated over A and hence a finite A-algebra, and $\mathfrak{q} \in \mathrm{LocIsom}(B/A'')$. The other inclusion follows from the first inclusion in (1).

To show now the inclusion $\mathrm{LocIsom}(B/A') \subseteq \mathrm{Isol}(B/A)$ in (1) we may assume by (5) that A' is finite over A. For $A = A'$ the inclusion is obvious and in general it holds because for every finite morphism $\mathrm{Spec}\,A' \to \mathrm{Spec}\,A$ all fibers are discrete.

Assertion (2) is clear, and the same is true for the inclusions "$\supseteq$" in (3). The inclusion $\mathrm{Isol}(S^{-1}B/S^{-1}A) \subseteq \mathrm{Isol}(B/A) \cap \mathrm{Spec}\,S^{-1}B$ holds because the fibers of $\mathrm{Spec}\,S^{-1}B \to \mathrm{Spec}\,S^{-1}(A)$ and of $\mathrm{Spec}\,B \to \mathrm{Spec}\,A$ over points of $\mathrm{Spec}\,S^{-1}(A)$ coincide.

Now let us prove the inclusion $\mathrm{LocIsom}(S^{-1}B/S^{-1}A) \subseteq \mathrm{LocIsom}(B/A) \cap \mathrm{Spec}\,S^{-1}B$: Let $\mathfrak{q}' \subseteq B$ be a prime ideal with $\mathfrak{q}' \cap S = \emptyset$ such that there exists an $f' \in S^{-1}A$ with $f' \notin S^{-1}\mathfrak{q}'$ and $(S^{-1}A)_{f'} = (S^{-1}B)_{f'}$. Then we find $s \in S$ such that $A_{f's} = B_{f's}$ (because B is a finitely generated A-algebra). As $s \notin \mathfrak{q}'$, we also have $f's \notin S^{-1}\mathfrak{q}'$.

Next we show that $\mathrm{Isol}(B/A)$ is stable under generization under the hypothesis that B is integral over $A[b]$. Denote, for a moment, by A' the integral closure of A in $A[b]$; since A' is integral over A, we have $\mathrm{Isol}(B/A) = \mathrm{Isol}(B/A')$, so that we may assume that A is integrally closed in $A[b]$. Let $\mathfrak{q} \in \mathrm{Isol}(B/A)$ and let $\mathfrak{q}' \subset B$ be a prime ideal with $\mathfrak{q} \supseteq \mathfrak{q}'$. Set $\mathfrak{p}' := A \cap \mathfrak{q}'$. Since B is integral over $A[b]$, $B/\mathfrak{q}'$ is integral over $A[b]/\mathfrak{p}'A[b]$. Assume $\mathfrak{q}' \notin \mathrm{Isol}(B/A)$, which means that $\mathfrak{q}'$ is not minimal or not maximal in the set of prime ideals over $\mathfrak{p}'$. We apply Lemma 12.74 to the homomorphism $A \to A[b]$ and the prime ideal $\mathfrak{p}'$. Since $\mathfrak{q}' \notin \mathrm{Isol}(B/A)$ and B is finite over $A[b]$, the only case of the lemma which can occur is case (iii), so $\mathfrak{q}'$ either maps to the generic point of $f^{-1}(\mathfrak{p}') \cong \mathbb{A}^1_{\kappa(\mathfrak{p}')}$, or to a closed point. In the first case, $(A/\mathfrak{p}')[T] \hookrightarrow B/\mathfrak{q}'$, $T \mapsto b$, is integral and injective. In the latter case, $\mathfrak{q}'$ must be maximal in the fiber $f^{-1}(\mathfrak{p}')$, because B is integral over $A[b]$. If it fails to be in $\mathrm{Isol}(B/A)$, as we assume, then it has a generization in the fiber over $\mathfrak{p}'$ which must map to the generic point of $\mathbb{A}^1_{\kappa(\mathfrak{p}')}$; we replace $\mathfrak{q}'$ by this generization.

So we may assume that $B/\mathfrak{q}'$ is an integral extension of $(A/\mathfrak{p}')[T]$. Replacing A by $A/\mathfrak{p}'$ and B by $B/\mathfrak{q}'$ we may assume that $\mathfrak{q}' = 0$ and that A and B are integral domains. Finally we may replace B by its integral closure in $\mathrm{Frac}\, B$ and A by its integral closure in B (using that for an integral ring extension $R \hookrightarrow S$ there are no inclusion relations between prime ideals of S lying over a given prime ideal of R; see Theorem B.56 (1)). Then A is integrally closed. Note that this implies that $A[T]$ is integrally closed (Proposition B.73 (7)).

However, in this situation $\mathrm{Isol}(B/A) \neq \emptyset$ implies that $\mathrm{Isol}(A[T]/A) \neq \emptyset$ because B is integral over $A[T]$ and $A[T]$ is integrally closed (using the going up and the going down property of integral extensions, Theorem B.56 (2) and (3)). But all fibers of $\mathrm{Spec}\, A[T] \to \mathrm{Spec}\, A$ are affine spaces of dimension 1 over a field and do not contain any isolated points. This is a contradiction, and we see that $\mathrm{Isol}(B/A)$ is stable under generization. $\qquad\square$

The following proposition is the local version of Zariski's Main Theorem:

Proposition 12.76. *Let B be a finitely generated algebra over a subring A, and let $\bar{A}$ be the integral closure of A in B. Then*

$$\mathrm{Isol}(B/A) = \mathrm{LocIsom}(B/\bar{A}).$$

Proof. We have already seen in Lemma 12.75 (1) that $\mathrm{LocIsom}(B/\bar{A}) \subseteq \mathrm{Isol}(B/A)$. Hence it remains to show the other inclusion. We prove this by induction on the number of generators n of the A-algebra B.

First step. Since $\mathrm{Isol}(B/A) \subseteq \mathrm{Isol}(B/\bar{A})$, it is enough to show that $\mathrm{Isol}(B/\bar{A}) = \mathrm{LocIsom}(B/\bar{A})$, so we can assume from now on that A is integrally closed in B. Then the case $n = 1$ follows from Lemma 12.74. Now let $n \geq 2$.

Second step. Write $B = A[b, b_2, \ldots, b_n]$ and let $\mathfrak{q} \in \mathrm{Isol}(B/A)$. By induction hypothesis and Lemma 12.75 (5) we find a finite $A[b]$-subalgebra A' of B and an element $f' \in A' \setminus \mathfrak{q}$ such that

$$(12.16.2) \qquad\qquad B_{f'} = A'_{f'},$$

in other words, there exists an open neighborhood U' of $\mathfrak{q}' := \mathfrak{q} \cap A'$ in $\mathrm{Spec}\, A'$ such that the preimage of U' in $\mathrm{Spec}\, B$ is isomorphic to U'. Hence $\{\mathfrak{q}'\}$ is open in its fiber over A and therefore $\mathfrak{q}'$ is isolated over A, i.e., $\mathfrak{q}' \in \mathrm{Isol}(A'/A)$, because A' is of finite type over A (Lemma 12.72). Assume that we have shown that $\mathfrak{q}' \in \mathrm{LocIsom}(A'/A)$, i.e., there exists an element $f_1 \in A \setminus \mathfrak{q}'$ with $A'_{f_1} = A_{f_1}$. Then the image of f' in A'_{f_1} is of the form g/f_1^m for some $g \in A$ and some $m \geq 1$. Replacing f' by $f' f_1^m$, we may assume that $f' \in A$. Then for $f := f' f_1$ we have $B_f = A_f$.

Therefore we may replace B by A' and we may (and do) assume from now on that B is a finite extension of $A[b]$ for some element $b \in B$.

Third step. We start with a preliminary remark which we use below. Let $R \subseteq S$ be a ring extension and let $s \in S^\times$ be a unit. Then s is integral over R if and only if $s \in R[s^{-1}]$.

Now assume there existed $\mathfrak{q} \in \mathrm{Isol}(B/A) \setminus \mathrm{LocIsom}(B/A)$. By Lemma 12.75 (4), $\mathrm{Isol}(B/A)$ is stable under generization, so we may assume that all prime ideals $\mathfrak{q}' \subsetneq \mathfrak{q}$ lie in $\mathrm{LocIsom}(B/A)$.

By Lemma 12.75 (3) we may assume that A is local with maximal ideal $\mathfrak{m} = \mathfrak{q} \cap A$. We claim that $B/\mathfrak{q}$ contains elements that are transcendental over $k := \kappa(\mathfrak{m})$ (which shows $\mathfrak{q} \notin \mathrm{Isol}(B/A)$).

Assume that every element of $B/\mathfrak{q}$ is integral over A. Then we find a monic polynomial $h \in A[T]$ such that $y := h(b) \in \mathfrak{q}$. As B is finite over $A[b]$ and $A[b]$ is finite over $A[y]$, B is finite over $A[y]$. By our choice of $\mathfrak{q}$, for every prime ideal $\mathfrak{r} \subsetneq \mathfrak{q}$ of B there exists an element $f_{\mathfrak{r}} \in A \setminus \mathfrak{r}$ such that

$$B_{f_{\mathfrak{r}}} = A[y]_{f_{\mathfrak{r}}} = A_{f_{\mathfrak{r}}}.$$

As B is finite over $A[y]$, we may even assume that $f_{\mathfrak{r}} B \subseteq A[y]$ (replacing $f_{\mathfrak{r}}$ by some power $f_{\mathfrak{r}}^N$ such that $f_{\mathfrak{r}}^N b_l \in A[y]$ for every generator b_l of B). Then we have

$$f_{\mathfrak{r}} \in \mathfrak{b} := \{\, g \in A[y] \; ; \; gB \subseteq A[y] \,\}.$$

Note that $\mathfrak{b}$ actually is an ideal of B. We have shown that it is not contained in any prime ideal $\mathfrak{r}$ of B with $\mathfrak{r} \subsetneq \mathfrak{q}$.

Consider the case that $\mathfrak{b} \subseteq \mathfrak{q}$. We claim that there exists an element $z \in B \setminus A[y]$ with $zy \in A[y]$. As the image of $\mathfrak{q}$ in $B/\mathfrak{b}$ is a minimal prime ideal, the image of y in $B/\mathfrak{b}$ is a zero-divisor. Thus we find an element $z' \in B \setminus \mathfrak{b}$ such that $z'y \in \mathfrak{b} \subseteq A[y]$. If $z' \notin A[y]$ we can set $z := z'$. If $z' \in A[y]$ then there exists $g \in B$ such that $gz' \notin A[y]$ (because $z' \notin \mathfrak{b}$) and we can set $z := gz'$.

Writing $zy = a_0 + a_1 y + \ldots a_m y^m \in A[y]$ and replacing z by $z - (a_1 + a_2 y + \cdots + a_m y^{m-1})$ we even find $z \in B$ with

$$z \notin A[y], \qquad zy \in A.$$

Then z is not nilpotent because A is integrally closed in B and $z \notin A$. Thus $B_z \neq 0$. Denote by $\tilde{A}$, $\tilde{y}$, and $\tilde{z}$ the images of A, y, and z in B_z. As $\tilde{z}$ is integral over $\tilde{A}[\tilde{y}]$, we find $\tilde{z} \in \tilde{A}[\tilde{y}, \tilde{z}^{-1}]$ by the preliminary remark above. As $\tilde{y}\tilde{z} \in \tilde{A}$, we have

$$\tilde{A}[\tilde{y}, \tilde{z}^{-1}] = \tilde{A}[\tilde{y}\tilde{z}, \tilde{z}^{-1}] = \tilde{A}[\tilde{z}^{-1}] \subseteq B_z.$$

This shows that $\tilde{z}$ is integral over $\tilde{A}$ (again by the preliminary remark above). It follows that z is integral over A and hence $z \in A$. This is a contradiction to $z \notin A[y]$.

Therefore we must have $\mathfrak{b} \not\subseteq \mathfrak{q}$. Let $f \in \mathfrak{b} \setminus \mathfrak{q}$. Then $B_f = A[y]_f$. We identify $\mathrm{Spec}\, B_f$ (resp. $\mathrm{Spec}\, A[y]_f$) with open subsets of $\mathrm{Spec}\, B$ (resp. $\mathrm{Spec}\, A[y]$). Via these identifications we find $\mathfrak{q} \notin \mathrm{LocIsom}(B_f/A) = \mathrm{LocIsom}(A[y]_f/A)$ and hence $\mathfrak{q}_0 := \mathfrak{q} \cap A[y]$ is not in $\mathrm{LocIsom}(A[y]/A)$. As $\mathfrak{q}$ is minimal in $\mathrm{Spec}\, B \setminus \mathrm{LocIsom}(B/A)$, the prime ideal $\mathfrak{q}_0$ will be minimal in $\mathrm{Spec}\, A[y] \setminus \mathrm{LocIsom}(A[y]/A)$. Thus Lemma 12.74 shows that $A[y]/\mathfrak{q}_0 \subseteq B/\mathfrak{q}$ contains elements that are transcendental over k. This contradicts our assumption that $B/\mathfrak{q}$ contains only elements that are integral over A and finishes the last step. $\square$

Let us rewrite the proposition in a more direct fashion using Lemma 12.75 (5), and state some corollaries.

Proposition 12.77. *Let B be a finitely generated algebra over a subring A. Let $\mathfrak{q} \subset B$ be a prime ideal that is isolated over A. Then there exists a finite subextension A' of A in B and an element $f \in A'$ with $f \notin \mathfrak{q}$ such that $B_f = A'_f$.*

Corollary 12.78. *Let B be a finitely generated algebra over a subring A such that A is integrally closed in B. Let $\mathfrak{q} \subset B$ be a prime ideal that is isolated over A. Then there exists an $f \in A$ with $f \notin \mathfrak{q}$ such that $B_f = A_f$.*

Corollary 12.79. *Let $f\colon X \to Y$ be a morphism locally of finite type and let $U \subseteq X$ be the set $\{\, x \in X \;;\; \dim_x f^{-1}(f(x)) = 0 \,\}$. Then U is open in X.*

Proof. We may assume that $X = \operatorname{Spec} B$ and $Y = \operatorname{Spec} A$ are affine. The ring homomorphism $\varphi\colon A \to B$ corresponding to f makes B into a finitely generated A-algebra. Replacing A by $A/\operatorname{Ker}(\varphi)$, we may assume that A is a subring of B. Let $x \in U$. By Proposition 12.77 there exists a finite A-subalgebra A' and (denoting by $f'\colon \operatorname{Spec} B \to \operatorname{Spec} A'$ the induced morphism) an open neighborhood U' of $f'(x)$ such that f' induces an isomorphism $f'^{-1}(U') \xrightarrow{\sim} U'$. Then $f'^{-1}(U')$ is an open neighborhood of x in U. $\qquad\square$

II: Proof of Zariski's main theorem.

To show the global version of Zariski's main theorem, we will need some facts from commutative algebra. A local ring A is called *henselian* if every finite A-algebra is isomorphic to a product of local rings. By Proposition B.46 (3) any complete local noetherian ring is henselian. The following result will be proved in Volume II (see also [EGAIV] (18.6.6)).

Lemma 12.80. *Let A be a local ring with maximal ideal $\mathfrak{m}_A$. Then there exists a faithfully flat local homomorphism $A \to A'$, where A' is a local henselian ring, such that $A'\mathfrak{m}_A$ is the maximal ideal of A'.*

If A is noetherian, then the Lemma 12.80 follows from basic commutative algebra (Corollary B.43) because we may choose A' as the completion of A.

We need another tool which will be proved (more generally) in Proposition 14.53 where we explain the notion of faithfully flat descent. This will allow us to apply a base change to a henselian ring as provided by Lemma 12.80.

Lemma 12.81. *Let $A \to A'$ be a flat local ring homomorphism of local rings and let $f\colon X \to Y$ be a morphism of A-schemes. Then f is an isomorphism if and only if its base change $f'\colon X \otimes_A A' \to Y \otimes_A A'$ is an isomorphism.*

We start with the following consequence of the local version of Zariski's main theorem.

Corollary 12.82. *Let A be a local henselian ring with maximal ideal $\mathfrak{m}$, let B be a finitely generated A-algebra, and let $f\colon \operatorname{Spec} B \to \operatorname{Spec} A$ be the corresponding morphism. Let $y \in \operatorname{Spec} A$ be the closed point and let $x \in f^{-1}(y)$ be a point which is isolated in $f^{-1}(y)$. Then $B_{\mathfrak{p}_x}$ is a finite A-algebra and $\operatorname{Spec} B_{\mathfrak{p}_x}$ is open in $\operatorname{Spec} B$.*

Proof. A nonzero quotient of a henselian ring is clearly again henselian. Replacing A by $A/\operatorname{Ker}(A \to B)$ we may assume that A is a subring of B. By Proposition 12.77 there exists a finite subextension $A' \subseteq B$ of A such that $B_{\mathfrak{p}_x}$ is a localization of A'. As A is henselian, $A' = \prod_{y'} A'_{\mathfrak{p}_{y'}}$, where y' runs through the finite set of closed points of $\operatorname{Spec} A'$. Thus $B_{\mathfrak{p}_x} = A'_{\mathfrak{p}_z} =: A''$, where z is the image of x in $\operatorname{Spec} A'$ under $\operatorname{Spec} B \to \operatorname{Spec} A'$. As A' is a finite A-algebra, the same holds for its quotient A'' which shows that $B_{\mathfrak{p}_x}$ is a finite extension of A. Let $e \in A'$ be the idempotent element such that $A'' = A'_e$. Then $e \notin \mathfrak{p}_x$ and $B_{\mathfrak{p}_x} = A'_e \subseteq B_e$. This shows $B_{\mathfrak{p}_x} = B_e$ which implies that $\operatorname{Spec} B_{\mathfrak{p}_x}$ is open in $\operatorname{Spec} B$. $\qquad\square$

Theorem 12.83. *Let* $f: X \to Y$ *be a separated morphism of finite type such that* $f^\flat: \mathscr{O}_Y \to f_*\mathscr{O}_X$ *is an isomorphism. Let* V *be the open set* $\{\, x \in X \;;\; \dim_x f^{-1}(f(x)) = 0 \,\}$ *in* X. *Then the restriction* $f_{|V}: V \to Y$ *is an open immersion and* $f^{-1}(f(V)) = V$.

The hypotheses of this theorem are satisfied for instance if X and Y are locally noetherian integral schemes, Y is normal with generic point η, f is proper, and the generic fiber $f^{-1}(\eta)$ is geometrically integral (see Exercise 12.29).

Proof. First step. Let $y \in f(V)$. We will first show that the induced morphism

$$(12.16.3) \qquad\qquad X^y := X \times_Y \operatorname{Spec} \mathscr{O}_{Y,y} \to \operatorname{Spec} \mathscr{O}_{Y,y}$$

is an isomorphism.

By Proposition 4.20 we may consider X^y as a subspace of X. And thus $x \in X^y$ is still isolated in its fiber over y. The properties of being separated and of finite type are stable under base change and the hypothesis $\mathscr{O}_Y = f_*\mathscr{O}_X$ is stable under flat base change $Y' \to Y$ by Proposition 12.6. Thus we may replace Y by $\operatorname{Spec} \mathscr{O}_{Y,y}$ and X by X^y and we can assume that $Y = \operatorname{Spec} A$, where A is a local ring, and that y is the closed point of Y. We will show that $X \to Y$ is an isomorphism.

Let $A \to A'$ be a henselian extension as in Lemma 12.80 and set $Y' := \operatorname{Spec} A'$ and $X' := X \times_Y Y'$. By Lemma 12.81 it suffices to show that the projection $f': X' \to Y'$ is an isomorphism. If $x \in V \cap f^{-1}(y)$ and $x' \in X'$ is any point lying over x, then x' is isolated in $(f')^{-1}(f'(x'))$. The condition $f_*\mathscr{O}_X = \mathscr{O}_Y$ is also preserved by the flat base change $A \to A'$, so we may assume that A is henselian.

Let $x \in V \cap f^{-1}(y)$, and let $U = \operatorname{Spec} B$ be an open affine neighborhood of x in X such that $U \cap f^{-1}(y) = \{x\}$. By Corollary 12.82 there exists an open neighborhood V' of x such that $V' = \operatorname{Spec} B_{\mathfrak{p}_x}$ and such that V' is finite over A. As f is separated, this implies that the open immersion $V' \hookrightarrow X$ is finite (Proposition 12.11) and hence closed (Proposition 12.12). Let $W \subseteq X$ be the open and closed complement of V'. Then $\Gamma(X, \mathscr{O}_X) = \Gamma(V', \mathscr{O}_X) \times \Gamma(W, \mathscr{O}_X)$. As $\Gamma(X, \mathscr{O}_X) = \Gamma(Y, f_*\mathscr{O}_X) = \Gamma(Y, \mathscr{O}_Y) = A$ is a local ring, this is only possible if W is empty. This shows that $V' = X$ is affine and thus $\Gamma(X, \mathscr{O}_X) = \Gamma(Y, \mathscr{O}_Y)$ implies that $X \to Y$ is an isomorphism.

Second step. Let $x \in V$ and $y := f(x)$. Considering $\operatorname{Spec} \mathscr{O}_{Y,y}$ as a subspace of Y, the underlying topological space of X^y can be identified with $f^{-1}(\operatorname{Spec} \mathscr{O}_{Y,y})$ by Proposition 4.20. Thus the first step shows that

$$(12.16.4) \qquad\qquad f_{|V} \text{ is injective}, \qquad f^{-1}(f(V)) = V.$$

Moreover X^y is the spectrum of a local ring and thus $X^y \cong \operatorname{Spec} \mathscr{O}_{X,x}$. Therefore $g := f_{|V}$ is injective, locally of finite type and it induces for all $x \in V$ isomorphisms

$$(12.16.5) \qquad\qquad \operatorname{Spec} \mathscr{O}_{X,x} \xrightarrow{\sim} X^{g(x)} \xrightarrow{\sim} \operatorname{Spec} \mathscr{O}_{Y,g(x)}.$$

We will show that g is an open immersion (using $\mathscr{O}_Y = f_*\mathscr{O}_X$ which holds by hypothesis). Note that if we knew that X and Y are schemes locally of finite presentation over some base S, and that g is an S-morphism, then we could simply apply Proposition 10.52 at this point.

We may assume that $Y = \operatorname{Spec} A$ is affine. It suffices to show that g induces a homeomorphism of V with an open subspace of Y. Let $x \in V$, set $y := g(x)$ and $S := A \setminus \mathfrak{p}_y$. Let $U = \operatorname{Spec} B$ be an open affine neighborhood of x in V. By (12.16.4) we have

(12.16.6) $$f^{-1}(f(U)) = U.$$

Then as an A-algebra B is generated by finitely many elements $b_1, \ldots, b_n$. By (12.16.5) we have $B_{\mathfrak{p}_x} = S^{-1}B = S^{-1}A$. Thus we find an $s \in S$ such that $b_i/1$ $(i = 1, \ldots, n)$ is in the image of $\varphi_s \colon A_s \to B_s$. Replacing Y by $D(s)$, X by $f^{-1}(D(s))$, and U by $\operatorname{Spec} B_s$, we may assume that $A \to B$ is surjective. Then U is an open neighborhood of x which is isomorphic to a closed subscheme $Z = \operatorname{Spec} A/\mathfrak{a}$ of Y (for some ideal $\mathfrak{a} \subseteq A$). Let $U' := f^{-1}(Y \setminus Z)$. By (12.16.6), X is the disjoint union of the open subsets U and U'. We find

$$A = \Gamma(X, \mathscr{O}_X) = \Gamma(U, \mathscr{O}_X) \times \Gamma(U', \mathscr{O}_X) \cong A/\mathfrak{a} \times \Gamma(U', \mathscr{O}_X),$$

where the first equality holds because $\mathscr{O}_Y = f_*\mathscr{O}_X$. This shows that Z is also an open subscheme of Y. Thus we have seen that every $x \in V$ has an open neighborhood which is homeomorphically mapped onto an open subspace of Y. As we already know that g is injective, this implies that g induces a homeomorphism of V with an open subspace of Y. $\qquad\square$

We now come to the proof of Zariski's main theorem Theorem 12.73. In the situation of Theorem 12.73 set $\mathscr{B} := f_*\mathscr{O}_X$. This is a quasi-coherent $\mathscr{O}_Y$-algebra (Proposition 10.10). Set $X' = \operatorname{Spec} \mathscr{B}$. Let $Z = \operatorname{Spec} \mathscr{C}$ be the integral closure of Y in $\mathscr{B}$ (Definition 12.41), let $(\mathscr{C}_\lambda)_\lambda$ be the filtered system of finitely generated $\mathscr{O}_Y$-subalgebras of $\mathscr{C}$, and set $Z_\lambda := \operatorname{Spec} \mathscr{C}_\lambda$. As $\mathscr{C}$ is integral over $\mathscr{O}_Y$, each subalgebra $\mathscr{C}_\lambda$ is a finite $\mathscr{O}_Y$-algebra. Therefore Z_λ is finite over Y, in particular Z_λ is qcqs. Moreover, $\mathscr{C}$ is the union of the $\mathscr{C}_\lambda$ (Corollary 10.51). The inclusions $\mathscr{C}_\lambda \hookrightarrow \mathscr{B} = f_*\mathscr{O}_X$ and $\mathscr{C} \hookrightarrow \mathscr{B}$ correspond to morphisms $X \to Z_\lambda$ and $X \to Z$ (Proposition 11.1).

Similarly, we approximate the Z-scheme X' by schemes of finite type: Let $f' \colon X' \to Z$ be the affine morphism corresponding to the inclusion $\mathscr{C} \to \mathscr{B}$. Then $f'_*\mathscr{O}_{X'}$ is a quasi-coherent $\mathscr{O}_Z$-algebra and $\operatorname{Spec} f'_*\mathscr{O}_{X'} = X'$. Again, $f'_*\mathscr{O}_{X'}$ is a filtered union of quasi-coherent $\mathscr{O}_Z$-algebras $\mathscr{D}_\mu$ of finite type. We set $X'_\mu := \operatorname{Spec} \mathscr{D}_\mu$ and let $f'_\mu \colon X'_\mu \to Z$ be the structure morphism.

For every quasi-compact open subset V' of $\operatorname{Isol}(f)$ we obtain the following diagram (where h, h_μ, g, and g_λ are defined by the condition that the diagram is commutative)

(12.16.7)

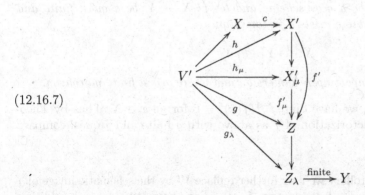

The following lemma shows that if g (resp. h) is an open immersion, then there exists an index λ (resp. an index μ) such that g_λ (resp. h_μ) is an open immersion.

Lemma 12.84. *Let S be a scheme, let $\mathscr{A}$ be a quasi-coherent $\mathscr{O}_S$-algebra which is the filtered union of quasi-coherent $\mathscr{O}_S$-subalgebras $\mathscr{A}_\nu$. Set $X := \operatorname{Spec} \mathscr{A}$ and $X_\nu := \operatorname{Spec} \mathscr{A}_\nu$.*

Let V be an S-scheme of finite type and let $g \colon V \to X$ be an open immersion. Then there exists an index ν such that the composition $V \xrightarrow{g} X \to X_\nu$ is an open immersion.

Proof. We may assume that $S = \operatorname{Spec} R$ is affine. Then $X = \operatorname{Spec} A$ and $X_\nu = \operatorname{Spec} A_\nu$, and A is the filtered union of the A_ν. As V is of finite type over S, it is quasi-compact. Hence g induces an isomorphism of V with a quasi-compact open subset W of X. Therefore we find elements $t_1, \ldots, t_n \in A$ with $\bigcup D_X(t_i) = W$. Set $f_i := g^*(t_i) \in \Gamma(V, \mathscr{O}_V)$. Then g induces an isomorphism $V_{f_i} \xrightarrow{\sim} D(t_i)$. Thus there exists an isomorphism $\varphi_i \colon B_i := \Gamma(V_{f_i}, \mathscr{O}_V) \xrightarrow{\sim} C_{t_i}$ of R-algebras for all $i = 1, \ldots, n$. As V is of finite type over S, the A-algebras B_i are generated by finitely many elements, say $s_{ij} \in B_i$. We write $\varphi_i(s_{ij}) = c_{ij}/t_i^{n_{ij}}$ for $c_{ij} \in C$. Choose an index ν such that all elements c_{ij} and t_i are in C_ν. Then φ_i induces an isomorphism $B_i \xrightarrow{\sim} (C_\nu)_{t_i}$ for all i. Hence g induces an isomorphism of V onto the open subscheme $\bigcup_i D_{X_\nu}(t_i)$. $\square$

Proof. (Zariski's Main Theorem 12.73) We use the notation of (12.16.7). By Theorem 12.83, the restriction of c to $\operatorname{Isol}(c)$ is an open immersion. As $\operatorname{Isol}(f) \subseteq \operatorname{Isol}(c)$ by (12.16.1), h is an open immersion. By Lemma 12.84 there exists an index μ such that h_μ is an open immersion and we may consider $\operatorname{Isol}(f)$ as an open subscheme of X'_μ. As $\mathscr{O}_Z$ is integrally closed in $f'_* \mathscr{O}_{X'}$, it is also integrally closed in the subalgebra $f'_{\mu,*} \mathscr{O}_{X'_\mu}$. As f'_μ is affine and of finite type we can apply Corollary 12.78 and see that there exists an open subscheme W of Z such that $f'^{-1}(W) = \operatorname{Isol}(f')$ and such that f'_μ induces an isomorphism $\operatorname{Isol}(f') \xrightarrow{\sim} W$. As we have $\operatorname{Isol}(f) \subseteq \operatorname{Isol}(f'_\mu)$, this shows that $\operatorname{Isol}(f) \to Z$ is an open immersion. In particular g is an open immersion. By Lemma 12.84 we deduce that there exists an index λ such that g_λ is an open immersion. This finishes the proof of Zariski's main theorem. $\square$

(12.17) Variants and applications of Zariski's main theorem.

The following corollary is also often called Zariski's main theorem.

Corollary 12.85. *Let Y be a qcqs scheme and let $f \colon X \to Y$ be a quasi-finite and separated morphism. Then there exists a factorization*

$$X \xrightarrow{j} Y' \xrightarrow{g} Y$$

of f, where j is a quasi-compact open immersion and where g is a finite morphism.

Proof. As f is quasi-finite, we have $\dim_x f^{-1}(f(x)) = 0$ for all $x \in X$. Thus by Theorem 12.73 there exists a factorization of f as above with g finite and j quasi-compact open immersion. $\square$

Remark 12.86. In the Corollary we may further replace Y' by the schematic image of j (Section (10.8)). Thus we find a factorization $f = g \circ j$ as above, where j is in addition schematically dominant.

Very often Corollary 12.85 is applied in the following form.

Corollary 12.87. *Let S be a noetherian scheme and let X and Y be schemes that are separated and of finite type over S. Then every morphism $f\colon X \to Y$ with finite fibers can be factorized as $f = g \circ j$ where j is an open immersion and g a finite morphism.*

Proof. As X and Y are both separated over S, any morphism between them is separated as well. As X and Y are of finite type over S, both are noetherian (Proposition 10.9) and in particular qcqs. Moreover, any morphism f between them is of finite type (Proposition 10.7). Thus by definition, f is quasi-finite if and only if it has finite fibers. $\qquad\square$

The following corollary is another version of Zariski's main theorem.

Corollary 12.88. *Let $f\colon X \to Y$ be a separated quasi-finite birational morphism of integral schemes and assume that Y is normal. Then f is an open immersion.*

Proof. We may assume that $Y = \operatorname{Spec} A$ is affine. As f is birational, it induces an isomorphism of function fields $K(Y) \overset{\sim}{\to} K(X)$. By Corollary 12.85 we find a factorization $f = g \circ j$, where $j\colon X \to Y'$ is an open immersion and $g\colon Y' \to Y$ is finite. Replacing Y' by the closure of $j(X)$, endowed with its reduced scheme structure, we may assume that Y' is integral and that j is dominant. In particular j induces an isomorphism of function fields $K(Y') \overset{\sim}{\to} K(X)$. This shows that g is birational as well. As g is finite, $Y' \cong \operatorname{Spec} A'$ where A' is a finite A-algebra such that $\operatorname{Frac} A = \operatorname{Frac} A'$. As A is integrally closed, we find $A = A'$ and g is an isomorphism. $\qquad\square$

The following characterization of finite morphisms is also a very useful application of Zariski's main theorem.

Corollary 12.89. *For a morphism $f\colon X \to Y$ of schemes the following assertions are equivalent.*
(i) *f is finite.*
(ii) *f is quasi-finite and proper.*
(iii) *f is affine and proper.*

Proof. "(ii) $\Rightarrow$ (i)". Let f be quasi-finite and proper. All properties are local on the target, thus we may assume that Y is affine (and in particular quasi-compact and separated). Thus we can apply Corollary 12.85 and find a factorization of f into an open immersion j followed by a finite morphism g. By Proposition 12.58, j is proper as well and hence a closed immersion. This implies that j is finite and hence $f = g \circ j$ is finite.

"(i) $\Rightarrow$ (iii)". By definition, any finite morphism is affine, and in Example 12.56 we have seen that any finite morphism is proper.

"(iii) $\Rightarrow$ (ii)". We have to check that all fibers of f are finite. Thus we may assume that $Y = \operatorname{Spec} k$, where k is a field. As f is affine and of finite type, we have $X = \operatorname{Spec} A$, where A is a finitely generated k-algebra. By Noether normalization we find a finite and surjective morphism $g\colon X \twoheadrightarrow \mathbb{A}_k^n$ of k-schemes. Composing g with an open embedding $\mathbb{A}_k^n \hookrightarrow \mathbb{P}_k^n$ we obtain a morphism $X \to \mathbb{P}_k^n$ which is proper by Proposition 12.58 (3) and in particular its image $\mathbb{A}_k^n$ is closed in $\mathbb{P}_k^n$. This shows $n = 0$ and hence that X is finite over k. $\qquad\square$

Corollary 12.90. *Let $f\colon X \to Y$ be a proper morphism and let V be the set of $y \in Y$ such that $f^{-1}(y)$ is a finite set. Then V is open in Y and the restriction $f^{-1}(V) \to V$ of f is a finite morphism.*

Proof. As f is of finite type, $f^{-1}(y)$ is of finite type over $\kappa(y)$ for $y \in Y$. Thus V consists of those points $y \in Y$ such that $\dim_x f^{-1}(y) = 0$ for all $x \in f^{-1}(y)$ (Proposition 5.20). Let $U \subseteq X$ be the set of $x \in X$ such that $\dim_x f^{-1}(f(x)) = 0$. This is an open subset by Corollary 12.79, and V is the complement of $f(X \setminus U)$ in Y. Thus V is open because f is closed. The restriction $f^{-1}(V) \to V$ is quasi-finite and proper and hence finite by Corollary 12.89. $\square$

Finally, we note the following corollary of Theorem 12.83. With the terminology introduced in Section (13.16) below it says that a separated quasi-finite morphism is quasi-affine.

Corollary 12.91. *Let $f\colon X \to S$ be a quasi-finite separated morphism. Then the canonical morphism $X \to \operatorname{Spec} f_*\mathcal{O}_X$ is a quasi-compact open immersion.*

(12.18) Criteria for a morphism to be a closed immersion.

Zariski's main theorem also allows us to deduce several criteria to check whether a morphism is a closed immersion. For the next corollary recall that a morphism $f\colon X \to Y$ is called a *monomorphism* if for all schemes T the induced map on T-valued points $X(T) \to Y(T)$ is injective. It is immediate that a morphism f is a monomorphism if and only if its diagonal Δ_f is an isomorphism.

Corollary 12.92. *A morphism of schemes is a closed immersion if and only if it is a proper monomorphism.*

Proof. Clearly a closed immersion is proper and a monomorphism. For the converse we may assume that $Y = \operatorname{Spec} A$ is affine. If $f\colon X \to Y$ is a proper monomorphism, the fiber $f^{-1}(y) \to \operatorname{Spec} \kappa(y)$ is a monomorphism for all $y \in Y$ (being monomorphism is stable under base change). Thus $f^{-1}(y)$ consists of at most one point. Therefore the proper monomorphism $f\colon X \to Y$ is finite by Corollary 12.89. We find $X = \operatorname{Spec} B$, where B is a finite A-algebra. Then $f^{-1}(y) = \operatorname{Spec} B_y$ with $B_y = B/\mathfrak{m}_y B$ a finite $\kappa(y)$-algebra. As the diagonal is an isomorphism, the multiplication $B_y \otimes_{\kappa(y)} B_y \to B_y$ is an isomorphism which is only possible if the vector space dimension $\dim_{\kappa(y)} B_y \leq 1$. Thus $B_y = \kappa(y)$ or $B_y = 0$ and we see that for all y, $A \to B$ is surjective modulo $\mathfrak{m}_y$. By Nakayama's lemma $A \to B$ is surjective, and f is a closed immersion. $\square$

This corollary will allow us to give a criterion when a morphism $f\colon X \to Y$ is a closed immersion in terms of the maps $f(T)\colon X(T) \to Y(T)$ on T-valued points; see Remark 15.13.

Proposition 12.93. *Let S be a scheme, let X and Y be S-schemes such that the structure morphisms $g\colon X \to S$ and $h\colon Y \to S$ are proper, and let $f\colon X \to Y$ be an S-morphism. Let $s \in S$ be a point and let $f_s = f \times \operatorname{id}_{\kappa(s)}\colon X \otimes_S \kappa(s) \to Y \otimes_S \kappa(s)$ be the induced morphism of fibers. If f_s is finite (resp. a closed immersion), then there exists an open neighborhood U of s such that $f_{|g^{-1}(U)}\colon g^{-1}(U) \to h^{-1}(U)$ is finite (resp. a closed immersion).*

Proof. We first claim that it suffices to show that for all $y \in h^{-1}(s)$ there exists an open neighborhood V_y of y such that the restriction $f^{-1}(V_y) \to V_y$ of f is finite (resp. a closed immersion). Indeed, if V denotes the union of the open sets V_y, then the restriction $f^{-1}(V) \to V$ is finite (resp. a closed immersion) because both properties are local on the target. As h is closed, there exists an open neighborhood U of s such that $h^{-1}(U) \subseteq V$ (Remark 12.57). This proves the claim.

The morphism f is proper by Proposition 12.58 (3). If f_s is finite, then the existence of an open neighborhood V_y of y for all $y \in h^{-1}(s)$ such that the restriction $f^{-1}(V_y) \to V_y$ of f is finite follows from Corollary 12.90.

Now let f_s be a closed immersion. As we already proved the corresponding assertion for finite morphism, we may assume that f is finite. Then $X = \operatorname{Spec} \mathscr{B}$, where $\mathscr{B} = f_* \mathscr{O}_X$ is a quasi-coherent $\mathscr{O}_Y$-algebra which is an $\mathscr{O}_Y$-module of finite type (Remark 12.10 (3)). The morphism f corresponds to $u := f^\flat \colon \mathscr{O}_Y \to \mathscr{B}$. For $y \in h^{-1}(s)$ let $u_y \colon \mathscr{O}_{Y,y} \to \mathscr{B}_y$ be the induced homomorphism on stalks. As f_s is a closed immersion, $u_y \otimes \operatorname{id}_{\kappa(s)} \colon \mathscr{O}_{Y,y}/\mathfrak{m}_s \mathscr{O}_{Y,y} \to \mathscr{B}_y/\mathfrak{m}_s \mathscr{O}_{Y,y}$ is a surjective homomorphism. As $\mathfrak{m}_s \mathscr{O}_{Y,y} \subseteq \mathfrak{m}_y$, Nakayama's lemma implies that u_y is surjective for all $y \in h^{-1}(s)$. As $\operatorname{Coker}(u)$ is an $\mathscr{O}_Y$-module of finite type, its support is closed in Y (Corollary 7.32). Thus for all $y \in h^{-1}(s)$ we find an open neighborhood V_y of y such that $u_{|V_y}$ is surjective. $\qquad\square$

If g is in addition flat, then the same result holds if we replace "closed immersion" by "isomorphism" in the statement; see Proposition 14.28.

The fiber f_s of f in Proposition 12.93 is a morphism between schemes over a field. For those morphisms there is the following geometric characterization of closed immersions.

Proposition 12.94. *Let k be a field, let X and Y be k-schemes of finite type, and let $f \colon X \to Y$ be a proper morphism of k-schemes. Then f is a closed immersion if and only if there exists an algebraically closed extension K of k such that the map $X(K) \to Y(K)$ induced by f is injective and such that for all $x \in X(K)$ the induced map on relative tangent spaces $T_x(X/k) \to T_{f(x)}(Y/k)$ is injective.*

Proof. The condition is clearly necessary. Let us prove that it is sufficient. Remark 12.16 shows that f is quasi-finite. As f is also proper, Corollary 12.89 shows that f is finite. To show that f is a closed immersion, we may assume that Y is affine, say $Y = \operatorname{Spec} A$. Then X is affine, say $X = \operatorname{Spec} B$. We have to show that the finite homomorphism $\varphi \colon A \to B$ corresponding to f is surjective. As $k \to K$ is faithfully flat, it suffices to show that $\varphi \otimes_k \operatorname{id}_K$ is surjective. Thus we may assume that $k = K$. In particular k is algebraically closed. Thus k-valued points of X (resp. Y) correspond to maximal ideals of B (resp. A) (Corollary 3.36). It suffices to show that for each maximal ideal $\mathfrak{m}$ of A, corresponding to $y \in Y(k)$, the homomorphism φ induces a surjective homomorphism $\varphi' \colon A' := A_\mathfrak{m} \to B' := B \otimes_A A_\mathfrak{m}$. The maximal ideals of B' correspond bijectively to the k-valued points of X lying over y. Thus by hypothesis, B' is either 0 or a local ring. We may assume that B' is local. Let $\mathfrak{n}$ be its maximal ideal and $x \in X(k)$ the corresponding point. By Remark 6.12 we have $T_x(X/k) = (\mathfrak{n}/\mathfrak{n}^2)^*$ and $T_{f(x)}(Y/k) = (\mathfrak{m}/\mathfrak{m}^2)^*$. The hypothesis on tangent spaces shows that φ' induces a surjective k-linear map $\mathfrak{m}/\mathfrak{m}^2 \to \mathfrak{n}/\mathfrak{n}^2$. Thus Nakayama's Lemma implies that $\mathfrak{n} = \varphi'(\mathfrak{m})B'$ and hence $B'/\varphi'(\mathfrak{m})B' = k$. Another application of Nakayama's Lemma shows that the A'-module B' is generated by a single element. Therefore φ' is surjective. $\qquad\square$

Remark 12.95. The hypothesis in Proposition 12.94 that f is injective on K-valued points and on tangent spaces in K-valued points may also expressed by saying that the map $X(K[\varepsilon]) \to Y(K[\varepsilon])$ induced by f is injective (where $K[\varepsilon] = K[T]/(T^2)$ is the ring of dual numbers over K). In Volume II we will study the tangent bundle of a scheme.

Then this injectivity means that f induces an injection on K-valued points of the tangent bundle.

Exercises

Exercise 12.1. Let $f\colon X \to Y$ be a morphism of schemes. Show that f is affine if and only if f_{red} is affine.

Exercise 12.2$\Diamond$. Let $f\colon X \to Y$ be a finite (resp. integral, resp. quasi-finite) morphism. Show that f_{red} possesses the same property.

Exercise 12.3. Let $f\colon X \to Y$ be a closed morphism, let $y \in Y$ be a point such that $f^{-1}(y)$ is contained in an open affine subscheme of X. Show that there exists an open neighborhood V of y such that the restriction $f^{-1}(V) \to V$ of f is affine. Deduce that every closed injective morphism is affine.

Exercise 12.4. Let R be a ring, $S = \operatorname{Spec} R$ and let $X \subseteq \mathbb{A}^1_S$ be a closed subscheme, $X = V(\mathfrak{a})$ where $\mathfrak{a} \subseteq R[T]$ is an ideal. Show that $X \to S$ is finite if and only if $\mathfrak{a}$ contains a monic polynomial.

Exercise 12.5. Let S be a scheme of characteristic $p > 0$ and let X be an S-scheme. Show that the relative Frobenius $F_{X/S}\colon X \to X^{(p)}$ is integral. Show that $F_{X/S}$ is finite if X is locally of finite type over S.

Exercise 12.6$\Diamond$. Let $Y = \operatorname{Spec} A$ be an integral affine scheme with generic point η and let $X \to Y$ be a qcqs morphism with generic fiber X_η. Let $\mathscr{F}$ be a quasi-coherent $\mathscr{O}_X$-module and let $\mathscr{F}_\eta$ be the induced $\mathscr{O}_{X_\eta}$-module. Show that there is a functorial isomorphism $\Gamma(X, \mathscr{F}) \otimes_A \operatorname{Frac} A \cong \Gamma(X_\eta, \mathscr{F}_\eta)$.

Exercise 12.7. Let k be an algebraically closed field and let $1 \le r \le n$ be integers. Let $J_r(k)$ be the set of matrices $A \in M_n(k)$ such that for every eigenvalue λ of A the number of Jordan blocks for the eigenvalue λ is $\le r$. Show that $J_r(k)$ is open and dense in $M_n(k)$.
   More generally, let M_n be the $\mathbb{Z}$-scheme of $(n \times n)$-matrices. Show that there exists a (necessarily) unique open subscheme $J_r \subseteq M_n$ such that the set of k-valued points of J_r is $J_r(k)$ for all algebraically closed fields k. Show that $J_r \times_\mathbb{Z} S$ is dense in $M_{n,S}$ for all schemes S.
   The open subscheme J_1 is called the scheme of *generic matrices*.

Exercise 12.8$\Diamond$. Let X be a scheme whose underlying topological space is finite and discrete. Show that any scheme morphism $X \to Y$ is affine.

Exercise 12.9. In the situation of Remark 12.28 assume that the order $\#G$ is a unit in R. Show that (12.7.1) is an isomorphism for all R-algebras B.

Exercise 12.10. Let $f\colon X \to Y$ be an integral morphism, let $\mathscr{E}$ be a locally free $\mathscr{O}_X$-module of finite constant rank n, and let $Z \subset Y$ be a finite set which is contained in an open affine subscheme V of Y. Show that there exists an open affine subscheme $V' \subseteq V$ with $Z \subseteq V'$ such that $\mathscr{E}_{|f^{-1}(V')} \cong \mathscr{O}^n_{X|f^{-1}(V')}$.
Hint: Reduce to the case that $X = \operatorname{Spec} B$ and $Y = \operatorname{Spec} A$ are affine. Use Exercise 10.33 to reduce to the case that B is a finite A-algebra.

Exercise 12.11. Let $K \supseteq k$ be a finite purely inseparable extension and let $\bar{k}$ be an algebraic closure of k. Let x (resp. y, resp. $\bar{y}$) be the unique point of $\operatorname{Spec} K$ (resp. $\operatorname{Spec} k$, resp. $\operatorname{Spec} \bar{k}$). Then $\operatorname{Spec} K \otimes_k \bar{k}$ consists of a single point $\bar{x}$. Show that $e_{x/y} = 1 = f'_{\bar{x}/\bar{y}}$ and $f'_{x/y} = [K : k] = e_{\bar{x}/\bar{y}}$.

Exercise 12.12. Determine for $\operatorname{Spec} \mathbb{Z}[T]/(T^2 + 1) \to \operatorname{Spec} \mathbb{Z}$ ramification and inertia indices in each point.
Hint: Use that -1 is not a square in $\mathbb{F}_p$ if and only if $p \equiv 3 \bmod 4$.

Exercise 12.13. Let k be a field of characteristic $\neq 2$, $n \geq 1$ an integer, let G be the group $\{\pm 1\}$ which acts on $k[T_1, \ldots, T_n]$ by $(-1) \cdot T_i := -T_i$ for all i. Show that $\mathbb{A}_k^n/G$ is smooth over k if and only if $n = 1$.

Exercise 12.14. Let $f: X \to Y$ be an affine morphism of schemes, and let $\mathscr{F}$ be a quasi-coherent $\mathscr{O}_X$-module and let T be an $\mathscr{F}$-torsor on X (Definition 11.10, considering $\mathscr{F}$ as a sheaf of abelian groups). Show that f_*T is an $f_*\mathscr{F}$-torsor on Y and that $T \mapsto f_*T$ defines an isomorphism $H^1(X, \mathscr{F}) \overset{\sim}{\to} H^1(Y, f_*\mathscr{F})$.
Hint: Use Proposition 12.5.

Exercise 12.15. Let X be a noetherian scheme. Show that X is affine if and only if each irreducible component of X (considered as a reduced subscheme) is affine.

Exercise 12.16. Let k be a field such that $\operatorname{char}(k) \neq 2$. Let $f \in k[T]$ be a monic polynomial which is not a square and set $X = \operatorname{Spec} k[T, U]/(U^2 - f)$.
(a) Show that X is geometrically integral. Moreover, show that X smooth over k (and in particular normal) outside the finite k-scheme $V(U) \cap X$.
(b) Show that X is normal if and only if f is square free.
(c) Determine the normalization of X.

Exercise 12.17.
(a) Let Z be an integral scheme and let $\pi: Z' \to Z$ be its normalization. Assume that π is of finite presentation (if Z is noetherian, this simply means that π is finite using Remark 12.10). Show that Z_{norm} is open in Z.
(b) Now let X be a locally noetherian scheme such that for each irreducible component Z of X (endowed with its reduced scheme structure) the normalization of Z is finite. Show that X_{norm} is open in X.

Exercise 12.18. Let X be a noetherian scheme and let U be an open subscheme such that the inclusion $j: U \hookrightarrow X$ is affine. Show that every irreducible component of $X \setminus U$ has codimension ≤ 1 in X.
Hint: Assume there exists a component Z of $X \setminus U$ with $\operatorname{codim}_X Z \geq 2$. Replacing X by $\operatorname{Spec} \mathscr{O}_{X,Z}$ and U by $U \cap \operatorname{Spec} \mathscr{O}_{X,Z}$ we may assume that $X = \operatorname{Spec} A$ for a local noetherian ring with $\dim A \geq 2$ and $U = X \setminus \{x\}$, where x is the closed point. Replacing A by a quotient $A/\mathfrak{p}$ for a minimal prime ideal $\mathfrak{p}$ we may assume that A is an integral domain. Replacing A by its normalization A' we may assume that A is normal. Then use Theorem 6.45 to get a contradiction.

Note that if A' is not finite over A (this can only happen if X is not quasi-excellent), then one should reduce first to the case that X is affine and of finite type over $\mathbb{Z}$ which ensures that X is excellent.

Exercise 12.19. Let $f: X \to Y$ be a morphism of schemes. Show that the following assertions are equivalent.

(i) f is integral.

(ii) f is affine and universally closed.

(iii) f is affine and the base-change $f_{\mathbb{A}_Y^1} \colon \mathbb{A}_X^1 \to \mathbb{A}_Y^1$ is closed.

Hint: Reduce to the case $X = \operatorname{Spec} B$ and $Y = \operatorname{Spec} A$ affine. To show that (iii) implies (i) consider for $b \in B$ the induced homomorphism $\varphi' \colon A' \to B_b$, where A' is the subring of B_b generated by the image of A and $1/b$. Consider the surjections $\alpha \colon A[T] \to A'$ and $B[T] \to B_b$ that send T to $1/b$ to show that $\operatorname{Spec} \varphi'$ is closed and surjective. Deduce that $1/b$ is invertible in A' and that therefore $b/1$ is in the image of α. This gives an integrality equation for b.

Exercise 12.20. Let $f \colon X \to Y$ be an integral birational morphism of integral schemes and assume that Y is normal. Show that f is an isomorphism.

Hint: This exercise does not use Zariski's main theorem.

Exercise 12.21. Let A and B be Dedekind rings, set $Y := \operatorname{Spec} A$ and $X := \operatorname{Spec} B$. Let $f \colon X \to Y$ be a finite dominant morphism. Let $C \subseteq B$ be any subring containing A and let $h \colon Z := \operatorname{Spec} C \to Y$ be the morphism corresponding to the inclusion $A \hookrightarrow C$.

(a) Show that h is surjective finite locally free of degree $[\operatorname{Frac} C : K]$ and that $\dim C = 1$.

(b) Form now on assume that C and B have the same field of fractions. Show that B is the integral closure of C and the inclusion $C \hookrightarrow B$ corresponds to a finite birational morphism $g \colon X \to Z$. Deduce that $\deg f = \deg h$.

(c) Let $x \in X$, $z := g(x)$. Show that $e_{x/z} = f_{x/z} = 1$ and deduce $g^{-1}(z) = \operatorname{Spec} \kappa(z)$.

(d) Show that g is a finite birational universal homeomorphism which induces isomorphisms on all residue fields.

(e) Give an example of the above situation where g is not an isomorphism.

Hint: E.g., consider $A = \mathbb{Z}$, $L = \mathbb{Q}[\sqrt{d}]$ where $d \neq 0, 1$ is a squarefree integer, and B the integral closure of A in L. Show that $C := \mathbb{Z}[\sqrt{d}]$ is not integrally closed if $d \equiv 1$ (mod 4) (($\sqrt{d}+1)/2$ is integral over $\mathbb{Z}$; its minimal polynomial is $T^2 - T + (d-1)/4$). *Remark*: It is not difficult to see that for $d \equiv 1$ (mod 4) the integral closure of $\mathbb{Z}$ in L is $\mathbb{Z}[(\sqrt{d}+1)/2]$. If $d \not\equiv 1$ (mod 4), the integral closure of $\mathbb{Z}$ in L is $\mathbb{Z}[\sqrt{d}]$.

Exercise 12.22. Let $f \colon X \to Y$ be an integral dominant morphism of integral schemes. Assume that Y is normal. Show that f is universally open.

Hint: Reduce to $Y = \operatorname{Spec} A$ and $X = \operatorname{Spec} B$ affine. Use Exercise 10.36 and Proposition B.73 (7) to show that it suffices to prove that f is open. For $b \in B$ let $p = T^n + a_{n-1}T^{n-1} + \cdots + a_0 \in A[T]$ be a monic polynomial with $p(b) = 0$ which has minimal degree. Show that $f(D(b)) = \bigcup_{i=0}^{n-1} D(a_i)$.

Exercise 12.23. Set $A := \mathbb{R}[U,T]/(U^2 + T^2)$, denote by u and t the image of U resp. T in A, and $X = \operatorname{Spec} A$.

(a) Show that the normalization X' of X is $\operatorname{Spec} A'$, where $A' = A[s]$ with $s = t/u$. Deduce that $A' \cong \mathbb{C}[S]$.

(b) Show that $X' \to X$ is a homeomorphism but it is not universally open.

Hint: Consider the base change $Y \to X$ with $Y = \operatorname{Spec} \mathbb{C}[U,T]/(U^2 + T^2)$.

Exercise 12.24◊. Show that the property of being "finite locally free" is stable under composition, stable under base change, and local on the target.

Exercise 12.25. Let $f \colon X \to Y$ be a finite surjective morphism and set $\mathscr{B} := f_* \mathcal{O}_X$ (this is a finite quasi-coherent $\mathcal{O}_Y$-algebra such that $X = \operatorname{Spec} \mathscr{B}$). Assume that X and Y are integral and that Y is normal. Let $L = K(X)$ and $K = K(Y)$ be the function fields, $n := [L : K]$, and let $N_{L/K} \colon L \to K$ be the norm map.

(a) Show that $N_{L/K}$ induces a homomorphism $N_{X/Y}\colon \mathscr{B}^{\times} \to \mathscr{O}_Y^{\times}$ of sheaves of abelian groups by considering $\Gamma(V, \mathscr{B})$ (resp. $\Gamma(V, \mathscr{O}_Y)$) as subrings of L (resp. of K). *Hint*: Theorem 5.13.

(b) Show that the homomorphism $H^1(Y, \mathscr{B}^{\times}) \to H^1(Y, \mathscr{O}_Y^{\times})$ induced by $N_{X/Y}$ yields a homomorphism of abelian groups

$$N_{X/Y}\colon \operatorname{Pic}(X) \to \operatorname{Pic}(Y).$$

(c) Show that $N_{X/Y}(f^*\mathscr{M}) = \mathscr{M}^{\otimes n}$ for all $\mathscr{M} \in \operatorname{Pic}(Y)$.

For every line bundle $\mathscr{L}$ on X we call $N_{X/Y}(\mathscr{L})$ the *norm of* $\mathscr{L}$.

Exercise 12.26. Let $f\colon X \to Y$ be a morphism of finite type and let $(X_i)_{i=1,\ldots,n}$ (resp. $(Y_i)_{i=1,\ldots,n}$) be a finite family of closed subschemes of X (resp. of Y) and let $f_i\colon X_i \to Y_i$ be morphisms such that for all i the following diagram commutes

$$
\begin{array}{ccc}
X_i & \longrightarrow & X \\
\downarrow{\scriptstyle f_i} & & \downarrow{\scriptstyle f} \\
Y_i & \longrightarrow & Y
\end{array}
$$

where the horizontal morphisms are the canonical closed immersions. Assume that X is the union of the X_i (as a set). Show that f is proper if and only if f_i is proper for all i.

Exercise 12.27$\Diamond$. Let $f\colon X \to Y$ be a morphism of finite type and let Y be noetherian. Show that f is proper if and only if for every irreducible component X' of X the restriction $f_{|X'_{\mathrm{red}}}$ is proper.
Hint: Exercise 12.26.

Exercise 12.28. Let k be a field, let X be a proper geometrically connected and geometrically reduced k-scheme. Let K be a field extension of k and let $p\colon X_K := X \otimes_k K \to X$ be the projection. Show that the induced map $p^*\colon \operatorname{Pic}(X) \to \operatorname{Pic}(X_K)$ is injective.
Hint: Use Proposition 12.66 to show that a line bundle $\mathscr{L}$ on X is trivial if and only if the canonical homomorphism $\Gamma(X, \mathscr{L}) \otimes_k \mathscr{O}_X \to \mathscr{L}$ is an isomorphism.

Exercise 12.29. Let $f\colon X \to Y$ be a proper morphism of locally noetherian integral schemes, and assume that Y is normal. Assume that the function field $K(Y)$ is algebraically closed in $K(X)$ (e.g., by Proposition 5.51, if the generic fiber of f is geometrically integral). Show that the natural homomorphism $\mathscr{O}_Y \to f_*\mathscr{O}_X$ is an isomorphism and that f has geometrically connected fibers (use Section (12.14)). Give an example of such a morphism which is birational but not finite.

Exercise 12.30$\Diamond$. Let S be a scheme, let X be a proper S-scheme and let Y be a separated S-scheme. Show that any quasi-finite S-morphism $f\colon X \to Y$ is finite.

Exercise 12.31. Let X, Y be integral schemes and let $f\colon X \to Y$ be a birational separated morphism of finite type. Let $y \in Y$ be a normal point such that there exists an isolated point in $f^{-1}(y)$. Show that there exists an open neighborhood V of y such that the restriction $f^{-1}(V) \to V$ of f is an isomorphism.

Exercise 12.32. Let $f\colon X \to Y$ be a morphism of finite type. Show that f is a universal homeomorphism if and only if f is purely inseparable, finite, and surjective.

Exercise 12.33. Let S be a scheme and let $f\colon X \to Y$ be a morphism of S-schemes. Assume that f is locally of finite type. Show that f is a monomorphism if and only if for all $s \in S$ the $\kappa(s)$-morphism on fibers $f_s\colon X_s \to Y_s$ is a monomorphism.

13   Projective morphisms

Contents

- Projective spectrum of a graded algebra
- Embeddings into projective space
- Blowing-up

The most important examples of schemes that are not affine are the schemes that are described in $\mathbb{P}^n_R$ by homogeneous equations $f_1, \ldots, f_r$ in $R[T_0, \ldots, T_n]$ (R some ring). As we have seen in Section (3.7), they are closed subschemes $V_+(f_1, \ldots, f_r)$ of $\mathbb{P}^n_R$. In this chapter we will formalize this construction by attaching to a graded R-algebra $A = \bigoplus_{d \geq 0} A_d$ an R-scheme $\operatorname{Proj} A$. For $A = R[T_0, \ldots, T_n]/(f_1, \ldots, f_r)$ this construction yields $V_+(f_1, \ldots, f_r)$. In particular we will find $\operatorname{Proj} R[T_0, \ldots, T_n] = \mathbb{P}^n_R$.

We can describe all quasi-coherent $\mathscr{O}_X$-modules on $X := \operatorname{Proj} A$ – at least if A is generated by finitely many elements in A_1. In this case every quasi-coherent $\mathscr{O}_X$-module $\mathscr{F}$ is associated to a graded A-module M. The isomorphism class of M is not uniquely determined but if we restrict to the full subcategory of so-called saturated graded A-modules we obtain an equivalence of this subcategory with the category of quasi-coherent $\mathscr{O}_X$-modules (Theorem 13.20). In particular we will see that every closed subscheme of $\mathbb{P}^n_R$ (being defined by a quasi-coherent ideal of $\mathscr{O}_{\mathbb{P}^n_R}$) is of the form $\operatorname{Proj} R[T_0, \ldots, T_n]/I$, where I is an ideal generated by homogeneous polynomials. Moreover, there exists a unique such ideal I that is saturated.

We then generalize the construction of the projective spectrum from graded R-algebras over a ring R to quasi-coherent graded $\mathscr{O}_S$-algebras over a scheme S which is necessary for many applications (e.g., the important construction of a blow-up in the last part). For a special kind of graded $\mathscr{O}_S$-algebras already encountered earlier, namely the symmetric algebra $\operatorname{Sym} \mathscr{E}$ of a quasi-coherent $\mathscr{O}_S$-module $\mathscr{E}$, we will prove that $\operatorname{Proj}(\operatorname{Sym} \mathscr{E})$ is nothing but the projective bundle $\mathbb{P}(\mathscr{E})$ defined in Section (8.8).

Having seen that we can describe closed subschemes of projective space $\mathbb{P}^n_S$ (S some scheme) and their quasi-coherent modules via graded algebras and graded modules, in the second part of this chapter we will study the question how to embed an arbitrary S-scheme X into $\mathbb{P}^n_S$ (or, more generally, into some projective bundle $\mathbb{P}(\mathscr{E})$). These embeddings will be defined by line bundles and the study of such embeddings will lead us to the important notion of (very) ample line bundles. Schemes that can be embedded as closed subschemes of projective bundles will be called projective. It is a central result that projective schemes are proper ("main theorem of elimination theory"). The converse does not hold. But in the last part we will also show that a proper morphism can be always precomposed with a birational projective morphism to obtain a projective morphism (Lemma of Chow).

The main topic of the last part will be a very useful construction: the blow-up of a scheme in a closed subscheme. It is given by an explicit description using the language of projective spectra developed in the first part of the chapter.

© Springer Fachmedien Wiesbaden GmbH, part of Springer Nature 2020
U. Görtz und T. Wedhorn, *Algebraic Geometry I: Schemes*, Springer Studium
Mathematik – Master, https://doi.org/10.1007/978-3-658-30733-2_14

Projective spectrum of a graded algebra

Projective spectra of graded algebras will be defined by a gluing process generalizing the construction of projective space by gluing copies of affine spaces.

(13.1) Graded rings and graded modules.

GRADED RINGS AND GRADED ALGEBRAS.

A *graded ring* is a ring A endowed with a direct sum decomposition $A = \bigoplus_{d \geq 0} A_d$ of abelian groups such that $A_d A_e \subseteq A_{d+e}$ for all $d, e \geq 0$. Then A is an A_0-algebra. If R is any ring, a graded ring A is called *graded R-algebra*, if A_0 is an R-algebra (equivalently, A is an R-algebra and $R A_d = A_d$ for all d).

If A and B are graded R-algebras, a *graded R-algebra homomorphism* $\varphi \colon A \to B$ is an R-algebra homomorphism such that $\varphi(A_d) \subseteq B_d$ for all $d \geq 0$.

For a graded R-algebra and for an integer $\delta \geq 0$ we set $A^{(\delta)} := \bigoplus_{d \geq 0} A_{d\delta}$. This is again a graded R-algebra.

GRADED MODULES AND HOMOGENEOUS SUBMODULES.

Let A be a graded ring. A *graded A-module* is an A-module M together with a decomposition $M = \bigoplus_{d \in \mathbb{Z}} M_d$ of abelian groups such that $A_d M_e \subseteq M_{d+e}$ for all $d \geq 0$ and $e \in \mathbb{Z}$.

The elements of A_d (or M_d) are called *homogeneous of degree d*. For a homogeneous element $0 \neq a \in A_d$ its degree d is denoted by $\deg a$. An arbitrary element $x \in A$ (resp. $x \in M$) has a unique expression as a finite sum $x = \sum_d x_d$, where $x_d \in A_d$ (resp. $x_d \in M_d$). The x_d are called the *homogeneous components of x*.

A *homomorphism $M \to N$ of graded A-modules* is an A-linear map $u \colon M \to N$ such that $u(M_d) \subseteq N_d$ for all d. More generally, for $k \in \mathbb{Z}$ we call an A-linear map $u \colon M \to N$ *homogeneous of degree k* if $u(M_d) \subseteq N_{d+k}$ for all d.

An A-submodule N of a graded A-module M is called *homogeneous* if the following equivalent conditions are satisfied.

(i) $N = \bigoplus_d (N \cap M_d)$.

(ii) N is generated by homogeneous elements of M.

(iii) For every $x \in N$ all its homogeneous components are again in N.

Then N (resp. M/N) is again a graded A-module with homogeneous summand $N_d = N \cap M_d$ (resp. $(M/N)_d = M_d/(N \cap M_d)$). The sum and the intersection of a family of homogeneous submodules is again a homogeneous submodule (for the sum use (ii), for the intersection (iii)). In particular, we may speak of the *homogeneous submodule generated by E*, where E is a subset of M: It is the intersection of all homogeneous submodules containing E or, equivalently, the submodule generated by all homogeneous components of all elements of E.

A *homogeneous ideal I of A* is a graded submodule of A. Then A/I is again a graded ring with $(A/I)_d = A_d/I_d$.

GRADED LOCALIZATION.

Let M be a graded A-module and let $f \in A$ be a homogeneous element. Then the localization M_f has a $\mathbb{Z}$-grading where the homogeneous elements of degree $d \in \mathbb{Z}$ are the elements of the form m/f^n where $m \in M$ is homogeneous such that $d = \deg m - n \deg f$. The subgroup of elements of degree zero of M_f is denoted by $M_{(f)}$. Note that $A_{(f)}$ is a subring of A_f which acts on $M_{(f)}$. Thus $M_{(f)}$ is an $A_{(f)}$-module.

Let $f \in A_d$ be homogeneous of degree $d > 0$. Then there is a ring isomorphism

$$(13.1.1) \qquad A_{(f)} \xrightarrow{\sim} A^{(d)}/(f-1)A^{(d)}, \qquad a/f^n \mapsto a \bmod (f-1),$$

where $A^{(d)}$ is the graded ring $\bigoplus_{e \geq 0} A_{ed}$. The inverse isomorphism is given by sending $a \in A_{ed}$ to a/f^e.

Example 13.1. Let R be a ring. The polynomial ring $A := R[X_0, \ldots, X_n]$ is a graded R-algebra where A_d is the R-submodule of homogeneous polynomials of degree d. We have $R[X_0, \ldots, X_n]_{(X_i)} = R[\frac{X_0}{X_i}, \ldots, \frac{X_n}{X_i}]$.

More generally, let M be any R-module. Then $A := \mathrm{Sym}_R(M)$ is a graded R-algebra.

RELEVANT PRIME IDEALS.

We set $A_+ := \bigoplus_{d \geq 1} A_d$. This is a homogeneous ideal. More generally, if $I \subseteq A$ is a homogeneous ideal, we set $I_+ := I \cap A_+$. This is again a homogeneous ideal.

A homogeneous prime ideal $\mathfrak{p} \subset A$ is called *relevant* if it does not contain A_+, i.e., if $\mathfrak{p}_+ \subsetneq A_+$.

Proposition 13.2. *Let A be a graded ring.*
(1) *Let $\mathfrak{p}, \mathfrak{p}' \subset A$ be relevant prime ideals. If $\mathfrak{p}_+ = \mathfrak{p}'_+$, then $\mathfrak{p} = \mathfrak{p}'$. A homogeneous ideal $I \subsetneq A_+$ is of the form $\mathfrak{p}_+$ for some relevant prime ideal $\mathfrak{p}$ of A if (and only if) for all homogeneous elements $a, b \in A_+ \setminus I$ one has $ab \notin I$.*
(2) *Let $S \subseteq A_+$ be a non-empty subset such that $s, t \in S$ implies $st \in S$ (hence $S \cup \{1\}$ is a multiplicative subset). Suppose that S does not contain every homogeneous element of A_+. Then the set of homogeneous ideals $I \subsetneq A_+$ with $S \cap I = \emptyset$ has maximal elements and each such maximal element I is of the form $\mathfrak{p}_+$ for a relevant prime ideal $\mathfrak{p}$.*
(3) *Let $I \subseteq A_+$ be a homogeneous ideal. Then $\mathrm{rad}(I)_+ = \mathrm{rad}(I) \cap A_+$ is again a homogeneous ideal. Moreover, $\mathrm{rad}(I)_+$ is the intersection of A_+ and of all relevant prime ideals containing I.*

Proof. (1). Let $I \subsetneq A_+$ be a homogeneous ideal and let $f \in A_+ \setminus I$ be homogeneous. If I is of the form $\mathfrak{p}_+$ we necessarily have

$$\mathfrak{p}_0 = \{ a \in A_0 \ ; \ f^r a \in I_{r \deg f} \text{ for all } r \geq 1 \}.$$

This already shows the uniqueness assertion in (1). It remains to show that $\tilde{I} := \mathfrak{p}_0 \oplus I$ is a prime ideal. It is immediate that $\tilde{I}$ is an ideal. Let $g, g' \in A \setminus \tilde{I}$. Write g and g' as sum of homogeneous elements $g = g_0 + \cdots + g_h$ and $g' = g'_0 + \cdots + g'_{h'}$. As $\tilde{I}$ is homogeneous, it suffices to show that $g_h g'_{h'} \notin \tilde{I}$. If $h \neq 0 \neq h'$, this follows from the hypothesis on I. If $h = 0$ (resp. $h' = 0$) we multiply g_h (resp. $g'_{h'}$) with a power of f and again we can apply the hypothesis on I.

(2). The existence of maximal elements follows from Zorn's lemma applied to the ordered set of ideal $I \subsetneq A_+$ with $I \cap S = \emptyset$. This set is non-empty because S does not contain every homogeneous element of A_+. To prove the second assertion in (2) we use (1). Let $a, b \in A_+ \setminus I$ be homogeneous elements. By the maximality of I, both $I + (a)$ and $I + (b)$ meet S. Thus we find $i, j \in I$ and $f, g \in A$ such that $h := (i + fa)(j + gb) \in S$. But if ab were in I, then $h \in I$ which is not possible because $I \cap S = \emptyset$.

(3). It suffices to show that $\mathrm{rad}(I)$ is the intersection $\mathfrak{a}$ of all relevant prime ideals containing I (then $\mathrm{rad}(I)$ and hence $\mathrm{rad}(I)_+$ are homogeneous). We may replace A by A/I and can therefore assume $I = 0$. Clearly we have $\mathrm{rad}\, I \subseteq \mathfrak{a}$. Conversely, if $f \notin \mathrm{rad}(I)$, then an ideal maximal among those properly contained in A_+ and not meeting $\{1, f, f^2, \dots\}$ is a relevant prime ideal by (2). Thus f is not contained in the intersection of all relevant prime ideals. $\qquad\square$

Remark 13.3. More generally, fix integers $k, \delta \geq 1$ and set $A_{++} := \bigoplus_{d \geq k} A_{d\delta}$. This is again a homogeneous ideal (depending on k and δ). A homogeneous prime ideal $\mathfrak{p} \subset A$ is relevant if and only if $\mathfrak{p}$ does not contain A_{++} (if $x \in A_+ \setminus \mathfrak{p}$ is homogeneous, then $x^{k\delta} \in A_{++} \setminus \mathfrak{p}$). A similar proof (using more complicated notation depending on k and δ) then shows that all the assertions of Proposition 13.2 remain true if we replace $(\ )_+$ everywhere by $(\ )_{++}$.

(13.2) Projective spectrum.

Let A be a graded ring. We attach to A a scheme $\mathrm{Proj}\, A$, called the *projective spectrum of A* as follows.

$\mathrm{Proj}\, A$ AS A TOPOLOGICAL SPACE.

Let $\mathrm{Proj}\, A$ be the set of relevant prime ideals $\mathfrak{p}$ of A (i.e., of homogeneous prime ideals of A that do not contain A_+). We endow $\mathrm{Proj}\, A$ with a topology by taking as closed sets the sets
$$V_+(I) := \{\mathfrak{p} \in \mathrm{Proj}\, A \ ; \ \mathfrak{p} \supseteq I\},$$
where $I \subseteq A_+$ is a homogeneous ideal of A. Clearly we have $\bigcap_i V_+(I_i) = V_+(\sum_i I_i)$, $V_+(I) \cup V_+(J) = V_+(I \cap J)$, $V_+(A_+) = \emptyset$, and $V_+(0) = \mathrm{Proj}\, A$. Therefore these sets form indeed the closed sets of a topology on $\mathrm{Proj}\, A$. For a homogeneous element $f \in A_+$ define the open set $D_+(f) := (\mathrm{Proj}\, A) \setminus V_+(f)$. In other words, $D_+(f)$ is the set of $\mathfrak{p} \in \mathrm{Proj}\, A$ with $f \notin \mathfrak{p}$.

By construction, $\mathrm{Proj}\, A$ is a subset of $\mathrm{Spec}\, A$. If $\mathfrak{a} \subseteq A$ is an arbitrary ideal, then $V(\mathfrak{a}) \cap \mathrm{Proj}(A) = V_+(\mathfrak{a}^h)$, where $\mathfrak{a}^h$ is the homogeneous ideal generated by $\mathfrak{a}$. This shows that $\mathrm{Proj}\, A$ carries the topology induced from $\mathrm{Spec}\, A$.

Using Proposition 13.2 the same arguments as for the topology on $\mathrm{Spec}\, A$ show that analogous results of Sections (2.1) and (2.2) hold for $\mathrm{Proj}\, A$ by considering only relevant prime ideals instead of arbitrary prime ideals and only homogeneous ideals contained in A_+ instead of arbitrary ideals. Some statements are collected in the following proposition.

Proposition 13.4. *Let A be a graded ring. For a subset $Y \subseteq \mathrm{Proj}\, A$ define*
$$I_+(Y) := (\bigcap_{\mathfrak{p} \in Y} \mathfrak{p}) \cap A_+.$$

(1) If $I \subseteq A_+$ is a homogeneous ideal, then $I_+(V_+(I)) = \mathrm{rad}(I)_+$. If $Y \subseteq \mathrm{Proj}\, A$ is a subset, then $V_+(I_+(Y)) = \overline{Y}$.

(2) The maps $Y \mapsto I_+(Y)$ and $I \mapsto V_+(I)$ define mutually inverse, inclusion reversing bijections between the set of homogeneous ideals $I \subseteq A_+$ such that $I = \mathrm{rad}(I)_+$ and the set of closed subsets of $\mathrm{Proj}\, A$.

Via this bijection, the closed irreducible subsets correspond to ideals of the form $\mathfrak{p}_+$, where $\mathfrak{p}$ is a relevant prime ideal.

(3) If $I \subseteq A_+$ is a homogeneous ideal, then $V_+(I) = \emptyset$ if and only if $\mathrm{rad}(I)_+ = A_+$. In particular $\mathrm{Proj}\, A = \emptyset$ if and only if every element in A_+ is nilpotent.

(4) The sets $D_+(f)$ for homogeneous elements $f \in A_+$ form a basis of the topology of $\mathrm{Proj}\, A$.

(5) Let $(f_i)_i$ be a family of homogeneous elements $f_i \in A_+$ and let I be the ideal generated by the f_i. Then $\bigcup_i D_+(f_i) = \mathrm{Proj}\, A$ if and only if $\mathrm{rad}(I)_+ = A_+$.

The open subspaces $D_+(f)$ (for $f \in A_+$ homogeneous) can be described as follows. Note that $D_+(f) = D(f) \cap \mathrm{Proj}\, A$. Let us prove that the continuous map $D(f) = \mathrm{Spec}\, A_f \to \mathrm{Spec}\, A_{(f)}$, $\mathfrak{p} \mapsto \mathfrak{p} \cap A_{(f)}$ restricts to a homeomorphism

$$(13.2.1) \qquad \psi_f \colon D_+(f) \xrightarrow{\sim} \mathrm{Spec}\, A_{(f)}, \qquad \mathfrak{p} \mapsto \mathfrak{p}_{(f)}$$

Indeed, if $\mathfrak{q} \subseteq A_{(f)}$ is a prime ideal, set $\mathfrak{p}_n := \{\, a \in A_n \;;\; a^{\deg f}/f^n \in \mathfrak{q} \,\}$ for $n \geq 0$ and define $\theta_f(\mathfrak{q}) := \bigoplus_n \mathfrak{p}_n$. Using Proposition 13.2 (1) it is immediate that θ_f is an inverse to ψ_f. It remains to show that ψ_f is open. Let $g \in A_+$ be a homogeneous element with $D_+(g) \subseteq D_+(f)$ (equivalently, $g \in \mathrm{rad}(f)_+$). Then for $\mathfrak{p} \in D_+(f)$ we have

$$(13.2.2) \qquad \begin{aligned} \mathfrak{p} \in D_+(g) &\Leftrightarrow \exists\, r,s \geq 1 : g^r/f^s \notin \mathfrak{p} A_f \\ &\Leftrightarrow \forall\, r,s \geq 1 : g^r/f^s \notin \mathfrak{p} A_f \\ &\Leftrightarrow g_{(f)} := g^{\deg f}/f^{\deg g} \notin \mathfrak{p}_{(f)}. \end{aligned}$$

This shows $\psi_f(D_+(g)) = D(g_{(f)})$ and in particular that ψ_f is open.

$\mathrm{Proj}\, A$ AS A SCHEME.

We define a sheaf of rings on $X := \mathrm{Proj}\, A$ by

$$(13.2.3) \qquad \Gamma(D_+(f), \mathscr{O}_X) := A_{(f)}$$

for $f \in A_+$ homogeneous. For $g \in A_+$ homogeneous with $D_+(g) \subseteq D_+(f)$ we let the restriction homomorphism be the canonical homomorphism $A_{(f)} \to (A_{(f)})_{g_{(f)}} = A_{(g)}$ (with $g_{(f)}$ defined as in (13.2.2)). This defines a ringed space which we denote again by $\mathrm{Proj}\, A$.

Proposition 13.5. *The ringed space* $X = \mathrm{Proj}\, A$ *is a separated scheme.*

Proof. By definition we have $(D_+(f), \mathscr{O}_{X | D_+(f)}) \cong \mathrm{Spec}\, A_{(f)}$ for every homogeneous element $f \in A_+$. This shows that $\mathrm{Proj}\, A$ is a scheme. To prove that $\mathrm{Proj}\, A$ is separated, it suffices to remark that $D_+(f) \cap D_+(g) = D_+(fg)$ is affine and that the multiplication $A_{(f)} \otimes A_{(g)} \to A_{(fg)}$ is surjective (Proposition 9.15). $\qquad\square$

Let R be a ring and let A be a graded R-algebra, i.e., we are given a ring homomorphism $\varphi\colon R \to A_0$. For a homogeneous element $f \in A_+$, the ring $A_{(f)}$ is an R-algebra and for $D_+(g) \subseteq D_+(f)$ the restriction homomorphisms $A_{(f)} \to A_{(g)}$ are R-algebra homomorphisms. This shows that $\operatorname{Proj} A$ is a scheme over $\operatorname{Spec} R$. In particular, $\operatorname{Proj} A$ is always a scheme over $\operatorname{Spec} A_0$.

Example 13.6. The fundamental example is of course the projective space. Let R be a ring and $A = R[X_0, \ldots, X_n]$ the polynomial ring graded by the degree of homogeneous polynomials. Then $\mathbb{P}_R^n = \operatorname{Proj} R[X_0, \ldots, X_n]$ with $D_+(X_i) = \operatorname{Spec} R[X_0/X_i, \ldots, X_n/X_i]$ for all $i = 0, \ldots, n$.

FUNCTORIALITY OF $\operatorname{Proj} A$.

Let A be a graded ring. The first kind of functoriality is the observation that we may "thin out" A and "change A_0" without changing the scheme $\operatorname{Proj} A$. More precisely:

Remark 13.7. Fix integers $k, \delta \geq 1$ and define a new graded ring A' by $A'_0 = \mathbb{Z}$, $A'_d := 0$ for $0 < d < k$ and $A'_d := A_{d\delta}$ for $d \geq k$. By Remark 13.3, $\mathfrak{p} \mapsto \mathfrak{p} \cap A'$ defines a bijection $\operatorname{Proj} A \to \operatorname{Proj} A'$. For any homogeneous element $f \in A_+$ we find $f^{k\delta} \in A'$. But it is clear that $D_+(f) = D_+(f^{k\delta})$ and it is easy to see that $A_{(f)} = A'_{(f^{k\delta})}$. Thus we have an isomorphism

$$\operatorname{Proj} A \overset{\sim}{\to} \operatorname{Proj} A'.$$

The second kind of functoriality is with respect to homomorphisms of graded rings. Let R be a ring. If $\varphi\colon A \to B$ is a homomorphism of graded R-algebras, the inverse image of a relevant prime ideal of B is in general not a relevant prime ideal of A. Thus $\operatorname{Proj} A$ is not functorial in A with respect to arbitrary homomorphisms of graded R-algebras. But there is a unique morphism of R-schemes

$$\operatorname{Proj} \varphi\colon G(\varphi) \to \operatorname{Proj} A,$$

where $G(\varphi) \subseteq \operatorname{Proj} B$ is the open subscheme

$$(13.2.4) \qquad G(\varphi) := \{\, \mathfrak{q} \in \operatorname{Proj}(B) \; ; \; \varphi^{-1}(\mathfrak{q}) \not\supseteq A_+ \,\} = \bigcup_{f \in A_+ \text{ homogeneous}} D_+(\varphi(f))$$

such that the restriction of $\operatorname{Proj} \varphi$ to $D_+(\varphi(f))$ (for $f \in A_+$ homogeneous) is the morphism $D_+(\varphi(f)) \to D_+(f)$ corresponding to the morphism $A_{(f)} \to B_{(\varphi(f))}$ induced by φ.

The following properties follow directly from the definitions.

Proposition 13.8. *Let* $\varphi\colon A \to B$ *be a graded R-algebra homomorphism and denote by* $^a\varphi := \operatorname{Proj} \varphi\colon G(\varphi) \to \operatorname{Proj} A$ *the associated morphism of R-schemes.*
(1) *Let* $f \in A_+$ *homogeneous. Then* $^a\varphi^{-1}(D_+(f)) = D_+(\varphi(f))$. *In particular,* $^a\varphi$ *is an affine morphism.*
(2) *Let* φ *be surjective. Then* $G(\varphi) = \operatorname{Proj} B$ *and* $\operatorname{Proj} \varphi\colon \operatorname{Proj} B \to \operatorname{Proj} A$ *is a closed immersion which induces an isomorphism* $\operatorname{Proj} B \overset{\sim}{\to} \operatorname{Proj} A/\operatorname{Ker}(\varphi)$.

(13.3) Properties of $\operatorname{Proj} A$.

Let R be a ring and let A be a graded R-algebra. We are interested in properties of the R-scheme $\operatorname{Proj} A$.

Lemma 13.9.
(1) A subset $E \subseteq A_+$ of homogeneous elements generates A_+ as an ideal if and only if E generates A as an A_0-algebra.
(2) The ideal A_+ is finitely generated if and only if A is an A_0-algebra of finite type.
(3) The ring A is noetherian if and only if A_0 is noetherian and A is an A_0-algebra of finite type.

Proof. (1). The condition is clearly sufficient. Conversely, we show by induction on d that A_d is the A_0-module generated by products of elements in E of degree $\leq d$. As E generates A_+ as an ideal, we have $A_d = \sum_{i=0}^{d-1} A_i(E \cap A_{d-i})$. Applying the induction hypothesis we are done.

(2). A finite generating set $S \subseteq A_+$ of the ideal A_+ (resp. of the A_0-algebra A) exists if and only if a finite generating set $E \subseteq A_+$ of homogeneous elements of the ideal A_+ (resp. of the A_0-algebra A) exists: We can take for E the set of all homogeneous components of all elements of S. Therefore (2) follows from (1).

(3). The condition is sufficient by Hilbert's basis theorem (Proposition B.34). If A is noetherian, $A_0 = A/A_+$ is noetherian and A_+ is finitely generated. Thus A is an A_0-algebra of finite type by (2). □

Lemma 13.10. *Let A be a graded ring that is a finitely generated A_0-algebra and let M be a graded A-module that is a finitely generated A-module.*
(1) *For all $n \in \mathbb{Z}$ the A_0-module M_n is finitely generated and there exists an integer n_0 such that $M_n = 0$ for all $n \leq n_0$.*
(2) *There exists an integer n_1 and an integer $d > 0$ such that $A_d M_n = M_{n+d}$ for all $n \geq n_1$.*
(3) *For every integer $\delta > 0$ the algebra $A^{(\delta)} := \bigoplus_{d \geq 0} A_{d\delta}$ is a finitely generated A_0-algebra.*

The proof will show that we can choose for d in (2) the least common multiple of $(\deg f_i)_i$ if $f_1, \ldots, f_r \in A_+$ generate A as A_0-algebra.

Proof. Let $f_1, \ldots, f_r \in A_+$ be homogeneous elements that generate A as A_0-algebra and set $d_i := \deg f_i > 0$. Let $x_1, \ldots, x_s \in M$ be homogeneous elements that generate the A-module M and set $n_j := \deg x_j$. Then every element in M_n is an A_0-linear combination of elements of the form

$$f_1^{\alpha_1} \ldots f_r^{\alpha_r} x_j, \qquad n_j + \sum_i \alpha_i d_i = n, \ \alpha_i \geq 0.$$

This shows (1). Let d be the least common multiple of the d_i and set $g_i := f_i^{d/d_i}$. Then $\deg g_i = d$ for all i. Let E be the finite set of elements of the form $f_1^{\beta_1} \ldots f_r^{\beta_r} x_j$ with $0 \leq \beta_i < d/d_i$ for all i. Let n_1 be the maximal degree of the elements in E. Then for $n \geq n_1$ every element in M_{n+d} is a linear combination of elements in E where the coefficients are monomials of positive degree in the g_i. This shows (2).

Similarly as in the proof of (1) one sees that the ideal $A_+^{(\delta)}$ is finitely generated. This implies (3) by Lemma 13.9. □

To study the properties of the R-scheme $\operatorname{Proj} A$ we may replace A_0 by R without changing the R-scheme $\operatorname{Proj} A$ (Remark 13.7).

Remark 13.11. Let R be a ring and let A be a finitely generated graded R-algebra with $A_0 = R$. Then there exists a finitely generated graded R-algebra A' with $A'_0 = R$ which is generated by finitely many elements of A'_1 such that the R-schemes $\operatorname{Proj} A$ and $\operatorname{Proj} A'$ are isomorphic: Indeed, applying Lemma 13.10 (2) to $M_j = A$ we find an integer $\delta > 0$ and an integer $d_1 \geq 1$ such that $A_{d\delta} = (A_\delta)^d$ for all $d \geq d_1$. We set $A' := R \oplus \bigoplus_{d \geq 1}(A_\delta)^d$. By Remark 13.7 we find $\operatorname{Proj} A \cong \operatorname{Proj} A'$.

Proposition 13.12. *Let A be a graded R-algebra with $A_0 = R$ such that A_+ is generated by finitely many elements. Then there exists a closed immersion of R-schemes $\operatorname{Proj} A \hookrightarrow \mathbb{P}^n_R$ for some $n \geq 0$. In particular $\operatorname{Proj} A$ is of finite type over R. If R is noetherian, $\operatorname{Proj} A$ is noetherian.*

Proof. Using Remark 13.11 we may in addition assume that A is generated as R-algebra by finitely many elements $f_0, \ldots, f_n \in A_1$. Then the homomorphism of R-algebras $R[X_0, \ldots, X_n] \to A$, $X_i \mapsto f_i$, is surjective and graded. Thus Proposition 13.8 shows that we obtain a closed immersion $\operatorname{Proj} A \hookrightarrow \operatorname{Proj} R[X_0, \ldots, X_n] = \mathbb{P}^n_R$. $\square$

With the notion of projectivity introduced below in Definition 13.68, Proposition 13.12 means that $\operatorname{Proj} A$ is projective over R.

(13.4) Quasi-coherent sheaves attached to graded modules.

Let A be a graded ring and set $X = \operatorname{Proj} A$. Similarly as in the affine case, where the category of quasi-coherent $\mathscr{O}_{\operatorname{Spec} R}$-modules is equivalent to the category of R-modules, we will define a functor $M \mapsto \tilde{M}$ from the category of graded A-modules to the category of quasi-coherent $\mathscr{O}_X$-modules. However, this functor is not an equivalence of categories (see Theorem 13.20, though).

Let M be a graded A-module. If $f, g \in A_+$ are homogeneous elements we have by Proposition 13.4

$$D_+(g) \subseteq D_+(f) \Leftrightarrow V_+(f) \subseteq V_+(g) \Leftrightarrow g \in \operatorname{rad}(Af)_+ \Leftrightarrow f/1 \in (A_g)^\times.$$

Thus there is a unique A_f-module homomorphism $M_f \to M_g$ extending the identity on M. This homomorphism is graded and thus induces an $A_{(f)}$-module homomorphism $\rho_{(g,f)} \colon M_{(f)} \to M_{(g)}$.

Proposition 13.13. *Let M be a graded A-module. There exists on $X = \operatorname{Proj} A$ a unique quasi-coherent $\mathscr{O}_X$-module $\tilde{M}$ such that $\Gamma(D_+(f), \tilde{M}) = M_{(f)}$ for every homogeneous element $f \in A_+$ and such that the restriction maps are given by $\rho_{(g,f)}$.*

Proof. The uniqueness is clear because the $D_+(f)$ form a basis of the topology. Moreover, if $\tilde{M}$ exists, $\tilde{M}_{|D_+(f)}$ is the quasi-coherent module $(M_{(f)})^\sim$ on $D_+(f) = \operatorname{Spec} A_{(f)}$ corresponding to the $A_{(f)}$-module $M_{(f)}$. But it is easy to check that these sheaves can be glued together. $\square$

In particular we find $\tilde{A} = \mathscr{O}_X$.

Remark 13.14. For any homogeneous element $f \in A_+$ we have a functor $M \mapsto (M_{(f)})^\sim$ from the category of graded A-modules to the category of quasi-coherent $\mathscr{O}_{D_+(f)}$-modules which is exact and commutes with direct sums, filtered inductive limits, and tensor products (Proposition 7.14). This shows that $M \mapsto \tilde{M}$ defines an exact covariant functor from the category of graded A-modules to the category of quasi-coherent $\mathscr{O}_{\operatorname{Proj} A}$-modules which commutes with direct sums, filtered inductive limits, and tensor products.

This functor is not faithful: Let M be a graded A-module such that there exists an $n \in \mathbb{Z}$ with $M_d = 0$ for all $d \geq n$. Then $M_{(f)} = 0$ for all $f \in A_+$ homogeneous and hence $\tilde{M} = 0$.

Of vital importance are the quasi-coherent $\mathscr{O}_X$-modules $\mathscr{O}_X(n)$, called *Serre's twisting sheaves*, that are defined as follows. For $n \in \mathbb{Z}$ and a graded A-module M we define a new graded A-module $M(n)$ by $M(n)_d := M_{n+d}$ for all $d \in \mathbb{Z}$. In particular we have the graded A-module $A(n)$. Define

$$(13.4.1) \qquad\qquad\qquad \mathscr{O}_X(n) := \widetilde{A(n)}.$$

Let $d > 0$ be an integer and $f \in A_d$. Then for any multiple $n = dk$ of d, the map $A_{(f)} = (A_f)_0 \to A(n)_{(f)} = (A_f)_n$ which is given by multiplication with f^k is an isomorphism. This shows that the multiplication with f^{-k} yields an isomorphism

$$(13.4.2) \qquad\qquad\qquad \mathscr{O}_X(n)_{|D_+(f)} \overset{\sim}{\to} \mathscr{O}_{X|D_+(f)}.$$

If A_+ is generated by A_1, we have $\operatorname{Proj} A = \bigcup_{f \in A_1} D_+(f)$. Therefore we find:

Proposition 13.15. *Let A be a graded ring such that A_+ is generated by A_1 and set $X = \operatorname{Proj} A$. Then $\mathscr{O}_X(n)$ is an invertible $\mathscr{O}_X$-module for all $n \in \mathbb{Z}$.*

From now on we will make for our study of quasi-coherent modules on $X = \operatorname{Proj} A$ the following assumption:

$$A \text{ is generated as an } A_0\text{-algebra by } A_1.$$

Equivalently, by Lemma 13.9, A_+ is generated as ideal by A_1. We say then briefly that A *is generated in degree* 1. We can also express this by saying that there is an isomorphism of graded A_0-algebras $A \cong A_0[(T_i)_{i \in I}]/\mathfrak{a}$, where $A_0[(T_i)_i]$ is graded by the usual degree of polynomials and where $\mathfrak{a}$ is a graded ideal. If the ideal A_+ is generated by finitely many elements in A_1, we can choose I to be finite. If A is a finitely generated A_0-algebra (which is very often the case), the hypothesis that A is generated by A_1 is harmless by Remark 13.11.

Set $\varepsilon_{f,g} = f^{-n} g^n \in \Gamma(D_+(f) \cap D_+(g), \mathscr{O}_X)^\times = A_{(fg)}^\times$ for $f, g \in A_1$. Then $(\varepsilon_{f,g})$ is a Čech 1-cocycle on the open affine covering $\mathcal{U} = (D_+(f))_{f \in A_1}$ of $\operatorname{Proj} A$. The associated line bundle is $\mathscr{O}_X(n)$ (Section (11.6)). This also shows that we have for all $n, m \in \mathbb{Z}$ isomorphisms

$$(13.4.3) \qquad\qquad \mathscr{O}_X(m) \otimes_{\mathscr{O}_X} \mathscr{O}_X(n) \overset{\sim}{\to} \mathscr{O}_X(n+m).$$

There are homomorphisms of A_0-modules

$$(13.4.4) \qquad\qquad \alpha_n \colon A_n \to \Gamma(X, \mathscr{O}_X(n)) \qquad \text{for all } n \in \mathbb{Z}$$

that are defined for $f \in A_1$ by $\alpha_n(a)_{|D_+(f)} := a/f^n$ (these elements of $\Gamma(D_+(f), \mathscr{O}_X)$ differ on $D_+(f) \cap D_+(g)$ by g^n/f^n and therefore glue together to a section of $\Gamma(X, \mathscr{O}_X(n))$). Using the isomorphism (13.4.2) the homomorphism α_n can also be defined by

$$\alpha_n(a)_{|D_+(g)} := a/1 \in A(n)_{(g)} = \Gamma(D_+(g), \mathscr{O}_X(n))$$

for every homogeneous element $g \in A_+$.

Example 13.16. Let R be a ring, $A = R[T_0, \ldots, T_d]$ and $X = \operatorname{Proj} A = \mathbb{P}_R^d$. Then $\mathcal{O}_X(n)$ is the line bundle associated to the cocycle $(T_i^{-n} T_j^n)_{i,j}$ on $(D_+(T_i))_i$. If $R = k$ is a field, we recover the line bundles $\mathcal{O}_X(n)$ defined in Example 11.45. The calculation of $\Gamma(X, \mathcal{O}_X(n))$ in that example did use only the cocycle description of $\mathcal{O}_X(n)$ and thus works for an arbitrary ring R. It shows that for an arbitrary ring R and all $n \in \mathbb{Z}$ the homomorphism α_n (13.4.4) is an isomorphism

$$(13.4.5) \qquad R[T_0, \ldots, T_d]_n \xrightarrow{\sim} \Gamma(\mathbb{P}_R^d, \mathcal{O}_{\mathbb{P}_R^d}(n)).$$

Example 13.17. Let R be an integral domain and let $f \in A := R[T_0, \ldots, T_d]$ be a nonzero homogeneous polynomial of degree $n > 0$. Let $V_+(f) \subset P := \mathbb{P}_R^d$ be the corresponding closed subscheme. Then the multiplication with f yields an injective homomorphism $A(-n) \to A$ of graded A-modules with cokernel $A/(f)$. Passing to the attached quasi-coherent modules we obtain an exact sequence of quasi-coherent $\mathcal{O}_P$-modules

$$(13.4.6) \qquad 0 \to \mathcal{O}_P(-n) \to \mathcal{O}_P \to \mathcal{O}_{V_+(f)} \to 0,$$

where we consider $\mathcal{O}_{V_+(f)}$ as a quasi-coherent $\mathcal{O}_P$-module with support in $V_+(f)$.

Remark 13.18. Recall that in Section (7.11) for an invertible sheaf $\mathcal{L}$ on a scheme X and for a section $s \in \Gamma(X, \mathcal{L})$ we defined the open subset $X_s(\mathcal{L})$ as the set of points $x \in X$ where $s(x) \neq 0$. If $X = \operatorname{Proj} A$ for a graded ring A and $f \in A_n$, then we have

$$(13.4.7) \qquad D_+(f) = X_{\alpha_n(f)}(\mathcal{O}_X(n)).$$

Indeed, let us call the right hand side X_f. We have $X = \bigcup_{g \in A_1} D_+(g)$. Therefore (13.4.7) is implied by the equality (for all $g \in A_1$)

$$X_f \cap D_+(g) = \{\, x \in \operatorname{Spec}(A_{(g)}) \; ; \; (f/g^n)(x) \neq 0 \,\} = D_+(f) \cap D_+(g).$$

(13.5) Graded modules attached to quasi-coherent sheaves.

We continue to assume that all graded rings A are generated in degree 1. Set $X = \operatorname{Proj} A$. For every $\mathcal{O}_X$-module $\mathcal{F}$ we define

$$(13.5.1) \qquad \mathcal{F}(n) := \mathcal{F} \otimes_{\mathcal{O}_X} \mathcal{O}_X(n).$$

Then (13.4.3) implies that $\mathcal{F}(n) \otimes \mathcal{O}_X(m) \cong \mathcal{F}(n+m)$. For each graded A-module M we then have $\widetilde{M}(n) = \widetilde{M(n)}$.

We now define a functor from the category of $\mathcal{O}_X$-modules to the category of graded A-modules by

$$(13.5.2) \qquad \Gamma_*(\mathcal{F}) := \bigoplus_{n \in \mathbb{Z}} \Gamma(X, \mathcal{F}(n)).$$

For $a \in A_d$ and $x \in \Gamma(X, \mathcal{F}(n))$ define $ax \in \Gamma(X, \mathcal{F}(n+d))$ as the image of $\alpha_d(a) \otimes x$ in $\Gamma(X, \mathcal{O}_X(d) \otimes \mathcal{F}(n)) = \Gamma(X, \mathcal{F}(n+d))$. This makes $\Gamma_*(\mathcal{F})$ into a graded A-module and we obtain the desired functor.

Remark 13.19. If A_+ is generated by finitely many elements of A_1, the projective spectrum $X = \operatorname{Proj} A$ is isomorphic to a closed subscheme of $\mathbb{P}^n_{A_0}$ (Proposition 13.12), in particular X is quasi-compact and separated. Let $f \in A_d$ and let $\mathscr{F}$ be a quasi-coherent $\mathscr{O}_X$-module. Then there is an isomorphism

$$(13.5.3) \qquad\qquad \sigma \colon \Gamma(D_+(f), \mathscr{F}) \xrightarrow{\sim} \Gamma_*(\mathscr{F})_{(f)}.$$

Indeed, set $s := \alpha_d(f) \in \Gamma(X, \mathscr{O}_X(d))$. Then we have $X_s(\mathscr{O}_X(d)) = D_+(f)$ (13.4.7) and for $t' \in \Gamma(X_s(\mathscr{O}_X(d)), \mathscr{F})$ there exists by Theorem 7.22 (2) an integer $n > 0$ such that $t' \otimes s^{\otimes n}$ can be extended to a section $t \in \Gamma(X, \mathscr{F}(nd))$. We set $\sigma(t') := t/f^n \in \Gamma_*(\mathscr{F})_{(f)}$. Then σ is well defined and injective by Theorem 7.22 (1). The surjectivity of σ is clear.

We will now define functorial homomorphisms $\alpha \colon M \to \Gamma_*(\tilde{M})$ and $\beta \colon \Gamma_*(\mathscr{F})^\sim \to \mathscr{F}$. We start with the definition of α for a graded A-module M. For this we generalize the homomorphism (13.4.4) to a homomorphism of graded A-modules

$$(13.5.4) \qquad\qquad \alpha = \alpha_M \colon M \to \Gamma_*(\tilde{M})$$

that is defined as follows. Fix $x \in M_n$. For all $f \in A_d$ define $\alpha^f_n(x) := x/1 \in (M_{(f)})_n = M(n)_{(f)} = \Gamma(D_+(f), \tilde{M}(n))$. It is easy to check that these local sections glue together to a section $\alpha_n(x) \in \Gamma(X, \tilde{M}(n))$. We obtain a homomorphism of abelian groups $\alpha_n \colon M_n \to \Gamma(X, \tilde{M}(n))$, and $\alpha = \bigoplus \alpha_n$ is a homomorphism of graded A-modules. Clearly, α is functorial in M.

To define β, let $\mathscr{F}$ be an $\mathscr{O}_X$-module and let M be the A-module $\Gamma_*(\mathscr{F})$. Let $f \in A_d$ ($d > 0$) and define a homomorphism of $A_{(f)}$-modules
$$(13.5.5)$$
$$\beta^f \colon \Gamma(D_+(f), \tilde{M}) = M_{(f)} \to \Gamma(D_+(f), \mathscr{F}), \quad x/f^n \mapsto (x_{|D_+(f)})(\alpha_{dn}(f^n)_{|D_+(f)})^{-1}.$$

Here we have $x \in M_{nd} = \Gamma(X, \mathscr{F}(nd))$ and $\alpha_{dn}(f^n)$ is a section in $\Gamma(X, \mathscr{O}_X(nd))$ whose restriction to $D_+(f) = D_+(f^n)$ is invertible by (13.4.7). It is easy to see that these homomorphisms are compatible with restrictions from $D_+(f)$ to $D_+(fg)$ for all $g \in A_e$ ($e > 0$) and we obtain a functorial homomorphism of $\mathscr{O}_X$-modules

$$(13.5.6) \qquad\qquad \beta = \beta_{\mathscr{F}} \colon \Gamma_*(\mathscr{F})^\sim \to \mathscr{F}.$$

We cannot expect α and β to be isomorphisms in general: Otherwise the functors $M \mapsto \tilde{M}$ and $\mathscr{F} \to \Gamma_*(\mathscr{F})$ would be equivalences of categories; but we have already seen in Remark 13.14 that $M \mapsto \tilde{M}$ will never be faithful (except in trivial cases). But if we restrict to a subcategory of the category of graded A-modules (and to the case that $X = \operatorname{Proj} A$ is quasi-compact) we obtain an equivalence. Thus call a graded A-module M *saturated* if $\alpha \colon M \to \Gamma_*(\tilde{M})$ is an isomorphism and let $(A\text{-GrMod}^{\mathrm{sat}})$ be the full subcategory of saturated graded A-modules. Let $(X\text{-QCoh})$ be the category of quasi-coherent $\mathscr{O}_X$-modules.

Theorem 13.20. *Let A be a graded ring such that A_+ is generated by finitely many elements of A_1 and let $X = \operatorname{Proj} A$. Then the functors*

$$(A\text{-GrMod}^{\mathrm{sat}}) \underset{\Gamma_*(\mathscr{F}) \,\leftarrow\!\shortmid\, \mathscr{F}}{\overset{M \mapsto \tilde{M}}{\rightleftarrows}} (X\text{-QCoh})$$

are mutually quasi-inverse and define therefore an equivalence of categories.

Proof. We first show that $\beta\colon \Gamma_*(\mathscr{F})^\sim \to \mathscr{F}$ is an isomorphism for all quasi-coherent $\mathscr{O}_X$-modules $\mathscr{F}$. But β^f (13.5.5) is by definition the inverse of the isomorphism (13.5.3) for all $f \in A_d$. This shows that β is an isomorphism.

It remains to show that $M := \Gamma_*(\mathscr{F})$ is a saturated A-module for every quasi-coherent $\mathscr{O}_X$-module $\mathscr{F}$. But it is immediate that the composition

$$ M \xrightarrow{\ \alpha\ } \Gamma_*(\tilde{M}) \xrightarrow{\ \Gamma_*(\beta)\ } M $$

is the identity. As we have already seen that $\Gamma_*(\beta)$ is an isomorphism, α is an isomorphism. $\qquad\square$

Remark 13.21. Assume that A is noetherian and that A_+ is generated by finitely many elements of A_1. Then $X = \operatorname{Proj} A$ is a noetherian scheme (Proposition 13.12). In Volume II we will see that one can describe the category of coherent $\mathscr{O}_X$-modules as follows. Let $\mathcal{C}$ be the abelian category of finitely generated graded A-modules and let $\mathcal{C}'$ be the full subcategory of modules in $\mathcal{C}$ which are finitely generated as A_0-modules. Then $M \mapsto \tilde{M}$ yields an equivalence between the quotient category $\mathcal{C}/\mathcal{C}'$ (in the sense of [Sch] 19.5.4)) and the abelian category of coherent sheaves on X.

Proposition 13.22. *Let A be a graded ring such that A_+ is generated by $f_1, \ldots, f_r \in A_1$, let $X = \operatorname{Proj} A$ and let $\mathscr{F}$ be a quasi-coherent $\mathscr{O}_X$-module of finite type. Then there exists an $n_0 \in \mathbb{Z}$ such that the following assertion holds. For all $n \geq n_0$ there exists a surjection $\mathscr{O}_X^k \to \mathscr{F}(n)$ for some $k \geq 0$ (dependent on n).*

Proof. For all $i = 1, \ldots, r$ we may choose finitely many sections $t'_{ij} \in \Gamma(D_+(f_i), \mathscr{F})$ that generate $\mathscr{F}_{|D_+(f_i)}$. By Theorem 7.22 we find an integer n_0 (which we may choose to be independent of i and j) such that $f_i^n t'_{ij}$ extends to a section $t_{ij} \in \Gamma(X, \mathscr{F}(n))$ for all $n \geq n_0$. These global sections define a homomorphism $u\colon \mathscr{O}_X^k \to \mathscr{F}(n)$ whose restriction to $D_+(f_i)$ is surjective for all i. Hence u is surjective. $\qquad\square$

(13.6) Closed subschemes of projective spectra.

Lemma 13.23. *Let A be a graded ring such that A_+ is generated by finitely many elements of A_1 and set $X = \operatorname{Proj} A$. Let M be a graded A-module and let $\mathscr{G} \subseteq \tilde{M}$ be a quasi-coherent $\mathscr{O}_X$-submodule. Then there exists a homogeneous submodule $N \subseteq M$ such that $\tilde{N} = \mathscr{G}$. If M is saturated, we can choose N to be saturated.*

Proof. Let $\alpha\colon M \to \Gamma_*(\tilde{M})$ be the homomorphism (13.5.4) and let P be its image. Then $\tilde{P}$ is the image of $\tilde{\alpha}$ which is an isomorphism $\tilde{M} \to \Gamma_*(\tilde{M})^\sim$ and hence $\tilde{P} = \tilde{M}$.

Set $N := \alpha^{-1}(\Gamma_*(\mathscr{G}))$. The restriction of α to N yields an injective homomorphism $\tilde{N} \to \Gamma_*(\mathscr{G})^\sim = \mathscr{G}$ whose image is $(\Gamma_*(\mathscr{G}) \cap P)^\sim = \mathscr{G} \cap \tilde{P} = \mathscr{G}$ because the functor $(\ )^\sim$ is exact.

If M is saturated, i.e., α is an isomorphism, then $N \to \Gamma_*(\mathscr{G}) = \Gamma_*(\tilde{N})$ is an isomorphism. $\qquad\square$

Proposition 13.24. *Let A be a graded ring such that A_+ is generated by finitely many elements of A_1, let $X = \operatorname{Proj} A$ and let $Z \subseteq X$ be a closed subscheme of X. Then there exists a homogeneous ideal $I_Z \subseteq A$ not containing A_+ such that $Z = \operatorname{Proj} A/I_Z$. If A is saturated as a graded A-module, we may choose I_Z to be saturated and then I_Z is the unique saturated homogeneous ideal of A such that $Z = \operatorname{Proj} A/I_Z$.*

Proof. Let $\mathscr{I} \subseteq \mathscr{O}_X$ be the quasi-coherent ideal defining Z and let $I_Z \subseteq A$ be the homogeneous ideal such that $\tilde{I} = \mathscr{I}$ constructed in Lemma 13.23. Then $Z = \operatorname{Proj} A/I_Z$ by Proposition 13.8. If A is saturated, then I_Z can be chosen to be saturated by Lemma 13.23. The uniqueness assertion follows from Theorem 13.20. $\qquad\qquad\qquad\qquad\square$

Remark 13.25. By going through the construction of the ideal I_Z in the proof of Proposition 13.24 one easily checks that I_Z has the following more explicit description. Let $f_1, \dots, f_n \in A_1$ be generators of A_+. Then $Z \cap D_+(f_i)$ is a closed subscheme of $D_+(f_i)$ and corresponds to an ideal $\mathfrak{a}_i \subseteq A_{(f_i)}$. For all $d \geq 0$ we set

$$I_d := \{ \, a \in A_d \; ; \; a/f_i^d \in \mathfrak{a}_i \text{ for all } i \, \}.$$

Then $I_Z = \bigoplus_d I_d$.

By far the most important examples are closed subschemes of projective space. Here we can say more and give some geometric interpretations:

Remark 13.26. For $A = R[T_0, \dots, T_n]$ for a ring R (and thus $\operatorname{Proj} A = \mathbb{P}_R^n$), then A is saturated as a graded A-module by Example 13.16. For any closed subscheme Z of $\mathbb{P}_R^n$ we have $Z = \operatorname{Proj} A/J$ where $J \subseteq A$ is a homogeneous ideal. By Proposition 13.24 there exists a unique saturated ideal $I = I_Z$ such that $Z = \operatorname{Proj} A/I$. It is called the *saturation of J*. It can be described explicitly as

$$I = J^{\mathrm{sat}} := \{ \, f \in A \; ; \; \exists N \geq 0 \; \forall i : T_i^N f \in J \, \}.$$

As I is saturated, α_I yields for all $d \in \mathbb{Z}$ an isomorphism

$$(13.6.1) \qquad\qquad\qquad I_d \overset{\sim}{\to} \Gamma(\mathbb{P}_R^n, \mathscr{I}(d)),$$

where $\mathscr{I} \subseteq \mathscr{O}_{\mathbb{P}_R^n}$ is the quasi-coherent ideal defining Z.

Consider now the case that $R = k$ is a field. Then we may express Proposition 13.24 by saying that any closed subscheme Z of $\mathbb{P}_k^n$ is the intersection of hypersurfaces. Moreover, the finite-dimensional k-vector space I_d consists of those homogeneous polynomials of degree d in $k[T_0, \dots, T_n]$ that vanish on Z. Thus a nonzero element of I_d defines a hypersurface of $\mathbb{P}_k^n$ of degree d containing Z and two such elements define the same hypersurface if and only if they differ by an element of $k^\times$. We obtain a bijection

$$(13.6.2) \qquad \left\{ \begin{matrix} \text{hypersurfaces of degree } d \\ \text{containing } Z \end{matrix} \right\} \leftrightarrow (\Gamma(\mathbb{P}_k^n, \mathscr{I}(d)) \setminus \{0\})/k^\times$$

(13.7) The projective spectrum of a quasi-coherent algebra.

Similarly as the construction of the (affine) spectrum Spec can be globalized from a ring to a quasi-coherent algebra (Section (11.2)), we will now globalize the construction of the projective spectrum Proj to graded quasi-coherent algebras. Then the previous results about Proj A and about quasi-coherent modules on Proj A can be generalized to this globalized situation. Usually we will state only the results but will not give details of the proofs.

Thus let S be a scheme. A *graded quasi-coherent $\mathscr{O}_S$-algebra* is a quasi-coherent $\mathscr{O}_S$-algebra $\mathscr{A}$ together with a decomposition into $\mathscr{O}_S$-submodules $\mathscr{A} = \bigoplus_{d \geq 0} \mathscr{A}_d$ such that $\mathscr{A}_d \mathscr{A}_e \subseteq \mathscr{A}_{d+e}$ for all integers $d, e \geq 0$.

We say that a graded quasi-coherent $\mathscr{O}_S$-algebra $\mathscr{A}$ is *of finite type* if there exists an open affine covering $(U_i)_i$ of S such that $\Gamma(U_i, \mathscr{A})$ is a finitely generated $\Gamma(U_i, \mathscr{O}_S)$-algebra for all i (equivalently, $\Gamma(U, \mathscr{A})$ is a finitely generated $\Gamma(U, \mathscr{O}_S)$-algebra for all open affine subschemes U of S). Moreover, $\mathscr{A}$ is said to be *generated by $\mathscr{A}_1$* if there exists an open affine covering $(U_i)_i$ of S such that the $\Gamma(U_i, \mathscr{O}_S)$-algebra $\Gamma(U_i, \mathscr{A})$ is generated by $\Gamma(U_i, \mathscr{A}_1)$ (equivalently, the $\Gamma(U, \mathscr{O}_S)$-algebra $\Gamma(U, \mathscr{A})$ is generated by $\Gamma(U, \mathscr{A}_1)$ for all open affine subschemes U of S).

Let $\mathscr{A}$ be an arbitrary graded quasi-coherent $\mathscr{O}_S$-algebra. Then for every open affine subscheme $U \subseteq S$,

$$\Gamma(U, \mathscr{A}) = \bigoplus_{d \geq 0} \Gamma(U, \mathscr{A}_d)$$

is a graded algebra over $\Gamma(U, \mathscr{O}_S)$ and we obtain a separated morphism of schemes $\pi_U \colon \operatorname{Proj} \Gamma(U, \mathscr{A}) \to U$. One checks that there is an S-scheme

$$\pi \colon \operatorname{Proj} \mathscr{A} \to S$$

together with U-isomorphisms $\eta_U \colon \pi^{-1}(U) \xrightarrow{\sim} \operatorname{Proj} \Gamma(U, \mathscr{A})$ for all open affine subschemes $U \subseteq S$. Moreover, the S-scheme $\operatorname{Proj} \mathscr{A}$ together with the isomorphisms η_U are unique up to unique isomorphism of S-schemes. We omit the details. The morphism π is always separated (as this can be checked locally on S). The S-scheme $\operatorname{Proj} \mathscr{A}$ is called the *projective spectrum of the graded quasi-coherent $\mathscr{O}_S$-algebra $\mathscr{A}$*.

Set $X := \operatorname{Proj} \mathscr{A}$. Again there is an exact functor $\mathscr{M} \mapsto \widetilde{\mathscr{M}}$ from the category of graded quasi-coherent $\mathscr{A}$-modules to the category of quasi-coherent $\mathscr{O}_X$-modules. In particular, we can define the $\mathscr{O}_X$-modules $\mathscr{O}_X(n) := \mathscr{A}(n)^\sim$.

Remark 13.27. The formation of $\operatorname{Proj} \mathscr{A}$ is compatible with base change: Let $g \colon S' \to S$ be a morphism of schemes. Then we have

(13.7.1) $$\operatorname{Proj}(g^* \mathscr{A}) \cong (\operatorname{Proj} \mathscr{A}) \times_S S'.$$

Let $g' \colon X' := \operatorname{Proj}(g^* \mathscr{A}) \to X := \operatorname{Proj} \mathscr{A}$ be the projection and let $\mathscr{M}$ be a graded quasi-coherent $\mathscr{A}$-module. Then $(g^* \mathscr{M})^\sim \cong g'^*(\widetilde{\mathscr{M}})$. In particular $\mathscr{O}_{X'}(n) \cong g'^* \mathscr{O}_X(n)$ for all $n \in \mathbb{Z}$.

Proposition 13.28. Let $\mathscr{L}$ be an invertible $\mathscr{O}_S$-module and let $\mathscr{A}_{\mathscr{L}}$ be the graded quasi-coherent $\mathscr{O}_S$-algebra $\bigoplus_{d \geq 0} \mathscr{A}_d \otimes \mathscr{L}^{\otimes d}$. Set $X_{\mathscr{L}} := \operatorname{Proj} \mathscr{A}_{\mathscr{L}}$ and let $\pi' \colon X_{\mathscr{L}} \to S$ be the structure morphism. There is an isomorphism of S-schemes

$$g_{\mathscr{L}} \colon X_{\mathscr{L}} \xrightarrow{\sim} X = \operatorname{Proj} \mathscr{A}.$$

such that $\mathscr{O}_{X_{\mathscr{L}}}(n) = g_{\mathscr{L}}^* \mathscr{O}_X(n) \otimes \pi'^*(\mathscr{L}^{\otimes n})$ for all $n \in \mathbb{Z}$.

Proof. If $\mathscr{L} = \mathscr{O}_S$, $g_{\mathscr{O}_S}$ is given by the natural isomorphism $\mathscr{A} \xrightarrow{\sim} \mathscr{A}_{\mathscr{L}}$. In general, we set $X' = X_{\mathscr{L}}$ and $\mathscr{A}' = \mathscr{A}_{\mathscr{L}}$. Let $(U_i)_i$ be an open covering of S such that there exist isomorphism $t_i \colon \mathscr{L}_{|U_i} \xrightarrow{\sim} \mathscr{O}_{S|U_i}$ for all i. Set $\mathscr{A}_i := \mathscr{A}_{|U_i}$ and $\mathscr{A}'_i := \mathscr{A}'_{|U_i}$. Identifying $\mathscr{L}_{|U_i}$ and $\mathscr{O}_{S|U_i}$ via t_i we obtain isomorphisms $\varphi_i \colon \mathscr{A}_i \xrightarrow{\sim} \mathscr{A}'_i$. Any other choice of isomorphisms $\mathscr{L}_{|U_i} \xrightarrow{\sim} \mathscr{O}_{S|U_i}$ differs from t_i by an invertible section $f_i \in \Gamma(U_i, \mathscr{O}_S^\times)$ and $\mathscr{L}$ is given by the Čech 1-cocycle $(f_i f_j^{-1})$. The corresponding isomorphism $\mathscr{A}_i \to \mathscr{A}'_i$ differs from φ_i by multiplication with f_i^d in degree d. This does not change the induced isomorphism $g_i \colon \operatorname{Proj} \mathscr{A}'_i \xrightarrow{\sim} \operatorname{Proj} \mathscr{A}_i$. Hence these local isomorphisms glue together to a global isomorphism $g_{\mathscr{L}}$. On the $\mathscr{O}_{X'}$-module $\mathscr{O}_{X'}(n)$ the change of the t_i induces multiplication with f_i^n. Thus $\mathscr{O}_{X'}(n)$ differs from $g_{\mathscr{L}}^* \mathscr{O}_X(n)$ by an invertible sheaf given by the cocycle $(\pi_i'^*(f_i^n)\pi_j'^*(f_j^{-n}))$, where $\pi_i' \colon \operatorname{Proj} \mathscr{A}'_i \to U_i$ is the structure morphism. This means $\mathscr{O}_{X_{\mathscr{L}}}(n) = g_{\mathscr{L}}^* \mathscr{O}_X(n) \otimes \pi'^*(\mathscr{L}^{\otimes n})$. $\qquad\square$

If $\mathscr{A}$ is a graded quasi-coherent $\mathscr{O}_S$-algebra generated by $\mathscr{A}_1$, set $X = \operatorname{Proj} \mathscr{A}$, and let $\pi \colon X \to S$ be the structure morphism. Then $\mathscr{O}_X(n)$ is invertible for all $n \in \mathbb{Z}$ and there is a homomorphism of $\mathscr{A}_0$-modules

$$(13.7.2) \qquad\qquad \alpha_n \colon \mathscr{A}_n \to \pi_*(\mathscr{O}_X(n))$$

which is a globalization of (13.4.4).

Again there is a functor $\mathscr{F} \mapsto \Gamma_*(\mathscr{F})$ from the category of quasi-coherent $\mathscr{O}_X$-modules to the category of graded $\mathscr{A}$-modules given by

$$(13.7.3) \qquad\qquad \Gamma_*(\mathscr{F}) := \bigoplus_{n \in \mathbb{Z}} \pi_* \mathscr{F}(n),$$

where $\mathscr{F}(n) = \mathscr{F} \otimes_{\mathscr{O}_X} \mathscr{O}_X(n)$.

The results of Section (13.5) can be globalized as follows.

Theorem 13.29. *Let S be a scheme and let $\mathscr{A}$ be a graded quasi-coherent $\mathscr{O}_S$-algebra generated by $\mathscr{A}_1$, where $\mathscr{A}_1$ is an $\mathscr{O}_S$-module of finite type. Set $X = \operatorname{Proj} \mathscr{A}$, and let $\pi \colon X \to S$ be the structure morphism.*

(1) *For every quasi-coherent $\mathscr{O}_X$-module $\mathscr{F}$ there exists a functorial isomorphism of $\mathscr{O}_X$-modules $\Gamma_*(\mathscr{F})^\sim \xrightarrow{\sim} \mathscr{F}$.*

(2) *Let S be quasi-compact and let $\mathscr{F}$ be a quasi-coherent $\mathscr{O}_X$-module of finite type. Then there exists an integer n_0 such that for all $n \geq n_0$ the canonical homomorphism $\pi^*(\pi_* \mathscr{F}(n)) \to \mathscr{F}(n)$ (7.8.10) is surjective.*

To relate Theorem 13.29 (2) and Proposition 13.22 we use the following result.

Proposition 13.30. *Let $f \colon X \to S$ be a qcqs morphism and let $\mathscr{F}$ be a quasi-coherent $\mathscr{O}_X$-module. Then the following assertions are equivalent.*

(i) *The canonical homomorphism $f^*(f_* \mathscr{F}) \to \mathscr{F}$ (7.8.10) is surjective.*

(ii) *There is a quasi-coherent $\mathscr{O}_S$-module $\mathscr{G}$ and a surjective homomorphism $f^* \mathscr{G} \twoheadrightarrow \mathscr{F}$.*

(iii) *For every open affine subscheme $U \subseteq S$, $\mathscr{F}_{|f^{-1}(U)}$ is generated by its global sections over $f^{-1}(U)$.*

If S is qcqs and if $\mathscr{F}$ is of finite type, these assertions are equivalent to

(ii') *There exists an $\mathscr{O}_S$-module $\mathscr{G}$ of finite presentation and a surjective homomorphism $f^* \mathscr{G} \twoheadrightarrow \mathscr{F}$.*

Proof. By Corollary 10.27, $f_*\mathscr{F}$ is quasi-coherent. Therefore (i) implies (ii). To show that (ii) implies (iii), we may assume that S is affine. Then $\mathscr{G}$ is generated by its global sections. Therefore $f^*\mathscr{G}$ and in particular $\mathscr{F}$ is generated by its global sections. To show that (iii) implies (i) we may assume that S is affine and that $\mathscr{F}$ is generated by its global sections. In other words we find a surjection $v\colon f^*(\mathscr{O}_S^{(I)}) = \mathscr{O}_X^{(I)} \twoheadrightarrow \mathscr{F}$. But v factors in $f^*(\mathscr{O}_S^{(I)}) \to f^*(f_*\mathscr{F}) \to \mathscr{F}$ which shows (i).

It remains to show that (ii) implies (ii') if S is qcqs and if $\mathscr{F}$ is of finite type. By Corollary 10.50 we can write $\mathscr{G}$ as filtered inductive limit of $\mathscr{O}_S$-modules $\mathscr{G}_i$ of finite presentation. Then Lemma 10.47 implies (ii'). $\qquad\square$

(13.8) Projective bundles as projective spectra.

We have
$$\mathbb{P}_S^n = \operatorname{Proj}\mathscr{O}_S[X_0,\dots,X_n] = \operatorname{Proj}\operatorname{Sym}((\mathscr{O}_S^{n+1})^\vee).$$

On the other hand, we also have seen in Section (8.8) that

$$\mathbb{P}_S^n = \mathbb{P}((\mathscr{O}_S^{n+1})^\vee).$$

We will generalize this identification now. Let $\mathscr{E}$ be a quasi-coherent $\mathscr{O}_S$-module. Its symmetric algebra $\operatorname{Sym}\mathscr{E}$ is a graded quasi-coherent $\mathscr{O}_S$-algebra which is generated in degree 1. Let $X := \operatorname{Proj}(\operatorname{Sym}\mathscr{E})$ be its projective spectrum and $\pi\colon X \to S$ the structure morphism.

The S-scheme $\operatorname{Proj}(\operatorname{Sym}\mathscr{E})$ is not changed if we tensor $\mathscr{E}$ with an invertible $\mathscr{O}_S$-module:

Lemma 13.31. *Let $\mathscr{L}$ be an invertible $\mathscr{O}_S$-module. Then there is an S-isomorphism $i\colon X := \operatorname{Proj}(\operatorname{Sym}\mathscr{E}) \xrightarrow{\sim} Y := \operatorname{Proj}(\operatorname{Sym}(\mathscr{E}\otimes\mathscr{L}))$ such that $i^*\mathscr{O}_Y(n) = \mathscr{O}_X(n)\otimes\pi^*(\mathscr{L}^{\otimes n})$ for all $n \in \mathbb{Z}$.*

In particular we have $\operatorname{Proj}(\operatorname{Sym}\mathscr{L}) \cong \operatorname{Proj}(\operatorname{Sym}\mathscr{O}_S) = S$.

Proof. There is a homomorphism $\operatorname{Sym}^n(\mathscr{E} \otimes \mathscr{L}) \to \operatorname{Sym}^n\mathscr{E} \otimes \mathscr{L}^{\otimes n}$ given on sections over an open subset U by

$$(x_1 \otimes t_1)\dots(x_n \otimes t_n) \mapsto (x_1 x_2 \cdots x_n) \otimes (t_1 \otimes \cdots \otimes t_n).$$

This is an isomorphism: We can work locally on S and assume that $\mathscr{L} = \mathscr{O}_S$; then this is clear. Therefore the Lemma follows from Proposition 13.28. $\qquad\square$

The homomorphism $\alpha_1\colon \operatorname{Sym}^1\mathscr{E} = \mathscr{E} \to \pi_*\mathscr{O}_X(1)$ (13.7.2) yields via adjointness (Proposition 7.11) a homomorphism

$$(13.8.1) \qquad\qquad \alpha_1^\sharp\colon \pi^*\mathscr{E} \to \mathscr{O}_X(1)$$

of $\mathscr{O}_X$-modules. We claim that $\alpha_1^\sharp$ is surjective. Indeed, let $u\colon \mathscr{E} \otimes_{\mathscr{O}_S} \operatorname{Sym}\mathscr{E} \to (\operatorname{Sym}\mathscr{E})(1)$ be the canonical homomorphism of graded $(\operatorname{Sym}\mathscr{E})$-modules. Then we have $\alpha_1^\sharp = \tilde{u}$. Clearly u is surjective. This proves our claim.

On the other hand, recall the S-scheme $p\colon \mathbb{P}(\mathscr{E}) \to S$ from Section (8.8): By definition there is a surjective homomorphism $p^*\mathscr{E} \twoheadrightarrow \mathscr{L}_{\mathrm{univ}}$ where $\mathscr{L}_{\mathrm{univ}}$ is the "universal quotient line bundle of $\mathscr{E}$". More precisely, for every S-scheme $f\colon T \to S$ the map

$$\mathrm{Hom}_S(T, \mathbb{P}(\mathscr{E})) \to \{\, \mathscr{H} \subset f^*\mathscr{E} \ \mathscr{O}_T\text{-submodule} \ ; \ f^*\mathscr{E}/\mathscr{H} \text{ is a line bundle} \,\},$$
$$u \mapsto \mathrm{Ker}(u^* p^* \mathscr{E} \twoheadrightarrow u^* \mathscr{L}_{\mathrm{univ}})$$

is bijective. In particular, we see that the surjective homomorphism (13.8.1) corresponds to a morphism

$$(13.8.2) \qquad\qquad\qquad r_{\mathscr{E}} \colon \mathrm{Proj}(\mathrm{Sym}\,\mathscr{E}) \to \mathbb{P}(\mathscr{E})$$

of S-schemes such that $r_{\mathscr{E}}^* \mathscr{L}_{\mathrm{univ}} = \mathscr{O}_X(1)$. We will show that this is an isomorphism:

Theorem 13.32. *Let $\mathscr{E}$ be a quasi-coherent $\mathscr{O}_S$-module. The morphism $r_{\mathscr{E}}$ (13.8.2) is an isomorphism.*

Proof. We define an inverse as follows. Let $v \colon p^*\mathscr{E} \twoheadrightarrow \mathscr{L} := \mathscr{L}_{\mathrm{univ}}$ be the universal quotient bundle on $P := \mathbb{P}(\mathscr{E})$. Functoriality of the symmetric algebra yields a surjective homomorphism of quasi-coherent $\mathscr{O}_P$-algebras

$$\varphi = \mathrm{Sym}(v) \colon \mathrm{Sym}(p^*\mathscr{E}) = p^*(\mathrm{Sym}\,\mathscr{E}) \to \mathrm{Sym}\,\mathscr{L} = \bigoplus_{d \geq 0} \mathscr{L}^{\otimes d}.$$

By (a globalization of) Proposition 13.8 we obtain a morphism

$$q \colon P = \mathrm{Proj}(\mathrm{Sym}\,\mathscr{L}) \xrightarrow{\ \mathrm{Proj}\,\varphi\ } \mathrm{Proj}\,p^*(\mathrm{Sym}\,\mathscr{E}) = \mathrm{Proj}(\mathrm{Sym}\,\mathscr{E}) \times_S P \to \mathrm{Proj}(\mathrm{Sym}\,\mathscr{E}),$$

where the first equality is the identification of Lemma 13.31 and where the last arrow is the projection. It is not difficult (albeit somewhat lengthy) to see that q and $r_{\mathscr{E}}$ are inverse to each other. $\qquad\square$

From now on we will usually identify $P := \mathbb{P}(\mathscr{E})$ and $\mathrm{Proj}(\mathrm{Sym}\,\mathscr{E})$ and denote the universal quotient line bundle on P by $\mathscr{O}_P(1)$.

We reformulate the theorem in a different form in the special case $\mathscr{E} = \mathscr{O}_S^{n+1}$ which will be useful when we study embeddings into projective space.

Corollary 13.33. *Let X be an S-scheme. There is a natural bijection between the set $\mathrm{Hom}_S(X, \mathbb{P}_S^n)$ of morphisms from X to $\mathbb{P}_S^n$ and the set of isomorphism classes of tuples $(\mathscr{L}, s_0, \dots, s_n)$, where $\mathscr{L}$ is a line bundle on X, and the $s_i \in \Gamma(X, \mathscr{L})$ are global sections of $\mathscr{L}$ which generate $\mathscr{L}$.*

Proof. We view $\mathbb{P}^{n+1} = \mathbb{P}(\mathscr{O}_S^{n+1})$ as the space of line bundle quotients of $\mathscr{O}_S^{n+1}$, using that the free sheaf $\mathscr{O}_S^{n+1}$ is self-dual. The corollary follows from the functorial description of $\mathbb{P}(\mathscr{O}_S^{n+1})$, because a tuple $(\mathscr{L}, s_0, \dots, s_n)$ as above is the same as a surjective homomorphism $\mathscr{O}_X^{n+1} \twoheadrightarrow \mathscr{L}$. In particular, given a morphism $X \to \mathbb{P}_S^n$, we obtain the line bundle $\mathscr{L}$ on X as the pull-back of $\mathscr{O}_{\mathbb{P}_S^n}(1)$, and the s_i as the pull-backs of the global sections of $\mathscr{O}_{\mathbb{P}_S^n}(1)$ given by the universal projection $\mathscr{O}_{\mathbb{P}_S^n}^{n+1} \twoheadrightarrow \mathscr{O}_{\mathbb{P}_S^n}(1)$. $\qquad\square$

Remark 13.34. If $S = \mathrm{Spec}\,k$, where k is a field, we can describe the map $X(k) \to \mathbb{P}_k^n(k)$ corresponding to the tuple $(\mathscr{L}, s_0, \dots, s_n)$ quite explicitly: Given $x \in X(k)$, there is an index i such that $s_i(x) \in \mathscr{L}_x \otimes_{\mathscr{O}_{X,x}} \kappa(x)$ does not vanish, because the s_i generate $\mathscr{L}$. Therefore for each j there exists a unique element $\alpha_j \in \kappa(x) = k$ such that $s_j(x) = \alpha_j s_i(x)$. We map x to $(\alpha_0 : \cdots : \alpha_n)$, and denote this point by $(s_0(x) : \cdots : s_n(x))$. It is clear that this point in $\mathbb{P}^n(k)$ does not depend on the choice of i.

Remark 13.35. Let S be a scheme, let $\mathscr{E}$ be a quasi-coherent $\mathscr{O}_S$-module and let $\mathscr{L}$ be an invertible $\mathscr{O}_S$-module. By Theorem 13.32 and Lemma 13.31 there exists an isomorphism

$$(13.8.3) \qquad i \colon P := \mathbb{P}(\mathscr{E}) \xrightarrow{\sim} Q := \mathbb{P}(\mathscr{E} \otimes \mathscr{L})$$

such that $i^* \mathscr{O}_Q(1) \cong \mathscr{O}_P(1) \otimes \pi^* \mathscr{L}$, where $\pi \colon P \to S$ is the structure morphism. If $f \colon T \to S$ is an S-scheme, then (13.8.3) is given on T-valued points by the restriction of the bijection

$$(13.8.4) \qquad \begin{aligned} \{\mathscr{O}_T\text{-submodules of } f^* \mathscr{E}\} &\xrightarrow{\sim} \{\mathscr{O}_T\text{-submodules of } f^* (\mathscr{E} \otimes \mathscr{L})\}, \\ \mathscr{U} &\mapsto \mathscr{U}_{\mathscr{L}} := \mathscr{U} \otimes f^* \mathscr{L}. \end{aligned}$$

Locally on S, the $\mathscr{O}_T$-modules $\mathscr{U}$ and $\mathscr{U}_{\mathscr{L}}$ are isomorphic. In particular $\mathscr{U}$ (resp. $f^* \mathscr{E}/\mathscr{U}$) is locally free of rank e if and only if $\mathscr{U}_{\mathscr{L}}$ (resp. $f^*(\mathscr{E} \otimes \mathscr{L})/\mathscr{U}_{\mathscr{L}}$) is locally free of rank e. Here $e \geq 0$ is some integer. Thus (13.8.4) induces also an isomorphism of S-schemes

$$(13.8.5) \qquad \operatorname{Grass}^e(\mathscr{E}) \xrightarrow{\sim} \operatorname{Grass}^e(\mathscr{E} \otimes \mathscr{L}).$$

Remark 13.36. Let $\mathscr{E}$ be a finite locally free $\mathscr{O}_S$-module and let $\pi \colon P := \mathbb{P}(\mathscr{E}) \to S$ be the structure morphism. Then the canonical homomorphism

$$\alpha_d \colon \operatorname{Sym}^d(\mathscr{E}) \to \pi_* \mathscr{O}_P(d)$$

is an isomorphism for all $d \in \mathbb{Z}$ (and in particular $\pi_* \mathscr{O}_P(d) = 0$ for $d < 0$). Indeed, the question is local on S and we may assume that $S = \operatorname{Spec} R$ is affine and that $\mathscr{E} = \tilde{E}$, where E is the R-module $(R^{n+1})^\vee$ for some n. Then $P = \mathbb{P}_R^n$ and the claim follows from Example 13.16.

(13.9) Affine cone.

In Section (1.21) we considered closed varieties $X(k) \subseteq \mathbb{P}^n(k)$ (k an algebraically closed field). We defined the affine cone $C(X)(k)$ of $X(k)$ as the closure of $\pi^{-1}(X(k))$ in $\mathbb{A}^{n+1}(k)$, where $\pi \colon \mathbb{A}^{n+1}(k) \setminus \{0\} \to \mathbb{P}^n(k)$ is the morphism $(x_0, \ldots, x_n) \mapsto (x_0 : \ldots : x_n)$. We will now explain how to generalize this construction using the language introduced above. We remark that the variety $X(k)$ corresponds to a scheme of the form $X = V_+(I) = \operatorname{Proj} A$, where $A = k[X_0, \ldots, X_n]/I$ for a homogeneous ideal I. Moreover, we have $A_0 = k$.

Let S be a scheme and let $\mathscr{A}$ be a graded quasi-coherent $\mathscr{O}_S$-algebra such that $\mathscr{A}_0 = \mathscr{O}_S$. Set

$$X := \operatorname{Proj} \mathscr{A}, \qquad C := C(\mathscr{A}) := \operatorname{Spec} \mathscr{A}.$$

The augmentation homomorphism $\mathscr{A} \to \mathscr{A}/\mathscr{A}_+ = \mathscr{A}_0 = \mathscr{O}_S$ is surjective and therefore defines a closed immersion $i \colon S = \operatorname{Spec} \mathscr{O}_S \hookrightarrow C$. We set

$$C^0 := C^0(\mathscr{A}) := C \setminus i(S).$$

We call C the *affine cone of* $\mathscr{A}$ or, by abuse of language, *of* X, $i(S)$ its *vertex* and the open subscheme C^0 the *pointed affine cone of* $\mathscr{A}$ or of X. The formation of X, C and hence of C^0 is compatible with arbitrary base change $T \to S$.

Proposition 13.37. *There exists a surjective affine S-morphism $\pi = \pi_{\mathscr{A}} \colon C^0 \to X$ which is compatible with base change (i.e., $\pi_{g^*\mathscr{A}} = \pi_{\mathscr{A}} \times_S \mathrm{id}_T$ for every morphism $g \colon T \to S$). Moreover, if $S = \operatorname{Spec} R$ is affine and $\mathscr{A} = \tilde{A}$, we have:*

(1) For every homogeneous element $f \in A_+$

$$\pi^{-1}(D_+(f)) = C_f$$

and the restriction of π to C_f corresponds to the injection $A_{(f)} \hookrightarrow A_f$.
(2) For every $f \in A_1$ the $D_+(f)$-scheme C_f is isomorphic to $D_+(f) \otimes_R R[T, T^{-1}]$.

Proof. Assume first that $S = \operatorname{Spec} R$ is affine and $\mathscr{A} = \tilde{A}$. For a homogeneous element $f \in A_+$ let $\pi_f \colon C_f \to D_+(f)$ be the morphism corresponding to the inclusion $A_{(f)} \hookrightarrow A_f$. For every prime ideal $\mathfrak{p} \subset A_{(f)}$, $\mathfrak{p}A_f$ is a prime ideal of A_f. This shows that π_f is surjective. Moreover, if $g \in A_+$ is a second homogeneous element, then $\pi_{f|C_{fg}}$ induces the morphism π_{fg}. Thus we can glue the π_f to an affine surjective morphism $\pi \colon C^0 \to X$ (note that the complement of $\bigcup_f C_f$ in C is $V(A_+)$ and hence $\bigcup_f C_f = C^0$).

If f is of degree 1, the homomorphism $A_{(f)}[T, T^{-1}] \to A_f$ of $A_{(f)}$-algebras, that sends T to $f/1$, is bijective: It is clearly surjective. Let $\sum_{d=-t}^u b_d(f/1)^d = 0$, where $b_d = a_d/f^m$ with $a_d \in A$ of degree m (we may choose $m > 0$ independent of d). Multiplying with a sufficiently high power of f we find an integer $k > t$ such that $\sum_d f^{d+k} a_d = 0 \in A$. As the degrees of the summands are all different, we have $f^{d+k} a_d = 0$ for all d and hence $b_d = 0$ in $A_{(f)}$. This shows (2).

The construction of π is clearly compatible with base change $T \to S$ and in particular with localization on S. Thus if S is arbitrary, we can glue the locally constructed morphisms to a global morphism $\pi_{\mathscr{A}}$. $\qquad\square$

Remark 13.38.
(1) If $\mathscr{A}$ is generated by $\mathscr{A}_1$ we thus find an open affine covering $(U_i)_i$ of X such that $\pi^{-1}(U_i) \cong \mathbb{A}^1_{U_i} \setminus \{0\}$. In particular π is surjective and smooth of relative dimension 1 in this case.
(2) If S is affine, $C^0 = \bigcup_f C_f$, where f runs through a set of elements in A_+ which generate an ideal whose radical is equal to A_+ (Proposition 13.4 (5)).

Example 13.39. In the case that $\mathscr{A} = \operatorname{Sym} \mathscr{E}$ for a quasi-coherent $\mathscr{O}_S$-module $\mathscr{E}$, we find $C = \mathbb{V}(\mathscr{E})$. Moreover, $i \colon S \to \mathbb{V}(\mathscr{E})$ is the zero section of $\mathbb{V}(\mathscr{E})$. Denote by $f \colon C \to S$ the structure morphism. By Proposition 11.4, the identity id_C corresponds to a homomorphism

$$(13.9.1) \qquad\qquad\qquad u \colon f^*\mathscr{E} \to \mathscr{O}_C$$

of $\mathscr{O}_C$-bundles and C^0 is the locus where u is surjective, i.e. $C^0 = C \setminus \operatorname{Supp}(\operatorname{Coker} u)$. The morphism $\pi \colon C^0 \to X = \mathbb{P}(\mathscr{E})$ corresponds via the universal property of $\mathbb{P}(\mathscr{E})$ (Section (8.8)) to the surjection $u_{|C^0}$.

Embeddings into projective space

Let S be a scheme. To study an S-scheme $f: X \to S$ it is often desirable to embed X into projective space $\mathbb{P}_S^n$. For instance, we have seen in Proposition 13.24 that closed subschemes of $\mathbb{P}_S^n$ are always given locally on S as vanishing schemes of homogeneous polynomials. We will start by proving the so-called fundamental theorem of elimination theory, namely that $\mathbb{P}_S^n$ is proper over S (Theorem 13.40). In particular we see that to embed X as a closed subscheme into $\mathbb{P}_S^n$ a necessary condition is that X is proper over S.

In Corollary 13.33 we have given a useful description of morphisms to projective space. We will use this description here and identify morphisms $X \to \mathbb{P}_S^n$ with tuples $(\mathscr{L}, s_0, \ldots, s_n)$. All the line bundles $\mathscr{L}$ appearing here obviously are generated by global sections (which is not true for an arbitrary line bundle!). However, if we are interested in closed embeddings of a proper S-scheme into projective space, much more must be true (Proposition 12.93 and Proposition 12.94): The resulting map must be injective on K-valued points (K some algebraically closed field over S), i.e., for any two distinct points $x, x' \in X(K)$, with $s_i(x), s_i(x') \neq 0$, say, there must be an index j such that $\frac{s_j(x)}{s_i(x)} \neq \frac{s_j(x')}{s_i(x')}$ – we say that the sections s_i must separate the points of X. Even this is not enough to obtain an immersion. The s_i have to separate tangent vectors as well; see Remark 13.55 below.

The upshot is that a line bundle which gives rise to an embedding into projective space must have very many global sections. Line bundles with this property are called *very ample*; see Definition 13.44 below.

More specifically, we will investigate the following questions:

(1) An embedding of X into some projective bundle $\mathbb{P}(\mathscr{E})$ is given by a surjection $u: f^*\mathscr{E} \twoheadrightarrow \mathscr{L}$, where $\mathscr{L}$ is some line bundle on X and where $\mathscr{E}$ is a quasi-coherent $\mathscr{O}_S$-module. This yields an S-morphism $r_u: X \to \mathbb{P}(\mathscr{E})$. From this point of view we can ask which conditions on $\mathscr{L}$ ensure that there exists a u such that r_u is an immersion. This leads us to the definition of a very ample line bundle and is the topic of Section (13.12).

(2) An embedding of a projective bundle $\mathbb{P}(\mathscr{E})$ into projective space $\mathbb{P}^n$ is given by a surjection $\mathscr{O}_S^{n+1} \twoheadrightarrow \mathscr{E}$. This yields a closed immersion $\mathbb{P}(\mathscr{E}) \hookrightarrow \mathbb{P}_S^n$. In fact, we may identify $\mathbb{P}(\mathscr{E})$ and $\mathbb{P}(\mathscr{E} \otimes \mathscr{K})$ for every line bundle $\mathscr{K}$ on S (Lemma 13.31). Thus it suffices to find a surjection $\mathscr{O}_S^{n+1} \twoheadrightarrow \mathscr{E} \otimes \mathscr{K}$ for some $\mathscr{K}$. This leads us to the definition of an ample line bundle and is the topic of Section (13.11).

As this terminology suggests, there is a relation between the notions of "ample" and of "very ample" line bundles which will be proved in Theorem 13.62.

We will call f a quasi-projective (resp. projective) morphism if there exists an immersion (resp. a closed immersion) $X \to \mathbb{P}_S^n$ (at least if S is "not too big", for arbitrary S this naive definition of quasi-projectivity and projectivity is too special) and study these properties in Section (13.15) below.

We conclude this part of the chapter with the study of schemes X that are quasi-compact and isomorphic to an open subscheme of an affine scheme. Equivalently (Proposition 13.80) every quasi-coherent $\mathscr{O}_X$-module is generated by its global sections in Section (13.16). These schemes will be called quasi-affine.

(13.10) Projective schemes are proper.

A crucial property of projective space, and hence of all closed subschemes of projective space, is that they are proper.

Theorem 13.40. *Let S be a scheme, and $n \geq 0$. Then the projective space $\mathbb{P}_S^n$ is proper over S.*

Proof. As $\mathbb{P}_S^n$ is separated and of finite type over S, we only have to show that $\mathbb{P}_S^n \to S$ is universally closed. We have $\mathbb{P}_T^n = T \times_S \mathbb{P}_S^n$ for every morphism $T \to S$. Thus it suffices to show that $\mathbb{P}_T^n \to T$ is closed for every scheme T. This question is local on T and we may assume that T is affine. Thus it suffices to show that $f\colon \mathbb{P}_S^n \to S$ is closed, where $S = \operatorname{Spec} R$ is affine.

Set $A := R[T_0, \ldots, T_n]$ and let $Z \subseteq \mathbb{P}_R^n$ be a closed subspace. We endow Z with some structure of a closed subscheme. Then $Z = V_+(I)$ for a homogeneous ideal $I \subset A$ (Proposition 13.24).

Let $s \in S \setminus f(Z)$. Write $B = \kappa(s)[T_0, \ldots, T_n]$ and let $\bar{I}$ be the image of I in B. By hypothesis, the intersection $f^{-1}(s) \cap V_+(I)$ is empty. This implies that the fiber product $f^{-1}(s) \times_{\mathbb{P}_R^n} V_+(I)$ is empty, and this fiber product can be identified with $\operatorname{Spec} \kappa(s) \times_R V_+(I) \cong V_+(\bar{I}) \subseteq \mathbb{P}_{\kappa(s)}^n$ by Remark 13.27. Using Proposition 13.4 (3), we see $B_+ \subseteq \operatorname{rad}(\bar{I})$. This implies that for d suitably large, we have $B_d = \bar{I}_d$, i.e., $(A_d/I_d) \otimes \kappa(s) = 0$.

Denote by $\mathfrak{p} \subset R$ the prime ideal corresponding to s. As A_d/I_d is a finitely generated R-module whose fiber over s is 0, the stalk $(A_d/I_d)_{\mathfrak{p}}$ vanishes by Nakayama's Lemma. As the support of the finitely generated R-module A_d/I_d is closed, we find an $h \in R \setminus \mathfrak{p}$ such that $h A_d \subseteq I_d$. In particular we find $h T_i^d \in I_d$ for all i, which shows that $D(h) \cap f(Z) = \emptyset$. As $h \in R \setminus \mathfrak{p}$, $D(h)$ is an open neighborhood of s. This proves that $S \setminus f(Z)$ is open. $\square$

This theorem is really a non-trivial result, and is sometimes called the fundamental theorem of elimination theory. As illustration and applications, we mention Exercise 13.14, Exercise 13.9, Section (13.17) on conic projections, and Theorem 14.132.

Corollary 13.41. *Let S be a scheme, let X be a closed subscheme of $\mathbb{P}_S^n$, and let Y be an S-scheme which is separated over S. Then every S-morphism $f\colon X \to Y$ is proper.*

Proof. As $\mathbb{P}_S^n$ is proper over S, every closed subscheme X is proper over S. Therefore f is proper by Proposition 12.58 (3). $\square$

Corollary 13.42. *Let S be a scheme and let $\mathscr{E}$ be a quasi-coherent $\mathscr{O}_S$-module of finite type. Then $\mathbb{P}(\mathscr{E})$ is proper over S.*

Proof. As properness can be checked locally on the target, we may assume that S is affine. Then there exists a surjective homomorphism $\mathscr{O}_S^n \twoheadrightarrow \mathscr{E}$. It corresponds to a closed immersion $\mathbb{P}(\mathscr{E}) \hookrightarrow \mathbb{P}(\mathscr{O}_S^n) \cong \mathbb{P}_S^{n-1}$. As $\mathbb{P}_S^{n-1}$ is proper, it follows that $\mathbb{P}(\mathscr{E})$ is proper. $\square$

(13.11) Ample line bundles.

Let X be a scheme and let $\mathscr{F}$ be a quasi-coherent $\mathscr{O}_X$-module of finite type. We would like to find criteria when it is possible to find an immersion $i\colon \mathbb{P}(\mathscr{F}) \to \mathbb{P}_X^n$ for some n. As $\mathbb{P}(\mathscr{F})$ is proper over X (Corollary 13.42), i is closed if it exists.

Remark 13.43. If X is quasi-compact and $\mathscr{F}$ is generated by its global sections, we find a surjection $\mathscr{O}_X^{n+1} \twoheadrightarrow \mathscr{F}$ (Lemma 10.47) and hence there exists a closed immersion $\mathbb{P}(\mathscr{F}) \hookrightarrow \mathbb{P}(\mathscr{O}_X^{n+1}) \cong \mathbb{P}_X^n$.

If X is affine, then every quasi-coherent $\mathscr{O}_X$-module is generated by its global sections.

The line bundle $\mathscr{O}_P(1)$ on a projective bundle $P = \mathbb{P}(\mathscr{E})$ has the property that for every finitely generated quasi-coherent $\mathscr{O}_P$-module $\mathscr{F}$ there exists $n \geq 0$ such that $\mathscr{F}(n) := \mathscr{F} \otimes \mathscr{O}_P(1)^{\otimes n}$ is generated by global sections (Proposition 13.22). This is a property of line bundles which is generally very interesting, and which leads us to the following definition (recall that the notions "invertible module" and "line bundle" are synonymous for us).

Definition 13.44. *Let X be a qcqs scheme. An invertible $\mathscr{O}_X$-module $\mathscr{L}$ is called* ample *if for every quasi-coherent $\mathscr{O}_X$-module $\mathscr{F}$ of finite type there exists an integer n_0 such that $\mathscr{F} \otimes \mathscr{L}^{\otimes n}$ is generated by its global sections for all $n \geq n_0$.*

Example 13.45. Let A be a graded ring such that A_+ is generated by finitely many elements in A_1 and set $X := \operatorname{Proj} A$. Then Proposition 13.22 shows that the line bundles $\mathscr{O}_X(d)$ are ample for $d \geq 1$.

Consider the particular case $X = \mathbb{P}_R^N$ for a ring R and for an integer $N > 0$. We claim that $\mathscr{O}_X(d)$ is not ample for $d \leq 0$. If $d < 0$, we have $\Gamma(X, \mathscr{O}_X(d)) = 0$ by Example 13.16. Therefore there exists no $n > 0$ such that $\mathscr{O}_X(d)^{\otimes n} = \mathscr{O}_X(nd)$ is generated by its global sections. In particular $\mathscr{O}_X(d)$ cannot be ample. If the structure sheaf $\mathscr{O}_X$ itself were ample, every quasi-coherent $\mathscr{O}_X$-module $\mathscr{F}$ of finite type would be generated by its global sections. But we have just seen that this is not the case for $\mathscr{F} = \mathscr{O}_X(-1)$.

Recall that if $f \in \Gamma(X, \mathscr{L})$ is a global section, we defined in Section (7.11) the open subset

$$X_f := X_f(\mathscr{L}) = \{\, x \in X \; ; \; f(x) \neq 0 \text{ in the fiber } \mathscr{L}(x) \,\}.$$

Remark 13.46. Let X be a scheme and let $\mathscr{L}$ be a line bundle on X.

(1) Let $f_1, \ldots, f_r \in \Gamma(X, \mathscr{L})$ be global sections and denote by $u \colon \mathscr{O}_X^r \to \mathscr{L}$ the corresponding homomorphism of $\mathscr{O}_X$-modules. Then $\bigcup_{i=1}^r X_{f_i} = X \setminus (\operatorname{Supp} \operatorname{Coker} u)$. In particular u is surjective (i.e., the f_i generate $\mathscr{L}$) if and only if $X = \bigcup_{i=1}^r X_{f_i}$.

(2) For every section $f \in \Gamma(X, \mathscr{L})$ let $u_f \colon \mathscr{O}_X \to \mathscr{L}$ be the corresponding homomorphism. Then Assertion (1) implies that the restriction of u_f to X_f is surjective. By Corollary 8.12 we obtain an isomorphism

$$(13.11.1) \qquad\qquad u_f|_{X_f} \colon \mathscr{O}_X|_{X_f} \xrightarrow{\sim} \mathscr{L}|_{X_f}.$$

(3) Let $\mathscr{M}$ be a second line bundle on X, and let $f \in \Gamma(X, \mathscr{L})$ and $g \in \Gamma(X, \mathscr{M})$ be global sections. Then $X_f(\mathscr{L}) \cap X_g(\mathscr{M}) = X_{f \otimes g}(\mathscr{L} \otimes \mathscr{M})$. In particular $X_f = X_{f \otimes d}$ for all $d \geq 1$.

(4) Via the isomorphism $X \cong \operatorname{Proj}(\bigoplus_{d \geq 0} \mathscr{L}^{\otimes d})$ (Lemma 13.31), X_f is identified with $D_+(f)$ for all $f \in \Gamma(X, \mathscr{L}^{\otimes d})$ and $d \geq 1$.

(5) For all $f \in \Gamma(X, \mathscr{L})$ the open immersion $X_f \hookrightarrow X$ is an affine morphism (Example 12.4 (4)). In particular, X_f is quasi-compact if X is quasi-compact.

If X is affine, then any invertible $\mathscr{O}_X$-module and in particular $\mathscr{O}_X$ is an ample line bundle. For $f \in \Gamma(X, \mathscr{O}_X)$ the open subset X_f is the principal open subset $D(f)$. The following proposition characterizes ample line bundles in terms of the associated "principal open sets". Similarly as the structure sheaf of an affine scheme, ample line bundles (and their tensor powers) have many sections, so that there are "many" associated principal open sets.

Proposition 13.47. *Let X be a qcqs scheme and let $\mathscr{L}$ be an invertible $\mathscr{O}_X$-module. Then the following assertions are equivalent.*

(i) *$\mathscr{L}$ is ample.*

(ii) *For every quasi-coherent ideal $\mathscr{I} \subseteq \mathscr{O}_X$ of finite type there exists an integer $n \geq 1$ such that $\mathscr{I} \otimes \mathscr{L}^{\otimes n}$ is generated by its global sections.*

(iii) *The open subsets X_f for $f \in \Gamma(X, \mathscr{L}^{\otimes n})$ and $n > 0$ form a basis of the topology of X.*

(iv) *There exists an integer $d \geq 1$ and finitely many sections $f_i \in \Gamma(X, \mathscr{L}^{\otimes d})$ such that X_{f_i} is affine for all i and such that $X = \bigcup_i X_{f_i}$.*

Proof. It is clear that (i) implies (ii).

(ii) $\Rightarrow$ (iii). Let $x \in X$ be a point and let U be an open affine neighborhood of x. Let $\mathscr{I}$ be a quasi-coherent ideal of $\mathscr{O}_X$ that defines a closed subscheme of X whose underlying subspace is $X \setminus U$. We may assume that $\mathscr{I}$ is of finite type (if X is noetherian, this is automatic, otherwise use Exercise 10.29). Then (ii) implies that there exists an integer $n > 0$ and $f \in \Gamma(X, \mathscr{I} \otimes \mathscr{L}^{\otimes n}) \subseteq \Gamma(X, \mathscr{L}^{\otimes n})$ such that $f(x) \neq 0$. Then $x \in X_f \subseteq U$.

(iii) $\Rightarrow$ (iv). Let $x \in X$ be a point and let U be an open affine neighborhood of x such that there exists an isomorphism $\eta \colon \mathscr{L}_{|U} \cong \mathscr{O}_U$. By (iii) we find an $f \in \Gamma(X, \mathscr{L}^{\otimes n})$ such that $x \in X_f \subseteq U$. But $X_f = D(\eta(f_{|U}))$ which shows that X_f is affine. Therefore there exists a finite open affine covering $(X_{f_i})_i$ of X with $f_i \in \Gamma(X, \mathscr{L}^{\otimes d_i})$. Replacing f_i by a suitable power (which does not change X_{f_i}) we may assume that all integers d_i are equal to the same d.

(iv) $\Rightarrow$ (i). Let $\mathscr{F}$ be a quasi-coherent $\mathscr{O}_X$-module of finite type. For $n \in \mathbb{Z}$ we set $\mathscr{F}(n) := \mathscr{F} \otimes \mathscr{L}^{\otimes n}$. By (iv) there exist an integer $d \geq 1$ and a finite open affine covering $(X_{f_i})_i$ of X with $f_i \in \Gamma(X, \mathscr{L}^{\otimes d})$. As X_{f_i} is affine, $\mathscr{F}_{|X_{f_i}}$ is generated by finitely many sections $g_{ij} \in \Gamma(X_{f_i}, \mathscr{F})$. By Theorem 7.22 there exists an integer m_0 such that $g_{ij} \otimes f_i^m$ can be extended to a section in $\Gamma(X, \mathscr{F}(dm))$ for all $m \geq m_0$. As there are only finitely many g_{ij}'s, we may choose m_0 independent of i and j. Therefore $\mathscr{F}(dm)$ is generated by its global sections for all $m \geq m_0$. Repeating the same argument for $\mathscr{F}(k)$ with $0 < k < d$ shows that there exists an integer m_k such that $\mathscr{F}(k)(md) = \mathscr{F}(k + md)$ is generated by its global sections for all $m \geq m_k$. This shows that we can find an integer n_0 such that $\mathscr{F}(n)$ is generated by its global sections for all $n \geq n_0$ (e.g., we can choose n_0 as the maximum of the dm_k). $\square$

In Volume II we will see cohomological characterizations of ample line bundles which generalize Serre's criterion 12.35 for affine schemes.

Any ample line bundle on X defines an open embedding of X into some projective spectrum: Let X be a qcqs scheme, set $R := \Gamma(X, \mathscr{O}_X)$, and let $q \colon X \to S := \operatorname{Spec} R$ be the canonical morphism (Remark 3.7). Let $\mathscr{F}$ be a quasi-coherent $\mathscr{O}_X$-module. By Corollary 10.27 its direct image $q_* \mathscr{F}$ is quasi-coherent. It corresponds to the R-module $\Gamma(S, q_* \mathscr{F}) = \Gamma(X, \mathscr{F})$. Thus via adjointness the identity $\Gamma(X, \mathscr{F})^{\sim} \to q_* \mathscr{F}$ corresponds to a homomorphism of $\mathscr{O}_X$-modules

(13.11.2) $$q^*(\Gamma(X,\mathscr{F})^\sim) \to \mathscr{F}.$$

Now let $\mathscr{L}$ be an invertible $\mathscr{O}_X$-module. We set $A_d := \Gamma(X, \mathscr{L}^{\otimes d})$ and $A = \bigoplus_{d\geq 0} A_d$. This is a graded R-algebra. The homomorphisms (13.11.2) for $\mathscr{L}^{\otimes d}$ and $d \geq 0$ then yield a homomorphism

(13.11.3) $$\varphi\colon q^*(\tilde{A}) \to \bigoplus_{d\geq 0} \mathscr{L}^{\otimes d} = \mathrm{Sym}(\mathscr{L}).$$

We have $\mathrm{Proj}\, q^*(\tilde{A}) = \mathrm{Proj}\, A \times_S X$ (13.7.1). By functoriality of Proj (Proposition 13.8) we obtain a morphism of S-schemes $r\colon G(\varphi) \to \mathrm{Proj}\, A$, where

$$G(\varphi) = \bigcup_{f\in A_+} X_f \subseteq X = \mathrm{Proj}(\mathrm{Sym}\,\mathscr{L}).$$

For $f \in A_+$ homogeneous we have $r^{-1}(D_+(f)) = X_f$ and the restriction of r to X_f yields a morphism $r_f\colon X_f \to D_+(f)$ which on global sections induces the isomorphism $A_{(f)} \xrightarrow{\sim} \Gamma(X_f, \mathscr{O}_X)$ given by Theorem 7.22. Therefore r_f is an isomorphism if X_f is affine.

Now assume that $\mathscr{L}$ is ample. Let E be the set of those homogeneous elements $f \in A_+$ such that X_f is affine. Then $(X_f)_{f\in E}$ is a covering of X (Proposition 13.47). Thus r yields an isomorphism

(13.11.4) $$X \xrightarrow{\sim} P^0 := \bigcup_{f\in E} D_+(f) \subseteq \mathrm{Proj}\, A$$

and P^0 is an open subscheme of $\mathrm{Proj}\, A$. In particular, r is an open immersion. In fact we have more precisely:

Proposition 13.48. *Let X be a qcqs scheme, let $\mathscr{L}$ be an ample invertible $\mathscr{O}_X$-module. Then the above morphism*

$$r\colon X \to \mathrm{Proj} \bigoplus_{d\geq 0} \Gamma(X, \mathscr{L}^{\otimes d})$$

is a quasi-compact schematically dominant open immersion.

In particular the existence of an ample line bundle on X implies that X is separated.

Proof. As $(D_+(f))_{f\in A_+}$ is an open affine covering of $\mathrm{Proj}\, A$ and as $r^{-1}(D_+(f)) = X_f$ is quasi-compact for all $f \in A_+$ by Remark 13.46 (5), r is quasi-compact. Therefore the restrictions r_f are open quasi-compact immersions as well for all $f \in A_+$. In particular $(r_f)_* \mathscr{O}_{X_f}$ is a quasi-coherent $\mathscr{O}_{D_+(f)}$-module (Corollary 10.27). It follows that $r_f^\flat\colon \mathscr{O}_{D_+(f)} \to (r_f)_* \mathscr{O}_{X_f}$ is a homomorphism of quasi-coherent modules on the affine scheme $D_+(f)$ which induces on global sections the isomorphism $A_{(f)} \xrightarrow{\sim} \Gamma(X_f, \mathscr{O}_X)$. Hence $r_f^\flat$ is an isomorphism and in particular r_f is schematically dominant for all $f \in A_+$. This shows that r is schematically dominant. $\square$

Proposition 13.49. *Let X be a qcqs scheme and let $\mathscr{L}$ be an ample $\mathscr{O}_X$-module. For every finite subset Z of X and for every open neighborhood U of Z there exist an integer $n > 0$ and a section $f \in \Gamma(X, \mathscr{L}^{\otimes n})$ such that X_f is an affine neighborhood of Z contained in U.*

Proof. Using Proposition 13.48 we consider X as an open subscheme of $P := \operatorname{Proj} A$ with $A = \bigoplus_{d \geq 0} \Gamma(X, \mathscr{L}^{\otimes d})$. Let Y be the complement of U in P. Then Y is of the form $V_+(I)$ for a homogeneous $I \subset A$ not containing A_+ (Proposition 13.24). The points of Z correspond to relevant prime ideals $\mathfrak{p}_1, \ldots, \mathfrak{p}_n$ of A which do not contain I. By prime ideal avoidance (Proposition B.2 (2)) there exists a homogeneous element $f \in I_+$ which is not contained in the union of the $\mathfrak{p}_i$. Then $Z \subset D_+(f) \subseteq U$ and $X_f = D_+(f)$. $\qquad\square$

Proposition 13.50. *Let X be a qcqs scheme, $\mathscr{L}$ and $\mathscr{L}'$ invertible $\mathscr{O}_X$-modules.*
(1) *Let $n > 0$ be an integer. Then $\mathscr{L}$ is ample if and only if $\mathscr{L}^{\otimes n}$ is ample.*
(2) *If $\mathscr{L}$ is ample, then there exists an integer n_0 such that $\mathscr{L}^{\otimes n} \otimes \mathscr{L}'$ is ample and generated by its global sections for all $n \geq n_0$.*
(3) *If $\mathscr{L}$ is ample and if there exists an integer $n' > 0$ such that $\mathscr{L}'^{\otimes n'}$ is generated by its global sections, $\mathscr{L} \otimes \mathscr{L}'$ is ample.*
(4) *If $\mathscr{L}$ and $\mathscr{L}'$ are ample, $\mathscr{L} \otimes \mathscr{L}'$ is ample.*

Assertion (3) can be further generalized, see Exercise 13.11.

Proof. Assertions (1) and (3) follow from (iii) of Proposition 13.47: For (1) it suffices to remark that $X_f = X_{f^{\otimes n}}$. Let us show (3). Let $x \in X$ be a point and let U be an open neighborhood of x. As $\mathscr{L}$ is ample, there exist $n > 0$ and $f \in \Gamma(X, \mathscr{L}^{\otimes n})$ with $x \in X_f \subseteq U$. As $\mathscr{L}'^{\otimes n'}$ is generated by its global sections, we find $f' \in \Gamma(X, \mathscr{L}'^{\otimes n'})$ such that $f'(x) \neq 0$. Set $g := f^{\otimes n'} \otimes f'^{\otimes n} \in \Gamma(X, (\mathscr{L} \otimes \mathscr{L}')^{\otimes nn'})$. Then $x \in X_g \subseteq X_f \subseteq U$ which shows that $\mathscr{L} \otimes \mathscr{L}'$ is ample. Assertion (4) follows from (3).

It remains to show (2). As $\mathscr{L}$ is ample, there exists an integer $m_0 > 0$ such that $\mathscr{L}' \otimes \mathscr{L}^{\otimes m}$ is generated by its global sections for all $m \geq m_0$. Therefore (3) shows that we can take $n_0 = m_0 + 1$. $\qquad\square$

Proposition 13.51. *Let X be a qcqs scheme and let $i \colon Z \to X$ be a quasi-compact immersion (e.g., if X is noetherian and i is an arbitrary immersion). If $\mathscr{L}$ is an ample $\mathscr{O}_X$-module, $i^*\mathscr{L}$ is an ample $\mathscr{O}_Z$-module.*

A much more general version of stability of ampleness under pull back will be proved in Proposition 13.83.

Proof. For $f \in \Gamma(X, \mathscr{L}^{\otimes n})$ set $f' := i^*(f) \in \Gamma(Z, (i^*\mathscr{L})^{\otimes n})$ (7.8.11). Then we have $X_f \cap Z = Z_{f'}$. Thus we conclude by Proposition 13.47 (iii). $\qquad\square$

(13.12) Immersions into projective bundles; very ample line bundles.

Let S be a scheme. To study an S-scheme $f \colon X \to S$ it is often desirable to embed X into the projective space $\mathbb{P}^n_S$ or, more generally, into a projective bundle $P := \mathbb{P}(\mathscr{E})$, where $\mathscr{E}$ is a quasi-coherent $\mathscr{O}_S$-module. Such an embedding i might not always exist, but if it does, we know that it is given by a surjection $f^*\mathscr{E} \twoheadrightarrow \mathscr{L}$, where $\mathscr{L}$ is an invertible $\mathscr{O}_X$-module (Section (8.8)). In this case we have $\mathscr{L} \cong i^*\mathscr{O}_P(1)$. This leads to the following definition.

Definition 13.52. Let $f\colon X \to S$ be a morphism of finite type. An invertible $\mathscr{O}_X$-module $\mathscr{L}$ is called very ample for f or over S if there exists an open covering (U_α) of S and for all α a U_α-immersion $i_\alpha\colon X_\alpha := f^{-1}(U_\alpha) \hookrightarrow P_\alpha := \mathbb{P}^{n_\alpha}_{U_\alpha}$ for some $n_\alpha \geq 0$ such that $\mathscr{L}_{|X_\alpha} \cong i_\alpha^* \mathscr{O}_{P_\alpha}(1)$.

In other words, we call the line bundle $\mathscr{L}$ very ample if it is locally on S the inverse image of $\mathscr{O}(1)$ under an immersion into $\mathbb{P}^n$.

Let $f\colon X \to S$ be a morphism of finite type. If there exists a very ample line bundle on X, then X can be embedded locally on S into $\mathbb{P}^n_S$. The converse does not hold in general: Even if X can be embedded locally on S into $\mathbb{P}^n_S$, these embeddings are not necessarily given by sections of a single line bundle.

Remark 13.53. Let $f\colon X \to S$ be a morphism of finite type, let $S' \to S$ be a morphism, and let $\mathscr{L}$ be an invertible $\mathscr{O}_X$-module. Set $X' := X \times_S S'$, let $f'\colon X' \to S'$ be the base change of f, and let $g\colon X' \to X$ be the projection. If $\mathscr{L}$ is very ample for f, then $g^*\mathscr{L}$ is very ample for f'. Indeed, as the question is local on S, we may assume that there exists an immersion $i\colon X \to P := \mathbb{P}^n_S$ such that $i^*\mathscr{O}_P(1) \cong \mathscr{L}$. Then $g^*\mathscr{L} \cong (i \times \mathrm{id}_{S'})^*\mathscr{O}_{P'}(1)$, where $P' = \mathbb{P}^n_{S'} = \mathbb{P}^n_S \times_S S'$. Thus $g^*\mathscr{L}$ is very ample.

Let $f\colon X \to S$ be a separated morphism of finite type and let $\mathscr{L}$ be an invertible $\mathscr{O}_X$-module. Let $(t_0, \ldots, t_n)$ be a family of global sections $t_i \in \Gamma(X, \mathscr{L})$ corresponding to a homomorphism $u\colon f^*\mathscr{O}_S^{n+1} = \mathscr{O}_X^{n+1} \to \mathscr{L}$. We set $\mathscr{E} := \mathscr{O}_S^{n+1}$. If the sections t_i generate $\mathscr{L}$ (i.e., if u is surjective), we obtain an S-morphism (not necessarily an immersion)

$$(13.12.1) \qquad r_u\colon X \to \mathbb{P}(\mathscr{E}) \cong \mathbb{P}^n_S.$$

If u is not surjective, we will still get a morphism from an open subscheme of X to $\mathbb{P}^n_S$: The induced homomorphism of graded $\mathscr{O}_X$-algebras $\varphi := \mathrm{Sym}(u)$ defines a morphism

$$\mathrm{Proj}\,\varphi\colon G(u) \to \mathbb{P}(f^*\mathscr{E}) = \mathbb{P}(\mathscr{E}) \times_S X,$$

where $G(u) := G(\varphi) \subseteq \mathbb{P}(\mathscr{L}) = X$ is (the globalization of) the open subscheme defined in (13.2.4). Denote the composition of $\mathrm{Proj}\,\varphi$ with the projection $\mathbb{P}(f^*\mathscr{E}) \to \mathbb{P}(\mathscr{E})$ by r.

If u is surjective, then $G(u) = X$ and $r = r_u$ is the corresponding morphism (13.12.1). In general, $G(u)$ is the locus where u is surjective, i.e., $G(u)$ is the complement of $\mathrm{Supp}(\mathrm{Coker}\,u)$.

We will make r more explicit on k-valued points (k some field over S). Replacing X by $X \times_S \mathrm{Spec}\,k$ and $\mathbb{P}^n_S$ by $\mathbb{P}^n_k$, r is given by (see Remark 13.34)

$$X(k) \ni x \mapsto (t_0(x) : \ldots : t_n(x)) \in \mathbb{P}^n(k),$$

where $(t_0(x) : \ldots : t_n(x))$ is defined as follows. Choose an open neighborhood U of x and an isomorphism $\eta_U\colon \mathscr{L}_{|U} \xrightarrow{\sim} \mathscr{O}_U$. Then $\eta_U(t_{i|U})(x) \in k$ and

$$(t_0(x) : \ldots : t_n(x)) := (\eta_U(t_{0|U})(x) : \ldots : \eta_U(t_{n|U})(x))$$

is a point in $\mathbb{P}^n(k)$ which is independent of the choice of η_U. In this case the set of k-valued points of $G(u)$ is the complement of $\{x \in X(k)\ ;\ t_i(x) = 0 \text{ for all } i\}$.

Let us return to the general situation. Even if u is surjective (i.e., the t_i generate $\mathscr{L}$), r_u will not necessarily be an immersion. There is the following criterion.

Lemma 13.54. *In the discussion above assume that $S = \operatorname{Spec} R$ is an affine scheme. Assume that there is a subset $J \subseteq \{0, \ldots, n\}$ such that*
(a) *$\bigcup_{j \in J} X_{t_j} = X$ (i.e., the sections t_j for $j \in J$ generate $\mathscr{L}$),*
(b) *X_{t_j} is affine for all $j \in J$,*
(c) *for all $j \in J$ the sections $(t_{i|X_{t_j}})(t_{j|X_{t_j}})^{-1}$ $(i = 0, \ldots, n)$ generate $\Gamma(X_{t_j}, \mathscr{L}) = \Gamma(X_{t_j}, \mathscr{O}_X)$ (13.11.1) as an R-algebra.*
Then the morphism r corresponding to $(t_0, \ldots, t_n)$ is an immersion. If J can be chosen as $\{0, \ldots, n\}$, then r is a closed immersion.

The criteria (a), (b), and (c) for $J = \{0, \ldots, n\}$ for r to be a closed immersion are also necessary (Exercise 13.10).

Proof. We identify $\mathbb{P}^n_S$ with $P := \operatorname{Proj} R[T_0, \ldots, T_n]$ such that $r^{-1}(D_+(T_i)) = X_{t_i}$ for all $i = 0, \ldots, n$. For all $j \in J$, the restriction $r_j \colon X_{t_j} \to D_+(T_j)$ of r is then a morphism of affine schemes which induces on global sections a homomorphism of R-algebras $\varphi_j \colon \Gamma(D_+(T_j), \mathscr{O}_P) \twoheadrightarrow \Gamma(X_{t_j}, \mathscr{O}_X) \cong \Gamma(X_{t_j}, \mathscr{L})$ with $\varphi_j(T_i/T_j) = (t_{i|X_{t_j}})(t_{j|X_{t_j}})^{-1}$. Therefore φ_j is surjective by condition (c) and r_j is a closed immersion. It follows that r factors into a closed immersion $X = \bigcup_{j \in J} X_{t_j} \to U := \bigcup_{j \in J} D_+(T_j)$ followed by the open immersion $U \hookrightarrow P$. In particular, r is an immersion which is closed if $J = \{0, \ldots, n\}$. $\square$

Remark 13.55. If X is a scheme that is proper over an algebraically closed field k, we also have the geometric characterization Proposition 12.94 of closed immersions. Thus a tuple $(\mathscr{L}, t_0, \ldots, t_n)$ yields a closed immersion $r \colon X \hookrightarrow \mathbb{P}^n_k$ if and only if the following three conditions are satisfied.
(1) The sections t_i must generate $\mathscr{L}$: For all $x \in X(k)$ there exists an i with $t_i(x) \neq 0$ (so that we obtain a morphism $r \colon X \to \mathbb{P}^n_k$).
(2) The t_i must separate points: Let $V \subseteq \Gamma(X, \mathscr{L})$ be the k-subvector space generated by the t_i. Then for all $x, x' \in X(k)$ there exists $t \in V$ such that $t(x) = 0$ and $t(x') \neq 0$, or vice versa (so that r is injective on k-valued points).
(3) The t_i must separate tangent vectors: For all $x \in X(k)$ the set $\{t \in V \; ; \; t(x) = 0\}$ generates $\mathfrak{m}_x \mathscr{L}_x / \mathfrak{m}_x^2 \mathscr{L}_x \cong \mathfrak{m}_x / \mathfrak{m}_x^2$ (so that the map induced by r on the dual of tangent spaces is surjective).

If S is "not too big", a very ample line bundle yields a global embedding into a projective bundle.

Proposition 13.56. *Let S be qcqs and let $f \colon X \to S$ be a morphism of finite type. Let $\mathscr{L}$ be an invertible $\mathscr{O}_X$-module. Then the following assertions are equivalent.*
(i) *$\mathscr{L}$ is very ample with respect to f.*
(ii) *$f_* \mathscr{L}$ is quasi-coherent, the canonical homomorphism $u \colon f^* f_* \mathscr{L} \to \mathscr{L}$ is surjective, and the corresponding S-morphism $r \colon X \to \mathbb{P}(f_* \mathscr{L})$ is an immersion.*
(iii) *There exists an $\mathscr{O}_S$-module $\mathscr{E}$ of finite presentation and a quasi-compact immersion $i \colon X \to P := \mathbb{P}(\mathscr{E})$ of S-schemes such that $\mathscr{L} \cong i^* \mathscr{O}_P(1)$.*
(iv) *There exists a graded quasi-coherent $\mathscr{O}_S$-algebra $\mathscr{A}$ such that $\mathscr{A}_1$ is an $\mathscr{O}_S$-module of finite type which generates $\mathscr{A}$ and there exists a quasi-compact open schematically dominant immersion $j \colon X \to P' := \operatorname{Proj} \mathscr{A}$ such that $\mathscr{L} \cong j^* \mathscr{O}_{P'}(1)$.*
(v) *For every open affine subscheme U of S there exists an integer $n \geq 0$ and an immersion $i \colon f^{-1}(U) \to \mathbb{P}^n_U$ of U-schemes such that $\mathscr{L}_{|f^{-1}(U)} \cong i^* \mathscr{O}_{\mathbb{P}^n_U}(1)$.*

As (i) and (ii) can both be checked locally on S, the equivalence of (i) and (ii) holds for an arbitrary scheme S.

Proof. As f is quasi-compact by hypothesis, any S-morphism $X \to Y$ to a separated S-scheme Y (e.g., $Y = \mathbb{P}(\mathscr{E})$ in (iii), or $Y = \operatorname{Proj}\mathscr{A}$ in (iv)) is quasi-compact (Proposition 10.3).

(v) $\Rightarrow$ (i). This is clear.

(i) $\Rightarrow$ (ii). All assertions in (ii) are local on S. Thus we may assume that S is affine and that there exists an S-immersion $r\colon X \hookrightarrow P := \mathbb{P}(\mathscr{O}_S^{n+1})$ such that $r^*\mathscr{O}_P(1) \cong \mathscr{L}$. Thus f is separated, and $f_*\mathscr{L}$ is quasi-coherent by Corollary 10.27. We denote by $u\colon f^*\mathscr{O}_S^{n+1} = \mathscr{O}_X^{n+1} \twoheadrightarrow \mathscr{L}$ the corresponding surjection. By adjunction of f^* and f_*, the homomorphism u factorizes into $\mathscr{O}_X^{n+1} \to f^*(f_*\mathscr{L}) \to \mathscr{L}$ and the second homomorphism is necessarily surjective. It yields an S-morphism $r'\colon X \to \mathbb{P}(f_*\mathscr{L}) = \operatorname{Proj}(\operatorname{Sym}\Gamma(X,\mathscr{L}))$. It remains to show that r' is an immersion. Now u corresponds to sections $t_0,\ldots,t_n \in \Gamma(X,\mathscr{L}) = \Gamma(S, f_*\mathscr{L})$ and we have $r'^{-1}(D_+(t_i)) = X_{t_i}$. We obtain a factorization of the immersion $r_{|X_{t_i}}$

$$X_{t_i} \xrightarrow{r'_{|X_{t_i}}} D_+(t_i) \to D_+(T_i)$$

for all i. Thus $r'_{|X_{t_i}}$ is an immersion. As the X_{t_i} cover X, we see that r' is an immersion.

(ii) $\Rightarrow$ (iii). By Corollary 10.50 we may write $f_*\mathscr{L}$ as filtered inductive limit of $\mathscr{O}_S$-modules $\mathscr{E}_\lambda$ of finite presentation. By Lemma 10.47 there exists an index λ_0 such that the composition $f^*\mathscr{E}_\lambda \to f^*f_*\mathscr{L} \twoheadrightarrow \mathscr{L}$ is surjective for all $\lambda \geq \lambda_0$. We obtain S-morphisms $r_\lambda\colon X \to \mathbb{P}(\mathscr{E}_\lambda)$. It suffices to show that there exists an index $\lambda_1 \geq \lambda_0$ such that r_λ is an immersion for all $\lambda \geq \lambda_1$.

As S is quasi-compact, this is a local question on S and thus we may assume that $S = \operatorname{Spec} R$ is affine. Then $\mathbb{P}(f_*\mathscr{L}) = \operatorname{Proj} A$, where $A = \operatorname{Sym}_R \Gamma(X,\mathscr{L})$. As f is of finite type, X is quasi-compact, and there exist finitely many $t_i \in \Gamma(X,\mathscr{L}) = \Gamma(S, f_*\mathscr{L})$ such that $(X_{t_i})_i$ is an open affine covering and such that

$$r_i := r_{|X_{t_i}}\colon X_{t_i} \to D_+(t_i)$$

is a closed immersion. It corresponds to a surjective R-algebra homomorphism

$$\varphi_i\colon \Gamma(D_+(t_i), \mathscr{O}_{\mathbb{P}(f_*\mathscr{L})}) \to B_i := \Gamma(X_{t_i}, \mathscr{O}_X)$$

As X is of finite type over S, the R-algebra B_i is generated by finitely many elements b_{ij}. Thus there exists an integer $m > 0$ and $s_{ij} \in A_m$ such that $\varphi_i(t_i^{-m}s_{ij}) = b_{ij}$ (we choose m sufficiently large such that it works for all i and j).

By hypothesis, A is the filtered inductive limit of graded algebras $A_\lambda := \operatorname{Sym}_R \Gamma(S, \mathscr{E}_\lambda)$. Thus there exists an index λ_1 and, for all $\lambda \geq \lambda_1$, homogeneous elements t_i^λ and s_{ij}^λ of A_λ that are sent to t_i and s_{ij}, respectively, under $A_\lambda \to A$. Then $r_\lambda^{-1}(D_+(t_i^\lambda)) = X_{t_i}$ and the restriction $X_{t_i} \to D_+(t_i^\lambda)$ of r_λ is a morphism of affine schemes which is surjective on global sections. Therefore r_λ is an immersion.

(iii) $\Rightarrow$ (iv). As remarked above, the immersion i is quasi-compact. Therefore there exists a factorization of i over an open schematically dominant immersion $j\colon X \to Z$, where $Z \subseteq \mathbb{P}(\mathscr{E})$ is a closed subscheme (Remark 10.31). By (a globalization of) Proposition 13.24, $Z = \operatorname{Proj}\mathscr{A}$ with $\mathscr{A} \cong \operatorname{Sym}\mathscr{E}/\mathscr{I}$ for a homogeneous quasi-coherent ideal $\mathscr{I}$. As $\operatorname{Sym}\mathscr{E}$ is generated by its degree 1 component $\mathscr{E}$, we see that $\mathscr{A}$ is of the desired form.

(iv) $\Rightarrow$ (v). We may assume that S is affine. Then we can find a surjection $\mathscr{E} \twoheadrightarrow \mathscr{A}_1$, where $\mathscr{E}$ is a free $\mathscr{O}_S$-module of finite type. The surjective homomorphism of graded $\mathscr{O}_S$-algebras $\operatorname{Sym}\mathscr{E} \twoheadrightarrow \operatorname{Sym}\mathscr{A}_1 \twoheadrightarrow \mathscr{A}$ yields a closed immersion $\operatorname{Proj}\mathscr{A} \hookrightarrow \mathbb{P}(\mathscr{E}) \cong \mathbb{P}_S^n$. This shows (v). $\qquad\square$

We also note the following compatibilities of very ample line bundles with respect to tensor products.

Proposition 13.57. *Let $f\colon X \to S$ be a morphism of finite type, and let $\mathscr{L}$ and $\mathscr{L}'$ be invertible $\mathscr{O}_X$-modules.*
(1) *Assume that $\mathscr{L}$ is very ample for f and that there exists locally on S a quasi-coherent $\mathscr{O}_S$-module $\mathscr{E}'$ and a surjection $u\colon f^*\mathscr{E}' \twoheadrightarrow \mathscr{L}'$. Then $\mathscr{L} \otimes \mathscr{L}'$ is very ample for f.*
(2) *If $\mathscr{L}$ and $\mathscr{L}'$ are very ample for f, $\mathscr{L} \otimes \mathscr{L}'$ is very ample for f.*

Proof. (1). We can assume that S is affine and that there is a surjection $u\colon f^*\mathscr{E}' \twoheadrightarrow \mathscr{L}'$. By Lemma 10.47 we may assume that $\mathscr{E}'$ is of finite type. Let $i'\colon X \to P' := \mathbb{P}(\mathscr{E}')$ be the S-morphism corresponding to u. As $\mathscr{L}$ is very ample, there exists also an S-immersion $i\colon X \to P := \mathbb{P}(\mathscr{E})$ for some finite free $\mathscr{O}_S$-module $\mathscr{E}$. Then $(i, i')_S\colon X \to P \times_S P'$ is an S-immersion. Composing $(i, i')_S$ with the Segre embedding $P \times_S P' \hookrightarrow Q := \mathbb{P}(\mathscr{E} \otimes \mathscr{E}')$ (8.8.6) we obtain an immersion $i''\colon X \hookrightarrow Q$. As we have $i^*\mathscr{O}_P(1) \cong \mathscr{L}$ and $i'^*\mathscr{O}_{P'}(1) \cong \mathscr{L}'$, the definition of the Segre embedding shows $i''^*\mathscr{O}_Q(1) \cong \mathscr{L} \otimes \mathscr{L}'$. Therefore $\mathscr{L} \otimes \mathscr{L}'$ is very ample.

(2). This follows from (1) because every very ample line bundle $\mathscr{L}'$ satisfies the condition on $\mathscr{L}'$ in (1) by Proposition 13.56. □

Remark 13.58. Let $\mathscr{E}$ be a quasi-coherent $\mathscr{O}_S$-module of finite type and let $P := \mathbb{P}(\mathscr{E})$ be the corresponding projective bundle. By Proposition 13.56 the line bundle $\mathscr{O}_P(1)$ is very ample. Therefore its d-fold tensor power $\mathscr{O}_P(d)$, $d > 0$ is also very ample by Proposition 13.57.

To conclude this section, we compare the notions of ample and very ample line bundles in the case of an affine base S.

Theorem 13.59. *Let $S = \operatorname{Spec} R$ be an affine scheme, let $f\colon X \to S$ be a morphism of finite type, and let $\mathscr{L}$ and $\mathscr{L}'$ be invertible $\mathscr{O}_X$-modules.*
(1) *If $\mathscr{L}$ is very ample for f, then $\mathscr{L}$ is ample.*
(2) *The following assertions are equivalent:*
   (i) *The line bundle $\mathscr{L}$ is ample.*
   (ii) *There exists an integer $n > 0$ such that $\mathscr{L}^{\otimes n}$ is very ample for f.*
   (iii) *For every line bundle $\mathscr{L}'$ on X there exists an integer $n_0 > 0$ (depending on $\mathscr{L}'$) such that $\mathscr{L}' \otimes \mathscr{L}^{\otimes n}$ is very ample for f for all $n \geq n_0$.*

In Section (13.14) we will define a relative notion of ampleness and generalize this theorem to arbitrary base schemes (Theorem 13.62).

Proof. (1). As $\mathscr{L}$ is very ample, we find an S-immersion $i\colon X \to P := \mathbb{P}^n_S$ such that $\mathscr{L} = i^*\mathscr{O}_P(1)$. As $\mathscr{O}_P(1)$ is ample by Example 13.45, $\mathscr{L}$ is ample by Proposition 13.51.

(2). Clearly (iii) implies (ii). If $\mathscr{L}^{\otimes d}$ is very ample, it is ample by (1). It follows that $\mathscr{L}$ is ample by Proposition 13.50 (1). This proves that (ii) implies (i).

It remains to prove that (i) implies (iii). The assumptions imply that X is qcqs. As a first step we show the implication "(i) $\Rightarrow$ (ii)". Since $\mathscr{L}$ is ample, replacing $\mathscr{L}$ by some power $\mathscr{L}^{\otimes d}$ for some $d \geq 1$ we may assume that there exist sections $f_i \in \Gamma(X, \mathscr{L})$ $(i = 1, \ldots, r)$ such that $(X_{f_i})_i$ is an open affine covering of X (Proposition 13.47). Set $A_i := \Gamma(X_{f_i}, \mathscr{O}_X) = \Gamma(X_{f_i}, \mathscr{L})$ (13.11.1). Let $\{\, a_{ij} \;;\; j = 1, \ldots, k_i \,\}$ be a set of generators of the R-algebra A_i. By Theorem 7.22 there exists for all i, j an integer $n_{ij} \geq 1$ such that $a_{ij} \otimes f_i^{\otimes n_{ij}}$ extends to a global section $b_{ij} \in \Gamma(X, \mathscr{L}^{\otimes n_{ij}})$. We may assume that all n_{ij} are equal to some sufficiently large integer n. Then by Lemma 13.54 the global sections $f_i^{\otimes n}$ ($i = 1, \ldots, r$) and b_{ij} ($i = 1, \ldots, r$; $j = 1, \ldots k_i$) define an immersion $i \colon X \hookrightarrow P := \mathbb{P}_S^N$ such that $\mathscr{L}^{\otimes n} \cong i^* \mathscr{O}_P(1)$.

Now let us show (iii) assuming (i). We have already shown that there exists an integer $d \geq 1$ such that $\mathscr{L}^{\otimes d}$ is very ample. As $\mathscr{L}$ is ample, there exists an integer m_0 such that $\mathscr{L}^{\otimes n} \otimes \mathscr{L}'$ is generated by its global sections for all $n \geq m_0$. Using Proposition 13.57 (1) we see that $\mathscr{L}^{\otimes n} \otimes \mathscr{L}'$ is very ample for all $n \geq d + m_0$. $\qquad\square$

(13.13) Linear systems.

Let R be a noetherian ring, and let $i \colon X \to \mathbb{P}_R^n$ be a closed immersion. We outline a slightly different (and more classical) view on this situation. Let $\mathscr{L} = i^* \mathscr{O}(1)$. We have the pull-back map (7.8.11)

$$ i^* \colon \Gamma(\mathbb{P}_R^n, \mathscr{O}(1)) \to \Gamma(X, \mathscr{L}). $$

Each regular section $H \in \Gamma(\mathbb{P}_R^n, \mathscr{O}(1))$ corresponds to a hyperplane in $\mathbb{P}_R^n$, and similarly, each regular section $s \in \Gamma(X, \mathscr{L})$ gives rise to an effective Cartier divisor in the linear equivalence class of divisors attached to $\mathscr{L}$ (cf. Corollary 11.30).

As explained in Remark 11.27, we can view effective Cartier divisors as subschemes (which locally are the vanishing scheme of one regular element). Let us denote by $D_H \subset \mathbb{P}_R^n$ and $D_s \subset X$ the closed subschemes corresponding to H and s as above, respectively. We have $X \subseteq H$ if and only if $i^*(H) = 0$, and otherwise $D_{i^*(H)} = i^{-1}(H)$, where the right hand side denotes the schematic inverse image of H, which we can also view as the schematic intersection $X \cap H$.

For the rest of the discussion let us assume that $R = k$ is a field, and that X is geometrically integral over k. In particular one has $\Gamma(X, \mathscr{O}_X) = k$ by Proposition 12.66. As X is integral, a section s of a line bundle $\mathscr{L}$ on X is regular if and only if it is non-zero. Further assume that i^* is injective, i.e., that X is not contained in a hyperplane of $\mathbb{P}_k^n$.

Two non-zero sections $s, s' \in \Gamma(X, \mathscr{L})$ give rise to the same closed subscheme if and only if they differ by a global section of $\mathscr{O}_X^\times$, i.e., $s' = \alpha s$, $\alpha \in \Gamma(X, \mathscr{O}_X^\times) = k^\times$ (Proposition 11.34). Therefore we can view the space of effective divisors in the linear equivalence class of $\mathscr{L}$ as the projective space of all lines in $\Gamma(X, \mathscr{L})$, i.e., as the set $\mathbb{P}(\Gamma(X, \mathscr{L})^\vee)(k)$.

The image of i^*, viewed as a linear subspace of $\mathbb{P}(\Gamma(X, \mathscr{L})^\vee)$, is called the *linear system attached to i*. In general, any linear subspace of $\mathbb{P}(\Gamma(X, \mathscr{L})^\vee)$ is called a *linear system of $\mathscr{L}$*. The whole space $\mathbb{P}(\Gamma(X, \mathscr{L})^\vee)(k)$ is called the *complete linear system* associated with $\mathscr{L}$. A linear system is called *base point free*, if the intersection $\bigcap_D \operatorname{Supp}(D)$, where D runs through all closed subschemes attached to the divisors in the linear system, is empty. The linear system attached to i is base point free.

Conversely, let L be a linear system of $\mathscr{L}$ corresponding to a subspace $V \subseteq \Gamma(X, \mathscr{L})$. Then L is base point free if and only if for every (or, equivalently, for one) choice of basis $t_0, \ldots, t_n$ of V the family $\underline{t} := (t_j)_{0 \leq j \leq n}$ generates $\mathscr{L}$, i.e. $\underline{t}$ defines a morphism $r_{\underline{t}} \colon X \to \mathbb{P}(V)$. If we do this for the linear system attached to i we have, up to a linear coordinate changes of $\mathbb{P}_k^n$, that $i = r_{\underline{t}}$.

We see that the effective Cartier divisors in the linear system attached to i are precisely the intersections of hyperplanes in $\mathbb{P}^n_k$ with X. On the other hand, the embedding i is "defined" by the choice of $n+1$ suitable linearly equivalent effective Cartier divisors (which become the inverse images of the coordinate hyperplanes in $\mathbb{P}^n_k$).

(13.14) Relatively ample line bundles.

We will now relate ample and very ample line bundles over arbitrary base schemes. We first define the following relative notion of an ample line bundle.

Definition 13.60. *Let $f\colon X \to S$ be a morphism of finite type. An invertible $\mathscr{O}_X$-module $\mathscr{L}$ is called (relatively) ample for f or f-ample or S-ample if $\mathscr{L}_{|f^{-1}(U)}$ is an ample line bundle for all open affine subschemes U of S.*

If we want to stress that $\mathscr{L}$ is ample in the previous sense (Definition 13.44), we will also say that $\mathscr{L}$ is *absolutely ample*.

Note that the existence of an f-ample line bundle implies that f is separated (Proposition 13.48).

Remark 13.61. Let $f\colon X \to S$ be a morphism of finite type and let $\mathscr{L}$ be an invertible $\mathscr{O}_X$-module.
(1) As on an affine scheme all line bundles are ample, we see that all line bundles are ample for f if f is an affine morphism.
(2) Proposition 13.51 shows that if $\mathscr{L}$ is ample for f and if $i\colon Z \to X$ is a quasi-compact immersion, then $i^*\mathscr{L}$ is ample for $f \circ i$.
(3) Let $\mathscr{L}$ be ample for f. The construction before Proposition 13.48 yields an open quasi-compact schematically dominant immersion of S-schemes

$$(13.14.1) \qquad\qquad r\colon X \hookrightarrow \operatorname{Proj}\bigoplus_{n\geq 0} f_*\mathscr{L}^{\otimes n}.$$

(4) Let f be quasi-separated and let $\mathscr{L}$ be absolutely ample. Then $\mathscr{L}$ is f-ample.
Indeed, then for every open affine subscheme U of S the inclusion $f^{-1}(U) \hookrightarrow X$ is quasi-compact (Remark 10.4) and thus $\mathscr{L}_{|f^{-1}(U)}$ is ample by Proposition 13.51.

We can now generalize Theorem 13.59 to the case where the base S is qcqs, but not necessarily affine.

Theorem 13.62. *Let $f\colon X \to S$ be a morphism of finite type, and let $\mathscr{L}$ and $\mathscr{L}'$ be invertible $\mathscr{O}_X$-modules.*
(1) *If $\mathscr{L}$ is very ample for f, then $\mathscr{L}$ is ample for f.*
(2) *Assume that S is quasi-compact. Then the following assertions are equivalent.*
   (i) *The line bundle $\mathscr{L}$ is ample for f.*
   (ii) *There exists an integer $n > 0$ such that $\mathscr{L}^{\otimes n}$ is very ample for f.*
   (iii) *For every line bundle $\mathscr{L}'$ on X there exists an integer $n_0 > 0$ (depending on $\mathscr{L}'$) such that $\mathscr{L}' \otimes \mathscr{L}^{\otimes n}$ is very ample for f for all $n \geq n_0$.*

Proof. (1). We may assume that S is affine, and the assertion follows from Theorem 13.59 (1).

(2). Again, everything is easily reduced to the case that S is affine, so it is enough to apply Theorem 13.59 (2). $\qquad\square$

As an interesting consequence of the theorem, we note

Proposition 13.63. *Let S be a scheme and let $f\colon X \to S$ be a morphism of finite type. Let $\mathscr{L}$ be an invertible $\mathscr{O}_X$-module. Then $\mathscr{L}$ is f-ample if and only if there exists an open covering (U_α) of S such that $\mathscr{L}_{|f^{-1}(U_\alpha)}$ is (absolutely) ample for all α.*

Proof. If $\mathscr{L}$ is f-ample, then the same is true for the restriction $\mathscr{L}_{|f^{-1}(V)}$ for any open subset $V \subseteq S$. Let $U \subseteq S$ be an open affine subset. We have to show that $\mathscr{L}_{|f^{-1}(U)}$ is ample. Replacing S by U and U_α by $U \cap U_\alpha$ we may assume that S is affine. Since S is quasi-compact, we may assume that the covering (U_α) is finite. Theorem 13.62 (2) implies that there exists $n > 0$ such that each $(\mathscr{L}^{\otimes n})_{|f^{-1}(U_\alpha)}$ is very ample for all α. Hence $\mathscr{L}^{\otimes n}$ is very ample for f, and hence $\mathscr{L}$ is f-ample (again by Theorem 13.62 (2)). $\qquad\square$

The proposition shows in particular that if S is affine, then $\mathscr{L}$ is ample if and only if $\mathscr{L}$ is ample for f.

Proposition 13.64. *Let $f\colon X \to S$ be a morphism of finite type and let $\mathscr{L}$ be an invertible $\mathscr{O}_X$-module. Let $g\colon S' \to S$ be a morphism of schemes. If $\mathscr{L}$ is ample for S, then $g'^*(\mathscr{L})$ is ample for the base change $f'\colon X' := X \times_S S' \to S'$ of f, where $g'\colon X' \to X$ is the projection.*

Proof. By Proposition 13.63 we may assume that S and S' are affine. Then g' is an affine morphism. Thus if $f \in \Gamma(X, \mathscr{L}^{\otimes d})$ $(d > 0)$ is a section such that X_f is affine, $g'^{-1}(X_f) = X'_{g'^*(f)}$ is affine, where $g'^*(f) \in \Gamma(X', g'^* \mathscr{L}^{\otimes n})$ is the pullback of f under g'. Now use Proposition 13.47 (iv). $\qquad\square$

If f is of finite presentation, then Proposition 13.63 can be strengthened as follows (Exercise 13.13). Let $s \in S$ be a point and let $j\colon X \times_S \operatorname{Spec} \mathscr{O}_{S,s} \to X$ be the canonical morphism. Then $j^* \mathscr{L}$ is ample if and only if there exists an open affine neighborhood U of s such that $\mathscr{L}_{|f^{-1}(U)}$ is ample (the "if"-part follows from Proposition 13.64).

In Volume II we will see that this result may be strengthened considerably if $f\colon X \to S$ is in addition proper: In this case $\mathscr{L}$ is ample over some open affine neighborhood U of s if and only if the restriction $\mathscr{L}_{|X_s}$ to the fiber over s is ample.

The property for a line bundle to be relatively ample is compatible with composition of morphisms in the following sense.

Proposition 13.65. *Let S be a quasi-compact scheme, let $f\colon X \to Y$ and $g\colon Y \to S$ be morphisms of finite type. Let $\mathscr{L}$ be an f-ample $\mathscr{O}_X$-module and let $\mathscr{K}$ be a g-ample $\mathscr{O}_Y$-module. Then there exists an integer n_0 such that $\mathscr{L} \otimes f^*(\mathscr{K})^{\otimes n}$ is ample for $g \circ f$ for all $n \geq n_0$.*

Proof. As we can cover S by finitely many affine schemes, we may assume that S is affine. The existence of $\mathscr{L}$ and of $\mathscr{K}$ implies that X is separated over Y and that Y is separated over S. Therefore X and Y are quasi-compact separated schemes.

We find an integer $d > 0$ and finitely many sections $t_i' \in \Gamma(Y, \mathscr{K}^{\otimes d})$ such that the $Y_{t_i'}$ form an open affine covering of Y (Proposition 13.47). Let $t_i \in \Gamma(X, f^*\mathscr{K}^{\otimes d})$ be the pull back of t_i' (7.8.11). Then $X_i := X_{t_i} = f^{-1}(Y_{t_i'})$ and the X_i cover X. As $\mathscr{L}_{|X_i}$ is ample (Proposition 13.51), there exist sections $s_{ij} \in \Gamma(X_i, \mathscr{L}^{\otimes m_i})$ such that the $(X_i)_{s_{ij}}$ form an affine covering of X_i. Replacing s_{ij} by a suitable power, we may also assume that all m_i

are equal to some integer $m > 0$. Using Theorem 7.22 we find an integer $N > 0$, which we may assume to be of the form $N = rm$, and sections

$$u_{ij} \in \Gamma(X, \mathscr{L}^{\otimes m} \otimes f^* \mathscr{K}^{\otimes dN}) = \Gamma(X, (\mathscr{L} \otimes f^* \mathscr{K}^{\otimes dr})^{\otimes m})$$

such that the restriction of u_{ij} to X_i is $s_{ij} \otimes t_i^{\otimes N}$. Then $X_{u_{ij}} = (X_i)_{s_{ij}}$ and therefore the $X_{u_{ij}}$ form an open affine covering of X. Setting $n_0 := dr$ we see that $\mathscr{L} \otimes f^* \mathscr{K}^{\otimes n_0}$ is ample by Proposition 13.47.

Moreover, there exists $k > 0$ such that $\mathscr{K}^{\otimes k}$ (and hence $f^* \mathscr{K}^{\otimes k}$) is generated by its global sections. Therefore $\mathscr{L} \otimes f^* \mathscr{K}^{\otimes n}$ is ample for all $n \geq n_0$ by Proposition 13.50 (3). This finishes the proof. $\square$

For finite locally free morphisms $X \to Y$ we can relate ample line bundles on X and on Y via the norm of line bundles (Remark 12.25). Compare also Proposition 14.58.

Proposition 13.66. *Let S be a scheme, let X and Y be S-schemes of finite type and let $f\colon X \to Y$ be a surjective finite locally free S-morphism.*
(1) Let $\mathscr{L}$ be an S-ample line bundle on X. Then $N_{X/Y}(\mathscr{L})$ is an S-ample line bundle on Y.
(2) Let $\mathscr{M}$ be a line bundle on Y. Then $\mathscr{M}$ is S-ample if and only if $f^* \mathscr{M}$ is S-ample.

Proof. For both assertions we may assume that S is affine. Then a line bundle is S-ample if and only if it is ample. Moreover X and Y are quasi-compact, and X is quasi-separated if and only if Y is quasi-separated (Proposition 10.25).

(1). We set $\mathscr{K} := N_{X/Y}(\mathscr{L})$. We have to show that the sets Y_t for $t \in \Gamma(Y, \mathscr{K}^{\otimes n})$ and $n \geq 1$ form a basis of the topology of Y. Let $y \in Y$ and V be an open affine neighborhood of y. As f is finite, $f^{-1}(y)$ is a finite set. As $\mathscr{L}$ is ample, there exists an $n \geq 1$ and $s \in \Gamma(X, \mathscr{L}^{\otimes n})$ such that X_s is an open affine neighborhood of $f^{-1}(y)$ contained in $f^{-1}(V)$ (Proposition 13.49). Let $t := N_{X/Y}(s) \in \Gamma(Y, N_{X/Y}(\mathscr{L}^{\otimes n})) = \Gamma(Y, \mathscr{K}^{\otimes n})$ (12.6.5). Considering s (resp. t) as a homomorphism $\mathscr{O}_X \to \mathscr{L}^{\otimes n}$ (resp. $\mathscr{O}_Y \to \mathscr{K}^{\otimes n}$), Proposition 12.26 shows that $f^{-1}(Y_t) = X_s$. Therefore Y_t is an open neighborhood of y contained in V.

(2). As f is affine, $f^* \mathscr{M}$ is ample if $\mathscr{M}$ is ample. Conversely, let $f^* \mathscr{M}$ be ample. We may assume that f has constant rank n. By (12.6.3) we have $N_{X/Y}(f^* \mathscr{M}) \cong \mathscr{M}^{\otimes n}$, and $\mathscr{M}^{\otimes n}$ is ample by (1). Therefore $\mathscr{M}$ is ample by Proposition 13.50 (1). $\square$

If X and Y are integral and Y is normal, then it suffices in Proposition 13.66 to assume that f is finite surjective (Exercise 13.18).

We conclude this section by giving a criterion of Grauert (without proof) for an invertible $\mathscr{O}_X$-module $\mathscr{L}$ to be ample. It says that $\mathscr{L}$ is ample if and only if the zero section of $\mathscr{L}$ can be contracted, more precisely:

Theorem 13.67. (Grauert's criterion for ampleness) *Let $f\colon X \to S$ be a separated morphism of finite type and let $\mathscr{L}$ be an invertible $\mathscr{O}_X$-module. Let $\mathbb{V}(\mathscr{L})$ be the corresponding geometric line bundle and let $z\colon X \to \mathbb{V}(\mathscr{L})$ be the zero section. Then $\mathscr{L}$ is ample for f if and only if there exist an S-scheme C, an S-section $\varepsilon\colon S \to C$, and an S-morphism $q\colon \mathbb{V}(\mathscr{L}) \to C$ satisfying the following conditions.*
(a) The diagram

is commutative.

(b) *The restriction* $\mathbb{V}(\mathscr{L}) \setminus z(X) \to C$ *of* q *is an open quasi-compact immersion whose image does not meet* $\varepsilon(S)$.

The proof of this theorem is quite lengthy but it uses only techniques explained in this chapter (see [EGAII] 8.8–8.10 for a proof). The hard part of the theorem is that this criterion is sufficient. For the converse, if one assumes that $\mathscr{L}$ is f-ample, one can choose $C = \operatorname{Spec} \mathscr{A}$, where $\mathscr{A} := \mathscr{O}_S \oplus \bigoplus_{d \geq 1} f_* \mathscr{L}^{\otimes d}$. Then the canonical homomorphism of $\mathscr{O}_X$-algebras $f^* \mathscr{A} \to \bigoplus_{d \geq 0} \mathscr{L}^{\otimes d} = \operatorname{Sym} \mathscr{L}$ defines an X-morphism $\mathbb{V}(\mathscr{L}) \to \operatorname{Spec} f^* \mathscr{A} = C \times_S X$, and we may take for q the corresponding S-morphism $\mathbb{V}(\mathscr{L}) \to C$. Using that the morphism (13.14.1) is a quasi-compact dominant immersion, it is not difficult to see that C, the section ε corresponding to the canonical projection $\mathscr{A} \to \mathscr{O}_S$, and q satisfy the conditions of the theorem.

Moreover, if f is proper, then one can prove that q is proper (and even projective if S is qcqs) and that its restriction yields an isomorphism $\mathbb{V}(\mathscr{L}) \setminus z(X) \overset{\sim}{\to} C \setminus \varepsilon(S)$.

(13.15) Quasi-projective and projective morphisms.

Definition and Proposition 13.68. *Let* $f \colon X \to S$ *be a morphism of finite type.*
(1) *The morphism* f *is called* projective *if there exist a quasi-coherent* $\mathscr{O}_S$-*module* $\mathscr{E}$ *of finite type and a closed* S-*immersion* $X \hookrightarrow \mathbb{P}(\mathscr{E})$.
(2) *The morphism* f *is called* quasi-projective *if there exists an invertible* $\mathscr{O}_X$-*module which is ample with respect to* f.
If S *is qcqs, then* f *is quasi-projective if and only if there exist a quasi-coherent* $\mathscr{O}_S$-*module* $\mathscr{E}$ *of finite type and an* S-*immersion* $X \hookrightarrow \mathbb{P}(\mathscr{E})$.

In general, it may happen that $f \colon X \to S$ is projective and there exists no immersion $X \hookrightarrow \mathbb{P}^n_S$ for some n. We refer to Summary 13.71 below for the implications between different possible notions of "(quasi-)projective". Here we follow [EGAII].

Proof. Assume that S is qcqs. If there exists an ample line bundle $\mathscr{L}$ on X, then some power of $\mathscr{L}$ is very ample and gives rise to an embedding $X \hookrightarrow \mathbb{P}(\mathscr{E})$ as desired; see Proposition 13.56. On the other hand, given such an embedding, the pull-back of $\mathscr{O}_{\mathbb{P}(\mathscr{E})}(1)$ to X is (very) ample on X by Proposition 13.51 and Proposition 10.3 (2). $\qquad\square$

Example 13.69. Let S be a scheme, let $\mathscr{E}$ be a quasi-coherent $\mathscr{O}_S$-module of finite type, and let $e \geq 0$ be an integer. Then the Grassmannian $\operatorname{Grass}^e(\mathscr{E})$ is projective over S. In particular, $\operatorname{Grass}^e(\mathscr{E})$ is proper over S.

Indeed, the Plücker embedding is a closed immersion $\operatorname{Grass}^e(\mathscr{E}) \hookrightarrow \mathbb{P}(\bigwedge^e \mathscr{E})$ (Remark 8.24). As $\bigwedge^e \mathscr{E}$ is again quasi-coherent of finite type, the claim follows.

We will give in Example 15.12 another proof that $\operatorname{Grass}^e(\mathscr{E})$ is proper using the valuative criterion.

Remark 13.70. Note that although the property for a line bundle on X to be S-ample is local on S, the existence of such a line bundle (and thus the property of being quasi-projective) is not local on S: Hironaka gave an example of a birational proper non-projective morphism $f\colon X \to S$ of proper schemes over $\mathbb{C}$, where X and S are smooth over $\mathbb{C}$ and irreducible of dimension 3, where S is in addition projective over $\mathbb{C}$ and where $S = U_1 \cup U_2$, where U_i is open and affine and such that the restriction $f^{-1}(U_i) \to U_i$ of f is projective for $i = 1, 2$ (see [Ha3] Appendix B, Example 3.4.1).

This also gives an example of a proper smooth scheme X over $\mathbb{C}$ which is not projective over $\mathbb{C}$. Moreover we obtain an example of two morphisms (namely f and $S \to \operatorname{Spec} k$) which are both locally on the target projective, but whose composition is not locally on the target projective. Cf. Exercise 13.6.

Note that X has to have dimension ≥ 3: In Theorem 15.18 we will prove that every smooth proper scheme over a field k of dimension 1 is projective (in fact this is true for arbitrary proper schemes of dimension 1). It also can be shown (e.g., [Ba] Theorem 1.28) that any smooth proper scheme over k of dimension 2 is projective (here smoothness is essential: there are non-smooth proper schemes of dimension 2 that are not projective).

Of course, if the base scheme S is the spectrum of a local ring (e.g., $S = \operatorname{Spec} k$ for a field k), there is no difference between the properties "(quasi-)projective locally on the target" and "(quasi-)projective". We summarize the results obtained so far as follows.

Summary 13.71. Let $f\colon X \to S$ be a morphism of finite type. Consider the following assertions.
(i) There exists an S-immersion (resp. a closed S-immersion) $X \hookrightarrow \mathbb{P}^n_S$ for some $n \geq 0$.
(ii) There exists an S-immersion (resp. a closed S-immersion) $X \hookrightarrow \mathbb{P}(\mathscr{E})$ for some quasi-coherent $\mathcal{O}_S$-module $\mathscr{E}$ of finite type.
(iii) There exists a quasi-compact open schematically dominant S-immersion (resp. an S-isomorphism) $X \hookrightarrow \operatorname{Proj} \mathscr{A}$, where $\mathscr{A}$ is a graded quasi-coherent $\mathcal{O}_S$-algebra such that $\mathscr{A}_1$ is of finite type and generates $\mathscr{A}$.
(iv) There exists a very ample line bundle for f on X (resp. there exists a very ample line bundle for f on X and X is proper over S).
(v) There exists an f-ample line bundle on X (resp. there exists an f-ample line bundle on X and X is proper over S).
(vi) f is quasi-projective (resp. f is projective).
Then we have the following implications.
(1) In general, the implications "(i) $\Rightarrow$ (ii) $\Leftrightarrow$ (iii) $\Rightarrow$ (iv) $\Rightarrow$ (v) $\Leftrightarrow$ (vi)" (resp. the implications "(i) $\Rightarrow$ (ii) $\Leftrightarrow$ (iii) $\Leftrightarrow$ (vi) $\Rightarrow$ (iv) $\Rightarrow$ (v)") hold.
(2) If the scheme S is qcqs (e.g., if S is noetherian), then the following implications hold: "(i) $\Rightarrow$ (ii) $\Leftrightarrow$ (iii) $\Leftrightarrow$ (iv) $\Leftrightarrow$ (v) $\Leftrightarrow$ (vi)".
(3) If S is qcqs such that there exists an ample line bundle $\mathscr{M}$ on S (e.g., if S is affine), then all the assertions are equivalent.
Indeed, all of these implications have already been shown or follow immediately from the definitions except for the implication "(ii) $\Rightarrow$ (i)" in (3). If $\mathscr{E}$ is a quasi-coherent $\mathcal{O}_S$-module of finite type, there exist integers $N, n \geq 0$ and a surjection $\mathcal{O}_S^{n+1} \twoheadrightarrow \mathscr{E} \otimes \mathscr{M}^{\otimes N}$ by the definition of "ample". It yields a closed immersion $\mathbb{P}(\mathscr{E}) = \mathbb{P}(\mathscr{E} \otimes \mathscr{M}^{\otimes N}) \hookrightarrow \mathbb{P}(\mathcal{O}_S^{n+1}) \cong \mathbb{P}^n_S$, where the first equality is given by Lemma 13.31. This proves "(ii) $\Rightarrow$ (i)".

All assertions in Summary 13.71 imply that locally on S there exists an S-immersion (resp. a closed S-immersion) $X \hookrightarrow \mathbb{P}^n_S$ for some $n \geq 0$. But Remark 13.70 shows that this condition does not imply in general that $X \to S$ is quasi-projective (resp. projective) even if S itself is projective over a field.

Corollary 13.72. *Let $f \colon X \to S$ be a morphism of schemes. If f is projective, then f is proper and quasi-projective. The converse holds if S is qcqs.*

Proof. If f is projective, f is proper by Theorem 13.40 and if $i \colon X \to P := \mathbb{P}(\mathscr{E})$ is a closed immersion (where $\mathscr{E}$ is a quasi-coherent $\mathscr{O}_S$-module of finite type), then $i^* \mathscr{O}_P(1)$ is very ample for f. Thus f is quasi-projective.

Conversely, assume f is proper and there exists an f-ample line bundle. If S is qcqs, there exists a very ample line bundle for f by Theorem 13.62 (2). Thus there exists an immersion $i \colon X \hookrightarrow \mathbb{P}(\mathscr{E})$ for some quasi-coherent $\mathscr{O}_S$-module of finite type by Proposition 13.56. As X is proper over S, the immersion i is closed. Therefore X is projective over S. $\qquad \square$

Remark 13.73. It follows from Proposition 13.64 that the property "quasi-projective" is stable under base change. If X is an S-scheme such that there exists a closed S-immersion $i \colon X \hookrightarrow \mathbb{P}(\mathscr{E})$ and if $g \colon S' \to S$ is a morphism, then the base change $i_{(S')}$ is a closed immersion of $X \times_S S'$ into $\mathbb{P}(\mathscr{E}) \times_S S' = \mathbb{P}(g^* \mathscr{E})$. Therefore the property "projective" is stable under base change.

Moreover it follows from Proposition 13.65 and Corollary 13.72 that they are also very often stable under composition: Let $f \colon X \to Y$ and $g \colon Y \to Z$ be quasi-projective (resp. projective) morphisms and assume that Z is quasi-compact (resp. qcqs). Then $g \circ f$ is quasi-projective (resp. projective).

Proposition 13.74. *Let $f \colon X \to S$ be a proper and quasi-projective morphism (e.g., if f is projective) and let $\mathscr{L}$ be an invertible $\mathscr{O}_X$-module that is ample for f. Then the open immersion (13.14.1) is an isomorphism of S-schemes*

$$r \colon X \xrightarrow{\sim} \operatorname{Proj} \bigoplus_{d \geq 0} f_* \mathscr{L}^{\otimes d}.$$

Proof. We already know that r is an open dominant immersion (Remark 13.61 (3)). As f is proper, r is also a closed immersion and hence surjective. $\qquad \square$

Corollary 13.75. *Let $S = \operatorname{Spec} R$ be an affine scheme, let $f \colon X \to S$ be a projective morphism and let $\mathscr{L}$ be an ample invertible $\mathscr{O}_X$-module. Then there exists an isomorphism of S-schemes*

$$X \cong \operatorname{Proj} \bigoplus_{d \geq 0} \Gamma(X, \mathscr{L}^{\otimes d}).$$

If X is quasi-projective over a field k, then there exists an immersion $X \hookrightarrow \mathbb{P}_k^N$ for some integer $N \geq 1$. It is a natural question how small N we can take to be. If X is projective and smooth over an infinite field k we will show in Theorem 14.132 that we may always take $N = 2 \dim(X) + 1$.

Proposition 13.76. *Let S be a scheme, let X and Y be S-schemes of finite type and let $f \colon X \to Y$ be a surjective finite locally free S-morphism.*
(1) *X is quasi-projective over S if and only if Y is quasi-projective over S.*
(2) *If S is qcqs, then X is projective over S if and only if Y is projective over S.*

Proof. Assertion (1) follows from Proposition 13.66. If S is qcqs, then "projective over S" is equivalent to "quasi-projective and proper over S" (Corollary 13.72). As finite morphisms are proper, X is proper over S if Y is proper over S. The converse holds by Proposition 12.59. Thus (2) follows from (1).   □

If $f\colon X \to S$ is an affine morphism of finite type, the structure sheaf $\mathscr{O}_X$ is very ample for f. In particular, f is quasi-projective. As a corollary we obtain the following characterization of finite morphisms.

Corollary 13.77. *A morphism of schemes $f\colon X \to S$ is finite if and only if it is affine and projective.*

Proof. Let f be finite. Then f is affine by definition. By the remark above, $\mathscr{O}_X$ is very ample and $f_*\mathscr{O}_X$ is an $\mathscr{O}_S$-module of finite type (Remark 12.10 (3)). By Proposition 13.56 there exists an immersion of S-schemes $X \hookrightarrow \mathbb{P}(f_*\mathscr{O}_X)$ which is necessarily a closed immersion because finite morphisms are proper. Hence f is projective.

Conversely, a projective morphism is proper (Theorem 13.40), and by Corollary 12.89 we know that an affine and proper morphism is finite.   □

(13.16) Quasi-affine morphisms.

If X is affine, then every line bundle on X is ample. The converse does not hold. This leads us to the following definition.

Proposition and Definition 13.78.
(1) *A qcqs scheme X is called* quasi-affine *if the following equivalent properties are satisfied*
   (i) *There exists an open embedding $X \hookrightarrow Y$ into an affine scheme Y.*
   (ii) *The structure sheaf $\mathscr{O}_X$ is ample.*
   (iii) *Every invertible $\mathscr{O}_X$-module is ample.*
(2) *A morphism $f\colon X \to S$ of schemes is called* quasi-affine *if there exists an open covering (U_α) by affine subschemes such that $f^{-1}(U_\alpha)$ is quasi-affine for all α.*

Proof. If there exists an embedding as in (i), then by Remark 13.61 (2), we have that $\mathscr{O}_X$ is ample, i. e. (ii). If (ii) holds, then every quasi-coherent $\mathscr{O}_X$-module of finite type is generated by its global sections, and therefore every invertible $\mathscr{O}_X$-module is ample. Clearly, (iii) implies (ii), and given (ii), Proposition 13.48 tells us that the morphism

$$X \to \operatorname{Proj} \bigoplus_{d \geq 0} \Gamma(X, \mathscr{O}_X^{\otimes d}) = \operatorname{Proj} \Gamma(X, \mathscr{O}_X)[T] = \operatorname{Spec} \Gamma(X, \mathscr{O}_X)$$

is an open immersion.   □

Remark 13.79.
(1) Any affine scheme is quasi-affine. Therefore any affine morphism is quasi-affine.
(2) Any quasi-affine morphism $f\colon X \to S$ of finite type is quasi-projective because $\mathscr{O}_X$ is an f-ample line bundle.
(3) Any quasi-compact immersion (e.g., an immersion of locally noetherian schemes) is quasi-affine by Remark 13.61 (2).

(4) More generally, we have seen in Corollary 12.91 that Zariski's main theorem implies that every separated quasi-finite morphism is quasi-affine.

(5) Let $R \neq 0$ be a ring and $X = \mathbb{A}_R^n \setminus \{0\}$ for an integer $n \geq 1$ (Example 4.5). Then X is an open quasi-compact subscheme of $\mathbb{A}_R^n$ and hence quasi-affine. It is affine if and only if $n = 1$.

(6) Let $f \colon X \to Y$ be a quasi-affine morphism of finite type, where Y is quasi-compact. As $\mathscr{O}_X$ is f-ample, Theorem 13.62 (2) shows that every invertible $\mathscr{O}_X$-module is very ample for f.

Proposition 13.80. *Let X be a qcqs scheme. The following assertions are equivalent.*

(i) *X is quasi-affine.*

(ii) *There exists an invertible $\mathscr{O}_X$-module $\mathscr{L}$ such that $\mathscr{L}$ and $\mathscr{L}^{-1}$ are ample.*

(iii) *Every quasi-coherent $\mathscr{O}_X$-module is generated by its global sections.*

(iv) *The canonical morphism $X \to \operatorname{Spec}\Gamma(X, \mathscr{O}_X)$ (Remark 3.7) is a quasi-compact open schematically dominant immersion.*

Proof. The implications "(iii) $\Rightarrow$ (i) $\Rightarrow$ (ii)" are clear, and (ii) implies that $\mathscr{O}_X$ is ample (and hence (i)) by Proposition 13.50 (4). If $\mathscr{O}_X$ is ample, then every quasi-coherent $\mathscr{O}_X$-module of finite type is generated by its global sections. As an arbitrary quasi-coherent $\mathscr{O}_X$-module is the inductive limit of its quasi-coherent submodules of finite type (Corollary 10.50), (iii) follows.

The equivalence of (i) and (iv) follows from the proof of Proposition 13.78. $\qquad\square$

Corollary 13.81. *A morphism $f \colon X \to Y$ of finite type is quasi-affine if and only if $\mathscr{O}_X$ is f-ample.*

Corollary 13.82. *A morphism of schemes $f \colon X \to Y$ is finite if and only if f is quasi-affine and proper.*

Proof. We have only to show that the condition is sufficient. The condition to be finite is local on the target and we may assume that Y is affine and that X is quasi-affine. Set $A := \Gamma(X, \mathscr{O}_X)$. By Section (3.3) f is the composition of the canonical morphism $j \colon X \to \operatorname{Spec} A$ followed by a morphism $g \colon \operatorname{Spec} A \to Y$. By Proposition 13.80, j is an open dominant immersion. As f is proper, j is also proper. Therefore j is an isomorphism. Thus X is affine and hence f is affine and proper. Thus f is finite by Corollary 12.89. $\quad\square$

Proposition 13.83. *Let S be a scheme, let X and Y be S-schemes of finite type and let $f \colon X \to Y$ be a quasi-affine S-morphism of finite type. If $\mathscr{L}$ is an S-ample $\mathscr{O}_Y$-module, then $f^*\mathscr{L}$ is an S-ample $\mathscr{O}_X$-module.*

Proof. We may assume that S is affine. Then a line bundle on X is S-ample if and only if it is ample. Corollary 13.81 shows that $\mathscr{O}_X$ is ample for f. Therefore there exists an integer $n > 0$ such that $f^*(\mathscr{L})^{\otimes n}$ is S-ample (Proposition 13.65). Thus $f^*(\mathscr{L})$ is ample by Proposition 13.50 (1). $\qquad\square$

We obtain the following criterion for ampleness.

Theorem 13.84. *Let S be a scheme, let $f\colon X \to S$ be a separated morphism of finite type, and let $\mathscr{L}$ be an invertible $\mathscr{O}_X$-module. Let $\mathscr{E}$ be a quasi-coherent $\mathscr{O}_S$-module of finite type, let $f^*\mathscr{E} \to \mathscr{L}$ be a surjective homomorphism, and let $r\colon X \to \mathbb{P}(\mathscr{E})$ be the corresponding morphism.*

(1) *If r is quasi-finite, then $\mathscr{L}$ is ample over S.*

(2) *Assume that f is proper. Then the following assertions are equivalent.*

    (i) *$\mathscr{L}$ is ample over S.*

    (ii) *r is finite.*

    (iii) *r is quasi-affine.*

Proof. As f is separated and of finite type and as $P := \mathbb{P}(\mathscr{E})$ is separated over S, the morphism r is separated and of finite type. If r is quasi-affine, then Proposition 13.83 implies that $\mathscr{L} = r^*\mathscr{O}_P(1)$ is ample. Thus Assertion (1) follows from Remark 13.79 (4).

In Assertion (2) the implication "(ii) $\Rightarrow$ (iii)" is clear and the implication "(iii) $\Rightarrow$ (i)" has already been proved. Assume that $\mathscr{L}$ is ample over S. As f is proper, r is proper as well, and it suffices to show that r is quasi-finite by Corollary 12.89. We only have to check that r has finite fibers. For this we may make a base change of the form $\operatorname{Spec}\kappa(s) \to S$ using that "relatively ample" is stable under base change (Proposition 13.64). Thus we may assume that S is the spectrum of a field k. By Remark 12.16 it suffices to check that $X_p := r^{-1}(p)$ is finite for all closed points $p \in P$. Then X_p is a closed subscheme of X and thus proper over k. By Proposition 13.51, the restriction $\mathscr{L}_{|X_p}$ is ample. On the other hand, $\mathscr{L}_{|X_p}$ is the inverse image of $\mathscr{O}_P(1)$ under the composition $X_p \to \operatorname{Spec}\kappa(p) \to P$. As any line bundle on $\operatorname{Spec}\kappa(p)$ is trivial, $\mathscr{L}_{|X_p} \cong \mathscr{O}_{X_p}$. Therefore $\mathscr{O}_{X_p}$ is ample, and X_p is quasi-affine. As X_p is also proper, X_p is finite by Corollary 13.82. $\qquad\square$

(13.17) Conic projections.

Let k be an algebraically closed field. Recall that we defined in Section (1.24) for a pair of complementary linear subspaces $C, D \subset \mathbb{P}^n(k)$ and for a projective variety $Z \subseteq D$ the cone $\overline{Z,C}$ of Z over C. We will now generalize this construction (and show that it suffices to assume that $Z \cap C = \emptyset$ instead of $Z \subseteq D$).

Let S be a scheme, let $\mathscr{E}$ be a finite locally free $\mathscr{O}_S$-module and let $C \subseteq \mathbb{P}(\mathscr{E})$ be a linear subbundle corresponding to a finite locally free quotient $q\colon \mathscr{E} \twoheadrightarrow \mathscr{F}$ of $\mathscr{E}$, cf. Section (8.8). Set $e := \operatorname{rk}(\mathscr{F}) - 1$. Thus $C = \mathbb{P}(\mathscr{F})$ is a linear subbundle of $\mathbb{P}(\mathscr{E})$ of rank e. For $m \geq 0$ let

$$L_m := \operatorname{LinSub}_m(\mathbb{P}(\mathscr{E})) = \operatorname{Grass}^{m+1}(\mathscr{E})$$

be the scheme of linear subbundles of rank m of $\mathbb{P}(\mathscr{E})$.

We denote by $\mathscr{G}$ the kernel of q. We first show that there exists a closed subscheme of L_m parameterizing linear subbundles that contain C:

Lemma 13.85. *For $m \geq e$ define a morphism on $(h\colon T \to S)$-valued points*

$$\iota\colon \operatorname{Grass}^{m-e}(\mathscr{G}) \to \operatorname{LinSub}_m(\mathbb{P}(\mathscr{E})),$$

$$(\mathscr{G}' \subseteq h^*\mathscr{G}) \mapsto \operatorname{Ker}(h^*\mathscr{E} \to h^*\mathscr{E}/\mathscr{G}').$$

Then ι is a closed immersion.

The image of ι (on T-valued points) consists of those linear subspaces of rank m which contain h^*C. Thus ι defines an isomorphism of $\operatorname{Grass}^{m-e}(\mathscr{G})$ onto a closed subscheme

(13.17.1) $$L_m^C := \mathrm{LinSub}_m^C(\mathbb{P}(\mathscr{E})) \subseteq \mathrm{LinSub}_m(\mathbb{P}(\mathscr{E}))$$

which parametrizes linear subbundles containing C.

Proof. Clearly ι is a monomorphism. Moreover it is proper (because it is a morphism between proper S-schemes by Example 13.69). Therefore ι is a closed immersion by Corollary 12.92. $\qquad\square$

For a subscheme X of $\mathbb{P}(\mathscr{E})$, we will also need a scheme that parametrizes pairs (x, Λ) with x a point in X and Λ a linear subbundle containing x:

Remark 13.86. Let $f\colon X \to S$ be an S-scheme and let $\xi\colon X \to \mathbb{P}(\mathscr{E})$ be an S-morphism. Its graph is an X-morphism $X \to \mathbb{P}(\mathscr{E}) \times_S X = \mathbb{P}(f^*\mathscr{E})$, in other words a linear subbundle of $\mathbb{P}(f^*\mathscr{E})$ of rank 0 which we again denote by ξ. For $m \geq 0$ we set

(13.17.2) $$H_m^\xi := \mathrm{LinSub}_m^\xi(\mathbb{P}(f^*\mathscr{E})).$$

Thus H_m^ξ is the X-scheme that parameterizes linear subbundles of $\mathbb{P}(f^*\mathscr{E})$ which contain ξ. Then H_m^ξ is a closed subscheme of $\mathrm{LinSub}_m(\mathbb{P}(f^*\mathscr{E})) = X \times_S \mathrm{LinSub}_m(\mathbb{P}(\mathscr{E}))$ and the projections restrict to morphisms

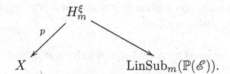

By Lemma 13.85, we have an isomorphism of X-schemes

(13.17.3) $$H_m^\xi \cong \mathrm{Grass}^m(\mathscr{G}),$$

where $\mathscr{G} \subseteq f^*\mathscr{E}$ is locally a direct summand of corank 1. In particular H_m^ξ is smooth and projective over X, and has geometrically integral fibers of dimension $m(\mathrm{rk}(\mathscr{E}) - 1 - m)$.

Viewed as an S-scheme, the $(h\colon T \to S)$-valued points of H_m^ξ are pairs (x, Λ), where Λ is a linear subbundle of $\mathbb{P}(h^*\mathscr{E})$ of rank m and where $x \in X_S(T)$ such that the linear subbundle $\xi \circ x$ of $\mathbb{P}(h^*\mathscr{E})$ of rank 0 is contained in Λ.

Example 13.87. Consider the case $S = \mathrm{Spec}\, k$, where k is a field. Then $\mathbb{P}(\mathscr{E}) \cong \mathbb{P}_k^n$. Assume that X is a closed subscheme of $\mathbb{P}_k^n$ (thus ξ is the inclusion). Then $H_m^\xi(k)$ consists of pairs (x, Λ) where $x \in X(k)$ and $\Lambda \subseteq \mathbb{P}_k^n$ is a linear subspace of dimension m containing x.

We return to the projection with center in C (Remark 8.18)

$$\pi_q\colon \mathbb{P}(\mathscr{E}) \setminus C \to \mathbb{P}(\mathscr{G}).$$

By Lemma 13.85 we may identify $\mathbb{P}(\mathscr{G})$ with the scheme L_{e+1}^C of subbundles of rank $e + 1$ of $\mathbb{P}(\mathscr{E})$ that contain C. With this identification, $\pi_q\colon \mathbb{P}(\mathscr{E}) \setminus C \to L_{e+1}^C$ sends a point x of $\mathbb{P}(\mathscr{E})$ (more precisely, a linear subbundle of $\mathbb{P}(\mathscr{E})$ of rank 0) to the smallest linear subbundle $\overline{x, C}$ containing x and C.

Now let $Z \subset \mathbb{P}(\mathscr{E})$ be a closed subscheme with $Z \cap C = \emptyset$. Then the restriction of π_q to Z is a morphism of proper S-schemes (Corollary 13.42). Therefore its image $\pi_q(Z)$ is a closed subspace of L_{e+1}^C and we may endow it with its structure as schematic image. We think of $\pi_q(Z)$ as the projection of X from the linear subbundle C to L_{e+1}^C.

We would like to define $\overline{Z,C}$ as the "union of all subbundles in $\pi_q(Z)$". This is then the cone with vertex C over Z. To do this in a rigorous way, we consider for $m := e + 1$ the closed subscheme $H_m := H_m^{\mathrm{id}}$ of $\mathbb{P}(\mathscr{E}) \times_S \mathrm{LinSub}_m(\mathbb{P}(\mathscr{E}))$ (13.17.2) together with its projections $p\colon H_m \to \mathbb{P}(\mathscr{E})$ and $r\colon H_m \to \mathrm{LinSub}_m(\mathbb{P}(\mathscr{E}))$. Define

$$\overline{Z,C} := p(r^{-1}(\pi_q(Z))).$$

As p is proper, this is a closed subspace of $\mathbb{P}(\mathscr{E})$ which we may view as a closed subscheme of $\mathbb{P}(\mathscr{E})$ by endowing it with the scheme structure given by the schematic image (Proposition 10.30). This concludes our construction of $\overline{Z,C}$ in the relative situation.

Locally on S, $\mathscr{F}$ is the direct sum of invertible $\mathcal{O}_S$-modules and we may factor π_q into projections from a point (i.e., into projections given by linear subbundles of rank 0). For those projections we have the following result.

Proposition 13.88. *Assume that C is a linear subbundle of rank 0 (i.e., $\mathscr{F}$ is a line bundle) and let Z be a closed subscheme of $\mathbb{P}(\mathscr{E})$ with $Z \cap C = \emptyset$. Then $\pi_{q|Z}\colon Z \to \mathbb{P}(\mathscr{G})$ is a finite morphism.*

Proof. As a morphism between proper S-schemes, π_q is proper. Moreover we have seen in Example 1.66 that π_q has finite geometric fibers and hence is quasi-finite (Remark 12.16). Therefore π_q is finite by Corollary 12.89. $\square$

This general result can be used to prove a projective variant of Noether's normalization theorem:

Theorem 13.89. *Let k be an infinite field, let X be a projective k-scheme and set $d := \dim X$. Then there exists a finite surjective morphism $\pi\colon X \twoheadrightarrow \mathbb{P}_k^d$.*

Proof. Once we have constructed a finite surjective morphism $\pi\colon X \twoheadrightarrow \mathbb{P}_k^n$ for some n, we automatically have $\dim X = \dim \mathbb{P}_k^n = n$ by Proposition 12.12.

As X is projective, there exists a finite k-morphism $f\colon X \to \mathbb{P}_k^m$ for some m (in fact even a closed immersion). We will prove the theorem by descending induction on m. If f is surjective, we are done. Otherwise, $\mathrm{Im}(f)$ is a proper closed subscheme of $\mathbb{P}_k^m$. As k is infinite, we find a k-valued point $q \in \mathbb{P}^m(k)$ which is not contained in $\mathrm{Im}(f)$ (because there exists no nonzero homogeneous polynomial that vanishes on all points of $\mathbb{P}^m(k)$). Let $\pi_q\colon \mathbb{P}_k^m \setminus \{q\} \to P' \cong \mathbb{P}_k^{m-1}$ be the projection with center q. Then $\pi_{q|\mathrm{Im}(f)}$ is finite by Proposition 13.88. Thus the composition $\pi_q \circ f\colon X \to \mathbb{P}_k^{m-1}$ is finite and we are done by the induction hypothesis. $\square$

Blowing-up

The procedure of blow-up, in the simplest case, replaces a point x in a (smooth) scheme X by its projectivized tangent space $\mathbb{P}(T_x X)$. For instance, if X is two-dimensional, then X is "blown-up" at x in the sense that x is replaced by a "sphere" $\mathbb{P}^1$ (think of the Riemann sphere $\mathbb{P}^1(\mathbb{C})$). Let us explain why this is a useful tool.

Let S be a scheme and let X and Y be two S-schemes. For simplicity, we will assume in this motivation that S is noetherian and that X and Y are separated of finite type over S. It is often the case that rather than by a morphism between them, X and Y are related by a rational map (cf. Section (9.6)), i.e., a map a priori only defined on the complement of a certain "bad locus" which is usually closed and nowhere dense. In other words there exists a morphism $f\colon U \to Y$, where U is an open dense subscheme of X. It is then a natural question whether it is possible to extend f to a morphism $X \to Y$. Note that by Proposition 9.27 there is a unique domain of definition of f, open inside X, such that f cannot be extended to a larger open subscheme. Here we are mainly interested in the question whether f can be extended after modifying X "in a controlled way".

There are two main obstructions to this. First, Y might have "holes": If Y is itself isomorphic to an open dense subscheme of an S-scheme Y', then at least it might be possible to extend f to a morphism $X \to Y'$. As a trivial example consider the case $X = \mathbb{P}^1_S$, $U = \mathbb{A}^1_S \subset X$, and $f = \mathrm{id}\colon U \to Y := \mathbb{A}^1_S$. Then Y has a "hole at infinity" but of course we can extend f to a morphism $X \to Y'$, where $Y' = \mathbb{P}^1_S$. By Nagata's compactification theorem (Theorem 12.70) we can always embed Y into a proper S-scheme $\bar{Y}$, and $\bar{Y}$ does not have "holes": If $j\colon \bar{Y} \to \bar{Y}'$ is an open immersion into a separated S-scheme with dense image, j is automatically proper and thus an isomorphism.

Therefore let us now assume that Y is proper over S. In Theorem 12.60 we have seen that if X is normal, then f can be extended at least to an open subscheme $\tilde{U} \subseteq X$ whose complement has codimension at least 2. Thus we might try the following. Fix a point $x \in Z := X \setminus U$. Let C be the germ of a normal curve in a point c (more precisely, C is the spectrum of a discrete valuation ring with closed point c) and let $\gamma\colon C \to X$ be a morphism with $\gamma(c) = x$ and $\gamma(C \setminus \{c\}) \subset U$ (we will see in Proposition 15.7 that such γ's always exist). Then $f \circ \gamma_{|C\setminus\{c\}}$ can be extended to a morphism $\gamma'\colon C \to Y$ and we set $f_\gamma(x) := \gamma'(c)$. This defines an "extension of f into x along γ". If X is reduced and f can be extended to a (necessarily unique) morphism on X, then all extensions along different γ must coincide because of continuity reasons. But usually $f_\gamma(x)$ will not be independent of the choice of γ. Thus we obtain different extensions of f for different choices of γ. This "indeterminacy" is the second obstruction to the extension of f: Compare Example 9.30.

To overcome this difficulty we might try to modify X: Find a morphism $\pi\colon \tilde{X} \to X$ which restricts to an isomorphism $\pi^{-1}(U) \overset{\sim}{\to} U$ and a morphism $\tilde{f}\colon \tilde{X} \to Y$ such that for all γ as above there exists a point $\tilde{x}_\gamma \in \pi^{-1}(x)$ with $\tilde{f}(\tilde{x}_\gamma) = f_\gamma(x)$. Moreover, we may even hope that the points in $\pi^{-1}(x)$ correspond to the different ways a curve γ may approach x from U, i.e., that there is an isomorphism of $\pi^{-1}(x)$ to the projective space of lines in the $\kappa(x)$-vector space $T_x(X)/T_x(Z)$ which we think of as the space of tangent vectors normal to Z (in Volume II we will give the definition of a normal sheaf of a closed subscheme).

We will see that it is indeed possible to find such a morphism $\pi\colon \tilde{X} \to X$ such that f can be extended to a morphism $\tilde{f}\colon \tilde{X} \to Y$. Moreover, one knows that π can be obtained by an explicit construction called a blow-up. But this construction depends on the choice of additional data and to find the right data is a very difficult problem in general. The additional hope about the shape of the fiber of π is in general too naive and will hold only if Z and X are smooth over S (or more generally, if Z is regularly immersed into X, a notion that will be studied in Volume II).

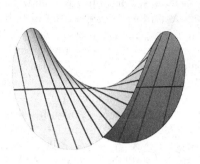

Figure 13.1: The picture shows $\mathbb{A}^2$ on the right, with a couple of lines marked, and (one chart of) the blow-up of the origin on the left. On the left, the horizontal line is the exceptional divisor, i.e., the inverse image of the origin, and the other lines are mapped to the marked lines on the right.

(13.18) An example of a blow-up.

We will start with a concrete example. Let k be an algebraically closed field, set $X := \mathbb{A}^2_k$ and let $f \colon U := \mathbb{A}^2_k \setminus \{0\} \to Y := \mathbb{P}^1_k$ be the morphism which is given on k-valued points by $(x, y) \mapsto (x : y)$. Thus we are in the situation described above. The tangent directions of curves γ through 0 in $\mathbb{A}^2_k$ are given by elements $(\lambda : \mu) \in T(k) := \mathbb{P}^1(k)$. In this case we may simply consider the line in $\mathbb{A}^2_k$ through 0 with direction $(\lambda : \mu)$ which is given on k-valued points by $\gamma \colon t \mapsto (\lambda t, \mu t)$. Then $f(\gamma(t)) = (\lambda t : \mu t) = (\lambda : \mu)$ is defined also for $t = 0$ and we have $f_\gamma(0) = (\lambda : \mu)$.

Thus we are looking for a morphism $\pi \colon \tilde{X} \to X$ whose restriction $\pi^{-1}(U) \to U$ is an isomorphism and such that $\pi^{-1}(0) \cong \mathbb{P}^1_k$. We define $\tilde{X}$ as the reduced k-scheme of finite type whose k-valued points are given by

$$\tilde{X}(k) := \{\, ((x, y), (\lambda : \mu)) \in X(k) \times T(k) \; ; \; \mu x = \lambda y \,\}$$

and π as the projection $((x, y), (\lambda : \mu)) \mapsto (x, y)$. We call $(\tilde{X}, \pi)$ (or simply $\tilde{X}$) the *blow-up of $\mathbb{A}^2_k$ at 0*. Then the morphism $j \colon (x, y) \mapsto ((x, y), (x : y))$ for $(x, y) \in \mathbb{A}^2 \setminus \{0\}$ defines an inverse isomorphism of $\pi^{-1}(U) \xrightarrow{\sim} U$. For the fiber over 0 we find $E := \pi^{-1}(0) \cong T$. We call E the *exceptional divisor* for reasons explained below. Then we have indeed a morphism $\tilde{f} \colon \tilde{X} \to \mathbb{P}^1_k$ given by $((x, y), (\lambda : \mu)) \mapsto (\lambda : \mu)$ which extends f.

We may describe $\tilde{X}$ locally as follows. Set $A := k[x, y]$ (and hence $\mathbb{A}^2_k = \operatorname{Spec} A$) and $A^x := k[x, y, y/x]$ which we consider as subalgebra of $k[x, y, x^{-1}]$ and similarly $A^y := k[x, y, x/y]$. Define $U^x := \operatorname{Spec} A^x$ and $U^y := \operatorname{Spec} A^y$. We have $A^x = k[x, y, \mu]/(\mu x = y)$ and hence $U^x = \tilde{f}^{-1}(D_+(\lambda))$ where $D_+(\lambda)(k) = \{\, (\lambda : \mu) \in \mathbb{P}^1(k) \; ; \; \lambda \neq 0 \,\}$. Similarly we find $U^y = \tilde{f}^{-1}(D_+(\mu))$ and thus U^x and U^y form an open covering of $\tilde{X}$. Moreover $E \cap U^x$ is defined in U^x by the single equation $x = 0$ and similarly $E \cap U^y = V(y)$ as subscheme of U^y. Thus E is a closed subscheme defined locally by a single regular element. In other words, E is an effective Cartier divisor (Remark 11.27). This explains the terminology for E introduced above. Globally, we can describe $\tilde{X}$ (with the projection to $\mathbb{P}^1$) as the geometric line bundle $\mathbb{V}(\mathscr{O}(1))$, see Exercise 13.24.

Now let $C \subset \mathbb{A}_k^2$ be the closed reduced k-subscheme of $\mathbb{A}_k^2$ whose k-valued points are $C(k) = \{ (x, y) \in \mathbb{A}^2(k) \; ; \; y^2 = x^2(x+1) \}$. This is a curve (i.e., a k-scheme of finite type of dimension 1) which is smooth at all points except 0. To simplify matters, we assume that $\operatorname{char}(k) \neq 2$. Then at 0 we find two tangent lines within $\mathbb{A}_k^2$ given by the equations $x = y$ and $x = -y$. Let $C' := \pi^{-1}(C)$, i.e.,

$$C'(k) := \{ ((x, y), (\lambda : \mu)) \in C(k) \times \mathbb{P}^1(k) \; ; \; \mu x = \lambda y \}.$$

Then C' has two irreducible components (considered as reduced schemes): the exceptional divisor E of the blow-up of $\mathbb{A}_k^2$ in 0 and the closure $\tilde{C}$ of $C \setminus \{0\} \cong \pi^{-1}(C \setminus \{0\})$ in $\tilde{X}$. We call $\tilde{C}$ the *strict transform of* C.

We set $B := k[x, y]/(y^2 - x^2(x+1))$ (and hence $C = \operatorname{Spec} B$) and $V^x := \operatorname{Spec} B[y/x]$ and $V^y := \operatorname{Spec} B[x/y]$. We have
(13.18.1)
$$U^x(k) \cap \tilde{C}(k) = \{ (x, y, \mu) \in \mathbb{A}^3(k) \; ; \; y^2 = x^2(x+1), \mu x = y, \mu^2 = x + 1 \} = V^x(k).$$

We find $U^x \cap \tilde{C} = V^x$. Similarly, we see $U^y \cap \tilde{C} = V^y$ with

(13.18.2) $\qquad V^y(k) = \{ (x, y, \lambda) \in \mathbb{A}^3(k) \; ; \; y^2 = x^2(x+1), x = \lambda y, \lambda^3 y + \lambda^2 = 1 \}.$

Thus we have an analogous local description of $\tilde{C}$ as for $\tilde{X}$ and we will call $\tilde{C}$ the *blow-up of* C *in* 0. In this case the blow-up separates the two tangent lines of C in 0: we have $E(k) \cap \tilde{C}(k) = \{((0, 0), (1 : 1)), ((0, 0), (1 : -1))\}$.

The equation $\lambda^3 y + \lambda^2 = 1$ in (13.18.2) shows that $\lambda \neq 0$. Therefore we find that $V^y \subset \tilde{f}^{-1}(D_+(\lambda)) = U^x$ which shows $V^y \subset V^x$ and hence $\tilde{C} = V^x$. The description of V^x in (13.18.1) then shows that the restriction of $\tilde{f}$ to $\tilde{C}$ given by $(x, y, \mu) \mapsto (1 : \mu)$ defines an isomorphism $\tilde{C} \xrightarrow{\sim} D_+(\lambda) \cong \mathbb{A}_k^1$. In particular $\tilde{C}$ is smooth over k. Thus the blow-up of C in its singular point yields a desingularization of C (it follows from Proposition 12.44 that in this case $\tilde{C}$ is simply the normalization of C). See Section (13.23) below about resolution of singularities for pointers towards vast generalizations of this phenomenon.

(13.19) Definition and universal property of blow-ups.

Recall from Remark 11.27 that we may consider effective Cartier divisors on a scheme X as closed subschemes Z of X such that the quasi-coherent ideal $\mathscr{I} \subset \mathscr{O}_X$ defining Z is an invertible $\mathscr{O}_X$-module, i.e., $\mathscr{I}$ is locally defined by a single regular element. Note that the empty subscheme is an effective Cartier divisor with this definition. Now the blow-up with center Z is defined by being universal with respect to transforming Z into an effective Cartier divisor:

Definition 13.90. *Let X be a scheme and let Z be a closed subscheme. A blow-up of X along Z is a scheme $\tilde{X}$ and a morphism $\pi \colon \tilde{X} \to X$ such that $\pi^{-1}(Z)$ is an effective Cartier divisor and which is universal with respect to this property: If $\pi' \colon \tilde{X}' \to X$ is any morphism such that $\pi'^{-1}(Z)$ is an effective Cartier divisor, then there is a unique morphism $g \colon \tilde{X}' \to \tilde{X}$ such that $\pi' = \pi \circ g$.*

Clearly, a blow-up $(\tilde{X}, \pi)$ of X along Z is unique up to unique isomorphism and we set $\operatorname{Bl}_Z(X) := \tilde{X}$. The effective Cartier-divisor $\pi^{-1}(Z)$ is called the *exceptional divisor* of the blow-up. The closed subscheme Z is called the *center* of the blow-up.

Before we show that a blow-up always exists, we first deduce formal consequences of the definition. First we remark that if Z is an effective Cartier divisor in X, then the universal property of the blow-up shows that $\operatorname{Bl}_Z(X) \to X$ is an isomorphism.

Proposition 13.91. *Let X be a scheme, let Z be a closed subscheme of X, and let $\pi\colon \mathrm{Bl}_Z(X) \to X$ be the blow-up of X along Z. Let $f\colon X' \to X$ be a morphism of schemes.*

(1) *There exists a unique morphism $\mathrm{Bl}_Z(f)\colon \mathrm{Bl}_{f^{-1}(Z)}(X') \to \mathrm{Bl}_Z(X)$ making the following diagram commutative*

(13.19.1)

$$
\begin{array}{ccc}
\mathrm{Bl}_{f^{-1}(Z)}(X') & \xrightarrow{\ \mathrm{Bl}_Z(f)\ } & \mathrm{Bl}_Z(X) \\
\downarrow & & \downarrow{\scriptstyle\pi} \\
X' & \xrightarrow{\quad f \quad} & X.
\end{array}
$$

(2) *If f is flat, then the diagram (13.19.1) is cartesian.*
(3) *Let U be the open subscheme $X \setminus Z$. Then the restriction of π to $\pi^{-1}(U)$ is an isomorphism $\pi^{-1}(U) \xrightarrow{\sim} U$.*
(4) *Let $\mathscr{I}$ be the quasi-coherent ideal of $\mathscr{O}_X$ corresponding to Z and suppose that for every affine open $V \subseteq X$, $\Gamma(V, \mathscr{I})$ contains a regular element (e.g., if X is integral and $\mathscr{I} \neq 0$), then π is birational.*

Proof. The existence and the uniqueness of $\mathrm{Bl}_Z(f)$ follow immediately from the definition. To show (2) set $\tilde{X}' := \mathrm{Bl}_Z(X) \times_X X'$ and let $p\colon \tilde{X}' \to \mathrm{Bl}_Z(X)$ be the projection which is flat because flatness is stable under base change. We have to show that the induced morphism $r\colon \mathrm{Bl}_{f^{-1}(Z)}(X') \to \tilde{X}'$ is an isomorphism. Let $E = \pi^{-1}(Z)$ be the exceptional divisor. By Corollary 11.51 the inverse image $E' := p^{-1}(E)$ is again an effective Cartier divisor. But E' is also the inverse image of $f^{-1}(Z)$ under the projection $\tilde{X}' \to X'$. Thus by the universal property of the blow-up there exists a unique morphism $r'\colon \tilde{X}' \to \mathrm{Bl}_{f^{-1}(Z)}(X')$. The universal property of the blow-up shows $r' \circ r = \mathrm{id}$ and the universal property of the fiber product shows $r \circ r' = \mathrm{id}$.

To show (3) we apply (2) to the open immersion $U \hookrightarrow X$ using that the universal property of the blow-up implies immediately that the blow-up in the empty subscheme is an isomorphism. Finally (4) follows from (3) because the hypothesis implies that U is schematically dense in X (Remark 9.24). Moreover the same remark shows that the complement of every effective Cartier divisor (and in particular $\pi^{-1}(U)$) is schematically dense in $\mathrm{Bl}_Z(X)$. $\qquad\square$

We will now construct the blow-up along a subscheme. Let X be a scheme, let Z be a closed subscheme, and let $\mathscr{I} \subseteq \mathscr{O}_X$ be the corresponding quasi-coherent ideal. For $d > 0$ let $\mathscr{I}^d$ be the d-th power of $\mathscr{I}$. We also set $\mathscr{I}^0 := \mathscr{O}_X$. Then $\mathscr{B} := \bigoplus_{d \geq 0} \mathscr{I}^d$ is a graded quasi-coherent $\mathscr{O}_X$-algebra which is generated in degree 1. We define $\tilde{X} := \mathrm{Proj}\,\mathscr{B}$ with structure morphism $\pi\colon \tilde{X} \to X$ and we set $E := \pi^{-1}(Z)$. We will show that $\tilde{X}$ is a blow-up of X along Z.

This can be checked locally on X. Thus we assume that $X = \mathrm{Spec}\,A$ is affine. Then $Z = V(I)$ for the ideal $I \subseteq A$ with $\tilde{I} = \mathscr{I}$ and $\mathscr{B} = \tilde{B}$, where $B := \bigoplus_{d \geq 0} B_d$ with $B_d = I^d$.

Let us describe $\tilde{X}$ locally: For $f \in I$ let $A[\frac{I}{f}]$ be the A-subalgebra of A_f which is generated by elements of the form x/f with $x \in I$. If we consider x and f as elements of $B_1 = I$, then x/f may be also considered as a degree zero element of B_f and we obtain an A-algebra isomorphism

$$
A[\frac{I}{f}] \xrightarrow{\sim} B_{(f)}
$$

whose inverse is given by considering y/f^d for $y \in I^d$ as an element of $A[\frac{I}{f}]$. Thus we see that if f runs through a generating set of I the $D_+(f) = \operatorname{Spec} A[\frac{I}{f}]$ form an open affine covering of $\tilde{X}$. If $(x_\alpha)_\alpha$ is a generating system of I, then we have surjective homomorphism of A-algebras

$$A[(T_\alpha)_\alpha]/((fT_\alpha - x_\alpha)_\alpha) \longrightarrow A[\frac{I}{f}], \qquad T_\alpha \mapsto x_\alpha/f$$

and the kernel are those elements that are annihilated by some power of f.

The ideal $IA[\frac{I}{f}]$ is generated by the image of f in $A[\frac{I}{f}]$ and this image is a regular element. Moreover, if $\varphi\colon A \to C$ is any A-algebra such that $\varphi(f)$ is regular and generates $\varphi(I)C$, then there exists a unique A-algebra homomorphism $A[\frac{I}{f}] \to C$ which sends x/f (for $x \in I$) to the unique element $c \in C$ such that $\varphi(f)c = \varphi(x)$. This shows that $\tilde{X} = \bigcup_{f \in I} D_+(f)$ is a final object in the category of all X-schemes $g\colon Y \to X$ such that $g^{-1}(\mathscr{I})\mathscr{O}_Y$ is locally generated by a regular element. Therefore $\tilde{X}$ is in fact a blow-up of X along Z and we have seen:

Proposition 13.92. *Let X be a scheme, $i\colon Z \hookrightarrow X$ be a closed subscheme and let $\mathscr{I} \subseteq \mathscr{O}_X$ be the corresponding quasi-coherent ideal. Then $\operatorname{Proj} \bigoplus_{d \geq 0} \mathscr{I}^d$ is a blow-up of X along Z.*

Example 13.93. Let us spell out the construction in the case of the blow-up of the origin in affine space $\mathbb{A}_R^n$ over some ring R. Write $A = R[T_1, \ldots, T_n]$, $I = (T_1, \ldots, T_n)$, and $B = \bigoplus B_n$, where $B_0 = A$, $B_n = I^n$. Note that in the ring B, we have to distinguish whether T_i is viewed as an element of B_0 or of B_1. We have a surjection of A-algebras $A[X_1, \ldots, X_n] \to B$ given by mapping X_i to $T_i \in I = B_1$. Its kernel is the ideal generated by $T_i X_j - T_j X_i$, $1 \leq i, j \leq n$. Therefore the blow-up of the origin, i.e., the closed subscheme $\operatorname{Spec} R = V(T_1, \ldots, T_n)$ inside $\mathbb{A}_R^n$ is given by the closed subscheme

$$\tilde{X} = V_+(T_i X_j - T_j X_i;\ 1 \leq i, j \leq n) \subset \mathbb{P}_R^{n-1} \times_R \mathbb{A}_R^n,$$

where the T_i are the coordinates of $\mathbb{A}_R^n$, and the $X_1, \ldots, X_n$ are the homogeneous coordinates on $\mathbb{P}_R^{n-1}$.

The same argument shows that we may identify the blow-up of $\mathbb{P}_R^n$ in the R-rational point $(1 : 0 : \ldots : 0)$ with the closed subscheme of $\mathbb{P}_R^{n-1} \times_R \mathbb{P}_R^n$ defined by the equations $T_i X_j - T_j X_i$ for $1 \leq i, j \leq n$, where $T_0, \ldots, T_n$ are the homogeneous coordinates of $\mathbb{P}_R^n$ and $X_1, \ldots, X_n$ those of $\mathbb{P}_R^{n-1}$.

See Exercises 13.24 and 13.25 for further explicit examples.

Remark 13.94. The exceptional divisor is given by

$$E := \operatorname{Proj} i^* \bigoplus_{d \geq 0} \mathscr{I}^d = \operatorname{Proj} \bigoplus_{d \geq 0} \mathscr{I}^d / \mathscr{I}^{d+1}$$

by Remark 13.27.

Example 13.95. Let X be a locally noetherian scheme and let $x \in X$ be a closed point such that $\mathscr{O}_{X,x}$ is regular of dimension d. Let $\pi\colon \tilde{X} \to X$ be the blow-up in $\{x\}$ (considered as a closed reduced subscheme). Then π is an isomorphism over $X \setminus \{x\}$ and for the exceptional divisor we have

$$\pi^{-1}(x) = \operatorname{Proj} \bigoplus_{i \geq 0} \mathfrak{m}_x^i / \mathfrak{m}_x^{i+1} \cong \mathbb{P}_{\kappa(x)}^{d-1},$$

where the last isomorphism is due to Proposition B.76 (ii).

In particular, if Z is a smooth closed point with residue class field k inside a k-scheme of finite type, then the exceptional divisor of the corresponding blow-up is the projective space of all lines inside the tangent space at the point Z. If Z is a not necessarily smooth point, then the exceptional divisor can be identified with the projectivization of the tangent cone, cf. Exercise 6.10.

The description of the blow-up in Proposition 13.92 yields the following properties.

Proposition 13.96. *Let X be a scheme, let Z be a closed subscheme, let $\mathscr{I} \subseteq \mathscr{O}_X$ be the corresponding quasi-coherent ideal, and let $\pi \colon \tilde{X} := \mathrm{Bl}_Z(X) \to X$ be the blow-up.*
(1) *Assume that $\mathscr{I}$ is of finite type (e.g., if X is locally noetherian). Then $\mathrm{Bl}_Z(X)$ is projective over X and $\mathscr{I}\mathscr{O}_{\tilde{X}} = \mathscr{O}_{\tilde{X}}(1)$ is very ample for π.*
(2) *If $i\colon Y \hookrightarrow X$ is a closed subscheme, the morphism $\mathrm{Bl}_Z(i)\colon \mathrm{Bl}_{Y\cap Z}(Y) \to \mathrm{Bl}_Z(X)$ (Proposition 13.91) is a closed immersion.*

In the situation of Assertion (2) we usually identify $\mathrm{Bl}_{Y\cap Z}(Y)$ with a closed subscheme of $\mathrm{Bl}_Z(X)$ which is called the *strict transform of Y under the blow-up of X along Z*. It can also be described as the schematic closure of $\pi^{-1}(Y \setminus Z)$ in $\mathrm{Bl}_Z(X)$ (Exercise 13.23).

Note that even if π is an isomorphism, the strict transform of Y is not necessarily isomorphic to Y (see Example 14.144). However this is the case if $Y \setminus Z$ is schematically dense in Y (e.g., if Y is integral and not contained in Z).

Proof. Assertion (1) follows from the Summary 13.71. To show (2) let $\mathscr{J} \subseteq \mathscr{O}_X$ be the quasi-coherent ideal defining Y. Then the closed subscheme $Y \cap Z$ of Y is defined by the quasi-coherent ideal $(\mathscr{I} + \mathscr{J})/\mathscr{J}$ of $\mathscr{O}_Y = \mathscr{O}_X/\mathscr{J}$. Thus $\mathrm{Bl}_Z(i)$ is the closed immersion corresponding to the surjective homomorphism $\bigoplus \mathscr{I}^d \twoheadrightarrow \bigoplus(\mathscr{I} + \mathscr{J}/\mathscr{J})^d$ of graded $\mathscr{O}_X$-algebras (Proposition 13.8). $\square$

Corollary 13.97. *Let X be an integral scheme, and let $\emptyset \neq Z \subsetneq X$ be a non-empty closed subscheme such that the defining quasi-coherent ideal is of finite type (e.g., if X is locally noetherian). Then $\mathrm{Bl}_Z(X)$ is an integral scheme and the structure morphism $\pi\colon \mathrm{Bl}_Z(X) \to X$ is birational, projective, and surjective.*

Proof. Set $U := X \setminus Z$. Then U is schematically dense in X because X is integral and $U \neq \emptyset$. Moreover $\pi^{-1}(U)$ is the complement of an effective Cartier divisor and therefore schematically dense in $\mathrm{Bl}_Z(X)$. Thus π is birational. It is projective by Proposition 13.96. In particular, π is closed (Corollary 13.72) and hence surjective. It remains to show that $\mathrm{Bl}_Z(X)$ is integral. As $\pi^{-1}(U) \cong U$ is irreducible and dense, $\mathrm{Bl}_Z(X)$ is irreducible. Let $\mathscr{I}$ be the quasi-coherent ideal defining Z. As X is integral, $B_V := \Gamma(V, \bigoplus_d \mathscr{I}^d)$ is an integral domain for every open affine subset $V \subseteq X$. This shows that $D_+(f) \subseteq \mathrm{Bl}_Z(X)$ is integral for every homogeneous element $f \in B_{V,+}$. As these open subschemes cover $\mathrm{Bl}_Z(X)$ (in fact, even form a base of the topology), $\mathrm{Bl}_Z(X)$ is integral. $\square$

We will discuss in Section (13.22) results that are a converse to this corollary.

(13.20) Blow-ups and extending rational maps.

Let $U \subseteq X$ be an open subscheme. We call a morphism $\pi\colon X' \to X$ a *U-admissible blow-up* if there exists a closed subscheme Z of X defined by a quasi-coherent ideal of

$\mathscr{O}_X$ of finite type such that $Z \cap U = \emptyset$ and such that $X' \cong \mathrm{Bl}_Z(X)$ (as schemes over X). Then π is projective by Proposition 13.96.

The next theorem says, roughly speaking, that if $f\colon X \dashrightarrow Y$ is a rational map of S-schemes and Y is proper over S, then there exists a blow-up $\tilde{X}$ of X which is an isomorphism over the domain of definition of f and a morphism $\tilde{X} \to Y$ of S-schemes which extends f, i.e., by blowing up X we can eliminate the indeterminacy which prevents extending f.

Theorem 13.98. *Let S be a qcqs scheme, let X be a qcqs S-scheme and let Y be a proper S-scheme. Let $U \subseteq X$ be a quasi-compact open dense subscheme and let $f\colon U \to Y$ be an S-morphism. Then there exists an U-admissible blow-up $\pi\colon \tilde{X} \to X$ and an S-morphism $\tilde{f}\colon \tilde{X} \to Y$ such that f is the composition of $U \overset{\sim}{\to} \pi^{-1}(U)$ and $\tilde{f}_{|\pi^{-1}(U)}$.*

For the proof we refer to [Co] Theorem 2.4 whose exposition is based on notes of Deligne. This proof – although difficult and lengthy – is quite elementary and uses only methods and results explained in this Volume. Another proof of Theorem 13.98 in the case that S is noetherian has been given by Lütkebohmert in [Lü], Lemma 2.2. Lütkebohmert's proof relies on a deep theorem of Raynaud and Gruson [RG] on "flattening" via blow-ups (see Section (14.34)).

Here we will show only a special but interesting case of Theorem 13.98. It is the following construction. Let $f\colon X \to S$ be a morphism of schemes. Let $\mathscr{E}$ be an $\mathscr{O}_S$-module of finite type, let $\mathscr{L}$ be an invertible $\mathscr{O}_X$-module, and let $u\colon f^*\mathscr{E} \to \mathscr{L}$ be a homomorphism of $\mathscr{O}_X$-modules. Set $U := X \setminus (\mathrm{Supp}\,\mathrm{Coker}\,u)$, i.e., U is the locus in X where u is surjective. By the universal property of $\mathbb{P}(\mathscr{E})$ there is a unique S-morphism $r\colon U \to \mathbb{P}(\mathscr{E}) =: P$ such that $r^*\mathscr{O}_P(1) \cong \mathscr{L}_{|U}$.

Proposition 13.99. *There exists a U-admissible blow-up $\pi\colon \tilde{X} \to X$ and a morphism $\tilde{r}\colon \tilde{X} \to \mathbb{P}(\mathscr{E})$ such that $\tilde{r}_{|\pi^{-1}(U)}$ is identified with r under the isomorphism $\pi^{-1}(U) \overset{\sim}{\to} U$ induced by π.*

Proof. Let $\mathscr{F} \subseteq \mathscr{L}$ be the image of u. Then $\mathscr{F}$ is a quasi-coherent $\mathscr{O}_X$-module of finite type. Let $\mathscr{I} := \mathscr{F} \otimes \mathscr{L}^{\otimes -1} \subseteq \mathscr{O}_X$ be the corresponding quasi-coherent ideal of finite type via the bijection (13.8.4). We have $\mathscr{I}_x = \mathscr{O}_{X,x}$ if and only if $\mathscr{F}_x = \mathscr{L}_x$, i.e., if and only if $x \in U$. Thus if we set $Z := V(\mathscr{I})$, then $X \setminus Z = U$.

Let $\pi\colon \tilde{X} \to X$ be the blow-up of X in Z which is U-admissible. By the definition of blow-up, the quasi-coherent ideal $(\pi^{-1}\mathscr{I})\mathscr{O}_{\tilde{X}}$ of $\pi^{-1}(Z)$ is an invertible ideal sheaf of $\mathscr{O}_{\tilde{X}}$ and thus the corresponding $\mathscr{O}_{\tilde{X}}$-submodule $\tilde{\mathscr{L}} := (\pi^{-1}\mathscr{I})\mathscr{O}_{\tilde{X}} \otimes \pi^*\mathscr{L}$ of $\pi^*\mathscr{L}$ is invertible as well. As $\tilde{\mathscr{L}}$ is the image of $\pi^*(f^*\mathscr{E}) \to \pi^*\mathscr{L}$, we have a surjection $(f \circ \pi)^*\mathscr{E} \twoheadrightarrow \tilde{\mathscr{L}}$. Thus we find a unique S-morphism $\tilde{r}\colon \tilde{X} \to \mathbb{P}(\mathscr{E})$ such that $\tilde{r}^*\mathscr{O}_P(1) \cong \tilde{\mathscr{L}}$. As $\tilde{\mathscr{L}}_{|\pi^{-1}(U)} = \pi^*\mathscr{L}_{|\pi^{-1}(U)}$, we see that the restriction of $\tilde{r}$ to $\pi^{-1}(U)$ corresponds to r via $\pi^{-1}(U) \overset{\sim}{\to} U$. $\qquad\square$

(13.21) Lemma of Chow.

The Lemma of Chow says that every separated morphism of finite type (resp. every proper morphism) becomes quasi-projective (resp. projective) after modification with a birational projective morphism, more precisely:

Theorem 13.100. *Let S be a qcqs scheme and let $f\colon X \to S$ be a separated morphism of finite type such that X has only finitely many irreducible components (which is automatic if S is noetherian).*

Then there exists a commutative diagram

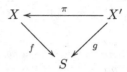

where g is quasi-projective and where π is surjective, projective, and there exists a quasi-compact open dense subscheme U of X such that $\pi^{-1}(U)$ is dense in X' and the restriction $\pi^{-1}(U) \to U$ of π is an isomorphism. Moreover:

(1) f is proper if and only if g is projective.

(2) If X is reduced (resp. irreducible, resp. integral), then X' can be chosen to be reduced (resp. irreducible, resp. integral).

It can be shown (see [Co] Corollary 2.6) that π can be chosen to be a blow-up.

Proof. Step 1: Reduction to the case that X is irreducible. Let $(X_i)_i$ be the finite family of irreducible components of X. We endow each X_i with the structure of a closed subscheme as follows. For all i let $V_i \subseteq X_i$ be a non-empty open quasi-compact subset such that V_i does not meet any other component X_j for $j \neq i$. Then V_i is open in X and we may consider V_i as an open subscheme of X. As X is quasi-separated and as V_i is quasi-compact, the inclusion $V_i \hookrightarrow X$ is a quasi-compact morphism. The schematic closure of V_i in X (Section (10.8)) is then a closed subscheme of X whose underlying topological space is X_i. In this way we consider X_i as a closed subscheme of X. If X is reduced, V_i and hence X_i is reduced.

Assume that there exists an irreducible S-scheme X_i', a morphism $\pi_i\colon X_i' \to X_i$ and an open dense subset U_i of X_i satisfying the desired properties for $X_i \to S$. Let X' be the disjoint union of the X_i'. Then X' is clearly quasi-projective over S. It is reduced if the X_i' are reduced. The morphism $\pi\colon X' \to X$ induced by the π_i is surjective. The closed immersion $X_i \hookrightarrow X$ is projective and thus the composition $X_i' \to X_i \hookrightarrow X$ is projective by Remark 13.73, and Remark 8.21 shows that $X' \to X$ is projective. Finally we may take for U the union of the open subsets $U_i \cap V_i$.

Step 2: Construction of π. By Step 1 we may assume that X is irreducible. Let $(S_j)_j$ be a finite open affine covering of S and let $(X_k)_{1 \leq k \leq n}$ be a finite covering of X by open affine non-empty subschemes X_k of X such that for all k the image $f(X_k)$ is contained in some S_j (depending on k). Then $X_k \to S_j$ is affine and of finite type and hence quasi-projective. Moreover the immersion $S_j \hookrightarrow S$ is quasi-compact and hence also quasi-projective. Thus $X_k \to S$ is quasi-projective for all k by Remark 13.73 and there exist open S-immersions $r_k\colon X_k \hookrightarrow P_k$, where P_k is a projective S-scheme.

Set $U := \bigcap_k X_k$. As X is irreducible, U is open and dense in X. The restrictions of r_k to U define a quasi-compact immersion of S-schemes

$$r\colon U \to P := P_1 \times_S \cdots \times_S P_n$$

and hence a quasi-compact immersion

$$h := (j, r)_S\colon U \hookrightarrow X \times_S P,$$

where $j\colon U \hookrightarrow X$ is the inclusion. Let X' be the schematic image of h in $X \times_S P$. Thus h factors into a schematically dominant open immersion $h'\colon U \hookrightarrow X'$ followed by a closed immersion $i\colon X' \hookrightarrow X \times_S P$. As U is irreducible, X' is irreducible. Define π as the composition

$$\pi\colon X' \xrightarrow{\ i\ } X \times_S P \xrightarrow{\ p_1\ } X,$$

where p_1 is the first projection. We set $U' := \pi^{-1}(U)$.

Step 3: Properties of π. The S-scheme P is projective because the P_i are projective over S. Therefore $X \times_S P$ is projective over X and thus π is projective.

The definition of X' implies that $U' = i^{-1}(U \times_S P)$ is the schematic closure of the subscheme $h(U)$ in $U \times_S P$. But the morphism $(\mathrm{id}_U, r_{|U})\colon U \hookrightarrow U \times_S P$ is the graph of the separated morphism r and hence a closed immersion. Thus h induces an isomorphism $U \xrightarrow{\sim} U'$ which is inverse to the restriction $U' \to U$ of π. This also shows that U' is schematically dense in X'. Finally, $\pi(X')$ contains U and is closed in X (because projective morphisms are closed). Therefore π is surjective.

Step 4: Reducedness. Let us show that X' is reduced if X is reduced. Step 1 shows that we may assume that X is integral. Then U is reduced and the schematic image X' of $h\colon U \to X \times_S P$ is reduced as well (Remark 10.32).

Step 5: X' is quasi-projective over S. It suffices to show that the composition

$$\iota\colon X' \xrightarrow{\ i\ } X \times_S P \xrightarrow{\ p_2\ } P$$

is an immersion (where p_2 is the second projection). We denote by $q_k\colon P \to P_k$ the k-th projection and set $W_k := q_k^{-1}(r_k(X_k))$ (which are open in P).

We claim that $\iota(X')$ is contained in $\bigcup_k W_k$. To show the claim we set $X'_k := \pi^{-1}(X_k)$. Then $(X'_k)_k$ is an open covering of X'. We have $r_k \circ \pi_{|U'} = q_k \circ \iota_{|U'}$ for all k. As U' is schematically dense in X' and thus in X'_k and as P_k is separated over S, it follows that $r_k \circ \pi_{|X'_k} = q_k \circ \iota_{|X'_k}$ (Proposition 9.19). This shows that $X'_k \subseteq \iota^{-1}(W_k)$ for all k. As the X'_k form a covering of X, this proves our claim.

Therefore to show that ι is an immersion, it suffices to show that the restriction $\iota_k\colon U'_k := \iota^{-1}(W_k) \to W_k$ of ι is an immersion for all k. Let Γ_k be the graph of the composition $u_k\colon W_k \to r_k(X_k) \xrightarrow{\sim} X_k \hookrightarrow X$ which is a closed subscheme of $W_k \times_S X$ because X is separated over S. The restrictions to U' of the two morphisms $u_k \circ p_2$ and p_1 from $W_k \times_S X$ to X are equal, thus $\mathrm{Ker}(u_k \circ p_2, p_1) = \Gamma_k$ majorizes U' and hence also U'_k. Therefore we can write ι_k as the composition $U'_k \hookrightarrow \Gamma_k \xrightarrow{\sim} W_k$ which shows that ι_k is an immersion.

Step 6: Proof of (1). If X' is projective over S, then X' is proper over S and hence X is proper over S by Proposition 12.59 because π is surjective. Conversely, if $X \to S$ is proper, then $X' \to X \to S$ is proper because properness is stable under composition. Thus $X' \to S$ is proper and quasi-projective and hence projective by Corollary 13.72. $\square$

The proof of Theorem 13.100 shows that if X is irreducible, we can choose U in such a way that $\pi^{-1}(U)$ is schematically dense in X'.

Corollary 13.101. *Let S be a locally noetherian scheme. Let $f\colon X \to S$ be a separated morphism of finite type. Then f is proper if and only if for all $n \geq 1$ the base change $f_{(\mathbb{A}_S^n)}\colon X \times_S \mathbb{A}_S^n \to \mathbb{A}_S^n$ is closed.*

Proof. The condition is necessary by definition. To prove that it is sufficient we may assume that $S = \operatorname{Spec} R$ is affine. We first show that $f_{(S')}$ is closed for every morphism $S' \to S$ of finite type. As this question is local on S', we may assume that S' is affine and hence that there exists a closed S-immersion $S' \hookrightarrow \mathbb{A}_S^n$ for some $n \geq 0$. By hypothesis $f_{(\mathbb{A}_S^n)}$ is closed and this implies that $f_{(S')}$ is closed.

By Chow's lemma (Theorem 13.100) there exists a commutative diagram

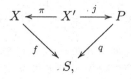

where π is projective and surjective, j is an open immersion, and q is projective. We may write j as the composition

$$j\colon X' \xrightarrow{\Gamma_j} X' \times_S P \xrightarrow{\pi \times \operatorname{id}_P} X \times_S P \xrightarrow{f_{(P)}} P.$$

As j is separated, Γ_j is a closed immersion. As π is projective (hence proper by Corollary 13.72), $\pi \times \operatorname{id}_P$ is closed. Finally $f_{(P)}$ is closed because P is of finite type over S. This shows that j is a closed immersion. Thus X' is proper and hence X is proper because π is surjective (Proposition 12.59). $\qquad\square$

(13.22) Blow-ups and birational morphisms.

Theorem 13.102. *Let X be an integral noetherian scheme such that there exists an ample line bundle on X (e.g., if X is quasi-projective over some affine scheme) and let X' be an integral scheme. Let $f\colon X' \to X$ be a projective birational morphism. Then there exists a closed subscheme Z of X and an isomorphism of X-schemes $X' \cong \operatorname{Bl}_Z(X)$.*

We will use in the proof the finiteness of direct images for coherent modules (Theorem 12.68) which will be proved in Volume II.

Proof. Let $\mathscr{L}$ be an f-ample invertible $\mathscr{O}_{X'}$-module. Then $X' \cong \operatorname{Proj} \mathscr{A}$ by Proposition 13.74, where $\mathscr{A} = \bigoplus_d f_*(\mathscr{L}^{\otimes d})$. By Theorem 12.68, $f_*(\mathscr{L}^{\otimes d})$ is for all d a coherent $\mathscr{O}_X$-module. As in Remark 13.11 and as X is quasi-compact, we can replace $\mathscr{L}$ by some positive power $\mathscr{L}^{\otimes \delta}$ to assume that $f_*(\mathscr{L})^d = f_*(\mathscr{L}^{\otimes d})$ for d sufficiently large. Then we have $X' \cong \operatorname{Proj} \bigoplus_d f_*(\mathscr{L})^d$.

Let $\mathscr{K}_X$ (resp. $\mathscr{K}_{X'}$) be the constant sheaf with value $K(X)$ (resp. $K(X')$). As f is birational, we have $f_* \mathscr{K}_{X'} = \mathscr{K}_X$. By Proposition 11.29 we can embed $\mathscr{L}$ into $\mathscr{K}_{X'}$ and we obtain an embedding $f_* \mathscr{L} \hookrightarrow \mathscr{K}_X$. Let $\mathscr{M}$ be an ample line bundle on X, which we may assume to be a submodule of the quasi-coherent $\mathscr{O}_X$-module $\mathscr{K}_X$ (cf. Section (11.12)). We claim that there exists an integer $n \geq 1$ such that $\mathscr{M}^{-n} f_* \mathscr{L} \subseteq \mathscr{O}_X$ (multiplication in $\mathscr{K}_X$). Let $\mathscr{T}$ be the coherent ideal $(\mathscr{O}_X : f_* \mathscr{L})$ (Remark 7.37) and let $n \geq 1$ be an integer such that $\mathscr{M}^n \mathscr{T}$ is generated by global sections. Thus we have $\mathscr{O}_X \subseteq \mathscr{M}^n \mathscr{T}$ and hence $\mathscr{M}^{-n} \subseteq \mathscr{T}$. This shows our claim.

We set $\mathscr{I} := \mathscr{M}^{-n} f_* \mathscr{L}$, let Z be the corresponding closed subscheme of X. Then $\operatorname{Bl}_Z(X) = \operatorname{Proj} \bigoplus_d (\mathscr{M}^{-dn}(f_* \mathscr{L})^d) \cong X'$ by Proposition 13.28. $\qquad\square$

We also mention the following result due to Conrad ([Co] 2.11 and 4.4) which shows that if we allow blow-ups on source and target, then any separated birational morphism of finite type can be made into an open immersion:

Theorem 13.103. *Let S be qcqs. Let $f: X \to S$ be a separated morphism of finite type such that there exists an open quasi-compact dense subscheme $U \subset S$ such that $f^{-1}(U)$ is dense in X and such that f induces an isomorphism $f^{-1}(U) \overset{\sim}{\to} U$. Then there exists a commutative diagram*

(13.22.1)
$$
\begin{array}{ccc}
\tilde{X} & \overset{j}{\longrightarrow} & \tilde{S} \\
\downarrow & & \downarrow \\
X & \overset{f}{\longrightarrow} & S,
\end{array}
$$

where $\tilde{S} \to S$ and $\tilde{X} \to X$ are U-admissible blow-ups and where j is an open immersion. If f is also proper, then (13.22.1) can be chosen such that j is an isomorphism.

More recently, Abramovich, Karu, Matsuki and Włodarczyk [AKMW], [Wlo] have proved the following "weak factorization theorem". See also Bonavero's survey [Bon].

Theorem 13.104. *Let k be an algebraically closed field of characteristic 0, and let X and Y be smooth and proper integral schemes over k. Then every birational map $\varphi: X \dashrightarrow Y$ factors as a sequence of blow-ups in closed subschemes that are themselves smooth schemes and inverses of such blow-ups, i.e., there is a sequence*

$$
X = X_0 \dashrightarrow X_1 \dashrightarrow \cdots \dashrightarrow X_{n-1} \dashrightarrow X_n = Y,
$$

of birational maps between smooth proper integral k-schemes, such that for each map ψ in the sequence, either ψ or ψ^{-1} is a blow-up morphism in a smooth center, and such that the composition of these birational maps is φ.

(13.23) Resolution of Singularities.

We conclude this chapter with some results on resolutions of singularities without giving any proofs. Let X be a reduced noetherian scheme. A *resolution of singularities* is a proper birational morphism $f: X' \to X$, where X' is regular. No general criterion is known for whether a scheme X admits a resolution of singularities. Grothendieck gave the following necessary criterion in [EGAIV] (7.9.5).

Proposition 13.105. *Let X be a reduced noetherian scheme such that every integral scheme Y, which is finite over X, admits a resolution of singularities. Then X is quasi-excellent (Definition 12.49).*

One can ask whether, conversely, every reduced (quasi-)excellent scheme has a resolution of singularities. This is an important question which is still wide open even for many special cases. For instance, it is not known whether all integral schemes of finite type over an algebraically closed field of positive characteristic have a resolution of singularities.

Here we mention a few positive results. The most important one is Hironaka's theorem (cf. [Hi]) on resolution of singularities in characteristic zero:

Theorem 13.106. *Let X be a reduced excellent scheme such that the residue fields $\kappa(x)$ of all points $x \in X$ have characteristic zero (e.g., if X is reduced and of finite type over a field of characteristic zero). Then X has a resolution of singularities.*

In fact Hironaka proves that for a scheme X as in Theorem 13.106 there exists a sequence

$$(13.23.1) \qquad \pi \colon X' = X_n \to X_{n-1} \to \cdots \to X_1 \to X_0 = X,$$

where X' is regular and where, for each i, $X_{i+1} \to X_i$ is a blow-up in a closed subscheme $D_i \subset X_i$ satisfying the following conditions.

(a) D_i is regular, and D_i is contained in the non-regular locus of X_i.

(b) X_i is normally flat along D_i, i.e., the $\mathscr{O}_{D_i}$-module $\bigoplus_{d \geq 0} \mathscr{I}_{D_i}^d / \mathscr{I}_{D_i}^{d+1}$ is flat (where $\mathscr{I}_{D_i} \subset \mathscr{O}_{X_i}$ is the quasi-coherent ideal sheaf defining D_i).

These conditions imply in particular that the restriction $\pi^{-1}(X_{\mathrm{reg}}) \to X_{\mathrm{reg}}$ of π is an isomorphism.

Hironaka's proof is extremely complicated. Since the middle of the 1990's, a number of substantial simplifications have been given. See Hauser's survey [Hau] and Kollár's book [Ko] for further details and references; see also Temkin's article [Te1] for a modern proof of Hironaka's result.

Combining this result with Nagata's compactification Theorem 12.70 we obtain:

Corollary 13.107. *Let S be an excellent noetherian $\mathbb{Q}$-scheme, let $f \colon X \to S$ be a separated morphism of finite type and assume that X is regular. Then there exists an open dominant immersion $j \colon X \to \bar{X}$ of S-schemes, where $\bar{X}$ is a regular proper S-scheme.*

It can be shown ([Te1]) that one can choose $\bar{X}$ in such a way that the boundary $\bar{X} \setminus j(X)$ is a divisor with particularly nice singularities (a so-called divisor with normal crossing).

In arbitrary characteristic the existence of a sequence as in (13.23.1) is known only for reduced excellent schemes X of dimension ≤ 2.

A weaker form of resolution of singularities has been shown by de Jong [dJ] for schemes of finite type over arbitrary fields:

Theorem 13.108. *Let k be a field and let X be an integral separated scheme of finite type over k. Then there exist k-morphisms*

$$\bar{X}_1 \xleftarrow{\;j\;} X_1 \xrightarrow{\;\pi\;} X,$$

where

(1) *$\bar{X}_1$ and X_1 are integral and regular, and $\bar{X}_1$ is projective over k,*

(2) *π is an alteration, i.e. π is proper, surjective, and there exists a non-empty open subset U of X such that $\pi^{-1}(U) \to U$ is finite.*

(3) *j is an open immersion such that the reduced closed subscheme $D := \bar{X}_1 \setminus j(X_1)$ is a strict normal crossing divisor (i.e., denoting by D_i for $i \in I$ the irreducible components of D, considered as reduced subscheme, then for all $J \subseteq I$ the schematic intersection $\bigcap_{i \in J} D_i$ is a regular closed subscheme of $\bar{X}_1$ of codimension $\#J$).*

There are plenty of variants and generalizations of the results stated above (e.g., embedded resolutions, or a variant of Theorem 13.108 for schemes over discrete valuation rings) which have many applications but go beyond the scope of this book.

Exercises

Exercise 13.1. Let $n \geq 1$ be an integer, R a ring, and let $a_0, \ldots, a_n \geq 1$ be integers. Endow $R[T_0, \ldots, T_n]$ with the structure of a graded R-algebra such that the degree of T_i is a_i. Then $\mathbb{P}_R(a_0, \ldots, a_n) := \operatorname{Proj} R[T_0, \ldots, T_n]$ is called *weighted projective space with weights* $(a_0, \ldots, a_n)$.

(a) Let $R[X_0, \ldots, X_n]$ be the polynomial ring with the standard grading. Show that the R-algebra homomorphism $R[T_0, \ldots, T_n] \to R[X_0, \ldots, X_n]$, $T_i \mapsto X_i^{a_i}$, yields a finite surjective morphism $\pi \colon \mathbb{P}_R^n \to \mathbb{P}_R(a_0, \ldots, a_n)$.

(b) Show that $\mathbb{P}_R(a_0, a_1) \cong \mathbb{P}_R^1$ for all $a_0, a_1 \geq 1$ (although π is usually not an isomorphism).

(c) Show that $\mathbb{P}_R(1, 1, 2) \cong V_+(X_1^2 - X_0 X_2) \subset \mathbb{P}_R^3$, where $X_0, \ldots, X_3$ denote homogeneous coordinates on $\mathbb{P}_R^3$. Deduce that $\mathbb{P}_R(1, 1, 2)$ is not smooth over R.

(d) Find for all triples (a_0, a_1, a_2) a graded R-algebra A such that A is generated by finitely many elements in A_1 and such that $\mathbb{P}(a_0, a_1, a_2) \cong \operatorname{Proj} A$.

Exercise 13.2. Let A be a graded ring and $X = \operatorname{Proj} A$. We say that a graded A-module M *satisfies condition* (TN) (resp. *satisfies condition* (TF)) if there exists an $n \in \mathbb{Z}$ such that $\bigoplus_{d \geq n} M_d = 0$ (resp. such that $\bigoplus_{d \geq n} M_d$ is a finitely generated A-module).

(a) Let $u \colon M \to N$ be a homomorphism of graded A-modules. Show that if $\operatorname{Ker}(u)$ (resp. $\operatorname{Coker}(u)$) satisfies (TN), then the associated homomorphism $\tilde{u} \colon \tilde{M} \to \tilde{N}$ of quasi-coherent $\mathscr{O}_X$-modules is injective (resp. surjective). We say that u is a *(TN)-isomorphism* if $\operatorname{Ker}(u)$ and $\operatorname{Coker}(u)$ satisfy (TN).

(b) Let M be a graded A-module that satisfies (TF). Show that $\tilde{M}$ is an $\mathscr{O}_X$-module of finite type.

(c) Let M be a graded A-module that satisfies (TF). Show that $\tilde{M} = 0$ if and only if M satisfies (TN).

Exercise 13.3. Let k be a field, $A = k[T_0, \ldots, T_n]$ and let M be a graded A-module that satisfies condition (TF) (Exercise 13.2). Show that M is saturated if and only if the following conditions hold.

(a) For $m \in M$ with $T_i m = 0$ for all i one has $m = 0$.

(b) For all homogeneous elements $m_i \in M$ $(i = 0, \ldots, n)$ with $T_i m_j = T_j m_i$ for all i, j there exists an $m \in M$ such that $m_i = T_i m$.

Exercise 13.4. Let A be a graded ring such that A_+ is generated by finitely many elements of A_1. For a graded A-module M define the *saturation* $M^{\mathrm{sat}} := \Gamma_*(\tilde{M})$. Show that M^{sat} is a saturated A-module and that we obtain a functor $(A\text{-GrMod}) \to (A\text{-GrMod}^{\mathrm{sat}})$ and that this functor is left adjoint to the inclusion functor $(A\text{-GrMod}^{\mathrm{sat}}) \to (A\text{-GrMod})$.

Exercise 13.5. Let R be a ring, $n_1, \ldots, n_r \geq 1$ integers and let $X \subset \mathbb{P}_R^{n_1} \times_R \cdots \times_R \mathbb{P}_R^{n_r}$ be a closed subscheme. Show that X is the vanishing scheme of polynomials $f_1, \ldots, f_m \in R[(T_{ji})_{1 \leq j \leq r, 0 \leq i \leq n_j}]$ which for all $j = 1, \ldots, r$ are separately homogeneous in the variable $(T_{ji})_{0 \leq i \leq n_j}$.

Exercise 13.6. We call a morphism $f \colon X \to S$ *locally quasi-projective* (resp. *locally projective*) if there exists an open affine covering $(U_\alpha)_\alpha$ of S and for all α a quasi-compact U_α-immersion (resp. a closed U_α-immersion) $f^{-1}(U_\alpha) \hookrightarrow \mathbb{P}_{U_\alpha}^{n_\alpha}$ for some integer $n_\alpha \geq 0$.

We remark that here "locally" means locally on the target, while for notions encountered earlier (like "morphisms locally of finite type") the "locally" meant locally on the source.

Prove the following assertions:
(a) Any quasi-compact immersion (resp. any closed immersion) is locally quasi-projective (resp. locally projective).
(b) Any locally quasi-projective morphism f is separated and of finite type.
(c) A morphism $f\colon X \to S$ is locally quasi-projective (resp. locally projective) if and only if there exists an open affine covering $(U_\alpha)_\alpha$ of S and for all α a quasi-compact U_α-immersion (resp. a closed U_α-immersion) $f^{-1}(U_\alpha) \hookrightarrow \mathbb{P}(\mathscr{E}_\alpha)$ for some quasi-coherent $\mathscr{O}_{U_\alpha}$-module $\mathscr{E}_\alpha$ of finite type.
(d) A scheme morphism $f\colon X \to S$ is locally projective if and only if f is proper and locally quasi-projective.

Exercise 13.7. Let A be a graded ring which is factorial (e.g., $A = k[T_0, \ldots, T_n]$ for a field k). Let $d, e > 0$ be integers and $f \in A_d$, $g \in A_e$ be nonzero homogeneous elements without a common prime factor. Set $X = \operatorname{Proj} A$ and $Z = V_+(f, g) = \operatorname{Proj} A/I$, where I is the homogeneous ideal generated by f and g. Show that the sequence

$$0 \to A(-d-e) \xrightarrow{\varphi} A(-d) \oplus A(-e) \xrightarrow{\psi} I \to 0,$$

with $\varphi(a) = (-ag, af)$ and $\psi(a, b) = af + bg$, is an exact sequence of graded A-modules. Deduce the existence of an exact sequence of $\mathscr{O}_X$-modules

$$0 \to \mathscr{O}_X(-d-e) \to \mathscr{O}_X(-d) \oplus \mathscr{O}_X(-e) \to \mathscr{I} \to 0,$$

where $\mathscr{I} \subseteq \mathscr{O}_X$ is the quasi-coherent ideal defining Z.

Exercise 13.8. In Exercise 1.30 we introduced the Veronese embedding. This exercise generalizes the construction there to projective bundles over schemes. Let S be a scheme, let $\mathscr{E}$ be a quasi-coherent $\mathscr{O}_S$-module of finite type, set $P := \mathbb{P}(\mathscr{E})$ and let $p\colon P \to S$ be the structure morphism. Fix an integer $d \geq 1$ and set $P(d) := \mathbb{P}(\operatorname{Sym}^d \mathscr{E})$. The surjection $p^*\mathscr{E} \twoheadrightarrow \mathscr{O}_P(1)$ yields a surjection $p^* \operatorname{Sym}^d(\mathscr{E}) = \operatorname{Sym}^d(p^*\mathscr{E}) \twoheadrightarrow \operatorname{Sym}^d(\mathscr{O}_P(1)) = \mathscr{O}_P(d)$. Thus we obtain a unique morphism $v_d\colon P \to P(d)$ such that $v_d^*(\mathscr{O}_{P(d)}(1)) = \mathscr{O}_P(d)$.

Show that v_d is a closed immersion. We call v_d the *d-fold Veronese embedding*. Show that if $S = \operatorname{Spec} k$ for an algebraically closed field, v_d is the embedding defined in Exercise 1.30.

Exercise 13.9. Let k be an algebraically closed field, let $n, d \geq 1$ be integers, and let $V = V_n^d$ be the k-vector space of homogeneous polynomials in indeterminates $T_0, \ldots, T_n$ of degree d. Let $P_n^d := \mathbb{P}(V^\vee)$ its projectivization (thus $P_n^d(k)$ consists of the elements of V modulo multiplication with nonzero scalars). Show that the set of reducible polynomials is closed in $P_n^d(k)$.
Hint: For $1 \leq e \leq d-1$ multiplication of polynomials yields a morphism $P_n^e \times P_n^{d-e} \to P_n^d$. Show that the image Z_e is closed and consider the union of the $Z_e(k)$.

Exercise 13.10. With the notation of Lemma 13.54 assume that r is a closed immersion. Show that then condition (a), (b), and (c) hold for $J = \{0, \ldots, n\}$.

Exercise 13.11. Let X be a qcqs scheme and let $\mathscr{L}$ and $\mathscr{L}'$ be invertible $\mathscr{O}_X$-modules. Assume that $\mathscr{L}$ is ample and that for all $x \in X$ there exists an integer $n > 0$ and a section $s' \in \Gamma(X, \mathscr{L}'^{\otimes n})$ such that $s'(x) \neq 0$. Show that $\mathscr{L} \otimes \mathscr{L}'$ is ample.

Exercise 13.12. Let $f\colon X \to S$ be a morphism of finite type and let $\mathscr{L}$ be an invertible $\mathscr{O}_X$-module.
(a) Let $\mathscr{K}$ be an invertible $\mathscr{O}_S$-module. Show that $\mathscr{L}$ is ample for f if and only if $\mathscr{L} \otimes f^*\mathscr{K}$ is ample for f.

(b) Let $g\colon X' \to X$ be a morphism of finite type and let $\mathscr{L}'$ be an invertible $\mathscr{O}_{X'}$-module. Assume that $\mathscr{L}'$ is very ample for g and that $\mathscr{L}$ is very ample for f. Show that $\mathscr{L}' \otimes f^*\mathscr{L}$ is very ample for $f \circ g$.

Exercise 13.13. Assume in the standard set up for schemes over inductive limits of rings (see Section (10.13) 1.–3.) that X_0 is of finite presentation. Let $\mathscr{L}_0$ be a line bundle on X_0 and set $\mathscr{L} := x_0^*\mathscr{L}_0$ and $\mathscr{L}_\lambda := x_{0,\lambda}^*\mathscr{L}_0$. Show that if $\mathscr{L}$ is ample over S (resp. very ample over S), then there exists an index λ_0 such that $\mathscr{L}_\lambda$ is ample over S_λ (resp. very ample over S_λ) for all $\lambda \geq \lambda_0$.

Exercise 13.14. Let k be an algebraically closed field. Let $m, n \geq 1$, and let $A = k[T_0, \ldots, T_m, S_0, \ldots, S_n]$. We identify $(\operatorname{Spec} A)(k)$ with the product of the space of monic polynomials of degree $m + 1$ and the space of monic polynomials of degree $n + 1$ over k, i. e., the indeterminates correspond to the coefficients of the polynomials. Prove that the subset of $(\operatorname{Spec} A)(k)$ consisting of pairs (f, g) (where f is a monic polynomial of degree $m + 1$, and g is a monic polynomial of degree $n + 1$) such that f and g have a common zero in k, is Zariski closed.

Exercise 13.15. Let $f\colon X \to S$ be a morphism of finite type, let $\mathscr{L}$ be an f-ample line bundle, and let $s \in \Gamma(X, \mathscr{L})$. Show that X_s is quasi-affine over S. If f is proper, show that X_s is affine over S.

Exercise 13.16. Let X be a qcqs scheme and set $A := \Gamma(X, \mathscr{O}_X)$. Show that the following assertions are equivalent.
(i) X is quasi-affine.
(ii) The canonical morphism $X \to \operatorname{Spec} A$ (Remark 3.7) is a homeomorphism of X onto a subspace of $\operatorname{Spec} A$.
(iii) The open subsets X_f for $f \in A$ form a basis of the topology of X.
(iv) Those open subsets X_f for $f \in A$ such that X_f is affine form a covering of X.

Exercise 13.17. Let k be a field and $S = \operatorname{Spec} k$. Let $n \geq 1$ be an integer and set $X := \mathbb{A}_k^{n+1} \setminus \{0\}$. Show that $\mathscr{O}_X$ is very ample over k and that the sections $s_i := T_i \in \Gamma(X, \mathscr{O}_X) = k[T_0, \ldots, T_n]$ for $i = 0, \ldots, n$ generate $\mathscr{O}_X$. Show that the corresponding morphism $X \to \mathbb{P}_k^n$ (13.12.1) is the canonical morphism (3.6.1).

Generalize these assertions to arbitrary schemes S.

Exercise 13.18. Let S be a scheme and let $f\colon X \to Y$ be a finite surjective morphism of quasi-compact S-schemes. Assume that X and Y are integral and that Y is normal. Let $N_{X/Y}\colon \operatorname{Pic}(X) \to \operatorname{Pic}(Y)$ be the norm homomorphism defined in Exercise 12.25.
(a) Let $\mathscr{L}$ be an invertible $\mathscr{O}_X$-module that is ample over S. Show that $N_{X/Y}(\mathscr{L})$ is ample over S.
(b) Let $\mathscr{K}$ be an invertible $\mathscr{O}_Y$-module. Show that $\mathscr{K}$ is ample over S if and only if $f^*\mathscr{K}$ is ample over S.
(c) Assume that Y is of finite type over S. Show that X is quasi-projective over S if and only if Y is quasi-projective over S.
(d) Assume that Y is of finite type over S and that S is qcqs. Show that X is projective over S if and only if Y is projective over S.

Exercise 13.19. Let X be a qcqs scheme and let $\mathscr{E}$, $\mathscr{E}_1$ and $\mathscr{E}_2$ be finite locally free $\mathscr{O}_X$-modules.
(a) Show that the following assertions are equivalent.

(i) For every quasi-coherent $\mathcal{O}_X$-module $\mathcal{F}$ of finite type there exists an integer n_0 such that $\mathrm{Sym}^n \mathcal{E} \otimes_{\mathcal{O}_X} \mathcal{F}$ is generated by its global sections for all $n \geq n_0$.

(ii) For every quasi-coherent $\mathcal{O}_X$-module $\mathcal{F}$ of finite type and for every $x \in X$ there exists an integer $n_0(x)$ such that for all $n \geq n_0(x)$ the global sections of $\mathrm{Sym}^n \mathcal{E} \otimes_{\mathcal{O}_X} \mathcal{F}$ generate the stalk $\mathrm{Sym}^n \mathcal{E}_x \otimes \mathcal{F}_x$ as $\mathcal{O}_{X,x}$-module.

The vector bundle $\mathcal{E}$ is called *ample* if it satisfies these equivalent conditions.

(b) Show that an invertible $\mathcal{O}_X$-module is ample in this sense if and only if it is ample in the sense of Definition 13.44.

(c) Show that any quotient vector bundle of an ample vector bundle is again ample. Show that a direct sum $\mathcal{E} = \mathcal{E}_1 \oplus \mathcal{E}_2$ is ample if and only if $\mathcal{E}_1$ and $\mathcal{E}_2$ is ample.

(d) Show that if $\mathcal{E}_1$ is ample and $\mathcal{E}_2$ is generated by its global sections, then $\mathcal{E}_1 \otimes \mathcal{E}_2$ is ample.

(e) Show that if $\mathcal{E}$ is ample, then there exists an $n_0 \geq 1$ such that $\mathrm{Sym}^n(\mathcal{E})$ is ample for all $n \geq n_0$.

Remark: In fact one can show that the tensor product of ample vector bundles is again ample. This implies that $\mathrm{Sym}^n(\mathcal{E})$ is ample for all $n \geq 1$.

(f) Show that if $\mathcal{E}$ ample, then $\det(\mathcal{E})$ is ample.

Exercise 13.20. Let $f : X \to Y$ be a closed morphism with finite fibers and assume that there exists an ample line bundle on X. Show that f is affine.
Hint: Exercise 12.3.

Exercise 13.21. Deduce Nagata's compactification Theorem 12.70 from Temkin's factorization Theorem 12.71.

Exercise 13.22. Let X be a scheme, let $\mathcal{I} \subseteq \mathcal{O}_X$ be a quasi-coherent ideal, and set $Z_n = V(\mathcal{I}^n)$ for all $n \geq 1$. Show that $\mathrm{Bl}_{Z_n}(X) \cong \mathrm{Bl}_{Z_1}(X)$ for all $n \geq 1$.

Exercise 13.23. Let X be a scheme, let Z be a closed subscheme, set $U := X \setminus Z$, and let $\pi : \mathrm{Bl}_Z(X) \to X$ be the blow-up (inducing an isomorphism $\pi^{-1}(U) \xrightarrow{\sim} U$). Let Y be a closed subscheme. Show that the schematic closure $\tilde{Y}$ (Remark 10.31) of $\pi^{-1}(U \cap Y)$ in $\mathrm{Bl}_Z(X)$ exists and that $\tilde{Y}$ is the strict transform of Y in $\mathrm{Bl}_Z(X)$.

Exercise 13.24. Let k be a field. Show that the blow-up $\tilde{X}$ of the origin in $\mathbb{A}_k^2$, viewed as a $\mathbb{P}_k^1$-scheme via $\tilde{X} \subset \mathbb{A}_k^2 \times \mathbb{P}_k^1 \to \mathbb{P}_k^1$, is the geometric line bundle $\mathbb{V}(\mathcal{O}(1))$. This is also an illustration of Theorem 13.67.

Exercise 13.25. Let k be a field, and let $X = V(XY - Z^2) \subset \mathbb{A}_k^3$. Let $\tilde{X} \to X$ be the blow-up of X in the origin $V(X, Y, Z) \subset \mathbb{A}_k^3$. Show that the morphism $X \setminus \{(0,0,0)\} \to \mathbb{P}_k^1$, $(x, y, z) \mapsto (x : z) = (z : y)$ extends to a morphism $\tilde{X} \to \mathbb{P}_k^1$, and identify $\tilde{X}$ as a $\mathbb{P}_k^1$-scheme with $\mathbb{V}(\mathcal{O}_{\mathbb{P}_k^1}(2))$. This is also an illustration of Theorem 13.67.

Exercise 13.26. Let X be a reduced scheme and let Z be a closed subscheme. Show that $\mathrm{Bl}_Z(X)$ is a reduced scheme.

Exercise 13.27. Let X be a scheme, let $Y_1, \ldots, Y_n$ be closed subschemes and let $\tilde{Y}_i$ be the strict transform of Y_i in $\mathrm{Bl}_{\cap Y_i}(X)$. Show that $\cap \tilde{Y}_i = \emptyset$.

Exercise 13.28. Let S be a scheme, $\mathcal{E}$ be a quasi-coherent $\mathcal{O}_S$-module, and let $V := \mathbb{V}(\mathcal{E})$ be the corresponding vector bundle. Let $f : V \to S$ be its structure morphism and let $z : S \to \mathbb{V}(\mathcal{E})$ be the zero section.

(a) Let $w: f^*\mathcal{E} \oplus \mathcal{O}_V \to \mathcal{O}_V$ the $\mathcal{O}_V$-module homomorphism whose restriction to $f^*\mathcal{E}$ is (13.9.1) and whose restriction to $\mathcal{O}_V$ is the identity. Show that w defines an affine open schematically dominant immersion of S-schemes

$$j := j_{\mathcal{E}} : : V \hookrightarrow \bar{V} := \mathbb{P}(\mathcal{E} \oplus \mathcal{O}_S).$$

and that $j_{\mathcal{E}}$ is functorial for surjective homomorphisms $\mathcal{E}' \twoheadrightarrow \mathcal{E}$.

(b) Show that the canonical projection $\mathcal{E} \oplus \mathcal{O}_S \to \mathcal{E}$ yields a closed immersion

$$i := i_{\mathcal{E}} : P := \mathbb{P}(\mathcal{E}) \hookrightarrow \bar{V}$$

such that $\bar{V} \setminus i(P) = j(V)$.

(c) Let $\bar{f}: \bar{V} \to S$ be the structure morphism, let $\bar{f}^*\mathcal{E} \oplus \mathcal{O}_{\bar{V}} = \bar{f}^*(\mathcal{E} \oplus \mathcal{O}_S) \twoheadrightarrow \mathcal{O}_{\bar{V}}(1)$ the universal quotient bundle, and let $\bar{u}: \bar{f}^*\mathcal{E} \to \mathcal{O}_{\bar{V}}(1)$ its restriction to the first direct summand. Show that the restriction of $\bar{u}$ to the open subscheme $\bar{V}^0 := \bar{V} \setminus j(z(S))$ is surjective and defines a surjective affine S-morphism

$$p: \bar{V}^0 \to P$$

such that $p \circ i = \mathrm{id}_P$ and such that $p_{|\bar{V}^0} = \pi$, where π is the projection of the pointed affine cone to P.

(d) Show that there exists an open affine covering $(U_i)_i$ of P such that $p^{-1}(U_i) \cong \mathbb{A}^1_{U_i}$ for all i.

(e) Describe all these morphisms on k-valued points (k some field) if $\mathcal{E} = (\mathcal{O}_S^{n+1})^\vee$ (and hence $V = \mathbb{A}_S^{n+1}$, $\bar{V} = \mathbb{P}_S^{n+1}$, and $P = \mathbb{P}_S^n$).

The scheme $\overline{\mathbb{V}(\mathcal{E})}$ is called the *projective closure* of $\mathbb{V}(\mathcal{E})$. The closed subscheme $i(\mathbb{P}(\mathcal{E}))$ is called the *locus at infinity of* $\mathbb{V}(\mathcal{E})$.

14 Flat morphisms and dimension

Contents

The main topic of this chapter is a detailed study of flat morphisms. Although it is impossible to fully capture the algebraic notion of flatness in geometric terms, there is a useful heuristic: If $f \colon X \to S$ is a flat morphism, then the fibers $f^{-1}(s)$ form a *continuously varying* family of varieties, as s varies in S. In simple situations, the converse to this statement is true, see Proposition 14.14, Theorem 14.34 and Theorem 14.128 below. One could say that flatness is the correct generalization of this naive point of view to the general case.

At the end, we take up again the study of the notion of dimension of a scheme. We now go beyond the case of schemes of finite type over a field (see Chapter 5), and in particular investigate the behavior of the dimensions of fibers of (flat) morphisms.

Flat morphisms

(14.1) First properties of flat morphisms.

Recall the definition of flatness (Definition 7.38):

Definition 14.1. *Let $f \colon X \to Y$ be a morphism of schemes, let $\mathscr{F}$ be an $\mathscr{O}_X$-module.*
(1) *Then $\mathscr{F}$ is called* flat over Y at x *or* f-flat at x *if $\mathscr{F}_x$ is a flat $\mathscr{O}_{Y,f(x)}$-module. It is called* flat over Y *or* f-flat *if $\mathscr{F}$ is flat over Y at all points $x \in X$.*
(2) *If $X = Y$ and $f = \mathrm{id}_X$, we simply say that $\mathscr{F}$ is flat at x if it is id_X-flat at x, i.e. if $\mathscr{F}_x$ is a flat $\mathscr{O}_{X,x}$-module. Similarly, $\mathscr{F}$ is called flat, if $\mathscr{F}_x$ is a flat $\mathscr{O}_{X,x}$-module for all $x \in X$.*
(3) *We say that f is* flat, *or that X is flat over Y, if $\mathscr{O}_X$ is flat over Y.*

Example 14.2. *Let Y be a scheme.*
(1) *If $Y = \operatorname{Spec} k$ for a field k, then every morphism of schemes $X \to Y$ is flat.*

© Springer Fachmedien Wiesbaden GmbH, part of Springer Nature 2020
U. Görtz und T. Wedhorn, *Algebraic Geometry I: Schemes*, Springer Studium
Mathematik – Master, https://doi.org/10.1007/978-3-658-30733-2_15

(2) The structure morphism $\mathbb{A}_Y^n \to Y$ is flat because polynomial rings are flat (Example B.19).

(3) As $\mathbb{P}_Y^n$ has an open cover by schemes that are flat over Y (more precisely, isomorphic to $\mathbb{A}_Y^n$), $\mathbb{P}_Y^n$ is flat over Y. More generally, for every finite locally free $\mathscr{O}_Y$-module $\mathscr{E}$ the projective bundle $\mathbb{P}(\mathscr{E})$ is flat over Y because locally on Y there exist isomorphisms $\mathbb{P}(\mathscr{E}) \cong \mathbb{P}_Y^n$.

Whereas in Chapter 7 we were mainly interested in flat $\mathscr{O}_X$-modules, in this chapter we will study flat morphisms between schemes in detail. We start by recording some simple properties.

Proposition 14.3.
(1) *A ring homomorphism $A \to B$ is flat if and only if the morphism $\operatorname{Spec} B \to \operatorname{Spec} A$ is flat in the sense of the above definition.*
(2) *Flatness is stable under base change and under composition of morphisms.*
(3) *Flatness is local on the source and on the target.*
(4) *Open immersions are flat morphisms.*
(5) *Let $f\colon X \to Y$ be a morphism. Then f is flat if and only if for every $y \in Y$, the pull-back $X \times_Y \operatorname{Spec} \mathscr{O}_{Y,y} \to \operatorname{Spec} \mathscr{O}_{Y,y}$ is flat.*

Proof. For part (1), see Proposition B.27: Flatness of a ring homomorphism $A \to B$ can be checked on the local rings. Part (2) follows from simple properties of the tensor product. Assertion (3) is clear from the definition, and this also implies (4). Finally, (5) follows from the definition because the local ring of $X \times_Y \operatorname{Spec} \mathscr{O}_{Y,y}$ in x is $\mathscr{O}_{X,x}$ by Proposition 4.20. $\square$

The property "flat" is also compatible with inductive limits of rings in the sense of Section (10.13), more precisely we have:

Proposition 14.4. *In the situation of Section (10.13), 1.–3., 5., and 6. we assume that X_0 and Y_0 are of finite presentation over S_0 and that $\mathscr{F}_0$ is an $\mathscr{O}_{X_0}$-module of finite presentation. Set $\mathscr{F} := x_0^* \mathscr{F}_0$ and $\mathscr{F}_\lambda := x_{0\lambda}^* \mathscr{F}_0$ for all λ. Then $\mathscr{F}$ is f-flat if and only if there exists a λ such that $\mathscr{F}_\lambda$ is f_λ-flat.*

This is a rather difficult result from commutative algebra which we do not show here (see [EGAIV] (11.2.6) or [RG] Part I (2.7) for a proof).

We recall that we have already seen in Corollary 10.85 that any morphism $f\colon X \to Y$ of finite presentation to an integral scheme Y is generically flat:

Theorem 14.5. *Let Y be an integral scheme and let be $f\colon X \to Y$ a quasi-compact morphism locally of finite presentation. Then there exists an open dense subset $V \subseteq Y$ such that $f_{|f^{-1}(V)}$ is flat.*

For later use, we note the following lemma:

Lemma 14.6. *Let $f\colon X \to Y$ be a quasi-compact morphism of S-schemes, and let $Z \subseteq Y$ be the schematic image of f (Section (10.8)). If $S' \to S$ is a flat morphism, then $Z' = Z \times_S S'$ is the schematic image of the base change $f'\colon X \times_S S' \to Y \times_S S'$.*

Proof. By Proposition 14.3 the base change $p\colon Y' := Y \times_S S' \to Y$ is again flat. Moreover $Z' = p^{-1}(Z)$ and thus we may assume that $Y = S$ and $Y' = S'$. By Proposition 10.30, the schematic image Z is the closed subscheme corresponding to the quasi-coherent ideal $\mathscr{K}_f := \mathrm{Ker}(f^\flat\colon \mathcal{O}_Y \to f_*\mathcal{O}_X)$. As $Y' \to Y$ is flat, we have $p^*\mathscr{K}_f = \mathrm{Ker}(f'^\flat)$. As f' is again quasi-compact, $Z' = V(\mathrm{Ker}(f'^\flat))$ is the schematic image of f'. $\square$

(14.2) Faithfully flat morphisms.

A concept of similar importance as *flatness* is the notion of *faithfully flat* morphism:

Definition 14.7. *A morphism* $f\colon X \to Y$ *is called* faithfully flat *if f is flat and surjective.*

If k is a field, then any morphism from a non-empty scheme X to $\mathrm{Spec}\, k$ is faithfully flat. For all schemes Y and $n \geq 0$ the structure morphisms $\mathbb{A}^n_Y \to Y$ and $\mathbb{P}^n_Y \to Y$ are faithfully flat.

Remark 14.8. The property "faithfully flat" is stable under base change, under composition of morphisms, and local on the target because this holds for the properties "flat" and "surjective".

We will study faithfully flat morphisms in detail in this chapter, and we will see that given a faithfully flat quasi-compact morphism $S' \to S$ and an S-scheme X, many properties of $X \times_S S'$ "descend" to X: if $X \times_S S'$ has a certain property, it follows that X has the same property.

The following lemma implies in particular that every flat local homomorphism between local rings gives rise to a faithfully flat morphism of their spectra.

Lemma 14.9. *Let* $f\colon X \to Y$ *be a flat morphism of schemes,* $x \in X$ *a point,* $y := f(x)$. *Let* $y' \in Y$ *be a generization of y (i.e.,* $y \in \overline{\{y'\}}$*). Then there exists a generization x' of x such that* $f(x') = y'$.

Proof. Set $A = \mathcal{O}_{Y,y}$ and $B = \mathcal{O}_{X,x}$ and $\varphi := f^\sharp_x\colon A \to B$. We have to show that the morphism $g\colon \mathrm{Spec}\, B \to \mathrm{Spec}\, A$ corresponding to the flat local homomorphism φ is surjective. As $\mathfrak{m}_y B \neq B$, B is a faithfully flat A-module. Therefore, $B \otimes_A \kappa(y') \neq 0$, or in other words: The fiber of $\mathrm{Spec}\, B \to \mathrm{Spec}\, A$ over $y' \in \mathrm{Spec}\, A$ is non-empty. $\square$

We can also express the statement of the lemma by saying that for every flat morphism $f\colon X \to Y$ and $x \in X$, we have $f(\mathrm{Spec}\, \mathcal{O}_{X,x}) = \mathrm{Spec}\, \mathcal{O}_{Y,f(x)}$ (where we consider the spectra of the stalks as subsets of X and Y, respectively). See Theorem 14.35 below for an important application of this fact: flat morphisms locally of finite presentation are open.

Now we relate the definition above to the notion of faithful flatness for modules over a ring (see Proposition B.17). As the terminology indicates, the two notions are equivalent in the following sense:

Proposition 14.10. *Let* $\varphi\colon A \to B$ *be a homomorphism of rings and let* $f\colon \mathrm{Spec}\, B \to \mathrm{Spec}\, A$ *be the corresponding morphism. Then f is faithfully flat if and only if B is a faithfully flat A-module.*

Proof. We already know that f is flat if and only if B is a flat A-module. Thus we may assume that f and B are flat. Then B is a faithfully flat A-module if and only if for every maximal ideal $\mathfrak{m} \subset A$ we have $\mathfrak{m}B \neq B$ (Proposition B.17 (iii)). If $\mathfrak{n} \subset B$ is any maximal ideal containing $\mathfrak{m}B$, $\varphi^{-1}(\mathfrak{n})$ is a prime ideal containing $\mathfrak{m}$ and hence equal to $\mathfrak{m}$. Therefore the flat A-module B is faithfully flat if and only if every closed point of $\operatorname{Spec} A$ lies in the image of f. But every point $y \in \operatorname{Spec} A$ is a generization (Definition 2.8) of a closed point of $\operatorname{Spec} A$. Thus the proposition follows from Lemma 14.9. $\qquad\square$

Using this equivalence, we can globalize the property of Definition B.17.

Proposition 14.11. *Let $f\colon X \to Y$ be a faithfully flat morphism of schemes, and let*

$$\mathscr{F}_\bullet: \qquad \mathscr{F}_1 \to \mathscr{F}_2 \to \mathscr{F}_3$$

be a sequence of quasi-coherent $\mathscr{O}_Y$-modules. Then the sequence $\mathscr{F}_\bullet$ is exact if and only if the pull-back $f^*\mathscr{F}_\bullet$ is an exact sequence of $\mathscr{O}_X$-modules.

Proof. Since the exactness can be checked on the stalks, hence in particular locally on the base, this follows from the previous proposition. $\qquad\square$

Proposition 14.11 immediately implies the following result.

Corollary 14.12. *Let $g\colon X \to S$ and $h\colon Y \to S$ be morphisms of schemes and let $f\colon X \to Y$ be a faithfully flat S-morphism. Then g is flat (resp. faithfully flat) if and only if h is flat (resp. faithfully flat).*

Proposition 14.13. *Let $S' \to S$ be a flat morphism. Let $f\colon X \to S$ be a morphism of schemes with the property that f maps every maximal point of X to a maximal point of S. Then the morphism $X \times_S S' \to S'$ has the same property.*

Proof. Let $x' \in X' := X \times_S S'$ be a maximal point, i.e., the generic point of an irreducible component of X'. As $S' \to S$ is flat, its base change $X' \to X$ is flat as well. By Lemma 14.9, the image x of x' in X is also a maximal point, so by assumption the same is true for the image $s := f(x)$ in S. Let s' denote the image of x' in S'. We want to show that it is a maximal point. After base change by $\operatorname{Spec} \mathscr{O}_{S,s} \to S$, we may assume that $S = \operatorname{Spec} \mathscr{O}_{S,s}$.

We may assume that S is reduced, so that $\mathscr{O}_{S,s}$ is a field because s is a maximal point. Therefore X is flat over S. In particular, $\mathscr{O}_{X',x'}$ is flat over $\mathscr{O}_{S',s'}$, and applying Lemma 14.9 once more, we obtain the desired conclusion. $\qquad\square$

(14.3) Example: Flat morphisms into Dedekind schemes.

Let R be a Dedekind domain. Recall that an R-module M is flat if and only if it is torsion-free (Proposition B.89). Moreover, recall that a Dedekind scheme is an integral noetherian scheme Y such that every local ring of Y is a discrete valuation ring or a field, see Section (7.13).

Proposition 14.14. *Let Y be a Dedekind scheme, let η be the generic point of Y and let $f\colon X \to Y$ be a morphism. Consider the following assertions.*
(i) *The morphism f is flat.*
(ii) *The scheme X is equal to the schematic closure (see Remark 10.31) of the generic fiber $f^{-1}(\eta)$ in X.*

(iii) *Every irreducible component of X dominates Y.*
Then the implications "(i) $\Leftrightarrow$ (ii) $\Rightarrow$ (iii)" hold. If X is reduced, all assertions are equivalent.

Proof. To prove "(i) $\Leftrightarrow$ (ii)", we may assume $Y = \operatorname{Spec} R$ for a Dedekind ring R and that $X = \operatorname{Spec} A$ is affine. Denote by K the field of fractions of R. The schematic closure of the generic fiber is $\operatorname{Spec} A/\mathfrak{a}$, where $\mathfrak{a} := \operatorname{Ker}(A \to A \otimes K) = A_{\mathrm{tors}}$. Therefore A is torsion-free if and only if the schematic closure of the generic fiber is X.

If there was an irreducible component X' of X that does not dominate Y, then its generic point θ is not contained in $f^{-1}(\eta)$ und thus also not in the closure of $f^{-1}(\eta)$. This is a contradiction to (ii).

Finally (iii) implies (ii) if X is reduced because a closed subscheme of X with the same underlying topological space is necessarily equal to X. $\square$

The proposition shows that it is easy to "flatten" a scheme X over a Dedekind scheme by replacing X by the schematic closure of its generic fiber (see Exercise 14.5 for more details).

Remark 14.15. In the situation of Proposition 14.14 let X be locally noetherian. Then (ii) (and hence the flatness of f) can be expressed by saying that all associated points of X lie in $f^{-1}(\eta)$ (Proposition 9.22).

Definition/Remark 14.16. Assume that $Y = \operatorname{Spec} R$ for a discrete valuation ring R. Then Y has two points, an open generic point η corresponding to the zero ideal and a closed point s, also called *special point*, corresponding to the maximal ideal (π), where $\pi \in R$ is a uniformizing element. Hence a morphism $f \colon X \to Y$ has two fibers, the *generic fiber* $X_\eta := f^{-1}(\eta)$ and the *special fiber* $X_s := f^{-1}(s)$. An R-module M is torsion-free if and only if it has no π-torsion. In particular, the following assertions are equivalent.
(i) The scheme X is flat over Y.
(ii) For all open affine subschemes U the R-module $\Gamma(U, \mathscr{O}_X)$ has no π-torsion.
(iii) For all $x \in X$, the R-module $\mathscr{O}_{X,x}$ has no π-torsion.
In (iii) it suffices to consider points x in the special fiber X_s since for all $x \in X_\eta$ the image of π in $\mathscr{O}_{X,x}$ is a unit.

Proposition 14.17. *Let R be a discrete valuation ring, let s be the special and η the generic point of $\operatorname{Spec} R$. Let $f \colon X \to \operatorname{Spec} R$ be a morphism. Assume that X is locally noetherian. Suppose that all maximal points of the special fiber $X_s := f^{-1}(s)$ lie in the closure of $X_\eta := f^{-1}(\eta)$ and that the special fiber X_s is reduced. Then X is flat over R.*

Proof. Let π be a uniformizing element of R. We show that for all $x \in X_s$, π is a regular element of $\mathscr{O}_{X,x}$. Hence we may assume that $X = \operatorname{Spec} A$ for a local noetherian ring A such that π is contained in its maximal ideal. Let $a \in A$ with $\pi a = 0$. We want to show that $a = 0$. Certainly $f^{-1}(\eta) = \operatorname{Spec} A \otimes K$ is contained in $V(a)$, because its points are prime ideals that do not contain π. Our assumption implies that $V(a) = \operatorname{Spec} A$ as sets, i.e., a is nilpotent. Since the special fiber is reduced, it follows that $a \in \pi A$, say $a = \pi a'$. We can apply the same reasoning to a', and by induction we get that $a \in \bigcap_i \pi^i A$. As π is contained in the maximal ideal of A, this intersection is 0 (see Corollary B.43). $\square$

In Theorem 14.34 we will prove a valuative criterion for flatness: A morphism $X \to Y$, locally of finite type, and with Y reduced and locally noetherian, is flat if every base change with respect to $\operatorname{Spec} R \to Y$, R a discrete valuation ring, is flat. This shows that understanding flatness over a discrete valuation ring is really the crucial case, at least as long as one considers reduced base schemes.

There is another way to read this: Flatness over a discrete valuation ring is easy to understand – it just says that the whole family equals the schematic closure of the generic fiber. By the valuative criterion, flatness over a reduced base is reasonably easy to understand in geometric terms, too. The great merit of the notion of flatness is that it is defined over any base, and gives us a means to use this notion over non-reduced base schemes, where it is harder to describe the geometric content.

(14.4) Examples of flat and of non-flat morphisms.

We start with examples of two classes of morphisms which are not flat, except in trivial cases. We will use the following lemma that will also be useful when we prove quasi-projectivity of curves in Theorem 15.18 below.

Lemma 14.18. *Let X be a scheme with finitely many irreducible components and let $f\colon X \to Y$ be a separated morphism. If there exists a covering $X = \bigcup_{i \in I} U_i$ by dense open subsets $U_i \subseteq X$ such that for all i, $f_{|U_i}$ is an immersion (resp. an open immersion), then f is an immersion (resp. an open immersion).*

Proof. It suffices to show that f is injective. We may assume that X and Y are reduced. Replacing Y by $\overline{f(X)}$, endowed with its reduced scheme structure, we may assume that f is dominant. As f is locally on X an immersion, $f_x^\sharp\colon \mathcal{O}_{Y,f(x)} \to \mathcal{O}_{X,x}$ is surjective for all $x \in X$ and hence bijective because f is dominant and Y is reduced. Thus f is flat and therefore the two projections $\operatorname{pr}_i\colon X \times_Y X \to X$ $(i = 1, 2)$ are flat. Thus $\operatorname{pr}_i(X \times_Y X)$ is stable under generization in X (Lemma 14.9).

Let $\eta_1, \ldots, \eta_n \in X$ be the maximal points of X. Since any U_i contains all of these points, the restriction of f to $\{\eta_1, \ldots, \eta_n\}$ is injective. As f is dominant, the images $\theta_i := f(\eta_i)$ are the maximal points of Y.

Let ζ be a maximal point of $X \times_Y X$. The image of ζ in Y is a maximal point of Y, say θ_j. As $\operatorname{pr}_i(X \times_Y X)$ is stable under generization, $\operatorname{pr}_1(\zeta)$ and $\operatorname{pr}_2(\zeta)$ are maximal points of X which both map to θ_j. Therefore $\operatorname{pr}_1(\zeta) = \operatorname{pr}_2(\zeta) = \eta_j$. Thus $\zeta \in Z := \operatorname{Spec}(\kappa(\eta_j) \otimes_{\kappa(\theta_j)} \kappa(\eta_j))$ by Lemma 4.28. But by the above, $\kappa(\theta_j) \to \kappa(\eta_j)$ is an isomorphism and hence $Z \cong \operatorname{Spec} \kappa(\eta_j) \cong \operatorname{Spec} \kappa(\theta_j)$. Therefore every maximal point of $X \times_Y X$ is in the image of the diagonal Δ_f. As f is separated, $\Delta_f(X)$ is closed. Therefore Δ_f is surjective.

Now let $x_1, x_2 \in X$ with $f(x_1) = f(x_2)$. Then there exists $z \in X \times_Y X$ such that $\operatorname{pr}_1(z) = x_1$ and $\operatorname{pr}_2(z) = x_2$. As Δ_f is surjective, there exists $x \in X$ with $\Delta_f(x) = z$ and hence $x_i = \operatorname{pr}_i(\Delta_f(x)) = x$ for $i = 1, 2$. $\square$

Proposition 14.19. *Let $f\colon X' \to X$ be a surjective separated morphism of integral schemes such that the induced extension of function field $K(X) \to K(X')$ is an isomorphism. Suppose that f satisfies moreover one of the following hypotheses.*
(a) *The morphism f is affine (e.g., if f is the normalization of X).*
(b) *The morphism f is locally of finite presentation (e.g., if f is the blow-up of a noetherian integral scheme X in a closed subscheme $Z \subsetneq X$).*
Then f is flat if and only if it is an isomorphism.

Proof. Suppose that f is flat and hence faithfully flat since f is surjective. To see that f is an isomorphism we may assume that $X = \operatorname{Spec} A$ for an integral domain A.

Suppose that f is affine. Then $X' = \operatorname{Spec} A'$, where A' is a faithfully flat A-algebra contained in the field of fractions of A. We claim that $A = A'$. Let $a' = a/s \in A'$ with $a, s \in A$ and $s \neq 0$. Then $a \in sA' \cap A = sA$, where the equality holds by Proposition B.22. Hence $a' \in A$. This shows the claim and that f is an isomorphism.

Now suppose that f is locally of finite presentation. For all $x' \in X'$ the local ring $\mathscr{O}_{X',x'}$ is a faithfully flat $\mathscr{O}_{X,f(x')}$-algebra (by Example B.18) contained in the field of fractions of $\mathscr{O}_{X,f(x')}$. Hence the same argument as above shows that $\mathscr{O}_{X,x} \to \mathscr{O}_{X,f(x')}$ is an isomorphism. Since f is locally of finite presentation, Proposition 10.52 shows that there exists an open neighborhood U' of x' such that $f_{|U'}$ is an open immersion. Therefore f is an open immersion by Lemma 14.18. As f is by hypothesis surjective, f is an isomorphism. $\qquad\square$

Proposition 14.20. *Let $f\colon X \to Y$ be a closed immersion. Then f is flat and of finite presentation if and only if it is an open immersion.*

Proof. A closed immersion is quasi-compact and separated. Thus the condition is sufficient because any open immersion is flat and locally of finite presentation. Now assume that f is flat and of finite presentation. As f is a closed immersion it yields a homeomorphism of X onto its image $f(X)$. Thus it suffices to show that for all $y \in f(X)$ there exists an open neighborhood V of y such that the restriction $f^{-1}(V) \to V$ is an isomorphism. This follows from the following lemma. $\qquad\square$

Lemma 14.21. *Let $f\colon X \to Y$ be a closed immersion of finite presentation which is flat at $x \in X$. Then there exists an open neighborhood V of $y := f(x)$ such that the restriction $f^{-1}(V) \to V$ of f is an isomorphism.*

Proof. Let $\mathscr{I} \subset \mathscr{O}_Y$ be the quasi-coherent ideal of the closed immersion. As f is of finite presentation, $\mathscr{I}$ is of finite type. By hypothesis the (nonzero) $\mathscr{O}_{Y,y}$-module $\mathscr{O}_{Y,y}/\mathscr{I}_y$ is flat and hence free by Proposition B.21. Thus $\mathscr{I}_y = 0$. As $\mathscr{I}$ is of finite type, its support is closed (Corollary 7.32) and there exists an open neighborhood V of y such that $\mathscr{I}_{|V} = 0$. $\qquad\square$

Now let us give some general positive results. First, "families of hypersurfaces" are flat:

Proposition 14.22. *Let S be a scheme, let $X \to S$ be a flat morphism, and let $Y \subseteq X$ be a closed subscheme of X, locally on X defined by a single equation, such that for every point $s \in S$, the inclusion of fibers $Y_s \subset X_s$ makes Y_s into an effective Cartier divisor of X_s. Suppose further that one of the following hypotheses is satisfied.*
(1) The schemes S and X are locally noetherian.
(2) The morphism $X \to S$ is locally of finite presentation.
Then Y is flat over S and Y is an effective Cartier divisor of X.

As Y is defined locally by one equation, the same is true for Y_s in X_s. Hence if X_s is integral, then Y_s is an effective Cartier divisor in X_s if and only if $Y_s \neq X_s$ since then every local section of $\mathscr{O}_X$ is regular if and only if it is locally non-zero.

We give the proof in the noetherian case and reduce in the locally of finite presentation case to a commutative algebra statement in [St].

Proof. All hypotheses and assertions are local on X and S. Hence we may assume that $S = \operatorname{Spec} A$ and $X = \operatorname{Spec} B$ are affine, where B is a flat A-algebra, and that Y is the vanishing locus of an element $f \in B$. We have to show that B/fB is flat over A and that f is regular in B. Both properties can be checked on local rings hence we can even assume that A and B are local rings where A und B are noetherian or B is the localization of an A-algebra of finite presentation. Then we can conclude by the following lemma in the noetherian case and by [St] 046Z in the locally of finite presentation case. $\qquad\square$

Lemma 14.23. *Let $A \to B$ be a local homomorphism of local noetherian rings and let $\mathfrak{m} \subset A$ be the maximal ideal. Let M be a finitely generated B-module and let f be an element of the maximal ideal of B. Then the following assertions are equivalent.*
(i) *The multiplication $M \to M$ with f is injective and M/fM is flat over A.*
(ii) *The multiplication $M/\mathfrak{m}M \to M/\mathfrak{m}M$ with f is injective and M is flat over A.*

Note that (ii) implies (i) trivially, if f is a unit in B.

Proof. We set $k := A/\mathfrak{m}$.

"(ii) $\Rightarrow$ (i)". Let $x \in M$ with $fx = 0$. Then $x \in \mathfrak{m}M$ by hypothesis. We have to show that $x = 0$. As $\bigcap_n \mathfrak{m}^n M = 0$ by Proposition B.42 it suffices to show that $x \in \mathfrak{m}^n M$ implies $x \in \mathfrak{m}^{n+1}M$ for all $n \geq 1$.

Let $(a_1, \ldots, a_r)$ be a generating system of $\mathfrak{m}^n$ whose image in $\mathfrak{m}^n/\mathfrak{m}^{n+1}$ is a k-basis. Let $v: A^{\oplus r} \to A$ be the linear map $(\alpha_i)_i \mapsto \sum_i \alpha_i a_i$. Let K be its kernel. As v is injective modulo $\mathfrak{m}$, we have $K \subseteq \mathfrak{m}^{\oplus r}$. As M is flat over A, one has an exact sequence

$$0 \longrightarrow K \otimes_A M \longrightarrow M^r \xrightarrow{v_M} M.$$

Write $x = \sum_i a_i y_i$ with $y_i \in M$. Then $(fy_i)_i \in \operatorname{Ker}(v_M)$ and hence $(fy_i)_i = \sum_\ell \beta_\ell \otimes z_\ell$ for $\beta_\ell \in K \subseteq \mathfrak{m}^{\oplus r}$ and $z_\ell \in M$. Hence the image of fy_i in $M/\mathfrak{m}M$ is zero and therefore by hypothesis $y_i \in \mathfrak{m}M$ for all i. This shows $x \in \mathfrak{m}^{n+1}M$.

To show that M/fM is a flat A-module it suffices to show the kernel C of the multiplication $\mathfrak{m} \otimes_A M/fM \to M/fM$ is zero by Theorem B.51. The snake lemma (Lemma B.6) applied to the commutative diagram with exact rows

$$
\begin{array}{ccccccc}
\mathfrak{m} \otimes M & \xrightarrow{\ f\ } & \mathfrak{m} \otimes M & \longrightarrow & \mathfrak{m} \otimes M/fM & \longrightarrow & 0 \\
\downarrow & & \downarrow & & \downarrow & & \\
0 \longrightarrow M & \xrightarrow{\ f\ } & M & \longrightarrow & M/fM & \longrightarrow & 0
\end{array}
$$

yields an exact sequence $0 \longrightarrow C \longrightarrow M/\mathfrak{m}M \xrightarrow{\ f\ } M/\mathfrak{m}M$. Hence $C = 0$ because the multiplication $M/\mathfrak{m}M \to M/\mathfrak{m}M$ with f is injective.

"(i) $\Rightarrow$ (ii)". By hypothesis, $0 \longrightarrow M \xrightarrow{\ f\ } M \longrightarrow M/fM \longrightarrow 0$ is an exact sequence. Applying Proposition B.16 (iii) shows that the multiplication with f is injective on $M \otimes_A N$ for every A-module N. Choosing $N = A/\mathfrak{m}$ shows that $M/\mathfrak{m}M \xrightarrow{\ f\ } M/\mathfrak{m}M$ is injective.

To show that M is flat over A it suffices to show that M/f^nM is flat over A for all n by Proposition B.50. We show this by induction on n. As the multiplication with f on M is injective we have an exact sequence

$$0 \longrightarrow M/f^nM \xrightarrow{\ f\ } M/f^{n+1}M \longrightarrow M/fM \longrightarrow 0.$$

Hence we conclude by Proposition B.16 (iv). $\qquad\square$

Theorem 14.24. *Let $f: X \to S$ be a smooth morphism of schemes. Then f is flat.*

Proof. Since flatness can be checked locally, it is enough to prove the assertion for a smooth morphism in "standard form": Hence we may assume that $S = \operatorname{Spec} R$ is affine and that $X = \operatorname{Spec} R[T_1, \ldots, T_n]/(f_1, \ldots, f_{n-d})$, where $f_1, \ldots, f_{n-d} \in R[T_1, \ldots, T_n]$ are polynomials such that for all $x \in X$, the Jacobian matrix $J_{f_1, \ldots, f_{n-d}}(x)$ has rank $n - d$.

For $i = 0, \ldots, n - d$ we set $X_i := \operatorname{Spec} R[T_1, \ldots, T_n]/(f_1, \ldots, f_i)$. Then we have a sequence of closed subschemes

$$X = X_{n-d} \subseteq X_{n-d-1} \subseteq \cdots \subseteq X_1 \subseteq X_0 = \mathbb{A}_R^n.$$

We show by induction that X_i is flat over R at all points of X. The assertion is clear for $i = 0$.

By Definition 6.14, X_i is smooth of relative dimension $n - i$ at all points of X. Using Proposition 14.22, it is enough to show that for every point $s \in \operatorname{Spec} R$, f_i is regular in $A_{i-1} := \kappa(s)[T_1, \ldots, T_n]/(f_1, \ldots, f_{i-1})$. As the fiber $X_{i-1,s} = \operatorname{Spec} A_{i-1}$ is smooth over the field $\kappa(s)$ at all point of X_s, Lemma 6.26 shows that all local rings $\mathcal{O}_{X_{i-1,s},x}$ for $x \in X_s$ are regular and in particular integral domains. Hence it suffices to show that the image $\bar{f}_i$ of f_i in $\mathcal{O}_{X_{i-1,s},x}$ is non-zero for all closed points x of X_s. But $\dim \mathcal{O}_{X_{i,s},x} = n - i$ by Theorem 5.22. As $\mathcal{O}_{X_{i,s},x} = \mathcal{O}_{X_{i-1,s},x}/(\bar{f}_i)$ this shows that $\bar{f}_i \neq 0$. □

Together with Corollary 6.32 we have now seen that a morphism locally of finite presentation is flat and its fibers are geometrically regular if it is smooth. In Volume II we will prove that the converse also holds.

Properties of flat morphisms

(14.5) The fiber criterion for flatness.

Let us consider the following situation: we fix a base scheme S and an S-morphism $f: X \to Y$. For $s \in S$ we denote by X_s and Y_s the fibers of X and Y, respectively. We want to investigate whether f is flat in terms of the fibers $f_s := f \times_S \operatorname{Spec} \kappa(s): X_s \to Y_s$. Obtaining a general result is only promising if we assume that X is flat over S (think of the case $Y = S$). It turns out that except for some mild finiteness conditions, this is enough to prove a "fiber criterion for flatness". In fact, we can even deal with the more general situation where we consider a quasi-coherent $\mathcal{O}_X$-module, as the theorem below shows. With notation as above, for an $\mathcal{O}_X$-module $\mathscr{F}$ we denote by $\mathscr{F}_s$ the pull-back to the fiber X_s.

Theorem 14.25. *Let S be a scheme, let $g: X \to S$ and $h: Y \to S$ be morphisms of schemes, and let $f: X \to Y$ be an S-morphism. Let $\mathscr{F}$ be a quasi-coherent $\mathcal{O}_X$-module. Let $x \in X$, $y = f(x)$, $s = h(y) = g(x)$, and suppose that the stalk $\mathscr{F}_x$ is not zero. Assume that one of the following two conditions is satisfied:*
(a) *The schemes S, X, and Y are locally noetherian, and $\mathscr{F}$ is coherent.*
(b) *The morphisms g and h are locally of finite presentation, and $\mathscr{F}$ is of finite presentation.*

Then the following are equivalent:
(i) $\mathscr{F}$ *is g-flat at x, and $\mathscr{F}_s$ is f_s-flat at x.*
(ii) *h is flat at y and $\mathscr{F}$ is f-flat at x.*

Theorem 14.44 below shows that under the assumption (b) the set of $x \in X$ satisfying (ii) is open in X.

Proof. First of all, note that (ii) implies (i): the first statement of (i) follows since $h = g \circ f$, and the second one because flatness is preserved by base change. It remains to prove the converse. We give the proof only in case assumption (a) is satisfied. The case of assumption (b) can be reduced to the noetherian case using the techniques of Chapter 10 and additional considerations concerning the compatibility of flatness and limits. See Proposition 14.4, [EGAIV] (11.3.10).

Since the flatness question can be checked on the local rings, the claim in the noetherian case follows from the following lemma. $\qquad\square$

Lemma 14.26. *Let $\varphi\colon A \to B$, $\psi\colon B \to C$ be local homomorphisms of local noetherian rings, let k be the residue class field of A, and let M be a finitely generated C-module. The following are equivalent:*
(i) *M is a flat A-module, and $M \otimes_A k$ is a flat $B \otimes_A k$-module.*
(ii) *B is a flat A-module and M is a flat B-module.*

Proof. Clearly, the second condition implies the first one. Now assume that (i) is satisfied.
Let $\mathfrak{m}$ be the maximal ideal of A. Consider the following commutative diagram

$$
\begin{array}{ccc}
\mathrm{gr}_0^{\mathfrak{m}}(M) \otimes_{\mathrm{gr}_0^{\mathfrak{m}}(A)} \mathrm{gr}^{\mathfrak{m}}(A) & \xrightarrow{\;\;v\;\;} & \mathrm{gr}^{\mathfrak{m}}(M) \\
\| & & \uparrow{\scriptstyle w} \\
\mathrm{gr}_0^{\mathfrak{m}}(M) \otimes_{\mathrm{gr}_0^{\mathfrak{m}}(B)} (\mathrm{gr}_0^{\mathfrak{m}}(B) \otimes_{\mathrm{gr}_0^{\mathfrak{m}}(A)} \mathrm{gr}^{\mathfrak{m}}(A)) & \xrightarrow{\;1 \otimes u\;} & \mathrm{gr}_0^{\mathfrak{m}}(M) \otimes_{\mathrm{gr}_0^{\mathfrak{m}}(B)} \mathrm{gr}^{\mathfrak{m}}(B),
\end{array}
$$

where $v = \gamma_M^{\mathfrak{m}}$, $u = \gamma_B^{\mathfrak{m}}$ and $w = \gamma_M^{\mathfrak{m}B}$ are the canonical surjective homomorphisms (B.9.1). We apply the local criterion for flatness, Theorem B.51 (with $I = \mathfrak{m}$). Since M is flat over A, the homomorphism v is bijective, and therefore all homomorphisms in the above diagram are isomorphisms. Applying the theorem again for B and $I = \mathfrak{m}B$, the flatness of $M \otimes_A k$ over $B \otimes_A k$ with the bijectivity of w shows that M is flat over B.

Since B is a local ring, M is even faithfully flat over B (see Example B.18), and it follows that $\mathrm{gr}_0^{\mathfrak{m}}(M) = M \otimes_A k$ is faithfully flat over $\mathrm{gr}_0^{\mathfrak{m}}(B) = B/\mathfrak{m}B$. Since $1 \otimes u$ in the diagram is an isomorphism, it follows that u is an isomorphism as well. Applying the local criterion for flatness once more, we obtain that B is flat over A. $\qquad\square$

The following special case of the theorem, where $\mathscr{F} = \mathscr{O}_X$, is by far the most important case:

Corollary 14.27. *Let S be a scheme, let $g\colon X \to S$ and $h\colon Y \to S$ be morphisms which are locally of finite presentation, and let $f\colon X \to Y$ be an S-morphism. The following are equivalent:*
(i) *g is flat and for every $s \in S$, $f_s\colon X_s \to Y_s$ is flat.*
(ii) *h is flat at all points of $f(X)$ and f is flat.*

As a first non-trivial application of the fiber criterion for flatness let us give a criterion when it suffices to check on fibers whether a morphism is an isomorphism:

Proposition 14.28. *Let S be a scheme and let $g\colon X \to S$ and $h\colon Y \to S$ be S-schemes. Let g and h be both proper and of finite presentation and let g be flat. Let $f\colon X \to Y$ be an S-morphism, let $s \in S$ be a point, and let $f_s = f \times \mathrm{id}_{\kappa(s)}\colon X \otimes_S \kappa(s) \to Y \otimes_S \kappa(s)$ be the induced morphism of fibers. If f_s is an isomorphism, then there exists an open neighborhood U of s such that $f_{|g^{-1}(U)}\colon g^{-1}(U) \to h^{-1}(U)$ is an isomorphism.*

Proof. First note that f is of finite presentation by Proposition 10.35 (3). By Proposition 12.93 we may assume that f is a closed immersion. The fiber criterion for flatness (Theorem 14.25) shows that f is flat at all $x \in g^{-1}(s)$. The same argument as in the first part of the proof of Proposition 12.93 shows that it suffices to find for all $y \in h^{-1}(s)$ an open neighborhood V_y such that the restriction $f^{-1}(V_y) \to V_y$ of f is an isomorphism. Thus the proposition follows from Lemma 14.21. $\square$

The proof shows that it suffices to assume that g is flat at all points $x \in g^{-1}(s)$. In Volume II we will show that if g and h are only locally of finite presentation and g is flat, then f is an isomorphism if and only if f_s is an isomorphism for all $s \in S$.

(14.6) The valuative criterion for flatness.

In this section, we will prove the valuative criterion for flatness that was already mentioned above. We start with a quite general result about "descent" of the property of being flat.

Proposition 14.29. *Let $(A, \mathfrak{m})$ be a local noetherian ring and let $A \to B$ be a local homomorphism of local noetherian rings. Let $\varphi\colon A \to A'$ be a homomorphism from A to a semi-local noetherian ring A' such that $\varphi^{-1}(\mathfrak{r}') = \mathfrak{m}$, where $\mathfrak{r}'$ is the Jacobson radical of A'. Furthermore, suppose that the induced homomorphism $\widehat{\varphi}\colon \widehat{A} \to \widehat{A'}$ between the completions (with respect to the $\mathfrak{m}$- and $\mathfrak{r}'$-adic topologies) is injective.*
    Then B is flat over A if and only if $B \otimes_A A'$ is flat over A'.

Proof. Since flatness is stable under base change, the condition is necessary. Therefore we assume that $B \otimes_A A'$ is flat over A'. To check that B is flat over A it is by Theorem B.51 enough to show that for all n, $B/\mathfrak{m}^n B$ is flat over $A/\mathfrak{m}^n$.

We first claim that for all $n \geq 1$ there exists an $N \geq 1$ such that $\varphi^{-1}(\mathfrak{r}'^N) \subseteq \mathfrak{m}^n$. The canonical homomorphisms $A \to \widehat{A}$ and $A' \to \widehat{A'}$ are faithfully flat and in particular injective and we consider them as inclusions. As $\widehat{\varphi}$ is injective, φ is also injective. We have $\widehat{\mathfrak{r}'^N} = (\widehat{\mathfrak{r}'})^N$ (Proposition B.41) and $(\widehat{\mathfrak{r}'})^N \cap A' = \mathfrak{r}'^N$ (Proposition B.42). Therefore we find $\varphi^{-1}(\mathfrak{r}'^N) = A \cap \widehat{\varphi}^{-1}((\widehat{\mathfrak{r}'})^N)$. As we also have $\widehat{\mathfrak{m}}^n \cap A = \mathfrak{m}^n$ it suffices to show for all $n \geq 1$ there exists an $N \geq 1$ such that $\widehat{\varphi}^{-1}((\widehat{\mathfrak{r}'})^N) \subseteq \widehat{\mathfrak{m}}^n$. But we have $\bigcap_N \widehat{\varphi}^{-1}((\widehat{\mathfrak{r}'})^N) = \widehat{\varphi}^{-1}(\bigcap_N (\widehat{\mathfrak{r}'})^N) = 0$ because $\widehat{\varphi}$ is injective. Thus Lemma B.45 implies the claim.

Therefore, if we set $\mathfrak{a}_N := \varphi^{-1}(\mathfrak{r}'^N)$, there exists a homomorphism $A/\mathfrak{a}_N \to A/\mathfrak{m}^n$. On the other hand, φ yields an injection $A/\mathfrak{a}_N \hookrightarrow A'/\mathfrak{r}'^N$ of Artinian rings and Lemma 14.30 below (applied to this injection and the $A/\mathfrak{a}_N$-module $B/\mathfrak{a}_N B$) shows that $B/\mathfrak{a}_N B$ is flat over $A/\mathfrak{a}_N$. By tensoring with $\otimes_{A/\mathfrak{a}_N} A/\mathfrak{m}^n$, we are done. $\square$

Lemma 14.30. *Let A be an Artinian ring, and let M be an A-module. Let $\varphi\colon A \to A'$ be an injective ring homomorphism such that $M' := M \otimes_A A'$ is flat over A'. Then M is flat over A.*

Proof. Fix a maximal ideal $\mathfrak{m}$ of A, and write $\mathfrak{I}_n = \varphi^{-1}(\mathfrak{m}^n A')$. Note that $\mathfrak{I}_1 = \mathfrak{m}$. The filtration $(\mathfrak{I}_n)_n$ satisfies the conditions of Corollary B.52, and therefore it is enough to show that the natural homomorphism in the upper row of the following commutative diagram is injective (we use the notation of Corollary B.52).

$$
\begin{array}{ccc}
\mathrm{gr}_0(M) \otimes_{\mathrm{gr}_0(A)} \mathrm{gr}_\bullet(A) & \longrightarrow & \mathrm{gr}_\bullet(M) \\
\downarrow & & \downarrow \\
\mathrm{gr}_0^{\mathfrak{m}A'}(M') \otimes_{\mathrm{gr}_0^{\mathfrak{m}A'}(A')} \mathrm{gr}_\bullet^{\mathfrak{m}A'}(A') & \longrightarrow & \mathrm{gr}_\bullet^{\mathfrak{m}A'}(M')
\end{array}
$$

Since M' is flat over A', the homomorphism in the lower row is injective, so it is enough to show that the left vertical arrow is injective. This arrow however can be written as

$$
\mathrm{gr}_0(M) \otimes_{\mathrm{gr}_0(A)} \bigoplus \mathfrak{I}_n/\mathfrak{I}_{n+1} \hookrightarrow M/\mathfrak{m}M \otimes_{A/\mathfrak{m}} \bigoplus \mathfrak{m}^n A'/\mathfrak{m}^{n+1} A'
$$
$$
= M'/\mathfrak{m}M' \otimes_{A'/\mathfrak{m}A'} \bigoplus \mathfrak{m}^n A'/\mathfrak{m}^{n+1} A' = \mathrm{gr}_0^{\mathfrak{m}A'} M' \otimes_{\mathrm{gr}_0^{\mathfrak{m}A'}(A')} \mathrm{gr}_\bullet^{\mathfrak{m}A'}(A'),
$$

where the arrow is injective because φ is injective and $M/\mathfrak{m}M$ is flat over $A/\mathfrak{m}A$. $\qquad\square$

The following lemma will be useful later when we want to apply Proposition 14.29.

Lemma 14.31. *Let $\varphi\colon A \to A'$ be an injective ring homomorphism of noetherian rings. Assume that A is local, and assume that in addition one of the following conditions is satisfied:*
(1) *The homomorphism $\varphi\colon A \to A'$ is finite.*
(2) *The ring A' is local, the completion $\widehat{A}$ of A is a domain, and φ is local, and essentially of finite type (Definition B.14).*
Then the completion $\widehat{\varphi}$ is also injective.

Proof. If φ is finite, then $\widehat{A'} = \widehat{A} \otimes_A A'$ by Proposition B.46, and the assertion follows from the injectivity of φ, because the homomorphism $A \to \widehat{A}$ is flat. This proves the lemma under Assumption (1).

Let us assume (2). Let $\mathfrak{m}$ be the maximal ideal of A, and denote by $\mathfrak{m}'$ the maximal ideal of A'. Looking at the definition of the completions as projective limits, we see that it suffices to show that for every $m \geq 0$, there exists $n \geq 0$ such that $\varphi^{-1}((\mathfrak{m}')^n) \subseteq \mathfrak{m}^m$. (This amounts to saying that the $\mathfrak{m}$-adic topology on A coincides with the topology induced by the $\mathfrak{m}'$-adic topology on A'.)

As $A \to \widehat{A}$ is injective, A is an integral domain. Since φ is injective, there exists a prime ideal $\mathfrak{q} \subset A'$ such that $\varphi^{-1}(\mathfrak{q}) = 0$, so that $A \to A'/\mathfrak{q}$ is still injective. Replacing A' by $A'/\mathfrak{q}$, we may assume that A' is a domain, as well.

Let $B = \widehat{A} \otimes_A A'$. Since A' is essentially of finite type over A, B is the localization of an $\widehat{A}$-algebra of finite type, too, and in particular noetherian. Since $\widehat{A}$ is flat over A, Proposition 14.13 shows that every irreducible component of $\operatorname{Spec} B$ dominates $\operatorname{Spec} \widehat{A}$. Let $x \in \operatorname{Spec} B$ be a point which projects to the closed points of $\operatorname{Spec} \widehat{A}$ and $\operatorname{Spec} A'$, and let $B' = \mathscr{O}_{\operatorname{Spec} B, x}$ be the corresponding local ring. We obtain a diagram of local homomorphisms of local noetherian rings. The map $A \to \widehat{A}$ is injective. Since Spec B' dominates Spec $\widehat{A}$ and $\widehat{A}$ is a domain, ψ is injective too. Therefore we can identify A and $\widehat{A}$ with subrings of B'. Let $\mathfrak{m}_{B'} \subset B'$ be the maximal ideal. Then $\bigcap (\mathfrak{m}_{B'})^i = 0$ (Corollary B.43), so the sequence of ideals $\psi^{-1}((\mathfrak{m}_{B'})^i)$ has intersection 0 in $\widehat{A}$. By Lemma B.45, for every $m \geq 0$, there exists n with $\psi^{-1}((\mathfrak{m}_{B'})^n) \subseteq \mathfrak{m}_{\widehat{A}}^m$. Intersecting with A, and writing χ for the map $A \to \widehat{A} \to B'$, we obtain $\chi^{-1}((\mathfrak{m}_{B'})^n) \subseteq \mathfrak{m}_A^m$. As $\mathfrak{m}' = \rho^{-1}(\mathfrak{m}_{B'})$ and thus $\mathfrak{m}'^n \subseteq \rho^{-1}(\mathfrak{m}_{B'}^n)$, we get

$$\varphi^{-1}((\mathfrak{m}')^n) \subseteq \chi^{-1}((\mathfrak{m}_{B'})^n) \subseteq \mathfrak{m}_A^m,$$

and the lemma is proved. $\qquad \square$

Combining the above results, we obtain:

Corollary 14.32. *Let $(A, \mathfrak{m})$ be a local noetherian ring, let $A \to B$ be a local homomorphism to a local noetherian ring B, and let $\varphi \colon A \to A'$ be a finite injective ring homomorphism.*

Then B is flat over A if and only if $B \otimes_A A'$ is flat over A'.

Proof. By Lemma 14.31 the completion $\widehat{\varphi}$ is injective. Under the hypothesis, A' is a semi-local ring by Proposition B.46. Moreover the inverse image of every maximal ideal of A' is $\mathfrak{m}$ because φ is finite. Thus the inverse image of the radical of A' is $\mathfrak{m}$, and we can apply Proposition 14.29. $\qquad \square$

Remark 14.33. Corollary 14.32 is a special case of descent of flatness: We say that a ring homomorphism $\varphi \colon A \to B$ *descends flatness* if every A-module M, such that $M \otimes_A B$ is a flat B-module, is flat over A. This condition has been examined in [RG] Part II, §1. We collect some statements (which we will not use in the sequel):
(1) (loc. cit. p. 55) If $\varphi \colon A \to B$ descends flatness and Spec A is connected, then φ is injective.
(2) (loc. cit. (1.2.10)) If an injective homomorphism $\varphi \colon A \to B$ descends flatness, then every A-module M, such that $M \otimes_A B = 0$, is 0. The converse holds if A is noetherian.
(3) (loc. cit. (1.2.4)) Every finite injective homomorphism $\varphi \colon A \to B$ descends flatness.

Now we can state, and prove (in many cases), the valuative criterion for flatness.

Theorem 14.34. *Let Y be a reduced, locally noetherian scheme, and let $f \colon X \to Y$ be a morphism locally of finite type. Then f is flat if and only if for every discrete valuation ring R and every morphism Spec $R \to Y$ the pull-back morphism $X \times_Y$ Spec $R \to$ Spec R is flat.*

The proof relies on the results on "descent of flatness" proved above. Proposition 14.29 allows us to check flatness over A after base change to some A-algebra A', but a crucial hypothesis is that the map $\widehat{A} \to \widehat{A'}$ must be injective. The only tool we have at our disposal to check this is Lemma 14.31 which deals with the situations where either A' is finite over A, or the completion $\widehat{A}$ is a domain.

We have Theorem 12.50 which says that if A is a local ring of a quasi-excellent scheme X (e.g., if X is of finite type over a field; see Theorem 12.51), and if A is a normal domain, then $\widehat{A}$ is a normal domain. We also know that the completion of a regular local ring is again regular, and hence a domain (see Proposition B.77 (6)). Therefore, we make *one* of the following additional assumptions

(1) Y is quasi-excellent (e.g., if Y is locally of finite type over a field).

(2) Y is regular.

Note that in Case (1) our proof still relies on Theorems 12.50 and 12.51 for which we did not give a proof. In Case (2) we use only the facts Proposition B.77 (1) and (6) from commutative algebra. The general case (and further generalizations) of Theorem 14.34 are proved in [EGAIV] (11.8.1) where a reduction in the spirit of the methods of Chapter 10 is used to reduce to the case that the local rings of Y are essentially of finite type over $\mathbb{Z}$ (and thus are quasi-excellent by Theorem 12.51).

Proof. Because flatness is stable under base change, the condition is necessary (regardless of our assumption). For the sufficiency, it is enough to show the flatness after base change to the local rings $\mathscr{O}_{Y,y}$ where y runs through the points of Y. Thus we may assume that $Y = \operatorname{Spec} A$, where A is a local noetherian ring. Under this additional assumption we will prove the theorem by induction on $\dim A$. Thus let $d \geq 0$ be an integer. By induction hypothesis we may assume that the theorem is already proved if $Y = \operatorname{Spec} A$ with $\dim A < d$ and Y is quasi-excellent or Y is regular. Now assume that $\dim A = d$ and that Y is quasi-excellent or regular.

(i). We will first prove the theorem under the additional assumption that the completion $\widehat{A}$ is an integral domain. As $A \to \widehat{A}$ is faithfully flat, X is flat over A if and only if $X \otimes_A \widehat{A}$ is flat over $\widehat{A}$. We have $\dim A = \dim \widehat{A}$. In any case $\widehat{A}$ is excellent (Theorem 12.51). If A is regular, so is $\widehat{A}$. Thus we may assume that $Y = \operatorname{Spec} A$, where A is a complete noetherian domain with $d = \dim A$. By Lemma 15.6 below there exists a discrete valuation ring R' which dominates A. Let R be its completion (which is again a discrete valuation ring). Then the composition $\varphi \colon A \hookrightarrow R' \hookrightarrow R$ is an injective local homomorphism of complete local noetherian rings. By hypothesis, we know that $X \otimes_A R$ is flat over R. Let $x \in X$. We want to show that X is flat over A in x. Let $y \in \operatorname{Spec} A$ be the image of x. We have to show that $X \otimes_A A_{\mathfrak{p}_y}$ is flat over $A_{\mathfrak{p}_y}$. Note that $A_{\mathfrak{p}_y}$ is quasi-excellent (resp. regular), if A is quasi-excellent (resp. regular) by definition (resp. by Proposition B.77 (1)). If y is not the closed point, then $\dim A_{\mathfrak{p}_y} < d$ and we are done by induction hypothesis. If y is the closed point, $A \to \mathscr{O}_{X,x}$ is a local ring homomorphism of local noetherian rings such that $\mathscr{O}_{X,x} \otimes_A R$ is flat over R. Then $\mathscr{O}_{X,x}$ is flat over A by Proposition 14.29.

(ii). If Y is regular, the assumption made in (i) holds as explained above. Thus we are done in this case. If Y is quasi-excellent, then we will reduce to the case that $\widehat{A}$ is a domain as follows.

Since A is reduced, it is a subring of the product $\prod_i A/\mathfrak{p}_i$, where the $\mathfrak{p}_i$ are the minimal prime ideals of A. This product is a semi-local noetherian ring, and the homomorphism $\widehat{A} \to \widehat{\prod_i A/\mathfrak{p}_i}$ between the $\mathfrak{m}$- and $\mathfrak{m}\prod_i A/\mathfrak{p}_i$-adic completions is injective, because completion is exact for finitely generated modules over a noetherian ring. By Corollary 14.32, we can replace A by the product, and we can then check the flatness by checking it over each factor.

Hence we may assume that A is an integral domain, and we denote by K the field of fractions of A. Let $\widetilde{A}$ be the normalization of A in K. This is a finite A-algebra by Theorem 12.50 (if A is essentially of finite type over a field we may also use Corollary 12.52), and in particular is a semi-local noetherian ring. We denote by φ the inclusion $A \to \widetilde{A}$.

We may replace A by $\widetilde{A}$, using Corollary 14.32 once more. Therefore we may assume that A is normal. Furthermore, to check flatness over A it is enough to check flatness over all localizations with respect to maximal ideals, so that we can assume that A is a local normal domain. As Y is quasi-excellent, $\widetilde{A}$ is again a local (normal) domain and we are done by step (i). $\qquad\square$

(14.7) Flat morphisms and open morphisms.

Theorem 14.35. *Let* $f\colon X \to Y$ *be a flat morphism which is locally of finite presentation. Then* f *is universally open.*

Proof. Since flatness and being locally of finite presentation are stable under base change, it is enough to show that f is open. This follows from the Lemma 14.9 and Corollary 10.72. $\qquad\square$

As smooth morphisms are locally of finite presentation and flat (Theorem 14.24), we obtain:

Corollary 14.36. *Every smooth morphism is universally open.*

We see that if $f\colon X \to Y$ is a faithfully flat morphism of schemes which is locally of finite presentation, then the topology on Y is the quotient topology with respect to f, i.e., a subset $U \subset Y$ is open if and only if $f^{-1}(U)$ is open in X.

We will show in Proposition 14.39 that this conclusion also holds if f is a quasi-compact faithfully flat morphism. On the other hand, it is easy to see that there exist flat and quasi-compact morphisms between noetherian schemes which are not open (Exercise 14.3). Similarly, it is easy to give examples of universally open morphisms (even universal homeomorphisms) which are not flat (Exercise 14.4). However, the following theorem shows that after "eliminating trivial counterexamples", there is a converse:

Theorem 14.37. *Let* Y *be a reduced locally noetherian scheme, let* $f\colon X \to Y$ *be a universally open morphism which is locally of finite type. Suppose that for every* $y \in Y$ *the fiber* $f^{-1}(y)$ *is a geometrically reduced* $\kappa(y)$*-scheme. Then* f *is flat.*

Proof. Since Y is reduced and locally noetherian, we can apply the valuative criterion for flatness, Theorem 14.34. So let $\operatorname{Spec} R \to Y$ be a morphism, where R is a discrete valuation ring, and let $X' = X \times_Y \operatorname{Spec} R$. We have to prove that the base change $f'\colon X' \to \operatorname{Spec} R$ is flat.

We apply Proposition 14.17. Let s denote the closed point of $\operatorname{Spec} R$, and let y be its image in Y. Since the fiber $f^{-1}(y)$ is geometrically reduced, we know that $f'^{-1}(s)$ is reduced. Therefore it remains to show that all maximal points of $f'^{-1}(s)$ lie in the closure of the generic fiber $f'^{-1}(\eta)$, where η denotes the generic point of $\operatorname{Spec} R$. Let $\xi \in f'^{-1}(s)$ be a maximal point. It is clearly enough to show that ξ has a generization in X' which lies in the generic fiber. But if this is not the case, then ξ could not have any proper generization in X', i.e., the closure of ξ would be an irreducible component of X'. But then we could find an open subset of this component which maps to $\{s\}$, a contradiction to the assumption that f' is open. $\qquad\square$

Theorem 14.38. *Let $f: X \to Y$ be a morphism of schemes where the underlying topological space of Y is discrete. Then f is universally open.*

Proof. We have to prove that for every morphism $Y' \to Y$, the base change $f': X' = X \times_Y Y' \to Y'$ is open. Because this question is local on Y, we can restrict to the case that Y, as a set, is a single point. By Proposition 4.34, we can replace f by f_{red}, the corresponding morphism of the underlying reduced schemes, so we may even assume that Y is the spectrum of a field k. In particular f is flat, so in order to apply Theorem 14.35 we would like to reduce to a situation where we have a morphism which is in addition locally of finite presentation.

Furthermore, given a covering $X = \bigcup U_i$ by open affines, and a morphism $Y' \to Y$, the fiber product $X \times_Y Y'$ is covered by the $U_i \times_Y Y'$. Therefore we may also assume that $X = \operatorname{Spec} B$ is affine. By a similar argument, we may assume that $Y' = \operatorname{Spec} A'$ is affine as well. Write $B' = B \otimes_k A'$.

After these reduction steps, we have to prove that for every $t \in B'$, the image V of $D(t)$ under the morphism $f': \operatorname{Spec} B' \to \operatorname{Spec} A'$ is open.

Now we write B as the inductive limit of its finitely generated k-subalgebras B_λ. We obtain $B' = \varinjlim(B_\lambda \otimes_k A')$, and the element t is the image of some $t_\lambda \in B_\lambda \otimes_k A'$ for suitable λ. Since the morphism $\operatorname{Spec} B_\lambda \otimes_k A' \to \operatorname{Spec} A'$ is flat and locally of finite presentation, Theorem 14.35 implies that it is open, so in particular the image V_λ of $D(t_\lambda)$ in $\operatorname{Spec} A'$ is open.

It is enough to show that $V = V_\lambda$. Since $D(t)$ is the inverse image of $D(t_\lambda)$ under the projection $\operatorname{Spec} B' \to \operatorname{Spec} B_\lambda \otimes_k A'$, we have $V \subseteq V_\lambda$. To prove the other inclusion, consider $y \in V_\lambda$. We have to show that $y \in V$, or in other words that $f'^{-1}(y) \cap D(t) \neq \emptyset$. We can rewrite the intersection $f'^{-1}(y) \cap D(t)$ as $g^{-1}(W)$, where

$$g: f'^{-1}(y) \to f_\lambda'^{-1}(y), \qquad W = f_\lambda'^{-1}(y) \cap D(t_\lambda).$$

The fibers are given by $f'^{-1}(y) = \operatorname{Spec} B' \otimes_{A'} \kappa(y) = \operatorname{Spec} B \otimes_k \kappa(y)$ and by $f_\lambda'^{-1}(y) = \operatorname{Spec}(B_\lambda \otimes_k A') \otimes_{A'} \kappa(y) = \operatorname{Spec} B_\lambda \otimes_k \kappa(y)$. Since $\kappa(y)$ is flat over k, the homomorphism $B_\lambda \otimes_k \kappa(y) \to B \otimes_k \kappa(y)$ is injective. Therefore g is dominant, and the inverse image of the non-empty open subset W under g must be non-empty. $\square$

(14.8) Topological properties of quasi-compact flat morphisms.

The goal of this section is to prove the following result.

Proposition 14.39. *Let $f: X \to Y$ be a quasi-compact faithfully flat morphism. Then the topology on Y is the quotient topology with respect to f (i.e., a subset Z of Y is closed if and only if $f^{-1}(Z)$ is closed).*

We introduce the following technical notion. A morphism $f: X \to Y$ of schemes is called *generizing* if for all $x \in X$, we have $f(\operatorname{Spec} \mathcal{O}_{X,x}) = \operatorname{Spec} \mathcal{O}_{Y,f(x)}$. We have seen in Lemma 14.9 that every flat morphism is generizing. Thus Proposition 14.39 immediately follows from the following more general assertion.

Lemma 14.40. *Let $f: X \to Y$ be a quasi-compact generizing morphism of schemes. Then the topology on the subspace $f(X)$ is the quotient topology with respect to f.*

For the proof we make the following definition: Let Y be a qcqs scheme. A subset $E \subset Y$ is called *pro-constructible* if there exists a morphism $f \colon \operatorname{Spec} A' \to Y$ from an affine scheme such that $f(\operatorname{Spec} A') = E$.

Remark 14.41. Let Y be a qcqs scheme.
(1) By Proposition 10.45, every constructible set of Y is pro-constructible.
(2) If $f \colon X \to Y$ is a quasi-compact morphism, $f(X)$ is pro-constructible: We may cover X by finitely many affine open subsets U_i and the composition $\coprod U_i \to X \to Y$ is a morphism of an affine scheme to Y with image $f(X)$.
(3) The intersection of two pro-constructible subsets Z and Z' of Y is pro-constructible (see also Exercise 14.12): We use (2). If $f \colon X \to Y$ and $f' \colon X' \to Y$ are quasi-compact morphisms with $f(X) = Z$ and $f'(X') = Z'$, then $X \times_Y X' \to Y$ is a quasi-compact morphism with image $Z \cap Z'$.

The key point is that for a generizing morphism to a qcqs scheme taking the inverse image of a pro-constructible subset is compatible with taking closures:

Lemma 14.42. *Let Y be a qcqs scheme, $Z \subseteq Y$ a pro-constructible set, and $f \colon X \to Y$ a generizing morphism of schemes. Then $\overline{f^{-1}(Z)} = f^{-1}(\overline{Z})$.*

Proof. The claim can be checked locally on Y. Moreover, if $V \subseteq Y$ is an open quasi-compact subset, $V \cap Z$ is again pro-constructible by Remark 14.41 (1) and (3). Thus we may assume that $Y = \operatorname{Spec} A$ is affine.

Clearly we have $\overline{f^{-1}(Z)} \subseteq f^{-1}(\overline{Z})$. Let $x \in U := X \setminus \overline{f^{-1}(Z)}$, then $\operatorname{Spec} \mathscr{O}_{X,x} \subseteq U \subseteq f^{-1}(Y \setminus Z)$ and hence $\operatorname{Spec} \mathscr{O}_{Y,f(x)} = f(\operatorname{Spec} \mathscr{O}_{X,x}) \subseteq Y \setminus Z$. We claim that this implies $y := f(x) \notin \overline{Z}$ (which proves the lemma).

Indeed let $Y' = \operatorname{Spec} A'$ be an affine scheme and $g \colon Y' \to Y$ a morphism such that $Z = g(Y')$. Let $\mathfrak{p} \subset A$ be the prime ideal corresponding to y. As $g^{-1}(\operatorname{Spec} \mathscr{O}_{Y,y}) = \emptyset$, we have

$$\varinjlim_{s \notin \mathfrak{p}} (A_s \otimes_A A') = A_{\mathfrak{p}} \otimes_A A' = 0$$

Thus there exists an $s \in A \setminus \mathfrak{p}$ such that $1 = 0$ in $A_s \otimes_A A'$, i.e., such that $D(s) \cap Z = \emptyset$. This shows $y \notin \overline{Z}$. $\square$

Proof. (of Lemma 14.40) The question is local on Y. We thus may assume that Y is affine. Let $Z \subseteq f(X)$ be a subset such that $F := f^{-1}(Z)$ is closed in X. We have to show that Z is closed in $f(X)$, i.e. $Z = \overline{Z} \cap f(X)$.

Endow F with its reduced subscheme structure, then $F \hookrightarrow X \to Y$ is a quasi-compact morphism and thus its image Z is pro-constructible. Thus we have

$$Z = f(f^{-1}(Z)) = f(\overline{f^{-1}(Z)}) = f(f^{-1}(\overline{Z})) = \overline{Z} \cap f(X),$$

where the third equality follows from Lemma 14.42. $\square$

A surjective morphism $f \colon X \to Y$ of schemes is also called *submersive* if the topology on Y is the quotient topology with respect to f. As "faithfully flat quasi-compact" is a property stable under base change, Proposition 14.39 shows:

Corollary 14.43. *A faithfully flat quasi-compact morphism is universally submersive.*

(14.9) Openness results.

We give some results about the openness of certain properties for flat morphisms.

Theorem 14.44. *Let $f\colon X \to Y$ be a morphism locally of finite presentation, and let $\mathscr{F}$ be an $\mathcal{O}_X$-module of finite presentation. Then the set U of points $x \in X$ such that $\mathscr{F}$ is f-flat at x is open in X.*

Proof. We sketch the proof. Applying the technique of elimination of noetherianness hypotheses (Section (10.18)), and a corresponding result about the compatibility of flatness and inductive limits of rings, one reduces to the case that Y is locally noetherian. See Proposition 14.4, [EGAIV] (11.3.1).

Since the question is local on X and Y, we may assume that $X = \operatorname{Spec} B$ and $Y = \operatorname{Spec} A$ are affine. Let $M = \Gamma(X, \mathscr{F})$. Similarly as in Lemma 14.9, one shows that U is stable under generization, so by Lemma 10.17 it is enough to prove that U is constructible. This amounts to showing (see Proposition 10.14) that for every closed irreducible subset $Z \subseteq X$ such that the generic point η of Z is in U, a non-empty open subset of Z is contained in U.

So let $Z \subseteq X$ be closed irreducible, with generic point η, and suppose $\eta \in U$. The point η corresponds to a prime ideal $\mathfrak{P} \subset B$, and the image of η in $\operatorname{Spec} A$ corresponds to a prime ideal $\mathfrak{p} \subset A$. Since η is in U, $M_{\mathfrak{P}}$ is flat over A. The local criterion for flatness then says that $(M/\mathfrak{p}M)_{\mathfrak{P}}$ is flat over $A/\mathfrak{p}A$, and that for each n, multiplication induces an isomorphism

$$(14.9.1) \qquad \mathfrak{p}^n/\mathfrak{p}^{n+1} \otimes_{A/\mathfrak{p}} (M/\mathfrak{p}M)_{\mathfrak{P}} \to \mathfrak{p}^n M_{\mathfrak{P}}/\mathfrak{p}^{n+1} M_{\mathfrak{P}}$$

of $B_{\mathfrak{P}}$-modules (Theorem B.51 (ii), where we consider each summand separately).

By Theorem 10.83, there exists $a \in A \setminus \mathfrak{p}$ such that $M_a/\mathfrak{p}M_a$ is a free $(A/\mathfrak{p})_a$-module, so if $\mathfrak{Q} \in D_B(a) \subseteq \operatorname{Spec} B$, then $M_{\mathfrak{Q}}/\mathfrak{p}M_{\mathfrak{Q}}$ is flat over $A/\mathfrak{p}$.

Furthermore, the homomorphism (14.9.1) is precisely the localization of the natural homomorphism

$$\mathfrak{p}^n/\mathfrak{p}^{n+1} \otimes_{A/\mathfrak{p}} M/\mathfrak{p}M \to \mathfrak{p}^n M/\mathfrak{p}^{n+1}M$$

at the prime ideal $\mathfrak{P}$. Since it is an isomorphism after localization at the prime ideal $\mathfrak{P}$, and since the source and the target are finitely generated modules over the noetherian ring B, it is an isomorphism after localization with respect to each prime ideal $\mathfrak{Q} \subset B$ in some open neighborhood V of $\mathfrak{P}$ in $\operatorname{Spec} B$ (see Proposition 7.27).

By the hard direction of the local criterion for flatness we get $V \cap D_B(a) \subseteq U$. $\square$

We now come back to results as discussed in Section (10.23). Let $\mathbf{P} = \mathbf{P}(Z, k)$ be a property of a scheme Z of finite type over a field k (e.g., the property of being geometrically integral or having a fixed dimension d). Then for a morphism $f\colon X \to S$ of finite presentation we studied the question whether

$$S_{\mathbf{P}} := \{\, s \in S \;;\; \mathbf{P}(f^{-1}(s), \kappa(s)) \text{ holds}\,\}$$

is a constructible subset of S. For instance we proved that for every integer d the set $S_{\mathbf{P}}$ is constructible if $\mathbf{P}(Z, k)$ is the property that the k-scheme Z has dimension d (Proposition 10.96).

We also discussed in Section (10.25) an example about the constructibility of local properties of fibers: Let $\mathbf{Q} = \mathbf{Q}(Z, z, k)$ be a property of a point z of a scheme Z locally of finite type over a field k. Then for a morphism $f\colon X \to S$ locally of finite presentation we may study the set

$$X_{\mathbf{Q}} := \{\, x \in X \; ; \; \mathbf{Q}(f^{-1}(f(x)), x, \kappa(f(x))) \text{ holds}\,\}.$$

We gave an example in Theorem 10.97, where we proved that for a fixed integer d the subset $X_{\mathbf{Q}}$ is constructible if $\mathbf{Q} = \mathbf{Q}(Z, z, k)$ is the property that $\dim_z(Z) = d$.

In many cases, the additional assumption that f is flat allows to prove that $X_{\mathbf{Q}}$ is not only constructible but open in X. Moreover, if f is flat and proper, then $S_{\mathbf{P}}$ is open in S for many properties $\mathbf{P}$.

We mention the following two results as examples without proof (see [EGAIV] (12.1.1) and (12.2.4) for a proof and Appendix E for a more exhaustive list):

Theorem 14.45. *Let $f \colon X \to S$ be a flat morphism locally of finite presentation. Then the set of $x \in X$ such that $f^{-1}(f(x))$ is geometrically reduced at x is open in X.*

Theorem 14.46. *Let $f \colon X \to S$ be a proper flat morphism of finite presentation. Then the following subsets of S are open:*
(1) *The set of points $s \in S$ such that $f^{-1}(s)$ is geometrically integral.*
(2) *The set of points $s \in S$ such that $f^{-1}(s)$ is geometrically reduced and equidimensional of a fixed dimension d.*

A last example is the following result that follows immediately from Proposition 14.28.

Proposition 14.47. *Let S be a scheme and let $g \colon X \to S$ and $h \colon Y \to S$ be S-schemes that are proper and of finite presentation over S. Assume that X is also flat over S. Let $f \colon X \to Y$ be an S-morphism and define*

$$U := \{\, s \in S \; ; \; f_s \colon X_s \to Y_s \text{ is an isomorphism}\,\}.$$

Then U is open in S and the restriction $g^{-1}(U) \to h^{-1}(U)$ of f is an isomorphism.

Faithfully flat descent

In this section we will discuss a large number of "descent" results of the following type: Given a faithfully flat quasi-compact morphism $S' \to S$, we will show that an S-scheme X (or an S-morphism $f \colon X \to Y$) satisfies a property $\mathbf{P}$ if the base change $X \times_S S'$ (or $f_{(S')} \colon X \times_S S' \to Y \times_S S'$, respectively) satisfies $\mathbf{P}$. Of course, depending on $\mathbf{P}$, additional hypotheses might be needed.

(14.10) Descent of properties of modules and algebras over a ring.

Proposition 14.48. *Let $A \to A'$ be a faithfully flat ring homomorphism, let M be an A-module and write $M' := M \otimes_A A'$.*
(1) *If M' is flat over A', then M is flat over A.*
(2) *If M' is of finite type over A', then M is of finite type over A.*
(3) *If M' is of finite presentation over A', then M is of finite presentation over A.*
(4) *If M' is locally free of rank n over A', then M is locally free of rank n over A.*

Proof. Part (1) is immediate from the definitions. To prove part (2), choose generators $e_i = \sum_j m_{ij} \otimes a_{ij} \in M'$, $m_{ij} \in M$, $a_{ij} \in A$. Denote by N the number of all m_{ij}, and denote by $\varphi \colon A^N \to M$ the morphism mapping the standard basis vectors to the m_{ij}. Clearly, $\varphi \otimes_A A'$ is surjective, and hence, because of the faithful flatness, so is φ, which shows that M is finitely generated over A.

Now part (3) follows easily: If M' is of finite presentation, then by (2), M is of finite type. Let N be the kernel of a surjection $A^n \to M$. By flatness, $N' = \mathrm{Ker}(A')^n \to M'$, so N' is of finite type (Proposition B.8), and applying (2) once more, we obtain that N is finitely generated, as desired. Since a module is locally free of finite rank if and only if it is flat and of finite presentation, and the rank can be checked after base change to any residue class field, (4) follows from (1) and (3). $\qquad\square$

Proposition 14.49. *Let $A \to A'$ be a faithfully flat ring homomorphism, let B be an A-algebra, and write $B' = B \otimes_A A'$.*
(1) *If B' is an A'-algebra of finite type, then B is of finite type over A.*
(2) *If B' is an A'-algebra of finite presentation, then B is of finite presentation over A.*

Proof. This is proved in the same way as parts (2) and (3) of the previous proposition (using $A[T_1, \ldots, T_n]$ instead of A^n and Proposition B.11 instead of Proposition B.8). $\quad\square$

(14.11) Descent of properties of morphisms of schemes.

Now we study the behavior of morphisms of schemes from this point of view. Clearly, whenever we expect to have results about descent, we must ask that the base change morphism be surjective. For many set-theoretical properties of morphisms, this is in fact sufficient, as the following proposition shows:

Proposition 14.50. *Let S be a scheme, and let $f \colon X \to Y$ be a morphism of S-schemes. Consider a surjective morphism $g \colon S' \to S$. Write $f' \colon X' := X \times_S S' \to Y' := Y \times_S S'$ for the morphism obtained by base change. If f' is*
(1) *surjective,*
(2) *injective,*
(3) *bijective, or*
(4) *a morphism with set-theoretically finite fibers,*
then so is f.

Proof. We have a cartesian diagram

(14.11.1)
$$\begin{array}{ccc} X' & \xrightarrow{\,f'\,} & Y' \\ {\scriptstyle u}\downarrow & & \downarrow{\scriptstyle v} \\ X & \xrightarrow{\,f\,} & Y, \end{array}$$

where the vertical arrows are surjective (Proposition 4.32). Part (1) follows immediately from this.

To prove parts (2) and (4), we must look a little more carefully at the fibers of f and f'. In fact, given $y' \in Y'$ with image y in Y, we have

$$X'_{y'} = X \times_Y Y' \times_{Y'} \operatorname{Spec} \kappa(y') \cong X_y \otimes_{\kappa(y)} \kappa(y').$$

In particular, the projection $X'_{y'} \to X_y$ is surjective, and this proves (2) and (4), because $Y' \to Y$ is a surjection. Together, (1) and (2) imply (3). $\qquad\square$

For properties that involve the topology of the schemes, or even the scheme structure, we must require that the base change morphism be faithfully flat quasi-compact. Under this assumption many properties of morphisms of schemes descend, and we list some examples. See Appendix C for a more complete list.

Proposition 14.51. *Let S be a scheme, and let $f \colon X \to Y$ be a morphism of S-schemes. Consider a faithfully flat quasi-compact morphism $g \colon S' \to S$. Write $f' \colon X' := X \times_S S' \to Y' := Y \times_S S'$ for the morphism obtained by base change. If f' is*
(1) *open,*
(2) *closed,*
(3) *a homeomorphism,*
(4) *quasi-compact,*
(5) *quasi-separated, or*
(6) *separated,*
then so is f.

Proof. We again consider the cartesian diagram (14.11.1). In this case the vertical arrows are quasi-compact and faithfully flat because both properties are stable under base change. The assertions (1)–(3) then follow immediately from Proposition 14.39, and, for part (3), Proposition 14.50.

For Part (4) it suffices to remark that under our hypotheses the vertical arrows in (14.11.1) are surjective and quasi-compact. Thus if $V \subseteq Y$ is quasi-compact, then $f^{-1}(V) = u(f'^{-1}(v^{-1}(V)))$ is quasi-compact.

The morphism f is separated (resp. quasi-separated) if and only if the diagonal Δ_f is a closed (resp. quasi-compact) morphism. Since $\Delta_{f'} = \Delta_f \times_Y Y'$, parts (5) and (6) follow from (2) and (4). $\qquad\square$

Remark 14.52.
(1) The proof of Proposition 14.51 uses only the property that g is universally submersive (Corollary 14.43) and quasi-compact. Thus the conclusions also hold if we assume for instance that g is surjective and proper.

One can also show that it also suffices to assume that g is surjective and universally open (Exercise 14.7), e.g., if g is faithfully flat and locally of finite presentation (Theorem 14.35).

(2) Let $\mathbf{Q}$ be a property of scheme morphisms that is stable under base change (e.g., if $\mathbf{Q}$ is the property "faithfully flat quasi-compact"). Let S be a scheme, let $f \colon X \to Y$ be a morphism of S-schemes, and $g \colon S' \to S$ a morphism possessing $\mathbf{Q}$. Write $f' \colon X' := X \times_S S' \to Y' := Y \times_S S'$ and assume that we have already shown that if f' has a property $\mathbf{P}$, then f has this property $\mathbf{P}$. Then the transitivity of the fiber product shows that if f' has $\mathbf{P}$ universally (Definition 4.31), then f has $\mathbf{P}$ universally.

In particular, we can add the properties "purely inseparable" (which means "universally injective") and "universally bijective" to the list in Proposition 14.50 and "universally open", "universally closed", and "universally homeomorphism" to the list in Proposition 14.51.

Next, we come to descent of scheme-theoretic properties of morphisms. Again we refer to Appendix C for a more exhaustive list.

Proposition 14.53. *Let S be a scheme, and let $f\colon X \to Y$ be a morphism of S-schemes. Consider a faithfully flat quasi-compact morphism $g\colon S' \to S$. Write $f'\colon X' := X \times_S S' \to Y' := Y \times_S S'$ for the morphism obtained by base change. If f' is*

(1) *(locally) of finite type, or (locally) of finite presentation,*

(2) *an isomorphism,*

(3) *a monomorphism,*

(4) *an open immersion, or a closed immersion, or a quasi-compact immersion*

(5) *proper,*

(6) *affine,*

(7) *finite, or*

(8) *quasi-finite,*

then so is f.

The hypothesis on g can be further weakened (Exercise 14.9).

Proof. Note that the projection $Y' \to Y$ is again faithfully flat and quasi-compact, so we can assume that $Y = S$.

The assertion for the properties "locally of finite type", and "locally of finite presentation" follow from Proposition 14.49, and together with Proposition 14.51 (4) and (6), we obtain (1).

Now assume that f' is an isomorphism. By Remark 14.52, f is a universal homeomorphism. We also see that f is quasi-compact and separated, and we have to show that the sheaf homomorphism $\theta\colon \mathscr{O}_Y \to f_*\mathscr{O}_X$ associated with f is an isomorphism. By assumption, the corresponding homomorphism $\theta'\colon \mathscr{O}_{Y'} \to f'_*\mathscr{O}_{X'}$ is an isomorphism, and this is the composition of the canonical isomorphism $g^*f_*\mathscr{O}_X \to f'_*\mathscr{O}_{X'}$ (see Proposition 12.6) and the pull-back $g^*\theta$. It follows that $g^*\theta$ is an isomorphism, and since g is faithfully flat, so is θ.

For the case of monomorphisms, note that $X \to Y$ is a monomorphism if and only if the diagonal morphism $X \to X \times_Y X$ is an isomorphism. Since the transition from a morphism to its diagonal morphism is compatible with base change, (3) follows from (2).

If f' is an open immersion, then $g^{-1}(f(X)) = f'(X')$ is open in Y', hence so is $f(X) \subseteq Y$ (Proposition 14.39). We can replace Y by $f(X)$, and then the assertion follows from (2).

Suppose that f' is a closed immersion. We can check that f is a closed immersion locally on Y, so we may assume that $Y = \operatorname{Spec} A$ is affine. Denote by $Z \subseteq Y$ the schematic image of f (see Section (10.8)). Lemma 14.6 shows that $Z' = Z \times_S S'$ can be naturally identified with the schematic image of f'. Since f' induces an isomorphism $X' \to Z'$ by assumption, part (2) implies that f factors through an isomorphism $X \to Z$, and hence is a closed immersion.

Let f' be a quasi-compact immersion. Then f is quasi-compact by Proposition 14.51. Let Z be the schematic image of f (Proposition 10.30). Then $Z' := Z \times_Y Y'$ is the schematic image of f' (Lemma 14.6) and the factorization of f' in $X' \to Z' \to Y'$ is an open immersion followed by a closed immersion. Thus by the assertions of part (4) which have already been proved, f is the composition of an open and a closed immersion and hence an immersion.

If f' is proper, f is separated by Proposition 14.51, universally closed by Remark 14.52, and of finite type by (1).

We now come to the case that f' is an affine morphism. In particular, f' is qcqs, and by Proposition 14.51 this also holds for f. We write $\mathscr{A} = f_*\mathscr{O}_X$, $\mathscr{A}' = f'_*\mathscr{O}_{X'}$ which are quasi-coherent by Corollary 10.27. By Proposition 12.6 we have $\mathscr{A}' = g^*\mathscr{A}$. Therefore we can identify the natural morphisms $X' \to \operatorname{Spec}\mathscr{A}' \to Y'$ with the base change of the morphisms $X \to \operatorname{Spec}\mathscr{A} \to Y$ with respect to $g\colon Y' \to Y$. The assumption that f' is affine is equivalent to saying that the morphism $X' \to \operatorname{Spec}\mathscr{A}'$ is an isomorphism, and it follows from (2) that the same is true for f, hence that f is affine.

If f' is finite, then f is affine and proper by the above, and hence is finite by Corollary 12.89. Finally, if f' is quasi-finite, then f is of finite type by (1) and has finite fibers by Proposition 14.50. $\square$

Definition 14.54. *Let* $\mathbf{P}$ *be a property of morphisms of schemes. We say that* $\mathbf{P}$ *is* stable under faithfully flat descent, *if the following condition holds. For every morphism of schemes* $f\colon X \to Y$ *and for every faithfully flat quasi-compact morphism* $Y' \to Y$ *such that the base change* $f'\colon X \times_Y Y' \to Y'$ *has property* $\mathbf{P}$, *the morphism* f *has property* $\mathbf{P}$.

Thus all the properties in the Propositions 14.50, 14.51, and 14.53 are stable under faithfully flat descent. The properties "projective" and "quasi-projective" are not stable under faithfully flat descent as they are not even local on the target (Remark 13.70); see however Proposition 14.57 below.

Example 14.55. Let k be a field, let $f\colon X \to Y$ be a morphism of k-schemes, and let k' be a field extension of k. Let $\mathbf{P}$ be a property of morphisms of schemes that is stable under faithfully flat descent. If $f \otimes \operatorname{id}_{k'}\colon X \otimes_k k' \to Y \otimes_k k'$ has property $\mathbf{P}$, then f has property $\mathbf{P}$: Indeed, as $\operatorname{Spec} k' \to \operatorname{Spec} k$ is clearly faithfully flat and quasi-compact, the same holds for its base change $Y \otimes_k k' \to Y$.

Thus if $\mathbf{Q}$ is a property which is stable under flat base change, then to prove that $\mathbf{Q}$ implies $\mathbf{P}$ for f, one can show this after base change to a field extension (e.g., to some algebraic closure).

Remark 14.56. Let $\mathbf{P}$ and $\mathbf{Q}$ be two properties of morphism of schemes $f\colon X \to S$ of finite presentation. We make the following hypotheses.
(a) $\mathbf{P}$ is stable under base change and compatible with inductive limits of rings.
(b) $\mathbf{Q}$ is local on the target, stable under base change and under faithfully flat descent, and compatible with inductive limits of rings. Moreover we assume that if $S = \operatorname{Spec} R$, where R is a complete local noetherian ring with maximal ideal $\mathfrak{m}$, then $X \to \operatorname{Spec} R$ satisfies $\mathbf{Q}$ if $X \otimes_R R/\mathfrak{m}^n \to \operatorname{Spec} R/\mathfrak{m}^n$ satisfies $\mathbf{Q}$ for all $n \geq 1$.
Then if $\mathbf{P}$ implies $\mathbf{Q}$ for all $f\colon X \to S$ such that S is the spectrum of a local Artinian ring, then $\mathbf{P}$ implies $\mathbf{Q}$ for all schemes S and all morphisms $f\colon X \to S$ of finite presentation.

Indeed, let $f\colon X \to S$ be a morphism of finite presentation with the property $\mathbf{P}$, where S is an arbitrary scheme. We want to show that f has the property $\mathbf{Q}$. As $\mathbf{P}$ is stable under base change and $\mathbf{Q}$ is local on the target, we may assume that $S = \operatorname{Spec} R$ is affine. Then R is the limit of a filtered inductive system of noetherian subrings. As $\mathbf{P}$ is compatible with inductive limits and $\mathbf{Q}$ is stable under base change, we may assume that R is noetherian. As $\mathbf{P}$ is stable under base change and as $\mathbf{Q}$ is local on the target and compatible with inductive limits of rings, we may even assume that R is a local noetherian ring. Then its completion $\hat{R}$ is a faithfully flat R-algebra. As $\mathbf{P}$ is stable under flat base change and $\mathbf{Q}$ is stable under faithfully flat descent, we may assume that R is a complete local noetherian ring with maximal ideal $\mathfrak{m}$. Then $R/\mathfrak{m}^n$ is local Artinian and by the last hypothesis on $\mathbf{Q}$ we may assume that R is local Artinian.

Proposition 14.57. *Let k be a field, let X be a k-scheme, and let k' be a field extension of k. Then $X \otimes_k k'$ is quasi-projective (resp. projective) over k' if and only if X is quasi-projective (resp. projective) over k.*

Proof. We know already that the condition is sufficient (Remark 13.73). As the property "proper" is stable under faithfully flat descent it suffices to show that X is quasi-projective over k if $X_{k'} := X \otimes_k k'$ is quasi-projective over k'. As $X_{k'}$ is of finite type over k', the k-scheme X is of finite type. By hypothesis there exists an immersion $X_{k'} \hookrightarrow \mathbb{P}^n_{k'}$. By the principle of finite field extension (Corollary 10.79) this implies the existence of an immersion $X_K \hookrightarrow \mathbb{P}^n_K$ over a finite extension K of k. Thus X_K is quasi-projective over K and also over k. As $\operatorname{Spec} K \to \operatorname{Spec} k$ is finite, the projection $X_K \to X$ is surjective finite locally free. Therefore X is quasi-projective by Proposition 13.76. $\qquad\square$

Proposition 14.58. *Let S be a scheme, and let $f: X \to Y$ be a morphism of S-schemes of finite type. Consider a faithfully flat quasi-compact morphism $g: S' \to S$. Write $f': X' := X \times_S S' \to Y' := Y \times_S S'$ for the morphism obtained by base change and $g': X' \to X$ for the projection. Let $\mathscr{L}$ be an invertible $\mathscr{O}_X$-module. Then $\mathscr{L}$ is ample (resp. very ample) for f if and only if $g'^*\mathscr{L}$ is ample (resp. very ample) for f'.*

Proof. By Remark 13.53 and Proposition 13.64 the condition is necessary. To show that it is sufficient, we may assume that $S = Y$ and that Y is affine. Then Y' is quasi-compact.

Assume that $g'^*\mathscr{L}$ is very ample for f'. The existence of an f'-ample line bundle implies that f' is separated. Therefore f is separated by Proposition 14.51. By Proposition 13.56 we have to show that (i) $f_*\mathscr{L}$ is quasi-coherent, (ii) the canonical homomorphism $f^*f_*\mathscr{L} \to \mathscr{L}$ is surjective, and (iii) the corresponding morphism $r: X \to \mathbb{P}(f_*\mathscr{L})$ is an immersion. As g is flat, we have

$$(14.11.2) \qquad\qquad g^*f_*\mathscr{L} \cong f'_*g'^*\mathscr{L}$$

by Proposition 12.6. Assertion (i) holds by Corollary 10.27. Assertion (ii) holds using Proposition 14.11 and (14.11.2) because the corresponding property of $g'^*\mathscr{L}$ holds and because g' is faithfully flat. Finally, as $g'^*\mathscr{L}$ is very ample, the base change of r via g is a quasi-compact immersion (using (14.11.2) to identify $\mathbb{P}(f_*\mathscr{L}) \times_Y Y'$ with $\mathbb{P}(f'_*g'^*\mathscr{L})$). Therefore r is a (quasi-compact) immersion by Proposition 14.53.

Now assume that $g'^*\mathscr{L}$ is ample for f'. As Y' is quasi-compact, there exists an integer $n > 0$ such that $g'^*\mathscr{L}^{\otimes n}$ is very ample (Theorem 13.62) for f'. Thus we have just seen that $\mathscr{L}^{\otimes n}$ is very ample for f and therefore $\mathscr{L}$ is ample for f. $\qquad\square$

(14.12) Descent of absolute properties of schemes.

Having considered properties of modules and of morphisms of schemes, we now look at properties of schemes and their behavior with respect to descent. Again, the notion of faithful flatness turns out to be the crucial ingredient which allows us to conclude that many interesting properties of schemes descend. Here are a few very simple examples: Let $X \to Y$ be faithfully flat. If X is reduced, then so is Y (because faithfully flat ring homomorphisms are injective); if X is irreducible, then so is Y (in fact, this is true whenever $X \to Y$ is surjective).

For the sake of completeness, we also note the following similar results about descent of absolute properties of schemes with respect to faithfully flat morphisms:

Proposition 14.59. *Let $f \colon X \to Y$ be a faithfully flat morphism of locally noetherian schemes, and let* **P** *denote the property of being*
(1) *reduced,*
(2) *normal, or*
(3) *regular.*
If X has the property **P***, then so has Y. If Y and all fibers $f^{-1}(y)$, $y \in Y$, have the property* **P***, then so has X.*

Proof. These statements are easily reduced to the corresponding results in commutative algebra (Proposition B.82 and Example B.80, Proposition B.73 (6), and Proposition B.77 (5)) $\qquad\qquad\qquad\qquad\qquad\qquad\qquad\qquad\qquad\qquad\qquad\qquad\quad$ $\square$

Corollary 14.60. *Let $f \colon X \to Y$ be a smooth surjective morphism of locally noetherian schemes. Then X is reduced (resp. normal, resp. regular) if and only if Y has this property.*

Proof. The morphism f is faithfully flat by Theorem 14.24. As smoothness is stable under base change, all fibers of f are smooth schemes over a field. Therefore all fibers are regular by Lemma 6.26 and in particular normal and reduced. Therefore Proposition 14.59 implies the corollary. $\qquad\qquad\qquad\qquad\qquad\qquad\qquad\qquad\qquad\qquad\qquad\qquad\qquad\qquad\quad$ $\square$

Remark 14.61. Proposition 14.59 and Corollary 14.60 also hold more generally for the properties (R_k) and (S_k) (k fixed) by Proposition B.82. In particular they hold for the property "Cohen-Macaulay".

Remark 14.62. Let S be a locally noetherian scheme and let $\mathscr{A}$ be a graded quasi-coherent $\mathscr{O}_S$-algebra such that $\mathscr{A}_0 = \mathscr{O}_S$ and such that $\mathscr{A}_1$ is a coherent $\mathscr{O}_S$-module that generates $\mathscr{A}$. Let $X = \operatorname{Proj}(\mathscr{A})$ be its projective spectrum, $C = \operatorname{Spec}(\mathscr{A})$ its affine cone and $C^0 \subset C$ its pointed affine cone (Section (13.9)). Let **P** be one of the properties "reduced", "normal", "regular", "irreducible", "(R_k)", or "(S_k)".
(1) If C has the property **P**, then C^0 and X have the property **P**.
(2) C^0 has property **P** if and only if X has property **P**.
Indeed, if C has property **P**, then the open subscheme C^0 has property **P**. Thus it suffices to show (2). The canonical projection $\pi \colon C^0 \to X$ is smooth and surjective (Remark 13.38). Thus for all properties except "irreducible" we are done by Corollary 14.60. If C^0 is irreducible, then X is irreducible because π is surjective. Conversely, let X be irreducible. As π is open by Corollary 14.36 and as for every $x \in X$ the fiber $\pi^{-1}(x) \cong \mathbb{A}^1_{\kappa(x)} \setminus \{0\}$ is irreducible, C^0 is irreducible by Proposition 3.24.

(14.13) Gluing of sheaves reconsidered.

Our next goal is to develop the theory of faithfully flat descent of quasi-coherent sheaves: We want to study, given a faithfully flat morphism $S' \to S$, which quasi-coherent sheaves on S' arise via pull-back from a quasi-coherent $\mathscr{O}_S$-module. We start with a reminder on gluing of morphisms of sheaves, and gluing of sheaves for the Zariski topology which uses a language which is close to the theory of descent which we are going to work out here. Let S be a scheme, let $S = \bigcup U_i$ be an open covering, and let $S' = \coprod U_i$ be the disjoint union of the U_i, so that we have a natural surjection $S' \to S$. We set $S'' = S' \times_S S' = \coprod_{i,j} U_i \cap U_j$, $S''' = S' \times_S S' \times_S S' = \coprod_{i,j,k} U_i \cap U_j \cap U_k$. We have the projections $p_1, p_2 \colon S'' \to S'$, $p_{12}, p_{23}, p_{13} \colon S''' \to S''$.

A sheaf $\mathscr{F}'$ on S' is just a tuple $(\mathscr{F}'_i)_i$ where each $\mathscr{F}'_i$ is a sheaf on U_i. Similarly, we view sheaves $\mathscr{F}''$ on S'' as tuples $(\mathscr{F}''_{ij})_{i,j}$. Then a gluing datum for a quasi-coherent sheaf $\mathscr{F}'$ on S' is exactly the same as an isomorphism $\varphi \colon p_1^* \mathscr{F}' \xrightarrow{\sim} p_2^* \mathscr{F}'$ which satisfies the "cocycle condition"

$$p_{23}^* \varphi \circ p_{12}^* \varphi = p_{13}^* \varphi.$$

Note that we are implicitly identifying certain pull-backs of $\mathscr{F}'$ to S''' (with respect to natural isomorphisms), e. g.

$$p_{12}^*(p_2^* \mathscr{F}') \cong (p_2 \circ p_{12})^* \mathscr{F}' = (p_1 \circ p_{23})^* \mathscr{F}' \cong p_{23}^*(p_1^* \mathscr{F}'),$$

so that the composition and the comparison of the left and the right hand sides make sense. We make these identifications here and in the sequel and usually omit checking that everything is compatible with them. For a more detailed account, see [Vis].

We obtain a category $\mathrm{QCoh}(S'/S)$ of quasi-coherent sheaves on S' together with a gluing datum, and a functor $\mathrm{QCoh}(S) \to \mathrm{QCoh}(S'/S)$ given by pull-back and the natural isomorphisms on intersections $U_i \cap U_j$. The statement that we can glue homomorphisms of quasi-coherent modules says precisely that this functor is fully faithful, while the statement that we can glue quasi-coherent modules translates to this functor being an equivalence of categories.

Our goal in this section is to generalize this result to a much larger class of morphisms $S' \to S$. The class of morphisms for which we can prove "gluing", or rather "descent" of quasi-coherent sheaves is the class of quasi-compact faithfully flat morphisms. This theory is due to Grothendieck. Heuristically, we still think of a quasi-compact faithfully flat morphism $S' \to S$ as a covering of S (in some "topology" which is much finer than the Zariski topology). In fact, the notion of *Grothendieck topology*, a generalization of the notion of topology, makes this heuristics precise. We will not need this notion, though, and therefore do not go into the details. Note that the class of all faithfully flat morphisms is too large from the point of view of a reasonable theory of descent.

Since an (infinite) Zariski-covering $X = \bigcup_i U_i$ in general will not give rise to a *quasi-compact* morphism $\coprod_i U_i \to X$, it is sometimes useful to combine these notions; see Exercise 14.8 and [Vis] 2.3.2 for a detailed discussion.

(14.14) Descent Data and the gluing functor.

Let $p \colon S' \to S$ be a quasi-compact faithfully flat morphism of schemes. We use the following notation:

$$S'' := S' \times_S S', \quad S''' := S' \times_S S' \times_S S',$$

and denote by

$$p_i \colon S'' \to S', \quad i \in \{1, 2\}, \qquad p_{ij} \colon S''' \to S'', \quad i, j \in \{1, 2, 3\}, \ i < j,$$

the projections on the ith factor and the (i, j)-th factors, respectively.

Definition 14.63. *Let $\mathscr{F}'$ be a quasi-coherent $\mathscr{O}_{S'}$-module. A descent datum on $\mathscr{F}'$ is a $\mathscr{O}_{S''}$-module homomorphism*

$$\varphi \colon p_1^*\mathscr{F}' \xrightarrow{\sim} p_2^*\mathscr{F}',$$

which satisfies the cocycle condition *$p_{23}^*\varphi \circ p_{12}^*\varphi = p_{13}^*\varphi$, i.e., more precisely, such that the following diagram commutes:*

(14.14.1)

$$
\begin{array}{ccccc}
p_{12}^*p_1^*\mathscr{F}' & \xrightarrow{\ p_{12}^*\varphi\ } & p_{12}^*p_2^*\mathscr{F}' = p_{23}^*p_1^*\mathscr{F}' & \xrightarrow{\ p_{23}^*\varphi\ } & p_{23}^*p_2^*\mathscr{F}' \\
\| & & & & \| \\
p_{13}^*p_1^*\mathscr{F}' & & \xrightarrow{\qquad p_{13}^*\varphi \qquad} & & p_{13}^*p_2^*\mathscr{F}'
\end{array}
$$

Let $(\mathscr{F}', \varphi)$, $(\mathscr{G}', \psi)$ be quasi-coherent $\mathscr{O}_{S'}$-modules with descent data. A morphism between these pairs is a homomorphism $\alpha \colon \mathscr{F}' \to \mathscr{G}'$ of $\mathscr{O}_{S'}$-modules which is compatible with the descent data, i.e., such that $p_2^*\alpha \circ \varphi = \psi \circ p_1^*\alpha$. We denote by $\mathrm{QCoh}(S'/S)$ the category of quasi-coherent $\mathscr{O}_{S'}$-modules with descent data.

There is a natural functor

$$\Phi_{S'/S} \colon \mathrm{QCoh}(S) \to \mathrm{QCoh}(S'/S)$$

which associates to each quasi-coherent $\mathscr{O}_S$-module $\mathscr{F}$ the pull-back $p^*\mathscr{F}$ together with its *canonical descent datum*

$$\varphi_{\mathrm{can}} \colon p_1^*(p^*\mathscr{F}) \cong (p \circ p_1)^*\mathscr{F} = (p \circ p_2)^*\mathscr{F} \cong p_2^*(p^*\mathscr{F}).$$

It is easy to check that φ_{can} satisfies the cocycle condition. We will show, Theorem 14.68, that for quasi-compact faithfully flat morphisms $S' \to S$ the functor $\Phi_{S'/S}$ is an equivalence of categories.

Right now, let us make the notion of descent datum more explicit in the affine case $S = \operatorname{Spec} R$, $S' = \operatorname{Spec} R'$, $S'' = \operatorname{Spec} R'' = \operatorname{Spec} R' \otimes_R R'$. A quasi-coherent $\mathscr{O}_{S'}$-module $\mathscr{F}'$ is given by an R'-module M', and $p_1^*\mathscr{F}'$, $p_2^*\mathscr{F}'$ correspond to the tensor products

$$M' \otimes_{R',\pi_1} (R' \otimes_R R'), \qquad M' \otimes_{R',\pi_2} (R' \otimes_R R')$$

where π_1 is given by $r' \mapsto r' \otimes 1$, and π_2 is given by $r' \mapsto 1 \otimes r'$. We can identify both tensor products with $M' \otimes_R R'$, *but* under these identifications in the first case the R''-module structure is the usual one, while in the second case it is given by $(a \otimes b)(m' \otimes r') = bm' \otimes ar'$. In particular, the identity homomorphism is not R''-linear, and hence does not give rise to a descent datum. The cocycle condition can also easily be translated into a condition on (tensor products of) modules.

Now assume that $M' = M \otimes_R R'$ arises as the pull-back of an R-module M. Under the identifications above, the canonical descent datum is given by

(14.14.2) $(M \otimes R') \otimes_R R' \xrightarrow{\sim} (M \otimes R') \otimes_R R', \quad m \otimes a \otimes b \mapsto m \otimes b \otimes a.$

As required, this map is R''-linear with respect to the R''-module structures explained above.

(14.15) Descent of homomorphisms.

We start with faithfully flat descent of homomorphisms of quasi-coherent sheaves. The following lemma will turn out to be crucial for all of the following. Let $R \to R'$ be a faithfully flat ring homomorphism, and let M be an R-module. For $n \geq 0$ let $R^{(n)}$ be the n-fold tensor product $R' \otimes_R \cdots \otimes_R R'$ (we set $R^{(0)} = R$, $R^{(1)} = R'$). For $i = 0, \ldots, n$ define maps $\varphi_i \colon M \otimes_R R^{(n)} \to M \otimes_R R^{(n+1)}$ by

$$\varphi_i(m \otimes r_1 \otimes \cdots \otimes r_n) = m \otimes r_1 \otimes \cdots \otimes r_i \otimes 1 \otimes r_{i+1} \otimes \cdots \otimes r_n$$

and

$$\delta^n \colon M \otimes_R R^{(n)} \to M \otimes_R R^{(n+1)}, \qquad x \mapsto \sum_{i=0}^{n} (-1)^i \varphi_i(x).$$

One checks easily that $\delta^n \circ \delta^{n-1} = 0$.

Lemma 14.64. *In this situation, the complex*

$$0 \to M \to M \otimes_R R' \to M \otimes_R R^{(2)} \to \cdots$$

is exact.

Proof. It is enough to check the exactness after a faithfully flat base change, so we may first tensor the whole sequence by $\otimes_R R'$. Then the exactness follows from the following formal argument (as a warm up, we have that the map $M \otimes_R R' \to M \otimes R' \otimes R'$ admits the retraction $m \otimes x_1 \otimes x_2 \mapsto m \otimes x_1 x_2$, and hence must be injective, which gives the exactness at the first position):

After tensoring by $\otimes_R R'$, the maps of our complex have the form

$$\delta^n_{R'} \colon M \otimes_R R^{(n+1)} \to M \otimes_R R^{(n+2)},$$

$$m \otimes x_1 \otimes \cdots \otimes x_{n+1} \mapsto \sum_{i=0}^{n} (-1)^i \varphi_i(m \otimes x_1 \otimes \cdots \otimes x_n) \otimes x_{n+1}.$$

Consider the maps

$$\gamma^n \colon M \otimes_R R^{(n+2)} \to M \otimes R^{(n+1)},$$

$$m \otimes x_0 \otimes \cdots \otimes x_{n+1} \mapsto \sum_{i=0}^{n} (-1)^i m \otimes x_1 \otimes \cdots \otimes x_i x_{i+1} \otimes \cdots \otimes x_{n+1},$$

for $n \geq 0$. By a straightforward computation, one sees that the map

$$\delta^{n-1}_{R'} \circ \gamma^{n-1} + \gamma^n \circ \delta^n_{R'} \colon M \otimes_R R^{(n)} \to M \otimes_R R^{(n)}$$

is the identity map. In analogy to the theory of homotopy, we can express this by saying that the family $\gamma^\bullet$ is a homotopy between the identity and the zero map.

It follows that the identity map and the zero map induce the same homomorphism on the cohomology groups $\operatorname{Ker} \delta^n / \operatorname{Im} \delta^{n-1}$ of this complex, which means that it must be exact. $\qquad \square$

Now we can describe which S'-morphisms between pull-backs $p^* \mathscr{F}$, $p^* \mathscr{G}$ of quasi-coherent $\mathscr{O}_S$-modules arise from $\mathscr{O}_S$-homomorphisms.

Proposition 14.65. *Let $p\colon S' \to S$ be a quasi-compact faithfully flat morphism of schemes. Then the functor $\Phi_{S'/S}$ is fully faithful. In other words, whenever $\mathscr{F}, \mathscr{G}$ are quasi-coherent $\mathscr{O}_S$-modules, then the sequence*

$$0 \to \operatorname{Hom}_S(\mathscr{F}, \mathscr{G}) \to \operatorname{Hom}_{S'}(p^*\mathscr{F}, p^*\mathscr{G}) \to \operatorname{Hom}_{S''}(q^*\mathscr{F}, q^*\mathscr{G}),$$

where $q = p \circ p_1 = p \circ p_2$, and where the last map is given by $\alpha \mapsto p_1^*\alpha - p_2^*\alpha$, is an exact sequence of abelian groups.

Proof. The morphism p being faithfully flat, the first map is injective (cf. Proposition 14.11). Furthermore, the composition $\operatorname{Hom}_S(\mathscr{F}, \mathscr{G}) \to \operatorname{Hom}_{S''}(q^*\mathscr{F}, q^*\mathscr{G})$ is clearly the zero map. It remains to check that every element of $\operatorname{Hom}_{S'}(p^*\mathscr{F}, p^*\mathscr{G})$ which is mapped to zero is the image of an element of $\operatorname{Hom}_S(\mathscr{F}, \mathscr{G})$.

By using gluing of morphisms with respect to a Zariski cover, we may assume that S is affine. Then in particular S' is quasi-compact, so that we can cover it as $S' = \bigcup U_i$ by finitely many open affine subschemes. Set $T' = \coprod_i U_i$, $T'' = T' \times_S T'$. Note that the morphism $r\colon T' \to S$ is again quasi-compact faithfully flat. We have a commutative diagram

$$
\begin{array}{ccccccc}
0 & \longrightarrow & \operatorname{Hom}_S(\mathscr{F}, \mathscr{G}) & \longrightarrow & \operatorname{Hom}_{S'}(p^*\mathscr{F}, p^*\mathscr{G}) & \longrightarrow & \operatorname{Hom}_{S''}(q^*\mathscr{F}, q^*\mathscr{G}) \\
& & \downarrow {\scriptstyle =} & & \downarrow & & \downarrow \\
0 & \longrightarrow & \operatorname{Hom}_S(\mathscr{F}, \mathscr{G}) & \longrightarrow & \operatorname{Hom}_{T'}(r^*\mathscr{F}, r^*\mathscr{G}) & \longrightarrow & \operatorname{Hom}_{T''}(s^*\mathscr{F}, s^*\mathscr{G})
\end{array}
$$

(Here s is the natural morphism $T'' \to S$.) Because the morphism $T' \to S$ is faithfully flat, the vertical arrows are injective. A simple diagram chase shows that the exactness assertion for the upper row follows from the corresponding assertion for the lower row. We may therefore replace S' by T', i.e., we may assume that S' is affine.

We have reduced the proof of the proposition to the affine situation $S = \operatorname{Spec} R$, $S' = \operatorname{Spec} R'$. Let us rephrase the situation in terms of R-modules. We denote by M, N the R-modules $\Gamma(S, \mathscr{F})$ and $\Gamma(S, \mathscr{G})$, resp. We are given an R'-module homomorphism $\varphi'\colon M \otimes_R R' \to N \otimes_R R'$ such that for all $m \otimes a \otimes b \in M \otimes_R R' \otimes_R R'$, when we write $\sum n_i \otimes c_i = \varphi'(m \otimes b)$, then

$$\varphi''(m \otimes a \otimes b) := \varphi'(m \otimes a) \otimes b = \sum n_i \otimes a \otimes c_i.$$

More conceptually, we can write $\varphi'' = p_1^*\varphi = p_2^*\varphi$. Denote by δ_M the homomorphism

$$\delta_M\colon M \otimes_R R' \to M \otimes_R R' \otimes_R R', \quad m \otimes a \to m \otimes 1 \otimes a - m \otimes a \otimes 1,$$

and correspondingly for δ_N. We have that $\varphi'' \circ \delta_M = \delta_N \circ \varphi'$, so $\varphi'(\operatorname{Ker} \delta_M) \subseteq \operatorname{Ker} \delta_N$.

Now Lemma 14.64 shows that $M = \operatorname{Ker} \delta_M$, $N = \operatorname{Ker} \delta_N$. Therefore φ', restricted to these kernels, induces a morphism $\varphi\colon M \to N$, and it is clear that $\varphi \otimes_R R' = \varphi'$. $\quad\square$

(14.16) Descent of quasi-coherent modules.

We keep the assumption that $p\colon S' \to S$ is quasi-compact and faithfully flat, and the notation introduced above. Our next goal is to show that the functor $\Phi_{S'/S}$ is an equivalence of categories, i.e., that it is also essentially surjective. We do this in several steps, and first establish its compatibility with base change.

Proposition 14.66. *Let* $f\colon T \to S$ *be a morphism, and denote by* $f'\colon T \times_S S' \to S'$, $f''\colon (T \times_S S') \times_T (T \times_S S') = T \times_S S'' \to S''$ *the projections. Then we have a commutative diagram of morphisms of functors:*

$$
\begin{array}{ccc}
\mathrm{QCoh}(S) & \xrightarrow{\;\Phi_{S'/S}\;} & \mathrm{QCoh}(S'/S) \\
\Big\downarrow{\scriptstyle f^*} & & \Big\downarrow \\
\mathrm{QCoh}(T) & \xrightarrow{\;\Phi_{T'/T}\;} & \mathrm{QCoh}(T \times_S S'/T)
\end{array}
$$

where the right column functor maps a pair $(\mathscr{F}', \varphi)$ *to its pull-back* $((f')^*\mathscr{F}', (f'')^*\varphi)$.

Proof. The proof consists of checking compatibilities of fiber products and is straightforward. $\qquad\square$

The next step is a result which says (according to the analogy with usual gluing of sheaves) that we can pass to a refinement of our covering.

Proposition 14.67. *Let* $f'\colon T' \to S'$ *be a morphism such that the composition* $p \circ f'\colon T' \to S$ *is quasi-compact and faithfully flat. Then the following diagram of morphisms of functors is commutative, and all three functors are fully faithful.*

$$
\begin{array}{ccc}
\mathrm{QCoh}(S) & \longrightarrow & \mathrm{QCoh}(S'/S) \\
\big\downarrow & \swarrow & \\
\mathrm{QCoh}(T'/S) & &
\end{array}
$$

Proof. It is easy to see that the diagram is commutative. We know by the above that the functors $\mathrm{QCoh}(S) \to \mathrm{QCoh}(S'/S)$, $\mathrm{QCoh}(S) \to \mathrm{QCoh}(T'/S)$ are fully faithful. Since we do not assume that f' is faithfully flat, we have to address this question for the remaining functor separately. We extend the above diagram to

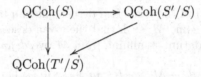

and since the projections from $S' \times_S T'$ to S' and to T' are quasi-compact and faithfully flat, our claim follows. $\qquad\square$

Theorem 14.68. *Let* $p\colon S' \to S$ *be a quasi-compact faithfully flat morphism. Then the functor*

$$
\Phi_{S'/S}\colon \mathrm{QCoh}(S) \to \mathrm{QCoh}(S'/S)
$$

is an equivalence of categories.

Proof. We know already (Proposition 14.65) that $\Phi_{S'/S}$ is fully faithful, so it remains to check the essential surjectivity.

We will prove this in three steps: We reduce to a statement that can be checked after faithfully flat base change; we apply the base change $S' \to S$, and hence may assume that $S' \to S$ has a section; using Proposition 14.67 we deal with the case where p has a section.

(i). Using the compatibility with base change, Proposition 14.66 and usual gluing, it is enough to prove the proposition locally on S, so we may assume that S is affine. Then S' is quasi-compact, and we find a quasi-compact faithfully flat morphism $T' \to S'$ with T' affine. Using Proposition 14.67 we may assume that S' is affine as well.

So we have $S = \operatorname{Spec} R$, $S' = \operatorname{Spec} R'$, say. Let M' be an R'-module with a descent datum φ. As we have seen already, in terms of modules the descent datum is an isomorphism

$$M' \otimes_{R'} R^{(2)} \stackrel{\sim}{\to} M' \otimes_{R'} R^{(2)}$$

which satisfies the cocycle condition. Lemma 14.64 shows that if M' descends to an R-module M, then we must have

$$M = \operatorname{Ker}(M \otimes R' \to M \otimes R^{(2)}) = \operatorname{Ker}(M' \to M' \otimes_R R'),$$

where the map is given, for $m \in M$, $a \in R'$, $m' := m \otimes a$, by

$$m \otimes a \mapsto m \otimes 1 \otimes a - m \otimes a \otimes 1 = \varphi_{\mathrm{can}}(m' \otimes 1) - m' \otimes 1,$$

see the description (14.14.2) of the canonical descent datum in terms of modules. Since we expect M' to come from some M, such that the given descent datum for M' coincides with the canonical descent datum stemming from M, we *define*

$$M := \operatorname{Ker}(M' \to M' \otimes_R R', \ m' \mapsto \varphi(m' \otimes 1) - m' \otimes 1).$$

We see that M' descends to an R-module if and only if the natural homomorphism $M \otimes_R R' \to M'$ is an isomorphism of R'-modules which is compatible with the descent data φ_{can} and φ.

(ii). This statement can be checked after a faithfully flat base change. We choose to apply base change by the morphism $p\colon S' \to S$. Because the morphism $p'\colon S' \times_S S' \to S'$ admits the diagonal $S' \to S' \times_S S'$ as a section, we may assume, after replacing p by p', that p has a section.

(iii). Now assume that $p\colon S' \to S$ has a section $s\colon S \to S'$. Then we may apply Proposition 14.67 with $T' = S$, $f' = s$. Clearly the functor $\operatorname{QCoh}(S) \to \operatorname{QCoh}(S/S)$ is an equivalence of categories, and it follows that all the functors in the resulting diagram are equivalences of categories. $\square$

(14.17) Descent of schemes.

Having proved faithfully flat descent for quasi-coherent sheaves, we can ask similar questions about other kinds of objects. The most interesting case is descent of schemes (equipped with descent data) for quasi-compact faithfully flat morphisms, generalizing the usual gluing of schemes (Section (3.5)). As a first step, we show that we have descent for morphisms as in the case of sheaves. However, descent for objects, i.e., for schemes, does not work in general; see Remark 14.73.

Definition 14.69. *Let* $p\colon S' \to S$ *be a morphism of schemes, and let* X' *be an* S'-*scheme. A descent datum* on X' *for* p *is an isomorphism*

$$\varphi\colon X' \times_{S',p_1} S'' \overset{\sim}{\to} X' \times_{S',p_2} S''$$

of S''-*schemes which satisfies the obvious cocycle condition.*

Note that for an S-scheme X we have, similarly as for $\mathscr{O}_S$-modules, a canonical isomorphism

$$(14.17.1) \qquad \varphi_{\mathrm{can}}\colon (X \times_S S') \times_{S',p_1} S'' = (X \times_S S') \times_{S',p_2} S''$$

which satisfies the cocycle condition, and which we call the canonical descent datum on $X \times_S S'$. Similarly as for quasi-coherent modules, we have the category $(\mathrm{Sch}/S'/S)$ of S'-schemes with descent data with respect to p, and a functor $p^*\colon (\mathrm{Sch}/S) \to (\mathrm{Sch}/S'/S)$.

Definition 14.70. *Let* $p\colon S' \to S$ *be a morphism of schemes. A descent datum* φ *for* p *on an* S'-*scheme* X' *is called* effective, *if the pair* (X',φ) *lies in the essential image of the functor* p^*.

Descent for schemes is much easier if the problem can be reduced to an affine situation, because affine scheme morphisms correspond to quasi-coherent algebras over the target, and for those we can apply the results for quasi-coherent modules proved above. Therefore the case where the descent datum is compatible with an affine covering is particularly approachable. The precise notion we use is this:

Definition 14.71. *Let* $p\colon S' \to S$ *be a quasi-compact faithfully flat morphism of schemes, and let* X' *be an* S'-*scheme equipped with a descent datum* φ. *We say that an open subscheme* $U' \subset X'$ *is* stable under φ, *if* φ *restricts to an isomorphism* $U' \times_{S',p_1} S'' \overset{\sim}{\to} U' \times_{S',p_2} S''$ *(which automatically is a descent datum on* U'*).*

We get the following theorem.

Theorem 14.72. *Let* $p\colon S' \to S$ *be a quasi-compact faithfully flat morphism.*
(1) *The functor*

$$(\mathrm{Sch}/S) \to (\mathrm{Sch}/S'/S), \quad X \mapsto (X \times_S S', \varphi_{\mathrm{can}})$$

from the category of S-*schemes to the category of* S'-*schemes with descent data with respect to* p *is fully faithful.*
(2) *A descent datum* φ *on an* S'-*scheme* X' *is effective if* X' *can be covered by open subschemes which are affine over* S' *and stable under* φ.

Proof. We first prove (1), so consider S-schemes X, Y. The question is local on S and Y, so we may assume that these are affine. Using the usual gluing of morphisms of schemes, one sees that one can also reduce to the case that X is affine. In that case, morphisms $X \to Y$ of S-schemes (and $X' \to Y'$ of S'-schemes, resp.) are nothing but homomorphisms of quasi-coherent $\mathscr{O}_S$-algebras (and $\mathscr{O}_{S'}$-algebras, resp.), so that the assertion follows from Proposition 14.65.

To prove (2), assume that X' with descent datum φ admits a covering $X' = \bigcup U_i'$ by φ-stable open subsets, each of which is affine over S'. Using a variant of Theorem 14.68 for quasi-coherent algebras instead of quasi-coherent modules, we see that all the affine schemes U_i' descend to affine S-schemes U_i. By descent of morphisms of schemes (i.e., part (1)), we see that the gluing datum for the family $(U_i')_i$ given by X' descends to a gluing datum for the family $(U_i)_i$. By usual gluing of schemes we obtain an S-scheme X, and this is the S-scheme we are looking for. $\qquad\square$

One can show that in part (2) of the theorem, it is enough to assume that X' admits a covering by quasi-affine open subschemes (see Section (13.16)) which are stable under φ; see [BLR] Thm. 6.1/6.

Remark 14.73. On the other hand, note that the descent of schemes is not effective in general; see [BLR] 6.7, or [Vis] for an example, a thorough discussion and further references. There are examples of quasi-compact faithfully flat morphisms $S' \to S$ and quasi-projective S'-schemes with descent data which do not descend to S.

On the other hand, under additional assumptions, descent can often be shown to be effective. We list some examples:

Remark 14.74. In the following cases, a descent datum φ on an S'-scheme X' for a quasi-compact faithfully flat morphism $p\colon S' \to S$ is effective:
(1) If $(X', \mathscr{L}')$ is a pair of a quasi-projective S'-scheme together with an S'-ample invertible sheaf $\mathscr{L}'$, equipped as a pair with a descent datum (in the obvious sense), then one can show that Theorem 14.72 can be applied so that the pair $(X', \mathscr{L}')$ descends to a pair $(X, \mathscr{L})$ of a quasi-projective S-scheme with an S-ample invertible sheaf $\mathscr{L}$. See [Vis], Thm. 4.38.
(2) If p is purely inseparable, every descent datum φ is effective (Exercise.14.19).
(3) Let p be finite locally free and assume that X' is quasi-projective over S'. Then every descent datum φ is effective (see [SGA1] VIII, Cor. 7.6).

(14.18) Representable functors are fpqc-sheaves.

Let S be a fixed scheme.

Definition 14.75. A functor $F\colon (\mathrm{Sch}/S)^{\mathrm{opp}} \to (\mathrm{Sets})$ is called an fpqc-sheaf over S if F is a Zariski-sheaf (see Section (8.3)), and if for every ring R over S and for every faithfully flat ring homomorphism $R \to R'$, the sequence

(14.18.1) $$F(\mathrm{Spec}\, R) \to F(\mathrm{Spec}\, R') \rightrightarrows F(\mathrm{Spec}\, R' \otimes_R R')$$

is exact.
A morphism of fpqc-sheaves is a morphism of functors.

As a consequence of Theorem 14.72 (1), we have:

Proposition 14.76. Let X be an S-scheme. Then the functor $h_X\colon (\mathrm{Sch}/S)^{\mathrm{opp}} \to (\mathrm{Sets})$, $h_X(T) = \mathrm{Hom}_S(T, X)$, is an fpqc-sheaf.

The Yoneda lemma 4.6 shows that we obtain a fully faithful functor of the category of schemes into the category of fpqc-sheaves.

Remark 14.77. There are variants of Definition 14.75: A Zariski-sheaf $F\colon (\mathrm{Sch}/S)^{\mathrm{opp}} \to (\mathrm{Sets})$ is called *fppf-sheaf* (resp. *sheaf for the étale topology*) if the sequence (14.18.1) is exact for every faithfully flat ring homomorphism $R \to R'$ such that R' is an R-algebra of finite presentation (resp. such that $\mathrm{Spec}\, R' \to \mathrm{Spec}\, R$ is étale). Every fpqc-sheaf is an fppf-sheaf, and every fppf-sheaf is a sheaf for the étale topology.

Many notions carry over directly from the "usual" sheaf setting to the theory of fpqc-sheaves (or fppf-sheaves, or sheaves for the étale topology), at least as long as no sheafification is required. For instance, fiber products (in the category of functors) of fpqc-sheaves (resp. fppf-sheaves, resp. sheaves for the étale topology) are again sheaves of the same kind.

If F is an fpqc-sheaf (or an fppf-sheaf, or a sheaf for the étale topology) on S, and $S' \to S$ is a morphism, then its base change $F \times_S S'$ is the functor $(F \times_S S')(T) := F(T)$ for every S'-scheme T (which we may also consider as S-scheme via composition with $S' \to S$). This is a sheaf of the same kind. Because the functor $F \times_S S' \colon (\text{Sch}/S')^{\text{opp}} \to (\text{Sets})$ is just the restriction of F to $(\text{Sch}/S')^{\text{opp}}$, we also write $F_{|S'}$ instead of $F \times_S S'$.

We note that one *can* sheafify every functor $(\text{Sch}/S)^{\text{opp}} \to (\text{Sets})$ to a sheaf for the étale topology or even to an fppf-sheaf (satisfying the usual universal properties), but this is more complicated than for usual sheaves, and we will not need it. In particular, one runs into problems of set theory. See Artin's notes [Art] for details.

(14.19) Torsors and H^1.

In Section (11.5), we have discussed torsors for the Zariski topology. There is a similar notion in the setting of the "fpqc-topology", the "fppf-topology", or the "étale topology". In this section we will concentrate on the "fppf-topology" and leave it to the reader to consider other "topologies".

Definition 14.78. *We call a morphism of schemes* fppf-morphism, *if it is faithfully flat and locally of finite presentation.*

The property "fppf-morphism" is stable under composition, stable under base change, local on the target, and stable under faithfully flat descent because this holds for the properties "faithfully flat" and "locally of finite presentation".

Lemma 14.79. *Let S be an affine scheme and let $f \colon S' \to S$ be an fppf-morphism. Then there exists a morphism $S'' \to S'$ such that the composition $g \colon S'' \to S$ is an affine fppf-morphism.*

Proof. Let (S_i') be an open affine covering of S'. Since f is open by Theorem 14.35 and S is quasi-compact, there exists a finite subset $I' \subseteq I$ such that the restriction $\bigcup_{i \in I'} S_i' \to S$ of f is still surjective. We set $S'' = \coprod_{i \in I'} S_i'$. Then S'' is affine and $S'' \to S$ is faithfully flat and of finite presentation. $\square$

Let S be a scheme. An *fppf-sheaf in groups* over S is a functor $(\text{Sch}/S)^{\text{opp}} \to (\text{Grp})$ such that its composition with the forgetful functor $(\text{Grp}) \to (\text{Sets})$ is an fppf-sheaf (Remark 14.77). Compare the definition of group scheme in Section (4.15). Proposition 14.76 implies that every group scheme over S gives rise to an fppf-sheaf in groups over S.

We fix an fppf-sheaf in groups G over S. There is the obvious notion of a G-action $G \times X \to X$ on an fppf-sheaf X over S and of a morphism of fppf-sheaves with G-action (cf. Section (11.5)).

Definition 14.80. *Let S be a scheme, and let G be an fppf-sheaf in groups over S. An* fppf-torsor *over S under G (or simply a G-fppf-torsor over S) is an fppf-sheaf X over S together with an action of G on X, such that the following two conditions are satisfied*

(a) *The morphism $G \times_S X \to X \times_S X$, $(g,x) \mapsto (gx, x)$ (on T-valued points) is an isomorphism (i.e., for all S-schemes T, the group $G(T)$ acts simply transitively on the set $X(T)$).*
(b) *There exists an fppf-morphism $S' \to S$ such that $X(S') \neq \emptyset$.*

We remark that condition (b) is automatic, if X itself is a scheme whose structure morphism $X \to S$ is fppf (by taking $S' = X$).

The G-fppf-torsors over S form a full subcategory of the category of fppf-sheaves with G-action. It follows from (a) that every morphism in this category is an isomorphism.

A G-fppf-torsor X is called the *trivial torsor* if it is isomorphic to G acting by multiplication on itself. In this case $X(S) \neq \emptyset$ because $G(S)$ is non-empty. Conversely, if the set $X(S)$ is non-empty, then the G-action induces a morphism $G \to X$ of G-fppf-torsors and thus X is trivial. We can read condition (b) as saying that after an fppf base change, i.e., "locally for the fppf topology", every torsor X is trivial.

Remark 14.81. Let G be an fppf-sheaf in groups over a scheme S and let X be an fppf-sheaf with G-action over S.

If X is an fppf-torsor under G, then for every morphism $S' \to S$ the base change $X_{|S'}$ is an fppf-torsor under $G_{|S'}$ over S'. By hypothesis (b) in Definition 14.80 there exists an fppf-morphism $S' \to S$ such that $X(S') \neq \emptyset$ which implies that $X_{|S'}$ is trivial.

Conversely, assume that there exists an fppf-morphism $S' \to S$ such that $X_{|S'}$ is a trivial fppf-torsor under $G_{|S'}$ over S'. Then X is an fppf-torsor under G. Indeed, we have to show that $\sigma\colon G \times_S X \to X \times_S X$, $(g, x) \mapsto (gx, x)$, is an isomorphism. As $G \times_S X$ and $X \times_S X$ are Zariski-sheaves we may work Zariski-locally on S and can thus assume that $S = \operatorname{Spec} R$ is affine. By Lemma 14.79 we may assume that there exists an fppf-morphism of the form $S' = \operatorname{Spec} R' \to S$, where R' is a faithfully flat R-algebra of finite presentation, such that $X_{|S'}$ is trivial. By the sheaf axioms for fppf-sheaves (14.18.1) we may check whether σ is an isomorphism after replacing S by S'. But then X is trivial and σ is an isomorphism.

Proposition 14.82. *Assume that G is an S-group scheme that is affine over S. Then every G-fppf-torsor X over S is representable by an S-scheme which is affine over S.*

Proof. Let X be a G-torsor over S. Using Zariski-gluing, we may work locally on S (Theorem 8.9), and hence assume that S is affine. By Lemma 14.79 there exists an affine fppf-morphism $f\colon S' \to S$ that trivializes X.

As the functor $X_{|S'} = X \times_S S'$ is the base change of the functor X, it is equipped with a descent datum which is then also a descent datum of schemes by Yoneda's lemma. Since $X_{|S'} \cong G_{|S'}$ is affine over S', the descent is effective, see Theorem 14.72. The S-scheme we obtain by descent is then isomorphic to X as an fppf-sheaf. In other words, X is representable. □

The hypothesis in Proposition 14.82 that G is affine over S can be replaced by other assumptions, e. g., it is enough to assume that G is quasi-projective over S and that the torsor X over S becomes trivial after restriction via a finite locally free surjective morphism $S' \to S$ (use Remark 14.74 instead of Theorem 14.72 in the proof of Proposition 14.82).

Remark 14.83. Assume that $G \to S$ has a property which is stable under base change, local on the target, and stable under faithfully flat descent. Then every G-fppf-torsor X that is an S-scheme has the same property: We may assume that S is affine and again by

Lemma 14.79 that X is isomorphic to G after base change with an affine fppf-morphism (which is in particular faithfully flat and quasi-compact).

One can develop the theory of fppf-torsors very much along the lines of our discussion of Zariski-torsors in Chapter 11. For instance, one can define (Čech) cohomology groups $H^1_{\mathrm{fppf}}(S' \to S, G)$ and $H^1_{\mathrm{fppf}}(S, G)$ which classify the isomorphism classes of G-torsors over S trivialized by $S' \to S$, and of all G-torsors over S, resp. These cohomology groups fit into a framework of derived functors of the global section functor for the fppf-topology. We will not go into the general theory here, but see the following Section (14.20) for a special case.

(14.20) Galois descent.

In this section we will study the particular case of faithfully flat descent, where the morphism $p\colon S' \to S$ is a "Galois covering", i.e., the analogue of a finite unramified covering map with a finite group Γ of deck transformations.

If Γ is any group and S is a scheme, we denote by $\Gamma \times S$ the disjoint union $\coprod_{\gamma \in \Gamma} S$. The canonical structure morphism $\Gamma \times S \to S$ is surjective and locally on the source an isomorphism. In particular it is faithfully flat. It is finite if Γ is finite. Equivalently, we can view the product $\Gamma \times S$ as the constant group scheme $\underline{\Gamma}_S$, see Example 4.43.

An action of Γ on an S-scheme S' by S-automorphisms then corresponds to a morphism of S-schemes $(\Gamma \times S) \times_S S' \to S'$ which is denoted on T-valued points (T an S-scheme) simply by $(\gamma, s') \mapsto \gamma s'$.

Definition 14.84. *Let Γ be a finite group. A* Galois covering *with* Galois group Γ *is a finite faithfully flat morphism $p\colon S' \to S$, together with an action of Γ on S' by S-automorphisms, such that the morphism*

$$\sigma\colon \Gamma \times S' \to S' \times_S S', \qquad (\gamma, s') \mapsto (s', \gamma s'),$$

given on T-valued points (T an S-scheme), is an isomorphism.

Note that a Galois covering $S' \to S$ is the same as a $\underline{\Gamma}_S$-torsor. We leave the easy proof of this statement as Exercise 14.22.

The prototype of a Galois covering is a morphism $\operatorname{Spec} k' \to \operatorname{Spec} k$, where k'/k is a finite Galois extension of fields. Since $k' \otimes_k k' \cong \prod_{\gamma \in \mathrm{Gal}(k'/k)} k'$, this is a Galois covering with Galois group $\Gamma = \mathrm{Gal}(k'/k)$.

As a warm-up, we note the following theorem, which describes Galois descent for vector spaces. For a k'-vector space V', we call an action of Γ on the abelian group V' an action on V' over the action of Γ on k', if for every $\sigma \in \Gamma$, $a \in k'$, $v \in V'$, we have $\sigma(av) = \sigma(a)\sigma(v)$.

Theorem 14.85. *Let k'/k be a finite Galois extension with Galois group Γ. The functor*

$$(k\text{-vector spaces}) \to (k'\text{-vector spaces with } \Gamma\text{-action over } k'), \qquad V \mapsto V \otimes_k k',$$

is an equivalence of categories with quasi-inverse $V' \mapsto (V')^\Gamma$, the k-vector space of invariants of V' under Γ.

As we will see below, the theorem is a special case of the general theorem on faithfully flat descent, but to illustrate it, we give a direct proof.

Proof. We denote by $k'[\Gamma]$ the twisted group algebra, i.e., the k'-vector space with basis $\{\gamma \in \Gamma\}$ and with multiplication

$$\Big(\sum_{\gamma \in \Gamma} a_\gamma \gamma\Big)\Big(\sum_{\delta \in \Gamma} b_\delta \delta\Big) := \sum_{\gamma, \delta} a_\gamma \gamma(b_\delta)\gamma\delta.$$

Then $k'[\Gamma]$ is a k-algebra and the category of k'-vector spaces with Γ-action over k' is the category of $k'[\Gamma]$-left modules.

Now $k'[\Gamma]$ acts k-linearly on k' by

$$\Big(\sum_\gamma a_\gamma \gamma\Big)c := \sum_\gamma a_\gamma \gamma(c).$$

This action defines a k-algebra homomorphism $\alpha\colon k'[\Gamma] \to \operatorname{End}_k(k')$, where $\operatorname{End}_k(k')$ denotes the k-linear endomorphisms of the k-vector space k'. Moreover, α is k'-linear and $\{\alpha(\gamma) \, ; \, \gamma \in \Gamma\}$ is linearly independent in the k'-vector space $\operatorname{End}_k(k')$ by Dedekind's theorem on linear independence of characters. Thus α is injective and counting dimensions we see that α is an isomorphism. Thus Morita equivalence (Section (8.12)) shows that $V \mapsto k' \otimes_k V$ defines the desired equivalence of categories.

To see that a quasi-inverse to this functor is given by $V' \mapsto (V')^\Gamma$ it now suffices to see that $(V \otimes_k k')^\Gamma = V$ which is clear. $\qquad\square$

To make the connection to the previous sections, note that we can express the notion of descent datum with respect to a Galois covering $S' \to S$ using the group action of Γ. Let X' be an S'-scheme. We call an action of Γ on X' *compatible with the action of Γ on S'* (or an *action over the Γ-action on S'*), if for every $\gamma \in \Gamma$ the diagram

(14.20.1)
$$\begin{array}{ccc} X' & \xrightarrow{\gamma} & X' \\ \downarrow & & \downarrow \\ S' & \xrightarrow{\gamma} & S' \end{array}$$

is commutative. Notice that, since γ is an automorphism, in this case the diagram is automatically cartesian, and that for $\gamma \neq \operatorname{id}$, the automorphisms of X' occurring here are *not* S'-morphisms.

By assumption, we have an isomorphism $\Gamma \times S' \xrightarrow{\sim} S'' := S' \times_S S'$, $(\gamma, s) \mapsto (s, \gamma(s))$. Therefore an isomorphism $S'' \times_{p_1, S'} X' \xrightarrow{\sim} S'' \times_{p_2, S'} X'$ of S''-schemes is the same as an isomorphism $\Gamma \times X' \xrightarrow{\sim} \Gamma \times X'$ such that for all γ the diagram (14.20.1) is commutative. It is tedious, but straightforward to check that the isomorphism $\Gamma \times X' \xrightarrow{\sim} \Gamma \times X'$, composed with the second projection, is a group action, i.e., satisfies the usual associativity condition, if and only if the corresponding isomorphism $S'' \times_{p_1, S'} X' \xrightarrow{\sim} S'' \times_{p_2, S'} X'$ is a descent datum, i.e., satisfies the cocycle condition.

An open subset $U' \subseteq X'$ is stable under the descent datum φ if and only if it is stable under all the automorphisms $\gamma \in \Gamma$. Assume that S and S' are affine. If $x \in X$ is a point whose Γ-orbit is contained in an open subscheme $U' \subseteq X'$, then the intersection $\bigcap_\gamma \gamma(U')$ is an open neighborhood of x which is stable under Γ. If X' is separated and U' is affine, then this intersection is again affine. By Proposition 13.49, if X' is quasi-projective over S', then every finite subset, so in particular every Γ-orbit, is contained in an open affine subset. Therefore, we obtain from Theorem 14.72:

Theorem 14.86. *Let S be an affine scheme, and let $S' \to S$ be a Galois covering with Galois group Γ. Then the natural functor*

(quasi-projective S-schemes) $\to$ (quasi-projective S'-schemes with comp. Γ-action)
$$X \mapsto (X \times_S S', \varphi_{\mathrm{can}})$$

is an equivalence of categories.

(14.21) Descent along torsors.

The Galois descent discussed in Section (14.20) is a special case of descent along torsors which we explain now. Let S be a scheme, let G be an S-group scheme which is quasi-compact and faithfully flat over S and let S' be an S-scheme with a G-action $a \colon G \times_S S' \to S'$ that makes S' into a G-torsor over S. Then $p \colon S' \to S$ is faithfully flat quasi-compact by Remark 14.83. We will now define G-actions on quasi-coherent $\mathcal{O}_{S'}$-modules over the G-action on S' and we will see that these actions correspond to descent data for $S' \to S$. In Section (14.20) we considered the case that $G = \underline{\Gamma}_S$, where Γ is a finite group.

DESCENT FOR QUASI-COHERENT MODULES.

Let $p_2 \colon G \times_S S' \to S'$ be the second projection. A *G-equivariant structure on a quasi-coherent $\mathcal{O}_{S'}$-module $\mathscr{F}'$* is an isomorphism of $\mathcal{O}_{G \times_S S'}$-modules

$$\theta \colon a^*\mathscr{F}' \overset{\sim}{\to} p_2^*\mathscr{F}'$$

satisfying the cocycle condition

(14.21.1) $p_{23}^*\theta \circ (\mathrm{id}_G \times a)^*\theta = (m \times \mathrm{id}_{S'})^*\theta.$

The pair $(\mathscr{F}', \theta)$ is then called a *G-equivariant quasi-coherent $\mathcal{O}_{S'}$-module*. A *homomorphism of G-equivariant quasi-coherent $\mathcal{O}_{S'}$-modules* $(\mathscr{F}', \theta) \to (\mathscr{G}', \eta)$ is an $\mathcal{O}_{S'}$-linear homomorphism $u \colon \mathscr{F}' \to \mathscr{G}'$ such that $\eta \circ a^*(u) = p_2^*(u) \circ \theta$. We obtain the category of G-equivariant $\mathcal{O}_{S'}$-modules. As $p \circ a = p \circ p_2$, the identity defines a G-equivariant structure θ_{can} of $p^*\mathscr{F}$ for every quasi-coherent $\mathcal{O}_S$-module $\mathscr{F}$.

Let $\sigma \colon G \times_S S' \to S' \times_S S'$ be the morphism given on T-valued points by $(g, s') \mapsto (gs', s')$ which is an isomorphism by hypothesis. As we have $a = p_1 \circ \sigma$ we can use σ to identify a G-equivariant structure θ on $\mathscr{F}'$ with an isomorphism $\varphi \colon p_1^*\mathscr{F}' \overset{\sim}{\to} p_2^*\mathscr{F}'$. It is straightforward to check that the cocycle condition (14.21.1) for θ and the cocycle condition (14.14.1) for φ are equivalent. Thus Theorem 14.68 shows:

Theorem 14.87. *The functor $\mathscr{F} \mapsto (p^*\mathscr{F}, \theta_{\mathrm{can}})$ yields an equivalence of categories*

$$\mathrm{QCoh}(S) \overset{\cong}{\longrightarrow} (G\text{-equivariant quasi-coherent } \mathcal{O}_{S'}\text{-modules}).$$

For a finite Galois extension $k' \supset k$, $S = \mathrm{Spec}\, k$, $S' = \mathrm{Spec}\, k'$, and $G = \underline{\Gamma}_S$, where $\Gamma = \mathrm{Gal}(k'/k)$, this theorem is equivalent to Theorem 14.85.

DESCENT FOR AFFINE SCHEMES.

Let $\pi' \colon X' \to S'$ be a morphism of schemes. We also view X as an S-scheme via $S' \to S$. As above we call an action $\tilde{a} \colon G \times_S X' \to X'$ of the S-group scheme G on the S-scheme X' *compatible with the action of G on S'* if the diagram

$$
\begin{array}{ccc}
G \times_S X' & \xrightarrow{\tilde{a}} & X' \\
{\scriptstyle \mathrm{id}_G \times \pi'} \downarrow & & \downarrow {\scriptstyle \pi'} \\
G \times_S S' & \xrightarrow{a} & S'
\end{array}
$$

(14.21.2)

is cartesian. The pair $(X', \tilde{a})$ consisting of the S'-scheme X' and the S-action $\tilde{a}$ is then also called a *G-equivariant S'-scheme*. A *morphism of G-equivariant S'-schemes* $(X', \tilde{a}) \to (Y', \tilde{b})$ is a morphism $f' \colon X' \to Y'$ of S'-schemes such that

$$
\begin{array}{ccc}
G \times_S X' & \xrightarrow{\tilde{a}} & X' \\
{\scriptstyle \mathrm{id}_G \times f'} \downarrow & & \downarrow {\scriptstyle f'} \\
G \times_S Y' & \xrightarrow{\tilde{b}} & Y'
\end{array}
$$

(14.21.3)

commutes. We obtain the category of G-equivariant S'-schemes. As in the section on Galois descent, we see that a G-equivariant structure on X' is the same as a descent datum on X' for $S' \to S$. It is effective by Theorem 14.72 if we can cover X' by open G-invariant subschemes which are affine over S'.

This is clearly possible if X' is affine over S'. In this case $X' \cong \operatorname{Spec} \mathscr{A}'$ for a quasi-coherent $\mathscr{O}_{S'}$-algebra $\mathscr{A}'$ (Corollary 12.2), and a G-equivariant structure on X' is the same as a G-equivariant structure on the quasi-coherent $\mathscr{O}_{S'}$-module $\mathscr{A}'$ that is an isomorphism of $\mathscr{O}_{G \times_S S'}$-algebras. We obtain:

Corollary 14.88. *The functor $X \mapsto X \times_S S'$ induces an equivalence of the category of S-schemes X such that $X \to S$ is affine and the category of G-equivariant S'-schemes X' such that $X' \to S'$ is affine.*

(14.22) Forms over fields.

Let us look at descent theory from a slightly different angle. We have seen already that there may exist non-isomorphic k-schemes X, Y such that $X \otimes_k k'$ and $Y \otimes_k k'$ are isomorphic. See in particular the section about Brauer-Severi schemes in Chapter 8. In other words, the k'-scheme $X \otimes_k k' \cong Y \otimes_k k'$ descends to different k-schemes. This is the phenomenon we will study here. For simplicity, we will restrict to the case of Galois extensions. We start with a general definition:

Definition 14.89. *Let k'/k be a field extension, and let X be a k-scheme. A k'-form of X is a k-scheme Y such that the k'-schemes $Y \otimes_k k'$ and $X \otimes_k k'$ are isomorphic.*

Now assume that k'/k is a finite Galois extension and set $\Gamma := \operatorname{Gal}(k'/k)$. We know already that the different ways how a given scheme X'/k' descends to k are given by descent data, or equivalently by Γ-actions on X'. We can however make this more manageable using the notion of Galois cohomology. In this way, we can in particular deal with the problem that different (but isomorphic) descent data, or Γ-actions, give rise to the same k-scheme.

Let A be a group on which Γ acts by group automorphisms (from the left). We denote the action of Γ on A by $(\gamma, a) \mapsto {}^{\gamma}a$. Let us define the first Galois cohomology group $H^1(\Gamma, A) =: H^1(k'/k, A)$ of Γ with coefficients in A.

A *cocycle* is a map $c\colon \Gamma \to A$, $\gamma \mapsto c(\gamma)$, such that for all $\gamma, \delta \in \Gamma$,

$$c(\gamma\delta) = c(\gamma)\,{}^{\gamma}c(\delta).$$

We denote by $Z^1(\Gamma, A)$ the set of cocycles. Cocycles c, c' are called equivalent, if there exists $a \in A$ with

$$c'(\gamma) = a^{-1}c(\gamma)\,{}^{\gamma}a \qquad \text{for all } \gamma \in \Gamma.$$

We denote by $H^1(\Gamma, A)$ (or by $H^1(k'/k, A)$) the set of equivalence classes. This is a pointed set: The distinguished point is the trivial cocycle $\gamma \mapsto 1$.

Now let X be a k-scheme, and let $X' = X \otimes_k k'$. The action of Γ on k' induces an operation of Γ on $X' = X \otimes_k k'$ by letting Γ act on the second factor. We denote the automorphism of X' given by $\gamma \in \Gamma$ again by γ. Note that γ is only an automorphism of k-schemes but not (in general) of k'-schemes. We obtain an action of Γ on the group $\mathrm{Aut}_{k'}(X')$ of automorphisms a of X' over k' by

$$(\gamma, a) \mapsto {}^{\gamma}a := \gamma \circ a \circ \gamma^{-1}.$$

Consider a k'-form Y of X. We fix an isomorphism $\alpha\colon Y \otimes_k k' \xrightarrow{\sim} X'$, and define a cocycle $c \in Z^1(\Gamma, \mathrm{Aut}_{k'}(X'))$ as follows. By its action on the second factor, Γ acts on $Y \otimes_k k'$, and we obtain a "twisted" action of Γ on X' by letting $\gamma \in \Gamma$ act on X' by

$$\gamma_Y := \alpha \circ \gamma \circ \alpha^{-1}\colon X' \xrightarrow{\sim} X'.$$

Note that this is an action over the action of Γ on $\mathrm{Spec}\,k'$; the automorphism γ_Y (just as the automorphism γ defined above) is not a k'-morphism. We define

$$c(\gamma) := \gamma_Y \circ \gamma^{-1} \in \mathrm{Aut}_{k'}(X').$$

It is easy to check that $\gamma \mapsto c(\gamma)$ is in fact a cocycle. Furthermore, replacing the isomorphism $\alpha\colon Y \otimes_k k' \xrightarrow{\sim} X'$ by a different isomorphism amounts to changing it by an automorphism of X', and it is easily seen that the cocycles obtained from different isomorphisms are equivalent. Therefore we can associate to Y an element of $H^1(\Gamma, \mathrm{Aut}_{k'}(X'))$ which is independent of the choice of isomorphism $Y \otimes_k k' \xrightarrow{\sim} X'$.

Theorem 14.90. *Let k'/k be a finite Galois extension, let X be a quasi-projective k-scheme, and denote by X' the base change $X \otimes_k k'$. The map defined above is a bijection between*
(i) *the set of isomorphism classes of k'-forms of X and*
(ii) *the set $H^1(k'/k, \mathrm{Aut}_{k'}(X'))$,*
where the distinguished point of $H^1(k'/k, \mathrm{Aut}_{k'}(X'))$ corresponds to the k'-form X.

Proof. Because we have the theory of Galois descent at our disposal, this is just a matter of translating back and forth between the description of a descent datum by an action of Γ over the action of Γ on k', and the cocycle description developed above. $\qquad\square$

We can deal with an arbitrary Galois extension by passing to the inductive limit:

Theorem 14.91. *Let k be a field, let k^{sep} be a separable closure of k, let X be a quasi-projective k-scheme, and set $X^{\mathrm{sep}} := X \otimes_k k^{\mathrm{sep}}$. There is a natural bijection between*
(i) *the set of isomorphism classes of k^{sep}-forms of X and*
(ii) *the set $H^1(k^{\mathrm{sep}}/k, \mathrm{Aut}_{k^{\mathrm{sep}}}(X^{\mathrm{sep}})) := \varinjlim_{k'} H^1(k'/k, \mathrm{Aut}_{k'}(X \otimes_k k'))$, where the limit runs over all finite Galois extensions of k inside k^{sep}.*

Proof. This follows from the above theorem and Corollary 10.80. $\square$

Compare Remark 11.13 for the analogous result for sheaves for the Zariski topology.

(14.23) Application to Brauer-Severi varieties.

Theorem 14.91 is particularly useful if the automorphism group of X' is well understood. For instance, we have shown that the automorphism group of projective space $\mathbb{P}^n_k$ over a field k is the projective linear group $\mathrm{PGL}_{n+1}(k)$, see Section (11.15). This allows us to classify Brauer-Severi varieties (or Brauer-Severi schemes) which we defined in Section (8.11).

We start with applying the descent results to Brauer-Severi varieties.

Proposition 14.92. *Let X be a Brauer-Severi variety over a field k. Then X is projective, smooth and geometrically integral over k. There exists a finite extension K of k such that $X \otimes_k K \cong \mathbb{P}^n_K$ for some n.*

Proof. By hypothesis there exists a field extension K of k such that $X \otimes_k K \cong \mathbb{P}^n_K$. Thus by Proposition 14.57 we know that X is projective over k. In particular, X is of finite type over k and we may apply the principle of finite extension (Corollary 10.80) to see that there exists a finite extension K of k such that $X \otimes_k K \cong \mathbb{P}^n_K$. By Corollary 5.54 X is geometrically integral because $\mathbb{P}^n_K$ is integral for every field K. Finally, as $\mathbb{P}^n_K$ is regular for every field K, the k-scheme X is smooth by Corollary 6.32. $\square$

A field extension K of k such that $X \otimes_k K \cong \mathbb{P}^n_K$ is also called a *splitting field of X*. To apply Theorem 14.91 we need to know that a separable closure k^{sep} is a splitting field (or, equivalently by Corollary 10.64, that there exists a splitting field that is a finite Galois extension of k). We will deduce this from the following result which is known as Châtelet's theorem.

Theorem 14.93. *Let X be a Brauer-Severi variety over the field k. The following are equivalent:*
(1) The set $X(k)$ of k-valued points is non-empty.
(2) The variety X is isomorphic to projective space over k.

Proof. Denote by $\bar{k}$ an algebraic closure of k. Clearly, (ii) implies (i). To show the converse, it is enough to prove that there exists a line bundle $\mathcal{K}$ on X such that the base change $\mathcal{K}_{\bar{k}}$ is $\mathcal{O}(1)$. In fact, then $\mathcal{K}_{\bar{k}}$ is generated by global sections. We know that $\Gamma(X_{\bar{k}}, \mathcal{K}_{\bar{k}}) = \Gamma(X, \mathcal{K}) \otimes_k \bar{k}$ by Corollary 12.8, so we can use Proposition 14.11 to show that $\mathcal{K}$ is generated by global sections, too, and hence defines a morphism $s \colon X \to \mathbb{P}(\Gamma(X, \mathcal{K}))$. Since s becomes an isomorphism after base change to $\bar{k}$, it is itself an isomorphism, and the theorem follows. To find $\mathcal{K}$, we have to give a construction of $\mathcal{O}(1)$ on $\mathbb{P}^n_{\bar{k}}$ which can be carried out over k. Since line bundles are associated with divisors rather than with points, we first pass to the blow-up of X in a k-rational point x (which exists by assumption).

So let $x \in X(k)$ be a k-rational point. We fix an identification $X_{\bar{k}} = \mathbb{P}^n_{\bar{k}}$ that identifies x with $(1 : 0 : \ldots : 0)$. Let $\pi \colon X' \to X$ be the blow-up of X in x and let $E \subset X'$ be the exceptional divisor. Then $E \cong \mathbb{P}^{n-1}_k$ (Example 13.95), and Example 13.93 shows

$$X'_{\bar{k}} = V_+(T_i X_j - T_j X_i; \ 1 \leq i, j \leq n) \subset \mathbb{P}^{n-1}_{\bar{k}} \times \mathbb{P}^n_{\bar{k}}$$

(with the notation introduced there) such that $\pi_{\bar{k}}$ is the restriction of the second projection. Let $\bar{\psi}\colon X'_{\bar{k}} \to \mathbb{P}^{n-1}_{\bar{k}}$ be the restriction of the first projection. Then all fibers of $\bar{\psi}$ are projective lines and $\bar{\psi}$ induces an isomorphism of $E_{\bar{k}}$ onto $\mathbb{P}^{n-1}_{\bar{k}}$.

We fix an ample line bundle $\mathscr{L}$ on X. Then $\mathscr{L}_{\bar{k}} \cong \mathcal{O}_{\mathbb{P}^n_{\bar{k}}}(d)$ for some integer $d > 0$. We let $\mathscr{M}$ be the line bundle $\pi^*\mathscr{L} \otimes \mathcal{O}_{X'}(E)^{\otimes -d}$ on X'. Then $\mathscr{M}_{\bar{k}}$ is generated by its global sections and thus $\mathscr{M}$ is generated by its global sections. Therefore $\mathscr{M}$ defines a morphism of k-schemes $r\colon X' \to \mathbb{P}(\Gamma(X', \mathscr{M})) \cong \mathbb{P}^N_k$.

Let $H \subset E$ be a hyperplane, set $D := \pi(r^{-1}(r(H)))$ and let $\mathscr{K} = \mathcal{O}_X(D)$. It is easily checked that $D_{\bar{k}}$ is a hyperplane of $\mathbb{P}^n_{\bar{k}}$ and thus $\mathscr{K}_{\bar{k}} \cong \mathcal{O}_{\mathbb{P}^n_{\bar{k}}}(1)$, as desired. $\qquad\square$

Corollary 14.94. *Let X be a Brauer-Severi variety over a field k. Then there exists a finite Galois extension k' of k such that $X_{k'} \cong \mathbb{P}^n_{k'}$ for some $n \geq 0$.*

Proof. As X is geometrically reduced, there exists a finite separable extension K of k such that $X(K) \neq \emptyset$ (Proposition 6.21). Thus $X_K \cong \mathbb{P}^n_K$ by Châtelet's theorem 14.93. Now we can choose k' as a normal hull of K over k. $\qquad\square$

Theorem 14.95. *The map* (8.13.2)

$$\left\{ \begin{matrix} central\ simple\ k\text{-}algebras \\ of\ degree\ n \end{matrix} \right\} \longrightarrow \left\{ \begin{matrix} isomorphism\ classes\ of\ Brauer\text{-}Severi \\ varieties\ over\ k\ that\ are\ forms\ of\ \mathbb{P}^{n-1}_k \end{matrix} \right\},$$

$$A \longmapsto \mathcal{BS}(A).$$

is bijective.

Proof. After the corollary and Property (iii) of central simple algebras in Section (8.13), it is enough to prove that the restriction of this map to the subsets of central simple algebras A, and forms X of $\mathbb{P}^{n-1}_k$ that split over a fixed Galois extension k'/k, i.e., $A \otimes k' \cong M_n(k')$, and $X \otimes_k k' \cong \mathbb{P}^{n-1}_{k'}$, resp., is a bijection. Write $\Gamma = \mathrm{Gal}(k'/k)$.

The idea of the proof is to show that on both sides, the objects over k are described by the same type of descent data on the split object over k'. We have seen above that forms of $\mathbb{P}^{n-1}_k$ are parameterized by $H^1(\Gamma, \mathrm{PGL}_n(k'))$.

Let us construct a map from the set of isomorphism classes of central simple k-algebras which split over k' to $H^1(\Gamma, \mathrm{PGL}_n(k'))$. Given such a central simple k-algebra A, choose an isomorphism $A \otimes_k k' \overset{\sim}{\to} M_n(k')$. The Galois action on the second factor of $A \otimes_k k'$ gives rise to an action of Γ on $M_n(k')$ over the action of Γ on k', and hence, in the same way as above, to a cocycle $\Gamma \to \mathrm{Aut}(M_n(k'))$. Now by a special case of the theorem of Skolem and Noether ([BouA8] 10.1, Corollaire de Théorème 1), the automorphism group of $M_n(k')$ is $\mathrm{PGL}_n(k')$, acting by conjugation. Therefore we get a cocycle $\Gamma \to \mathrm{PGL}_n(k')$. Changing the isomorphism $A \otimes_k k' \overset{\sim}{\to} M_n(k')$ gives rise to an equivalent cocycle, and we obtain the desired map from the set of isomorphism classes of central simple k-algebras which split over k' to $H^1(\Gamma, \mathrm{PGL}_n(k'))$. One shows that this map is a bijection, the main ingredient of the proof being Galois descent for vector spaces (Theorem 14.85).

We have constructed a triangle

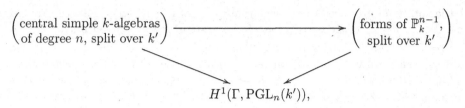

$$H^1(\Gamma, \mathrm{PGL}_n(k')),$$

and it remains to show that this triangle is commutative. But under Morita equivalence (Section (8.12)) the action of PGL_n on the set of submodules of k'^n corresponds to the action of PGL_n on $\mathrm{End}(k'^n) = M_n(k')$ by conjugation. $\qquad\square$

Dimension and fibers of morphisms

In the remainder of this chapter, we come back to the notion of dimension which we now want to discuss in a more general setting than in Chapter 5. For the definition of the dimension of a topological space, and some very general observations, see Section (5.3). Also recall the definition of the dimension of X in a point $x \in X$ (Section (5.7)) and in particular that for a scheme X locally of finite type over a field and a closed point $x \in X$ one has

$$\dim_x X = \dim \mathscr{O}_{X,x} = \max_Z \dim Z,$$

where Z runs through the set of irreducible components of X containing x.

(14.24) Dimension of fibers of morphisms of locally noetherian schemes I.

The main topic below will be the study of the dimension of the fibers of a morphism of finite type. We start with the following result:

Lemma 14.96. *Let $f: X \to Y$ be a morphism, locally of finite type. Let $x \in X$, $y = f(x)$. Then*

$$\dim_x f^{-1}(y) = \dim \mathscr{O}_{X,x} \otimes_{\mathscr{O}_{Y,y}} \kappa(y) + \operatorname{trdeg}_{\kappa(y)} \kappa(x).$$

In particular, if x is a closed point of $f^{-1}(y)$, then

$$\dim_x f^{-1}(y) = \dim \mathscr{O}_{X,x} \otimes_{\mathscr{O}_{Y,y}} \kappa(y).$$

Proof. Let Z be the closure of the point x inside $f^{-1}(y)$. The latter is a $\kappa(y)$-scheme locally of finite type, and therefore

$$\dim_x f^{-1}(y) = \dim Z + \operatorname{codim}_{f^{-1}(y)} Z.$$

by Proposition 5.30 (2), applied to all irreducible components of $f^{-1}(y)$ which contain x. Now $\dim Z = \operatorname{trdeg}_{\kappa(y)} \kappa(x)$ (Theorem 5.22), and $\operatorname{codim}_{f^{-1}(y)} Z = \dim \mathscr{O}_{f^{-1}(y),x} = \dim \mathscr{O}_{X,x} \otimes_{\mathscr{O}_{Y,y}} \kappa(y)$, where the first equality is (5.8.1), and the second one holds because the two rings are isomorphic, as is seen from the definition of the fiber $f^{-1}(y)$ as a scheme (cf. Remark 4.21). $\qquad\square$

Rewriting Proposition B.68 in terms of schemes, we obtain:

Corollary 14.97. *Let $f\colon X \to Y$ be a morphism between locally noetherian schemes. Let $x \in X$ and $y = f(x)$. Then*

$$(14.24.1) \qquad \dim \mathscr{O}_{X,x} \le \dim \mathscr{O}_{Y,y} + \dim(\mathscr{O}_{X,x} \otimes_{\mathscr{O}_{Y,y}} \kappa(y)),$$

and if f is flat at x, then we have equality.

Remark 14.98. The proof of Proposition B.68 in [Mat] Theorem 15.1 shows that to have equality in (14.24.1) it suffices to assume that f is generizing (e.g., if f is open); cf. Section (14.8).

Lemma 14.99. *Let R be a noetherian ring, let $\mathfrak{p} \subset R$ be a prime ideal, and let $\mathfrak{q} = \mathfrak{p}R[T]$. Then $\mathfrak{q}$ is a prime ideal of $R[T]$ and $\mathfrak{q} \cap R = \mathfrak{p}$. There exist infinitely many prime ideals $\mathfrak{q}' \subset R[T]$ with $\mathfrak{q}' \cap R = \mathfrak{p}$, $\mathfrak{q}' \ne \mathfrak{q}$, and there are no inclusions between them. Furthermore, for any such $\mathfrak{q}'$, we have*

$$\dim R[T]_{\mathfrak{q}'} = \dim R[T]_{\mathfrak{q}} + 1 = \dim R_{\mathfrak{p}} + 1.$$

Proof. Identifying $R[T]/\mathfrak{q}$ with $(R/\mathfrak{p})[T]$, we see that the prime ideals $\mathfrak{q}'$ correspond bijectively to the non-zero prime ideals of $\mathrm{Frac}(R/\mathfrak{p})[T]$. This proves all the assertions except for the dimension equalities. It also follows from this reasoning that $\dim R[T]_{\mathfrak{q}'} \ge \dim R[T]_{\mathfrak{q}} + 1$.

On the other hand, we have

$$\dim R[T]_{\mathfrak{q}'} \le \dim R_{\mathfrak{p}} + 1$$

by Proposition B.68, applied to the homomorphism $R_{\mathfrak{p}} \to R[T]_{\mathfrak{q}'}$, because $R[T]_{\mathfrak{q}'} \otimes_{R_{\mathfrak{p}}} R_{\mathfrak{p}}/\mathfrak{p} \cong (R_{\mathfrak{p}}/\mathfrak{p})[T]_{\mathfrak{q}'}$ is a discrete valuation ring and hence has dimension 1.

Now it remains to show that $\dim R[T]_{\mathfrak{q}} \ge \dim R_{\mathfrak{p}}$. But if $\mathfrak{p}_0 \subsetneq \cdots \subsetneq \mathfrak{p}_n = \mathfrak{p}$ is a chain of prime ideals in R, then $\mathfrak{p}_0 R[T] \subset \cdots \subset \mathfrak{p}_n R[T] = \mathfrak{q}$ is a chain of prime ideals, and all inclusions are strict, by the above. $\qquad\square$

Theorem 14.100. *Let R be a noetherian ring, and let $n \ge 0$. Then*

$$\dim R[X_1, \ldots, X_n] = \dim R + n.$$

Proof. By induction, we only have to show that $\dim R[T] = \dim R + 1$. It is easy to see that $\dim R[T] \ge \dim R + 1$, cf. (5.3.1). To show the converse inequality, let $\mathfrak{q}'$ be a maximal ideal of $R[T]$, and let $\mathfrak{p} = \mathfrak{q}' \cap R$, $\mathfrak{q} = \mathfrak{p}R[T]$. We then are in the situation of Lemma 14.99, and obtain

$$\dim R[T]_{\mathfrak{q}'} = \dim R_{\mathfrak{p}} + 1 \le \dim R + 1.$$

Letting $\mathfrak{q}$ vary over all maximal ideals of $R[T]$, this proves the claim. $\qquad\square$

As an immediate corollary to the theorem, we get:

Corollary 14.101. *Let X be a locally noetherian scheme, and let $n \ge 0$. Then*

$$\dim \mathbb{A}_X^n = \dim \mathbb{P}_X^n = \dim X + n.$$

Note that the noetherian hypothesis is needed here: there exist rings R of dimension 1 such that $\dim R[T] = 3 = \dim R + 2$. On the other hand, it is always true that $\dim R + 1 \le \dim R[T] \le 1 + 2\dim R$, [BouAC] VIII §2 no. 2, Cor. 2.

(14.25) Universally catenary schemes.

In general the class of locally noetherian schemes is too large to expect a good behavior of the notion of dimension. Certain useful dimension formulas hold only for the smaller class of so-called universally catenary schemes. Note that the results below show that most schemes arising "in practice" are universally catenary.

Definition 14.102.
(1) *A ring R is called* catenary, *if for every two prime ideals $\mathfrak{p} \subset \mathfrak{p}' \subset R$, there exists a chain $\mathfrak{p} = \mathfrak{p}_0 \subsetneq \mathfrak{p}_1 \subsetneq \cdots \subsetneq \mathfrak{p}_n = \mathfrak{p}'$ which does not admit a refinement, and all such chains have the same length.*
(2) *A ring R is called* universally catenary, *if R is noetherian, and every finitely generated R-algebra is catenary.*
(3) *A locally noetherian scheme X is called* (universally) catenary, *if all of its local rings are (universally) catenary.*

Remark 14.103.
(1) Obviously, every quotient and every localization of a catenary ring is again catenary.
(2) Thus in the definition of a universally catenary ring it is enough to check that every polynomial ring $R[T_1, \ldots, T_n]$ over R is catenary. One can show that it is even enough that $R[T_1]$ is catenary ([Mat] Thm. 31.7, Cor. 1).
(3) Combining (1) and (2) we see that every localization of a universally catenary ring is universally catenary.
(4) Theorem 14.106 (2) below implies that conversely if for a noetherian ring A all localizations $A_\mathfrak{p}$ ($\mathfrak{p} \subset A$ prime ideal) are universally catenary, then A is catenary. In particular, A is universally catenary if and only if $\operatorname{Spec} A$ is universally catenary.
(5) By (4) and (3) we see that if Y is a universally catenary scheme and $f \colon X \to Y$ is a morphism locally of finite type, then X is a universally catenary scheme.
(6) Even if R is a catenary integral domain, it does not follow that all maximal chains of prime ideals in R have the same length (see Exercise 5.7).

Proposition 14.104. *Let k be a field, and let X be a k-scheme locally of finite type. Then X is universally catenary.*

Proof. It suffices to show that every finitely generated k-algebra R is catenary (Remark 14.103 (3)). So let us show that all chains between prime ideals $\mathfrak{p} \subset \mathfrak{p}' \subset R$ which do not admit a refinement have the same length. Dividing by $\mathfrak{p}$, we may assume that $\mathfrak{p} = (0)$. Given maximal chains between $\mathfrak{p}$ and $\mathfrak{p}'$, we obtain maximal chains in $\operatorname{Spec} R$ by choosing a maximal chain between $\mathfrak{p}'$ and a maximal ideal of R containing $\mathfrak{p}'$ (Theorem 5.19 (2)), and the claim follows from Proposition 5.30 (1). □

But in fact much more is true:

Proposition 14.105. *Let R be a regular ring (or, more generally, a Cohen-Macaulay ring). Then every scheme locally of finite type over $\operatorname{Spec} R$ is universally catenary.*

To prove this result it clearly suffices to show that every regular ring (or, more generally, every Cohen-Macaulay ring) is universally catenary. For this we refer to [Mat] Thm. 17.9.

Finally we will use the following dimension formula and characterization of universally catenary rings (see [Mat], Theorems 15.5 and 15.6).

Theorem 14.106.

(1) *Let $R \subseteq B$ be integral domains where R is noetherian. Let $\mathfrak{p} \subset R$ be a prime ideal, and let $\mathfrak{P} \subset B$ be a prime ideal with $\mathfrak{P} \cap R = \mathfrak{p}$. Then*

$$(14.25.1) \qquad \dim R_\mathfrak{p} + \operatorname{trdeg}_{\operatorname{Frac} R} \operatorname{Frac}(B) \geq \dim B_\mathfrak{P} + \operatorname{trdeg}_{\kappa(\mathfrak{p})} \kappa(\mathfrak{P}).$$

(2) *A noetherian ring A is universally catenary if and only if one has equality in (14.25.1) for $R = A/\mathfrak{p}$ and B, for every $\mathfrak{p} \in \operatorname{Spec} A$, and every finitely generated extension ring $B \supset A/\mathfrak{p}$ which is an integral domain.*

(14.26) Dimension of fibers.

We start with the simple observation that the fiber dimension of a morphism of finite type between quasi-compact schemes is bounded:

Proposition 14.107. *Let $f \colon X \to Y$ be a morphism of finite type, and suppose that Y is quasi-compact. Then there exists $n \in \mathbb{Z}_{\geq 0}$ such that for every $y \in Y$, $\dim f^{-1}(y) \leq n$.*

Proof. We may assume that $Y = \operatorname{Spec} A$ and $X = \operatorname{Spec} B$ are affine. Then B is a finitely generated A-algebra, and if there exists a system of n generators, then for every $y \in Y$, the $\kappa(y)$-algebra $B \otimes_A \kappa(y)$ admits a system of n generators as a $\kappa(y)$-algebra. This implies that $\dim f^{-1}(y) = \dim \operatorname{Spec} B \otimes_A \kappa(y) \leq n$. $\qquad\square$

The dimension formula Theorem 14.106 has the following consequences for morphisms of schemes:

Lemma 14.108.

(1) *Let X, Y be irreducible locally noetherian schemes, let $f \colon X \to Y$ be a dominant morphism locally of finite type, and denote by ξ and η, resp., the generic points of X and Y. Let $e = \dim f^{-1}(\eta) = \operatorname{trdeg}_{\kappa(\eta)} \kappa(\xi)$ be the dimension of the generic fiber. Let $x \in X$ and $y = f(x)$. Then*

$$(14.26.1) \qquad e + \dim \mathscr{O}_{Y,y} \geq \operatorname{trdeg}_{\kappa(y)} \kappa(x) + \dim \mathscr{O}_{X,x}$$

(2) *If in addition Y is universally catenary (e. g., if Y is locally of finite type over a field), then equality holds in (14.26.1).*

(3) *If Y is universally catenary and x is closed in $f^{-1}(y)$, then*

$$\dim \mathscr{O}_{X,x} = \dim \mathscr{O}_{Y,y} + e.$$

Proof. We may assume that X and Y are affine (hence f is of finite type), and reduced. Because f is dominant, it corresponds to an injection $A \to B$ of integral domains, and the morphism $\mathscr{O}_{Y,y} \to \mathscr{O}_{X,x}$ induced on the stalks is injective, too.

Now we apply the dimension formula Theorem 14.106 (1) to $\mathscr{O}_{Y,y}$, B, and the prime ideal of B corresponding to x. This gives us

$$\dim \mathscr{O}_{Y,y} + \operatorname{trdeg}_{\mathscr{O}_{Y,y}} B \geq \dim \mathscr{O}_{X,x} + \operatorname{trdeg}_{\kappa(y)} \kappa(x).$$

Since $e = \dim f^{-1}(\eta) = \operatorname{trdeg}_{\operatorname{Frac}(\mathscr{O}_{Y,y})} \operatorname{Frac}(B \otimes_A \operatorname{Frac}(\mathscr{O}_{Y,y})) = \operatorname{trdeg}_{\operatorname{Frac}(\mathscr{O}_{Y,y})} \operatorname{Frac}(B)$, this is the same as (14.26.1).

If Y is universally catenary, by Theorem 14.106 (2) we even have equality. If in addition x is closed in the fiber $f^{-1}(y)$, then $\kappa(x)$ is a finite extension of $\kappa(y)$ (because $f^{-1}(y)$ is a $\kappa(y)$-scheme locally of finite type), and therefore $\operatorname{trdeg}_{\kappa(y)} \kappa(x) = 0$. $\qquad\square$

Proposition 14.109. *Let Y be a noetherian scheme, and let $f\colon X \to Y$ be a morphism of finite type.*

(1) Let $n \in \mathbb{Z}_{\geq 0}$ be such that $\dim f^{-1}(y) \leq n$ for all $y \in Y$, then

$$\dim X \leq \dim Y + n.$$

(2) Assume in addition that X and Y are irreducible and that Y is universally catenary, and that f is surjective. Then

$$\dim X = \dim Y + \dim f^{-1}(\eta),$$

where η is the generic point of Y.

Proof. As every point in the noetherian scheme X has a specialization that is a closed point in X, we find

$$(14.26.2) \qquad\qquad \dim X = \sup_{x \text{ closed}} \dim \mathscr{O}_{X,x}.$$

After this remark let us prove (1). Let X_i, $i = 1, \ldots, n$, be the irreducible components of X, viewed as closed integral subschemes of X. For each i, let Z_i be the schematic image of X_i, i.e., the integral closed subscheme of Y whose underlying topological space is the closure of $f(X_i)$, so that f induces dominant morphisms $X_i \to Z_i$.

The restrictions $X_i \to Z_i$ of f again satisfy the condition on the fiber dimension, so it follows from Lemma 14.108 (1), applied to the closed points $x \in X_i$, using (14.26.2), that

$$\dim X_i \leq n + \dim Z_i \leq n + \dim Y.$$

Therefore $\dim X \leq n + \dim Y$.

To prove (2) set $e := \dim f^{-1}(\eta)$. We find by Lemma 14.108 (3)

$$e + \dim Y = e + \sup_{y \in Y} \dim \mathscr{O}_{Y,y} = \sup_{y \in Y} \sup_{\substack{x \in f^{-1}(y) \\ \text{closed}}} \dim \mathscr{O}_{X,x}$$

The right hand side is $\leq \dim X$ and $\geq \sup_{x \text{ closed}} \dim \mathscr{O}_{X,x}$ because if x is closed in X, then x is closed in $f^{-1}(f(x))$. Thus by (14.26.2) the right hand side is equal to $\dim X$. $\square$

Using Exercise 5.5 one can prove the proposition under the more general assumption that Y is locally noetherian and that f is locally of finite type.

Remark 14.110. The proof shows that Proposition 14.109 (2) still holds if f is only dominant and $\dim Y = \sup_{y \in f(X)} \dim \mathscr{O}_{Y,y}$. This is for instance the case for all dominant morphisms if Y is of finite type over a field: As $f(X)$ is a constructible and dense, it contains a subset V that is dense and open in Y and by Theorem 5.22 (3) one has $\dim V = \dim Y$.

(14.27) Semi-continuity of fiber dimensions.

Now we prove that the fiber dimension is an upper semi-continuous function (on the *source* of a morphism), a result due to Chevalley. We start with a lemma:

Lemma 14.111. *Let Y be an irreducible locally noetherian scheme, let X be an irreducible scheme, and let $f\colon X \to Y$ be a dominant morphism of finite type. Denote by η the generic point of Y, and let $e = \dim f^{-1}(\eta)$. Then for every point $x \in X$, all irreducible components of $f^{-1}(f(x))$ have dimension $\geq e$.*

We will prove this lemma only if Y is in addition universally catenary. Note that the proof of the main result in this section (Theorem 14.112) uses the lemma only in this case. The general case can be deduced from this using a limit argument (as in Theorem 10.69) to reduce to the case that $Y = \operatorname{Spec} A$, where A is a finitely generated $\mathbb{Z}$-algebra (see [EGAIV] (13.1.1) for the details). Then Y is universally catenary by Proposition 14.105.

Proof. (if Y is universally catenary) Let $y \in f(X)$. As the local ring $\mathcal{O}_{Y,y}$ is universally catenary, we can apply Lemma 14.108 (2). We let Z be an irreducible component of $f^{-1}(y)$, and denote by x the generic point of Z. Then (14.26.1) reads

$$e + \dim \mathcal{O}_{Y,y} = \operatorname{trdeg}_{\kappa(y)} \kappa(x) + \dim \mathcal{O}_{X,x}.$$

Since x corresponds to a minimal prime ideal of $\mathcal{O}_{X,x} \otimes \kappa(y)$, we have $\dim \mathcal{O}_{X,x} \otimes \kappa(y) = 0$, and hence $\dim \mathcal{O}_{X,x} \leq \dim \mathcal{O}_{Y,y}$ by Proposition B.68. Furthermore, we have $\dim Z = \operatorname{trdeg}_{\kappa(y)} \kappa(x)$ (see Theorem 5.22). The Lemma is proved. $\qquad\square$

The next theorem is Chevalley's theorem about the semi-continuity of fiber dimensions: the fiber dimension of a morphism (locally of finite type) jumps up on closed subsets.

Theorem 14.112. *Let $f\colon X \to Y$ be a morphism locally of finite type. Then the function*

$$X \to \mathbb{Z}_{\geq 0}, \qquad x \mapsto \dim_x f^{-1}(f(x))$$

is upper semi-continuous, i.e., for every n the set

$$F_n = F_n(X) = \{\, x \in X \;;\; \dim_x f^{-1}(f(x)) \geq n \,\}$$

is closed.

We prove the result under the assumption that f is locally of finite presentation. The general case can be reduced to this case using an approximation argument in the spirit of Chapter 10, see [EGAIV] (13.1.3).

Proof. (if f is locally of finite presentation) Obviously, we may assume that $Y = \operatorname{Spec} A$ and $X = \operatorname{Spec} B$ are affine. We know, see Section (10.18), that we can find a subring $A_0 \subseteq A$, finitely generated over $\mathbb{Z}$, and an A_0-algebra B_0 of finite type, such that $B = A \otimes_{A_0} B_0$. Denote by $g\colon \operatorname{Spec} A \to \operatorname{Spec} A_0$, $g'\colon \operatorname{Spec} B \to \operatorname{Spec} B_0$ the canonical morphisms. We denote by $f_0\colon \operatorname{Spec} B_0 \to \operatorname{Spec} A_0$ the morphism induced from the A_0-algebra structure of B_0.

Let $y \in Y = \operatorname{Spec} A$, $y_0 = g(y)$. Then $f^{-1}(y) = f_0^{-1}(y_0) \times_{\operatorname{Spec} \kappa(y_0)} \operatorname{Spec} \kappa(y)$. Let $x \in f^{-1}(y)$, $x_0 = g'(x)$. Then $\dim_x f^{-1}(y)$ is the supremum over the dimensions of the irreducible components of $f^{-1}(y)$ containing x. Since the projection $f^{-1}(y) \to f_0^{-1}(y_0)$ is open, every such irreducible component Z dominates an irreducible component of $f_0^{-1}(y_0)$ (which obviously contains x_0), which has the same dimension as Z by Proposition 5.38. This shows that $\dim_x f^{-1}(y) = \dim_{x_0} f_0^{-1}(y_0)$, and hence $F_n(X) = g'^{-1}(F_n(\operatorname{Spec} B_0))$.

Therefore now we assume that A is a finitely generated $\mathbb{Z}$-algebra and B is an A-algebra of finite type. In particular A is universally catenary by Proposition 14.105. Since the dimension is a topological property, we may assume that $X = \operatorname{Spec} B$ and $Y = \operatorname{Spec} A$ are reduced, and by noetherian induction we may assume that the statement is true for all morphisms with target Y' a proper closed subscheme of Y.

Write $X_1, \ldots, X_n$ for the irreducible components of X, provided with the reduced scheme structure. Then $F_n(X) = \bigcup_i F_n(X_i)$, and therefore we may assume that X is irreducible. Since the morphism f factors through its schematic image, a closed subscheme of Y, we may also assume that f is dominant, and that Y is irreducible. Denote by η the generic point of Y.

Now if $n \leq \dim f^{-1}(\eta)$, then by Lemma 14.111 we have $F_n(X) = X$, and we are done. On the other hand, if $n > \dim f^{-1}(\eta)$, there exists an open neighborhood $U \subseteq Y$ of η with $F_n \subseteq f^{-1}(Y \setminus U)$ by Proposition 10.95. Replacing the morphism f by its restriction $f^{-1}(Y \setminus U) \to Y \setminus U$, we are done by noetherian induction. $\qquad\square$

Since a point x is isolated in its fiber $f^{-1}(f(x))$ if and only if $\dim_x f^{-1}(f(x)) = 0$, we immediately get a new proof of Corollary 12.79:

Corollary 14.113. *Let $f\colon X \to Y$ be a morphism locally of finite type. Then the set of points $x \in X$ which are isolated in their fiber, is open in X.*

Remark 14.114. The proof of Theorem 14.112 shows that if f is locally of finite presentation, then for every n the open immersion $\{\, x \in X \;;\; \dim_x f^{-1}(f(x)) \leq n \,\} \hookrightarrow X$ is quasi-compact.

For proper morphisms the fiber dimension is also upper semi-continuous on the target:

Corollary 14.115. *Let $f\colon X \to Y$ be a closed morphism locally of finite type. Then for every n the subset*
$$Y_n := \{\, y \in Y \;;\; \dim f^{-1}(y) \geq n \,\}$$
is closed in Y.

Proof. The set Y_n is the image under f of the closed set $F_n(X)$ of Theorem 14.112. As f is closed, Y_n is closed. $\qquad\square$

For general morphisms $f\colon X \to Y$ of finite type the function $y \mapsto \dim f^{-1}(y)$ is not necessarily upper semi-continuous, even if $Y = \operatorname{Spec} R$ for a discrete valuation ring R and f is faithfully flat (Exercise 14.24).

Theorem 14.116. *Let $f\colon X \to Y$ be an open morphism locally of finite type. Suppose that Y is universally catenary, irreducible and noetherian, that $\dim Y < \infty$, and that X is equidimensional. Assume that the following condition is satisfied.*
  (D) *For every irreducible component X' of X one has*
$$\dim Y = \sup_{y \in f(X')} \dim \mathcal{O}_{Y,y}.$$
Then for all $y \in f(X)$ the fiber $f^{-1}(y)$ is equidimensional and
$$\dim X = \dim Y + \dim f^{-1}(y).$$

Any flat morphism locally of finite type between locally noetherian schemes is open (Theorem 14.35).

Remark 14.117. Under the other hypotheses of Theorem 14.116 Condition (D) is satisfied in the following cases.

(1) The restriction of f to every irreducible component of X is surjective: This is clear because $\dim Y = \sup_{y \in Y} \dim \mathcal{O}_{Y,y}$ (Lemma 5.7 (4)).

   In particular, Condition (D) holds if f is surjective and X is irreducible.

(2) f is proper (or, more generally, closed): As f is open, the restriction of f to each irreducible component is dominant. Therefore it is surjective if f is closed. Now use (1).

(3) Y is of finite type over a field (in this case Y is automatically universally catenary, noetherian and of finite dimension): As explained in (2), for each irreducible component X' its image $f(X')$ is constructible and dense and thus contains a subset V that is dense and open in Y. Then $\dim V \le \sup_{y \in f(X')} \dim \mathcal{O}_{Y,y} \le \dim Y$ and we conclude because $\dim V = \dim Y$ by Theorem 5.22 (3).

(4) Y is of finite type over a scheme S that satisfies the following two conditions.
   (a) S is a universally catenary locally noetherian Jacobson scheme (Exercise 10.16).
   (b) For every irreducible component S' of S and every closed point $s \in S'$ one has $\dim \mathcal{O}_{S',s} = \dim S'$.

   Indeed, then Y also satisfies conditions (a) and (b) ([EGAIV] (10.6.1)) and one can argue as in (3) using [EGAIV] (10.6.2) instead of Theorem 5.22 (3).

   If S is a noetherian scheme of dimension ≤ 1 such that every irreducible component of dimension 1 has infinitely many points, then S satisfies conditions (a) and (b) ([EGAIV] (10.7.1)). In particular, Condition (D) of the theorem is satisfied if Y is of finite type over $\operatorname{Spec} \mathbb{Z}$.

In general, Condition (D) cannot be omitted, even if f is faithfully flat and X is connected (Exercise 14.24).

Proof. *(i).* Let $\eta \in Y$ be the generic point. Let X' be an irreducible component of X and let θ be its generic point. As f is open, $f(\theta) = \eta$. Thus $\{\theta\}$ is dense in $f^{-1}(\eta) \cap X'$ which is therefore irreducible. As $f^{-1}(\eta) \cap X'$ is of finite type over the field $\kappa(\eta)$, we have $\dim(f^{-1}(\eta) \cap X') = \dim_\theta f^{-1}(\eta)$. By Condition (D) und using Remark 14.110 we can apply Proposition 14.109 (2) to $f_{|X'}$ and obtain

$$\dim Y + \dim_\theta f^{-1}(\eta) = \dim X' = \dim X,$$

where the second equality holds because X is equidimensional. Letting X' vary we obtain that $\dim X < \infty$ and because of the natural bijection between the sets of irreducible components of X and of $f^{-1}(\eta)$, $f^{-1}(\eta)$ is equidimensional of dimension $\dim X - \dim Y$.

(ii). Let $x \in X$ and write $y = f(x)$. By (i) it suffices to show

$$\dim_x f^{-1}(y) = \dim_\theta f^{-1}(\eta)$$

for some maximal point θ of X. Let $X_1, \dots, X_r$ be the irreducible components of X that contain x. Then $\dim \mathcal{O}_{X,x} = \max_i \dim \mathcal{O}_{X_i,x}$ and we can choose an irreducible component X' containing x such that $\dim_{X',x} = \dim \mathcal{O}_{X,x}$. Let θ be its generic point.

From Theorem 14.112 we have $\dim_x f^{-1}(y) \ge \dim_\theta f^{-1}(\eta)$, so it remains to prove the other inequality. From Lemma 14.108 (1), we have

$$\dim \mathcal{O}_{X',x} - \dim \mathcal{O}_{Y,y} + \operatorname{trdeg}_{\kappa(y)} \kappa(x) \le \dim_\theta f^{-1}(\eta).$$

By our assumption on X' it holds for $\mathscr{O}_{X,x}$ instead of $\mathscr{O}_{X',x}$, as well. The openness implies that $\dim \mathscr{O}_{Y,y} + \dim \mathscr{O}_{X,x} \otimes_{\mathscr{O}_{Y,y}} \kappa(y) = \dim \mathscr{O}_{X,x}$ (Remark 14.98), so that we can rewrite the left hand side as $\dim \mathscr{O}_{X,x} \otimes_{\mathscr{O}_{Y,y}} \kappa(y) + \mathrm{trdeg}_{\kappa(y)} \kappa(x)$. Therefore the claim follows from Lemma 14.96. $\square$

We will see later (Corollary 14.130 below) that conversely, under a regularity assumption the dimension equality of the theorem (for all y) implies that f is flat. Note that we get a new proof of Theorem 14.100 about the dimension of affine space $\mathbb{A}_Y^n$ (which however also relies on Proposition B.68, which was the main ingredient of our first proof).

Corollary 14.118. *Let $f\colon X \to Y$ be a dominant morphism of finite type. Assume that X is irreducible and that Y is noetherian, universally catenary, irreducible, and of finite dimension. There exists an open dense subscheme $V \subseteq Y$ contained in $f(X)$ such that the following assertions hold.*

(1) For all $y \in V$ the fiber $f^{-1}(y)$ is equidimensional and $\dim f^{-1}(V) = \dim f^{-1}(y) + \dim V$.

(2) For all $y \in Y$ and every irreducible component Z of $f^{-1}(y)$ one has $\dim Z + \dim V \geq \dim f^{-1}(V)$.

Proof. We may assume that Y is integral. Let η be its generic point. By generic flatness (Corollary 10.85) there exists an open dense subscheme $V \subseteq Y$ such that the restriction $f_V\colon f^{-1}(V) \to V$ of f is flat. As $f(X)$ is constructible and dense, we may shrink V such that $V \subseteq f(X)$. Then f_V is faithfully flat and (1) follows from Theorem 14.116.

To show (2) note that (1) in particular implies $\dim f^{-1}(V) = \dim f^{-1}(\eta) + \dim V$. Hence (2) follows from Lemma 14.111. $\square$

Remark 14.119. Let $f\colon X \to Y$ be a dominant morphism of finite type of irreducible schemes. If Y is of finite type over a field k, then Y satisfies all the assumptions in Corollary 14.118. Moreover, X is also of finite type over k and therefore $\dim V = \dim Y$ and $\dim f^{-1}(V) = X$ by Theorem 5.22 (3).

More generally, we have $\dim V = \dim Y$ and $\dim f^{-1}(V) = X$ if Y is of finite type over a scheme S satisfying the conditions in Remark 14.117 (4) using [EGAIV] (10.6.2). This is the case if Y is of finite type over $\mathbb{Z}$.

Corollary 14.120. *Let Y be a scheme of finite type over a field and let $r \geq 0$ be an integer. Let $f\colon X \to Y$ be a morphism of finite type such that $\dim f^{-1}(\eta) \geq r$ for every maximal point η of Y. Then $\dim X \geq r + \dim Y$.*

Proof. Let Y' be an irreducible component of Y such that $\dim Y' = \dim Y$. Then the restriction $f^{-1}(Y') \to Y'$ is dominant and $\dim f^{-1}(Y') \leq \dim X$. Thus we may assume that Y is irreducible with generic point η. Let $X_1, \ldots, X_r$ be the irreducible components of X. Removing all irreducible components that do not meet $f^{-1}(\eta)$, we may assume that each X_i dominates Y. If we had $\dim X_i < \dim Y + r$ for all i, then Corollary 14.118 (1) and Remark 14.119, applied to the restriction $X_i \to Y$ of f, show that $f^{-1}(\eta) \cap X_i$ has dimension $< r$ for all i which contradicts the hypothesis $\dim f^{-1}(\eta) \geq r$. $\square$

Corollary 14.121. *Let Y be a scheme of finite type over a field, and let $f\colon X \to Y$ be a dominant morphism of finite type such that all non-empty fibers of f have dimension r. Then $\dim X = r + \dim Y$.*

Proof. By Proposition 14.109 (1) we have $\dim X \leq r + \dim Y$. As f is dominant, all fibers of maximal points of Y are non-empty. Hence the equality follows from Corollary 14.120. $\square$

Remark 14.122. Again Corollary 14.120 and Corollary 14.121 hold more generally if Y is of finite type over a scheme S satisfying the conditions in Remark 14.117 (4), e.g., if Y is of finite type over $\mathbb{Z}$.

Proposition 14.123. *Let $f \colon X \to Y$ be a proper flat morphism of finite presentation and set $V := \{\, y \in Y \;;\; \dim f^{-1}(y) = 0 \,\}$. Then V is open and closed in Y and the restriction $f^{-1}(V) \to V$ of f is finite locally free.*

Proof. This question is local on Y and thus we may assume that Y is affine. Since f is of finite presentation and all properties of f are compatible with inductive limits of rings, we may assume, by Theorem 10.69, that Y and hence X is noetherian. Since f is open and closed, we can restrict to the case that f is surjective.

Set $W := Y \setminus V$. Then W is the image of the set $F_1(X)$ of Theorem 14.112, so by that theorem and since f is proper, it is closed in Y. Thus V is open in Y.

Let us show that V is also closed. By Lemma 10.17 it is enough to show that it is closed under specialization. By Proposition 15.7 below, it is enough to show the following statement: Let T be the spectrum of a discrete valuation ring, $\eta \in T$ the generic point, $t \in T$ the closed point, and $f \colon Z \to T$ a proper flat morphism such that the fiber $f^{-1}(\eta)$ has dimension 0. Then the special fiber $f^{-1}(t)$ has dimension 0, too. In fact, this is easily checked directly. We give an argument using our previous results, instead: Since Z_{red} is again flat over T (use Proposition 14.14), and since the dimensions do not change, we may assume that Z is reduced. Applying Proposition 14.14 again, we see that under this assumption all irreducible components of Z are also proper and flat over T. Therefore we can apply Theorem 14.116 to each of these and we see that $\dim f^{-1}(t) = \dim f^{-1}(\eta) = 0$. This proves that V is closed.

The restriction $f^{-1}(V) \to V$ is proper, flat, and of finite presentation. By definition it is quasi-finite and hence finite by Corollary 12.89. Thus it is finite locally free (Proposition 12.19). $\square$

Dimension and regularity conditions

(14.28) Cohen-Macaulay schemes.

We have seen (Theorem 14.116) that the dimension of the fibers of a flat morphism is locally constant. In the next section, we will show that under a suitable hypothesis, e.g., if the source and the target are regular, the converse is true: Then we can check flatness just by looking at the dimensions of the fibers. To prove this result, one uses an induction during which non-regular schemes arise. Therefore we introduce the notion of *Cohen-Macaulay* schemes which is better adapted to this purpose. We refer to Definition B.79 for the definition of Cohen-Macaulay rings.

Definition 14.124. *A locally noetherian scheme X is called* Cohen-Macaulay, *if all the local rings $\mathcal{O}_{X,x}$, $x \in X$, are Cohen-Macaulay.*

Since fields are obviously Cohen-Macaulay, Proposition B.86 shows that every regular ring is Cohen-Macaulay, so we get:

Proposition 14.125. *Every regular scheme is Cohen-Macaulay.*

It is hard to fully capture the condition of being Cohen-Macaulay in geometric terms, but we note the following consequences:

Proposition 14.126. *Let X be a locally noetherian Cohen-Macaulay scheme.*
(1) *The scheme X is universally catenary.*
(2) *If X is generically reduced, then X is reduced. More precisely, X does not have embedded components.*
(3) *Assume that X is connected. Then X is connected in codimension 2, i.e., if $Z \subseteq X$ is a closed subscheme of codimension $\operatorname{codim}_X Z \geq 2$, then $X \setminus Z$ is connected.*
(4) *If in addition X is locally of finite type over a field k, then it is equidimensional.*

Proof. It follows from Proposition 14.105 that X is universally catenary. To prove (2), assume that $A = \mathcal{O}_{X,x}$ is a local ring of X. By assumption it is generically reduced, i.e., all localizations with respect to minimal prime ideals are reduced. Let $\mathfrak{p}$ be an associated prime ideal of A (Section (B.11)), and let $\mathfrak{q} \subseteq \mathfrak{p}$ be a minimal prime ideal of A. Then

$$\dim A/\mathfrak{p} \geq \operatorname{depth} A/\mathfrak{p} = \operatorname{depth} A = \dim A \geq \dim A/\mathfrak{q},$$

so $\mathfrak{q} = \mathfrak{p}$, and we see that all associated prime ideals are minimal. Exercise 9.16 then implies that A is reduced.

Let us prove (3). Assume that $X \setminus Z$ is the union of two disjoint open and closed subsets X_1 and X_2. Let $X_i' \subset X$ be a closed subscheme such that $X_i = X_i' \setminus Z$, $i = 1, 2$. We may assume that $Z = X_1' \cap X_2'$ (set-theoretically) by enlarging the X_i', if necessary. Let $z \in X_1' \cap X_2'$, and $A = \mathcal{O}_{X,z}$. Denote by $\mathfrak{a}_i \subset A$ the ideal corresponding to X_i'. We then have $A/(\mathfrak{a}_1 + \mathfrak{a}_2) = \mathcal{O}_{Z,z}$, and $\mathfrak{a}_1 \cap \mathfrak{a}_2$ is nilpotent. We would like to show that $\dim \mathcal{O}_{Z,z} \geq \dim \mathcal{O}_{X,z} - 1$ which by Proposition B.84 (2) means that we want to show that $\operatorname{depth}(\mathfrak{a}_1 + \mathfrak{a}_2) \leq 1$. But this follows from Hartshorne's connectedness theorem, Proposition B.85.

Now let us assume that X is of finite type over a field k. Because every closed point of an irreducible component $Z \subseteq X$ has codimension $\dim Z$ in Z, it is enough to show that for every local ring $\mathcal{O}_{X,x}$, all $\operatorname{Spec} \mathcal{O}_{X,x}/\mathfrak{p}$, where $\mathfrak{p} \subset \mathcal{O}_{X,x}$ is a minimal prime ideal, have the same dimension. But it follows from the reasoning we used to prove (2) that all these dimensions agree with $\dim \mathcal{O}_{X,x}$. $\square$

The third property shows that the union of two planes which intersect in a single point is not Cohen-Macaulay.

(14.29) Equidimensional morphisms between regular schemes are flat.

Lemma 14.127. *Let $\varphi \colon A \to B$ be a local homomorphism of local noetherian rings, let k denote the residue class field of A. Assume that A is regular, that B is Cohen-Macaulay, and that $\dim B = \dim A + \dim B \otimes_A k$. Then φ is flat.*

Proof. We prove the lemma by induction on $\dim A$. If $\dim A = 0$, then A is a field, so B is flat over A. Now assume that $\dim A > 0$. Denote by $\mathfrak{m}$ the maximal ideal of A, and let $x \in \mathfrak{m} \setminus \mathfrak{m}^2$. Note that A/xA is regular of dimension $\dim A - 1$ by Proposition B.77 (3). By Proposition B.68, we have

$$\dim B/xB \leq \dim A/xA + \dim B \otimes_A k = \dim A - 1 + \dim B \otimes_A k = \dim B - 1.$$

Furthermore, $\dim B/xB \geq \dim B - 1$ by Corollary B.64, so we have equality here, and hence $\dim B/xB = \dim A/xA + \dim(B/xB) \otimes_{A/xA} k$. In particular, x is a regular element of B, hence B/xB is Cohen-Macaulay by Proposition B.86, and by induction hypothesis B/xB is flat over A/xA. Now it follows from Lemma 14.23 that B is flat over A, as desired. $\qquad\square$

Theorem 14.128. *Let $f\colon X \to Y$ be a morphism of locally noetherian schemes which is locally of finite type, and let $x \in X$, $y = f(x)$. Assume that $\mathscr{O}_{Y,y}$ is regular, $\mathscr{O}_{X,x}$ is Cohen-Macaulay, and that*

$$\dim \mathscr{O}_{X,x} = \dim \mathscr{O}_{Y,y} + \dim \mathscr{O}_{X,x} \otimes_{\mathscr{O}_{Y,y}} \kappa(y).$$

Then f is flat at x.

Proof. Apply Lemma 14.127 to the local rings of Y and X. $\qquad\square$

Under additional assumptions, the conditions of the theorem are particularly easy to check, or even automatically satisfied:

Corollary 14.129. *Let $f\colon X \to Y$ be a finite dominant morphism between regular schemes. Then f is flat, and hence finite locally free.*

Proof. We can apply Theorem 14.128: For every $x \in X$ and $y := f(x)$, the induced morphism on stalks $f_x^\sharp\colon \mathscr{O}_{Y,y} \to \mathscr{O}_{X,x}$ is injective because f is dominant and Y is reduced (Corollary 2.11). As f is finite, we have $\dim \mathscr{O}_{X,x} = \dim \mathscr{O}_{Y,y}$, and $\dim \mathscr{O}_{X,x} \otimes_{\mathscr{O}_{Y,y}} \kappa(y) = 0$ by Theorem B.56. It follows that f is flat, and by Proposition 12.19 that f is finite locally free. $\qquad\square$

Corollary 14.130. *Let k be a field, let X, Y be k-schemes locally of finite type, and let $f\colon X \to Y$ be a morphism. Assume that Y is regular, and that X is Cohen-Macaulay (e.g., if X is regular). Assume that X is equidimensional of dimension e, that Y is equidimensional of dimension d, and that all fibers $f^{-1}(y)$, $y \in Y$, are equidimensional of dimension $e - d$. Then f is flat.*

Proof. By Proposition 10.7 (3), f is locally of finite type. We apply the theorem to the closed points $x \in X$. The assumptions of the theorem are satisfied, because $y = f(x)$ is again closed, so $\dim \mathscr{O}_{X,x} = \dim X$, $\dim \mathscr{O}_{Y,y} = \dim Y$, and because $\dim \mathscr{O}_{X,x} \otimes_{\mathscr{O}_{Y,y}} \kappa(y) = \dim_x f^{-1}(y)$ (Lemma 14.96). $\qquad\square$

If we make weaker assumptions on the singularities of X and Y, we cannot expect f to be flat. But we still have the following theorem, essentially due to Chevalley:

Theorem 14.131. *Let X and Y be irreducible locally noetherian schemes and assume that Y is normal with generic point η. Let $f\colon X \to Y$ be a dominant morphism locally of finite type such that $\dim_x f^{-1}(f(x)) = \dim f^{-1}(\eta)$ for all $x \in X$. Then f is universally open. If in addition, all fibers of f are geometrically reduced, then f is flat.*

We will not give a proof here. The first assertion is proved in [EGAIV] (14.4.4). The second then follows from Theorem 14.37.

We conclude this part of the chapter with two applications to projective schemes over fields.

(14.30) Embeddings into low-dimensional projective spaces.

A scheme X over a field k is projective if and only if X is isomorphic to a closed subscheme of $\mathbb{P}_k^N$. The following theorem gives an upper bound for how small N can be chosen if X is smooth over k.

Theorem 14.132. *Let k be an infinite field, let X be a k-scheme which is projective and smooth over k, and let $d := \dim X$. Then there exists a closed immersion $X \hookrightarrow \mathbb{P}_k^{2d+1}$.*

Proof. We may assume that X is a closed subscheme of $\mathbb{P}_k^N$ for some N. Let $H \subset \mathbb{P}_k^N$ be a linear subspace of dimension $N - 1$ and let U be the open subscheme $\mathbb{P}_k^N \setminus (X \cup H)$. The theorem is proved if for $N > 2d + 1$ we can find a point $q \in U(k)$ such that the restriction $\pi_{q|X} \colon X \to H = \mathbb{P}_k^{N-1}$ of the projection with center q (see Example 1.66) is a closed immersion. In fact we will show that there exists a non-empty open subscheme V of U such that $\pi_{q|X}$ is a closed immersion for all $q \in V(k)$ (this implies the theorem because $V \neq \emptyset$ implies $V(k) \neq \emptyset$ for an open subscheme V of $\mathbb{P}_k^N$ if k is infinite).

Let $\bar{k}$ be an algebraic closure of k. For $q \in U(\bar{k})$ the restriction of $\pi_q \colon \mathbb{P}_{\bar{k}}^N \setminus \{q\} \to H \otimes_k \bar{k}$ to $X \otimes_k \bar{k}$ is again denoted by π_q. We will show that there exist closed subschemes $Y_1, Y_2 \subset U$ with $\dim Y_i < \dim U$ such that the following conditions hold.
(a) Let $q \in U(\bar{k})$ be such that the projection $\pi_q(\bar{k}) \colon X(\bar{k}) \to H(\bar{k})$ is not injective. Then $q \in Y_1(\bar{k})$.
(b) Let $q \in U(\bar{k})$ be such that the map $d\pi_{q,x} \colon T_x(X/k) \to T_{\pi_q(x)}(H/k)$ on tangent spaces induced by the projection π_q is not injective for some $x \in X(\bar{k})$. Then $q \in Y_2(\bar{k})$.
Then Proposition 12.94 shows that we may take $V := U \setminus (Y_1 \cup Y_2)$.

Let C be the reduced subscheme of $U \times_k X \times_k X$ such that $C(\bar{k})$ consists of those triples (u, x, y) with $x \neq y$ and which are collinear as points in $\mathbb{P}^N(\bar{k})$, see Example 4.41. Denote by $\mathrm{pr}_1 \colon U \times_k X \times_k X \to U$ the projection and let Y_1 be the closure of $\mathrm{pr}_1(C)$ (viewed as a closed reduced subscheme of U). Then Y_1 satisfies condition (a). Let us show that $\dim Y_1 < \dim U$. It suffices to show that $\dim C < \dim U$ (apply Theorem 5.22 to the irreducible components of Y_1). Let $\mathrm{pr}_2 \colon C \to X \times_k X$ be the restriction to C of the projection to the last two factors. For $(x, y) \in X(\bar{k}) \times X(\bar{k})$ with $x \neq y$ the $\bar{k}$-valued points of the fiber $\mathrm{pr}_2^{-1}(x, y)$ can be identified with $U(\bar{k}) \cap \overline{xy}$, where $\overline{xy}$ denotes the line through x and y. Thus $\dim(\mathrm{pr}_2^{-1}(x, y)) \leq 1$ for all $(x, y) \in X(\bar{k}) \times X(\bar{k})$ (note that by definition of C the fiber $\mathrm{pr}_2^{-1}(x, x)$ over points of the diagonal is empty). By Proposition 14.109 (1), we see that $\dim C \leq 2d + 1 < N = \dim U$.

Let $T \subset U \times_k X$ be the closed subscheme such that $T(\bar{k})$ consists of pairs (u, x) such that $u \in T_x(X \subset \mathbb{P}^N/k)$, the projective tangent space in x, see Section (6.7). It is not difficult to see that the set of these pairs indeed is the set of $\bar{k}$-valued points of a closed subscheme of $U \times_k X$. As X is proper over k, the projection $\mathrm{pr}_1 \colon U \times X \to U$ is closed and thus $Y_2 := \mathrm{pr}_1(T)$ is closed and we may view Y_2 as a closed reduced subscheme of U. If there exists an $x \in X(\bar{k})$ such that $d\pi_{q,x}$ is not injective, then $q \in Y_2(\bar{k})$. Thus Y_2 satisfies condition (b). To see that $\dim Y_2 < \dim U = N$ we consider again the second projection $\mathrm{pr}_2 \colon T \to X$. For all $x \in X(\bar{k})$ the fiber $\mathrm{pr}_2^{-1}(x)$ is contained in $T_x(X \subset \mathbb{P}^N/k)$. Thus we see that the fibers have at most dimension $\dim_{\bar{k}} T_x(X \subset \mathbb{P}^N/k) = \dim_{\bar{k}} T_x(X/k) = d$

because we assumed that X is smooth over k. Again Proposition 14.109 shows that $\dim T \leq 2d$ and hence $\dim Y_2 \leq \dim T < N$. $\qquad\square$

Remark 14.133. The proof shows the following more general result. Let X be a projective scheme, $d := \dim X$ and $\delta := \sup_{x \in X(\bar{k})} \dim T_x(X/k)$. Then there exists a closed immersion $X \hookrightarrow \mathbb{P}_k^n$ with $n := \max(2d+1, d+\delta)$.

Conversely, for every $d > 0$ there exist smooth projective k-varieties of dimension d which cannot be embedded into $\mathbb{P}_k^{2d}$.

(14.31) Degree of subschemes of projective space.

Let k be a field, let $n \geq 1$ an integer, and let $X \subseteq \mathbb{P}_k^n$ be a closed subscheme of dimension $d \geq 0$. If X is a hypersurface, $X = V_+(f)$ for a non-constant homogeneous polynomial, we defined the degree of X to be the degree of f, see Section (5.11). We now want to define the degree of X for an arbitrary closed subscheme of $\mathbb{P}_k^n$. Roughly speaking, we define the degree as follows: Since X has dimension d, we expect that a sufficiently general $(n-d)$-dimensional linear subspace has 0-dimensional intersection with X, i. e. the intersection is just a finite number of points. Furthermore, it turns out that the number of points (in a "general" situation) depends only on X, and $\deg X$ will be defined as this number. The concept of generic points is the perfect tool to make this precise without requiring a big effort.

Fix an integer m with $0 \leq m \leq n$ and let $L_m := \operatorname{LinSub}_m(\mathbb{P}_k^n)$ be the scheme of linear subspaces of $\mathbb{P}_k^n$ of dimension m (Section (8.8)). Recall that $L_m \cong \operatorname{Grass}_{m+1,n+1}$ is geometrically integral, projective and smooth of relative dimension $(m+1)(n-m)$ over k. In Remark 13.86 we defined the incidence scheme H_m^X whose k-valued points are pairs (x, Λ), where $x \in X(k)$ and $\Lambda \in L_m(k)$ such that $x \in \Lambda$. We denote by $p \colon H_m^X \to X$ and by $q \colon H_m^X \to L_m$ the two projections.

The fiber of q over $\Lambda \in L_m(k)$ is then the k-scheme $X \cap \Lambda$. The "generic linear subspace of dimension m" is simply the canonical morphism $\operatorname{Spec} \kappa(\eta) \to L_m$, where η is the generic point of L_m. Therefore we consider the $\kappa(\eta)$-scheme $X_m := q^{-1}(\eta)$ as the "intersection of X with a generic linear subspace of dimension m".

Proposition 14.134. *The morphism* $p \colon H_m^X \to X$ *is surjective, projective, smooth of relative dimension* $m(n-m)$ *with geometrically integral fibers. If X is irreducible, then H_m^X is irreducible and X_m is either empty or irreducible.*

Proof. The assertions about p follow from (13.17.3). Let X be irreducible and assume that X_m is non-empty. As p is smooth, p is open (Corollary 14.36). Thus p is surjective, open, and with irreducible fibers, and hence Proposition 3.24 shows that H_m^X is irreducible. Then X_m must contain the generic point of H_m^X, in particular it is irreducible itself. $\qquad\square$

Proposition 14.135. *The $\kappa(\eta)$-scheme X_m has dimension $d + m - n$.*

For $d + m < n$ the assertions means that X_m is empty.

Proof. Let X' be an irreducible component of X of dimension $d' \leq d$, considered as closed integral subscheme. By construction, we have $H_m^{X'} = p^{-1}(X')$. Thus $H_m^{X'}$ is irreducible, and X_m' is irreducible or empty by Proposition 14.134. If $X_m' \neq \emptyset$, then its closure in H_m^X

is $H_m^{X'}$. This shows that if $X'' \neq X'$ is another irreducible component of X with $X_m'' \neq \emptyset$ we cannot have $X_m' \subseteq X_m''$ because otherwise $H_m^{X'} \subseteq H_m^{X''}$ and hence $X' \subseteq X''$. We see that, as a topological space, the irreducible components of X_m are those X_m' that are non-empty. In particular we may assume that X (and hence H_m^X) is irreducible.

By Corollary 14.121 applied to p we see that $\dim H_m^X = d + m(n - m)$ because all fibers of p have dimension $m(n - m)$ by Proposition 14.134. If $X_m \neq \emptyset$, then $H_m^X \to L_m$ is dominant and Remark 14.110 shows

$$\dim X_m = \dim H_m^X - \dim L_m = d + m(n - m) - (m + 1)(n - m) = d + m - n.$$

This also shows that for $d + m < n$ we must have $X_m = \emptyset$. $\square$

Remark 14.136. As having a fixed dimension is a constructible property (Proposition 10.96) and as every constructible subset containing the generic point of an irreducible scheme also contains an open dense subset, Proposition 14.135 implies that there exists an open dense subset V of L_m such that $\dim q^{-1}(y) = d + m - n$ for all $y \in V$.

The Lemma shows that for $m = n - d$ the $\kappa(\eta)$-scheme X_m is of dimension zero and hence finite over $\kappa(\eta)$ (Proposition 5.20).

Definition 14.137. *The positive integer* $\deg X := \dim_{\kappa(\eta)} \Gamma(X_{n-d}, \mathcal{O}_{X_{n-d}})$ *is called the* degree *of* $X \subseteq \mathbb{P}_k^n$.

We will see in Volume II that for hypersurfaces this definition agrees with the definition given in Section (5.11). Note that for hypersurfaces in $\mathbb{P}_k^2$ (i.e., plane curves) this follows from Bézout's theorem.

Hilbert schemes

(14.32) Definition of Hilbert schemes.

In this section we discuss the notion of Hilbert schemes (basically without giving any proofs). Hilbert schemes are schemes parameterizing all closed subschemes of a given scheme. They are defined in terms of the functor they represent, so we start by defining the Hilbert functor:

Definition 14.138. *Let* S *be a scheme.*
(1) *Let* X *be an* S*-scheme. Define*

$$\mathrm{Hilb}(X/S) := \{ Z \subseteq X \ \text{closed subscheme such that}$$
$$Z \to S \ \text{is proper, flat, and of finite presentation} \}$$

(2) *Let* X *be an* S*-scheme. The* Hilbert functor $\mathrm{Hilb}_{X/S}$ *is the functor*

$$(\mathrm{Sch}/S)^{\mathrm{opp}} \to (\mathrm{Sets}), \qquad T \mapsto \mathrm{Hilb}(X \times_S T/T).$$

For a morphism $T' \to T$ the fiber product $- \times_T T'$ induces a map $\mathrm{Hilb}(X \times_S T/T) \to \mathrm{Hilb}(X \times_S T'/T')$, so this really defines a functor. Grothendieck has proved that the Hilbert functor is representable in a very general setting; we have

Theorem 14.139. *Let S be a noetherian scheme and $f \colon X \to S$ a (quasi-)projective morphism. Then the Hilbert functor* $\mathrm{Hilb}_{X/S}$ *is representable; the representing functor is called the* Hilbert scheme. *All of its connected components are (quasi-)projective.*

We will not prove this difficult result here. See [FGA] for a sketch of Grothendieck's proof for projective schemes and the articles by Nitsure [FGAex] and by Altman and Kleiman [AK] for more details. Altman and Kleiman also prove a variant of the theorem where the hypothesis that S is noetherian is replaced by a stronger assumption on f.

The assumption that X is quasi-projective over S is essential. There exist proper smooth schemes X of dimension 3 over $S = \mathrm{Spec}\,\mathbb{C}$ such that $\mathrm{Hilb}_{X/S}$ is not representable by a scheme.

(14.33) Hilbert schemes of points.

Let us discuss a special, relatively simple, variant of the Hilbert scheme introduced above:

Definition and Proposition 14.140. *Let S be a noetherian scheme, let $f \colon X \to S$ be a (quasi-)projective morphism, and fix $n \in \mathbb{Z}_{\geq 0}$. The* Hilbert functor of 0-dimensional subschemes of length n of X over S, $\mathrm{Hilb}_{X/S}^n$, *is the open and closed subfunctor of* $\mathrm{Hilb}_{X/S}$ *defined by*

$$\mathrm{Hilb}_{X/S}^n(T) = \{\, Z \in \mathrm{Hilb}(X \times_S T/T) \;;\; Z \text{ is finite locally free of rank } n \text{ over } T \,\}.$$

We also write $X^{[n]}$ for $\mathrm{Hilb}_{X/S}^n$.

Proof. We have to prove that the subfunctor we defined is open and closed inside the Hilbert functor. Let $Z \in \mathrm{Hilb}(X \times_S T/T)$, so by definition Z is proper, flat, and of finite presentation over T. Denote by $f \colon Z \to T$ the morphism making Z into a T-scheme. By Proposition 14.123 the set $V := \{\, t \in T \;;\; \dim f^{-1}(t) = 0 \,\}$ is open and closed in T and the restriction $f^{-1}(V) \to V$ is finite locally free. The rank of the fibers is then a locally constant function on V and our claim is proved. $\square$

We will vastly generalize this proposition in Volume II when we discuss Hilbert polynomials.

Example 14.141. Let $S = \mathrm{Spec}\,k$, where k is an algebraically closed field, and let C be a connected smooth projective k-scheme of dimension 1. Let $n \geq 1$. What are the k-valued points $C^{[n]}(k)$ of the Hilbert scheme of 0-dimensional subschemes of length n? They correspond to 0-dimensional subschemes of C, and each such subscheme Z can be written as a disjoint union

$$Z = \coprod_{z \in Z} \mathrm{Spec}\,\mathscr{O}_{Z,z}$$

(see Proposition 5.11). Furthermore, each local ring $\mathscr{O}_{Z,z}$ is a quotient of the local ring $\mathscr{O}_{C,z}$, a discrete valuation ring, and hence has the form $\mathscr{O}_{C,z}/\mathfrak{m}_z^{n_z}$, where $\mathfrak{m}_z \mathscr{O}_{C,z}$ denotes the maximal ideal; we have $\sum n_z = n$. In particular, Z is completely determined by the tuple $(z_1, \ldots, z_n)$, where we count each $z \in Z$ with multiplicity n_z, and consider such tuples up to permutation. In other words, we obtain a natural bijection

$$C^{[n]}(k) \overset{\sim}{\to} C^n(k)/S_n.$$

One can show that this bijection gives rise to an isomorphism $C^{[n]} \cong C^n/S_n$ of k-schemes (cf. Section (12.7)). See [FGAex] 7.3 for a more detailed discussion.

Example 14.142. Let k be an algebraically closed field, and let X be a smooth projective variety of dimension d. What are the 0-dimensional subschemes of X of length 2, or in other words, the k-valued points of $X^{[2]}$? We can have either two distinct points $\{x_1, x_2\}$, or a subscheme Z of length 2 concentrated in a single point x. Then Z is determined by an ideal $\mathfrak{a} \subset \mathscr{O}_{X,x}$ such that $\dim_k \mathscr{O}_{X,x}/\mathfrak{a} = 2$. In particular,

$$\mathfrak{m}^2 \subseteq \mathfrak{a} \subseteq \mathfrak{m},$$

where $\mathfrak{m} \subset \mathscr{O}_{X,x}$ denotes the maximal ideal. In other words, the set $X^{[2]}(k)$ is the disjoint union of $(X \times X \setminus \Delta)(k)/(\mathbb{Z}/2)$ (where Δ denotes the diagonal and $\mathbb{Z}/2$ acts by permuting the factors) and the projective space of hyperplanes in $\mathfrak{m}/\mathfrak{m}^2$, or equivalently, of lines in the tangent space $T_{X,x} = (\mathfrak{m}/\mathfrak{m}^2)^*$.

This gives us the following natural candidate for a scheme representing the functor $X^{[2]}$: Let $\tilde{X}$ be the blow-up of $X \times X$ along the diagonal Δ. The group $\mathbb{Z}/2$ acts on $X \times X$ by exchanging the two factors, and since the diagonal is fixed under this action, the action extends naturally to $\tilde{X}$. Let H be the quotient of $\tilde{X}$ by this $\mathbb{Z}/2$-action, cf. Section (12.7). By what we have seen above, we have a natural bijection

$$H(k) \xrightarrow{\sim} X^{[2]}(k),$$

and one can show that this bijection gives rise to an isomorphism $H \cong X^{[2]}$ of k-schemes. Again we refer to [FGAex] 7.3 for further details.

(14.34) Flatness by blow-ups.

Raynaud and Gruson have developed in [RG] a useful "flattening technique" via blow-ups:

Theorem 14.143. *Let S be a qcqs scheme, let $U \subseteq S$ be a quasi-compact open subscheme, and let $f : X \to S$ be a morphism of finite presentation such that $f^{-1}(U)$ is flat over U. Then there exists a U-admissible blow-up (Section (13.20)) $\pi : S' \to S$ such that the strict transform $\tilde{X}$ of X under $\pi_{(X)} : X' := S' \times_S X \to X$ (in other words, $\tilde{X}$ is the schematic closure of $\pi_{(X)}^{-1}(f^{-1}(U))$ in X') is flat over S'.*

This difficult theorem is proved in [RG] (5.2.2). Here we will give only a plausibility argument, why such a result could hold: If S is noetherian and X is projective over S, then the Hilbert scheme $\mathrm{Hilb}_{X/S}$ exists and the flat U-scheme $f^{-1}(U)$ (considered as a closed subscheme of itself) corresponds to a morphism $s : U \to \mathrm{Hilb}_{X/S}$. Let S' be the schematic image and let $Z \subset X \times_S S'$ be the closed S'-flat subscheme corresponding to the closed immersion $S' \hookrightarrow \mathrm{Hilb}_{X/S}$. Then it is not difficult to check that $\pi : S' \to S$ is projective, that π is an isomorphism over U, and that $\pi^{-1}(U)$ is schematically dense in S'. Thus π is close to being a blow-up (Section (13.22)). Moreover, one can show that Z is the schematic closure of $\pi_{(X)}^{-1}(f^{-1}(U))$ in X' and is flat over S'. Thus Z would be the strict transform of X if π is a blow-up.

Example 14.144. Let $S = \mathrm{Spec}\, R$, where R is a discrete valuation ring with maximal ideal $\mathfrak{m}$, let $\eta \in S$ be the generic point, and set $U := \{\eta\}$. Let $f : X \to S$ be a morphism of finite type. Then $X_\eta := f^{-1}(U)$ is flat over $U = \mathrm{Spec}\, \kappa(\eta)$. In this case the blow-up in $\mathfrak{m}$ is an isomorphism because $\mathfrak{m}$ is generated by a regular element. The strict transform of X is the schematic closure of X_η in X and thus is flat over R by Proposition 14.14.

Exercises

Exercise 14.1. Use Proposition 14.4 to give a new proof of Theorem 10.84.

Exercise 14.2. Let S be a Dedekind scheme and let $f\colon X \to S$ be a morphism locally of finite type. Show that the following assertions are equivalent.
(i) f is open.
(ii) f is universally open.
(iii) f_{red} is flat.
(iv) Every irreducible component of X dominates S.

Exercise 14.3. Let Y be a noetherian integral scheme with generic point η such that $\{\eta\}$ is not open in Y (e.g., $Y = \operatorname{Spec} \mathbb{Z}$ or $Y = \mathbb{P}^1_k$ for a field k). Let $f\colon X = Y \amalg \operatorname{Spec} \kappa(\eta) \to Y$ be the canonical morphism. Show that f is a quasi-compact faithfully flat morphism between noetherian schemes which is not open.

Exercise 14.4. Show that the following morphisms f are universal homeomorphism which are not flat.
(a) The inclusion $f\colon X_{\mathrm{red}} \hookrightarrow X$, where X is a noetherian non-reduced scheme.
(b) Let k be a field, $X = \operatorname{Spec} k[T, U]/(TU, U^2)$ and $f\colon X \to \operatorname{Spec} k[T]$ the morphism corresponding to the inclusion $k[T] \hookrightarrow k[T, U]/(TU, U^2)$. Describe in this case also the schematic closure of the generic fiber of f.

Exercise 14.5. Let S be a Dedekind scheme, let η be its generic point, let $f\colon X \to S$ be a morphism of schemes and denote by $X_\eta := f^{-1}(\eta)$ its generic fiber. Let $i\colon X_\eta \hookrightarrow X$ be the inclusion. For a closed subscheme Z^0 of X_η denote by $\overline{Z^0}$ the schematic closure of Z^0 in X.
(a) Show that $Z \mapsto i^{-1}(Z)$ yields a bijection

$$\left\{ \begin{array}{l} \text{closed subschemes } Z \subseteq X \\ \text{such that } Z \text{ is flat over } S \end{array} \right\} \leftrightarrow \{ \text{closed subschemes } Z^0 \subseteq X_\eta \}$$

whose inverse is given by $Z^0 \mapsto \overline{Z^0}$.
(b) Let Y be a second S-scheme, and $Z^0 \subseteq X_\eta$ and $W^0 \subseteq Y_\eta$ be closed subschemes. Show that
$$\overline{Z^0 \times_{\kappa(\eta)} W^0} = \overline{Z^0} \times_S \overline{W^0}.$$

Exercise 14.6◊. Show that every faithfully flat quasi-compact morphism of schemes is an epimorphism in the category of schemes.

Exercise 14.7. Let $f\colon X \to Y$ be a morphism of S-schemes and let $g\colon S' \to S$ be a universally submersive locally quasi-compact surjective morphism (Exercise 10.3). Show that if the base change $f \times \operatorname{id}_{S'}\colon X \times_S S' \to Y \times_S S'$ satisfies one of the properties in Proposition 14.51, then f satisfies the same property.
Hint: Note that all properties are local on the target.
   Show that every surjective universally open morphism is universally submersive and locally quasi-compact surjective.

Exercise 14.8. A morphism $g\colon Y' \to Y$ is called an *fpqc-morphism* if it is faithfully flat and if every quasi-compact open subset of Y is the image of a quasi-compact open subset of X (see also Exercise 10.3).

(a) Show that the property "fpqc-morphism" is local on the target, stable under composition and stable under base change.
(b) Show that every fpqc-morphism is universally submersive.
(c) Show that every faithfully flat quasi-compact morphism is fpqc. Show that every faithfully flat open morphism is fpqc.

Exercise 14.9. Let $f\colon X \to Y$ be a morphism of S-schemes and let $g\colon S' \to S$ be an fpqc-morphism (Exercise 14.8). Show that if the base change $f \times \mathrm{id}_{S'}\colon X \times_S S' \to Y \times_S S'$ satisfies one of the properties in Proposition 14.53, then f satisfies the same property.

Exercise 14.10. Let $f\colon X \to Y$ be a faithfully flat morphism.
(a) Let Z_1 and Z_2 be subschemes of Y. Show that Z_1 majorizes Z_2 if and only if $f^{-1}(Z_1)$ majorizes $f^{-1}(Z_2)$. Deduce that $Z \mapsto f^{-1}(Z)$ yields an injection from the set of subschemes of Y to the set of subschemes of X.
(b) Assume that f is in addition quasi-compact. Show that a subscheme Z of Y is closed (resp. open) if and only if $f^{-1}(Z)$ is a closed (resp. open) subscheme of X.

Exercise 14.11. Let $f\colon X \to Y$ be a faithfully flat morphism, let D be a divisor on Y and let f^*D its inverse image. Show that $f^*D \geq 0$ if and only if $D \geq 0$. Deduce that the pullback $f^*\colon \mathrm{Div}(Y) \to \mathrm{Div}(X)$ is injective.

Exercise 14.12. Let Y be a qcqs scheme.
(a) Show that a subset E of Y is pro-constructible if and only if E is the intersection of constructible subsets. Deduce that any intersection of pro-constructible sets is again pro-constructible.
(b) Give an example of a pro-constructible subset (e.g., of $\mathrm{Spec}\,\mathbb{Z}$) which is not constructible.
(c) Show that any finite subset of Y is pro-constructible.

Exercise 14.13◊. Let Y be a qcqs scheme and let $f\colon X \to Y$ be a flat morphism. Show that $\overline{f^{-1}(y)} = f^{-1}(\overline{\{y\}})$ for every point $y \in Y$.

Exercise 14.14. Let $f\colon X \to Y$ be a quasi-compact dominant morphism and $g\colon Y' \to Y$ a generizing morphism. Show that the base change $f_{(Y')}\colon X \times_Y Y' \to Y'$ is quasi-compact and dominant.
Hint: Exercise 10.1.

Exercise 14.15. Let S be a locally noetherian scheme and let X and Y be S-schemes of finite type. Let $f, g\colon X \to Y$ be two morphisms of S-schemes.
(a) Show that $f = g$ if and only if for all S-schemes of the form $\mathrm{Spec}\,R$, where R is a local Artinian ring, the induced maps $f(R)$ and $g(R)$ on R-valued points are equal.
(b) Let $S = \mathrm{Spec}\,k$ for a field k. Then show that $f = g$ if and only if $f(R) = g(R)$ for all finite local k-algebras R.
(c) Let $S = \mathrm{Spec}\,k$ for a field k and assume that X is geometrically reduced. Then show that $f = g$ if and only if there exists an algebraically closed extension Ω of k such that $f(\Omega) = g(\Omega)$.
Hint: Use Exercise 14.6 to show that in Exercise 10.23 one can replace $\mathscr{O}_{X,x}$ by its completion. Then use Exercise 3.20.

Exercise 14.16. Let $p\colon S' \to S$ be a submersive morphism (e.g., if f is faithfully flat quasi-compact). Show that

$$\mathrm{Ouv}(S) \to \mathrm{Ouv}(S') \rightrightarrows \mathrm{Ouv}(S' \times_S S')$$

is exact, where $\mathrm{Ouv}(X)$ denotes the set of open subsets of a topological space X.

Exercise 14.17. Let S be a scheme, let $S' \to S$ be a quasi-compact faithfully flat morphism, X, Y be S-schemes. Assume that Y is separated over S. Let $U \subseteq X$ be a schematically dense open subset, and let $f\colon X \times_S S' \to Y \times_S S'$ be a morphism whose restriction to $U \times_S S'$ descends to S (i.e., $f_{|U \times_S S'}$ arises by base change from a morphism $U \to Y$). Show that f descends to a morphism $X \to Y$.

Exercise 14.18. Let $p\colon S' \to S$ be a faithfully flat quasi-compact morphism, let X and Y be S-schemes, let $f\colon X \to Y$ be a quasi-compact morphism of S-schemes, and let Z be a closed subscheme of Y. Show that f factors through Z if and only if $f \times \mathrm{id}_{S'}\colon X \times_S S' \to Y \times_S S'$ factors through $Z \times_S S'$.
Hint: Apply Exercise 14.10 to Z and the schematic image of f.

Exercise 14.19. Let $p\colon S' \to S$ be a faithfully flat quasi-compact purely inseparable morphism. Show that for every S'-scheme X' every descent datum φ on X' for p is effective.
Hint: Show that every open subset of X' is stable under φ.

Exercise 14.20. A morphism $f\colon X \to S$ of schemes is called *locally free* if there exists an open affine covering $(S_i)_{i \in I}$ of S and for all $i \in I$ an affine and faithfully flat S_i-scheme $S_i' = \mathrm{Spec}(R_i')$ and an open affine covering (X_{ij}) of $X \times_S S_i'$, say $X_{ij} = \mathrm{Spec}\, A_{ij}$, such that A_{ij} is a free R-module (i.e., "locally for the fpqc-topology X is the spectrum of a free $\mathscr{O}_S$-algebra").
(a) If S is the spectrum of a field, any morphism $X \to S$ is locally free.
(b) Show that every locally free morphism if flat. Show that the converse holds, if $S = \mathrm{Spec}\, R$, where R is an Artinian ring.
    Hint: Proposition B.37.
Remark: It can be shown ([RG] Part I, (3.3.1), (3.3.5), and (3.3.12)) that any smooth morphism $f\colon X \to S$ with geometrically integral fibers is locally free.

Exercise 14.21. Let $f\colon X \to S$ be a locally free morphism (Exercise 14.20) and let Y be a closed subscheme of X. Show that there exists a closed subscheme $\mathrm{Eq}_{Y,X}$ of S such that a morphism $f\colon S' \to S$ factors through $\mathrm{Eq}_{Y,X}$ if and only if $Y \times_S S' = X \times_S S'$.

Exercise 14.22. Prove that a Galois covering $S' \to S$ is the same as a $\Gamma \times S$-torsor, where Γ is viewed as the constant group scheme $\coprod_{\gamma \in \Gamma} \mathrm{Spec}\,\mathbb{Z}$.

Exercise 14.23. Let S be a scheme, let $g\colon X \to S$ and $h\colon Y \to S$ be flat morphisms locally of finite presentation, and let $f\colon X \to Y$ be a morphism of S-schemes. Assume that g is proper. Show that there exists an open subscheme $S' \subseteq S$ such that a morphism $T \to S$ factors through S' if and only if $f_{(T)}\colon X_{(T)} \to Y_{(T)}$ is flat.
Hint: Let X' be the points of $x \in X$ where f is flat and set $S' = S \setminus g(X \setminus X')$. Use the fiber criterion for flatness.

Exercise 14.24. Let R be a discrete valuation ring, $K = \mathrm{Frac}\, R$, and $\pi \in R$ a uniformizing element. Define an R-algebra $A := R[T, T_1, T_2]/(T_1(\pi T - 1), T_2(\pi T - 1))$ and let $f\colon X := \mathrm{Spec}\, A \to S := \mathrm{Spec}\, R$ be the corresponding morphism of schemes, let $\eta \in S$ the generic and $s \in S$ the closed point.

(a) Show that X is reduced, that the homomorphism $A \to K[T_1, T_2]$, $T \mapsto \pi^{-1}$, and $A \to R[T]$, $T_i \mapsto 0$, yield closed immersions $X_1 := \operatorname{Spec} K[T_1, T_2] \hookrightarrow X$ and $X_2 := \operatorname{Spec} R[T] \hookrightarrow X$ which identify X_1 and X_2 with the irreducible components of X. Deduce that X is equidimensional of dimension 2.

(b) Show that S is universally catenary and that f is faithfully flat. Show that $f^{-1}(s) \cong \mathbb{A}^1_{\kappa(s)}$ and that $f^{-1}(\eta)$ is connected with two irreducible components, one isomorphic to $\mathbb{A}^1_K$ and the other one isomorphic to $\mathbb{A}^2_K$.

(c) Show that the schematic intersection $X_1 \cap X_2$ consists of a single closed point $x \in X$ and that $\dim \mathcal{O}_{X_1,x} = 2$ and $\dim \mathcal{O}_{X_2,x} = 1$.

Exercise 14.25. Let $n \geq 1$ be an integer. Let C be the scheme of finite type over $\mathbb{Z}$ such that
$$C(R) = \{ (A, B) \in M_n(R) \times M_n(R) \; ; \; AB = BA \}, \qquad (R \text{ a ring}).$$
Let $p_1, p_2 \colon C \to M_n$ be the two projections.

(a) Show that for $x \in M_n$ the fiber $p_i^{-1}(x)$ is an affine space over $\kappa(x)$ of dimension $\geq n$. Show that if x lies in the locus J_1 of generic matrices (Exercise 12.7), then $p_i^{-1}(x) \cong \mathbb{A}^n_{\kappa(x)}$.

(b) Set $U_i := p_i^{-1}(J_1)$. Show that $U_2 \cap p_1^{-1}(x) \neq \emptyset$ for all $x \in M_n$. Deduce that $U_2 \cap p_1^{-1}(x)$ is dense in $p_1^{-1}(x)$ for all $x \in M_n$.
   Hint: It suffices to show that for any matrix $A \in M_n(k)$ with coefficients in an algebraically closed field k there exists a generic matrix $B \in M_n(k)$ such that $AB = BA$.

(c) Let k be a field and set $X_k := X \otimes_{\mathbb{Z}} k$ for a scheme X over $\mathbb{Z}$. Show that $U_{1,k}$ and $U_{2,k}$ are open dense irreducible subschemes of C_k and deduce that C_k is irreducible.
   Remark: To the authors' knowledge it is an open problem whether C_k is reduced.

(d) Show that $\dim(U_{i,k}) = n^2 + n$ and deduce that $\dim C_k = n^2 + n$.

(e) Show that C is irreducible and $\dim C = n^2 + n + 1$.

(f) Show that $C \to \operatorname{Spec} \mathbb{Z}$ is universally open.
   Hint: Exercise 14.2.
   Remark: If we knew that C_k is reduced, then it would be easy to see that C is flat over $\mathbb{Z}$ (Why?).

Exercise 14.26. Let S be a locally noetherian scheme. A functor $F \colon (\operatorname{Sch}/S)^{\operatorname{opp}} \to (\operatorname{Sets})$ is called *formally proper* if for every discrete valuation ring R with field of fractions K and every morphism $\operatorname{Spec} R \to S$ the canonical map $F(\operatorname{Spec} K) \to F(\operatorname{Spec} R)$ is bijective (cf. Theorem 15.9).

Let X be a proper S-scheme. Show that $\operatorname{Hilb}_{X/S}$ is formally proper.

15   One-dimensional schemes

Contents

In this chapter we will apply the results obtained so far to noetherian schemes of dimension one. Arbitrary one-dimensional noetherian schemes will be called *absolute curves*. Examples for absolute curves are rings of integers in number fields (i.e., finite extensions of $\mathbb{Q}$) or schemes of finite type over a field k of pure dimension one. The latter we will call *curves over k*.

Morphisms into and from one-dimensional schemes

Noetherian schemes X of dimension one are particularly simple: Constructible sets are open or closed and thus images of scheme morphisms $Y \to X$ of finite type are easy to describe; X is normal if and only if X is regular and thus the normalization of X yields already a desingularization (under the mild hypothesis that the normalization is finite); rational morphisms of normal curves into proper schemes are everywhere defined.

(15.1) Absolute curves.

Proposition 15.1. *Let X be a non-empty noetherian scheme and let X_i $(1 \le i \le n)$ be the irreducible components. Suppose that no irreducible component consists of a single point. The following assertions are equivalent.*
(i) For every closed point $x \in X$, $\dim \mathcal{O}_{X,x} = 1$ (i.e., $\mathrm{codim}_X(\{x\}) = 1$).
(ii) The closed irreducible subsets of X are the X_i and the closed points of X.
(iii) The scheme X is of pure dimension 1 (i.e., $\dim X_i = 1$ for all i).

We call a scheme satisfying these equivalent conditions an *absolute curve*.

Proof. As X is quasi-compact, every closed irreducible set contains a closed point of X. This shows the equivalence of all assertions. □

© Springer Fachmedien Wiesbaden GmbH, part of Springer Nature 2020
U. Görtz und T. Wedhorn, *Algebraic Geometry I: Schemes*, Springer Studium
Mathematik – Master, https://doi.org/10.1007/978-3-658-30733-2_16

Examples 15.2.
(1) Let k be a field. Then $\mathbb{A}_k^1$ and $\mathbb{P}_k^1$ are absolute curves. Every plane curve over k is an absolute curve, cf. Section (5.14).
(2) A scheme X is a Dedekind scheme (Section (7.13)) of dimension $\neq 0$ if and only if X is a normal connected absolute curve.

Clearly, if X is an absolute curve, every irreducible component of X is again an absolute curve. For irreducible absolute curves we have the following corollary of Proposition 15.1.

Corollary 15.3. *The constructible sets of an irreducible absolute curve X are the finite sets of closed points and the complements of finite sets of closed points. In particular, every constructible set is open or closed.*

Proof. As X is noetherian, any closed subspace of X has only finitely many irreducible components. Therefore the closed subsets are X itself and finite sets of closed points. If we intersect an open subset U with such a closed subset it is either U itself or a finite set of closed points. Therefore any locally closed subset is open or closed. Moreover the union of a non-empty open subset and a closed subset is open again. Therefore all constructible sets are open or closed. $\square$

Proposition 15.4. *Let Y be an irreducible absolute curve, let $X \neq \emptyset$ be a connected scheme, and let $f \colon X \to Y$ be a morphism of finite type.*
(1) *Then $f(X)$ is either an open dense subset of Y or a closed point of Y.*
(2) *If f closed, then f is either surjective or $f(X)$ consists of a single closed point.*
(3) *If X is integral and Y is normal, then f is flat if and only if $f(X)$ does not consist of a single closed point.*

Proof. (1) follows from Corollary 15.3 and from Chevalley's theorem (Theorem 10.20) that the image of a morphism of finite type is constructible, and (2) is implied by (1). If $f(X)$ does not consist of a single closed point, then f is dominant by (1) and hence flat by Proposition 14.14. Conversely, if f is flat, then f is open (Theorem 14.35). Thus $f(X)$ cannot consist of a single closed point. $\square$

NORMALIZATION OF ABSOLUTE CURVES.

One aspect that makes absolute curves much easier to study than higher-dimensional schemes is the fact that an absolute curve is normal if and only if it is regular (Proposition 6.40). We have seen that we may attach to every integral scheme X its normalization X'. In general, X' is not finite over X. It can even happen that X' is not noetherian even if X is noetherian.

But for absolute curves such pathologies cannot occur, as the Theorem of Krull and Akizuki (Theorem 12.54) shows: the normalization $\pi \colon X' \to X$ of an integral absolute curve X in any finite extension of its function field is a Dedekind scheme, and for every proper closed subscheme $Z \subsetneq X$, the restriction $\pi^{-1}(Z) \to Z$ is finite.

Even in the case of integral absolute curves, π need not be finite. On the other hand, we have seen that if X is of finite type over a field or over $\mathbb{Z}$ (or, more generally, quasi-excellent), then π is always finite, see Section (12.12). Moreover, in this case the locus X_{norm} of normal points is open and dense. Therefore the normalization $X' \to X$ is a desingularization in the sense of Section (13.23).

EXTENSION OF RATIONAL MAPS FROM NORMAL CURVES.

For curves, Theorem 12.60 specializes to the following statement:

Proposition 15.5. *Let S be a scheme, let X be a normal noetherian S-scheme of dimension 1, and let Y be a proper S-scheme. Then every rational S-map $f\colon X \dashrightarrow Y$ extends to a morphism $X \to Y$.*

Of course the condition that $\dim X = 1$ is crucial in the proposition, which will turn out to be one of the essential ingredients in our study of proper curves over a field below.

Valuative criteria

In this section we prove a kind of converse to Proposition 15.5 which says, roughly speaking, that we can detect whether a morphism $Y \to S$ is proper by checking whether every morphism from a pointed curve to Y extends to the whole curve. In fact, instead of honest curves, it suffices to work with "germs of curves", i.e., with spectra of discrete valuation rings. This gives a geometric meaning to the notion of properness which should be compared with the notion of a complete Riemannian manifold in differential geometry, cf. [He] I §10.

(15.2) Morphisms of spectra of valuation rings to schemes.

Recall that if A and B are local rings (both contained in some other ring), we say that B *dominates* A if $A \subseteq B$ and $\mathfrak{m}_B \cap A = \mathfrak{m}_A$ (where $\mathfrak{m}_R$ denotes the maximal ideal of a local ring R). For the definition and basic properties of valuation rings we refer to Definitions/Propositions B.69, B.71.

Lemma 15.6. *Let A be a local integral domain, $K = \operatorname{Frac} A$ and let K' be an extension of K. There exists a valuation ring A' with $\operatorname{Frac}(A') = K'$ that dominates A. If A is in addition noetherian and $K' \supseteq K$ is finitely generated, then we can find a discrete valuation ring A' with $\operatorname{Frac}(A') = K'$ that dominates A.*

Proof. We may assume that A is not a field. We first consider the case $K = K'$. By definition, a valuation ring is a maximal element in the set of local subrings of K ordered by domination. Using Zorn's lemma it is easy to see that A is dominated by a valuation ring A' with $\operatorname{Frac} A' = K$.

Now let A be noetherian with maximal ideal $\mathfrak{m}$, set $X = \operatorname{Spec} A$ and let $\pi\colon X' \to X$ be the blow-up of X in the closed set $Y := \{\mathfrak{m}\}$. As $\mathfrak{m}$ is finitely generated, π is of finite type (even projective). Therefore X' is noetherian. Let Z be an irreducible component of the exceptional divisor $\pi^{-1}(Y)$ and let $B := \mathscr{O}_{X',Z}$ be its local ring. Then B is a local ring of dimension 1 which is noetherian and has K as field of fractions (because π is birational and of finite type). Its integral closure A'' in K is then a Dedekind ring by Theorem 12.54 and for any maximal ideal of A'' the localization of A'' in this maximal ideal is a discrete valuation ring dominating A.

For a general K' we therefore can assume that A is already a valuation ring (resp. a discrete valuation ring in the noetherian case) with valuation v. Let $(T_i)_{i \in I}$ be a transcendence basis of $K' \supset K$. Choosing a well ordering on I, using Proposition B.72 (1) and transfinite induction on I we can extend v to a valuation on $K((T_i)_{i \in I})$ such that the value group is not changed. Thus we may assume for the proof that $K' \supseteq K$ is algebraic. But it is always possible to extend a valuation v to an algebraic extension of K (loc. cit. (2)), and this extension is discrete if v is discrete and $K' \supseteq K$ is finite (loc. cit. (3)). This shows the lemma. $\qquad\square$

Proposition 15.7. *Let Y be a scheme, and let $y \neq y'$ be two points such that y' is a specialization of y. Let $\iota \colon \kappa(y) \to K$ be a field extension. Then there exists a valuation ring A with $\operatorname{Frac} A = K$ and a morphism $g \colon \operatorname{Spec} A \to Y$ with $g(s) = y'$ and $g(\eta) = y$ such that the induced morphism $\kappa(y) \to \kappa(\eta) = \operatorname{Frac} A$ is given by ι. Here $s \in \operatorname{Spec} A$ is the closed point and $\eta \in \operatorname{Spec} A$ is the generic point.*

Moreover, if Y is locally noetherian and K is a finitely generated extension of $\kappa(y)$, we may in addition assume that A is a discrete valuation ring.

Proof. The canonical morphism $\operatorname{Spec} \mathscr{O}_{Y,y'} \to Y$ is a homeomorphism onto its image, and this image contains y which we can therefore consider as a point of $\operatorname{Spec} \mathscr{O}_{Y,y'}$. We may replace Y by $\operatorname{Spec} A$ with $A := \mathscr{O}_{Y,y'}/\mathfrak{p}_y$. Therefore we can assume that Y is the spectrum of a local integral domain A and that y is its generic and y' its closed point. Moreover, if Y is locally noetherian, then A is noetherian. Now the proposition follows from Lemma 15.6. $\qquad\square$

(15.3) Valuative criteria.

To formulate the valuative criteria we introduce the following notion. Consider a commutative diagram of schemes of the following form

(15.3.1)
$$\begin{array}{ccc} \operatorname{Spec} K & \xrightarrow{\ u\ } & X \\ {\scriptstyle j}\downarrow & & \downarrow{\scriptstyle f} \\ \operatorname{Spec} A & \xrightarrow{\ v\ } & Y. \end{array}$$

A *lift of v* is a morphism $\tilde{v} \colon \operatorname{Spec} A \to X$ such that $\tilde{v} \circ j = u$ and $f \circ \tilde{v} = v$.

Theorem 15.8. *(Valuative criterion; general version) Let $f \colon X \to Y$ be a morphism of schemes. Then the following assertions are equivalent.*
(i) *f is separated (resp. universally closed, resp. proper).*
(ii) *f is quasi-separated (resp. quasi-compact, resp. quasi-separated and of finite type) and for all diagrams (15.3.1) where A is a valuation ring and $K = \operatorname{Frac} A$ there exists at most one (resp. at least one, resp. a unique) lift of v.*

For many applications the following noetherian version is useful.

Theorem 15.9. *(Valuative criterion; noetherian version) Let Y be a locally noetherian scheme and let $f \colon X \to Y$ be a morphism of finite type. Then the following assertions are equivalent.*
(i) *f is separated (resp. proper).*

(ii) *For all diagrams* (15.3.1) *where A is a discrete valuation ring and $K = \operatorname{Frac} A$ there exists at most one (resp. a unique) lift of v.*

We will prove only the noetherian variant Theorem 15.9. The implication "(i) $\Rightarrow$ (ii)" in Theorem 15.8 has already been proved in Lemma 12.61. We refer to Exercise 15.1 (or to [EGAInew] 5.5) for a proof of the other implication (note that the valuative criterion for a morphism to be proper follows from the valuative criteria for a morphism to be separated and to be universally closed).

Proof. (of Theorem 15.9) We first prove the valuative criterion for morphisms to be separated. If f is separated, Corollary 9.9 shows that there can be at most one lift.

For the converse we can work locally on Y and assume that Y is noetherian. Let $\Delta_f \colon X \to X \times_Y X$ be the diagonal. We have to show that its image is closed. As Δ_f is a morphism of finite type between noetherian schemes, its image is constructible and it suffices to show that it is stable under specialization by Lemma 10.17. Let $z \in \Delta(X)$ and let $z' \neq z$ be a specialization of z in $X \times_Y X$. By Proposition 15.7 there exists a discrete valuation ring A and a morphism $v' \colon \operatorname{Spec} A \to X \times_Y X$ with $v'(\eta) = z$ and $v'(s) = z'$ (where η is the generic point and s the closed point of $\operatorname{Spec} A$). Denote by p_1 (resp. p_2) the first (resp. second) projection $X \times_Y X \to X$. As $z \in \Delta(X)$, the restrictions of $p_1 \circ v'$ and $p_2 \circ v'$ to $\operatorname{Spec} \kappa(\eta)$ are equal. Denote this restriction by u. If we set $\tilde{v}_1 := p_1 \circ v'$ and $\tilde{v}_2 := p_2 \circ v'$, we have $f \circ \tilde{v}_1 = f \circ \tilde{v}_2$. Thus $\tilde{v}_1$ and $\tilde{v}_2$ are both lifts of v and hence $\tilde{v}_1 = \tilde{v}_2$ which shows that $z' \in \Delta(X)$.

Now we show the valuative criterion for morphisms to be proper. If f is proper, the existence of a unique lift of v follows from Proposition 15.5 applied to the rational Y-map $u \colon \operatorname{Spec} A \dashrightarrow X$.

Thus we assume that (ii) holds. We have already seen that (ii) implies that f is separated. It suffices to prove that the base change $f_{(Y')} \colon X \times_Y Y' \to Y'$ is closed for every morphism $Y' \to Y$ of finite type (Corollary 13.101). Then $f_{(Y')}$ is separated and of finite type and Y' is locally noetherian. If f satisfies (ii), the same is true for $f_{(Y')}$. Therefore it suffices to prove that (ii) implies that f is closed.

Let $Z \subseteq X$ be a closed subset which we endow with its reduced subscheme structure. If (ii) holds for f, it also holds for $f_{|Z}$. Thus it suffices to show that the image $f(X)$ is closed. It is constructible by Chevalley's theorem (Theorem 10.20) and we have to show that it is stable under specialization. Thus let $y \in f(X)$ and let $y' \in Y$ be a specialization. Fix a point $x \in X$ with $f(x) = y$. As f is of finite type, $\kappa(x)$ is a finitely generated field extension of $\kappa(y)$. By Proposition 15.7 there exists a discrete valuation ring A with $\operatorname{Frac} A = \kappa(x)$ and a morphism $v \colon \operatorname{Spec} A \to Y$ such that $v(\eta) = y$ and $v(s) = y'$, where η (resp. s) is the generic (resp. closed) point of $\operatorname{Spec} A$. By (ii) we find a lift $\tilde{v} \colon \operatorname{Spec} A \to X$ of v. This shows that $y' = f(\tilde{v}(s))$ is in the image of f. $\qquad\square$

The proof shows that in the noetherian case the valuative criterion for separatedness also applies to morphisms *locally* of finite type.

Combining Proposition 15.5 and the valuative criterion we can give a more geometric variant of the valuative criterion:

Corollary 15.10. *Let Y be a locally noetherian scheme and let $f \colon X \to Y$ be a morphism of finite type. Then f is proper if and only if for all normal absolute curves C which are schemes over Y, for every closed point $c \in C$, and for every Y-morphism $g \colon C \setminus \{c\} \to X$ there exists a morphism $\bar{g} \colon C \to X$ such that $\bar{g}_{|C \setminus \{c\}} = g$.*

Remark 15.11. For applications it is helpful to restrict to particularly simple discrete valuation rings to apply the valuative criterion. The proof shows that if we want to use the valuative criterion to prove that a certain morphism is separated or proper we may always replace a valuation ring A by a valuation ring A' that dominates A (this holds also for the general version of the valuative criterion). It can be shown that if A is a valuation ring and k' is any extension of the residue field $k = A/\mathfrak{m}_A$ of A, there exists a complete valuation ring A' with residue field k' that dominates A ([EGAInew] **0** (6.5.6)). If A is a discrete valuation ring, then there exists such an A' which is in addition discrete and such that $\mathfrak{m}_A A' = \mathfrak{m}_{A'}$ ([EGAInew] **0** (6.8.3)). In particular we see:

(1) In Theorem 15.8 it suffices to verify (ii) for complete valuation rings with algebraically closed residue field.

(2) In Theorem 15.9 it suffices to verify (ii) for complete discrete valuation rings with algebraically closed residue field.

Note that much is known about the structure of complete discrete valuation rings (e.g., [Se2] II, §4–§6):

(1) Let A be a complete discrete valuation ring whose residue field k and fraction field have the same characteristic. Then $A \cong k[\![T]\!]$.

(2) Let A be a complete discrete valuation ring whose residue field k is perfect of characteristic $p > 0$ and whose field of fractions has characteristic 0, let v be its normalized valuation, and let $e := v(p)$ be the absolute ramification index of A. Let $W(k)$ be the ring of Witt vectors of k. Then there exists a unique injective ring homomorphism $\iota\colon W(k) \to A$ inducing the identity on residue fields. It makes A into a free $W(k)$-module of rank e and there exists a uniformizing element π of A such that $A = W(k)[\pi]$ and such that π satisfies an Eisenstein equation $\pi^e + b_{e-1}\pi^{e-1} + \cdots + b_0 = 0$ with $b_i \in W(k)$ divisible by p and b_0 not divisible by p^2.

For instance this shows that if we want to show that a morphism of schemes of finite type over a field k is separated or proper it suffices to verify the valuative criterion for valuation rings of the form $K[\![T]\!]$, where K is an algebraically closed extension of k.

Example 15.12. Let $n \geq 1$ and $0 \leq d \leq n$ be integers. Let us give a new proof of the fact (Theorem 13.40) that $\mathrm{Grass}_{d,n} \to \mathrm{Spec}\,\mathbb{Z}$ is proper using the valuative criterion Theorem 15.9.

By the valuative criterion we have to show that for every discrete valuation ring A with field of fractions K the canonical map $\mathrm{Grass}_{d,n}(A) \to \mathrm{Grass}_{d,n}(K)$ is bijective. But this map is by definition of $\mathrm{Grass}_{d,n}$ simply the map

$$\left\{\begin{matrix}\text{direct summands } U \text{ of } A^n \\ \text{of rank } d\end{matrix}\right\} \to \left\{\begin{matrix}\text{subvector spaces } W \text{ of } K^n \\ \text{of dimension } d\end{matrix}\right\}$$
$$U \mapsto U \otimes_A K,$$

which is bijective with inverse map $W \mapsto W \cap A^n$.

In particular we obtain a new proof that $\mathbb{P}^n_{\mathbb{Z}}$ is proper over $\mathbb{Z}$ and hence that $\mathbb{P}^n_Y$ is proper over Y for all schemes Y. In this special case, we can also check the valuative criterion in terms of homogeneous coordinates. A morphism $u \in \mathbb{P}^n(K)$, is given by homogeneous coordinates $(u_0 : \cdots : u_n)$, $u_i \in K$, not all zero (see Exercise 3.19). Multiplying all the u_i by a suitable scalar $\lambda \in K^\times$, we may assume that all $u_i \in A$, and that there exists i_0 with $u_{i_0} \in A^\times$ (in fact, we might even arrange things so that $u_{i_0} = 1$). Furthermore, under this condition the tuple $(u_i)_i$ is uniquely determined up to multiplication by a unit of A. To simplify the notation, let us assume that $i_0 = 0$. Then $\left(\frac{u_1}{u_0}, \ldots, \frac{u_n}{u_0}\right)$ defines an A-valued point of $\mathbb{A}^n$, and hence, embedding $\mathbb{A}^n$ as the usual chart $U_0 \subset \mathbb{P}^n$, an A-valued point $\widetilde{u}$ of $\mathbb{P}^n$. Using the assumption that $\operatorname{Spec} A$ is a Y-scheme, $\widetilde{u}$ gives rise to an A-valued point of $\mathbb{P}_Y^n = \mathbb{P}^n \times_{\operatorname{Spec} \mathbb{Z}} Y$, whose base change to K coincides with u by construction. This is the desired lift.

Remark 15.13. Combining the valuative criterion with Corollary 12.92 we obtain a characterization for a morphism being a closed immersion via its induced map on points. We formulate a noetherian version.

Let S be a locally noetherian scheme and let $f \colon X \to Y$ be an S-morphism of S-schemes of finite type. Then f is a closed immersion if and only if the following two conditions hold.
(a) For every S-scheme T the induced map $f(T) \colon X_S(T) \to Y_S(T)$ on T-valued points is injective.
(b) For every Y-scheme T of the form $T = \operatorname{Spec} R$ for a discrete valuation ring R with field of fractions K the canonical map $\operatorname{Hom}_Y(\operatorname{Spec} R, X) \to \operatorname{Hom}_Y(\operatorname{Spec} K, X)$ is bijective.

As the property "closed immersion" is local on the target and as we can glue morphisms (Proposition 3.5), it suffices to check condition (a) for affine schemes $T = \operatorname{Spec} B$ that are schemes over some open affine subscheme $V = \operatorname{Spec} A$ of S. As f is of finite presentation (Remark 10.36), we may even assume that B is a finitely generated A-algebra by Theorem 10.63. Again one can use Remark 15.11 to reduce to checking condition (b) only for special discrete valuation rings.

Curves over fields

A scheme of finite type over a field k that is of pure dimension 1 is called a curve over k. There are a number of results and constructions which rely on the assumption of dimension 1 and which in higher dimensions are either false or much more complicated:
- All separated curves are automatically quasi-projective.
- Normal (= regular) curves can be easily compactified by a regular curve.
- Proper normal integral curves are birationally equivalent if and only if they are isomorphic.

(15.4) Curves and morphism between curves.

Definition 15.14. *Let k be a field. A non-empty k-scheme X of finite type is called a* curve *over k if it satisfies the following conditions (which are equivalent by Theorem 5.22).*
(i) *X is an absolute curve.*
(ii) *$\operatorname{trdeg}_k \kappa(\eta) = 1$ for every generic point η of an irreducible component of X.*

Remark 15.15. Let k be a field, and let C be a curve over k.
(1) Every non-empty open subscheme of C is again a curve over k.
(2) If C is integral, then its normalization $\pi \colon C' \to C$ is finite and birational (see the discussion of normalization of absolute curves in Section (15.1)).

(3) A point x of C is normal if and only if it is regular (Proposition 6.40).
(4) If C is generically reduced (i.e., there exists an open dense subscheme U of C that is reduced), then the set $\mathrm{Sing}(C)$ of non-regular points of C is finite.

Indeed, we may assume that C is irreducible. As the complement of U is finite (Corollary 15.3), we may also assume that C is integral. As C is of finite type over a field, the normal locus C_{norm} is open and dense in C (Section (12.12)). Thus its complement $\mathrm{Sing}(C)$ is finite (again by Corollary 15.3).

Proposition 15.16. *Let $f\colon X \to Y$ be a morphism of irreducible curves over a field k. Assume that f is not constant (i.e., $f(X)$ does not consist of a single point).*
(1) *The morphism f is dominant and quasi-finite.*
(2) *If X is proper over k and Y is separated, then f is finite and surjective and Y is proper over k.*
(3) *If X is proper over k and integral and Y is separated and normal, then f is surjective finite locally free (and Y is proper).*

Proof. (1). The image $f(X)$ is constructible by Theorem 10.20 and it is irreducible because X is irreducible. Hence if it is not a point, then Proposition 15.4 shows that it is dense in Y. Saying that f is quasi-finite is equivalent to saying that all fibers of f have dimension 0. If f had a fiber of dimension 1, this fiber would have to be all of X, so that f had to be constant.

(2). If X is proper and Y is separated, then f is proper by Proposition 12.58. Being proper and quasi-finite, f is finite by Corollary 12.89. As proper morphisms are closed, f is surjective by (1). As images of proper schemes are proper (Proposition 12.59), Y is proper.

(3). By Proposition 15.4 (3) the morphism f is flat and hence finite locally free by (2).								$\square$

Let $f\colon X \to Y$ be a non-constant morphism of integral curves over k. The proposition shows that the generic fiber is non-empty and finite. Thus f induces a finite extension of function fields $K(Y) \hookrightarrow K(X)$ and $[K(X):K(Y)]$ is called the *degree of f*. It is denoted by $\deg f$.

If f is finite locally free we can in particular apply the results on inertia and ramification indices, cf. Section (12.6).

(15.5) Projectivity of curves.

Unlike higher dimensional varieties, all separated curves over a field are quasi-projective. To prove this (at least for generically reduced curves), we start with the following lemma:

Lemma 15.17. *Let A be a Dedekind domain. Then every open subscheme of $\mathrm{Spec}\,A$ is affine.*

Proof. Write $X = \mathrm{Spec}\,A$. By induction, it is enough to show that for every closed point $x \in X$, the complement $U = X \setminus \{x\}$ is affine. Denote by $\mathfrak{m} \subset A$ the maximal ideal corresponding to x. Let $s \in A$ be an element such that $s\mathscr{O}_{X,x} = \mathfrak{m}_x\mathscr{O}_{X,x}$, and let $\{x_1 = x, x_2, \ldots, x_n\} = V(s)$ (set-theoretically). Denote by $\mathfrak{m}_i$ the maximal ideal in A corresponding to x_i. For $i = 2, \ldots, n$, let $s_i \in \mathfrak{m}_i \setminus \mathfrak{m}$.

For d_i sufficiently large, the element $t := s^{-1}\prod_{i=2}^n s_i^{d_i}$ lies in $\bigcap_{u \in U}\mathscr{O}_{X,u} = \mathscr{O}_X(U) \subset \mathrm{Frac}(A)$. On the other hand $t \notin A$.

The A-submodule of $\mathrm{Frac}(A)$ generated by 1 and t^{-1} is a locally free A-module of rank 1, and hence is equal to the A-module of global sections of a line bundle $\mathscr{L}$ on X. The global sections 1, t^{-1} give rise to a morphism $i\colon X \to \mathrm{Proj}\,A[X_0, X_1] = \mathbb{P}^1_X$ (see Corollary 13.33). This is a morphism of X-schemes, i.e., i is a section of id_X, and hence is a closed immersion by Example 9.12. By construction, $i^{-1}(D_+(X_1))$ is the locus in X where t^{-1} does not vanish, i.e., $i^{-1}(D_+(X_1)) = U$. This means that U can be identified with a closed subscheme of $D_+(X_1) \cong \mathbb{A}^1_X$ and therefore is affine. $\qquad\square$

Note that in the setting of the lemma, U is *not* in general a principal open subset of $\mathrm{Spec}\,A$. In fact, there exist Dedekind rings A with a closed point $x \in X := \mathrm{Spec}\,A$ such that $\mathscr{O}([x])$ has infinite order as an element of the Picard group $\mathrm{Pic}(X)$. If $X \setminus \{x\}$ had the form $D(s)$, then some power of $\mathscr{O}([x])$ would correspond to the principal ideal (s).

Theorem 15.18. *Let k be a field, and let C be a separated curve over k. Then C is quasi-projective over k.*

At this point, we will prove the theorem under the additional assumption that C is generically reduced. We will come back to the general case in Volume II, where we will study ampleness of line bundles from a cohomological point of view.

Proof. We may assume that C is connected (Remark 8.21).

(i). We first prove the theorem under the following assumption:

The curve C has an open affine covering $C = \bigcup_{i \in I} U_i$ such that every U_i contains the finite set $\mathrm{Sing}\,C$ of non-normal points of C.

We may assume that the index set I is finite and that each $U_i \neq \emptyset$. Then U_i meets every irreducible component of C because among the singular points are in particular those points which lie on more than one irreducible component. Thus U_i is open and dense in C and the complement consists of finitely many closed points all of whose local rings are normal. As U_i is affine and of finite type over k, it admits an immersion into some affine space and hence also an immersion $g_i\colon U_i \hookrightarrow \mathbb{P}^{n_i}_k$. By Proposition 15.5 we can extend g_i to a morphism $\bar{g}_i\colon C \to \mathbb{P}^{n_i}_k$ which is separated because C is separated.

Combining all the $\bar{g}_i$, we obtain a morphism $g\colon C \to \prod_i \mathbb{P}^{n_i}_k \to \mathbb{P}^N_k$ using the Segre embedding. Since each $\bar{g}_{i|U_i}$ is an immersion, by assumption, the restrictions $g_{|U_i}$ are immersions. It follows from Lemma 14.18 that g is an immersion, so C is quasi-projective.

(ii). The first step proves the theorem in particular for every normal curve. Now let C be an arbitrary generically reduced curve, let $C_1, \ldots, C_n$ be its irreducible components, and let $\tilde{C}$ be the disjoint union of the normalizations $\tilde{C}_i$ of the $(C_i)_{\mathrm{red}}$. We obtain a finite surjective map $f\colon \tilde{C} \to C$, see Section (12.12), and $\tilde{C}$ is quasi-projective.

We will show that C satisfies the assumption above. Let $x \in C$. We have to find an open affine neighborhood U of x which also contains $\mathrm{Sing}\,C$. By Chevalley's criterion for affineness, Theorem 12.39, it is enough to find an open subset $U \subseteq C$ such that $f^{-1}(U)$ is affine and contains the finite set $Z := f^{-1}(\mathrm{Sing}\,C) \cup f^{-1}(x)$. As $\tilde{C}$ is quasi-projective, we find an affine open subset $V \subseteq \tilde{C}$ which contains Z (Proposition 13.49). Write $V = \tilde{C} \setminus \{x_1, \ldots, x_m\}$, the x_i being closed points of $\tilde{C}$. Then

$$\tilde{U} := V \setminus \bigcup_i f^{-1}(f(x_i))$$

has the form $f^{-1}(U)$, is affine by Lemma 15.17, and contains Z. $\qquad\square$

Using Theorem 14.132 we obtain the following result.

Theorem 15.19. *Let C be a smooth proper curve over an infinite field k. Then there exists a closed immersion $C \hookrightarrow \mathbb{P}^3_k$.*

(15.6) Normal completion of curves.

For curves it is easy to see that every normal curve admits a unique normal compactification, more precisely:

Theorem 15.20. *Let C be a normal separated curve over a field k. Then there exists a projective normal curve $\hat{C}$ and an open dominant k-immersion $j \colon C \to \hat{C}$ satisfying the following universal property. For every k-morphism $f \colon C \to X$ to a proper k-scheme X there exists a unique k-morphism $\bar{f} \colon \hat{C} \to X$ such that $\hat{f} \circ j = f$.*

A curve (or, more generally, a scheme of finite type) over k is sometimes called *complete* if it is proper over k. We call $(\hat{C}, j)$ a *normal completion* of C. Clearly it is unique up to unique isomorphism.

Proof. We may assume that C is connected and hence irreducible. To show the existence of $\hat{C}$, we can assume that C is a subscheme of $\mathbb{P}^n_k$ by Theorem 15.18. Let $\bar{C}$ be the closure of C in $\mathbb{P}^n_k$, endowed with its reduced subscheme structure. Then $\bar{C}$ is an integral projective curve over k. Let $\pi \colon \hat{C} \to \bar{C}$ be the normalization of $\hat{C}$. Then π is finite (Corollary 12.52) and $\hat{C}$ is again projective. As C is normal, the restriction of π to $\pi^{-1}(C)$ is an isomorphism. Thus we can consider C as an open dense subscheme of $\hat{C}$.

The universal property of $\hat{C}$ follows from Proposition 15.5 $\qquad\qquad\qquad\qquad\square$

(15.7) Curves and extensions of transcendence degree 1.

Theorem 15.21. *Let k be a field. There is a contravariant equivalence between the categories of*
 (i) *normal proper integral curves over k (with non-constant morphisms),*
 (ii) *extension fields K of k, finitely generated and of transcendence degree 1 (with k-homomorphisms),*
given by mapping a curve C as in (i) to its function field $K(C)$.

Proof. Mapping a curve C to its function field defines a functor between the two categories of the theorem; for objects this is clear, and for morphisms we use that every non-constant morphism is dominant, Proposition 15.16. We define a quasi-inverse as follows: Let K/k be a finitely generated extension field with $\operatorname{trdeg}_k K = 1$. This means that for some $T \in K$, transcendental over k, the extension $K/k(T)$ is finite. We denote by C the normalization of the projective line $\mathbb{P}^1_k$, whose function field we identify with $k(T)$, in K (Section (12.11)). This is a normal integral scheme, finite over $\mathbb{P}^1_k$ (see Corollary 12.52), and hence a normal proper integral curve. Furthermore, C has function field $K(C) = K$.

It remains to show that every field homomorphism $\alpha \colon K(C) \to K(C')$ between the function fields of normal proper integral curves C, C' gives rise to a morphism $C' \to C$ of algebraic curves. But Lemma 9.33 shows that α is induced from a dominant rational map $C' \dashrightarrow C$. Since C is proper over k and C' is normal, this rational map extends uniquely to a morphism $C' \to C$ (see Proposition 15.5), as desired. $\qquad\qquad\square$

Recall also that a proper integral curve C is geometrically integral, if and only if k is algebraically closed in the function field $K(C)$ and the extension $K(C)/k$ is separable, Proposition 5.51, so that this property can easily be checked on the function field side.

Compare the theorem with Proposition 9.35 which shows that the category of finitely generated field extensions of k is equivalent to the category of integral k-schemes of finite type with dominant rational maps as morphisms. Any dominant rational map of normal proper connected curves is everywhere defined and a finite locally free morphism. Any birational map of normal proper connected curves is an isomorphism by Corollary 12.88.

Corollary 15.22. *Let C be a normal connected curve over a field k and let $\hat{C}$ be its normal completion. Then there exist canonical bijections between the following three sets.*
(a) $\{\, u \in K(C) \;;\; u \text{ transcendental over } k \,\}$.
(b) $\{\, f \colon C \to \mathbb{P}^1_k \;;\; f \text{ dominant} \,\}$.
(c) $\{\, f \colon \hat{C} \to \mathbb{P}^1_k \;;\; f \text{ dominant} \,\}$.

Note that every dominant morphism $\hat{C} \to \mathbb{P}^1_k$ is surjective finite locally free by Proposition 15.16.

Proof. The function field of $\mathbb{P}^1_k$ is the purely transcendental extension $k(T)$. Thus the bijection between (a) and (c) is given by the equivalence of categories in Theorem 15.21. The bijection between (b) and (c) is given by the universal property of $\hat{C}$ (Theorem 15.20). $\square$

Remark 15.23. Let C be a normal proper integral curve over k. Mapping each closed point of C to its local ring (viewed as a subring of $K(C)$), we obtain a bijection between the set of closed points of C and the set of discrete valuation rings $A \subset K(C)$ with $k^\times \subset A^\times$. We leave the proof of this fact as Exercise 15.10.

Definition 15.24. *Let k be a field, and let K be a finitely generated extension field of k with $\operatorname{trdeg}_k K = 1$. We call the complete normal curve C over k with $K(C) = K$ the complete normal model of K, or of any curve C' with function field $K(C') = K$.*

For every normal curve C over k, the complete model of $K(C)$ is the normal completion $\hat{C}$ of C (Theorem 15.20).

Divisors on curves

(15.8) Divisors on absolute curves.

Let C be an absolute curve. If C is connected and normal, C is a Dedekind scheme and divisors on C are particularly simple because Cartier divisors are the same as Weil divisors (Theorem 11.40 (2)) and Weil divisors are just formal linear combinations of closed points on C with integer coefficients (see Example 11.47 and Section (7.13)).

In general we can describe divisors on arbitrary (absolute) curves as follows.

Proposition 15.25. *Let C be an absolute curve, let C^1 be the set of points $x \in C$ such that $\dim \mathscr{O}_{C,x} = 1$ and let $j_x\colon \operatorname{Spec} \mathscr{O}_{C,x} \to X$ be the canonical morphism of $x \in C^1$. Then we have an isomorphism of abelian groups*

$$(15.8.1) \qquad \operatorname{Div}(X) \overset{\sim}{\to} \bigoplus_{x \in C^1} \operatorname{Div}(\mathscr{O}_{C,x}), \qquad D \mapsto \sum_{x \in C^1} j_x^*(D).$$

Note that $\operatorname{Div}(\mathscr{O}_{C,x}) = (\operatorname{Frac} \mathscr{O}_{C,x})^\times / \mathscr{O}_{C,x}^\times$, where $\operatorname{Frac} \mathscr{O}_{C,x}$ is the total ring of fractions of $\mathscr{O}_{C,x}$ (see also Exercise 11.18). If C is normal, then $\mathscr{O}_{C,x}$ is a discrete valuation ring for all $x \in C^1$ and the normalized valuation defines an isomorphism $\operatorname{Div}(\mathscr{O}_{C,x}) \overset{\sim}{\to} \mathbb{Z}$. In this case (15.8.1) is simply the isomorphism $\operatorname{Div} X \to Z^1(X)$ identifying Cartier divisors with Weil divisors.

Proof. Let $D \in \operatorname{Div}(X)$. For $x \notin \operatorname{Supp}(D)$, D is given in a neighborhood of x by a local equation which is a unit, thus $j_x^*(D) = 0$. The complement of $\operatorname{Supp}(D)$ is (even schematically) dense in C. Thus $\operatorname{Supp}(D)$ consists of only finitely many points because C is an absolute curve. This shows that (15.8.1) is well defined. As local equations f_i for a divisor are uniquely determined by their germs in points $x \in C^1$, the map (15.8.1) is injective. To show the surjectivity we first recall that $\operatorname{Frac} \mathscr{O}_{C,x} = \mathscr{K}_{C,x}$ and that $\Gamma(U, \mathscr{K}_C) = \operatorname{Frac} \Gamma(U, \mathscr{O}_C)$ for every open affine subset U of C (Remark 11.25). Thus it suffices to show that for $f \in (\operatorname{Frac} \mathscr{O}_{C,x})^\times$ there exists an open affine neighborhood U of x and $g \in (\operatorname{Frac} \Gamma(U, \mathscr{O}_C))^\times$ such that $g_x = f$ and such that $g_y \in \mathscr{O}_{C,y}^\times$ for all $y \in U$, $y \neq x$. But in the beginning of the proof we saw that for U and g as above the support of $\operatorname{div}(g)$ is finite. By shrinking U we may assume that $\operatorname{Supp}(\operatorname{div}(g)) = \{x\}$. □

Corollary 15.26. *Let C be an absolute curve. Then every divisor D on C is the difference $D = D_1 - D_2$ of two effective divisors such that $\operatorname{Supp} D_i \subseteq \operatorname{Supp} D$ for $i = 1, 2$.*

Corollary 15.27. *Let C be an absolute curve without points that are open in C and let U be an open dense subset of C. Then there exists an effective divisor D on C such that $\operatorname{Supp} D$ is contained in U and meets every irreducible component of C.*

A curve over a field does not have open points.

Proof. For each irreducible component C_i of C ($i = 1, \ldots, n$) choose a closed point $x_i \in C_i$ which is contained in U but in no other irreducible component (this is possible because C does not contain open points). As the number of associated points of C is finite, we may assume that x_i is not an embedded associated point. Thus in $\mathscr{O}_{C,x_i}$ there exists a regular element f_i that is not a unit. Let $D_i = (f_i) \in \operatorname{Div}(\mathscr{O}_{C,x_i})$ be the attached effective divisor and let $D \in \operatorname{Div}(C)$ be the divisor corresponding to $(D_1, \ldots, D_n)$ via (15.8.1). □

(15.9) Degree of a divisor on a curve.

If C is an absolute curve, then a Weil divisor is a formal finite $\mathbb{Z}$-linear combination $\sum_x n_x[x]$ of points $x \in C^1$, where C^1 is the set of points $x \in C$ such that $\dim \mathscr{O}_{C,x} = 1$. Equivalently, C^1 is the set of closed points of C.

Definition 15.28. *Let C be a curve over a field k. Let D be a divisor on C and let* $\mathrm{cyc}(D) = \sum_{x \in C^1} n_x[x]$ *be the corresponding Weil divisor on C. The degree* $\deg D = \deg_k D$ *of D is defined as*

$$\deg D = \sum_{x \in C^1} n_x[\kappa(x) : k].$$

If D is effective, we may consider D as closed subscheme that is locally defined by a regular equation (Remark 11.27). For all $x \in C^1$ choose local equations $f_x \in \mathscr{O}_{C,x}$ of D. Then $n_x = \mathrm{lg}(\mathscr{O}_{C,x}/f_x)$ and hence $n_x[\kappa(x) : k] = \dim_k \mathscr{O}_{C,x}/f_x$ by (12.5.2). The underlying topological space of the scheme D is the finite set $\mathrm{Supp}\, D$ and thus it is finite over k (Proposition 5.20). As a scheme

$$D = \mathrm{Spec} \prod_{x \in \mathrm{Supp}\, D} \mathscr{O}_{C,x}/f_x.$$

Thus we see that for effective divisors D one has

(15.9.1) $$\deg D = \dim_k \Gamma(C, \mathscr{O}_D) = \dim_k \Gamma(D, \mathscr{O}_D).$$

Remark 15.29. Let C be a curve over a field k.
(1) The map $\deg \colon \mathrm{Div}(C) \to \mathbb{Z}$ is a homomorphism of abelian groups: It is the composition of the homomorphism $\mathrm{cyc}\colon \mathrm{Div}(C) \to Z^1(C)$ with the homomorphism $Z^1(C) \to \mathbb{Z}$ which sends $\sum n_x[x]$ to $\sum n_x[\kappa(x) : k]$.
(2) Let k' be a field extension and let $p \colon C \otimes_k k' \to C$ be the projection. Then the flat pullback $p^*(D)$ of a divisor D on C (Proposition 11.50) is a divisor on the curve $C \otimes_k k'$ over k'. We have

$$\deg_k D = \deg_{k'} p^* D.$$

Indeed, as $p^* \colon \mathrm{Div}(C) \to \mathrm{Div}(C \otimes_k k')$ is a homomorphism of groups and as every divisor on C is a difference of effective Cartier divisors (Corollary 15.26), it suffices to show the equality if D is effective. But then $p^*D = p^{-1}(D) = D \otimes_k k'$ if we consider D as a closed subscheme, and we have by (15.9.1)

$$\deg_k D = \dim_k \Gamma(D, \mathscr{O}_D) = \dim_{k'} \Gamma(p^{-1}(D), \mathscr{O}_{p^{-1}(D)}) = \deg_{k'} p^* D.$$

Let us investigate the pullback of divisors for curves in a special case. Let $f \colon C_1 \to C_2$ be a finite morphism of integral curves over a field k, where C_2 is normal, i.e., a Dedekind scheme. Then f is surjective finite locally free by Proposition 15.4 and Proposition 15.16. Since pullback of divisors is a homomorphism between the divisor groups, it is enough to understand the pullback of divisors of the form $[x]$, where $x \in C_2$ is a closed point. Then $f^*[x]$ is the effective Cartier divisor corresponding to the closed subscheme $f^{-1}(x)$ of C_1 (Exercise 15.12) and we obtain

$$\deg_k f^*[x] = [\kappa(x) : k] \dim_{\kappa(x)} \Gamma(f^{-1}(x), \mathscr{O}_{f^{-1}(x)})$$
$$= [\kappa(x) : k][K(C_1) : K(C_2)] = [K(C_1) : K(C_2)] \deg_k[x],$$

where the second equality holds because f is finite locally free, compare Proposition 12.21. Recalling $\deg f = [K(C_1) : K(C_2)]$ and extending this result by linearity to arbitrary divisors, we obtain:

Proposition 15.30. *Let* $f: C_1 \to C_2$ *be a finite morphism of integral curves over a field* k. *Assume that* C_2 *is normal. Let* D *be a divisor on* C_2. *Then*

$$\deg f^* D = \deg f \deg D.$$

The proposition also holds without the assumption that C_2 is normal.

The following theorem is an analogue in algebraic geometry of the fact that all meromorphic functions on a compact Riemann surface (e.g., the Riemann sphere) have the same number of zeros and poles, if counted with multiplicities.

Theorem 15.31. *Let* C *be a proper curve over a field* k *and let* D *and* D' *be linearly equivalent divisors. Then* $\deg D = \deg D'$.

We will give two proofs. The second one uses finiteness of cohomology for coherent sheaves on a proper scheme (Theorem 12.68 which will be proved in Volume II). The first one is elementary but works only under the additional hypothesis that C is normal.

Since the degree map is a group homomorphism, it is enough to show that all divisors $\mathrm{div}(f)$, $f \in \Gamma(C, \mathscr{K}_C^\times)$, have degree 0. To show this, we may pass to the individual connected components of C, and hence we may assume that C is connected.

Proof. (if C is normal) Since C is normal and connected, we are considering $f \in K(C)^\times$. If f is algebraic over k, its divisor is trivial. Thus we may assume that f is transcendental, so f defines a surjective finite locally free morphism $\tilde{f}: C \to \mathbb{P}_k^1$ by Corollary 15.22, such that $\mathrm{div}(f) = \tilde{f}^*([0]) - \tilde{f}^*([\infty])$. Since the divisor $[0] - [\infty]$ on $\mathbb{P}_k^1$ obviously has degree 0, the result follows from Proposition 15.30.   □

Let us start with the preparation for the second proof of Theorem 15.31. Let C be a proper curve over k and let $\mathscr{F}$ be a coherent $\mathscr{O}_C$-module. We will see in Volume II that $\Gamma(C, \mathscr{F})$ and $H^1(C, \mathscr{F})$, the cohomology group defined in Section (11.5), are finite-dimensional k-vector spaces. Thus the following definition makes sense.

Definition 15.32. *Let* C *be a proper curve over a field* k, *and let* $\mathscr{F}$ *be a coherent* $\mathscr{O}_C$-*module. The number*

$$\chi(\mathscr{F}) = \dim \Gamma(C, \mathscr{F}) - \dim H^1(C, \mathscr{F})$$

is called the Euler characteristic *of* $\mathscr{F}$.

In Volume II we will see that $H^1(C, -)$ is the first right-derived functor of $\Gamma(C, -)$, and that on the curve C all higher derived functors of Γ vanish. Therefore we can also write $H^0(C, \mathscr{F}) := \Gamma(C, \mathscr{F})$ and view the Euler characteristic as the alternating sum over the dimensions of all cohomology groups of C with coefficients in $\mathscr{F}$. The number $p_a(C) = 1 - \chi(\mathscr{O}_C)$ is called the *arithmetic genus of the curve* C, and is an important invariant of C. The genus should be seen as a "topological" invariant of C; if C is a normal proper connected curve over the complex numbers, then its genus is equal to the genus of the associated Riemann surface, which measures the size of the fundamental group (in somewhat rough terms, the genus is the number of holes, the Riemann sphere having no holes, a one-dimensional complex torus having one hole, etc.)

Let D be an effective divisor on an arbitrary curve C over k. Then we have an exact sequence

$$0 \to \mathscr{O}_C(-D) \to \mathscr{O}_C \to \mathscr{O}_D \to 0$$

of $\mathscr{O}_C$-modules. We obtain an exact cohomology sequence (11.5.7) of k-vector spaces

$$(15.9.2) \quad \begin{aligned} 0 &\to \Gamma(C, \mathscr{O}_C(-D)) \to \Gamma(C, \mathscr{O}_C) \to \Gamma(C, \mathscr{O}_D) \\ &\to H^1(C, \mathscr{O}_C(-D)) \to H^1(C, \mathscr{O}_C) \to H^1(C, \mathscr{O}_D) = 0 \end{aligned}$$

where the last equality follows from Proposition 12.32 because D is affine. We obtain from (15.9.2):

Proposition 15.33. *Let C be a proper curve over a field k and let D be an effective divisor on C. Then*

$$\deg D = \dim_k \Gamma(C, \mathscr{O}_D) = \chi(\mathscr{O}_C) - \chi(\mathscr{O}_C(-D)).$$

This is part of the Riemann-Roch theorem (see Theorem 15.35) below. We can use it to show Theorem 15.31:

Proof. (of Theorem 15.31) By Proposition 15.33 we see that $\deg D$ depends only on its linear equivalence class if D is an effective divisor. Now if $\mathrm{div}(f)$ is any principal divisor, we can write $\mathrm{div}(f) = D - D'$ where D and D' are effective (Corollary 15.26). As D and D' are linearly equivalent by definition, we see that $\deg D = \deg D'$ and hence $\deg \mathrm{div}(f) = 0$. This completes the second proof of Theorem 15.31. $\square$

Therefore if C is a proper curve over a field k, we obtain a group homomorphism

$$(15.9.3) \quad \deg \colon \operatorname{Pic}(C) = \operatorname{DivCl}(C) \to \mathbb{Z}.$$

In particular, we can speak of the degree of a line bundle $\mathscr{L}$ on C. We have

$$\deg \mathscr{L} = \chi(\mathscr{O}_C) - \chi(\mathscr{L}^{-1}).$$

and $\deg D = \deg \mathscr{O}_C(D)$ for every divisor D on C.

If C has a normal k-rational point, then the degree map (15.9.3) is surjective.

Remark 15.34. Let C be a proper connected normal curve over k. If $\mathscr{L} = \mathscr{O}(D)$ for a divisor D, and $f \in \Gamma(C, \mathscr{L})$, $f \neq 0$, then by definition of the line bundle $\mathscr{O}(D)$, $D + \mathrm{div}(f)$ is an effective divisor, and in particular $\deg D = \deg(D + \mathrm{div}(f)) \geq 0$. In other words, if D is a divisor of degree < 0, then $\Gamma(C, \mathscr{O}(D)) = 0$.

(15.10) Theorem of Riemann-Roch.

Given a line bundle $\mathscr{L}$ on a curve C over k, it is an interesting and important problem to determine the dimension $\dim \Gamma(C, \mathscr{L})$. This is called the Riemann-Roch problem, and its answer is provided by the Theorem of Riemann and Roch. We state a weak version of the theorem:

Theorem 15.35. (Riemann-Roch) *Let C be a normal proper curve over a field k, and let $g := p_a(C)$ be the genus of C. There exists a divisor class $K \in \operatorname{DivCl}(C)$, called the* canonical divisor class, *with $\deg K = 2g - 2$, such that for every divisor class $D \in \operatorname{DivCl}(C)$, we have*

$$\dim \Gamma(C, \mathscr{O}(D)) - \dim \Gamma(C, \mathscr{O}(K - D)) = \chi(\mathscr{O}(D)) = \deg D + \chi(\mathscr{O}_C).$$

The first equality which allows us to rewrite the Euler characteristic as a difference of two dimensions of spaces of local sections is a special case of the *Serre Duality Theorem*. The second equality is Proposition 15.33. The term $\dim \Gamma(C, \mathcal{O}(K - D))$ should be seen as a correction term. In particular, if $\deg D > 2g-2$, then $\deg K - D < 0$, so by Remark 15.34, $\dim \Gamma(C, \mathcal{O}(K - D)) = 0$.

The same statement holds in a slightly modified form if one drops the assumption that C be normal. In the general case, one has to replace the term $\mathcal{O}(K - D)$ by $\omega_{C/k} \otimes \mathcal{O}(-D)$, where $\omega_{C/k}$ is a certain coherent sheaf on C, the "dualizing sheaf for the morphism $C \to \operatorname{Spec} k$". If C is normal, then $\omega_{C/k}$ is a line bundle, and hence of the form $\mathcal{O}(K)$ for some divisor class K.

We will prove the theorem in Volume II using cohomological methods. For an elementary proof, see [Fu1] Chap. 8. The Theorem of Riemann and Roch is an indispensable tool in the study of algebraic curves. Its generalizations, the Theorem of Hirzebruch-Riemann-Roch, and the Theorem of Grothendieck-Riemann-Roch, are of similar importance for the understanding of algebraic varieties of higher dimension.

The projective line $\mathbb{P}^1$ has genus 0, and the canonical line bundle is $\mathcal{O}(-2)$. It is easy to check, using Example 13.16, that the equality stated in the theorem holds in this case. (See Exercise 15.13).

Exercises

Exercise 15.1. The goal is to prove that the non-noetherian version of the valuative criterion in Theorem 15.8 is sufficient for a morphism to be separated or universally closed.

(a) Show that for quasi-separated morphisms $f: X \to Y$ the valuative criterion for separatedness is sufficient.
  Hint: Apply Exercise 10.2 to the diagonal of f.

(b) Show that for quasi-compact morphisms $f: X \to Y$ the valuative criterion for a morphism to be universally closed is sufficient.
  Hint: Show that it suffices to show that f is closed. Then use again Exercise 10.2.

Exercise 15.2. Let Y be a locally noetherian scheme, let $f: X \to Y$ be a morphism locally of finite type. Show that f is universally open if and only if for all morphisms $g: Y' \to Y$, where Y' is the spectrum of a discrete valuation ring, the base change $f_{(Y')}: X \times_Y Y' \to Y'$ is open.
Hint: Exercise 14.2.

Exercise 15.3. Let S be a Dedekind scheme with function field K and let $X \to S$ be a proper morphism. Show that the canonical map $X(S) \to X(\operatorname{Spec} K)$ is bijective.

Exercise 15.4. Let Y be a locally noetherian integral scheme, X an integral scheme, and $f: X \to Y$ a dominant morphism of finite type. Show that f is proper if and only if every discrete valuation ring of $K(X)$ that dominates a local ring of Y also dominates a local ring of X.

Exercise 15.5. Let C be an integral curve over a field k. Show that C is proper over k if and only if its normalization C' is proper over k.

Exercise 15.6. Let Y be an integral curve over a field k and let $f: X \to Y$ be a finite morphism. Show that there exists an open dense subscheme V of Y such that the restriction $f^{-1}(V) \to V$ of f is finite locally free.

Exercise 15.7$\Diamond$. Let C be an integral curve over a field k of characteristic $p > 0$ and let $F_{C/k}: C \to C^{(p)}$ be the relative Frobenius. Show that $C^{(p)}$ is an integral curve and that $\deg F_{C/k} = p$.

Exercise 15.8. Let k be a perfect field of characteristic $p > 0$ and let $f: C \to C'$ be a non-constant morphism of integral proper smooth curves over k. Assume that the associated extension $K(C') \hookrightarrow K(C)$ of function fields is purely inseparable. Show that there exists an integer $r \geq 1$ and an isomorphism $g: C' \xrightarrow{\sim} C^{(p^r)}$ such that $g \circ f$ is the r-th power of the relative Frobenius $F_{C/k}$.

Exercise 15.9. Let k be an algebraically closed field. In $\mathbb{P}_k^2$ with coordinates X, Y, Z define the plane curve $C := V_+(Z^2Y^2 - X^4 + Z^4)$.
(a) Show that C is an integral curve and that $(0 : 1 : 0)$ is the only singular point of C.
(b) In $\mathbb{P}_k^3$ with coordinates $T_0, \ldots, T_3$ define $C' = V_+(T_3T_0 - T_1^2, T_0^2 + T_2^2 - T_3^2)$. Show that C' is smooth over k.
(c) Show that $C' \cap D_+(T_0) \cong C_{\mathrm{sm}}$ and deduce that C' is a normalization of C.

Exercise 15.10. Let C be a normal proper integral curve over the field k. Show that mapping each closed point of C to its local ring (viewed as a subring of $K(C)$) induces a bijection between the set of closed points of C and the set of discrete valuation rings $A \subset K(C)$ with $k^\times \subset A^\times$.
Hint: Exercise 15.4.

Exercise 15.11. Let C be a connected smooth proper curve over a field k. Show that the following assertions are equivalent.
(i) C is isomorphic to $\mathbb{P}_k^1$.
(ii) The function field $K(C)$ is isomorphic to $k(T)$ (T an indeterminate).
(iii) The (arithmetic) genus of C is 0 and $C(k) \neq \emptyset$.
(iv) There exists $p \in C(k)$ such that $\dim \Gamma(C, \mathscr{O}_C([p])) \geq 2$.
(v) There exist $p, q \in C(k)$, $p \neq q$, such that the divisors $[p], [q]$ are linearly equivalent.
Hint: Show "(ii) $\Leftrightarrow$ (i) $\Rightarrow$ (iii) $\Rightarrow$ (iv) $\Rightarrow$ (v) $\Rightarrow$ (ii)". Use the Theorem of Riemann and Roch.

Exercise 15.12. Let $f: C_1 \to C_2$ be a finite morphism of integral curves over a field k, and assume that C_2 is normal. Let $x \in C_2$ be a closed point. We consider the divisor $[x]$ as a Cartier divisor on C_2. Show that its pull-back $f^*[x]$ is the effective Cartier divisor on C_1 corresponding to the closed subscheme $f^{-1}(x) \subset C_1$.

Exercise 15.13. Let k be a field, and let D be a divisor on $\mathbb{P}_k^1$. Prove the Riemann-Roch equation

$$\dim \Gamma(\mathbb{P}_k^1, \mathscr{O}(D)) - \dim \Gamma(\mathbb{P}_k^1, \mathscr{O}(D) \otimes \mathscr{O}(-2)) = \deg D + 1.$$

Hint: Use Example 13.16.

Exercise 15.14. Let k be an algebraically closed field, $C := V_+(Y^2Z - X^3 - X^2Z) \subset \mathbb{P}_k^2$. Show that the group $\mathrm{DivCl}^0(C) := \{ [D] \in \mathrm{DivCl}(C) ; \deg D = 0 \}$ is naturally isomorphic to $k^\times$.

Exercise 15.15. Let C be a proper normal connected curve over a field k with arithmetic genus 1. Let D be a divisor on C with $\deg D > 0$. Show that $\dim_k \Gamma(C, \mathscr{O}_C(D)) = \deg D$.

16 Examples

Contents

In this chapter we consider several examples. Each example is given in such a way that it progresses along the theory introduced in the book and that it is possible to study the examples in parallel to the main text. We indicate in the section titles up to which chapter definitions and results are used in that particular section.

Determinantal varieties

In this example we study determinantal varieties which are – roughly speaking – the schemes of matrices of rank $\leq r$ for some fixed r.

We fix integers $m, n \geq 1$ and an integer $r \geq 0$ which is usually assumed to be $\leq \min(n, m)$.

(16.1) Determinantal varieties (Chapter 1).

Let k be an algebraically closed field. We consider the space $M_{n \times m}(k)$ of all $(n \times m)$-matrices with coefficients in k as an affine variety, isomorphic to $\mathbb{A}^{nm}(k)$. It is the variety corresponding to the polynomial ring

$$R_{n \times m} := k[(T_{ij})_{1 \leq i \leq n, 1 \leq j \leq m}].$$

For an integer r with $0 \leq r \leq \min(n, m)$ we consider the subset

(16.1.1) $$\Delta^r_{n \times m}(k) := \{ A \in M_{n \times m}(k) \; ; \; \mathrm{rk}(A) \leq r \}.$$

To see that this is an affine algebraic set we introduce some notation. Let R be any ring and let $A \in M_{n \times m}(R)$ be a matrix. Recall that an r-*minor of A* is the determinant of an $(r \times r)$-submatrix of A. We denote by $I_r(A)$ the ideal in R generated by all r-minors of A. Now let $A_{\mathrm{univ}} \in M_{n \times m}(R_{n \times m})$ be the matrix whose (i, j)-th coefficient is the indeterminate T_{ij}. As a matrix with coefficients in a field has rank $\leq r$ if and only if all its $(r + 1)$-minors vanish, we have an equality of affine algebraic sets

© Springer Fachmedien Wiesbaden GmbH, part of Springer Nature 2020
U. Görtz und T. Wedhorn, *Algebraic Geometry I: Schemes*, Springer Studium
Mathematik – Master, https://doi.org/10.1007/978-3-658-30733-2_17

$$\Delta_{n\times m}^r(k) = V(I_{r+1}(A_{\text{univ}})).$$

By Proposition 1.12 we thus have $I(\Delta_{n\times m}^r(k)) = \text{rad}(I_{r+1}(A_{\text{univ}}))$. In fact one can show (e.g., [BV] Theorem 5.3 and Theorem 5.7)

(16.1.2) $$\text{rad}(I_{r+1}(A_{\text{univ}})) = I_{r+1}(A_{\text{univ}}).$$

This is proved in loc. cit. by defining on $B_{n\times m}^r := R_{n\times m}/I_{r+1}(A_{\text{univ}})$ an additional structure, called "straightening law" which ensures that $B_{n\times m}^r$ is reduced. Although the proof is elementary, it is not easy and quite lengthy, and it would lead us to far astray. In some special cases this can be shown easily (see Exercise 16.1).

Remark 16.1. Transposition of matrices defines an isomorphism $M_{n\times m}(k) \xrightarrow{\sim} M_{m\times n}(k)$ inducing an isomorphism of $\Delta_{n\times m}^r(k)$ with $\Delta_{m\times n}^r(k)$. Thus if we are interested in the structure of $\Delta_{n\times m}^r(k)$ we may always assume that $m \leq n$.

Next we will see that $\Delta_{n\times m}^r(k)$ is irreducible and hence an affine variety in the sense of Definition 1.46:

Proposition 16.2. $\Delta_{n\times m}^r(k)$ *is irreducible.*

We call these affine varieties $\Delta_{n\times m}^r(k)$ *determinantal varieties*.

Proof. The proof is based on the simple fact that images of irreducible spaces are again irreducible (Corollary 1.16) and that a matrix $A \in M_{n\times m}(k)$ is of rank $\leq r$ if and only if it can be written as a product of matrices $A = BC$ with $B \in M_{n\times r}(k)$ and $C \in M_{r\times m}(k)$. We consider pairs (B,C) of such matrices as elements of the (irreducible) variety $\mathbb{A}^{nr+rm}(k)$. Matrix multiplication $(B,C) \mapsto BC$ is clearly defined by polynomials and thus defines a morphism of affine algebraic sets $\mathbb{A}^{nr+rm}(k) \to M_{n\times m}(k)$ whose image is $\Delta_{n\times m}^r(k)$. This shows the claim. $\square$

We obtain a sequence of closed affine subvarieties of $M_{n\times m}(k)$:

$$\{0\} = \Delta_{n\times m}^0(k) \subset \Delta_{n\times m}^1(k) \subset \cdots \subset \Delta_{n\times m}^{\min(n,m)}(k) = M_{n\times m}(k)$$

Remark 16.3. By Proposition 1.20 and using (16.1.2) we see that $I_{r+1}(A_{\text{univ}})$ is a prime ideal.

There is also a projective variant of $\Delta_{n\times m}^r(k)$: As $I_{r+1}(A_{\text{univ}})$ is by definition generated by homogeneous polynomials $f_1, \ldots, f_s$ of degree $r+1$ (the $(r+1)$-minors of A_{univ}), we may consider

(16.1.3) $$\Delta_{+,n\times m}^r(k) := V_+(f_1, \ldots, f_s)$$

as a closed subset of the projective space of lines in $M_{n\times m}(k)$ (isomorphic to $\mathbb{P}^{nm-1}(k)$).

Then $\Delta_{n\times m}^r(k)$ is the affine cone of $\Delta_{+,n\times m}^r(k)$ (Proposition 1.63) and we have a surjection $\Delta_{n\times m}^r(k) \setminus \{0\} \to \Delta_{+,n\times m}^r(k)$ which shows that $\Delta_{+,n\times m}^r(k)$ is irreducible and hence a projective variety in the sense of Definition 1.62. We call these projective varieties *projective determinantal varieties*.

(16.2) Determinantal varieties as schemes (Chapter 3).

We now can define determinantal varieties as affine schemes and extend coefficients from algebraically closed fields to arbitrary rings. More precisely, we define for a ring R the R-algebra

$$B^r := B^r_{n \times m, R} := R[(T_{ij})_{1 \leq i \leq n, 1 \leq j \leq m}]/I_{r+1}.$$

where I_{r+1} is the ideal generated by the $(r+1)$-minors of the $(n \times m)$-matrix whose (i,j)-th component is T_{ij}. We define

(16.2.1) $\Delta^r := \Delta^r_{n \times m, R} := \operatorname{Spec} B^r_{n \times m, R}$

We call this R-scheme a *determinantal scheme*.

We can also define a projective analogue of the determinantal schemes. As all the minors generating I_{r+1} are homogeneous polynomials we obtain by the construction in Section (3.7) a closed subscheme

(16.2.2) $\Delta^r_+ := \Delta^r_{n \times m, +, R} := V_+(I_{r+1})$

of $\mathbb{P}^{nm-1}_R$. We call these R-schemes *projective determinantal schemes*.

If $R = k$ is an algebraically closed field, then I_{r+1} is equal to its radical (16.1.2). Thus the affine scheme $\Delta^r_{n \times m, k}$ and its projective variant $\Delta^r_{n \times m, +, k}$ are reduced schemes. Via Theorem 3.37 they correspond to the affine variety $\Delta^r_{n \times m}(k)$ (resp. projective variety $\Delta^r_{n \times m, +}(k)$) defined in Section (16.1). Let us collect the properties of the schemes $\Delta^r_{n \times m, k}$ and $\Delta^r_{n \times m, +, k}$ we have seen so far.

Proposition 16.4. *Let k be an algebraically closed field. The determinantal schemes $\Delta^r := \Delta^r_{n \times m, k}$ and $\Delta^r_+ := \Delta^r_{n \times m, +, k}$ are integral schemes of finite type over k. These schemes form chains of closed affine subschemes*

$$\operatorname{Spec} k \cong \Delta^0 \subsetneq \Delta^1 \subsetneq \cdots \subsetneq \Delta^{\min(n,m)} \cong \mathbb{A}^{nm}_k$$

and closed projective subschemes

$$\emptyset = \Delta^0_+ \subsetneq \Delta^1_+ \subsetneq \cdots \subsetneq \Delta^{\min(n,m)}_+ \cong \mathbb{P}^{nm-1}_k.$$

(16.3) Points of determinantal varieties (Chapter 4).

Let R be a ring. The formation of the determinantal R-schemes $\Delta^r_{n \times m, R}$ and its projective variant $\Delta^r_{n \times m, +, R}$ commute with base change: Let $R \to R'$ be a homomorphism of rings. Then we have by (4.11.1) and (4.12.4)

(16.3.1) $\Delta^r_{n \times m, R} \otimes_R R' = \Delta^r_{n \times m, R'}, \qquad \Delta^r_{n \times m, +, R} \otimes_R R' = \Delta^r_{n \times m, +, R'}.$

In particular, $\Delta^r_{n \times m, R} = \Delta^r_{n \times m, \mathbb{Z}} \otimes_{\mathbb{Z}} R$ and $\Delta^r_{n \times m, +, R} = \Delta^r_{n \times m, +, \mathbb{Z}} \otimes_{\mathbb{Z}} R$. For an arbitrary scheme S we define

$$\Delta^r_{n \times m, S} := \Delta^r_{n \times m, \mathbb{Z}} \times_{\mathbb{Z}} S, \qquad \Delta^r_{n \times m, +, S} := \Delta^r_{n \times m, +, \mathbb{Z}} \times_{\mathbb{Z}} S.$$

Let $M_{n \times m, S}$ be the S-scheme, isomorphic to $\mathbb{A}^{nm}_S$, whose T-valued points for an S-scheme T is the set $M_{n \times m}(\Gamma(T, \mathscr{O}_T))$. Then we have a chain of closed subschemes

$$S = \Delta^0_{n \times m, S} \subset \Delta^1_{n \times m, S} \subset \cdots \subset \Delta^{\min(m,n)}_{n \times m, S} = M_{n \times m, S}.$$

We define an open subscheme of $\Delta^r_{n \times m, S}$

(16.3.2) $$ {}^0\Delta^r_{n \times m, S} := \Delta^r_{n \times m, S} \setminus \Delta^{r-1}_{n \times m, S}.$$

To describe the T-valued points of $\Delta^r_{n \times m, S}$ and ${}^0\Delta^r_{n \times m, S}$ we need the following notion.

Definition 16.5. *Let R be a ring. A matrix $A \in M_{n \times m}(R)$ is said to have* rank $\leq r$ *if all $(r+1)$-minors of A are zero. It is said to have* rank r *if it is of rank $\leq r$ and if the ideal in R generated by all r-minor of A is equal to R.*

We will see in Lemma 16.17 that A has rank r if and only if the image of the linear map $R^m \to R^n$ defined by A is a direct summand of R^n of rank r.

Proposition 16.6. *If $T \to S$ is an S-scheme, then the T-valued points of $\Delta^r_{n \times m, S}$ (resp. of ${}^0\Delta^r_{n \times m, S}$) are matrices $A \in M_{n \times m}(\Gamma(T, \mathscr{O}_T))$ whose rank is $\leq r$ (resp. $= r$).*

Proof. The description of the T-valued points of $\Delta^r = \Delta^r_{n \times m, S}$ is the definition of the determinantal scheme. To describe the T-valued points of ${}^0\Delta^r = {}^0\Delta^r_{n \times m, S}$ consider the morphism of S-schemes $d_r \colon M_{n \times m, S} \to \mathbb{A}^N_S$ which is defined on T-valued points by sending a matrix $A \in M_{n \times m}(\Gamma(T, \mathscr{O}_T))$ to the tuple of all r-minors of A (thus $N = \binom{n}{r}\binom{m}{r}$). Then $\Delta^{r-1} = d_r^{-1}(0)$, where $0 \in \mathbb{A}^N_S$ denotes the closed subscheme defined by the zero section (Remark 3.15). Therefore ${}^0\Delta^r$ is the inverse image of $\mathbb{A}^N_S \setminus \{0\}$ under $d_{r|\Delta^r}$. Thus the description of ${}^0\Delta^r(T)$ follows from the description of the T-valued points of $\mathbb{A}^N_S \setminus \{0\}$ in Example 4.5. $\qquad\square$

For every scheme S we have an action of the S-group scheme $\mathrm{GL}_{n,S} \times \mathrm{GL}_{m,S}$ on $\Delta^r_{n \times m, S}$ which we define on T-valued points for any S-scheme T. By Corollary 4.7 it suffices to do this for S-schemes which are affine, i.e. of the form $T = \mathrm{Spec}\, R$, and we set:

(16.3.3)
$$(\mathrm{GL}_n(R) \times \mathrm{GL}_m(R)) \times \Delta^r_{n \times m, S}(R) \to \Delta^r_{n \times m, S}(R),$$
$$((U, V), A) \mapsto U A V^{-1}$$

Example 16.7. Assume that $n, m \geq 2$ and $r = 1$. Let K be any field. A matrix A in $M_{n \times m}(K)$ has rank ≤ 1 if and only if it is the product $A = BC$ of matrices $B \in M_{n \times 1}(K)$ and $C \in M_{1 \times m}(K)$. Thus Remark 4.9 shows that for every ring R matrix multiplication defines a morphism of R-schemes

$$\mathbb{A}^n_R \times \mathbb{A}^m_R \cong M_{n \times 1, R} \times M_{1 \times m, R} \to M_{n \times m, R} \cong \mathbb{A}^{nm}_R$$

with (set-theoretic) image $\Delta^1_{n \times m, R}$. As explained in Section (4.14), this morphism induces a morphism

$$\mathbb{P}^{n-1}_R \times \mathbb{P}^{m-1}_R \to \mathbb{P}^{nm-1}_R$$

with (set-theoretic) image $\Delta^1_{n \times m, +, R}$. This is nothing but the Segre embedding (4.14.1). In particular it is a closed immersion.

If $R = k$ is a field, then $\Delta^1_{n \times m, +, k}$ is reduced and the Segre embedding yields an isomorphism $\mathbb{P}^{n-1}_k \times \mathbb{P}^{m-1}_k \xrightarrow{\sim} \Delta^1_{n \times m, +, k}$. We will see in Example 16.36 that this is true over an arbitrary ring R.

(16.4) Dimension of determinantal varieties (Chapter 5).

Let k be a field, and let $\Delta_k^r = \Delta_{n \times m, k}^r$ and $\Delta_{+, k}^r = \Delta_{n \times m, +, k}^r$ be the determinantal schemes over k.

Remark 16.8. For every algebraically closed field extension K of k the K-scheme $\Delta_K^r = \Delta_k^r \otimes_k K$ is integral by Proposition 16.4. Therefore Δ_k^r is geometrically integral by Corollary 5.54. The same argument shows that $\Delta_{k, +}^r$ is geometrically integral.

We want to determine the dimension of Δ_k^r. For this we make the following construction. Let V, W be finite-dimensional k-vector spaces, $m = \dim V$, $n = \dim W$. We consider the affine space of linear maps $\underline{\mathrm{Hom}}_k(V, W)$. More precisely, $\underline{\mathrm{Hom}}_k(V, W)$ is the k-scheme, isomorphic to $\mathbb{A}_k^{mn}$, whose R-valued points for some k-algebra R are given by

$$\underline{\mathrm{Hom}}_k(V, W)(R) := \mathrm{Hom}_R(V \otimes_k R, W \otimes_k R).$$

Let $U \subseteq V$ be a k-subvector space of dimension r. Then we have for any k-algebra R maps

$$(16.4.1) \qquad \rho_U : \underline{\mathrm{Hom}}_k(V, W)(R) \to \underline{\mathrm{Hom}}_k(U, W)(R), \qquad f \mapsto f_{|U}$$

which are functorial in R. Hence they define a morphism of k-schemes $\underline{\mathrm{Hom}}_k(V, W) \to \underline{\mathrm{Hom}}_k(U, W)$ by Yoneda's lemma.

We choose bases $(v_1, \ldots, v_m)$ of V and $(w_1, \ldots, w_n)$ of W such that $(v_1, \ldots, v_r)$ is a basis of U. These choices define isomorphisms of k-schemes $\underline{\mathrm{Hom}}_k(V, W) \xrightarrow{\sim} M_{n \times m, k}$ with the k-scheme of $(n \times m)$-matrices and $\underline{\mathrm{Hom}}_k(U, W) \xrightarrow{\sim} M_{n \times r, k}$. Recall that we defined in (16.3.2) the open subscheme ${}^0\Delta_{n \times r, k}^r \subset M_{n \times r, k}$ of matrices of rank r. Let $Y_U \subset \Delta_{n \times m, k}^r$ be the open subscheme which is the inverse image (Section (4.11)) of ${}^0\Delta_{n \times r, k}^r$ under the morphism

$$\Delta_{n \times m, k}^r \hookrightarrow M_{n \times m, k} \cong \underline{\mathrm{Hom}}_k(V, W) \xrightarrow{\rho_U} \underline{\mathrm{Hom}}_k(U, W) \cong M_{n \times r, k}.$$

In other words, Y_U is the open subscheme of $\Delta_{n \times m, k}^r$ of linear maps whose restriction to U is of rank r.

Lemma 16.9. *The k-scheme Y_U is isomorphic to an open subscheme of $\mathbb{A}_k^{r(m+n-r)}$.*

Proof. There are bijections, functorial in a k-algebra R

$$Y_U(R) = \{ A = (B, B') \in M_{n \times r}(R) \times M_{n \times (m-r)}(R) = M_{n \times m}(R) ; \ \mathrm{rk}(B) = \mathrm{rk}(A) = r \}$$
$$\xrightarrow{\sim} \{ (B, C) \in M_{n \times r}(R) \times M_{r \times (m-r)}(R) ; \ \mathrm{rk}(B) = r \}$$
$$= {}^0\Delta_{n \times r, k}^r(R) \times M_{r \times (m-r), k}(R),$$

where the isomorphism is given by $(B, B') \mapsto (B, B^{-1}B')$. This makes sense as B yields an isomorphism onto its image which contains the image of B'. The inverse isomorphism is given by $(B, C) \mapsto (B, BC)$. Thus Y_U is isomorphic to an open subscheme of $M_{n \times r, k} \times_k M_{r \times (m-r), k} \cong \mathbb{A}_k^{r(n+m-r)}$. $\qquad \square$

Remark 16.10. The union of the open subschemes Y_U, where U runs through the set of r-dimensional subspaces of k^m, is the scheme ${}^0\Delta_{n \times m, k}^r$ of $(n \times m)$-matrices of rank r.

This shows in particular that ${}^0\Delta^r_{n\times m,k}$ is reduced (without using the reducedness of $\Delta^r_{n\times m,k}$ for which we did not give a proof).

Proposition 16.11. *For the determinantal schemes one has* $\dim \Delta^r_{n\times m,k} = r(m+n-r)$ *and, for* $r > 0$, $\dim \Delta^r_{n\times m,+,k} = r(m+n-r)-1$.

Proof. Let U be any r-dimensional subvector space of k^m. Then $Y_U \subset \Delta^r_{n\times m,k}$ is open and non-empty. As $\Delta^r_{n\times m,k}$ is irreducible by Remark 16.8, Y_U is a dense. Therefore we have $\dim \Delta^r_{n\times m,k} = \dim Y_U$ by Theorem 5.22 (3). Using Lemma 16.9, we see that $\dim Y_U = \dim \mathbb{A}^{r(m+n-r)}_k = r(m+n-r)$. The second equality follows from Lemma 5.39 because $\Delta^r_{n\times m,k}$ is the affine cone of $\Delta^r_{n\times m,+,k}$. $\qquad\square$

(16.5) Smooth locus of determinantal varieties (Chapter 6).

Let k be a field. We now study the question where the determinantal k-scheme $\Delta = \Delta^r_{n\times m,k}$ is smooth over k. For $r = \min(m,n)$ we have $\Delta = M_{n\times m,k} \cong \mathbb{A}^{nm}_k$ and thus Δ is smooth over k in this case. For $r = 0$ we find $\Delta = \operatorname{Spec} k$ which is again smooth over k. In general recall the open subscheme ${}^0\Delta = {}^0\Delta^r_{n\times m,k}$ of matrices of rank $= r$ of Δ defined in (16.3.2).

Proposition 16.12. *Assume* $0 < r < \min(n,m)$. *The locus of smooth points (over k) of* Δ *is given by*

$$\Delta_{\mathrm{sm}} = {}^0\Delta.$$

Proof. By Remark 16.10, ${}^0\Delta$ has an open covering by the subschemes Y_U, where U runs through the set of r-dimensional k-subvector spaces of k^n. The k-schemes Y_U are isomorphic to open subschemes of some affine space over k (Lemma 16.9), in particular Y_U is smooth over k. Therefore ${}^0\Delta$ is smooth over k. This shows ${}^0\Delta \subseteq \Delta_{\mathrm{sm}}$.

To show $\Delta_{\mathrm{sm}} \subseteq {}^0\Delta$ it suffices to show by Theorem 6.28 that for every closed point $x \in \Delta \setminus {}^0\Delta$ we have $\dim_{\kappa(x)} T_x(\Delta/k) > \dim \Delta$. We consider x as a $\kappa(x)$-valued point of $(\Delta \setminus {}^0\Delta) \otimes_k \kappa(x)$. Then the relative tangent space $T_x(\Delta/k)$ is the Zariski tangent space $T_x(\Delta \otimes_k \kappa(x))$ (Remark 6.12). Now $\Delta \otimes_k \kappa(x)$ is the determinantal scheme $\Delta^r_{n\times m,\kappa(x)}$ and Proposition 5.38 shows that $\dim \Delta = \dim \Delta \otimes_k \kappa(x)$. Replacing Δ by $\Delta \otimes_k \kappa(x)$ and k by the finite extension $\kappa(x)$ we may therefore assume that x is a k-rational point.

We use Example 6.5 to show that for every k-rational point $x \in \Delta^{r-1}_{n\times m,k} = \Delta \setminus {}^0\Delta$ one has $\dim T_x(\Delta) = nm$ ($>\dim\Delta$ by Proposition 16.11). For subsets $J \subseteq \{1,\dots,m\}$ and $I \subseteq \{1,\dots,n\}$ with $\#I = \#J = r+1$ we set $d_{I,J} := \det(T_{I,J})$, where $T_{I,J}$ is the submatrix of the $(n\times m)$-matrix of indeterminates (T_{ij}) whose rows (resp. columns) are those indexed by elements of I (resp. of J). Thus by definition $\Delta = V(I_{r+1})$ where I_{r+1} is the ideal generated by all elements $d_{I,J}$ and Example 6.5 shows that

$$T_x(\Delta) = \operatorname{Ker}(A(x)\colon k^{nm} \longrightarrow k^N),$$

where $A(x)$ is the matrix

$$A(x) = \left(\frac{\partial d_{I,J}}{\partial T_{ij}}(x)\right)_{\substack{I,J \\ i,j}} \in M_{N\times nm}(k)$$

and where $N = \binom{n}{r+1}\binom{m}{r+1}$ is the number of pairs of subsets I and J as above. We write $I = \{i_1 < \dots < i_{r+1}\}$ and $J = \{j_1 < \dots < j_{r+1}\}$. Then

$$d_{I,J} = \sum_{\pi \in S_{r+1}} \mathrm{sgn}(\pi) T_{i_1,j_{\pi(1)}} \cdots T_{i_{r+1},j_{\pi(r+1)}}$$

and for $1 \leq i \leq n$ and $1 \leq j \leq m$ we obtain as partial derivative

$$\frac{\partial d_{I,J}}{\partial T_{ij}} = \begin{cases} 0, & \text{if } i \notin I \text{ or } j \notin J; \\ \sum\limits_{\substack{\pi \in S_{r+1} \\ \pi(s)=t}} \mathrm{sgn}(\pi) T_{i_1,j_{\pi(1)}} \cdots \widehat{T}_{i_s,j_{\pi(s)}} \cdots T_{i_{r+1},j_{\pi(r+1)}}, & \text{if } i = i_s \text{ and } j = j_t . \end{cases}$$

All these partial derivatives are elements of the ideal I_r. Therefore for $x \in \Delta_{n \times m,k}^{r-1} = V(I_r)$ we see that $A(x) = 0$. $\square$

Remark 16.13. As Δ^r is a closed subscheme of the k-scheme $M_{n \times m,k}$, the tangent space $T_A(\Delta^r)$ in a point $A \in \Delta^r(k)$ is a subvector space of $T_A(M_{n \times m,k}) = M_{n \times m}(k)$. The proof of Proposition 16.12 shows that for $0 < r < \min(n,m)$ and $A \notin {}^0\Delta^r$ one has $T_A(\Delta^r) = M_{n \times m}(k)$. For a smooth point A one can again use Example 6.5 to calculate $T_A(\Delta^r)$ as a subvector space of $M_{n \times m}(k)$. Here we state the result only in the case $n = m = r + 1$ where Δ^r is simply the vanishing scheme of the determinant. In this case an easy calculation shows that for a matrix $A \in {}^0\Delta^r(k)$

$$T_A(\Delta^r) = \{ B \in M_m(k) \; ; \; (\tilde{A}, B) = 0 \},$$

where $\tilde{A}$ is the adjoint matrix of A and $(\ , \ )$ is the perfect bilinear form on $M_n(k)$ given by $((a_{ij}), (b_{ij})) = \sum_{i,j} a_{ij} b_{ij}$.

Proposition 16.12 shows that for $0 < r < \min(n,m)$ the singular locus Δ_{sing} of non-smooth points in $\Delta = \Delta_{n \times m,k}^r$ is the closed subset $\Delta_{n \times m,k}^{r-1}$. Thus by Proposition 16.11 we have

(16.5.1) $\dim \Delta - \dim \Delta_{\mathrm{sing}} = n + m - 2r + 1 \geq 3$

Remark 16.14. As Δ is irreducible and as smooth points are regular, we see that all points $x \in \Delta$ with $\dim \mathcal{O}_{\Delta,x} \leq 2$ are regular. In particular, Δ is regular in codimension 1. One can show that Δ is always normal (in fact it is Cohen-Macaulay [BV] (5.14), in particular S_2 and hence normal by Serre's criterion Proposition B.81). In the case $n = m = r + 1$, where Δ is a hypersurface in $M_{n \times m,k}$, this follows from Proposition 6.41.

(16.6) Determinantal schemes of linear maps (Chapter 8).

Until now we considered the locus of matrices of rank $\leq r$, i.e., of linear maps between free modules with a chosen basis. The language of quasi-coherent modules over schemes will allow us to redefine (and generalize) determinantal varieties in a basis-free way. This uses systematically the exterior product.

Proposition 7.48 shows that a linear map u between finitely generated free modules has rank $\leq r$ (i.e., all $r + 1$-minors vanish) if and only if $\bigwedge^{r+1}(u) = 0$. Thus we may generalize the determinantal schemes as follows.

Let S be a scheme, let $\mathscr{E}$ and $\mathscr{F}$ be quasi-coherent $\mathcal{O}_S$-modules, and let $r \geq 0$ be an integer. We define a functor

$$\Delta^r_{\mathscr{E} \to \mathscr{F}} \colon (\mathrm{Sch}/S)^{\mathrm{opp}} \to (\mathrm{Sets}),$$

(16.6.1)
$$(f \colon T \to S) \mapsto \{\, u \in \mathrm{Hom}_{\mathscr{O}_T}(f^*\mathscr{E}, f^*\mathscr{F}) \;;\; \bigwedge^{r+1}(u) = 0 \,\}.$$

If $\mathscr{E} = \mathscr{O}_S^m$ and $\mathscr{F} = \mathscr{O}_S^n$, then $\Delta^r_{\mathscr{E} \to \mathscr{F}}$ is the functor h_Δ attached to the scheme $\Delta = \Delta^r_{n \times m, S}$ (Section (4.1)).

Remark 16.15. For arbitrary quasi-coherent $\mathscr{O}_S$-modules $\mathscr{E}$ and $\mathscr{F}$ and for any morphism $f \colon T \to S$ (yielding a morphism $h_T \to h_S$ of functors) we have by definition

(16.6.2)
$$\Delta^r_{\mathscr{E} \to \mathscr{F}} \times_{h_S} h_T = \Delta^r_{f^*\mathscr{E} \to f^*\mathscr{F}},$$

where the left hand side is the fiber product of functors (Example 8.3).

In general, the functor $\Delta^r_{\mathscr{E} \to \mathscr{F}}$ will not be representable. But we have the following result.

Lemma 16.16. *Assume that $\mathscr{E}$ and $\mathscr{F}$ are finite locally free. Then the functor $\Delta^r_{\mathscr{E} \to \mathscr{F}}$ is representable by an S-scheme.*

We will denote the representing S-scheme again by $\Delta^r_{\mathscr{E} \to \mathscr{F}}$. We call schemes of this type again *determinantal schemes*.

It can be shown (using the construction of quasi-coherent bundles in Chapter 11) that $\Delta^r_{\mathscr{E} \to \mathscr{F}}$ is representable if $\mathscr{E}$ is an arbitrary quasi-coherent $\mathscr{O}_S$-module (Exercise 16.2).

Proof. If $f \colon T \to S$ is any S-scheme and (U_i) is a Zariski covering, then for every homomorphism of $\mathscr{O}_T$-modules $u \colon f^*\mathscr{E} \to f^*\mathscr{F}$ the exterior power $\bigwedge^{r+1}(u)$ vanishes if and only if $\bigwedge^{r+1}(u_{|U_i}) = \bigwedge^{r+1}(u)_{|U_i} = 0$ for all i (the first equality is the functoriality of the isomorphism (7.20.6), where the scheme morphism is the inclusion $U_i \hookrightarrow T$). This shows that $\Delta^r_{\mathscr{E} \to \mathscr{F}}$ is a Zariski sheaf (for any $\mathscr{E}$ and $\mathscr{F}$). Thus by Theorem 8.9 the question whether $\Delta^r_{\mathscr{E} \to \mathscr{F}}$ is representable is Zariski-local on S. Using (16.6.2) we may therefore assume that $\mathscr{E} = \mathscr{O}_S^m$ and $\mathscr{F} = \mathscr{O}_S^n$ for some integers $n, m \geq 0$. But in this case $\Delta^r_{\mathscr{E} \to \mathscr{F}}$ is representable by $\Delta^r_{n \times m, S}$. $\square$

In $\Delta^r_{n \times m}$ we defined the open subscheme ${}^0\Delta^r_{n \times m}$ of matrices whose rank is equal to r. To understand this condition in basis free terms we use the following result.

Lemma 16.17. *Let R be a ring and let $u \colon M \to N$ be a homomorphism of R-modules. Assume that M is free of m and N is free of rank n and let A be the matrix of u with respect to some bases of M and N. Let $1 \leq r \leq n$ be an integer. Then the following assertions are equivalent.*

(i) *The image of u is a direct summand of N of rank r.*
(ii) *All $(r+1)$-minors of A are zero and the ideal generated by the r-minors of A is R itself.*
(iii) *The image of $\bigwedge^r(u)$ is a direct summand of $\bigwedge^r N$ of rank 1.*

Proof. We use the chosen bases to identify M with R^m and N with R^n. All assertions can be checked after localization. Thus we may in addition assume that R is a local ring. Let k be its residue field. The image of u is generated by the column vectors $v_1, \ldots, v_m \in R^n$ of A. Then the image of $\bigwedge^r(u)$ is generated by the vectors v_I for $I \in \mathcal{F}_r(m)$ (with the notation introduced in Section (7.20)). We also denote the image of v_i in $R^n \otimes_R k$ by $\bar{v}_i$.

"(i) $\Rightarrow$ (iii)". Let $\text{Im}(u)$ be a direct summand of rank r. Then $\text{Im}(u)$ is a free R-module. After renumbering we may assume the images of $v_1, \ldots, v_r$ in $R^n \otimes_R k$ generate $\text{Im}(u) \otimes_R k$. Then Nakayama's lemma implies that $v_1, \ldots, v_r$ generate $\text{Im}(u)$ and hence form a basis of $\text{Im}(u)$ which can be completed to a basis of N. Then $v_1 \wedge \cdots \wedge v_r$ is a basis of $\text{Im}(\bigwedge^r u)$ which can be completed to a basis of $\bigwedge^r N$.

"(iii) $\Rightarrow$ (ii)". Assertion (iii) implies that there exists some $I \in \mathcal{F}_r(m)$ such that v_I is a free generator of $\text{Im}(\bigwedge^r u)$ and that all v_J for $J \in \mathcal{F}_r(m)$ are a multiples $a_J v_I$ of v_I. We claim that $v_K = 0$ for all $K \in \mathcal{F}_{r+1}(m)$. Indeed, write $K = J \cup \{l\}$ with $J \in \mathcal{F}_r(m)$. Then $v_K = \pm v_J \wedge v_l = \pm a_J(v_I \wedge v_l)$. If $l \in I$ we obtain $v_K = 0$. Otherwise fix $i \in I$, and set $I' := I \setminus \{i\}$ and $J' := I' \cup \{l\} \in \mathcal{F}_r(m)$. Then $v_I \wedge v_l = \pm v_{I'} \wedge v_l \wedge v_i = \pm a_{J'} v_I \wedge v_i = 0$.

The claim shows that all $(r+1)$-minors of A are zero. As $\text{Im}(\bigwedge^r u)$ is a direct summand we see that the image $\bar{v}_I$ of v_I in $(\bigwedge^r N) \otimes_R k$ is nonzero, i.e., the matrix in $M_{n \times r}(k)$ with column vectors $\bar{v}_i$ for $i \in I$ has rank r. Therefore it has a nonzero r-minor. The corresponding r-minor of A is then invertible.

"(ii) $\Rightarrow$ (i)". Let $I \in \mathcal{F}_r(m)$ such that the matrix A_I with column vectors v_i, $i \in I$, has an invertible r-minor, say the determinant of $A_{I,J}$ for some $J \in \mathcal{F}_r(n)$. As all $(r+1)$-minors vanish, $\text{Im}(u)$ is generated by $\{v_i \; ; \; i \in I\}$ and hence $\text{Im}(u) = \text{Im}(u_I)$, where $u_I \colon R^r \to R^n$ is the linear map defined by A_I. Then the submodule of R^n generated by the standard basis vectors e_l for $l \notin J$ is a complement of $\text{Im}(u_I)$. This shows that $\text{Im}(u)$ has a complement of rank $n - r$. $\qquad\square$

Let $\mathscr{E}$ and $\mathscr{F}$ be finite locally free $\mathcal{O}_S$-modules. We assume that $\mathscr{E}$ and $\mathscr{F}$ have constant rank m and n, respectively. Finally let r be an integer with $0 \le r \le \min(n, m)$.

Define a subfunctor ${}^0\Delta^r_{\mathscr{E} \to \mathscr{F}}$ of $\Delta^r_{\mathscr{E} \to \mathscr{F}}$ whose value on an S-scheme $f \colon T \to S$ is given by

$$(16.6.3) \qquad {}^0\Delta^r_{\mathscr{E} \to \mathscr{F}}(T) := \{\, u \in \Delta^r_{\mathscr{E} \to \mathscr{F}}(T) \; ; \; \text{Coker}(u) \text{ is locally free of rank } n - r \,\}$$

By Proposition 8.10, $\text{Coker}(u)$ is locally free of rank $n - r$ if and only if $\text{Im}(u)$ is locally a direct summand of $f^*\mathscr{F}$ of rank r.

Proposition 16.18. *The inclusion* ${}^0\Delta^r_{\mathscr{E} \to \mathscr{F}} \hookrightarrow \Delta^r_{\mathscr{E} \to \mathscr{F}}$ *is representable by an open immersion, i.e.,* ${}^0\Delta^r_{\mathscr{E} \to \mathscr{F}}$ *is representable by an open subscheme of the S-scheme* $\Delta^r_{\mathscr{E} \to \mathscr{F}}$.

Proof. As the property "open immersion" is local on the target and in particular local on S, we may assume that $S = \text{Spec } R$ is affine and that $\mathscr{E} = \tilde{M}$ and $\mathscr{F} = \tilde{N}$, where $M = R^m$ and $N = R^n$. But in this case Lemma 16.17 shows that ${}^0\Delta^r_{\mathscr{E} \to \mathscr{F}}(T) = {}^0\Delta^r_{n \times m, S}(T)$ (16.3.2) by Proposition 16.6. $\qquad\square$

The proof shows that if $\mathscr{E}$ and $\mathscr{F}$ are free modules, ${}^0\Delta^r_{\mathscr{E} \to \mathscr{F}}$ is the scheme ${}^0\Delta^r_{n \times m, S}$.

(16.7) Separatedness and smooth bundles of determinantal schemes (Chapter 9).

From now on we fix the following notation. Let S be a scheme, $\mathscr{E}$ and $\mathscr{F}$ finite locally free $\mathcal{O}_S$-modules, and let $r \ge 0$ be an integer. Let $\Delta = \Delta^r_{\mathscr{E} \to \mathscr{F}}$ be the corresponding determinantal scheme over S.

Remark 16.19. There exists an open affine covering $(U_i)_i$ of S such that $\mathscr{E}_{|U_i} \cong \mathscr{O}_S^{m_i}$ and $\mathscr{F}_{|U_i} \cong \mathscr{O}_S^{n_i}$. Thus $\Delta \times_S U_i \cong \Delta^r_{n_i \times m_i, U_i}$ is affine for all i. In particular $\Delta \times_S U_i$ is separated over U_i by Rémark 9.8. As "separated" is local on the target (Proposition 9.13), we see that $\Delta^r_{\mathscr{E} \to \mathscr{F}}$ is separated over S.

We will now define a basis-free variant of the locus where a certain minor is invertible. To do this we start with a general construction. For all quasi-coherent $\mathscr{O}_S$-modules $\mathscr{G}$ and $\mathscr{H}$ we define functors

$$(16.7.1) \quad \begin{aligned} \underline{\mathrm{Hom}}(\mathscr{G}, \mathscr{H}) &: (\mathrm{Sch}/S)^{\mathrm{opp}} \to (\mathrm{Sets}), & (h \colon T \to S) &\mapsto \mathrm{Hom}_{\mathscr{O}_T}(h^* \mathscr{G}, h^* \mathscr{H}), \\ \underline{\mathrm{Isom}}(\mathscr{G}, \mathscr{H}) &: (\mathrm{Sch}/S)^{\mathrm{opp}} \to (\mathrm{Sets}), & (h \colon T \to S) &\mapsto \mathrm{Isom}_{\mathscr{O}_T}(h^* \mathscr{G}, h^* \mathscr{H}). \end{aligned}$$

Lemma 16.20. *Let $\mathscr{G}$ and $\mathscr{H}$ be finite locally free $\mathscr{O}_S$-modules. Then $\underline{\mathrm{Hom}}(\mathscr{G}, \mathscr{H})$ and $\underline{\mathrm{Isom}}(\mathscr{G}, \mathscr{H})$ are represented by S-schemes which are smooth and affine over S. $\underline{\mathrm{Isom}}(\mathscr{G}, \mathscr{H})$ is an open subscheme of $\underline{\mathrm{Hom}}(\mathscr{G}, \mathscr{H})$ which is schematically dense if $\mathrm{rk}(\mathscr{G}) = \mathrm{rk}(\mathscr{H})$.*

In fact, one can show that $\underline{\mathrm{Hom}}(\mathscr{G}, \mathscr{H})$ is representable if $\mathscr{H}$ is finite locally free and $\mathscr{G}$ is an arbitrary quasi-coherent $\mathscr{O}_S$-module; see Exercise 11.8.

Proof. Clearly both functors are Zariski-sheaves. Thus all assertions are local on S and we may assume that $S = \mathrm{Spec}\, R$ is affine and that $\mathscr{G} = \mathscr{O}_S^m$ and $\mathscr{H} = \mathscr{O}_S^n$. But then $\underline{\mathrm{Hom}}(\mathscr{G}, \mathscr{H})$ is isomorphic to the R-scheme $M_{n \times m, R} \cong \mathbb{A}_R^{nm}$. In particular it is smooth and affine over S. The scheme $\underline{\mathrm{Isom}}(\mathscr{G}, \mathscr{H})$ is either empty (if $m \neq n$) or isomorphic to $\mathrm{GL}_{n,R}$ (if $m = n$). In the second case, $\mathrm{GL}_{n,R}$ is the principal open set of $M_{n \times n, R} \cong \mathrm{Spec}\, R[(T_{ij}); 1 \leq i, j \leq n]$ defined by the determinant det. Hence $\mathrm{GL}_{n,R}$ is also affine. As one coefficient of the polynomial det is not a zero divisor, det is not a zero-divisor (in fact, all coefficients of det are ± 1). Thus $\mathrm{GL}_{n,R}$ is schematically dense in $M_{n \times n, R}$ by Remark 9.24. $\qquad \square$

Now let $\mathscr{U} \subseteq \mathscr{E}$ be a locally direct summand of rank r and let $\mathscr{V} \subseteq \mathscr{F}$ be a locally direct summand such that $\mathscr{F}/\mathscr{V}$ is locally free of rank r. We obtain a morphism of S-schemes, defined on T-valued points for an S-scheme $h \colon T \to S$ by:

$$(16.7.2) \quad \begin{aligned} \sigma_{\mathscr{U}, \mathscr{V}} \colon \Delta^r_{\mathscr{E} \to \mathscr{F}} &\to \underline{\mathrm{Hom}}(\mathscr{U}, \mathscr{F}/\mathscr{V}), \\ (u \colon h^* \mathscr{E} \to h^* \mathscr{F}) &\mapsto (h^* \mathscr{U} \hookrightarrow h^* \mathscr{E} \xrightarrow{u} h^* \mathscr{F} \twoheadrightarrow h^*(\mathscr{F}/\mathscr{V})) \end{aligned}$$

We set

$$(16.7.3) \quad Y_{\mathscr{U}, \mathscr{V}} := \sigma_{\mathscr{U}, \mathscr{V}}^{-1}(\underline{\mathrm{Isom}}(\mathscr{U}, \mathscr{F}/\mathscr{V})).$$

If $\mathscr{E} = \mathscr{O}_S^m$, $\mathscr{F} = \mathscr{O}_S^n$, $\mathscr{U} = \mathscr{O}_S^J$ for $J \in \mathcal{F}_r(m)$, and $\mathscr{V} = \mathscr{O}_S^{\{1,\dots,n\}\setminus I}$ for $I \in \mathcal{F}_r(n)$, then $\sigma_{\mathscr{U}, \mathscr{V}}$ is the morphism which sends a matrix A to the $(r \times r)$-submatrix $A_{I,J}$. In this case

$$(16.7.4) \quad Y_{I,J} := Y_{\mathscr{U}, \mathscr{V}}$$

is the locus of matrices where $\det(A_{I,J})$ is invertible.

The formation of $Y_{\mathscr{U}, \mathscr{V}}$ is compatible with base change: If $g \colon S' \to S$ is a morphism of schemes, then we have an identity of open subschemes of $\Delta^r_{\mathscr{E} \to \mathscr{F}} \times_S S' = \Delta^r_{g^* \mathscr{E} \to g^* \mathscr{F}}$:

$$(16.7.5) \quad Y_{\mathscr{U}, \mathscr{V}} \times_S S' = Y_{g^* \mathscr{U}, g^* \mathscr{V}}.$$

The morphism $\sigma_{\mathcal{U},\mathcal{V}}$ is the composition of the following two morphisms, defined on $(h\colon T \to S)$-valued points,

$$(16.7.6) \qquad \begin{aligned} \rho_{\mathcal{U}} &\colon \Delta^r_{\mathcal{E}\to\mathcal{F}} \to \underline{\mathrm{Hom}}(\mathcal{U},\mathcal{F}), & u &\mapsto u_{|h^*\mathcal{U}}, \\ \pi_{\mathcal{V}} &\colon \underline{\mathrm{Hom}}(\mathcal{U},\mathcal{F}) \to \underline{\mathrm{Hom}}(\mathcal{U},\mathcal{F}/\mathcal{V}), & v &\mapsto (h^*\mathcal{U} \xrightarrow{v} h^*\mathcal{F} \twoheadrightarrow h^*\mathcal{F}/\mathcal{V}). \end{aligned}$$

Remark 16.21. The morphism $\rho_{\mathcal{U}}$ and $\pi_{\mathcal{V}}$ are surjective on R-valued points for any affine S-scheme $h\colon \mathrm{Spec}\,R \to S$ because in this case one can choose complements of $h^*\mathcal{U}$ in $h^*\mathcal{E}$ and of $h^*\mathcal{V}$ in $h^*\mathcal{F}$. Therefore $\sigma_{\mathcal{U},\mathcal{V}}$ is surjective on R-valued points. In particular $\rho_{\mathcal{U}}$, $\pi_{\mathcal{V}}$, and $\sigma_{\mathcal{U},\mathcal{V}}$ are surjective (Proposition 4.8).

For every $(h\colon T \to S)$-valued point $w \in \underline{\mathrm{Hom}}(\mathcal{U},\mathcal{F}/\mathcal{V})(T)$ the additive group of $\mathrm{Hom}_{\mathcal{O}_T}(h^*\mathcal{U}, h^*\mathcal{V})$ acts by addition simply transitively on $\pi_{\mathcal{V}}^{-1}(w)$. This action is functorial in T. Thus if $T = \mathrm{Spec}\,R$ is affine, $\pi_{\mathcal{V}}^{-1}(w)$ is non-empty and any choice of an element in $\pi_{\mathcal{V}}^{-1}(w)$ yields an isomorphism of the T-scheme $\underline{\mathrm{Hom}}(h^*\mathcal{U}, h^*\mathcal{V})$ with the fiber product of

$$\begin{array}{c} \underline{\mathrm{Hom}}(\mathcal{U},\mathcal{F}) \\ \downarrow{\scriptstyle \pi_{\mathcal{V}}} \\ T \xrightarrow{w} \underline{\mathrm{Hom}}(\mathcal{U},\mathcal{F}/\mathcal{V}) \end{array}$$

In particular we see that for every open affine subscheme T of $\underline{\mathrm{Hom}}(\mathcal{U},\mathcal{F}/\mathcal{V})$ we have an isomorphism of T-schemes

$$(16.7.7) \qquad \pi_{\mathcal{V}}^{-1}(T) \cong \underline{\mathrm{Hom}}(h^*\mathcal{U}, h^*\mathcal{V}) = \underline{\mathrm{Hom}}(\mathcal{U},\mathcal{V}) \times_S T.$$

Similarly, for every $(h\colon T \to S)$-valued point $v \in \underline{\mathrm{Hom}}(\mathcal{U},\mathcal{F})(T)$ we have the following action of the additive group $\mathrm{Hom}_{\mathcal{O}_T}(h^*(\mathcal{E}/\mathcal{U}), h^*\mathcal{U})$, which we consider as a subgroup of $\mathrm{Hom}_{\mathcal{O}_T}(h^*\mathcal{E}, h^*\mathcal{E})$:

$$\mathrm{Hom}_{\mathcal{O}_T}(h^*(\mathcal{E}/\mathcal{U}), h^*\mathcal{U}) \times \rho_{\mathcal{U}}^{-1}(v) \to \rho_{\mathcal{U}}^{-1}(v),$$
$$(a, u) \mapsto u + u \circ a.$$

This is an action because for all $a, a' \in \mathrm{Hom}_{\mathcal{O}_T}(h^*(\mathcal{E}/\mathcal{U}), h^*\mathcal{U}) \subset \mathrm{Hom}_{\mathcal{O}_T}(h^*\mathcal{E}, h^*\mathcal{E})$ we have $a \circ a' = 0$. In fact $\rho_{\mathcal{U}}$ is the basis-free variant of the morphism (16.4.1). Now assume that v has rank r, i.e. $v \in {}^0\Delta^r_{h^*\mathcal{U}\to h^*\mathcal{V}}$. Then the same argument as in Lemma 16.9 shows that this action is simply transitive. As for $\pi_{\mathcal{F}}$ above we deduce for all open affine subschemes T of ${}^0\Delta^r_{\mathcal{U}\to\mathcal{F}} \subset \underline{\mathrm{Hom}}(\mathcal{U},\mathcal{F})$ the existence of an isomorphism of T-schemes

$$(16.7.8) \qquad \rho_{\mathcal{U}}^{-1}(T) \cong \underline{\mathrm{Hom}}(h^*(\mathcal{E}/\mathcal{U}), h^*\mathcal{U}) = \underline{\mathrm{Hom}}(\mathcal{E}/\mathcal{U},\mathcal{U}) \times_S T.$$

We define the locus of homomorphisms u whose restriction to $\mathcal{U}$ have rank r:

$$(16.7.9) \qquad Y_{\mathcal{U}} := \rho_{\mathcal{U}}^{-1}({}^0\Delta^r_{\mathcal{U}\to\mathcal{F}}).$$

Using Lemma 16.20 and Remark 16.21 we deduce from (16.7.7) and (16.7.8) the following result.

Proposition 16.22. *The morphisms*

$$\pi_{\mathcal{V}}\colon \underline{\mathrm{Hom}}(\mathcal{U},\mathcal{F}) \to \underline{\mathrm{Hom}}(\mathcal{U},\mathcal{F}/\mathcal{V}),$$
$$\rho_{\mathcal{U}}\colon Y_{\mathcal{U}} \to {}^0\Delta^r_{\mathcal{U}\to\mathcal{F}}$$

are surjective, affine, and smooth of relative dimension $r(\mathrm{rk}(\mathscr{F}) - r)$ *and* $(\mathrm{rk}(\mathscr{E}) - r)r$, *respectively.*

In fact, one can deduce from (16.7.7) and (16.7.8) the stronger assertion that $\pi_{\mathscr{Y}}$ and $\rho_{\mathscr{U}|Y_{\mathscr{U}}}$ can be endowed with the structure of a geometric vector bundles in the sense of Definition 11.6.

Corollary 16.23. *Assume that $\mathscr{E}$ and $\mathscr{F}$ are of constant rank $m \geq 1$ and $n \geq 1$, respectively and that $0 \leq r \leq \min(m, n)$. The S-schemes $Y_{\mathscr{U},\mathscr{V}}$, $Y_{\mathscr{U}}$, and ${}^0\Delta^r_{\mathscr{E} \to \mathscr{F}}$ are all smooth of relative dimension $r(m + n - r)$ over S.*

Corollary 16.24. *The locus of points of $\Delta^r_{\mathscr{E} \to \mathscr{F}}$ that are smooth over S is ${}^0\Delta^r_{\mathscr{E} \to \mathscr{F}}$.*

Proof. The locus of points of $\Delta^r_{\mathscr{E} \to \mathscr{F}}$ that are smooth over S contains ${}^0\Delta^r_{\mathscr{E} \to \mathscr{F}}$ by Corollary 16.23. Assume there existed a smooth point $x \in \Delta^r_{\mathscr{E} \to \mathscr{F}} \setminus {}^0\Delta^r_{\mathscr{E} \to \mathscr{F}}$. Let s be its image in S. As smoothness is stable under base change, x is a point of the fiber $(\Delta^r_{\mathscr{E} \to \mathscr{F}} \setminus {}^0\Delta^r_{\mathscr{E} \to \mathscr{F}}) \otimes_S \kappa(s)$ which is smooth over $\kappa(s)$ but this is a contradiction to Proposition 16.12. $\qquad \square$

Remark 16.25. It can be shown that the open subschemes $Y_{\mathscr{U},\mathscr{V}}$ are schematically dense in $\Delta^r_{\mathscr{E} \to \mathscr{F}}$. Indeed, in Proposition 16.34 we will see that $\Delta^r_{\mathscr{E} \to \mathscr{F}}$ is flat over S. As formation of $Y_{\mathscr{U},\mathscr{V}}$ is compatible with base change by (16.7.5), we may assume that $S = \mathrm{Spec}\, k$ for a field k (Remark 9.25). But in this case $Y_{\mathscr{U},\mathscr{V}}$ is open and non-empty in an integral scheme (Remark 16.8) and thus schematically dense.

In particular, the open subschemes ${}^0\Delta^r_{\mathscr{E} \to \mathscr{F}}$ and $Y_{\mathscr{U}}$ are schematically dense in $\Delta^r_{\mathscr{E} \to \mathscr{F}}$ because they contain $Y_{\mathscr{U},\mathscr{V}}$.

Remark 16.26. (Birational equivalence class of determinantal schemes) Let R be a ring and $S = \mathrm{Spec}\, R$. Assume $\mathscr{E} = \mathscr{O}_S^m$ and $\mathscr{F} = \mathscr{O}_S^n$. With the notation of (16.2.1) we then have

$$\Delta^r_{\mathscr{E} \to \mathscr{F}} \cong \Delta^r_{n \times m, R} = \mathrm{Spec}\, B^r_{n \times m, R}.$$

We set $\Delta = \Delta^r_{n \times m, R}$ and $B = B^r_{n \times m, R}$. To exclude trivial cases we assume that $1 \leq r \leq \min(m, n)$.

Fix subsets $J \subseteq \{1, \ldots, m\}$ and $I \subseteq \{1, \ldots, n\}$ with $\#I = \#J = r$. Let

$$(16.7.10) \qquad\qquad d_{I,J} \colon \Delta \to \mathbb{A}^1_R$$

be the morphism of R-schemes which is given on R'-valued points (R' some R-algebra) by sending a matrix $A \in \Delta^r_{n \times m, R}(R')$ to the determinant of the submatrix $A_{I,J}$. We consider $d_{I,J}$ as an element of B. Then $Y_{I,J}$ (16.7.4) is the locus in $\Delta^r_{n \times m, R}$, where $d_{I,J}$ is invertible, i.e., $Y_{I,J} = \mathrm{Spec}\, B[1/d_{I,J}]$.

It is a special case of Lemma (6.4) of [BV] that there is an isomorphism

$$(16.7.11) \qquad\qquad \varphi \colon B[1/d_{I,J}] \cong R[T_1, \ldots, T_s][\delta^{-1}]$$

of R-algebras, where $s = r(m + n - r)$ and

$$(16.7.12) \qquad\qquad \delta = \varphi(d_{I,J}) \in R[T_1, \ldots, T_s].$$

If R is an integral domain, φ can be chosen such that δ is a prime element in $R[T_1, \ldots, T_s]$. In this case $\Delta^r_{n \times m, R}$ is birationally isomorphic to $\mathbb{A}^s_R$.

(16.8) Finiteness properties of determinantal schemes (Chapter 10).

Proposition 16.27. *The structure morphism* $\Delta^r_{\mathscr{E} \to \mathscr{F}} \to S$ *is of finite presentation.*

Proof. The property "of finite presentation" is local on the target. Thus we may assume that $S = \operatorname{Spec} R$ is affine and that $\mathscr{E} = \mathcal{O}_S^m$ and $\mathscr{F} = \mathcal{O}_S^n$. But then $\Delta^r_{\mathscr{E} \to \mathscr{F}} = \Delta^r_{n \times m, R}$ is an affine R-scheme which is defined in $M_{n \times m, R} \cong \mathbb{A}_R^{nm}$ be finitely many equations. Thus $\Delta^r_{\mathscr{E} \to \mathscr{F}}$ is of finite presentation. $\qquad\square$

Remark 16.28. If $S = \operatorname{Spec} R$ is affine and R is an R_0-algebra for some ring R_0 (e.g., $R_0 = \mathbb{Z}$), then R is the inductive limit of the finitely generated R_0-subalgebras R_λ. Let $S_0 = \operatorname{Spec} R_0$ and $S_\lambda = \operatorname{Spec} R_\lambda$ and $s_\lambda \colon S \to S_\lambda$ be the canonical morphism. By Theorem 10.60 there exists an index λ and finite locally free $\mathcal{O}_{S_\lambda}$-modules $\mathscr{E}_\lambda$ and $\mathscr{F}_\lambda$ such that $s_\lambda^* \mathscr{E}_\lambda \cong \mathscr{E}$ and $s_\lambda^* \mathscr{F}_\lambda \cong \mathscr{F}$. Thus we have an isomorphism of S-schemes

$$\Delta^r_{\mathscr{E}_\lambda \to \mathscr{F}_\lambda} \times_{S_\lambda} S \cong \Delta^r_{\mathscr{E} \to \mathscr{F}}.$$

(16.9) Determinantal schemes, Flattening stratification, and Fitting stratification (Chapter 11).

Let S be a scheme. The flattening stratification of an $\mathcal{O}_S$-module $\mathscr{H}$ can also be expressed using determinantal schemes – at least if $\mathscr{H}$ is of finite presentation.

Let $\mathscr{H}$ be an $\mathcal{O}_S$-module and assume that there exists an exact sequence of $\mathcal{O}_S$-modules

$$(16.9.1) \qquad\qquad \mathscr{E} \xrightarrow{\;u\;} \mathscr{F} \to \mathscr{H} \to 0,$$

where $\mathscr{E}$ and $\mathscr{F}$ are finite locally free $\mathcal{O}_S$-modules of constant rank. This implies that $\mathscr{H}$ is of finite presentation and in particular quasi-coherent. Conversely if $\mathscr{H}$ is of finite presentation, then such a sequence exists whenever S is affine.

The morphism u corresponds to an S-scheme morphism $i_u \colon S \to \underline{\operatorname{Hom}}(\mathscr{E}, \mathscr{F})$ (16.7.1). Let $0 \leq r \leq \operatorname{rk}(\mathscr{F}) =: n$ be an integer. We consider the determinantal schemes $\Delta^{n-r}_{\mathscr{E} \to \mathscr{F}}$ and ${}^0\Delta^{n-r}_{\mathscr{E} \to \mathscr{F}}$ as subschemes of finite presentation of $\underline{\operatorname{Hom}}(\mathscr{E}, \mathscr{F})$. We set

$$F_{\geq r}(u) := i_u^{-1}(\Delta^{n-r}_{\mathscr{E} \to \mathscr{F}}), \qquad F_{=r}(u) := i_u^{-1}({}^0\Delta^{n-r}_{\mathscr{E} \to \mathscr{F}}).$$

Then $F_{\geq r}(u)$ is a closed subscheme of finite presentation of S. Moreover $F_{=r}(u) := F_{\geq r}(u) \setminus F_{\geq r+1}(u)$ is open in $F_{\geq r}(u)$. We think of $F_{\geq r}(u)$ (resp. $F_{=r}(u)$) as the locus in S, where $\operatorname{Coker}(u) = \mathscr{H}$ has rank $\geq r$ (resp. $= r$).

Let us make this more precise. By the definition of ${}^0\Delta^r_{\mathscr{E} \to \mathscr{F}}$ (16.6.3), a morphism $f \colon T \to S$ of schemes factors through $F_{=r}(u)$ if and only if the cokernel of $f^*(u)$ is locally free of rank r. But $\operatorname{Coker}(f^*(u)) = f^* \mathscr{H}$. Thus $F_{=r}(u)$ is part of the flattening stratification for $\mathscr{H}$ (Section (11.8))

$$F_{=r}(u) = F_{=r}(\mathscr{H}).$$

In particular, the subscheme $F_{=r}(u)$ depends only on $\mathscr{H}$ and not on the choice of the presentation (16.9.1). The same assertion is true for the closed subschemes $F_{\geq r}(u)$:

Lemma 16.29. *Let $\mathscr{E}' \xrightarrow{u'} \mathscr{F}' \to \mathscr{H} \to 0$ be a second presentation of $\mathscr{H}$ by finite locally free modules. Then $F_{\geq r}(u) = F_{\geq r}(u')$.*

Proof. The equality of subschemes can be checked locally and thus we may assume that $S = \operatorname{Spec} R$, $\mathscr{H} = \tilde{P}$, $\mathscr{E} = \tilde{M}$, $\mathscr{F} = \tilde{N}$, $\mathscr{E}' = \tilde{M}'$, $\mathscr{F}' = \tilde{N}'$, where $M = R^m$, $N = R^n$, $M' = R^{m'}$, $N' = R^{n'}$ are free R-modules. Then u and u' are given by linear maps A and A' with $A \in M_{n \times m}(R)$ and $A' \in M_{n' \times m'}(R)$, where we identify linear maps between standard free modules and matrices. We obtain exact sequences of R-modules

$$R^m \xrightarrow{A} R^n \xrightarrow{g} P \to 0, \qquad R^{m'} \xrightarrow{A'} R^{n'} \xrightarrow{g'} P \to 0.$$

Let $B \colon R^{n'} \to R^n$ and $B' \colon R^n \to R^{n'}$ be linear maps with $g' = gB$ and $g = g'B'$. Then

$$C := \begin{pmatrix} I_n & 0 \\ B' & I_{n'} \end{pmatrix} \begin{pmatrix} I_n & -B \\ 0 & I_{n'} \end{pmatrix}$$

defines an automorphism of $R^n \oplus R^{n'}$ such that the following diagram commutes

$$
\begin{array}{ccccccc}
R^m \oplus R^{n'} & \xrightarrow{A \oplus I_{n'}} & R^n \oplus R^{n'} & \xrightarrow{(g,0)} & P & \longrightarrow & 0 \\
& & \downarrow{\scriptstyle C} & & \| & & \\
R^n \oplus R^{m'} & \xrightarrow{I_n \oplus A'} & R^n \oplus R^{n'} & \xrightarrow{(0,g')} & P & \longrightarrow & 0.
\end{array}
$$

The ideal generated by the $(n-r)$-minors of A is the same as the ideal generated by the $(n+n'-r)$-minors of $A \oplus I_{n'}$ and hence $F_{\geq r}(A) = F_{\geq r}(A \oplus I_{n'})$. Similarly, we see that $F_{\geq r}(A') = F_{\geq r}(I_n \oplus A')$. Moreover, the ideal generated by the $(n+n'-r)$-minors of $A \oplus I_{n'}$ is also the ideal generated by all $(n+n'-r)$-minors of all matrices $(d_{ij})_{1 \leq i \leq n+n'-r, 1 \leq j \leq n+n'}$, where $d_i = (d_{i,1}, \dots, d_{i,n+n'})$ are vectors of $\operatorname{Ker}(g, 0)$ for $i = 1, \dots, n+n'-r$. Similarly, the ideal generated by the $(n+n'-r)$-minors of $I_n \oplus A'$ can be described via $(n+n'-r)$-tuples of vectors in $\operatorname{Ker}(0, g')$. As C induces an isomorphism $\operatorname{Ker}(g, 0) \xrightarrow{\sim} \operatorname{Ker}(0, g')$, we find $F_{\geq r}(A \oplus I_{n'}) = F_{\geq r}(I_n \oplus A')$. $\square$

For every morphism $f \colon T \to S$ we have $f^{-1}(F_{\geq r}(u)) = F_{\geq r}(f^* u)$ because the formation of determinantal schemes is compatible with base change. In particular we have for every open subscheme U of S an equality $F_{\geq r}(u) \cap U = F_{\geq r}(u_{|U})$ of closed subschemes of U. This shows the following result.

Proposition 16.30. *Let S be a scheme and let $\mathscr{H}$ be an $\mathscr{O}_S$-module of finite presentation.*

(1) *For every integer $r \geq 0$ there exists a unique closed subscheme $F_{\geq r}(\mathscr{H})$ of S such that for all open subschemes U of S and for all exact sequences $\mathscr{E} \xrightarrow{u} \mathscr{F} \to \mathscr{H}_{|U} \to 0$ of $\mathscr{O}_U$-modules, where $\mathscr{E}$ and $\mathscr{F}$ are finite locally free, one has $F_{\geq r}(\mathscr{H}) \cap U = F_{\geq r}(u)$.*

(2) *One has a chain of closed subschemes of finite presentation*

(16.9.2) $\qquad \dots \subseteq F_{\geq r}(\mathscr{H}) \subseteq F_{\geq r-1}(\mathscr{H}) \subseteq \dots \subseteq F_{\geq 0}(\mathscr{H}) = S$

and $F_{=r}(\mathscr{H}) = F_{\geq r}(\mathscr{H}) \setminus F_{\geq r+1}(\mathscr{H})$.

(3) *For every morphism $f \colon T \to S$ of schemes one has*

$$f^{-1}(F_{\geq r}(\mathscr{H})) = F_{\geq r}(f^* \mathscr{H}).$$

For $r \geq 0$ we also write $F_{>r}(\mathscr{H})$ instead of $F_{\geq r+1}(\mathscr{H})$. The quasi-coherent ideal of $\mathscr{O}_S$ corresponding to the closed subscheme $F_{>r}(\mathscr{H})$ is called the r-th *Fitting ideal* of $\mathscr{H}$ and is denoted by $\mathrm{Fitt}_r \, \mathscr{H}$. If $S = \mathrm{Spec}\, R$ is affine and $\mathscr{H} = \tilde{P}$ for an R-module of finite presentation, $\mathrm{Fitt}_r \, P$ denotes the ideal of R corresponding to $\mathrm{Fitt}_r \, \mathscr{H} \subseteq \mathscr{O}_S$. We call the chain of closed subschemes (16.9.2) the *Fitting stratification of S with respect to $\mathscr{H}$*. There is an equality of closed subsets of S

$$F_{>r}\mathscr{H} = \{\, s \in S \;;\; \dim_{\kappa(s)} \mathscr{H}(s) > r \,\} = V(\mathrm{Ann}(\bigwedge^{r+1} \mathscr{H})).$$

Thus the radicals of the ideals $\mathrm{Fitt}_r \, \mathscr{H}$ and $\mathrm{Ann}(\bigwedge^{r+1} \mathscr{H})$ are equal. In fact, it can be shown (e.g., [BE]) that

$$\mathrm{Ann}(\bigwedge^{r+1} \mathscr{H})^{n-r+1} \subseteq \mathrm{Fitt}_r \, \mathscr{H} \subseteq \mathrm{Ann}(\bigwedge^{r+1} \mathscr{H}),$$

where $n \geq 0$ is an integer such that there exists an open covering (U_i) of S and a surjection $\mathscr{O}_{U_i}^n \twoheadrightarrow \mathscr{H}_{|U_i}$ for all i (such an n need not exist if S is not quasi-compact; in this case we set $\mathrm{Ann}(\bigwedge^{r+1} \mathscr{H})^{n-r+1} = 0$).

Remark 16.31. The fact that $F_{>r-1}\mathscr{H} \setminus F_{>r}\mathscr{H}$ is $F_{=r}\mathscr{H}$ of the flattening stratification can be expressed as follows. For every morphism $f \colon T \to S$ the inverse image $f^*\mathscr{H}$ is locally free of rank r if and only if $f^*(\mathrm{Fitt}_r \, \mathscr{H}) = \mathscr{O}_T$ and $f^*(\mathrm{Fitt}_{r-1} \, \mathscr{H}) = 0$.

For finitely generated modules M over a principal ideal domains the Fitting ideals determine the elementary divisors of M and hence the isomorphism class of M (Exercise 16.4). Over more general rings this does not hold: All finitely generated projective modules of a fixed rank have the same Fitting ideals. But even if R is a ring, where all finitely generated projective modules are free, the Fitting ideals of M do not determine M up to isomorphism in general.

(16.10) Divisor class group of determinantal schemes (Chapter 11).

Let k be a field, let n, m and $0 < r < \min(n, m)$ be integers, and let $\Delta = \Delta_{n \times m, k}^r$ be the corresponding determinantal scheme. Fix subsets $J \subseteq \{1, \ldots, m\}$ and $I \subseteq \{1, \ldots, n\}$ with $\#I = \#J = r$. We use the notation of Remark 16.26. Let $d_{I,J} \colon \Delta \to \mathbb{A}_k^1$ be the function on Δ that sends a matrix A to the determinant of the submatrix $A_{I,J}$. Then the principal open subscheme $\Delta[d_{I,J}^{-1}]$ of Δ, where $d_{I,J}$ is invertible, is $Y_{I,J}$. By Remark 16.26 there is an isomorphism $\Delta[d_{I,J}^{-1}] \cong \mathrm{Spec}\, k[T_1, \ldots, T_s][\delta^{-1}]$ which sends $d_{I,J}$ to δ. As $k[T_1, \ldots, T_s][\delta^{-1}]$ is a factorial ring (Proposition B.75 (1)), we have by Example 11.44

$$\mathrm{Pic}(\Delta[d_{I,J}^{-1}]) = \mathrm{Cl}(\Delta[d_{I,J}^{-1}]) = 0.$$

All irreducible components $X_1, \ldots, X_t$ of the closed subscheme $V(d_{I,J})$ of Δ have codimension 1 (Theorem 5.32). Thus Proposition 11.42 tells us that there is a surjection

$$\pi \colon \bigoplus_{i=1}^{t} \mathbb{Z}[X_i] \twoheadrightarrow \mathrm{Cl}(\Delta).$$

We want to determine t and the kernel of π. Before giving the answer to this in general, we calculate an explicit example.

Example 16.32. Consider the case $m = n = 2$ and $r = 1$. Then $\Delta = \operatorname{Spec} B$ with $B := k[a, b, c, d]/(ad - bc)$. We choose $J = I = \{1\}$ and thus $d_{I,J} = a$. Thus $V(d_{I,J}) = \operatorname{Spec} B/(a) = \operatorname{Spec} k[b, c, d]/(bc)$ and we see that $V(d_{I,J})$ has two irreducible components, namely the vanishing scheme $X_1 := V(a, b)$ of the ideal generated by the 1-minors in the same row as a and the vanishing scheme $X_2 := V(a, c)$ of the 1-minors in the same column as a. Each of them is isomorphic to $\mathbb{A}_k^2$. In $Z^1(\Delta)$ we have

$$[X_1] + [X_2] = \operatorname{cyc}(d_{I,J})$$

which is zero in $\operatorname{Cl}(\Delta)$. This is the only relation: Assume that $\lambda_1[X_1] + \lambda_2[X_2] = \operatorname{cyc}(f)$ in $Z^1(\Delta)$ for $\lambda_1, \lambda_2 \in \mathbb{Z}$ and some $f \in K(\Delta)^\times$. As $[X_1]$ and $[X_2]$ are in the kernel of $Z^1(\Delta) \to Z^1(\Delta[a^{-1}])$, the image of $\operatorname{cyc}(f)$ in $Z^1(\Delta[a^{-1}])$ is zero, i.e., f is a unit in $B[1/a]$. But in this case the isomorphism (16.7.11) takes the form

$$\varphi\colon B[1/a] \overset{\sim}{\to} k[a, b, c][1/a]$$

Thus (after possibly multiplying f with some nonzero element in k) we find $f = a^\mu$ for some $\mu \in \mathbb{Z}$. Therefore we have in $Z^1(\Delta)$

$$\lambda_1[X_1] + \lambda_2[X_2] = \operatorname{cyc}(f) = \mu \operatorname{cyc}(a) = \mu([X_1] + [X_2])$$

and hence $\lambda_1 = \lambda_2 = \mu$. This shows $\operatorname{Cl}(\Delta) = \mathbb{Z}[X_1] = \mathbb{Z}[X_2] \cong \mathbb{Z}$.

For arbitrary m, n, $0 < r < \min(m, n)$, and I, J we obtain a similar result (although the calculations are much more involved, see [BV] §8): Let X_1 be the vanishing scheme in Δ where all r-minors of the $(r \times m)$-submatrix with rows in I vanish, and let X_2 be the vanishing scheme in Δ where all r-minors of the $(n \times r)$-submatrix with columns in J vanish. Then X_1 and X_2 are integral subschemes of Δ of codimension 1, and they are the irreducible components of $V(d_{I,J})$. There is an exact sequence of free abelian groups

(16.10.1) $$0 \to \mathbb{Z}([X_1] + [X_2]) \longrightarrow \mathbb{Z}[X_1] \oplus \mathbb{Z}[X_2] \overset{\pi}{\longrightarrow} \operatorname{Cl}(\Delta) \to 0.$$

In particular $\operatorname{Cl}(\Delta) \cong \mathbb{Z}$.

(16.11) Affineness of Δ and ${}^0\Delta$ (Chapter 12).

Let S be a scheme, let $\mathscr{E}$ and $\mathscr{F}$ be finite locally free $\mathscr{O}_S$-modules and let $r \geq 0$ be an integer. For every S-scheme $h\colon T \to S$ there are functorial bijections

$$\operatorname{Hom}_S(T, \mathbb{V}(\mathscr{E} \otimes \mathscr{F}^\vee)) \overset{(11.3.4)}{=} \Gamma(T, h^*(\mathscr{E}^\vee \otimes \mathscr{F})) \overset{(7.5.4)}{=} \operatorname{Hom}_{\mathscr{O}_T}(h^*\mathscr{E}, h^*\mathscr{F}).$$

Thus we can identify the S-scheme $\underline{\operatorname{Hom}}(\mathscr{E}, \mathscr{F})$ (16.7.1) with the quasi-coherent bundle $\mathbb{V}(\mathscr{E} \otimes \mathscr{F}^\vee)$. As $\mathscr{E} \otimes \mathscr{F}^\vee$ is finite locally free, $\underline{\operatorname{Hom}}(\mathscr{E}, \mathscr{F})$ is a vector bundle over S.

In particular, $\underline{\operatorname{Hom}}(\mathscr{E}, \mathscr{F})$ and its closed subscheme $\Delta^r_{\mathscr{E} \to \mathscr{F}}$ are affine over S. Thus

$$\Delta^r_{\mathscr{E} \to \mathscr{F}} \cong \operatorname{Spec}(\operatorname{Sym}(\mathscr{E} \otimes \mathscr{F}^\vee)/\mathscr{I}_{r+1}),$$

where $\mathscr{I}_{r+1}$ is a quasi-coherent ideal of the quasi-coherent $\mathscr{O}_X$-algebra $\operatorname{Sym}(\mathscr{E} \otimes \mathscr{F}^\vee)$.

The open subscheme ${}^0\Delta^r_{\mathscr{E}\to\mathscr{F}}$ is only affine over S in trivial cases. If $\mathscr{E}$ and $\mathscr{F}$ are of constant rank m and n, respectively, and $0 < r < \min(m, n)$, then ${}^0\Delta^r_{\mathscr{E}\to\mathscr{F}}$ is not affine over S. Indeed if this was the case, then the open immersion ${}^0\Delta^r_{\mathscr{E}\to\mathscr{F}} \hookrightarrow \Delta^r_{\mathscr{E}\to\mathscr{F}}$ would be affine by Proposition 12.3 (3). As the property "affine" is stable under base change $\operatorname{Spec} k \to S$, where k is a field, and as the formation of the determinantal schemes are compatible with base change, this would imply that ${}^0\Delta := {}^0\Delta^r_{n\times m,k} \hookrightarrow \Delta := \Delta^r_{n\times m,k}$ is affine and hence that ${}^0\Delta$ is an open affine subscheme of Δ. But the complement of ${}^0\Delta^r_{n\times m,k}$ in $\Delta^r_{n\times m,k}$ is $\Delta^{r-1}_{n\times m,k}$ which is closed and irreducible of codimension ≥ 3 (16.5.1). As Δ is normal (Remark 16.14), the restriction map $\Gamma(\Delta, \mathscr{O}_\Delta) \to \Gamma({}^0\Delta, \mathscr{O}_\Delta)$ is an isomorphism by Theorem 6.45. Thus if ${}^0\Delta$ was affine, the inclusion ${}^0\Delta \hookrightarrow \Delta$ would be an isomorphism; contradiction. In fact, it is not necessary to use that Δ is normal, to obtain a contradiction (see Exercise 12.18).

(16.12) Projective determinantal schemes (Chapter 13).

If $\mathscr{E}$ and $\mathscr{F}$ are (globally) free modules over a scheme S of constant rank, we defined in Section (16.3) "projective variants" $\Delta^r_{n\times m,+,S}$ of $\Delta^r_{n\times m,S}$. We will now globalize this construction to arbitrary finite locally free $\mathscr{O}_S$-modules.

As remarked in Section (16.11) we have $\Delta^r_{\mathscr{E}\to\mathscr{F}} = \operatorname{Spec}\mathscr{A}$, where $\mathscr{A}$ is the quotient of the graded quasi-coherent $\mathscr{O}_S$-algebra $\operatorname{Sym}(\mathscr{E}\otimes\mathscr{F}^\vee)$ by a quasi-coherent ideal $\mathscr{I}_{r+1}$. We claim that $\mathscr{I}_{r+1}$ is homogeneous. Indeed, this can be checked locally on S and thus we may assume $S = \operatorname{Spec} R$ affine and $\mathscr{E} = \mathscr{O}_S^m$, $\mathscr{F} = \mathscr{O}_S^n$ for integers $n, m \geq 0$. Identifying $\operatorname{Sym}\mathscr{E}\otimes\mathscr{F}^\vee$ with the quasi-coherent graded algebra corresponding to the R-algebra $R[(T_{ij})_{1\leq i\leq n, 1\leq j\leq m}]$, the quasi-coherent ideal $\mathscr{I}_{r+1}$ corresponds to the ideal I_{r+1} generated by the $(r+1)$-minors of the $(n\times m)$-matrix whose (i,j)-th component is T_{ij} (16.2.1). This shows the claim. Thus $\mathscr{A}$ is a graded quasi-coherent $\mathscr{O}_S$-algebra. Clearly, $\mathscr{A}$ is generated by $\mathscr{A}_1$, and $\mathscr{A}_1$ is an $\mathscr{O}_S$-module of finite type (because the same assertions hold for $\operatorname{Sym}(\mathscr{E}\otimes\mathscr{F}^\vee)$). We set

$$(16.12.1) \qquad\qquad \Delta_+ := \Delta^r_{\mathscr{E}\to\mathscr{F},+} := \operatorname{Proj}\mathscr{A}.$$

Then, using the language introduced in Section (13.9), Δ is the affine cone of Δ_+. We call these scheme *projective determinantal schemes*.

The surjection $\operatorname{Sym}(\mathscr{E}\otimes\mathscr{F}^\vee) \twoheadrightarrow \mathscr{A}$ yields a closed immersion $\Delta_+ \hookrightarrow \mathbb{P}(\mathscr{E}\otimes\mathscr{F}^\vee)$. Thus we see:

Proposition 16.33. *The S-scheme Δ_+ is projective over S. In particular it is proper over S.*

(16.13) Flatness of determinantal schemes (Chapter 14).

Let S be a scheme, let $\mathscr{E}$ and $\mathscr{F}$ be finite locally free $\mathscr{O}_S$-modules and let $r \geq 0$ be an integer. Let $\Delta := \Delta^r_{\mathscr{E}\to\mathscr{F}}$ be the corresponding determinantal scheme and $\Delta_+ := \Delta^r_{\mathscr{E}\to\mathscr{F},+}$ be its projective variant.

Proposition 16.34. *The schemes Δ and Δ_+ are flat over S.*

Proof. We first show that Δ is flat over S. We may work locally on S and can thus assume that $\Delta = \Delta^r_{n\times m,R}$ for a ring R. As $\Delta^r_{n\times m,R} = \Delta^r_{n\times m,\mathbb{Z}} \otimes_\mathbb{Z} R$ and flatness is compatible with base change, we may assume $R = \mathbb{Z}$. By Proposition 14.3 (5) we may replace $\mathbb{Z}$

by $\mathbb{Z}_{(p)}$ for some prime number p. Thus we may assume that R is a discrete valuation ring. Let $s \in S = \operatorname{Spec} R$ be the special point and $\eta \in S$ be the generic point and let Δ_s and Δ_η be the special and the generic fiber, respectively. The special fiber Δ_s is integral (Remark 16.8). Thus by Proposition 14.17 it suffices to show that the generic point ξ of Δ_s lies in the closure of Δ_η. But ξ lies in the open subscheme ${}^0\Delta^r_{n \times m, R}$ which is smooth (Corollary 16.23) and in particular flat (Theorem 14.24). Thus ξ is even in the closure of the generic fiber of ${}^0\Delta^r_{n \times m, R}$.

The S-scheme Δ is the affine cone of Δ_+. As Δ is flat over S, the pointed affine cone Δ^0 is also flat over S. The projection $\Delta^0 \to \Delta_+$ is smooth and surjective (Remark 13.38) and in particular faithfully flat (Theorem 14.24). Therefore Δ_+ is flat over S by Corollary 14.12. $\qquad\square$

In fact, a more constructive study of determinantal schemes shows that $\Delta^r_{n \times m, R} = \operatorname{Spec} B$, where B is an R-algebra which is free as R-module (and in particular flat over R); see [BV] Theorem (5.3).

Corollary 16.35. *Let S be locally noetherian. If S is reduced (resp. irreducible, resp. normal, resp. Cohen-Macaulay), then $\Delta^r_{\mathcal{E} \to \mathcal{F}}$ and $\Delta^r_{\mathcal{E} \to \mathcal{F}, +}$ are reduced (resp. irreducible, resp. normal, resp. Cohen-Macaulay).*

Proof. If the affine cone $\Delta = \Delta^r_{\mathcal{E} \to \mathcal{F}}$ has any of the above properties, the same holds for $\Delta^r_{\mathcal{E} \to \mathcal{F}, +}$ by Remark 14.62. The structure morphism $f \colon \Delta \to S$ is flat and of finite presentation and hence open. Replacing S by its image, we may assume that f is faithfully flat. All fibers of f are normal, Cohen-Macaulay and geometrically integral (Remark 16.14 and Remark 16.8). Thus the corollary follows from Proposition 14.59 and Remark 14.61. $\qquad\square$

Example 16.36. Let $\sigma \colon \mathbb{P}(\mathcal{E}) \times_S \mathbb{P}(\mathcal{F}^\vee) \to \mathbb{P}(\mathcal{E} \otimes \mathcal{F}^\vee)$ be the Segre embedding (Section (8.8.6) and Remark 8.19). Clearly σ factors through $\Delta^1_{\mathcal{E} \to \mathcal{F}, +}$. In fact σ yields an isomorphism $\sigma' \colon \mathbb{P}(\mathcal{E}) \times_S \mathbb{P}(\mathcal{F}^\vee) \xrightarrow{\sim} \Delta^1_{\mathcal{E} \to \mathcal{F}, +}$. Indeed, we know the result already if S is the spectrum of a field (Example 16.7). As the formation of σ and the determinantal scheme is compatible with base change, this implies that σ' is an isomorphism on all fibers over S. But $\mathbb{P}(\mathcal{E}) \times_S \mathbb{P}(\mathcal{F}^\vee)$ is proper and flat over S (Corollary 13.42 and Example 14.2) and $\Delta^1_{\mathcal{E} \to \mathcal{F}, +}$ is proper over S (Proposition 16.33). Thus σ' is an isomorphism by Proposition 14.28.

Cubic surfaces and a Hilbert modular surface

In this part of the example chapter we study some surfaces: cubic surfaces (starting with the Clebsch cubic surface) and an example of a Hilbert modular surface. See [Ha3] V.4, [Bea] Chapter IV, [Ge], [Hu].

In this example k will always denote a field. For simplicity we assume that k is algebraically closed of characteristic zero although most arguments are also valid for fields of characteristic > 5.

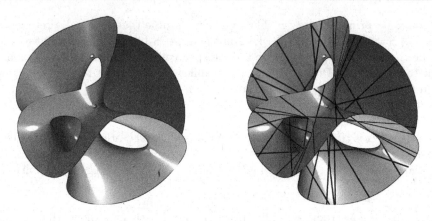

Figure 16.1: A picture of the (points in $\mathbb{R}^3$ of) the Clebsch cubic surface. On the right, the 27 lines are marked.

(16.14) The Clebsch cubic surface (Chapter 1).

Let

$$\mathcal{C} := V_+(X_0 + X_1 + X_2 + X_3 + X_4,\ X_0^3 + X_1^3 + X_2^3 + X_3^3 + X_4^3) \subset \mathbb{P}^4(k).$$

The variety $\mathcal{C}$ is named the *Clebsch cubic surface* surface after A. Clebsch who studied it around 1870. Since $V_+(X_0 + X_1 + X_2 + X_3 + X_4) \cong \mathbb{P}^3(k)$, we can also consider $\mathcal{C}$ as a closed subset, defined by one homogeneous polynomial of degree 3, in $\mathbb{P}^3(k)$:

$$\mathcal{C} \cong V_+(X_0^3 + X_1^3 + X_2^3 + X_3^3 - (X_0 + X_1 + X_2 + X_3)^3) \subset \mathbb{P}^3(k).$$

The open subsets $D_+(X_0 + X_1 + X_2 + X_3 - X_i)$, $i = 0, \dots, 3$, cover $\mathbb{P}^3(k)$, so to check that $\mathcal{C}$ is a projective variety (i.e., that $\mathcal{C}$ is irreducible), it is enough to check, for each i, that

$$\mathcal{C} \cap D_+(X_0 + X_1 + X_2 + X_3 - X_i) \subset D_+(X_0 + X_1 + X_2 + X_3 - X_i) \cong \mathbb{A}^3(k)$$

is an affine variety. We check that it is defined by an irreducible polynomial. By symmetry, it is enough to consider the case $i = 0$. (As coordinates on $D_+(X_1 + X_2 + X_3) \cong \mathbb{A}^3(k)$ we take $\frac{X_0}{X_1+X_2+X_3}$, $\frac{X_1}{X_1+X_2+X_3}$, $\frac{X_2}{X_1+X_2+X_3}$.) The dehomogenization amounts to setting $X_1 + X_2 + X_3 = 1$, so the polynomial we have to consider is

$$f_0 = X_0^3 + X_1^3 + X_2^3 + (1 - X_1 - X_2)^3 - (X_0 + 1)^3 = -3\left((X_0 + \tfrac{1}{2})^2 - g\right)$$

where

$$g = (X_2 - \tfrac{1}{2})^2 - (X_2 - 1)X_1^2 - (X_2^2 - 1)^2 X_1 \in k[X_1, X_2].$$

Since g is not a square in $k[X_1, X_2]$, it follows that f_0 is irreducible.

Let $\zeta \in k$ be a primitive 5-th root of unity. The k-variety $\mathcal{C}$ contains the following 27 lines in $\mathbb{P}^4(k)$ (and similarly, via the identification $V_+(X_0 + X_1 + X_2 + X_3 + X_4) \cong \mathbb{P}^3(k)$, contains 27 lines of $\mathbb{P}^3(k)$):

(1) The line $V_+(X_0, X_1+X_2, X_3+X_4)$ and all its images under the action of the symmetric group S_5 by permutation of the coordinates. These are 15 lines altogether.

(2) The line through $(1 : \zeta : \zeta^2 : \zeta^3 : \zeta^4)$ and $(1 : \zeta^{-1} : \zeta^{-2} : \zeta^{-3} : \zeta^{-4})$, and all its images under the action of the symmetric group S_5 by permutation of the coordinates. These are 12 lines altogether.

These containments are easily checked in terms of the explicit equations of $\mathcal{C}$. In fact, the lines in this list are the only ones on $\mathcal{C}$:

Theorem 16.37. *The Clebsch cubic surface $\mathcal{C}$ contains precisely these 27 lines.*

See also Corollary 16.42 which shows that there are only finitely many lines on any "smooth" cubic surface. For the notion of smoothness, we refer to Chapter 6. Compare also the discussion in Section (16.22). Note that the situation is much different for quadrics in $\mathbb{P}^3(k)$: Every quadric in $\mathbb{P}^3(k)$ contains infinitely many lines, see Exercise 4.19.

One checks easily that there exist precisely 10 points in $\mathcal{C}$ which lie on 3 of the lines. These points are called *Eckardt points*.

(16.15) A Hilbert modular surface (Chapter 5).

Another example that we study is a special case of a Hilbert modular surface. We will explain this terminology in Section (16.33). Here we define $\mathcal{H}$ simply via equations:

$$\mathcal{H} := V_+(\sigma_2, \sigma_4) \subset \mathbb{P}_k^4,$$

where σ_i denotes the i-th elementary symmetric polynomial in $X_0, \ldots, X_4$ (which we use as homogeneous coordinates on $\mathbb{P}_k^4$), i.e.,

$$\sigma_2 = \sum_{0 \le i < j \le 4} X_i X_j$$

$$\sigma_4 = X_{1234} + X_{0234} + X_{0134} + X_{0124} + X_{0123},$$

where we use X_{1234} as a short-hand notation for $X_1 X_2 X_3 X_4$ etc.

If σ_2 were reducible, then it would be the product of two linear polynomials, and it is easy to see that this is not the case. Therefore $V_+(\sigma_2)$ is integral and has dimension 3. By Proposition 5.40, it follows that every irreducible component of $\mathcal{H}$ has dimension 2. In fact, we have

Proposition 16.38. *The k-scheme $\mathcal{H}$ is irreducible.*

Proof. We denote by $C = V(\sigma_2, \sigma_4) \subset \mathbb{A}_k^5$ the cone over $\mathcal{H}$. The morphism $C \setminus \{0\} \to \mathcal{H}$ induces a bijection of the sets of irreducible components. It follows that every irreducible component of C is 3-dimensional, and that it is enough to show that C is irreducible.

Consider the morphism

$$f : C \to \mathbb{A}^3, \qquad (x_0, \ldots, x_4) \mapsto (\sigma_1(x), \sigma_3(x), \sigma_5(x)).$$

The morphism f is surjective, and all fibers of f are finite sets. In fact, more is true: The theory of elementary symmetric polynomials shows that the k-algebra extension $k[\sigma_1, \ldots, \sigma_5] \subset k[X_0, \ldots, X_4]$ is finite, and hence the homomorphism $k[\sigma_1, \sigma_3, \sigma_5] \to k[X_0, \ldots, X_4]/(\sigma_2, \sigma_4)$ which is precisely the homomorphism corresponding to f, is finite too. It follows that for every irreducible component $C' \subseteq C$, the homomorphism $k[\sigma_1, \sigma_3, \sigma_5] \to \Gamma(C')$ is finite, so that $\dim f(C') = 3$ by Proposition 5.12. In other words, the generic point of C' maps to the generic point η of $\mathbb{A}_k^3$.

It is therefore enough to show that the generic fiber $f^{-1}(\eta)$ of f is irreducible. This fiber is the spectrum of the ring

$$k(T_1, T_2, T_3) \otimes_{k[T_1,T_2,T_3]} k[X_0, \ldots, X_4]/(\sigma_2, \sigma_4),$$

where the homomorphism $k[T_1, T_2, T_3] \to k[X_0, \ldots, X_4]/(\sigma_2, \sigma_4)$ is the one corresponding to f, i.e., $T_1 \mapsto \sigma_1$, $T_2 \mapsto \sigma_3$, $T_3 \mapsto \sigma_5$.

This ring is a localization of the ring $k[X_0, \ldots, X_4]/(\sigma_2, \sigma_4)$, and in the localization, we can rewrite the equations $\sigma_2 = \sigma_4 = 0$ as

$$X_0 = -(\sigma_1')^{-1}\sigma_2', \qquad X_0 = -(\sigma_3')^{-1}\sigma_4',$$

where σ_i' denotes the i-th elementary symmetric polynomial in $X_1, \ldots, X_4$. This shows that the ring in question is isomorphic to a localization of

$$k[X_1, X_2, X_3, X_4]/(\sigma_3'\sigma_2' - \sigma_1'\sigma_4')$$

Thus it is sufficient to show that the latter ring is a domain, or in other words that the polynomial $\sigma_3'\sigma_2' - \sigma_1'\sigma_4'$ is irreducible. Viewing it as a quadratic polynomial in X_1, this irreducibility can be checked without too much difficulty. $\square$

(16.16) The Clebsch cubic surface as a smooth k-scheme (Chapter 6).

The computations in Section (16.14) show that the Clebsch cubic surface $\mathcal{C} = V_+(X_0 + X_1 + X_2 + X_3 + X_4, X_0^3 + X_1^3 + X_2^3 + X_3^3 + X_4^3) \subset \mathbb{P}_k^4$ is given in $\mathbb{P}_k^3$ as the vanishing scheme of one irreducible homogeneous polynomial. Therefore $\mathcal{C}$ is an integral k-scheme and it follows from Proposition 5.40 that $\mathcal{C}$ has dimension 2.

Then $\mathcal{C}$ is smooth over k: In fact, the points where the smoothness condition is violated are the points in $\mathcal{C} \cap V_+(3X_i^2 - 3X_j^2; i \neq j)$, and this set is empty; in fact, even $V_+(X_0 + \cdots + X_4) \cap V_+(3X_i^2 - 3X_j^2; i \neq j) = \emptyset$.

(16.17) Lines on smooth cubic surfaces (Chapter 6).

A *cubic surface* is a hypersurface of $\mathbb{P}_k^3$ defined by a homogeneous polynomial $f \neq 0$ of degree 3.

Proposition 16.39. *Let $S \subset \mathbb{P}_k^3$ be a smooth cubic surface. Let h be a linear polynomial in the homogeneous coordinates on $\mathbb{P}_k^3$, defining a hyperplane $H = V_+(h)$. Then the schematic intersection $V_+(f, h)$ of S and H is reduced. In other words, it is defined (inside $H \cong \mathbb{P}_k^2$) by a cubic polynomial which is either irreducible, or has a linear and an irreducible quadratic factor, or has three distinct linear factors.*

Proof. Let W, X, Y, Z be the homogeneous coordinates on $\mathbb{P}_k^3$. The intersection $V_+(f, h)$ is given, inside H, by a polynomial of degree 3. If this polynomial is not reduced, then it must have a multiple linear factor. This linear factor corresponds to a line $L \subset V_+(f, h)$. After a change of coordinates, we may assume that $h = W$, and that $L = V_+(W, X)$. Write $f = g_1(X, Y, Z) + W g_2(W, X, Y, Z)$. The above condition translates to the condition that X^2 divides g_1.

In this situation, one easily checks that all points of $V_+(W, X, g_1)$ are singular points of S. This is a contradiction to the assumption that S is smooth over k, because $V_+(W, X, g_1)$ is the zero locus of a homogeneous polynomial of degree 2 in $V_+(W, X) \cong \mathbb{P}_k^1$, and hence is non-empty. $\square$

Corollary 16.40. *Let $S \subset \mathbb{P}_k^3$ be a smooth cubic surface. Let $s \in S$ be a closed point of S. Then there exist at most 3 lines of $\mathbb{P}_k^3$ which contain s and are contained in S. If there are two or three such lines, then they are coplanar, i.e., they all lie in some hyperplane of $\mathbb{P}_k^3$.*

Proof. Let $T = T_x(S \subset \mathbb{P}_k^3)$ be the projective tangent space of S in the point s (Section (6.7)). Because S is smooth of dimension 2, this is a 2-dimensional linear subspace of $\mathbb{P}_k^3$. Every line containing s and contained in S gives rise to a line in the hyperplane $T \subset \mathbb{P}_k^3$. This proves that all such lines are coplanar, and Proposition 16.39 shows that there are at most 3 lines in $S \cap T$. $\square$

Lemma 16.41. *Let $S \subset \mathbb{P}_k^3$ be a smooth cubic surface. Let $L \subset S$ be a line. Then there exist precisely 5 hyperplanes $H \subset \mathbb{P}_k^3$ such that $L \subset H$ and that $H \cap S$ is the union of three different lines.*

Proof. To ease the notation, we use W, X, Y, Z as homogeneous coordinates on $\mathbb{P}_k^3$. After a change of coordinates, we may assume that the line L is given as $L = V_+(Y, Z)$. Let f be a homogeneous polynomial with $S = V_+(f)$. Since $L \subset S$ by assumption, Y or Z occur in every monomial in f, so we can write

$$f = AW^2 + 2BWX + CX^2 + 2DW + 2EX + F,$$

where $A, B, C, D, E, F \in k[Y, Z]$ and A, B, C are linear, D, E are quadratic, and F is cubic.

All hyperplanes $H \subset \mathbb{P}_k^3$ containing L are of the form $H = V_+(\alpha Y - \beta Z)$. Let us first assume that $\beta \neq 0$; then we can even assume that $\beta = 1$, i.e., $H = V_+(\alpha Y - Z) \cong \mathbb{P}_k^2$, where we can use W, X, Y as homogeneous coordinates in $\mathbb{P}^2$, and $S \cap H$ is defined inside H by the cubic polynomial

$$
\begin{aligned}
f_H :=\ & A(Y, \alpha Y)W^2 + 2B(Y, \alpha Y)WX + C(Y, \alpha Y)X^2 \\
& + 2D(Y, \alpha Y)W + 2E(Y, \alpha Y)X + F(Y, \alpha Y) \\
=\ & Y\big(A(1, \alpha)W^2 + 2B(1, \alpha)WX + C(1, \alpha)X^2 \\
& + 2D(1, \alpha)WY + 2E(1, \alpha)XY + F(1, \alpha)Y^2\big).
\end{aligned}
$$

Since the polynomials $A, \ldots, E$ are homogeneous, we can conclude that for $(\alpha : \beta) \in \mathbb{P}^1(k)$ the intersection $H \cap S$ consists of three lines if and only if the quadric $V_+(q)$ is the union of two lines, where

$$q = A(\beta, \alpha)W^2 + 2B(\beta, \alpha)WX + C(\beta, \alpha)X^2 + 2D(\beta, \alpha)WY + 2E(\beta, \alpha)XY + F(\beta, \alpha)Y^2.$$

This statement even holds without the assumption that $\beta \neq 0$. By Proposition 16.39, $V_+(q)$ is reduced, so it is singular if and only if it is the union of two distinct lines. Now q is singular if and only if the matrix of the corresponding bilinear form does not have full rank, cf. Example 6.16 (3). This means that the set of points $(\alpha : \beta) \in \mathbb{P}^1(k)$ such that the intersection $H \cap S$ is singular is precisely the zero set of the polynomial

$$\Delta = \det \begin{pmatrix} A & B & D \\ B & C & E \\ D & E & F \end{pmatrix} \in k[Y, Z].$$

Note that Δ is homogeneous of degree 5. Thus we have to show that $V_+(\Delta) \subset \mathbb{P}_k^1$ consists of precisely 5 points, or in other words, that Δ does not have a double zero. To do so, we have to use the assumption that S is smooth.

Let us consider a zero of the polynomial Δ. After another change of coordinates, we may assume that the zero is given by $Z = 0$, i.e., $(\beta : \alpha) = (1 : 0)$, $H = V_+(Z)$. We have to show that Δ is not divisible by Z^2. Let $P \in \mathbb{P}^3(k)$ be the singular point of $V_+(q) \subset H \subset \mathbb{P}_k^3$. First assume that $P \in L$, i.e., all the three lines which make up $H \cap S$ go through the point P. We can then change coordinates such that $q = X^2 + Y^2$ (so $P = (1 : 0 : 0 : 0)$). This means that A, B, D, and E are all divisible by Z, but C and F are not. We obtain that

$$\Delta \equiv ACF \not\equiv 0 \bmod Z^2.$$

Similarly, if $P \notin L$, then we may assume that $q = WX$ (so $P = (0 : 0 : 1 : 0)$), and

$$\Delta \equiv B^2 F \bmod Z^2.$$

Here B is not divisible by Z, and since P is a smooth point of S, we have that F is not divisible by Z^2. The lemma is proved. $\square$

With the notation of the lemma, we obtain that there exist exactly 10 lines contained in S which meet L (and are different from L). They can be numbered as 5 pairs (L_i, L_i') of lines, $i = 1, \ldots, 5$, such that for all i the lines L, L_i, L_i' are coplanar (i.e., the intersection of the i-th hyperplane of the lemma is $L \cup L_i \cup L_i'$), and that for $i \neq j$ we have

$$(L_i \cup L_i') \cap (L_j \cup L_j') = \emptyset.$$

From this point, by a mostly combinatorial argument, one can prove that S contains exactly 27 lines; see [Bea] Ch. IV. We content ourselves with the following result:

Corollary 16.42. *Let $S \subset \mathbb{P}_k^3$ be a smooth cubic surface. Then S contains at most finitely many lines of $\mathbb{P}_k^3$.*

Proof. Assume that L_1 is a line contained in S. (If no such line exists, then the statement of the corollary is satisfied. But in fact, we will see in Theorem 16.45 that every smooth cubic hypersurface contains a line.) Lemma 16.41 implies that there exists a hyperplane $H \subset \mathbb{P}_k^3$ such that the intersection $H \cap S$ is the union of three distinct lines L_1, L_2, L_3. Every line of $\mathbb{P}_k^3$ meets H, therefore every line $L \subset S$ meets at least one of L_1, L_2, L_3. It is therefore enough to show that there exist only finitely many lines in S which meet L_1. But we know from Lemma 16.41 that there are only finitely many hyperplanes in $\mathbb{P}_k^3$ which contain L_1 and another line of S. $\square$

(16.18) Singularities of the Hilbert modular surface $\mathcal{H}$ (Chapter 6).

We want to describe the singular points of $\mathcal{H} = V_+(\sigma_2, \sigma_4) \subset \mathbb{P}_k^4$, where σ_i denotes the i-th elementary symmetric polynomial in $X_0, \ldots, X_4$. We introduce some further notation:

$$\sigma_2^i = \sum_{\substack{0 \le a < b \le 4 \\ a,b \notin \{i\}}} X_a X_b, \qquad \sigma_2^{ij} = \sum_{\substack{0 \le a < b \le 4 \\ a,b \notin \{i,j\}}} X_a X_b$$

To determine the singular points of $\mathcal{H}$, we compute the partial derivatives

$$\frac{\partial \sigma_2}{\partial X_i} = \sigma_1 - X_i$$

$$\frac{\partial \sigma_4}{\partial X_i} = \sigma_3 - X_i \sigma_2^i$$

and the (2×2)-minors of the Jacobian matrix. The minor corresponding to the columns i, j is

$$\begin{aligned}
M_{ij} &= \frac{\partial \sigma_2}{\partial X_i}\frac{\partial \sigma_4}{\partial X_j} - \frac{\partial \sigma_2}{\partial X_j}\frac{\partial \sigma_4}{\partial X_i} \\
&= (\sigma_1 - X_i)(\sigma_3 - X_j \sigma_2^j) - (\sigma_1 - X_j)(\sigma_3 - X_i \sigma_2^i) \\
&= (X_j - X_i)(\sigma_3 - \sigma_1 \sigma_2^{ij} - X_{ij}(\sigma_1 - X_i - X_j)) \\
&= (X_j - X_i)((X_j - \sigma_1)X_i^2 + (X_j - \sigma_1)^2 X_i + \sigma_3 - \sigma_1 \sigma_2^j),
\end{aligned}$$

where we have written $\sigma_2^i = \sigma_2^{ij} + X_j(\sigma_1 - X_i - X_j)$, and similarly for σ_2^j, to get from the second to the third line. Note that by Example 6.16 (2) we can in fact compute the singular locus by using the homogeneous equations defining $\mathcal{H}$.

The singular locus of $\mathcal{H}$ is the set $V_+(M_{ij}, i, j) \cap \mathcal{H}$. Now consider $x_1, \ldots, x_4 \in k$ such that $x = (x_0 : \cdots : x_4) \in X(k)$ is a singular point of $\mathcal{H}$. Write $\sigma_1(x) := \sigma_1(x_0, \ldots, x_4) = \sum x_i$; with analogous notation we have $\sigma_3(x)$, and $\sigma_2(x) = 0$. Because $\sigma_2(x) = 0$, the x_i cannot be all equal; in particular, there exists j with $x_j \neq \sigma_1(x)$, and by symmetry we may assume that $j = 0$. Since the chosen point is assumed to be singular, in particular all the minors M_{i0}, $i = 1, \ldots, 4$ vanish. Now if $i \in \{1, \ldots, 4\}$ with $x_i \neq x_0$, then $M_{i0}(x) = 0$ implies that x_i is a zero of the quadratic polynomial

$$(x_0 - \sigma_1(x))T^2 + (x_0 - \sigma_1(x))T + \sigma_3(x) - \sigma_1(x)\sigma_2^0(x) \in k[T].$$

This polynomial can have at most two distinct zeros, and we conclude that the set $\{x_0, x_1, x_2, x_3, x_4\}$ has at most three elements.

Now we distinguish cases: If three of the coordinates of x are equal, then it is easy to see that we must have $x \in S$, where

$$S := \{(1:0:0:0:0), (0:1:0:0:0), (0:0:1:0:0), (0:0:0:1:0), (0:0:0:0:1)\},$$

and these five points are easily seen to lie in $V_+(M_{ij}, i, j) \cap \mathcal{H}$, so they are singular points of $\mathcal{H}$.

Otherwise, we may assume, after possibly renumbering again, that $x_0 = x_1$, $x_2 = x_3$. One shows by an explicit computation that $V_+(X_0 - X_1, X_2 - X_3) \cap V_+(M_{ij}, i, j) \cap \mathcal{H} = \emptyset$, so this case does not contribute any singular points. Therefore we see that

$$\mathcal{H}_{\text{sing}} := \mathcal{H} \setminus \mathcal{H}_{\text{sm}} = S.$$

In particular, we see that the singular locus of $\mathcal{H}$ is 0-dimensional. In other words, $\mathcal{H}$ is regular in codimension 1, and we can use this to show that $\mathcal{H}$ is a normal variety. In fact, we know that $\mathcal{H}$ is of dimension 2 and is defined by 2 equations in $\mathbb{P}_k^4$. For the local rings of $\mathcal{H}$, this means that they are quotients of a 4-dimensional regular local ring by an ideal generated by 2 elements. These 2 elements form a regular sequence because of dimension reasons. By Proposition B.86, all the local rings are Cohen-Macaulay rings. Therefore it follows from Serre's criterion for normality, Proposition B.81, that $\mathcal{H}$ is normal.

(16.19) Smooth cubic surfaces are rational (Chapter 9).

Lemma 16.43. *Let $S \subset \mathbb{P}_k^3$ be a smooth cubic surface which contains a line of $\mathbb{P}_k^3$. Then S contains two disjoint lines of $\mathbb{P}_k^3$.*

Proof. Let $L \subset S$ be a line. By Lemma 16.41 we find two hyperplanes $H_1, H_2 \subset \mathbb{P}_k^3$ which contain L and intersect S in three lines. Pick lines $L_i \subset H_i \cap S$, $i = 1, 2$, which are lines different from L. Because of Proposition 16.39 we have $L \cap L_1 \cap L_2 = \emptyset$. Furthermore, for $i \in \{1, 2\}$, L and any point of $L_i \setminus L$ spans H_i, so the assumption that $H_1 \neq H_2$ implies that $L_1 \cap L_2 = \emptyset$. $\square$

Proposition 16.44. *Let $S \subset \mathbb{P}_k^3$ be a smooth cubic surface. Then S is rational, i.e., S and $\mathbb{P}_k^2$ are birationally equivalent.*

Proof. Let $L_1, L_2 \subset S$ be disjoint lines of $\mathbb{P}_k^3$. Clearly it is enough to prove that S and $L_1 \times L_2 \cong \mathbb{P}_k^1 \times \mathbb{P}_k^1$ are birationally equivalent (Exercise 9.18). We begin by constructing a rational map $\varphi\colon S \dashrightarrow L_1 \times L_2$. For every closed point $x \in \mathbb{P}_k^3 \setminus (L_1 \cup L_2)$, the plane generated by x and L_2 meets L_1 in a unique point $\varphi_1(x)$, and symmetrically, the plane generated by x and L_1 meets L_2 in a unique point $\varphi_2(x)$. Computing these intersection points in terms of equations of S, L_1, L_2, one sees that their coordinates are given by rational functions, so that we obtain a morphism

$$\mathbb{P}_k^3 \setminus (L_1 \cup L_2) \to L_1 \times L_2, \quad x \mapsto (\varphi_1(x), \varphi_2(x)).$$

By restriction we get the desired rational map $\varphi\colon S \dashrightarrow L_1 \times L_2$.

Now we construct a rational map $\psi\colon L_1 \times L_2 \dashrightarrow S$. Given $p_1 \in L_1(k)$, $p_2 \in L_2(k)$, let $L \subset \mathbb{P}_k^3$ be the unique line through p_1 and p_2. There exist only finitely many lines contained in S (Corollary 16.42) and therefore except for finitely many pairs (p_1, p_2), the line L is not contained in S. This means that $L \cap S$ is given inside $L \cong \mathbb{P}_k^1$ as the zero set of a non-vanishing homogeneous polynomial of degree 3, i.e., L meets S in three points p_1, p_2 and $\psi(p_1, p_2)$. (We count the intersection points with multiplicities.)

It is clear that φ and ψ are inverse to each other where they are defined. $\square$

The domain of definition of $\mathbb{P}_k^3 \dashrightarrow L_1 \times L_2$ is $\mathbb{P}_k^3 \setminus (L_1 \cup L_2)$, but one can show that the map $\varphi\colon S \dashrightarrow L_1 \times L_2$ extends to a morphism $S \to L_1 \times L_2$.

(16.20) Picard group of the Clebsch cubic surface (Chapter 11).

Each of the 27 lines on the Clebsch cubic surface $\mathcal{C}$ is a divisor on $\mathcal{C}$, and we obtain a map $\mathbb{Z}^{27} \to \mathrm{Pic}(\mathcal{C})$. One can show that this homomorphism is surjective. It is not injective, however: Assume that $H_1, H_2 \subset \mathbb{P}_k^3$ are different hyperplanes which each intersect $\mathcal{C}$ in the union of three lines L_{1j} resp. L_{2j}, $j = 1, 2, 3$ (compare Proposition 16.39). In the case at hand, we could take $H_1 = V_+(X_0)$, $H_2 = V_+(X_1)$, for instance, but the same argument applies to every smooth cubic surface. Then the divisors $[L_{11}] + [L_{12}] + [L_{13}]$ and $[L_{21}] + [L_{22}] + [L_{23}]$ are linearly equivalent, because the divisors $[H_1]$ and $[H_2]$ of $\mathbb{P}_k^3$ are linearly equivalent.

One can show that $\mathrm{Pic}(\mathcal{C}) \cong \mathbb{Z}^7$, see [Ha3], Proposition V.4.8. See also the sketchy discussion in Section (16.21).

(16.21) Cubic surfaces in $\mathbb{P}_k^3$ as blow-ups (Chapter 13).

Let us make a remark, without any proofs, about Proposition 16.44 where we constructed, given a smooth cubic surface S over k, a birational map $S \to \mathbb{P}_k^1 \times_k \mathbb{P}_k^1$. As stated there, this rational map actually extends to a morphism $S \to \mathbb{P}^1 \times \mathbb{P}^1$ (of course, its inverse does not extend to a morphism). Now one can show that this morphism $S \to \mathbb{P}^1 \times \mathbb{P}^1$ can actually be identified with the blow-up of 5 points in $\mathbb{P}^1 \times \mathbb{P}^1$.

Furthermore, it is not hard to check that the k-scheme obtained as the blow-up of a point in $\mathbb{P}^1 \times \mathbb{P}^1$ is isomorphic to the scheme obtained by blowing up 2 points in $\mathbb{P}_k^2$. As a consequence, we can also view S as the blow-up of 6 points in $\mathbb{P}^2$. Conversely, whenever $p_1, \ldots, p_6 \in \mathbb{P}^2(k)$ are such that no three p_i lie on a line of $\mathbb{P}_k^2$, and not all six lying on a quadric, then the blow-up of $\mathbb{P}_k^2$ in the points $p_1, \ldots, p_6$ is isomorphic to a smooth cubic surface in $\mathbb{P}_k^3$.

Using the description of S as a blow-up $\pi \colon S \to \mathbb{P}_k^2$, we can outline why $\mathrm{Pic}(S) \cong \mathbb{Z}^7$. In fact, we have an exceptional divisor over each of the blown-up points, each of which gives rise to an element of $\mathrm{Pic}(S)$, and also the pull-back $\pi^* \mathcal{O}(1)$. These seven elements can be shown to be a $\mathbb{Z}$-basis of $\mathrm{Pic}(S)$. See [Ha3] V.4, in particular Proposition V.4.8, and [Bea] Chapter IV.

(16.22) Lines on cubic surfaces (Chapter 14).

A dimension argument now allows us to show that any (not necessarily smooth) cubic surface contains a line.

Theorem 16.45. *Let $X \subset \mathbb{P}_k^3$ be a cubic surface, i.e., $X = V_+(f)$ for some homogeneous polynomial f of degree 3. Then X contains a line of $\mathbb{P}_k^3$.*

Proof. Let P be the projective space of homogeneous polynomials in $T_0, \ldots, T_3$ of degree 3 over k, up to scalars, i.e., $P \cong \mathbb{P}^{19}$. Let $G = \mathrm{Grass}_{2,4}$ be the Grassmannian of lines in $\mathbb{P}_k^3$. We know that $\dim G = 4$ (see Corollary 8.15).

We attach to each k-valued point $p \in P(k)$ its zero set $V_+(p) \subset \mathbb{P}_k^3$. Then

$$Z(k) := \{(\ell, p) \in G(k) \times P(k); \; \ell \subset V_+(p)\}$$

is a closed subvariety (in the sense of Chapter 1) of $(G \times P)(k)$, i.e., it is the set of k-valued points of a unique reduced closed subscheme $Z \subset G \times P$.

The projections give rise to morphisms $\alpha \colon Z \to G$, $\beta \colon Z \to P$. We first study the fibers of α. It is clear that all fibers of α over closed points are isomorphic (formally, the group $\mathrm{PGL}_4(k)$ acts on $\mathbb{P}_k^3$, and hence on G, P and Z, the action on $G(k)$ is transitive and α is equivariant with respect to this action). So let us compute the fiber over the line $\ell = V_+(T_0, T_1) \subset \mathbb{P}_k^3$. This line is contained in a cubic $V_+(p)$ if and only if

$p(0,0,T_2,T_3) \in k[T_2,T_3]$ is the zero polynomial, i.e., if and only if the coefficients of T_2^3, $T_2^2 T_3$, $T_2 T_3^2$ and T_3^3 vanish. Therefore all fibers of α over closed points are isomorphic to the projective space of dimension $\dim P - 4$. It follows, for instance from Theorem 14.112, that all fibers of α have dimension $\dim P - 4$. Using Corollary 14.121, we obtain that $\dim Z = \dim P$.

Now consider the morphism $\beta \colon Z \to P$. The statement of the theorem amounts to the surjectivity of β. If β is not surjective, then its image, which is closed because β is a proper morphism, and irreducible, has codimension ≥ 1. By Proposition 14.109 (2) and the semi-continuity of fiber dimensions (Theorem 14.112), we would find that all non-empty fibers of β have dimension ≥ 1.

Therefore it suffices to find a single point $p \in P(k)$ such that $\beta^{-1}(p)$ is non-empty and finite. Take p to be the point corresponding to a smooth cubic surface (e.g., the Clebsch cubic) and use Corollary 16.42. $\qquad\square$

Note that the morphism β in the proof of the theorem also has fibers which are not finite: for instance the singular cubic $V_+(T_0 T_1 T_2)$ which is just the union of three hyperplanes contains infinitely many lines. But for smooth cubic surfaces, the existence of a line implies the following result, as indicated after Lemma 16.41.

Theorem 16.46. *Let $X \subset \mathbb{P}_k^3$ be a smooth cubic surface. Then X contains exactly 27 different lines of $\mathbb{P}_k^3$.*

Cyclic quotient singularities

In this part of the chapter we will study some quotients of surfaces by cyclic groups. In this example k will always denote an algebraically closed field of characteristic zero.

(16.23) A cyclic quotient singularity (Chapter 1).

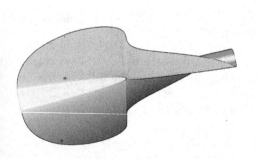

Let $A = k[X^3, Y^3, XY]$ considered as a k-subalgebra of $B = k[X,Y]$, and let $f \colon \mathbb{A}^2(k) \to Z$ be the corresponding morphism of schemes. Writing A as the quotient $k[S,T,U]/(U^3 - ST)$ (via $S \mapsto X^3$, $T \mapsto Y^3$, $U \mapsto XY$), we can write $Z \cong V(U^3 - ST) \subset \mathbb{A}^3(k)$. Note that the polynomial $U^3 - ST$ is irreducible, so that Z is in fact an affine variety.

We continue to view Z as a closed subvariety of $\mathbb{A}^3(k)$. For $z = (0,0,0) \in Z$, the fiber $f^{-1}(z)$ is the single point $(0,0)$, but for all other points $z \in Z$, the fiber $f^{-1}(z)$ consists of 3 different points.

Figure 16.2: The solution set (in $\mathbb{R}^3$) of the equation $U^3 - ST = 0$.

In fact, if we let the group G of third roots of unity in $k^\times$ (so $G \cong \mathbb{Z}/3$) act on $\mathbb{A}^2(k)$ by $\zeta \cdot (x,y) = (\zeta x, \zeta^2 y)$, then the fibers of f are exactly the G-orbits. So we can consider Z as the "quotient" $\mathbb{A}^2(k)/G$; compare Sections (12.7) and (16.28).

(16.24) A resolution of a cyclic quotient singularity (Chapter 3).

From now on we consider Z as the scheme

$$Z := \operatorname{Spec} k[S, T, U]/(U^3 - ST) \cong \operatorname{Spec} k[X^3, Y^3, XY]$$

and we think of Z as the quotient of $\mathbb{A}_k^2$ by the group of third roots of unity (this will be made more precise in Section (16.28)). We will see that Z has an "isolated singularity" in 0 (Section (16.26)) and that there is a natural way to "resolve this singularity" (Section (16.29)).

Here we will define this resolution by a gluing process. Consider the following gluing datum: We set $U_1 = U_2 = U_3 = \mathbb{A}_k^2$, where as coordinates on U_i we use u_i, v_i. We glue

$$U_1 \supset D(u_1) \cong D(u_2) \subset U_2 \text{ via } u_2 = u_1^{-1},\ v_2 = u_1^2 v_1,$$

$$U_1 \supset D(v_1) \cong D(v_3) \subset U_3 \text{ via } v_3 = v_1^{-1},\ u_3 = v_1^2 u_1,$$

$$U_2 \supset D(u_2 v_2) \cong D(u_3 v_3) \subset U_3 \text{ via } u_3 = u_2^3 v_2^2,\ v_3 = u_2^{-2} v_2^{-1},$$

and denote the scheme obtained by gluing by $\tilde{Z}$. We have maps

$$U_1 \to Z, \qquad (u_1, v_1) \mapsto (u_1 v_1^2, u_1^2 v_1, u_1 v_1),$$

$$U_2 \to Z, \qquad (u_2, v_2) \mapsto (u_2^3 v_2^2, v_2, u_2 v_2),$$

$$U_3 \to Z, \qquad (u_3, v_3) \mapsto (u_3, u_3^2 v_2^3, u_3 v_3),$$

which glue to a morphism

$$\pi \colon \tilde{Z} \to Z.$$

One checks that the restriction $\pi^{-1}(Z \setminus \{(0,0,0)\}) \to Z \setminus \{(0,0,0)\}$ is an isomorphism. The closed subscheme $\pi^{-1}((0,0,0))$ (with the reduced scheme structure) can be identified with the union (inside a $\mathbb{P}_k^2$) of two projective lines intersecting in a single point.

(16.25) Fibers of a cyclic quotient singularity (Chapter 4).

Let $A = k[XY, X^3, Y^3] \cong k[U, T, S]/(U^3 - TS)$, and $Z = \operatorname{Spec} A$. Let $f \colon \mathbb{A}_k^2 \to Z$ be the morphism corresponding to the inclusion $A \hookrightarrow k[X, Y]$. We compute the fibers $f^{-1}(z)$ as k-schemes for the closed points $z \in Z(k)$. Consider Z as $V(U^3 - TS) \subset \mathbb{A}_k^3$, and write $z = (s, t, u)$ accordingly. By definition, we have

$$f^{-1}(z) = \operatorname{Spec} k[X, Y] \otimes_A A/(U - u, T - t, S - s) = k[X, Y]/(XY - u, X^3 - t, Y^3 - s).$$

If $z = (0, 0, 0)$ (with respect to the coordinates S, T, U), then

$$f^{-1}(0, 0, 0) = \operatorname{Spec} k[X, Y]/(XY, X^3, Y^3),$$

a scheme which topologically has a single point, but whose affine coordinate ring is a 5-dimensional k-vector space (with basis $1, X, Y, X^2, Y^2$).

On the other hand, let $z = (s, t, u) \neq (0, 0, 0)$, with $t \neq 0$ (otherwise $s \neq 0$, which leads to a symmetric situation). Then it is easy to see that $f^{-1}(z) \cong \operatorname{Spec} k[X]/(X^3 - t)$ and thus $f^{-1}(z) \cong \operatorname{Spec} k^3$ because the polynomial $X^3 - t$ has three distinct zeros. Therefore $f^{-1}(z)$ is the union of three copies of $\operatorname{Spec} k$ in this case. In particular, its affine coordinate ring has dimension 3. So this dimension (which is the number of points in the fiber counted with multiplicity, cf. Section (5.14)), jumps at the point $(0, 0, 0)$. This reflects the fact that the morphism f is not flat, see Section (16.30).

(16.26) Singularities of a cyclic quotient singularity (Chapter 6).

We will now study the singularities of $Z = \operatorname{Spec}[S, T, U]/(U^3 - ST)$. The partial derivatives are

$$\frac{\partial(U^3 - ST)}{\partial S} = -T, \quad \frac{\partial(U^3 - ST)}{\partial T} = -S, \quad \frac{\partial(U^3 - ST)}{\partial U} = 3U^2,$$

and the only point (of Z) where they all vanish is $(0,0,0)$. Therefore we find that the smooth locus of Z is $Z_{\mathrm{sm}} = Z \setminus \{(0,0,0)\}$. The tangent space at the singular point $(0,0,0)$ has dimension 3.

Recall the isomorphism $A := k[S, T, U]/(U^3 - ST) \cong k[X^3, Y^3, XY] \subset k[X, Y]$ (see Section (16.23)). It shows that we can identify A with the subring of polynomials invariant under the transformation $X \mapsto \zeta X$, $Y \mapsto \zeta^2 Y$, where $\zeta \in k$ is a fixed primitive 3rd root of unity. One checks that the field of fractions of A admits an analogous description as a subfield of $k(X, Y)$. Using this, it is easy to see that A is integrally closed in its field of fractions, i.e., that Z is normal.

The scheme $\tilde{Z}$ constructed in Section (16.24) has by definition an open covering of schemes isomorphic to $\mathbb{A}_k^2$. In particular $\tilde{Z}$ is smooth over k.

(16.27) Separatedness of the resolution (Chapter 9).

Proposition 9.15 allows us to check that the scheme $\tilde{Z}$ defined in Section (16.24) is separated. For instance, with the notation of Section (16.24), $U_1 \cap U_2$ is the affine open $D(u_1) \subset U_1$, and its coordinate ring $k[u_1, u_1^{-1}, v_1]$ is obviously generated by u_1, v_1, $u_2 = u_1^{-1}$, v_2.

In Section (16.29) we will see that $\tilde{Z}$ is a blow-up (Chapter 13) of Z in its singular point. This also shows that $\tilde{Z}$ is separated.

(16.28) Cyclic quotient surface singularities (Chapter 12).

Let $V = \mathbb{A}_k^2$, and fix natural numbers $0 < a < r$. Let G be the group of r-th roots of unity in $k^\times$, so $G \cong \mathbb{Z}/r$ non-canonically. We consider the G-action on V given by

$$\zeta \cdot v = \begin{pmatrix} \zeta & \\ & \zeta^a \end{pmatrix} v.$$

Correspondingly, $\zeta \in G$ acts on $k[X, Y]$ (according to the normalization in Section (12.7)) by $\zeta \cdot X = \zeta^{-1} X$ and $\zeta \cdot Y = \zeta^{-a} Y$. Let $Z := Z_{a,r} := V/G$ be the quotient of this action, i.e., $Z = \operatorname{Spec} k[X, Y]^G$, and let $p \colon \mathbb{A}_k^2 \to Z$ be the finite surjective quotient morphism corresponding to the inclusion $k[X, Y]^G \subset k[X, Y]$. Clearly Z is integral and $\dim Z = \dim \mathbb{A}_k^2$ because p is finite and surjective (Proposition 12.12). By Example 12.48, Z is normal.

The normality of Z implies that the singular locus of Z has codimension ≥ 2 (Proposition 6.40). As $\dim Z = 2$, there exist at most finitely many singular points in Z. If $p(x, y) \in Z(k)$ be a singular point, it is easy to convince oneself that then all points $p(\alpha x, \alpha y)$, $\alpha \in k$, are singular as well. Therefore the only point of Z which is possibly singular is $p(0, 0)$. The singularity at this point is called a cyclic quotient singularity.

In all non-trivial cases the point $p(0, 0)$ is in fact a singular point (because the group G, seen as a subgroup of $\mathrm{GL}_2(\mathbb{C})$, is not generated by pseudo-reflections unless it is trivial). See the references in Remark 12.31.

Proposition 16.47. *The monomials*

$$X^i Y^j, \quad 0 \le i, j \le n, \ (i,j) \ne (n,n), \ i + aj \equiv 0 \bmod n,$$

are a generating system of the k-algebra $k[X,Y]^G$.

Proof. It is clear that *all* monomials $X^i Y^j$ with $i + aj \equiv 0 \bmod n$ are a k-basis of $k[X,Y]^G$. Furthermore, we can write every monomial $X^i Y^j \in k[X,Y]^G$ with $i > n$, or $j > n$, or $i = j = n$, as a product of X^n, Y^n and other monomials of the form $X^i Y^j$ with $0 \le i, j \le n$, $(i,j) \ne (n,n)$ and $i + aj \equiv 0 \bmod n$. $\square$

Example 16.48. The simplest example is $r = 2$, $a = 1$, in which case we have $G \cong \mathbb{Z}/2$, $k[X,Y]^G = k[X^2, XY, Y^2] \cong k[S,T,U]/(U^2 - ST)$, the final isomorphism being given by $S \mapsto X^2$, $T \mapsto Y^2$, $U \mapsto XY$. The spectrum of this ring can be identified with the "nilpotent cone", i.e., the variety of nilpotent matrices of trace 0 in $M_{2 \times 2, k}$.

Example 16.49. Now let us consider the case $r = 3$, $a = 2$. Here $G \cong \mathbb{Z}/3$, and

$$k[X,Y]^G = k[X^3, XY, Y^3] \cong k[S,T,U]/(U^3 - ST),$$

the isomorphism being given by $S \mapsto X^3$, $T \mapsto Y^3$, $U \mapsto XY$. This is exactly the example considered above in Sections (16.23), (16.24), (16.25), (16.26).

Example 16.49 shows that the generating system given in the proposition is not minimal in general. One can find a minimal generating system using the Hirzebruch-Jung algorithm which relates the invariant ring to the development of $\frac{r}{r-a}$ as a continued fraction $a_1 - 1/(a_2 - 1/\dots)$. See [Ri].

(16.29) The resolution as blow-up (Chapter 13).

We continue the study of the cyclic quotient singularity $Z = \operatorname{Spec} k[U,T,S]/(U^3 - TS)$. As we have seen before, the singular locus of Z is $\{(0,0,0)\} = V(S,T,U)$. Let us compute the blow-up $\tilde{Z}$ of Z with respect to the ideal (S,T,U).

Similarly as described in Example 13.93, we can cover $\tilde{Z}$ by 3 charts D_U, D_S, D_T, corresponding to the three generators U, T, S of the ideal under consideration:

(1) To describe D_U, we introduce new variables $U' = U$, $T' = \frac{T}{U}$, $S' = \frac{S}{U}$, and obtain

$$D_U = \operatorname{Spec} k[U', T', S']/((U')^3 - (U'T')(U'S'), U'\text{-torsion})$$
$$= \operatorname{Spec} k[U', T', S']/(U' - T'S') \cong \operatorname{Spec} k[T', S'] = \mathbb{A}_k^2.$$

Inside this chart, the exceptional divisor is given by $U = T = S = 0$, i.e., in terms of the right hand side, as $V(T'S') \subset \mathbb{A}_k^2$. Geometrically, it is the union of the coordinate axes.

(2) To describe D_T, we introduce new variables $U' = \frac{U}{T}$, $T' = T$, $S' = \frac{S}{T}$, and obtain

$$D_T = \operatorname{Spec} k[U', T', S']/((U'T')^3 - S'(T')^2, T'\text{-torsion})$$
$$= \operatorname{Spec} k[U', T', S']/((U')^3 T' - S') \cong k[U', T'],$$

where the exceptional divisor is $V(T')$.

(3) Since everything is symmetric with respect to T and S, the third case is analogous to the second one, and we obtain $D_S \cong k[U', S']$, and the exceptional divisor is $V(S')$. Analyzing how the three charts are glued, we see that $\pi\colon \tilde{Z} \to Z$ is precisely the morphism constructed in Section (16.24). In particular, we find that the exceptional divisor is the union of two projective lines, intersecting "transversally" (see Exercise 6.7) in a single point. Note in particular that the exceptional divisor is not irreducible.

By Corollary 13.97 the morphism $\pi\colon \tilde{Z} \to Z$ is birational projective surjective and an isomorphism over the smooth locus of Z. As $\tilde{Z}$ is smooth over k, the morphism π is a resolution of singularities in the sense of Section (13.23).

(16.30) Non-flatness of quotients (Chapter 14).

The computation of the schematic fibers of the quotient morphism $f\colon \mathbb{A}_k^2 \to Z = \operatorname{Spec} k[X^3, Y^3, XY]$ shows that the finite morphism f is not flat (otherwise f would be finite locally free and its degree (12.6.1) would be constant). Note that we cannot apply Corollary 14.129 because Z is not regular.

The resolution of singularities $\pi\colon \tilde{Z} \to Z$ is an isomorphism of the smooth locus of Z but the fiber over the singular point has dimension 1. Thus π cannot be open by Theorem 14.116 and a fortiori π is not flat (Theorem 14.35).

Abelian varieties

In this example we will touch upon some basic properties of projective varieties with a group structure which are called abelian varieties. We will also make some remarks on "moduli spaces of abelian varieties" (without any proofs).

(16.31) Smoothness of algebraic group schemes (Chapter 6).

Let k be a field. For group schemes (Section (4.15)) over k smoothness is easy to check (see also Exercise 16.8).

Proposition 16.50. *Let G be a group scheme locally of finite type over a field k. Then G is smooth over k if and only if G is geometrically reduced.*

Proof. The condition is clearly necessary. To show that it is sufficient, we may assume that k is algebraically closed (Remark 6.30 (2)). As G is reduced, Theorem 6.19 shows that the smooth locus of G is non-empty. Furthermore, the group $G(k)$ acts by scheme automorphisms on G via the multiplication map, and the induced action on $G(k)$ is clearly transitive. Thus we see that every closed point of G is smooth over k. As the closed points are very dense, G is smooth over k. $\qquad\square$

It can be shown (e.g., [DG] II, §6, 1.1) that if $\operatorname{char}(k) = 0$, then every group scheme locally of finite type over k is automatically smooth. In positive characteristic there are plenty of examples of non-smooth group schemes (see for instance Exercise 16.6).

Proposition 16.51. *Let k be a perfect field, G a group scheme locally of finite type over k. Then G_{red} is closed k-subgroup scheme of G which is smooth over k.*

Proof. As k is perfect, the k-scheme G_{red} is geometrically reduced (Corollary 5.57) and $G_{\text{red}} \times G_{\text{red}}$ is reduced (Proposition 5.49). Therefore the multiplication $m\colon G \times_k G \to G$ induces a morphism $m_{\text{red}}\colon G_{\text{red}} \times G_{\text{red}} \to G_{\text{red}}$. The inversion $i\colon G \to G$ and the unit $e\colon \operatorname{Spec} k \to G$ induce k-morphisms $i_{\text{red}}\colon G_{\text{red}} \to G_{\text{red}}$ and $e_{\text{red}}\colon \operatorname{Spec} k \to G_{\text{red}}$. The functoriality of $(\ )_{\text{red}}$ (Proposition 3.51) implies that $(G_{\text{red}}, m_{\text{red}}, i_{\text{red}}, e_{\text{red}})$ is a group scheme over k. It is smooth by Proposition 16.50. $\square$

Corollary 16.52. *Let G be a group scheme locally of finite type over a field k. Then G is geometrically irreducible if and only if G is connected.*

Proof. We only have to show that G is geometrically irreducible if G is connected. By definition $G(k) \neq \emptyset$. This implies that G is geometrically connected (Exercise 5.23). Therefore we may assume that k is algebraically closed and it suffices to show that G is irreducible. Replacing G by G_{red} we may assume that G is smooth over k (Proposition 16.51). Then G is regular (Theorem 6.28) and in particular all local rings of G are integral domains. Therefore G is connected if and only if G is irreducible (Exercise 3.16). $\square$

(16.32) Abelian varieties (Chapter 12).

Proper connected smooth group schemes are called abelian varieties:

Definition 16.53. *Let k be a field. An* abelian variety *over k is a k-group scheme which is connected, geometrically reduced and proper over k.*

Remark 16.54. An abelian variety is automatically smooth and geometrically integral over k (Proposition 16.50 and Corollary 16.52).

Proposition 16.55. *(Rigidity Lemma) Let k be a field, and let X be a geometrically reduced, geometrically connected proper k-scheme such that $X(k) \neq \emptyset$. Let Y be an integral k-scheme, and let Z be a separated k-scheme. Let $f\colon X \times Y \to Z$ be a morphism such that for some $y \in Y(k)$, $f_{|X \times \operatorname{Spec} \kappa(y)}$ factors through a k-valued point $z \in Z(k)$. Then f factors through the projection $p_2\colon X \times Y \to Y$.*

Proof. Let $x \in X(k)$, viewed as a morphism $\operatorname{Spec} k \to X$, and consider the morphisms f and $g := f \circ (x \times \operatorname{id}_Y) \circ p_2$ from $X \times Y$ to Z. It suffices to show that $f = g$ or, equivalently, that the subscheme $\operatorname{Eq}(f, g)$ of $X \times Y$, where these morphisms coincide, is equal to $X \times Y$. Let $U \subset Z$ be an open affine neighborhood of z. Because X is proper over k, the projection $p_2\colon X \times Y \to Y$ is closed. By hypothesis $p_2^{-1}(y) \subseteq f^{-1}(U)$, and therefore there exists an open neighborhood V of y in Y such that $p_2^{-1}(V) \subseteq f^{-1}(U)$ (Remark 12.57). Let $y' \in V$. Then the restriction of f to $X \times_k \operatorname{Spec} \kappa(y') \subset X \times Y$ yields a morphism $X \otimes_k \kappa(y') \to U \otimes_k \kappa(y')$ of $\kappa(y')$-schemes, which factors through a $\kappa(y')$-valued point by Corollary 12.67. This shows that $\operatorname{Eq}(f, g)$ contains all points of $X \times V$, and hence contains the dense open subset $X \times V$. Moreover $\operatorname{Eq}(f, g)$ is closed because Z is separated (see Definition 9.7 (ii)). Because $X \times Y$ is reduced (Proposition 5.49), $\operatorname{Eq}(f, g) = X \times Y$ (Corollary 9.9). $\square$

The proposition remains true without the condition that $X(k) \neq \emptyset$. One can prove this by applying a base change to a suitable extension field of k first, and then using the descent techniques of Chapter 14.

Corollary 16.56.
(1) *Let A, B be abelian varieties over the field k, and let $f\colon A \to B$ be a morphism of k-schemes, such that f maps the unit element $e_A \in A(k)$ of A to the unit element e_B of B. Then f is a homomorphism of group schemes (Definition 4.42).*
(2) *Let A be an abelian variety over the field k. Then for every k-scheme S, the group $A(S)$ is commutative. Equivalently, if $s\colon A \times A \to A \times A$ denotes the morphism $(x,y) \mapsto (y,x)$, then A is commutative in the sense that $m \circ s = m$.*

Proof. Consider the morphism

$$g\colon A \times A \longrightarrow B \times B \xrightarrow{\ m_B\ } B,$$

where the morphism $A \times A \to B \times B$ is defined as $(f \circ m_A) \times (i_B \circ m_B \circ (f \times f))$. In terms of S-valued points, g maps (a_1, a_2) to $f(a_1 a_2)(f(a_1)f(a_2))^{-1}$, so the statement of part (1) is equivalent to $g(A \times A) = \{e_B\}$. By the group axioms, we have $g(\{e_A\} \times A) = g(A \times \{e_A\}) = \{e_B\}$, so applying Proposition 16.55 twice we find that g factors through the first, and through the second projection $A \times A \to A$. This implies that g must be constant, as desired.

The second part now follows immediately, because the first part shows that the inversion morphism $i\colon A \to A$ mapping each element to its inverse is a group scheme homomorphism. $\qquad\square$

(16.33) Moduli spaces of abelian varieties (Chapter 13).

In Definition 16.53 we have defined the notion of abelian variety over a field k. In the following paragraphs we go through a number of technical definitions which enable us to state a theorem which relates a "parameter space" of abelian varieties with additional structure to the surfaces $\mathcal{H} = V_+(\sigma_2, \sigma_4) \subset \mathbb{P}^4_k$ and, less directly, to the Clebsch cubic surface $\mathcal{C}$.

The rough underlying idea of these parameter spaces, or *moduli spaces*, is to define a scheme $\mathcal{M}$ such that $\mathcal{M}(k)$ is the set of isomorphism classes of abelian varieties over k. However, there are several technical problems with this idea. First of all, an equality of sets alone is not a very interesting statement, so it is more appropriate to define $\mathcal{M}$ as a functor on the category of k-schemes. To this end, one introduces the notion of *abelian scheme* which is the correct notion of a family of abelian varieties. An abelian scheme over a scheme S is a proper smooth group scheme A over S such that all fibers of the structure morphism $A \to S$ are abelian varieties, i.e., are geometrically connected. Now we could consider the functor

$$S \mapsto \{\text{isomorphism classes of abelian schemes over } S\}.$$

Unfortunately, this functor is not representable. One of the main problems is that abelian schemes always admit non-trivial automorphisms. These automorphisms allow us to define abelian schemes (over suitable S) which are isomorphic locally on S, but are not isomorphic. This means that the above functor is not a Zariski sheaf, and a fortiori is not representable (Section (8.3)). The situation becomes better if we consider abelian schemes with additional data (where isomorphisms have to respect the additional data, so that in the end there are fewer isomorphisms). We will sketch the ingredients one uses for the definition of representable moduli functors of abelian varieties, but only for abelian varieties over a field, rather than for abelian schemes.

One can show that every abelian variety over k admits an ample line bundle, and hence is a projective k-scheme. Let us assume for simplicity that k is algebraically closed. Let A be an abelian variety over k. A *polarization* of A is an equivalence class of ample line bundles on A, where we call line bundles $\mathscr{L}$, $\mathscr{L}'$ equivalent, if the line bundle $\mathscr{L} \otimes \mathscr{L}^{-1}$ is translation invariant, i.e., $t_x^*(\mathscr{L} \otimes \mathscr{L}^{-1}) \cong \mathscr{L} \otimes \mathscr{L}^{-1}$ for every $x \in A(k)$. Here t_x denotes the (right) multiplication by x, i.e., $t_x \colon A \to A$, $a \mapsto ax$. While our definition is specific for abelian varieties, the resulting equivalence relation is so-called algebraic equivalence and can be defined for more general varieties.

For an abelian variety A over k, we denote by $\mathrm{End}(A)$ its ring of endomorphisms of A, i.e., of all scheme morphisms $A \to A$ which respect the group law. Addition in $\mathrm{End}(A)$ is induced from the group law on A, and multiplication is given by composition of homomorphisms. If k has characteristic 0, then "generically" abelian varieties have endomorphism ring $\mathbb{Z}$.

Here we are interested in abelian varieties with more endomorphisms. For the situation we have in mind, fix a real quadratic field $K := \mathbb{Q}(\sqrt{d})$, $d > 1$ a square-free integer, and denote by $\mathcal{O}_K$ its ring of integers, i.e., the integral closure of $\mathbb{Z}$ in K.

Definition 16.57. *Let k be an algebraically closed field of characteristic 0. A polarized abelian variety with real multiplication by $\mathcal{O}_K$ over k is an abelian variety A over k together with a polarization and a ring homomorphism $\iota \colon \mathcal{O}_K \to \mathrm{End}(A)$, such that for some ample line bundle $\mathscr{L}$ in the equivalence class corresponding to the polarization, the following condition holds:*

For every $f \in \iota(\mathcal{O}_K)$ and every $x \in A(k)$,

$$f^* t_x^* \mathscr{L} \otimes \mathscr{L} \cong t_{f(x)}^* \mathscr{L} \otimes f^* \mathscr{L}.$$

While polarizations are essential for the representability, the tool to eliminate automorphisms is the notion of level structure. For $n \in \mathbb{Z}$ we have the morphism $[n] \colon A \to A$, $x \mapsto x + \cdots + x = nx$, i.e., we add x to itself n times. As A is a commutative group scheme, $[n]$ is a homomorphism of group schemes. Its kernel is denoted by $A[n]$. It is a closed subgroup scheme of A. See Definition 4.45.

Definition 16.58. *Let A be an abelian variety over an algebraically closed field k of characteristic 0. A full level n-structure on A is an isomorphism*

$$A[n](k) \overset{\sim}{\to} (\mathbb{Z}/n\mathbb{Z})^{2 \dim A}.$$

of abstract groups.

It is not clear that for a given A there exists a level structure (but this is true). If k has characteristic $p > 0$, and p divides n, then there never exists a level structure in the above sense, so one has to resort to more advanced techniques.

Theorem 16.59. *Let $K = \mathbb{Q}(\sqrt{5})$. Let $X = V_+(\sigma_2, \sigma_4) \subset \mathbb{P}^4$, and let $X_0 \subset X$ be the smooth locus, i.e., the complement of the 5 points $(1:0:0:0:0)$, $(0:1:0:0:0)$, ..., $(0:0:0:0:1)$ (see Section (16.18)). Then there is a natural bijection*

$$X_0(\mathbb{C}) = \left\{ \begin{array}{c} \text{polarized abelian varieties with} \\ \text{real multiplication by } \mathcal{O}_K, \\ \text{of dimension 2,} \\ \text{and with a full level 2-structure} \end{array} \right\} / \cong$$

This is a very deep theorem, which involves complex analytic methods as well as arithmetic ingredients (more precisely, Hilbert modular forms). See [Ge], VIII, Theorem (2.1). As mentioned above, the key point here is to make proper sense of the word *natural* above. One way to do so is the functorial point of view which amounts to defining a moduli functor of polarized abelian schemes with real multiplication and level structure. In the situation at hand, there is a second interesting method: Via the embedding $X_0 \subset \mathbb{P}^4_{\mathbb{C}}$, we can view $X_0(\mathbb{C})$ as a complex manifold. One can show that there is a surjective holomorphic map $\mathbb{H}^2 \to X_0(\mathbb{C})$, where $\mathbb{H} = \{ z \in \mathbb{C} \; ; \; \operatorname{Im} z > 0 \}$ is the complex upper half plane. Furthermore, via this morphism one can identify $X_0(\mathbb{C})$ with the quotient of $\mathbb{H}^2$ under the action of a certain discrete group.

The theorem admits vast generalizations. Restricting ourselves to the setup above, called the case of Hilbert modular surfaces, one finds the following general picture, independently of K and of the choice of level structure. In fact, there are many more choices for the level structure than just the full level n-structures defined above. The moduli space X_0 obtained from the corresponding moduli problem is a surface over k. Unlike in the case above, X_0 is not smooth in general. It may have cyclic quotient singularities, but no other singularities. There is a natural compactification $X \supset X_0$ which is normal, and such that $X \setminus X_0$ is a finite set of closed points. These points are singular in X. By blowing up the singular points in a suitable way, one obtains a resolution of singularities $\tilde{X} \to X$.

Let us go back to the case discussed above, i.e., fix $K = \mathbb{Q}(\sqrt{5})$ and the level structure as above. In this case there exists a morphism $\tilde{X} \to \mathcal{C}$, where $\mathcal{C} \subset \mathbb{P}^3_k$ denotes the Clebsch cubic surface, which identifies $\tilde{X}$ with the blow-up of $\mathcal{C}$ in the ten Eckardt points, i.e., the ten points on $\mathcal{C}$ in which three of the 27 lines intersect. It is a typical phenomenon that $\tilde{X}$ is the blow-up of some other smooth surface, i.e., $\tilde{X}$ is not "minimal".

(16.34) Inflexion points (Chapter 6).

Let k be an algebraically closed field.

Definition 16.60. *Let $C \subset \mathbb{P}^2_k$ be a one-dimensional closed subscheme. A point $x \in C(k)$ is called an* inflexion point *(or* flex*), if C is smooth at x and the projective tangent space $L := T_x(C \subset \mathbb{P}^2/k)$ of C in x, i.e., the line in $\mathbb{P}^2_k$ tangent to C in x, has intersection multiplicity $i_x(C, L) \geq 3$.*

Example 16.61. Assume that C is a cubic curve, i.e., $C = V_+(f)$ where f is of degree 3. Let $x \in C(k)$ be an inflexion point of C. After a change of coordinates we may assume that $x = (0:1:0)$ and that the line L as in the definition is $L = V_+(Z)$. These two facts translate to the condition that the monomials Y^3 and XY^2 do not occur in f, and that Y^2Z occurs with a coefficient $\neq 0$ which we may assume to be $= 1$ (where we use X, Y, Z as homogeneous coordinates on $\mathbb{P}^2_k$). Identifying $L \cong \mathbb{P}^1_k$ (with homogeneous coordinates X, Y), the scheme-theoretic intersection $C \cap L$ is given by the polynomial $f(X, Y, 0)$ which by the above has the form $aX^3 + bX^2Y$ $(a, b \in k)$. We have to check whether this polynomial, or more precisely its dehomogenization obtained by setting $Y = 1$, has (at least) a triple zero at $X = 0$, and obviously this is the case if and only if $b = 0$. In other words, if f is a homogeneous polynomial of degree 3, then $(0 : 1 : 0)$ is a flex if and only if the monomials Y^3, XY^2 and X^2Y do not occur in f, and Y^2Z does occur with a non-zero coefficient.

To handle the case of a general curve $C = V_+(f) \subset \mathbb{P}^2_k$, one looks at the Hessian matrix

$$H_f = \left(\frac{\partial^2 f}{\partial X_i \partial X_j} \right)_{i,j=0,1,2} \in M_{3 \times 3}(k[X_0, X_1, X_2]),$$

where we now use homogeneous coordinates labeled by X_0, X_1, X_2. We then have

Proposition 16.62. *Assume that* char $k \neq 2$. *A point* $x \in C = V_+(f) \subset \mathbb{P}^2_k$ *is a flex of C if and only if x is a smooth point of C and* $\det H_f(x) = 0$.

The proof of the proposition is not particularly hard, but because we will apply it only for cubic curves, we restrict to that case.

Proof. [if $\deg f = 3$] In Example 16.61 we have seen that the point $x = (0 : 1 : 0)$ is a flex of C if and only certain monomials in the homogeneous coordinates do, resp. do not, occur. It is a straightforward computation to check that this is equivalent to the condition that x is a smooth point of C and that $\det H_f(0, 1, 0) = 0$.

Now the proposition follows from the compatibility of the Hessian matrix with change of coordinates in $\mathbb{P}^2_k$. In fact, if $A \in \mathrm{GL}_3(k)$, then one checks using the chain rule for differentiation that, denoting by A the induced automorphism of $\mathbb{P}^2_k$ as well,

$$H_{f \circ A}(x) = A^t \cdot H_f(Ax) \cdot A,$$

so in particular we see that a point x is a flex of C if and only if $A(x)$ is a flex of $A(C)$. $\square$

Corollary 16.63. *Assume that* char $k \neq 2$. *Let* $C = V_+(f) \subset \mathbb{P}^2_k$ *be a smooth curve of degree* $\deg f \geq 3$. *Then C has a flex.*

Proof. Since all points of C are smooth, it is enough to show that $C \cap V_+(\det H_f) \neq \emptyset$. This follows from Bézout's theorem (Theorem 5.61) because our assumption on the degree of f ensures that $\det H_f$ is not constant. $\square$

(16.35) Elliptic Curves (Chapter 15).

Let k be an algebraically closed field. Let $E \subset \mathbb{P}^2_k$ be an integral curve of degree 3, i.e., a curve defined by an irreducible cubic polynomial, and assume that E has a flex (Definition 16.60). As Corollary 16.63 shows, this is always true if E is smooth and char $k \neq 2$. (One can show using the theorem of Riemann and Roch, Theorem 15.35, that the assumption on the characteristic of k can be dropped.)

After a change of variables, we may assume that $E = V_+(f)$, where f is a cubic homogeneous polynomial of the special form ("Weierstrass form")

$$f = Y^2Z + a_1XYZ + a_3YZ^2 - X^3 - a_2X^2Z - a_4XZ^2 - a_6Z^3,$$

i.e., that the monomials Y^3, X^2Y, XY^2 do not occur, and the coefficients of Y^2Z and X^3 are scaled to 1. See Example 16.61 and note that the irreducibility of f ensures that the monomial X^3 must occur, since otherwise f would be divisible by Z.

From now on, we always assume that $E = V_+(f)$ with f of the above form. We then have $E \cap V_+(Z) = \{(0:1:0)\}$ (as sets).

Let us compute the smooth locus of E in the special case that char $k \neq 2$, and that $a_1 = a_3 = 0$, i.e.,

$$f = Y^2Z - X^3 - a_2X^2Z - a_4XZ^2 - a_6Z^3.$$

In fact, under the assumption char $k \neq 2$, one can always arrange that $a_1 = a_3 = 0$ by a suitable change of variables. The partial derivatives of the above polynomial are

$$\frac{\partial f}{\partial X} = -3X^2 - 2a_2XZ - a_4Z^2, \quad \frac{\partial f}{\partial Y} = 2YZ, \quad \frac{\partial f}{\partial Z} = Y^2 - a_2X^2 - 2a_4XZ - 3a_6Z^2.$$

One sees easily that the point $(0:1:0)$ is not singular. On the other hand, consider $E \cap D_+(Z)$. To study this affine piece of E we set $Z = 1$. By Lemma 6.1 (or by passing to the dehomogenized equations) we see that the singular locus of $X \cap D_+(Z)$ is the common zero locus of f and $\frac{\partial f}{\partial X}$, $\frac{\partial f}{\partial Y}$, where we always plug in $Z = 1$. The equation $2Y = 0$ implies that $y = 0$ at every singular point $(x:y:1)$ of E. The remaining equations then say precisely that x must be a multiple root of the polynomial $X^3 + a_2X^2 + a_4X + a_6$. So we see that E is smooth if and only if this polynomial does not have a multiple zero. (Furthermore, since a cubic polynomial has at most one double zero, we also see that a curve E as above has got at most one singular point, cf. Exercise 6.25.)

Now we come back to the general case, and we assume that E is smooth. As a side remark, we note that Bézout's Theorem 5.61 shows that every curve in $\mathbb{P}^2_k$ is connected, and hence every smooth curve is irreducible. We denote by o the point $(0:1:0) \in E(k)$.

Every line $L \subset \mathbb{P}^2_k$ intersects E in three points (counted with multiplicities), again by Bézout's theorem. We define a map $m\colon E(k) \times E(k) \to E(k)$ as follows: Given $p, q \in E(k)$, let L_1 be the unique line in $\mathbb{P}^2_k$ through p and q. (In case $p = q$, we let L_1 be the tangent line to E in this point, which is uniquely determined because of the smoothness condition.) Let r be the third point of intersection of E and L_1. If the intersection multiplicities of E and L_1 are not all $= 1$, then r will be one of the points p, q (in fact, we may even have $r = p = q$). Similarly, let L_2 be the line through o and r, and define $m(p, q)$ to be the third intersection point of L_2 and E. See Figure 16.3. This gives rise to a map $m\colon E(k) \times E(k) \to E(k)$ and writing down the construction in terms of an equation for E, it is easy to see that m comes from a morphism $m\colon E \times_k E \to E$.

For every $p \in E(k)$, we have $m(p, o) = m(o, p) = p$, and for any $p_1, p_2 \in E(k)$ we have $m(p_1, p_2) = m(p_2, p_1)$. If $p \in E(k)$, and q is the third point of intersection of E and the line through p and o, then $m(p, q) = o$. In fact, mapping $p \mapsto q$ defines a morphism $i\colon E \to E$. Altogether, we have almost proved that m, i and o define the structure of a group scheme on E, the only missing piece being the associativity of m. It is true that m is associative, but this is harder to prove. An elegant proof can be given using the Riemann-Roch theorem. We will take up this theme again in Volume II. Altogether we see that E is a projective integral group scheme over Spec k, in other words, E is an abelian variety.

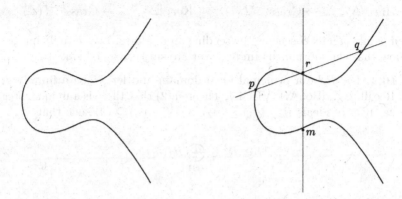

Figure 16.3: The ($\mathbb{R}$-valued points of the) elliptic curve given in $\mathbb{P}^2$ by the homogeneous equation $Y^2Z - X^3 - X^2Z - \frac{1}{10}Z^3 = 0$. On the right, the construction of the group law is illustrated. The point $m = m(p,q)$ is the result of adding p and q. The point o lies on the line of infinity of $\mathbb{P}^2$ and therefore is not visible in the picture. The line through a point r and through o is the vertical line passing through r.

Conversely, one can show that every one-dimensional abelian variety over k is isomorphic to a smooth curve of degree 3 in $\mathbb{P}_k^2$. These curves are called *elliptic curves*. This name originates from a (not very close) relationship to ellipses: Computing the arc length of an ellipse leads to the computation of "elliptic integrals". These integrals cannot be solved in terms of elementary functions, but only by "elliptic functions". These are, in some sense, rational functions on elliptic curves.

For a detailed treatment of elliptic curves and their arithmetic, see [Kn], [Si].

Exercises

Exercise 16.1. Let k be a field and $n \geq 1$ an integer. Show that

$$\det := \sum_{\sigma \in S_n} \operatorname{sgn}(\sigma) T_{1\sigma(1)} \cdots T_{n\sigma(n)}$$

is an irreducible polynomial in $R := k[(T_{ij})_{1 \leq i,j \leq n}]$. Deduce that $R/(\det)$ is an integral domain.

Exercise 16.2. Let S be a scheme, let $\mathscr{E}$ be a quasi-coherent $\mathscr{O}_S$-module, and let $\mathscr{F}$ be a finite locally free $\mathscr{O}_S$-module. Show that the functor $\Delta^r_{\mathscr{E} \to \mathscr{F}}$ (16.6.1) is representable by a closed subscheme of $\underline{\operatorname{Hom}}(\mathscr{E}, \mathscr{F})$ and that the closed immersion $\Delta^r_{\mathscr{E} \to \mathscr{F}} \hookrightarrow \underline{\operatorname{Hom}}(\mathscr{E}, \mathscr{F})$ is of finite presentation.
Hint: Exercise 11.8.

Exercise 16.3. Let S be a scheme, let $\mathscr{E}$ and $\mathscr{F}$ be finite locally free $\mathscr{O}_S$-modules of constant rank m and n, respectively, and let r be an integer with $1 \leq r \leq \min(n,m)$. Define morphisms of S-schemes

$$\text{Im}\colon {}^{0}\Delta^{r}_{\mathscr{E}\to\mathscr{F}} \to \text{Grass}^{n-r}(\mathscr{F}), \qquad \text{Ker}\colon {}^{0}\Delta^{r}_{\mathscr{E}\to\mathscr{F}} \to \text{Grass}^{m-r}(\mathscr{E})$$

on T-valued points (T any S-scheme) by sending an $\mathscr{O}_T$-linear map u to its image (resp. its kernel). Show that the morphisms Im and Ker are surjective and smooth.

Exercise 16.4. Let R be a principal ideal domain and let M be a finitely generated R-module. Recall (e.g., [BouAII] VII, 4.4, Theorem 2) that there is a unique integer $n \geq 0$ and a unique chain of ideals $R \supsetneq (a_1) \supseteq (a_2) \supseteq \cdots \supseteq (a_s) \supsetneq (0)$ such that

$$M \cong R^n \oplus \bigoplus_{i=1}^{s} R/(a_i).$$

Show that

$$\text{Fitt}_r\, M = \begin{cases} 0, & \text{if } r < n; \\ (a_1 a_2 \cdots a_{r-n+1}), & \text{if } n \leq r < n+s; \\ R, & \text{if } r \geq n+s. \end{cases}$$

and deduce that the isomorphism class of M is uniquely determined by the Fitting ideals of M.

Exercise 16.5. What are the inertia and the ramification degrees at all points of the quotient morphism $f\colon \mathbb{A}^2_k \to Z = \text{Spec}\, k[X^3, Y^3, XY]$?

Exercise 16.6. Let k be a field, $p \geq 0$ its characteristic, and let $n \geq 1$ be an integer. Let μ_n be the group scheme of n-th root of unity over k, i.e. $\mu_n(R)$ is the group $\{\, a \in R^\times \;;\; a^n = 1 \,\}$ for every k-algebra R. Its underlying scheme is $\text{Spec}\, k[T]/(T^n - 1)$. Show that μ_n is smooth over k if and only if p does not divide n.

Exercise 16.7. Let k be a field and let G be a k-group scheme locally of finite type. Let G' be a connected component of G.
(1) Show that G' is irreducible and of finite type over k. Show that G' is geometrically irreducible if $G'(k) \neq \emptyset$.
(2) Show that G is equidimensional.
(3) Show that the geometric number of connected components of G' is equal to the geometric number of irreducible components of G'.
(4) Let $k = \mathbb{Q}$ and $G = \mu_p$ the group scheme of p-th root of unity for a prime number p (Exercise 16.6). Show that G has two connected components and that the geometric number of connected components is p.

Exercise 16.8. Let k be a field, let G be a k-group scheme locally of finite type, and let $e \in G(k)$ be its unit section. Show that G is smooth over k if and only if $\dim_k T_e(G) = \dim G$.

A   The language of categories

We assume that the reader is familiar with the concept of a category and of a functor. In this appendix we briefly fix some notation and recall some results which are used in Volume I. Everything in this appendix can be found in any of the standard text books to category theory, such as [McL], [Mit], or [Sch]. We avoid any discussion of set-theoretical difficulties and refer for this to the given references.

(A.1) Categories.

Recall that a category $\mathcal{C}$ consists of (1) a collection of objects, (2) for any two objects X and Y a collection $\operatorname{Hom}_{\mathcal{C}}(X,Y) = \operatorname{Hom}(X,Y)$ of morphisms from X to Y, (3) for every object X an element $\operatorname{id}_X \in \operatorname{Hom}(X,X)$, (4) for any three objects X, Y, Z a composition map $\circ\colon \operatorname{Hom}(X,Y) \times \operatorname{Hom}(Y,Z) \to \operatorname{Hom}(X,Z)$, such that composition of morphisms is associative, and the elements id_X are neutral elements with respect to composition. We write $f\colon X \to Y$ if f is a morphism from X to Y.

Remark A.1. As we stated above, we are deliberately vague here regarding set-theoretic foundations, specifically regarding the question what is meant by a *collection*. As is well known, a too naive view on this leads to contradictions quickly – Russell's paradoxon concerning the collection of all collections which do not contain themselves being a well-known one. Another obvious problem is that one would like to attach a cardinality to each set (possibly infinite, of course), but the collection of all sets, say, obviously cannot have a cardinality in a meaningful way. The usual way out of these problems is to formalize the notion of set by a system of axioms (like the system given by Zermelo and Fraenkel, plus the axiom of choice (ZFC)) which formulates certain minimal existence requirements (specifically, one requires the existence of a set which is infinite) and certain constructions that can be used to construct further sets from given ones. The collection of all sets, for instance, is then an object which cannot be accessed within this system of axioms, and is called a *class* (or a *proper class* when one wants to stress that it is a class, but not a set).

With a framework like this in place, the notion of collection above could be replaced by the notion of class to obtain a very general notion of category. There are obvious variants: Allowing the collection of objects to be a class, but requiring that all $\operatorname{Hom}(X,Y)$ are sets, one arrives at the notion of *locally small category*. For instance, the category of sets is locally small. Restricting further, a *small category* is a category where the objects as well as the morphisms form sets.

Since constructions based on proper classes rather than sets are "dangerous" (more formally: not covered by the axiomatic framework of set theory), it is sometimes necessary to artificially restrict to locally small, or even small categories. There are (at least) two standard ways to do so. One is Grothendieck's notion of *universe* (see e.g. [KS]) where one, roughly speaking, fixes a (very large) set, called the universe, and then considers only objects contained in the given universe. For certain constructions, it might be necessary to switch to a larger universe. The other, similar, approach is to fix a large cardinal, and then to only consider sets of cardinality less than the given cardinal. Cf. [St] 0007, 000H.

© Springer Fachmedien Wiesbaden GmbH, part of Springer Nature 2020
U. Görtz und T. Wedhorn, *Algebraic Geometry I: Schemes*, Springer Studium
Mathematik – Master, https://doi.org/10.1007/978-3-658-30733-2

Both approaches work well for a theory like algebraic geometry because typically one is only interested in sets (groups, rings, schemes, ...) with bounded cardinality. On the other hand, both approaches have the draw-back that they involve a choice of some kind and that it might sometimes be painful to check (or: not be true) that all constructions are independent of this choice.

If $f\colon X \to Y$ is a morphism, then a morphism $g\colon Y \to X$ is called a *section* (resp. a *retraction*) if $f \circ g = \mathrm{id}_Y$ (resp. if $g \circ f = \mathrm{id}_X$).

A morphism $f\colon X \to Y$ is called an *isomorphism* if there exists a morphism $g\colon Y \to X$ such that $f \circ g = \mathrm{id}_Y$ and $g \circ f = \mathrm{id}_X$. We often write $f\colon X \overset{\sim}{\to} Y$ to indicate that f is an isomorphism. We also write $X \cong Y$ and say that X and Y are *isomorphic* if there exists an isomorphism $X \overset{\sim}{\to} Y$.

A *subcategory* of a category $\mathcal{C}$ is a category $\mathcal{C}'$ such that every object of $\mathcal{C}'$ is an object of $\mathcal{C}$ and such that $\mathrm{Hom}_{\mathcal{C}'}(X', Y') \subseteq \mathrm{Hom}_{\mathcal{C}}(X', Y')$ for any pair (X', Y') of objects of $\mathcal{C}'$, compatibly with composition of morphisms and identity elements. The subcategory $\mathcal{C}'$ is called *full* if $\mathrm{Hom}_{\mathcal{C}'}(X', Y') = \mathrm{Hom}_{\mathcal{C}}(X', Y')$ for all objects X' and Y' of $\mathcal{C}'$.

For every category $\mathcal{C}$ the *opposite category*, denoted by $\mathcal{C}^{\mathrm{opp}}$, is the category with the same objects as $\mathcal{C}$ and where for two objects X and Y of $\mathcal{C}^{\mathrm{opp}}$ we set $\mathrm{Hom}_{\mathcal{C}^{\mathrm{opp}}}(X, Y) := \mathrm{Hom}_{\mathcal{C}}(Y, X)$ with the obvious composition law.

A morphism $f\colon X \to Y$ in a category $\mathcal{C}$ is called *monomorphism* (resp. *epimorphism*) if for every object Z left composition with f is an injective map $\mathrm{Hom}(Z, X) \to \mathrm{Hom}(Z, Y)$ (resp. right composition with f is an injective map $\mathrm{Hom}(Y, Z) \to \mathrm{Hom}(X, Z)$). We often write $f\colon X \hookrightarrow Y$ (resp. $f\colon X \twoheadrightarrow Y$) to indicate that f is a monomorphism (resp. epimorphism). Every isomorphism is a monomorphism and an epimorphism. The converse does not hold in general.

An object Z in $\mathcal{C}$ is called *final* (resp. *initial*), if $\mathrm{Hom}_{\mathcal{C}}(X, Z)$ (resp. $\mathrm{Hom}_{\mathcal{C}}(Z, X)$) has exactly one element for all objects X. For any two final (resp. initial) objects Z and Z' there is a unique isomorphism $Z \overset{\sim}{\to} Z'$.

Notation A.2. Throughout this book we use the following notation for specific categories:

  (Sets) the category of sets,
  $\widehat{\mathcal{C}}$ the category of all contravariant functors of $\mathcal{C}$ to the category of sets (see also Section (4.2)),
  (Ring) the category of (commutative) rings,
  (A-Mod) the category of A-modules for a ring A,
  (R-Alg) the category of (commutative) R-algebras for a ring R,
  (Sch) the category of schemes (see Section (3.1)),
  (Sch/S) the category of S-schemes for a scheme S (if $S = \mathrm{Spec}\,R$ is affine we also write (Sch/R)) (see Section (3.1)).
  (Aff) the category of affine schemes (see Section (2.11)).
  ($\mathscr{O}_X$-Mod) the category of $\mathscr{O}_X$-modules for a sheaf of rings $\mathscr{O}_X$ (see Section (7.1)).

(A.2) Functors, equivalence of categories, adjoint functors.

Given categories $\mathcal{C}$ and $\mathcal{D}$, a (covariant) functor $F\colon \mathcal{C} \to \mathcal{D}$ is given by attaching to each object C of $\mathcal{C}$ an object $F(C)$ of $\mathcal{D}$, and to each morphism $f\colon C \to C'$ in $\mathcal{C}$ a morphism $F(f)\colon F(C) \to F(C')$, compatibly with composition and identity elements. A contravariant functor from $\mathcal{C}$ to $\mathcal{D}$ is by definition a covariant functor $F\colon \mathcal{C}^{\mathrm{opp}} \to \mathcal{D}$, where C^{opp} is the opposite category of C. Sometimes we use the notation $F\colon C \to D$ for a contravariant functor, in which case we explicitly state that F is contravariant.

If F is a functor from C to D and G a functor from $D \to \mathcal{E}$, we write $G \circ F$ for the composition.

For two functors $F, G\colon C \to D$ we call a family of morphisms $\alpha(S)\colon F(S) \to G(S)$ for every object S of C *functorial in S* or a *morphism of functors* if for every morphism $f\colon T \to S$ in C the diagram

$$
\begin{array}{ccc}
F(T) & \xrightarrow{\ \alpha(T)\ } & G(T) \\
{\scriptstyle F(f)}\downarrow & & \downarrow{\scriptstyle G(f)} \\
F(S) & \xrightarrow{\ \alpha(S)\ } & G(S)
\end{array}
$$

commutes. With this notion of morphism we obtain the category of all functors from C to D. We denote by $\widehat{C}$ the category of all contravariant functors of C into the category of sets.

A functor $F\colon C \to D$ is called *faithful* (resp. *fully faithful*) if for all objects X and Y of C the map $\text{Hom}_C(X, Y) \to \text{Hom}_D(F(X), F(Y))$, $f \mapsto F(f)$ is injective (resp. bijective). The functor F is called *essentially surjective* if for every object Y of D there exists an object X of C and an isomorphism $F(X) \cong Y$.

A functor $F\colon C \to D$ is called an *equivalence of categories* if it is fully faithful and essentially surjective. This is equivalent to the existence of a *quasi-inverse functor* G, i.e., of a functor $G\colon D \to C$ such that $G \circ F \cong \text{id}_C$ and $F \circ G \cong \text{id}_D$. Similarly, considering contravariant functors, we obtain the notion of an *anti-equivalence of categories*.

Let $F\colon C \to D$ and $G\colon D \to C$ be functors. Then G is said to be *right adjoint* to F and F is said to be *left adjoint* to G if for all objects X in C and Y in D there exists a bijection

$$\text{Hom}_C(X, G(Y)) \cong \text{Hom}_D(F(X), Y)$$

which is functorial in X and in Y. If a functor F (resp. a functor G) has a right adjoint functor (resp. a left adjoint functor), this adjoint functor is unique up to isomorphism. If F is an equivalence of categories, then a quasi-inverse functor is right adjoint and left adjoint to F.

(A.3) Inductive and projective limits.

Ordered sets.

Let I be a set.
(1) A relation $\leq$ on I is called *partial preorder* or simply *preorder*, if $i \leq i$ for all $i \in I$ and $i \leq j$, $j \leq k$ imply $i \leq k$ for all $i, j, k \in I$.
(2) A preorder $\leq$ is called *partial order* or simply *order* if $i \leq j$ and $j \leq i$ imply $i = j$ for all $i, j \in I$.
(3) A preorder $\leq$ is called *filtered* if for all $i, j \in I$ there exists a $k \in I$ with $i \leq k$ and $j \leq k$.
(4) A partial order $\leq$ is called *total order* if for all $i, j \in I$ one has $i \leq j$ or $j \leq i$.

Every preordered set I can be made into a category, again denoted by I, whose objects are the elements of I and for two elements $i, j \in I$ the set of morphisms $\text{Hom}_I(i, j)$ consists of one element if $i \leq j$ and is empty otherwise. There is a unique way to define a composition law in this category.

PROJECTIVE LIMITS AND PRODUCTS.

Let I be a preordered set and let $\mathcal{C}$ be a category. A *projective system* in $\mathcal{C}$ indexed by I is a functor $I^{\mathrm{opp}} \to \mathcal{C}$. In other words, it is a tuple $(X_i)_{i \in I}$ of objects in $\mathcal{C}$ together with morphisms $\varphi_{ij} \colon X_j \to X_i$ for all $i, j \in I$ with $i \leq j$ such that $\varphi_{ii} = \mathrm{id}_{X_i}$ and $\varphi_{ij} \circ \varphi_{jk} = \varphi_{ik}$ for all $i \leq j \leq k$.

A *projective limit* (or simply *limit*) of a projective system $((X_i)_{i \in I}, (\varphi_{ij})_{i \leq j})$ in $\mathcal{C}$ is an object X in $\mathcal{C}$ together with a family of morphisms $(\varphi_i \colon X \to X_i)_{i \in I}$ such that $\varphi_i = \varphi_{ij} \circ \varphi_j$ for all $i \leq j$ which is final with this property, i.e., if $(Y, (\psi_i)_{i \in I})$ is a tuple of an object Y and morphisms $\psi_i \colon Y \to X_i$ such that $\psi_i = \varphi_{ij} \circ \psi_j$ for all $i \leq j$, then there exists a unique morphism $\psi \colon Y \to X$ such that $\varphi_i \circ \psi = \psi_i$ for all $i \in I$. A projective limit $(X, (\varphi_i)_i)$ need not exist, but if it exists, then it is uniquely determined up to unique isomorphism, and is denoted by

$$\varprojlim_{i \in I} X_i.$$

Example A.3.
(1) If I is endowed with the discrete order, i.e., $i \leq j$ if and only if $i = j$, then

$$\prod_{i \in I} X_i := \varprojlim_{i \in I} X_i$$

is called the *product* of the family $(X_i)_{i \in I}$.
(2) If I consists of three elements j_1, j_2, k whose nontrivial order relations are $k \leq j_1$ and $k \leq j_2$, then

$$X_{j_1} \times_{X_k} X_{j_2} := \varprojlim_{i \in \{j_1, j_2, k\}} X_i$$

is called the *fiber product*; see also Section (4.4).
(3) For $I = \emptyset$, the notion of projective limit is the same as that of final object of $\mathcal{C}$.

Example A.4.
(1) Let $\mathcal{C}$ be the category of sets. Then $\prod_{i \in I} X_i$ is the usual cartesian product, and the final object is the singleton, i.e., the set consisting of one element. For an arbitrary preordered set I the projective limit of a projective system $((X_i)_{i \in I}, (\varphi_{ij})_{i \leq j})$ exists and is given by

$$X = \varprojlim_{i \in I} X_i = \{ (x_i)_{i \in I} \in \prod_{i \in I} X_i \; ; \; \varphi_{ij}(x_j) = x_i \text{ for all } i \leq j \},$$

where the maps $\varphi_i \colon X \to X_i$ are the restrictions of the projections $\prod_i X_i \to X_i$.
(2) Let $\mathcal{C}$ be the category of groups, the category of rings, or the category of left or right R-modules for a fixed (not necessarily commutative) ring R. Then projective limits always exist and can be constructed as follows: The underlying set of the projective limit of a projective system in $\mathcal{C}$ is the projective limit in the category of sets. Multiplication, addition, or scalar multiplication is defined componentwise. The existence of inverses and neutral elements (as appropriate) is clear.

Inductive Limits and Sums.

Now consider the dual notion of projective limits, i.e., the notion obtained by "reversing all arrows": Let I be a preordered set and let $\mathcal{C}$ be a category. An *inductive system* in $\mathcal{C}$ indexed by I is a projective system in $\mathcal{C}^{\mathrm{opp}}$. In other words, it is a tuple $(X_i)_{i \in I}$ of objects in $\mathcal{C}$ together with morphisms $\varphi_{ji} \colon X_i \to X_j$ for all $i, j \in I$ with $i \leq j$ such that $\varphi_{ii} = \mathrm{id}_{X_i}$ and $\varphi_{kj} \circ \varphi_{ji} = \varphi_{ki}$ for all $i \leq j \leq k$.

An *inductive limit* (or *colimit*) of an inductive system $((X_i)_{i \in I}, (\varphi_{ji})_{i \leq j})$ in $\mathcal{C}$ is an object X in $\mathcal{C}$ together with a family of morphisms $(\varphi_i \colon X_i \to X)_{i \in I}$ such that $\varphi_i = \varphi_j \circ \varphi_{ji}$ for all $i \leq j$ which is initial with this property, i.e., if $(Y, (\psi_i)_{i \in I})$ is a tuple of an object Y and morphisms $\psi_i \colon X_i \to Y$ such that $\psi_i = \psi_j \circ \varphi_{ji}$ for all $i \leq j$, then there exists a unique morphism $\psi \colon X \to Y$ such that $\psi \circ \varphi_i = \psi_i$ for all $i \in I$. An inductive limit $(X, (\varphi_i)_i)$ is uniquely determined up to unique isomorphism (if it exists) and it is denoted by

$$\varinjlim_{i \in I} X_i.$$

Example A.5.
(1) If I is endowed with the discrete order, i.e., $i \leq j$ if and only if $i = j$, then

$$\coprod_{i \in I} X_i := \varinjlim_{i \in I} X_i$$

is called the *sum* or the *coproduct* of the family $(X_i)_{i \in I}$.
(2) If $I = \emptyset$, then the inductive limit is the same as an initial object of $\mathcal{C}$.

Example A.6.
(1) Let $\mathcal{C}$ be the category of sets. Then $\coprod_{i \in I} X_i$ is the disjoint union of the X_i, and the initial object is the empty set. In the category of sets, arbitrary inductive limits exist.
(2) If $\mathcal{C}$ is the category of topological spaces, then $\coprod_{i \in I} X_i$ is the disjoint union of X_i endowed with the unique topology such that the induced topology on X_i is the given one and such that X_i is open and closed in $\coprod_{i \in I} X_i$ for all $i \in I$.
(3) Let $\mathcal{C}$ be the category of groups, the category of rings, or the category of left or right R-modules for a fixed (not necessarily commutative) ring R. Let $((X_i)_i, (\varphi_{ji})_{i \leq j})$ be an inductive system in $\mathcal{C}$ such that I is a filtered preordered set. Then the underlying set of the inductive limit in $\mathcal{C}$ is the disjoint union $\coprod_{i \in I} X_i$ of sets modulo the following equivalence relation: For $x \in X_i$ and $x' \in X_{i'}$ we say $x \sim x'$ if there exists an element $j \geq i, i'$ such that $\varphi_{ji}(x_i) = \varphi_{ji'}(x')$. This is an equivalence relation because I is filtered. Multiplication resp. addition of two equivalence classes $[x]$ and $[x']$ is done by choosing representatives $x \in [x]$ and $x' \in [x']$ such that $x, x' \in X_i$ (which is possible because I is filtered) and then multiply resp. add x and x' in X_i. Scalar multiplication is defined by scalar multiplication on representatives.
(4) Let R be a commutative ring and let $\mathcal{C}$ be the category of R-algebras. Then the sum in $\mathcal{C}$ of two R-algebras A and B is the tensor product $A \otimes_R B$.

Limits and adjoint functors.

Let $\mathcal{C}$ and $\mathcal{D}$ be two categories and let $F \colon \mathcal{C} \to \mathcal{D}$ be a functor. If F admits a left adjoint functor G (i.e., F is right adjoint to G), then F commutes with projective limits: For every projective system $((X_i)_i, (\varphi_{ij})_{i \leq j})$ in $\mathcal{C}$ such that its projective limits exists, the projective limit of $((F(X_i))_i, (F(\varphi_{ij}))_{i \leq j})$ exists and the canonical morphism

$$F(\varprojlim_{i \in I} X_i) \to \varprojlim_{i \in I} F(X_i)$$

is an isomorphism. Dually, a functor which admits a right adjoint commutes with inductive limits.

(A.4) Additive and abelian categories.

ADDITIVE CATEGORIES.

An *additive category* is a category $\mathcal{C}$ such that for any pair (X, Y) of objects the set $\mathrm{Hom}_{\mathcal{C}}(X, Y)$ is endowed with the structure of an abelian group such that
(a) the composition law is bilinear,
(b) there exists a final object in $\mathcal{C}$,
(c) for any pair (X, Y) of objects there exists the coproduct of X and Y which is then usually denoted by $X \oplus Y$ with morphisms $i: X \to X \oplus Y$ and $j: Y \to X \oplus Y$ and called the *direct sum of X and Y*; in other words for every object Z the homomorphism

$$\mathrm{Hom}_{\mathcal{C}}(X \oplus Y, Z) \to \mathrm{Hom}_{\mathcal{C}}(X, Z) \times \mathrm{Hom}_{\mathcal{C}}(Y, Z), \qquad u \mapsto (u \circ i, u \circ j)$$

is bijective.

In an additive category $\mathcal{C}$ the final object is also an initial object. Generally an object which is initial and final is called a *zero object* and is denoted by 0. In an additive category, all finite direct sums exist. Moreover, for every finite family $(X_i)_i$ objects in $\mathcal{C}$ the product $\prod_i X_i$ exists and the canonical morphism $\bigoplus_i X_i \to \prod_i X_i$ is an isomorphism.

If $\mathcal{C}$ is an additive category, the opposite category $\mathcal{C}^{\mathrm{opp}}$ is again additive.

Let $\mathcal{C}$ and $\mathcal{D}$ be additive categories. A functor $F: \mathcal{C} \to \mathcal{D}$ is called *additive* if for all objects X and Y the map $F: \mathrm{Hom}_{\mathcal{C}}(X, Y) \to \mathrm{Hom}_{\mathcal{D}}(F(X), F(Y))$ is a homomorphism of abelian groups.

KERNEL, COKERNEL AND ABELIAN CATEGORIES.

Let $\mathcal{C}$ be an additive category and let $u: X \to Y$ be a morphism. An object $\mathrm{Ker}(u)$ together with a morphism $i: \mathrm{Ker}(u) \to X$ (resp. an object $\mathrm{Coker}(u)$ together with a morphism $p: Y \to \mathrm{Coker}(u)$) is called *kernel of u* (resp. *cokernel of u*) if $u \circ i = 0$ (resp. $p \circ u = 0$) and if for every object Z the map

$$\mathrm{Hom}_{\mathcal{C}}(Z, \mathrm{Ker}(u)) \to \{ f \in \mathrm{Hom}_{\mathcal{C}}(Z, X) \; ; \; u \circ f = 0 \}, \qquad g \mapsto i \circ g$$
$$(\text{resp.} \quad \mathrm{Hom}_{\mathcal{C}}(\mathrm{Coker}(u), Z) \to \{ f \in \mathrm{Hom}_{\mathcal{C}}(Y, Z) \; ; \; f \circ u = 0 \}, \qquad g \mapsto g \circ p)$$

is bijective. Note that $\mathrm{Ker}(u) = 0$ (resp. $\mathrm{Coker}(u) = 0$) is equivalent to saying that u is a monomorphism (resp. an epimorphism). If u has a cokernel $p: Y \to \mathrm{Coker}(u)$ and p has a kernel, we call the kernel of p the *image of u* and denote it by $\mathrm{Im}(u) \to Y$. Similarly, if u has a kernel $i: \mathrm{Ker}(u) \to X$ and i has a cokernel, we call this cokernel the *coimage of u* and denote it by $X \to \mathrm{Coim}(u)$. If $\mathrm{Coim}(u)$ and $\mathrm{Im}(u)$ exist, there exists a unique morphism $\bar{u}: \mathrm{Coim}(u) \to \mathrm{Im}(u)$ such that

$$
\begin{array}{ccc}
X & \xrightarrow{\;u\;} & Y \\
\downarrow & & \uparrow \\
\mathrm{Coim}(u) & \xrightarrow{\;\bar{u}\;} & \mathrm{Im}(u)
\end{array}
$$

commutes.

An additive category $\mathcal{C}$ is called *abelian* if for any morphism $u\colon X \to Y$ the kernel and the cokernel of u exist and the induced morphism $\bar{u}\colon \mathrm{Coim}(u) \to \mathrm{Im}(u)$ is an isomorphism.

EXACT SEQUENCES.

Let $\mathcal{C}$ be an abelian category. A sequence of morphism

$$X \xrightarrow{u} Y \xrightarrow{v} Z$$

is called *exact* if $v \circ u = 0$ and the induced morphism $\mathrm{Im}(u) \to \mathrm{Ker}(v)$ is an isomorphism. More generally, any sequence of morphism is called *exact* if any successive pair of morphisms is exact.

Let $\mathcal{C}$ and $\mathcal{D}$ be abelian categories. An additive functor $F\colon \mathcal{C} \to \mathcal{D}$ is *left exact* (resp. *right exact*) if for any exact sequence $0 \to X' \to X \to X''$ (resp. $X' \to X \to X'' \to 0$) the sequence $0 \to F(X') \to F(X) \to F(X'')$ (resp. $F(X') \to F(X) \to F(X'') \to 0$) is exact. The functor F is *exact* if it is right exact and left exact. A functor F is exact if and only if for all exact sequences $X \xrightarrow{u} Y \xrightarrow{v} Z$ the sequence

$$F(X) \xrightarrow{F(u)} F(Y) \xrightarrow{F(v)} F(Z)$$

is exact.

If the additive functor F has a left adjoint (resp. a right adjoint) functor $G\colon \mathcal{D} \to \mathcal{C}$, then F is left exact (resp. right exact). In particular, any additive equivalence of abelian categories is exact.

B   Commutative Algebra

In this appendix we collect some results from commutative algebra that are needed in our exposition. We will not give any proofs (except if we were not able to find a suitable textbook reference in which case we give a proof at the end of the appendix). Definitions and results are ordered by topic but not necessarily by logical dependence. Often references are given for whole sections supplemented to references within the section if the result is not contained in the main reference. Sometimes results follow immediately from the definitions and in this case there are no references.

We assume that the reader is familiar with the following basic notions.
- rings, modules, and their homomorphisms;
- basic constructions with modules: submodules and quotient modules (in particular ideals and quotient rings), sum and intersection of submodules, direct sum, product, tensor product, inductive limit, projective limit;
- localization of rings and of modules, prime ideals, maximal ideals;
- complexes and exact sequences;
- algebras over a ring.

All rings are commutative with 1, unless explicitly stated otherwise, and all ring homomorphisms send 1 to 1.

(B.1) Basic definitions for rings.

As a reminder we collect in this section some basic definitions about rings. The main reference is [AM] Chapter 1. Let A be a ring.

An element $a \in A$ is called
(1) *zero divisor*, if there exists $0 \neq b \in A$ with $ab = 0$;
(2) *regular*, if a is not a zero divisor;
(3) *unit*, if there exists $b \in A$ with $ab = 1$;
(4) *nilpotent*, if there exists an integer $n \geq 1$ with $a^n = 0$;
(5) *irreducible*, if it is not a unit and if $a = bc$ implies that b or c is a unit;
(6) *prime*, if the principal ideal (a) is a prime ideal (equivalently, a is not a unit and if a divides bc, then a divides b or c);
(7) *idempotent*, if $a^2 = a$.
The units of A form an abelian group under multiplication, denoted by $A^\times$.

A ring A is called
(1) *(integral) domain* or simply *integral*, if $A \neq 0$ and A has no zero divisors $\neq 0$;
(2) *field*, if $A \neq 0$ and every element $\neq 0$ is a unit;
(3) *local*, if A has exactly one maximal ideal;
(4) *semi-local*, if A has finitely many maximal ideals;
(5) *reduced*, if A has no nilpotent elements $\neq 0$;
(6) *principal ideal domain*, if A is an integral domain and every ideal of A is generated by a single element.

© Springer Fachmedien Wiesbaden GmbH, part of Springer Nature 2020
U. Görtz und T. Wedhorn, *Algebraic Geometry I: Schemes*, Springer Studium
Mathematik – Master, https://doi.org/10.1007/978-3-658-30733-2

An ideal $\mathfrak{a} \subset A$ is called

(1) *prime ideal*, if $\mathfrak{a} \neq A$ and if $a, b \notin \mathfrak{a} \Rightarrow ab \notin \mathfrak{a}$ for all $a, b \in A$ (equivalently, $A/\mathfrak{a}$ is an integral domain);

(2) *maximal ideal*, if $\mathfrak{a} \neq A$ and there exists no ideal $\mathfrak{b}$ with $\mathfrak{a} \subsetneq \mathfrak{b} \subsetneq A$ (equivalently, $A/\mathfrak{a}$ is a field);

(3) *nilpotent*, if there exists an integer $n \geq 1$ such that $\mathfrak{a}^n = 0$;

(4) *principal ideal*, if $\mathfrak{a}$ can be generated by one element.

Every proper ideal $\mathfrak{a} \subsetneq A$ is contained in a maximal ideal of A. The set of prime ideals of A is denoted by $\operatorname{Spec} A$.

Proposition B.1. (Chinese remainder theorem) *Let* $\mathfrak{a}_1, \ldots, \mathfrak{a}_n$ *ideals of A such that* $\mathfrak{a}_i + \mathfrak{a}_j = A$ *for all* $i \neq j$. *Then* $\bigcap_i \mathfrak{a}_i = \prod_i \mathfrak{a}_i$ *and the projections* $A \to A/\mathfrak{a}_i$ *yield an isomorphism of rings*

$$A / \prod_i \mathfrak{a}_i \xrightarrow{\sim} \prod_i A/\mathfrak{a}_i.$$

For an ideal $\mathfrak{a} \subseteq A$ we call

(B.1.1) $$\operatorname{rad} \mathfrak{a} := \{\, f \in A \;;\; \exists r \in \mathbb{Z}_{\geq 0} : f^r \in \mathfrak{a} \,\} = \bigcap_{\mathfrak{p} \supseteq \mathfrak{a} \text{ prime}} \mathfrak{p}$$

the *radical of $\mathfrak{a}$*. An ideal $\mathfrak{a}$ is called *radical ideal* if $\mathfrak{a} = \operatorname{rad} \mathfrak{a}$ (equivalently, $A/\mathfrak{a}$ is reduced). For an ideal $\mathfrak{a}$ of A the following implications hold

$$\mathfrak{a} \text{ maximal ideal} \Rightarrow \mathfrak{a} \text{ prime ideal} \Rightarrow \mathfrak{a} \text{ radical ideal.}$$

The radical of the zero ideal consists of all nilpotent elements and is called the *nilradical of A*, denoted by $\operatorname{nil}(A)$. If $\operatorname{nil}(A)$ is generated by finitely many elements, it is nilpotent. We call

$$\mathfrak{r}(A) := \{\, a \in A \;;\; \forall b \in A : 1 - ab \in A^\times \,\} = \bigcap_{\mathfrak{m} \subset A \text{ maximal}} \mathfrak{m}$$

the *Jacobson radical of A*.

If A is a local ring, we denote by $\mathfrak{m}_A$ its maximal ideal and by $\kappa(A) = A/\mathfrak{m}_A$ its *residue field*. A homomorphism of local rings $\varphi \colon A \to B$ is called *local*, if $\varphi(\mathfrak{m}_A) \subseteq \mathfrak{m}_B$. As a shortcut for "let A be a local ring with maximal ideal $\mathfrak{m}$", we sometimes write "let $(A, \mathfrak{m})$ be a local ring".

Proposition B.2.

(1) *Let $\mathfrak{p}$ be a prime ideal containing a (finite) product $\prod_i \mathfrak{a}_i$ of ideals $\mathfrak{a}_i$ of A. Then $\mathfrak{p} \supseteq \mathfrak{a}_i$ for some i. If $\mathfrak{p} = \bigcap_i \mathfrak{a}_i$ or $\mathfrak{p} = \prod_i \mathfrak{a}_i$, then $\mathfrak{p} = \mathfrak{a}_i$ for some i.*

(2) *(prime ideal avoidance) Let $\mathfrak{p}_1, \ldots, \mathfrak{p}_r$ be prime ideals and let $\mathfrak{a}$ be an ideal with $\mathfrak{a} \subseteq \bigcup_i \mathfrak{p}_i$, then $\mathfrak{a} \subseteq \mathfrak{p}_i$ for some i.*

ALGEBRAS.

An *A-algebra* B is a ring B together with a ring homomorphism $\varphi \colon A \to B$. We usually suppress the homomorphism φ from the notation and write ab instead of $\varphi(a)b$ for $a \in A$ and $b \in B$. Multiplication with elements of A makes every A-algebra B into an A-module. Any ring A is a $\mathbb{Z}$-algebra in a unique way. We denote the A-algebra of polynomials in a family $(T_i)_{i \in I}$ of indeterminates by $A[(T_i)_i]$. Most often we will only use the case that $I = \{1, \ldots, n\}$ is finite and then we write $A[T_1, \ldots, T_n]$.

Let B be an A-algebra. The A-subalgebra B generated by a family $(b_i)_i$ of elements of B is denoted by $A[(b_i)_i]$ (and again we write $A[b_1,\dots,b_n]$ if $I = \{1,\dots,n\}$). There is a unique A-algebra homomorphism $\beta\colon A[(T_i)_i] \to A[(b_i)_i]$ with $\beta(T_i) = b_i$ for all $i \in I$. We call the family $(b_i)_{i\in I}$ *algebraically independent* if β is injective.

The ring of formal power series in variables $T_1,\dots,T_n$ with coefficients in A is denoted by $A[\![T_1,\dots,T_n]\!]$.

(B.2) Basic definitions for modules.

The main reference is [AM] Chapter 2. Let A be a ring.

Let M be an A-module and N, N' be submodules of M. Then

$$(N : N') := \{\, a \in A \; ; \; aN' \subseteq N \,\}$$

is an ideal of A. We call $\mathrm{Ann}(M) := (0 : M)$ the *annihilator of M*. An A-module is called *faithful* if $\mathrm{Ann}(M) = 0$.

We recall different versions of Nakayama's lemma.

Proposition B.3. (Nakayama's lemma) *Let M be an A-module and let $\mathfrak{a} \subseteq A$ be an ideal.*

(1) *Assume that M is finitely generated and $\mathfrak{a}M = M$. Then there exists $x \equiv 1 \,(\mathrm{mod}\,\mathfrak{a})$ such that $xM = 0$.*

(2) *(see also [BouAC] II, 3.2, Corollary 1 of Proposition 4) Let $u\colon N \to M$ be an A-module homomorphism. Assume that $\mathfrak{a}$ is nilpotent or that M is finitely generated and $\mathfrak{a}$ is contained in the Jacobson radical of A. Then u is surjective if and only if $u \otimes \mathrm{id}_{A/\mathfrak{a}}\colon N/\mathfrak{a}N \to M/\mathfrak{a}M$ is surjective.*

(3) *Assume A is local with maximal ideal $\mathfrak{m}$ and that M is finitely generated. Then $m_1,\dots,m_r \in M$ generate M if and only if their images generate the $(A/\mathfrak{m})$-vector space $M/\mathfrak{m}M$.*

Corollary B.4. ([Mat] Theorem 2.4) *Let M be a finitely generated A-module and let $u\colon M \to M$ be a surjective endomorphism. Then u is bijective.*

Proposition B.5. (Five Lemma, [Mat] Appendix B) *Consider a commutative diagram of A-modules with exact rows*

$$
\begin{array}{ccccccccc}
M_1 & \longrightarrow & M_2 & \longrightarrow & M_3 & \longrightarrow & M_4 & \longrightarrow & M_5 \\
\downarrow{\scriptstyle u_1} & & \downarrow{\scriptstyle u_2} & & \downarrow{\scriptstyle u_3} & & \downarrow{\scriptstyle u_4} & & \downarrow{\scriptstyle u_5} \\
N_1 & \longrightarrow & N_2 & \longrightarrow & N_3 & \longrightarrow & N_4 & \longrightarrow & N_5.
\end{array}
$$

Assume that u_2 and u_4 are isomorphisms, u_1 is surjective and u_5 is injective. Then u_3 is an isomorphism.

Lemma B.6. (Snake lemma, [Mat] Appendix B) *Let*

$$
\begin{array}{ccccccc}
M & \xrightarrow{\ f\ } & N & \xrightarrow{\ g\ } & P & \longrightarrow & 0 \\
\downarrow{\scriptstyle \alpha} & & \downarrow{\scriptstyle \beta} & & \downarrow{\scriptstyle \gamma} & & \\
0 & \longrightarrow & X & \xrightarrow{\ u\ } & Y & \xrightarrow{\ v\ } & Z
\end{array}
$$

be a commutative diagram of A-modules whose rows are exact sequences. Then there is an exact sequence

$$\operatorname{Ker}\alpha \xrightarrow{\;f'\;} \operatorname{Ker}\beta \xrightarrow{\;g'\;} \operatorname{Ker}\gamma \xrightarrow{\;\delta\;} \operatorname{Coker}\alpha \xrightarrow{\;u'\;} \operatorname{Coker}\beta \xrightarrow{\;v'\;} \operatorname{Coker}\gamma$$

where f', g', u', v' are the maps induced by f, g, u, and v, and where the "boundary map" δ is defined as follows: Let $c \in \operatorname{Ker}\gamma$, let $b \in N$ with $g(b) = c$, let $x \in X$ with $u(x) = \beta(b)$, and define $\delta(c)$ as the image of x in $\operatorname{Coker}\alpha$.

FREE MODULES.

An A-module M if called *free* if there exists an isomorphism $A^{(I)} := \bigoplus_{i\in I} A \xrightarrow{\sim} M$ for some set I. If $A \neq 0$, the cardinality of I is uniquely determined by M and called the *rank* of M. In $A^{(I)}$ the family $(e_i)_{i\in I}$, where $e_i \in A^{(I)}$ is the element whose j-th entry is δ_{ij} for all $i, j \in I$, is a basis, called the *standard basis* of $A^{(I)}$.

LENGTH OF A MODULE ([AM] Chap. 6).

Am A-module $M \neq 0$ is called *simple* if it has no submodules other than 0 and M itself (equivalently, $M \cong A/\mathfrak{m}$, where $\mathfrak{m}$ is a maximal ideal of A). A *composition series* of an A-module M is a chain
$$0 = M_0 \subset M_1 \subset \cdots \subset M_l = M$$
of submodules such that M_i/M_{i-1} is simple for all $i = 1, \ldots, l$.

If M has a composition series the tuple of the isomorphism classes of M_i/M_{i-1} depends up to order only on M. In particular the integer l depends only on M. It is called the *length of M* and denoted by $\lg_A(M)$. If M does not have a composition series, we set $\lg_A(M) := \infty$.

If $0 \to M' \to M \to M'' \to 0$ is an exact sequences of A-modules, then M is of finite length if and only if M' and M'' are of finite length. In this case $\lg_A(M) = \lg_A(M') + \lg_A(M'')$.

TORSION OF A MODULE.

Let A be an integral domain and let M be an A-module. Then

$$M_{\text{tors}} := \{\, m \in M \;;\; \exists\, 0 \neq a \in A : am = 0 \,\}$$

is a submodule of M, called *torsion module of M*. The module M is called *torsion-free*, if $M_{\text{tors}} = 0$.

(B.3) Finiteness conditions for modules and algebras.

Definition B.7. *Let M be an A-module.*
(1) *M is called* finitely generated *or of finite type if there is a surjection $A^n \twoheadrightarrow M$ for some integer $n \geq 0$.*
(2) *M is called of finite presentation if there is an exact sequence $A^m \to A^n \to M \to 0$ for some integers $n, m \geq 0$.*

Every A-module of finite presentation is of finite type. The converse holds if A is noetherian.

Proposition B.8. *Let* $0 \to M' \to M \to M'' \to 0$ *be an exact sequence of A-modules.*
(1) *([Mat] Exercise 2.5) If M' and M'' are of finite type (resp. of finite presentation), then M is of finite type (resp. of finite presentation).*
(2) *([Mat] Theorem 2.6) If M is of finite type and M'' is of finite presentation, then M' is of finite type.*
(3) *([Mat] Exercise 2.5) If M is of finite presentation and M' is of finite type, then M'' is of finite presentation.*

Definition B.9. *Let B be an A-algebra.*
(1) *B is called* finitely generated *or* of finite type *if there exists a surjective homomorphism of A-algebras $\pi \colon A[T_1, \dots, T_n] \to B$.*
(2) *B is called* of finite presentation *if there exists π as above such that $\operatorname{Ker}(\pi)$ is a finitely generated ideal.*

Every A-algebra of finite presentation is of finite type. The converse holds, if A is noetherian by Hilbert's basis theorem (Proposition B.34).

Lemma B.10. *(see Section (B.17) below) Let R be a ring, and let $(A_\lambda)_{\lambda \in \Lambda}$ be a filtered inductive system of R-algebras. Let $A = \varinjlim A_\lambda$. Furthermore, let B be an A-algebra of finite presentation. Then there exists $\lambda \in \Lambda$ and an A_λ-algebra B_λ of finite presentation such that $B \cong B_\lambda \otimes_{A_\lambda} A$.*

Proposition B.11. *(see Section (B.17) below) Let $A \xrightarrow{\varphi} B \xrightarrow{\pi} C$ be ring homomorphisms.*
(1) *If B is an A-algebra of finite presentation and C is a B-algebra of finite presentation, then C is an A-algebra of finite presentation.*
(2) *Let C be an A-algebra of finite presentation and let B be a finitely generated A-algebra. Then C is a B-algebra of finite presentation. If $\pi \colon B \to C$ is surjective, then $\operatorname{Ker}(\pi)$ is a finitely generated ideal of B.*

Definition B.12. *An A-algebra B is called* finite *if B is finitely generated as an A-module.*

Clearly, any finite A-algebra is finitely generated.

Proposition B.13. *(see Section (B.17) below) A finite A-algebra B is of finite presentation as A-module if and only if B is of finite presentation as A-algebra.*

Definition B.14. *An A-algebra B is called* essentially of finite type, *if it is isomorphic to the localization of an A-algebra of finite type.*

(B.4) Projective, flat and faithfully flat modules.

Let A be a ring.

Definition/Proposition B.15. *([Mat] Appendix B) An A-module P is called* projective *if it satisfies the following equivalent conditions.*
(i) *The functor $\operatorname{Hom}(P, -)$ is right exact, i.e., for every surjective homomorphism of A-modules $p \colon M \twoheadrightarrow M'$ and every homomorphism $u' \colon P \to M'$ there exists a homomorphism $u \colon P \to M$ with $p \circ u = u'$.*
(ii) *For every surjection $p \colon M \to P$ there exists a homomorphism $i \colon P \to M$ such that $p \circ i = \operatorname{id}_P$.*
(iii) *P is a direct summand of a free A-module.*

Definition/Proposition B.16. ([BouAC] I, 2.3, Proposition 1 and Proposition 4) *An A-module M is called* flat *if it satisfies the following equivalent conditions.*
(i) *For every exact sequence $N' \to N \to N''$ of A-modules the tensored sequence $N' \otimes M \to N \otimes M \to N'' \otimes M$ is exact.*
(ii) *For every finitely generated ideal $\mathfrak{a} \subset A$ the homomorphism $\mathfrak{a} \otimes M \to M$, $a \otimes m \mapsto am$ is injective.*
(iii) *For every exact sequence $0 \to E \to F \to M \to 0$ of A-modules and for every A-module N the tensored sequence $0 \to E \otimes N \to F \otimes N \to M \otimes N \to 0$ is exact.*
(iv) *For every exact sequence $0 \to E \to F \to M \to 0$ of A-modules, E is flat if and only if F is flat.*

Definition/Proposition B.17. ([Mat] Theorem 7.2) *An A-module M is called* faithfully flat *if it satisfies the following equivalent conditions.*
(i) *A sequence $N' \to N \to N''$ of A-modules is exact if and only if the tensored sequence $N' \otimes M \to N \otimes M \to N'' \otimes M$ is exact.*
(ii) *M is flat and for every A-module N the equality $M \otimes_A N = 0$ implies $N = 0$.*
(iii) *M is flat and for every maximal ideal $\mathfrak{m}$ of A we have $M \neq \mathfrak{m}M$.*

Example B.18. Let $A \to B$ be a local homomorphism of local rings and let M be a nonzero finitely generated B-module which is flat over A. Then M is a faithfully flat A-module. Indeed, by Proposition B.17 it suffices to show that $\mathfrak{m}M \neq M$, where $\mathfrak{m}$ is the maximal ideal of A. But by hypothesis $\mathfrak{m}B$ is contained in the maximal ideal of B and thus $\mathfrak{m}M = M$ would imply $M = 0$ by Nakayama's lemma (Proposition B.3).

An A-algebra B is called *flat* (resp. *faithfully flat*) if B is flat (resp. faithfully flat) as an A-module.

Example B.19. For $n \geq 0$ the polynomial ring $A[T_1, \ldots, T_n]$ is a faithfully flat A-algebra (because every nonzero free module is faithfully flat). If A is noetherian, the ring of formal power series $A[\![T_1, \ldots, T_n]\!]$ is a faithfully flat A-algebra ([BouAC] III, 3.4, Corollary 3 of Theorem 3) and for a maximal ideal $\mathfrak{m}$ of A one has $A[\![T_1, \ldots, T_n]\!]/\mathfrak{m}A[\![T_1, \ldots, T_n]\!] = (A/\mathfrak{m})[\![T_1, \ldots, T_n]\!]$.

The following permanence properties follow immediately from the definitions.

Proposition B.20. *Let $\mathbf{P}$ be one of the properties "of finite type", "of finite presentation", "free", "projective", "flat", "faithfully flat".*
(1) *If M and N are A-modules with the property $\mathbf{P}$, then $M \otimes_A N$ has property $\mathbf{P}$.*
(2) *Let $\varphi \colon A \to B$ be a ring homomorphism and let M be an A-module having the property $\mathbf{P}$. Then the B-module $B \otimes_A M$ has property $\mathbf{P}$.*
(3) *Every filtered inductive limit of flat modules is again flat.*
(4) *If $(M_i)_{i \in I}$ is a family of A-modules, then $\bigoplus_i M_i$ is projective (resp. flat) if and only if M_i is projective (resp. flat) for all i.*

Every free module is projective, and every projective module is flat. In particular, if $A = k$ is a field, every k-vector space is flat. Every flat module over an integral domain is torsion-free.

Proposition B.21. ([BouAC] II, 5.3, Proposition 5 with Proposition B.29) *Let A be a semi-local ring and let M be a finitely presented flat module such that the dimensions $\dim_{\kappa(\mathfrak{m})} M/\mathfrak{m}M$ are equal for all maximal ideals $\mathfrak{m}$. Then M is free.*

Proposition B.22. ([Mat] Theorem 7.5) *Let* $\varphi\colon A \to B$ *be a faithfully flat ring homomorphism. Then* φ *is injective and* $\mathfrak{a} = \varphi^{-1}(\varphi(\mathfrak{a})B)$ *for every ideal* $\mathfrak{a}$ *of* A.

(B.5) Localization.

The main reference is [AM] Chapter 3. Let A be a ring and M an A-module. Let $S \subseteq A$ be a multiplicative subset (i.e., $1 \in S$ and $a, b \in S \Rightarrow ab \in S$). We denote the localization of A (resp. M) with respect to S by $S^{-1}A$ (resp. $S^{-1}M$). Elements in $S^{-1}A$ are denoted by $\frac{a}{s}$ for $a \in A$ and $s \in S$. Two elements $\frac{a}{s}, \frac{a'}{s'} \in S^{-1}A$ are equal if and only if there exists $t \in S$ with $tas' = ta's$. Analogous notation is used for $S^{-1}M$. The usual formulas for adding and multiplying fractions define on $S^{-1}A$ a ring structure and on $S^{-1}M$ the structure of an $S^{-1}A$-module.

If $f \in A$ and $\mathfrak{p} \subset A$ is a prime ideal, we set

(B.5.1)
$$A_f := S_f^{-1}A, \qquad M_f := S_f^{-1}M,$$
$$A_{\mathfrak{p}} := (A \setminus \mathfrak{p})^{-1}A, \qquad M_{\mathfrak{p}} := (A \setminus \mathfrak{p})^{-1}M$$

where $S_f = \{\, f^r \; ; \; r \geq 0 \,\}$ with the convention $f^0 := 1$.

The ring homomorphism

$$\iota = \iota_S \colon A \to S^{-1}A, \qquad a \mapsto \frac{a}{1}$$

is called the canonical homomorphism. We have $\iota(S) \subseteq (S^{-1}A)^{\times}$ and ι is universal with this property: If $\varphi\colon A \to B$ is any ring homomorphism with $\varphi(S) \subseteq B^{\times}$, then there exists a unique ring homomorphism $\psi\colon S^{-1}A \to B$ such that $\psi \circ \iota = \varphi$. This characterizes the pair $(S^{-1}A, \iota)$ uniquely up to unique isomorphism.

The following assertions follow immediately from the definition.
(1) $\iota\colon A \to S^{-1}A$ is an isomorphism if and only if $S \subseteq A^{\times}$.
(2) ι is injective if and only if S consists only of regular elements. In this case two elements $\frac{a}{s}, \frac{a'}{s'} \in S^{-1}A$ are equal if and only if $as' = a's$.
(3) $S^{-1}A = 0$ if and only if S contains a nilpotent element of A ($\Leftrightarrow 0 \in S$).

If $u\colon M \to N$ is a homomorphism of A-modules, the map $S^{-1}u\colon S^{-1}M \to S^{-1}N$ sending $\frac{m}{s}$ to $\frac{u(m)}{s}$ is a homomorphism of $S^{-1}A$-modules. Therefore $M \mapsto S^{-1}M$ yields a functor from the category of A-modules to the category of $S^{-1}A$-modules.

Proposition B.23. *Let* S *be a multiplicative set of a ring* A.
(1) *The homomorphism* $S^{-1}A \otimes_A M \to S^{-1}M$, $\frac{a}{s} \otimes m \mapsto \frac{am}{s}$ *is an isomorphism of* $S^{-1}A$-*modules for every* A-*module* M.
(2) *The* A-*algebra* $S^{-1}A$ *is flat.*
(3) *The functor* $M \mapsto S^{-1}M$ *is exact and commutes with direct sums, filtered inductive limits, and tensor products.*

Every ideal of $S^{-1}A$ is of the form $S^{-1}\mathfrak{a}$ for some ideal $\mathfrak{a} \subseteq A$, and the map $\mathfrak{p} \mapsto S^{-1}\mathfrak{p}$ defines a bijection

(B.5.2) $\{\mathfrak{p} \subset A \text{ prime ideal} \; ; \; \mathfrak{p} \cap S = \emptyset\} \leftrightarrow \{\text{prime ideals of } S^{-1}A\}$.

Write $S \leq T$ if for every prime ideal $\mathfrak{p}$ of A, $\mathfrak{p} \cap T = \emptyset$ implies $\mathfrak{p} \cap S = \emptyset$. This defines a partial preorder on the set of multiplicative subsets. Using that the group of units in a ring is the complement of the union of all prime ideals and the universal property of the localization, we obtain:

Lemma B.24. *Let $S, T \subseteq A$ be multiplicative subsets. There exists a (necessarily unique) A-algebra homomorphism $\iota_{T,S} \colon S^{-1}A \to T^{-1}A$ if and only if $S \leq T$. In particular, this is the case if $S \subseteq T$.*

In particular, we have for a prime ideal $\mathfrak{p}$ and an element $f \in A \setminus \mathfrak{p}$ a homomorphism $A_f \to A_{\mathfrak{p}}$ and the induced homomorphism

$$(B.5.3) \qquad \varinjlim_{f \in A \setminus \mathfrak{p}} A_f \to A_{\mathfrak{p}}$$

is an isomorphism of A-algebras. Using Proposition B.23 we also obtain for all A-modules M an isomorphism, functorial in M,

$$(B.5.4) \qquad \varinjlim_{f \in A \setminus \mathfrak{p}} M_f \xrightarrow{\sim} M_{\mathfrak{p}}.$$

The set S of regular elements in A is multiplicative and we call $\operatorname{Frac} A := S^{-1}A$ the *ring of total fractions*. If A is an integral domain (and hence $S = A \setminus \{0\}$), $\operatorname{Frac} A$ is a field, called the *field of fractions* of A.

For every prime ideal $\mathfrak{p}$ of A the canonical homomorphism $A \to A_{\mathfrak{p}}/\mathfrak{p}A_{\mathfrak{p}}$ induces an isomorphism

$$(B.5.5) \qquad \operatorname{Frac}(A/\mathfrak{p}) \xrightarrow{\sim} A_{\mathfrak{p}}/\mathfrak{p}A_{\mathfrak{p}}$$

and $\kappa(\mathfrak{p}) := A_{\mathfrak{p}}/\mathfrak{p}A_{\mathfrak{p}}$ is called the *residue field at $\mathfrak{p}$*.

(B.6) Local-global principles.

The main reference is [AM] Chapter 3.

Proposition B.25. ([BouAC] II, 3.3, Theorem 1) *For a sequence $M \to N \to P$ of A-modules the following assertions are equivalent.* ·
(i) *The sequence $M \to N \to P$ is exact.*
(ii) *For all prime ideals $\mathfrak{p}$ of A the sequence $M_{\mathfrak{p}} \to N_{\mathfrak{p}} \to P_{\mathfrak{p}}$ is exact.*
(iii) *For all maximal ideals $\mathfrak{m}$ of A the sequence $M_{\mathfrak{m}} \to N_{\mathfrak{m}} \to P_{\mathfrak{m}}$ is exact.*

In particular, a morphism $u \colon M \to N$ of A-modules is injective (resp. surjective) if and only if $u_{\mathfrak{p}} \colon M_{\mathfrak{p}} \to N_{\mathfrak{p}}$ is injective (resp. surjective) for all prime ideals $\mathfrak{p}$ of A. Applying this to $N = 0$ one obtains:

Corollary B.26. $M = 0$ *if and only if $M_{\mathfrak{m}} = 0$ for all maximal ideals $\mathfrak{m}$ of A.*

Proposition B.27. ([Mat] Theorem 7.1) *Let $\varphi \colon A \to B$ be a homomorphism of rings and let N be a B-module (also considered as A-module via φ). The following assertions are equivalent.*
(i) *N is a flat A-module.*
(ii) *For every prime ideal $\mathfrak{q}$ of B the localization $N_{\mathfrak{q}}$ is a flat $A_{\varphi^{-1}(\mathfrak{q})}$-module.*
(iii) *For every maximal ideal $\mathfrak{n}$ of B the localization $N_{\mathfrak{n}}$ is a flat $A_{\varphi^{-1}(\mathfrak{n})}$-module.*

Proposition B.28. (Lemma 3.20 in the main text; [BouAC] II, 5.1, Corollary of Proposition 3) *Let $f_1, \ldots, f_r \in A$ be elements that generate the unit ideal and let M be an A-module. Then M is finitely generated (resp. of finite presentation) if and only if for all i the A_{f_i}-module M_{f_i} is finitely generated (resp. of finite presentation).*

Proposition B.29. ([Mat] Theorem 7.12 and Corollary) *For an A-module M the following assertions are equivalent.*
(i) M *is finitely generated and projective.*
(ii) M *is of finite presentation and $M_{\mathfrak{p}}$ is a free $A_{\mathfrak{p}}$-module for all prime ideals $\mathfrak{p}$ of A.*
(iii) M *is of finite presentation and $M_{\mathfrak{m}}$ is a free $A_{\mathfrak{m}}$-module for all maximal ideals $\mathfrak{m}$ of A.*
(iv) M *is flat and of finite presentation.*

Let $n \geq 0$ be an integer. A finitely generated projective A-module M is said to be of *rank n* if $\mathrm{rk}_{A_{\mathfrak{p}}} M_{\mathfrak{p}} = n$ for all prime ideals $\mathfrak{p}$ of A. For a discussion of the properties of projective and flat modules, respectively, in the non-finitely generated case, we refer to [RG]; see also [Dr] and [St] 058B which fixes an error in [RG].

Let A be an integral domain. For every A-module M one has $S^{-1}(M_{\mathrm{tors}}) = (S^{-1}M)_{\mathrm{tors}}$. This implies:

Proposition B.30. *For an A-module M the following assertions are equivalent.*
(i) M *is torsion-free.*
(ii) $M_{\mathfrak{p}}$ *is torsion-free for all prime ideals $\mathfrak{p}$ of A.*
(iii) $M_{\mathfrak{m}}$ *is torsion-free for all maximal ideals $\mathfrak{m}$ of A.*

(B.7) Noetherian and Artinian modules and rings.

Let A be a ring.

Definition/Proposition B.31. ([AM] Proposition 6.1 and Proposition 6.2) *An A-module M is called* noetherian, *if it satisfies the following equivalent conditions.*
(i) *Any ascending chain $N_0 \subseteq N_1 \subseteq \dots$ of submodules becomes stationary (i.e., there exists an integer $r \geq 0$ such that $N_i = N_r$ for all $i \geq r$).*
(ii) *Any non-empty subset of submodules of M has a maximal element.*
(iii) *Every submodule of M is finitely generated.*

An A-module M is called *Artinian* if any non-empty subset of submodules of M has a minimal element (equivalently, every descending chain $N_0 \supseteq N_1 \supseteq \dots$ of submodules becomes stationary).

Proposition B.32. ([AM] Proposition 6.3) *Let $0 \to M' \to M \to M'' \to 0$ be an exact sequence of A-modules. Then M is noetherian (resp. Artinian) if and only if M' and M'' are noetherian (resp. Artinian).*

An A-module M is noetherian and Artinian if and only if M is of finite length ([AM] Proposition 6.8).

The ring A is called *noetherian* (resp. *Artinian*) if A as an A-module is noetherian (resp. Artinian). Thus a ring is noetherian if and only if every ideal is finitely generated. Every Artinian ring is noetherian (an analogous statement for modules does not hold).

Proposition B.33.
(1) ([AM] Proposition 6.5) *Every finitely generated module over a noetherian (resp. Artinian) ring is itself noetherian (resp. Artinian).*
(2) ([AM] Proposition 7.3; Proposition B.36 below) *If the ring A is noetherian (resp. Artinian) and $S \subseteq A$ is a multiplicative subset, then $S^{-1}A$ is noetherian (resp. Artinian).*

Proposition B.34. (Hilbert's basis theorem) ([AM] Corollary 7.7) *Let A be a noetherian ring. Then any finitely generated A-algebra B is again noetherian.*

Proposition B.35. ([BouAC] III, 2.6, Corollary 6 of Theorem 2) *Let A be a noetherian ring. Then for all $n \geq 0$ the ring of formal power series $A[\![T_1, \ldots, T_n]\!]$ is noetherian.*

The following proposition gives a characterizations of Artinian rings.

Proposition B.36. ([AM] Chapter 8) *Let A be a ring. The following are equivalent:*
  (i) *The ring A is Artinian, i. e. it satisfies the descending chain condition for ideals.*
  (ii) $\lg_A(A) < \infty$.
  (iii) *The ring A is noetherian, and every prime ideal is a minimal prime ideal.*
  (iv) *The ring A is noetherian, and the natural map*

$$A \to \prod_{\mathfrak{p} \in \operatorname{Spec} A} A_{\mathfrak{p}}$$

is an isomorphism.
  (v) *The ring A is noetherian and for all prime ideals $\mathfrak{p}$ of A, the localization $A_{\mathfrak{p}}$ is Artinian.*
If A is Artinian, A has only finitely many prime ideals.

Proposition B.37. ([BouAC] II, 3.2, Corollary 2 of Proposition 5) *Let A be a local Artinian ring. Then an A-module is flat if and only if it is free.*

(B.8) Completion.

The main reference is [Mat] §8. Let A be a ring and let $\mathfrak{a} \subseteq A$ be an ideal.

Definition/Remark B.38. *Let M be an A-module. The $\mathfrak{a}$-adic topology on M is the coarsest topology on M such that for all $m \in M$ the sets $m + \mathfrak{a}^n M$ for $n \geq 0$ form a fundamental system of open neighborhoods of m.*
  (1) *The $\mathfrak{a}$-adic topology on A makes A into a topological ring (i.e., addition and multiplication are continuous). Moreover, M with its $\mathfrak{a}$-adic topology is a topological A-module over the topological ring A (i.e., addition and scalar multiplication are continuous).*
  (2) *The $\mathfrak{a}$-adic topology on M is Hausdorff if and only if $\bigcap_{n \geq 0} \mathfrak{a}^n M = 0$.*
  (3) *Let $\varphi \colon A \to B$ be an A-algebra, let $\mathfrak{b} \subseteq B$ be an ideal and assume that there exists an integer $r \geq 1$ such that $\varphi(\mathfrak{a}^r) \subseteq \mathfrak{b}$. Then φ is continuous with respect to the $\mathfrak{a}$-adic topology on A and the $\mathfrak{b}$-adic topology on B.*
  (4) *In the situation of (3), let N be a B-module endowed with the $\mathfrak{b}$-adic topology. Then every A-linear homomorphism $u \colon M \to N$ is continuous. In particular (for $B = A$, $\mathfrak{b} = \mathfrak{a}$, and $\varphi = \operatorname{id}_A$), every homomorphism of A-modules is continuous with respect to the $\mathfrak{a}$-adic topology.*

For an A-module M we endow $M/\mathfrak{a}^n M$ with the discrete topology (which is also the quotient topology with respect to the $\mathfrak{a}$-adic topology) and call the topological A-module

$$\hat{M} := \hat{M}^{\mathfrak{a}} := \varprojlim_n M/\mathfrak{a}^n M$$

the *completion* of M. For $M = A$ we obtain a topological ring $\hat{A}$, and $\hat{M}$ is a topological $\hat{A}$-module.

The projections $M \to M/\mathfrak{a}^n M$ yield a continuous homomorphism $i\colon M \to \hat{M}$ of A-modules, called the *canonical homomorphism*. The canonical homomorphism $A \to \hat{A}$ is a continuous ring homomorphism.

Every homomorphism $u\colon M \to N$ of A-modules induces an $\hat{A}$-linear homomorphism $\hat{u}\colon \hat{M} \to \hat{N}$ on the $\mathfrak{a}$-adic completions. Thus we obtain a functor $M \mapsto \hat{M}$ from the category of A-modules to the category of $\hat{A}$-modules.

Definition B.39. *An A-module M is called* $\mathfrak{a}$-adically complete *if the canonical homomorphism $M \to \hat{M}$ is an isomorphism. In particular, A is said to be $\mathfrak{a}$-adically complete if the canonical homomorphism $A \to \varprojlim_n A/\mathfrak{a}^n A$ is an isomorphism.*

If A is a local ring, we usually endow A (and every A-module M) with the $\mathfrak{m}$-adic topology, where $\mathfrak{m}$ is the maximal ideal of A. If A (or M) is $\mathfrak{m}$-adically complete, we simply say that A (or M) is *complete*.

Let $\mathfrak{a}$ be again an arbitrary ideal of A. If $\iota\colon N \to M$ is the inclusion of an A-submodule N of M, then ι is continuous with respect to the $\mathfrak{a}$-adic topology, but the topology on N induced by the $\mathfrak{a}$-adic topology on M may be strictly coarser than the $\mathfrak{a}$-adic topology on N. The Artin-Rees lemma asserts that this cannot happen if A is noetherian and M is finitely generated:

Proposition B.40. (Artin-Rees lemma, [Mat] Theorem 8.5) *Let A be noetherian, $\mathfrak{a} \subseteq A$ an ideal, M a finitely generated A-module, and $N \subseteq M$ a submodule. Then there exists an integer $r > 0$ such that $\mathfrak{a}^n M \cap N = \mathfrak{a}^{n-r}(\mathfrak{a}^r M \cap N)$ for all $n > r$. In particular, the $\mathfrak{a}$-adic topology on M induces the $\mathfrak{a}$-adic topology on N.*

Proposition B.41. *Let A be noetherian, $\mathfrak{a} \subseteq A$ an ideal, and M a finitely generated A-module.*

(1) *The $\mathfrak{a}$-adic completion $\hat{A}$ is noetherian and the canonical homomorphism $A \to \hat{A}$ is flat.*

(2) *The canonical homomorphism $\hat{A} \otimes_A M \to \hat{M}$ is an isomorphism.*

(3) *The topology on $\hat{M}$ is the $\mathfrak{a}$-adic topology as an A-module and the $\mathfrak{a}\hat{A}$-adic topology as an $\hat{A}$-module, $\hat{M}$ is complete with this topology, and for all $n \geq 1$ we have ([BouAC] III, 2.12, Corollary 2 of Proposition 16)*

$$\widehat{\mathfrak{a}^n M} = \widehat{\mathfrak{a}^n}\hat{M} = \mathfrak{a}^n \hat{M}.$$

(4) *The canonical homomorphism $i\colon M \to \hat{M}$ induces isomorphisms*

$$\mathfrak{a}^n M/\mathfrak{a}^{n+1}M \xrightarrow{\sim} \mathfrak{a}^n \hat{M}/\mathfrak{a}^{n+1}\hat{M}, \qquad M/\mathfrak{a}^n M \xrightarrow{\sim} \hat{M}/\mathfrak{a}^n \hat{M}.$$

(5) *For every exact sequence $M' \to M \to M''$ of finitely generated A-modules, the induced sequence $\hat{M}' \to \hat{M} \to \hat{M}''$ of $\hat{A}$-modules is exact.*

Proposition B.42. ([BouAC] III, 3.3, Proposition 6 and III, 3.5, Proposition 9 and its Corollary 1) *Let A be noetherian, $\mathfrak{a} \subseteq A$ an ideal, and $\hat{A}$ its $\mathfrak{a}$-adic completion. The following assertions are equivalent.*

(i) *$\mathfrak{a}$ is contained in the Jacobson radical of A.*

(ii) *For every finitely generated A-module M one has $\bigcap_{n \geq 0} \mathfrak{a}^n M = 0$.*

(iii) *For every finitely generated A-module M every submodule of M is closed in the $\mathfrak{a}$-adic topology.*

(iv) *$\hat{A}$ is a faithfully flat A-module.*

If these equivalent conditions are satisfied, then for every finitely generated A-module M and for every submodule N one has

$$N = \hat{N} \cap M = (\hat{A}N) \cap M.$$

Corollary B.43. ([BouAC] III, 3.4, Corollary of Proposition 8) *Let A be a local noetherian ring. Then $\hat{A}$ is a complete local noetherian ring, $\hat{A}$ is a faithfully flat A-algebra, $\mathfrak{m}_{\hat{A}} = \mathfrak{m}_A \hat{A}$, and for every ideal $\mathfrak{b} \subsetneq A$ we have*

$$\bigcap_{n \geq 0} \mathfrak{b}^n = 0 \qquad \text{and} \qquad \mathfrak{b} = \hat{\mathfrak{b}} \cap A.$$

Corollary B.44. *Let A be a noetherian ring which is complete for the $\mathfrak{a}$-adic topology. Then every finitely generated A-module is complete for the $\mathfrak{a}$-adic topology.*

Lemma B.45. ([BouAC] III, 2.7, Proposition 8) *Let A be a complete local noetherian ring with maximal ideal $\mathfrak{m}$, and let $\mathfrak{a}_N \subset A$, $N \geq 0$, be a descending sequence of ideals with $\bigcap_N \mathfrak{a}_N = 0$. Then for every $n \geq 0$, there exists N with $\mathfrak{a}_N \subseteq \mathfrak{m}^n$.*

Proposition B.46. ([BouAC] III, 2.13, Proposition 17; III, 3.4, Proposition 8; V, 2.1, Corollary 3 of Theorem 1; V, 2.1, Proposition 5) *Let A be a semi-local noetherian ring with maximal ideals $\mathfrak{m}_1, \ldots, \mathfrak{m}_n$, let $\mathfrak{r}(A) = \mathfrak{m}_1 \cdots \mathfrak{m}_r$ be its Jacobson radical, and let B be a finite A-algebra.*
(1) *The $\mathfrak{r}$-adic completion $\hat{A}$ of A is isomorphic to $\prod_{i=1}^n \widehat{A_{\mathfrak{m}_i}}$, and the $\widehat{A_{\mathfrak{m}_i}}$ are complete local noetherian rings.*
(2) *B is semi-local and noetherian, and the $\mathfrak{r}(B)$-adic topology on B coincides with the $\mathfrak{r}(A)$-adic topology.*
(3) *If A is complete, B is complete. In this case B is a finite product of complete local noetherian rings.*

Examples B.47.
(1) Let $A = R[T_1, \ldots, T_n]$ be a polynomial ring over some ring R and let $\mathfrak{a} = (T_1, \ldots, T_n)$. Then $\hat{A} \cong R[\![T_1, \ldots, T_n]\!]$, and the topology on $\hat{A}$ is the $(T_1, \ldots, T_n)$-adic topology. Thus, if R is noetherian, $R[\![T_1, \ldots, T_n]\!]$ is noetherian as well.
(2) Let A be noetherian and $\mathfrak{a} = (a_1, \ldots, a_n)$. Then the $\mathfrak{a}$-adic completion $\hat{A}$ is isomorphic to $A[\![T_1, \ldots, T_n]\!]/(T_1 - a_1, \ldots, T_n - a_n)$.
(3) Every local Artinian ring is complete.

Let A be a ring, $\mathfrak{a} \subseteq A$ be an ideal. For every A-module we set

$$\mathrm{gr}^{\mathfrak{a}}(M) := \bigoplus_{n \geq 0} \mathrm{gr}_n^{\mathfrak{a}} M, \qquad \text{with } \mathrm{gr}_n^{\mathfrak{a}}(M) := \mathfrak{a}^n M / \mathfrak{a}^{n+1} M.$$

Then $\mathrm{gr}^{\mathfrak{a}}(A)$ is a graded ring (see Section (13.1)), called the *associated graded ring*, and $\mathrm{gr}^{\mathfrak{a}}(M)$ is a graded $\mathrm{gr}^{\mathfrak{a}}(A)$-module, called the *associated graded module*. The construction yields a functor $M \mapsto \mathrm{gr}^{\mathfrak{a}}(M)$ from the category of A-modules to the category of graded $\mathrm{gr}^{\mathfrak{a}}(A)$-modules.

Example B.48. ([BouAC] III, 2.3, Examples (2)) Let R be a ring, $A = R[\![T_1, \ldots, T_n]\!]$ and $\mathfrak{a} = (T_1, \ldots, T_n)$. Then $\mathrm{gr}^{\mathfrak{a}}(A) \cong R[T_1, \ldots, T_n]$.

Proposition B.49. ([AM] Lemma 10.23) *Let $(A, \mathfrak{a})$ and $(B, \mathfrak{b})$ be two rings with ideals, let M be an A-module and N a B-module, and let $u \colon M \to N$ be a homomorphism of abelian groups such that $u(\mathfrak{a}^n M) \subseteq \mathfrak{b}^n N$ for all $n \geq 1$. We equip M and N with the $\mathfrak{a}$- and $\mathfrak{b}$-adic topologies. Let $\hat{u} \colon \hat{M} \to \hat{N}$ and $\mathrm{gr}(u) \colon \mathrm{gr}^{\mathfrak{a}}(M) \to \mathrm{gr}^{\mathfrak{b}}(N)$ be the induced homomorphisms. If $\mathrm{gr}(u)$ is surjective (resp. injective), then $\hat{u}$ is surjective (resp. injective).*

(B.9) Local criterion for flatness.

Let A be a ring.

Proposition B.50. ([Mat] Theorem 22.1) *Let B be a noetherian A-algebra, let $\mathfrak{b} \subset B$ be an ideal that is contained in the Jacobson radical of B, and let M be a finitely generated B-module. Suppose that $M/\mathfrak{b}^n M$ is a flat A-module for all $n \geq 1$, then M is a flat A-module.*

Let $\mathfrak{a} \subset A$ be an ideal and M an A-module. For all $n \geq 0$ multiplication induces a surjective homomorphism $\gamma_n \colon \mathfrak{a}^n/\mathfrak{a}^{n+1} \otimes_{A/\mathfrak{a}} M/\mathfrak{a}M \twoheadrightarrow \mathfrak{a}^n M/\mathfrak{a}^{n+1}M$. Thus we obtain a surjective homomorphism of $\mathrm{gr}^{\mathfrak{a}}(A)$-modules

$$(B.9.1) \qquad \gamma_M^{\mathfrak{a}} \colon \mathrm{gr}^{\mathfrak{a}}(A) \otimes_{\mathrm{gr}_0^{\mathfrak{a}}(A)} \mathrm{gr}_0^{\mathfrak{a}}(M) \twoheadrightarrow \mathrm{gr}^{\mathfrak{a}}(M).$$

Theorem B.51. ([Mat] Theorem 22.3) *Let A be a ring, and let $\mathfrak{a} \subseteq A$ be an ideal. Let M be an A-module, and suppose that $\mathfrak{a}$ is nilpotent, or that there exists a noetherian A-algebra B, such that $\mathfrak{a}B$ is contained in the Jacobson radical of B and such that the A-module structure on M is induced from a B-module structure which makes M a finitely generated B-module. The following are equivalent:*
(i) *The A-module M is flat.*
(ii) *The $A/\mathfrak{a}$ module $M/\mathfrak{a}$ is flat and the natural homomorphism (B.9.1) is an isomorphism.*
(iii) *The $A/\mathfrak{a}$ module $M/\mathfrak{a}$ is flat and the multiplication $\mathfrak{a} \otimes_A M \to M$ is injective.*
(iv) *For all $n \geq 1$, the $A/\mathfrak{a}^n$-module $M/\mathfrak{a}^n M$ is flat.*

Note that in the statement of the equivalent conditions in the theorem, B does not play a role; it just serves to impose a certain finiteness condition on M. More precisely, it is enough to require that M is an $\mathfrak{a}$-adically ideal-separated A-module, i.e., $\bigcap \mathfrak{a}^n(\mathfrak{b} \otimes M) = 0$ for every finitely generated ideal $\mathfrak{b}$ of A.

By induction one deduces easily the following corollary.

Corollary B.52. *Let A be a ring with a finite filtration of ideals*

$$0 = \mathfrak{I}_{N+1} \subset \mathfrak{I}_N \subset \cdots \subset \mathfrak{I}_1 \subset \mathfrak{I}_0 = A,$$

such that $\mathfrak{I}_m \mathfrak{I}_n \subseteq \mathfrak{I}_{n+m}$ for all n, m. Let M be an A-module, and denote by $\mathrm{gr}\, A = \bigoplus_{n=0}^{N} \mathfrak{I}_n/\mathfrak{I}_{n+1}$, $\mathrm{gr}\, M = \bigoplus_{n=0}^{N} \mathfrak{I}_n M/\mathfrak{I}_{n+1}M$ the associated graded ring and module, respectively. Suppose that $M \otimes_A (A/\mathfrak{I}_1)$ is a flat $A/\mathfrak{I}_1$-module and that the natural homomorphism

$$\mathrm{gr}_0(M) \otimes_{\mathrm{gr}_0(A)} \mathrm{gr}(A) \to \mathrm{gr}(M)$$

is injective. Then M is a flat A-module.

(B.10) Integral ring homomorphisms.

The main reference is [AM] Chapter 5. Let A be a ring and let $\varphi \colon A \to B$ be an A-algebra.

Definition/Proposition B.53.
(1) *An element $b \in B$ is called* integral *over A if it satisfies the following equivalent conditions.*
  (i) *There exists a monic polynomial $f \in A[T]$ such that $f(b) = 0$.*
  (ii) *The A-subalgebra $A[b]$ of B is a finite A-algebra.*
  (iii) *There exists a finite A-subalgebra C of B that contains b.*
(2) *The subset A' of elements of B that are integral over A forms an A-subalgebra of B.*

The ring homomorphism $\varphi \colon A \to B$ is called *integral* if all elements of B are integral over A.

An A-algebra is finite if and only if it is integral and of finite type.

Proposition B.54. *Let B be a finite (resp. integral) A-algebra.*
(1) *If C is a finite (resp. integral) B-algebra, then C is a finite (resp. integral) A-algebra.*
(2) *If A' is any A-algebra, then $A' \otimes_A B$ is a finite (resp. integral) A'-algebra.*
(3) *(see Section (B.17) below) Let A_0 be a ring, and let $A = \varinjlim_\lambda A_\lambda$ be the filtered inductive limit of A_0-algebras A_λ. Let B_0 be an A_0-algebra of finite presentation such that $B := A \otimes_{A_0} B_0$ is a finite A-algebra. Then there exists an index λ such that the A_λ-algebra $B_\lambda := A_\lambda \otimes_{A_0} B_0$ is finite.*

Let B be an A-algebra. Then

$$(\text{B.10.1}) \qquad\qquad A' := \{\, b \in B \;;\; b \text{ is integral over } A \,\}$$

is a A-subalgebra of B which is called the *integral closure* of A in B. If $A = A'$, then A is called *integrally closed in B*. If A is a domain, A is called *integrally closed* if A is integrally closed in its fraction field.

Proposition B.55. *Let B be an A-algebra and let A' be the integral closure of A in B. Then for every multiplicative set $S \subset A$, the integral closure of $S^{-1}A$ in $S^{-1}B$ is $S^{-1}A'$.*

Theorem B.56. *Let $\varphi \colon A \to B$ be an integral injective ring homomorphism.*
(1) *Let $\mathfrak{q}_0 \subsetneq \mathfrak{q}_1$ be prime ideals of B. Then $\varphi^{-1}(\mathfrak{q}_0) \subsetneq \varphi^{-1}(\mathfrak{q}_1)$.*
(2) *(going up) Let $\mathfrak{p}_1 \subsetneq \cdots \subsetneq \mathfrak{p}_r$ be a chain of prime ideals of A and let $\mathfrak{q}_1 \subsetneq \cdots \subsetneq \mathfrak{q}_m$ (for some $0 \le m < r$) be a chain of prime ideals of B such that $\varphi^{-1}(\mathfrak{q}_i) = \mathfrak{p}_i$ for all $i = 1, \ldots, m$. Then this chain can be completed to a chain $\mathfrak{q}_1 \subsetneq \cdots \subsetneq \mathfrak{q}_r$ of prime ideals of B such that $\varphi^{-1}(\mathfrak{q}_i) = \mathfrak{p}_i$ for all $i = 1, \ldots, r$.*
(3) *(going down) Let A and B be integral domains and assume that A is integrally closed. Let $\mathfrak{p}_1 \subsetneq \cdots \subsetneq \mathfrak{p}_r$ be a chain of prime ideals of A and let $\mathfrak{q}_n \subsetneq \cdots \subsetneq \mathfrak{q}_r$ (for some $1 < n \le r$) be a chain of prime ideals of B such that $\varphi^{-1}(\mathfrak{q}_i) = \mathfrak{p}_i$ for all $i = n, \ldots, r$. Then this chain can be completed to a chain $\mathfrak{q}_1 \subsetneq \cdots \subsetneq \mathfrak{q}_r$ of prime ideals of B such that $\varphi^{-1}(\mathfrak{q}_i) = \mathfrak{p}_i$ for all $i = 1, \ldots, r$.*
(4) *([Mat] Exercise 9.2) One has $\dim A = \dim B$.*

Proposition B.57. *Let A be a noetherian integrally closed domain, let L be a finite separable extension of $\operatorname{Frac} A$, and let B be the integral closure of A in L. Then B is a finite A-algebra.*

Theorem B.58. (Noether normalization theorem, [BouAC] V,3.1, Theorem 1) *Let k be a field and let $A \neq 0$ be a finitely generated k-algebra.*
(1) *There exist elements $t_1, \ldots, t_d \in A$ such that the homomorphism of k-algebras $\varphi \colon k[T_1, \ldots, T_d] \to A$, $T_i \mapsto t_i$, is injective and finite.*
(2) *If $\mathfrak{a}_0 \subset \mathfrak{a}_1 \subset \cdots \subset \mathfrak{a}_r \subsetneq A$ is a chain of ideals in A ($r \geq 0$), then the t_i in (1) can be chosen such that $\varphi^{-1}(\mathfrak{a}_i) = (T_1, \ldots, T_{h(i)})$ for all $i = 0, \ldots, r$ and suitable $0 \leq h(0) \leq h(1) \leq \cdots \leq h(r)$.*

(B.11) Associated Primes.

The main reference is [Mat] §6. Let A be a ring and let M be an A-module. A prime ideal $\mathfrak{p}$ of A is called *an associated prime ideal of M* if there exists an element $m \in M$ such that $\mathfrak{p} = \mathrm{Ann}(m) = \{\, a \in A \; ; \; am = 0 \,\}$. The set of associated prime ideals of M is denoted by $\mathrm{Ass}\, M$ or $\mathrm{Ass}_A M$. By definition, $\mathrm{Ass}\, M$ consists of those prime ideals $\mathfrak{p}$ such that M contains a submodule that is isomorphic to $A/\mathfrak{p}$.

Proposition B.59. *Let A be a noetherian ring and let M be an A-module.*
(1) *If $M \neq 0$, then $\mathrm{Ass}\, M \neq \emptyset$.*
(2) *One has $\bigcup_{\mathfrak{p} \in \mathrm{Ass}\, M} \mathfrak{p} = \{\, a \in A \; ; \; \exists\, 0 \neq m \in M : am = 0 \,\}$.*
(3) *For every multiplicative subset S of A, we have*

$$(\text{B.11.1}) \qquad \mathrm{Ass}(S^{-1}M) = \mathrm{Spec}(S^{-1}A) \cap \mathrm{Ass}(M),$$

where $\mathrm{Spec}(S^{-1}A) = \{\, \mathfrak{p} \subset A \text{ prime ideal} \; ; \; \mathfrak{p} \cap S = \emptyset \,\}$.
(4) *If M is finitely generated, then $\mathrm{Ass}\, M$ is finite,*

$$\mathrm{Ass}\, M \subseteq \mathrm{Supp}\, M := \{\, \mathfrak{p} \subset A \text{ prime ideal} \; ; \; M_{\mathfrak{p}} \neq 0 \,\},$$

and $\mathrm{Ass}\, M$ and $\mathrm{Supp}\, M$ have the same minimal elements.
In particular $\mathrm{Ass}\, A$ is finite, it contains the minimal prime ideals of A, and $\bigcup_{\mathfrak{p} \in \mathrm{Ass}\, A} \mathfrak{p}$ is the set of zero divisors of A.

(B.12) Regular sequences and dimension.

Let A be a ring and let M be an A-module.

Definition B.60. *A finite sequence $(a_1, \ldots, a_n)$ of elements $a_i \in A$ is called* regular *for M or M-regular if for all $i = 1, \ldots, n$ the multiplication with a_i on the A-module $M/(a_1 M + \cdots + a_{i-1} M)$ is injective and if $M/(a_1 M + \cdots + a_n M) \neq 0$.*

Proposition/Definition B.61. ([BouAC10] 1.4, Corollaire 1 de Théorème 2) *Let A be a local noetherian ring with maximal ideal $\mathfrak{m}$, and let M be a finitely generated A-module. Let $r \geq 0$ be an integer. Then the following assertions are equivalent.*
(i) *The M-regular sequences $(a_1, \ldots, a_r)$ of elements $a_i \in \mathfrak{m}$ of length r are precisely those that are maximal among the M-regular sequences of elements in $\mathfrak{m}$.*
(ii) *For every M-regular sequence $(a_1, \ldots, a_r)$ of elements $a_i \in \mathfrak{m}$ of length r the A-module $M/(a_1 M + \cdots + a_r M)$ has a nonzero element which is annihilated by $\mathfrak{m}$.*
(iii) *There exists an M-regular sequence $(a_1, \ldots, a_r)$ of elements $a_i \in \mathfrak{m}$ of length r such that $M/(a_1 M + \cdots + a_r M)$ has a nonzero element which is annihilated by $\mathfrak{m}$.*

Such an integer always exists. It is called the depth of M and denoted by $\mathrm{depth}_A(M)$ or simply $\mathrm{depth}(M)$.

There is also the following more general version of depth.

Proposition/Definition B.62. ([Mat] Theorem 16.7) *Let A be a noetherian ring, M a finite A-module and $\mathfrak{a} \subseteq A$ an ideal such that $\mathfrak{a}M \neq M$. Then all maximal M-regular sequences of elements contained in $\mathfrak{a}$ have the same length. This number is called the $\mathfrak{a}$-depth of M and denoted by $\mathrm{depth}(\mathfrak{a}, M)$.*

Let A be a ring. Then

$$\dim A := \sup\{\, l \in \mathbb{N}_0 \;;\; \exists \mathfrak{p}_0 \subsetneq \mathfrak{p}_1 \subsetneq \cdots \subsetneq \mathfrak{p}_l \text{ chain of prime ideals of } A\,\}$$

is called the *Krull dimension* or simply the *dimension* of the ring A. For a prime ideal $\mathfrak{p}$ of A we call

$$\mathrm{ht}(\mathfrak{p}) := \dim(A_{\mathfrak{p}})$$

the *height of* $\mathfrak{p}$. It is the supremum of the lengths of chains of prime ideals of A that have $\mathfrak{p}$ as their maximal element. For an arbitrary ideal $\mathfrak{a} \subsetneq A$ we set

$$\mathrm{ht}(\mathfrak{a}) := \inf\{\, \mathrm{ht}(\mathfrak{p}) \;;\; \mathfrak{a} \subseteq \mathfrak{p} \text{ prime ideal}\,\}.$$

Proposition B.63. (Krull's principal ideal theorem) ([Mat] Theorem 13.6) *Let A be a noetherian ring. If $f \in A$, and $\mathfrak{p} \subset A$ is a prime ideal which is minimal among all prime ideals containing f, then $\mathrm{ht}\,\mathfrak{p} \leq 1$. Conversely, if $\mathfrak{p} \subset A$ is a prime ideal of height ≤ 1, then there exists $f \in A$ such that $\mathfrak{p}$ is minimal among all prime ideals containing f.*

Corollary B.64. *Let A be a local noetherian ring and $a \in A \setminus A^{\times}$. Then $\dim A/(a) \geq \dim A - 1$. If a is regular, then $\dim A/(a) = \dim A - 1$.*

Corollary B.65. (Theorem of Artin-Tate) (see Section (B.18) below) *Let A be a noetherian integral domain. Then A is semi-local of dimension ≤ 1 if and only if there exists an $f \in A$ such that A_f is a field.*

Proposition B.66. ([Mat] Theorem 17.2, Theorem 13.4) *Let A be a local noetherian ring. Then $\mathrm{depth}\,A \leq \dim A < \infty$.*

Proposition B.67. ([Mat] Theorem 13.9) *Let A be a local noetherian ring, and let $\hat{A}$ be the completion with respect to its maximal ideal. Then $\dim A = \dim \hat{A} < \infty$.*

Proposition B.68. ([Mat] Theorem 15.1) *Let $\varphi\colon A \to B$ be a local homomorphism of noetherian local rings, and denote by $\mathfrak{m}$ the maximal ideal of A. Then*

$$\dim B \leq \dim A + \dim B/\mathfrak{m}B,$$

and if φ is flat, then equality holds.

(B.13) Valuation rings.

Let A and B local rings with $A \subseteq B$. Then we say that B *dominates* A if $\mathfrak{m}_B \cap A = \mathfrak{m}_A$. Given a field K, the set of local subrings of K is inductively ordered with respect to domination order.

Definition/Proposition B.69. ([BouAC] VI, 1.2, Theorem 1; §3.2) *Let K be a field and let A be a subring of K. The ring A is called a* valuation ring *of K if it satisfies the following equivalent properties.*
(i) *For every $a \in K^\times$ one has $a \in A$ or $a^{-1} \in A$.*
(ii) *Frac $A = K$ and the set of ideals of A is totally ordered by inclusion.*
(iii) *A is local and a maximal element in the set of local subrings of K with respect to the domination order.*
(iv) *There exists a totally ordered abelian group G and a map $v \colon K \to G \cup \{\infty\}$ with*
    (a) *$v(ab) = v(a) + v(b)$ (with $g + \infty := \infty + g := \infty$ for $g \in G$).*
    (b) *$v(a + b) \geq \min\{v(a), v(b)\}$ (where $\infty > g$ holds by definition for all $g \in G$).*
    (c) *$v(a) = \infty \Leftrightarrow a = 0$.*
    such that $A = \{ a \in K \; ; \; v(a) \geq 0 \}$.

A map $v \colon K \to G \cup \{\infty\}$ satisfying (a)–(c) is called a *valuation for A on K*, or a *valuation on K with valuation ring A*. The totally ordered abelian group $v(K^\times) \subseteq G$ is called the *value group of A*. It is isomorphic to the totally ordered abelian group $K^\times / A^\times$ and in particular unique up to isomorphism.

Corollary B.70. ([Mat] Theorem 10.2) *Let K be a field, $R \subseteq K$ be a subring, $\mathfrak{p}$ a prime ideal of R. Then there exists a valuation ring A of K such that $R \subseteq A$ and $\mathfrak{m}_A \cap R = \mathfrak{p}$.*

Proposition B.71. ([Mat] Theorem 11.1, Theorem 11.2) *Let A be a ring which is not a field. Then A is called a* discrete valuation ring, *if it satisfies the following equivalent properties.*
(i) *A is a local principal ideal domain.*
(ii) *A is a local one-dimensional integrally closed noetherian ring.*
(iii) *A is a noetherian valuation ring.*
(iv) *A is noetherian, local, $\dim A > 0$, and its maximal ideal $\mathfrak{m}$ is a principal ideal.*
(v) *There is a valuation $v \colon \operatorname{Frac} A \to \mathbb{Z} \cup \{\infty\}$ such that $A = \{ a \in \operatorname{Frac} A \; ; \; v(a) \geq 0 \}$.*

If A is a discrete valuation ring, a generator π of the unique maximal ideal $\mathfrak{m}$ of A is called a *uniformizing element*. Every nonzero ideal of A is of the form (π^n) for some $n \geq 0$. Moreover, we have

(B.13.1)
$$A^\times = \{ a \in A \; ; \; v(a) = 0 \},$$
$$\mathfrak{m} = \{ a \in A \; ; \; v(a) > 0 \}.$$

A valuation v for A as in (v) is called *normalized* if v is surjective (equivalently, $v(\pi) = 1$ for every uniformizing element π).

Proposition B.72. *Let A be a valuation ring, $K = \operatorname{Frac} A$, and let $v \colon K \to G \cup \{\infty\}$ be a valuation for A with values in some totally ordered abelian group G.*
(1) ([BouAC] VI, 10.1, Proposition 1) *There exists an extension of v to $K(T)$ (T transcendental) with values in the same group G.*
(2) ([BouAC] VI, 1.2, Cor. to Thm. 2) *If $L \supseteq K$ is a field extension, there exists an extension of v to a valuation on L.*
(3) ([BouAC] VI, 8.1, Corollary 3 of Proposition 1) *If $L \supseteq K$ is a finite extension and v is discrete, then any extension of v to L is a discrete valuation.*

(B.14) Normal, factorial, regular and Cohen-Macaulay rings.

A ring A is called *normal* if $A_{\mathfrak{p}}$ is an integrally closed integral domain for all prime ideals $\mathfrak{p}$.

Proposition B.73. *Let A be a ring.*
(1) ([AM] Proposition 5.13) *A is integrally closed if and only if A is normal and an integral domain.*
(2) *If A is normal, then $S^{-1}A$ is normal for every multiplicative subset $S \subset A$.*
(3) ([Mat] Theorem 11.5, Theorem 12.3) *Let A be a noetherian integral domain. Then A is normal if and only if*
  (a)
$$A = \bigcap_{\substack{\mathfrak{p} \subset A\ prime \\ \mathrm{ht}\ \mathfrak{p}=1}} A_{\mathfrak{p}}$$

  (b) *For all prime ideals $\mathfrak{p}$ of height one, $A_{\mathfrak{p}}$ is a discrete valuation ring.*
(4) ([Mat] Theorem 12.4) *Let A be a normal noetherian integral domain, $a \in (\mathrm{Frac}\,A)^{\times}$. Then there exist only finitely many prime ideals $\mathfrak{p}$ of A of height one with $a \notin A_{\mathfrak{p}}^{\times}$ (in other words, such that $v_{\mathfrak{p}}(a) \neq 0$, where $v_{\mathfrak{p}}$ is a valuation for the discrete valuation ring $A_{\mathfrak{p}}$).*
(5) ([Mat] Theorem 10.3) *Every valuation ring is integrally closed.*
(6) ([Mat] Theorem 23.9) *Let $A \to B$ be a faithfully flat morphism of noetherian rings. Then B is normal if and only if A is normal and if for all prime ideals $\mathfrak{p} \subset A$, the ring $B \otimes_A \kappa(\mathfrak{p})$ is normal.*
(7) ([BouAC] V, 1.3, Corollary 2 of Proposition 13) *If A is normal, then $A[T_1, \dots, T_n]$ is normal.*

Proposition/Definition B.74. ([Ei] 0.2) *An integral domain A is called* factorial *or* unique factorization domain *if it satisfies the following equivalent conditions.*
 (i) *Every nonzero element f of A can be written as a product $f = f_1 f_2 \cdots f_r$ of irreducible elements, and the f_i are uniquely determined up to order and up to multiplication by a unit u_i.*
 (ii) *Every nonzero element f of A can be written as a product $f = f_1 f_2 \cdots f_r$ of irreducible elements, and every irreducible element is a prime element.*
 (iii) *Every irreducible element is a prime element, and every ascending chain of principal ideals becomes stationary.*

Proposition B.75. *Let A be an integral domain.*
(1) ([BouAC] VII, 3.4, Proposition 3; VII, 3.5, Theorem 2) *If A is factorial, then the polynomial ring $A[T_1, \dots, T_n]$ is factorial and the localization $S^{-1}A$ is factorial for every multiplicative subset $S \subset A$.*
(2) ([Mat] Theorem 20.1) *Let A be noetherian. Then A is factorial if and only if every prime ideal of height 1 is a principal ideal.*
(3) ([Mat] Theorem 20.2) *Let A be noetherian and let $S \subset A$ be a multiplicative set generated by prime elements. Then A is factorial if and only if $S^{-1}A$ is factorial.*

Definition/Proposition B.76. ([AM] Theorem 11.22) *A local noetherian ring $(A, \mathfrak{m})$ with residue field k is called* regular *if the following equivalent conditions are satisfied.*
 (i) $\dim A = \dim_k \mathfrak{m}/\mathfrak{m}^2$.
 (ii) $\mathrm{gr}^{\mathfrak{m}}(A) \cong k[T_1, \dots, T_d]$.

(iii) $\mathfrak{m}$ *can be generated by* $\dim A$ *elements.*
A noetherian ring A is called regular, *if for all maximal ideals $\mathfrak{m} \subset A$ the localization $A_\mathfrak{m}$ is a regular local ring.*

For every noetherian local ring A as above one has $d := \dim A \le \dim_k \mathfrak{m}/\mathfrak{m}^2$. Thus A is regular if and only if $\mathfrak{m}$ can be generated by a sequence $(a_1, \dots, a_d)$ of d elements. Such a sequence is then automatically A-regular and the homomorphism of k-algebras $k[T_1, \dots, T_d] \to \mathrm{gr}^\mathfrak{m}(A), T_i \mapsto (a_i \bmod \mathfrak{m}^2) \in \mathrm{gr}_1^\mathfrak{m}(A)$ is an isomorphism.

Proposition B.77. *Let $(A, \mathfrak{m})$ be a regular local ring with residue field k.*
(1) ([Mat] Theorem 19.3) *For every prime ideal $\mathfrak{p} \subset A$, the localization $A_\mathfrak{p}$ is regular.*
(2) (Theorem of Auslander-Buchsbaum, [Mat] Theorem 20.3) *The ring A is a unique factorization domain (and in particular an integral domain).*
(3) ([Mat] Theorem 14.2) *Let $a_1, \dots, a_r \in \mathfrak{m}$. Then the images of the a_i in $\mathfrak{m}/\mathfrak{m}^2$ are linearly independent over k if and only if the quotient $A/(a_1, \dots, a_r)$ is regular and has dimension $\dim A - r$.*
(4) ([Mat] Theorem 19.5) *The polynomial ring $A[T]$ and the ring of formal power series $A[\![T]\!]$ are regular.*
(5) ([Mat] Theorem 23.7) *Let $A \to B$ be a flat local ring homomorphism of local noetherian rings A and B. If B is regular, then A is regular. If A is regular and $B/\mathfrak{m}_A B$ is regular, then B is regular.*
(6) (by (5)) *A local noetherian ring R is regular if and only if $\widehat{R}$ is regular.*

Let R be a local noetherian ring. If $\dim R = 0$, then R is regular if and only if R is a field. If $\dim R = 1$, then R is regular if and only if its maximal ideal is a principal ideal, i.e., if and only if R is a discrete valuation ring (Proposition B.71).

Remark B.78.
(1) Every local regular ring is factorial (Theorem of Auslander-Buchsbaum). Every factorial ring is integrally closed. In particular, every regular ring is normal.
(2) If R is normal and noetherian, then by Proposition B.73 (3) (b) the localization $R_\mathfrak{p}$ is regular for all prime ideals $\mathfrak{p}$ of R of height ≤ 1 (as $R_\mathfrak{p}$ is always a domain, $R_\mathfrak{p}$ is a field if the height of $\mathfrak{p}$ is 0). In particular we see that if R is a noetherian ring of dimension ≤ 1, then R is normal if and only if R is regular.

There exist rings R with $\dim R \ge 2$ that are normal but not regular; see Exercise 6.23.

Definition B.79. *Let A be a noetherian ring and let $k \ge 0$ be an integer.*
(1) *We say that A satisfies (R_k) if $A_\mathfrak{p}$ is regular for all prime ideal $\mathfrak{p}$ with $\mathrm{ht}(\mathfrak{p}) \le k$.*
(2) *We say that A satisfies (S_k) if $\mathrm{depth}\, A_\mathfrak{p} \ge \min(\mathrm{ht}(\mathfrak{p}), k)$ for all prime ideals $\mathfrak{p}$.*
(3) *A is called* Cohen-Macaulay *if $\mathrm{depth}\, A_\mathfrak{p} = \mathrm{ht}\,\mathfrak{p}$ for all prime ideals $\mathfrak{p}$.*

Example B.80. Let A be a noetherian ring.
(1) A is Cohen-Macaulay (resp. regular) if and only if A satisfies (S_k) (resp. (R_k)) for all $k \ge 0$.
(2) Every ring satisfies (S_0). A ring satisfies (S_1) if and only if every associated prime ideal is a minimal prime ideal.
(3) The ring A is reduced if and only if A satisfies (S_1) and (R_0).

Proposition B.81. (Serre's normality criterion, [Mat] Theorem 23.8) *A noetherian ring is normal if and only if it satisfies (R_1) and (S_2).*

Proposition B.82. ([Mat] Theorem 23.9) *Let $A \to B$ be a flat local homomorphism of local noetherian rings, and let $k \geq 0$ be an integer.*
(1) *If B satisfies (R_k) (resp. (S_k)), then A satisfies (R_k) (resp. (S_k)).*
(2) *If A and $B \otimes_A \kappa(\mathfrak{p})$ satisfy (R_k) (resp. (S_k)) for every prime ideal $\mathfrak{p}$ of A, then B satisfies (R_k) (resp. (S_k)).*

Proposition B.83. ([Mat] Theorem 17.6) *Let A be a noetherian ring. Then A is Cohen-Macaulay if and only if the unmixedness property holds for A, i.e., for every ideal $\mathfrak{a} \subset A$ of height $r \geq 0$ generated by r elements, all minimal prime ideals of A containing $\mathfrak{a}$ have the same height.*

Proposition B.84. *Let A be a local Cohen-Macaulay ring.*
(1) ([Mat] Theorem 17.3) *All associated prime ideals of A have the same height (i.e., Spec A has no embedded components, see Section (9.5), and is equidimensional, see Section (5.3)).*
(2) ([Mat] Theorem 17.4) *Let $\mathfrak{a} \subsetneq A$ be an ideal. Then $\dim A/\mathfrak{a} + \operatorname{depth}(\mathfrak{a}, A) = \dim A$.*

Proposition B.85. (Hartshorne's connectedness theorem, [Ei] Theorem 18.12) *Let A be a noetherian local ring, let $\mathfrak{a}_1, \mathfrak{a}_2 \subset A$ be ideals such that $\mathfrak{a}_1 \cap \mathfrak{a}_2$ is nilpotent, and that there is no inclusion relation between the radicals of $\mathfrak{a}_1$ and $\mathfrak{a}_2$. Then $\operatorname{depth}(\mathfrak{a}_1 + \mathfrak{a}_2, A)) \leq 1$.*

Proposition B.86. ([Mat] Theorem 17.3) *Let A be a local noetherian ring and let $\mathfrak{a}$ be an ideal generated by an A-regular sequence. Then A is Cohen-Macaulay if and only if $A/\mathfrak{a}$ is Cohen-Macaulay. In particular, every regular ring is Cohen-Macaulay.*

(B.15) Dedekind domains and principal ideal domains.

Definition/Proposition B.87. ([Mat] 11.6; [AM] Theorem 9.3) *An integral domain A is called a* Dedekind domain *or* Dedekind ring, *if it satisfies the following equivalent properties.*
(i) *A is either a field or a noetherian ring such that $A_\mathfrak{m}$ is a discrete valuation ring for every maximal ideal $\mathfrak{m}$.*
(ii) *A is normal, noetherian, and $\dim A \leq 1$.*
(iii) *A is regular and $\dim A \leq 1$.*
(iv) *Every non-zero ideal of A can be written as a finite product of prime ideals.*
(v) *Every non-zero ideal of A is a projective A-module of rank 1.*
If A is a Dedekind domain, then the factorization in (iv) into prime ideals is unique up to order.

Proposition B.88. ([Mat] Theorem 20.7) *A ring A is a principal ideal domain if and only if A is factorial and a Dedekind ring.*

Proposition B.89. *Let M be an A-module.*
(1) ([BouAII] VII, §3, Corollary 3 of Theorem 1) *Let A be a principal ideal domain. Then M is free if and only if M is projective.*
(2) ([BouAII] VII, 4.4, Corollary 2 of Theorem 2) *Let A be a principal ideal domain, and let M be finitely generated. Then M is free if and only if M is torsion-free.*
(3) ([BouAC] I, 2.4, Proposition 3; Proposition B.30 and Proposition B.27) *Let A be a Dedekind domain. Then M is flat if and only if M is torsion-free.*

(4) (by (2), Proposition B.30 and Proposition B.29) *Let A be a Dedekind domain and let M be finitely generated. Then M is projective if and only if M is torsion-free.*

Proposition B.90. (Krull-Akizuki theorem, [Mat] Theorem 11.7) *Let A be a noetherian integral domain of dimension 1, let L be a finite extension of $\mathrm{Frac}\,A$, and let B be an A-subalgebra of L. Then B is noetherian, $\dim B \leq 1$, and $\lg_A(B/\mathfrak{b}) < \infty$ for every nonzero ideal $\mathfrak{b} \subseteq B$.*

(B.16) Field extensions.

A ring homomorphism $\iota\colon k \to K$ of fields is called a *field extension*. Then ι is automatically injective. Very often we will consider ι as inclusion and omit it from the notation.

If $(T_i)_{i \in I}$ is a family of indeterminates, the polynomial ring $k[(T_i)_{i \in I}]$ is an integral domain. Its field of fractions is denoted by $k((T_i)_{i \in I})$. It is a field extension of k.

The dimension of the k-vector space K is denoted by $[K : k]$ and is also called the *degree* of $k \to K$. A *subextension of* $k \to K$ is a k-subalgebra of K which is a field. The intersection of a family of subfields is again a subfield. If $(a_i)_{i \in I}$ is a family of elements a_i of K, we denote by $k((a_i)_{i \in I})$ the smallest subextension which contains all elements a_i.

Definition B.91. *A field extension $\iota\colon k \to K$ is called*
(1) finite, *if ι is finite (Definition B.12).*
(2) algebraic, *if ι is integral (Definition B.53). An element $a \in K$ is called* algebraic *over k if it is integral over k.*
(3) transcendental, *if it is not algebraic.*
(4) purely transcendental, *if there exists an algebraically independent family $(t_i)_{i \in I}$ of elements $t_i \in K$ such that $K = k((t_i)_{i \in I})$ (equivalently $K \cong k((T_i)_{i \in I})$ for a family $(T_i)_{i \in I}$ of indeterminates).*
(5) finitely generated, *if there exist finitely many elements $a_1, \ldots, a_n \in K$ such that $K = k(a_1, \ldots, a_n)$.*
(6) separable, *if for every field extension $k \to L$ the tensor product $K \otimes_k L$ is reduced. An element $a \in K$ is called* separable *over k if it is a root of an irreducible polynomial $f \in k[T]$ that has only simple roots in an algebraically closed extension of k.*

Separable field extensions are not necessarily algebraic but separable elements are by definition always algebraic.

Let $\iota\colon k \to K$ be a field extension. The set

$$k^{\mathrm{alg}} := \{\, a \in K \;;\; a \text{ is algebraic over } k \,\}$$
$$(\text{resp. } k^{\mathrm{sep}} := \{\, a \in K \;;\; a \text{ is separable (and algebraic) over } k \,\})$$

is a subextension of K and called the *algebraic closure of k in K* (resp. *separable closure of k in K*). We say that k is *algebraically closed in K* (resp. *separably closed in K*) if $k = k^{\mathrm{alg}}$ (resp. $k = k^{\mathrm{sep}}$).

Definition B.92. *An algebraic field extension $\iota\colon k \to K$ is called*
(1) inseparable, *if it is not separable.*
(2) purely inseparable, *if k is separably closed in K.*
(3) normal, *if for every extension $K \to \Omega$ with Ω algebraically closed and for every homomorphism $\sigma\colon K \to \Omega$ of k-algebras we have $\sigma(K) \subseteq K$.*
(4) Galois, *if $k \to K$ is normal and separable.*

Let $k \to K$ be an algebraic extension and let k^{sep} be the separable closure of k in K. Then $k^{\mathrm{sep}} \to K$ is purely inseparable. We set

(B.16.1) $$[K : k]_{\mathrm{sep}} := [k^{\mathrm{sep}} : k], \qquad [K : k]_{\mathrm{insep}} := [K : k^{\mathrm{sep}}]$$

and call $[K : k]_{\mathrm{sep}}$ (resp. $[K : k]_{\mathrm{insep}}$) the *separability degree* (resp. *inseparability degree*) of $k \to K$.

Remark B.93.
(1) ([BouAII] V, 2.2, Theorem 1; V, 3.3, Proposition 3; V, 14.7, Corollary 3 of Proposition 7) Let $k \to K$ and $K \to L$ be field extensions. Then the composition $k \to L$ is finite (resp. algebraic, resp. finitely generated) if and only if the extension $k \to K$ and $K \to L$ have the same property.
(2) ([BouAII] V, 3.2, Theorem 2) An extension is finite if and only if it is algebraic and finitely generated.
(3) ([BouAII] V, 15.3, Proposition 6) Every purely transcendental extension is separable.

Definition B.94. *A field k is called*
(1) algebraically closed, *if there exists no nontrivial algebraic extension of k.*
(2) separably closed, *if there exists no nontrivial separable algebraic extension of k.*
(3) perfect, *if every field extension of k is separable (equivalently, $\mathrm{char}(k) = 0$, or $\mathrm{char}(k) = p > 0$ and the Frobenius $k \to k$, $x \mapsto x^p$ is surjective).*

A field is algebraically closed if and only if it is perfect and separably closed.

Let k be a field. A field extension $k \to K$ is called an *algebraic closure* (resp. *separable closure*, resp. *perfect closure*) of k if K is algebraically closed (resp. separably closed, resp. perfect) and if for every extension $k \to L$, where L is algebraically closed (resp. separably closed, resp. perfect), there exists a k-homomorphism $K \to L$.

Proposition B.95. ([BouAII] V, 4.3, Theorem 2; V, 7.8, Proposition 14; V, 5.2, Proposition 3) *An algebraic closure (resp. separable closure, resp. perfect closure) of k always exists and is an algebraic (resp. separable algebraic, resp. purely inseparable algebraic) extension of k. An algebraic closure (resp. separable closure) of k is unique up to isomorphism of k-algebras; a perfect closure of k is unique up to unique isomorphism of k-algebras.*

Let $k \to \Omega$ be an algebraically (resp. separably) closed extension of k. Then for every algebraic (resp. separable algebraic) extension $k \to K$ there exists a k-algebra homomorphism $K \to \Omega$.

Proposition/Definition B.96. *Let $k \to K$ be a field extension.*
(1) ([BouAII] V, 14.2, Theorem 1) *There exists an algebraically independent family $(t_i)_{i \in I}$ of elements $t_i \in K$ such that K is an algebraic extension of $k((t_i)_{i \in I})$. Such a family is called a* transcendence basis *of the field extension $k \to K$.*
(2) ([BouAII] V, 14.3, Theorem 3) *Any two transcendence bases of $k \to K$ have the same cardinality. This cardinality is called the* transcendence degree *and denoted by $\mathrm{trdeg}_k(K)$.*
(3) ([BouAII] V, 14.3, Corollary of Theorem 4) *If $K \to L$ is a second field extension, we have $\mathrm{trdeg}_k(L) = \mathrm{trdeg}_k(K) + \mathrm{trdeg}_K(L)$.*
(4) ([BouAII] V, 14.7, Proposition 17) *The extension $k \to K$ is finitely generated if and only if there exists a finite transcendence basis $(t_1, \ldots, t_n)$ such that K is a finite extension of $k(t_1, \ldots, t_n)$.*

Therefore a field extension $k \to K$ is algebraic if and only if $\operatorname{trdeg}_k(K) = 0$.

To characterize separable extensions we recall that if A is a ring and M an A-module, we call a $\mathbb{Z}$-linear map $d \colon A \to M$ a *derivation*, if $d(aa') = a\,d(a') + a'\,d(a)$ for all $a, a' \in A$.

Proposition B.97. ([BouAII] V, 15.4, Theorem 2; V, 15.6, Theorem 4; V, 16.4, Theorem 3; V, 16.7, Corollary of Theorem 5; V, 7.3, Proposition 6) *Let $k \to K$ be a field extension. The following assertions are equivalent.*

(i) $k \to K$ *is separable.*

(ii) *For every finite purely inseparable extension L of k, the ring $K \otimes_k L$ is reduced.*

(iii) *There exists a perfect extension k' of k such that $K \otimes_k k'$ is reduced.*

(iv) *For every algebraically closed extension Ω of K and for all k-linearly independent elements $a_1, \ldots, a_n \in K$ there exist k-automorphisms $\sigma_1, \ldots, \sigma_n$ of Ω such that $\det(\sigma_i(a_j))_{i,j} \neq 0$.*

(v) *For every K-vector space V and every derivation $d \colon k \to V$ (considering V as a k-vector space via $k \to K$) there exists a derivation $D \colon K \to V$ which extends d.*

If $k \to K$ is algebraic, Assertions (i) – (v) are equivalent to

(vi) *Every $a \in K$ is separable over k.*

If $k \to K$ is finitely generated, Assertions (i) – (v) are equivalent to

(vii) *There exists a transcendence basis $(t_1, \ldots, t_n)$ of K over k such that K is a finite separable extension of $k(t_1, \ldots, t_n)$.*

Proposition B.98. (Primitive element theorem, [BouAII] V, 7.4, Theorem 1) *Every finite separable field extension is generated by a single element.*

Remark B.99. ([BouAII] V, 7.9) Let $k \to K$ be an algebraic field extension.

(1) We have
$$[K : k] = [K : k]_{\mathrm{sep}} [K : k]_{\mathrm{insep}}.$$

(2) Let Ω be an algebraically closed extension of k. Then $[K : k]_{\mathrm{sep}}$ is finite if and only if the set $\operatorname{Hom}_{k\text{-Alg}}(K, \Omega)$ of k-embeddings $K \to \Omega$ is finite, and in this case we have

$$[K : k]_{\mathrm{sep}} = \# \operatorname{Hom}_{k\text{-Alg}}(K, \Omega).$$

(3) If $K \to L$ is a second field extension,

$$[L : k]_{\mathrm{sep}} = [L : K]_{\mathrm{sep}} [K : k]_{\mathrm{sep}},$$
$$[L : k]_{\mathrm{insep}} = [L : K]_{\mathrm{insep}} [K : k]_{\mathrm{insep}}.$$

Lemma B.100. (see Section (B.19) below) *For two field extensions $k \to K$ and $k \to L$ we have*

$$\dim(K \otimes_k L) = \min(\operatorname{trdeg}_k(K), \operatorname{trdeg}_k(L))$$

(If $\operatorname{trdeg}_k(K)$ and $\operatorname{trdeg}_k(L)$ are infinite cardinals, the assertion is $\dim(K \otimes_k L) = \infty$.)

Proposition B.101. (see Section (B.19) below) *For a field extension $k \to K$ the following assertions are equivalent.*

(i) k *is separably closed in K.*

(ii) *For every field extension $k \to L$, the ring $K \otimes_k L$ has only one minimal prime ideal.*

(iii) *For every finite separable extension $k \to L$, the ring $K \otimes_k L$ has only one minimal prime ideal.*

(iv) *There exists a separably closed extension $k \to \Omega$ such that the ring $K \otimes_k \Omega$ has only one minimal prime ideal.*

Corollary B.102. (see Section (B.19) below) *For a field extension $k \to K$ the following assertions are equivalent.*

(i) *$k \to K$ is purely inseparable (and in particular algebraic).*

(ii) *For every extension L of k there exists at most one k-embedding $K \to L$.*

(iii) *For every extension L of k the ring $K \otimes_k L$ has a single prime ideal.*

In the following sections we give proofs for those of the above results for which we do not know of any easily available textbook reference.

(B.17) Proofs of results on algebras of finite presentation.

Note that if B is an A-algebra of finite presentation and $A \to A'$ is a ring-homomorphism, then $A' \otimes_A B$ is an A'-algebra of finite presentation.

Proof. (of Lemma B.10) We can write $B = A[\underline{T}]/(f_1, \ldots, f_m)$ with $A[\underline{T}] = A[T_1, \ldots, T_n]$. Since Λ is filtered, we find $\lambda \in \Lambda$ such that all the coefficients of the f_j are contained in the image of the canonical homomorphism $\varphi_\lambda \colon A_\lambda \to A$. We choose preimages $f_{\lambda j}$ of the f_j under the homomorphism $A_\lambda[\underline{T}] \to A[\underline{T}]$. Then $A_\lambda[\underline{T}]/(f_{\lambda 1}, \ldots, f_{\lambda m}) \otimes_{A_\lambda} A = B$, so we can set $B_\lambda = A_\lambda[\underline{T}]/(f_{\lambda 1}, \ldots, f_{\lambda m})$. $\square$

Remark B.103. As we can write every ring as the filtered union of its finitely generated $\mathbb{Z}$-subalgebras, we see that if B is an A-algebra of finite presentation there exists a finitely generated $\mathbb{Z}$-subalgebra A_0 of A and an A_0-algebra B_0 of finite type such that $B \cong A \otimes_{A_0} B_0$.

Proof. (of Proposition B.11) Assertion (1) follows at once from Remark B.103 using Lemma B.10 and the fact that if B_0 is a finitely generated A_0-algebra and C_0 is a finitely generated B_0-algebra, then C_0 is a finitely generated A_0-algebra.

Let us show (2). As C is a finitely generated A-algebra, C is also a finitely generated B-algebra and we find a surjection of B-algebras $\beta \colon B[\underline{T}] := B[T_1, \ldots, T_n] \twoheadrightarrow C$. To show the first assertion of (2) it suffices to prove that $\mathrm{Ker}(\beta)$ is a finitely generated ideal. Now by assumption there also exists a surjection of A-algebras $A[\underline{X}] := A[X_1, \ldots, X_m] \twoheadrightarrow B$. Tensoring with $A[\underline{T}]$ we obtain a surjection

$$\alpha \colon A[\underline{X}, \underline{T}] \twoheadrightarrow B[\underline{T}] \xrightarrow{\ \beta\ } C.$$

If $\mathrm{Ker}(\alpha)$ is finitely generated, then its image $\mathrm{Ker}(\beta)$ in $B[\underline{T}]$ is finitely generated. Therefore it suffices to show the second assertion of (2) where $\pi \colon B = A[Y_1, \ldots, Y_r] \twoheadrightarrow C$ is a surjection of a polynomial algebra.

Let A_0 be a finitely generated $\mathbb{Z}$-subalgebra of A and C_0 an A_0-algebra of finite type such that $C \cong A \otimes_{A_0} C_0$. If A_0 is sufficiently large, then we find $y_i \in C_0$ such that $1 \otimes y_i = \pi(Y_i)$. If $c_1, \ldots, c_t$ generate C_0 as A_0-algebra we find polynomials P_j with coefficients in A such that $1 \otimes c_j = P_j(1 \otimes y_1, \ldots, 1 \otimes y_r)$. If A_0 is sufficiently large, the P_j have coefficients in A_0 and we find, after enlarging A_0 once more if necessary, that

$c_j = P_j(y_1, \ldots, y_r)$. In particular, the y_i generate C_0. Then $\pi_0 \colon A_0[Y_1, \ldots, Y_r] \to C_0$, $Y_i \mapsto y_i$, is a surjection whose kernel is a finitely generated ideal because A_0 is noetherian. As $\mathrm{id}_A \otimes \pi_0 = \pi$, the kernel of π is finitely generated because $\mathrm{Ker}(\pi)$ is the image of $A \otimes_{A_0} \mathrm{Ker}(\pi_0)$ in $A[Y_1, \ldots, Y_r]$. $\qquad\square$

Proof. (of Proposition B.13) Assume that the finite A-algebra B is of finite presentation as a module. Thus we find an exact sequence of A-modules $A^m \xrightarrow{u} A^n \xrightarrow{p} B \to 0$. The matrix defining $u \colon A^m \to A^n$ has coefficients in some finitely generated $\mathbb{Z}$-subalgebra A_0 of A and thus $u = \mathrm{id}_A \otimes u_0$ for an A_0-linear map $u_0 \colon A_0^m \to A_0^n$ and $B_0 := \mathrm{Coker}(u_0)$ is an A_0-module such that $A \otimes_{A_0} B_0 \cong B$. Let $m \colon B \otimes_A B \to B$ be the multiplication. For $i = 1, \ldots, n$ set $b_i := p(e_i)$, where $(e_i)_i$ is the standard basis of A^n. Then

$$m(b_j \otimes b_k) = \sum_i a_{ijk} b_i$$

with $a_{ijk} \in A$. Adjoining these elements to A_0 we may assume that $a_{ijk} \in A_0$ such that $m = \mathrm{id}_A \otimes m_0$ for a homomorphism $m_0 \colon B_0 \otimes_{A_0} B_0 \to B_0$. Enlarging A_0 further we may assume that m_0 is associative, commutative and unital. Then B_0 is a finite A_0-algebra which is of finite presentation as A_0-algebra because A_0 is noetherian. Therefore $B \cong A \otimes_{A_0} B_0$ is an A-algebra of finite presentation.

Conversely assume that B is of finite presentation as A-algebra. Choose finitely many generators b_i of B as an A-module. Arguing as in the last part of the proof of Proposition B.11 we find a finitely generated $\mathbb{Z}$-subalgebra A_0 of A, an A_0-algebra B_0 such that $B \cong A \otimes_{A_0} B_0$ and a surjection $\pi_0 \colon A_0[\underline{T}] \twoheadrightarrow B_0$ such that $1 \otimes b_{0i} = b_i$, where $b_{0i} := \pi_0(T_i)$. As B is a finite A-algebra, each b_i is a zero of a monic polynomial with coefficients in A. If A_0 is sufficiently large, we may assume that also b_{i0} is integral over A_0. Then B_0 is generated by finitely many integral elements and thus is a finite A_0-algebra (Proposition B.53). As A_0 is noetherian, B_0 is of finite presentation as A_0-module. Therefore B is of finite presentation as A-module. $\qquad\square$

Proof. (of Proposition B.54 (3)) One can use the same arguments as in the second part of the previous proof. $\qquad\square$

(B.18) Proof of a theorem by Artin and Tate.

Proof. (of Corollary B.65) Assume that A is semi-local of dimension ≤ 1. Then there are only finitely many non-zero prime ideals $\mathfrak{m}_1, \ldots, \mathfrak{m}_r$ and all of them are maximal. As A is integral, there exists $0 \neq f \in \prod_i \mathfrak{m}_i$. The localization A_f is then an integral domain whose only prime ideal is the zero ideal. Thus A_f is a field.

Conversely, let $f \in A$ such that A_f is a field. Then every non-zero prime ideal contains f. Let $\mathfrak{p}_1, \ldots, \mathfrak{p}_r$ be those prime ideals which are minimal with this property (there are only finitely many because $A/(f)$ is noetherian). Assume there existed a maximal ideal $\mathfrak{m}$ which contains one of the $\mathfrak{p}_i$ properly. By prime ideal avoidance (Proposition B.2), there exists $g \in \mathfrak{m} \setminus \bigcup_i \mathfrak{p}_i$. Let $\mathfrak{q}$ be a prime ideal that is minimal among those that contain g. By Krull's principal ideal (Proposition B.63), $\mathfrak{q}$ has height 1. Therefore $\mathfrak{q} \neq 0$ and thus $f \in \mathfrak{q}$. Moreover $\mathfrak{q}$ must be one of the $\mathfrak{p}_i$ because otherwise $\mathrm{ht}(\mathfrak{q}) > 1$. This is a contradiction.

Therefore such a maximal ideal $\mathfrak{m}$ cannot exist and the $\mathfrak{p}_i$ must be the maximal ideals. This shows that A is semi-local of dimension ≤ 1. $\qquad\square$

(B.19) Proofs of results on field extensions.

Proof. (of Lemma B.100) We may assume that $\operatorname{trdeg}_k(K) \leq \operatorname{trdeg}_k(L)$. Let $\underline{T} = (T_i)_{i \in I}$ be a transcendence basis of K over k. Set $K' := k(\underline{T})$. Then $K' \otimes_k L \hookrightarrow K \otimes_k L$ is an integral extension and thus $\dim(K' \otimes_k L) = \dim(K \otimes_k L)$ by Theorem B.56. Thus we may assume $K = k(\underline{T})$. Then $K \otimes_k L$ is a localization of the ring $k[\underline{T}] \otimes_k L = L[\underline{T}]$.

As K is a purely transcendental extension with $\operatorname{trdeg}_k(K) \leq \operatorname{trdeg}_k(L)$, we find a k-embedding $K \hookrightarrow L$ and hence an L-algebra homomorphism $K \otimes_k L \to L$. Its kernel is a maximal ideal $\mathfrak{m}$ of $K \otimes_k L$ with residue field L. Moreover $\mathfrak{m} \cap L[\underline{T}]$ is a maximal ideal with residue field L. It is thus generated by a family $(T_i - b_i)_{i \in I}$ for $b_i \in L$. Thus $\operatorname{ht}(\mathfrak{m}) = \operatorname{ht}(\mathfrak{m} \cap L[\underline{T}]) \geq \#I$ (where $\#I := \infty$ if I is infinite) and hence $\dim(K \otimes_k L) \geq \#I$. This proves the lemma if $\operatorname{trdeg}_k(K)$ is infinite. If $n := \operatorname{trdeg}_k(K) < \infty$, then $\dim K \otimes_k L \leq \dim L[T_1, \dots, T_n] = n$ because $K \otimes_k L$ is a localization of $L[T_1, \dots, T_n]$. $\qquad\square$

Proof. (of Proposition B.101) We first claim that if $k \hookrightarrow L \hookrightarrow M$ are field extensions and $K \otimes_k M$ has only one minimal prime ideal, then $K \otimes_k L$ has only one minimal prime ideal. Indeed, let $\iota \colon K \otimes_k L \to K \otimes_k M$ be the induced homomorphism. Let $\mathfrak{p}_1$ and $\mathfrak{p}_2$ be minimal prime ideals of $K \otimes_k L$. There exist prime ideals $\mathfrak{q}_1$ and $\mathfrak{q}_2$ of $K \otimes_k M$ such that $\iota^{-1}(\mathfrak{q}_i) = \mathfrak{p}_i$ for $i = 1, 2$. Let $\mathfrak{q}$ be the unique minimal prime ideal of $K \otimes_k M$. Then for $i = 1, 2$ we find $\mathfrak{q} \subseteq \mathfrak{q}_i$ and hence $\iota^{-1}(\mathfrak{q}) \subseteq \iota^{-1}(\mathfrak{q}_i) = \mathfrak{p}_i$ which implies $\iota^{-1}(\mathfrak{q}) = \mathfrak{p}_i$ by the minimality of $\mathfrak{p}_i$. Thus $\mathfrak{p}_1 = \mathfrak{p}_2$.

In more geometric terms, we could say that the morphism $\operatorname{Spec} K \otimes_k M \to \operatorname{Spec} K \otimes_k L$ is surjective as the base change of a surjective morphism, and that the source of this morphism is irreducible by assumption. Therefore the target is irreducible.

Assertion (i) implies (ii) by [BouAII] V, 17.2, Corollary of Proposition 1 and clearly (ii) implies (iv). Let L be a finite separable extension of k. There exists a k-embedding $L \hookrightarrow \Omega$ and therefore (iv) implies (iii) by the claim above.

It remains to show that (iii) implies (i). Let $L \supseteq k$ be a finite separable subextension of K. By the claim above we know that $L \otimes_k L$ has a unique minimal prime ideal. As L is separable, $L \otimes_k L$ is also reduced (Proposition B.97) and hence an integral domain. As it is a finite k-algebra, it is a field. As any nonzero ring homomorphism with a field as domain is injective, this implies that the multiplication $L \otimes_k L \to L$, $a \otimes b \mapsto ab$ must be bijective. This shows that $L = k$ because otherwise $\dim_k L \otimes_k L > \dim_k L$. $\qquad\square$

Proof. (of Corollary B.102) "(iii) $\Rightarrow$ (ii)". We have $\operatorname{Hom}_k(K, L) = \operatorname{Hom}_L(K \otimes_k L, L)$ and thus two different k-embeddings would define different maximal ideals of $K \otimes_k L$.

"(i) $\Rightarrow$ (iii)". Lemma B.100 implies that $K \otimes_k L$ has dimension 0 and Proposition B.101 then shows that $K \otimes_k L$ has only one prime ideal.

"(ii) $\Rightarrow$ (i)". Apply (ii) to an algebraic closure L of K. $\qquad\square$

(B.20) The resultant and the discriminant.

The reference for this section is [BouAII] IV §6 no. 6, no. 7.

Let R be a ring. Let m, n be positive integers, and consider polynomials

$$f = a_m X^m + \cdots + a_0, \quad g = b_n X^n + \cdots + b_0$$

in $R[X]$ (we do not require $a_m, b_n \neq 0$ at this point).

Definition B.104. *The resultant of the polynomials f, g (for degrees m, n) is the determinant of the $((m+n) \times (m+n))$-matrix*

$$\begin{pmatrix}
a_m & & & & b_n & & & & \\
a_{m-1} & a_m & & & \vdots & \ddots & & & \\
\vdots & a_{m-1} & \ddots & & \vdots & & \ddots & & \\
\vdots & \vdots & \ddots & & b_0 & & & \ddots & \\
a_0 & \vdots & & a_m & & \ddots & & & \ddots \\
& a_0 & & a_{m-1} & & & \ddots & & b_n \\
& & \ddots & \vdots & & & & \ddots & \vdots \\
& & & \ddots & \vdots & & & & \vdots \\
& & & & a_0 & & & & b_0
\end{pmatrix}$$

where empty entries are understood as 0. We denote the resultant by $\mathrm{res}_{m,n}(f,g)$. If $m = \deg(f)$, $n = \deg(g)$, then we also write $\mathrm{res}(f,g)$ instead.

Proposition B.105. *Let $f, g \in R[X]$. Assume that f is monic and that $g \neq 0$. The following are equivalent.*
 (i) $\mathrm{res}(f,g) \in R^\times$,
 (ii) there exist $u, v \in R[X]$ with $uf + vg = 1$,
 (iii) the residue class of g in $R[X]/(f)$ lies in $(R[X]/(f))^\times$.
If $R = K$ is a field, then the above conditions are also equivalent to
 (iv) For every field extension L/K, the polynomials f and g have no common zero in L.

Definition B.106. *Let $f \in R[X]$ be a monic polynomial of degree m. The discriminant of f is*

$$\mathrm{disc}(f) = (-1)^{m(m-1)/2}\,\mathrm{res}(f, f') \in R.$$

This notion is related to the notion of discriminant of a finite free R-algebra as follows: The discriminant $\mathrm{disc}(f)$ equals the discriminant of $A = R[X]/(f)$ for the basis $1, X, \ldots, X^{m-1}$, i.e., the determinant of the matrix $(\mathrm{tr}_{A/R}(X^{i+j}))_{i,j=1,\ldots,m}$.

Proposition B.107. *Let $m \geq 1$. There is a unique polynomial $\Delta_m \in \mathbb{Z}[A_0, \ldots, A_{m-1}]$ such that for every ring R and every monic polynomial*

$$f = X^m + \sum_{i=0}^{m-1} a_i X^i \in R[X]$$

of degree m we have

$$\mathrm{disc}(f) = \Delta_m(a_0, \ldots, a_{m-1}).$$

Moreover, Δ_m has degree $2m - 2$, and is homogeneous if A_i is assigned degree $m - i$.

Corollary B.108.

(1) Let $\varphi \colon R \to S$ be a ring homomorphism, let $f \in R[X]$ be a monic polynomial, and let $^\varphi f$ be the polynomial in $S[X]$ obtained by applying φ to the coefficients of f. Then $\varphi(\mathrm{disc}(f)) = \mathrm{disc}(^\varphi f)$.

(2) If $R = K$ is a field and $f \in K[X]$ is a monic polynomial, then $\mathrm{disc}(f) \neq 0$ if and only if f has no multiple zeros in an algebraic closure of K.

C   Permanence for properties of morphisms of schemes

Here we collect which properties of morphisms of schemes satisfy certain permanence properties. Let **P** be a property of morphisms of schemes. Recall that we made the following definitions.

(LOCS) **P** satisfies (LOCS) if **P** is local on the source, i.e., for all morphisms of schemes $f\colon X \to Y$ and for every open covering $(U_i)_i$ of X the morphism f satisfies **P** if and only if its restriction $U_i \to Y$ satisfies **P** for all i.

(LOCT) **P** satisfies (LOCT) if **P** is local on the target, i.e., for all morphisms of schemes $f\colon X \to Y$ and for every open covering $(V_j)_j$ of Y the morphism f satisfies **P** if and only if its restriction $f^{-1}(V_j) \to V_j$ satisfies **P** for all j.

(COMP) **P** satisfies (COMP) if **P** is stable under composition, i.e., for all morphisms of schemes $f\colon X \to Y$ and $g\colon Y \to Z$ satisfying **P**, the composition $g \circ f$ also satisfies **P**.

(BC) **P** satisfies (BC) if **P** is stable under base change, i.e., if for all morphisms of schemes $f\colon X \to Y$ satisfying **P** and for an arbitrary morphism $Y' \to Y$ the base change $f_{(Y')}\colon X \times_Y Y' \to Y'$ satisfies **P**.

(CANC) **P** satisfies (CANC) if for all morphisms of schemes $f\colon X \to Y$ and $g\colon Y \to Z$ such that $g \circ f$ satisfies **P**, the morphism f satisfies **P**.

(IND) **P** satisfies (IND) if **P** is compatible with limits of projective systems of schemes with affine transition maps. More precisely, let S_0 be a quasi-compact and quasi-separated scheme, let $(S_\lambda)_\lambda$ be a filtered projective system of S_0-schemes that are affine over S_0, and let S be the projective limit. Let X_0 and Y_0 be S_0-schemes which are of finite presentation over S_0, let $f_0\colon X_0 \to Y_0$ be a morphism of S_0-schemes. For all λ we set $X_\lambda := X_0 \times_{S_0} S_\lambda$, $Y_\lambda := Y_0 \times_{S_0} S_\lambda$, and $f_\lambda := f_0 \times_{S_0} \mathrm{id}_{S_\lambda}$. We finally define $X := X_0 \times_{S_0} S$, $Y := Y_0 \times_{S_0} S$, and $f := f_0 \times_{S_0} \mathrm{id}_S$. Then we say that **P** satisfies (IND) if for all data $(S_0, (S_\lambda), f_0)$ as above the morphism f possesses **P** if and only if there exists an index λ_0 such that f_λ possesses **P** for all $\lambda \geq \lambda_0$.

(DESC) **P** satisfies (DESC) if **P** is stable under faithfully flat descent, i.e., given a morphism of schemes $f\colon X \to Y$ and a faithfully flat quasi-compact morphism $Y' \to Y$ such that the base change $f' := f_{(Y')}\colon X \times_Y Y' \to Y'$ possesses **P**, then f possesses **P**.

For each property for morphisms $f\colon X \to Y$ of schemes that we defined in the text we list whether it satisfies the permanence properties above together with references. Sometimes these permanences hold only under additional hypotheses. In that case we use the notation introduced above. Permanence properties not listed do not hold (or are not known to us to hold). Every entry starts with stating which immersions satisfy the given property.

© Springer Fachmedien Wiesbaden GmbH, part of Springer Nature 2020
U. Görtz und T. Wedhorn, *Algebraic Geometry I: Schemes*, Springer Studium
Mathematik – Master, https://doi.org/10.1007/978-3-658-30733-2

quasi-separated: [EGAII] (5.1.10); (IND): [EGAIV] (8.10.5); (DESC): [EGAIV] (2.7.1).

quasi-compact. Closed immersions, immersions of locally noetherian schemes; (LOCT), (COMP), (BC): Proposition 10.3; (CANC) if g is quasi-separated (e.g., if Y is locally noetherian): Remark 10.4; (IND): by definition; (DESC): Proposition 14.51.

quasi-compact and dominant. (COMP), (BC) if the base change $Y' \to Y$ is flat; (DESC): [EGAIV] (2.6.4)

quasi-compact immersion. Closed immersions, immersions of locally noetherian schemes; (LOCT), (COMP), (BC), (CANC) if g is quasi-separated (e.g., if Y is locally noetherian): holds for "quasi-compact" and "immersion"; (IND): by definition; (DESC): Proposition 14.53.

quasi-finite. Quasi-compact immersions; (LOCT), (COMP), (BC), (CANC) if g is quasi-separated: Proposition 12.17; (IND): [EGAIV] (8.10.5); (DESC): Proposition 14.53.

quasi-projective. Quasi-compact immersions (e.g., immersions of locally noetherian schemes); (COMP) if Z is quasi-compact, (BC), (CANC) if g is quasi-separated: [EGAII] (5.3.4); (IND): [EGAIV] (8.10.5).

quasi-separated. Immersions; (LOCT), (COMP), (BC), (CANC): Proposition 10.25; (IND): [EGAIV] (8.10.5); (DESC): Proposition 14.51.

separated. Universally injective morphisms (and hence immersions); (LOCT), (COMP), (BC), (CANC): Proposition 9.13; (IND): [EGAIV] (8.10.5); (DESC): Proposition 14.51.

smooth. Open immersions; (LOCS), (LOCT), (COMP), (BC): Proposition 6.15; (CANC) if g is unramified: Volume II; (IND): [EGAIV] (17.7.8); (DESC): [EGAIV] (17.7.3).

surjective. (LOCT), (COMP), (BC): Proposition 4.32; (CANC) if g is injective; (IND): [EGAIV] (8.10.5); (DESC).

universally bijective. (LOCT), (COMP), (BC); (DESC): Remark 14.52.

universally closed. Closed immersions; (LOCT), (COMP), (BC); (CANC) if g is separated: Remark 9.11; (DESC): Remark 14.52.

universal homeomorphism. (LOCT), (COMP), (BC); (DESC): Remark 14.52.

universally injective. See "purely inseparable".

universally open. Open immersions; (LOCT), (COMP), (BC); (CANC) if g is unramified: Volume II; (DESC): Remark 14.52.

D Relations between properties of morphisms of schemes

In the figure we illustrate the most important implications between properties of a scheme morphism $f\colon X \to Y$. We use the following abbreviations:

- qc means *quasi-compact*; qcqs as usual means *quasi-compact and quasi-separated*;
- ft means *of finite type*, fp means *of finite presentation*, lfp means *locally of finite presentation*.

The dotted arrows are conditional (i.e., they hold if the source is noetherian, or if the target is qcqs, respectively, as indicated). We left out a few implications which follow directly from the definitions (such as "faithfully flat + qc ⇒ qc").

References for the equivalences in the diagram:

- A universal homeomorphism of finite type is purely inseparable (clear, because this is the same as being universally injective). In particular the diagonal morphism is surjective (Exercise 9.9), so the morphism is separated, and being universally closed and of finite type, it is proper. Now Corollary 12.89 shows that is it finite, and of course it is surjective. Conversely, a purely inseparable surjective morphism is universally bijective, and if it is finite (hence universally closed), then it is a universal homeomorphism (Exercise 12.32).
- A closed immersion obviously is a proper monomorphism, and the converse holds by Corollary 12.92.
- Clearly, a finite morphism is (quasi-)affine and proper (see Example 12.56 (3)). For the converse, see Corollary 13.82.
- Every integral morphism is affine by definition and universally closed by Proposition 12.12, Proposition 12.11 (2). Conversely, as outlined in Exercise 12.19, every affine universally closed morphism is integral.

References for the implications with bold arrows:

- Every faithfully flat quasi-compact morphism is universally submersive by Corollary 14.43.
- Faithfully flat morphisms locally of finite presentation are open by Theorem 14.35.
- Smooth morphisms are flat by Theorem 14.24.
- Purely inseparable morphisms are separated, because their diagonal morphism is surjective (Exercise 9.9).
- Every separated quasi-finite morphism is quasi-affine by Corollary 12.91 (and, by definition, of finite type).
- By Corollary 13.77, every finite morphism is projective.
- Corollary 13.72 shows that every projective morphism is proper and quasi-projective, and that the converse holds if the target is qcqs.

© Springer Fachmedien Wiesbaden GmbH, part of Springer Nature 2020
U. Görtz und T. Wedhorn, *Algebraic Geometry I: Schemes*, Springer Studium
Mathematik – Master, https://doi.org/10.1007/978-3-658-30733-2

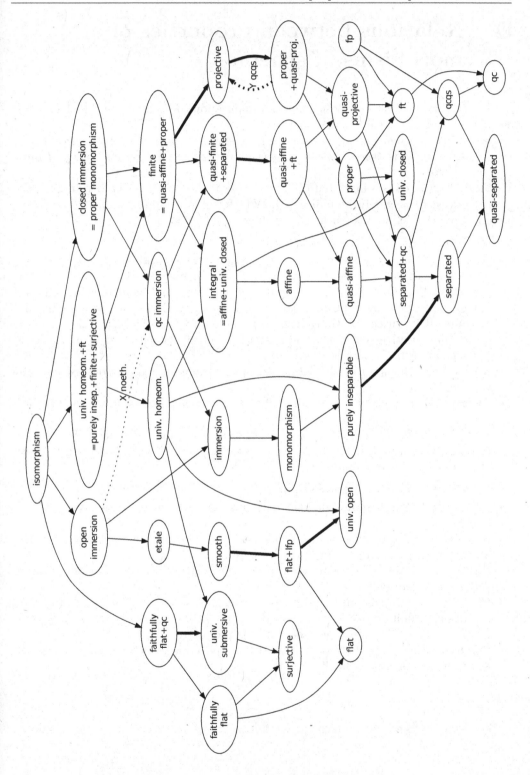

E   Constructible and open properties

In this appendix we collect some (ind-)constructible properties. For the convenience of the reader we recall (and refine) some definitions and properties for constructible subsets. Note that we use the terminology of [EGAInew], which differs from that of [EGAIV] §1.

(1) Let S be a scheme and let E be a subset of S.
   - Let S be noetherian. Then E is constructible if and only if it is the finite union of locally closed sets.
   - Let S be qcqs. Then E is constructible if and only if it is the finite union of sets of the form $U \cap (S \setminus V)$, where $U, V \subseteq S$ are quasi-compact open subsets of S.
   - Let S be an arbitrary scheme. Then E is constructible if and only if for every qcqs open subscheme $U \subseteq S$ the intersection $E \cap U$ is constructible.

(2) We define a more general notion for a subset E of a scheme S.
   - Let S be noetherian. Then E is called *ind-constructible* if and only if it is the union of locally closed sets.
   - Let S be qcqs. Then E is called *ind-constructible* if and only if it is the union of sets of the form $U \cap (S \setminus V)$, where $U, V \subseteq S$ are quasi-compact open subsets of S.
   - Let S be an arbitrary scheme. Then E is called *ind-constructible* if and only if for every qcqs open subscheme $U \subseteq S$ the intersection $E \cap U$ is constructible.

(3) A subset E of a scheme is constructible if and only if E and $S \setminus E$ are ind-constructible ([EGAInew] (7.2.9)).

(4) Let E be an ind-constructible subset of an irreducible scheme S. If E contains the generic point of S, then E contains an open dense subset of S ([EGAInew] (7.2.8)).

(5) Let E be a constructible subset of an irreducible scheme S. Then either E or $S \setminus E$ contains an open dense subset of S (follows from (3) and (4); see also Proposition 10.14).

(6) Let $f \colon S' \to S$ be a morphism of schemes and let $E \subseteq S$ be a constructible (resp. ind-constructible) subset of S. Then $f^{-1}(E)$ is constructible (resp. ind-constructible) in S' (Proposition 10.43, [EGAInew] (7.2.3) (vi)).

Notation E.1. In this appendix S will always denote a scheme. If X is an S-scheme, we will denote by X_s the fiber of X over a point $s \in S$. If $f \colon X \to Y$ is a morphism of S-schemes, then $f_s \colon X_s \to Y_s$ is the induced morphism on fibers. For a quasi-coherent $\mathscr{O}_X$-module $\mathscr{F}$, we set $\mathscr{F}_s := \mathscr{F} \otimes_{\mathscr{O}_S} \kappa(s)$ which is a quasi-coherent $\mathscr{O}_{X_s}$-module. For every homomorphism $u \colon \mathscr{F} \to \mathscr{G}$ of quasi-coherent $\mathscr{O}_X$-modules we denote by $u_s \colon \mathscr{F}_s \to \mathscr{G}_s$ the induced homomorphism of $\mathscr{O}_{X_s}$-modules. If $Z \subseteq X$ is a subset, then we set $Z_s := Z \cap X_s$.

© Springer Fachmedien Wiesbaden GmbH, part of Springer Nature 2020
U. Görtz und T. Wedhorn, *Algebraic Geometry I: Schemes*, Springer Studium
Mathematik – Master, https://doi.org/10.1007/978-3-658-30733-2

(E.1) Constructible and open properties on the target.

CONSTRUCTIBLE PROPERTIES OF SCHEMES.

Let X be an S-scheme of finite presentation and set

$$E_{\mathbf{P}}(X/S) := \{\, s \in S \; ; \; \mathbf{P}(X_s, \kappa(s)) \text{ holds} \,\},$$

where $\mathbf{P} = \mathbf{P}(Z, k)$ is one of the following properties of a scheme Z of finite type over a field k.

(1) $\{\, s \in S \; ; \; X_s \text{ is empty} \,\}$ is constructible in S (Proposition 10.96). It is open if X is proper and flat over S (by (8) below).

(2) $\{\, s \in S \; ; \; X_s \text{ is finite} \,\}$ is constructible in S (Proposition 10.96). It is open if X is proper and flat over S (by (8) below).

(3) $\{\, s \in S \; ; \; X_s \text{ is purely inseparable over } \kappa(s) \,\}$ is constructible in S ([EGAIV] (9.2.6)).

(4) $\{\, s \in S \; ; \; \dim X_s \in \Phi \,\}$ is constructible in S, where Φ is a given subset of $\mathbb{N}_0 \cup \{-\infty\}$ (Proposition 10.96).

(5) $\{\, s \in S \; ; \; \{\, \dim Z \; ; \; Z \text{ irreducible component of } X_s \,\} \subseteq \Phi \,\}$ is constructible in S, where Φ is a given subset of $\mathbb{N}_0 \cup \{-\infty\}$ ([EGAIV] (9.8.5)).

(6) $\{\, s \in S \; ; \; \{\, \dim Z \; ; \; Z \text{ irreducible component of } X_s \,\} \supseteq \Phi \,\}$ is constructible in S, where Φ is a given finite subset of $\mathbb{N}_0 \cup \{-\infty\}$ ([EGAIV] (9.8.5)). It is open if X is proper and flat over S ([EGAIV] (12.2.1)).

(7) $\{\, s \in S \; ; \; \dim X_s < n \,\}$ is open in S if $X \to S$ is proper, where $n \geq 0$ is a fixed integer (Corollary 14.115). It is open and closed in S if S is locally noetherian and $X \to S$ is open and closed ([EGAIV] (14.2.5)).

(8) $\{\, s \in S \; ; \; \{\, \dim Z \; ; \; Z \text{ associated component of } X_s \,\} \subseteq \Phi \,\}$ is constructible in S, where Φ is a given subset of $\mathbb{N}_0 \cup \{-\infty\}$ ([EGAIV] (9.8.5)). It is open if X is proper and flat over S ([EGAIV] (12.2.1)).

(9) The set of $s \in S$, where the geometric number of irreducible components of X_s is contained in Φ, is constructible, where $\Phi \subseteq \mathbb{N}_0$ is a given subset ([EGAIV] (9.7.9)).

(10) The set of $s \in S$, where the geometric number of connected components of X_s is contained in Φ, is constructible, where $\Phi \subseteq \mathbb{N}_0$ is a given subset ([EGAIV] (9.7.9)). It is open and closed in S if S is locally noetherian, and f is proper, universally open, and with geometrically reduced fibers. It is also open and closed in S if f is finite and étale.

(11) $\{\, s \in S \; ; \; X_s \text{ is geometrically reduced} \,\}$ is constructible in S ([EGAIV] (9.7.7)). It is open in S, if X is proper and flat over S ([EGAIV] (12.2.1)).

(12) $\{\, s \in S \; ; \; X_s \text{ is geometrically integral} \,\}$ is constructible in S ([EGAIV] (9.7.7)). It is open in S, if X is proper and flat over S ([EGAIV] (12.2.1)).

(13) The set of $s \in S$, where X_s is geometrically reduced and the geometric number of connected components of X_s is contained in Φ, is constructible, where $\Phi \subseteq \mathbb{N}_0$ is a given subset (by (10) and (11)). It is open in S, if X is proper and flat over S ([EGAIV] (12.2.4)).

(14) $\{\, s \in S \; ; \; X_s \text{ does not have embedded components} \,\}$ is constructible in S ([EGAIV] (9.8.1)).

(15) $\{\, s \in S \; ; \; X_s \text{ is Cohen-Macaulay} \,\}$ is constructible in S ([EGAIV] (9.9.3)). It is open in S, if X is proper and flat over S ([EGAIV] (12.2.1)).

(16) $\{\, s \in S \; ; \; X_s \text{ satisfies } (S_r) \,\}$ is constructible in S, where $r \geq 1$ is a fixed integer ([EGAIV] (9.9.3)). It is open in S, if X is proper and flat over S ([EGAIV] (12.2.1)).

(17) $\{s \in S \,;\, X_s$ satisfies (S_r) and is equidimensional$\}$ is constructible in S, where $r \geq 1$ is a fixed integer ([EGAIV] (9.9.3) and [EGAIV] (9.8.5)). It is open in S, if X is proper and flat over S ([EGAIV] (12.2.1)).

(18) $\{s \in S \,;\, X_s$ is smooth over $\kappa(s)\}$ is constructible in S ([EGAIV] (9.9.5) and Corollary 6.32). It is open in S, if X is proper and flat over S ([EGAIV] (12.2.4)).

(19) $\{s \in S \,;\, X_s$ is étale over $\kappa(s)\}$ is constructible in S (by (18) and (4))

(20) $\{s \in S \,;\, X_s$ is geometrically normal over $\kappa(s)\}$ is constructible in S ([EGAIV] (9.9.5)). It is open in S, if X is proper and flat over S ([EGAIV] (12.2.4)).

(21) $\{s \in S \,;\, X_s$ satisfies (R_k) geometrically$\}$ is constructible in S, where $r \geq 0$ is a fixed integer ([EGAIV] (9.9.5)).

CONSTRUCTIBLE PROPERTIES OF MORPHISMS OF SCHEMES.

Let $f \colon X \to Y$ be a morphism of S-schemes of finite presentation and for $s \in S$ let $f_s \colon X_s \to Y_s$ be the induced morphism on fibers.

(1) $\{s \in S \,;\, f_s$ is surjective$\}$ is constructible ([EGAIV] (9.3.2)).

(2) $\{s \in S \,;\, f_s$ is quasi-finite$\}$ is constructible ([EGAIV] (9.3.2)).

(3) $\{s \in S \,;\, f_s$ is purely inseparable$\}$ is constructible ([EGAIV] (9.3.2)).

(4) $\{s \in S \,;\, f_s$ is dominant$\}$ is constructible ([EGAIV] (9.6.1)).

(5) $\{s \in S \,;\, f_s$ is separated$\}$ is constructible ([EGAIV] (9.6.1)).

(6) $\{s \in S \,;\, f_s$ is proper$\}$ is constructible ([EGAIV] (9.6.1)).

(7) $\{s \in S \,;\, f_s$ is finite$\}$ is constructible ([EGAIV] (9.6.1)). If X and Y are proper over S, then this set is open (Proposition 12.93).

(8) $\{s \in S \,;\, f_s$ is an immersion$\}$ is constructible ([EGAIV] (9.6.1)).

(9) $\{s \in S \,;\, f_s$ is a closed immersion$\}$ is constructible ([EGAIV] (9.6.1)). If X and Y are proper over S, then this set is open (Proposition 12.93).

(10) $\{s \in S \,;\, f_s$ is an open immersion$\}$ is constructible ([EGAIV] (9.6.1)).

(11) $\{s \in S \,;\, f_s$ is an isomorphism$\}$ is constructible ([EGAIV] (9.6.1)). If X is proper and flat over S and Y is proper over S, then this set is open in S (Proposition 14.28).

(12) $\{s \in S \,;\, f_s$ is a monomorphism$\}$ is constructible ([EGAIV] (9.6.1)).

(13) $\{s \in S \,;\, f_s$ is flat$\}$ is constructible ([EGAIV] (11.2.8)).

(14) $\{s \in S \,;\, f_s$ is smooth$\}$ is constructible ([EGAIV] (17.7.11)).

(15) $\{s \in S \,;\, f_s$ is étale$\}$ is constructible ([EGAIV] (17.7.11)).

(16) $\{s \in S \,;\, f_s$ is affine$\}$ is ind-constructible ([EGAIV] (9.6.2)).

(17) $\{s \in S \,;\, f_s$ is quasi-affine$\}$ is ind-constructible ([EGAIV] (9.6.2)).

(18) $\{s \in S \,;\, f_s$ is projective$\}$ is ind-constructible ([EGAIV] (9.6.2)).

(19) $\{s \in S \,;\, f_s$ is quasi-projective$\}$ is ind-constructible ([EGAIV] (9.6.2)).

CONSTRUCTIBLE PROPERTIES OF EXISTENCE OF MORPHISMS OF SCHEMES.

Let X and Y be schemes of finite presentation over a scheme S and let $\mathbf{P}$ be a property of morphisms of schemes that is compatible with inductive limits of rings. Then the set of $s \in S$ such that there exists a field extension $k' \supseteq \kappa(s)$ and a morphism $X_{k'} \to Y_{k'}$ that has property $\mathbf{P}$ is an ind-constructible subset of S ([EGAIV] (9.3.5)).

CONSTRUCTIBLE PROPERTIES OF MODULES.

Let $g \colon X \to S$ be a morphism of finite presentation and let $\mathscr{F}$ be an $\mathscr{O}_X$-module of finite presentation.

(1) $\{s \in S \,;\, \mathscr{F}_s$ locally free $\mathscr{O}_{X_s}$-module of rank $n\}$ is constructible ([EGAIV] (9.4.7)).

(2) $\{s \in S \; ; \; \mathscr{F}_s \xrightarrow{u_s} \mathscr{G}_s \xrightarrow{v_s} \mathscr{H}_s$ is exact $\}$ is constructible, where $\mathscr{F} \xrightarrow{u} \mathscr{G} \xrightarrow{v} \mathscr{H}$ is a sequence of $\mathcal{O}_X$-modules of finite presentation ([EGAIV] (9.4.4)).

(3) $\{s \in S \; ; \; \mathscr{F}_s$ is a torsion-free $\mathcal{O}_{X_s}$-module $\}$ is constructible if X_s is a geometrically integral $\kappa(s)$-scheme for all $s \in S$ ([EGAIV] (9.4.8)).

(4) $\{s \in S \; ; \; \mathscr{F}_s$ is a torsion $\mathcal{O}_{X_s}$-module $\}$ is constructible if X_s is a geometrically integral $\kappa(s)$-scheme for all $s \in S$ ([EGAIV] (9.4.8)).

(5) $\{s \in S \; ; \; \mathscr{F}_s$ is a Cohen-Macaulay $\mathcal{O}_{X_s}$-module $\}$ is constructible in S ([EGAIV] (9.9.3)). It is open if X is proper and $\mathscr{F}$ is f-flat ([EGAIV] (11.2.1)).

(6) $\{s \in S \; ; \; \mathscr{L}_s$ is ample over $\kappa(s)\}$ is ind-constructible in S, if $\mathscr{L}$ is an invertible $\mathcal{O}_X$-module ([EGAIV] (9.6.2)). It is open if f is proper and then $\mathscr{L}_{|g^{-1}(E)}$ is relatively ample over E ([EGAIV] (9.6.4)).

(7) $\{s \in S \; ; \; \mathscr{L}_s$ is f_s-ample (resp. f_s-very ample) $\}$ is ind-constructible in S, where $f \colon X \to Y$ is a morphism of S-schemes of finite presentation ([EGAIV] (9.6.2)). If f is proper, then this set is constructible ([EGAIV] (9.6.3)).

(8) $\{s \in S \; ; \; \{\dim Z \; ; \; Z$ irreducible component of $\mathrm{Supp}(\mathscr{F}_s)\} \subseteq \Phi\}$ is constructible, where $\Phi \subseteq \mathbb{N}_0 \cup \{-\infty\}$ is a fixed subset ([EGAIV] (9.8.5)).

(9) $\{s \in S \; ; \; \mathscr{F}_s$ is f_s-flat $\}$ is constructible, where $f \colon X \to Y$ is a morphism of S-schemes of finite presentation ([EGAIV] (11.2.8)).

CONSTRUCTIBLE PROPERTIES OF SUBSETS.

Let X be an S-scheme of finite presentation and let $Z, Z' \subseteq X$ be two constructible subsets. Then the following subsets E are constructible in S.

(1) $E := \{s \in S \; ; \; Z_s \neq \emptyset\}$ ([EGAIV] (9.5.1)).

(2) $E := \{s \in S \; ; \; Z_s \subseteq Z'_s\}$ ([EGAIV] (9.5.2)).

(3) $E := \{s \in S \; ; \; Z_s$ dense in $Z'_s\}$ if $Z \subseteq Z'$ ([EGAIV] (9.5.3)).

(4) $E := \{s \in S \; ; \; Z_s$ open (resp. closed, resp. locally closed) in $X_s\}$ ([EGAIV] (9.5.4)).

(5) $E := \{s \in S \; ; \; \{\dim Z' \; ; \; Z' \subseteq Z_s$ irreducible component $\} \subseteq \Phi\}$ if Z_s is locally closed in X_s for all $s \in S$. Here $\Phi \subseteq \mathbb{Z} \cup \{-\infty\}$ is a given subset ([EGAIV] (9.5.4)).

(E.2) Constructible and open properties on the source.

CONSTRUCTIBLE PROPERTIES OF SCHEMES.

Let $f \colon X \to S$ be a morphism locally of finite presentation.

(1) $\{x \in X \; ; \; \mathcal{O}_{X_{f(x)},x}$ is equidimensional $\}$ is constructible in X ([EGAIV] (9.9.1)).

(2) $\{x \in X \; ; \; \dim_x(X_{f(x)}) < n\}$ is open in X, where $n \geq 0$ is a fixed integer (Theorem 14.112). It is open and closed in X if f is smooth ([EGAIV] (17.10.2)).

(3) $\{x \in X \; ; \; X_{f(x)}$ geometrically integral in $x\}$ is constructible in X (see Exercise 5.21) ([EGAIV] (9.9.4)). It is open in X if X is flat over S ([EGAIV] (12.1.1)).

(4) $\{x \in X \; ; \; X_{f(x)}$ geometrically reduced in $x\}$ is constructible in X ([EGAIV] (9.9.4)). It is open in X if X is flat over S ([EGAIV] (12.1.1)).

(5) $\{x \in X \; ; \; X_{f(x)}$ smooth in $x\}$ is constructible in X ([EGAIV] (9.9.4)). It is open in X if X is flat over S ([EGAIV] (12.1.6)).

(6) $\{x \in X \; ; \; X_{f(x)}$ geometrically normal in $x\}$ is constructible in X ([EGAIV] (9.9.4)). It is open in X if X is flat over S ([EGAIV] (12.1.6)).

(7) $\{x \in X \; ; \; \mathcal{O}_{X_{f(x)},x}$ is Cohen-Macaulay $\}$ is constructible in X ([EGAIV] (9.9.2)). It is open in X if X is flat over S ([EGAIV] (12.1.1)).

(8) $\{x \in X \; ; \; \mathcal{O}_{X_{f(x)},x}$ satisfies $(S_k)\}$ is constructible in X, where $k \geq 1$ is a fixed integer ([EGAIV] (9.9.2)). It is open in X if X is flat over S ([EGAIV] (12.1.6)).

(9) $\{x \in X \; ; \; \dim \mathscr{O}_{X_{f(x)},x} - \operatorname{depth} \mathscr{O}_{X_{f(x)},x} \le k\}$ is constructible in X, where $k \ge 0$ is a fixed integer ([EGAIV] (9.9.4)). It is open in X if X is flat over S ([EGAIV] (12.1.1)).

(10) $\{x \in X \; ; \; \dim_x Z_{f(x)} \in \Phi\}$ is constructible in X, where $Z \subseteq X$ is a constructible subset such that Z_s is closed in X_s for all $s \in S$ and $\Phi \subset \mathbb{N}_0 \cup \{-\infty\}$ is a fixed finite subset ([EGAIV] (9.9.1)).

(11) $\{x \in X \; ; \; x$ is not contained in an embedded component of $X_{f(x)}\}$ is constructible in X ([EGAIV] (9.9.2)). It is open in X if X is flat over S ([EGAIV] (12.1.1)).

(12) $\{x \in X \; ; \; \dim Z \in \Phi$, where Z is an associated component of $X_{f(x)}$ with $x \in Z\}$ is constructible in X, where Φ is a given finite subset of $\mathbb{N}_0 \cup \{\pm\infty\}$ ([EGAIV] (9.9.2)). It is open in X if X is flat over S ([EGAIV] (12.1.1)).

(13) $\{x \in X \; ; \; f$ is flat in $x\}$ is open in X (Theorem 14.44).

CONSTRUCTIBLE PROPERTIES OF MODULES.

Let $f \colon X \to S$ be a morphism locally of finite presentation and let $\mathscr{F}$ be an $\mathscr{O}_X$-module of finite presentation. Then the following subsets E are constructible in X.

(1) $E := \{x \in X \; ; \; \mathscr{F}_{f(x)}$ is Cohen-Macaulay in $x\}$ ([EGAIV] (9.9.2)).

(2) $E := \{x \in X \; ; \; \mathscr{F}_{f(x)}$ satisfies (S_k) in $x\}$, where $k \ge 1$ is a fixed integer ([EGAIV] (9.9.2)).

(3) $E := \{x \in X \; ; \; \mathscr{F}_{f(x)} \to \mathscr{G}_{f(x)} \to \mathscr{H}_{f(x)}$ is exact in $x\}$, where $\mathscr{F} \to \mathscr{G} \to \mathscr{H}$ is a complex of $\mathscr{O}_X$-modules of finite presentation ([EGAIV] (9.9.6)). It is open if $\mathscr{H}$ is f-flat ([EGAIV] (12.3.3)).

(4) $\{x \in X \; ; \; \mathscr{F}$ is f-flat in $x\}$ is open in X (Theorem 14.44).

(5) $\{x \in X \; ; \; (u \otimes \operatorname{id}_{\kappa(f(x))})_x \colon (\mathscr{F}_{f(x)})_x \to (\mathscr{G}_{f(x)})_x$ is injective$\}$
$= \{x \in X \; ; \; (\operatorname{Ker} u)_x = 0$ and $\operatorname{Coker} u$ is f-flat in $x\}$ is open in X if $\mathscr{G}$ is f-flat, where $u \colon \mathscr{F} \to \mathscr{G}$ is a homomorphism of $\mathscr{O}_X$-modules of finite presentation ([EGAIV] (11.3.7)).

Bibliography

[AHK] A. Altman, R. Hoobler, S. Kleiman, *A note on the base change map for cohomology*, Compositio Math. **27** (1973), 25–38.

[AK] A. Altman, S. Kleiman, *Compactifying the Picard scheme*, Adv. in Math. **35** (1980), 50–112.

[AKMW] D. Abramovich, K. Karu, K. Matsuki, J. Włodarczyk, *Torification and factorization of birational maps*, J. Amer. Math. Soc. **15** (2002), no. 3, 531–572.

[AM] M.F. Atiyah, I.G. MacDonald, *Introduction to Commutative Algebra*, Westview Press (1994).

[An] M. André, *Localisation de la lissité formelle*, manuscripta math. **13** (1974), 297–307.

[Art] M. Artin, *Grothendieck topologies*, Mimeogr. notes at Harvard University (1962).

[Ba] L. Bădescu, *Algebraic Surfaces*, Universitext, Springer (2001).

[BE] D.A. Buchsbaum, D. Eisenbud, *What annihilates a module?*, J. Algebra **47** (1977), 231–243.

[Bea] A. Beauville, *Surfaces algébriques complexes*, Astérisque **54**, 3rd ed. (1982).

[BH] W. Bruns, J. Herzog, *Cohen-Macaulay rings*, Cambridge Univ. Press, (1993).

[BLR] S. Bosch, W. Lütkebohmert, M. Raynaud, *Néron models*, Erg. der Mathematik und ihrer Grenzgeb. (3) **21**, Springer (1990).

[Bon] L. Bonavero, *Factorisation faible des applications birationelles [d'après Abramovich, Karu, Matsuki, Wlodarczyk et Morelli]*, Sém. Bourbaki **880** (2000/01), in: Astérisque **282** (2002), 1–37.

[BouAI] N. Bourbaki, *Algebra*, Chapters 1–3, Springer (1989).

[BouAII] N. Bourbaki, *Algebra*, Chapters 4–7, Springer (1988).

[BouA8] N. Bourbaki, *Algèbre*, Chapitre 8, Hermann (1958).

[BouAC] N. Bourbaki, *Commutative algebra*, Chapters 1–7, 2nd printing, Springer (1989).

[BouAC10] N. Bourbaki, *Algèbre commutative*, Chapitre 10, Masson, Paris (1998).

[BouGT] N. Bourbaki, *General topology*, Chapters 1–4, 2nd printing, Springer (1989).

[BouLie] N. Bourbaki, *Lie groups and Lie algebras*, Chapters 4–6, 2nd printing, Springer (2009).

[BV] W. Bruns, U. Vetter, *Determinantal rings*, Lecture Notes in Math. **1327**, Springer (1988).

© Springer Fachmedien Wiesbaden GmbH, part of Springer Nature 2020
U. Görtz und T. Wedhorn, *Algebraic Geometry I: Schemes*, Springer Studium
Mathematik – Master, https://doi.org/10.1007/978-3-658-30733-2

[Ch] Sém. Claude Chevalley, vol. I (1956–58), http://www.numdam.org/actas/SCC/

[Co] B. Conrad, *Deligne's notes on Nagata compactifications*, J. Ramanujan Soc. **22** (2007) no. 3, 205–257, Erratum: **24** (2009) no. 4, 427–428.

[DG] M. Demazure, P. Gabriel, *Groupes algébriques. Tome I: Géométrie algébrique, généralités, groupes commutatifs*, Masson, North-Holland (1970).

[dJ] A.J. de Jong, *Smoothness, semi-stability and alterations*, Inst. Hautes Études Sci. Publ. Math. **83** (1996), 51–93.

[Dr] V. Drinfeld, *Infinite-dimensional vector bundles in algebraic geometry (an introduction)*, in: The unity of mathematics, Progr. Math. **244**, Birkhäuser 2006, 263–304.

[Du] A. Ducros, *Les espaces de Berkovich sont excellents*, Annales de l'institut Fourier **59** (2009), 1407–1516.

[EGAI] A. Grothendieck, J.A. Dieudonné, *Eléments de Géométrie Algébrique I*, Publ. Math. IHÉS **4** (1960), 5–228.

[EGAInew] A. Grothendieck, J.A. Dieudonné, *Eléments de Géométrie Algébrique I*, Springer (1971).

[EGAII] A. Grothendieck, J.A. Dieudonné, *Eléments de Géométrie Algébrique II*, Publ. Math. IHÉS **8** (1961), 5–222.

[EGAIII] A. Grothendieck, J.A. Dieudonné, *Eléments de Géométrie Algébrique III*, Publ. Math. IHÉS **11** (1961), 5–167; **17** (1963), 5–91.

[EGAIV] A. Grothendieck, J.A. Dieudonné, *Eléments de Géométrie Algébrique IV*, Publ. Math. IHÉS **20** (1964), 5–259; **24** (1965), 5–231; **28** (1966), 5–255; **32** (1967), 5–361.

[EH] D. Eisenbud, J. Harris, *The geometry of schemes*, Graduate Texts in Math. **197**, Springer (2000).

[Ei] D. Eisenbud, *Commutative algebra with a view toward algebraic geometry*, Graduate Texts in Math. **150**, Corrected 3rd printing, Springer (1999).

[FGA] A. Grothendieck, *Fondements de la géométrie algébrique*, Sém. Bourbaki, 1957–1962.

[FGAex] B. Fantechi et. al., *Algebraic Geometry: Grothendieck's FGA Explained*, Mathematical Surveys and Monographs **123**, AMS 2005.

[Fu1] W. Fulton, *Algebraic Curves: An Introduction to Algebraic Geometry*, Addison Wesley 1974; Third edition available at http://www.math.lsa.umich.edu/~wfulton/CurveBook.pdf

[Fu2] W. Fulton, *Intersection theory*, 2nd edition, Springer (1998).

[Ge] G. van der Geer, *Hilbert modular surfaces*, Erg. der Mathematik und ihrer Grenzgeb. (3) **16**, Springer (1988).

[GH] P. Griffiths, J. Harris, *Principles of algebraic geometry*, Wiley- Interscience (1978), reprinted 1994.

[Gi] J. Giraud, *Cohomologie non abélienne*, Grundlehren der math. Wiss. **179**, Springer (1971).

[Gl] S. Glaz, *Commutative coherent rings*, Lecture Notes in Math. **1371**, Springer (1989).

[GS] P. Gille, T. Szamuely, *Central simple algebras and Galois cohomology*, Cambridge Studies in Adv. Math. **101**, Cambridge Univ. Press (2006).

[Ha] G. Harder, *Lectures on Algebraic Geometry I*, Aspects of Math. **E 35**, Vieweg (2008).

[Ha1] R. Hartshorne, *Ample subvarieties of algebraic varieties*, Lecture Notes in Math. **156**, Springer (1970).

[Ha2] R. Hartshorne, *Varieties of small codimension in projective space*, Bull. Amer. Math. Soc. **80** (1974), 1017–1032.

[Ha3] R. Hartshorne, *Algebraic Geometry*, corrected 8th printing, Graduate Texts in Math. **52**, Springer (1997).

[Har] J. Harris, *Algebraic Geometry: a first course*, Graduate Texts in Math. **133**, Springer (1995).

[Hau] H. Hauser, *The Hironaka theorem on resolution of singularities (or: A proof we always wanted to understand)*, Bull. A. M. S. (N.S.) **40** (2003), no. 3, 323–403.

[He] S. Helgason, *Differential Geometry, Lie Groups, and Symmetric Spaces* Graduate Studies in Mathematics, AMS (2001).

[Hi] H. Hironaka, *Resolution of singularities of an algebraic variety over a field of characteristic zero: I-II*, Ann. of Math. **79** (1964), 109–326.

[HN] G. Harder, M.S. Narasimhan, *On the cohomology groups of moduli spaces of vector bundles on curves*, Math. Ann. **212** (1974/75), 215–248.

[Ho] G. Horrocks, *Vector bundles on the punctured spectrum of a local ring*, Proc. London Math. Soc. **14** (1964), 689–713.

[HoMu] G. Horrocks, D. Mumford, *A rank 2 vector bundle on $\mathbb{P}^4$ with 15,000 symmetries*, Topology **12** (1973), 63–81.

[Hu] B. Hunt, *The Geometry of some special Arithmetic Quotients*, Lecture Notes in Math. **1637**, Springer (1996)

[Jo] J.-P. Jouanolou, *Théorèmes de Bertini et applications*, Progr. Math. **42**, Birkhäuser (1983).

[Kl] S. Kleiman, *Misconceptions about K_X*, Enseign. Math. (2) **25** (1979), no. 3-4, 203–206.

[Kn] A. Knapp, *Elliptic Curves*, Princeton Univ. Press 1992.

[Ko] J. Kollár, *Lectures on Resolution of Singularities*, Annals of Math. Studies **166**, Princeton Univ. Press 2007.

[KS] M. Kashiwara, P. Schapira, *Categories and Sheaves*, Grundlehren der math. Wiss. **332**, Springer (2006).

[Ku1] H. Kurke, *Einige Eigenschaften von quasi-endlichen Morphismen von Präschemata*, Monatsberichte der deutschen Akademie der Wissenschaften zu Berlin **9** (1967), 248–257.

[Ku2] H. Kurke, *Über quasi-endliche Morphismen von Präschemata*, Monatsberichte der deutschen Akademie der Wissenschaften zu Berlin **10** (1968), 389–393.

[La] S. Lang, *Algebra*, Graduate Texts in Math. **211**, 4th ed., Springer (2004).

[Liu] Q. Liu, *Algebraic geometry and arithmetic curves*, Oxford Graduate Texts in Math. **6**, Oxford Univ. Press, 2002.

[Lü] W. Lütkebohmert, *On compactifications of schemes*, manuscripta mathematica **80** (1993), 95–111.

[Mat] H. Matsumura, *Commutative ring theory*, Cambridge studies in advanced mathematics **8**, Cambridge Univ. Press (1989).

[McL] S. Mac Lane, *Categories for the working mathematician*, 2nd edition, Springer (1998).

[Mit] B. Mitchell, *Theory of categories*, Academic Press (1965).

[MO] MathOverflow, `https://mathoverflow.net/` (append the specified reference to this URL).

[Mu1] D. Mumford, *The red book of varieties and schemes*, 2nd edition, Lecture Notes in Math. **1358**, Springer (1999).

[Mu2] D. Mumford, *Algebraic geometry I : complex projective varieties*, Grundlehren der math. Wiss. **221**, Springer (1976).

[Na1] M. Nagata, *Local rings*, Interscience Publishers (1960).

[Na2] M. Nagata, *Imbedding of an abstract variety in a complete variety*, J. Math. Kyoto **2** (1962), 1–10.

[Na3] M. Nagata, *A generalization of the imbedding problem*, J. Math. Kyoto **3** (1963), 89–102.

[Nau] N. Naumann, *Representability of* Aut_F *and* End_F, Comm. Algebra **33** (2005), no. 8, 2563–2568.

[Neu] J. Neukirch, *Algebraic number theory*, Springer (1999).

[NSW] J. Neukirch, A. Schmidt, K. Wingberg, *Cohomology of number fields*, Second edition. Grundlehren der math. Wiss. **323**, Springer (2008).

[Per] D. Perrin, *Algebraic geometry. An introduction*, Springer (2008).

[Pes] C. Peskine, *An algebraic introduction to complex projective geometry. 1. Commutative algebra*, Camb. Studies in Adv. Math. **47**, Cambridge Univ. Press (1996).

[Pey] A. Perry, *Faithfully flat descent for projectivity of modules*, arxiv:1001.0038 (2010)

[Po] D. Popescu, *General Néron desingularization and approximation*, Nagoya Math. J. **104** (1986), 85–115.

[Re] R. Remmert, *Local theory of complex spaces*, Chapter 1 of *Several complex variables VII: sheaf-theoretical methods in complex analysis*, Encyclopaedia of mathematical sciences **74**, Springer (1994).

[Rei] M. Reid, *Undergraduate Algebraic Geometry*, London Math. Soc. Student Texts **12**, 2nd ed. (1990)

[RG] M. Raynaud, L. Gruson, *Critères de platitude et projectivité, Techniques de "platification" d'un module*, Inventiones Math. **13** (1971), 1–89.

[Ri] O. Riemenschneider, *Deformationen von Quotientensingularitäten (nach zyklischen Gruppen)*, Math. Ann. **209** (1974), 211–248.

[Sch] H. Schubert, *Categories*, Springer (1972).

[Se1] J.-P. Serre, *Faisceaux algébriques cohérents*, Ann. of Math. **61** (1955), 197–278.

[Se2] J.-P. Serre, *Local Fields*, Graduate Texts in Math. **67**, 2nd corrected printing, Springer (1995).

[SGA1] A. Grothendieck, M. Raynaud, *Revêtements étales et groupe fondamental*, Documents Mathématiques 3, Société Mathématique de France 2003.

[Sh] I.R. Shafarevich, *Basic Algebraic Geometry*, 2nd edition (in two volumes) Grundlehren der math. Wiss. **213**, Springer (1994).

[Si] J. Silverman, *The arithmetic of elliptic curves*, Graduate Texts in Math. **106**, Springer (1986).

[Sp] M. Spivakovsky, *A new proof of D. Popescu's theorem on smoothing of ring homomorphisms*, J. of the American Mathematical Society **12** (1999), 381–444.

[St] Stacks project collaborators, *Stacks project*, https://stacks.math.columbia.edu/.

[Surf] Surf, by Stephan Endrass and others, http://surf.sourceforge.net/

[Te1] M. Temkin, *Desingularization of quasi-excellent schemes in characteristic zero*, Advances Math. **219** (2008), 488–522.

[Te2] M. Temkin, *Relative Riemann-Zariski spaces*, Israel J. Math. **185** (2011), 1–42.

[TT] R.W. Thomason, T. Trobaugh, *Higher Algebraic K-Theory of Schemes and of Derived Categories*, in *The Grothendieck Festschrift, Volume III* (ed. by P. Cartier et. al.), Progress in Mathematics **88**, Birkhäuser Boston (1990), 247–435.

[Vis] A. Vistoli, *Grothendieck topologies, fibered categories and descent theory*, in: [FGAex], 1–104.

[We] A. Weil, *Basic number theory*, 3rd edition, Springer (1974).

[Wlo] J. Włodarczyk, *Combinatorial structures on toroidal varieties and a proof of the weak factorization theorem*, Invent. Math. **154** (2003), no. 2, 223–331.

Detailed List of Contents

© Springer Fachmedien Wiesbaden GmbH, part of Springer Nature 2020
U. Görtz und T. Wedhorn, *Algebraic Geometry I: Schemes*, Springer Studium
Mathematik – Master, https://doi.org/10.1007/978-3-658-30733-2

Index of Symbols

© Springer Fachmedien Wiesbaden GmbH, part of Springer Nature 2020
U. Görtz und T. Wedhorn, *Algebraic Geometry I: Schemes*, Springer Studium
Mathematik – Master, https://doi.org/10.1007/978-3-658-30733-2

Index

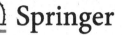

Printed in the United States
By Bookmasters